On a Tree Fallen Across The Road

The tree the tempest with a crash of wood
Throws down in front of us is not to bar
Our passage to our journey's end for good
But just to ask us who we think we are

Robert Frost, American Poet, 1874–1963

Metallic Powders for Additive Manufacturing

Metallic Powders for Additive Manufacturing

Science and Applications

Enrique J. Lavernia
Texas A&M University
College Station, USA

Kaka Ma
Colorado State University
Fort Collins, USA

Julie M. Schoenung
Texas A&M University
College Station, USA

James F. Shackelford
University of California
Davis, USA

Baolong Zheng
University of California
Irvine, USA

WILEY

Published by John Wiley & Sons, Inc., Hoboken, New Jersey.
Published simultaneously in Canada.

Library of Congress Cataloging-in-Publication Data Applied for:

Hardback ISBN: 9781119908111

Cover Design: Wiley
Cover Images: © MarinaGrigorivna/Shutterstock; Courtesy of Dr. Ramesh Subramanian, Siemens Energy Inc.;
Lavernia Group, Texas A&M University

Set in 9.5/12.5pt STIXTwoText by Straive, Chennai, India
Printed and bound by CPI Group (UK) Ltd, Croydon, CR0 4YY

C9781119908111_310124

Contents

About the Authors

Dr. Enrique J. Lavernia, currently holds the title of Professor and M. Katherine Banks Chair, in the Departments of Materials Science and Engineering and Mechanical Engineering at Texas A&M University. Prior to that, he was a Distinguished Professor in the Department of Materials Science and Engineering in the Samueli School of Engineering (2015–2023) and the Provost and Executive Vice Chancellor for the University of California, Irvine from 2015–2020. Dr. Lavernia was previously engineering dean (2002–2008 and 2010–2015) at UC Davis and a Distinguished Professor of Chemical Engineering and Materials Science. He also served as UC Davis' provost and executive vice chancellor (2008–2010).

In 2022, Prof. Lavernia was recognized with an *R&D 100 Award* for his work with Dr. Todd Monson (Sandia National Laboratories, NM, USA) on Iron Nitride Soft Magnetics. Also in 2022, Dr. Lavernia was recognized with ASM International, *Albert Sauveur Achievement Award* "for *sustained and pioneering studies on the fundamental mechanisms that govern the interrelationship between processing, microstructure and mechanical behavior of structural materials.*" He was elected as a Foreign Member of the Chinese Academy of Engineering in 2020. He also received the Acta Materialia Gold Medal in 2020. In 2019, Dr. Lavernia was awarded a *Doctor of Science in Technology, honoris causa,* by Aalto University, Helsinki, Finland. In 2018, he received the Distinguished Engineering Educator Award from the National Engineers' Council, and in 2017, he became a member of the National Academy of Inventors. Dr. Lavernia was elected to become a Fellow of the National Academy of Inventors in November 2016. Also in 2016, he received the Alexander von Humboldt Foundation Research Award as well as the Leadership Award from the TMS Society. In 2015, he was inducted into the Hispanic Hall of Fame by the HEENAC Great Minds in STEM. In 2014, he was awarded the TMS Fellows Award Class of 2014 by the Minerals, Metals and Materials Society. Elected to the National Academy of Engineering in 2013, Dr. Lavernia is also a fellow of the Minerals, Metals and Materials Society, the Materials Research Society, the American Society of Mechanical Engineers, the American Association for the Advancement of Science, and ASM International. He is the recipient of the 2013 Edward DeMille Campbell Memorial Lectureship and the 2013 ASM International Gold Medal Award. Named Presidential Young Investigator by the National Science Foundation, Dr. Lavernia also received a Young Investigator Award from the Office of Naval Research. In 2011, he received the Hispanic Engineer National Achievement Award and the Society for the Advancement of Chicanos and Native Americans in Science Distinguished Scientist Award.

Dr. Lavernia has published over 670 journal papers, 231 conference papers, 1 book, 11 edited books/journals, and 21 book chapters and has been awarded 11 patents on topics ranging from nano-materials to aluminum alloys. His research interests include the synthesis and behavior of nanostructured, high entropy, and multi-scale materials with particular emphasis on processing fundamentals and physical behavior; high temperature-high pressure atomization processes; and additive manufacturing. He earned his BSc from Brown University in 1982 and his MSc and PhD degrees in 1984 and 1986, respectively, from the Massachusetts Institute of Technology.

Dr. Kaka Ma is an Associate Professor in the Department of Mechanical Engineering and the School of Materials Science and Engineering at Colorado State University (CSU). She earned her B.S. degree in materials physics from the University of Science and Technology of China in June 2006, and then PhD in materials science and engineering from the University of California (UC), Davis in December 2010. Following several years of postdoctoral research and some part-time instructor experiences at UC Davis and UC Irvine, she joined CSU as a tenure-track Assistant Professor in August 2016. Ma's research interests sit at the interface of materials science, mechanical engineering, and sustainability. Her research is focused on powder metallurgy for both metals and ceramics, with the overarching goal of discovering new processing-structure-properties correlation to sustainably develop materials for next-generation structural, electronic, and energy components. Her research group mainly utilizes spark plasma sintering and laser-directed energy deposition to develop functionally graded materials, thermionic materials, and structural materials and investigate the sustainability issues associated with these powder technologies. She has published more than 50 peer-reviewed journal papers, given about 20 invited talks, and authored more than 70 presentations at technical conferences, workshops, and at other universities. She is a member of the Minerals, Metals & Materials Society (TMS), Material Research Society (MRS), and America Makes. Ma is the founding editor and Editor-in-chief of Results in Materials. She has also been serving on the reviewer/editorial board for Materials Science & Engineering A, Journal of the American Ceramics Society, Journal of Materials Science, Additive Manufacturing, and so forth.

Professor Julie M. Schoenung currently holds the title of Professor and Wofford Cain Chair III in the Departments of Materials Science and Engineering and Mechanical Engineering at Texas A&M University. She is also the Founding Co-Director of the Lincoln Dynamic Foundation World Institute for Sustainable Development of Materials (WISDOM). Prior to joining Texas A&M, Professor Schoenung served as the Founding Department Chair in the Department of Materials Science and Engineering at the University of California, Irvine (UCI) and held the title of Distinguished Professor. She also held the Daniel G. Aldrich, Jr. endowed chair position at UCI. Prior to her positions at UCI, Professor Schoenung held faculty appointments UC Davis and at Cal Poly Pomona (California State Polytechnic University, Pomona).

Professor Schoenung is a member of the National Academy of Engineering (NAE), and a Fellow of the American Association for the Advancement of Science (AAAS), the National Academy of Inventors (NAI), the Minerals, Metals and Materials Society (TMS), the Materials Research Society (MRS), ASM International, the American Ceramic Society, and the Alpha Sigma Mu Honor Society. Professor Schoenung received her PhD and MS in materials engineering from the Massachusetts Institute of Technology, and her BS in ceramic engineering from the University of Illinois, Urbana-Champaign. Professor Schoenung conducts research into structure-processing-property mechanistic relationships in a variety of materials systems. Her current research focus is on high entropy ceramics and additive manufacturing of ceramics, cermets, composites, and metals, including the application of alternative feedstock materials generated from waste products. Professor Schoenung is also a pioneer in the field of sustainable development of materials, with many years of experience studying the materials-selection process in a variety of applications. She conducts research into the analysis of factors that guide the materials-selection decision-making process, such as economics, environmental impact and toxicity, cost-performance trade-offs, policy, and sustainability standards.

James F. Shackelford has BS and MS degrees in Ceramic Engineering from the University of Washington and a PhD in Materials Science and Engineering from the University of California, Berkeley. Following a postdoctoral fellowship at McMaster University in Canada, he joined the University of California, Davis, where he is currently a Distinguished Professor Emeritus in the Department of Materials Science and Engineering and is a Visiting Professor Emeritus at UC Irvine. For many years, he served as the Associate Dean for Undergraduate Studies in the College of Engineering at UC Davis and later as the Director of the University Honors Program that serves students from a wide spectrum of majors. He teaches and conducts research in the structural characterization and processing of materials. His current focus in teaching is doing so through online technologies. A member of the American Ceramic Society and ASM International, he was named a Fellow of the American Ceramic Society in 1992, was named a Fellow of ASM International in 2011, and received the Outstanding Educator Award of the American Ceramic Society in 1996 and the Albert Easton White Distinguished Teacher Award from ASM International in 2019. In 2016, Professor Shackelford received the Inaugural Award for Outstanding Contributions to Materials Education at the North American Materials Education Symposium (NAMES) held at the University of California, Berkeley. He has published over 150 archived papers and books including *Introduction to Materials Science for Engineers* now in its 9th Edition and which has been translated into Chinese, German, Italian, Japanese, Korean, Portuguese, and Spanish. He is also the coauthor of the *CRC Materials Science and Engineering Handbook* now in its 4th Edition and *The Glass of Wine* covering the intersection of the glass and wine industries published by The American Ceramic Society and Wiley.

Dr. Baolong Zheng holds a Bachelor of Science degree (1987) and a Master of Science degree (1989) in Metallurgy and Materials Engineering from the University of Science and Technology Beijing (USTB) in China. He also obtained a PhD in Materials Science and Engineering from the University of California (UC) Davis in the USA in 2006. Following the completion of his PhD, Dr. Zheng worked as a Postdoctoral Researcher and Research Scientist at UC Davis (2006–2015) and UC Irvine (2016–2024), where he specialized in research on gas atomization (GA), additive manufacturing (AM) deposition, and field assisted sintering technologies (FAST). He worked on developing Al-, Ti-, Ni-, Mg-, and Fe-based alloys and metal matrix composites (MMCs) using these technologies, as well as Al-Ni metallic foam materials. He has expertise in the synthesis and characterization of non-equilibrium nanostructured and amorphous Mg-, Fe-, and Al-based alloys, high entropy alloys, soft magnetic materials, and MMCs, achieved through various powder metallurgy and thermo-mechanical processing techniques including GA, high-energy milling/cryomilling, AM, SPS, CIP/HIP, extrusion, ECAP, HPT, and rolling.

Dr. Zheng has an extensive background in academia and research. He was an R&D Engineer (1990–1993) at the National Engineering Research Center for Adv. Rolling Technology of China (NERCAR), an Assistant Professor (1993–1996) and Associate Professor (1997–2000) at USTB, and a Research Assistant (2000–2001) at UC Berkeley.

Dr. Zheng has published over 90 journal papers, more than 100 conference presentations, and 8 books or chapters, showcasing his contributions to the field of metallurgy and materials engineering.

Preface

Motivation

In recent years, additive manufacturing (AM) has evolved as an important manufacturing approach that has the potential to revolutionize the fabrication of both advanced and conventional materials by simplifying the number of process steps that are required to attain a final product, particularly with complex three-dimensional geometries. This is consistent with the data shown in Figure 1, which summarizes the statistical data of articles on the topic of additive manufacturing published up to 2019 as well as the number of additive manufacturing papers under specific material term searches.

It is evident from a review of the published literature as well as from extensive industrial interests in additive manufacturing that many existing metallic additive manufacturing technologies require powder as the raw input material, and mostly rely on atomized metal powders as the source material. The quality of starting powder, such as chemical composition, particle size, and morphology, is critical to successful attempts to produce functional additive manufacturing components. It is also evident that although atomization has existed for several hundred years, the complex interrelationships that exist between the characteristics of the metal powders and the requirements of additive manufacturing remain largely unexplored. Moreover, it is also evident based on a large body of literature that perhaps one of the most critical elements that is hindering progress in additive manufacturing is in fact a lack of knowledge of the precise role that metal powders play in determining both the processing as well as the final performance of additively manufactured parts.

The motivation for this book is to introduce the reader to both the science and technology of atomized metal powders and how these relate to the process of additive manufacturing. The book seeks to set a foundation for the underlying science that governs the formation and microstructure of atomized metallic droplets. This information is then correlated with the behavior of metallic powders during additive manufacturing. We seek to establish the fundamental relationships that exist among the chemistry, microstructure, and morphology of atomized metallic powders and their behavior during additive manufacturing.

It was an important goal of the authors to deliver a book that is comprehensive in its coverage of all topics related to the atomization of metal powders and additive manufacturing. Concepts are introduced at a fundamental level, placing particular emphasis on information that relates to practical engineering problems. As such, the book is useful to the novice but also serves as a reference book for those who are active in the practice of metal powder atomization and/or additive manufacturing. We hope that the book will also play a role in education, and as such, we envision this book to be useful as a reference book in graduate or advanced undergraduate courses, for example.

Organization

In this book, we begin with a description of the fundamentals of powder atomization to familiarize the reader with the mechanisms that govern microstructure evolution in atomized metal powders. We describe the variety of processes that currently exist for the atomization of metal powders. We then establish the relationships that exist between the characteristics of atomized powders and the performance requirements of additively manufactured parts. We also discuss alternative feedstock materials for metal additive manufacturing other than atomized powder, such as powders produced from ball milling of machining chips, and how these alternative feedstock materials affect

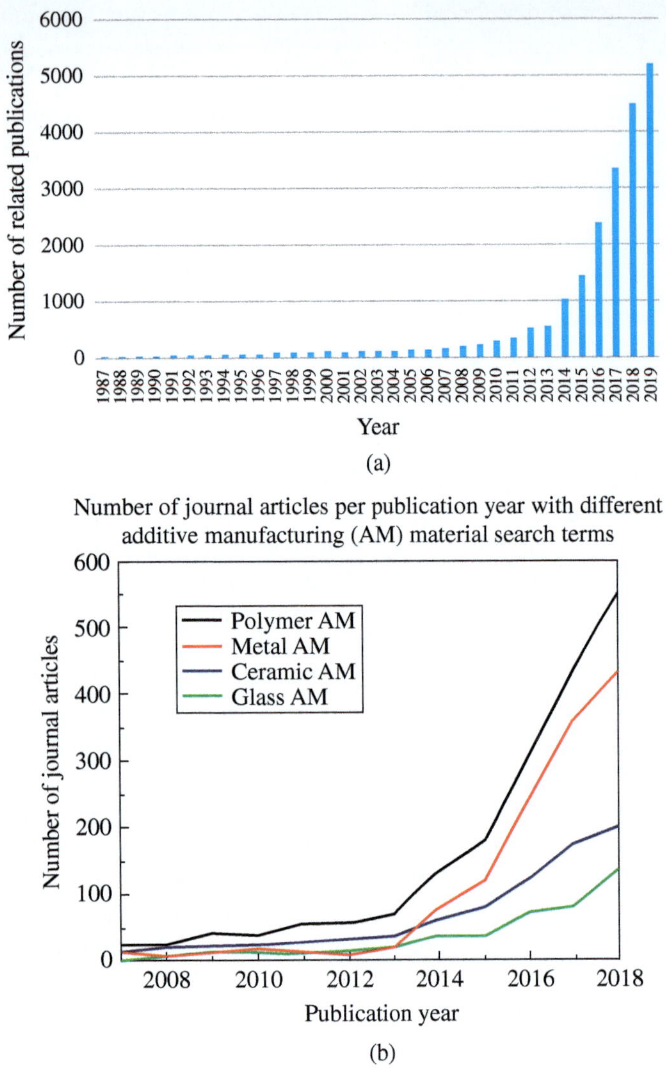

Figure 1 (a) Statistical data of articles on the topic of additive manufacturing published in 2019, based on the database of Web of Science Core Collection in July 2020 [1], and (b) number of additive manufacturing papers under specific material term searches [2].

the choices of additive manufacturing processing windows (parameters) and the microstructure and properties of the final additive manufacturing parts. We finish with a perspective on the potential influence of metal powder atomization and additive manufacturing on environmental impact and economic viability, which is less commonly addressed in the current literature, and, we provide a brief outlook on the short-term needs and long-term future directions of metal powders for additive manufacturing.

In Part I, we begin with an introduction to the process of atomization describing the various methods that exist to generate metallic droplets. We proceed to describe the various theories that exist to describe the formation of droplets, the physics that govern their morphology, and their ultimate overall microstructure evolution. We discuss the parameters that exercise some degree of control over atomization paying, particular attention to ones that directly affect the particle size distribution and how these are related to the various processing variables. We discuss the various models that exist to describe the mechanisms that govern atomization. The book then addresses the heat transfer and solidification of droplets, paying particular attention to the critical thermal and solidification conditions that govern the evolution of microstructure.

In Part II, we introduce the various processes that are used for additive manufacturing and the relationship between metal powder characteristics and additive manufacturing processing, and ultimately part quality. This section of the book begins with an overview of the various techniques that are used to produce additively manufactured components. The next chapter covers powder-laser-molten-pool interactions, followed by the influence of powder production methods and chemistry on additive manufacturing. We then discuss the influence of powder on microstructure evolution, defect formation, and residual stress, respectively, and finally present a discussion on the influence of powder on the physical and chemical behavior of additively manufactured parts. Our final two chapters describe societal issues (such as economic viability and sustainability) followed by perspectives on future directions and challenges.

January 2024

Enrique J. Lavernia
Kaka Ma
Julie M. Schoenung
James F. Shackelford
Baolong Zheng

References

1 Liu, G., Zhang, X., Chen, X. et al. (2021). Additive manufacturing of structural materials. *Mater. Sci. Eng., R* 145: 100596.

2 Wachtel, P.F., Lindberg, G.P., Musgraves, J.D. et al. (2019). A comparison study of additive manufacturing techniques applied to chalcogenide glass. In: *Proceedings SPIE 11004, Thermosense: Thermal Infrared Applications XLI*. Baltimore, MD, USA: SPIE.

Acknowlegments

Enrique J. Lavernia

I sincerely thank my family, whose steadfast support and boundless affection have been my pillars, particularly in challenging circumstances. I am deeply thankful to my spouse and soulmate, Julie Schoenung, whose enduring patience, love, and encouragement have been instrumental in my life. My amazing children, Alejandro Carlos and Laura Kathryn, who consistently bring me happiness, pride, and unwavering backing. The spirit and love of my brother, Carlos J. Lavernia, is a remarkable guardian over me and our family.

I also thank my wonderful MSE colleagues and dedicated staff for their care, integrity, and hard work. My long-time friend and colleague, Jeff Lefkoff, has always been there for me on our regular beach walks and for lending a listening ear. My countless friends and colleagues locally, nationally, and globally, including Diran Apelian, Michael Stamos, Horst Hahn, and many others whose friendship and care have sustained me.

I am grateful to over 40 PhD students, many MSc candidates, postdoctoral scientists, undergraduate students, visiting scholars, and colleagues. I extend special appreciation to Dr. Yizhang Zhou and Dr. Baolong Zheng, the coauthor of this book. Their intellectual contributions and the meaningful friendships we have developed over time have significantly enriched my life and inspired me during adversity. My colleague and friend, Darryl Mack, consistently demonstrates technical excellence and patience, which are instrumental in ensuring the safety and success of our continually evolving and expanding state-of-the-art experimental facilities.

My heartfelt thanks to Aaron May, Brian Sun, John Littrell, and Alisa Chandler, whose personal care, energy, and vast legal knowledge have helped sustain me through the past several years.

Finally, Jim, Julie, Kaka, and Baolong, I cannot express enough gratitude for your unwavering commitment and hard work in our collaborative writing of this book. Despite the many hours spent in virtual meetings and countless discussions about content, references, keywords, etc., you never faltered in your dedication. I am humbled by your steadfast support and willingness to spend endless hours working on various chapters. Thank you for being an integral part of this journey and making it an unforgettable experience.

Kaka Ma

I would like to start by thanking Dr. Enrique Lavernia for inviting me to this exciting opportunity to discuss and share experiences, opinions, and visions of the impactful and rapidly expanding field of additive manufacturing, focusing on the topic of metallic powders. I cannot express enough gratitude to my former PhD advisor, Dr. Julie M. Schoenung, who not only trained me with professional attributes and encouraged me to my current career stage but also became a good friend, always lending her ears and sharing her wisdom when I face challenges in work or life. I am deeply grateful to this awesome team of coauthors, Enrique, Julie, Jim, and Baolong, for the fruitful discussions along the way and their understanding and patience when things slow down.

I am always grateful to my family for the constant love and support. I appreciate my husband, Dr. Xiaochuan Tang, for his understanding and support of my commitment to writing this book in addition to my regular faculty working loads. He not only took care of extra household chores, but also provided valuable insights and suggestions, given his background of applied physics and his expertise in computational materials science when I was writing this book. My kids, Erika Tang and Ethan Tang, always cheer me up with joy and pride, as well as endless childish yet inspiring questions. I also appreciate the resources and support provided by Colorado State University that allow me to advance my research in the field of advanced manufacturing. I am thankful to my colleagues at CSU, who have supported me in my career path in many ways.

Julie M. Schoenung

I wish to thank my coauthors for their hard work, collegiality, and patience while writing this book together. I've learned a great deal from them, both technically and professionally. We have worked together very well as a team, with the persistent encouragement of each other and especially our team leader, Dr. Enrique J. Lavernia. I also wish to acknowledge the significant contributions of Dr. Xin Wang. Her efforts expanded the scope of the book beyond what was originally planned. I thank Dr. Jim Shackelford for being the first person to ever invite me to contribute a chapter to a book. I thank Kaka Ma, Xin Wang, and more than twenty other PhD and master's students for keeping me inspired to learn new things and for giving me the opportunity to relish in their career growth and success. I thank Baolong Zheng, Yizhang Zhou, Darryl Mack, Alex Dupuy, Amy Ricks, and the MSE staff for constant technical and administrative support to keep my group and the department running while I spent time working on this book. Their loyalty is immeasurable. I thank my faculty colleagues and collaborators for always encouraging me to think outside the box, and to not be afraid to venture into new fields of research. Working at the intersection of disciplines has kept me motivated and has allowed me and my colleagues to have a greater impact and to hopefully move the needle, at least slightly. I wish to thank my family for their love and support throughout my life, my career, and while working on this book. To my husband, Enrique, and to our children, Alejandro and Laura, my most sincere gratitude for always believing in me and supporting me despite my frequent moments of doubt. To my parents, Alois and Rita Schoenung, for believing in the value of education, and investing in their daughters despite limited resources. To my sisters, Chris, Marji, Suzi, and Kathy, for letting me be the baby of the family, yet nurturing my independence, as well as my personal, spiritual, and intellectual growth. A special thank you to Suzi for forging the path into both engineering and graduate studies. Finally, to so many other role models, including the countless unnamed and unrecognized women scientists and engineers who opened the doors of opportunity for the rest of us to follow.

James F. Shackelford

I begin by acknowledging Enrique Lavernia for inviting me to join him and an excellent group of coauthors to cover this important and timely discussion of the rapidly emerging field of additive manufacturing of metals. My family has been more than "patient and understanding" during this and other writing projects. Penelope has been an occasional coauthor as well as family matriarch, master chef and sommelier. Our son Scott and his family, Megumi, Mia, and Toki are a constant source of joy and pride. I am also deeply grateful to colleagues at the University of California, Davis and the University of California, Irvine who continue to encourage and support my ongoing explorations in the endlessly fascinating material world.

Baolong Zheng

I would like to express my deepest gratitude and appreciation to my former PhD advisor, current mentor and boss, Professor Enrique J. Lavernia for his advice, guidance, inspiration, and support in helping me complete my dissertation, papers, and presentations, and various research projects. His erudition and insights, enthusiasm, and optimism provided me with an endless source of knowledge and energy. Thanking him for bringing me to work in these rich scientific content and vibrant fields of gas atomization and additive manufacturing (my dissertation: Synthesis and behavior of metallic glasses via gas atomization and laser deposition). At that time, Prof. Lavernia's group was one of earliest research labs working on directed energy deposition (DED) (laser engineering net shipping [LENS]) of additive manufacturing in the world. Special words of thanks also go to Dr. Yizhang Zhou, Darryl Mack and all the other colleagues. My very special thanks to my wife, Xiaoting, and my twin daughters, Penny and Annie, for their profound love, patience, understanding, and unconditional support throughout my life, studies, and work in the USA.

Part I

Atomization of Metallic Powder

1

Overview of Atomization Techniques

1.1 History of Metallic Powder and Atomization Techniques

1.1.1 Metal Powders

The high surface-to-volume ratio (i.e. volume per surface area) of a sphere renders it an ideal geometry for handling, pressing, forming, and/or reacting particles into a bulk form. It is hence not surprising that **metal powders**, in their most elementary form, emerged as useful "engineering" materials, even in ancient times. **Metal powders** can be produced using either chemical/**mechanical methods** or **fluid atomization**. Mechanical methods involve the physical breakdown of a large particle into a smaller one, whereas **chemical methods** typically involve a solid-state process in which a metal oxide is reduced into a metal. For example, the reduction of iron, also called sponge iron, occurs when iron ore is exposed to a reducing gas. Powders produced using chemical reduction methods are not widely used in **additive manufacturing (AM)** and hence are not covered in this chapter. There is some preliminary research on the use of powders produced by mechanical methods in AM, and these will be discussed in a subsequent chapter.

Powder metallurgy, as the practice of working with metal powders is widely known, has a rich history, from the Incas using powder techniques to work precious metals to ancient Egypt, Africa, and India (see Table 1.1). In fact, the use of elevated temperature metals can be traced back about 7 000 years, long before furnaces capable of reaching the temperature required for melting them. Hence, it has been suggested that objects made of iron about 5 000 years ago were likely fabricated using powder metals. The first application of a powder metal product was a bronze ball bearing, capable of self-lubrication and used for automotive applications in 1927 [1].

The Oxford dictionary defines the word ***atomize*** as to reduce something into atoms or into very small pieces. Clearly, based on this definition, one can hypothesize that the term atomization has been loosely applied to the disintegration of any bulk material into smaller components. One can envision the disintegration of water as it emerges from a nozzle or waterfall into droplets or similarly the disintegration of rock by mechanical means into very small pieces. Fluid atomization, however, is generally described as a process where a liquid metal is disrupted by a high-velocity fluid such as air, nitrogen, argon, helium, or water in some cases. The actual process of atomization occurs when there is a transfer of **kinetic energy** from the atomizing medium to the metal being disintegrated. There are different forces in action during atomization that depend on the fluid being used for the energy transfer. In the case of water, for example, it is typically the pressure of the medium that dictates the efficiency of the atomization process and thereby the resultant distribution of droplets that emerges during disintegration. In the case of **gas atomization (GA)**, however, it is typically found that the gas-to-metal ratio dominates the resulting distribution of droplets, and various mathematical relationships have been developed to capture this relationship. In the case of gases, for example, increases in pressure that exceed 0.1 MPa (this is the pressure at which a **sonic velocity** is reached) result in only very small increments in gas velocity. In comparison, to reach sonic velocity with a water jet, for example, a pressure of nearly 40 MPa is needed in the case of water, and velocity continues to increase as the square root of the pressure.

There are various methods that can be effectively used to produce metal powders from valuable metals and alloys such as **nickel**, **cobalt**, copper, titanium, aluminum, and stainless steel base alloys. However, GA is generally viewed as the method of choice due to its capacity for large production quantities, the ability to control chemistry, and the geometrical properties of the powder particles that can be produced. These characteristics render atomization as the preferred technique to manufacture powders for AM, although, as we will discuss in subsequent chapters, alternative methods for powder preparation are being explored.

Metallic Powders for Additive Manufacturing: Science and Applications, First Edition.
Enrique J. Lavernia, Kaka Ma, Julie M. Schoenung, James F. Shackelford, and Baolong Zheng.

Table 1.1 Major Historical Developments in Powder Metallurgy.

Date	Development	Origin
3000 BCE	"Sponge iron" for making tools	Egypt, Africa, India
1200 CE	Cementing platinum grains	South America (Incas)
1781	Fusible platinum-arsenic alloy	France, Germany
1790	Production of platinum-arsenic chemical vessels commercially	France
1822	Platinum powder formed into a solid ingot	France
1826	High-temperature sintering of platinum powder compacts on a commercial basis	Russia
1829	Wollaston method of producing compact platinum from platinum sponge (basis for the modern PM technique)	England
1830	Sintering compacts of various metals	Europe
1859	Platinum fusion process	
1870	Patent for bearing materials made from metal powders (the forerunner of self-lubricating bearings)	United States
1878–1900	Incandescent lamp filaments	United States
1915–1930	Cemented carbides	Germany
Early 1900s	Composite metals	United States
	Porous metals and metallic filters	United States
1920s	Self-lubricating bearings (used commercially)	United States
1940s	Iron powder technology	Europe
1950s and 1960s	Powder metallurgy wrought and dispersion-strengthened products, including powder forgings	United States
1970s	**Hot isostatic pressing**, PM tool steels, and superplastic superalloys	United States
1980s	Rapid solidification, powder injection molding technology, and binder-treated ferrous premixes	United States and Europe
1990s	Intermetallics, metal–matrix composites, spray forming, nanoscale powders, water-atomized pre-alloyed ferrous powders with **molybdenum** as the principal alloying element, and warm compaction	United States and United Kingdom
2000s	Warm-die compaction, additive manufacturing (3D printing) on a commercial basis	United States and Europe

Source: Samal and Newkirk [1] / ASM International.

1.1.2 Atomizer Designs

A critical aspect of atomization technology involves **atomizer design**. The atomizer is used to disintegrate the molten materials into a dispersion of droplets, which eventually solidify into powders, and therefore different designs result in various distributions of droplet (and hence powder) sizes. Since size distribution is an important variable that governs the solidification behavior of droplets, atomizer design directly influences the resultant microstructure of the powder, and hence the final properties of the consolidated material. Historically, a variety of atomizer designs have been developed to various degrees of success. There is, however, no general agreement among the scientific community as to the optimal atomizer design; more likely, the design used depends on the desired powder characteristics and the ultimate intended application of the powder. Moreover, and as will be addressed in a later chapter, optimal processing parameters and atomizer design are often material-specific (these can change even for slight compositional variations).

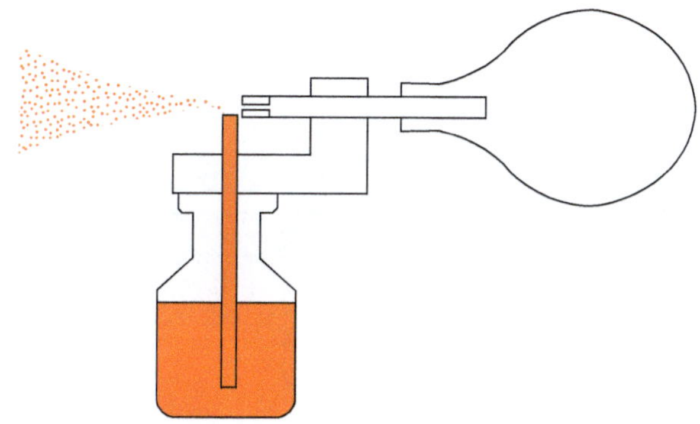

Figure 1.1 Schematic diagram showing design of first "atomizer". Source: De Vilbiss [2] / United States Patent and Trademark office / Public Domain.

The design of an atomizer is based on airflow dynamics; as horizontal air travels over a vertical tube, it creates a suction effect and aspirates the air and liquid in the tube. In essence, an atomizer uses Bernoulli's principle to disintegrate a liquid into a mist of fine droplets. The first atomizer nozzle is described as an **aspirator nozzle** and is credited to Dr. Thomas DeVilbliss of Toledo, Ohio, in the late nineteenth century. He invented this device to spray medicine in the back of patients' throats [2, 3], as shown in Figure 1.1. Not long after Dr. DeVilbliss used the device shown schematically in Figure 1.1 for medical purposes, the first **perfume atomizers** emerged in 1907, and a new industry was born.

In the case of metals, the first atomizer design is credited to Hall, who used it to atomize aluminum powders in the late 1920s [4]. *In the design proposed by Hall, compressed air is used as the atomization media,* and it is delivered via an **annular orifice** at a predetermined **convergent angle** (see Figure 1.2). As the gas flows through the **annular orifice**, it creates an aspiration effect that suctions the molten metal into the nozzle cavity. The amount of gas can be changed by changing the size of the air gap or the temperature and pressure of the gas.

The **Hall nozzle** design falls into a category of nozzle designs known as confined nozzles or close-coupled nozzles. This nozzle design involves closely coupling the liquid metal stream with the gas jets to maximize the transfer of energy from the atomization medium (gas) to the metal being disintegrated. A typical **closed-type atomizer** is illustrated schematically in Figure 1.3 [5]. It normally contains 10–18 gas jets symmetrically arranged around a cavity designed to contain the metal delivery tube. Typically, the gas nozzles are positioned at 1.0–1.5 diameters apart, depending on the desired powder characteristics. The geometry of the gas nozzles is designed to provide maximum

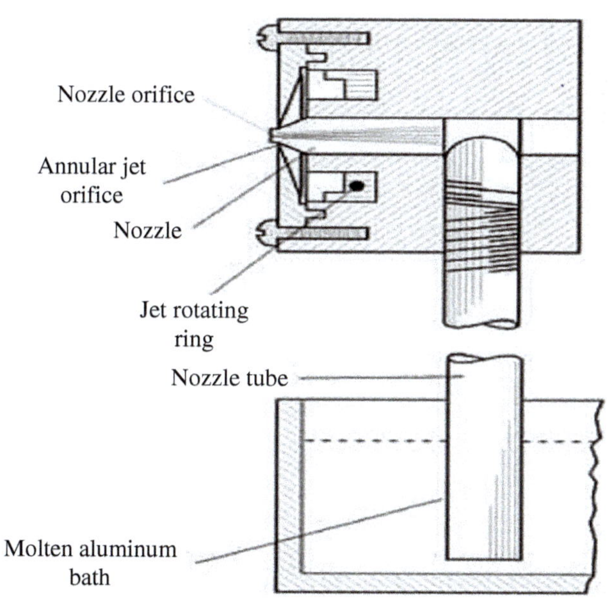

Figure 1.2 Aspirating nozzle design developed by E.J. Hall in 1920s. Source: Hall [4] / United States Patent and Trademark office / Public Domain.

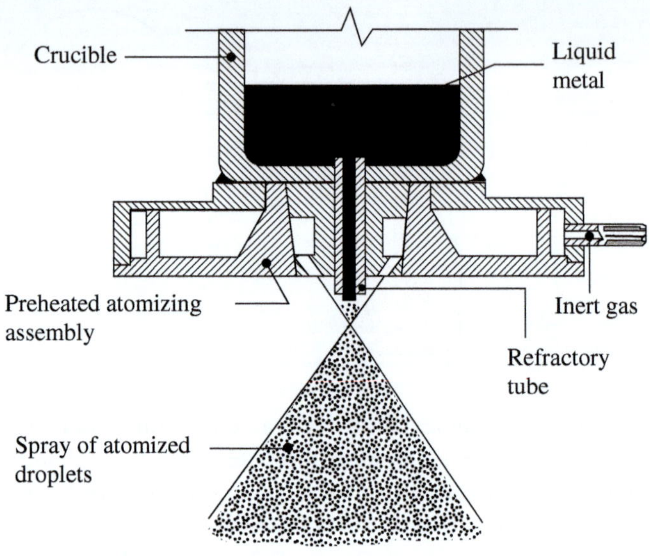

Figure 1.3 Closed-type atomizer used for spray deposition processing. Source: Singer [5].

kinetic energy to the atomization gas, thereby leading to an "efficient" atomization. As shown in Figure 1.3, the primary advantage of a closed-type atomizer is that the molten metal is delivered directly to the atomization region, thereby avoiding instability of the metal stream and producing a stable atomized droplet spray. In confined designs, the molten metal travels a very short distance before being impinged upon by the high-energy atomization gas jets. The major problem that is associated with this type of atomizer is that it is very sensitive to metal freeze-up, a condition that occurs when the liquid metal at the end of the delivery tube solidifies prematurely because of the rapid heat extraction that occurs in this region. Accordingly, careful control of atomization parameters must be exercised when this type of atomizer is utilized [6]. During atomization with confined nozzles, it is frequently reported that the metal travels on the surface of the metal delivery tube, such that when it interacts with the atomization gas, it is effectively a thin film. This filming effect is beneficial for the efficient breakup of the liquid, and its formation is related to the gas flow dynamics, as illustrated in Figure 1.4.

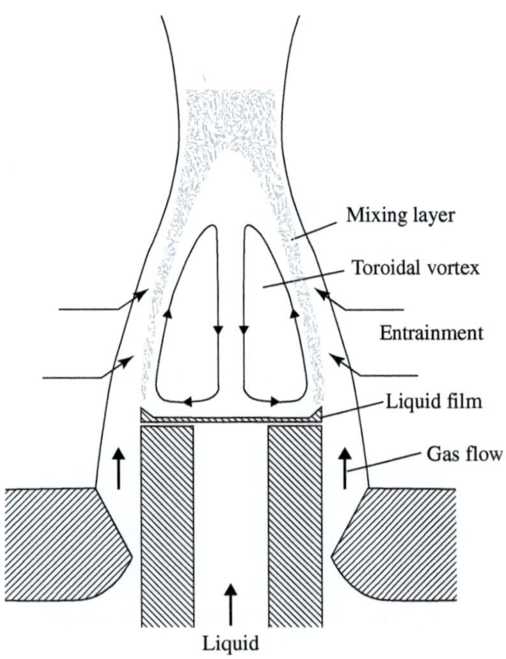

Figure 1.4 **Fluid flow pattern** at the tip of a confined nozzle during atomization. Source: Unal [7] / Taylor & Francis.

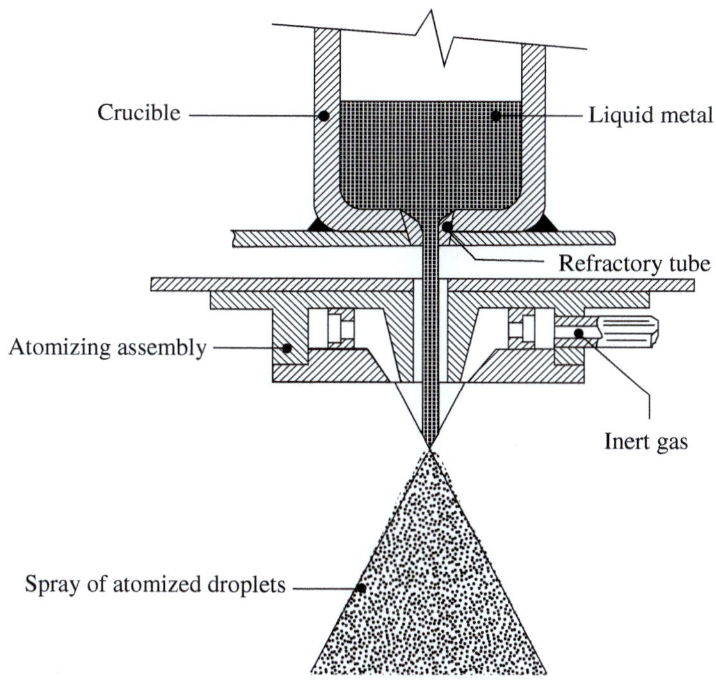

Figure 1.5 Open-type atomizer used for spray deposition processing. Source: Singer [5].

The toroidal recirculation vortexes that form in confined nozzles originate from fluid entrainment that occurs on the inner boundaries of an annular jet [7, 8]. As a jet exits an annular cavity, there is a mixing zone that forms as the gas jet "shears" the molten metal; there, the jet pulls fluid from the mainstream jet. An important consequence of this vortex is the formation of a low-pressure zone at the tip of the nozzle; this can aspirate metal during atomization, thereby leading to changes in the gas-to-metal flow ratio.

In view of the difficulties associated with the **metal freeze-up phenomenon**, many researchers use an open-type atomizer [5, 9, 10]. A schematic diagram of such an atomizer is illustrated in Figure 1.5. An open-type atomizer is also referred to in the literature as having a free-fall design. In the open-type atomizer, liquid metal is allowed to fall freely to the atomization point (i.e. the focal point of the atomization gas jets). Since the metal delivery nozzle tube is not in direct contact with the atomization gas jets, nozzle freeze-up is avoided. The principal problem that is associated with open-type atomizers, however, is the instability of the molten metal stream, which leads to oscillation and asymmetry of the atomization cone. This instability occurs because the free-falling molten metal stream is easily deflected by the turbulence in the surrounding gas envelope.

In the 1980s, researchers used a so-called **ultrasonic atomizer** [11, 12] for GA. The **USGA nozzle** makes use of the theory that a high-frequency (>100 Hz) pulsed gas stream with high-amplitude (primary and secondary) **shock waves** will promote the efficient atomization of molten metals [13, 14] (see Figure 1.6). Accordingly, a narrow distribution of droplets may be produced with relatively small average droplet size. The design of an ultrasonic GA

Figure 1.6 Schematic of USGA nozzle design. Source: Gerking [15].

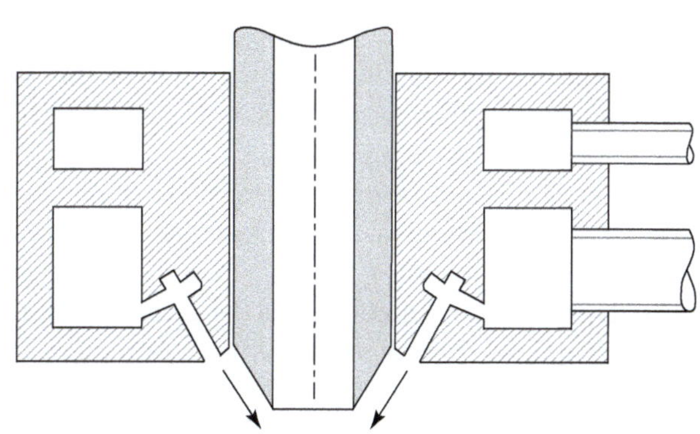

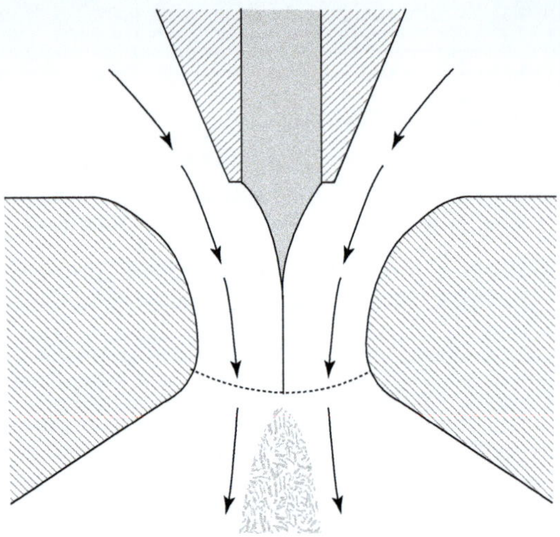

Figure 1.7 Schematic of Nanoval nozzle design. Source: Miller et al. [20].

nozzle incorporates a **Hartmann resonant chamber** to generate high-frequency and high-velocity pulsed gas jets [13]. When operating under proper atomization conditions, the USGA atomizer can generate gas jets of supersonic velocity superimposed with ultrasonic pulses [14]. Hence, it has been argued, albeit not without controversy, that the atomization efficiency that may be achieved with this atomizer design exceeds that of conventional atomizers, thereby allowing close control over the **droplet** size distribution [14, 16, 17].

Other atomizer designs include the National Institute of Science and Technology (NIST) design, which is like the USGA nozzle but does not include the Hartmann cavities [18, 19], and the **Nanoval design** (see Figure 1.7), in which molten metal is delivered directly to the throat of a converging–diverging nozzle in the form of a thin stream [15].

1.2 Melt Atomization

Melt atomization is perhaps the most promising and economic technique for the fabrication of fine and spherical metallic powders used for conventional powder metallurgy and metal AM. The mechanism for melt atomization involves the disintegration of a melt film or a stream into droplets by the application of energy with high-velocity fluids, such as gases (air, **nitrogen**, **argon**, and helium), water, and synthetic oils, and even combinations of two fluids have been used to fabricate fiber powders. Inert gas atomization (IGA), also referred to as GA, has become extremely popular and is increasingly being used to produce fine, high-purity spherical powders in large quantities for the **stock powder** used in AM.

There are other atomization techniques that are not based on the principle of melt disintegration by impingement with a second fluid, such as the use of **centrifugal forces** in **centrifugal shot casting**, **vibrational energy**, **electrical fields**, **melt extraction**, **rotating consumable electrode**, plasma micro-atomization, **melt saturation** and **explosion**, and capillary and **standing wave atomization** [21]. These processes are generally being used to produce special powders for refractory or reactive metals used in advanced materials.

The primary advantage of the melt atomization process lies in its flexibility in terms of producing the desired alloy chemistry and particle size. The process can produce small quantities of experimental powders and can also be scaled up to produce large quantities of powder for commercial applications. The average particle size (ranging 5–250 μm) and cooling rates (ranging 10^2–10^6 K s^{-1}) can be controlled depending on the processing parameters for tailoring the microstructure of produced powder.

Atomization is suitable for producing powders of metals and alloys where the melting point of the material is not too high. High-temperature materials like **tungsten**, **hafnium**, **molybdenum**, and **ceramic** are not usually produced by melt atomization, except by specialized processes such as **plasma atomization**.

Contamination may occur due to the crucible (usually a graphite or ceramic crucible) in which the atomization melt is contained. This problem becomes extremely acute for processing high-temperature and high-performance materials, such as Ti alloys. One way to avoid that problem is to produce a thin "skull" of the same material that will be atomized within the crucible. The "skull" can effectively insulate the molten materials from the crucible and thus reduce contamination to a great extent. Since the crucibles, in which the "skull" is formed, need to be cooled, they are made of highly thermally conductive metals, like **copper**. Another technique is to have no crucible at all and use the materials to be atomized as an electrode that is gradually melted and atomized. In addition, another alternative is to use ceramic filters to filter out large ceramic particles from the melt before atomization.

The basic concept in the common atomization processes is to deliver energy to the liquid melt via the impinging fluid. Variations in the fluid focusing system and the type of fluid used determine the final powder characteristics. Subsequent sections of this chapter will briefly describe various melt atomization processes that are presently in use and some of the new processes that are being developed to produce finer and spherical powders. The requirements of the powder used for AM are also providing a large impetus to the melt atomization processes to provide fine and spherical powders that are suitable for AM process.

1.3 Gas Atomization (GA)

Melt atomization is the dominant method used commercially to produce metal and alloy powders from Al, Cu, Fe, Ti, Ni, and other alloys, because high production rates favor economies of scale. Atomization, as the name implies, involves the energetic disintegration of a liquid into micrometer-sized droplets, as shown in Figure 1.1. The principle of melt atomization is ancient and formed the basis for a British patent in 1872 to produce **lead powders** by drawing off and spraying molten lead using a stream injector [21]. Since the first large-scale production of atomized iron powder during World War II, melt atomization technology has been steadily implemented and improved, partly due to the widespread application of pre-alloyed powders.

Air atomization is where both atomization and melting are carried out in the air. Compressed air jets are used to break up the molten metal, and often cooling is achieved by sucking large volumes of cooling air through the equipment. This method has been widely used since ~1900 to make tin, lead, **aluminum**, and zinc powders, but also for some copper and copper alloys. It was used for processing Fe–C alloys in the middle of the twentieth century but is now obsolete. As would be expected, oxygen contents of air-atomized powders are relatively high, typically in the range of 1000 ppm to 1%. Particle shapes range from irregular for Al and Zn to spheroidal for copper, gold, and silver. Particle sizes for Ni and Fe alloys are ~50–200 μm and for Sn and Pb are ~10–30 μm.

IGA is where the melt is broken up by inert gas, normally nitrogen or **argon**, and thus protected from oxidation. This was developed in the post-World War II period with melting in air and used for a very wide range of alloys, ranging from Sn and Pb to Au, Ag, Cu, Co, Ni, and Fe alloys. The major fields of application for this technique have been powders for **thermal spraying**, **hard facing**, **brazing**, **dentistry**, and more recently, **hot isostatic pressing (HIP)**, **metal injection molding (MIM)**, and AM. Oxygen contents can be far lower than in air-atomized powders, typically in the range of 100–500 ppm, but highly alloy-dependent. Particle morphologies are typically spherical, but agglomeration and satellite formation can occur. Median particle sizes are typically 30–200 μm for Fe, Ni, and Co alloys with the upper limit dictated by the large vessel sizes needed to avoid splatting of still-molten larger droplets on the walls.

As a type of melt atomization, GA is the process where the molten metal is disrupted by a high-velocity gas. Subsonic, supersonic, and ultrasonic GA utilize different gas velocity jets, as implied by the names used to describe the techniques. During GA, a high-pressure atomization gas, such as air, nitrogen, argon, or helium, is discharged from the pressure reservoir of the atomizer into a chamber, which is typically maintained at a low environmental pressure. The compressible atomization gas is accelerated to a high velocity as it expands from the high reservoir pressure into a low-pressure chamber. The energetic impingement of a high-velocity gas jet causes the molten metal steam to deform and disintegrate into fine droplets. The **gas-to-metal ratio** is an important factor that governs particle size for GA, rather than being dominated by the pressure of the medium like in the case of water atomization.

GA involves the disintegration of an unconfined stream of liquid metal into fine particles by a focused jet or jets of atomizing gas. The design of the nozzle and the liquid-metal feed mechanism often varies. The gas jet can be directed through several discrete openings or through an **annular-ring** type of opening. There are two types of atomization nozzle design – "free-fall" and "confined." A "free-fall" configuration is often used in the conventional GA technique. In this technique, the gas jet(s) strikes the metal stream at a point that is some distance away from the point where the liquid metal leaves the crucible (i.e. flow becomes unconfined). Problems arise from the stream instability when very high gas velocities and sharp jet angles are used to try to produce finer particles. This conventional type of GA can produce powders with a mean particle size usually greater than 50 μm.

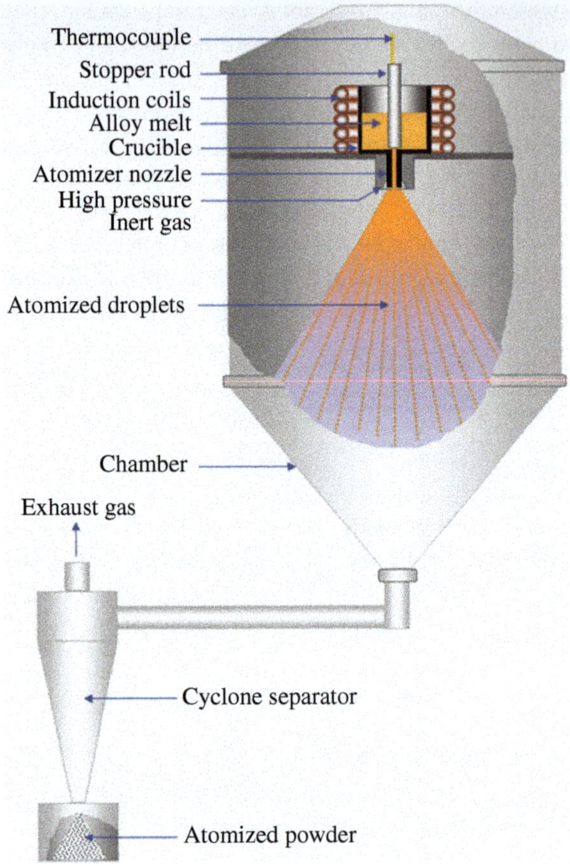

Figure 1.8 Schematic diagram of gas atomization structure and processing. Source: Adapted from Zheng et al. [22].

The liquid stream is disintegrated in a horizontal or vertical direction during GA processing. Most GA processes rely on the vertical direction of liquid metal flow. The horizontal gas atomizers can normally be used for low-temperature metals and alloys. The molten metal is siphoned up through a feed tube and disintegrated by high-velocity gas. Vertical GA is normally used for high-temperature materials, with inert gases being used as the atomization fluid. A schematic of the conventional vertical type of gas atomizer is shown in Figure 1.8.

The chamber size should be large enough so that the particles produced can have free flight through the chamber and solidify prior to reaching the collection chamber wall. In this process, the goal is to generate fine particle sizes (around the 10 μm range). Finer particles can be generated by higher gas flow rates, higher melt temperatures, and steep angles between the gas jet and the melt stream. The **thermal conductivity** of the atomizing gas, the metal surface energy, and the atomizer nozzle design all play important roles in determining the final powder characteristics.

The commonly accepted mechanism for conventional GA involves the formation of a stable sheet that later becomes wavy in nature, although as will be discussed in the next chapter, many questions remain. Cylindrical ligaments are torn off from the sheet, and these in turn break up into number of small, elongated droplets that then become spherical. In fact, the process of GA is very complex and far from being completely understood. Nevertheless, it is well known that the powder particle size distribution depends on a variety of different parameters. One set of parameters is related to the physical properties of the melt, e.g. density and temperature, which have an influence on viscosity and surface tension. Another set of parameters is related to the gas flow field, which depends, e.g. on the geometry of the gas nozzle, the type of atomization gas, and the gas pressure.

To achieve smaller average particle sizes and rapid solidification (RS) rates, close-coupled (also known as confined or direct jet) GA is the preferred route [23]. In this case, the gas jet impinges on the stream of molten metal at the point where it emerges from the melt guide tube, atomization nozzle. This results in the most efficient transfer of the gas kinetic energy to the liquid melt, which is the key to the formation of fine powders. Little of the gas energy is lost in turbulence as in the case of an open-die or **free-fall arrangement**. Also, a higher angle of impingement is possible, thus improving the fine particle yield. The major challenges are due to the proximity of the gas jet to the molten metal guide tube.

High-pressure gas atomization (HPGA) was a natural evolution of the normal GA process. The necessity to develop finer powders and produce fast cooling rates resulted in the development of the HPGA process. It was reported that using 573 K (300 °C) superheat, 4.5 MPa argon atomizing pressure, and a 2.5-mm-diameter pour tube was optimum for atomizing copper [24]. HPGA of copper, amalgam, and a lead–tin alloy exhibited mean particle sizes of 25, 20, and 15 μm, respectively. From the point of view of the large demand for finer powders, this is also a very important development.

With increasing demand for finer powders, the powder producers are adopting new techniques that could yield a higher percentage of finer powders. Research and development effort in this area has therefore increased tremendously. The need for producing fine particle size with a narrow size distribution has resulted in the development of ultrasonic GA. This form of atomization uses high-frequency, supersonic gas jets as the atomizing medium. The production of the ultrasonic waves is accomplished by accelerating high-pressure gas through a resonance cavity. The passage of the highly pressurized gas such as nitrogen, argon, or **helium** through the resonance cavities results in the attainment of ultrasonic frequencies of 80–100 kHz and supersonic gas velocities often reaching Mach 1.7 – 2.5 [25, 26]. A typical ultrasonic gas atomizer uses around 16–20 resonating cavities that direct the jets to a circle of small radius surrounding the liquid metal stream. Various investigators have used the ultrasonic GA process to produce rapidly solidified powders of different metals and alloys.

Given the high velocity of the gas pulse, the mechanism for this form of atomization does not follow the usual ligament formation mechanism that has been proposed for conventional GA. Instead, the liquid metal stream behaves like an impacted low-shear-strength solid. Despite some progress, it is apparent that more research is needed to complete our understanding of ultrasonic GA.

Among the methods of powder production, the proportion of powder produced by GA has accounted for 80% of the total. The GA method can be divided into the following types according to the different heating elements of the equipment: **vacuum induction melting inert gas atomization (VIGA), plasma melting induction gas atomization (PIGA)**, **plasma torch atomization (PA)**, and **electrode induction melting gas atomization (EIMGA**/EIGA).

In the VIGA process, the melting materials may easily be contaminated by the crucible, which is particularly challenging for high-purity reactive metals such as Ti. EIGA can avoid the contamination from the crucible by using ceramic-free electrode induction, but it is limited to small-size feedstocks and low-melt flow rates. PIGA can be used to prepare Ti and Ti alloys. However, considerable electric powder is required in this technique; thus, complex safety precautions are required due to much more liquid volume involved as compared to EIGA. The plasma rotational electrode process (PREP) technique can also be used to prepare the spherical Ti powder, but the powders generally have a wide PSD [26]. Generally, the above methods can be used to prepare Ti powders with good sphericity. However, it remains challenging for these methods to obtain fine Ti powders with a narrow particle size distribution (PSD) in a single batch.

1.4 Vacuum Induction Gas Atomization (VIGA)

Vacuum inert gas atomization (VIGA) involves the melting and pouring of an alloy prior to atomization inside of a vacuum chamber to allow the production of the most oxidation-sensitive and reactive alloys, such as Fe-, Ni-, and Co-based alloys containing Al, Ti, and rare earths. This includes superalloys, such as **IN718**, **maraging steels**, and **M-Cr-Al-Y alloys**. This technique was developed during the 1950s and 1960s when there was a push to explore the potential benefits of RS to allow the production of more highly alloyed superalloys for aerospace and defense applications [27]. This proved to be a very challenging field of application, but, after several decades of development, there are now many thousands of tons per year of VIGA-produced superalloy powders. This intensive development has meant that the technology lends itself well to producing powders for HIP, MIM, and AM. Oxygen contents in the 50–200 ppm range are achievable. Particle morphology is generally spherical, and the size distribution is comparable to that obtained by IGA.

The VIGA atomization system generally consists of three production stages: (i) melting in vacuum induction melting furnace; (ii) transfer of the molten metal or alloy to a crucible; and (iii) powder atomization by pressurized gas resulting in a fine dispersion of powders. In the VIGA system, a vacuum induction melting unit is integrated with an IGA unit. The starting raw materials are heated and melted by frequency induction furnaces using electromagnetic induction, which couples electrical power into the crucible/material under vacuum atmosphere. This design

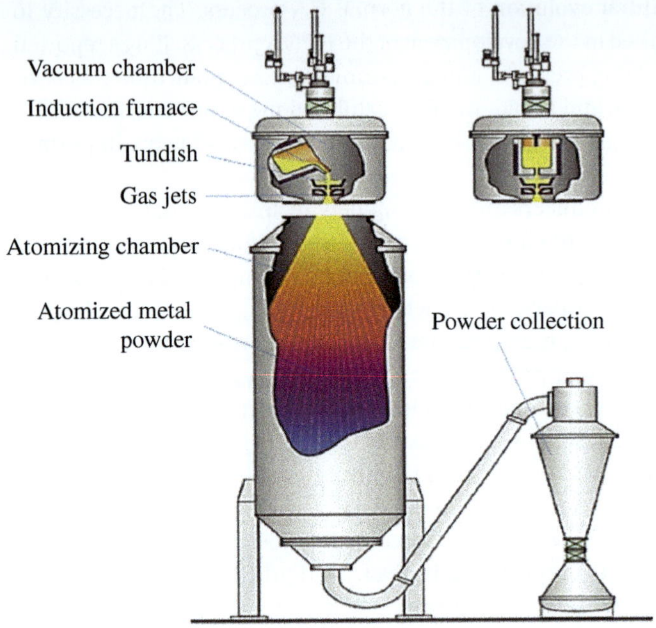

Figure 1.9 Schematic diagram of vacuum induction gas atomization (VIGA) processing. Source: ALD Vacuum Technologies AG [28].

allows the metal alloys to attain the lowest oxygen content possible. Once the desired melt homogeneity and chemical composition have been achieved, the material is poured into a tundish by crucible tilting, as shown in Figure 1.9. The equipment directs the liquid metal to focus the gas jet. The fine metal stream, flowing from the tundish orifice into the atomization nozzle system, is subjected to a high-pressure, inert gas jet and then atomized. The combination of molten metal and gas jet impacts, atomizes, breaks up molten metal into small droplets, and creates a spray of micron-sized droplets that solidify in the atomization chamber located directly underneath the atomization nozzle, and the droplets eventually solidify into spherical morphology. The VIGA process provides excellent control over alloy powder composition and oxygen content.

Two main atomization stages, including primary atomization and secondary atomization, are generally involved in vacuum induction melting GA. The primary atomization refers to the cylindrical flow of high-temperature molten alloy flowing out of the delivery tube, which interacts with the high-speed gas flow field, disperses, forms a liquid film, and then breaks into an initial droplet dispersion. Secondary atomization, also known as the second crushing, refers to the process that occurs as the droplets that are initially formed are entrained by the airflow into the turbulent layer, that is, the process of secondary crushing and additional crushing into smaller droplets.

Due to the limitations of the hardware equipment and crucible, the heating temperature of vacuum induction melting is limited to the range of 1500–1600 °C. And due to the influence of the ceramic crucible and the guide nozzle, impurities will end up in the alloy melt, which will affect the purity of the prepared metal powder. Powders produced using VIGA typically fall in the 0–150 μm size range, and the gases that are used in the nozzle system are N_2, Ar, or He.

VIGA is one of the leading powder-making processes for the production of large volumes of a variety of high-performance spherical, high-grade metal powders, and essential for high-quality manufacturing of Ni-based superalloys, Fe-, Al-, Co-, Cr-based and other special alloy powders, as well as high-purity metal powders with low oxygen and low or adjusted nitrogen concentrations for use in AM, laser cladding, thermal spray, powder metallurgy, HIP, and other advanced manufacturing fields.

1.5 Electrode Induction Melting Gas Atomization (EIMGA)

High-quality metallic powders are in high demand to produce advanced materials that are required by many industries, including the automotive, medical, and aerospace industries, for example. In some cases, performance requirements include behavior at elevated temperatures, and reactive and refractory metals, intermetallic compounds, and superalloys are examples of such high-temperature, high-performance materials that demand super-clean, low-oxygen-containing, rapidly solidified powders. IGA is well established as the most efficient and cost-effective approach for meeting the stringent requirements of these powders and producing them in the required production quantities. There are, however, some challenges associated with the production of atomized powders, including: ceramic impurity pickup due to the ceramic-lined crucible in which the material is melted prior to IGA and powder yield for a particular size requirement. Ceramic inclusions in these high-performance materials are a major cause of their failure during operating conditions and hence represent an important challenge. The production of reactive, high-melting-temperature metal powders requires a contactless technique avoiding direct contact with refractory crucible walls. The industry has adopted various ingenious approaches to address these challenges. A straightforward approach to address the ceramic inclusion problem is to eradicate the cause of the problem, which is the ceramic crucible. This approach leads to the concept of ceramic-crucible-free or even crucible-free melting coupled with IGA. Direct melting of an electrode by induction heating is a contactless technique suitable for **reactive high melting temperature alloys**, such as Ti, Zr, and Nb. The technique has found application in the production of Ti alloy powders by GA and is widely described as EIMGA [29, 30]. The EIMGA involves the advantageous features of induction melting and atomization of the master alloy without the use of melting crucible.

The EIMGA process can be described as essentially an induction drip-melting atomization process that utilizes a rotating electrode that gradually moves down through a specifically designed induction melting coil. The EIMGA process is a versatile method for converting a wide variety of metals and special alloy electrodes into metal powders. Notably, all metals and alloys with melting points up to 2500 °C can be converted into powders with the EIMGA process. Electrodes with up to 150 mm diameter and up to 1000 mm length with charge weights from 5 to 2000 kg can be used as feedstock.

The EIMGA technique requires a pre-alloyed rod. The desired material is first formed into an electrode and does not require a ceramic crucible. The lower end of the rod is dropped into a conical induction coil, and continuous melting occurs by melting from the surface to the center of the electrode. The end of the rod is heated up, and the melt drops into the center of a gas nozzle, where it is atomized by Ar gas. An electric drive controls the movement of the consuming electrode into the induction coil. The melt flow rate can be precisely controlled by the induction power applied and the downward velocity of the rod. The melt flow rates achievable for y-TiAl alloys can reach to around 50 $kg\,h^{-1}$ [31]. To guarantee a symmetrical melting of the tip, the rod rotates at a low frequency of 5 rpm. In contrast to rotating electrode processes (REP) [32, 33], where a high rotational speed is used, for EIMGA the rods need not to be balanced.

The electrode bottom is immersed in an induction coil, where it melts and produces dripping or streaming liquid metal, as shown in Figure 1.10a,b. The dynamics of the melting process are complex, involving the dynamic adjustment of the electromagnetic field due to propagation of the melt front and motion relative to the coil. The melt flow is controlled by the melting rate, the gravity force in the liquid film, and the electromagnetic forces. The high-frequency coil current (100–200 kHz) attains the desired energy penetration on the surface of the electrode and creates the induced current layer in the electrode, which is spatially nonuniform along the surface. The arrangement of the induction coil is also of great importance. The special arrangement provides the desired superheat to the melt, which flows down the front-end section and is fed directly into the atomizing nozzle. The current flows in a thin skin and concentrates on the inner surfaces of the coil due to confinement of magnetic flux. The liquid film is confined by an electromagnetic "pinching" force, but the nonuniformity of this force along the surface creates a rather strong electromagnetically driven flow. This type of flow can, in some cases, oppose the dripping of the film flow due to gravity and lead to an interruption of the melt stream. A quasi-steady state is desirable, avoiding unstable dripping and maintaining a conically symmetrical shape at the electrode bottom. To stabilize the circumferential uniformity, the ingot is rotated at low angular velocities of 5–12 turns per minute. This means that angular momentum transport in the conical melt film flow must be included in the model.

The EIMGA system design is generally based on a vertically oriented, rotating electrode that is continuously melted at one end in an inert gas atmosphere without a ceramic crucible using controlled motion into a specially designed conical induction coil. The design of the induction coil and the frequency of the generator are optimized to achieve the necessary heat, causing the melt to drop from one end of the electrode. With proper controls, a continuous stream of molten material falls directly into the center of an annular slit. Then, the molten metal is transformed with an inert gas nozzle using a high-velocity gas stream to **atomize** the melt, as shown in Figure 1.10c. The atomization is carried

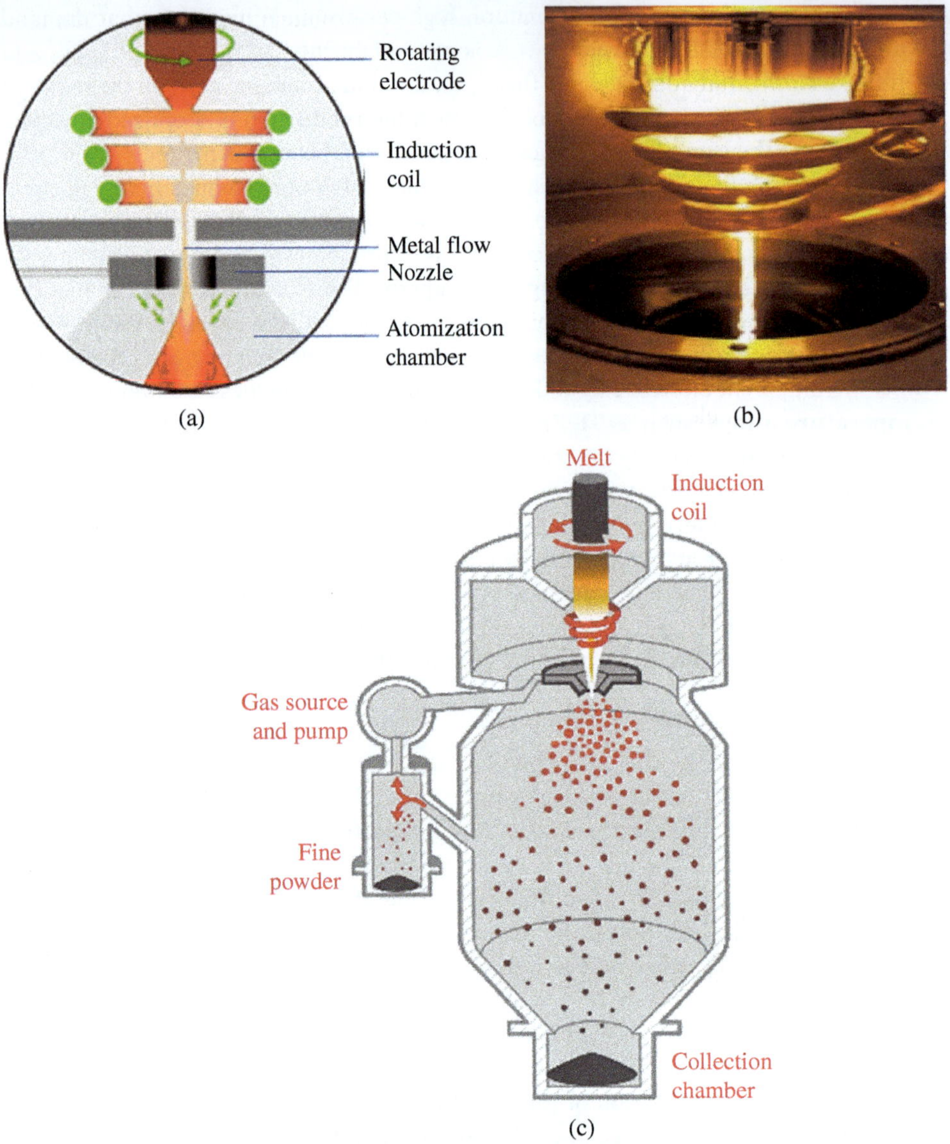

Figure 1.10 (a) Schematic of the electrode induction melting gas atomization (EIMGA) process for the production of metal powders. Source: LPW Technology, UK [34]. (b) Picture of melt liquid in an EIMGA process before atomization. Source: Ref. [35] / Arcast. (c) Schematic diagram of the EIMGA. Source: Lalwani [36].

out by a high-velocity inert gas stream. The micro-droplets spray solidifies while traveling down the atomization tower and forms spherically shaped fine powders which are collected in a vacuum-tight powder container.

Atomization of the molten metal is realized by the spraying action, resulting from injection of inert gas. Unlike other technologies, the EIMGA processing route is devoid of the use of ceramic crucible during melting of alloy, in addition to other advantages such as lower power consumption and simpler equipment setup. The superalloy powders prepared using such a crucible-free technology can prevent the introduction of Al_2O_3, SiO_2, and other ceramic impurities into the powder preparation process. It is also expected to reduce the quantity of inclusions in the powders, thereby improving the high-temperature performances of subsequently formed powder metallurgy superalloys. A further advantage of this technique is the small electric power required, since the liquid volumes are very small. This is also the reason why this technique is considered as very safe. A disadvantage is seen in the use of pre-alloyed rods.

The disadvantage associated with this process is the need to prefabricate electrodes from the desired materials. The cost of alloy prefabricated bars is higher than that of alloy ingots, and the uniformity of the bar alloy has a greater impact on the chemical composition of the powder after atomization. Segregation in the length of the rod can result in chemically heterogeneous alloy powder. It is difficult to control the melting rate of metal bars. Be sure that it is easy to cause the flow of liquid to be cut off, and the "broken rod" will block the draft tube.

Powders produced with EIMGA systems typically have the following characteristics: spherical shape, good rheological flow characteristics, particle sizes ranging from 40 μm ≤ d50 ≤ 100 μm, high purity, high flow ability, low O, N, and N concentrations, and rapidly solidified and homogeneous microstructure. The ceramic-free material processing and the outstanding process simplicity and robustness are the inherent advantages of the EIMGA process. The application-specific combination of melting and atomization techniques can enable high-quality manufacturing of a wide range of super-clean metal powders, including superalloys, precious metals, refractory and reactive metal powders, including Ti and Ti-alloys (e.g. CP-Ti, $TiAl_6V_4$, TiAl), Zr-, Nb-, Ta-base alloys, precious metals and their alloys, ultrapure reactive copper and aluminum alloys, and other specialty metals and alloys. Powders produced by the EIMGA process can be used in numerous applications. The main materials and application areas of powders produced with EIMGA systems are as follows: feedstock for AM technologies, aviation industry, power engineering, sputter targets, feedstock for MIM and HIP, electronics and chemical industries, biomedical engineering, and precious metal industry.

1.6 Plasma Rotating Electrode Process (PREP)

In this class of atomization technique, **centrifugal atomization** is perhaps the most notable. In centrifugal atomization, the breakup of the liquid occurs by thinning the metal into a film using centrifugal forces. This stands in contrast to conventional methods, where disintegration of a molten stream occurs via direct gas impingement. Centrifugal atomization, also sometimes described as **rotary atomization**, involves the delivery of metal onto a rapidly spinning disk that rotates at very high velocities. This technique has been widely used to prepare pre-alloyed powders of aluminum, zirconium, **titanium**, and iron, and various superalloys have been produced on a commercial scale.

PREP is also a centrifugal atomization process where a consumable metal bar is melted while being rotated rapidly about its longitudinal axis. As a first step, material of the target powder composition is fabricated into the geometry of a rod electrode that serves as an anode. This consumable rod electrode is then rotated at velocities that range in the 100's of revolutions per second and gradually moved in as the electrode is consumed. This displacement of the anode is carefully governed by external drives to ensure that the rapidly moving rod electrode remains spinning concentrically. As the metal is ejected from the rapidly spinning electrode, it travels in an inert atmosphere as it spheroidizes under the action of surface tension and hydrodynamic forces. The consumable electrode can be melted using a variety of energy sources. For example, in the case of titanium-alloyed powders in the United States, helium plasma is generally used [37] to melt the end of a rapidly rotating bar, and molten droplets are spun off and solidify in flight in a helium atmosphere. The melting and subsequent atomization are conducted in a circular stainless steel tank (such as 2.5 m in diameter) maintained at a positive pressure of helium. A plasma arc is used to melt the end of a rapidly rotating rod of metal material inside a chamber. Because the melting and solidification processes occur within an inert gas environment, they produce uniform metal particles that contain almost no impurities such as oxygen.

The high-speed rotating electrode (raw material) can also be melted using a plasma arc under the protection of the high-purity inert atmosphere. As the molten metal is ejected and travels in the inert gas environment, sometimes it impinges on the cold wall chamber prior to full spheroidization. The melting of the electrode at one end can also be accomplished using a stationary tungsten electrode. This, however, can result in small amounts of tungsten contamination in the powders.

During PREP processing, the metal or alloy is first manufactured into the self-consuming electrode bars. Then, a plasma arc is used to melt the end of the rapidly rotating bar, and molten droplets are spun off and solidify in flight in an insert atmosphere. The electrodes are pre-alloyed bars nominally of Φ60–65 mm in diameter and are rotated at speeds of up to 15 000 rpm. Particle size depends on the alloy composition, the electrode diameter, and the rotation speed. The typical sizes for Ti-6Al-4V powders are between 100 and 300 μm with a median diameter of about 175 μm. The PREP powder is spherical while having good flow characteristics and a tap density of about 65% solid substance. Various process modifications have been studied and implemented to produce finer powder [38]. In particular, the REP has been combined with GA.

The REP is also used by similar technologies that are available in Russia, China, and Japan. The PREP and REP powders are used to produce PM components for aerospace, gas turbine engineering, environment control, and

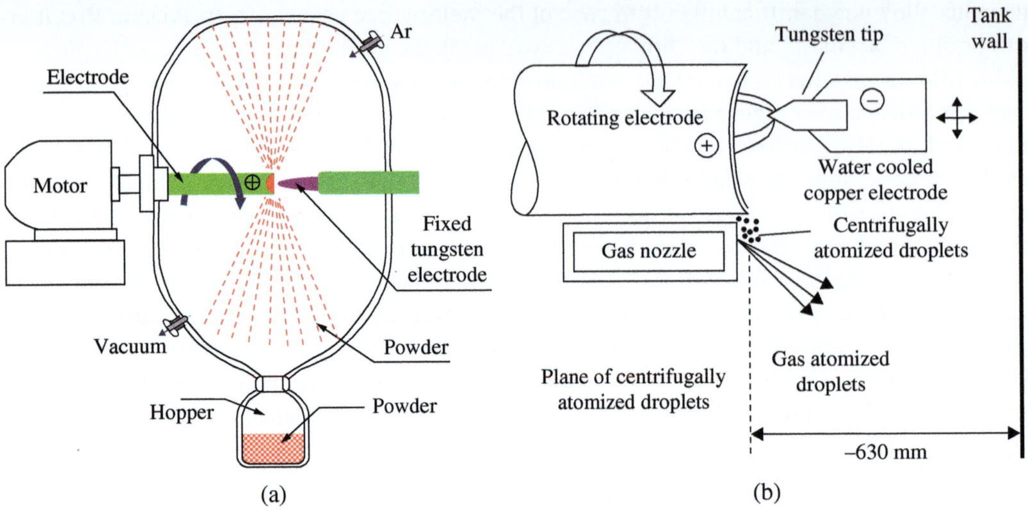

Figure 1.11 Conceptual schematic of (a) PREP principle. Source: Chen et al. [40] / with permission of Elsevier and (b) gas-assisted, rotating electrode process. Source: Nachtrab et al. [38].

biomedical applications. PREP has been developed to produce Ni-based alloy powders, Ti alloy powders, stainless steel alloy powders, and other refractory metal powders. The high-quality metal powders produced by PREP are widely used for electron-selective melting, laser melting, cladding, coating, HIP, etc.

Powders produced by the REP are generally spherical in nature. Compared to gas-atomized powder, PREP powder is relatively coarse. Various process modifications have been investigated to produce finer powder. A typical average size of the powder particle produced by the REP is 150–250 μm, and the particle size distribution is usually narrow [39]. It is interesting to note that the use of a large-diameter disk electrode, rather than a bar electrode, has been reported to decrease the mean particle size by more than 40% and to reduce the cost of input material. In another process modification, the metal droplets can be quenched into liquid argon. It was reported that the results of this approach were a major reduction in powder particle size [38]. In another modification, GA was coupled with the REP, as shown in Figure 1.11a. The REP powder process is similar to PREP, except that a tungsten arc is used instead of a plasma to melt the end of the rotating electrode [41]. As shown in Figure 1.11b, the yield of smaller than 44 μm titanium powder increased from less than 1% with regular REP to greater than 20% with GA assisted REP.

The characteristics of metal powders produced by PREP include low oxygen and other impurities (high purity), perfectly spherical and essentially satellite-free, narrow particle size distribution, high flowability and high apparent/packing density, and low internal porosity. Importantly, this process has been used extensively to produce powders of reactive and refractory materials.

The metal or alloy is made into self-consuming electrode bar stock, and the tip/end face of the electrode rotating at high speed is melted by argon or nitrogen plasma arc, and the centrifugal force generated by the electrode rotating at high speed throws out the melted metal liquid to form small droplets due to the high centrifugal force of the rotation of the rod. The droplets are cooled at high speed in the inert gas and solidified into spherical powder particles.

An inherent limitation of PREP, however, is that it is difficult to obtain finer powders because there are limitations to the maximum rotational speed that can be achieved in the electrode rod. In addition, a small quantity of coarse residual inclusions is commonly found in PREP powders [42]. To further reduce the occurrence of inclusions, research into the development of new preparation technologies for ultra-pure fine powders is essential.

1.7 Spark Plasma Discharge Spheroidization (SPDS)

Spark plasma discharge spheroidization (SPDS) is a metal powder preparation method that was originally developed to reduce the inclusions and refine the particle size of the powder. The **spark erosion technique** was proposed by Svedberg in 1924 [43] and was used to produce colloidal suspensions in the early years. Nowadays, it is widely used in various industries for the fabrication of hard or tough materials, wire-cutting, and powder production. American materials and electrochemical research (MER) has used the plasma discharge spark process to produce ultrafine Ti and tool steel powders with controlled size distribution ranging from 1 to 10 μm [44]. In addition, Ti–Ni–Zr and Ti–Ni–Hf powders have been obtained via spark erosion in liquid argon from melted shape memory master alloys [45].

SPDS process primarily takes place with the spark discharge between the pre-alloyed charges that are immersed in a dielectric fluid. The molten and subsequently evaporated materials are ejected from the electrodes and quenched in the dielectric medium after a complex plasma creation and breakdown. These materials ultimately condense as powder in the dielectric medium. The powder is quenched in the dielectric fluid and experiences extremely rapid cooling. Compared with the atomization process, like GA or PREP, this process can produce superfine powder at a much higher cooling rate up to $\sim 10^7$ K s^{-1}.

The major steps of SPDS processing include the dielectric breakdown and the formation of a discharge channel, followed by the energy distribution and heat transmission, RS, and ejection and transfer of powder. A schematic diagram of the SPDS used in the preparation of superalloy powders out of the plates/bars is shown in Figure 1.12.

During SPDS processing, a repetitively pulsed spark discharge between the electrodes immersed in dielectric fluid is maintained. The forming mechanism of particles can be divided into four steps: (i) when the gap between the electrodes becomes narrow enough, a plasma channel is generated between the electrodes, that are immersed in the dielectric fluid (such as water, kerosene, or liquid inert gas) and a pulsed power source, while the plasma channel is surrounded by a sheath of vaporized dielectric; (ii) due to the repetitive spark discharge, a super-high temperature, as high as 10,000 K, develops at local of the electrode, which leads to localized melting or vaporization of the electrode; (iii) the molten metal droplets and vapors are inserted into an insulating medium; (iv) finally, the powders can be obtained by cooling of the molten droplets or condensation and cooling of the vapors in the dielectric liquid. SPDS can avoid the introduction of ceramic during the powder preparation process.

As an example, fine spherical particles of PM Superalloy FGH4096 with sizes less than 40 μm were obtained via the SPDS process [46]. The sphericities and particle sizes of the powders for three different pilot dielectric liquids of kerosene, alcohol with liquid argon, and pure alcohol were reported to be different. FGH4096 powders show higher sphericity and finer particle sizes with alcohol/liquid argon used as the dielectric medium, as shown in Figure 1.13. Experiments on a spark erosion facility showed that spherical powders with particle sizes tuned from several nm to 40 μm can be fabricated. The smooth surface and uniformity in particle size were also obtained by appropriately adjusting the processing parameters.

Interestingly, the solidification microstructure at the surface of the SPDS powder shows no dendritic and cellular morphologies and instead reveals a solidification cell structure. The main causes for the resultant microstructure included (i) the particle sizes were much smaller than those obtained by the traditional processes (GA and PREP), and (ii) the solidification speed and cooling rate of the powder were also relatively higher than those obtained with

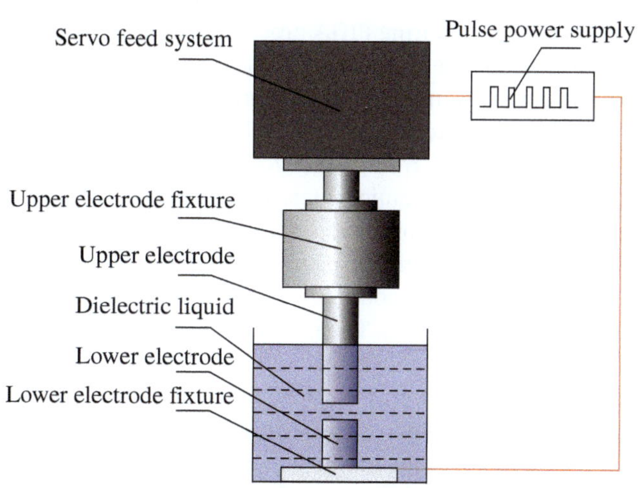

Figure 1.12 Schematic diagram of SPDS apparatus used for the preparation of powders. Source: Jia et al. [46] / IOP Publishing.

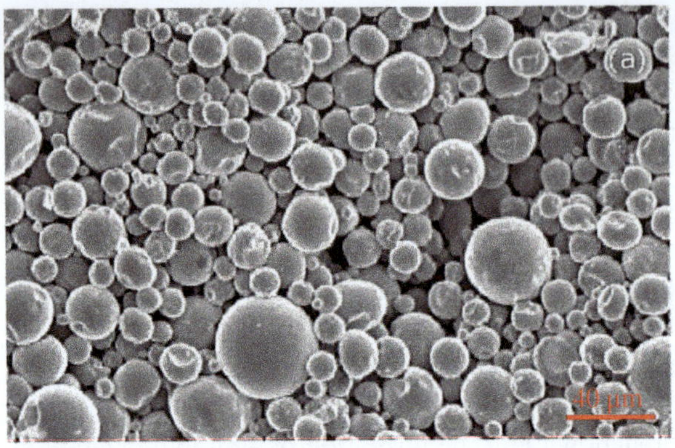

Figure 1.13 Morphology images of the SPDS powders prepared in alcohol with liquid argon. Source: Jia et al. [46] / IOP Publishing.

traditional processes (about 1 order of magnitude higher). Therefore, the heat transfer conditions limit the formation of dendrites, and a cell structure evolves in the powder.

The SPDS powder products have a broad potential market. Besides the preparation of PM superalloys for making turbine disks, the method can also be used for injection molding, rapid laser prototyping technologies, and AM. In short, superalloy powders prepared by SPDS processing have the advantages of high sphericity, small particle size, uniform microstructure, and less inclusion contents.

1.8 Plasma Induction Gas Atomization (PIGA)

The production of high-temperature and reactive materials, such γ-TiAl-based alloy powder, is driven by the need for these materials in many engineering applications. Ideally, the process should be ceramic-free to avoid possible contamination at the elevated temperatures that are required for reactive metals (around 1450 °C). In the molten state, these materials aggressively react with nearly any crucible materials. In addition, the atomized droplets and the hot powder particles are susceptible to reactions with oxygen and nitrogen as well as carbon pickup. For these reasons, only very specialized melting techniques that are crucible-free [32, 47] or that involve cold crucible melting [48, 49] can be employed. An example of a crucible-free technique is the EIMGA technique, as shown in Section 1.3. In addition, the leakage rate of the respective atomization facility must be very low, such that fully inert conditions can be guaranteed throughout the entire process. For subsequent powder handling, it is required that the powder tank can be demounted and transferred into a system of glove boxes operated under argon gas without any breach of the inert gas line.

An example of a cold crucible technique is the plasma melting induction PIGA technique [50]. The main features of the PIGA unit are illustrated schematically in Figure 1.14 [41, 51]. Plasma has been used extensively in fabricating various metal powders. The plasma process may be divided into two categories. One is that the plasma is used to form a melt pool that is then atomized by some conventional process, as applied during PIGA processing. In this case, the plasma essentially serves as the source of thermal energy. In the second case, the plasma is used to form the molten particles from either agglomerates or other starting powders and can be atomized by impacting the molten droplets traveling at very high velocities against a moving or stationary substrate material, as described in Section 1.4, PREP.

This technique has been especially developed for Ti and Ti alloys and is currently operated as a pilot plant at the GKSS Research Center, Geesthacht, Germany. In this technology, the use of ceramic materials is completely avoided. By means of a He-operated plasma torch, a melt bath is generated in a water-cooled copper crucible. In this process, a thin skull of the desired material is formed and prevents the contamination of the actual material in the crucible. At the tap hole of the crucible, an induction heated water-cooled copper funnel is mounted. It forms a thin melt stream and guides it into the center of a gas nozzle, where the melt is atomized by Ar gas. The new guiding system provides additional heating of the molten material if desired. The melt flow rate for TiAl alloys can be about 250 kg h^{-1}. Because of the melt bath formation and the induction guiding, considerable electric power is required. Since the liquid volume is much higher if compared to the EIGA technique, complex safety precautions are required. An advantage of this technique is seen in the chemically homogeneous melt bath, which serves as the source for the melt stream.

The PIGA method uses a plasma heat source to improve the stability and efficiency of the heat source, especially for high-temperature metals. Because the PIGA method uses a water-cooled copper crucible, when the metal stream is in

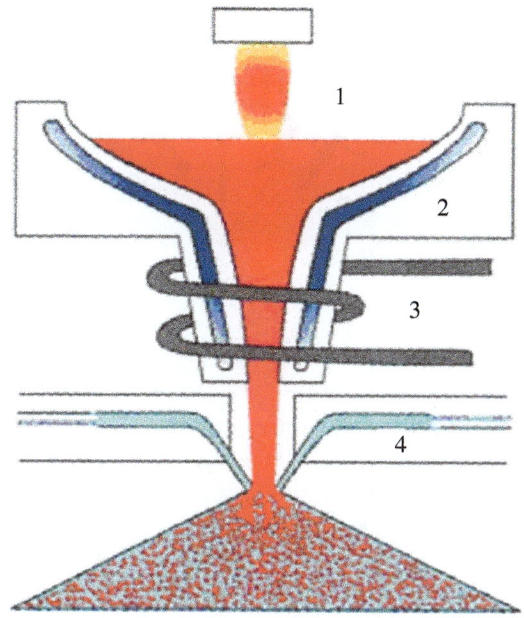

Figure 1.14 Schematic illustration of the plasma melting induction guiding gas atomization (PIGA) technique. (1) Plasma torch, (2) cold copper crucible, (3) induction-heated cold copper funnel, and (4) gas nozzle. Source: Gerling et al. [41] / John Wiley & Sons.

contact with the water-cooled copper crucible, a layer of parent metal is formed on the surface of the crucible, which isolates the subsequent metal stream from direct contact with the copper crucible wall and improves the preparation of metal powder's purity. PIGA is generally suitable for manufacturing reactive metals and high-temperature alloys, such as Ti, TiAl, FeGd, FeTb, Zr, and Cr, which cannot be processed using a ceramic crucible. It is also suitable for manufacturing high-purity metals, oxides, carbides, and nitrides, as well as ultrafine nanopowders.

The process has the advantage of utilizing scrap, granules, bars, etc., as feedstock material, and additional alloying can be accomplished during melting. The new high-power inert gas plasma torch is also suitable for the handling of reactive metal alloys containing volatile constituents. The disadvantages include the use of very high-power plasma torches exhibiting a high degree of erosion of the electrode, which can contaminate the metal powder.

1.9 Plasma-Atomized Wire (PAW)

The fabrication of metal powders using plasma technology is known as plasma atomization. In the **plasma-atomized wire (PAW) process**, plasma torches are used to melt a wire and produce the powder. The droplets are cooled and collected in a manner like that in GA. PAW powders are very clean, since the chemistry comes from a solid metal wire and the powder is spherical with excellent flow rate.

The plasma atomization technology initially went into commercial production in 1998 to enable the production of a fine particle distribution of high-purity reactive metal powders such as titanium. Plasma atomization technology can produce premium-quality spherical powders of reactive and high melting point materials, such as titanium, **nickel**, zirconium, **molybdenum**, niobium, tantalum, tungsten, and their alloys. The PAW process is primarily used for titanium alloys due to the challenges of handling molten titanium in GA. Plasma atomization has been widely developed, resulting in a high-purity spherical titanium powder in a wide variety of sizes. Atomization of plasma has the capacity of producing various sizes of titanium powder from very fine (−25 μm) to coarse (+125 μm) [51].

Powders with a spherical morphology have the benefit of having the properties of good flowability and hence capable of producing high density. In powder manufacturing processes, powders with good flowability require fewer additives than powders with low flowability. Moreover, powders with good flowability will also sinter to a high density when consolidated using HIP, for example.

Figure 1.15 shows a plasma atomization process scheme using a three-plasma arc, with titanium wire as a material feed [52]. The plasma arc is utilized as a high-enthalpy heat source that will melt and spray molten titanium wire, resulting in a titanium powder that has a very high-purity round shape.

Plasma atomization uses wire as a feedstock, which is ideal for traceability from the ingot to the final powder product. Tight quality control of the wire chemistry ensures batch-to-batch consistency. As opposed to conventional

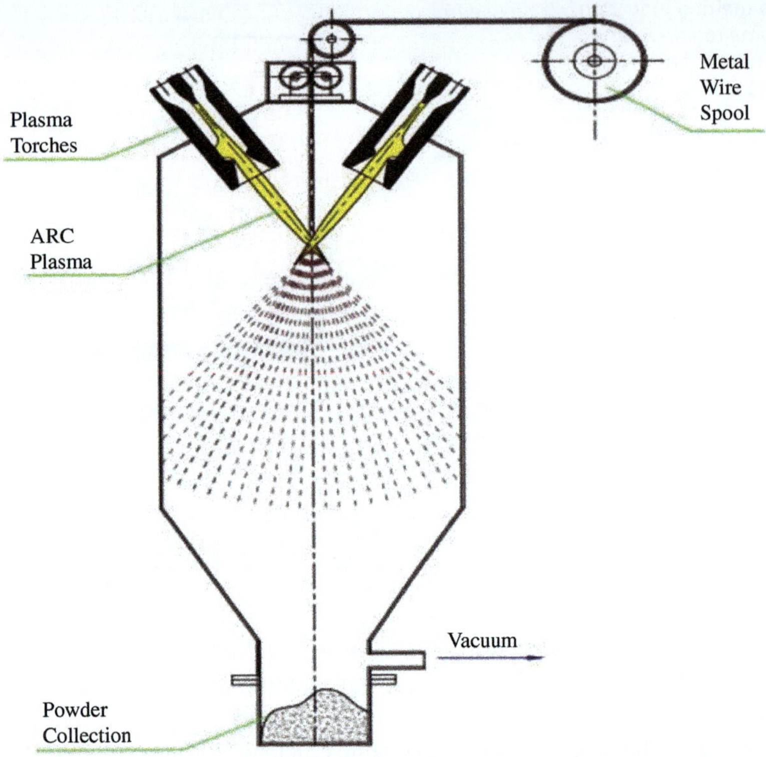

Figure 1.15 Schematic representation of the plasma atomization process to produce Ti powders. Source: Grenier [52].

GA, plasma atomization uses three plasma torches to instantaneously melt and atomize the wire in a single step. The intense heat of the plasma maintains the droplets well above their melting point throughout multiple phases of breakup, which increases the yield of very fine particles.

Once droplets are formed, they require some time to assume the most stable shape, a sphere. The cooling tower provides a low-velocity and low-turbulence environment that allows the particle to spheroidize under ideal conditions.

PAW provides a unique opportunity for producing fine, spherical, rapidly solidified powders of practically any material. The process offers the advantage of container-less melting and thus reduces the chances of contamination from the crucible. PAW process offers the highest purity possible with highly spherical particles and minimal satellite content. PAW powder thus exhibits exceptional flowability and packing properties, is well renowned and extensively used in AM, MIM, hot and **cold isostatic pressing** (HIP and CIP) and thermal and cold **spray coatings**. The main challenges with the PAW process are the high manufacturing costs, due to low production rates, and limited alloy flexibility, as the starting material must be available as wire.

Another related PAW method is PA. The principle of the PA method is that the metal wire is melted by adding a plasma heat source through a straightening machine at a certain rate, and then the molten metal droplets are atomized by high-pressure gas. The powder prepared by the PA method also has the characteristics of high sphericity, high purity, and low oxygen content. However, because the raw material is wire, the cost of raw materials is increased, and the types of metal powders prepared with it are limited.

1.10 Water Atomization (WA)

The process of producing metal powders by the disintegration of the molten metal stream using water is commonly referred to as **water atomization (WA)**. This is one of the most popular techniques for producing nonreactive metal and alloy powders. The WA technique presents a cost-effective and efficient approach to producing metallic powders and is currently used to produce low-alloy iron and steel powders for pressing and sintering.

WA is being broadly used since the middle twentieth century. The use of an atomizing chamber with inert gas has allowed the production of large tonnages of Fe powder and of a wide range of alloy steels, including stainless and tool steels. It is also widely used for Au, Ag, Zn, and Cu alloys and for some types of Ni and Co alloys. Oxygen contents are highly alloy-dependent, ranging from 500 ppm for some self-fluxing Ni-Cr-B-Si alloys and Cu alloys to 1% for high manganese steels. The morphology of WA powders is generally somewhat irregular but is greatly affected by alloy composition and atomizing conditions such that apparent densities can range from as low as 20% to ~50% of solid. Particle sizes for steels range from ~30 to 1000 μm [53], as the rapid quenching of the water jets allows larger particles to freeze quickly.

The rate of heat extraction during the WA process is quite high. The apparatus consists of water jets (usually coming through multiple jets or an **annular ring**) directed onto a molten stream of metal or alloy. The key variables that influence the process are the melt superheat and water pressure, which controls the water velocity. High water pressure results in finer powder particle sizes. Other important parameters are the melt stream diameter, metal flow rate, and water jet apex angle. The nozzle-to-melt distance is not as critical as GA since water has lower compressibility and higher density than gases and thus its energy does not diminish as rapidly. However, this distance should be minimized to maintain the melt superheat.

Due to the rapid extraction of heat, the final particle shape is quite irregular, as the particles have less time to spheroidize compared to the GA process. Thus, it is often necessary to have large superheat values for control of the particle morphology. The nature of the environment results in the final powder being slightly contaminated by a surface oxide layer, which under certain circumstances can be reduced by a subsequent hydrogen reduction step.

A schematic view of the WA process is shown in Figure 1.16 [54]. Water-atomized powders have excellent compressibility, but their apparent density is normally lower than that of gas-atomized powders due to the particle morphology. The water pressure has been reported to be the prime controlling factor for the WA process. It was reported that water-atomized steel powder using a water pressure of 1.7 MPa resulted in powders with a mean particle size of 117 μm compared to a mean particle size of 42 μm when the water pressure was 13.8 MPa [55].

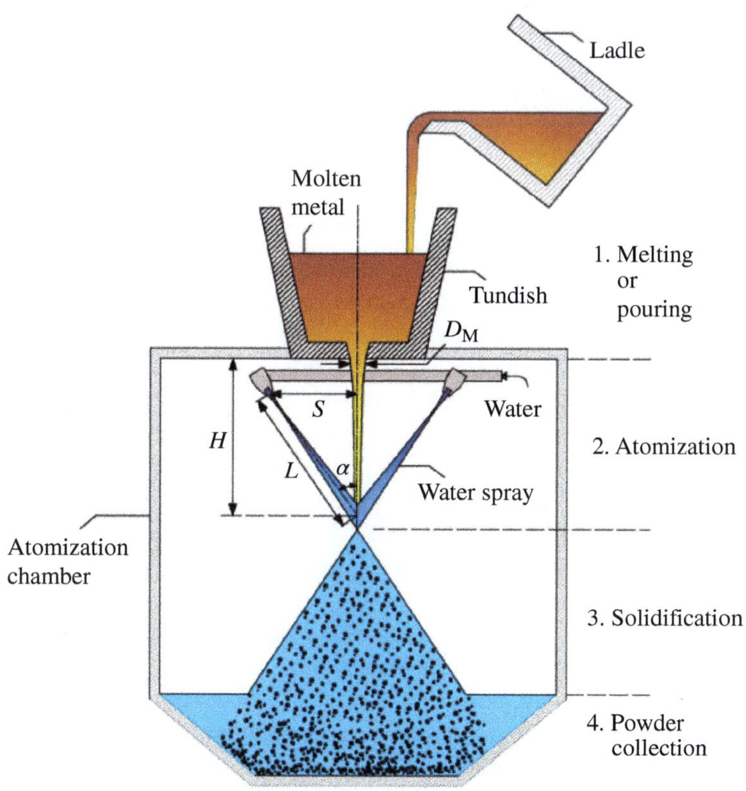

Figure 1.16 Schematic view of the water atomization process. Source: Asgarian et al. [54].

The need for finer powders (in the 10 μm range) has resulted in the high-pressure water atomization (HPWA) process in which water pressures up to 70 MPa have been used to produce powders with an average particle size of around 10 μm [56]. In the same report, it has been shown that the water jet nozzle configuration also influences the powder particle size. The authors have reported that a V-jet nozzle configuration produces finer powders than the commonly used cone nozzle configuration using similar atomizing pressures.

The process of high-pressure WA is an important development in the direction of producing atomized powders in the 10 μm range. The demand for these finer powders is increasing, especially due to the emergence of powder injection molding as an important fabrication route. Thus, it is expected that high-pressure WA could partially fill the gap between the supply and demand for reasonably inexpensive powders in the 10 μm particle size range. However, the powder morphology could be a problem for these powders.

High-pressure WA can provide fine powders, as discussed above. The process also can produce a wide range of powder particle size, size distribution, and powder apparent density. With increasing atomization water pressure, the particle size tends to become finer, and the particle size distribution also tends to become broader. The atomizing water pressure, which varied from 10 to 150 MPa, was found to yield finer average particle sizes with increasing water pressure with water-atomized 316-L stainless steel powders [57].

Ultra-high-pressure water atomization (UHPWA) involves the use of very high water pressures, typically in the range of 100–200 MPa. The water jets are generally supersonic, moving at velocities of 400–500 m s^{-1}, and atomization is partly due to the shockwaves around the water jets at such pressures. It is normally carried out in an inert gas-purged atomizing chamber and allows the production of very fine metallic powders with median particle sizes less than 10 μm. Due to the high surface area of such fine powders and the oxidation potential of the water, the oxygen contents are typically in the range of 2000–5000 ppm. The particle shape can range from somewhat irregular to spherical, allowing high tap densities of ~4.5 g ml^{-1} for MIM grades.

Few studies have been carried out on the mechanisms that govern the WA process. It has been suggested that the water jet is usually broken up into minute fragments before it impinges on the metal stream. Thus, very small water droplets impact the molten liquid stream, causing small appendages to appear on the molten liquid surface (like small waves) from which a metal droplet is ejected. The water droplet after striking the liquid film immediately evaporates. This is termed the "scrape" mechanism that operates during WA events.

In general, WA can produce powders that have an average particle size of around 200 μm. The particle morphology is generally irregular, and the powders usually have a wide size distribution. However, progress in high-pressure WA has led to the successful production of large tonnages of fine powders.

1.11 Summary

The material summarized in this first chapter reveals that powder metallurgy, as the practice of working with metal powders, has a deep and rich history. In some ways, the use of powder metals can be traced back several thousand years. Moreover, it is also evident that powder metals find applications in numerous areas of technology and engineering. A review of early **atomizer designs** reveals that the fundamental idea of breaking a liquid into a dispersion of fine droplets can be accomplished with very simple atomizer designs, such as the first atomizer design created by Hall, who used it for aluminum powder, to much more complex atomizer designs that involve intricate use of fluid flow phenomena to enhance the efficiency of atomization. In this chapter, we also reviewed the various techniques that are being used to produce many types of metal powders. In general, the breakup of metal powders can be accomplished using energetic gas jets, and, depending on the chemistry of the metal being atomized, gas chemistry is selected and used under various pressure conditions. It is also apparent that in many cases, it is imperative to control the environment in the chamber where the metal is being prepared to avoid deleterious surface reactions. Modern atomizer designs include sophisticated melting approaches that allow the handling and processing of reactive metals such as titanium. In general, desired powder morphology and particle size can be achieved through the identification of optimal atomization parameters. Furthermore, powder qualities are known to affect powder flowability, which can improve the injection-ability and spread-ability of powder particles for AM process. Generally, good injection-ability and spread-ability are associated with good powder feeding and packing fractions, which can lead to high-quality AM components. Therefore, in the following chapters, the parametric information for the GA is presented.

Nomenclature

AM	additive manufacturing
CIP	cold isostatic pressing
EIGA	electrode induction melting gas atomization
GA	gas atomization
HIP	hot isostatic pressing
HPGA	high-pressure gas atomization
HPWA	high-pressure water atomization
IGA	inert gas atomization
MIM	metal injection molding
PIGA	plasma melting induction gas atomization
PA	plasma torch atomization
PAW	plasma atomized wire
PIGA	plasma induction gas atomization
PREP	plasma rotational electrode process
RS	rapid solidification
REP	rotating electrode processes
SPDS	spark plasma discharge spheroidization
VIGA/VIMGA	vacuum induction melting inert gas atomization
UHPWA	ultra-high-pressure water atomization
WA	water atomization

References

1 Samal, P.K. and Newkirk, J.W. (2015). *ASM Handbook, Volume 07 – Powder Metallurgy*, 3. ASM International.
2 DeVilbiss, T.A. (1909). Atomizer, vol U.S. Patent #938,648.
3 Harrington, E.L. (1927). *J. Opt. Soc. Am.* 14 (1): 87–97. https://doi.org/10.1364/JOSA.14.000087.
4 Hall, E.J. (1928). Process for Disintegrating Metal, U.S. Patent #1,659,291.
5 Singer, A.R.E. (1970). *Met. Mater.* 426.
6 Lavernia, E.J., Srivatsan, T.S., and Rangel, R.H. (1992). *Atomiz. Spr.* 2: 253.
7 Unal, A. (1987). *Mater. Sci. Technol.* 3: 1029–1039.
8 Unal, A. (1988). *Mater. Sci. Technol.* 4: 909–915.
9 Singer, A.R.E. (1972). *J. Inst. Met.* 100: 185.
10 Brooks, R.G., Leatham, A.G., Coombs, J.S., and More, C. (1977). *Metall. Met. Forming* 9: 157.
11 Lavernia, E.J. 1984. Liquid dynamic compaction of a rapidly solidified 7075 aluminum alloy modified with 1% Ni and 0.8% Zr, thesis, Massachusetts Institute of Technology, Cambridge, MA.
12 Lavernia, E.J. and Grant, N.J. (1988). *J. Mater. Sci. Eng.* 98: 381–394.
13 Hueter, T.F. and Bolt, R.H. (1962). *Sonics*. New York: John Wiley & Sons.
14 Lavernia, E.J. and Grant, N.J. (1986). *Met. Powder Rep.* 41: 255.
15 Gerking, L. (1992). *Powder Met. Int.* 25: 59–65.
16 Mir, J.M. (1980). *J. Acoust. Soc. Am.* 67: 201.
17 Clyne, T.W., Ricks, R.A., and Goodhew, P.J. (1984). *Proceedings, 5th Conference on Rapidly Quenched Metals (R.Q.5)* (ed. S. Steeb and H. Warliment), 903. Wurzburg, Germany: North Holland Physics Publishing Amsterdam.
18 Ridder, S.D., Osella, S.A., Espina, P.I., and Biancaniello, F.S. (1992). *Int. J. Powder Metall.* 28: 133–147.
19 Ridder, S.D. and Biancaniello, F.S. (1989). *Mater. Sci. Eng., A* 98: 47–51.
20 Miller, S.A., Wojcik, L.A., Miller, R.S. et al. (1995). *Advances in Powder Metallurgy and Particulate Materials*, vol. 1 (ed. M. Phillips and J. Porter), 31–40. Metal Powder Industries Federation/APMI.
21 Marriott, W. (1872). vol patent 3322 British.
22 Zheng, B., Lin, Y., Zhou, Y., and Lavernia, E.J. (2009). *Metall. Mater. Trans. B* 40B: 768–778.
23 Lavernia, E.J., Srivatsan, T.S., and Rangel, R.H. (1992). *J. Atomiz. Spr.* 2: 253–274.

24 Lin, C.S., Jih, G.J., Chang, C.S., and Chiou, S.Y. (1989). *Advances in Powder Metallurgy*, vol. 2 (ed. T.G.G.a.W.F. Jandeska), 57. Princeton, NJ: Metal Powder Industries Federation.

25 Veistinen, M.K., Lavernia, E.J., Baram, J.C., and Grant, N.J. (1989). *Int. J. Powder Metall* 25: 89.

26 Mohanty, T., Tripathi, B., Mahata, T., and Sinha, P. (2014). Arc plasma assisted rotating electrode process for preparation of metal pebbles. *International Symposium on Discharges and Electrical Insulation in Vacuum (ISDEIV)* 741–744.

27 Samal, P.K. and Newkirk, J.W. (ed.) (2015). *ASM Handbook: Powder Metallurgy, ASM International*, 907. USA: Materials Park.

28 ALD Vacuum Technologies AG, Metal Powder Technology, http://www.metamak.com.tr/UserFiles/File/demir%20disi/vakum%20metalurjisi/powder%20metallurgy/Metal_powder.pdf.

29 Gerling, R., Schimansky, F.P., Wegmann, G., and Zhang, J.X. (2002). *Materials Sci. Eng., A* 326: 73–78.

30 Franz, H., Plochl, L., and Schimansky, F.P. (2008). Recent advances of titanium alloy powder production by ceramic-free inert gas atomization. *Proc. Titanium 2008, International Titanium Association* 1–4.

31 Gerling, R., Schimansky, F.P., and Clemens, H. (2003). *Gamma Ti- tanium Aluminides* (ed. Y.-W. Kim, H. Clemens, and A. Rosenberger). Warrendale, PA: TMS.

32 Preauchat, B., Popoff, F., and Thomas, M. (2002). *Adv. Eng. Mater.* 4: 133.

33 Grewe, H., Potschke, J., and Weilin, W.G. (1986). *ZwF* 81: 413.

34 Dawes, J., Bowerman, R., and Trepleton, R. (2015). *Johnson Matthey Technol. Rev.* 59: 243–256.

35 http://www.arcastinc.com/news.htm, 2021.

36 Lalwani, G. (2020). https://bonezonepub.com/2020/06/09/titanium-powder-quality-key-to-additive-manufacturing-success/.

37 Froes, F.H. and Eylon, D. (1990). *Int. Mater. Rev.* 35: 162–182.

38 Nachtrab, W.T., Roberts, P.R., and Newborn, H.A. (1993). *Key Eng. Mater.* 77–78: 115–140.

39 Bose, A. (1995). *Advances in Particulate Materials*. Elsevier.

40 Chen, G., Zhao, S.-y., Tan, P. et al. (2017). *Trans. Nonferrous Met. Soc. China* 27: 2647–2655.

41 Gerling, R., Clemens, H., and Schimansky, F.P. (2004). *Adv. Eng. Mater.* 6 (1-2): 23–38. https://doi.org/10.1002/adem.200310559.

42 Zhang, M.J., Li, F.G., Wang, S.Y., and Liu, C.Y. (2011). *Mater. Sci. Eng., A* 528: 4030–4039.

43 Svedberg, T. (1924). *Colloid Chemistry, Wisconsin Lectures*. New York: Chemical Catalog Co.).

44 Newell, K. (1999). *Met. Powder Rep.* 54: 22.

45 Monastyrsky, G., Ochin, P., Wang, G.Y. et al. (2011). *Chem. Metals Alloys* 4: 188.

46 Jia, C., Ge, C., and Yan, Q. (2016). *Chin. Phys. B* 25: 026103.

47 Hohmann, M., Diemar, W., and Ludwig, N. (1992). *Advances in Powder Metallurgy & Particulate Materials*, vol. 1 (ed. J.M. Capus and R.M. German), 27. Princrton, NJ: MPIF.

48 Gerling, R., Schimansky, F.P., and Wagner, R. (1992). *Advances in Powder Metallurgy & Particulate Materials*, vol. 1 (ed. J.M. Capus and R.M. German), 215. Princeton, NJ: MPIF.

49 Yolton, C.F. and Beckman, J.P. (1995). *Mater. Sci. Eng., A* A192: 597.

50 Stenzel, O.W., Sick, G., and Hohmann, M. (1990). DE4011 392A1, German Patent.

51 Alagheband, A. and Brown, C. (1998). *Met. Powder Rep.* 53: 26–28.

52 Grenier (1997). *Met. Powder Rep.* 52: 34–37.

53 Dunkley, J.J. (2019). *Johnson Matthey Technol. Rev.* 63: 226–232.

54 Asgarian, A., Tang, Z., Bussmann, M., and Chattopadhyay, K. (2020). *Powder Metall.* 63: 288–299.

55 German, R.M. (1984). *Powder Metallurgy Science*. Princeton, NJ: Metal Powder Industries Federation.

56 Seki, Y., Okamoto, S., Takigawa, H., and Kawai, N. (1990). *Met. Powder Rep.* 45: 38.

57 Kimura, A., Nakabayashi, K., and Shimura, T. (1990). *Met. Powder Rep.* 45: 106.

2

Atomization

2.1 Introduction

Early interest in atomization was derived from the use of **powder metallurgy** (PM) processing for the production of fully dense bodies as an alternative **manufacturing** route to conventional **ingot metallurgy** and forging routes, as described in Chapter 1. Today, PM techniques are being widely applied to a comprehensive spectrum of engineered materials including high-strength **aluminum alloys**, iron-based alloys, **superalloys**, **titanium alloys**, **ceramics**, **composites**, and effectively most materials of commercial and academic interest [1, 2]. The various processes that are generally used for the production of **powders** have been classified under the following categories [3, 4]:

(a) *Atomization* which may be used, at least in principle, for any metal or alloy system that can be melted,
(b) ***Chemical reduction*** as in the production of sponge iron from scale or iron oxides,
(c) ***Mechanical crushing*** or attrition which is widely used for brittle materials such as beryllium, antimony, and permanent magnetic materials,
(d) ***Vapor condensation*** which is used for making nanocrystalline powders, and
(e) ***Electrolysis*** which is used for the preparation of high-purity powders.

The most widely used processes for tonnage production of alloy powders are atomization and **chemical reduction** [5]. Atomization, however, is widely recognized as the most flexible technique. It allows for the production of a broad range of alloy compositions with extensive control over resulting powder characteristics and properties.

In most cases, atomization involves the impingement of a liquid stream by a **high-velocity fluid**, which may be either gas or liquid. During the atomization of molten metals, the droplets that are produced will eventually solidify. In the literature, the terms "droplet" and "powder" are used interchangeably. In this chapter as well as in other chapters in this book, however, "droplet" will be used to describe an atomized material that is in the liquid or semiliquid state, whereas "powder" will be used to describe an atomized material that is in the solid state.

The principle of atomization is ancient and, in its most rudimentary state, formed the basis for a **British patent** granted in 1872 to produce **lead powders** by drawing-off and spraying molten lead using a steam injector [6]. Subsequent years saw the arrival of two American patents granted in 1924 and 1925 describing an atomizing nozzle in which compressed gas is released from an **annulus** concentric with an orifice [7]. Currently, most atomization nozzles continue to operate based on the principle of disintegrating a molten stream of metal using a high-velocity fluid. Gas atomization is often preferred over other atomization methods for the production of pre-alloyed powders since a careful degree of control can be exercised on

(a) **powder chemistry**,
(b) powder cleanliness, and
(c) powder size distribution and shape [8, 9].

In additive manufacturing, it is essential that the atomized droplets meet specifications with respect to yield (in terms of particle size range), mean droplet size, and size distribution. These intrinsic characteristics of the droplets, along with shape properties, critically influence the microstructure in the final product and the consolidation of metal powders in PM materials [10, 11].

The objective of this chapter is to examine the physical process of atomization in which a molten metal stream is disintegrated and droplets are produced. In particular, the relevant physical parameters which influence the process

Metallic Powders for Additive Manufacturing: Science and Applications, First Edition.
Enrique J. Lavernia, Kaka Ma, Julie M. Schoenung, James F. Shackelford, and Baolong Zheng.

of atomization are examined and discussed in an attempt to elucidate their effects on droplet size and size distribution. Section 2.2 will discuss the methods that are commonly used for the atomization of metals. Section 2.3 summarizes the various phenomenological mechanisms that govern liquid breakup and droplet formation. Section 2.4 addresses the proper control of atomization parameters. Section 2.5 briefly introduces the methods of powder characterization. Section 2.6 focuses on the influences of processing parameters on powder size distribution, and Section 2.7 reviews the various theoretical models that have been developed in an effort to understand liquid breakup and atomization. Finally, Section 2.8 summarizes empirical relationships that relate powder size to processing variables.

2.2 Atomization Technology

2.2.1 Energy Consumption During Atomization

During atomization, the bulk liquid is disintegrated into fine droplets (generally in the micrometer-sized range) which exhibit a much larger surface-to-volume ratio as compared to the starting material. Hence, the atomization of molten metals inherently involves the creation of a large amount of **surface area**. During atomization, energy is transferred from the atomization media, such as a gas, a fluid, or a mechanical surface, to the molten metal. The driving force for atomization is generally provided by the kinetic energy of the atomization media. In atomization, molten metal has to be deformed and disintegrated into fine droplets. From an energy conservation standpoint, the kinetic energy imparted by the atomization media on the molten metal is partially dissipated in two important processes: to overcome the viscous forces of the molten metal that resist deformation and to overcome the surface energy forces that resist free surface creation. The kinetic energy that remains in the atomization media is eventually dissipated into the environment. The energy that is transferred from the atomization media to the molten metal in the form of surface energy for unit mass of powders produced, ΔE_{s}, may be estimated from a simple formulation:

$$\Delta E_{\mathrm{s}} = \sigma_1(\Sigma S_{\mathrm{d}} - \Sigma S_{\mathrm{b}}) \tag{2.1}$$

where σ_1 is the surface energy of the molten metal; ΣS_{d} and ΣS_{b} are the total surface areas of the droplets and that of the bulk material, respectively, for the unit mass of powders produced. When compared to a collection of droplets of equivalent mass, the surface energy of a bulk liquid is normally negligible, so that the second term in Eq. (2.1) may be safely ignored. It is useful to illustrate this **energy consumption** by considering the atomization of fine aluminum droplets. For typical atomization conditions, the average size of the droplets that are produced during atomization ranges from 60 to 120 μm. Table 2.1 summarizes the total surface area that would be created. For example, for an average droplet size of 80 μm, the total surface area that is created during atomization is 27.8 $\mathrm{m^2\ kg^{-1}}$. The corresponding increase in surface energy is 25.6 $\mathrm{kJ\ kg^{-1}}$ of aluminum powders. It is worth noting that due to the low efficiency of energy transfer from the atomization media to the molten metal, the energy calculated from Eq. (2.1) contributes only a small fraction of the overall energy consumption during the atomization process.

The **efficiency** of atomization may be defined as the ratio of the total surface energy created to the total energy consumed during atomization [12]. Then for perfectly spherical droplets of identical size, d, the efficiency is given by

$$\zeta = \frac{\Delta E_{\mathrm{s}}}{\Delta E_{\mathrm{t}}} = \frac{6\sigma_1}{\rho_1 d \Delta E_{\mathrm{t}}} \tag{2.2}$$

where ζ is the atomization efficiency; ΔE_{t} is the total energy consumed by the atomization media for each unit mass of powder produced; d is the average droplet diameter; and ρ_1 is the density of the molten metal. Alternatively, Eq. (2.2) may be expressed in terms of a power ratio as

$$\zeta = \frac{P_{\mathrm{s}}}{P_{\mathrm{t}}} = \frac{6\sigma_1 \dot{m}_1}{\rho_1 d P_{\mathrm{t}}} \tag{2.3}$$

where P_{s} and P_{t} represent the power associated with surface energy creation and energy consumption, respectively, and $\dot{m}_1$ is the metal flow rate during atomization. When Eq. (2.3) is applied to gas atomization, the power consumed by the atomization gas can be calculated from the atomization pressure and the gas flow rate by assuming isothermal gas expansion. This power, P_{t}, is expressed as

$$P_{\mathrm{t}} = \frac{\dot{m}_{\mathrm{g}}}{\Omega_{\mathrm{g}}} RT \ln\left(\frac{p_{\mathrm{o}}}{p_{\mathrm{en}}}\right) \tag{2.4}$$

Table 2.1 Surface energy increase during atomization of Al [from Eq. (2.1)].

Droplet size (μm)	Total surface area ΣS_d (m^2)	Surface energy increase ΔE ($kJ\,kg^{-1}$)
60	37.0	34.1
80	27.8	25.6
100	22.2	20.4
120	18.5	17.0

where Ω_g is the molar weight of the atomization gas; R is the gas constant; p_{en} is the environmental pressure, and $\dot{m}_g$, T, and p_o are the mass flow rate, temperature, and initial pressure of the atomization gas, respectively. For gas atomization, the overall energy efficiency is relatively low, being on the order of ~3%. Accordingly, inert gas atomization in the absence of **gas recycling** may become an expensive process [13], which when combined with additive manufacturing may lead to high manufacturing costs.

2.2.2 Molten Metal Atomization Methods

The energy that is required for atomization can be imparted onto the molten metal in a variety of ways. As previously discussed, it is generally supplied in the form of kinetic energy. Depending on the mode in which the energy is supplied, atomization processes can be classified into the following major categories:

(a) twin fluid atomization,
(b) **vacuum atomization,**
(c) **centrifugal atomization**, and
(d) **mono-sized droplet generation**.

For the production of metal droplets/powders, the fluid being atomized in the **twin fluid atomization** process is molten metal. A secondary fluid is used as the atomization media to breakup the molten metal into droplets. During atomization, jets of the secondary fluid are formed and accelerated using a stream injector. These jets are then focused onto a stream of molten metal to cause disintegration. This secondary fluid is known as the **atomization fluid.** In view of the significant difference in the physical properties between the fluid atomizing media and the molten metal, **twin-fluid atomization** is also referred to as **two-phase atomization.**

In the process of **vacuum atomization**, the starting molten metal contains dissolved gases. During atomization, the molten metal is injected into a low-pressure/vacuum environment. On exposure to the vacuum, the entrapped gases expand and are released into the low-pressure atmosphere causing the deformation and disintegration of the molten metal [1].

In centrifugal atomization, the molten metal is formed directly on, or injected from, a source onto a surface that is rotating at high speeds. The metal is stretched into liquid sheets under the combined action of the frictional force developed at the liquid–rotor interface and the centrifugal force. Under the action of centrifugal force, the sheets escape from the fringe of the rotor. The liquid metal sheets become unstable and eventually breakup into droplets during their flight.

The twin fluid atomization methods can be further classified into various categories according to the physical properties and flow characteristics of the atomization fluid. These categories are summarized as follows:

(a) water atomization,
(b) oil atomization,
(c) subsonic gas atomization,
(d) supersonic gas atomization, and
(e) ultrasonic gas atomization.

The selection of a particular atomization method involves the consideration of a number of factors. Such factors may include economic considerations, scale of production, the physical and chemical properties of the fluid to be atomized, the physical and chemical properties of the droplets/powders to be produced and the morphology of the droplets/powders desired.

Water and oil atomization methods involve the use of water or oil as the atomization media. During atomization, the atomization fluid is accelerated using a fluid injector. The atomization fluid is generally released as discrete jets. Atomizers that release the atomization media as sheets have also been successfully applied [1]. During impingement, the molten metal deforms and disintegrates under the action of the high-energy atomization fluid. Separation and desiccation are normally necessary to collect the metallic powders from the slurry of atomization fluid and powders. Water and oil atomization methods are generally used for the high-tonnage production of metallic powders when a certain degree of contamination can be tolerated.

In additive manufacturing processes **powder chemistry** will affect the quality of the fabricated component. Accordingly, contamination from the atomization media should be minimized to the extent possible. In view of this particular requirement, water and oil atomization methods are not widely used for the preparation of powders to be used in additive manufacturing processes. In the sections that follow subsonic gas atomization, supersonic gas atomization, ultrasonic gas atomization, centrifugal atomization, and monosized droplet generation will be discussed in detail because either they are being used to prepare powders for **additive manufacturing**, or are actively being studied for this purpose. It should be noted, however, that the fundamental mechanisms that govern the formation of metallic droplets are similar for all twin-fluid atomization methods, regardless of the particular atomization fluid used.

Subsonic, supersonic, and ultrasonic gas atomization techniques are referred to collectively as **gas atomization**. The gas atomization is most frequently used to produce metallic droplets for **additive manufacturing**. Generally, the droplets produced form a conical or linear spray as shown in an earlier chapter.

2.2.3 Subsonic Gas Atomization

Subsonic gas atomizers employ subsonic gases as the atomization fluid. A schematic diagram of a typical conical gas atomizer is shown in Figure 2.1, along with the key operating parameters. The gas atomizer consists of a gas inlet tube, a **rotrusion distance (or reservoir)**, and **gas nozzles** (i.e. the gas outlets of the atomizer). Depending on each particular design, the **gas nozzle** may be a series of discrete holes or may be an annular **ring-shaped slit**. The gas nozzles are designed to be inclined conically and are focused in front of the orifice of a delivery tube at an **apex angle**, α. One-half of the apex angle is referred to as the **attack angle** or **impingement angle**. As shown in Figure 2.1b, a typical gas atomizer generally consists of 12–18 gas orifices positioned at equal distances on the circumference of a circle. The atomization gas is fed into the atomizer from a gas inlet line which is connected to a pressurized gas

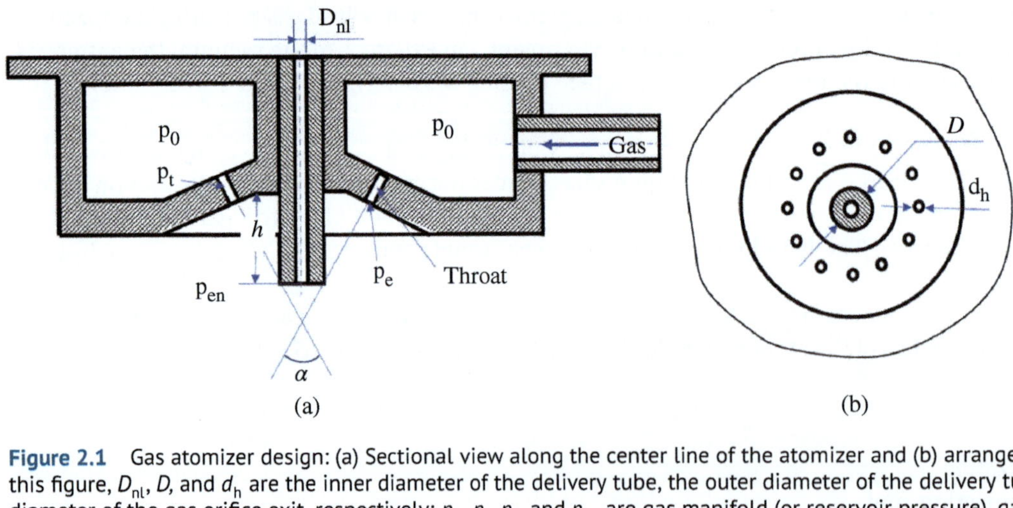

Figure 2.1 Gas atomizer design: (a) Sectional view along the center line of the atomizer and (b) arrangement of gas orifices. In this figure, D_{nl}, D, and d_h are the inner diameter of the delivery tube, the outer diameter of the delivery tube, and the inner diameter of the gas orifice exit, respectively; p_o, p_t, p_e, and p_{en} are gas manifold (or reservoir pressure), gas orifice throat pressure, gas exit pressure and environmental pressure, respectively; h is delivery tube protrusion distance; and α is apex angle.

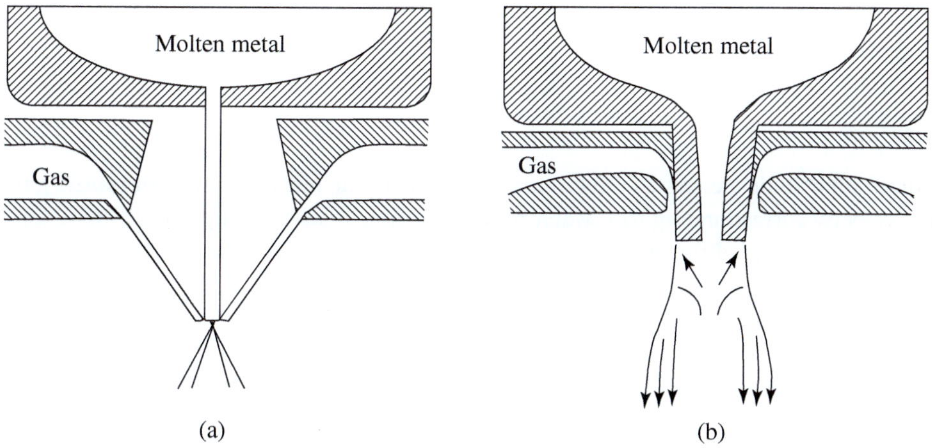

Figure 2.2 Atomizer designs: (a) free fall design and (b) confined design. Source: Adapted from Klar and Shafer [4].

source such as a compressor or a gas tank. The molten metal is typically supplied and directed into the atomizer using a **ceramic delivery tube**.

During atomization, a high-pressure atomization gas is discharged from the pressurizing manifold (or reservoir) of the atomizer (at a pressure p_0) into a chamber that is typically maintained at low environmental pressure (at a pressure p_{en}). The compressible atomization gas accelerates to a high velocity as it expands from the high reservoir pressure into the low-pressure chamber. During this process, the molten metal stream is deformed and disintegrated into fine droplets by the energetic impingement of high-velocity atomization gas jets. The disintegration process of molten metal is the topic of Section 2.3.

The gas atomizers can be classified into two basic types according to the relative positions between the atomization gas jets and the molten metal stream. These are **closed-type atomizers** and **open-type atomizers**. The schematic diagrams of such atomizers are shown in Figure 2.2 [2, 14]. In view of the flow characteristics of the molten metal, the closed-type atomizer is also referred to as a **confined design**. The open-type atomizer is also termed a **free-fall design** [2, 4]. In free-fall designs, the metal is allowed to fall under the action of gravitational forces for a certain distance (2–20 cm [2, 4]) prior to interacting with the atomization gas jets. In confined designs, the metal travels a very short distance or is pre-filmed before being impinged upon by the high-energy gas jets. An important characteristic of the confined designs is that the atomization jets are generally tangential to the edge of the molten metal delivery tube (see Figure 2.2). Therefore, the flow rate of the molten metal stream is significantly affected by the **aspiration pressure** generated by the high-velocity gas jets at the **orifice** of the metal delivery tube. In addition, aspiration essentially produces a pre-filming effect in which the molten metal leaving the delivery tube is stretched into films under the action of the aerodynamic forces in the low-pressure region. The formation of an aspiration zone and the pre-filming effect will be discussed in detail in Section 2.4. One of the advantages of confined atomizers is their higher atomization efficiency relative to that associated with free-fall atomizers. In addition, confined atomizers produce a relatively stable spray which is very important for powder quality. Confined atomizers, however, are extremely sensitive to metal freeze-up, a condition that occurs when the liquid metal at the end of the delivery tube solidifies as a result of the combined effects of **pre-filming** and rapid heat extraction that occur at the orifice of the delivery tube. As such process parameters must be optimized for individual alloy compositions to avoid premature solidification during atomization.

In addition to the aforementioned atomizer designs (i.e. confined and freefall), the configuration of the atomization gas orifices has also been the subject of extensive research. There are three general types of orifice designs that have been employed: converging, converging/diverging, and constant cross-section designs. As shown in Figure 2.2a, in the **converging design**, the cross-sectional diameter of the gas nozzles decreases gradually along the nozzle orifice [4]. In this type of design, the velocity of the atomization gas jets is confined to the subsonic regime regardless of the pressurizing conditions of the atomizer. In the **converging/diverging design**, as shown in Figure 2.2b, the cross-sectional diameter of the nozzles decreases initially, reaches a minimum value, and then increases towards the nozzle orifice [4]. In this design, the narrowest region is referred to as the **nozzle throat** [2, 4, 14–17]. The gas jets accelerated by a convergent/divergent nozzle can either be in the subsonic or the supersonic velocity regime, depending on the pressurization conditions and the geometry of the atomizer. In the constant cross-section design, the diameter of the gas nozzles remains constant.

2.2.4 Supersonic Gas Atomization

Supersonic gas atomization utilizes supersonic gas jets as the atomization media. The terminology **supersonic gas jets** implies gas jets with a flow velocity corresponding to a Mach number greater than one. Supersonic atomizers are primarily closed-type atomizers. The design of the molten metal delivery system and the attack angle are similar in supersonic and subsonic atomizers as evident from the data summarized in Table 2.2 [16–21].

The gas nozzle ducts in supersonic atomizers are designed to achieve supersonic flow. A typical supersonic atomizer is shown schematically in Figure 2.2b. A supersonic atomizer incorporates convergent/divergent gas nozzles. Such convergent/divergent design is also known as a **De Laval Nozzle**. Under proper pressurizing conditions and atomizer geometry, the gas jet flowing through the converging/diverging section of the nozzle is **overexpanded** and the resulting jet that emerges from the nozzle achieves supersonic velocity [15, 22]. Table 2.3 compares the reported gas velocities that are associated with various types of atomizers utilized for the atomization of molten metals. As can be seen from the results, supersonic gas velocities are achieved in the De Laval-type atomizers operating under pressurizing conditions that are comparable to those used in subsonic atomizers. Figure 2.3 shows the measured Ar gas velocity as a function of atomization pressure (the inlet gas pressure of the atomizer) [22]. The corresponding atomizer used in obtaining the gas velocity data that is shown in this figure was a De Laval design. In this case, supersonic flow develops downstream of the **nozzle throat** area at an inlet pressure of approximately 120 kPa. It should be noted

Table 2.2 Characteristics of various atomizer designs.

Atomizer	Type	Number of jets	Jet diameter or gap (mm)	Delivery tube diameter (mm)	Attack angle (°)	Application	References
NIST	Closed	18	0.8	2.5	22.5	Atom.	[18]
GE–CCN	Closed	Ring	0.5	3.8	22.0	Atom.	[18]
DJ–CF	USGA	18	—	—	22.5	Atom.	[19]
Ames	HPGA	20	—	—	22.5	Atom.	[21]
Ames	HPGA	20	—	—	31.5	Atom.	[21]
—	Sonic	4	4.5	4.4	27.5	Atom.	[20]
—	Sonic	4	4.5	4.4	34.5	Atom.	[20]
IN3/ON7	De Laval	Ring	1.4	4.0	—	Atom.	[16, 17]

NIST, atomizer designed at National Institute of Standards and Technology, Gaithersburg, MD, USA; GE-CCN, General Electric, Schenectady, NY, USA, close-coupled nozzle; DJ-CF, discrete jet confined-feed atomizer; USGA, ultrasonic gas atomizer; HPGA, high-pressure gas atomizer; Ames, atomizer designed in Ames Laboratory, Ames, IA, USA; -not available.

Table 2.3 Atomization gas velocity determined for various atomizer designs.

Atomizer	Design	Measurement technique	Atomization pressure (kPa)	Maximum gas velocity (ms^{-1})	Mach number	Atomization gas	References
INEL	De Laval	—	120	490	—	Ar	[22]
IN3/ON7	De Laval	—	2120	—	2.6	N_2	[19]
IN3/ON7	De Laval	—	3200	—	3.0	Ar	[19]
—	USGA	Pitot tube	—	—	1.7	Ne	[23]
Osprey	Subsonic	Pitot tube	560	85	—	N_2	[23]
Osprey	Subsonic	Pitot tube	1040	120	—	N_2	[23]

INEL, atomizer designed at Idaho National Engineering Laboratory, Idaho Falls, ID, USA; USGA, ultrasonic gas atomizer; IN3/ON7, serial number for a confined atomizer; —, not reported.

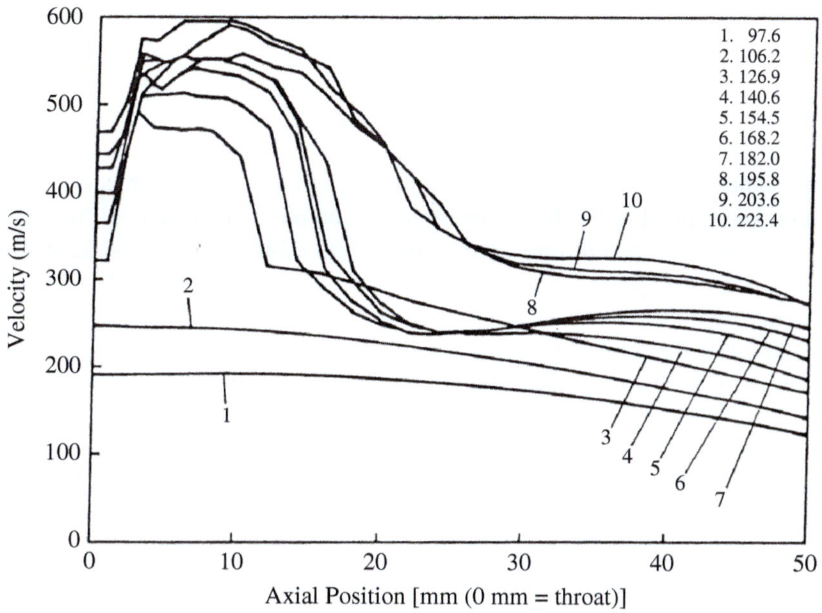

Figure 2.3 Gas velocity as a function of operating pressure and the distance from the throat of a De Laval atomizer. Source: McHugh and Key [22] / U.S. Department of Energy / Public Domain.

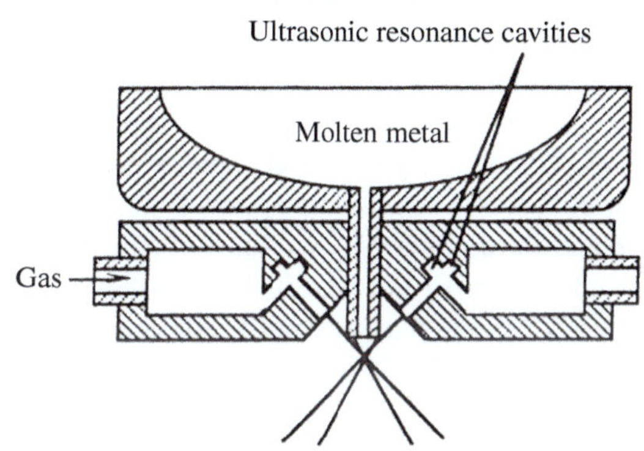

Figure 2.4 An ultrasonic atomizer design. Source: Veistinen et al. [24] / APMI International.

that as the gas jets exit the nozzle orifices, their velocity decays rapidly, mostly due to energy dissipation. As a result, the initially supersonic flow degrades into subsonic flow only a short distance from the orifices. The length of the supersonic regime in the gas jets depends on the pressurizing conditions of the atomizer. For example, the work of McHugh and Key [22] shows that for a low pressure of 120 kPa, the supersonic gas flow zone is essentially confined inside the gas nozzle orifice. The velocity of the gas falls in the subsonic regime as the gas exits the orifices. McHugh and Key noted that the length of the supersonic flow regime increases with increasing gas pressure. When the gas inlet pressure exceeds 233 kPa, the atomizing gas remains supersonic even after it exits the orifice [22].

It is important to note, however, that supersonic flow velocities have also been reported in nozzles that do not incorporate converging/diverging designs [21, 24–33]. This suggests that under certain conditions involving high pressures and complex orifice geometries, supersonic gas flow may be achieved, although the associated mechanisms remain the subject of controversy. An atomizer of this type is shown schematically in Figure 2.4 and is discussed in detail below.

2.2.5 Ultrasonic Gas Atomization (USGA)

A technique that has been widely used in the atomization of fine powders is **ultrasonic gas atomization** [21, 24–33]. The fundamental difference between ultrasonic gas atomization and other gas atomization techniques, such as subsonic and supersonic atomization, is that the atomization gas jets formed by an ultrasonic atomizer carry shock waves with frequencies beyond the audible range (i.e. 20 –100 kHz) [21, 24, 25, 27–32, 34]. The ultrasonic waves are generated as the gas flow through two resonant cavities, as shown in Figure 2.4 [24]. In principle, the shock waves are meant to increase the efficiency of the molten metal breakup process, although there is no general agreement as to the effectiveness of this approach. There are several other approaches that may be utilized to superimpose ultrasonic waves on atomizing gas jets. For example, attaching an electroacoustic transducer unit onto a gas atomizer should produce a similar effect since the ultrasonic pulses are transmitted to the atomizing media. The transducers that have been employed for atomization include both magnetostrictive and piezoelectric types [30, 34]. Ultrasonic waves have also been successfully superimposed onto the atomizing gas jets using acoustic transducers. Unfortunately, such devices can only be used in the atomization of the relatively low melting point materials, which are not typically used in additive manufacturing.

Figure 2.4 Shows a typical ultrasonic gas atomizer that is used for the atomization of molten metals. The gas nozzles are designed to incorporate resonance cavities (or resonance chambers). The design of the resonance chambers involves a mechanism that is similar to that of the Hartman tube, that is, high pressure coupled with flow interference. When a high-pressure gas stream passes through a resonance chamber, a portion of the gas is rapidly accelerated creating a sequence of ultrasonic shock waves which are superimposed onto the gas jets exiting the nozzle orifice.

As shown in Figure 2.4, the nozzles of most ultrasonic gas atomizers are not designed in a converging/diverging fashion. The gas jets generated from such nozzles are, however, generally characterized by a supersonic flow velocity. The supersonic nature of the gas jets generated from ultrasonic gas atomizers has been studied both theoretically and experimentally [23, 24, 33]. Liu et al. [33], for example, studied the gas flow characteristics of an ultrasonic gas atomizer using three distinct experimental methods, namely, **Schlieren photography**, pitot tube, and hot wire anemometry measurements. The experiments were conducted using helium gas with an ultrasonic gas atomizer operating at various manifold pressures. The gas flow was also visualized using a Schlieren photographic technique. In the pitot tube method, an overpressure is established in a pitot tube positioned inside of a gas flow field. The overpressure is then converted into a gas flow rate. One example of such an experiment is shown in Figure 2.5 where the measured pitot tube overpressure is presented as a function of the downstream distance from the nozzle exit. The results reveal a very high overpressure near the nozzle exit. The pressure decreases as the distance from the nozzle exit increases. The measured gas flow velocities are summarized in Figure 2.6, along with the calculated velocity of the atomized powders. This figure shows that the flow velocity of the gas jets is supersonic within a considerable distance from the nozzle exit. The supersonic flow characteristics were also confirmed by Schlieren's photography. Figure 2.7 shows that the length of the supersonic flow is about 125 mm when the atomizer is operating at a manifold pressure of 7.6 MPa.

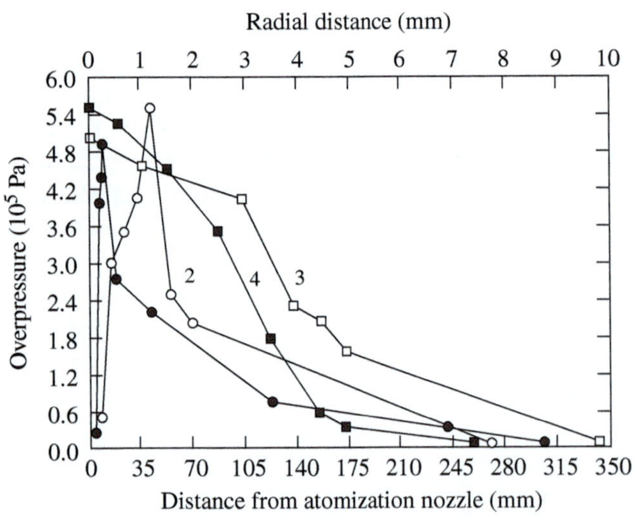

Figure 2.5 Pitot tube measurements showing the overpressure as a function of the distance from the atomization gas nozzle (curves 1 and 2) and of the radial distance from the melt axis at two gas converging points (curve 3, $L = 9$ mm without delivery tube; curve 4, $L = 42$ mm with delivery tube) dynamic atomization gas pressure, 6.9 MPa. Source: Reproduced by permission of Elsevier Science from Liu et al. [33].

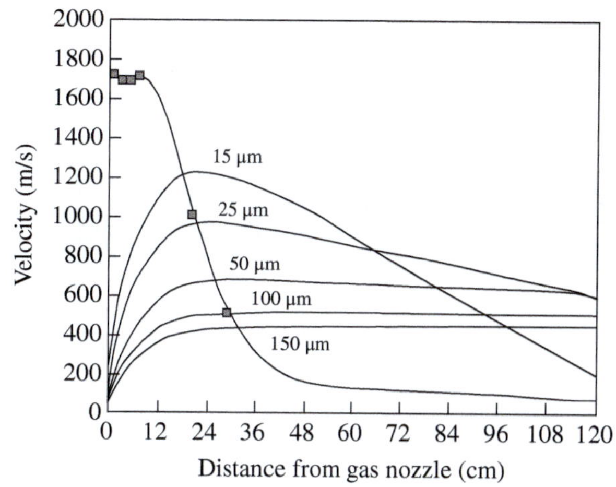

Figure 2.6 Measured gas velocity and calculated velocity of Al–5 wt% Mn–2.5 wt%Cr powder particle of various diameters as a function of the distance from the atomization nozzle: solid square, pitot tube results; open square, hot wire results. Source: Reproduced by permission of Elsevier Science from Liu et al. [33].

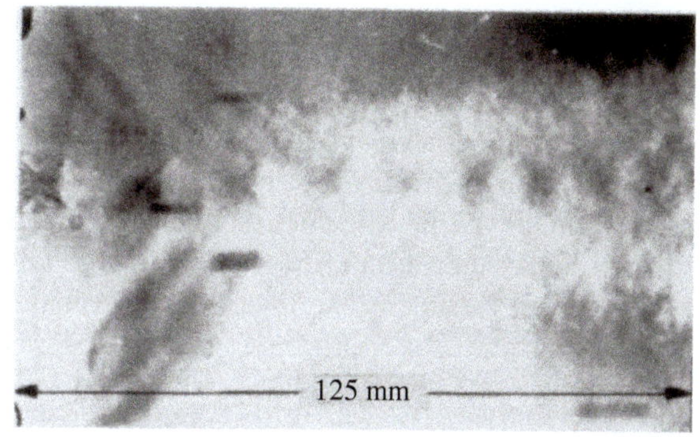

Figure 2.7 Schlieren photograph of ultrasonic helium gas atomization at a dynamic gas pressure of 7.6 MPa. The length of the supersonic flow region is about 125 mm. No melt was present during the measurements. Source: Liu et al. [33] / Reproduced with permission from Elsevier.

In related studies, exit gas velocities greater than Mach 2 have been reported by Backmark et al. [35].

Due to the flow complexity, the mechanism of supersonic gas flow from an ultrasonic atomizer remains unclear. One mechanism, however, was proposed by Veistinen et al. [24] to explain the supersonic flow behavior associated with ultrasonic atomizers based on the gas flow mechanism of a De Laval tube. They proposed that the growth of a boundary layer in the exit section of the ultrasonic nozzle reduces the effective area for gas flow. The behavior of the gas flowing through the reduced section may effectively lead to the formation of a throat in the constant cross-section gas nozzle. The phenomenon of generating an effective throat in a constant cross-section duct is known as the **vena contracta effect** and is shown schematically in Figure 2.8 [24]. In this case, if the ratio of pressure at the effective

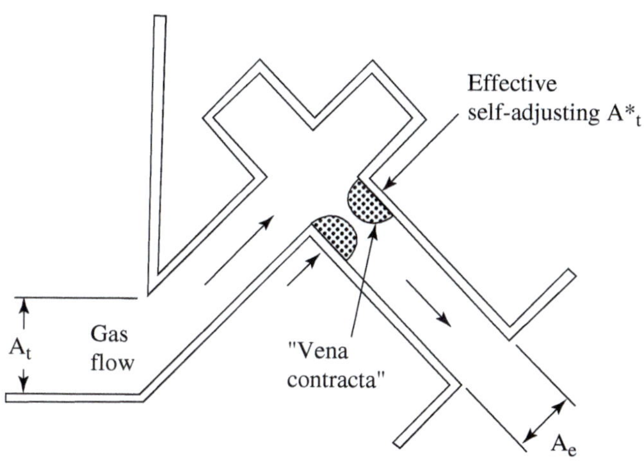

Figure 2.8 Resonance cavities and effective throat and vent areas in a USGA nozzle (not to scale). Reduction in throat area, yielding an effective A_t, is shown (highly exaggerated). Source: Reproduced by permission of APMI International [24].

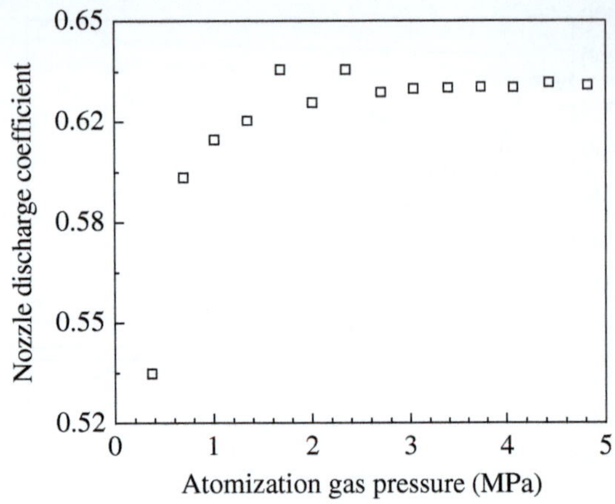

Figure 2.9 Experimentally determined nozzle discharge coefficient as a function of the atomization gas pressure (manifold pressure). Source: Reproduced by permission of APMI International [24].

throat area to that in the manifold satisfies the condition $p_t/p_o = p_t^*/p_o$, the supersonic flow may occur. Veistinen et al. verified the presence of a reduced throat section by measuring the effective gas flow rate. They argued that if an effective throat area is present, the flow rate of the atomization gas should be less than the theoretical flow rate calculated on the basis of the geometrical cross-section of the gas nozzles. Accordingly, they determined the ratio of the effective throat area to that of the normal nozzle duct area by determining the ratio of the measured gas mass flow rate to the theoretically calculated one. In this study, the measured gas mass flow rate was characterized by a duct discharge coefficient which is defined as the ratio of the experimentally determined gas mass flow rate to the theoretically calculated one under frictionless conditions. The gas discharge coefficients, as determined by Veistinen et al. [24], are summarized in Figure 2.9 for Ar gas. As shown in the figure, for low manifold pressures of less than 0.3 MPa, the discharge coefficient is 0.535. According to the proposed mechanism involving an effective throat, such a coefficient corresponds to a 53.5% reduction in the duct area at the effective throat, relative to the normal nozzle duct area.

2.2.6 Centrifugal Atomization

Centrifugal atomization differs from the aforementioned gas atomization techniques in that the atomization media is a mechanical rotation device. The energy imparted to the molten metal in the centrifugal atomization process is in the form of kinetic energy generated by means of a rotating surface [36–42]. Hence, centrifugal atomization is also referred to as **rotary atomization**. The most commonly used centrifugal atomization technique is the **rotating disc method**. Figure 2.10 illustrates that centrifugal atomization systems consist generally of a melting unit, a rotating media, and a collection unit, that is, a chamber or a substrate [37]. In the example shown in Figure 2.10, the melting unit consists of a crucible and an induction heating coil. The rotating media is simply a metallic disc placed at a distance underneath the crucible. The design of collection unit depends on the final product desired. When the end product powders, the collection unit is generally an environmental chamber with a relatively large diameter. The large diameter allows the droplets to have sufficient time to solidify into powders before they reach the chamber wall. In the Pratt Whitney (USA) design, as shown in Figure 2.11, inert gas is flushed down the chamber in order to ensure rapid cooling [43].

The advantage of centrifugal atomization Is that the atomizing conditions and cooling rate of the droplets may be controlled independently [37, 41]. The powders produced from centrifugal atomization are normally spherical and few satellites are attached. However, erosion, dissolution, oxidation, and creep of the rapidly spinning substrate material present some practical problems. For example, erosion and dissolution of the spinner may become severe when high melting temperature materials are atomized.

There are a variety of other atomization configurations that incorporate interesting modifications to the melting mechanism and rotation media. These are consumable electrode atomization, rotating crucible atomization, and rotating ingot atomization [43]. Figure 2.12 illustrates the various configurations that are available for centrifugal atomization. In the **consumable electrode method** process, as shown in Figure 2.12a [44], the starting metal/alloy is first fabricated into consumable electrodes. The electrode is heated and melted using a high-energy source. The molten metal then falls down onto a rapidly rotating disc or cup where it is atomized. The high energy source may be

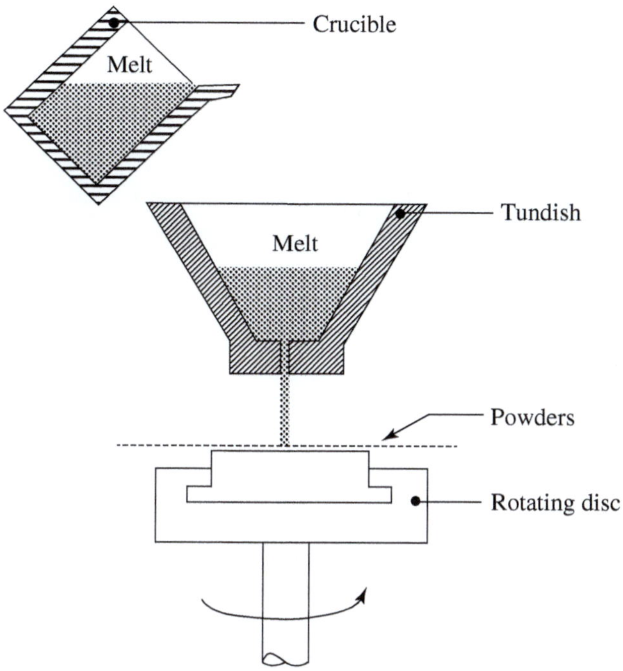

Figure 2.10 The principle of rotating disc centrifugal atomization. Source: Adapted from Kato et al. [37].

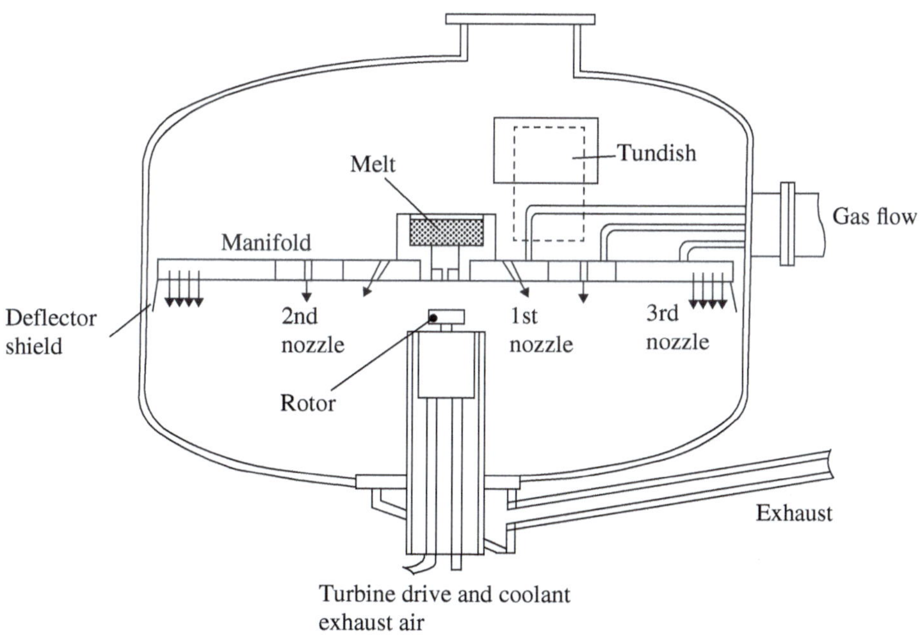

Figure 2.11 The Pratt Whitney design for centrifugal atomization. Source: [43] / Claitor's Publishing.

an electron beam or a plasma beam [44–46]. Such a method is suitable for reactive and high-temperature materials; examples are nickel-based superalloys and **titanium alloys**. In the **rotating crucible method**, the starting material is placed in a crucible and is melted using a high-energy source as illustrated in Figure 2.12b [47]. The crucible (or ingot mold) rotates at a high angular speed during atomization. Under the action of centrifugal forces, the molten metal moves along the crucible wall and is ejected into the crucible lip where it is atomized. In the **rotating ingot method**, the starting material ingot is first fabricated into a rod that rotates at high angular speeds during centrifugal atomization. The ingot is melted continuously during atomization by an electric arc generated between the ingot and an electrode. The molten metal is ejected from the rotating ingot and is atomized as shown in Figure 2.12c [48].

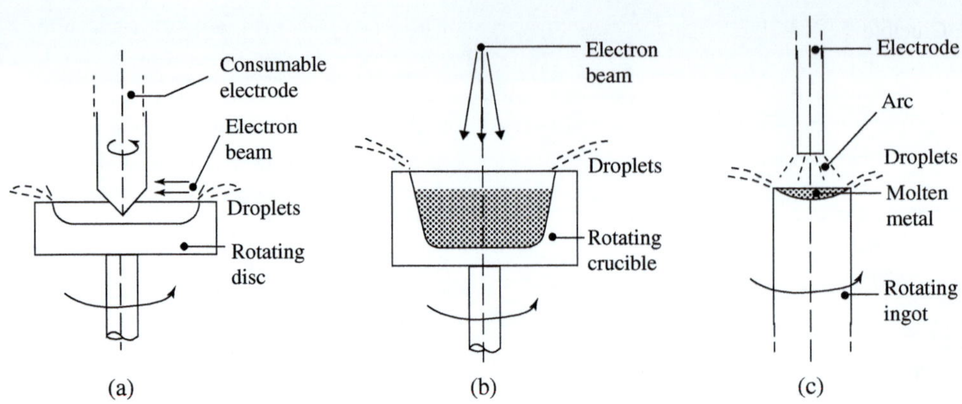

Figure 2.12 Centrifugal atomization methods: (a) consumable electrode. Source: Stephan [44], (b) rotating crucible. Source: [43] / Claitor's Publishing, and (c) rotating ingot. Source: Friedman [48].

2.2.7 Mono-sized Droplet Atomization

Mono-sized droplet atomization refers to the generation of uniformly sized droplets from liquid streams [49–55]. The principle of mono-sized droplet atomization is based on the Rayleigh mechanism of liquid column breakup. When the length-to-diameter ratio of a liquid stream exceeds a certain limiting value the liquid column becomes

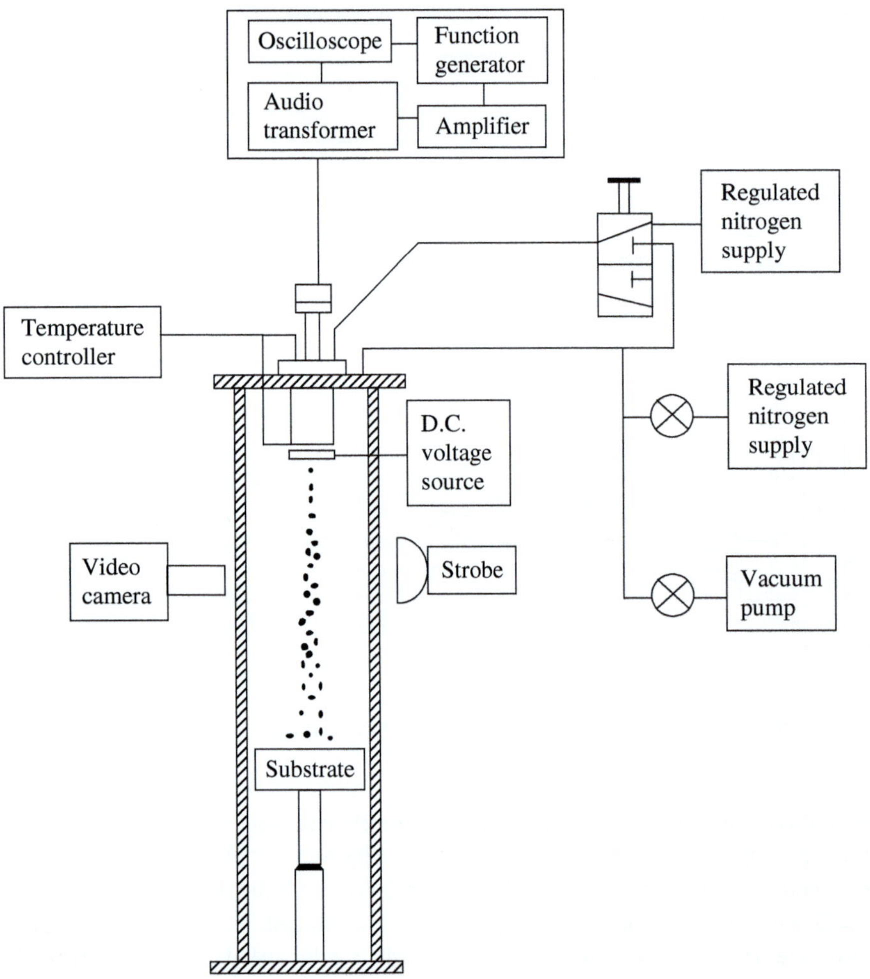

Figure 2.13 The mono-sized droplet generation device and the generation of mono-sized droplets. Source: Passow et al. [50] / Springer Nature.

unstable under the action of surface tension forces. Consequently, it will breakup into droplets by the growth of swollen and contracting waves which exhibit wavelengths greater than the circumference of the column.

Details of the Rayleigh breakup mechanism will be discussed in Sections 2.3 and 2.7.

Figure 2.13 schematically illustrates a mono-sized droplet generation device [50]. In this device, a liquid is contained in an airtight container that has a bottom pouring delivery tube leading to a large chamber. During atomization, the liquid container is pressurized so that the liquid may flow out of the container to form a free liquid stream in the chamber. The diameter of the liquid stream that flows out of the crucible can be controlled by changing the inner diameter of the delivery tube. Mechanical vibration is induced into the liquid stream by attaching a vibrational device to the bottom or the top of the liquid container in order to control the breakup of the liquid stream [49–55]. During the free fall of the liquid stream, the vibrational waves, sinusoidal or other more complex types, grow on the liquid and eventually cause the liquid stream to breakup. The vibrational frequency generally used in this method is in the 5–300 kHz range **which** is selected on the basis of the liquid stream diameter and the size of droplets desired. As illustrated in Figure 2.13, since the wavelength of the vibrational wave is uniform, the droplets thus generated exhibit a uniform size distribution. The size of the droplets may be controlled by changing the diameter of the liquid stream or by changing the vibrational frequency. However, since this method of atomization relies on the Rayleigh breakup, the size of the droplets generated is generally not less than the diameter of the original liquid stream.

Figure 2.14 shows the uniform-sized droplets that are generated using this method [51, 52]. Depending on the particular droplet generation mechanism, the important processing variables in the mono-sized droplet atomization method are the diameter of the liquid stream, the vibrational frequency, and the liquid stream velocity. One of the most important liquid properties that influence droplet size is surface tension. The mono-sized droplet atomization techniques have been successfully used for the production of droplets of uniform sizes [49–55]. As a result of the need to use vibration-generating devices, such as transducers which typically operate at low temperatures, this method is presently being used particularly for the generation of droplets from low-temperature liquids such as water, organic liquids, and low melting temperature metals. Moreover, the efficiency of this method is limited by the diameter of the droplets produced and by the need to use a large number of liquid nozzles when large mass flow rates are desired.

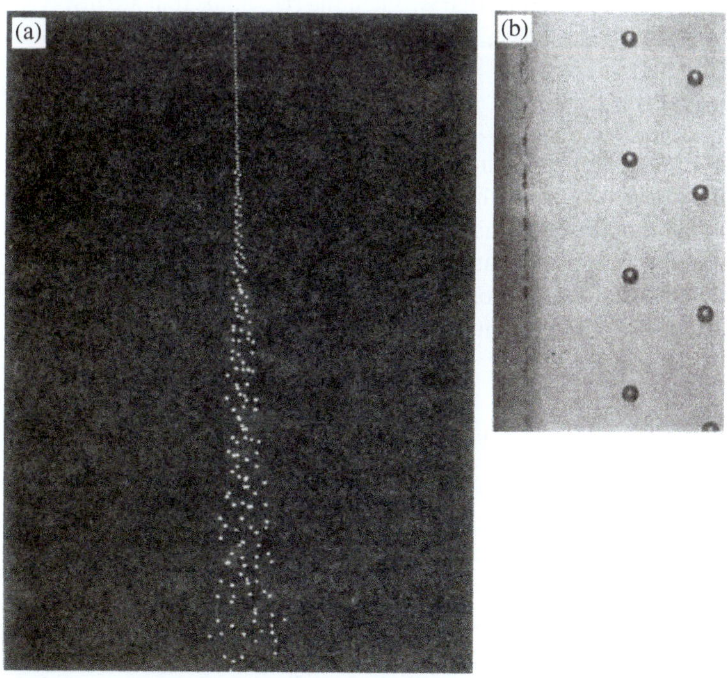

Figure 2.14 Uniform-sized droplet generated utilizing mono-sized droplet generation devices: (a) low magnification. Source: Choi [51] / American Institute of Physics (b) high magnification. Source: Ikawa et al. [56] / Reproduced with permission from The Iron and Steel Institute of Japan.

2.3 Formation of Droplets

2.3.1 Regimes of Liquid Breakup

A detailed understanding of the mechanisms that govern droplet formation during atomization is essential for optimal atomizer design and proper control of atomization conditions, and hence powder quality. Consequently, the mechanisms that govern the formation of droplets during atomization have been the subject of numerous research studies. Inspection of the available literature reveals that some comprehensive reviews are available in this area [1, 2, 10, 13, 57–60] and hence, only the principal findings will be described here. The atomization of a bulk liquid into droplets relies on the mechanical disturbances which an atomizer imposes on a liquid. It is important to note, however, that as a result of the various atomizer designs and the large number of processing variables involved, the actual breakup process of liquid into droplets may differ from case to case. There are, however, considerable similarities in the mechanisms of droplet formation from a disturbed liquid. A review of the various mechanisms that have been proposed to explain breakup mechanisms shows that most of the mechanisms of droplet formation involve two fundamental processes: the formation of an initial disturbance wave in the liquid with increasing amplitude, and the breakdown of ligaments from the disturbed liquid as the wave amplitude reaches a certain critical value. It should be noted that the term **formation of a droplet** is used hereafter to describe the process in which droplets form from a volume of disturbed liquid **which** could be in the form of a liquid film (sheet), a stream, or a large droplet.

The breakup of a liquid is intimately coupled to the interactions that occur between the liquid and the surrounding environment or atomization media. The relative velocity between the liquid and the gas is considered to be the most important factor that affects the liquid breakup process. The physics of atomization and the classical mechanisms believed to be responsible for liquid jet atomization were recently reviewed by Chigier [60]. On the basis of this work, the classical breakup regimes of a liquid may be classified according to the relative velocity between the atomizing and atomized fluids. These regimes include the Rayleigh regime, the first and second **wind** regimes, and the atomization regime. In the Rayleigh regime, the liquid is at negligible relative velocity with the gas environment. In this case, **surface tension** is primarily responsible for the breakup. In the first and second **wind** regimes, the relative velocity between the atomization and atomized fluid is significant. In this case, the aerodynamic interactions dominate the breakup process in the liquid. In the atomization regime, the two fluids interact at a high relative velocity. Moreover, liquid jet disruption occurs close to the nozzle exit plane. The following discussion will focus on the mechanisms of liquid breakup and the formation of droplets in the atomization regime, that is, when the two fluids that are involved in atomization are at a high relative velocity, because of its theoretical and practical importance.

2.3.2 Mechanisms of Atomization

The starting point in considering the formation of droplets from a volume of liquid is to understand how the liquid becomes unstable under the presence of some initial disturbances. In the literature, a somewhat oversimplified but classical picture of droplet formation from a disturbed liquid sheet was proposed by Dombrowski and Johns [61]. They considered the instability growth within a sheet of liquid moving at a relative velocity with the surrounding gas environment. Figure 2.15 shows the proposed mechanisms of droplet formation from a liquid sheet [61]. According to the classical schematic diagram shown in Figure 2.15, three important stages are responsible for the disintegration of a liquid sheet into droplets. These stages are

(a) formation and growth of disturbance waves in the liquid,
(b) disruption of liquid sheet into fragments, and
(c) formation of droplets by the further breakup of fragments.

First, **waves** initiate on the liquid sheet as a result of the disturbances that are imposed by the ambient atmosphere or the atomization media. The variation of air pressure and the shear force generated by the relative velocity at the gas–liquid interface causes some of these waves to grow in amplitude as a function of time and the distance from the nozzle exit. Second, when the most rapidly growing waves reach some critical amplitude the sheet of liquid with ripples and protuberances becomes unstable. At this point, fragments are torn off from the liquid sheet at the crests and troughs. Third, liquid fragments become unstable under the aerodynamic and surface tension forces and further break down into ligaments. Finally, droplets are formed by the spheroidization of ligaments under the action of surface tension forces.

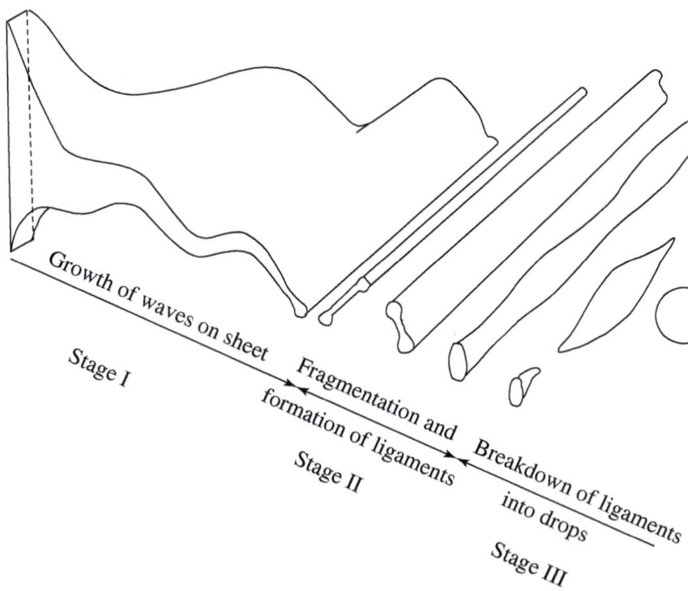

Figure 2.15 Mechanism of disintegration of a liquid sheet. Source: Dombrowski [61] / with permisison of Elsevier.

Although the simplistic view shown in Figure 2.15 was proposed to rationalize the formation of droplets from a liquid sheet, it may be applicable in the case of other forms of liquids such as cylindrical liquid jets or droplets of size greater than a certain critical value. A critical droplet size may be defined as the smallest size of droplets which is unstable under the pressure and shear forces that are imposed by the atomization environment.

Recent results suggest, however, that the **twin fluid atomization** of liquids may be more complex than the model proposed by Dombrowski and Johns (see Figure 2.16) [61]. To that effect, Klar and Fesko [1] suggested that the conventional atomization of molten metal may consist of five distinct stages. These stages are illustrated schematically in Figure 2.16 and are described as follows:

(a) formation of waves from initial disturbances,
(b) formation of fragments from the liquid,
(c) formation of droplets by the disintegration of fragments,
(d) breakup of large droplets, and
(e) coalescence of droplets during collision.

In stage I, waves form in the liquid as a result of the initial disturbances that are present on the surface of the liquid. There are several proposed origins of the initial disturbances. For example, it has been suggested that the most important disturbances are generated inside the delivery tube [60]. **Turbulence** can be generated inside a delivery tube by flow separation at sharp corners, wall roughness, boundary layers, and shear flow [60]. In stage II, the amplitude of the growing waves reaches a critical value. In this stage fragments with relatively large aspect ratios are torn off from the disturbed liquid. In stage III, the fragments that were torn off from the liquid during stage II become unstable and droplets are formed by the subsequent breakup and spheroidization of liquid fragments. Essentially, these first three stages are similar to the classical liquid breakup process described in Figure 2.15 [61]. Moreover, it has been proposed that the droplets formed during stage III may be subjected to further breakup when their size is greater than a certain critical value [1]. Accordingly, for droplets that are large than the critical size, the aerodynamic forces that are generated due to the relative velocity between gas and droplets may exceed the restoring force arising from surface tension [62–67]. In stage IV, the droplets that are greater than a critical size experience further deformation, becoming first flat and then bowl-shaped. The resultant liquid with a bowl-like geometry eventually bursts into droplets of smaller sizes [66, 68]. Since in this stage, new droplets are formed by the breakup of coarse droplets that were formed earlier, this particular stage is also referred to as **secondary atomization**. In stage V, collisions occur between droplets which formed during the early stages, leading under certain conditions, to droplet coalescence.

In earlier work, the presence of the various atomization stages that were suggested by Klar and Fesko [1] was identified indirectly during the study of the atomization behavior of molten metals. Figure 2.17 shows the presence of

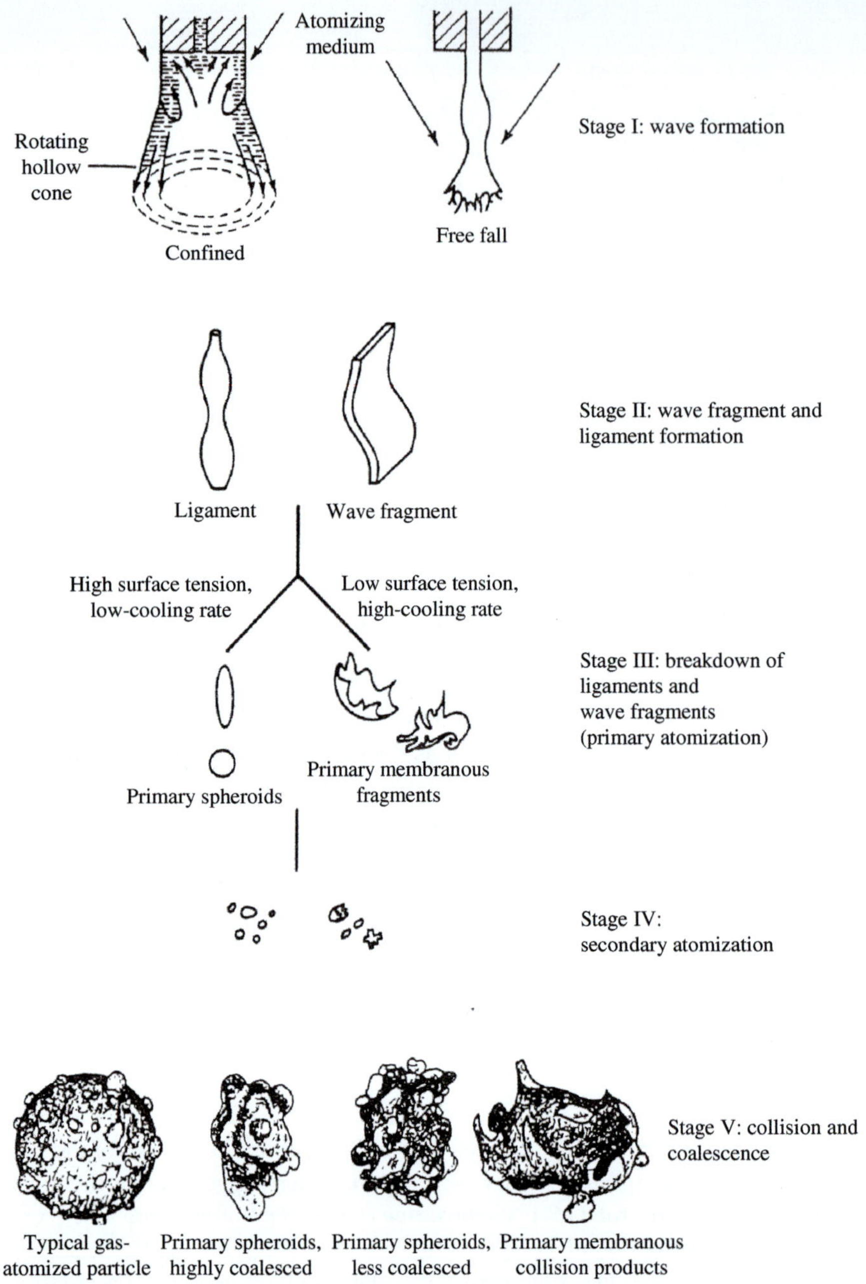

Figure 2.16 Droplet formation stages during twin fluid atomization. Source: Reproduced by permission of ASM International [1].

ligament-shaped droplets that were obtained during water and gas atomization [69]. The formation of liquid membranes during atomization is evidenced by the morphology of the water-atomized droplets shown in Figure 2.18 [66]. The collision of droplets during atomization is a well-documented phenomenon. Droplet collision is frequently inferred from the observation of satellite powder structures that are commonly present in twin fluid atomized powders (see Figure 2.19) [25]. The experimental evidence shown in Figure 2.20 also indicates the occurrence of collisions during atomization. In Figure 2.20, liquid droplets were presumably impinged upon by powders during atomization [70]. As a consequence, the deformed liquid droplets formed caps on solid droplets or **solidified** into an irregular morphology [70, 71]. However, the presence of the aforementioned five stages of atomization has been documented using in situ techniques, such as high-speed photography [17, 18, 20, 39, 60, 72–75]. The experimentally observed

Figure 2.17 Formation of ligament-shaped powders in tin atomized using N_2 gas at 350°C (metal flow rate = 50 gs^{-1} and gas flow rate = 20 gs^{-1}), powder size = −60 + 80 mesh. Source: See and Johnston [69] / Reproduced with permission from Elsevier.

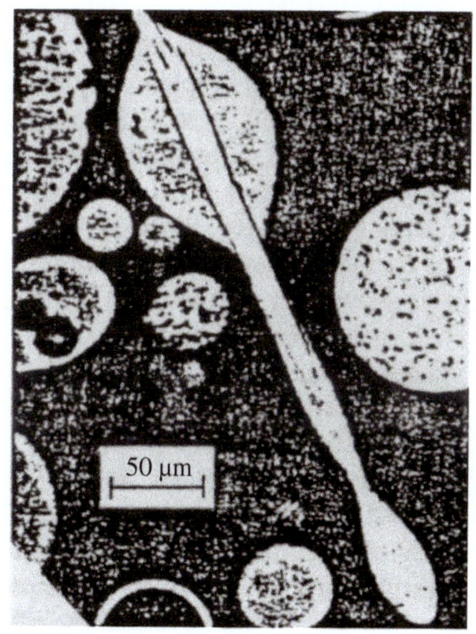

Figure 2.18 Morphology of water-atomized powders showing evidence of membranous fragments. Source: Lane et al. [66] / Reproduced with permission from American Chemical Society.

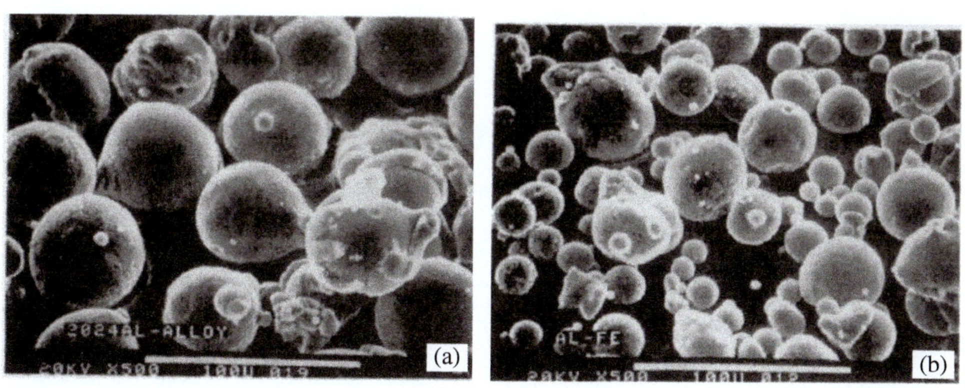

Figure 2.19 Satellite structures formed during atomization of Al alloys: (a) 2024 A1 alloy and (b) Al–Fe alloy. Source: Rai et al. [25] / Reproduced with permission from Elsevier.

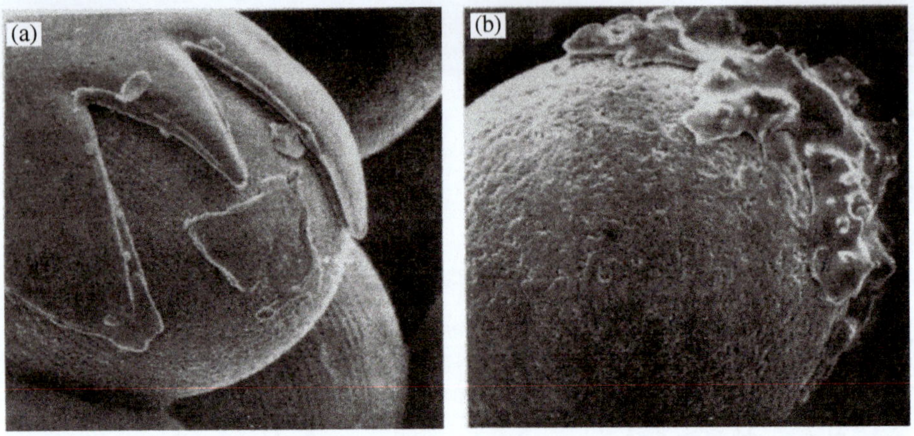

Figure 2.20 Evidence of particle collision and coalescence during atomization: (a) caps formed by splattered droplets, and (b) irregular shape caps formed by splattered droplets. Source: Kear et al. [70] / Reproduced with permission from Springer Nature.

atomization phenomena are, however, far more complex than the mechanisms suggested by Figures 2.15 and 2.16. Details of these observed atomization phenomena will be discussed in Sections 2.3.3–2.3.6.

It should be noted that the atomization of metals differs from that of other liquids, in that the solidification events that are associated with metals may affect the breakup processes and the resultant size of droplets. For example, considering stage IV, if the critical size of the droplets which may be subjected to secondary atomization is sufficiently small, droplets may already be solid or partially solid prior to secondary atomization. In this case, there may not be sufficient time for secondary atomization to occur. Alternatively, in stage V, if the droplets are completely solid, collisions may not necessarily lead to coalescence.

An atomization mechanism that is analogous to that of Klar and Fesko [1], was proposed by See and Johnston [69] to account for the experimental observations that are frequently reported for the atomization of metals. This mechanism is shown schematically in Figure 2.21. According to the mechanism proposed the atomization of liquid metals may be divided into three important stages:

(a) primary atomization,
(b) secondary atomization, and
(c) solidification.

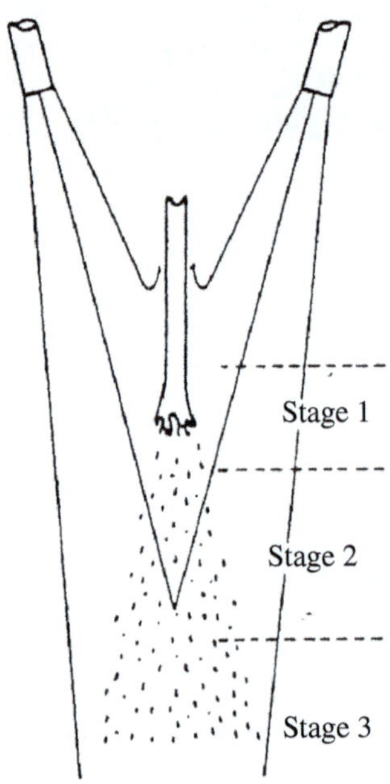

Figure 2.21 The three-stage atomization mechanism as proposed by See and Johnston. Source: Reproduced by permission of Elsevier Science from See and Johnston [69].

The primary atomization stage proposed by See and Johnston [69] is essentially similar to the three-stage breakup of a liquid proposed by Dombrowski and Johns [61] (see Figure 2.16). See and Johnston, however, argued that primary atomization occurs in the atomization system (gas–liquid system) before the maximum relative velocity is attained. During the secondary atomization stage, droplets formed during primary atomization enter a higher relative velocity regime. Under these conditions, droplets that are formed during primary atomization may undergo further disintegration. The stable size of the droplets that are formed during this stage is determined by a balance between the restoring surface tension force and the aerodynamic force. See and Johnston suggested that droplets formed during secondary atomization regain their spherical shape in the third stage. Two important processes occur during the solidification stage: restoration of droplets into a low-energy morphology (spherical) and solidification. On the basis of this model, See and Johnston proposed that the competition between spheroidization and solidification determines the final morphology of the powders produced during atomization.

The aforementioned discussion covered three fundamental physical mechanisms that are associated with the atomization process and the formation of droplets from a bulk liquid. These mechanisms were originally developed on the basis of experimental studies on the atomization behavior of a variety of liquids. Accordingly, such mechanisms may be generalized to account for the atomization behavior of liquids under a variety of conditions. In principle, the atomization process of liquids is far more complex than the simple events that are described by currently available models. Accordingly, droplet formation mechanisms may differ from experiment to experiment depending on factors such as the geometry of the liquid (cylindrical rod, sheet, or film) and atomization conditions employed. In the sections that follow, the experimental observations on the atomization behavior of liquid streams (cylindrical rods) and sheets will be discussed in detail to illustrate the similarities and differences that are present in the breakup of these liquid geometries.

2.3.3 Atomization of Cylindrical Liquids

The formation of droplets from a liquid stream (cylindrical jet) during gas atomization is dependent on the relative velocity between the liquid stream and the atomization gas (when a gas is used to promote atomization) or the surrounding air (when a liquid is simply injected into the air). At low relative velocities, the formation of droplets appears to be in general agreement with the **Rayleigh behavior** (the breakup phenomenon predicted by Rayleigh's theory [76]). At very high relative velocities, however, the droplet formation mechanism falls into the atomization regime. Detailed experimental studies of the atomization behavior of liquid streams by gas jets have been conducted by Nukiyama and Tanasawa [74] and Chigier [60]. Nukiyama and Tanasawa [74] studied the atomization of liquid streams under various kinetic conditions. In this study, the breakup process of the bulk liquid and the ensuing formation of droplets was monitored using high-speed photographic techniques. Figure 2.22 shows the dependence of

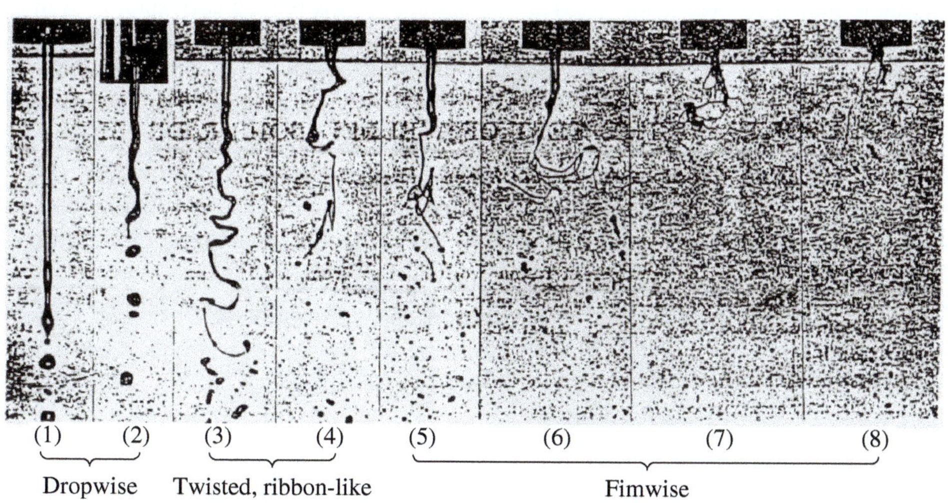

Figure 2.22 Atomization of liquid streams by gas jets and formation of droplets. Source: Nukiyama et al. [74] / Reproduced with permission from Elsevier.

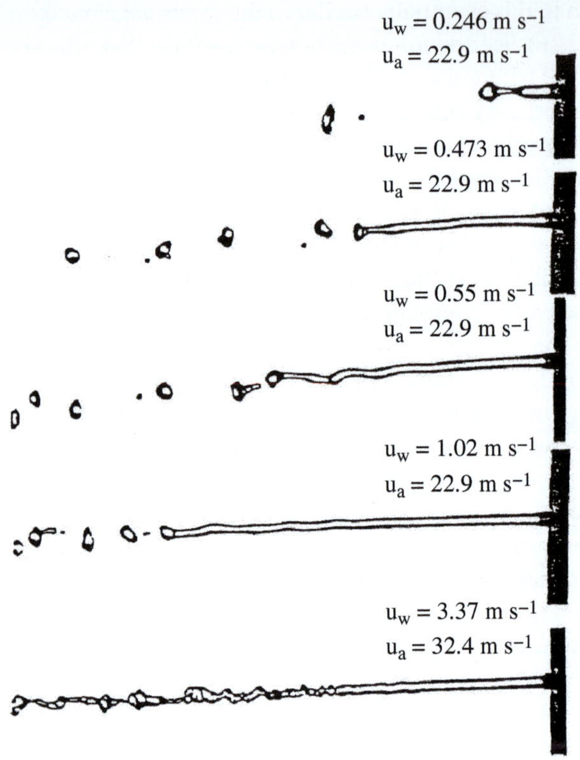

Figure 2.23 Axisymmetric Rayleigh disintegration of water jets in a coaxial air stream (jet diameter = 1.0 mm) u_w and u_a are the velocities of water and air, respectively. Source: Chigier [60] / U.S. Department of Commerce / Public Domain.

the liquid breakup process on the relative velocity between the liquid stream and the gas [74]. On the basis of phenomenological observations, Nukiyama and Tanasawa classified the atomization behavior of liquid streams into three important regimes, depending on the relative velocity between the liquid and the atomization gas [74]:

(a) dropwise atomization,
(b) twisted ribbon atomization, and
(c) filmwise atomization.

At very low relative velocities, atomization is characterized by dropwise atomization. In this velocity regime, a liquid stream becomes unstable with increasing distance from the nozzle. Swelling and contractions of growing amplitude form on the liquid stream and the stream eventually breakup into separate droplets. The diameter of the droplets that are formed during this regime is similar to that of the original liquid stream. This regime corresponds to the Rayleigh breakup regime where **surface tension** dominates the breakup process [74]. In related work, Chigier [60] observed a similar breakup mechanism by injecting jets of liquid through an air blast atomizer (but without airflow). Figure 2.23 shows photographs illustrating the breakup of water jets in a coaxial air stream corresponding to the Rayleigh breakup regime [60].

In the intermediate relative velocity regime, atomization is characterized by twisted ribbon-like atomization [74]. In this regime, initial disturbances grow in amplitude as the liquid stream leaves the nozzle (delivery tube) and the original stream is twisted into liquid ribbons. The twisting action increases with increasing distance from the atomizer and continues until ribbons finally break free and form fine liquid ligaments. The **ligaments** further disintegrate into droplets as the waves that form on their surface reach some critical amplitude. This regime of atomization concurs with Castleman's theory [77] in which aerodynamics plays an important role in the process of breakup. A small proportion of the droplets that are formed in this velocity regime are of a diameter that is smaller than that of the original liquid stream.

At high relative velocities, atomization is characterized by film-wise atomization. In this regime, the vigorous impingement of gas jets causes the twisted ribbons to flatten and to form bowl-shaped, thin liquid films. The films eventually burst into fine droplets when the surface tension force can no longer resist the aerodynamic force. The droplets formed in this atomization regime are much smaller in diameter relative to the original liquid stream [74]. The observations of Chigier et al. [60, 78, 79] on the breakup process in the high relative velocity regime reveal more

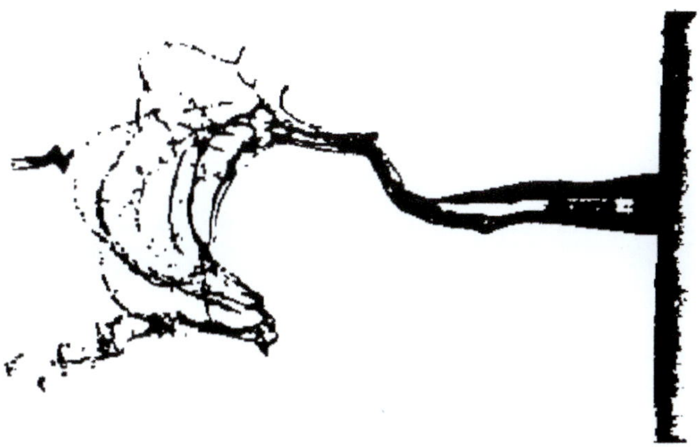

Figure 2.24 Disintegration of coaxial water jets. Source: Chigier [60] / Reproduced with permission from National Institute of Standards and Technology (NIST).

information than that described above. For example, the photographs shown in Figure 2.24 [60] reveal the disintegration of coaxial water jets during air atomization using an air blast atomizer. In these particular experiments, the liquid velocity was 1.1 ms^{-1}, whereas the air velocity was 45.8 ms^{-1}. The photographs shown in Figure 2.24 reveal a sequence of events during atomization. Figure 2.24 reveals that the original cylindrical liquid jet is first distorted and curled up into thin membranes. The **membranes** subsequently stretch between liquid rims of thicker dimension. Under the impingement of air, the membranes burst into a shower of fine droplets. The rims then breakup into droplets with diameters that are significantly larger than those formed from membranes [60].

2.3.4 Atomization of Liquid Sheets

Despite the significant differences in geometry, the experimentally observed breakup of liquid sheets is remarkably similar to those of large droplets and cylindrical liquid jets. The experimental work of Chigier [60] and Mansour and Chigier [78, 79] is useful in understanding the fundamental process of breakup of a liquid sheet during atomization. Chigier [60] and Mansour and Chigier [78, 79] used the airblast atomizer shown in Figure 2.25 to study the formation of droplets from a two-dimensional liquid sheet (water). Figure 2.26 shows the morphological development of a liquid sheet during the atomization process [78]. From this figure, it is evident that waves initially form on the sheets and that their amplitude increases with time. The growth of waves is accompanied by the formation of cells along the

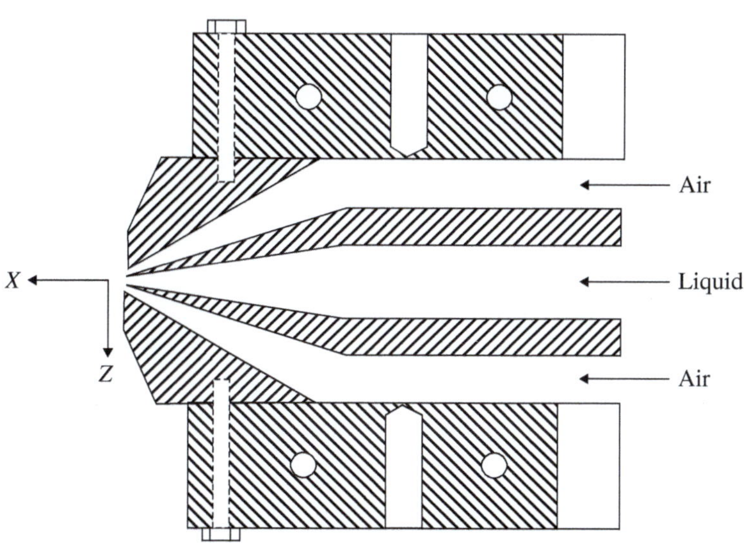

Figure 2.25 Two-dimensional variable geometry air blast atomizer. Source: Reproduced by permission of the American Institute of Physics [79].

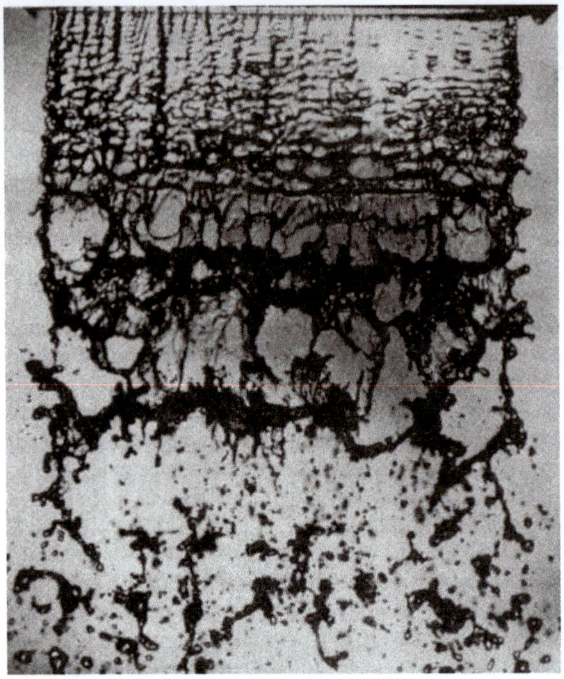

Figure 2.26 Cell structures formed on the liquid sheet surface during atomization. Source: Mansour and Chigier [78] / Reproduced with permission from American Institute of Physics.

length direction of the sheet. The cells are defined by rims of much greater thickness along their boundaries. During the breakup process, cells explode into fine droplets in a similar manner as the bagging of large droplets described in the previous sections. Rims, however, breakup at a later time than membranes forming droplets with concomitant larger diameters [78].

The close resemblance between the atomization behavior of cylindrical jets and liquid sheets under high-velocity conditions is based on the formation and breakup of curled liquid membranes and rims. Liquid membranes shatter into finer droplets and rims break into coarser droplets. Similarities between the breakup behavior of liquid sheets and cylindrical liquid streams can be clearly seen by comparing the side view of the atomized liquid sheet, as shown in Figure 2.27 [78], with that shown in Figure 2.24 for a cylindrical liquid stream [60].

Figure 2.27 Side view of a liquid sheet during atomization showing perforations which expand under the action of surface tension (same conditions as those for Figure 2.25). Source: Mansour and Chigier [78] / Reproduced with permission from American Institute of Physics.

2.3.5 Droplet Formation Under Conventional Gas Atomization Conditions

The breakup mechanisms that were discussed in the previous section represent the general cases when a volume of liquid experiences disturbances from an atomization environment. The actual breakup process that a liquid will experience under real atomization conditions will depend on a number of variables, for example, on atomizer design. This is illustrated by the two atomizer designs that are described in a previous section, that is, the **free fall design** and the confined design. The different mechanisms of liquid breakup that are associated with these two atomizer designs are illustrated in Figure 2.16. In the free fall atomizer design, except for the formation of surface disturbances, the liquid maintains its integrity prior to impingement by the gas jets. Accordingly, the droplet formation mechanism is similar to that described earlier in Figures 2.22 and 2.23. In this case, the breakup process depends on the relative velocity between the liquid and the atomization gas, as well as on other factors such as gas flow rate and liquid flow rate. The precise influence that these factors have on droplet size will be discussed subsequently in Section 2.5. The breakup process that is experienced by a liquid during confined gas atomization is slightly different from that present during free fall atomization. In a confined atomizer, the bulk liquid forms a film prior to impingement by the gas jets. The evolution of this pre-filming effect is discussed as follows. During confined gas atomization, multiple gas jets impinge the delivery tube tangentially prior to merging at a focal point located in front of the orifice of the delivery tube. The rapid flow of atomization gas passing the edge of the delivery tube results in a **low-pressure region** in front of the delivery tube orifice. This low-pressure region is shown schematically in Figure 2.28 [17] for a confined, annular atomizer. The aspiration effect that is associated with the low-pressure region causes the liquid exiting the delivery tube to expand and flow along the surface of the low-pressure zone forming an envelope of liquid film right above the focal point of the gas jets. This phenomenon is generally referred to in the literature as **pre-filming**. Under these

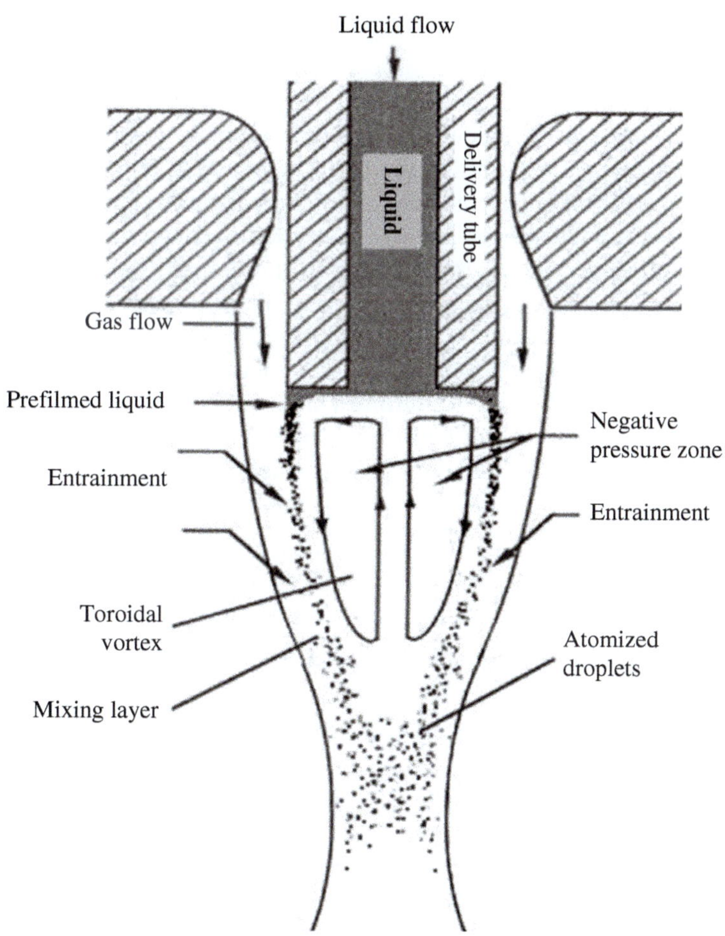

Figure 2.28 Gas and liquid flow patterns in an annular gas atomizer showing the presence of negative pressure zone in front of the delivery tube orifice and the pre-filming of liquid prior to atomization. Source: Adapted from Ünal [17].

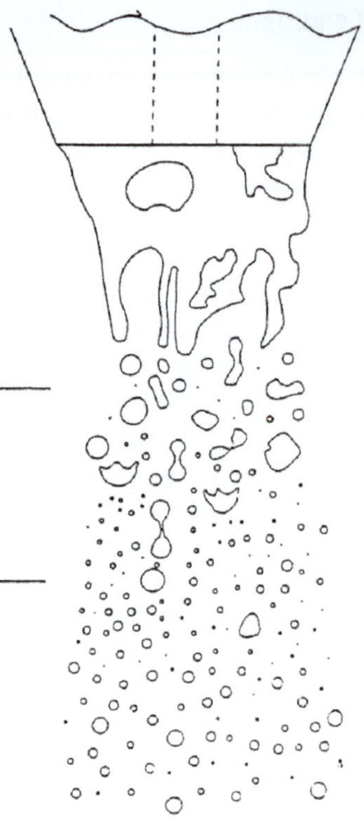

Figure 2.29 The formation of droplets during atomization using a confined atomizer: stage 1, formation of ligaments and primary droplets; stage 2, secondary atomization; and stage 3, coalescence and solidification of droplets. Source: Reproduced by permission of TMS and ASM International [18].

conditions, the liquid being atomized is essentially a film instead of a cylindrical jet. Figure 2.29 shows the mechanism of droplet formation from a confined atomizer as suggested by Ridder et al. [18] In this study, they proposed a three-stage atomization mechanism following **pre-filming**, namely, primary atomization, secondary atomization, and solidification and coalescence, as shown in Figure 2.29. These stages coincide with the schematic diagram shown in Figure 2.21 and the breakup behavior of each stage has been discussed in detail in previous sections, and hence will not be repeated here. Figure 2.30 shows the pre-filming effect during the atomization of water using an annular atomizer with De Laval-type **gas nozzles** [17]. As illustrated schematically in Figure 2.28, in this type of atomizer the liquid is delivered to the atomization zone by means of a delivery tube. As is evident from the photograph shown in Figure 2.30, instead of flowing directly to the focal point of the atomization gas in front of the delivery tube as a cylindrical liquid rod, the liquid is stretched along the surface of the delivery tube into a film which flows towards the edge of the delivery tube. Then the liquid film is atomized at the edge of the delivery tube as it is impinged directly by the atomization gas. A similar pre-filming effect was observed by Unal [80] during the atomization of molten metals

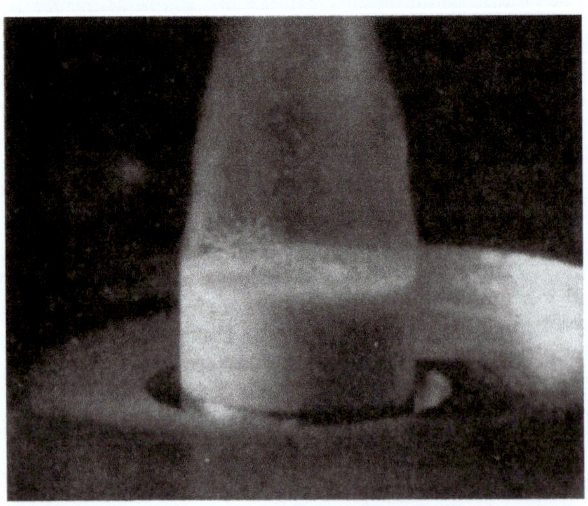

Figure 2.30 Prefilming of water during atomization using an annular atomizer. Source: Ünal [17] / Reproduced with permission from Springer Nature.

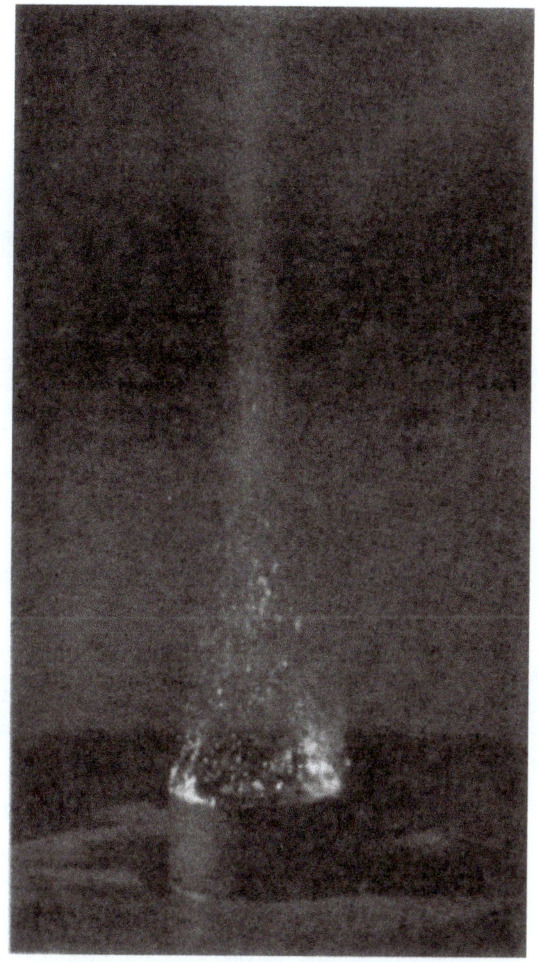

Figure 2.31 Prefilming of metal during gas atomization using a confined atomizer (metal flow rate = 0.45 kg min^{-1}). Source: Ünal [80] / Informa UK Limited.

using the same type of confined gas atomizer. Figure 2.31 shows a photograph taken during the atomization of 2014 A1 from this study.

2.3.6 Droplet Formation During Centrifugal Atomization

During centrifugal atomization, a centrifugal force is imparted by a rotating surface onto a liquid. The breakup mechanisms that are present during centrifugal atomization are influenced by the opposing effects of the metal flow rate and the rotational velocity of the atomization media. In early studies, Fraser et al. [38] investigated the breakup phenomena of liquids, such as oil and kerosene, as a function of feeding rate using rotary cups. In the rotary cup method, a liquid is fed into a rapidly rotating cup as shown in Figure 2.32. During centrifugal rotation, the liquid spreads along the inner wall of the cup and is subsequently ejected from the cup rim into the environment. Some of the results obtained from these early studies provide interesting insight into the liquid breakup mechanisms that are present during centrifugal atomization. For example, as shown in Figure 2.32, there are three different types of breakup phenomena present during centrifugal atomization, depending primarily on liquid flow rate [38]. At very low liquid flow rates, that is, less than 16.7 g min^{-1}, droplets form directly at the rim of the rotating cup (see Figure 2.22). At intermediate liquid flow rates, that is, approximately 133 g min^{-1}, the liquid ejected from the rim of the cup first degenerates into ligaments which further disintegrate into droplets during flight (see Figure 2.32b) [38]. At high liquid flow rates, that is, more than 750 g min^{-1}, the liquid first spreads into a thin film under the action of frictional and centrifugal forces [38]. Interestingly, and as shown in Figure 2.32c, the breakup mechanisms that are experienced by the liquid film thus formed are remarkably similar to that experienced by liquid sheets during gas atomization (see Figure 2.26).

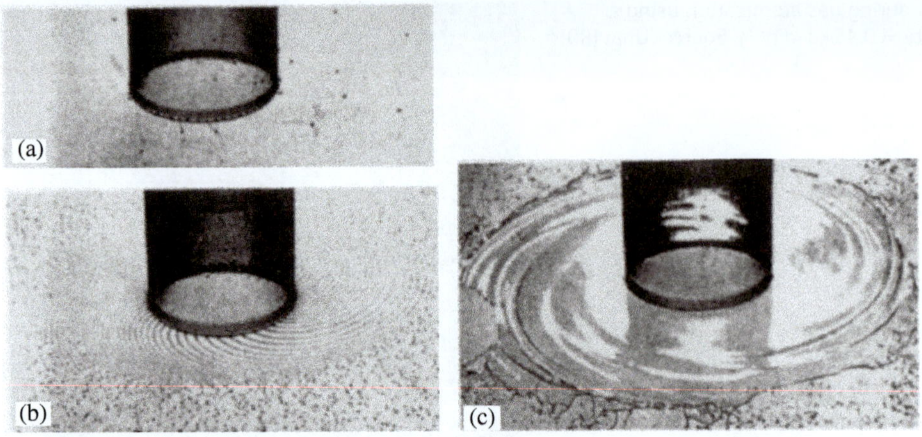

Figure 2.32 The breakup of liquids during centrifugal atomization with liquid flow rates of: (a) 16.7 g min^{-1}, (b) 133 g min^{-1}, and (c) 750 g min^{-1}. Source: Fraser et al. [38] / Reproduced with permission from Elsevier.

2.4 Control of Atomization Parameters

2.4.1 Classification of Processing Variables

For a particular melt composition, there are numerous factors that may affect the distribution of droplet sizes that are obtained during atomization. These factors may be broadly classified into two general categories:

(a) factors related to atomization conditions and atomizer design, and
(b) factors related to the physical properties of molten metals and the atomization media.

The first category includes factors such as atomizer geometry, metal stream dimensions, metal superheat, overpressure of crucible, outer diameter of delivery tube, protrusion length of delivery tube, atomization gas pressure, vibrational frequency (if present), and gas velocity. The second category includes **surface tension**, **viscosity**, melting temperature, and density of the material being atomized; and density, **viscosity**, sonic velocity, and thermal characteristics of the atomization gas. The factors that are encompassed in the first category are of interest, since control of process parameters may be readily utilized to produce desired droplet size distributions. Moreover, several of these processing parameters are interrelated, and some of them influence droplet size indirectly. The important atomization variables that govern droplet size may be identified as those related to the metal and those related to the atomization gas. The metal-related processing variables are

(a) liquid flow rate, and
(b) metal superheat.

The important atomization gas-related processing variables are

(a) gas flow rate,
(b) gas velocity,
(c) vibrational frequency,
(d) gas viscosity, and
(e) gas temperature.

These variables can be controlled to some extent by varying the processing parameters or by varying atomizer design. The control of metal superheat during atomization is straightforward. The following sections will discuss the factors that influence metal flow rate, gas flow rate, and gas velocity and how they are related to atomization parameters that may be readily controlled.

2.4.2 Factors Affecting Metal Flow Rate

Metal flow rate during gas atomization is governed directly by three important factors: gravitational force, crucible overpressure, p_c, and orifice pressure at the exit of the metal delivery tube, p_{ori}. The magnitude of the gravitational

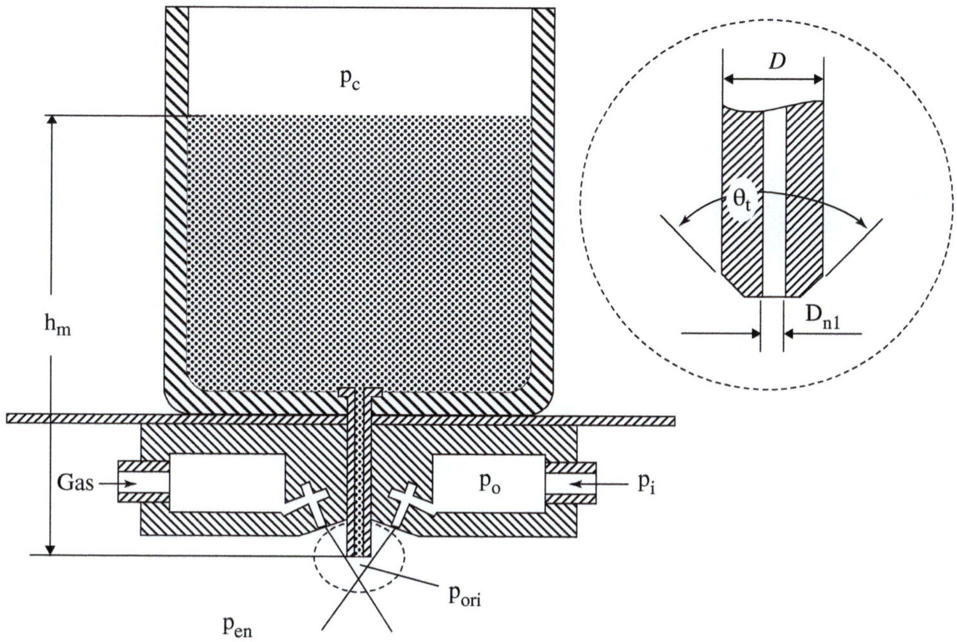

Figure 2.33 Factors affecting metal flow. In this figure, h_m is the distance between the metal surface and the orifice of the delivery tube; θ_t is the tip angle of the delivery tube; D_{nl} and D are the inner and outer diameters of the delivery tube; and p_c, p_i, p_o, $p_{ori,}$ and P_{en} are crucible overpressure, gas inlet pressure, manifold pressure, delivery tube orifice pressure, and environmental pressure, respectively.

force depends on the distance between the surface of the liquid metal inside the crucible and the orifice of the delivery tube, h_m (see Figure 2.33). Crucible overpressure is deliberately applied, whenever necessary, to control the flow rate of the molten metal. The orifice pressure, p_{ori}, at the exit of the delivery tube is generated as a result of the tangential flow of high-pressure **atomization** gas and it is subjected to changes that depend on atomizer design and atomization conditions. For commonly used atomization conditions, the orifice pressure is generally lower than the environmental pressure inside the atomization chamber. Accordingly, an aspiration effect is produced which enhances the flow of molten metal from the delivery tube. The pressure field at the exit of the delivery tube is sometimes characterized throughout the literature by **aspiration pressure**. Hence, **aspiration pressure**, p_a, is defined as the pressure differential between the orifice pressure, p_{ori}, and the environmental pressure, p_{en}

$$p_a = p_{ori} - p_{en} \tag{2.5}$$

Experimentally, the orifice pressure or, alternatively, the aspiration pressure has been identified to be closely related to the following factors [21, 24, 27, 80, 81]:

(a) atomization pressure,
(b) apex angle of gas jets,
(c) angle of a **tapered delivery tube**, θ_t (see Figure 2.33),
(d) **protrusion distance** of the delivery tube, and
(e) gas type (e.g., N_2, Ar, He, etc.).

When all other parameters remain constant, the orifice pressure varies significantly with gas atomization pressure. In the absence of crucible overpressure, when the orifice pressure is lower than the environmental pressure, the aspiration pressure assumes a negative value and the pressure field at the orifice enhances the melt flow rate. However, when the orifice pressure is greater than the environmental pressure, the pressure field at the delivery tube exit diminishes melt flow rate. In the extreme case, when the magnitude of the orifice pressure is sufficiently high, it leads to backflow of the melt and the occurrence of bubbling.

The latter phenomenon often leads to premature solidification of the metal and thus is generally avoided during atomization. Figure 2.34 shows an example of the influence of atomization pressure on orifice pressure [21, 82].

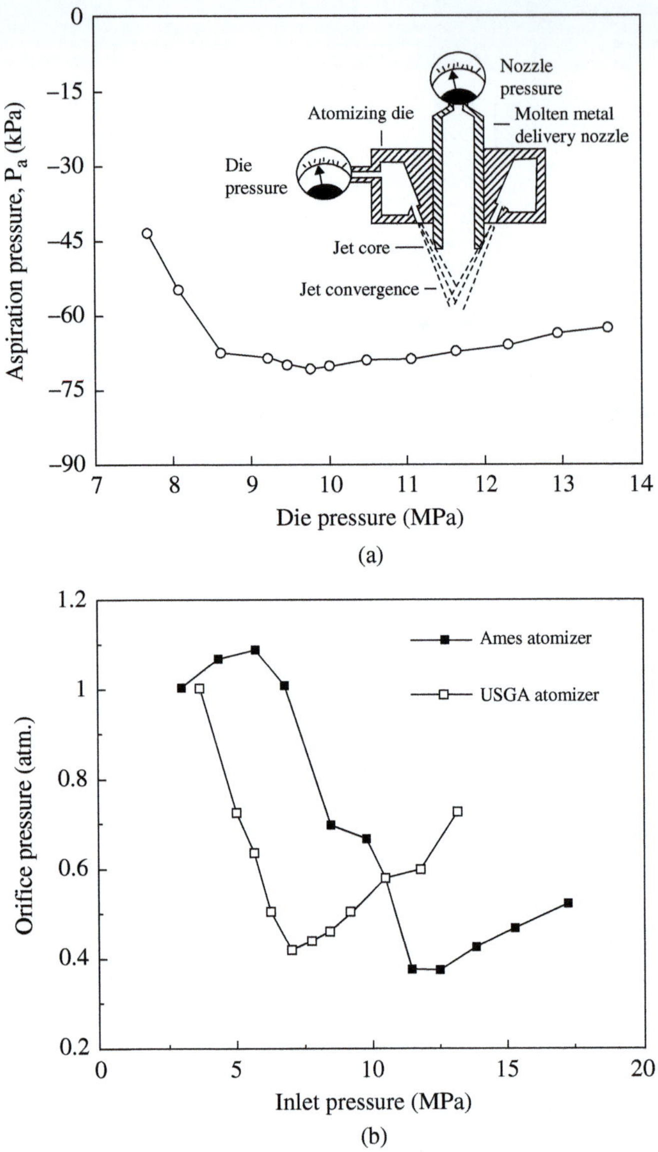

Figure 2.34 (a) Influence of atomization (die) pressure on the aspiration pressure at the exit of the delivery tube for atomization using Ar gas. Source: Reproduced by permission of Elsevier Science from Ridder and Biancaniello [82]. and (b) Influence of atomization pressure on the magnitude pressure at the delivery tube exit (orifice pressure) for atomization using Ar gas. Source: Reproduced by permission of Elsevier Science from Anderson et al. [21].

In Figure 2.34a the experimentally determined orifice pressure is shown in terms of aspiration pressure which is directly measurable using a simple pressure gauge [82]. The negative values shown in this figure are a result of the fact that the pressure at the nozzle exit is lower than the environmental pressure. In view of the fact that the environmental pressure is generally maintained at a constant value during atomization, the results shown in Figure 2.34a reveal that the aspiration effect increases with increasing atomization pressure. Moreover, a general trend that is evident in this figure is the initial decrease in orifice pressure with increasing atomization pressure (die pressure). This is followed by a slight increase in orifice pressure with further increases in atomization pressure. Figure 2.34b illustrates the orifice pressure as a function of atomization gas inlet pressure for two different atomizer designs: the Ames and the USGA atomizers [21]. The relationships between nozzle exit pressure and atomization pressure follow a similar trend for both types of atomizers. However, the minimum orifice pressures corresponding to these two different types of atomizers occur at very different atomization pressures, that is, at 7 MPa for the USGA atomizer and at 12.5 MPa for the Ames atomizer [24].

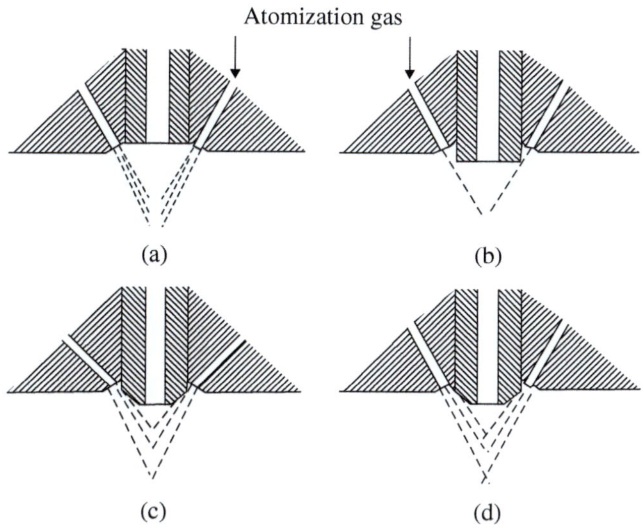

Figure 2.35 Nozzle tip designs: (a) retracted tip; (b) extended square tip; (c) tapered extended tip with 63° apex angle and (d) tapered extended tip with 45° apex angle. Source: Reproduced by permission of Elsevier Science from Anderson et al. [21].

Atomizer design, delivery tube geometry, and the delivery tube protrusion critically influence the pressure field in the vicinity of the delivery tube and consequently, the melt flow rate. Anderson et al. [21] investigated the effect of various atomizer designs on the delivery tube orifice pressure. The various atomizer designs used in this study are illustrated in Figure 2.35. In design a, the protrusion length of the atomizer is made sufficiently short so that there is no direct contact between the atomization gas jets and the delivery tube. Design b is generally used in commercial atomizers where a square-edged end of the delivery tube is extended into the gas flow field. In designs c and d, the delivery tube is tapered at a 45° angle. The primary difference between designs c and d involves the apex angle. Design c incorporates an apex angle of 63°, whereas design d incorporates an apex angle of 45°. Figure 2.36 [21] shows the ratio between orifice pressure and environmental pressure as a function of atomization pressure for these four atomizer designs. It is evident that designs a and c do not generate an aspiration effect. This is evidenced by pressure ratios of greater than unity corresponding to designs a and c for atomization pressures in the 2.5–20 MPa range. As shown in Figure 2.36, Designs b and d produce similar aspiration patterns [21]. It is worth noting that a maximum

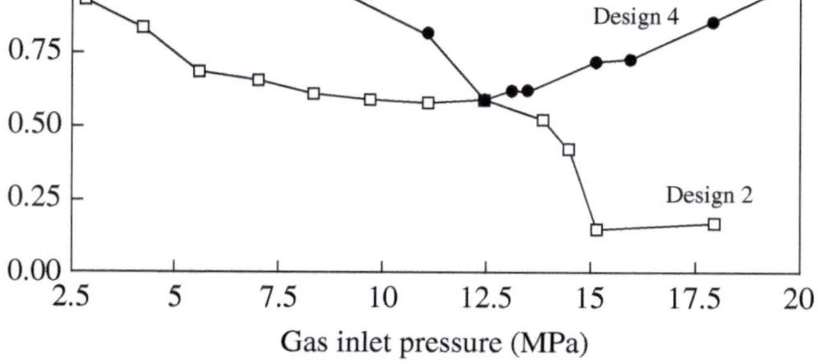

Figure 2.36 Summary of static pressure tests using the tip designs corresponding to Figure 2.32 (labeled 1–4) and a USGA atomizer. Source: Reproduced by permission of Elsevier Science from Anderson et al. [21].

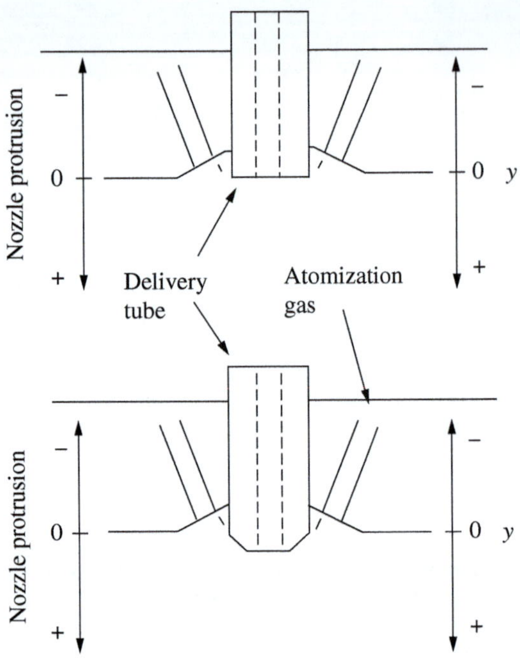

Figure 2.37 The protrusion distance for two atomizer geometries (y = nozzle protrusion distance). Source: Reproduced by permission of Elsevier Science from Veistinen et al. [81].

aspiration is produced at some intermediate atomization pressure range. Accordingly, a very high apex angle is not desirable from the standpoint of melt flow rate. In fact, the apex angle that is employed in most commercial atomizers that are used for producing metal powders is 45°. Moreover, apex angles in the 20–22.5° range are also being used for atomization.

The influence of the delivery tube protrusion distance on the aspiration effect was studied by Veistinen et al. [81] using an ultrasonic atomizer. In this study, the relative position between the nozzle exit and the atomization gas was defined using a protrusion distance, y, as shown schematically in Figure 2.37 where y is defined to have a value of zero at the lower surface of the atomizer [81]. The experimental results of Veistinen et al. are summarized in Figure 2.38 in terms of aspiration pressure, p_a, that is, from Figure 2.38, it is evident that when the delivery tube is in a retracted position (negative y), the pressure condition at the delivery tube exit becomes relatively unstable exhibiting a value that may be either greater or less than atmospheric pressure. A positive protrusion distance leads

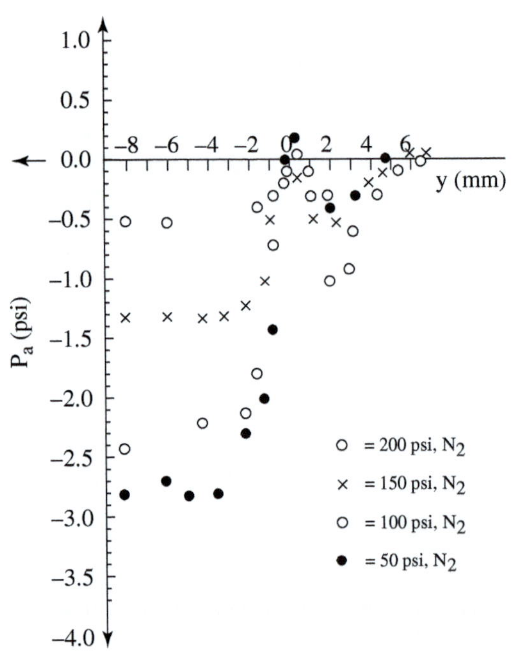

Figure 2.38 Pressure differentials at the exit of a square-ended nozzle relative to the environment as a function of nozzle protrusion distance, y. Source: Reproduced by permission of Elsevier Science from Veistinen et al. [81].

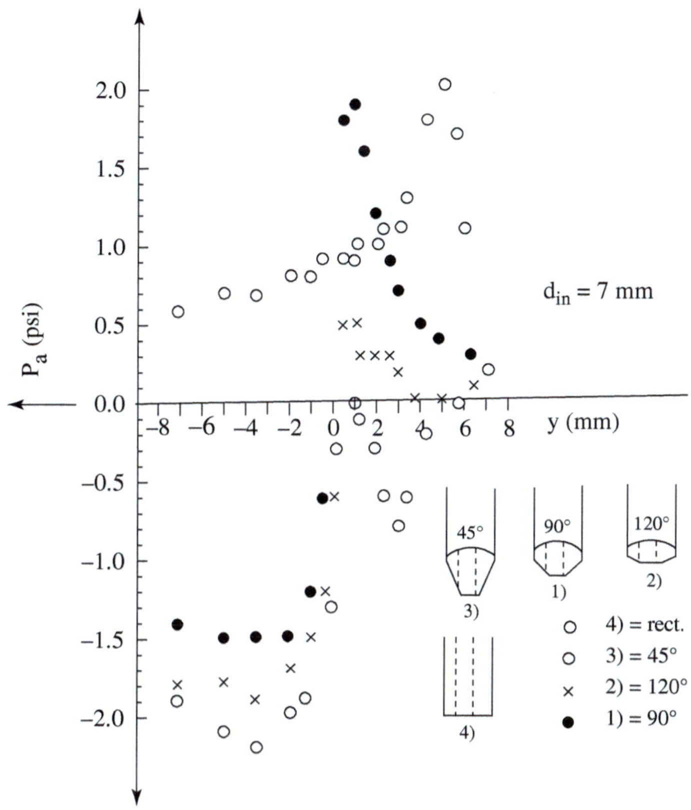

Figure 2.39 The influence of nozzle geometry on pressure field at the delivery tube exit for Ar atomization gas at 150 psi (1.0 MPa). Source: Reproduced by permission of Elsevier Science from Veistinen et al. [81].

to a subatmospheric pressure field at the orifice of the delivery tube. Consequently, the aspiration pressure is negative. As shown in Figure 2.38, in the positive protrusion distance regime, the pressure at the delivery tube exit decreases with increasing protrusion distance. This decrease in orifice pressure is followed by a saturation regime where any further increase in protrusion distance has a marginal effect on nozzle exit pressure. The pressure phenomenon at the exit of the melt delivery tube, when properly managed, may be readily utilized to control the melt flow rate. For example, it is possible to maintain a relatively constant melt flow rate, even as the metallostatic pressure level decreases with time, by selecting a nozzle position and geometry that promote a stable aspiration pressure condition.

The morphology of the nozzle also exerts a significant influence on the gas flow field and the concomitant metal flow rate. This phenomenon has been the subject of several systematic studies aimed at improving atomization efficiency [21, 27, 81]. Figure 2.39 shows the influence of nozzle geometry on the orifice pressure [81]. The results show that when all other atomization parameters are maintained constant, nozzle geometry determines the protrusion distance at which the maximum aspiration pressure or overpressure occurs. For example, as shown in Figure 2.39 when the tapered angle of the nozzle tip is less than 45°, the orifice pressure is always greater than the environmental pressure. Consequently, the pressure field at the orifice of the delivery tube diminishes metal flow. When the tapered angle is greater than 90°, aspiration occurs at large protrusion distances. Increasing the angle of the tapered nozzle leads to an increase in the protrusion that is necessary for producing maximum aspiration pressure. Moreover, when the tapered angle increases, as shown in Figure 2.39, the magnitude of the maximum aspiration pressure increases. Therefore, these results suggest that it is not necessary to utilize a tapered nozzle in order to generate a stable aspiration condition at the delivery tube exit.

2.4.3 Metal Flow Rate

The driving force for melt flow during atomization arises from three sources: crucible overpressure, metallostatic pressure due to the vertical distance, h_m (see Figure 2.33), and aspiration pressure. If one assumes that the crucible contains a limited amount of metal (i.e. no material is added during atomization), then the height of the melt in the crucible, h_m will decrease during atomization. A simple mass balance condition requires that the decrease in mass of the melt contained in the crucible must equal the mass of melt flowing out of the delivery tube [83]

$$A_c\dot{h}_m + A_n V_1 = 0 \tag{2.6}$$

or

$$-\dot{h}_m = \beta V_1 \tag{2.7}$$

where A_c and A_n are the inner cross-sectional areas of the crucible and that of the delivery tube, respectively; V_1 is the velocity of the melt stream flowing out of the nozzle; h_m is defined in Figure 2.33, β is the area ratio of A_n to A_c. Moreover, the conservation of momentum requires that [83]

$$\frac{1}{2}\rho_1 V_1^2 + p_{ori} = \frac{1}{2}\rho_1 \dot{h}_m^2 + p_c + \rho_1 g h_m \tag{2.8}$$

where ρ_1 is the density of the melt; g is the acceleration due to gravity; p_{ori} is the nozzle orifice pressure; and p_c is the crucible overpressure. Combining Eq. (2.7) with Eq. (2.8) yields an expression for the velocity of the liquid stream emanating from the nozzle as

$$V_1 = \left[\frac{2(p_c - p_{ori} + \rho_1 g h_m)}{\rho_1(1-\beta^2)}\right]^{1/2} \tag{2.9}$$

The mass flow rate, $\dot{m}_1$, may be calculated from

$$\dot{m}_1 = \rho_1 A_n V_1 \tag{2.10a}$$

$$\dot{m}_1 = A_n\left[\frac{2\rho_1(p_c - p_{ori} + \rho_1 g h_m)}{(1-\beta^2)}\right]^{1/2} \tag{2.10b}$$

In the more general case, the flow characteristics of a liquid metal can be incorporated into Eq. (2.10a) and (2.10b) by using a discharge coefficient, C_d. Eq. (2.10b) then becomes

$$\dot{m}_1 = A_n C_d\left[\frac{2\rho_1(p_c - p_{ori} + \rho_1 g h_m)}{(1-\beta^2)}\right]^{1/2} \tag{2.11}$$

The discharge coefficient, C_d, in Eq. (2.11) given by

$$C_d = \left(\frac{1}{\alpha} + e_f\right)^{-(1/2)} \tag{2.12}$$

where α and e_f are parameters that depend on the Reynolds number at the exit of the delivery tube and the geometry of the crucible/nozzle assembly [84].

The simplest case for melt flow occurs with a free fall atomizer when the crucible is not pressurized (open-top crucible). Under this condition, both p_c and p_{ori} equal the environmental pressure. Accordingly, Eq. (2.11) reduces to

$$\dot{m}_1 = A_n C_d \rho_1\left[\frac{2gh_m}{(1-\beta^2)}\right]^{1/2} \tag{2.13}$$

According to Eq. (2.13), the metal flow rate, $\dot{m}_1$, is uniquely governed by the gravitational force. Since h_m decreases continuously during atomization, the corresponding metal flow rate decreases with time. In order to obtain relatively stable atomization conditions, the metal flow rate should be controlled by either feeding molten metal continuously into the crucible or by pressurizing the crucible.

In the case of closed-type atomizers, it is generally reported that the magnitude of the aspiration pressure is significantly higher than the metallostatic pressure. Hence, under these conditions, the influence of h_m on flow rate may be safely ignored. The simplest case for metal flow occurs when an open top crucible is used and when the cross-sectional area of the crucible is much greater than that of the metal delivery tube. In this case, p_c in Eq. (2.11) equals the environmental pressure and β can be ignored. Eq. (2.11) then reduces to

$$\dot{m}_1 = A_n C_d(-2\rho_1 p_a)^{1/2} \tag{2.14}$$

where p_a is given by Eq. (2.5). The relationship between aspiration pressure and metal flow rate was determined experimentally by Oh et al. [85] in a study of the pressurization and flow behavior during gas atomization. Their

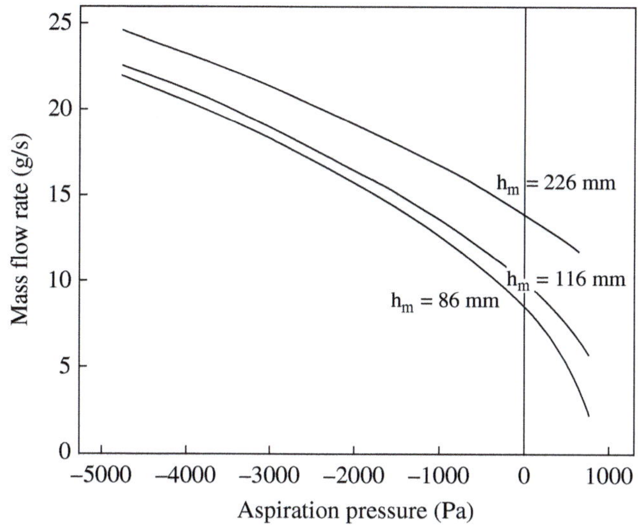

Figure 2.40 The effect of aspiration pressure, P_a, and metallostatic pressure head, h_m, on metal flow rate. Source: Oh et al. [85].

results, summarized in Figure 2.40, reveal that the mass flow rate is proportional to $(-p_a)^{1/2}$. Close inspection of Figure 2.40 indicates that the metallostatic pressure head, h_m, does affect the mass flow rate. Moreover, the influence of metallostatic pressure becomes increasingly important with decreasing magnitude of aspiration pressure [85].

2.4.4 Gas Flow Rate and Velocity

The primary factors that govern gas flow rate during atomization are gas nozzle area, nozzle geometry (i.e. straight, convergent, or divergent designs), inlet pressure, and gas chemistry. It is worth noting that gas inlet pressure is often used, instead of atomizer manifold pressure throughout the literature, although these two pressures may differ slightly under practical atomization conditions. The former is much easier to determine experimentally than the latter. Some results are available, regarding the measurement of gas flow rate during atomization. Backmark et al. [86], for example, measured the gas flow of an ultrasonic gas atomizer as a function of gas inlet pressure. In this work, bottles of gas were used as a gas source with a maximum static pressure of 10.4 MPa. The gas inlet pressure and the gas flow rate were measured during atomization. The results, summarized in Figure 2.41, indicate that the inlet

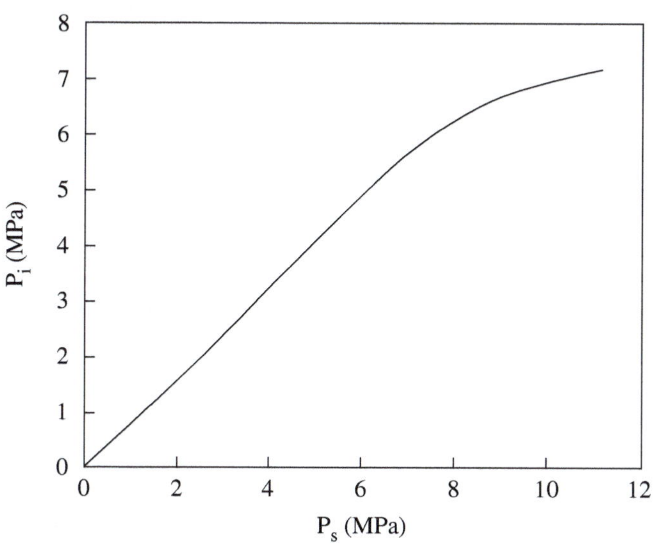

Figure 2.41 Influence of static gas pressure, P_s on dynamic gas inlet pressure, P_i. Source: Reproduced by permission of Verlag Schmid [86].

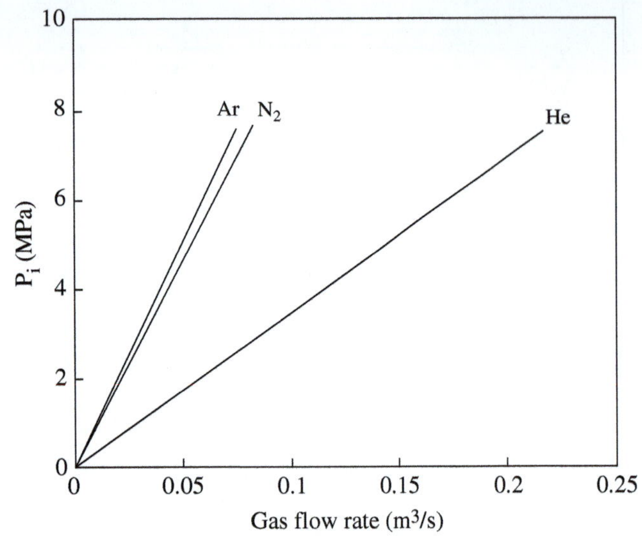

Figure 2.42 Relationship between gas inlet pressure, P_i, and gas flow rate for three different gases: Ar, N_2, and He. Source: Reproduced by permission of Verlag Schmid [86].

pressure is substantially lower than the static tank pressure [86]. Moreover, the results in Figure 2.42 show that the gas flow rate depends linearly on gas inlet pressure [86]. The results also reveal that for the same gas inlet pressure, the resultant gas flow rate is the highest for He and decreases in the sequence of He, N_2, and Ar. A similar dependence of gas flow rate on inlet pressure has been documented by various other investigators [24, 86]. It should be noted, however, that the measured gas flow rate also depends on other factors such as gas nozzle geometry and area, as well as the pressure of the environment into which the atomization gas is discharged. The differences in gas flow rate for atomizers of different designs are illustrated clearly in Figure 2.43 [87]. It is evident from these results that the two ultrasonic atomizers operating at identical atomization pressures exhibit very different gas flow rates due to differences in total gas nozzle area.

Alternatively, the flow rate of the atomization gas may be readily determined on the basis of fluid mechanics principles. It is necessary, however, to establish atomization conditions and atomizer design before such calculations can be performed. These atomization conditions are the manifold or atomization pressure, the density of the atomization gas in the manifold, and the environmental pressure in the atomization chamber. The atomizer design parameters include the geometry and the total area of the gas nozzle orifices. The relevant fluid mechanics principles that are necessary for the calculation of gas flow rate and velocity are available in the literature [88, 89]. In particular, only

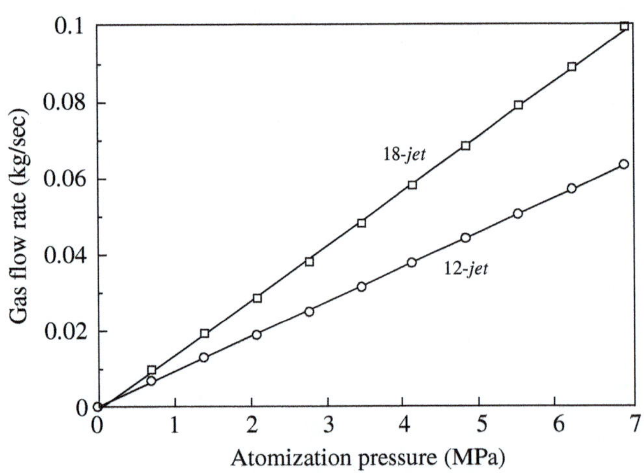

Figure 2.43 Influence of inlet pressure on gas flow rate for two different ultrasonic gas atomizers. Source: Reproduced by permission of Elsevier Science from Liang and Lavernia [87].

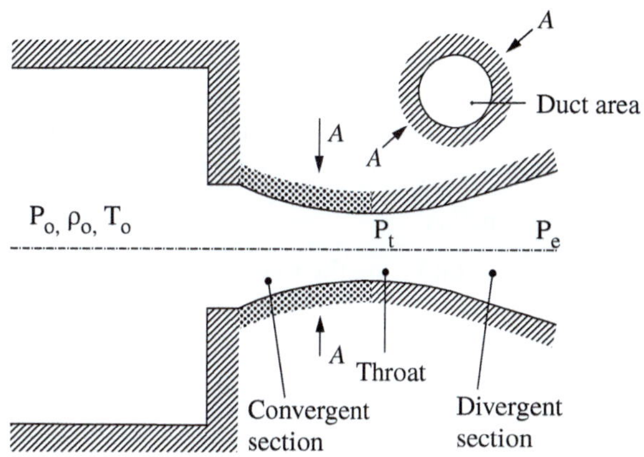

Figure 2.44 A convergent–divergent tube, where p_o, ρ_o, T_o are the stagnant pressure, density, and temperature, p_t is the throat pressure and p_e is the exit pressure. Source: Reproduced with permission of McGraw-Hill, Inc. from F. White [88].

simple examples will be considered here to illustrate the influence of processing parameters and atomizer design on the characteristics of gas flow. For example, on the basis of fluid mechanics, the calculation of gas velocity may be treated as the general case when a gas flows from a high-pressure reservoir to a low-pressure reservoir through a contracted orifice. In the discussion that follows, some of the terminologies that are generally used in fluid mechanics will be used to describe the flow rate and velocity of gas jets during atomization. Therefore, a duct is simply the tube that connects the high-pressure and low-pressure reservoirs, which can be considered to be the nozzle orifice in the case of an atomizer. The stagnant pressure is the pressure of the gas reservoir that is connected to the duct, which can be considered to be the manifold pressure. Finally, the stagnant gas density is defined as the density of the atomization gas in the manifold.

There are perhaps three basic gas duct geometries that are generally used in gas atomizers. These are **straight duct**, **convergent duct**, and **convergent–divergent duct** (the De Laval tube) geometries. The straight and convergent duct designs may be considered to be the limiting cases of the convergent–divergent duct design. A typical convergent–divergent duct geometry is illustrated schematically in Figure 2.44 and consists of a convergent section, a throat where the duct area is minimum and a divergent section. In the case of an ideal gas and assuming adiabatic and isentropic conditions, the relationship between the variation in duct area and gas velocity change is given by [88]

$$\frac{dA}{A}\frac{1}{(Ma^2-1)} = -\frac{dp}{\rho_g V_g^2} \tag{2.15}$$

where A is the area of the duct; Ma is the Mach number; p and ρ are the gas pressure and density, respectively; and V_g is the gas velocity. It follows from Eq. (2.15) that when the Mach number is less than unity, in the convergent section, a decrease in duct area is accompanied by an increase in gas velocity and a decrease in gas pressure. A sonic velocity occurs when Ma reaches unity. According to Eq. (2.15), sonic flow is possible only at the throat area when the change in duct area, dA/A, equals zero. In the divergent section, however, when the Mach number is greater than unity, the velocity of the gas increases with increasing duct area. Hence, the acceleration of the gas is accompanied by an expansion. The mechanism of gas acceleration through a convergent–divergent duct may be described as follows. First, the gas accelerates as it passes through the convergent section, eventually achieving sonic velocity at the throat area when certain critical pressure conditions are satisfied. Second, the sonic gas expands and accelerates further to a supersonic velocity as it passes through the divergent section.

Assuming isentropic conditions, the relationship between stagnation pressure, p_o (see Figure 2.33), and gas pressure in the duct, p, (see Figure 2.44) is given

$$\frac{p_0}{p} = \left[1 + \frac{1}{2}(\gamma - 1)Ma^2\right]^{\gamma/(\gamma-1)} \tag{2.16}$$

where $\gamma = c_p/c_v$; and c_p and c_v are the constant pressure and constant volume heat capacity of the gas, respectively. The flow rate is described as having achieved a critical condition when the velocity of the gas at the throat area is sonic, that is, when Ma equals unity. From Eq. (2.16), the critical gas pressure at the throat area, p_t^*, corresponding to this critical flow condition may be expressed in terms of the stagnant pressure, p_o, as

$$\frac{p_t^*}{p_o} = \left(\frac{2}{\gamma + 1}\right)^{\gamma/(\gamma-1)} \tag{2.17}$$

Therefore if one considers the schematic diagram of the duct geometry shown in Figure 2.44, supersonic flow is possible only when the throat pressure p_t equals the critical value p_t^*.

In the atomization literature, the gas flow rate is generally defined to be the mass of the atomization gas that passes through the duct (nozzle orifices) per unit time. Therefore, the mass flow rate may be calculated from the product of density and velocity of the gas, and the total cross-sectional area of the duct as

$$\dot{m}_g = \rho_g V_g A \tag{2.18}$$

where $\dot{m}_g$, ρ_g, and V_g are the mass flow, density, and velocity of the atomization gas, respectively; and A is the total duct area. The gas flow rate divided by the total duct area is defined as the mass flux. It then follows that the mass flux of a gas passing through any section of a duct is inversely proportional to the area ratio [88]:

$$\frac{\rho V_g}{\rho_t V_{gt}} = \frac{A_t}{A} \tag{2.19}$$

where A_t is the area of the throat as shown in Figures 2.44 and 2.45. Figure 2.45 illustrates the influence of the area ratio, A/A_t, on the gas Mach number for an ideal gas. The minimum value of A/A_t or the maximum mass flux occurs at an area ratio of unity and for sonic conditions (i.e. $Ma = 1$). This implies that the maximum gas flow rate occurs when the gas flow in the throat reaches sonic conditions or when the gas flow in a convergent-divergent duct reaches supersonic conditions. It then follows that when an atomizer reaches supersonic conditions, the gas flow rate will be constant if stagnant conditions (stagnant pressure, p_o, stagnant density ρ_o, and stagnant temperature, T_o) are imposed. In this case, the atomizer is said to be "choked." A simplified description of "choked" behavior is offered as follows. Consider, a duct with stagnation conditions of p_o, ρ_o, and T_o. Initially, the exit pressure, p_e, of the duct is the same as the stagnation pressure, p_o. When the exit pressure, p_e, is reduced below p_o the gas begins to flow out of the duct and the flow rate is governed by both p_o and p_e. Accordingly, it is anticipated that the gas flow rate will increase with increasing p_o and decreasing p_e. When p_e is less than a certain value the throat pressure attains a critical pressure p_t^* and the gas flow rate reaches its maximum value. From this point on, any further decrease in p_e will not result in a further increase in gas flow rate. Alternatively, when the throat pressure p_t attains the critical value, p_t^*, the gas flow rate through the duct is independent of the downstream gas pressure p_e; it is governed solely by the stagnation pressure of the reservoir or atomizer.

In general, the gas flow rate can be calculated using Eq. (2.18). The flow rates for subsonic and supersonic flow conditions, however, need to be considered separately. For an ideal gas under adiabatic and isentropic conditions, the gas density that appears in Eq. (2.18) is related to the stagnation density, ρ_0, by [88]

$$\frac{\rho_0}{\rho_g} = \left[1 + \frac{1}{2}(\gamma - 1)Ma^2\right]^{1/(\gamma-1)} \tag{2.20}$$

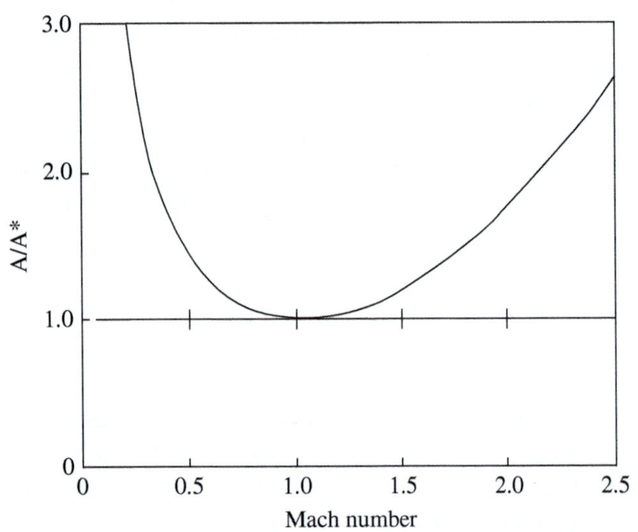

Figure 2.45 Influence of area ratio, A/A^* (where A is the duct area and A^* is the duct area in the throat) on Mach number for ideal isentropic gas flow behavior. Source: Reproduced with permission of McGraw-Hill, Inc. from F. White [88].

The gas flow rate may be determined by combining Eqs. (2.17), (2.18), and (2.20). Since as described above, the duct is choked under supersonic conditions, for a supersonic flow the gas flow rate can be calculated by computing the flow rate at the throat area using the critical conditions $\left(Ma = 1, \rho_t = \rho_t^*, \text{and } V_{gt} = V_{gt}^*\right)$ as

$$\dot{m}_g = \rho_t^* V_{gt}^* A_t \tag{2.21}$$

$$\dot{m}_g = A_t \left[\gamma \rho_o p_o \left(\frac{2}{\gamma + 1}\right)^{\gamma+1/\gamma-1}\right]^{1/2} \tag{2.22}$$

According to Eq. (2.22), for supersonic flow, gas flow rate is a function of the gas characteristics (e.g. γ) and stagnation conditions only.

For subsonic flow conditions, the gas flow rate can also be calculated by combining Eqs. (2.17), (2.18), and (2.20). In this case, however, the Mach number is less than unity and may be substituted by the pressurizing conditions of the atomizer (i.e. ρ_o, p_o, and p_e) as

$$\dot{m}_g = A_e \left(\frac{p_e}{p_o}\right)^{1/\gamma} \left\{\rho_o p_o \frac{2\gamma}{(\gamma - 1)} \left[1 - \left(\frac{p_e}{p_o}\right)^{\gamma-1/\gamma}\right]\right\}^{1/2} \tag{2.23}$$

The condition for subsonic flow is

$$\frac{p_t^*}{p_o} > \left(\frac{2}{\gamma + 1}\right)^{\gamma/\gamma-1} \tag{2.24}$$

Inspection of Eq. (2.23) shows that subsonic gas flow is governed by both the upstream stagnation pressure, p_o, and the downstream exit pressure, p_e.

The velocity of the atomization gas can be determined by calculating the Mach number of the gas when the geometry of the gas duct is known:

$$\frac{1}{Ma}\left[\frac{2 + (\gamma - 1)Ma^2}{(\gamma + 1)}\right]^{(\gamma+1)/2(\gamma-1)} = \frac{A}{A_t} \tag{2.25}$$

Therefore, the Mach number of the atomization gas can be directly calculated from the area ratio A/A_t. The actual velocity of the gas can be computed from the Mach number as

$$V_g = Ma(\gamma RT)^{1/2} \tag{2.26}$$

where R is the gas constant and T is the gas temperature. Eq. (2.25) has two possible solutions for each value of the area ratio, A/A_t. One solution is valid for subsonic conditions and the other one is valid for supersonic conditions. The determination of sonic or supersonic flow relies on knowledge of the throat pressure. It should be noted that Eqs. (2.22) and (2.26) are derived for the assumption of ideal gas under adiabatic and isentropic conditions. Hence, the results calculated using these equations may incur in certain error in both gas flow rate and gas velocity. By using Eq. (2.26), the velocity of the atomization gas can be calculated at the exit of the **gas nozzle** which can be considered to be the initial velocity of the atomization gas. The velocity of a gas, however, decreases rapidly as soon as it leaves the atomizer. Details of the decaying behavior of the atomization gas velocity will be discussed in Chapter 4.

2.5 Powder Size Distribution

In order to evaluate the influence of processing parameters on atomization behavior, one must first establish the important physical characteristics of powders or droplets. Inspection of the atomization literature reveals that powders produced by atomization techniques generally exhibit a wide distribution of powder sizes and morphologies. Therefore, the characterization of a collection of powders normally requires parameters such as average size, **size distribution**, and shape. Characterization of powders is a well-established area and a number of excellent books are available [14, 90–93]. Accordingly, this section will be limited to a description of those parameters that will be used extensively throughout this book, including average powder size and powder size distribution.

2.5.1 Powder Size

One approach to represent the **size distribution** of a population of powders is to graph the frequency of equally spaced size categories, $g(d_i)$, as a function of powder size in a **histogram** (see Figure 2.46) [90]. Alternatively, the **size distribution** may be represented in terms of a cumulative frequency, $G(d_i)$, which is defined as the fraction of powders that fall in the size range that is smaller than d_i (see Figure 2.47b) [93]. As illustrated in Figure 2.46, the three most important parameters that are used to characterize a powder size distribution are the **mode**, the **mean**, and the **median**. The mode is defined as the powder size that corresponds to the maximum frequency. The median is defined as the powder size that corresponds to the 50% cumulative frequency. Finally, the mean is defined as the weighted average size of the powders. It is important to note that distribution functions, such as those shown in Figure 2.46, may be constructed on the basis of different powder characteristics such as mass, number, volume, and **surface area**. Two of these characteristics that are most frequently used to represent a population of powders are mass and number, which correspond to distribution functions that are expressed in terms of weight and number fractions. The average powder sizes that are used throughout this book include mass median diameter, mass mean diameter, volume mean diameter, and Sauter mean diameter (SMD) [90–93]. The mass median diameter is defined

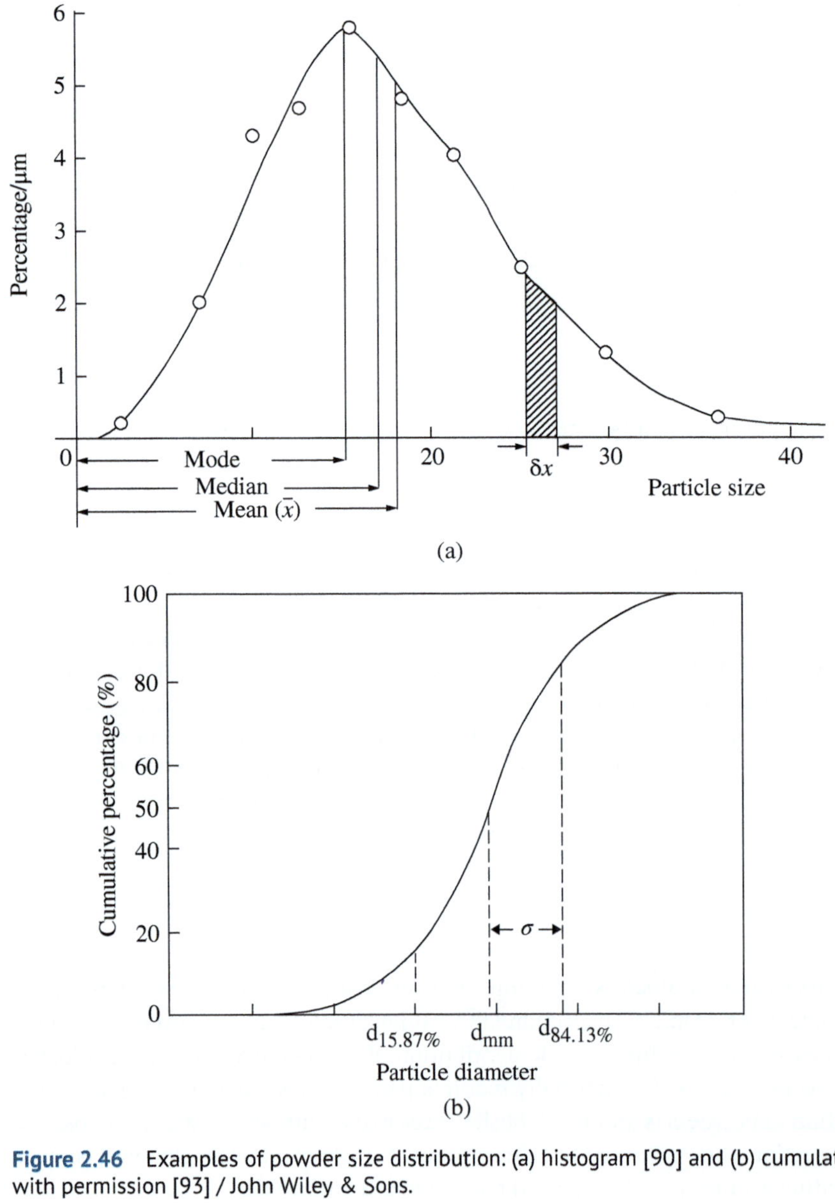

Figure 2.46 Examples of powder size distribution: (a) histogram [90] and (b) cumulative frequency curve. Source: Reproduced with permission [93] / John Wiley & Sons.

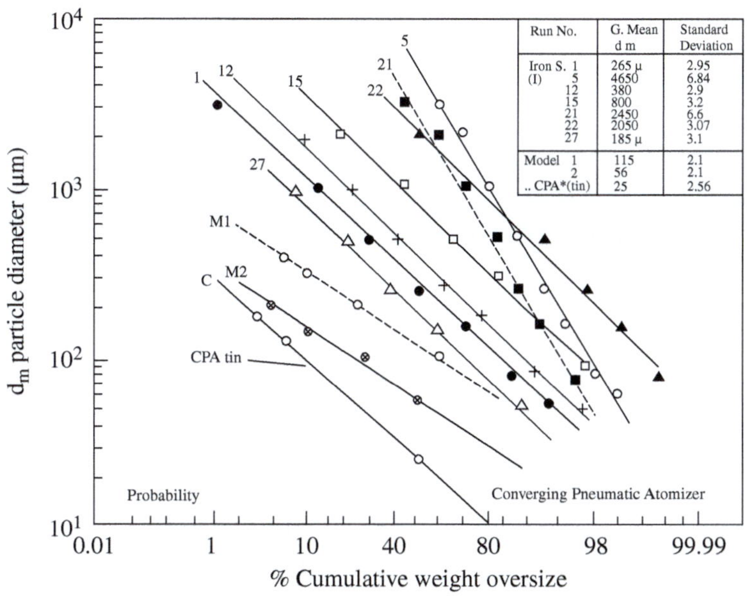

Figure 2.47 Weight percent cumulative size distribution of metal powders. Source: Reproduced with permission [73] / The Minerals, Metals and Materials Society.

as the median value as determined from a powder size distribution curve established on the basis of weight fractions. The mass mean diameter, defined as d_m, is determined on the basis of a weighted average as

$$d_m = \sum_{1}^{n} d_i g_w(d_i) \tag{2.27}$$

where $g_w(d_i)$ is the weight fraction of powders in the size range of $d_i \sim d_{i+1}$; n is the total number of size categories into which the population of powders is divided; and d_i is the powder size in the ith size category. The volume mean powder diameter is defined as:

$$d_v = \left[\sum_{1}^{n} d_i^3 g_n(d_i) \right]^{1/3} \tag{2.28}$$

where $g_n(d_i)$ is the number fraction of powders that fall in the size range corresponding to $d_i \sim d_{i+1}$, and d_v is the volume mean diameter. The SMD is defined as the volume-to-surface mean diameter and is calculated from the following equation:

$$d_{Saut} = \frac{\sum_1^n d_i^3 g_n(d_i)}{\sum_1^n d_i^2 g_n(d_i)} \tag{2.29}$$

where d_{Saut} is the SMD.

2.5.2 Size Distribution

The discussion of powder characteristics has been centered so far around the average powder diameter. The spread of the powder size distribution is an important characteristic that will determine the yield of powders that lie within a specified size range. The distribution of powder sizes that are typically generated during atomization is described frequently using a logarithmic normal function, although other distribution functions (e.g. normal) are sometimes used [90–93]. An important characteristic of a logarithmic normal distribution is that the mode, median and mean values are equal [90–93]. Moreover, the spread of a logarithmic size distribution of powders is readily characterized using a geometric standard deviation (designated as σ in Figure 2.46b) [90]. To illustrate this point, Figure 2.47 shows several examples of typical size distribution curves obtained from metal powders that were produced using a converging pneumatic atomizer. For a logarithmic normal powder size distribution, cumulative frequency as a function of powder size falls in a straight line in a double logarithmic graph. Therefore, a logarithmic normal size distribution can be characterized using two parameters: the mass median size, d_{mm} (or d_{50}), and the geometric standard deviation, σ_g. The logarithmic normal probability distribution function, $g(d)$ is given by [93]

$$g(d) = (2\pi)^{1/2}\ln(\sigma_g)\exp\left\{-\frac{(\ln d - \overline{\ln d})^2}{2\ln^2\sigma_g}\right\} \tag{2.30}$$

The geometric standard deviation, σ_g, in a logarithmic normal distribution may be calculated directly from

$$\sigma_g = \frac{d_{84}}{d_{50}} \tag{2.31a}$$

or

$$\sigma_g = \frac{d_{50}}{d_{16.1}} \tag{2.31b}$$

where d_{16} and d_{84} are the 16% and 84% cumulative weight undersize, respectively. All of these numbers are readily obtainable from a logarithmic probability plot.

Available experimental evidence suggests that the geometric standard deviation of atomized powders decreases with decreasing powder size. Consequently, the dependence of σ_g on d_{mm} (mass median diameter) may be described using a power law relationship in the form [4, 73, 94]:

$$\sigma_g = a d_{mm}^b \tag{2.32}$$

where a and b are constants. For example, for an Al–Li alloy Weldalite™049 [95], the values of a and b were reported to be 0.527 and 0.377, respectively. Hence, any variations in mass median diameter resulting from variations in the atomization parameters will affect also the spread of the distribution.

2.6 Effect of Processing Variables

2.6.1 Important Atomization Variables

In gas atomization, metal droplets are generated as a result of the dynamic interactions that occur between the gas jets and the molten metal. For fixed melt composition and atomizer design, the variables that directly influence droplet size distribution include metal superheat, metal flow rate, gas flow rate and velocity, the chemical and physical properties of the molten metal, and the chemical and physical properties of the atomization gas. In centrifugal atomization, the variables that affect the droplet size distribution include metal flow rate, rotor velocity (i.e. rotor diameter and angular speed), metal superheat, the chemical and physical properties of the molten metal, and the presence of cooling gas. A comprehensive list of these variables is summarized in Figures 2.48 and 2.49 for twin fluid atomization and centrifugal atomization, respectively. The relevant powder characteristics are also illustrated in Figures 2.48 and 2.49; the connecting lines illustrate the fact that a single atomization variable may influence several powder characteristics.

2.6.2 Atomization Pressure

Atomization pressure is the single most critical processing parameter that affects the **size distribution** of droplets that are produced by atomization. It has been consistently observed and documented by numerous investigators that when the atomization pressure increases, with other variables being kept constant, the resultant mass median diameter of the powders decreases [1, 3, 5, 12, 20, 73, 74, 96–103]. Figure 2.50 illustrates the influence of static atomization pressure on the **size distribution** of **lead powders** obtained using compressed air [58]. The results in this figure show that gas exit velocity increases with increasing static atomization pressure, with a concomitant decrease in particle size. Moreover, the results shown in Figure 2.50 support the premise that logarithmic normal statistics may be used to characterize the powder size distribution. Although atomization gas pressure is an important variable, its effect on the average powder size is indirect. In fact, gas pressure influences powder size through changes to the following three variables: mass flow rate of the metal, $\dot{m}_1$ [e.g. by producing aspiration, see Eq. (2.14)], gas flow rate, $\dot{m}_g$

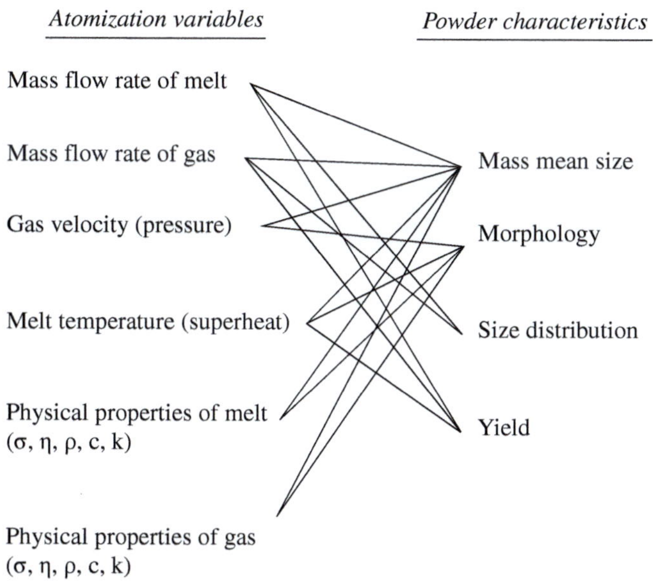

Figure 2.48 Factors affecting powder characteristics in twin fluid atomization of metals, where σ is surface energy, η is viscosity, ρ is density, c is heat capacity and k is thermal conductivity. Source: Adapted from Kato et al. [37].

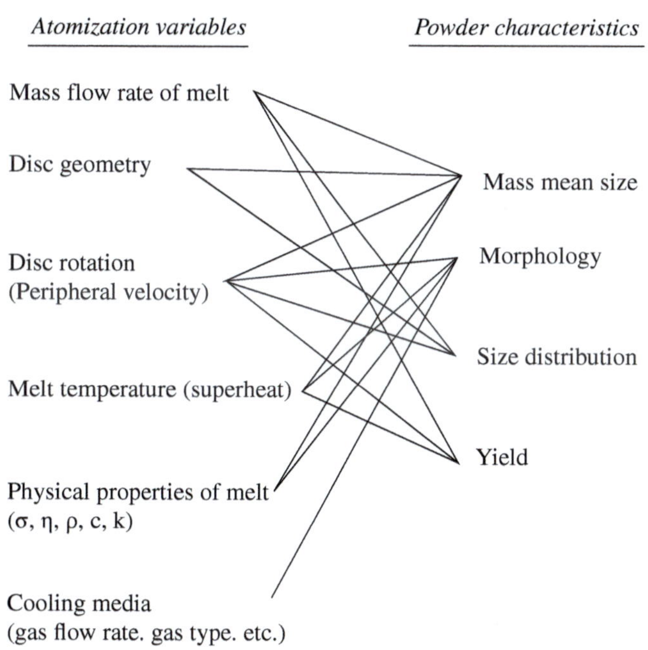

Figure 2.49 Factors affecting powder characteristics in centrifugal atomization of metals, where σ is surface energy, η is viscosity, ρ is density, c is heat capacity and k is thermal conductivity. Source: Adapted from Kato et al. [37].

[see Eq. (2.23)], and gas velocity, V_g [see Eqs. (2.23)–(2.26)]. The precise influence of atomization pressure on powder size is a complex phenomenon because average powder size decreases with increasing gas flow rate and gas velocity but increases with increasing metal flow rate. This is illustrated in Figure 2.51 where the cumulative frequency of powders produced by an ultrasonic atomizer is presented as a function of atomization pressure [20]. The results in this figure show that an increase in atomization pressure in the range between 10 and 12.5 MPa results in a refinement in powder size. However, increasing the atomization pressure from 12.5 to 17.3 MPa leads to an increase in powder size [21]. The precise influence of atomization pressure on the distribution of powder sizes is also intimately coupled to atomizer design.

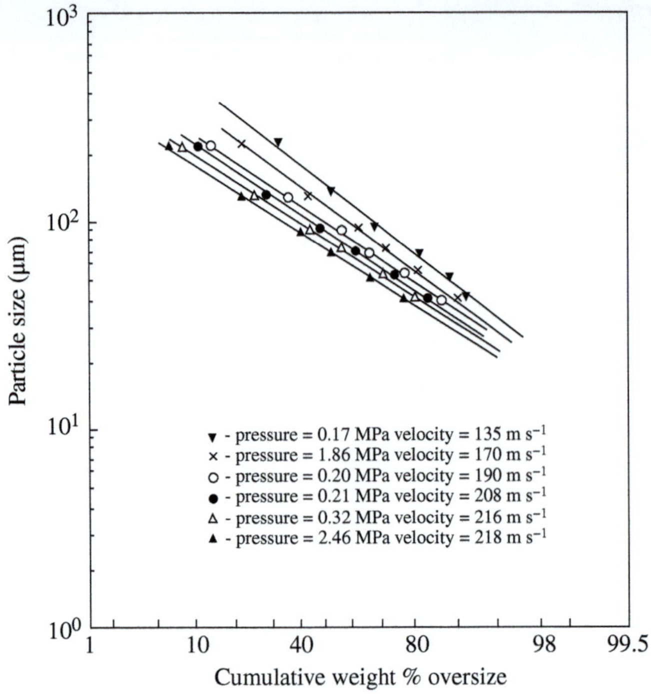

Figure 2.50 Size distribution of lead powders as a function of atomization pressure and gas velocity (atomized using compressed air). Source: Reproduced with permission of Verlag Schmid [58].

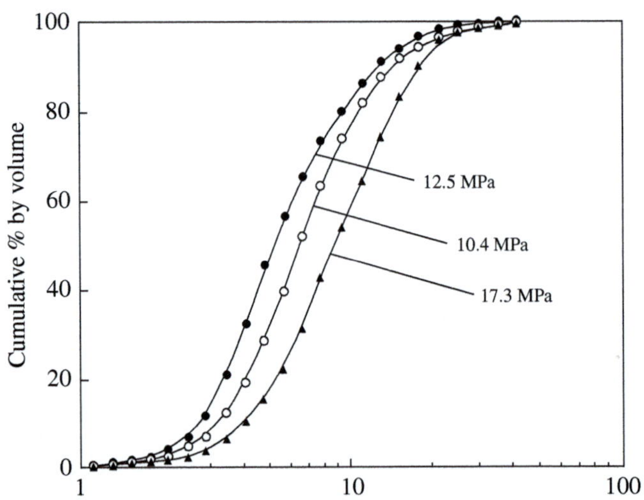

Figure 2.51 Size distribution of Sn–5 wt%Pb powders atomized using Ar gas for various atomization pressures. Source: Reproduced with permission of Elsevier Science from Anderson et al. [21].

2.6.3 Liquid Flow Rate

As discussed in Section 2.4.3, gas pressure influences the metal flow rate by promoting an aspiration or pressurization effect at the orifice of the metal delivery tube [24, 26, 27, 104]. In related studies, Couper and Singer [26] documented a series of "flow pressure maps" to demonstrate how metal flow rate could be increased by aspiration pressure (negative pressure) or retarded by bubbling (positive pressure). Moreover, Thompson [105] reported an increase in average powder size with increasing atomization gas pressure and rationalized this behavior on the basis of an increase in metal flow rate as a result of aspiration.

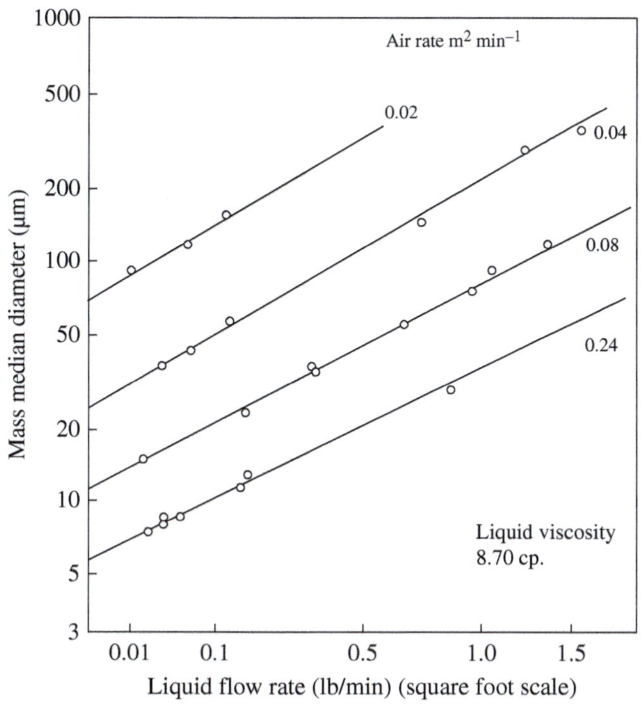

Figure 2.52 Effect of liquid mass flow rate (wax) on spray droplet size. Source: Reproduced by permission of the American Institute of Chemical Engineers. [12].

The influence of liquid mass flow rate on the mass median size of droplets was systematically studied by Kim and Marshall [12] using a pneumatic atomizer. The liquids used in this particular study were wax–polyethylene (Epolene N) mixtures. The results of this study, as summarized in Figure 2.52, indicate a general trend of increasing mass median diameter of droplets with increasing liquid mass flow rate [12]. Moreover, Kim and Marshall noted that the mass median droplet size obeyed a power law relationship with the square root of the liquid mass flow rate [12]. A similar trend of increasing droplet size with increasing liquid mass flow rate has also been documented for metals [73, 80].

2.6.4 Gas Velocity

The influence of gas velocity on atomization behavior is relatively well documented. In general, an increase in gas velocity results in a decrease in the average droplet size [74, 99–101, 106, 107]. It should be noted, however, that it is the **relative velocity** between the atomization gas and the atomized media that governs the breakup process during atomization and the concomitant size distribution of droplets. The influence of relative velocity on atomization behavior has been systematically studied by Weiss and Worsham [99] for a molten wax, Acrawax C. The results of this study, which are summarized in Figure 2.53, reveal a continuous decrease in the mass median diameter of atomized droplets with increasing relative velocity [99]. Weiss and Worsham reported that the mass median size of atomized droplets follows a power law relationship with relative velocity. Similar relationships between the size of atomized droplets and the relative velocity during atomization have also been reported by a number of investigators working with wax, oil, and molten metals [74, 100, 101, 107].

The results of numerous experimental studies have led to the development of a general empirical relationship between the mass median diameter of atomized droplets, d_{50}, and the relative velocity between the liquid and the atomization gas (absolute value), V, as

$$d_{50} \propto V^{-a} \tag{2.33}$$

The empirical exponent, a, in Eq. (2.33) has been reported to fall in the range of 1.0 – 1.68 for different atomizer designs and materials [74, 99, 101, 107]. The experimentally determined droplet size is sometimes reported as a function of the aerodynamic force in terms of the kinetic energy of the gas, $\frac{1}{2}V_g^2\rho_g$. Kim and Marshall [12], for example, noted that as

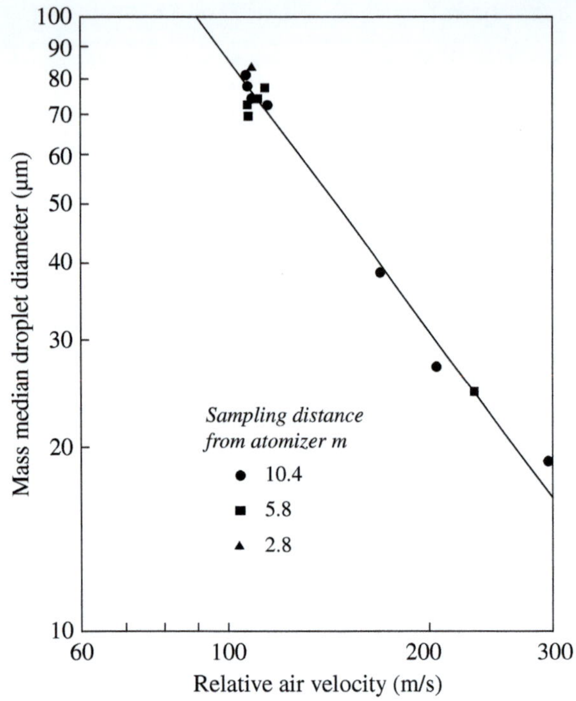

Figure 2.53 Effect of relative velocity between gas and liquid (Acra wax C) on atomized droplet size. Copyright © 1959 AIAA – Source: Reprinted with permission [99].

the **gas-to-liquid** ratio becomes very large, the mass median diameter of the droplets approaches an asymptotic value which was proposed to be the limiting diameter that can be achieved using a particular atomizer design. They used this concept of a limiting mass median diameter as a parameter to investigate the effect of the aerodynamic force on atomization. Figure 2.54 summarizes the results obtained in this study for air-atomized wax-polyethylene (Epolene N) using various pneumatic atomizer designs [12]. The results in this figure support the premise that average droplet size depends on the aerodynamic force during atomization [12]. For example, as shown in Figure 2.54, the limiting droplet diameter decreases with the increasing magnitude of aerodynamic force. Moreover, the limiting droplet sizes

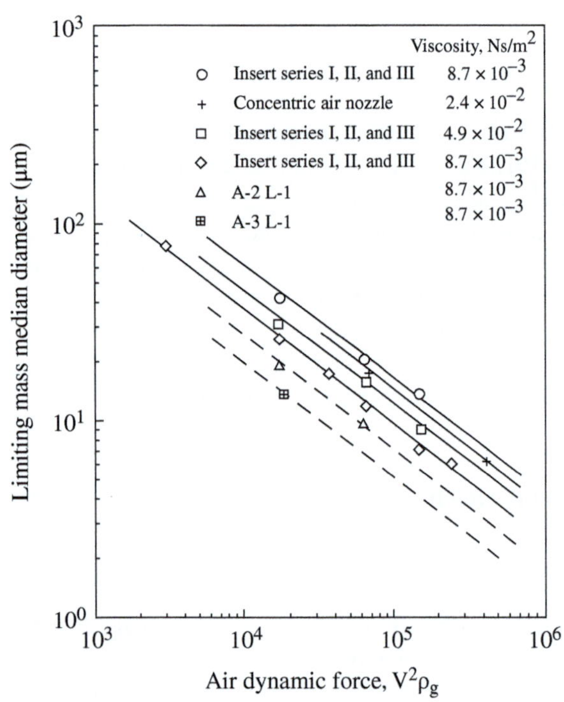

Figure 2.54 Effect of air dynamic force on atomized droplet size of wax–polyethylene mixtures for various atomizer designs (insert series, A-2 L-l and A3 L-1 refer to different atomizer designs) where $\frac{1}{2}V^2\rho_g$ is the aerodynamic force exerted by the gas. Source: Reproduced by permission of the American Institute of Chemical Engineers. © 1971 AIChE. All rights reserved [12].

fall into a power law relationship with the magnitude of the aerodynamic force. The exponent of the aerodynamic force, as determined from Figure 2.54, is –0.57 for a single air pneumatic atomizer [12].

The reported refinement in droplet size with increasing relative velocity has been rationalized in terms of secondary droplet disintegration mechanisms. In principle, a droplet that is formed during primary disintegration of a bulk liquid may be subjected to additional pressure forces when it travels at a relative velocity with the surrounding atomization fluid. The droplet will undergo further disintegration if the aerodynamic forces that develop from the surrounding fluid exceed the restoring force due to **surface tension**. If the aerodynamic force is defined in terms of kinetic energy and the surface tension force is defined in terms of the surface energy of droplets, this condition, termed the maximum stability criterion, is given by [69]

$$\frac{1}{2}\rho_g V^2 = C\frac{\sigma_1}{d_{max}} \tag{2.34}$$

where C is a constant, σ_1 is the liquid surface tension, and d_{max} is the maximum size of a droplet that is stable in the gas field.

The relationship between gas velocity and size distribution of atomized droplets is complex because of the large number of interrelated variables that influence atomization. For example, the relative velocity between gas and fluid used in Eq. (2.33) is generally defined to be the exit velocity of the atomization gas. In fact, the actual gas velocity and the concomitant relative velocity between the atomization gas and the liquid do not remain constant during atomization. Experimentally, it is generally observed that as the atomization gas leaves the nozzle orifice, it spreads and slows down during its travel to the metal contact point [14, 108, 109]. In related studies, Rai et al. [25] observed that, for a given gas pressure, the powder size is strongly dependent on the d_t/R_s ratio, where d_t is the throat diameter of each gas jet and R_s is the distance over which spreading of the gas jet occurs. This phenomenon is illustrated in Figure 2.55 [25] which shows a rapid increase in average powder size with decreasing d_t/R_s ratio.

2.6.5 Gas Flow Rate

Gas flow rate influences not only the distribution of droplet sizes but also the economic viability of the atomization processes. Since the gases that are generally used for atomization of high-quality metal powders are Ar, N_2, and He, an elevated gas flow rate translates into a high cost per unit amount of powder produced. The effect of gas flow rate on droplet size is relatively well established [12, 20, 80, 99]. Generally, available experimental results reveal a trend of decreasing droplet size with increasing gas flow rate. This trend is clearly indicated in Figure 2.52 for wax droplets

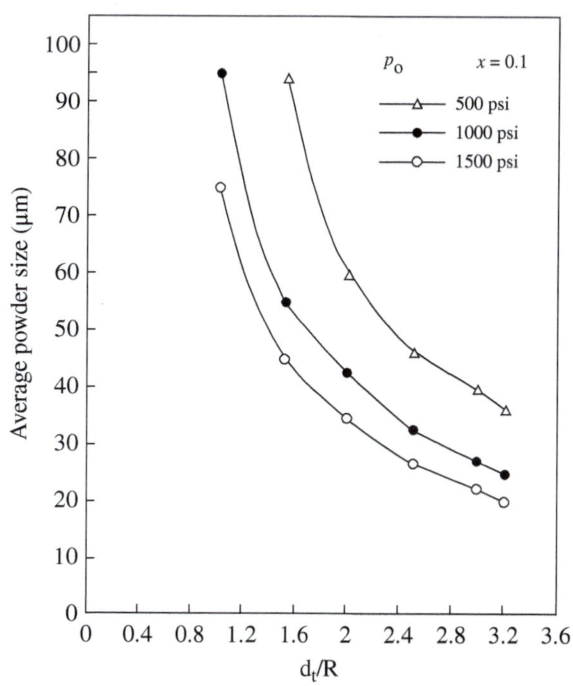

Figure 2.55 Average powder size as a function of the ratio between gas jet diameter (d_t) and spread distance (R) for three atomization pressures (p_o) (x = metal-to-gas flow rate) Source: Reproduced with permission from Rai et al. [25].

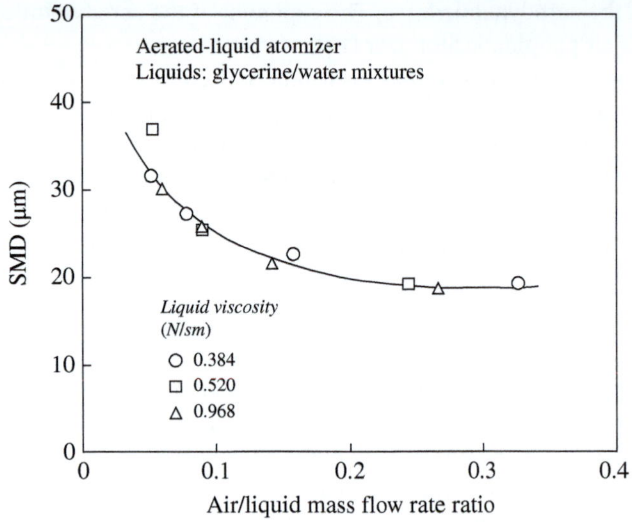

Figure 2.56 Influence of air/liquid ratio on the Sauter mean diameter (SMD) for liquids of high viscosities [110] (data from [112]).

produced by gas atomization [12]. As shown in this figure, corresponding to each liquid flow rate, the mass median droplet diameter of wax droplets produced using a gas flow rate of 0.24 m^3 min^{-1} is much finer than that of droplets produced using a gas flow rate of 0.02 m^3 min^{-1}.

It should be noted that in atomization studies, it is a common practice to present droplet size distribution data as a function of the **gas-to-liquid** mass flow rate ratio. As expected on the basis of the previous discussion, an increase in the mass flow rate ratio leads to a refinement in droplet size. This trend is shown in Figure 2.56 [110, 111] for air atomized water–glycerin mixtures. A similar behavior has been reported for various other liquids, including atomized water, wax, and liquid metals [12, 20, 73, 80, 99, 110].

There are some limiting cases that should be highlighted regarding the influence of the gas and liquid flow rates on atomization behavior. For a constant gas flow rate, a decrease in liquid flow rate is necessarily accompanied by an increase in the gas-to-liquid mass flow rate ratio. However, the mass median droplet size generally reaches a limiting value when the gas-to-liquid flow rate becomes very large. Accordingly, further increases in the gas-to-liquid flow rate ratio will not result in a significant refinement of the droplet size. This behavior has been documented for air atomized wax–polyethylene (Epolene N) [12] and air–water–glycerin mixtures [110, 111]. Lefebvre [110] and Beck *et al.* [111], for example, studied the atomization behavior of water–glycerin mixtures as a function of the air/liquid flow rate ratio. The results of this study, presented in Figure 2.56, show that the atomized droplet size of water–glycerin decreases with increasing air/liquid flow rate ratio. At flow rate ratios greater than 0.25, however, the average droplet size becomes insensitive to changes in air/liquid flow rate ratio. Consequently, further increases in the air/liquid ratio beyond this value promote only marginal refinement of the droplet size. For very small gas-to-liquid flow rate ratios, the droplet size versus flow rate ratio curves, corresponding to various gas flow rates, converge and the spray deteriorates. In this case, the velocity condition falls out of the atomization regime and the liquid cannot be effectively atomized (see Section 2.3). Kim and Marshall [12], for example, studied the atomization behavior of wax–polyethylene (Epolene N) using pneumatic atomizers operated with compressed air. They noted that the limiting gas-to-liquid flow rate ratio beyond which the wax spray deteriorated is 0.1.

2.6.6 Mechanical Disturbances

In an effort to promote droplet refinement, vibrational waves, and other mechanical disturbances have been superimposed on atomization gas jets and/or the fluid being atomized. The effectiveness of this approach, however, remains to be validated since often it is difficult to ascertain the individual influence of multiple disintegration mechanisms during gas atomization [21, 24–32, 34]. The important role played by gas dynamics in the determination of atomizing characteristics has led to the development of complex atomizer designs. Ultrasonic gas atomizers offer an example of

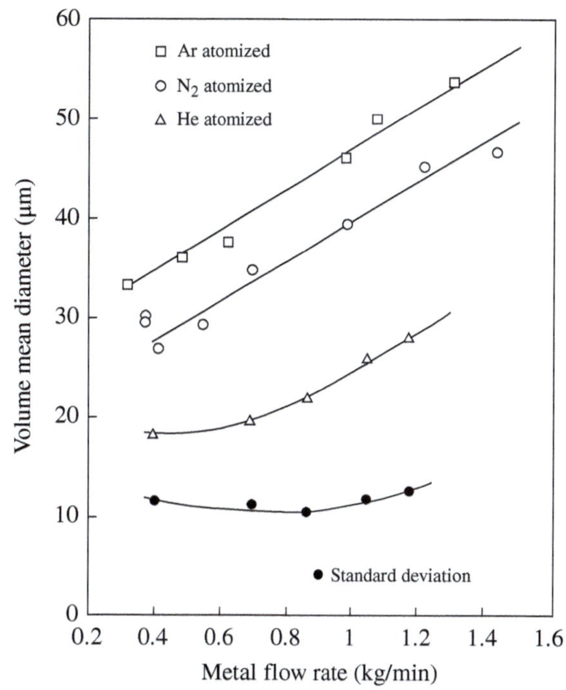

Figure 2.57 Influence of atomization gases on powder size (2014 Al): atomization pressure = 1.56 MPa; atomization temperature = 875°C. Source: Ünal [80] / Taylor & Francis.

such a design [21, 24–32, 34]. The USGA atomizer design employs a high gas pressure coupled with a flow interference mechanism, similar to that in a conventional whistle (Hartmann whistle), to generate high energy supersonic gas flow velocities. Exit gas velocities greater than Mach 2 have been reported during USGA atomization [35].

2.6.7 Physical Properties of Atomization Gas

There is limited fundamental understanding on the precise influence of the physical properties of the atomization gas on the atomization behavior of liquids. In a study of the atomization behavior of 2014 Al, Ünal [80] compared the volume mean diameter of powders produced by atomization using Ar, N_2 and He gases. The results of this study are summarized in Figure 2.57 [80] for various metal flow rates and an atomization pressure of 1.56 MPa. The results in this figure indicate that for constant atomization pressure and metal flow rate, the average powder size produced is finest for He and coarsest for Ar. Moreover, Ünal calculated the various physical properties of these three different gases at the nozzle exit and reported that the most important differences that are present between these three gases are gas velocity, gas flow rate, and thermal conductivity. For example, at the same atomization pressure, the mass flow rate of He is 3.1 times lower than that of Ar, while its velocity is approximately 3.1 times greater than that of Ar. The mass flow rate of the atomization gases as measured by Ünal for He, Ar, and N_2 were 1.47, 3.48, and 3.41 kg min^{-1}, respectively. The corresponding gas velocities calculated for He, Ar, and N_2 by assuming isentropic flow and using a gas pressure of 1.56 MPa are 1459, 462, and 575 ms^{-1}, respectively. Since a lower gas flow rate would result in a larger average powder size, Ünal [80] attributed the observed differences in atomized powder size primarily to the differences in gas exit velocity.

2.6.8 Liquid Viscosity

In the atomization and fluid mechanics literature, two **viscosity** nomenclatures are commonly used, namely, **dynamic viscosity**, η, and **kinematic viscosity**, ν ($\nu = \eta/\rho$ where ρ is density). In the discussion that follows the terminology "**viscosity**" will be used to describe dynamic viscosity, unless otherwise mentioned. The precise influence of liquid viscosity on atomization behavior depends on atomizer design and on atomization conditions, and consequently, reported results often appear to be contradictory. For example, whereas in some atomization studies, the mean size of droplets was reported to be strongly dependent on liquid viscosity [12, 113–116], in other studies droplet size was reported to be relative insensitive to variations in liquid viscosity [111, 117, 118]. In atomization

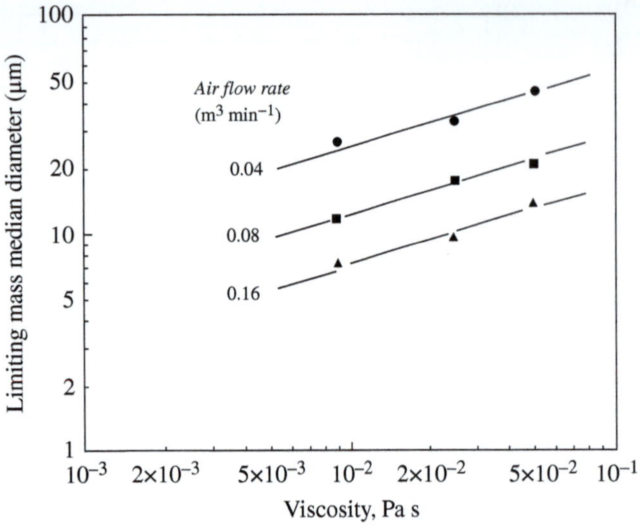

Figure 2.58 Limiting mass median diameter as a function of liquid viscosity [mixtures of wax and polyethylene (Epolene N)]. Source: Reproduced by permission of the American Institute of Chemical Engineers. [12].

studies that involve the use of co-flowing airblast atomizers, it is often hypothesized that the breakup of the liquid follows the classical wavy-sheet mechanism (see Figure 2.15) in which waves are formed on the liquid surface as a result of the aerodynamic force caused by the flow fields of the gas and liquid [60, 77, 119–121]. In this case, the **viscosity** of the liquid exerts a significant influence on the average size of atomized droplets. The average size of droplets decreases with decreasing liquid viscosity [12]. Kim and Marshall [12], for example, studied the influence of **viscosity** on the breakup of liquids, using a pneumatic gas atomizer. In their study, the **viscosity** of the liquid was systematically varied by mixing wax with polyethylene (Epolene N) in preselected ratios. Since the density and surface tension of the liquid mixtures were practically insensitive to the proportions of wax and polyethylene used, the observed changes in droplet size in their study were rationalized on the basis of the variations in liquid viscosity. Kim and Marshall [12] used the limiting mass mean diameter as a parameter to investigate the effect of **viscosity** on atomization. The limiting droplet sizes that were measured in this study are presented in Figure 2.58 [12] as a function of liquid viscosity for different air flow rates. The results in this figure clearly show that a decrease in **viscosity** leads to a decrease in the limiting droplet size. The experimentally observed trend of refinement in droplet size with decreasing liquid viscosity may be attributed to two separate effects, namely, **pre-filming** and wave growth. First, it is easier to deform a liquid with a low viscosity relative to one with a high viscosity. Therefore, a decrease in liquid viscosity will promote a decrease in liquid sheet thickness at the atomization edge, hereby promoting the formation of fine droplets. Second, a low liquid viscosity facilitates the development and growth of instabilities in the liquid sheet which ultimately causes refinement in droplet size [110].

In atomizer designs in which the gas and the liquid are not co-flowing, liquid disintegration relies primarily on the energetic impingement of the gas jets on the liquid stream. It has been suggested by Lefebvre [110] that for such atomizers the liquid sheet leaving the nozzle does not have ample time to develop a wavy structure. It is torn down directly into small fragments by the vigorous impingement of gas jets. In this case, the mean size of atomized droplets is less sensitive to variations in liquid viscosity. Experimental support for this suggestion is provided by the results obtained by Beck et al. [111] on the atomization of liquid mixtures of glycerin and water. Their results were summarized by Lefebvre [110] and are presented in Figure 2.59 for different liquid viscosities and relative velocities [110, 111]. The results in Figure 2.59 are shown in terms of the SMD. From the results in this figure, it is evident that an increase in **viscosity** by a factor of thirteen promotes only a 6% increase in SMD [111]. In related work, Sattelmeyer and Wittig [117] studied the effect of liquid viscosity on atomization using a pre-filming airblast atomizer. In their study, the impingement of the air jets on the liquid films was reported to be very close to the atomizing edge. Their results failed to reveal an influence of liquid viscosity on SMD. In related work, Buckner et al. [112] studied the atomization behavior of high-viscosity mixtures of glycerin and water using an effervescent atomizer. Their results are summarized in Figure 2.56 [110, 112] for different liquid viscosities. In agreement with the results of Sattelmeyer and Wittig, the results of Buckner et al. indicate that when atomization takes place very rapidly, liquid viscosity has

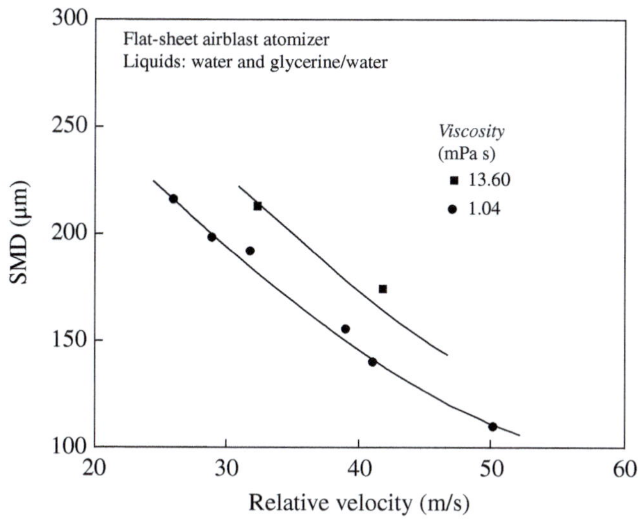

Figure 2.59 Influence of liquid viscosity on Sauter mean diameter (SMD) of water–glycerine droplets atomized using air [110] data from ref. [111].

only a marginal influence on SMD. As shown in Figure 2.56, the SMD data of liquids of various viscosities essentially fall on the same curve [110, 112].

The atomization behavior of metals differs slightly from that of other liquids in that at a relatively low temperature, the **viscosity** of melt increases and solidification is likely to occur, thereby limiting the breakup process. Putimtsev [114], for example, atomized a number of different metals and alloys and reported that both **surface tension** and liquid metal viscosity influence the morphology of atomized powders. In this study, it was observed that the tendency to form spherical particles was strongly favored by a low melt viscosity and a high melt surface tension. Putimtsev [115, 116] also studied the effect of **viscosity** on powder size for a number of **iron alloys** (Fe, Fe–Al, Fe–Si) and concluded that as melt viscosity increases it becomes more progressively difficult to obtain finer powders. While the results from this and other investigations [108, 122] suggest that melt viscosity is more important than **surface tension** in controlling powder size distribution, Klar and Shafer [123] concluded otherwise. Results from the work of Klar and Shafer on the atomization of copper alloys revealed that, while surface tension played an important role on determining powder size, the effect of melt viscosity was negligible.

2.6.9 Liquid Surface Tension

Fundamentally, atomization is a process that generates a large amount of **surface area** as a bulk liquid is disintegrated into fine droplets. Accordingly, an increase in surface tension is expected to increase the difficulty of liquid breakup. Inspection of the scientific literature reveals that a number of experiments have been conducted in an effort to isolate the effects of surface tension on atomization [108, 110, 117, 118, 123]. Klar and Shafer [123], for example, varied the surface tension of pure copper melt by changing the oxygen content in the molten copper. Their results revealed an increase in **viscosity** and a decrease in surface tension of molten copper with increasing oxygen content. Moreover, an increase in oxygen content in molten copper was accompanied by a refinement in atomized powder size. Klar and Shafer [123] argued that since an increase in liquid viscosity will tend to increase powder size, the observed refinement in copper powder size with increasing oxygen content must be attributed to surface tension effects. The observed trend of decreasing mean powder size with decreasing surface tension agrees in general with the results from many other investigations [108, 110, 117, 118, 122, 123].

The influence of liquid surface tension on the size distribution of atomized droplets depends on the particular liquid breakup mechanisms that are present as well as on the physical properties of the liquid. In the atomization of liquid metals, for example, when the temperature is close to the liquidus temperature, the **viscosity** of the melt may be sufficiently high to resist rapid deformation of the liquid. In this case, droplet size may depend primarily on viscosity. Moreover, in some atomizer designs, in which atomization relies on the energetic impingement of gas jets on the liquid stream, flow instabilities do not have sufficient time to develop. Lefebvre [124] postulated that when

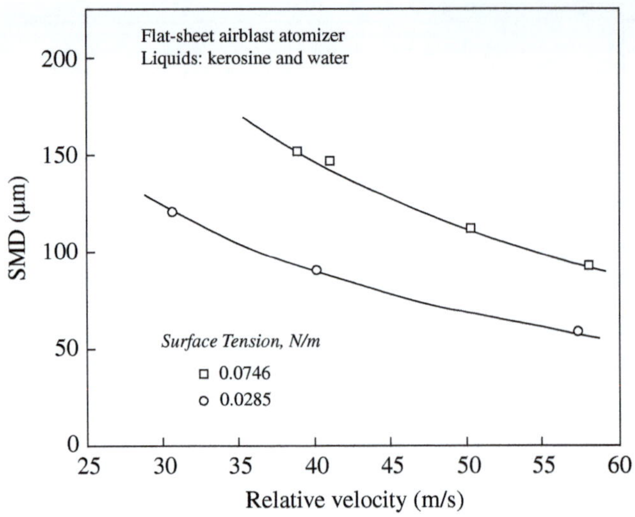

Figure 2.60 Influence of surface tension on Sauter mean diameter (SMD) of atomized droplets of kerosene and water mixtures [110] data from ref. [112].

atomization occurs rapidly the breakup process and the concomitant mean droplet size are governed primarily by the ratio of the **kinetic energy** of the atomization gas and the surface energy of the liquid. This results in a formulation that relates the mean droplet size with the kinetic factors of the gas and the physical properties of the liquid as [124]

$$d_{\text{Saut}} = \frac{3}{\left[1/t_1 + \xi\rho_1 V_g^2/4\sigma_1(1 + \dot{m}_1/\dot{m}_g)\right]} \tag{2.35}$$

where t_1 is the liquid sheet thickness; ξ is the atomization efficiency; ρ_1 is the liquid density; V_g is the gas velocity; σ_1 is the liquid surface tension, and; $\dot{m}_1$ and $\dot{m}_g$ are the mass flow rates of the liquid and the gas, respectively. Accordingly, the influence of surface tension is dependent on the gas velocity that is used for atomization. For high gas velocities, the second term in the denominator of Eq. (2.35) becomes overwhelmingly important. In this case, the mean droplet size is directly proportional to the surface tension of the liquid. Very significant effects of liquid surface tension on mean droplet size have been observed by a number of authors using relatively vigorous atomization conditions [82, 108, 111, 117, 118]. One example is provided by the results obtained by Beck et al. [111] on atomized mixtures of kerosene and water. The results of this study, which are summarized in Figure 2.60, reveal that an increase in surface tension from 0.0285 to 0.0746 Nm^{-1} resulted in an increase in average droplet size that exceeds 30% [106, 111].

2.6.10 Fluid Temperature

The influence of fluid temperature on atomization behavior has been the topic of many investigations [5, 6, 69, 80, 97, 99, 116, 123, 125, 126]. Figure 2.61 shows the results obtained by Weiss and Worsham [99] on the atomization behavior of wax. In this study, Weiss and Worsham [99] avoided the complexities that arise from the freezing of wax droplets during atomization, by atomizing the wax into an environment that was maintained at the same temperature as that of the molten wax. Their results indicate a refinement of droplet size with increasing wax temperature during atomization. Moreover, Weiss and Worsham [99] also measured the physical properties of the wax at the corresponding atomization temperatures. The results of their measurements are summarized in Table 2.4, where it is evident

Table 2.4 Physical properties of dyed wax.

Temperature (°C)	149	177	204
Specific gravity ($g\,m^{-3}$)	0.828	0.811	0.806
Surface tension (dyn. cm^{-1})	22.0	20.1	18.2
Viscosity (centipoise)	11.3	5.86	3.25

Source: Weiss and Worsham [99] / American Institute of Aeronautics and Astronautics.

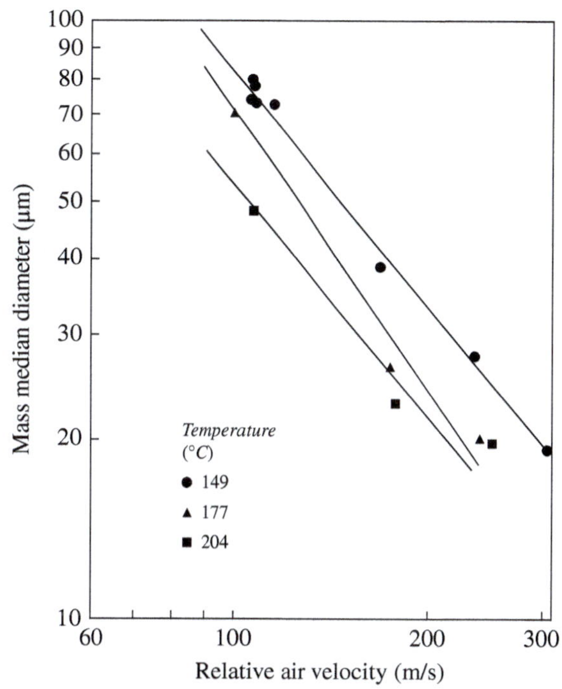

Figure 2.61 Effects of atomization temperature on mass median droplet diameter of molten wax. Copyright © 1959 AIAA – Source: Reprinted with permission [99].

that the physical property of wax that is most critically affected by temperature variations is liquid viscosity. Weiss and Worsham attributed the observed temperature dependence of the mean droplet size, primarily to the decrease in wax viscosity that accompanies an increase in atomization temperature.

Extensive research results are available [5, 6, 69, 80, 97, 99, 116, 123, 125, 126] regarding the Influence of temperature on the atomization of metals. The extent to which molten metal is heated above its melting point (superheat) has a significant influence on the resulting atomization characteristics. The effect of temperature on the atomization of molten metals can generally be evaluated from the characteristics of the atomized powders, for example, powder morphology and average size. There exists ample experimental evidence suggesting that an increase in melt superheat promotes powder size refinement [5, 6, 69, 80, 97, 99, 116, 123, 125, 126]. One example is shown in Figure 2.61 [80] for atomized 2014 A1 alloy. It is evident from the results in this figure that an increase in melt superheat temperature leads to a refinement in powder size. Interestingly, the experimental results shown in Figure 2.62 also reveal that a decrease in the width of the size distribution (geometric standard deviation) accompanies an increase in superheat temperature [80].

The influence of superheat temperature on atomization is related to the synergistic dependence of the physical properties and the solidification behavior of molten metals on temperature. In general, the experimentally observed refinement in powder size with increasing superheat temperature can be attributed to one or a combination of the following:

(a) decrease in melt viscosity,
(b) decrease in melt surface tension,
(c) decrease in melt density, and
(d) occurrence of solidification.

Thompson [105], for example, suggested that the observed trend of decreasing powder size with increasing superheat temperature is due to the dependence of melt density on temperature. He argued that as the temperature of a metal increases, while maintaining the actual volume of metal that passes through the pouring tube constant, the resultant mass flow rate decreases. Hence, this decrease in melt density results in an increase in the gas-to-metal mass flow rate ratio, which in turn promotes a finer powder size. Thompson's explanation, however, fails to account for the variations in other physical properties with temperature, such as melt viscosity and surface tension, which may play a more important role.

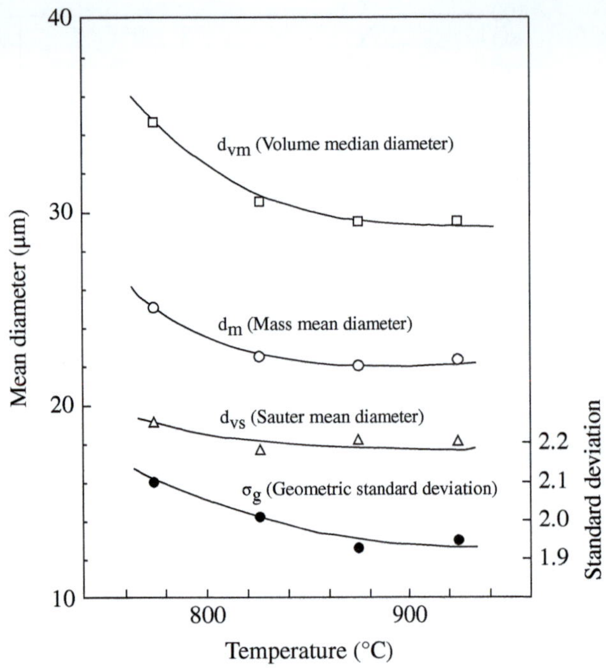

Figure 2.62 Influence of melt temperature on powder size of nitrogen-atomized 2014 Al: gas pressure = 1.56 MPa; metal flow rate = 3.42 kg min^{-1}. Source: Ünal [80] / Taylor & Francis.

Experimental results regarding the temperature dependence of surface tension and **viscosity** are available for some metals. The results are, however, generally limited to temperatures above the liquidus temperature. For example, above the liquidus temperature, the surface tension of Al depends on temperature as [127]

$$\sigma_1 = 0.866 - 1.5 \times 10^{-4}(T - T_1) \tag{2.36}$$

where σ_1 is the liquid surface tension in Nm^{-1}, T is the temperature of the molten metal in K and T_1 is the melting temperature. It is interesting to note that the temperature dependence of surface tension is not as significant as one may expect. For example, according to Eq. (2.36), an increase in temperature of 100 K above the melting temperature of Al will reduce the surface tension by only 2%.

For metals, at temperatures above the liquidus, an increase in melt temperature leads generally to a reduction in melt viscosity. The temperature dependence of the viscosity of molten metals may be expressed as [127]

$$\eta = \eta_0 \exp\left(-\frac{\Delta G}{RT}\right) \tag{2.37}$$

where η is dynamic viscosity, η_0 is a constant, ΔG is the activation energy of mass transport and R is the gas constant. In the case of molten Al, the constants η_0 and ΔG have values of 1.492×10^2 Pa s and 16.5 kJ mol^{-1}, respectively. Hence, the temperature dependence of viscosity appears to be stronger than that of surface tension. For example, an increase in temperature of 100 K above the liquidus temperature of Al corresponds to a decrease in viscosity, as calculated on the basis of Eq. (2.37), of 19%. This decrease in viscosity may account for the reduction in average powder size in atomized 2014 Al corresponding to temperatures above 825°C, as reported by Ünal [80] (see Figure 2.62).

2.6.11 Solidification Event

A marked dependence of powder size distribution on atomization temperature has been documented by a number of authors working with metallic materials [80, 105]. One example is provided by the results shown in Figure 2.63 [80]. The results suggest that for temperatures lower than 825°C, the temperature dependence of powder size in this figure appears to be more significant relative to that above 825°C. Ünal [80] attributed this phenomenon to the occurrence of solidification events during atomization at temperatures below 825°C. Another example of the possible influence of solidification on atomization behavior is provided by the results obtained by See and Johnston [69] on lead and tin.

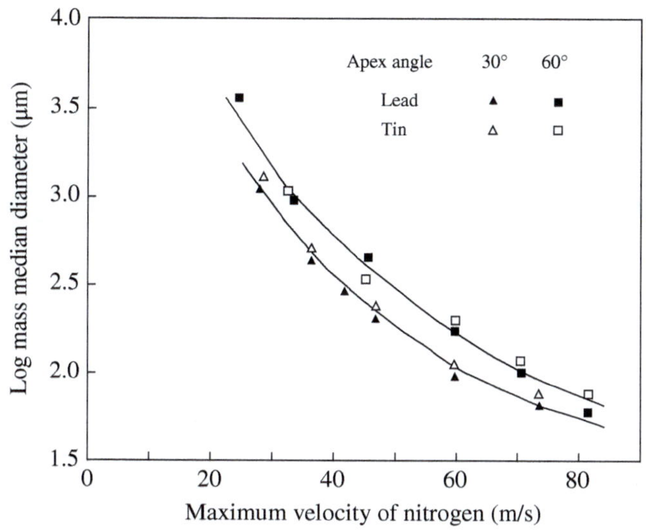

Figure 2.63 Relationship between mass median particle diameter of lead and tin, and maximum velocity of nitrogen. Mass flow rates: lead = 90 gs^{-1} and tin = 50 gs^{-1}. Metal nozzle diameter, 3.17 mm. Source: Reproduced by permission of the Institute of Materials [69].

Table 2.5 Approximate percentage of ligaments in atomized powders.

Temp (°C)	Percentage of ligaments in sieve fractions				
	+35	−35 + 45	−45 + 60	−60 + 80	−80 + 120
275	40	65	75	75	70
350	2	15	30	70	75
450	—	10	20	30	35

Atomization of tin; metal nozzle diameter = 0.317 mm; gas impingement angle = 30°; N_2 flow rate = 20 g s^{-1}.
Source: Adapted from See and Johnston [69].

They reported that lowering the atomization temperature hinders the complete development of a spherical droplet morphology. Interestingly, low atomization temperatures and vigorous atomization conditions were noted to promote the retention of morphologies that developed during the various stages of a liquid breakup. To that effect, See and Johnston determined quantitatively the percentage of **solidified** (e.g. frozen-in) ligaments as a function of atomization temperature. Their results, summarized in Table 2.5, suggest that an increase in atomization temperature generally leads to a decrease in the percentage of **prematurely solidified** ligaments [69]. See and Johnston [69], however, did not observe an appreciable refinement in powder size by increasing atomization temperature.

The premature solidification of molten metal ligaments during atomization is the result of a competition between spheroidization and solidification events. Molten metal ligaments tend to spheroidize under the driving force of surface tension. Simultaneously, however, the ligaments may solidify due to the vigorous thermal energy exchange that takes place between the metal and the atomization gas. In the literature, attempts have been made to predict the occurrence of **ligament-shaped powders** by comparing the time required for spheroidization to that required for solidification. The work of Nichiporenko and Naida [128], Rao [94], and that of See and Johnston [69] have led to the following simplified representation of a characteristic **cooling time** and **solidification time** as:

$$t_1 = \frac{d_{\text{lig}}\rho_1}{6h_c}\left[c_{\text{pl}}\ln\left(\frac{T_i - T_1}{T_1 - T_g}\right)\right] \tag{2.38a}$$

and

$$t_{\text{sol}} = \frac{d_{\text{lig}}\rho_1}{6h_c}\left[c_{\text{pl}}\ln\left(\frac{T_i - T_1}{T_1 - T_g}\right) - \left(\frac{\Delta H_f}{T_1 - T_g}\right)\right] \tag{2.38b}$$

where, t_1 and t_{sol} are the times required for the ligament to reach the liquidus temperature and for the completion of solidification, respectively; d_{lig} is the diameter of the ligament; ρ_1, c_{pl}, T_1 and ΔH_f are the density, heat capacity, liquidus temperature and heat of fusion of the metal, respectively; T_i the initial temperature of the ligament; and T_g is the temperature of the atomization gas. The variable h_c in Eq. (2.38) is the heat transfer coefficient at the ligament/gas interface, which may be calculated from [128]

$$h_c = \frac{k_g}{d_{lig}}(2.0 + 0.6Re^{0.5}Pr^{0.33}) \tag{2.39}$$

where k_g is the thermal conductivity of the atomization gas; and *Re* and *Pr* are the Reynolds number of the ligaments and the Prandtl number of the gas, respectively [128]. It is worth noting that h_c in Eq. (2.39) is dependent on the relative velocity between the ligaments and the gas. There are a number of formulations that have been developed in an effort to predict accurately the spheroidization time of ligaments [69, 94, 128]. The simplest one is that proposed by Nichiporenko and Naida [128] which is given by

$$t_{sph} = \frac{3\pi^2\eta_1}{4V_{lig}\sigma_1}\left(r^4 - r_{lig}^4\right) \tag{2.40}$$

where t_{sph} is the **spheroidization time** or **residence time**; η_1 and σ_1 are the viscosity and surface tension of the liquid, respectively; V_{lig} is the volume of the ligament; and r_{lig} and r are the diameters of the ligament prior to and after spheroidization, respectively. Accordingly, if the time required for solidification [Eq. (2.38b)] is less than that required for spheroidization (**residence time**), the resultant powder morphology is likely to be affected by atomization temperature and by other relevant atomization parameters such as atomization pressure, gas-to-melt flow rate ratio and the physical properties of the atomization gas.

2.6.12 Apex Angle

The atomization apex angle has been reported to affect the resultant powder size distribution [69, 72, 108, 129, 130]. See and Johnston [69], for example, studied the influence of apex angle on the atomization behavior of lead and tin using a free-fall atomizer and N_2 gas. The range of apex angles studied in their investigation included 30, 60, and 90° with the geometric impingement point of the gas jets on the metal stream being maintained 20 cm below the orifice of the metal **delivery tube**. According to the results of See and Johnston [69], there are two notable effects of apex angle on atomization behavior. First, an increase in apex angle in the range between 30 and 60° was noted to promote an increase in atomized powder size. This phenomenon is illustrated in Figure 2.63, where the logarithmic mass median diameter of atomized powders is presented as a function of gas velocity [69]. There is, however, some difficulty in interpreting these results due to the intrinsic variations in atomization parameters that are coupled with different atomizer designs. Second, for large values of the apex angle, a significant gas flow upwards from the atomization zone through the liquid metal delivery tube was noted. See and Johnston observed that when the apex angle exceeds 90°, a considerable amount of liquid metal was ejected upwards during atomization, thereby hindering metal flow.

2.6.13 Variables in Centrifugal Atomization

Centrifugal atomization has been successfully utilized for the atomization of rapidly solidified powders including **superalloys**, **titanium alloys**, specialty steels, and **aluminum alloys** [41, 44, 46–48, 131]. This technique has been used also in the study of atomization fundamentals using liquids, such as oil, water, and wax [39, 43, 132]. Figure 2.49 shows that the important variables which govern **droplet** size during centrifugal atomization are the physical properties of the liquid, liquid temperature (superheat for molten metals), liquid flow rate, disc diameter, and rotational speed. The influence of the physical properties and the temperature of the liquid on atomization behavior is similar for both gas atomization and centrifugal atomization, and hence will not be repeated here. The most important processing parameters that affect droplet size during centrifugal atomization are the melt flow rate, the angular speed, and the diameter of the rotating device. It is generally observed that the atomized droplet size increases with increasing liquid flow rate [39, 132], consistent with the previous discussion on twin fluid atomization. Moreover, when other atomization conditions remain constant, the average size of the atomized droplets decreases with increasing rotor diameter and rotational speed. Fraser et al. [38], for example, examined the atomization behavior of oil using a rotary cup atomization device in which the liquid was fed from the center of the cup and sprayed at the lip. The results of this study, shown in Figure 2.64, indicate considerable droplet refinement as the cup peripheral speed is

Figure 2.64 Mean droplet size as a function of rotary cup peripheral speed. Source: Adapted from Fraser et al. [38].

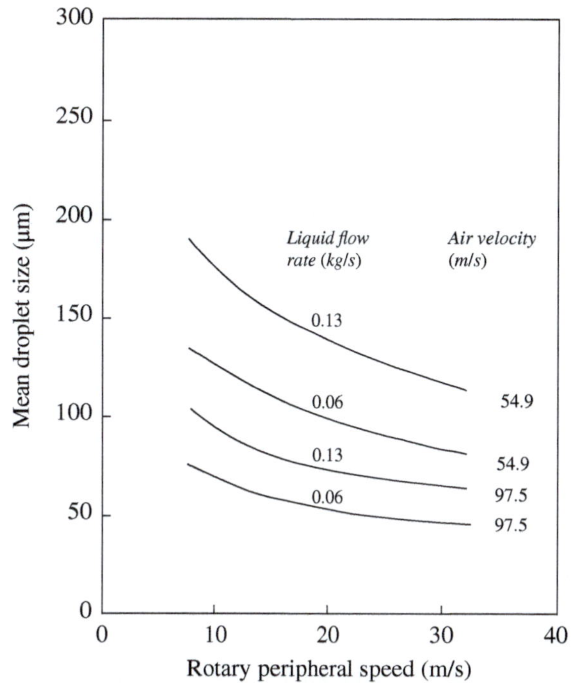

increased. Another notable feature that is evident in Figure 2.64 is the increasing droplet refinement with decreasing the mass flow rate of the liquid, consistent with the previous discussion on twin fluid atomization.

The observed trend of decreasing average droplet size with increasing rotational velocity is related to the liquid breakup mechanisms that are present during centrifugal atomization. During rotary cup atomization, a liquid is spread and pre-filmed on the cup's surface as a result of (a) the frictional forces that develop between the liquid and the surface of the cup, and (b) the centrifugal force due to rotation. During atomization, disturbance waves develop due to the presence of aerodynamic forces at the liquid–air interface. Some of these disturbances grow and become unstable. These disturbances eventually promote the detachment of liquid ligaments that ultimately disintegrate and spheroidize into droplets. Hence, on the basis of this simple mechanism, the size distribution of the droplets should be closely related to the volume of the ligaments, and therefore to the thickness of the liquid film that forms on the surface of the cup. A decrease in film thickness is expected to promote droplet refinement during rotary cup atomization. This premise is supported by the results shown in Figure 2.65 taken from the work of Fraser et al. [38], which reveal that increasing rotary cup speed (rpm) and diameter both lead to a thinning of the oil film during rotation.

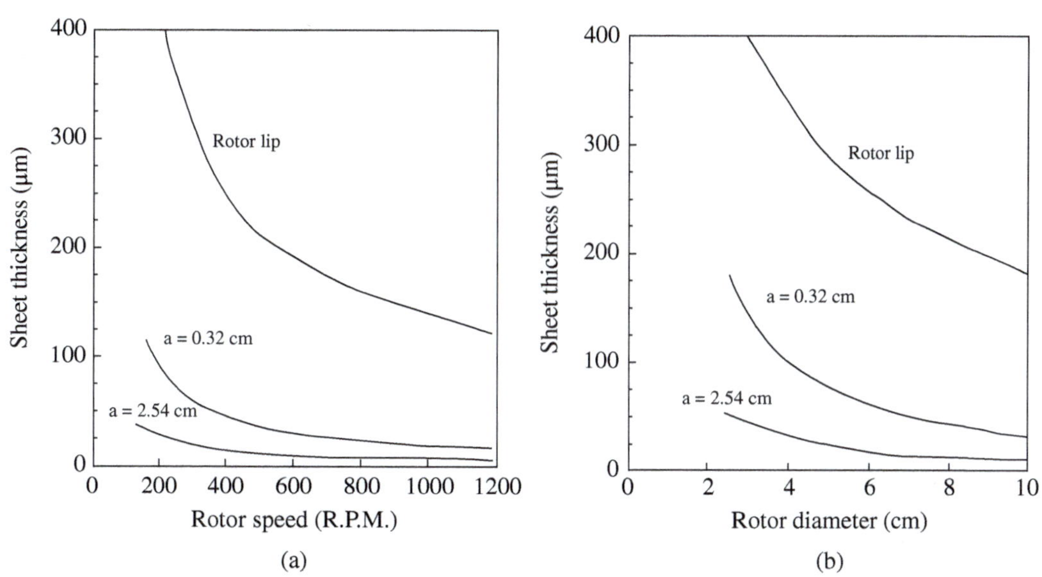

Figure 2.65 Sheet thickness as a function of operating variables during rotary cup atomization: (a) effect of rotary speed; (b) effect of rotary diameter where *a* is the radial distance from the cup lip [38].

Theoretically, it has been proposed by Jones [42] that the size of a droplet formed at the **rim** of a rotating disc depends on the balance that is established between the centrifugal forces and the surface tension forces. On the basis of this work, the following equation for the average droplet diameter, d_{mm} has been developed [42]

$$d_{mm} = \frac{a}{\omega}\left(\frac{\sigma_1}{\rho_1 D_d}\right)^{1/2} \tag{2.41}$$

where a is an experimental constant that is generally assigned a value of 4 [42], ω and D_d are the angular speed and diameter of the disc, respectively, and ρ_1 and σ_1 are liquid density and surface tension, respectively.

2.7 Theoretical Models of Atomization

The previous sections addressed the influence of processing parameters and material properties on atomization behavior. Many of the results discussed were obtained on the basis of phenomenological observations that were made during experimental studies. In an effort to elucidate fundamental phenomena, a number of investigators have studied the problem of atomization using theoretical analysis [133–136]. Typically, such theoretical analyses involve the formulation of models to predict **droplet size** from fundamental considerations of the growth of disturbances that are induced in the liquid during atomization. These models provide useful insight into the fundamental mechanisms that govern droplet formation and therefore play a critical role in the control and optimization of atomization processes. In the sections that follow, available models of atomization are described in detail, paying particular attention to the limiting assumption that was invoked in the formulation.

2.7.1 Breakup of Liquid Rods or Fragments

In analytical models, it is often assumed that the first step during atomization involves the breakup of an unstable liquid into ligaments as illustrated in Figure 2.15. If the aspect ratio of a ligament is sufficiently large it will subsequently become unstable under the combined influence of aerodynamic and surface tension forces. Accordingly, the ligament will breakup further into fragments of smaller aspect ratios. In the absence of secondary atomization, the fragments subsequently spheroidize into droplets of identical volume. Hence, in order to predict the average droplet size, one has to address two fundamental problems: the relationship between the diameter of the ligaments and the atomization variables; and the relationship between the size of the fragments and that of the ligaments. Hence, if the size of the fragments can be calculated, **droplet size** can be predicted on the basis of a mass balance consideration.

Fundamentally, the breakup of a liquid ligament, as shown in Figure 2.15, can be visualized by considering a long liquid rod. When the length of an inviscid liquid rod exceeds its circumference, it becomes geometrically unstable. In this case, some of the disturbances that are present in the liquid may grow in amplitude with time. When the amplitude of such disturbances reaches a certain critical value, the liquid rod will breakup into fragments of smaller length. **Rayleigh** [76] analyzed the growth of instabilities in a long liquid rod of diameter d_{lig} by considering the formation of fragments of the same diameter and with a length of $L_{lig} = sd_{lig}$ (where s is the aspect ratio of fragments). In this study, Rayleigh [76] proposed that there are numerous disturbances that are present on a liquid rod; however, only those fastest-growing disturbances are responsible for the breakup of the liquid rod. By applying Lagrange's method, Rayleigh [76] obtained a relationship between disturbance amplitude and time as

$$\alpha_d = \alpha_{do} \exp(qt) \tag{2.42}$$

where α_d is the disturbance amplitude, α_{do} is its initial value and t is time. The parameter q in Eq. (2.42) is given by [76]:

$$q = F\left(\frac{8T}{\sigma_1 d_{lig}^3}\right)^{1/2} \tag{2.43}$$

where F is a function of the rod aspect ratio only, T is a turbulence parameter, and d_{lig} is the rod diameter.

The general relationships described by Eqs. (2.42) and (2.43) were then used by Castleman [77] in the study of the breakup of liquid ligaments into droplets during atomization. Castleman assumed that each fragment produced by the growth of instabilities on ligaments ultimately develops into a droplet under the action of surface tension. He then

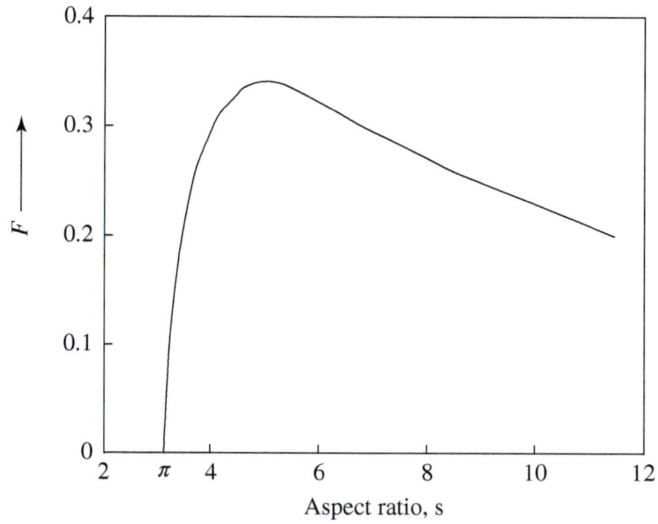

Figure 2.66 The dependence of the instability function, F, on the aspect ratio of ligaments (or liquid rods), s. Source: Reproduced with permission from Castleman [77].

used a mass balance approach to develop a relationship between the diameter of the droplet formed, d, the diameter, d_{lig}, and the aspect ratio, s, of the fragment as [77]:

$$d = \left(\frac{3s}{2}\right)^{1/3} d_{lig} \tag{2.44}$$

Since the fastest growing disturbances, that is, those with the largest q, are considered to be responsible for the formation of fragments, the most probable disturbances are those with a maximum F value [see Eq. (2.43)]. The results of Castleman's analysis [77] are shown in Figure 2.66 for F as a function of fragment aspect ratio, s. The results in this figure indicate that the most probable fragments correspond to an aspect ratio value of $s = 4.5$. Therefore, on the basis of Castleman's analysis [77], the resultant droplet size may be calculated in terms of the fragment size (e.g. diameter and aspect ratio) as

$$d = 1.89d_{lig} \tag{2.45a}$$

or alternatively,

$$d_{lig} = 0.53d \tag{2.45b}$$

where d and d_{lig} are the diameters of the droplets and that of the original liquid rod, respectively.

Tyler [121] suggested that the length of the resultant fragments should equal the wavelength, λ, of the swellings and contractions that are developed on the ligaments, and that the diameter of the fragments will equal that of the ligaments, $d_{lig}(s = \lambda/d_{lig})$. Tyler [121] further assumed that each fragment will eventually spheroidize into a single droplet of identical volume and calculated the size of the droplet from a mass balance equation [121]

$$\frac{\lambda}{d_{lig}} = s = \frac{2}{3}\left(\frac{d}{d_{lig}}\right)^3 \tag{2.46}$$

where λ is the wavelength of the most probable disturbance waves. Tyler [121] experimentally determined the aspect ratio of fragments by measuring the diameter of liquid ligaments and the size of droplets formed during atomization. His results indicate that the aspect ratio of ligaments falls in the range of 4.3–5.3, with an average value of 4.66. This average value of the fragment aspect ratio is in very good agreement with the theoretical value of 4.5 as was calculated by Castleman [77].

The importance of Eqs. (2.45) and (2.46) lies in the fact that they provide the starting point for the development of various atomization models. In many of these models a planar liquid geometry, rather than a cylindrical one, is assumed, presumably as a result of two reasons. First, the problem posed by the growth of waves is made more tractable by assuming a two-dimensional liquid geometry. Second, in many atomizers, liquid sheets are produced as a result of the pre-filming effect, even though the original liquid exiting the nozzle orifice is cylindrical (see, e.g. Figures 2.28 and 2.30). Regarding the origin of the initial disturbances, it has been suggested that when a liquid sheet exits an orifice, several intrinsic disturbances are present [61, 119, 120]. It is agreed generally that the responsible phenomena for the initial disturbances include the large velocity discontinuity that is present between the liquid jet and the atomizing fluid; fluctuations in back pressure; and **turbulence** in the atomizing fluid at the boundary condition discontinuity of the liquid as it crosses the nozzle exit plane [11, 60].

2.7.2 Formation of Droplets by Sheet Breakup

The work of Bradley [119, 120] provides useful insight into the fundamentals of droplet formation from a liquid sheet during atomization. Using the concept of the fastest-growing wavelength of a disturbance, Bradley developed a mathematical model for liquid breakup. This model considers the problem posed by the breakup of a two-dimensional liquid sheet when a high-velocity gas is flowing over its surface. To simulate the conditions that are present during atomization with high-velocity gases, Bradley included in his model the compressibility of the gas and the viscosity of the liquid. In order to calculate the wave number (or wavelength) of the fastest-growing disturbances, Bradley considered the growth of disturbances on the surface of a liquid with infinite depth which occupies a space corresponding to $y < 0$. He assumed that gas flows at a uniform velocity, V, over the liquid surface in the x direction as defined by the coordinates x and z. The fastest-growing wave number was determined for this semi-infinite liquid and was then applied to calculate the breakup of a liquid sheet. Bradley assumed that the initial motion of a liquid with an interfacial disturbance may be described by a set of linearized differential equations as [119]

$$\frac{\partial u_1}{\partial t} = -\frac{1}{\rho_1}\frac{\partial P_1}{\partial x} + \nu_1\left(\frac{\partial^2 u_1}{\partial x^2} + \frac{\partial^2 u_1}{\partial y^2}\right) \tag{2.47a}$$

$$\frac{\partial v_1}{\partial t} = -\frac{1}{\rho_1}\frac{\partial P_1}{\partial y} + \nu_1\left(\frac{\partial^2 v_1}{\partial x^2} + \frac{\partial^2 v_1}{\partial y^2}\right) - g \tag{2.47b}$$

$$\frac{\partial u_1}{\partial x} + \frac{\partial v_1}{\partial y} = 0 \tag{2.47c}$$

where u_1 and v_1 are the disturbance velocities of the liquid in the direction parallel and perpendicular to the undisturbed liquid surface, respectively; P_1 is the perturbation pressure; g is the gravitational acceleration; and ρ_1 and ν_1 are the density and the kinematic viscosity of the liquid, respectively. Bradley [119, 120] further assumed that the wave function for the disturbances may be described by the following equation:

$$\bar{y} = Re[\alpha_d(k, t)\exp(ikx)] \tag{2.48}$$

where $\bar{y}$ is the disturbance; α_d is its amplitude; k is the wave number; t is time, and x and y are the coordinates within and perpendicular to the surface of the undisturbed liquid. By superimposing relevant boundary conditions and considering the pressure exerted by the compressible gas flowing over the liquid surface, Bradley calculated the amplitude of the disturbances as a function of time. On the basis of Bradley's analysis, the amplitude of the disturbances may be expressed as a function of time as

$$\alpha_d = A\exp(\lambda_1 t) + B\exp(\lambda_2 t) \tag{2.49}$$

where λ_1 and λ_2 are functions of the wave number and are given by the following expressions:

$$\lambda_1, \lambda_2 = -\nu_1 k^2 \pm \left[\nu_1^2 k^4 - \frac{k\sigma_1}{\rho_1}(k - k_1)(k - k_2)\right]^{1/2} \tag{2.50a}$$

$$k_1, k_2 = \frac{\rho_g V^2}{2\sigma_1(1 - Ma^2)^{1/2}} \mp \left(\frac{\rho_g^2 V^4}{4\sigma_1^2(1 - Ma^2)} - \frac{\rho_1 g}{\sigma_1}\right)^{1/2} \tag{2.50b}$$

where ρ_g is the density of the gas, V is the **relative velocity** between the gas and the liquid, and Ma is the Mach number of the gas. According to Eq. (2.50), for disturbances with a wave number in the range of $k < k_1$ and $k > k_2$, λ_1 and λ_2 are either negative or complex numbers with a negative part. Hence, such disturbances will diminish with time. However, when k falls in the range of $k_1 < k < k_2$, λ_1 is real and positive, and in this case, the wave amplitude grows exponentially with time. Since λ_1 governs the growth rate of the wave amplitude, it is referred to as the growth factor by Bradley [119].

Bradley [119] made several additional assumptions in order to relate the growth factor and wave number as given by Eq. (2.50). First, he argued that the fastest growing waves, that is, the waves with maximum λ_1 value, are primarily responsible for the breakup of the liquid. Second, he assumed that the ligaments that are torn off from a sheet of liquid during atomization are a fraction of the wavelength of the most probable disturbances. Third, torn ligaments will breakup further into fragments of the same diameter, but with a shorter length. Hence, each fragment will form one droplet of identical volume. Fourth, the relationship between the diameter of the droplets and that of the ligaments may be described by Eq. (2.45). On the basis of the aforementioned assumptions, the size of the droplets is then related to the most probable wavelength by [119]

$$d = 1.89\varepsilon\lambda_{\mathrm{m}} \tag{2.51}$$

where ε is a fractional factor and λ_{m} is the wavelength of the most probable disturbances. Now the problem of predicting the size of the atomized droplets is narrowed down to the determination of the disturbance wavelength corresponding to the maximum growth rate. Since wavelength is related to wave number by $\lambda = 2\pi/k$, Eq. (2.51) can be rewritten as

$$d = \frac{11.8\varepsilon}{k_{\mathrm{m}}} \tag{2.52}$$

Bradley [119] determined the most probable wave number by plotting the growth factor, λ_1, as a function of wave number, k. It follows that the most probable wave number k_{m} corresponds to $\lambda_{1,\mathrm{max}}$. An example of this analysis is shown in Figure 2.67 for gas-atomized water. The size of the droplets formed during atomization can be computed from the calculated k_m using Eq. (2.51). Bradley [119] predicted the size of atomized droplets using Eq. (2.52) and assuming a fractional factor with a magnitude of $\varepsilon = 1/4$. Figure 2.68 compares the calculated and measured droplet sizes. The observed order of magnitude agreement between the measured and calculated droplet sizes is encouraging, if one considers the large number of interrelated variables that influence atomization.

In a related investigation, Dombrowski and Johns [61] developed an important theoretical model in which the liquid is not assumed to be inviscid. In considering the atomization of a liquid sheet, Dombrowski and Johns [61] assumed that the principal cause of instabilities in the liquid sheet occurs as a result of interactions with an atomizing media, whereby rapidly growing waves are imposed on the sheet.

Similar to Bradley's approach, Dombrowski and Johns [61] treated the problem posed by the atomization of a liquid sheet as follows:

(a) disintegration of the liquid sheet into a series of long ligaments as a result of interactions with the atomization gas.
(b) disintegration of ligaments into fragments, followed by the contraction of fragments into droplets of identical volume.

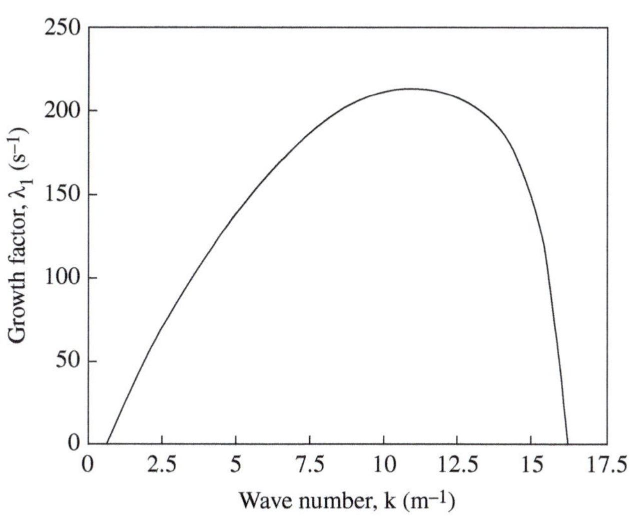

Figure 2.67 Growth factor, λ_1, as a function of the wave number, k (for air-atomized water with an air velocity of 10 ms^{-1}). Source: Reproduced with permission from Bradley Copyright (1973) The American Physical Society [119].

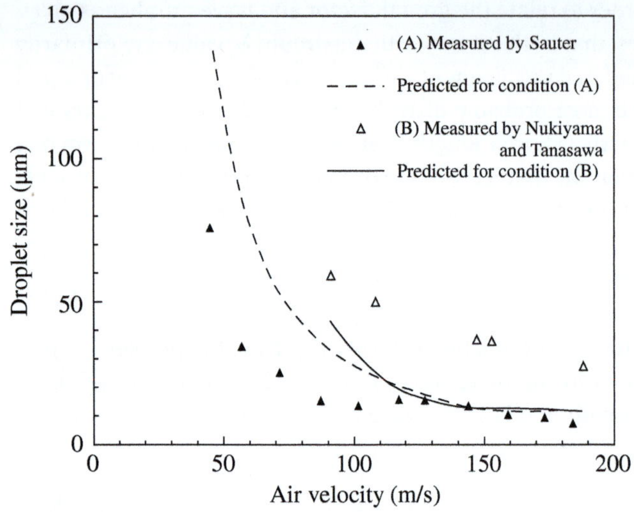

Figure 2.68 The measured (triangles) and calculated [dashed and solid lines, Eq. (2.52) droplet size for air/water systems. Data from Tables 1 and 2 of Ref. [119]. Source: Reproduced with permission from Bradley Copyright (1973) The American Physical Society.

For the breakup of the liquid sheet into fragments, they assumed that disintegration occurs when the wave amplitude reaches a critical value and ligaments are torn off from the sheet. In their model, Dombrowski and Johns [61] assumed that the disturbance waves may be described by a sinusoidal function with a time-dependent wave amplitude as

$$y = \alpha_\mathrm{d} \sin(kx + \varpi) \tag{2.53}$$

$$\alpha_\mathrm{d} = \alpha_\mathrm{do} \exp(\lambda_1) \tag{2.54}$$

where y is the displacement of the neutral axis of the liquid sheet from its undisturbed position in a direction normal to the surface, ϖ is the phase angle, x is the coordinate along the length direction of the undisturbed sheet and λ_1 is a growth factor. Moreover, it is important to note that Dombrowski and Johns [61] made the following assumptions in considering liquid sheet disintegration. First, the fastest-growing waves are primarily responsible for disintegration. Accordingly, the width of the ligaments torn off from the liquid sheet equals half of the wavelength of the fastest-growing waves. Second, the thickness of the liquid sheet decreases with time. Third, the ligaments that are torn off from the liquid sheet contract into cylindrical rods of length and volume that are equal to those of the original ligaments. From these assumptions, it follows that [61]

$$d_\mathrm{lig} = \left(\frac{4h_\mathrm{t}}{k_\mathrm{m}}\right)^{1/2} \tag{2.55}$$

where h_t is the thickness of the liquid sheet.

The diameter of the cylindrical ligaments can be determined if the wave number of the most probable waves and the thickness of the sheet during disintegration can be established. Dombrowski and Johns [61] developed a model to formulate the growth behavior of sinusoidal waves assuming the form given by Eqs. (2.53) and (2.54). In this model, the growth in wave amplitude is related to the wave numbers by considering a balance between the various forces that are present when a high-velocity gas flows over a liquid

$$F_\mathrm{p} + F_\sigma + F_\mathrm{I} + F_\eta = 0 \tag{2.56}$$

where F_p is the pressure force resulting from the interaction between the atomization gas and the liquid, F_σ is the surface tension force, F_I is the inertial force and F_η is the viscous force. The components of these forces in a direction normal to the undisturbed liquid sheet are given by [60]

$$F_\mathrm{p} = 2k\rho_\mathrm{g} V^2 yz\mathrm{d}x \tag{2.57a}$$

$$F_\sigma = \frac{\partial}{\partial x}\left(2\sigma_1 z \frac{\partial y}{\partial x}\right) \mathrm{d}x \tag{2.57b}$$

$$F_I = -\rho_1 \left(h_t \frac{\partial^2 y}{\partial t^2} + \frac{\partial h_t}{\partial t}\frac{\partial y}{\partial t} \right) z\mathrm{d}x \tag{2.57c}$$

$$F_\eta = \frac{\partial}{\partial x}\left(\eta_1 z h_t \frac{\partial^2 y}{\partial t \partial x} \right) \mathrm{d}x \tag{2.57d}$$

where x and z are the coordinates lying in the neutral surface of the undisturbed sheet with x in the direction of liquid flow; y is the coordinate normal to the neutral surface of the liquid sheet; and h_t is the thickness of the liquid sheet. By combining Eqs. (2.52, 2.54, 2.56, and 2.57) and making some simplifying assumptions, Dombrowski and Johns [61] developed a formulation of the growth factor as

$$\lambda_1 = \left\{ \frac{2}{\rho_1}(\rho_g V^2 k - \sigma_1 k^2) \right\}^{1/2} \int_0^t h_t^{-1/2} \mathrm{d}t \tag{2.58}$$

Equation (2.58) may be used to determine the most probable wave number corresponding to the fastest-growing waves. By combining Eqs. (2.55) and (2.58) Dombrowski and Johns [61] obtained the following expression which relates the diameter of the cylindrical ligaments to the atomization variables and the physical properties of the atomization gas and the liquid:

$$d_{\mathrm{lig}} = 0.9614 \left(\frac{K^2 \sigma_1^2}{\rho_g \rho_1 V^4} \right)^{1/6} \left(1 + 2.60 \eta_1^3 \left[\frac{K \rho_g^4 V^7}{72 \rho_1^2 \sigma_1^5} \right]^{1/2} \right)^{1/5} \tag{2.59}$$

where K is the product of liquid sheet thickness and length along its traveling direction, that is, $K = h_t x$, and is a parameter that is dependent on the operational conditions of the atomizers. Hence, the magnitude of K has to be determined for each type of atomizer under various operating conditions.

The cylindrical ligament diameter, d_{lig}, in Eq. (2.59) can be used to calculate the size of droplets which are produced from the subsequent breakup of the fragments. In principle, this calculation can be accomplished by applying Eq. (2.45a). Dombrowski and Johns [61], however, adopted an alternative approach. They considered the further disintegration of ligaments by the development and growth of symmetrical waves. Dombrowski and Johns assumed that the wave number of the fastest growing waves is k'_m, and that each wave eventually breaks up into a droplet. A mass balance formulation then leads to the following equation relating the droplet size with the wave number and the ligament diameter as

$$d^3 = \frac{2\pi}{k'_m} d_{\mathrm{lig}}^2 \tag{2.60}$$

Since the diameter of the ligaments is known from Eq. (2.59), the droplet size can be calculated from Eq. (2.60), if the wave number k'_m is known. It is worth noting that this wave number is defined for waves that develop on the ligaments and hence it differs from the wave number that develops on a liquid sheet as defined in Eqs. (2.55) and (2.58).

In calculating the wave number of symmetrical waves which develop on the surface of ligaments, Dombrowski and Johns [61] applied the formulation obtained by Weber [137]. Weber's treatment of the instability of liquid ligaments relates the wave number, k'_m, of the fastest-growing waves to the ligament diameter, d_{lig}, by [137]

$$k'_m d_{\mathrm{lig}} = \frac{1}{\sqrt{2}} \left[1 + \frac{3\eta_1}{(\rho_1 \sigma_1 d_{\mathrm{lig}})^{1/2}} \right]^{1/2} \tag{2.61}$$

It should be noted that Eq. (2.61) was derived for the case when surface tension governs liquid breakup. Moreover, Dombrowski and Johns assumed that the ligaments move transversely through the atmosphere, and that the effect of the surrounding atmosphere on the wave number is minimal and surface tension controls breakup. Accordingly, they combined Eqs. (2.60) and (2.61) and related the droplet size with the diameter of fragments as [61]

$$d = \left(\frac{3\pi}{\sqrt{2}} \right)^{1/3} d_{\mathrm{lig}} \left[1 + \frac{3\eta_1}{(\rho_1 \sigma_1 d_{\mathrm{lig}})^{1/2}} \right]^{1/6} \tag{2.62}$$

Equation (2.62) together with Eq. (2.59) may be used to predict atomized droplet size from atomization variables and the physical properties of the liquid and the atomization gas for liquid sheet atomization.

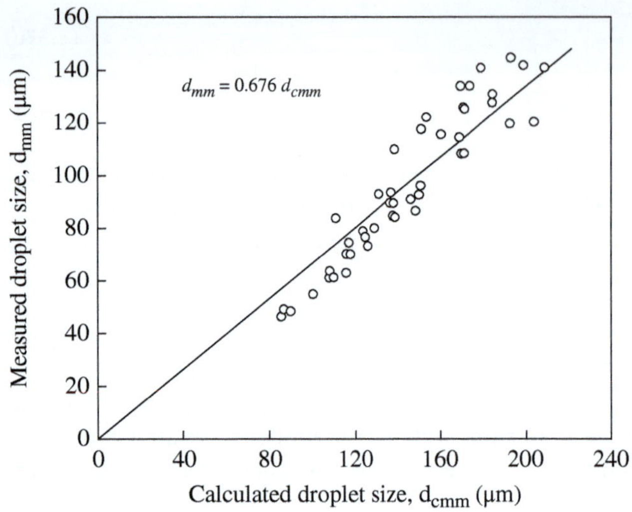

Figure 2.69 Comparison of the measured and calculated wax droplet sizes using Eq. (2.62) where d_{mm} is the measured mass median diameter and d_{cmm} is the calculated mass median diameter. Source: Reproduced from Dombrowski [61] / with permisison of Elsevier.

Dombrowski and Johns [61] sought to verify the applicability of their model by comparing the droplet sizes predicted on the basis of Eqs. (2.59) and (2.62) to those obtained experimentally using a **fan-spray** type atomizer. Accordingly, they studied the aerodynamic instability and disintegration of viscous sheets having a wide range of viscosities (0.1–1000 centipoise). The results of this study, which are summarized in Figure 2.69, reveal that the droplet diameters predicted from Eqs. (2.59) and (2.62) are generally larger than those determined experimentally [61]. Moreover, the results reveal that the measured and the calculated droplet sizes follow a linear relationship of $d_{mm} = 0.676 d_{cmm}$, where d_{mm} and d_{cmm} are the measured and calculated droplet sizes, respectively.

In addition to the oversimplifying assumptions that are associated with currently available models, there are two other phenomena that deserve mentioning since they hinder the accurate prediction of droplet size. The first one is premature solidification of ligaments and fragments during metal disintegration and droplet formation. The second phenomenon which is often overlooked in the development of atomization models is secondary atomization. The effect of solidification on droplet size has been addressed in Section 2.4. The phenomenon of secondary atomization has been addressed by a number of authors [63, 69]. The maximum stable droplet size derived on the basis of aerodynamic forces, surface tension forces, viscous forces, and inertial forces is given by Eq. (2.34).

The two aforementioned examples of atomization models (Bradley, Dombrowski, and Johns) provide a fundamental framework which may be effectively used for droplet size prediction in simple air/liquid systems. It must be realized, however, that these and other models developed so far generally assume a two-dimensional breakup process. The classical description presented by Dombrowski and his co-workers indicates that a planar liquid sheet distorts two-dimensionally according to the Kelvin–Helmholtz mode and will breakup into transverse ligaments which eventually disintegrate into droplets according to the Rayleigh instability mode. The Rayleigh or surface tension instability is attributable to the force imbalance that is associated with two radii of curvature present in the perturbed ligaments. However, recent experimental evidence [78, 138, 139] indicates that this two-dimensional description is indeed too simplistic. Only under extremely well-controlled conditions will the sheet distort two-dimensionally. In practice, three-dimensional distortion occurs as described in Figure 2.27; a liquid sheet may curl up during atomization under the action of aerodynamic forces accompanied by the formation of cells and rims. In this case, droplets may be produced by the breakup of cells and rims when they exceed their stability limit.

2.8 Empirical Models

The previous sections reveal that although the available analytical models are very useful in evaluating fundamental physical phenomena during atomization, their usefulness is limited by a large number of simplifying assumptions that are made. It has been discussed in Section 2.3, that results from direct experimental observations suggest that the breakup of a liquid under practical atomization conditions is far more complex than that proposed by the available physical models [1, 2, 10, 13, 57–60]. Often, the discrepancies between the droplet sizes predicted using available analytical models and the experimentally measured ones are significant. As a result, investigators have employed empirical approaches in an effort to develop relationships between droplet size and atomization variables. The formulation of empirical relationships for atomization generally involves two separate but closely related stages. In the first stage, the particular effect that one or more parameters have on the resultant powder (droplet) characteristics (e.g. size distribution, morphology, etc.) is determined experimentally. This is often accomplished by conducting a large number of experiments in which the atomization variables of interest are systematically varied. In the second stage, the experimental data are mathematically analyzed using, for example, regression techniques, and the particular influence of an atomization variable, or a group of variables, is established. Inspection of the available scientific literature reveals that this empirical approach has been extensively used and as a result, there are a variety of correlations available. It is worth noting, however, that the relationships developed by various investigators depend closely on the selection of experimental conditions and atomizer design. Accordingly, the applicability of these relationships is often limited to a narrow range of atomization conditions and a particular atomizer design. Some of the most widely used correlations are discussed in detail in the following sections.

2.8.1 Nukiyama and Tanasawa Analysis

A correlation for the prediction of the mean droplet diameter obtained from converging pneumatic atomizers was developed by Nukiyama and Tanasawa [74, 100, 101]. The equations in this study were developed on the basis of results obtained from several hundred experiments conducted under different experimental conditions and using various liquids, such as oil and water. The atomization nozzles used in this study were operated using compressed air. The Nukiyama and Tanasawa correlation for Sauter mean droplet diameter is given by

$$d_{\mathrm{Saut}} = \frac{585}{V}\left[\frac{\sigma_1}{\rho_1}\right]^{1/2} + 597\left[\frac{\eta_1}{(\rho_1\sigma_1)^{1/2}}\right]^{0.45}\left[\frac{1000\dot{v}_1}{\dot{v}_g}\right]^{1.5} \tag{2.63}$$

where d_{Saut} is the Sauter mean droplet diameter (in μm), and $\dot{v}_1$ and $\dot{v}_g$ are the volume flow rates of the atomization gas and that of the liquid, respectively. It is worth noting that the constants in this empirical relationship were determined on the basis of the following unit system: ρ_1 = liquid density, g cm^{-3}; η_1 = liquid viscosity, poise; σ_1 = liquid surface tension, dynes cm^{-1}; and V = relative velocity, ms^{-1}. An interesting prediction of Eq. (2.63) is the dependence of droplet size on surface tension and viscosity of the liquid. Accordingly, at large gas-to-liquid volume flow rate ratio values, the second term on the right-hand side of Eq. (2.63) is of minor importance. In this case, droplet size is governed by the factor of $\sigma_1/V(\rho_1)^{1/2}$ and the influence of surface tension becomes important. For small gas-to-liquid volume flow rate ratio values, the first term on the right-hand side is less significant. In this case, the droplet size is governed by the factor of $\left(\eta_1/(\sigma_1\rho_1)^{1/2}\right)^{0.45}((1000\dot{v}_1)/\dot{v}_g)^{1.5}$ and the influence of liquid viscosity becomes important. The validity of the above correlation developed by Nukiyama and Tanasawa was established and confirmed by several other investigators [102, 103]. For example, in an independent study, Lewis et al. [102] used three Venturi atomizers with different throat diameters to atomize water using high-pressure air. The droplet diameters determined from this study correlated well with those predicted on the basis of Eq. (2.63). In an experimental investigation performed under similar conditions, Bitron [103] extended the applicability of the Nukiyama–Tanasawa relation to the supersonic gas flow regime. It is important to note, however, that the applicability of the Nukiyama–Tanasawa equation is limited to fluids whose physical properties fall in the range of those for oils and aqueous liquids [74, 100, 101]. Moreover, the validity of this relationship has not been confirmed for atomizer designs that differ from the one used in the original investigation. It should be noted that this particular correlation does not contain atomizer parameters, such as the dimensions of the liquid. Moreover, it is worth noting that Eq. (2.63) is dimensionally inconsistent.

2.8.2 Wigg Analysis

Wigg [140] analyzed the average droplet sizes of waxes of various viscosities produced using air-blast twin fluid atomizers (see Figure 2.70). The experimental data obtained in this study reveal that the average size of wax droplets is inversely related to the mass flow rate ratio between the molten wax and the atomization gas. On the basis of energy

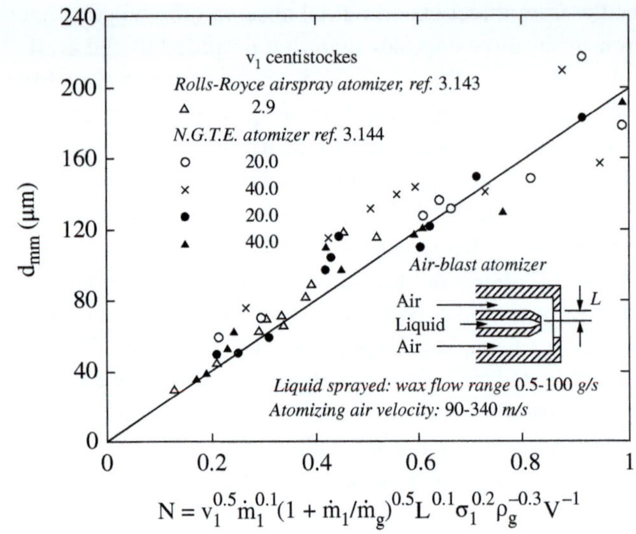

Figure 2.70 Droplet size correlation for wax sprays. Variables are defined in the nomenclature. Source: Wigg [140].

and momentum considerations, Wigg defined the initial velocities of the molten wax and that of the gas as V_1 and V_g, respectively. He proposed that following breakup, the atomized droplets and the gas approached at the same velocity, V_2. It then follows that the net loss of kinetic energy from the gas, ΔE_g, may be determined from [140]

$$2\Delta E_g = \dot{m}_1 V_1^2 + \dot{m}_g V_g^2 - (\dot{m}_1 + \dot{m}_g) V_2^2 \tag{2.64}$$

Moreover, if momentum is conserved during atomization, it follows that [140]

$$\dot{m}_1 V_1 + \dot{m}_g V_g = (\dot{m}_1 + \dot{m}_g) V_2 \tag{2.65}$$

If the relative velocity between the liquid and the gas at the beginning of atomization is $V = Vg - V_1$, by eliminating V_2 from Eqs. (2.64) and (2.65) this yields the formulation for the loss of kinetic energy in the gas as $\dot{m}_1 \dot{m}_g V^2/(\dot{m}_1 + \dot{m}_g)$. Wigg argued that the size of the droplets produced during atomization should be inversely proportional to the energy loss in the atomization gas. Accordingly, the mean size of the droplets is predicted to be a function of $(1 + \dot{m}_1/\dot{m}_g)/V^2$. Wigg analyzed the results obtained by Wood [141] on air-atomized molten waxes and concluded that the size of droplets produced during atomization is actually a function of the square root of the above factor as

$$d_m \propto \frac{\dot{m}_1^{0.1}}{V} \left(1 + \frac{\dot{m}_1}{\dot{m}_g}\right)^{0.5} = F \tag{2.66}$$

Wigg [140] argued that the rate of deformation of the liquid is high if it is atomized by a compressible fluid. Accordingly, a considerable amount of kinetic energy contained in the atomization fluid must be dissipated to overcome the viscous forces. Hence, Wigg used the F function given in Eq. (2.66) as a factor to obtain empirical relationships by plotting the results obtained on molten waxes of different viscosities. He noted that the curves corresponding to different liquid viscosities fall on separate straight lines. Moreover, the slope of these straight lines was noted to be proportional to the square root of viscosity [140, 142]. Accordingly, Wigg [140] further modified Eq. (2.66) to include liquid viscosity as an additional factor:

$$d_m \propto \frac{\eta_1^{0.5} \dot{m}_1^{0.1}}{V} \left(1 + \frac{\dot{m}_1}{\dot{m}_g}\right)^{0.5} \tag{2.67}$$

Wigg postulated that the average size of droplets should also be a function of other parameters, such as the dimension of the atomizer, L (see Figure 2.71), liquid and gas densities, and liquid surface tension. Accordingly, Wigg included these factors in Eq. (2.67) as

$$d_{mm} \propto \frac{\eta_1^{0.5} m_1^{0.1}}{V} \left(1 + \frac{\dot{m}_1}{\dot{m}_g}\right)^{0.5} L^m \sigma_1^n \eta_g^b \rho_g^c \rho_1^e \tag{2.68}$$

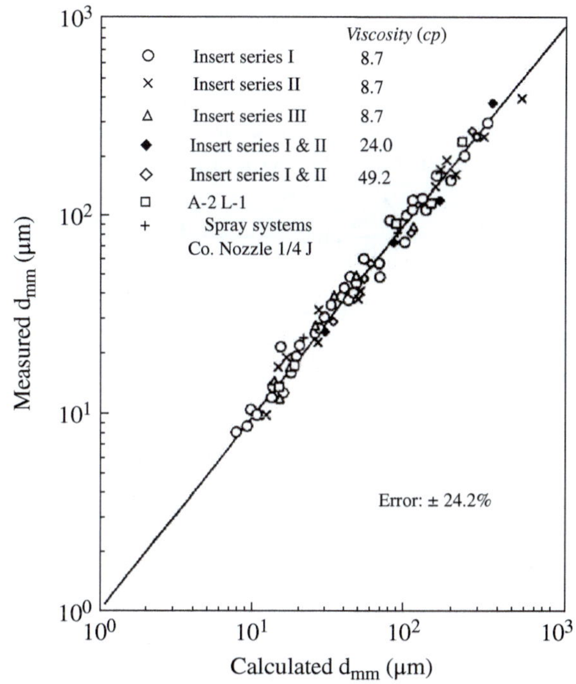

Figure 2.71 Correlation for convergent-type pneumatic atomizers [12]. Insert series I, II, III, and A-2, L1 all correspond to different atomizer designs. Source: Reproduced by permission of the American Institute of Chemical Engineers. © 1971 AIChE. All rights reserved.

where L is the atomizer dimension (height of air **annulus** as illustrated in Figure 2.70). The empirical relationship between the average droplet size and the atomization parameters may then be obtained by determining the exponents m, n, b, c, and e. Wigg further assumed that the effect of gas viscosity on atomization is negligible, so that b in Eq. (2.68) equals zero. Then in order to make Eq. (2.68) dimensionally correct, he assigned different values to the exponential constants, m, n, c, and e. Eq. (2.68) then becomes

$$d_{\mathrm{mm}} \propto \frac{\nu_1^{0.5} \dot{m}_1^{0.1}}{V}\left(1+\frac{\dot{m}_1}{\dot{m}_{\mathrm{g}}}\right)^{0.5} L^{0.1}\sigma_1^{0.2}\rho_{\mathrm{g}}^{-0.3} = N \tag{2.69}$$

To explore the validity of Eq. (2.69) Wigg plotted the droplet sizes obtained experimentally by Clare and Radcliffe [142] as well as those obtained by Fraser and Eisenklam [143] for waxes against the function N as defined in Eq. (2.69). The results, shown in Figure 2.70, indicate an almost linear relationship between the experimentally determined sizes of wax droplets and the function N, with a slope of approximately 200. Accordingly, the relationship between atomized droplet size and atomization parameters is given by

$$d_{\mathrm{mm}} = 200N \tag{2.70}$$

In the same study, Wigg analyzed the size distribution of atomized water droplets in which the coalescence of droplets is considered to be significant. He suggested that the coalescence depends primarily on the number of droplets formed per unit volume, as well as on the range of droplet sizes and velocities. Moreover, Wigg concluded that when droplet coalescence is significant the N function has to be modified in order to obtain an acceptable correlation between the calculated droplet and the measured droplet sizes. According to Wigg, when coalescence is present the average size of the droplets obtained during atomization may be calculated from

$$d_{\mathrm{mm}} = 200N\left[1+2.5\left(\frac{\dot{m}_{\mathrm{L}}}{\dot{m}_{\mathrm{g}}}\right)^{0.6}\dot{m}_1^{0.1}\right] \tag{2.71}$$

It is important to note, however, that Eq. (2.71) does not correlate well with the results obtained by Nukiyama and Tanasawa [74, 100, 101] for atomized droplets of glycerin/water mixtures. In particular, when tested against the experimental results obtained by Nukiyama and Tanasawa [74, 100, 101], the effect of liquid viscosity predicted by Eq. (2.71) is greater than that of the measured one. Moreover, the empirical relationship defined by Eq. (2.71), though correlated well with the experimental data in the case of droplet coalescence, is dimensionally incorrect. It may be improved by taking into consideration the physical mechanisms that govern coalescence.

2.8.3 Kim and Marshall Analysis

In view of the observation that the atomization behavior of liquids is affected by the synergistic interaction of several phenomena, including fluid dynamics, heat flow, and mass transfer, Kim and Marshall [12] employed dimensional analysis in an attempt to correlate droplet size with experimental parameters. In this approach the variables which influence atomization were combined into dimensionless groups:

$$\frac{d_{mm}}{T} = f\left(\frac{d_h}{D_{nl}}, We, \frac{\dot{m}_g}{\dot{m}_1} \frac{\eta_1}{[D_{nl}\rho_1\sigma_1]^{1/2}}, \frac{\eta_1}{\eta_g}, \frac{\rho_1}{\rho_g}\right) \tag{2.72}$$

where T is a gas turbulence parameter that has a dimension of length; D_{ng} and D_{nl} are the diameters of gas and liquid nozzles, respectively; and We is the Weber number which is defined as the ratio between aerodynamic and surface tension forces:

$$We = \frac{D_{nl}V^2\rho_g}{\sigma_1} \tag{2.73}$$

Kim and Marshall investigated the atomization characteristics of convergent, pneumatic atomizers. The droplet sizes determined in this study were correlated with the dimensionless groups to obtain an empirical relationship. The relation obtained by Kim and Marshall for a concentric single-air pneumatic atomizer [12] is as follows:

$$d_{mm} = \frac{a\sigma_1^{0.41}\eta_1^{0.32}}{[V^2\rho_g]^{0.57}A^{0.36}\rho_1^{0.16}} + \frac{b}{V^{0.54}}\left(\frac{\eta_1^2}{\rho_1\sigma_1}\right)^{0.17}\left[\frac{\dot{m}_g}{\dot{m}_1}\right]^m \tag{2.74}$$

where d_{mm} is the mass median diameter; m is an experimental constant; the values for a and b are 249 and 1260, respectively for single air pneumatic atomizers; and A is the cross-sectional area of the **gas nozzle**. Equation (2.74) indicates that the most important variables in atomization are (a) the dynamic forces of the atomization gas and (b) and the mass flow ratio of air to liquid. A notable feature of Eq. (2.74) is its dependence on atomization conditions. For example, the exponent, m, in Eq. (2.74) varies with the mass flow rate ratio, with $m = -1$ for $\dot{m}_g/\dot{m}_1 \leq 3$ and $m = -0.5$ for $\dot{m}_g/\dot{m}_1 > 3$. This correlation offers a satisfactory prediction of droplet size for convergent-type pneumatic atomizers. To that effect, Figure 2.71 compares experimental results on mixtures of wax and polyethylene (Epolene N) obtained by Kim and Marshall to the droplet sizes that are predicted by Eq. (2.74) [12]. It is evident from Figure 2.71 that the mass median diameters calculated using Eq. (2.74) agree reasonably well with the experimental ones.

It should be noted that the applicability of Eq. (2.74) depends on the specific atomizer design. For example, Kim and Marshall have also developed a parallel empirical relationship for concentric double-air pneumatic atomizers [12] in which the mass median diameter of droplets is related to the experimental parameters as

$$d_{mm} = \frac{a\sigma_1^{0.41}\eta_1^{0.32}}{(V^2\rho_g)_{av}^{0.72}\rho_1^{0.16}} + \frac{b}{V_{av}^{0.54}}\left(\frac{\eta_1}{\rho_1\sigma_1}\right)^{0.17}\left(\frac{\dot{m}_g}{m_1}\right)^m \tag{2.75}$$

where

$$(V^2\rho_g)_{av} = f(V^2\rho_g)_{pri} + (1-f)(V^2\rho_g)_{sec} \tag{2.76a}$$

and

$$V_{av} = fV_{pri} + (1-f)V_{sec} \tag{2.76b}$$

where the subscripts "pri" and "sec" refer to primary and secondary air nozzles, respectively; / is the weight fraction of total airflow from the primary gas nozzle; and the other variables have the same meaning as defined previously. The exponent, m, in Eq. (2.75) has the same value for each range of mass flow rate ratios as that used in Eq. (2.74). The magnitudes of the parameters a and h used in Eq. (2.75) for a double air nozzle atomizer are 8140 and 1240, respectively, compared to 249 and 1260 for a single air nozzle atomizer. It should be noted that these constants were obtained using the following unit system: surface energy = dyn cm^{-1}; viscosity = cp; velocity = ft s^{-1}; density = lb ft^{-3}; and area = ft^2. A noticeable difference between the equations obtained for a single and a double air nozzle atomizer is the exponential factor of air dynamic force. Accordingly, Kim and Marshall [12] determined that the values of this exponent are 0.57 and 0.72 for single and double-air atomizers, respectively.

2.8.4 Schmitt Analysis

A theoretical approach similar to the one developed and implemented by Kim and Marshall [12] was proposed by Schmitt [98] for the disintegration of molten metals using high-velocity gases. Schmitt formulated a relationship for the maximum possible droplet size obtainable under given atomization conditions by considering the various forces that are present. The forces are grouped into the following three basic non-dimensional parameters:

$$Z_1 = \frac{\text{forces on surface}}{\text{forces on inertia}} = \frac{\sigma_1}{\rho_1 V^2 d_{\text{max}}} \tag{2.77a}$$

$$Z_2 = \text{viscosity effect} = \frac{\eta_1^2}{\sigma_1 \rho_1 d_{\text{max}}} \tag{2.77b}$$

$$Z_3 = \frac{\text{inertial forces of atomizing gas}}{\text{inertia of liquid}} = \frac{\rho_g}{\rho_1} \tag{2.77c}$$

where d_{max} is the maximum size of droplets. Based on the three non-dimensional parameters Z_1, Z_2, and Z_3, Schmitt derived an equation that correlates the maximum powder size, atomization gas velocity, and the physical properties of the gas and the molten metal as

$$d_{\text{max}} = \left(\frac{10^5 \sigma_1}{4.8 \rho_1 V_g^2}\right)^{12/13} \left(\frac{10^6 \eta_1^2}{\sigma_1 \rho_1}\right)^{1/13} \tag{2.78}$$

where V_g is the velocity of the atomization gas. It is worth noting that Eq. (2.78) does not contain parameters related to atomizer design, such as the diameter of the liquid stream. This particular equation can only be used to predict the maximum size of droplets on the basis of limited atomization parameters, such as liquid density, surface tension, viscosity, and relative velocity between the gas and the liquid. The maximum droplet size predicted on the basis of Eq. (2.78), although important, provide less useful information about a collection of powders when compared, for example, to average droplet size and droplet size distribution. Schmitt tested the validity of Eq. (2.78) using the experimental data obtained from gas-atomized IN100 Ni-based superalloy. He found that the maximum powder size obtained from atomized IN100 alloy correlated well with those predicted using Eq. (2.78). Moreover, further experimental work is a need in order to confirm the validity of Eq. (2.78) for other liquids or metals.

2.8.5 Weiss and Worsham Analysis

Weiss and Worsham [99] studied the gas atomization behavior of a synthetic wax, Acrawax C. This particular wax was selected for atomization studies because it has a low melting temperature (i.e. 140–143°C) and is not volatile. In this study, Weiss and Worsham atomized the molten wax by injecting it into a high-velocity hot air stream using liquid injection tubes (or delivery tubes) with inner diameters of 1.2, 2.4, and 4.8 mm. The direction of liquid injection was either co-stream or contrastream with the direction of airflow. Hot air was utilized for atomization in order to maintain the temperature constant during atomization and to systematically study the influences of viscosity and surface tension. In their experiments, Weiss and Worsham studied the effects of atomization parameters, such as the relative velocity between the two fluids; the surface tension, viscosity, density, and stream diameter of the molten wax; and the viscosity and density of the atomization gas, on the mass median droplets size to obtain a relationship between the atomization conditions and the droplet size. One example is the effect of relative velocity. On the basis of their experimental work, Weiss and Worsham proposed that the relationship between relative velocity and atomized droplet size may be expressed as

$$d_{\text{mm}} \propto V^a \tag{2.79a}$$

where a is an exponential constant. By plotting the experimentally determined mass median droplet size, d_{mm}, and relative velocity, V, in a double logarithmic scale, Weiss and Worsham determined the magnitude of a to be –1.33. It then follows that

$$d_{\text{mm}} \propto V^{-1.33} \tag{2.79b}$$

By assuming $\rho_1 = 1000\rho_g$ and systematically establishing the individual influences of each critical parameter, Weiss and Worsham obtained the following relationship between the mass median droplet diameter and the atomization parameters:

$$d_{\text{mm}} = K_1 V^{-1.33} \sigma_1^{0.41} \eta_1^{0.34} \eta_g^{0.09} V_1^{0.08} D_{\text{nl}}^{0.16} \rho_1^{-0.84} \rho_g \left(1 + \frac{10^3 \rho_g}{\rho_1}\right) \tag{2.80}$$

where V_1 is the velocity of the liquid stream, and K_1 is a proportionality constant. Weiss and Worsham further rearranged Eq. (2.80) into a dimensionless form which provides physical insight into the particular effects of each group of parameters on the breakup of a liquid and on droplet formation [99]:

$$\left(\frac{d_{\text{mm}} \rho_g V^2}{\sigma_1}\right) = K_2 \left[\frac{V \eta_1}{\sigma_1}\right]^{2/3} \left[1 + \frac{10^3 \rho_g}{\rho_1}\right] \left[\frac{\dot{m}_1 \rho_1 \sigma_1 \eta_g}{\eta_1^4}\right]^{1/12} \tag{2.81}$$

Weiss and Worsham investigated the validity of Eq. (2.81), by plotting the left-hand side term of Eq. (2.81) versus the right-hand side, without the constant K_2. From the results presented in Figure 2.72, the proportionality constant K_2 was determined to be 0.61 for the wax/gas system. It should be noted that in this equation, it is assumed that $\rho_1 = 1000\,\rho_g$. Weiss and Worsham observed that the data corresponding to large liquid delivery tube diameters correlated well with Eq. (2.81) with standard deviations between the measured and correlated droplet diameters being 9.3% and 9.6% for delivery tube diameters of 4.8 and 2.4 mm, respectively. When injecting liquid using a small diameter delivery tube and atomizing at low relative velocities (i.e. less than 107 ms^{-1}), Eq. (2.81) failed to provide a good correlation between the measured and the calculated droplet diameters. As shown in Figure 2.72, in this case, the droplet sizes predicted using Eq. (2.45) tend to be greater than the measured ones. Moreover, it is worth noting that in this analysis it is implicitly assumed that the atomization gas is incompressible. The liquid stream diameter, D_{nl} velocity, V_1, and density, ρ_1, that appear in Eq. (2.80) were combined into the liquid flow rate term in Eq. (2.81). This rearrangement, however, eliminated a very useful practical parameter, that is, the diameter of the liquid stream which under some conditions may affect droplet size.

2.8.6 Lubanska Analysis

A correlation applicable to the gas atomization of liquid metals based on the Nukiyama–Tanasawa analysis [see Eq. (2.63)], and the Wigg analysis [see Eqs. (2.70) and (2.71)] was put forth by Lubanska [73]. Lubanska studied the atomization behavior of molten iron using a spray ring-type atomizer. This type of atomizer is similar to those illustrated in Figures 2.1 and 2.2 and consists of a ring of discrete gas nozzles symmetrically surrounding a liquid

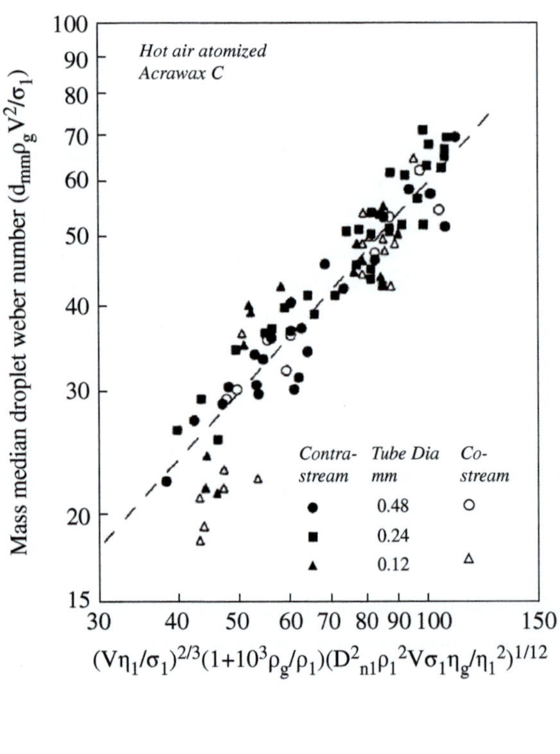

Figure 2.72 Correlation of mass median droplet diameters obtained from coaxial tube injector atomizers where contrastream = air and liquid flow in opposite directions, costream = air and liquid flow in the same direction, and tube dia. = inner diameter of the liquid injector. Copyright © 1959 AIAA – Source: Reproduced with permission Weiss [99] / American Institute of Aeronautics and Astronautics.

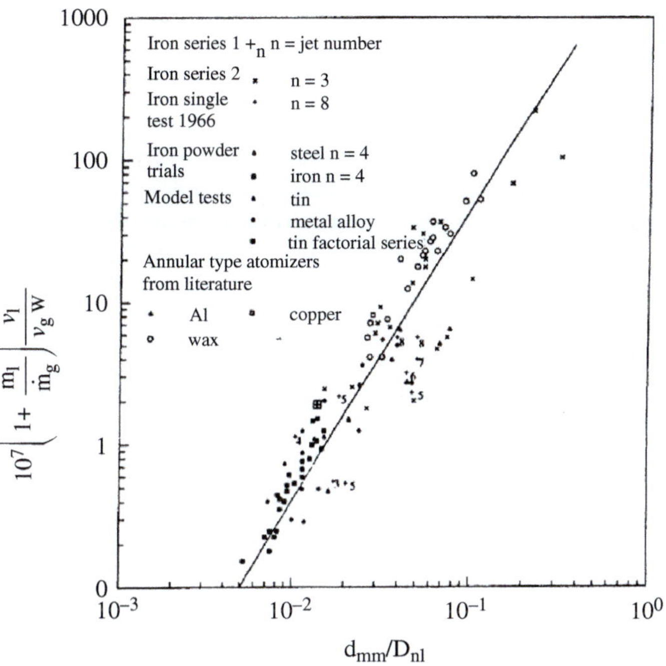

Figure 2.73 Correlation of atomized data for liquid metals. Source: Lubanska [73] / The Minerals, Metals and Materials Society.

delivery tube. In order to develop a relationship between average powder size and atomization conditions, Lubanska conducted two series of pilot plant atomization experiments (defined as series 1 and series 2 in Figure 2.73) in which the relevant atomization variables were systematically changed. When plotting the results of iron powders against the function, N [see Eq. (2.70)], in Wigg's formula, Lubanska did not obtain a satisfactory correlation. The data obtained on atomized iron powders deviated from Wigg's formula for large N values. Lubanska [73] analyzed the available empirical relationships deduced from experimental results obtained from various atomizers and liquid–gas systems and noted that many of the equations contain the function, $(1 + (\dot{m}_1/\dot{m}_g))$. Lubanska argued that this function was atomizer independent and could be generally applied to various liquids and atomizer designs. In addition, Lubanska noted the relative importance of the viscosities of liquid and gas on the basis of the results obtained by Wigg [140]. Accordingly, Lubanska developed a formula which retained the function, $(1 + (\dot{m}_1/\dot{m}_g))$, and the ratio of kinetic viscosities of liquid and gas, and the Weber number. The so-called Lubanska Equation in dimensionless form is

$$\frac{d_{mm}}{D_{nl}} = K\left[\frac{v_1}{v_g We}\left(1+\frac{\dot{m}_1}{\dot{m}_g}\right)\right]^{1/2} \tag{2.82a}$$

or

$$\frac{d_{mm}}{D_{nl}} = KF^{1/2} \tag{2.82b}$$

where K is a proportionality constant. Experimental results obtained for nitrogen-atomized iron powders were used to verify the applicability of Eq. (2.82) and to determine the magnitude of the constant K. Lubanska presented the factor, d_{mm}/D_{nl} versus the function F and obtained a good power-law relationship (see Figure 2.73). Lubanska suggested that the magnitude of the constant K is in the range of 40–50 and should be determined for each particular atomizer used.

On the basis of the previous discussion, it is evident that the applicability of the available equations is severely hindered by the large number of concurrent variables affecting the atomization process. In most cases, the usefulness of an equation is limited to particular experimental parameters and property ranges such as (a) fluid type, (b) pressure range, (c) atomizer type, and (d) gas type. Even theoretical studies attempting to establish universal curves for drop-sized prediction [119] await confirmation by means of more extensive experimental studies. Nevertheless, the primary importance of the investigations that have been conducted lies in enhancing our understanding of the key parameters that affect the atomization process.

2.9 Summary

In Chapter 2, we discussed the physical process of atomization which involves the disintegration of a molten metal stream and the subsequent generation of droplets. Particularly, we described the numerous physical parameters which influence atomization and discussed how these affect the final droplet size and distribution. Chapter 2 also presents a summary of the various methods that are used for the atomization of metals, and includes a discussion of atomizer design, droplet generation, and the influence of process and **physical parameters** on the resultant powder size distribution. This chapter also presents a thorough discussion of the phenomena and associated mechanisms that govern the actual liquid breakup and eventual formation of droplets. The information reviewed in this chapter reveals that atomization is an inherently stochastic process that is in its nature random and very difficult to analyze. A number of investigators over the years have developed correlations that can be used to predict powder size distribution although their accuracy is often limited by a narrow process parameter space and metal selection. Nevertheless, these correlations are useful in designing experiments and providing fundamental insight into the process of atomization. This chapter also reviews the various theoretical models that have been developed in an effort to understand liquid breakup and atomization. It is apparent that, despite progress from a fundamental perspective, there are many important questions that elude investigators, particularly in terms of an accurate prediction of an average power size and its corresponding distribution.

Nomenclature

Variable	Description	Units
A_c	Inner area of the crucible	m^2
A_e	Area of gas nozzle exit	m^2
A_n	Inner area of the liquid delivery tube	m^2
A_t	Area of nozzle throat	m^2
C_d	Liquid discharge coefficient	—
c_p	Constant pressure heat capacity	$J\,mol^{-1}\,K^{-1}$
C_{pl}	Constant pressure heat capacity of liquid	$J\,mol^{-1}\,K^{-1}$
c_v	Constant volume heat capacity	$J\,mol^{-1}\,K^{-1}$
d	Droplet size	μm
d_{16}	16% cumulative undersize	μm
d_{50}	50% cumulative undersize	μm
d_{84}	84% cumulative undersize	μm
d_h	Inner diameter of the gas orifice exit	m
d_{lig}	Diameter of liquid ligament	m
d_{Saut}	Sauter mean diameter	μm
d_m	Mass mean diameter	μm
d_{max}	Maximum stable droplet size	μm
d_{mm}	Mass median diameter	μm
d_v	Volume mean diameter	μm
D	Outer diameter of the delivery tube	m
D_{nl}	Inner diameter of the delivery tube	m
h_c	Heat transfer coefficient	$Wm^{-2}\,K^{-1}$
h_m	Height of liquid	m
h_t	Thickness of liquid sheet	m
ΔH_f	Heat of fusion	$J\,mol^{-1}$

Variable	Description	Units
ΔE_s	Total surface energy per unit mass of powder	$J\,kg^{-1}$
ΔE_t	Total energy consumption per unit mass of powder	$J\,kg^{-1}$
g	Gravitational acceleration	ms^{-2}
k_g	Thermal conductivity of gas	$Wm^{-1}\,K^{-1}$
k_m	Wave number of the fastest growing waves in a liquid sheet	m^{-1}
k'_m	Wave number of the fastest growing waves in a liquid rod	m^{-1}
L_{lig}	Length of fragments	m
Ma	Mach number	—
$\dot{m}_l$	Liquid flow rate	$kg\,s^{-1}$
$\dot{m}_g$	Gas flow rate	$kg\,s^{-1}$
p_a	Aspiration pressure	MPa
p_c	Crucible overpressure	MPa
p_e	Gas nozzle exit pressure	MPa
p_{en}	Environmental pressure	MPa
p_o	Atomization pressure or manifold pressure	MPa
p_{ori}	Delivery tube orifice pressure	MPa
Pr	Prandtl number	—
p_t	Throat pressure	MPa
p_t^*	Critical throat pressure	MPa
r	Diameter of ligaments after spheroidization	m
r_{lig}	Diameter of ligaments prior to spheroidization	m
R	Gas constant	$J\,mol^{-1}\,K^{-1}$
Re	Reynolds number	—
R_s	Spreading distance of gas jets	m
We	Weber number	—
t_{sol}	Solidification time	s
t_{sph}	Spheroidization time	s
T	Temperature	K
T_l	Melting temperature or liquidus temperature	K
T_g	Gas temperature	K
V	Relative velocity between gas and liquid	$m\,s^{-1}$
V_g	Velocity of atomization gas	$m\,s^{-1}$
V_{gt}	Velocity of atomization gas at the nozzle throat	$m\,s^{-1}$
V_{gt}^*	Critical velocity of atomization gas at the nozzle throat	$m\,s^{-1}$
V_l	Velocity of the liquid stream	$m\,s^{-1}$
y	Liquid delivery tube protrusion distance	m
α_d	Disturbance amplitude	m
α_{do}	Initial disturbance amplitude	m
α	Apex angle	degree
γ	c_p/c_v	—
η	Dynamic viscosity	Pa s
ν	Kinematic viscosity	$m^2\,s^{-1}$
ρ_o	Stagnant gas density	$kg\,m^{-3}$

Variable	Description	Units
ρ_t^*	Critical gas density at the throat	$kg\,m^{-3}$
ρ_1	Liquid density	$kg\,m^{-3}$
ΣS_d	Total surface area per unit mass of droplets	$m^2\,kg^{-1}$
ΣS_b	Total surface area per unit mass of bulk material	$m^2\,kg^{-1}$
σ_g	Geometric standard deviation	—
σ_1	Liquid surface tension	$N\,m^{-1}$
ξ	Atomization efficiency	—
$\overline{\omega}$	Phase angle	degree
ω	Angular velocity	$rad\,s^{-1}$
ω_g	Molar volume of the gas	$m^3\,mol^{-1}$

References

1 Klar, E. and Fesko, J.W. (1984). *Powder Metallurgy, Metals Handbook*. Materials Park, Ohio: American Society for Materials.
2 Lawley, A. (1983). *Rapid Solidification Technology Source Book* (ed. R.L. Ashbrook), 47. Ohio: ASM, Materials Park.
3 Hirschhorn, J.S. (1969). *Introduction to Powder Metallurgy*. New Jersey: American Powder Metallurgy Institute.
4 Klar, E. and Shafer, W.M. (1972). *Powder Metallurgy for High Performance Applications* (ed. J.J.B.A.B. Weiss), 57. Syracuse, New York: Syracuse University Press.
5 Gummeson, P.U. (1972). *Powder Metall.* 15: 67.
6 Marriott, W. (1872). British Patent 3322.
7 Hall, E.J. (1917). US Patents, No. 1, 501,449 and 1, 545,253.
8 Lavernia, E.J., Srivatsan, T.S., and Rangel, R.H. (1992). *J. Atomization Sprays.* 2: 253.
9 Grant, P.S. (1995). *Prog. Mater. Sci.* 39: 497.
10 Bauckhage, K. *Proceedings of ICLASS-91, 15–18 July 1991, National Institute of Standards and Technology*, 1991, 33. Gaithersburg, Maryland: US Department of Commerce.
11 Lavernia, E.J., Ayers, J.A., and Srivatsan, T.S. (1992). *Int. Mater. Rev.* 37: 1.
12 Kim, K.Y. and Marshall, W.R. (1971). *AICHE J.* 17: 575.
13 Lawley, A. (1977). *Int. J. Powder Met. Powder Tech.* 13: 169.
14 Lawley, A. (1992). *Atomization, the Production of Metal Powders, Monographs in P/M series.* Princeton, New Jersey: MPIF.
15 Biancaniello, F.S., Espina, P.I., Mattingly, G.E., and Ridder, S.D. (1989). *Mater. Sci. Eng. A.* 119: 161.
16 Ünal, A. (1989). *Metall. Trans. B.* 20B: 833.
17 Ünal, A. (1989). *Metall. Trans. B.* 20B: 613.
18 Ridder, S.D., Biancaniello, F.S., Espina, P.I., and Miller, S.A. (1989). *Synthesis and Analysis in Materials Processing: Characterization and Diagnostics of Ceramic and Metal Particulate Processing* (ed. E.J. Lavernia, H. Henein, and I. Anderson), 73. Warrendale, Pennsylvania: TMS.
19 Ünal, A. (1986). *Proceedings of Aluminum Technology '86* (ed. T. Sheppard), 673. London: Institute of Metals.
20 See, J.B., Runkle, J.C., and King, T.B. (1973). *Metall. Trans. A.* 4A: 2669.
21 Anderson, I.E., Figliola, R.S., and Morton, H. (1991). *Mater. Sci. Eng.* A148: 101.
22 McHugh, K.M. and Key, J.F. (1993). *Performance aspect of De Laval spray-forming nozzles, the Fluids Engineering Conference, 20–24 June 1993*, 91–93. New York, NY: ASME.
23 Liu, J., Arnberg, L., Backstrom, N. et al. (1988). *Powder Metall. Int.* 20: 17–22.
24 Veistinen, M.K., Lavernia, E.J., Baram, J.C., and Grant, N.J. (1989). *Int. J. Powder Metall.* 25: 89–92.
25 Rai, G., Lavernia, E.J., and Grant, N.J. (1984). *J. Meteorol.* 37: 22.

26 Couper, M.J. and Singer, R.F. (1984). *Proceedings of the 5th International Conference on Rapidly Quenched Metals* (ed. S. Steeb and H. Warlimont). North Holland, Amsterdam: NORTH_HOLLAND.

27 Baram, J.C., Veistinen, M.K., Lavernia, E.J. et al. (1988). *J. Mater. Sci.* 23: 2457.

28 Kumar, M. (1990). *Int. J. Powder Metall.* 26: 307.

29 Nilsson, E.O.F., Nilsson, S.I., and Hagelin, E.G. (1961). US Patent No. 2997245.

30 Mizutani, Y., Uga, Y., and Nishimoto, T. (1972). *Bull. Jpn. Soc. Mech. Eng.* 15: 620.

31 Domalavage, P., Ashdown, C., O'Handley, R.C., and Grant, N.J. (1983). *Mater. Sci. Eng. Letts.* 57: L1.

32 Anand, V., Kaufman, A.J., and Grant, N.J. (1980). *Rapid Solidification Processing: Principles and Technologies* (ed. R. Mehrabian, B. Kear, and M. Cohen), 273. Baton Rouge, Louisiana: Claitor's Publishing Division.

33 Liu, J., Arnberg, L., Bäckström, N. et al. (1988). *Mater. Sci. Eng. A.* 98: 43.

34 Tove, P.A. (1964). US Patent No. 3302892.

35 Backmark, U., Bäckström, N., Arnberg, L. (December 1985). Production of metal powder by ultrasonic gas atomization, Swedish Institute for Metals Research, Report No. IM 2051.

36 Young, J.M., Itoh, G., Jacobs, M.H., and Johnson, T.P. (1993). *Powder Metallurgy in Aerospace, Defense and Demanding Applications* (ed. F.H. Froes), 139. Princeton, New Jersey: MPIF.

37 Kato, T. and Kusaka, K. (1990). *Mater. Trans. JIM.* 31: 363.

38 Fraser, R.P., Dombrowski, N., and Routley, J.H. (1963). *J. Inst. Fuel.* 36: 316.

39 Dogan, C. and Saritas, S. (1994). *Int. J. Powder Metall.* 30: 419.

40 Cooper, K.P. and Ayers, J.D. (1994). *Int. J. Powder Metall.* 30: 77.

41 Halada, K. and Suga, H. (1990). *J. Jpn. Soc. Powder Metall.* 37: 492.

42 Jones, H. (1984). *J. Mater. Sci.* 19: 1043.

43 Cox, A.R., Moore, J.B., and Reuth, E.C.V. (1976). *Superalloys: Metallurgy and Manufacture* (ed. B.H. Kear and et al.), 45. Baton Rouge, Louisiana: Claitor's Publishing Division.

44 Stephan, H. (1976). *Advanced Fabrication Techniques in Powder Metallurgy and Their Economic Implications, 42nd Meeting of the Structures and Materials Panel (AGARD).* Ottawa, Canada.

45 Nikolaev, A.V. and Vilenskaya, M.L. (1991). *Phys. Chem. Mater. Treat.* 25: 160.

46 Grant, N.J. (1972). *Powder Metallurgy for High Performance Applications* (ed. J.J. Burke and B. Weiss), 85. Syracuse, New York: Syracuse University Press.

47 Sutcliffe, P.W. and Morton, P.H. (1976). *Advanced Fabrication Techniques in Powder Metallurgy and Their Economic Implications, 42nd Meeting of the Structures and Materials Panel (AGARD).* Fort Belvoir: Defense Technical Information Center.

48 Friedman, G. (1976). *Advanced Fabrication Techniques in Powder Metallurgy and Their Economic Implications, 42nd Meeting of the Structures and Materials Panel (AGARD).* Fort Belvoir: Defense Technical Information Center.

49 Warnica, W.D., van Reenen, M., Renksizbulut, M., and Strong, A.B. (1991). *Rev. Sci. Instrum.* 62: 3037.

50 Passow, C.H., Chun, J.H., and Ando, T. (1993). *Metall. Trans. A.* 24A: 1187.

51 Choi, K.J. and Delcorio, B. (1990). *Rev. Sci. Instrum.* 61: 1689.

52 Ström, L. (1969). *Rev. Sci. Instrum.* 40: 778.

53 Orme, M. and Muntz, E.P. (1987). *Rev. Sci. Instrum.* 58: 279.

54 Orme, M. and Muntz, E.P. (1990). *Phys. Fluids J. A.* 2: 1124.

55 Orme, M., Willis, K., and Nguyen, V. (1993). *Phys. Fluids J. A.* 5: 80.

56 Ikawa, Y., Itami, T., Kumagai, K. et al. (1990). *ISIJ Intern.* 30: 756.

57 Mehrotra, S.P. (1981). *Powder Metall. Int.* 13: 80.

58 Shinde, S.L. and Tenkoldar, G.S. (1977). *Powder Metall. Int.* 9: 180.

59 Johnston, G.H. and See, J.B. (1974). From unpublished work, Massachusetts Institute of Technology.

60 Chigier, N.A. (1991). *Proceedings ICLASS-91, 15–18 July 1991, National Institute of Standards and Technology*, 1. Gaithersburg, Maryland: US Department of Commerce.

61 Dombrowski, N. and Johns, W.R. (1963). *Chem. Eng. Sci.* 18: 203.

62 Orr, J.C. (1961). *Particulate Tech.* New York: MacMillan Co.

63 Gordo, G.D. (1959). *J. Appl. Phys.* 30: 1759.

64 Naida, Y.I. (1971). *Sovt. Powder Metall. Met. Ceram.* 10: 21.

65 Nichiporenko, O.S. (1974). *Sovt. Powder Metall. Met. Ceram.* 13: 433.

66 Lane, W.R. (1951). *Ind. Eng. Chem.* 43: 1312.
67 Magarvey, R.H. and Taylor, B.W. (1956). *J. Appl. Phys.* 27: 1129.
68 Hinze, J.O. (1948). *Appl. Sci. Res.* A1: 273.
69 See, J.B. and Johnston, G.H. (1978). *Powder Technol.* 21: 119.
70 Kear, B.H., Holiday, P.R., and Fox, A.R. (1983). *Rapid Solidification Technology, Source Book* (ed. E.R.L. Ashbrook), 66. Ohio: ASM, Metals Park.
71 Klar, E. and Fesko, J. (1981). *Prog. Powder Metall.* 37: 47.
72 Fraser, R.P., Dombrowski, N., and Routley, J.H. (1963). *Chem. Eng. Sci.* 18: 339.
73 Lubanska, H. (1970). *J. Meteorol.* 22: 45.
74 Nukiyama, S. and Tanasawa, Y. (1940). *Trans. Soc. Mech. Engrs. (Japan)* 6: S7.
75 Fritsching, U., Uhlenwinkel, V., and Bauckhage, E. (1992). *Phoenics J. Comp. Fluid Dynam. Appl.* 5: 81.
76 Rayleigh, L. (1879). *Proc. Lond. Math. Soc.* 10: 4.
77 Castleman, R.A. Jr. (1931). *US Bur. Stand. J. Res.* 6: 369.
78 Mansour, A. and Chigier, N. (1990). *Phys. Fluids A.* 2: 706.
79 Mansour, A. and Chigier, N. (1991). *Phys. Fluids A.* 3: 2971.
80 Ünal, A. (1987). *Mater. Sci. Technol.* 3: 1029.
81 Veistinen, M.K., Lavernia, E.J., Abinante, M., and Grant, N.J. (1987). *Mater. Lett.* 5: 373.
82 Ridder, S.D. and Biancaniello, F.S. (1988). *Mater. Sci. Eng. A.* A98: 47.
83 Upadhyay, R.K. and Spacil, H.S. (1989). *Surf. Coat. Technol.* 37: 379.
84 Ando, T., Tsao, C.Y., Wahlroos, J., and Grant, N.J. (1990). *Int. J. Powder Metall.* 26: 311.
85 Oh, B.W., Hong, S.P., and Lee, Z.H. (1992). *J. Korean Inst. Met. Mater.* 30: 1056.
86 Backmark, U., Bäckström, N., and Arnberg, L. (1986). *Powder Metall. Int.* 18: 338.
87 Liang, X. and Lavernia, E.J. (1993). *Mater. Sci. Eng. A.* A161: 221.
88 White, F.M. (1979). *Fluid Mechanics*. New York: McGraw-Hill, Inc.
89 Shapiro, A.H. (1953). *The Dynamics and Thermodynamics of Compressible Fluid Flow*. New York: The Ronald Press Co.
90 Allen, T. (1981). *Particle Size Measurement*, 3e. London, New York: Chapman and Hall Ltd.
91 Beddow, J.K. (1980). *Particulate Science and Technology*. New York: Chemical Publishing Co. Inc.
92 Fayed, M.E. and Otten, L. (1984). *Handbook of Powder Science and Technology*. New York: Van Nostrand Reinhold Co.
93 Murphy, C.H. (1984). *Handbook of Particle Sampling and Analysis Methods*. Verlag Chemie Intern. Inc.
94 Rao, P. (1973). Shape and Other Properties of Atomized Metal Powders, vol PhD Dissertation. Drexel University. Philadelphia, Pennsylvania.
95 Marinkovich, J., Mohamed, F.A., Pickens, J.R., and Lavernia, E.J. (1989). *J. Meteorol.* 41: 36.
96 Beddow, J.K. (1978). *The Production of Metal Powders*. Philadelphia, Pennsylvania: Heyden and Sons, Ltd.
97 Singer, A.R.E., Coombs, J.S., and Leatham, A.G. (1974). *Modern Developments in Powder Metallurgyvol 8* (ed. H.H. Housner and W.E. Smith), 263. New York: Plenum.
98 Schmitt, H. (1979). *Powder Metall. Int.* 11: 68.
99 Weiss, M.A. and Worsham, C.H. (1959). *ARS J.* 30: 252.
100 Nukiyama, S. and Tanasawa, Y. (1938). *Trans. Soc. Mech. Engrs. (Japan)* 4: 86.
101 Nukiyama, S. and Tanasawa, Y. (1938). *Trans. Soc. Mech. Engrs. (Japan)* 4: 138.
102 Lewis, H.C., Edwards, D.G., Goglia, M.J. et al. (1948). *Ind. Eng. Chem.* 40: 67.
103 Bitron, M.D. (1955). *Ind. Eng. Chem.* 47: 23.
104 Baram, J. (1988). *Mater. Sci. Eng. A.* 98: 65.
105 Thompson, J.S. (1948). *J. Inst. Met.* 74: 101.
106 Kim, M.H., H. Jones, Proceeding of the 4th International Conference on Rapidly Quenched Metals, Sendai, Japan, 1981.
107 Marshall, W.R. Jr., (1984). *Chem. Eng. Prog. Monogr. Ser.* 15: 141–146.
108 Tallmadge, J.A. (1978). *New Powder Metallurgy Processing Techniques and Analyses* (ed. H. Kuhn and A. Lawley), 1. New York: Academic Press.
109 Fritsching, U., Liu, H., and Bauckhage, K. (1991). *Proceedings ICLASS-91* (ed. H.G. Semerjian), 491. Gaithersburg, Maryland: National Institute of Standards and Technology.

110 Lefebvre, A.H. *Proceedings ICLASS-91, 15–18 July 1991, National Institute of Standards and Technology*, vol. 1991, 49. Gaithersburg, Maryland: US Department of Commerce.
111 Beck, J.E., Lefebvre, A.H., and Koblish, T.R. (1991). *J. Propuls. Power.* 7: 207.
112 Buckner, H.N., Sojka, P.E., and Lefebvre, A.H. (1989). *Aerated Atomization of High Viscosity Newtonian Liquids, Proceedings of 3rd Annual Conference on Liquid Atomization and Spray Systems*, 56. Irvine, California, USA.
113 Lefebvre, A.H. (1989). *Atomization and Sprays.* New York: Hemisphere Pub. Co.
114 Putimtsev, B.N. (1968). *Sov. Powder Metall. Met. Ceram.* 7: 839.
115 Putimtsev, B.N. (1972). *Sov. Powder Metall. Met. Ceram.* 11: 171.
116 Putimtsev, B.N. (1972). *Sov. Powder Metall. Met. Ceram.* 11: 863.
117 Sattelmayer, T. and Wittig, S. (1986). *J. Eng. Gas Turbines Power.* 108: 465.
118 Kennedy, J.B. (1985). High weber number SMD correlations for pressure atomizers, ASME Paper 85-GT-37.
119 Bradley, D.J. (1973). *Phys. D: Appl. Phys.* 6: 1724.
120 Bradley, D.J. (1973). *Phys. D: Appl. Phys.* 6: 2267.
121 Tyler, E. (1993). *Philos. Mag.* 16: 504.
122 Hall, E.J. (1924). US Patents, No. 1, 501,449 and 1, 545,253.
123 Klar, E. and Shafer, W.M. (1975). *Int. J. Powder Met. Powder Tech.* 11: 241.
124 Lefebvre, A.H. (1992). *J. Eng. Gas Turbines Power.* 114: 89.
125 Szekely, J. and Themelis, N.J. (1971). *Rate Phenomena in Process Metallurgy.* New York: Wiley.
126 Small, S. and Bruce, T.J. (1968). *Int. J. Powder Metall.* 4: 7.
127 Weat, R.C. and Astle, M.J. (ed.) (1981). *CRC Handbook of Physics and Chemistry*, 62e. Boca Raton, Florida: CRC Press.
128 Nichiporenko, O.S. and Naida, Y.I. (1968). *Sov. Powder Metall. Met. Ceram.* 7: 509.
129 Naida, Y.I., Nichiporenko, O.S., and Fen'Ko, M.A. (1975). *Sov. Powder Metall. Met. Ceram* 14: 442.
130 Y.I. Naida, 10 (1971) 187.
131 Decuurs, J., Devillard, J. and, Sainfort, G. (1976). Advanced fabrication techniques in powder metallurgy and their economic implications, 42nd Meeting of the Structures and Materials Panel (AGARD), Ottawa, Canada.
132 Walton, W.H. and Prewett, W.C. (1949). *Proc. Phys. Soc.* 62: 341.
133 Mottern, J.W. (1978). Reaction of velocity of profiles in laminar jets. *AIChe. Symp. Ser.* 75.
134 Bogy, D.B. (1979). *Annu. Rev. Fluid Mech.* 11: 207.
135 Naida, Y.I., Ivanov, V.S., Fedorov, S.F., and Manasyan, R.B. (1980). Transl. Poroshkovaya Metallurgiya 208 1.
136 Lin, S.P. and Kang, D.J. (1987). *Phys. Fluids A.* 7: 2000.
137 Weber, C.Z. (1931). *Angew. Math. Mech.* 11: 136.
138 Stapper, B.E., Sowa, W.A., and Samuelsen, G.S. (1992). *J. Eng. Gas Turbines Power-Trans ASME.* 114: 39.
139 Rangel, R.H. and Hess, C.F. (1990). Nonlinear spacial instability of a fluid sheet, 28th Aerospace Science Meeting and Exhibition AIAA-90-0118 Reno, Nevada.
140 Wigg, L.D. (1964). *J. Inst. Fuel.* 37: 500.
141 Wood, R. (July 1954). Unpublished work at Thronton Research Center.
142 Clare, H. and Radcliffe, A. (1954). *J. Inst. Fuel.* 27: 510.
143 Fraser, R.P. and Eisenklam, P. (1956). *Trans. Inst. Chem. Eng.* 34: 294.

3

Heat Transfer and Solidification of Droplets

3.1 Introduction

Atomization is a complex process that involves the generation of droplets from molten metal. The details associated with atomization, including the mechanisms that govern the **breakup** of a liquid, have been addressed in Chapter 2. Accordingly, the dispersion of micrometer-sized droplets, which are generated during atomization, travels a predetermined path, before undergoing a phase transition into solid powders. As discussed in Chapter 2, droplets that are generated during atomization are typically characterized by a broad size distribution. The occurrence and extent of solidification in droplets depend on factors such as droplet size, superheat temperature, physical properties of the melt, physical properties of the atomization gas, and relative velocity between the droplets and the atomization gas.

The transfer of thermal energy occurs as soon as the molten metal departs the inductively heated region along the path to the metal delivery tube. As shown in Figure 3.1, heat transfer and atomization occur simultaneously, as illustrated. To maintain the heat transfer problem tractable, it is generally assumed that heat transfer and the concomitant solidification events commence immediately following the formation of the droplets (e.g. the spheroidization of molten metal ligaments; see Chapter 2). Moreover, it is assumed that at the time immediately following droplet formation, the droplets are exactly at the same superheat temperature as that of the molten metal in the crucible. These limiting assumptions, and others that will be discussed in the present chapter, are necessitated by the complexity of the problem posed by a molten metal during atomization.

Atomization is only one of several engineering processes that necessitate a fundamental understanding of droplet heat transfer and solidification to achieve optimum combinations of microstructure and properties [1–12]. For example, considerable interest in droplet solidification was prompted by the development of atomization as a rapid solidification technology. Processes that involve the heat transfer and solidification of droplets include:

(a) **twin fluid atomization** [13–16],
(b) **rotary atomization** [17, 18],
(c) **electrohydrodynamic atomization** (EHD) [19–21],
(d) **drop tube solidification** [22, 23],
(e) **magnetic levitation** [24, 25],
(f) solidification of droplets embedded in metal matrices [26, 27],
(g) **droplet emulsion** [28, 29],
(h) **controlled drop size atomization** [30, 31], and
(i) thermal and plasma **thermal spray process** [32, 33].

Powder atomization has been extensively used as an effective method to achieve refined microstructures through rapid solidification [1–21]. It is well established that the thermal and solidification conditions that are present during the atomization of droplets promote the formation of rapidly solidified microstructures that are characterized by one or a combination of the following:

(a) refined microstructures, such as grain size and **dendrite arm spacings** [34–39],
(b) reduced microsegregation of alloying elements [19, 40],
(c) formation of **metastable phases** [37, 39],
(d) extended **metastable solid solubility** [19, 39–41], and
(e) modified phase morphologies [35, 37, 38].

Metallic Powders for Additive Manufacturing: Science and Applications, First Edition.
Enrique J. Lavernia, Kaka Ma, Julie M. Schoenung, James F. Shackelford, and Baolong Zheng.

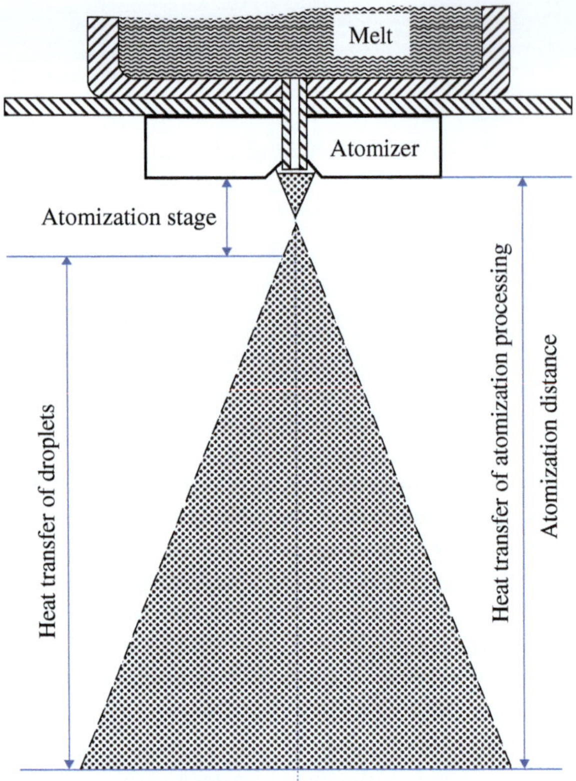

Figure 3.1 The various stages associated with heat transfer and solidification during spray deposition.

The heat transfer and solidification events that are present during the atomization stage critically influence microstructure formation in the powders. Figure 3.2 shows atomized powders and the various solidification microstructures that can form.

In particular, the heat transfer and solidification characteristics of the droplet dispersion influence the resultant microstructure and properties through the following effects [42]:

(a) determining the amount of thermal energy contained in the droplet spray during solidification,
(b) producing solid phases that will subsequently act as nucleation or growth centers for solidification,
(c) producing a partially liquid, partially solid spray of droplets that contains predetermined proportion of liquid and solid phases, and
(d) promoting the formation of solid phases with concomitant rapidly solidified characteristics because of the high cooling rates (10^3–10^5 K s^{-1}) that are associated with this stage.

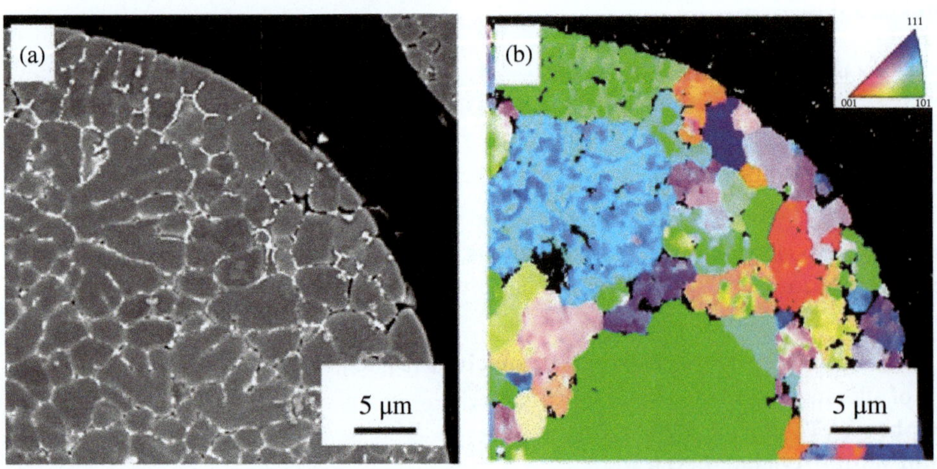

Figure 3.2 Solidification microstructure of atomized powder particles of Al7075 Alloy: (a) BSE contrast; (b) EBSD micrograph. Source: Molnarova et.al (2016)/MDPI [42].

The total amount of thermal energy that is released during atomization is finite. It then follows that the thermal energy that is dissipated during droplet travel determines the evolution of **microstructure** in the powders. Accordingly, the thermal and solidification conditions that are experienced by the powders are intimately coupled to the heat transfer and solidification environment in the spray [1–12].

This chapter will discuss the heat transfer and solidification phenomena that occur during droplet flight and the associated **microstructural development** in atomized droplets. Section 3.2 discusses the important thermal and solidification conditions; superheat, undercooling, **hypercooling**, **dynamic solidification**, and kinetic solidification. Section 3.3 focuses on the heat transfer mechanisms that govern droplet solidification; radiative and convective heat transfer phenomena. Section 3.4 deals with the nucleation behavior; **homogeneous nucleation**, **heterogeneous nucleation**, distribution of **nucleates**, and undercooling. Section 3.5 addresses the solidification behavior of droplets; temperature distribution, cooling rate, **solidification time**, and interfacial velocity. Section 3.5 reviews the **microstructural development**, solidification morphology, **dendrite arm spacing**, microstructural refinement, phase selection, and solute redistribution.

3.2 Important Thermal and Solidification Conditions

3.2.1 Thermal Conditions

The disintegration of material by high-energy gas jet atomization, as described in Chapter 3, is facilitated by the low viscosities that are associated with molten metals and alloys. Accordingly, prior to atomization, it is first necessary to provide sufficient thermal energy to increase the temperature of the material above its liquidus temperature. The difference between the temperature of molten metal and its equilibrium liquidus temperature is defined as **superheat**:

$$\Delta T_s = T - T_l \tag{3.1}$$

where ΔT_s is the superheat, T is the temperature of the liquid (above the liquidus temperature), and T_l is the liquidus temperature. Hence, a liquid that is at a temperature above the liquidus temperature is said to be **superheated**. As discussed in Chapter 3, superheating the material prior to atomization is necessary to avoid premature solidification of the molten metal in the delivery tube and to ensure complete spheroidization of the droplets.

During atomization, a droplet exchanges thermal energy with the atomization environment, and eventually cools down to the liquidus temperature. Thermodynamic restrictions dictate that solidification does not start immediately when the temperature of a droplet reaches the liquidus temperature. Instead, and depending on factors such as cooling rate, the potency and distribution of nucleation agents, to name a few, nucleation of solid phases commences at a certain temperature below the liquidus temperature. The difference between the equilibrium liquidus temperature and the nucleation temperature is referred to as **undercooling**:

$$\Delta T = T_l - T_N \tag{3.2}$$

where ΔT is the undercooling, and T_N is the nucleation temperature. Hence, a liquid which is at a temperature below the liquidus temperature is referred to as **an undercooled liquid**. If nucleation of the solid phases is triggered at extraneous sources, such as impurities or other nucleation sites, the undercooling achieved is **heterogeneous nucleation undercooling**. In atomization, it is very difficult to suppress heterogeneous nucleation because of the presence of droplet surfaces and other potential nucleation agents. Consequently, nucleation of the solid is generally triggered by heterogeneous nucleation mechanisms, in all but the smallest droplet (<10 μm). Ideally, if no heterogeneous nucleation reagents are present or active, the nucleation temperature of a droplet will depend on the physical properties of the molten metal, the droplet size, and the cooling conditions that are present. In this case, the undercooling achieved is defined as **homogeneous nucleation undercooling**.

Nucleation is followed by the growth of solid phases at the expense of the liquid. The thermal energy that is released during the liquid-to-solid phase transformation is defined as the latent heat of fusion which is dissipated into the undercooled liquid that is at a lower temperature. Consequently, a droplet experiences a redistribution of thermal energy because of two competing processes. First, thermal energy is released at the solid–liquid interface because of the liquid-to-solid phase transformation. Second, thermal energy is dissipated at the surface of the droplet into the atomization environment. The temperature of an undercooled droplet will hence depend on the balance between the "heating" and the "cooling" effects. When solidification occurs in an undercooled droplet, the initial solidification rate

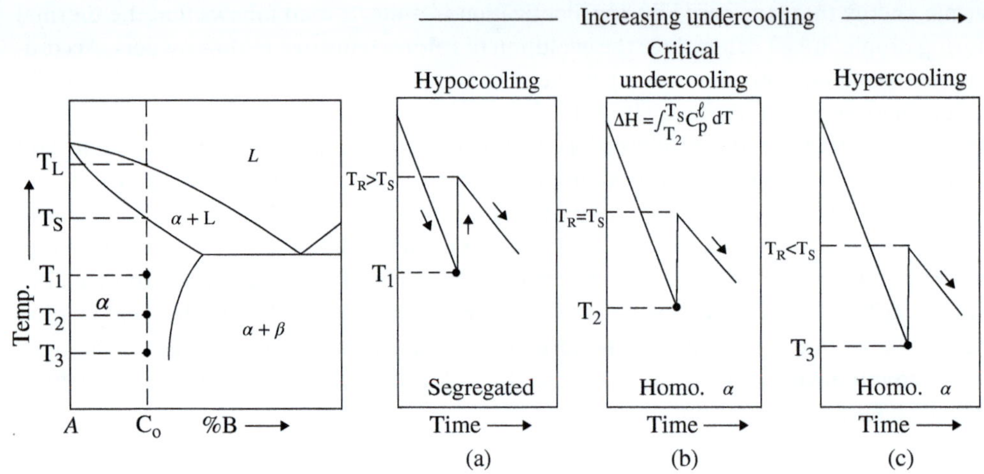

Figure 3.3 Concept of critical undercooling and hypercooling, T_1, T_2, and T_3 are the undercooling temperatures of alloy, C_0; and T_R is the recalescence temperature [43].

(i.e. the increase in the fraction of the solid as a function of time) is extremely high. Consequently, the rate of thermal energy release from the solid–liquid interface into the undercooled liquid is much faster than the rate of thermal energy dissipation from the droplet surface into the environment. As a result, the temperature of the droplet increases, a phenomenon that is referred to as **recalescence**.

The concept of undercooling can be illustrated by means of the schematic diagram shown in Figure 3.3 [43]. In this figure, an alloy of composition C_0 is undercooled into the single-phase region, a. With progressive increase in temperature during **recalescence**, droplet temperature returns to the two-phase ($\alpha + L$) region. **Recalescence** helps bring back the temperature to precisely the solidus temperature (T_S); this condition is termed as **critical undercooling** ($T_R = T_S$). When recalescence brings the temperature into the region below the solidus temperature, this condition is termed **hypercooling** ($T_R < T_S$). When undercooling is sufficiently high, solidification is sufficiently rapid such that the thermal energy dissipation from the droplet surface can be ignored as compared with the energy released from phase transformation [44]. Hence, in the absence of significant heat transfer to the surroundings, the solid–liquid phase transition is solely responsible for the temperature increase in an undercooled droplet during solidification. In this case, the thermal energy that is released during solidification equals the thermal energy consumed as the temperature of the droplet increases. Under the critical undercooling condition, **recalescence** will bring the temperature of the droplet from the undercooled temperature to the solidus temperature at the completion of solidification. If one assumes that the heat capacities of the solid and liquid phases are equal and that both latent heat capacity and heat of fusion are independent of temperature, an energy balance may be used to define the above-mentioned critical undercooling condition for an alloy [41, 45–47]:

$$\Delta T_{cr} = \frac{\Delta H_f}{c_{p,l}} + (T_1 - T_S) \tag{3.3}$$

where ΔH_f is the latent heat of fusion; $c_{p,l}$ is the molar heat capacity of the liquid phase; and T_1 and T_S are the liquidus and solidus temperatures, respectively. For a pure metal, the liquidus and solidus temperatures are equal, and the critical condition defined by Eq. (3.3) then reduces to [48]

$$\Delta T_{cr} = \frac{\Delta H_f}{c_{p,l}} \tag{3.4}$$

Moreover, when the undercooling exceeds the critical value defined by Eqs. (3.2) and (3.3), i.e. when

$$\Delta T \geq \Delta T_{cr} \tag{3.5}$$

the material is said to be hypercooled [45–47], During **hypercooling**, solidification may proceed to completion in the absence of external heat extraction and the recalescence temperature remains below the solidus temperature.

3.2.2 Solidification Considerations

A simple but very useful diagram describing the solidification history of atomized droplets was introduced by Levi and Mehrabian [43, 49, 50]. They described the solidification history of droplets using various paths in an enthalpy–temperature diagram. One example of such diagrams is shown in Figure 3.4 for an Al droplet. In this figure, the solidification history of the droplet is presented in terms of dimensionless enthalpy, ψ, and dimensionless temperature, θ. The dimensionless enthalpy and dimensionless temperature are given by

$$\psi = \frac{H - H_{\mathrm{sf}}}{\Delta H_{\mathrm{f}}} \tag{3.6}$$

and

$$\theta = c_{\mathrm{p,l}} \frac{T - T_{\mathrm{l}}}{\Delta H_{\mathrm{f}}} \tag{3.7}$$

where H is the enthalpy of the droplet at a temperature T, f is the fraction of solid, and H_{sf} is the enthalpy of the solid at the melting temperature. This diagram can be understood by considering the relative magnitudes of ψ and θ when a droplet contains different amounts of solid phases. The dimensionless enthalpy in Eq. (3.6) assumes a value of zero when a droplet is completely solidified ($f = 1$) and is at the melting temperature. The dimensionless enthalpies of superheated, undercooled, and hypercooled droplets are greater than unity, between zero and unity, and less than zero, respectively. When a droplet is completely solidified and is below the melting temperature, its dimensionless enthalpy is less than zero. The dimensionless temperature is zero at the melting temperature, greater than zero for a superheated droplet, between zero and minus unity for an undercooled droplet, and less than minus unity for a hypercooled droplet. There are five distinctive curves shown in Figure 3.4. The two inclined straight lines represent the cooling paths of pure liquid and pure solid. A pure liquid or solid droplet losses enthalpy at a rate that depends primarily on heat capacity. In this case, enthalpy is a linear function of temperature. There are three very important solidification paths that are illustrated in Figure 3.4; these are represented by the curves labeled (1)–(3). The vertical path 1 shown in Figure 3.4 represents the cooling behavior that is associated with **equilibrium solidification**. Equilibrium solidification is an ideal condition in which a solid–liquid transformation proceeds isothermally at the melting temperature. There is no temperature change during equilibrium solidification of a pure material, this process is defined by a vertical line in the ψ–θ diagram. Since the only driving force of equilibrium solidification is external heat extraction; the rate of enthalpy release during phase transformation equals the rate of thermal energy removal from the droplet. The horizontal solidification path (curve 2) represents the ideal condition of **adiabatic**

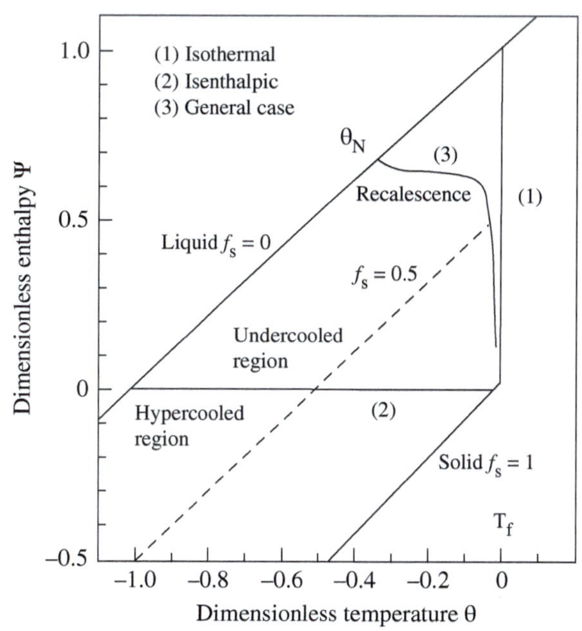

Figure 3.4 Enthalpy (ψ)–temperature (θ) diagram showing possible solidification "paths" where $\psi = (H - H_{\mathrm{sf}})/\Delta H_{\mathrm{f}}$ is the dimensionless enthalpy; $\theta = c_{\mathrm{pl}}(T - T_1)/\Delta H_{\mathrm{f}}$ is the dimensionless temperature; and θ_{n} is the dimensionless nucleation temperature. Source: Levi and Mehrabian [49] / Reproduced with permission from Springer Nature.

solidification. In **adiabatic solidification** there is no thermal energy dissipation from the droplet into the environment. The thermal energy released during the solid–liquid phase transformation is transferred into the **undercooled liquid** which surrounds the solid phase. The process is inherently **isenthalpic** because of the absence of external thermal energy extraction. The temperature of the droplet, however, increases with increasing extent of solidification because of **recalescence**. Therefore, isenthalpic solidification is represented by a horizontal line in the ψ–θ diagram. For the ideal process shown in Figure 3.4, curve 2, the liquid is hypercooled, such that **recalescence** causes the droplet temperature to increase, approaching the liquidus temperature upon the completion of solidification.

The two aforementioned solidification paths represent extreme conditions: **isothermal** solidification and **isenthalpic** solidification. Under practical atomization conditions, it is generally difficult to achieve such ideal solidification conditions. For example, the presence of a low initial undercooling will alter the path of ideal isothermal solidification. Moreover, when external thermal energy removal may not be neglected, the solidification of a hypercooled droplet will depart from ideal isenthalpic conditions. As a result, the solidification path of an undercooled droplet deviates from a horizontal line in the ψ–θ diagram. Under practical atomization conditions, there is a certain amount of undercooling present, and the rate of external thermal energy removal cannot be neglected when compared to the rate of energy released during the phase transformation. In this case, the droplet solidification is described by a mixture of path 1 and path 2. Such a general solidification path is illustrated by curve 3 in Figure 3.4 for an undercooled droplet. Curve 3 is characterized by two distinct solidification regimes. The first regime is close to a horizontal line, while the second regime is close to a vertical line in the ψ–θ diagram. In the first regime, solid phases nucleate and grow at the expense of the **undercooled liquid**. The phenomenon of solidification in an **undercooled liquid** is known as **dynamic solidification**. As a result of the rapid rate of thermal energy release from the solid–liquid interface the temperature of the droplet increases. Since external thermal energy extraction also influences the temperature history of a droplet during **dynamic solidification**, the **dynamic solidification** conditions illustrated by curve 3 differ from those of adiabatic solidification as illustrated by curve 2. The slope of the dynamic solidification curve is normally smaller than that corresponding to adiabatic solidification because the rate of **recalescence** is generally low when the rate of external thermal energy extraction is high. The second regime illustrated by curve 3 resembles that present during isothermal solidification. In this solidification regime, the temperature of a droplet reaches a value close to the melting temperature because of **recalescence**. Solidification is driven primarily by the dissipation of thermal energy to an external source, i.e. the surrounding environment. Although in this regime the solidification conditions differ from those present at equilibrium, it is often referred to as equilibrium solidification to simplify the description.

Mehrabian [50] calculated the cooling and solidification history of Al droplets by assuming various initial undercoolings for a heat transfer condition that is close to adiabatic. The results of this calculation are summarized in the ψ–θ diagrams shown in Figure 3.4. As shown in this figure, the solidification history of a droplet

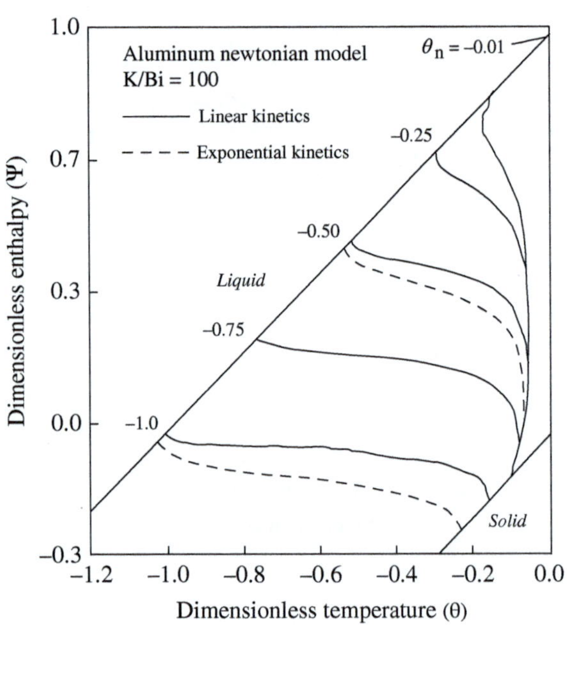

Figure 3.5 Effect of nucleation temperature on the thermal history of undercooled Al droplets calculated from Newtonian model where θ_n is the dimensionless nucleation temperature. Source: Mehrabian [50] / Reproduced with permission from Taylor & Francis.

is governed by the combined effects of initial **undercooling** and heat transfer, i.e. by the competition between dynamic solidification events and equilibrium solidification events. It is worth noting that at relatively low initial undercoolings (e.g. $\theta_n = -0.01$), the effect of equilibrium solidification becomes significant. The ψ–θ curves are almost vertical, indicating the importance of **external heat transfer** on solidification history. At relatively high initial undercoolings (e.g. $\theta_n = -1.0$), undercooling governs the solidification history of the droplet. The solidification paths, as defined by the ψ–θ curves, become increasingly horizontal as the initial undercooling is increased. As the initial undercooling increases to that corresponding to **hypercooling**, the ψ vs. θ curves become nearly horizontal. In this case, solidification is dominated by dynamic solidification (Figure 3.5).

3.3 Heat Transfer

Heat transfer is a process during which **thermal energy** is released from a thermal source (high-temperature region of a system) to a **thermal sink** (normally the low-temperature region of a system). In view of the complex conditions that are present during droplet solidification, it is useful to classify heat transfer into two general categories: **internal heat transfer** and **external heat transfer**. As shown in Figure 3.6, during internal heat transfer the solidifying phases may be the thermal sources, whereas the **undercooled liquid** represents the **thermal sink**. Accordingly, the energy that is released during the solid–liquid phase transformation is dissipated in part into the undercooled liquid. This internal heat transfer mechanism causes thermal energy redistribution within the droplet. External heat transfer, however, involves the dissipation of thermal energy from the droplet into the environment.

3.3.1 Heat Transfer Mechanisms

The dissipation of thermal energy from the surface of a droplet into the surrounding environment involves two important mechanisms: radiative and convective heat transfer. If one assumes that the environmental temperature equals the temperature of the atomization gas, the radiative heat flux is given by

$$J_r = \sigma_r \varepsilon_r \left(T^4 - T_g^4\right) \tag{3.8}$$

where J_r is the radiative heat flux; σ_r is the Stefan–Boltzmann constant; ε_r is the surface emissivity; and T_g is the temperature of the atomization gas. In a study of droplet solidification, Levi [48] defined a radiative heat transfer coefficient such that the radiative heat flux can be expressed in a form analogous to that of a convective heat flux. Accordingly, the radiation heat transfer coefficient becomes [48]

$$h_r = \sigma_r \varepsilon_r \frac{T^4 - T_g^4}{T - T_g} \tag{3.9}$$

By applying this formulation, the radiative thermal energy flux across the surface of a droplet may be conveniently expressed in the form:

$$J_r = h_r (T - T_g) \tag{3.10}$$

The radiative heat transfer coefficient in this case is, however, a function of droplet temperature.

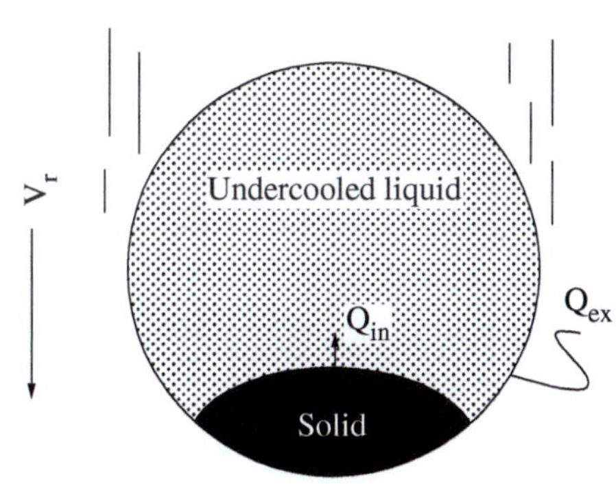

Figure 3.6 Internal and external heat transfer during droplet solidification. In this diagram V_r is the relative velocity between the droplet and the atomization gas; Q_{in} represents internal heat transfer and Q_{ex} represents external heat transfer.

In convective heat transfer, the thermal energy flux across the **droplet/gas interface** is generally defined using a simple formulation as

$$J_c = h_c(T - T_g) \tag{3.11}$$

where J_c is the thermal energy flux across the **droplet/gas interface**, and h_c is the convective heat transfer coefficient. In atomization, this seemingly simple equation is made complex by the fact that the magnitude of the conductive heat transfer coefficient is a function of many interrelated atomization variables. For example, the convective heat transfer coefficient is critically related to the physical properties of the gas, such as density, viscosity, and conductivity at the droplet/gas interfacial region. Moreover, such properties vary significantly during droplet flight. More importantly, when a droplet travels in an atomization environment, the relative velocity between the gas and the droplet governs the magnitude of the convective heat transfer coefficient. As will be discussed in Section 3.3.2, this relative velocity may vary from a large positive value through zero to a large negative value.

Experimental determination of the magnitude of the convective heat transfer coefficient poses a challenging task. Even in the case of solid–solid contact, such as the conditions that are normally encountered in die casting, permanent mold casting, and splat quenching, the experimentally determined heat transfer coefficients vary significantly, depending on experimental conditions [43, 51–57]. Some of the experimentally determined heat transfer coefficients have been summarized by Mehrabian [50] as shown in Table 3.1. It should be noted that the **heat transfer coefficient** of a droplet varies significantly during atomization. One approach that is used to evaluate **heat transfer** is to use an average heat transfer coefficient which may be determined using empirical methods. For example, the magnitude of the heat transfer coefficient may be calculated from experimental measurements of dendrite arm spacings in solidified droplets [57, 58]. In this approach, the measured dendrite arm spacings are converted into an average cooling rate based on a predetermined empirical relationship between dendrite arm spacing and average cooling rate. The heat transfer coefficient is then computed from the converted average cooling rate [57, 58]. Some values of average heat transfer coefficients that were determined using this method are summarized in Table 3.2 [57]. A notable feature in this table is the difference in the value of the convective heat transfer coefficient between gas atomization and EHD atomization. For example, as shown in Table 3.2, the heat transfer coefficient for Ar-atomized droplets is 9.6×10^3 Wm^{-2} K^{-1}, whereas that corresponding to EHD-atomized droplets is 1.6×10^3 W m^{-2} K^{-1}. This difference is related to the thermal environments that are present in these two processes. In the gas atomization process, droplets are exposed to a high relative velocity (e.g. 10–1000 ms^{-1}) as they travel in the atomization gas. Accordingly, convective heat transfer dominates the overall dissipation of thermal energy. In the EHD atomization, however, droplets are generated by exposing molten metal to an electrostatic force, and the process is conducted in a vacuum environment of up to 10^{-5} mm Hg. Under such high vacuum conditions, radiative heat transfer governs the overall thermal energy loss, and the effective heat transfer coefficient is relatively low.

The above-described empirical approach used to determine a heat transfer coefficient is useful in that it offers a simple technique that may be used to estimate the relative importance of certain experimental parameters such as droplet size and atomization environment, etc. Unfortunately, however, this approach fails to provide any detailed information regarding the variation of heat transfer coefficient during the droplet cooling history. Inspection of the available literature reveals that it is also possible to estimate the magnitude of a heat transfer coefficient during atomization using the available analytical and **numerical models** [2, 3, 45, 48, 59–64]. These are described in Section 3.3.2.

Table 3.1 Experimentally determined heat transfer coefficients, *h*.

Condition	Thickness (mm)	h (Wm^{-2} K^{-1})	References
Al splat on Ni substrate	10^{-3}	1.1×10^5–2.8×10^5	[57]
Al drop spreading on Fe substrate	0.15	1.7×10^4–1.8×10^5	[58]
Al in cast in steel mold	10	3.1×10^3	[59]
Al in pressure cast in steel mold (89.7 MPa)	10	3.3×10^4	[59]
Al die cast in steel mold	1.6	7.94×10^4	[60]

Source: Adapted from Mehrabian [50].

Table 3.2 Experimentally determined heat transfer coefficients, h, inferred from dendrite arm spacings.

Condition	Powder size (μm)	Dendrite arm spacing (μm)	Cooling rate (Ks^{-1})	h ($Wm^{-2}\ K^{-1}$)
Argon atomized Maraging 300 steel	75	~2	2.1×10^4	9.6×10^3
REP Maraging 300 steel	170	~3	5.5×10^3	5.4×10^4
EHD atomized Maraging 300 steel	650	~6.5	4.2×10^2	1.6×10^3

Source: Mehrabian [57] / Reproduced with permission from Claitor's Publishing Division.

3.3.2 Heat Transfer Coefficient

The empirical correlations that are available to estimate the value of the convective heat transfer coefficient generally involve the use of dimensionless numbers. Accordingly, it is useful to define the Reynolds number, *Re*, the Prandtl number, *Pr*, and the Nusselt number, *Nu*. The Reynolds number is defined as

$$Re = \frac{(V_g - V_d)d}{v_g} \tag{3.12}$$

where v_g is the kinematic viscosity of the atomization gas. The Prandtl number is the ratio between the momentum diffusivity and the thermal diffusivity of the atomization gas. It is related to the heat capacity, heat conductivity, and viscosity of the gas by

$$Pr = \frac{c_{p,g}\eta_g}{k_g} \tag{3.13}$$

where $c_{p,g}$, k_g, and η_g are heat capacity per unit mass, thermal conductivity, and dynamic viscosity of the environmental gas, respectively. The Nusselt number defines the ratio of the thermal conductance of the **droplet/gas interface** to that of the atomization gas. It is related to the convective heat transfer coefficient at the **droplet/gas interface** and to the thermal conductivity of the gas by

$$Nu = \frac{h_c d}{k_g} \tag{3.14}$$

where d is droplet size. In most heat transfer coefficient formulations, the convective heat transfer coefficient across a **droplet/gas interface** is approximated utilizing the well-established Ranz–Marshall correlation [65]:

$$Nu = 2 + 0.6\mathrm{Re}^{1/2}Pr^{1/3} \tag{3.15}$$

Hence, combining Eqs. (3.12)–(3.15), the convective heat transfer coefficient across a **droplet/gas interface** can be determined as follows:

$$h_c = \frac{k_g}{d}(2 + 0.6\mathrm{Re}^{1/2}Pr^{1/3}) \tag{3.16a}$$

or

$$h_c = \frac{k_g}{d}\left\{2 + 0.6\left[\frac{(V_g - V_d)d}{v_g}\right]^{1/2}\left(\frac{c_{p,g}\eta_g}{k_g}\right)^{1/3}\right\} \tag{3.16b}$$

From Eqs. (3.14–3.16) the magnitude of the convective heat transfer coefficient of atomized droplets can be calculated from a knowledge of the gas and droplet velocities as well as the droplet size and the physical properties of the gas.

In numerical studies of atomization, the heat transfer coefficient across droplet/gas interfaces is often calculated using Eq. (3.16a,b) based on the gas and droplet velocities. It is useful to determine the convective heat transfer coefficient of droplets as a function of flight distance, or equivalently, as a function of flight time. It is worth noting that, despite differences in initial conditions and other processing variables, the results published in the literature consistently reveal a similar trend of convective heat transfer coefficient as a function of flight distance for atomized droplets [62–64, 66–69]. One example is shown in Figure 3.7 from the work of Liu et al. [62] for Ni_3Al droplets

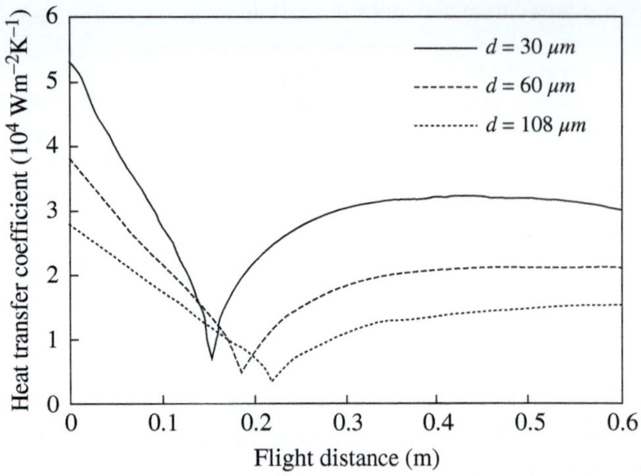

Figure 3.7 Calculated heat transfer coefficient for Ni_3Al droplets as a function of flight distance and droplet size (atomization gas = N_2). Source: Liu et al. [62].

atomized under supersonic conditions. As shown in Figure 3.7, the magnitude of the convective heat transfer coefficient reaches maximum value during the initial stages (e.g. close to the atomization region) [62]. It decreases rapidly to a minimum value which is followed by an increase with increasing flight distance. At relatively large flight distances, an increase in flight distance is accompanied by a decrease in the magnitude of the convective heat transfer coefficient [62]. Another important feature that is evident from Figure 3.7 is that the magnitude of the convective heat transfer coefficient corresponding to each flight distance is inversely related to droplet size. The only exception to this trend occurs for flight distances corresponding to a minimum value of the convective heat transfer coefficient.

There is an important heat transfer condition that is present when the differential velocity between the atomization gas and the droplets approaches a value of zero. In this case, the convective heat transfer coefficient reaches a minimum value. The transfer of thermal energy that occurs when droplets are at zero relative velocity with the atomization gas corresponds to **static heat transfer**. Accordingly, the corresponding heat transfer coefficient is known as a **static heat transfer coefficient**. The static heat transfer coefficient may be derived easily by setting the differential velocity between the droplet and the gas in Eq. (3.16) to zero. It then follows that

$$h_{cs} = \frac{2k_g}{d} \tag{3.17}$$

where h_{cs} is the static convective heat transfer coefficient.

It is not difficult to infer from Figure 3.7 that the magnitude of the static heat transfer coefficient corresponds to about one-third of that of the maximum convective heat transfer coefficient under normal atomization conditions [62]. In a related study, Grant et al. [63] compared the static heat transfer coefficient with the average convective heat transfer coefficient of droplets (e.g. under non-static conditions). The results of this study revealed that in the droplet size regime of 10–200 μm, the magnitude of the static heat transfer coefficient is between one-half and one-eighth of the heat transfer coefficient under convective conditions. Moreover, the differences between static and average heat transfer coefficients increased with increasing droplet size [63].

The theoretical importance of Eq. (3.17) stems from the fact that it provides a lower bound to the magnitude of the convective heat transfer coefficient. It is not difficult to infer from Eq. (3.16) that when droplet size decreases, the differences in magnitude between the static and convective heat transfer coefficients diminish. Accordingly, when droplet size approaches a limiting value of zero the second term in the right-hand side of Eq. (3.16) approaches zero. For very small droplet sizes, this implies that the relative motion between droplet and atomization gas does not critically influence the magnitude of the convective heat transfer coefficient. Accordingly, in the literature, Eq. (3.17) is sometimes used to approximate the convective heat transfer coefficient for fine droplets [48]. Although the use of Eq. (3.17) instead of (3.16) in the heat transfer formulation can cause a considerable error in the calculated results, the simplicity of Eq. (3.17) sometimes renders it possible to derive analytical solutions for undercooling and solidification problems. The insight provided by an analytical solution is valuable to the understanding of the physical process of solidification. Mehrabian et al. [58] estimated the magnitude of the convective heat transfer coefficient for atomization using Ar gas at a velocity of Mach one. The heat transfer coefficient estimated for droplets of 75 and 25 μm were 5.9×10^3 and 1.1×10^4 Wm^{-2} K^{-1}, respectively. Mehrabian et al. noted that the heat transfer coefficients calculated for higher conductivity gases and finer droplet sizes are greater than the above values but are generally less than 10^5 Wm^{-2} K^{-1}.

3.3.3 Gas Velocity

The magnitude of the convective heat transfer coefficient is critically influenced by the velocity differential between droplet and atomization gas (see Eq. 3.16). Therefore, the calculation of the rate of thermal energy extraction from a droplet necessitates a knowledge of the velocity profiles of both atomization gas and droplet. The determination of the initial velocity of the atomization gas has been described in detail in Chapter 3. It is worth noting, however, that during atomization gas velocity decays rapidly as the distance from the atomizer increases. The decay behavior of the gas velocity has been experimentally studied by several investigators using hot wire, pitot tube, and laser Doppler anemometry (LDA) techniques for initial gas velocities that fall in both subsonic and supersonic atomization regimes [64, 70–73]. Bewlay and Cantor [64], for example, determined the three-dimensional flow field of an atomization gas exiting a closed-type atomizer using LDA. Their results are summarized in Figure 3.8 as a two-dimensional map which sections the atomization cone along the center line [64]. The measured gas velocities are presented in terms of velocity vectors. The results shown in Figure 3.8 indicate that for a given initial gas velocity the velocity of the gas in the atomization cone is a function of both the distance from the atomizer and the distance from the center line of the atomization cone (i.e. radial distance). To simplify experimentation and theoretical analysis, in most studies the gas velocities along the centerline axis are employed as representative of the gas flow behavior [64, 70–72]. The results of these and other experimental measurements suggest that the gas velocity decays rapidly with increasing distance from the atomizer [64, 70–72]. Accordingly, various empirical relationships have been developed to describe the decay behavior [6, 63, 70, 71]. Grant et al. [63, 70], for example, plotted the logarithm of the velocity as a function of the axial distance from the atomizer. The results, shown in Figure 3.9, reveal that the decrease in gas velocity with distance from the atomizer can be approximated using an exponential decay function of the following form [63, 70]:

$$V_g = V_{gi} \exp\left(-\frac{Z}{\lambda_v}\right) \tag{3.18}$$

where V_{gi} is the initial velocity of the gas, Z is the axial distance from the atomizer and λ_v is an exponential decay coefficient. It is worth noting that the data obtained from different atomization experiments, as shown in Figure 3.9, fall

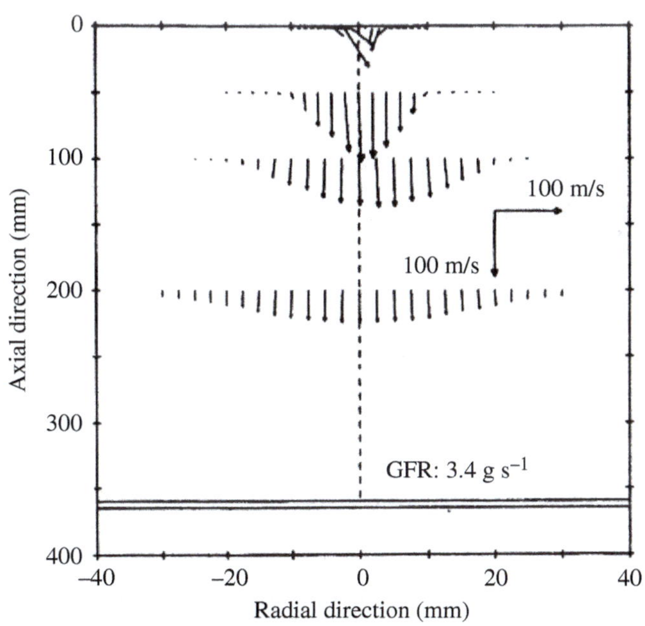

Figure 3.8 Laser Doppler anemometer map of atomization gas velocity for an atomizer-substrate distance of 360 mm and a gas flow rate (GSR) of 3.4 g s^{-1}. Source: Bewlay and Cantor [64] / Reproduced with permission from Springer Nature.

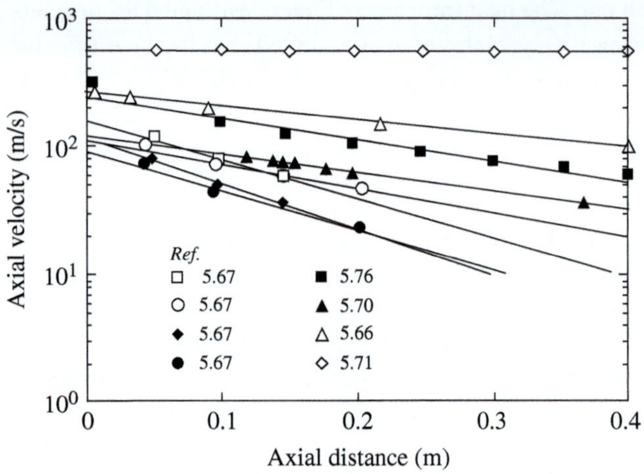

Figure 3.9 Exponential decay behavior of axial gas velocities during atomization. Source: Grant et al. [63].

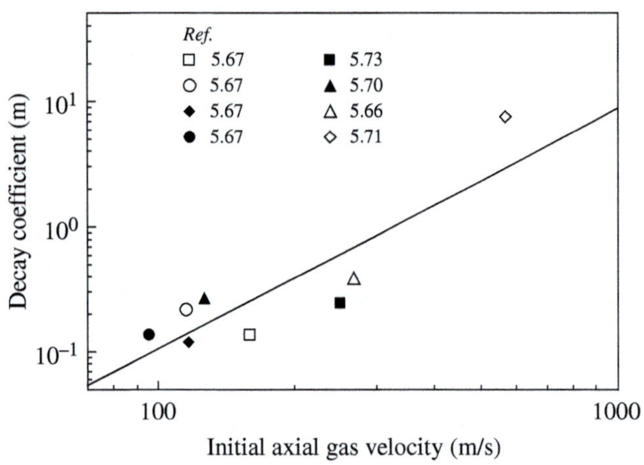

Figure 3.10 Decay coefficient as a function of initial gas velocity. Source: Grant et al. [63].

on separate lines. Grant et al. [63, 70] rationalized this phenomenon by arguing that the exponential decay coefficient is also a function of the initial velocity of the atomization gas. By plotting the decay coefficient as a function of the initial gas velocity on a double logarithmic scale (see Figure 3.10), Grant et al. derived a power law empirical relationship between the decay coefficient and the initial gas velocity:

$$\lambda_{\mathrm{v}} = a_1 V_{\mathrm{gi}}^n \tag{3.19}$$

where a_1 and n are experimental constants. The most accurate representation of the data was obtained with $a_1 = 3.04 \times 10^{-4}$ ($\mathrm{s}^{1.24}\ \mathrm{m}^{-0.24}$) and $n = 1.24$.

3.3.4 Droplet Velocity

The velocity of a droplet during atomization is initially at some minimum value. The drag force that is exerted by the atomization gas on the droplet accelerates and then decelerates the droplet as it travels toward the deposition substrate. In the literature, both theoretical and experimental approaches have been successfully employed to determine the velocity of droplets in a gas velocity field. Among the experimental techniques that are available, LDA is generally preferred because of its high spatial resolution and ability to measure gas and droplet velocities simultaneously. A more detailed description of this experimental technique is provided in Chapter 9. In related studies, Bewlay and Cantor [64] measured the velocity field of droplets in an atomization cone using LDA. As an example, some of their results are summarized in Figure 3.11 for a flight distance of up to 200 mm [64]. The results shown in Figure 3.11 are presented in terms of velocity vectors. The magnitude of the velocity is indicated on the

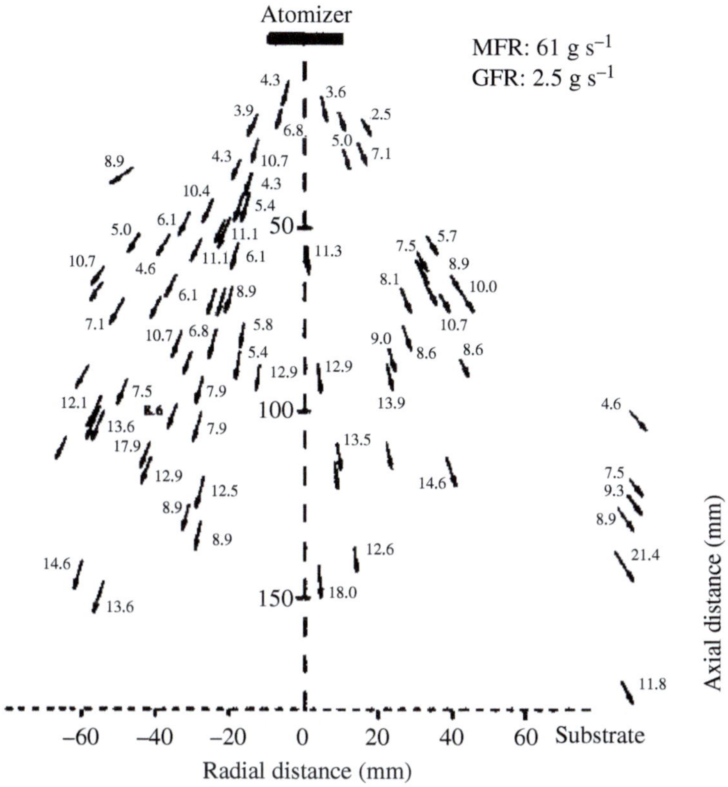

Figure 3.11 Droplet velocity vectors of Sn–38 wt%Pb as determined from laser Doppler anemometer measurements [metal flow rate (MFR) = 61 g s^{-1} and N_2 atomization gas flow rate (GFR) = 2.5 g s^{-1}. Source: Bewlay and Cantor [64] / Reproduced with permission from Springer Nature.

graph adjacent to each velocity vector. According to these results, although the velocity vectors are predominantly vertical, there is a considerable radial component in some cases. The radial velocity components become increasingly important toward the fringes of the atomization cone as illustrated by the droplet trajectories. In the central portion of the atomization cone, the radial velocity components are minimum. Another trend that is illustrated in Figure 3.11 is an increase in droplet velocity with flight distance for a flight distance of less than 200 mm [64].

Since the droplets produced during atomization exhibit a wide range of size distribution and that droplet velocity is affected by their size, it is very difficult to establish an accurate relationship between droplet velocity and flight distance for a given droplet in a single atomization experiment. Alternatively, Bewlay and Cantor [64] measured the mean droplet velocity (average velocity of different droplet sizes) as a function of flight distance during N_2 gas atomization of an Sn–38 wt%Pb alloy. Their results, shown in Figure 3.12, indicate that the mean droplet velocity increases with increasing flight distance [64]. The velocity reaches a maximum value at a flight distance of approximately 300 mm. To illustrate the various velocity trajectories that are associated with different droplet sizes, Bewlay and Cantor injected atomized Sn–38 wt%Pb powders of preselected sizes into an atomization gas field. The measured velocities of the injected powders are also shown in Figure 3.12 whereby they are compared to those of droplets formed directly from atomization [64]. Interestingly, the differences in velocity of different droplet sizes as determined using LDA measurements are relatively small (see Figure 3.12).

The experimental determination of droplet trajectories and the associated velocities provides useful guiding information that may be used to optimize process parameters. It is important to note, however, that the usefulness of these measurements is limited by several factors. First, and as has been addressed in previous sections, there is a wide distribution of droplet sizes that are produced during atomization. Consequently, it is difficult to establish a precise relationship between droplet size and velocity experimentally, for each droplet size of interest. Second, droplet velocity and trajectory are critically dependent on the atomization conditions that are employed during the actual measurements. Accordingly, results obtained from one set of experimental conditions may not be directly applicable to other experimental conditions. This factor limits the universal applicability of experimental data.

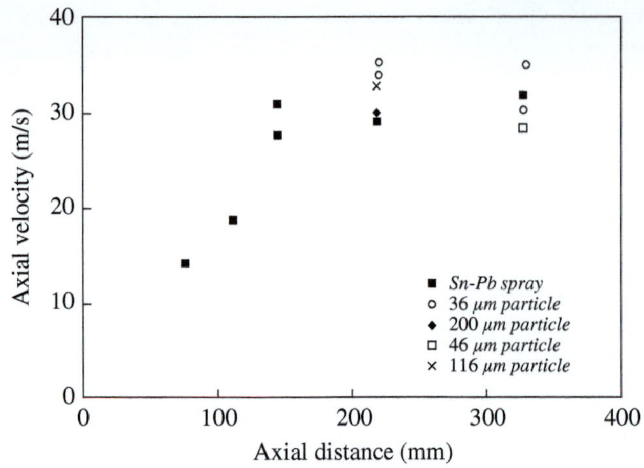

Figure 3.12 Droplet velocities determined from laser Doppler anemometer experiments using Sn–38 wt%Pb and pre-atomized Sn–38 wt%Pb powders. Source: Bewlay and Cantor [64] / Reproduced with permission from Springer Nature.

Third, experimentally, it is difficult to trace the trajectory of a particular droplet. The data obtainable with most LDA and photographic techniques are the position and velocity of a collection of droplets at a particular time, as shown in Figure 3.11. Hence, employing data such as the one shown in Figure 3.12 incurs a fundamental assumption. That is, the velocity profile of a droplet at different times and spatial positions may be statistically represented by the velocities and positions of a collection of droplets at a given time. This assumption is valid only when many droplets are measured and when all of the droplets involved in such measurements are of identical size.

Alternatively, droplet velocity may be determined from theoretical considerations of the momentum fields of droplets and gas. In this approach, the velocities of the atomization gas and the droplets are calculated based on fundamental fluid mechanics principles. Inspection of the atomization literature reveals that this approach has been extensively used in determining the velocity history of atomized droplets [2–6, 9, 15, 55, 60–63, 66, 70, 71, 74, 75]. The fundamental formulations that are generally used are discussed in detail as follows.

The velocity of a droplet in a gas field may be calculated from **Newton's second law**: force = mass × acceleration. Accordingly, there are three separate forces that must be considered in calculating the velocity history of a droplet: a gravitational force, F_g, a fluid drag force, F_d, and a buoyant force, F_b. These forces are combined into a formulation of Newton's second law as follows:

$$m_d \frac{dV_d}{dt} = F_g + F_d - F_b \tag{3.20}$$

where m_d is the mass of the droplet, t is time, and V_d is the velocity of the droplet. Since under commonly used spray conditions, a droplet is directed downward from an atomizer to a deposition surface, the gravitational force generally causes acceleration of the droplet along its moving direction and is expressed as

$$F_g = m_d g \tag{3.21}$$

where g is the gravitational acceleration. Following the discussion of Szekely [75], the effect of fluid drag force depends on the relative velocity between the gas and the droplet. Initially, when the velocity of a droplet is less than that of the atomization gas, fluid drag force is in the same direction as the gravitational force. This force changes its sign when droplet velocity exceeds gas velocity, a condition that occurs in the later stages of droplet flight. According to Szekely, fluid drag force can be calculated from

$$F_d = \frac{A_d}{8} C_{drag} \rho_g (V_g - V_d)|(V_g - V_d)| \tag{3.22}$$

where A_d is the surface area of the droplet; C_{drag} is the fluid drag coefficient; ρ_g is the density of the atomization gas; and V_g is the velocity of the atomization gas. The buoyant force acts in a direction that opposes the motion of the droplet and may be expressed as

$$F_b = \frac{\rho_g}{\rho_d} m_d g \tag{3.23}$$

where ρ_d is the density of the droplet. A differential equation for droplet velocity may then be derived by combining Eqs. (3.20)–(3.23). This results in [71, 75]

$$\frac{dV_d}{dt} = \left(1 - \frac{\rho_g}{\rho_d}\right) g + C_{drag} \frac{3\rho_g}{4d\,\rho_d} (V_g - V_d)|V_g - V_d| \tag{3.24}$$

The non-vectorial form of Eq. (3.24) implies that only movement along a single axis is considered. This axis is typically selected to lie along the symmetric axis of the atomization cone. The drag coefficient, C_{drag}, in Eq. (3.24) is a function of the Reynolds number, *Re*, and is given by [71, 75]:

$$C_{drag} = 0.28 + \frac{6}{Re^{0.5}} + \frac{21}{Re} \tag{3.25}$$

In principle, the gas velocity at any time or spatial location within the center line of the atomization cone may be calculated from Eq. (3.18). Accordingly, velocity or acceleration can be calculated from Eq. (3.18)–(3.20). These equations are applicable within a wide range of Reynolds numbers: $0.1 < Re < 4000$ [71, 76].

Figure 3.13 shows an example of the calculated velocities of the atomization gas and atomized droplets of various sizes, as a function of the distance from the atomizer [62]. The results shown in this figure indicate that for each droplet size, the velocity of a droplet is initially smaller than that of the atomization gas. The velocity of the atomization gas, however, decays rapidly while the droplets are being accelerated. Because of the acceleration of the droplets and the simultaneous deceleration of the gas, the gas velocity becomes smaller than that of the droplets at certain spatial distance. The drag force, however, increases rapidly to a magnitude that is equal to the magnitude of the gravitational force. At this instant, the droplet begins to experience deceleration. Moreover, from the results shown in Figure 3.13, for a given gas velocity field, droplet velocity critically depends on droplet size. Accordingly, fine droplets experience faster acceleration and deceleration relative to coarse droplets. During the initial stages of flight, the velocity of fine droplets always exceeds that of coarse ones.

The aforementioned formulation has been used extensively throughout the atomization literature for the calculation of droplet velocity. There are, however, alternative approaches that may be utilized to calculate droplet velocity. For example, for conditions when the initial gas velocity is relatively high, the gravitational and buoyant forces in Eq. (3.20) may be ignored without leading to significant errors. Moreover, the fluid drag force that the atomization gas imparts on the droplets may be estimated by invoking Stoke's law. Under this condition, the fluid drag force is simplified into [63]

$$F_d = 3\pi d \eta_g (V_g - V_d) \tag{3.26}$$

where η_g is the viscosity of the atomization gas. Grant et al. [63] tested the validity of Eq. (3.26) relative to that of Eq. (3.22). Their results indicate that Eq. (3.26) predicts a much lower fluid drag relative to that calculated from Eq. (3.22). Consequently, droplet velocities that are calculated utilizing Eq. (3.26) are significantly smaller relative to those calculated utilizing Eq. (3.22). The magnitude of the convective heat transfer coefficients that are calculated

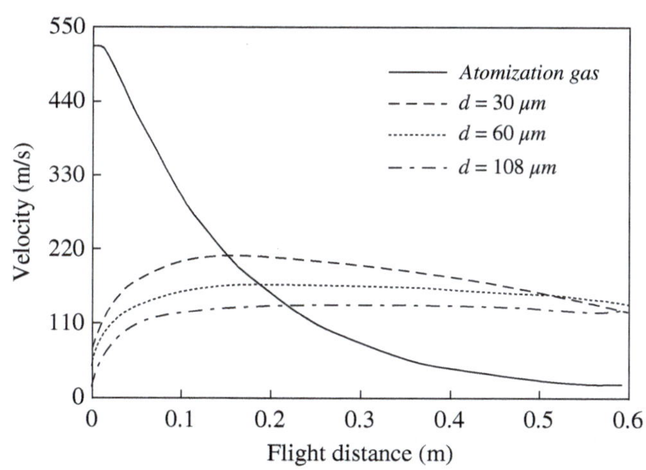

Figure 3.13 Calculated droplet velocity profile along the spray axis as a function of droplet size for N_2-atomized Ni_3Al droplets. Source: Liu et al. [62].

Table 3.3 Average heat transfer coefficient, cooling rate and fight time of Al–4 wt%Cu droplets corresponding to a 40 cm flight distance calculated utilizing different fluid drag models.

	Mean heat transfer coefficient (10^3 W m^{-2} K^{-1})		Mean cooling rate (10^3 Ks^{-1})		Flight time (ms)	
Diameter (μm)	RM relation	Stoke's law	RM relation	Stoke's law	RM relation	Stoke's law
10	8.3	9.1	146	88	11	11
20	6.0	6.5	111	64	8.6	9
30	3.9	5.0	82	50	7.5	11
50	3.6	3.6	37	36	7.7	16
80	2.7	2.7	11	23	8.9	25
120	2.1	2.2	10	15	10	36
200	1.5	1.8	9.4	8	13	58

RM relation, Ranz–Marshall relation.
Source: Adapted from Grant et al. [63].

utilizing these equations is, however, similar. This phenomenon is not surprising, if one considers that the magnitude of the convective heat transfer coefficient is governed primarily by the differential velocity between the gas and the droplets instead of by the absolute velocity of the droplets. As an example, the droplet velocity, residence time, and convective heat transfer coefficient, as calculated by Grant et al. [63], are summarized in Table 3.3 for a flight distance of 40 cm. As shown in this table, since Stoke's law [Eq. (3.26)] yields a fluid drag that is smaller than that calculated based on Eq. (3.22), the flight time calculated utilizing Stoke's law for each flight distance is much longer than that calculated utilizing Eq. (3.22). The magnitude of the convective heat transfer coefficients calculated from Stoke's law Eq. (3.26) is, however, slightly greater than those calculated utilizing Eq. (3.22). Regarding droplet velocity, Eq. (3.22) may provide a more accurate prediction than Eq. (3.26) since the former equation invokes fewer simplifying assumptions. However, because of its simplicity, Eq. (3.26) may be preferred when only an estimation of fluid drag, or droplet velocity is required.

It is evident from the aforementioned discussion that the velocity fields of the atomization gas and the droplets may be readily calculated using available correlations. The gas velocity may be calculated from empirical relationships like that represented by Eq. (3.12). Moreover, and despite differences in numerical assumptions and experimental correlations, inspection of available literature reveals several interesting observations. First, measured and calculated droplet velocities and trajectories exhibit similar behavior as a function of flight distance [6, 60, 62–64, 66, 77]. Second, the measured and calculated velocities are of the same order of magnitude [6, 62–64, 77]. Finally, it is important to note that there are various phenomena whose influence on droplet velocity and trajectory are not fully understood and that merit detailed investigation. These phenomena include droplet rotation, droplet-droplet interactions, and the effect of non-spherical droplet geometries. Clearly, the implications of these phenomena and others on droplet behavior during atomization must be established if a complete phenomenological understanding of droplet evolution is to be attained.

3.4 Nucleation

The thermal environment that is present during atomization generally leads to rapid solidification kinetics. For example, the high cooling rates that are experienced by the droplets during atomization often promote extended undercooling levels. There are two important effects that undercooling may have on the ensuing solidification microstructure. First, at relatively high undercooling levels **dynamic solidification** is normally characterized by a **solid–liquid interfacial velocity** that is much higher than that present during equilibrium solidification. Consequently, the **microstructure** that evolves under conditions of high undercooling may be dramatically different from that present under equilibrium conditions. Second, when certain conditions are present equilibrium solid phases are absent altogether. One extreme example of this effect is the formation of an **amorphous phase** from an undercooled liquid. To illustrate the influence of undercooling on solidification, the present section will address the phenomena of homogeneous and heterogeneous nucleation.

3.4.1 Homogeneous Nucleation

The problem posed by the phenomenon of undercooling is complex and has been studied extensively. However, precise calculation of the degree of undercooling experienced by a material prior to solidification remains limited to relatively ideal conditions and simple alloy systems. These difficulties arise in part from lack of a comprehensive mathematical formulation of undercooling as well as from the uncertainties that are associated with the required thermal–physical parameters. Accordingly, in the sections that follow, our current understanding of the phenomenon of undercooling is presented, with the relevant simplifying assumptions.

As has been addressed in the previous sections, when a droplet travels or resides in a gas environment, it losses thermal energy through convective and radiative heat transfer. Since thermal energy is extracted across the surface of a droplet, significant temperature gradients may develop across the diameter of the droplet. This condition renders it very difficult to calculate the **nucleation rate** and undercooling experienced by a droplet using an analytical approach. Fortunately, when droplet size is small, i.e. <300 μm, heat conduction inside of a droplet occurs at a rate that is comparatively faster than heat extraction across a droplet surface. In this case, the resistance to thermal energy loss is governed by heat transfer across the droplet/gas boundary. Accordingly, in many of the theoretical studies of droplet solidification behavior, the temperature within the droplets is assumed to be uniform [34, 35, 40, 41, 60, 63, 66]. Under the assumption of a uniform temperature profile, a droplet temperature is conveniently defined. The validity of the uniform temperature assumption will be discussed in detail in Section 3.5 in reference to the temperature gradient that develops within a droplet.

3.4.1.1 Free Energy of Nucleation

The **free energy changes** that are associated with the nucleation of a solid phase may be derived from classical nucleation theory [78–80]. It follows that when the temperature of a liquid falls below the equilibrium melting temperature, embryos of the thermodynamically stable solid phases will form because of local atomic fluctuations. If the molar volume difference between solid and liquid phases is neglected, the formation of an embryo from a homogenous liquid involves the replacement of a volume of liquid by an identical volume of the solid phase and the creation of a solid–liquid interface. Accordingly, the **free energy change** that is associated with the formation of an embryo during **homogeneous nucleation** is expressed as the summation of the changes in interfacial free energy and in volume free energy, i.e.

$$\Delta G = \Delta G_{\mathrm{v}} \frac{V_{\mathrm{e}}}{\Omega} + \sigma_{\mathrm{s1}} S_{\mathrm{e}} \tag{3.27}$$

where ΔG is the free energy of formation of an **embryo**; ΔG_{V} and $\sigma_{\mathrm{S}}1$ are the difference in molar free energy between the solid and liquid phases, and the solid–liquid interfacial energy, respectively; V_{e} and S_{e} are the volume and surface area of the embryo, respectively; and Ω is the molar volume. To maintain the problem as tractable, it is generally assumed that the geometry of the embryo is spherical. Accordingly, for a spherical embryo of radius, r_{e}, Eq. (3.27) reduces into:

$$\Delta G = \frac{4\pi}{3\Omega} r_{\mathrm{e}}^3 \Delta G_{\mathrm{v}} + 4\pi r_{\mathrm{e}}^2 \sigma_{\mathrm{s1}} \tag{3.28}$$

According to this equation, the free energy that is associated with the formation of an embryo depends critically on embryo size. As illustrated schematically in Figure 3.14, the change in free energy ΔG assumes a maximum value, ΔG^*, corresponding to some critical size, r_{e}^* [81]. Hence, embryos may grow or disappear from the liquid depending on the free energy conditions. For an embryo of size smaller than r_{e}^*, the decrease in volume free energy incurred during its growth is less than the corresponding increase in surface energy. Accordingly, the growth of such embryos will cause a net increase in the free energy of the system. Such embryos are not energetically favorable and will eventually dissolve in the liquid. For an embryo of size greater than r_{e}^*, further growth will reduce the free energy of the system. Such embryos will grow and develop into nuclei. Mathematical solution of Eq. (3.28) for a critical embryo size leads to the well-known expression for the free energy of nucleation [78–80]:

$$\Delta G^* = \frac{16\pi\sigma_{\mathrm{s1}}^3 \Omega^2}{3\Delta G_{\mathrm{V}}^2} \tag{3.29}$$

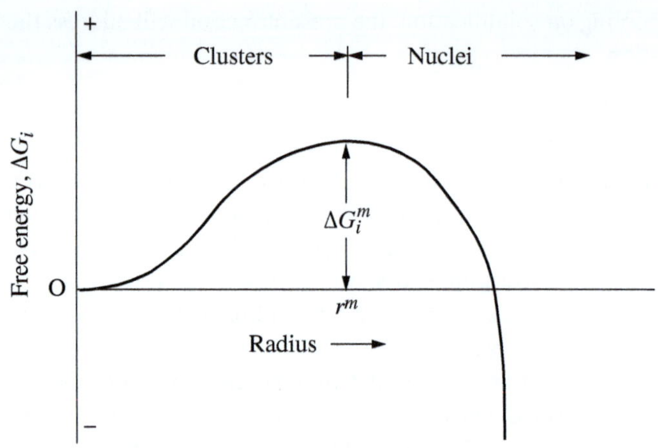

Figure 3.14 Free energy as a function of embryo size. Source: Flemings [81].

The only two variables in Eq. (3.29) are the solid–liquid interfacial energy and the volumetric free energy difference between the solid and liquid phases. The solid–liquid interfacial energy is generally determined from nucleation experiments where the interfacial energy is deduced from experimentally measured nucleation rates. Since in such experiments nuclei form in a temperature range, the interfacial energy determined using this approach is considered to be an average value. Accordingly, the temperature dependence of the interfacial energy is generally not available.

The volume free energy term in Eq. (3.29) can be readily expressed as a function of temperature by making certain assumptions. The simplest condition involves a first-order assumption which is often made in calculating the free energy of nucleation as a function of temperature [47, 48]. Following the second law of thermodynamics, at any temperature, the volume free energy term in Eqs. (3.28) and (3.29) can be written as

$$\Delta G_v = \Delta H - T\Delta S \tag{3.30}$$

where ΔH and ΔS are the molar enthalpy and entropy differences between the solid and the liquid phases, respectively. For a pure metal at precisely the melting temperature, T_1, the free energies of the liquid and of the solid are equal. In this case, ΔG_v is zero. It follows from Eq. (3.30) that

$$\Delta S_f = \frac{\Delta H_f}{T_1} \tag{3.31}$$

where ΔS_f is the molar entropy of fusion. When $T \neq T_1$ (i.e. not at the melting temperature), Eq. (3.30) yields a non-zero value. However, when the extent of undercooling is relatively low (i.e. the actual temperature is close to the melting temperature), ΔG_v in Eq. (3.30) can be estimated using a first-order approximation. In such an approximation, ΔH and ΔS are relatively insensitive to temperature and are replaced by their values at the melting temperature. With this assumption, and by combining Eqs. (3.30) and (3.31) the volume free energy can be expressed in the form of

$$\Delta G_v = \frac{\Delta H_f \Delta T}{T_f} \tag{3.32}$$

This is the linear relationship originally developed by Turnbull [82] by assuming that the differential heat capacity between the solid and the undercooled liquid phase is zero. Furthermore, combining Eqs. (3.29) and (3.32) yields

$$\Delta G^* = \frac{16\pi\sigma_{s1}^3 T_f^2 \Omega^2}{3\Delta H_f^2 \Delta T^2} \tag{3.33}$$

Equation (3.33) has been repeatedly used in the calculation of the free energy of nucleation [81, 83], the rate of nucleation [47], undercooling [47, 48], and to derive the value of interfacial energy from experimentally determined nucleation rates [44, 82].

It should be noted, however, that Eq. (3.32) is valid only when the degree of undercooling is relatively low and/or when the difference in heat capacity between the solid and the undercooled liquid phases is negligible [84, 85]. Available experimental evidence shows that the differential heat capacity between solid and liquid is non-zero even at the melting temperature and is a function of undercooling [86–88]. As an example, Figure 3.15 shows

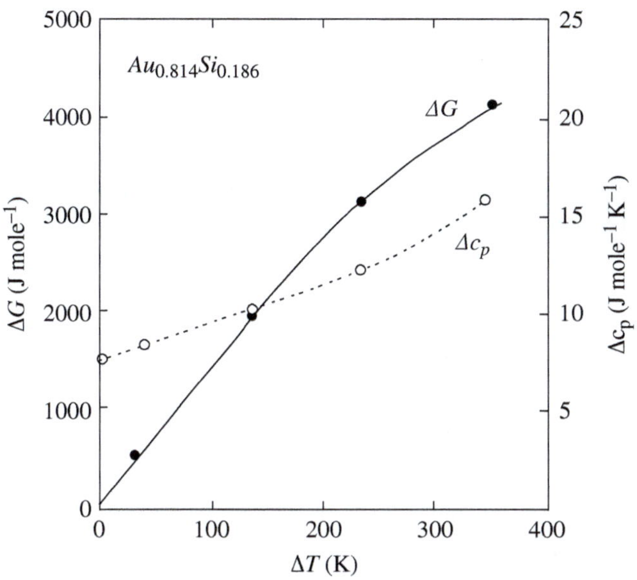

Figure 3.15 Temperature dependence of Δc_p and Gibbs' free energy difference between the undercooled liquid and the equilibrium solid phases in $Al_{0.814}Si_{0.186}$. Source: Dubey and Ramachandrarao [88].

the differential heat capacity between the equilibrium solid phases and the undercooled liquid as a function of undercooling for an amorphous-forming alloy $Au_{0.814}Si_{0.186}$ [88]. The free energy difference between solid and undercooled liquid is also included in this figure for comparison. It is clear from the results shown in Figure 3.15 that the difference in heat capacity between solid and liquid increases in magnitude with increasing undercooling. As a result, the dependence of free energy on temperature becomes non-linear at high undercoolings [88].

Calculation of the temperature dependence of ΔG_v requires a precise knowledge of the magnitude of the heat capacity of the liquid at undercooled temperatures. In particular, the differential heat capacity between the solid and the liquid, Δc_p. Such information is not readily available due to the difficulties that are associated with the measurement of the heat capacity of undercooled liquids. In the literature, various attempts have been made to formulate ΔG_v–T relationships that incorporate physical parameters such as T_1 and ΔH_f that are more easily determined experimentally [89–91]. Hoffman [89], for example, assumed that the differential heat capacity is non-zero but is independent of temperature. Based on this assumption, the temperature dependence of ΔG_v becomes non-linear, and can be approximated using the readily available melting temperature, T_1, as follows:

$$\Delta G_v = \frac{\Delta H_f \Delta T}{T_1}\left(\frac{T}{T_1}\right) \tag{3.34}$$

Jones and Chadwick [90] employed a similar constant Δc_p assumption. In this study, the enthalpy and entropy terms in Eq. (3.30) were expressed as a function of the temperature as follows:

$$\Delta H = \Delta H_f + \int_{T_1}^{T} \Delta c_p \mathrm{d}T \tag{3.35a}$$

$$\Delta S = \Delta S_f + \int_{T_1}^{T} \frac{\Delta c_p}{T} \mathrm{d}T \tag{3.35b}$$

Since Δc_p is assumed to be constant, ΔS and ΔH can be obtained by simple integration from T_1 to T. Combination of Eq. (3.30) with (3.35a,b) yields

$$\Delta G_v = \frac{\Delta H_f \Delta T}{T_1} - \frac{\Delta c_p \Delta T^2}{(T_1 - T)} \tag{3.36}$$

In related work, Dubey and Ramachandrarao [88] developed an expression for ΔG_v as a function of temperature based on the fundamental thermodynamic formulations represented by Eqs. (3.30) and (3.35). Instead of assuming a constant differential heat capacity, they employed a physical model for the undercooled liquid that was originally developed for glasses in which heat capacity is expressed as a function of temperature [92–96]. Accordingly, Dubey and Ramachandrarao [88] obtained an expression for ΔG_v as a function of temperature as follows

$$\Delta G_v = \frac{\Delta H_f \Delta T}{T_1} - \frac{\Delta c_p^f \Delta T^2}{2T}\left[1 - \frac{1}{3}\gamma\delta\frac{\Delta T}{T} + \sum_{n=4}^{\infty}(-1)^n\left(\frac{\gamma\delta\Delta T}{T}\right)^{n-2}\frac{2}{n!}\right] \tag{3.37}$$

where Δc_p^f is the differential heat capacity at the melting temperature, and δ and γ are functions whose values depend on the material's properties. It was argued that the product of δ and γ varies only slightly for many substances and is less than unity [88]. Noting that the higher-order terms in Eq. (3.37) diminish rapidly with increasing undercooling, and if the product of δ and γ equals 0.5, Dubey and Ramachandrarao [88] reduced Eq. (3.37) into a simplified form:

$$\Delta G_v = \frac{\Delta H_f \Delta T}{T_1} - \frac{\Delta c_p^f \Delta T^2}{2T}\left[1 - \frac{\Delta T}{6T}\right] \tag{3.38}$$

In Eq. (3.38) the difference in the volume of Gibbs' free energy between the solid and the undercooled liquid phase, ΔG_v, is expressed in terms of measurable parameters at the melting temperature. The physical properties required for the calculation of ΔG_v are ΔH_f, T_1, and the differential heat capacity at the melting temperature, Δc_p^f. It should be noted that when the degree of undercooling is relatively small, the higher-order terms in Eq. (3.38) can be further ignored without causing a considerable error. Under this condition, Eq. (3.38) reduces to the linear function originally obtained by Turnbull [82]. However, when the extent of undercooling is significant, the second and third terms need to be taken into consideration.

In general, the differential heat capacity between an undercooled solid and a liquid can be expressed by an empirical relationship of the type:

$$\Delta c_p = \Delta a + \Delta bT + \Delta cT^{-2} \tag{3.39}$$

where Δa, Δb, and Δc are experimental constants. Hence, whatever the precise empirical relationship of Eq. (3.39) may be, the ΔG_v–T relationship can always be calculated using Eqs. (3.30), (3.35), and (3.39).

3.4.1.2 Nucleation Rate

The simplest expression available for nucleation rate is perhaps the steady-state nucleation rate. The nucleation rate of a crystalline solid from a liquid depends on the rate of formation of embryos (or clusters). Only embryos of size greater than a certain critical value are energetically favorable and may eventually develop into nuclei. Under steady-state conditions, the size distribution of embryos present is assumed to be constant. Therefore, the rate of nucleation is constant at a given temperature. Based on classical nucleation theory, the general expression for steady-state nucleation rate is [47]

$$I(T) = A\exp\left[-\frac{\Delta G^*}{kT}\right] \tag{3.40a}$$

ΔG^* can be calculated using various approaches as discussed in the previous sections. The simplest approach, however, involves the assumption that ΔG_v is a linear function of undercooling. Consequently, by expressing ΔG^* in terms of measurable physical parameters, as given by Eq. (3.33), Eq. (3.40a) may be expressed as

$$I(T) = A\exp\left[-\frac{16\pi\sigma_{s1}^3\Omega^2 T_1^2}{3kT\Delta H_f^2\Delta T^2}\right] \tag{3.40b}$$

where $I(T)$ is the nucleation rate which has the units of number of nuclei per unit volume per unit time. The pre-exponential factor, A, represents the effect of atomic attachment of liquid atoms to embryos of critical size. In nucleation studies, several expressions have been developed for the calculation of this pre-exponential factor, A [36, 47, 48, 84, 97–110]. Perhaps the simplest expression is in the form [84, 99, 101]:

$$I(T) = \frac{k_v}{\eta_1}\exp\left[-\frac{16\pi\sigma_{s1}^3\Omega^2 T_1^2}{3kT\Delta H_f^2\Delta T^2}\right] \tag{3.41}$$

where k_v is a constant which is relatively insensitive to temperature. The magnitude of k_v is on the order of 10^{30} [36, 99]. The factor η_1 is the liquid viscosity which can be expressed as a function of temperature from [102]

$$\eta_1 = B\exp\left[\frac{C}{(T - T_\infty)}\right] \tag{3.42}$$

where B, C, and T_∞ are constants. The difficulties that are associated with the application of Eq. (3.42) arise from the fact that most of the experimental values available for B, C, and T_∞, are determined for superheated melts. Hence, the extrapolation of such data to the undercooled regime may lead to significant errors.

In related work, Hirth [47] used an alternative expression for the pre-exponential factor, A, in Eq. (3.40a,b). Based on this work, the formulation of the resultant nucleation rate becomes [47]

$$I(T) = 10^{-2} 4\pi \left(\frac{r_e^*}{a}\right)\left(\frac{a}{\Omega}\right) D_l \exp\left[-\frac{16\pi\sigma_{sl}^3 \Omega^2 T_l^2}{3kT\Delta H_f^2 \Delta T^2}\right] \tag{3.43}$$

where a is the interatomic distance and D_l is the liquid diffusivity. Although there are several formulations available that may be used to calculate the pre-exponential factor, A, in Eq. (3.40a), there is considerable discrepancy between theoretical and experimental values. The magnitude of the pre-exponential factor, A, determined based on theoretical considerations is on the order of 10^{41} m^{-3}s^{-1} [84, 103], whereas experimental results reveal that A is generally on the order of 10^{46} m^{-3}s^{-1} [84, 85, 104]. Hence, the reported discrepancy represents a few orders of magnitude. Fortunately, as will be discussed in subsequent sections, the effect of this seemingly large factor on the calculated undercooling is far less important than that of the solid–liquid interfacial energy. In the literature, both theoretical and experimental values for the pre-exponential factor A have been used in undercooling calculations [47].

Equations (3.40)–(3.43) represent expressions for the steady-state nucleation rate. It is worth noting that non-steady-state nucleation occurs during the transient time period prior to the establishment of steady state. During non-steady-state nucleation, the number of embryos increases with increasing time until steady-state conditions are reached. In this case, the rate of nucleation is a function of both temperature and time. It is anticipated that in the case of rapid solidification conditions, such as those present during atomization or melt spinning, non-steady-state nucleation may have an important influence only on the onset temperature of nucleation.

3.4.1.3 Homogeneous Undercooling

Undercooling was defined in an earlier section as the difference between the equilibrium melting temperature and the temperature at the onset of nucleation. Theoretically, nucleation temperature is the temperature at which one solid nucleus successfully forms from a liquid. As has been addressed in the previous sections, under steady-state conditions, nucleation is a statistical process in which embryos continuously form and dissolve in the liquid. Only the embryos which exhibit a size that is greater than the critical size may develop into nuclei (i.e. $r_e > r_e^*$ in Figure 3.14).

In view of the statistical nature of this process, nucleation is a time-, temperature-, and volume-dependent phenomenon. Under the assumption of steady-state nucleation, the number of nuclei that form during a continuous cooling process, N, can then be obtained from an integration of the nucleation rate over the residence time, t_N, spent by the liquid between the melting temperature and nucleation temperature [47, 48, 101, 105]:

$$N = v_d \int_0^{t_N} I(T)\mathrm{d}t \tag{3.44a}$$

where v_d is the volume of liquid. Alternatively, the integration can be completed over the temperature regime between the melting and nucleation temperatures for a continuous cooling process [12, 13, 34, 36, 40, 49]:

$$N = v_d \int_{T_l}^{T_N} \frac{I(T)}{\dot{T}}\mathrm{d}T \tag{3.44b}$$

where N is the number of nuclei, T_N is the nucleation temperature, $\dot{T} = \mathrm{d}T/\mathrm{d}t$ is the cooling rate, and $I(T)$ is the nucleation rate. At the nucleation temperature, at least one nucleus has formed from the liquid. This condition is specified as

$$v_d \int_{T_l}^{T_N} \frac{I(T)}{\dot{T}}\mathrm{d}T = 1 \tag{3.45}$$

Hence, Eq. (3.45) may be used to calculate the onset temperature for nucleation. In Eqs. (3.44a,b) and (3.45) the steady-state nucleation rate is used to calculate the onset temperature of nucleation. In a typical rapid solidification process, such as atomization, steady-state nucleation conditions may never be achieved. Since under non-steady-state conditions the nucleation rate increases with increasing time, with the highest possible nucleation rate corresponding to steady-state conditions, the use of a steady-state nucleation rate in Eq. (3.45) will, therefore, lead to an overestimation of the nucleation rate and the nucleation temperature.

In principle, Eq. (3.45) can be readily used to calculate the nucleation temperature and thus the undercooling. The difficulties that are associated with the integration of the relevant expression can always be circumvented by using numerical approaches [48, 106]. To develop an analytical expression for undercooling, Hirth [47] expanded the nucleation rate in Eq. (3.45) around the nucleation temperature using the Taylor series. Consequently, the integration of Eq. (3.45) is reduced into a simplified expression as [47]

$$\frac{0.01I(T_N)v_d\Delta T}{\dot{T}} \cong 1 \tag{3.46}$$

By combining Eq. (3.46) with Eq. (3.43) Hirth obtained an analytical expression for undercooling as follows:

$$\Delta T^2 = \frac{16\pi\sigma_{s1}^3\Omega^2T_1^2}{3kT_N\Delta H_f^2 \ln\left[10^{-3}d^3\left(r_e^*/a\right)^2(a/\Omega)D_1\Delta T/\dot{T}\right]} \tag{3.47}$$

In a related study of droplet solidification behavior, Levi [48] obtained a similar equation for undercooling in the form of

$$\Delta T^2 = \frac{F_1(T_N)}{\log[F_2(T_N)d^3/\dot{T}} \tag{3.48}$$

where the values of F_1 and F_2 depend on material properties and the nucleation temperature.

Using Eqs. (3.47) and (3.48) the effect of various factors, such as cooling rate, droplet size, and solid–liquid interfacial energy on nucleation undercooling can be evaluated. In these equations, cooling rate and droplet size are embedded in a logarithmic term. Accordingly, their effect on undercooling is less significant than that of interfacial energy. To compare the influence of the various factors on undercooling, Levi [48] conducted theoretical calculations for the achievable **homogeneous nucleation undercooling** in Ni and Al droplets (see Figure 3.16). In Figure 3.16 the influence of droplet size on reduced undercooling, i.e. $\Delta T/T_1$, is illustrated. The results reveal that the extent of **homogeneous nucleation undercooling** increases with decreasing droplet size. For example, as illustrated for Al droplets, when the droplet size decreases from 100 to 10 μm, the reduced undercooling increases from 0.22 to about 0.28. This corresponds to a difference in absolute undercooling of approximately 55 K. Since the cooling rate increases with decreasing droplet size, the droplet size effect illustrated in Figure 3.16 encompasses the dual effects of both liquid volume (size) and cooling rate.

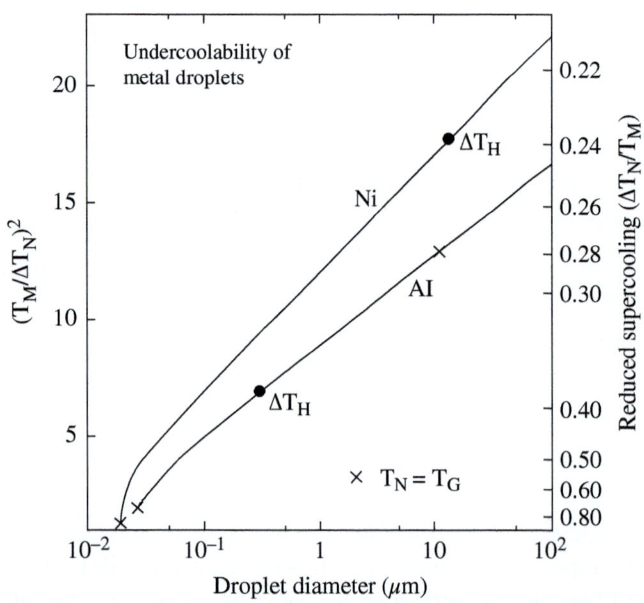

Figure 3.16 Achievable undercoolings prior to homogeneous nucleation in Ni and Al droplets cooled by He gas at 100 ms^{-1} where ΔT_n is the undercooling; T_g is the gas temperature; T_m is the nucleation temperature; and T_N is the nucleation temperatures. Source: Levi [48] / Reproduced with permission from Springer Nature.

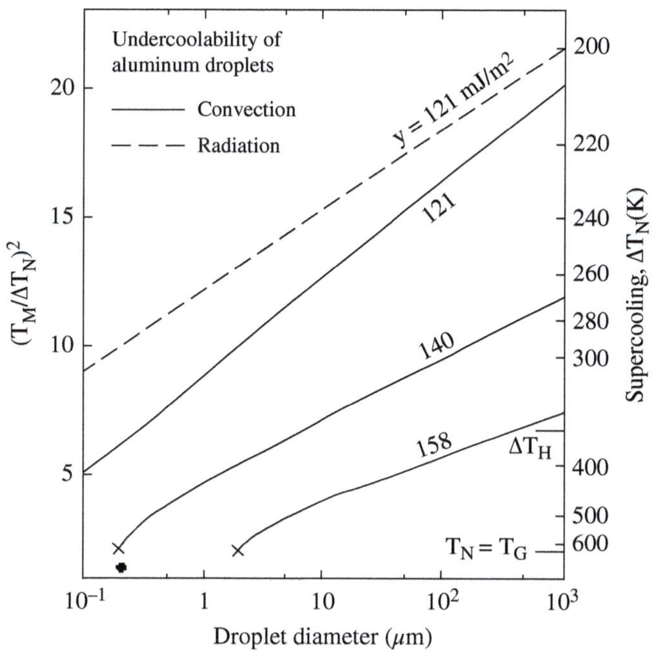

Figure 3.17 Effect of cooling efficiency and solid–liquid interfacial energy on the predicted undercoolablity of Al droplets for cooling in He at 100 ms^{-1} where δTn is the undercooling; T_g is the gas temperature, T_m is the melting temperature and T_N is the nucleation temperature. Source: Levi et al. [48] / Reproduced with permission from Springer Nature.

Figure 3.17 shows the reduced undercooling as a function of droplet size as calculated for pure Al [48]. Two distinct effects are illustrated in this figure: cooling rate and interfacial energy. For example, for the same droplet size and interfacial energy, convective heat transfer (high cooling rate) leads to a higher undercooling relative to that associated with radiative heat transfer (low cooling rate). The most significant effect shown in Figure 3.17 arises from interfacial energy. Accordingly, a slight increase in interfacial energy leads to significant increase in the calculated undercooling. For example, for a 10 μm Al droplet, an increase in interfacial energy from 0.121 to 0.158 J m^{-1} would increase the undercooling from 261 to 468 K [48].

It is important to note that Eqs. (3.47) and (3.48) have been derived for pure metals. However, these equations are often employed in the calculation of homogenous nucleation undercoolings for multicomponent homogeneous systems, such as alloys [47, 104, 107]. This simplifying assumption is often made due to the lack of alternative robust formulations that incorporate the effect of alloying additions. Since the presence of alloying elements is expected to influence the free energy and the interfacial energy of alloys as compared to that of pure elements, a degree of error will be inherently associated with the calculated results.

Inspection of the scientific literature reveals that some investigations have addressed the effect of alloy composition on the achievable undercoolings in simple binary alloy systems such as Sn—Bi and Cu—Ni [108, 109]. Figures 3.18 and 3.19 summarize the experimentally determined undercoolings as a function of alloy composition for the Sn—Bi and Cu—Ni systems [109]. It is worth noting from the results shown in these figures that for both Sn—Bi and Cu—Ni the undercooling versus composition curves follow a trend that is like that of the liquidus temperatures. Although the undercoolings shown in these figures may not be strictly defined as homogeneous nucleation undercoolings, the trend of composition on undercooling is clear. For more complex systems the phase diagrams are sophisticated and such systematic determination of undercooling is not available. However, it is expected that when alloying can induce changes to the solid–liquid interfacial energy or can promote heterogeneous nucleation, its effect on undercooling will be significant.

Thompson and Spaepen [109] developed a simple theoretical model which considers the undercooling–composition relationship for simple binary systems such as Sn—Bi and Cu—Ni. They modified the original formulation developed for the free energy difference between undercooled liquid and solid phases, as well as that of the interfacial energy. Thompson and Spaepen assumed that a binary alloy of components A and B form a regular solution in the liquid state and a dilute or ideal solution in the solid state. They further assumed that the solid phase that is likely

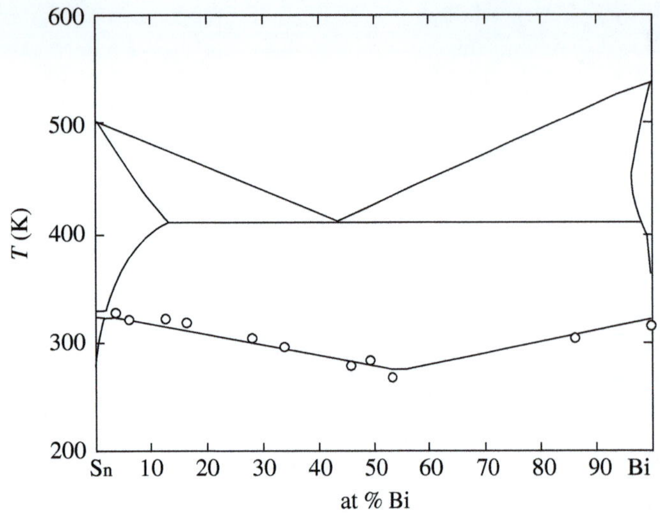

Figure 3.18 The experimental (open circles) and theoretical (solid line on open circles) undercoolings in the binary Sn–Bi system are shown in the phase diagram. Source: Thompson and Spaepen [109].

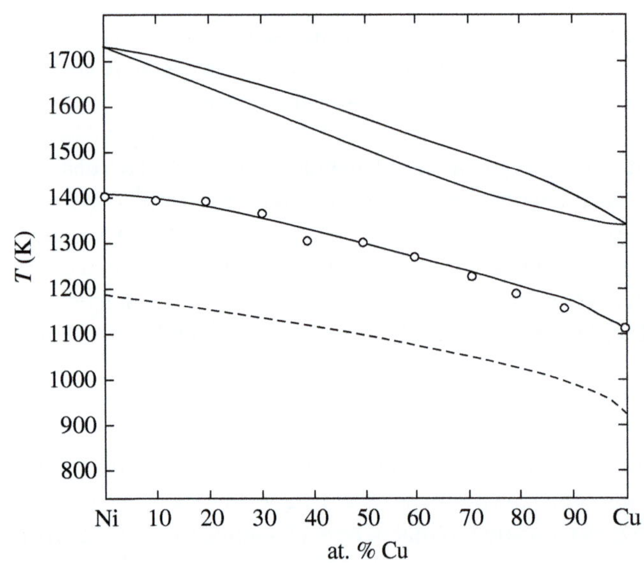

Figure 3.19 Phase diagram for Cu–Ni showing the undercooling data of Cech and Turnbull [111]. The dashed curve is the theoretical prediction for $\alpha_m = 0.86$. The line through the data is the theoretical prediction using a heterogeneous nucleation reduction factor $f(\theta) = 0.16$ (from ref. [112]). Source: Thompson and Spaepen [109].

to **nucleate** from an alloy is the one which leads to the largest decrease in free energy. This condition is expressed as [109]

$$\Delta G_v \overline{\Omega} = \Delta \mu_A = \Delta \mu_B \tag{3.49}$$

where $\overline{\Omega}$ is the average molar volume of solid which is assumed to follow a rule of mixture behavior between the molar volumes of pure components A and B, and $\Delta \mu_A$ and $\Delta \mu_B$ are the chemical potential differences between the liquid and solid phases for components A and B. The free energy difference between the undercooled liquid and solid alloy phases may then be approximated from

$$\Delta G_v = \frac{1}{\overline{\Omega}} \left[\Delta T \Delta S_f^A + RT \ln\left(\frac{x_1^A}{x_s^A}\right) - RT_1 \ln\left(\frac{x_1^A}{x_{s,eq}^A}\right) \right] \tag{3.50}$$

where x_l^A is the composition of A in the liquid phase, x_s^A is the composition of A in the solid phase, and $x_{s,eq}^A$ is the equilibrium composition of A in the solid phase as determined from the equilibrium phase diagram. To correlate interfacial free energy, σ_{sl}, with alloy composition, Thompson and Spaepen [109] employed the model of interfacial energy for pure species with close-packed surfaces in the bod-centered cubic (b.c.c.), fac-centered cubic (f.c.c.), and hexagonal close-packed (h.c.p.) crystal structure [110, 113]:

$$\sigma_{sl} = \frac{\alpha_m \Delta S_f}{(N_A \overline{\Omega})^{1/3}} T \tag{3.51}$$

where α_m is a geometrical factor and N_A is Avogadro's constant. The compositional dependence of σ_{sl} may then be related to the compositional dependence of the molar volume and entropy as follows:

$$\overline{\Omega} = x_s^A \Omega^A + \left(1 - x_s^A\right) \Omega^B \tag{3.52a}$$

$$\Delta S_f = x_s^A \Delta S_f^A + \left(1 - x_s^A\right) \Delta S_f^B \tag{3.52b}$$

where Ω^A and Ω^B are the molar volumes of components A and B, respectively, and ΔS_f^A and ΔS_f^B are the molar entropies of A and B, respectively.

In principle, undercooling in simple binary alloy systems may be estimated by combining Eq. (3.29), (3.40), and (3.45) with Eq. (3.49)–(3.51), (3.52a,b). To that effect, Thompson and Spaepen [109] compared the calculated homogeneous undercooling values with the maximum undercoolings measured by Perepezko [114] for the Sn—Bi system. The results are summarized in Figure 3.18 as a function of alloy composition where the open circles represent the experimentally determined maximum undercoolings and the solid line along the circles represents the calculated undercoolings. The results in this figure demonstrate relatively good agreement between the experimental and theoretical undercooling values. When compared with the undercooling values measured for the Cu—Ni system, as shown in Figure 3.19, the theoretically calculated undercoolings appear to be much higher (dashed line) [109, 111, 112]. However, when the free energy of nucleation is multiplied by a factor of $f(\theta) = 0.16$, there is excellent agreement between the theoretical calculations and the experimental results. This phenomenon may be an indication of heterogeneous nucleation in the Cu—Ni system. The details of heterogeneous undercooling will be addressed in Section 3.4.2.

It should be noted that the derivation of Eqs. (3.49)–(3.52) invokes considerable simplifying assumptions, i.e. regular solution in the liquid phase and dilute or ideal solution in the solid phase. For more complex systems the derivation of ΔG_v requires sophisticated thermodynamic formulations. A good agreement between experimental and theoretical results in this area has not been achieved, primarily due to the lack of both theoretical model and reliable experimental results. A noteworthy attempt, however, was reported by Thompson and Spaepen [109] in their calculation of undercooling in systems in which the solid phase that forms from the liquid may be considered dilute. The results obtained from such a formulation are shown in Figure 3.20 for the Pb—Sn system [109, 115]. In this system, in the Pb-rich side of the phase diagram, Sn exhibits a considerable solid solubility in Pb. Clearly, this behavior represents a departure from the assumption of a dilute solid solution that is associated with the derivation of Eqs. (3.48)–(3.52). Although in this case, the theoretical and experimental results exhibit similar trends with alloy composition, the agreement is less satisfactory [109]. In the atomization literature, considerable efforts have been made to estimate undercooling, homogeneous or heterogeneous, of complex alloy systems using equations that have been derived primarily for pure metal or simple multicomponent systems. In view of the discussion, there are numerous simplifying assumptions that are inherent to the available undercooling formulations. Accordingly, it is important to exercise great care when interpreting numerical results that involve the calculation of undercooling, and experimental verification should always be incorporated whenever possible.

3.4.2 Heterogeneous Nucleation

In the previous sections, it has been shown that the maximum achievable homogeneous undercooling is critically restricted by the physical and chemical properties of the liquid and solid phases as well as by the cooling efficiency of the system. In most practical cases, however, the maximum achievable undercooling is limited by the presence of nucleation reagents (or nucleants). The influence of heterogeneous nucleation reagents is perhaps best illustrated by the frequently reported experimental observations that the extent of undercooling increases with increasing melt

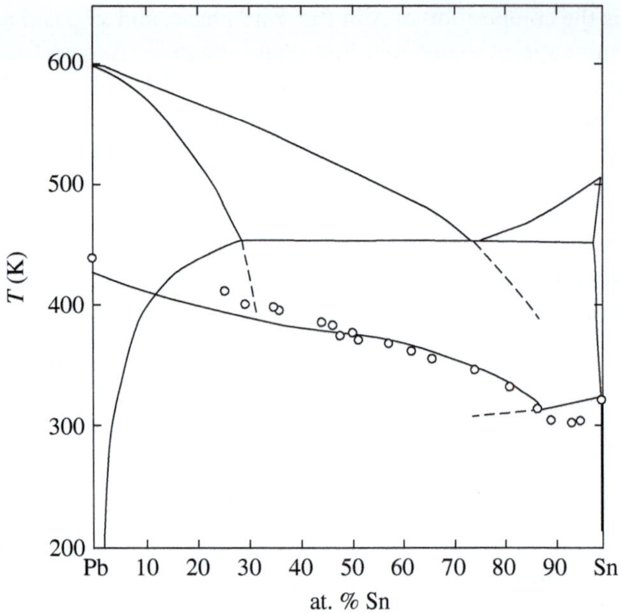

Figure 3.20 Phase diagram for Pb—Sn showing the undercooling data of Cooper et al. [115]. The line through the data points represents the theoretical prediction. The dashed portion corresponds to the extension of the liquidus and solidus lines on the Pb-rich side. Source: Thompson and Spaepen [109].

purity. In such studies, the maximum achievable undercoolings are often estimated either by post-solidification microstructural analysis [35] or more reliably, by direct thermal analysis [43, 85, 104, 108, 109, 116–119]. The maximum undercooling values determined, however, are often closely related to the experimental conditions used by the investigators. For example, the undercooling of Ga originally determined by Turnbull [119] was 70 K. For the same material, however, the maximum undercooling determined in later experiments by Miyazawa and Pound [85] was 101 K.

The experiments conducted by Willnecker et al. [110] perhaps best illustrate the important effect of nucleation reagents. In this study, undercooling measurements were made on a simple Ni—Cu binary alloy system, in which the oxide phase becomes a potential heterogeneous nucleation reagent. Accordingly, Willnecker et al. flushed a He—H_2 gas mixture into their experimental facility to reduce the copper oxides that formed on the surface of droplets. Figure 3.21 summarizes the maximum achievable undercooling in Ni—Cu droplets as a function of reduction time (gas flushing time) [110]. The results shown in this figure reveal a transient increase in undercooling with increasing reduction time. This transient stage is followed by the saturation of undercooling whereby any further increases in reduction time do not influence undercooling. Based on these experimental results, Willnecker et al. concluded that copper oxides located on the surface of droplets are responsible for the observed heterogeneous nucleation behavior. Therefore, increasing the time exposure to a reducing He—H_2 atmosphere, with concomitant reduction of surface oxide phases, led to an increase in the achievable undercoolings.

3.4.2.1 Heterogeneous Nucleants

A review of the scientific literature indicates that there are numerous substances that may act as potential nucleation sites. During ingot solidification, for example, **mold surfaces**, liquid/air interfaces, and **impurity inclusions** are all potential nucleation sites. Sometimes inoculating reagents are deliberately introduced into the liquid to promote nucleation and thereby increase grain refinement. In the case of droplet solidification, there are two basic types of potential nucleation reagents: surface nucleants and volume-distributed nucleants. Surface nucleants are normally oxides or other reaction products that form because of interactions with a gaseous environment (as in atomization) or a liquid (as in **droplet emulsion**) that may catalyze the nucleation of solid phases from the parent liquid [108, 110]. The nucleation potency of surface nucleants is related to their physical properties, as well as to the physical properties of the liquid and solid phases. This observation is illustrated by the results obtained by Rasmussen and Loper [108] with Bi and In emulsified with air and other chemicals. In this study, they found that when Bi droplets were

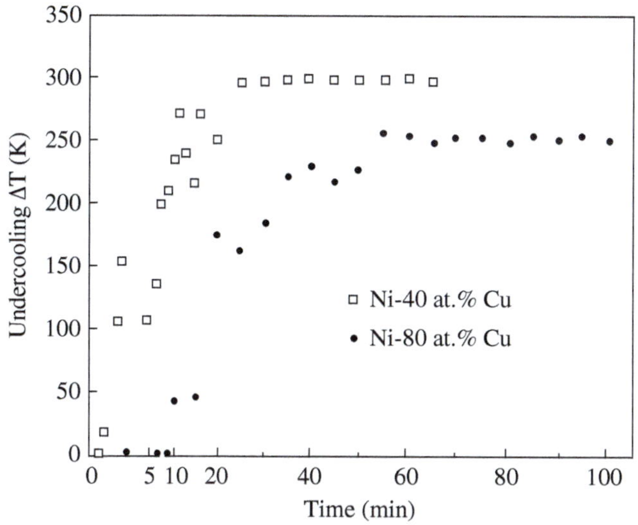

Figure 3.21 Undercooling behavior of Ni–Cu alloys as a function of reducing time during exposure to a He–H_2 atmosphere. Source: Reproduced by permission of Elsevier Science from R. Willnecker, D. M. Herlach and B. Feuerbacher (1988) *Mater. Sci. Eng.*, **98**, 85–88 [110].

emulsified with both air and di-tertiary-butyl peroxide a substantial undercooling of up to 160 K was achieved. When the In droplets were emulsified in air, however, an undercooling of only a few degrees was achieved. Hence, the results suggest that oxide is a strong catalyst for solid In. Moreover, the frequent observation of nucleation events originating at surfaces in atomized droplets supports the premise that surface nucleation is an important phenomenon that influences droplet solidification in the presence of surface nucleants. Two examples of surface nucleation events are shown in Figure 3.22 [37, 120] for Fe—Ni and Al—Fe powders. In the Fe—Ni powder shown in Figure 3.22a, for example, nucleation originated in one clearly defined region on the surface of the droplet. Following nucleation, the growth front stretched radially into the entire droplet [120]. In the Al—Fe droplet shown in Figure 3.22b, solidification appears to have started on the surface and formed a shell. In this case, the growth front subsequently moved radially inward into the interior of the droplet [37].

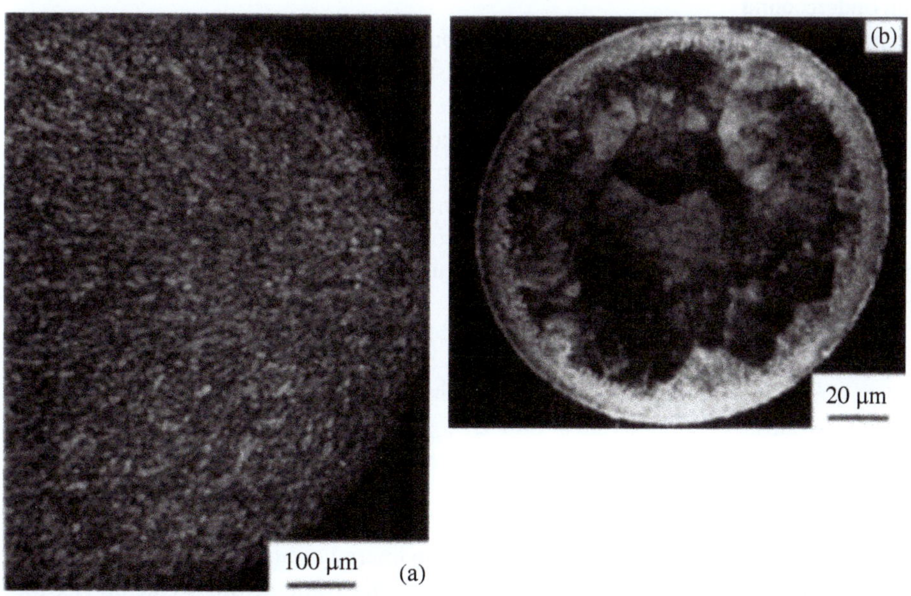

Figure 3.22 Microstructure of droplets showing surface-induced nucleation events: (a) a $Fe_{65}Ni_{35}$ droplet produced using the drop tube method. Source: Cochrane et al. [120] / Elsevier and (b) an atomized **Al**–8 wt%Fe droplet . Source: Cochrane et al. [37] / Reproduced with permission from Elsevier.

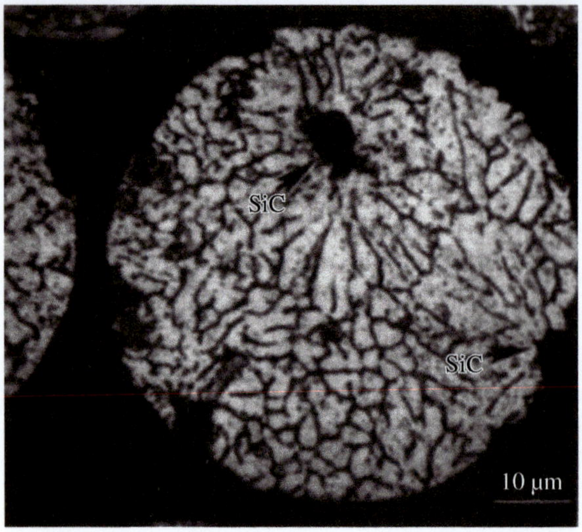

Figure 3.23 Microstructure of an Al–4 wt%Si/SiC droplet produced by spray atomization and co-injection showing what appears to be SiC particle-induced nucleation. Source: Wu and Lavernia [123] / Reproduced with permission from Springer Nature.

The volume-distributed nucleants may include impurity inclusions, primary phases formed during peritectic reactions, and even extraneous particles, depending on experimental conditions. Impurity inclusions are thought to be the most common nucleants that are frequently present. Their presence is difficult to detect directly using microscopic techniques. Available theoretical approaches that have been used to determine the distribution of such nucleants will be discussed in a later section. Moreover, primary phase-induced nucleation of secondary phases normally occurs in peritectic alloy systems [114, 121]. Regarding peritectic reactions, recent work has demonstrated that these reactions essentially involve the undercooling of a metastable liquid containing primary phase particles [114]. Hence, in these reactions, secondary phases nucleate heterogeneously on the surface of primary phases [114]. It is worth noting that heterogeneous nucleation on extraneous particles is frequently observed in metal matrix composites (MMCs) where, under certain conditions, the reinforcement phases may catalyze solidification [122, 123]. In the case of spray atomization of MMCs, the reinforcing particles may be introduced into a molten metal prior to atomization (see Chapter 4) [122] or co-injected into the atomized droplet dispersion [123]. An example of particle-induced nucleation is illustrated in Figure 3.23 for an Al–wt%Si MMC droplet reinforced with SiC particles.

3.4.2.2 Heterogeneous Nucleation Undercooling

There are several important factors that govern heterogeneous nucleation in droplets. These include available nucleation sites, nucleation potency, and nucleant distribution. The distribution of nucleants is particularly important in droplet solidification when heterogeneous nucleation occurs on volume-distributed catalytic sites. In this section, the phenomenon of heterogeneous nucleation will be discussed under the basic assumption that nucleants of identical potency are always present. The presence of a distribution of nucleants and its effect on nucleation will be discussed in the next section. Following simple classical nucleation theory [78, 84, 100, 124, 125], the heterogeneous nucleation rate can be formulated using a parallel approach as that used in the formulation of homogenous nucleation. Accordingly, the nucleation rate, $I(T)$, may be calculated as

$$I(T) = A_{\mathrm{i}} \exp\left[-\frac{\Delta G^{*}_{\mathrm{het}}}{kT}\right] \tag{3.53}$$

where the subscript of the prefactor A_{i} may be "v" to stand for nucleation induced by volume-distributed catalytic sites and "s" for nucleation induced by surface catalytic sites, and $\Delta G^{*}_{\mathrm{het}}$ is the free energy of heterogeneous nucleation. The free energy of heterogeneous nucleation in Eq. (3.53) is given by [124]

$$\Delta G^{*}_{\mathrm{het}} = \Delta G^{*} f(m, x) \tag{3.54}$$

where $f(m, x)$ is a catalytic potency function of the nucleants; ΔG^{*} is the free energy of homogeneous nucleation; and m and x are functions of the wetting angle and the size of the catalyst. The value of the prefactor, A_{v}, for volume-distributed nucleants depends on the number density and size of the nucleants. Following the simple approach employed by Turnbull and Fisher [126], the pre-exponential factor for heterogeneous nucleation differs from that used in homogenous nucleation only by a factor of n_{s}/n_{1}. Where n_{s} is the number of catalytic sites per unit volume of liquid and n_1 is the total number of liquid atoms per unit volume. It follows that the pre-exponential factor in Eq. (3.53) can be expressed in an analogous form as that used in Eq. (3.41):

$$I(T) = \frac{n_s}{n_1}\frac{k_v}{\eta_1}\exp\left[-\frac{\Delta G^*_{het}}{kT}\right] \tag{3.55}$$

For surface-induced nucleation, as in the case of oxide phases on the surface of a droplet, the pre-exponential factor in Eq. (3.55) is generally multiplied by a factor of $\phi(\phi \le 1)$ to take into account the fact that only a fraction of the surface sites will catalyze nucleation [36]. In this case, n_s in Eq. (3.55) is defined as the number of surface atoms per unit volume of liquid in the droplet. Inspection of the available literature reveals that different expressions have been used to calculate the magnitude of the pre-exponential factor, A_s, for surface-induced heterogeneous nucleation. For example, the form of A_i used by Turnbull and Fisher [126] is $A_s = 10^{33} n_s/n_1$ cm^{-3} s^{-1}. Boettinger and Perepezko [36], however, argued that it is more appropriate to use a surface nucleation rate whereby $A_s = \phi 10^{22}/\eta_1$ cm^{-2} s^{-1}. It is worth noting, however, that both expressions are used interchangeably by converting from volume to surface and vice versa.

The function $f(m, x)$, introduced in Eq. (3.54), refers to the catalytic potency of the nucleants. The catalytic potency of a nucleant depends on both the wetting characteristics between the catalyst and the solid phase, and the size and geometry of the nucleant. The simplest case that may be considered occurs when the size of the nucleant is much greater than that of the critically sized embryo. This condition can be idealized into the formation of an embryo on an infinitely large flat substrate (i.e. the size and curvature effects are ignored). An expression for the f function was derived by Turnbull and Fisher [126] for this case as

$$f(\theta_w) = \frac{1}{4}(2 + m)(1 - m)^2 \tag{3.56}$$

where $m = \cos\theta_w$, and θ_w is the wetting angle between the solid phase and the nucleant as shown in Figure 3.24. This equation is applicable, for example, when solidification initiates on a flat surface, such as the surface of a chill mold. Eq. (3.56) may be used when the surface of a droplet is covered by a monolayer of oxide phases in which the size of the catalytic substrate is much greater than that of the critically sized nuclei. When the size of the nucleation reagent is comparable to that of the embryos, however, the influence of nucleant size and geometry on nucleation potency must be taken into consideration [124, 125].

Fletcher [124, 125] conducted theoretical studies to elucidate the influence of nucleant size on nucleation potency. He considered the special case of nucleation events originating on the surface of a spherical nucleant. A schematic diagram illustrating such a nucleation mechanism is shown in Figure 3.24 [124, 125]. In this case, the free energy of nucleation may be calculated from [124, 126]

$$\Delta G = \Delta G_v \frac{V_e}{\Omega} + \sigma_{sl} S_{sl} + (\sigma_{sc} - \sigma_{lc}) S_{sc} \tag{3.57}$$

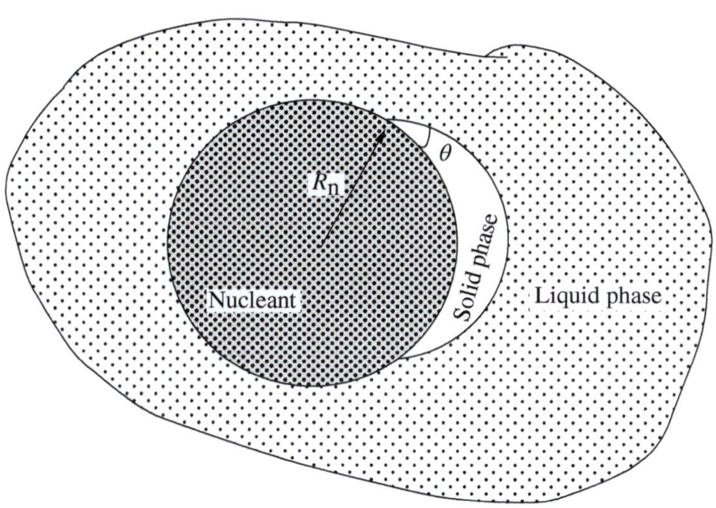

Figure 3.24 The nucleation of a solid phase on a nucleant. In the diagram, R_n is the radius of the nucleant and θ is the wetting angle between the nucleant and the solid phase. Source: Adapted from Fletcher [124].

where σ_{sc} and σ_{1c} are the interfacial energies between the solid phase and the nucleant, and between the liquid phase and the nucleant, respectively; and S_{s1} and S_{sc} are the contact areas between the solid phase and the liquid phase and between the solid and the nucleant, respectively. Using a formulation that is analogous to that used in the case of homogeneous nucleation, Fletcher [124, 125] derived an expression for the free energy of heterogeneous nucleation as

$$\Delta G^*_{het} = \frac{16\pi\sigma^3_{s1}\Omega^2}{3\Delta G^2_v} f(m, x) \tag{3.58}$$

where $m = \cos\theta = (\sigma_{1c} - \sigma_{sc})/\sigma_{s1}$; $x = R_n/r^*$ is the reduced nucleant size; R_n is the radius of the nucleant; and r^* is the size of the critical nuclei. According to Fletcher [124] the $f(m, x)$ function has the following form

$$f(m,x) = \frac{1}{2} + \frac{1}{2}\left(\frac{1-mx}{\zeta}\right)^3 + \frac{3}{2mx^2}\left(\frac{x-m}{\zeta} - 1\right) + \frac{1}{2x^3}\left[2 - \frac{3(x-m)}{\zeta} + \left(\frac{x-m}{\zeta}\right)^3\right] \tag{3.59}$$

where

$$\zeta = (1 - 2mx + x^2)^{1/2} \tag{3.60}$$

Figure 3.25 illustrates the effect of the parameters x and m on the nucleation potency function, $f(m, x)$ [124]. Since the free energy for heterogeneous nucleation is equal to that of homogeneous nucleation multiplied by a factor $f(m, x)$, a smaller $f(m, x)$ in this figure implies a lower energy barrier and therefore a higher nucleation potency. As shown in Figure 3.25, the nucleation potency of nucleant particles increases with increasing particle size. Moreover, when the size ratio between the particles and the critical nuclei exceeds a value of 10, the nucleation potency reaches a saturation value and there is negligible change in nucleation potency with further increases in particle size. When the corresponding size ratio is less than unity, however, the nucleation potency decreases dramatically. In this case, as the size ratio decreases, the $f(m, x)$ function approaches unity, a condition that approaches homogenous nucleation [124].

3.4.2.3 Distribution of Nucleants

In the aforementioned discussion on heterogeneous nucleation, it was assumed that nucleants are always present. Whereas this assumption is likely to be valid in the case of solidification of bulk liquids, where it is highly improbable that the liquid is completely devoid of nucleants, it may not always be applicable in the case of atomization.

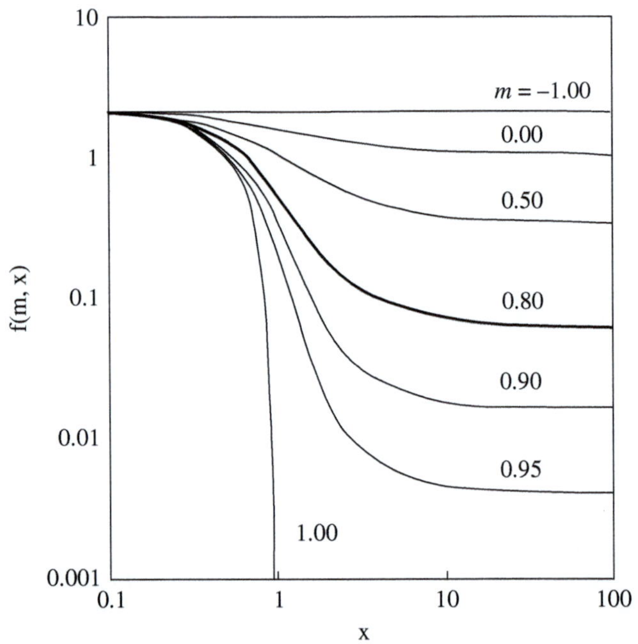

Figure 3.25 The geometrical factor $f(m, x)$ in terms of the ratio $x = R_n/r^*$. Source: Reproduced by permission of the American Institute of Physics [124].

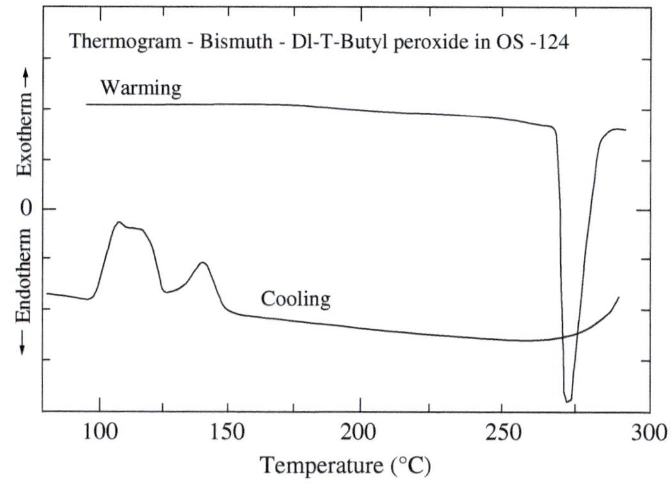

Figure 3.26 Thermogram for pure Bi. Emulsion formed by di-*t*-butyl peroxidation in the presence of isophthalic acid in OS-124. Source: Rasmussen and Loper [108].

In atomization, the total number of nucleants present in the bulk liquid is distributed among many droplets of various sizes. Hence, if the total number of nucleants present in the bulk liquid is less than the number of droplets, a fraction of the droplets must be free of nucleants [127, 128].

To evaluate the influence of nucleant distribution on undercooling it is necessary to take into consideration the following three factors: the presence of different nucleants (different θ), the size distribution of nucleants (same θ but different size), and the distribution of the nucleants among droplets. When different nucleants are present, the undercoolings achieved by various droplets prior to nucleation may be different. Figure 3.26 shows an example of **multinucleation** kinetics in emulsified droplets [108]. In the emulsion/undercooling experiments, a collection of emulsified droplets is heated to a temperature above the melting point; then it is cooled until solidified. The undercoolings experienced by these droplets are determined collectively using differential thermal analysis (DTA). In the DTA **thermogram**, the onset of a peak in the cooling curve represents the nucleation temperature, and that in the heating curve represents the melting temperature. The difference between the onset temperatures of the peaks in the heating and the cooling curves therefore represents the measure of undercooling. Moreover, the width of the cooling peak is indicative of a range of undercooling temperatures. The two separate peaks shown in Figure 3.26 suggest the presence of two different nucleation mechanisms. In this figure, the broad range of temperatures that are exhibited by the peaks suggests very different nucleation potencies associated with nucleation mechanism. As a comparison, a single nucleation mechanism with similar nucleation potency is shown in Figure 3.27 for liquid tin when the liquid alloy is properly emulsified. In this graph, there is a single, sharp nucleation peak indicating a narrow range of undercooling.

The influence of the size distribution of nucleants on undercooling is difficult to evaluate experimentally. Theoretically, some trends may be deduced based on Eq. (3.55)–(3.60). For example, if one considers a single nucleation mechanism, when the nucleants present in various droplets exhibit a distribution of sizes, the undercooling experienced by these droplets will be different. To that effect, according to Eq. (3.55)–(3.60) the nucleation potency increases with increasing nucleant size. Therefore, as undercooling increases, large nucleants will become active first. Hence, only at relatively high undercoolings will smaller nucleants become active and influence nucleation events. This phenomenon will lead to a range of undercoolings even for droplets under identical cooling rate conditions.

The influence of the distribution of nucleants among droplets on undercooling is of practical importance. To that effect, indirect experimental evidence suggests that the undercooling experienced by droplets of identical size might differ considerably [39, 49, 128, 129]. These large differences in undercooling are normally considered to be a result of the spatial distribution of nucleants in droplets. The differences in undercooling behavior are often deduced from microstructural studies conducted with droplets of similar sizes [39, 49, 128, 129]. This phenomenon is illustrated in Figure 3.28 based on results obtained by Boettinger [39] with atomized Al—4Cu powders. As shown in Figure 3.28, although the two powders are of similar size, the microstructural features that are present are significantly different. The **microstructure** shown in Figure 3.28a is coarse dendritic and therefore is likely to have been solidified under

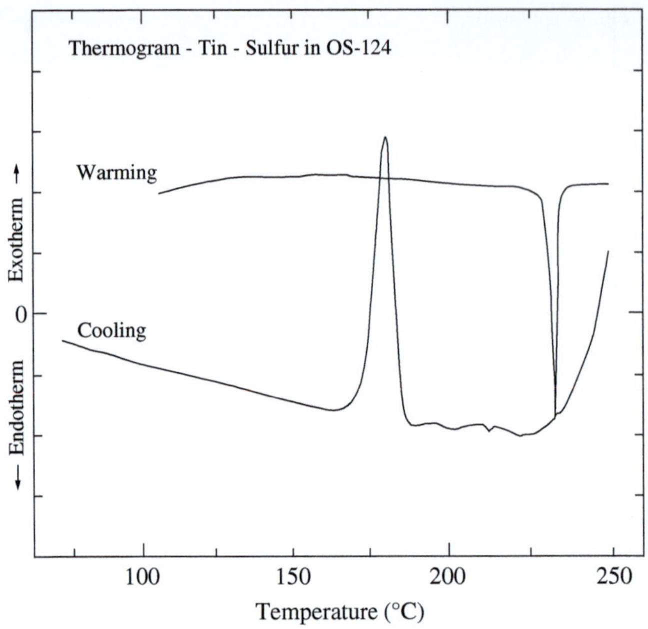

Figure 3.27 Thermogram for pure tin. Emulsion formed by di-*t*-butyl peroxidant in OS-124 in the presence of isophthalic acid. Source: Rasmussen and Loper [108].

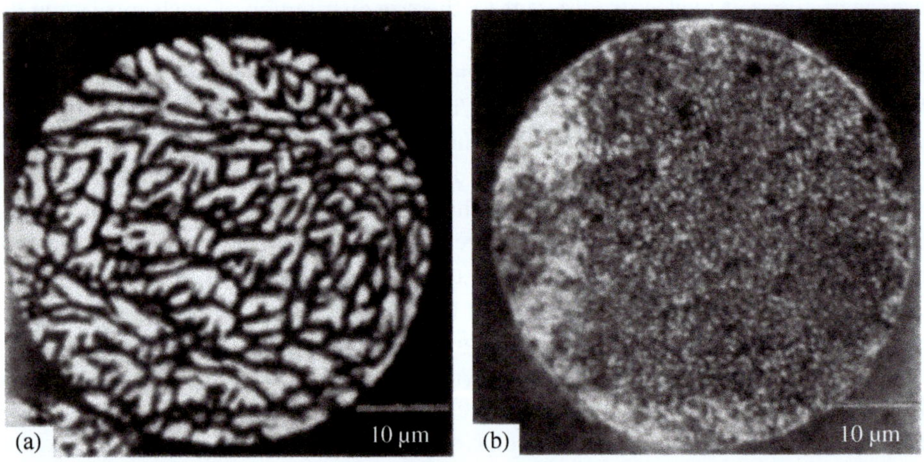

Figure 3.28 Al–4 wt%Cu powders of the same size showing differences in solidification microstructure: (a) solidified with marginal undercooling and (b) solidified with high undercooling. Source: Boettinger [39] / Reproduced with permission from Elsevier.

low undercooling conditions. The microstructure in Figure 3.28b, however, is free of catalytic nucleation centers and probably formed under relatively high undercooling conditions. Like the case of Al—4Cu, very different solidification microstructures have been observed in atomized Fe—Ni powders of similar sizes. Droplet solidification in the Fe—Ni system has been studied extensively [127–131]. Thoma and Perepezko [131], for example, studied the solidification behavior of Fe—Ni droplets using a drop tube method. Their results indicated that in the same powder size range of 30–44 μm, some of the powders exhibited a strain-free cellular/dendritic b.c.c. microstructure (δ-phase), some of the powders exhibited a martensite structure, while the rest of the powders were characterized by an equiaxed dendrite structure of the f.c.c. (γ-phase) [131]. Since in the drop tube technique droplets of similar sizes experience similar cooling rates, the observed differences in solidification microstructure were attributed to differences in undercooling behavior [128, 131].

Regarding the distribution of nucleants in droplets during atomization, some theoretical results are available [36, 132, 133]. To maintain the problem as tractable, available formulations often assume that the nucleants are of identical size and catalytic potency. The influence of nucleant distribution on undercooling relies critically on the isolation of small volumes of liquid in droplet production techniques. If one assumes that the number of nucleants contained in a bulk liquid is finite, when the bulk liquid is subdivided into fine droplets, the nucleants are necessarily distributed among the droplets. Hence, it is generally assumed that in each volume of liquid, the nucleants that are present will be randomly distributed. The highest possible degree of dispersion of nucleants among droplets may then be described by a Poisson distribution function [36, 132, 133]. Following this assumption, the probability that a volume of liquid is free of nucleants, P, may be expressed as [36]

$$P = \exp(-N_v v_d) \tag{3.61}$$

where N_v is the number density of nucleants, and v_d is the volume of liquid. When this formulation is applied to a collection of droplets of identical diameter, d, it yields the following expression for the fraction of droplets, f_n, that is free of nucleants [133]:

$$f_n = \exp\left[-\frac{\pi N_v d^3}{6}\right] \tag{3.62a}$$

or

$$f_n = \exp\left[-\left(\frac{d}{d_0}\right)^3\right] \tag{3.62b}$$

where d_0 is an exponential constant, the value of which can be determined from curve-fitting experimental results

In related studies, Drehman and Turnbull [133] applied the concept of a Poisson distribution function to investigate the phenomenon of surface-induced heterogeneous nucleation. They obtained the following relationship for the fraction of droplets that is free of nucleants [133]

$$f_n = \exp[-\pi N_s d^2] \tag{3.63a}$$

or

$$f_n = \exp\left[-\left(\frac{d}{d_0}\right)^2\right] \tag{3.63b}$$

where N_s is the number density of surface nucleants. Eqs. (3.62a,b) and (3.63a,b) can be used to calculate the fraction of nucleant-free droplets for a given nucleant density. These equations have been used, together with experimental results, to evaluate percentage of nucleant-free droplets and droplet diameter [133]. Boettinger and Perepezko [36], for example, estimated the nucleant density that is present in Al–4.5 wt%Cu droplets produced using gas atomization. Based on experimental observations, Boettinger and Perepezko argued that whereas undercooling has some effect on the solidification of droplets that are smaller than 100 μm, undercooling critically influences the solidification behavior of droplets that are smaller than 10 μm. Under these conditions, and if heterogeneous nucleation will occur in a droplet in the presence of a single nucleant, the nucleant density should range between 10^6 and 10^9 cm^{-3}. Boettinger and Perepezko calculated the fraction of nucleant-free droplets by assuming various nucleant densities. The results of their analysis are shown in Figure 3.29, for various droplet diameters [36]. Conceptually, these results demonstrate the importance of isolating the liquid into small volumes if a large degree of undercooling is desired. For example, when the nucleant density is 2×10^7 m^{-3}, for a droplet size of 40 μm about 40% of the droplets will be nucleant free. Alternatively, if the droplet size is reduced to 20 μm, more than 80% of the droplets will be nucleant free.

Eqs. (3.61)–(3.63) have been used to rationalize the experimentally determined fraction of cellular microstructure in Al—Cr alloy by Adkins and Tsakiropoulos [132] and the fraction of glassy droplets in Pd—Si droplets by Drehman and Turnbull [133]. Figure 3.30 compares the experimentally determined fraction of glassy Pd—Si droplets to the results calculated based on Eq. (3.63). The solid and open circles in this figure illustrate the experimentally determined fraction of glassy droplets and the solid lines represent the results predicted from Eq. (3.63) using an experimentally determined constant, d_0. The results show that Eq. (3.63) yields results that are in good agreement with the experimentally determined fraction of glassy droplets. Moreover, Drehman and Turnbull [133] compared experimental results with theoretical predictions for volume nucleant-induced heterogeneous nucleation and homogeneous nucleation. Their experimental results were in good agreement with those predicted based on Eq. (3.62), an observation that was interpreted to be suggestive of surface nucleation. This reported agreement between theory and experiment suggests the following two important points. First, the assumption of a Poisson function to describe

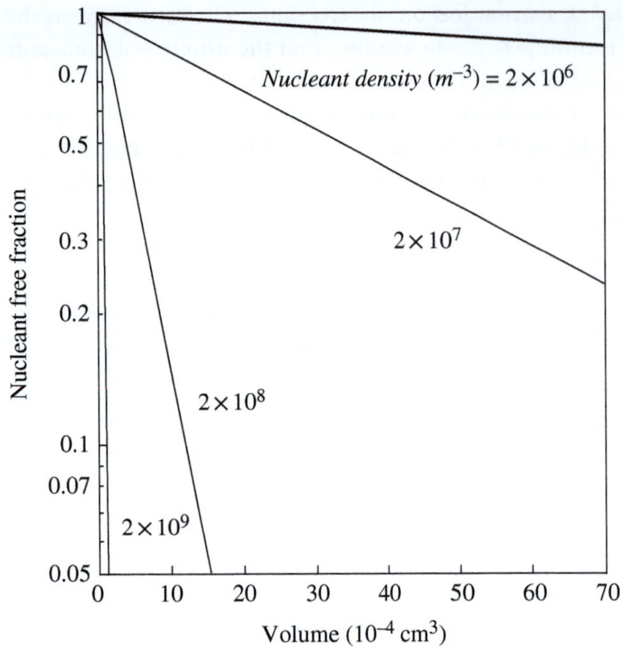

Figure 3.29 Nucleation-free fraction in different powder sizes for several nucleant densities. Source: Boettinger and Perepezko [36] / Reproduced with permission of The American Institute of Mining, Metallurgical and Petroleum Engineers.

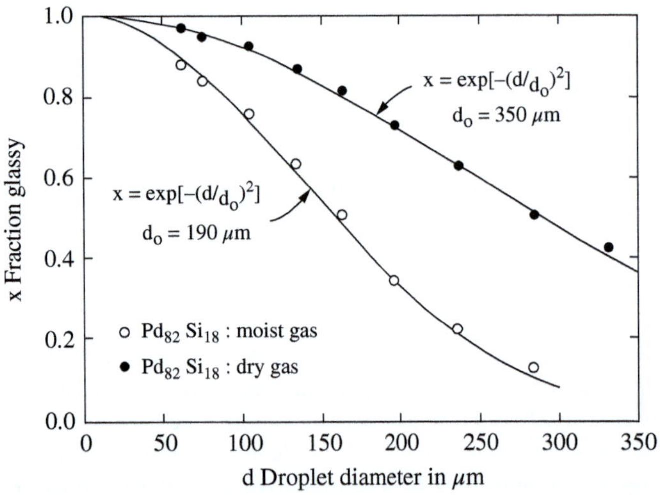

Figure 3.30 The droplet diameter dependence of the fraction of glassy droplets for $Pd_{82}Si_{18}$ in dried and undried He (80 vol.%) H_2 (20 vol.%); the data were fit by $x = \exp.[-(d/d_0)^2]$. Source: Drehman and Turnbull [133].

the distribution of nucleants is a valid one. Second, surface heterogeneous nucleation may be the most critical factor limiting the attainment of large undercooling levels.

3.5 Solidification of Droplets

The ultimate objective of the study of droplet solidification is to understand the fundamental mechanisms that govern the evolution of microstructure. To that effect, there is ample experimental evidence which shows that the formation of solidification microstructure (e.g., phase composition and morphology) is intimately related to cooling rate and undercooling conditions [34–39, 48, 49, 58, 132–142]. The influence of cooling rate and undercooling is associated with two basic processes: nucleation and growth of the solid phases. Initial undercooling is responsible for phase selection in which phases that are energetically favorable will **nucleate** first from the melt. The influence of growth on microstructure formation may be understood by simply considering the two extreme cases that occur during droplet solidification: adiabatic solidification and equilibrium solidification. In adiabatic solidification, the extent of undercooling governs the **solid–liquid interfacial velocity**. In **equilibrium solidification**, the movement of the solidification interface is controlled by external thermal energy extraction. In the sections that follow, the important factors that govern the formation of microstructure during **atomization** such as temperature distribution inside of droplets, cooling rate, **solidification time**, and **solid–liquid interfacial velocity** will be discussed in detail.

3.5.1 Temperature Distribution in Droplets

The formulation of models to calculate the thermal and solidification history of droplets has been a topic of considerable interest [43, 48, 58, 143, 144]. The models that are available to predict the thermal and solidification history of droplets can be grouped into two separate categories: analytical and **numerical models**. The **analytical models** normally invoke considerable simplifying assumptions but provide a large degree of physical insight into the problem. **Numerical models**, on the other hand, provide details about the temperature distribution as a function of time. Fundamentally, the thermal history of a droplet can be calculated by solving the appropriate heat transfer equations. To describe the thermal behavior of an atomized droplet, the simplest approach is to use spherical coordinates. The governing equation of heat transfer for a single-phase material in which heat transfer occurs primarily in the radial direction (i.e., θ and z components can be safely ignored) is

$$\frac{\partial T}{\partial t} = \alpha_{\mathrm{i}}\left(\frac{\partial^2 T}{\partial r^2} + \frac{2}{r}\frac{\partial T}{\partial r}\right) \tag{3.64}$$

where r is the radial coordinate, α is the thermal diffusivity, and the subscript i stands for 'l' for liquid phase or "s" for solid phase. The thermal diffusivity is related to the physical properties of single phases as

$$\alpha_{\mathrm{i}} = \frac{k_{\mathrm{i}}}{\rho_{\mathrm{i}} c_{\mathrm{p,i}}} \tag{3.65}$$

where k, ρ, and c_{p} are the thermal conductivity, density, and heat capacity per unit mass, respectively. There are two separate boundary conditions that are associated with this heat transfer problem. Symmetry restrictions at the center of the droplet require that

$$\frac{\partial T}{\partial r} = 0 \tag{3.66}$$

At the droplet/gas interface, the boundary condition is more complex. However, if the interface is a layer and the gas temperature is assumed to be constant, an interfacial heat transfer coefficient can be defined. In this case, the boundary condition at the droplet gas interface is

$$k\frac{\partial T}{\partial r} = -h(T - T_{\mathrm{g}}) \tag{3.67}$$

where h is interfacial heat transfer coefficient. Accordingly, the temperature distribution inside of a droplet may be readily calculated by numerically solving Eqs. (3.64)–(3.67).

The resistance to thermal energy loss from a single-phase droplet includes heat conduction within the droplet, and convective and radiative heat transfer across the surface of the droplet. To simplify the present discussion, in the text that follows only convective heat transfer at a droplet's surface will be considered. The effect of radiative heat transfer, if important, may be combined into a convective heat transfer coefficient using a formula as described in Eq. (3.9). In heat transfer problems of this type, a dimensionless number, the Biot number, is normally introduced to facilitate comparison of the relative importance of heat conduction and heat convection. As an example, one may consider the simple case of a one-dimensional system, such as a thin film of liquid of thickness $2L$ that loses thermal energy from both surfaces. The conductivity of the liquid is defined as k_1 and the convective heat transfer coefficient is defined as h_{c}. To further facilitate analysis let us assume that the temperature distribution in the liquid film is linear. In this case, the thermal energy flux passing through the body of the film is approximated by

$$J_{\mathrm{k}} = k_1\frac{\Delta T_{\mathrm{ci}}}{L} \tag{3.68}$$

where J_k is the conductive heat flux, and ΔT_{ci} is the temperature difference between the center and surface of the film. Since the temperature in the film is governed by a differential equation that is like Eq. (3.64) for the one-dimensional case, the assumption of a linear temperature profile is not strictly correct. When the film thickness dimension is small and the conductivity is high, however, a linear temperature profile assumption offers a close approximation to the actual temperature distribution. The thermal energy flux across the film/gas interface is limited by convective heat transfer and is given by

$$J_c = h_c(T_i - T_g) \tag{3.69}$$

where J_c is the convective heat flux, and T_i is the surface temperature of the liquid film. Under the assumption of a linear temperature profile, the magnitudes of J_k and J_c should be equal. This condition is expressed as

$$k_1 \frac{\Delta T_{ci}}{L} = h_c(T_i - T_g) \tag{3.70a}$$

or

$$\Delta T_{ci} = Bi(T_i - T_g) \tag{3.70b}$$

where Bi is defined as the Biot number and is given by

$$Bi = \frac{h_c L}{k_1} \tag{3.71}$$

In Eq. (3.70a,b), the temperature difference within the liquid is proportional to Bi. Hence, when the magnitude of Bi is very small, the temperature difference in the liquid becomes negligible.

In the heat transfer literature, the dimensionless number Bi, as introduced in the aforementioned simple example, is often generalized into a characteristic parameter. Accordingly, it is not necessary to make any assumptions regarding the temperature distribution in the liquid. For any volume of material, a Biot number can be defined with the parameter L being a characteristic dimension of the material under consideration. The Biot number provides a measure of the relative importance of heat conduction in the liquid volume versus heat convection at the interface. According to Eq. (3.71), for small values of Biot numbers, the primary resistance to heat transfer occurs at the droplet/gas interface. In this case, the heat conduction within the material occurs relatively fast and the temperature difference between different regions of the material volume is relatively small.

When applied to droplets, the Biot number becomes

$$Bi = \frac{h_c r_d}{k_1} \tag{3.72}$$

where r_d is droplet radius. Mehrabian et al. [58], used numerical heat transfer techniques to calculate the temperature distribution inside of atomized droplets. The results of their study, as shown in Figure 3.31, can be used to visualize the relative influence of the Biot number on the temperature profile inside of a droplet. The results in this figure are presented in terms of a dimensionless temperature, $(T - T_g)/(T_f - T_g)$, and as a function of dimensionless distance from the droplet center r/r_d. Figure 3.31 shows the temperature distribution in a liquid Fe droplet at the instant when the surface of the droplet reaches the melting temperature. The initial **superheat** temperature for this calculation was assumed to be $(T - T_g)/(T_f - T_g) = 1.3$; this corresponds to a droplet temperature of 2265 K. As shown in Figure 3.31, the influence of the Biot number on the temperature distribution inside of the droplet is significant. The temperature differential across the interior of the droplet is negligible only when the Biot number is less than 0.01. For larger values of the Biot number, the temperature difference within the droplet becomes significant [58].

3.5.2 Newtonian Solidification

Based on the aforementioned discussion, the temperature distribution inside of a droplet can be calculated using very basic governing differential heat transfer equations along with some initial and boundary conditions. The discussion also showed that when the Biot number is small, the temperature differential across a droplet may be safely neglected. Such **thermal conditions** are referred to as Newtonian cooling conditions. Hence, under the assumption of Newtonian cooling conditions, heat extraction from a droplet is governed by heat convection at the surface. Moreover, at any instant, the temperature distribution within the droplet is nearly uniform.

In droplet cooling, the magnitude of the error introduced by a Newtonian cooling assumption depends on the values of the heat transfer coefficient and the conductivity. Moreover, inspection of Eq. (3.72) reveals that the magnitude

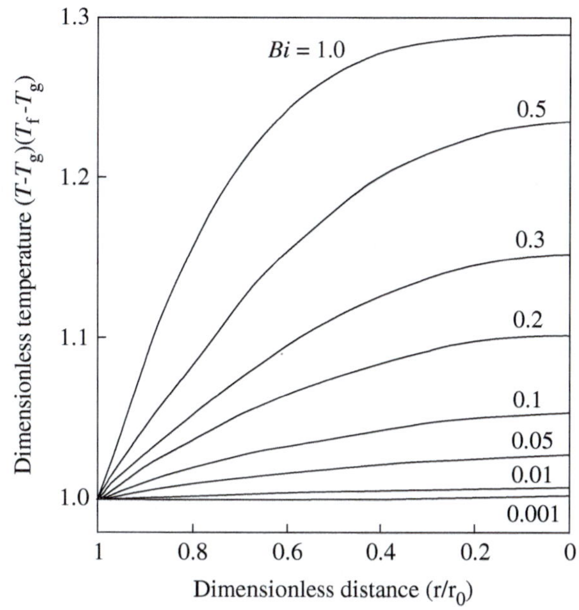

Figure 3.31 Dimensionless temperature distribution in a liquid iron droplet when its surface reaches the melting temperature T_1 for different Biot numbers (dimensionless initial superheat $(T - T_g)/(T_1 - T_g) = 1.3$. Source: Reproduced by permission of Plenum Press [58].

Table 3.4 Typical thermal conductivity and Biot number for atomized droplets.

Material	Conductivity† liquid ($Wm^{-1}\ K^{-1}$)	Conductivity† solid ($Wm^{-1}\ K^{-1}$)	Bi (liquid) (50 μm diameter)	Bi (solid) (50 μm diameter)
Al	100	200	0.013	0.0063
Cu	165	244	0.0076	0.0051
Mg	90	130	0.014	0.0096
Ni	30	60	0.042	0.021

Calculated for a heat transfer coefficient of $5 \times 10^4\ Wm^{-2}\ K^{-1}$.
Source: Adapted from Clyne [106].

of the error associated with this assumption decreases with decreasing droplet size. Generally, the assumption of Newtonian cooling is considered to be acceptable when the Biot number is less than 10^{-2} [43, 58]. Taking Al as an example, if the average heat transfer coefficient is taken as $5 \times 10^4\ Wm^{-2}\ K^{-1}$, the droplet size that satisfies the condition $Bi < 0.01$ is 50 μm. In the literature, the assumption of Newtonian cooling has been applied to droplet sizes of several hundred micrometers. Although a Newtonian assumption in the case of large droplets leads to some degree of uncertainty, the accompanying analysis provides useful information on droplet cooling and solidification. The Biot numbers of several materials are calculated by assuming a heat transfer coefficient of $5 \times 10^4\ W\ m^{-2}\ K^{-1}$ for a droplet size of 50 μm. The results, summarized in Table 3.4, demonstrate the important influence of thermal conductivity on the Biot number. For example, for the same material, the magnitude of the Biot number is smaller in the solid state relative to that corresponding to the liquid state because of the greater conductivity of solids.

3.5.3 Cooling Rate

There are two important simplifications that arise in the analysis of droplet cooling as a direct result of the Newtonian cooling assumption. First, since the temperature within a droplet is assumed to be uniform, a droplet temperature may be defined. Second, an instantaneous droplet cooling rate can also be calculated. Considering a single-phase droplet and assuming Newtonian cooling conditions, the temperature of a droplet can be calculated directly from an energy conservation standpoint, since the temperature decrease in a droplet is a direct result of the thermal energy loss that occurs at the droplet's surface. It then follows that a cooling rate may be calculated as [48, 70, 145, 146]

$$\dot{T} = \frac{3\Omega h_c}{c_{pi} r_d}(T - T_g) \tag{3.73}$$

In related studies, Mehrabian et al. [58] expressed the average cooling rate of a droplet into a parallel dimensionless form as

$$\Theta = 3Bi(\theta + Ste) \tag{3.74}$$

In Eq. (3.74), Θ is a dimensionless cooling rate, θ is a dimensionless temperature and *Ste* is a dimensionless Stefan number. These three dimensionless numbers are defined as follows [58]:

$$\Theta = \frac{r_d^2 c_{ps}}{\alpha_i \Delta H_f}\dot{T} \tag{3.75a}$$

$$\theta = \frac{c_{ps}(T - T_f)}{\Delta H_f} \tag{3.75b}$$

$$Ste = \frac{c_{ps}(T_f - T_g)}{\Delta H_f} \tag{3.75c}$$

The aforementioned Newtonian analysis has been used extensively in calculating the temperature history of droplets during atomization [2, 4, 5, 13, 43, 60, 61, 63, 69, 72, 143, 145]. A useful parameter that can be obtained from such formulations is the average cooling rate that the droplets experience prior to the onset of solidification. Such a cooling rate is an indication of the effectiveness of thermal energy extraction during atomization. Theoretical and experimental work in droplet atomization generally reveals an average cooling rate that falls in the range of 10^2–10^6 Ks^{-1}. The precise magnitude of the cooling rate depends on several factors, such as the cooling conditions that are present and the droplet size involved [13]. A theoretical estimation of droplet cooling rate was reported by Mehrabian et al. [58] using an average heat transfer coefficient of 10^5 Wm^{-2} K^{-1}. Accordingly, the average cooling rates obtained for droplet diameters of 100 and 10 μm are 8.3×10^5 and $8.3\times10^6 Ks^{-1}$, respectively [58]. Detailed calculations of droplet cooling rate are plentiful throughout the atomization literature [2, 34, 60, 63, 69]. Table 3.5 summarizes some examples of calculated cooling rates as a function of droplet size [34, 63]. The results reported by Samuel [34], for example, were obtained for liquid (Al-Li-Co) droplets assuming an average heat transfer coefficient of 5×10^4 Wm^{-2} K^{-1}. The results of Grant et al. [63], however, were calculated using a variable heat transfer coefficient as given by Eq. (3.25). In the calculations of Grant et al., the cooling rate was averaged over a flight distance of 40 cm. In this case, the **recalescence** experienced by the undercooled droplets was incorporated into the calculations. Consequently, the calculated cooling rate values are relatively lower than those obtained by Samuel [34, 63]. The results from both calculations, however, reveal the same trend of increasing cooling rate with decreasing droplet size.

Experimentally, the relationship between droplet size and cooling rate has been determined for several alloy systems. A standard approach that is frequently used to determine the droplet size–cooling rate relationship is to invoke the relationship between **secondary dendrite arm spacing** (SDAS) and droplet diameter. The relationship between SDAS and cooling rate, which is determined from separate experiments, is then used to establish a relationship between cooling rate and droplet size [16, 34, 147, 148]. For example, using the experimentally determined SDAS–cooling rate equation determined by Matyja et al. [16], Samuel [34] estimated the cooling rate of atomized Al—3Li powders. These results indicated a cooling rate in the range of 10^4–10^6 Ks^{-1} for droplets in the size range of 5–200 μm. In a separate study, Webster et al. [147] determined the cooling rate of Al—Li droplets in the Li compositional range of 3–5 wt%. Their results indicated that the cooling rate of droplets in the size range of 5–200 μm is between 10^2 and 10^4 Ks^{-1}. The discrepancy between these results indicates that droplet cooling rate determined

Table 3.5 Calculated droplet cooling rates.

Droplet size (μm)	5	10	15	20	50	80	100	120	125	200
$\dot{T}$, Al-3Li(Ks^{-1})[a]	1.7×10^7	–	5.6×10^7	–	1.7×10^6	1.1×10^6	8.5×10^5	–	6.8×10^5	4.2×10^5
$\dot{T}$, Al—4Cu (Ks^{-1})[b]	–	1.5×10^5	–	1.1×10^5	3.7×10^4	1.1×10^4	–	1.0×10^4	–	9.4×10^3

a) By Samuel [34] using a constant heat transfer coefficient of 5×10^4 W m^{-2} K^{-1} for liquid droplets.
b) By Grant et al. [63] using an initial gas velocity of 150 m s^{-1} and a variable heat transfer coefficient with values averaged over a flight distance of 40 cm.
$\dot{T}$, cooling rate.

using empirically established SDAS–cooling relationships is accurate only to one or two orders of magnitude. This may be rationalized if one considers the following. On one hand, droplet cooling rate depends on the experimental atomization conditions that are present. Hence, extrapolation of a dendrite-cooling rate relationship, as determined from one particular set of experiments inherently overlooks differences in experimental droplet cooling conditions. Moreover, experimentally, it has been reported that the chemistry of the atomization gas has a significant influence on the droplet size–cooling rate relationship. As an example, Figure 3.32 shows the cooling rate determined for Al droplets as a function of droplet diameter [2]. In this figure, the droplets were produced using ultrasonic gas atomization with Ar, He, and N_2 at an atomization pressure of 80 kg cm^{-2}. It is evident that the type of atomization gas used critically influences cooling rate. Accordingly, cooling rate is the highest for He-atomized droplets and lowest for Ar-atomized droplets [2]. On the other hand, alloy composition normally has a significant influence on the resultant dendrite arm spacing. Therefore, extrapolation of an SDAS–cooling rate relationship determined for one alloy composition to another may lead to significant errors. The effect of alloy composition on dendrite arm spacing is illustrated in Figure 3.33 for N_2-atomized Al—Si droplets. The results clearly reveal a decrease in dendrite arm spacing with increasing Si content [149].

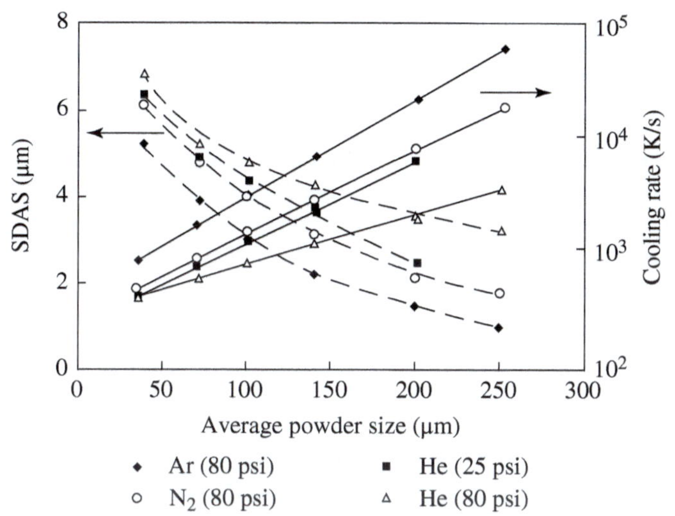

Figure 3.32 The variation in dendrite arm spacing and cooling rate with powder size for various gas atomization pressures. Source: Lavernia [2].

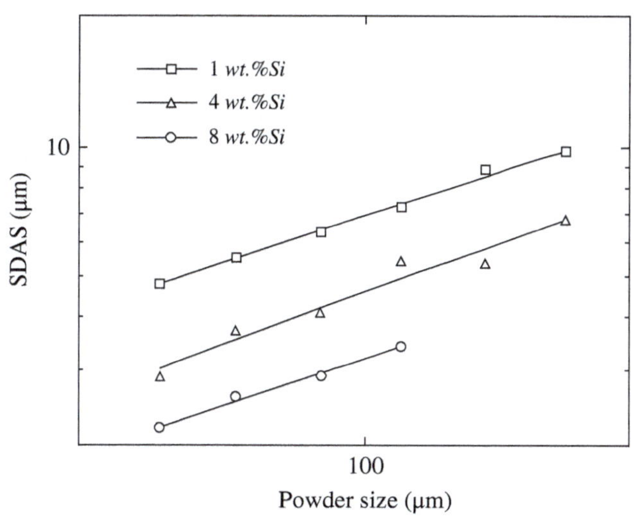

Figure 3.33 Secondary dendrite arm spacing (SDAS) as a function of powder size as measured for N_2-atomized Al—Si powders of various compositions. Source: Wu and Lavernia [149].

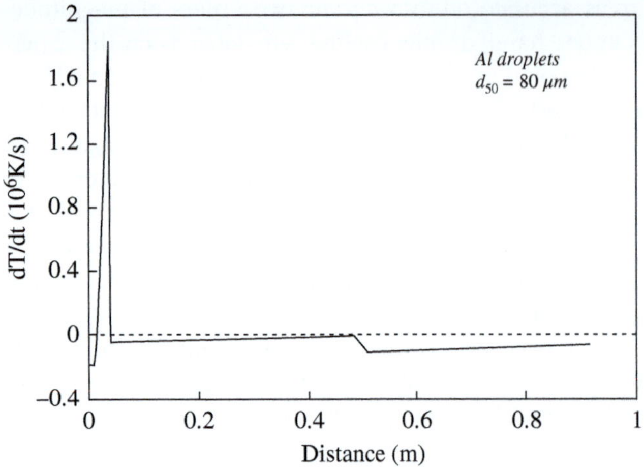

Figure 3.34 Calculated time rate of change in temperature for an Al droplet d_{50} as a function of the flight distance. Source: Gutierrez-Miravete et al. [60] / Reproduced with permission from Springer Nature.

In the aforementioned experimental studies, an average cooling rate can be inferred from the microstructural analysis. Moreover, details of droplet cooling history may be calculated on the basis of simple heat transfer considerations. For example, Gutierrez-Miravete et al. [60] calculated the cooling rate of Al droplets atomized using N_2 gas. An average droplet size of 80 μm was selected in their calculations which corresponds to the d_{50} (mass mean diameter) value for ultrasonic gas atomization of Al. The results obtained by Gutierrez-Miravete et al. are summarized in Figure 3.34 as a function of droplet flight distance. According to the results shown in Figure 3.34, a droplet experiences both cooling and heating during flight. In the cooling regime, the cooling rate varies in the range of 10^4–10^6 K s^{-1}.

3.5.4 Solidification Time

A useful parameter that can be derived from solidification models is the **solidification time** which is defined as the time elapsed between the onset and the completion of solidification. Solidification time, therefore, represents the time period during which the microstructure is exposed to coarsening in the mushy state (i.e. liquid and solid). Assuming Newtonian solidification conditions, solidification time can be readily calculated from energy conservation considerations. Solidification time can be obtained by integrating the fraction of solid phase from zero to unity. For **equilibrium solidification**, an equation for the calculation of solidification time is then given by

$$t_f = \frac{r_d \Delta H_f}{3h\Omega(T_1 - T_g)} \tag{3.76}$$

where t_f is the solidification time. It should be noted that the validity of this equation is limited to equilibrium solidification, and the presence of undercooling leads to a more complex expression. However, in view of the fact that solidification in the presence of undercooling always occurs at a rate that is faster than that in the absence of undercooling, Eq. (3.76) offers an upper bound for solidification time. In related work, Mehrabian et al. [58] derived a dimensionless form of Eq. (3.76) for Newtonian cooling conditions and equilibrium solidification as

$$\frac{\alpha_1 t_f}{r_d^2} = \left[3BiSte\frac{c_{pl}}{c_{ps}}\right]^{-1} \tag{3.77}$$

Inspection of Eqs. (3.76) and (3.77) reveal that solidification time can be minimized by reducing droplet size and/or by increasing Biot number, i.e. by increasing the magnitude of the heat transfer coefficient. For example, for a Fe droplet of 50 μm in diameter, a heat transfer coefficient of 5×10^4 W m^{-2} K^{-1} and a Biot number of 0.031, the solidification time estimated by Mehrabian et al. is on the order of 10^{-4}, whereas that corresponding to a 500 μm Fe droplet is on the order of 10^{-3}. Samuel [34] calculated the solidification time for Al—3Li droplets using the same heat transfer coefficient (i.e. 5×10^4 Wm^{-2} K^{-1}). According to his results for a droplet size range of 5–200 μm, the solidification time falls in the range of 1.9×10^{-5} and 8×10^{-3} seconds.

3.5.5 Interfacial Velocity

For conventional solidification conditions, the **solid–liquid interfacial velocity** can be determined relatively accurately. For example, transparent organic substances are sometimes used to visually monitor the movement of the solid–liquid interface. Although in metals, visual observation of the movement of the solid–liquid interface is precluded as a result of their non-transparent optical characteristics, **interfacial velocity** may be determined through clever experimental designs. One example is the Bridgeman method in which solidification is induced by displacing a liquid metal through a cold zone at a controlled velocity. Hence, the interfacial velocity can be readily controlled by changing the speed at which the metal passes though the cold zone. In droplets, the interfacial velocity is difficult to measure directly. Accordingly, in many cases estimation of interfacial velocities during atomization relies on theoretical modeling and computation [2, 34, 36, 43, 63, 106, 143].

3.5.5.1 Equilibrium Solidification

The calculation of interfacial velocity under equilibrium solidification conditions may be simplified further by assuming Newtonian heat transfer conditions and considering the solidification of a pure metal. Moreover, in the case of equilibrium solidification, it is generally assumed that the liquid-to-solid transformation proceeds at the melting temperature. Accordingly, the latent heat of fusion that is released during the liquid-to-solid phase transformation is equal to the amount of thermal energy that is dissipated into the environment. There is, however, one additional assumption that is necessary in order to complete the formulation. Since the magnitude of the interfacial velocity is influenced by the number of nucleation events and the morphology of the interface, a single nucleation event with a smooth solid–liquid interface is often assumed to maintain the problem as tractable. Hence, an energy balance formulation on a concentric spherical nucleus that is growing radially from the center of the droplet toward the surface yields the following equation:

$$V_{\mathrm{s1}} = \frac{r_{\mathrm{d}}^2 h(T_1 - T_{\mathrm{g}})\Omega}{r^2[(c_{\mathrm{ps}} - c_{\mathrm{p1}})T_1 + \Delta H_{\mathrm{f}}]} \tag{3.78}$$

where $V_{\mathrm{sl}} = \mathrm{d}r/\mathrm{d}t$ is the instantaneous interfacial velocity; r is the instantaneous interfacial position; and r_{d} is the radius of the droplet. Using similar assumptions, Mehrabian et al. [58] derived a formulation for the calculation of interfacial velocity in a dimensionless form as follows:

$$\frac{r_{\mathrm{d}}}{\alpha_1}V_{\mathrm{s1}} = \left(\frac{1}{\varphi^*}\right)^2 Bi\frac{c_{\mathrm{p1}}}{c_{\mathrm{ps}}}Ste \tag{3.79}$$

where $\varphi^* = r/r_{\mathrm{d}}$, the dimensionless interfacial position. It is interesting to note that Eq. (3.79) is essentially the same as Eq. (3.77) with the exception that the latter equation assumes $c_{\mathrm{pl}} = c_{\mathrm{ps}}$.

It is useful to understand the influence of the Biot number on interfacial velocity and thereby assess the error introduced by the assumption of Newtonian heat transfer into the calculations. To that effect, the interfacial velocities calculated by Mehrabian et al. [58] using a non-Newtonian approach are summarized in Figure 3.34. The results in this figure, which are shown for concentric isotherms, are calculated for different interfacial positions, $\Phi^* = r/r_{\mathrm{d}}$, and for various Biot numbers. As shown in Figure 3.35, interfacial velocity increases with increasing Biot number and decreasing solid fraction (i.e. with decreasing Φ^*). In the case of Newtonian solidification, inspection of Eq. (3.79), reveals that for a constant value of interfacial position, Φ^*, the dimensionless interfacial velocity follows a linear relationship with the Biot number. The curves shown in Figure 3.35, however, deviate from straight lines. The results of Mehrabian et al. demonstrate that the ratio between interfacial velocities calculated using non-Newtonian and Newtonian assumptions is close to unity at the beginning of solidification, i.e. for low Φ^* values. Accordingly, the deviation between non-Newtonian and Newtonian behavior increases with increasing fraction of solid and Biot number. In related studies, Levi and Mehrabian [143] reported detailed calculations of the interfacial profile, position, and velocity during atomization using a numerical approach. In this study, solidification was assumed to initiate at one location on the droplet surface and the interface moved toward the interior of the droplet during solidification. The results of these calculations shown are in Figure 3.36 where the solid lines represent the instantaneous positions of the interface, the factor f_{s} by the lines indicates the fraction solidified, and V_{sl} in the bracket represents instantaneous interfacial velocity. The interfacial velocities shown in Figure 3.36 were calculated assuming a dimensionless initial undercooling of $\theta = c_{\mathrm{p,l}}(T - T_1)/\Delta H_{\mathrm{f}} = -0.5$, a K_m/Bi value of 400, and an interfacial kinetic coefficient of 2 cm K^{-1} s^{-1}. The results in this figure demonstrate that solid–liquid interfacial

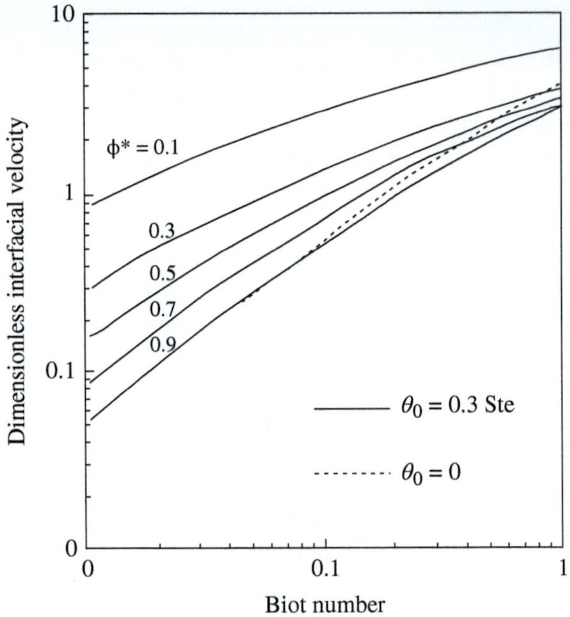

Figure 3.35 Dimensionless solid–liquid interfacial velocity during solidification of iron droplets as a function of Biot number and fractional position of the interface $\Phi^* = r/r_d$, and $\theta_0 = c_{p,1}(T - T_1)/\Delta H_f$. Source: Reproduced by permission of Plenum Press [58].

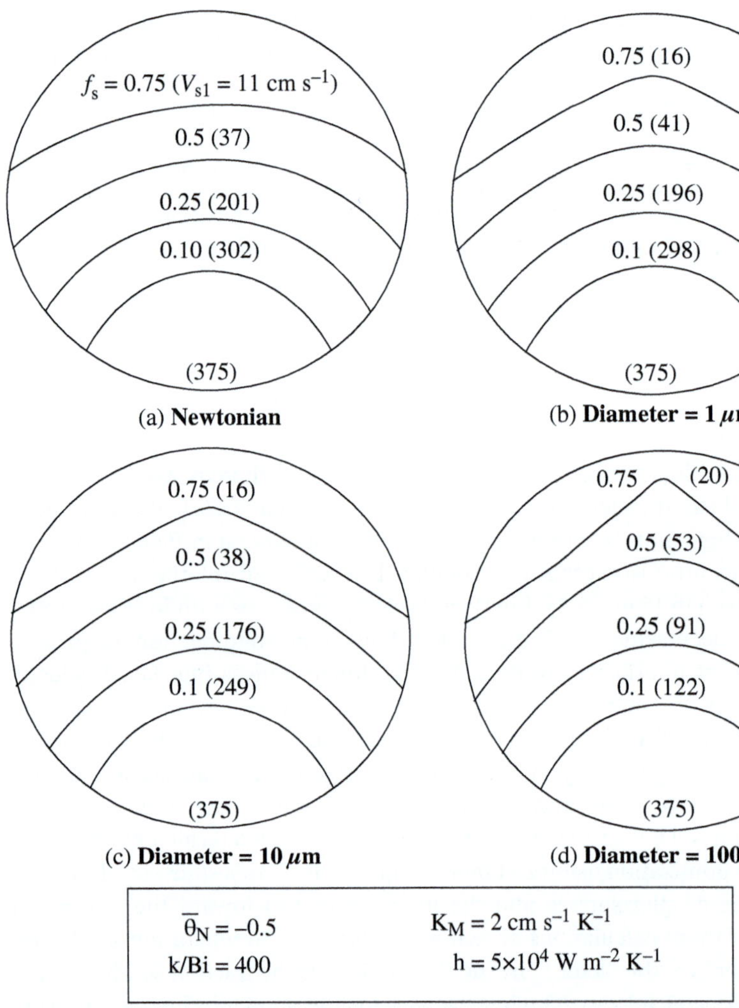

Figure 3.36 Interfacial profiles and velocities predicted from Newtonian [solid lines in (a) and dashed lines in (b), (c), and (d)] and non-Newtonian models [solid lines in (b), (c), and (d)] [$\theta = c_{p,l}(T - T_1)/\Delta H_f$]. The deviations from the Newtonian velocity limit increase with droplet size. Source: Levi and Mehrabian [143] / Reproduced with permission from Springer Nature.

velocity decreases with increasing droplet size and that the deviation of the solid–liquid interfacial behavior from Newtonian conditions increases with increasing droplet diameter or with increasing Biot number [143]. It is worth noting that the magnitude of the **solid–liquid interfacial velocity** that is calculated for equilibrium solidification is significantly smaller than 1 ms^{-1}. For example, Mehrabian et al. [58] calculated the interfacial velocity for Fe droplets at a solid fraction of 0.5 and assuming a heat transfer coefficient of 5×10^4 Wm^{-2} K^{-1}. The result of this study showed that the interfacial velocities for droplet sizes of 50 and 500 μm were 0.11 and 0.05 ms^{-1}, respectively [58].

3.5.5.2 Dynamic Solidification

In **dynamic solidification**, the velocity of the solid–liquid interface depends on the magnitude of undercooling present. Accordingly, the calculation of interfacial velocity requires physical models that relate interfacial velocity with undercooling. The physical models that relate interfacial velocity with undercooling generally invoke one of two types of growth mechanisms: continuous growth and **stepwise growth**; the presence of one or the other depends on the degree of undercooling. For example, it has been proposed that stepwise growth of a solid is favored when there is a low driving force present, i.e. at low undercoolings, whereas continuous growth is favored at high undercoolings. During solidification the release of the latent heat of fusion at the solid–liquid interface alters the thermal profile ahead of the interface. As a result, the magnitude of the local undercooling at the solid–liquid interface differs from the bulk undercooling in the liquid. Since the growth of the solid is a transport phenomenon that occurs at the local solid–liquid interface, the undercooling that governs the associated interfacial growth kinetics should be interfacial undercooling, rather than bulk undercooling. Hence, it is intuitive to define an interfacial undercooling, ΔT_i, to distinguish the undercooling at the solid–liquid interface from the bulk undercooling in the liquid. It has been suggested that the two growth regimes, i.e. stepwise growth and continuous growth, are separated by a transitional regime where the driving force depends on the roughness of the interface at the atomic level [150, 151]. Cahn [150] developed criteria to predict the presence of a particular growth mechanism, i.e. stepwise or continuous. The criteria involve expressing the driving force for growth of the interface in terms of the magnitude of the **free energy change** per unit volume as a result of growth. Since under a first-order approximation, the magnitude of **free energy change** during growth is proportional to undercooling [see Eqs. (3.34–3.38)], Cahn [150] concluded that the transition from stepwise growth to continuous growth occurs at some critical interfacial undercooling. The magnitude of the critical free energy proposed for the transition from a stepwise growth regime to the transitional regime is [150]

$$\Delta G_v^* = \frac{\sigma_{sl}\kappa}{a} \tag{3.80}$$

where ΔG_v^* is the critical volume free energy difference between the solid and the liquid phase, κ is a function of surface roughness with a magnitude of less than unity, and a is the interatomic distance [151]. Accordingly, the **critical undercooling** for the stepwise-to-continuous growth transition may be estimated from the critical free energy as defined by Eq. (3.80) and (3.32) as

$$\Delta T_i^* = \frac{\sigma_{sl}\kappa\Omega T_l}{a\Delta H_f} \tag{3.81}$$

Cahn [150] proposed that with increasing undercooling, transition from stepwise growth to continuous growth begins at $\Delta T_i > \Delta T_i^*$; however, complete continuous growth actually occurs at undercoolings that are equivalent to $\Delta T_i \geq \pi\Delta T_i^*$. It then follows that the transitional growth regime falls in the undercooling range of $\Delta T_i^* < \Delta T_i < \pi\Delta T_i^*$. The two growth regimes are discussed in detail below.

Continuous Growth The continuous growth mechanism assumes a diffuse solid–liquid interface [150–152]. The origin of this mechanism may be traced to the problem posed by the vapor deposition of crystalline phases [150–152]. In addressing this particular problem, it was postulated that at elevated temperatures, such as those close to the melting point, the solid–liquid interface is a rough surface. Moreover, the roughness of such a diffuse interface was proposed to consist of several atomic layers. Hence, across this diffuse solid–liquid interface, the characteristic parameters of the two phases, i.e. solid and liquid, vary continuously. In the continuous growth mechanism, the solid–liquid interface is assumed to maintain its roughness during growth so that the energy barrier during crystal growth is minimized. The solid–liquid interface then advances in a more or less continuous manner normal to its plane.

On the basis of classical continuous crystal growth theories, the rate of interfacial advance is proportional to a rudimentary step height and atomic attachment frequency [153] as

$$V_{s1} = av \tag{3.82}$$

where a is the step height (atomic distance in this case) and v is the frequency at which atoms in the liquid jump across the solid–liquid interface. In a crystal growth process, atoms will jump back and forth across the interface. Hence, when the rate of forward jump from the liquid to the solid exceeds that of backward jump from the solid to the liquid, the interface will grow at the expense of the liquid phase. When net jumps are considered, the growth rate may be expressed in terms of the associated free energy change as [153]:

$$V_{s1} = av_0 \exp\left(-\frac{E_a}{RT_i}\right)\left[1 - \exp\left(-\frac{\Delta G_v}{RT_i}\right)\right] \tag{3.83}$$

where v_0 is the atomic vibration frequency, E_a is the activation energy for atomic transport, and ΔG_v is the molar free energy difference between the liquid and the solid phases. This equation is valid when the interfacial configuration is such that all of the sites are equally favorable for attachment. Hence, Eq. (3.83), in combination with the value of the free energy change, ΔG_v, may be used to estimate the interfacial velocity. Eq. (3.83) is, however, often simplified by assuming a first-order approximation in the calculation of the free energy [see Eq. (3.34)]. In this case, Eq. (3.83) can be expressed in a more general form as [99]

$$V_{s1} = \phi V_0 \left[1 - \exp\left(-\frac{\Delta H_f \Delta T_i}{RT_f T_i}\right)\right] \tag{3.84}$$

where V_0 is a kinetic constant proportional to the interfacial rearrangement frequency and ϕ is the fraction of sites on the solid–liquid interface that are available for atomic attachment. For a diffuse solid–liquid interface ϕ is approximately unity.

The rationale for the presence of a velocity constant, V_0, in Eq. (3.84) is based on the two approaches that have been used to develop this equation. One approach assumes that the attachment kinetics of the atoms are limited by the collision of liquid atoms on the solid–liquid interface [99, 154]. In this case, V_0 can be approximated by an upper bound value, that is, the velocity of sound in the liquid phase [99, 154]. A second approach is to consider that the transport process across the interface is analogous to diffusion. In this case, V_0 is limited by diffusional processes and may be calculated as [151]:

$$V_0 = \beta \frac{D_1}{a} \tag{3.85a}$$

and

$$\beta = \left(\frac{a}{\delta}\right)^2 \left(\frac{v_i}{v_1}\right) \tag{3.85b}$$

where δ is the mean jump distance in the liquid and v is the jump frequency, with the subscripts "i" and "1" corresponding to interface and liquid, respectively. Levi and Mehrabian [143] compared the magnitude of V_0 for the cases of collision-limited and diffusion-limited attachment. If V_0 is assumed to be the velocity of sound, its value is on the order of 10^3 ms^{-1} in liquid metals. For a diffusion-limited process, the dimensionless factor β in Eq. (3.38b) is considered to be on the order of 1–10^2 [151], whereas those of D_1 and a in liquid metals can be assumed to be $D_1 \sim 10^{-9}$ m^{2}s^{-1} and $a \sim 10^{-10}$ m. In this case, the magnitude of the velocity constant V_0 would be on the order of 10^1–10^3 ms^{-1} [143]. Hence, the selection of V_0 can cause significant differences in the calculated solid–liquid interfacial velocities. In view of the fact that at high undercooling levels, atomic transport becomes sluggish, an assumption of liquid diffusion-controlled interfacial growth may be more practical. Therefore, in most crystal growth studies the diffusion-limited velocity constants determined on the basis of Eq. (3.85a,b) are utilized [48, 143].

In continuous growth, a limiting condition is applicable when the undercooling level is relatively small as compared to the melting temperature, i.e. $\Delta T_i \ll T_1$ or $T_i \approx T_1$. The crystal growth rate under these conditions is relatively well established. In this case, assuming $\phi = 1$, the interfacial velocity of Eq. (3.84) can be approximated by [151]

$$V_{s1} \approx \beta\left(\frac{D_1}{a}\right)\left(\frac{\Delta H_f}{RT_1^2}\right)\Delta T_i \tag{3.86a}$$

where D_l is the self-diffusion coefficient in the liquid at the melting temperature. Eq. (3.86a) implies that the interfacial velocity may be approximated as a linear function of undercooling. For simplicity, Eq. (3.86a) is often expressed as

$$V_{sl} = K_m \Delta T_i \tag{3.86b}$$

where K_m is defined as the linear interface kinetic coefficient and may be calculated from Eq. (3.86a). Eq. (3.86b) is widely used in the calculation of **solid–liquid interfacial velocity** in undercooled materials [41, 44, 48, 51, 61, 99, 155], There is, however, some degree of uncertainty in the calculation of the linear kinetic coefficient, K_m. For example, Levi and Mehrabian [143] proposed a lower bound of K_m when $\beta = 1$. In the case of Al, the diffusion coefficient is $D_l \sim 10^{-5}$ cm^2 s^{-1}; the calculated value of K_m is 2 cm s^{-1} K^{-1}. In a related study, Lavernia et al. [41] calculated the value of K_m for an Mg—Al alloy. The diffusion coefficient at 873 K was assumed to be 1.1×10^{-8} cm^{2}s^{-1}. By assuming an interfacial step equivalent to the lattice parameter of Mg and approximating the heat of fusion to be equivalent to that of pure Mg, they obtained a value of $K_m = 3 \times 10^{-2}$ cm s^{-1} K^{-1} for $\beta = 10^2$. Hence for values of β in the range of 1–10^2, the calculated K_m fell in the 3×10^{-4} to 3×10^{-2} cm s^{-1} K^{-1} range [41].

3.5.5.3 Stepwise Growth

In the case of low undercoolings, the advance of the solid–liquid interface occurs by stepwise growth. In the stepwise growth mechanisms, the solid–liquid interface is assumed to be flat at the atomic level. In such mechanisms atoms attach to the solid–liquid interface by jumping from the liquid to any edges or kinks that may be present on the flat solid–liquid interface. Hence, the attachment of atoms from the liquid phase onto the interface leads to the advance of edges. The direction of advance of the edges is, however, lateral, i.e. within the plane of the interface. Continuous attachment of atoms to the edges results in a new crystal layer that sweeps across the interface. The solid–liquid interface advances one step each time the growth front sweeps across the interface. The kinetics of stepwise growth depends on the nature of the edges. Three mechanisms have been proposed to be active in the stepwise growth regime, including

(a) **lateral growth** at available interfacial steps [151],
(b) dislocation-assisted growth [151, 156], and
(c) nucleation-assisted growth [157–159].

In the case of **lateral growth**, atoms attach directly to any available surface steps. In related work, Cahn et al. [151] argued that the edges of a flat crystal surface are diffuse and that in **lateral growth** the atoms attach onto the crystal edges in the same manner as in continuous growth on a diffuse interface. The relationship between growth velocity and undercooling, as described by Eq. (3.86a,b), is therefore applicable for the advance of interfacial steps. Cahn et al. thus modified Eq. (3.86) to calculate the lateral velocity of a diffuse interfacial step as

$$V_{step} \approx \beta \left(\frac{D_{l,f} \Delta H_f \Delta T_i}{a R T_l^2} \right) (2 + \kappa^{-1/2}) \tag{3.87}$$

where κ is a factor which describes the degree of diffusion of an interfacial step, and V_{step} is the velocity at which interfacial steps sweep across the solid–liquid interface. In the case of sharp steps, κ equals unity; in the case of very diffuse steps, κ is less than unity so that the step velocity becomes greater. Hence, the interfacial velocity can be calculated from this step velocity if the step distance on the interface is known. If the step distance is λ, the frequency, ν_{step}, at which the interfacial steps pass a certain point of the interface is

$$\nu_{step} = \frac{V_{step}}{\lambda} \tag{3.88}$$

The advancing velocity of the solid–liquid interface in this stepwise growth mechanism is therefore the product of step frequency and step height [151]

$$V_{sl} = \beta \left(\frac{D_{l,f} \Delta H_f \Delta T_i}{\lambda R T_l^2} \right) (2 + \kappa^{-1/2}) \tag{3.89}$$

This expression is obtained by assuming that the step height is equal to the interatomic spacing. For step heights that are greater than the interatomic spacing, the velocity should be greater than that predicted by Eq. (3.89). Since the step distance is normally much greater than the interatomic distance, the velocity given by Eq. (3.89) should be smaller than the continuous growth velocity as calculated by Eq. (3.86).

In the dislocation growth mechanism, the **lateral growth** steps are provided by the crystal edges that are created by screw dislocations intersecting the solid–liquid interface. In this case, Frank [156] suggested that the advancing velocity of a solid–liquid interface can be calculated from the step velocity and the spacing between steps as the spacing between steps as

$$V_{\text{s1}} = \frac{aV_{\text{step}}}{4\pi r_{\text{c}}} \tag{3.90}$$

where r_{c} is the radius of a critical nucleus. The radius of a critical nucleus, r_{c}, is a measure of the spacing between steps originating from screw dislocations. When r_{c} is expressed in terms of undercooling, by combining Eqs. (3.89) and (3.90) the interfacial growth velocity in a dislocation growth mechanism becomes [151]

$$V_{\text{s1}} = \beta\left(\frac{D_1\Delta H_{\text{f}}^2}{4\pi RT_1^3\sigma_{\text{s1}}\Omega}\right)\left(\frac{1+2\kappa^{1/2}}{\kappa}\right)\Delta T_{\text{i}}^2 \tag{3.91}$$

It is worth noting that in this equation V_{sl} is proportional to the square of the interfacial undercooling.

In nucleation-assisted growth, the solid–liquid interface is assumed to be crystallographically perfect. In this case, direct attachment of individual atoms to a two-dimensional, sharp solid–liquid interfacial interface is energetically unfavorable. Hence, two-dimensional nucleation on the atomically smooth solid–liquid interface is required to create crystal steps on the solid–liquid interface. The problem posed by the two-dimensional nucleation of a solid on a solid–liquid interface has been studied by a number of investigators [157–159]. The free energy associated with the nucleation of a two-dimensional step of height a is

$$\Delta G^* = -\frac{\pi\varepsilon_{\text{s1}}\sigma_{\text{s1}}^2}{\Delta G_{\text{v}}} \tag{3.92}$$

where ε_{sl} is the free energy per unit length of a step of height, a. Hilling [159] suggested that the **nucleation rate** of interfacial steps per unit area may be approximated by

$$I = V_{\text{step}}\frac{N_{\text{A}}}{\Omega}\left(\frac{\Delta H_{\text{f}}\Delta T_{\text{i}}}{RT_1^2}\right)^{1/2}\exp\left(-\frac{\Delta G^*}{RT_{\text{i}}}\right) \tag{3.93}$$

where N_A is Avogadro's constant. Two limiting cases have been identified for nucleation-assisted growth: step velocity-limited growth and nucleation-limited growth [151, 159]. For simplicity, one may assume that the area of a solid–liquid interface is A, and that the maximum and minimum length of straight steps are L_{max} and L_{min}, respectively. Thus $L_{\text{max}}/V_{\text{step}}$ is the time for an interfacial step to sweep through the entire solid–liquid interface and $1/(IA)$ is the time interval between surface nucleation events. Hilling [159] and Cahn et al. [151] suggested that when $L_{\text{max}}/V_{\text{step}} \ll 1/(IA)$ the advance of the solid–liquid interface is limited by nucleation. In this case, each step may have enough time to sweep across the entire interface and interfacial velocity becomes insensitive to the step-growth velocity [151]:

$$V_{\text{sl}} = aIA \tag{3.94}$$

In the other extreme, $L_{\text{max}}/V_{\text{step}} \gg 1/(IA)$, and advance of the interface is limited by the step-growth rate. In step velocity-limited growth, it is assumed that multiple nucleation events may occur on the flat interface and that the growth front from a nucleus will be annihilated when it collides with the front from another nucleus. In this case, the advance rate of the solid–liquid interface derived by Hilling [159] is given by

$$V_{\text{s1}} = V_{\text{step}}\left(\frac{\Delta H_{\text{f}}\Delta T_{\text{i}}}{RT_1^2}\right)^{1/6}\exp\left(-\frac{\Delta G^*}{3kT_{\text{i}}}\right) \tag{3.95}$$

In the presence of high undercooling conditions, and assuming that linear growth kinetics are applicable, it is possible to evaluate the relative importance of **recalescence** and heat transfer on the temperature of a droplet. For simplicity, the linear growth kinetics as expressed by Eq. (3.86b) may be converted into a dimensionless form as [143]

$$\overline{V}_{\text{s1}} = \overline{K}_{\text{m}}\Delta\overline{T}_{\text{i}} \tag{3.96}$$

where $\overline{V}_{\text{s1}} = V_{\text{s1}}d/(2\alpha_1)$ is the dimensionless interfacial velocity;
$\overline{K}_{\text{m}} = K_{\text{m}}(\Delta H_{\text{f}}r)/(\alpha_1 c_{\text{p,1}})$ is the dimensionless interfacial kinetic coefficient; and
$\Delta\overline{T}_{\text{i}} = c_{p,1}\Delta T_{\text{i}}/\Delta H_{\text{f}}$ is the dimensionless interfacial undercooling.

3.5.5.4 Experimentally Determined Interfacial Velocities

Inspection of the available literature reveals that various investigators have attempted to experimentally determine the **solid–liquid interfacial velocity** in undercooled liquids. These experiments, however, are often conducted using pure metals and under strictly controlled experimental conditions. For example, a number of authors have studied the interfacial velocity–undercooling relationship in pure Ni using thermal field measurement techniques [138, 140, 160–163]. In such experimental studies, pure Ni is generally melted using an electromagnetic levitation melting technique. The levitated melt is then undercooled by exposing the sample to mixtures of Ar and H_2 gases. Undercooling is thus controlled by artificially promoting nucleation at desired undercooling levels. This is accomplished, for example, by touching undercooled liquid with a pyrex whisker. Owing to the non-transparent nature of the liquid metal, the interfacial velocity is determined by measuring the thermal field that is generated during **recalescence**. In this approach, the thermal field is measured using two photodiodes located at a preselected distance. The interfacial velocity is then calculated by dividing the distance between the photodiodes by the time interval between which the two photodiodes record recalescence [140]. The extent of undercooling is thus determined from direct temperature measurement of the liquid. In view of this particular experimental arrangement, the undercooling recorded experimentally represents bulk undercooling, rather than interfacial undercooling.

In a related study, Colligan and Bayles [160] used high-speed motion photographs to monitor the progress of solidification in undercooled liquid Ni. They determined interfacial velocity by using film timing marks. As an example, Figure 3.37 shows the interfacial velocity–bulk undercooling relationship as determined by Colligan and Bayles [160] for undercooled pure Ni. From the results shown in Figure 3.37, in the undercooling range of 20–220 K, the interfacial velocity increases with increasing undercooling. The experimental results shown in Figure 3.37 exhibit a certain degree of scatter when the undercooling level exceeds 0.1 T_1 (i.e. 173 K). A similar trend of increasing interfacial velocity with undercooling was also reported by Walker [161] for undercooled Ni and Co melts. The data from both experiments fit a power law function as suggested by Jackson [164] in the form of

$$V_{s1} = A(\Delta T)^n \tag{3.97}$$

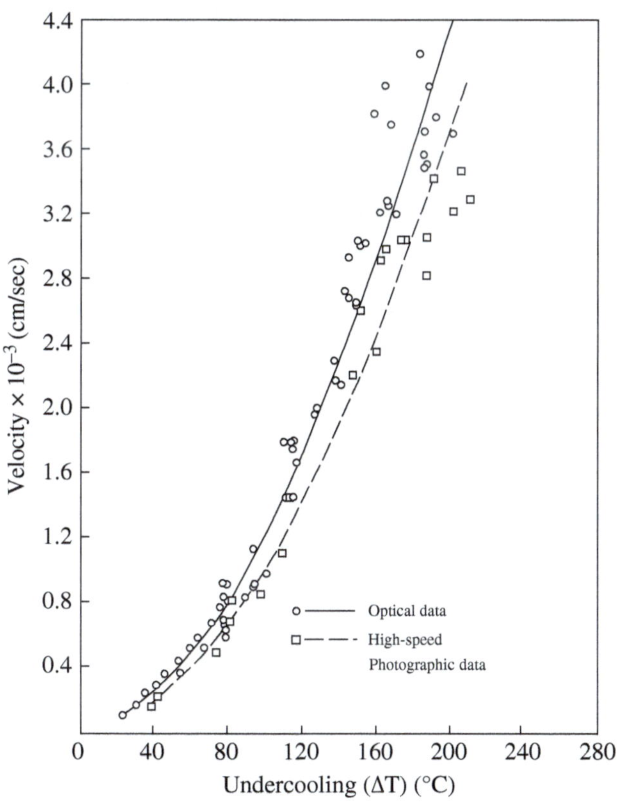

Figure 3.37 Solid–liquid interfacial velocity of pure Ni as a function of bulk undercooling. Source:. Colligan and Baylers [160].

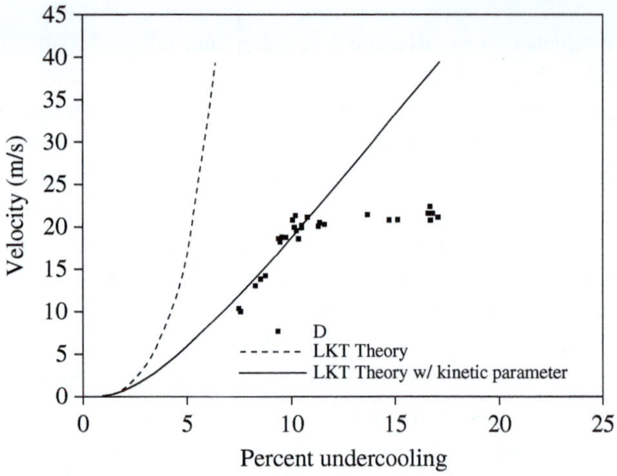

Figure 3.38 The experimentally determined interfacial velocity in undercooled Ni as a function of reduced undercooling. Source: Bassler et al. [140] / Reproduced with permission from Springer Nature.

where A and n are experimentally determined constants. On the basis of available experimental results, a parabolic function with $n = 2$ yields optimal curve fit. Accordingly, a parabolic empirical relationship is present between interfacial velocity and undercooling as [164]

$$V_{\mathrm{sl}} = A(\Delta T)^2 \tag{3.98}$$

When the experimental results obtained by Walker [161] are compared to Eq. (3.98) the constant, A, assumes a value of 0.142 for Ni and 0.17 for Co. It is worth noting, however, that the results from both Walker [161] and Colligan and Bayles [160], deviate from those predicted by Eq. (3.98) when undercooling exceeds 0.1 T_1. The maximum interfacial velocity determined by Walker for which the curves remain parabolic is approximately 50 m s^{-1} for both Ni and Co. In a related study, Bassler et al. [140] studied the solidification of undercooled Ni by measuring the interfacial thermal fields. Their results are summarized in Figure 3.38 for an undercooling in the range of 133 K (0.077 T_1) to 292 K (0.169 T_1). From this figure, there are two clear stages in the interfacial velocity versus undercooling curves. At low undercoolings, the interfacial velocity may be approximated by a parabolic function. This is consistent with the findings of Colligan and Bayles [160]. At high undercoolings, the interfacial velocities fall on a horizontal line, suggesting an upper limit for interfacial velocities, consistent with the results of Walker [161] and those of Hofmeister et al. [165]. According to the results of Bassler et al. [140] the transition between these two zones occurs at an undercooling of 0.1 T_1. This trend is consistent with the findings of Walker [161] for Ni. The highest interfacial velocities observed by Hofmeister et al. [165] and Bassler et al. [140] were on the order of 20 m s^{-1}. This value is significantly lower than the interfacial velocities observed by Colligan and Bayles [160], Walker [161], and Schleip et al. [166] (i.e. 50 m s^{-1}), in their earlier work. This discrepancy suggests that considerable care must be exercised when interpreting experimentally determined interfacial velocities.

The experimental observation of an upper limit in interfacial velocity at high undercoolings is inconsistent with the theoretical prediction of the continuous growth theory. To rationalize this discrepancy, Bassler et al. [140] argued that the interfacial velocity at high undercoolings is limited by the rate of diffusional atomic attachment. To support this argument, they estimated the speed at which short-range diffusion begins to limit solidification velocity. The highest interfacial velocity proposed by Bassler et al. was D_L/a. Hence, using a self-diffusion coefficient of 3.85×10^{-9} m^2 s^{-1} and the closest interatomic spacing, they estimated a limiting interfacial velocity of 14.5 m s^{-1} for the case when the **solute** trapping is ignored.

Willnecker et al. [138] studied the solidification behavior of undercooled binary $Cu_{70}Ni_{30}$ and ternary $Cu_{69}Ni_{30}B_1$ alloys. The ternary composition was selected on the basis of binary $Cu_{70}Ni_{30}$ with the addition of 1.0 wt% of B to reduce atomic mobility in the undercooled liquid. Willnecker et al. observed two distinct regimes in experimentally determined interfacial velocity/undercooling curves. The results of their study are summarized in Figure 3.39 for various degrees of undercooling [138]. Similar to the earlier results reported for pure Ni [160] in the initial regime, i.e. at low undercoolings, the interfacial velocity obeys a power law relationship with undercooling. Beyond

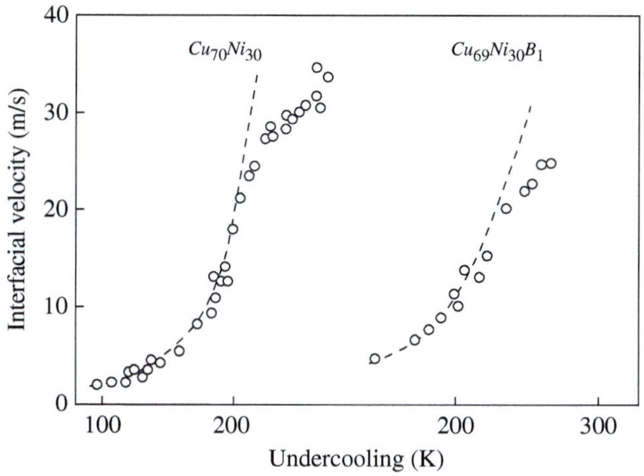

Figure 3.39 Solid–liquid interfacial velocities measured for $Cu_{70}Ni_{30}$ and $Cu_{69}Ni_{30}B_1$ alloys as a function of undercooling. Source: Reproduced by permission of the American Institute of Physics [138].

a critical degree of undercooling the interfacial velocity deviates from a power law relationship and follows a linear relationship with undercooling. The transition between these two regimes occurs at an interfacial velocity corresponding to $V_{s1} = D_l/a = 19\ \text{m s}^{-1}$ for binary $Cu_{70}Ni_{30}$ [138]. According to Willnecker et al., for any given undercooling, the interfacial velocity for ternary alloy $Cu_{69}Ni_{30}B_1$ is lower relative to that of binary alloy $Cu_{70}Ni_{30}$. The critical temperature at which the transition occurs between a power law and a linear relationship was 32 K higher for the ternary alloy relative to that for the binary alloy [138]. Willnecker et al. attributed the observed low interfacial velocity and high critical transition temperature in $Cu_{69}Ni_{30}B_1$ to a decrease in atomic mobility, i.e. in growth kinetic coefficient, as a result of the addition of B. The estimated kinetic growth coefficient, K_m, was $1.6\ \text{m K}^{-1}\ \text{s}^{-1}$ for $Cu_{70}Ni_{30}$ and $0.8\ \text{m K}^{-1}\ \text{s}^{-1}$ for $Cu_{69}Ni_{30}B_1$ [138].

In the aforementioned experimental studies, interfacial velocity appears to follow a parabolic relationship with undercooling, at relative high undercoolings (i.e. up to 0.1 T_1). Such results may not be suggestive of a dislocation growth mechanism for two reasons. First, the dislocation mechanism is operative at very low undercoolings. Second, the experimentally determined undercoolings often correspond to bulk undercoolings, whereas in the theoretical relationships described by Eq. (3.84)–(3.95) the undercoolings represent local interfacial undercoolings. The observed discrepancy between experimental measurements and theory may be attributed to the fact that at the micrometer level the solid–liquid interface is not flat and that the thermal and solutal fields at the interface may cause significant differences between the local undercooling at the solid–liquid interface and that in the bulk liquid. To that effect, solidification studies on undercooled metals frequently report that the morphology of the growth front is dendritic instead of planar [138, 140, 160–166].

A dendrite growth model was developed by Lipton et al. [167] to describe solidification at low undercoolings (i.e. low interfacial velocity). In related work, Trivedi and Kurz showed that for large Péclet numbers (i.e. high interfacial velocity) the complete stability analysis must be modified to predict the growth behavior of a solid–liquid interface. Subsequently, Lipton et al. [168] expanded the dendrite growth model developed by Lipton et al. [167] to encompass conditions of large Péclet numbers (the LKT theory). In the dendrite growth model, the growth velocity of isolated dendrites is related to bulk undercooling in the liquid instead of the interfacial undercooling. In the case that the advance of the solid–liquid occurs with local interfacial equilibrium, Lipton et al. [168] suggested that the bulk undercooling in a liquid may be expressed as the summation of the individual contributions from thermal, solutal, and curvature contributions. This summation in dimensionless form is

$$\Delta\overline{T} = \Delta\overline{T}_t + \Delta\overline{T}_c + \Delta\overline{T}_r \tag{3.99}$$

where $\Delta\overline{T}$, $\Delta\overline{T}_t$, $\Delta\overline{T}_c$, and $\Delta\overline{T}_r$ are the bulk undercooling, thermal undercooling, solutal undercooling, and curvature undercooling, respectively. The bar over the symbols indicates that the variables are dimensionless. The dimensionless undercooling is related to the dimensional undercooling, ΔT, by a general expression of the form:

$$\Delta\overline{T} = \frac{\Delta T c_{p,1}}{\Delta H_f} \tag{3.100}$$

When the growth kinetics of the interface become important, the assumption of local equilibrium at the interface may not be valid. In the LKT theory, the interfacial equilibrium condition is relaxed by using a kinetic relationship between interfacial velocity and interfacial undercooling. The dimensional form of this relationship is given by Eq. (3.86b). Moreover, interfacial undercooling, in dimensionless form, is given in Eq. (3.96). Hence, when interfacial kinetics are incorporated the bulk undercooling of Eq. (3.99) is expressed as

$$\Delta\overline{T} = \Delta\overline{T}_{\mathrm{t}} + \Delta\overline{T}_{\mathrm{c}} + \Delta\overline{T}_{\mathrm{r}} + \Delta\overline{T}_{\mathrm{i}} \tag{3.101}$$

where $\Delta\overline{T}_{\mathrm{i}}$ is dimensionless interfacial undercooling and is given by Eq. (3.96). Each of the undercooling terms in Eq. (3.101) can be expressed separately in terms of the dendrite radius and the solid–liquid interfacial velocity. For an isolated dendrite of radius R_{d} and growth velocity V_{sl}, solution of the thermal interface fields yields the thermal undercooling [169]:

$$\Delta\overline{T}_{\mathrm{t}} = Iv(P_{\mathrm{t}}) \tag{3.102}$$

where $Iv(p)$ is the Ivantsov function, and $P_{\mathrm{t}} = (V_{\mathrm{sl}}R_{\mathrm{d}})/(2\alpha_1)$ is the thermal Péclet number [169]. Moreover, solution of the solutal fields at the interface leads to an expression for **constitutional undercooling** [168]:

$$\Delta\overline{T}_{\mathrm{c}} = \frac{C_0|m|c_{\mathrm{p,l}}}{\Delta H_{\mathrm{f}}}\left[\frac{1}{1-(1-k_0)Iv(P_{\mathrm{c}})} - 1\right] \tag{3.103}$$

where m is the slope of the liquidus curve, k_0 is the equilibrium partition coefficient, and $P_{\mathrm{c}} = (V_{\mathrm{sl}}R_{\mathrm{d}})/(2D_1)$ is the solutal Péclet number. The curvature-induced undercooling of an isolated parabolic dendrite is related to the dendrite tip radius by [168]:

$$\Delta\overline{T}_{\mathrm{r}} = \frac{2d_0}{R_{\mathrm{d}}} \tag{3.104}$$

where d_0 is the capillarity length. It then follows that combination of Eqs. (3.100)–(3.104) yields an expression for bulk undercooling:

$$\Delta\overline{T} = Iv(P_{\mathrm{t}}) + \frac{C_0|m|c_{\mathrm{p,l}}}{\Delta H_{\mathrm{f}}}\left[\frac{1}{1-(1-k_0)Iv(P_{\mathrm{c}})} - 1\right] + \frac{2d_0}{R_{\mathrm{d}}} + \frac{\overline{V}_{\mathrm{sl}}}{\overline{K}_{\mathrm{m}}} \tag{3.105}$$

Eq. (3.105) contains three interrelated unknown variables: bulk undercooling, ΔT (or $\Delta\overline{T}$), dendrite tip radius, R_{d}, and interfacial velocity, V_{sl}. Accordingly, to obtain a relationship between interfacial velocity and bulk undercooling, the dendrite radius in Eq. (3.105) needs to be eliminated from this expression. Eliminating R_{d} from Eq. (3.105) requires another independent relationship between undercooling and dendrite radius or between interfacial velocity and dendrite radius.

Lipton et al. [167] developed a relationship between R_{d} and V_{sl} based on an **interfacial stability criterion** developed by Mullins and Sekerka [170], and Trivedi and Kurz [171]. In this formulation, the tip radius of dendrites is the shortest wavelength which is stable under local thermal, solutal, and interfacial velocity conditions. Then, for an isothermal dendrite tip, Lipton et al. [167] obtained the following expression:

$$\overline{V}_{\mathrm{sl}} = \frac{P_{\mathrm{t}}^2}{4\pi^2}\left[\xi_{\mathrm{t}} + 2\Lambda\xi_{\mathrm{c}}\frac{C_0(1-k_0)|m|\alpha_1 c_{\mathrm{p,l}}}{D_1\Delta H_{\mathrm{f}}}\right] \tag{3.106}$$

where ξ_{t} and ξ_{c} are thermal and solutal stability parameters. The functions Λ, and the parameters ξ_{t} and ξ_{c} in Eq. (3.106) are given by

$$\Lambda = [1-(1-k_0)Iv(P_{\mathrm{c}})]^{-1} \tag{3.107a}$$

$$\xi_{\mathrm{t}} = 1 - \frac{1}{\left(1+4\pi^2/P_{\mathrm{t}}^2\right)^{1/2}} \tag{3.107b}$$

$$\xi_{\mathrm{c}} = 1 + \frac{2k_0}{1-2k_0-\left(1+4\pi^2/P_{\mathrm{c}}^2\right)^{1/2}} \tag{3.107c}$$

With Eqs. (3.105) and (3.106) the interfacial velocity is singularly related to the bulk undercooling. For each given bulk undercooling the solution of the interfacial velocity is obtained by deriving a relationship between dendrite radius and interfacial velocity from Eq. (3.105). When the relationship is combined with Eq. (3.106) an interfacial velocity can be obtained. The advantage of using Eqs. (3.105) and (3.106) lies in the fact that the interfacial velocity is directly related to the bulk undercooling ΔT which may be determined experimentally or may be calculated theoretically.

Bassler et al. [140] compared their experimental results with the predictions of the LKT theory [168] for dendrite growth in highly undercooled liquids, as predicted based on Eqs. (3.105) and (3.106). The theoretical predictions obtained based on the LKT theory are designated in Figure 3.38 by the solid and dashed lines for models with [see Eq. (3.99)] and without a kinetic factor K_m [see Eq. (3.101)]. Bassler et al. noted that the LKT model predictions in the absence of interfacial kinetics were not in good agreement with those obtained experimentally. However, when the interfacial kinetics are included in the LKT model, the theoretical predictions are in close agreement with experimental results obtained by Bassler et al. as well as the results obtained by other investigators for undercoolings less than 0.1 T_1 (i.e. <173 K).

3.6 Microstructural Development

The microstructure that evolves during droplet solidification is a result of a combination of several interrelated variables, such as undercooling, **nucleation rate**, and solid–liquid interfacial velocity. Inspection of the available literature consistently reveals that the microstructure of atomized droplets generally incorporates one or more of the following characteristics:

(a) phase morphologies that are different from those associated with equilibrium solidification [19, 44, 138, 172, 173],
(b) refinement in dendrite arm spacing and grain size [172, 174–176],
(c) formation of non-equilibrium crystalline phases because of competitive nucleation and growth mechanisms [35, 101, 128, 131], and
(d) solute trapping and partitionless solidification [19, 44, 49, 138, 173, 177–180].

These characteristics will be discussed in detail in the text that follows.

3.6.1 Solidification Morphology

The evolution of **solidification morphology** may occur within a single-phase regime as the solidification conditions change. For example, for a given distribution of droplets, the **solidification morphology** is critically dependent on droplet size [57, 181–183]. Samuel [34] studied the microstructure of gas-atomized Al—3Li, and Al-3Li-Co powders in the powder size range of 5–200 μm. It was reported that for Al—3Li in the powder size range of less than 25 μm, the microstructure was primarily a **dendritic** type, whereas, in powder sizes greater than 200 μm, the microstructure was characterized by a mixture of elongated and equiaxed dendrites. The equiaxed dendrite structure was a result of slow solidification conditions. This microstructure was characterized by the absence of directional growth features and was different from the cellular or **dendritic** structures that are commonly observed in castings [34]. Furthermore, Samuel noted that the addition of Co caused a change in **solidification morphology**. For example, when 0.4 wt%Co was added to the alloy, the microstructure of powders that were smaller than 25 μm remained primarily **dendritic**. With increasing powder size, however, the solidification microstructure became **cellular**. Although a well-defined powder size range for the dendritic-to-cellular transition was not observed, the proportion of cellular structure was found to increase with increasing powder size.

Levi and Mehrabian [49] studied the microstructure of EHD-atomized Al—Si powders. In this study, it was reported that in the submicrometer size regime, large powders were characterized by a mixture of cellular and featureless microstructures. The proportion of featureless microstructure increased with decreasing powder size. For extremely fine powders, i.e. in the size range of less than 200 μm, the cellular features disappeared and the microstructures were completely featureless [49]. In general, in the powder size range normally encountered in atomization, as powder size decreases, the **solidification morphology** changes accordingly. The available experimental evidence seems to suggest the following sequential change in **solidification morphology** with refinement in powder size or with increasing solidification rate:

$$\text{Cellular} \Rightarrow \text{Dendritic} \Rightarrow \text{Cellular} \Rightarrow \text{Featureless (Planar)} \tag{3.108}$$

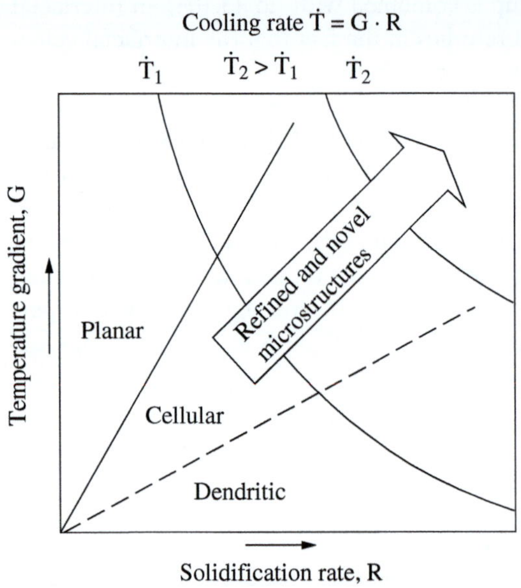

Figure 3.40 Representation of change of solidification morphology as a function of thermal gradient and interfacial velocity. Source: Mehrabian [50] / Reproduced with permission from Taylor & Francis.

It should be noted that the cellular solidification microstructure in the slow solidification rate regime (i.e. in large powders) is seldom observed in atomized powders. Moreover, featureless microstructures are generally observed in very fine powders, mostly in the submicrometer size regime. Therefore, the most frequently observed change in **solidification morphology** in atomized powders may be summarized as follows (with decreasing powder size):

$$\text{Dendritic} \Rightarrow \text{Cellular} \Rightarrow \text{Planar} \tag{3.109}$$

Since the rate of thermal energy extraction and the degree of undercooling both increase with decreasing powder size, the evolution in **solidification morphology** described by Eq. (3.109) should be closely coupled to the magnitude of the interfacial velocity. In addition, it has also been suggested that the thermal gradient ahead of the solid–liquid interface is an important factor that promotes changes to the solidification morphology [98]. Figure 3.40 is frequently used to illustrate the transition in solidification morphology as a function of interfacial velocity and thermal gradient in a velocity regime that falls below the absolute stability limit [43]. The results in this figure show that for constant thermal gradient, increasing the interfacial velocity is likely to cause transition in solidification morphology in the following sequence:

$$\text{Planar} \Rightarrow \text{Cellular} \Rightarrow \text{Dendritic} \tag{3.110}$$

The analysis of microstructure is further complicated by the fact that even powders of identical size may exhibit significant variations in solidification morphology, depending primarily on undercooling [38, 41, 141, 142]. In fact, it is not uncommon to observe several solidification morphologies in a single powder size range [35, 39, 49]. One example is the microstructure of Al—4Cu that is shown in Figure 3.28 from the work of Boettinger [39]. This figure illustrates the dramatic differences in scale and morphology of the microstructures that can be present in the same powder size range. Since for the same powder size and experimental conditions, the differences in cooling rate are unlikely to be significant, the observed differences in microstructure can only be rationalized in terms of the undercooling. In related studies, Boettinger et al. [35] noted the breakdown of interfacial morphology from microcellular to coarse dendritic in an Al–8 wt% Fe powder.

Theoretically, the transition in solidification morphology has been rationalized in terms of the thermal and solutal fields ahead of the interface and the interfacial velocity. The equilibrium composition of a solid that forms from a liquid differs from that of the liquid depending on the phase diagram of interest. As an example, one may consider solidification of the α – phase in a portion of a binary (of elements A and B) phase diagram shown in Figure 3.41a. At a temperature, T, in the solid–liquid two-phase regime, the concentration of solute B in the solid phase is C_s which is less than that in the liquid phase with which it is in equilibrium, C_1. Hence, under equilibrium conditions, the composition of a solid phase and that of the liquid are related by a partition coefficient as

$$C_s = k_o C_1 \tag{3.111}$$

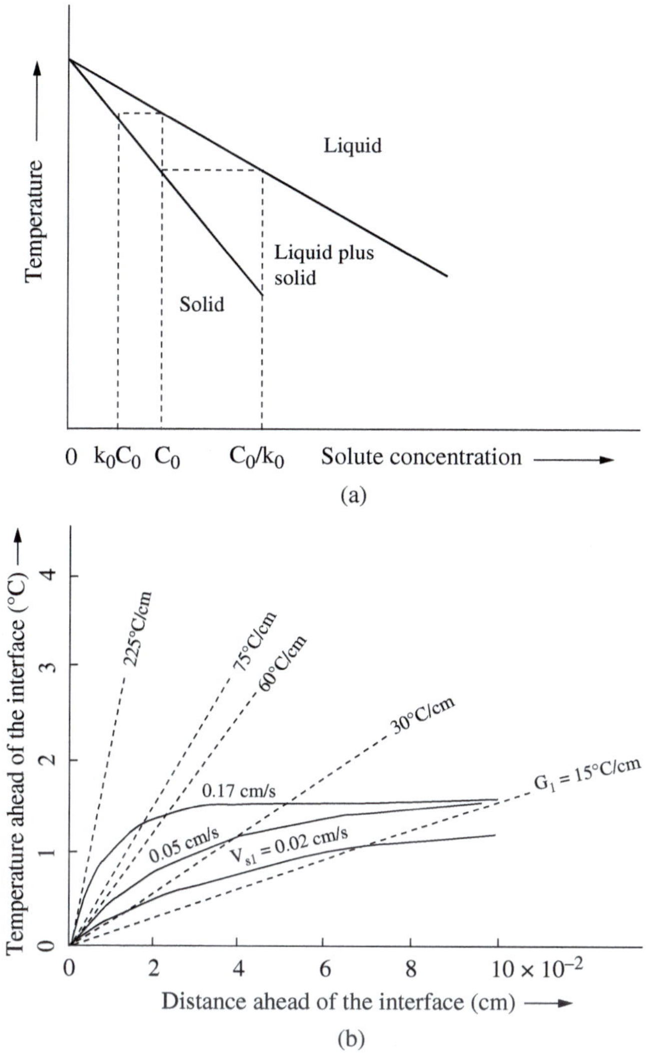

Figure 3.41 (a) a portion of a binary (A − B) phase diagram and (b) the presence of constitutional undercooling. Source: Tiller et al. [184].

where C_s and C_l are the composition of the solid phase and liquid phases at equilibrium, and k_o is the equilibrium partition coefficient. This thermal equilibrium condition requires that the solute atoms be rejected from the solid–liquid interface into the liquid phase during solidification. At very low interfacial velocities, the solute atoms rejected into the liquid phase have ample time to diffuse from the interface into the bulk liquid. Accordingly, the interface may proceed with a planar morphology. At higher interfacial velocities, the competition between interfacial movement and solute diffusion results in the buildup of an interfacial solute-rich layer in the liquid ahead of the solid–liquid interface. Therefore, the liquidus temperature in the liquid ahead of the solid–liquid interface is dictated by the solutal distribution. When the temperature gradient imposed by the environment is less than that of the constitutional temperature profile, the liquid ahead of the interface is effectively undercooled. This condition is shown schematically in Figure 3.41 and was defined as **constitutional undercooling** by Tiller et al. [184]. The **constitutional undercooling** ahead of the interface favors the growth of a solid with a cellular or dendritic morphology. This transition from a planar to a cellular interface generally occurs in the low solidification regime [81, 184]. In the rapid solidification rate regime, due to the high interfacial velocities involved, there is only limited time available so that the lateral segregation of solute can occur only for perturbations of very short wavelength. This effect results in a decrease in cellular spacing with increasing interfacial velocity. In the limit, the solidification velocity is sufficiently high so that the increase in interfacial free energy because of curvature effects overrides the decrease in free energy that is associated with solute segregation.

In this case, interfacial perturbations become energetically unfavorable and the solidification front proceeds with a planar morphology.

In the literature, various theoretical models have been developed to rationalize the observed transition between planar and cellular microstructures at low interfacial velocities [170, 184, 185] and from cellular to planar morphologies at high interfacial velocities [170, 186, 187]. In the low interfacial velocity regime, Tiller et al. [184] considered the effect of **constitutional undercooling** on the transition of solidification morphology. As shown in Figure 3.41, under steady-state conditions, from solute diffusion considerations, the constitutional temperature gradient at the interface, G_{1c}, is obtained by solving the solute diffusion field ahead of the interface [184]:

$$G_{1c} = -\frac{V_{s1} m_1 C_o (1 - k_o)}{k_o D_1} \tag{3.112}$$

where C_o is the alloy composition and *ml* is the slope of the liquidus line. Tiller et al. argued that when the liquid ahead of the interface is constitutionally undercooled, any solid phase that grows into the undercooled region will become energetically preferred. Under this condition, the original planar solid–liquid interface will break down into a cellular or dendritic morphology. The condition of **constitutional undercooling**, i.e., $G_{1s} \geq G_1$, therefore becomes an **interfacial stability criterion**. Hence, a planar interface is stable when [184, 185]

$$\frac{G_1}{V_{s1}} \geq -\frac{m_1 C_o (1 - k_o)}{k_o D_1} \tag{3.113}$$

where G_1 is the thermal gradient which is imposed by the environment. Eq. (3.113) predicts a transition from a planar to a cellular/dendritic solidification morphology as the solid–liquid interface velocity increases. The condition specified by Eq. (3.113) is defined as the **constitutional stability criterion**.

Although Eq. (3.113) successfully predicts the transition from a planar to a cellular/dendritic solidification morphology, it fails to predict the stability of an interface at high interfacial velocities because it only considers the effect of **constitutional undercooling**. At high interfacial velocities, corresponding to a very small dendrite tip radius, capillarity effects become more important and must be taken into consideration. A thorough analysis of interfacial stability during directional solidification was documented by Mullins and Sekerka [170]. One of the most important accomplishments of the Mullins–Sekerka analysis is the prediction of the absolute stability of a planar interface at high interfacial velocities. Under directional **solidification conditions**, the thermal gradient in the liquid phase is always positive and hence the movement of the solid–liquid interface is constrained by the cooling conditions that are imposed on the system. Mullins and Sekerka assumed a solid–liquid interface with superimposed initial perturbations. In this study, they derived a condition for interfacial stability based on the fundamental differential equations that govern the thermal and solutal fields. In the Mullins–Sekerka analysis, the stability of a solid–liquid interface is evaluated by studying the behavior of initial perturbations that are present on a solid–liquid interface. Mullins and Sekerka argued that an interface is stable if the perturbations diminish with time. According to Mullins and Sekerka, a solid–liquid interface becomes unstable when its velocity satisfies the condition:

$$V_c < V_{sl} < V_a \tag{3.114}$$

where V_c and V_a are the critical solid–liquid interfacial velocities for the constitutional stability and absolute stability criteria, respectively. Accordingly, a planar interface is stable at both low and high interfacial velocities. At very high interfacial velocities, capillarity effects become so important that a planar interface will be more stable than a cellular or dendritic interface. This condition is known as the **absolute stability** condition. The **constitutional stability criterion** shown in Eq. (3.114) can be expressed separately as

$$V_{s1} < V_c = \frac{\overline{G} D_1}{\Delta T_o} \tag{3.115}$$

with

$$\overline{G} = \frac{k_s G_s + k_1 G_1}{k_s + k_1} \tag{3.116a}$$

and

$$\Delta T_o = -\frac{m_1 C_o (1 - k_o)}{k_o} \tag{3.116b}$$

where ΔT_o is the freezing temperature range; G_s is the temperature gradient in the solid phase; and k_1 and k_s are thermal gradients in the liquid and solid phases, respectively. Eq. (3.115) reduces to the **constitutional stability criterion** given by Eq. (3.113) when G_s approaches G_1. The absolute stability criterion is given by [170]:

$$V_{s1} > V_a = \frac{D_1 \Delta T_o}{\Gamma k_o} \tag{3.117}$$

where $\Gamma = \Omega\sigma_{s1}/\Delta S_f$ is the Gibbs–Thompson factor.

The limitation of the Mullins–Sekerka analysis is the assumption that the thermal diffusion length is much greater than the wavelength of the perturbations [170]. As a result, the stability criterion applies strictly only to the **constrained growth** of a solid–liquid interface. Hence, in the presence of thermal undercooling, a solid–liquid interface will always be unstable. Experimental observations, however, suggest that a planar interface may be stable in undercooled alloys if the interfacial velocity exceeds a certain critical value [39, 49]. Trivedi and Kurz [171] re-examined the Mullins–Sekerka analysis and extended it to the large thermal Péclet number regime. They showed that a planar interface may also be stable in undercooled liquids. For an undercooled liquid with $G_s = 0$ and $k_s = k_1$, Trivedi and Kurz concluded that the absolute stability criterion should contain two terms, one from the contribution of solutal diffusion and the other from the contribution of thermal diffusion [171]:

$$V_a = \frac{D_1 \Delta T_o}{\Gamma k_o} + \frac{\alpha_1 \Delta T_h}{\Gamma} \tag{3.118}$$

The first term in Eq. (3.118) is the same as the absolute stability criterion derived by Mullins and Sekerka. The second term in this equation denotes the thermal absolute velocity, with $\Delta T_h = \Delta H_f / c_{p1}$ being the **hypercooling** limit. The second term can be rewritten in a more explicit form as

$$\frac{\Gamma c_{p1}}{\Delta H_f} = \frac{\alpha_1}{V_{s1}} \tag{3.119}$$

Eq. (3.119) states that when the capillarity length is greater than the thermal diffusion length, a planar interface is stable. It should be noted that the thermal stability predicted based on Eqs. (3.115)–(3.118) is derived based on the assumption of an equilibrium partition coefficient k_o. At high interfacial velocity conditions, the partition coefficient is likely to be different from its equilibrium value.

3.6.2 Microstrutural Refinement

3.6.2.1 Dendrite Arm Spacing

The scale of the microstructure that is observed following solidification is influenced by several processes: **nucleation**, growth of the solid phase, and coarsening of the solid phase. The factors that govern the scale of the solidification microstructure in atomized droplets have been studied extensively [16, 34, 57, 120, 147, 181, 182, 188–191]. However, since the evolution of microstructure during droplet solidification is inferred primarily from post-solidification microstructural observations, the most direct parameter that is used in solidification microstructure behavior studies is powder size. As discussed in Section 3.6.1, in large powders the morphology of the solidification microstructure is frequently dendritic and remains relatively unchanged with powder size. The scale of the microstructure, however, becomes finer with decreasing powder size. The most convenient parameters that are used to characterize the microstructural scale are secondary SDAS and grain size (GS). The former factor represents a typical segregation distance while the latter parameter denotes a typical dislocation slip distance. Primary dendrite arm spacing (PDAS) is sometimes used to characterize the scale of the microstructure. PDAS is, however, experimentally difficult to determine and is not as representative of segregation distance as the SDAS. The SDAS generally decreases with decreasing powder size. As an example, Figure 3.42 shows the powder size dependence of SDAS in some Ni-based superalloys [189]. Based on the results obtained on René 95 superalloy powders, Domingue et al. [181] proposed a power law empirical relationship between SDAS and powder size as

$$\text{SDAS} = k_p d^m \tag{3.120}$$

where k_p and m are experimentally determined constants. The constants a and m determined for argon-atomized René 95 were 0.139 and 0.484, respectively, and those for centrifugally atomized powders of identical composition were 0.209 and 0.435, respectively [181].

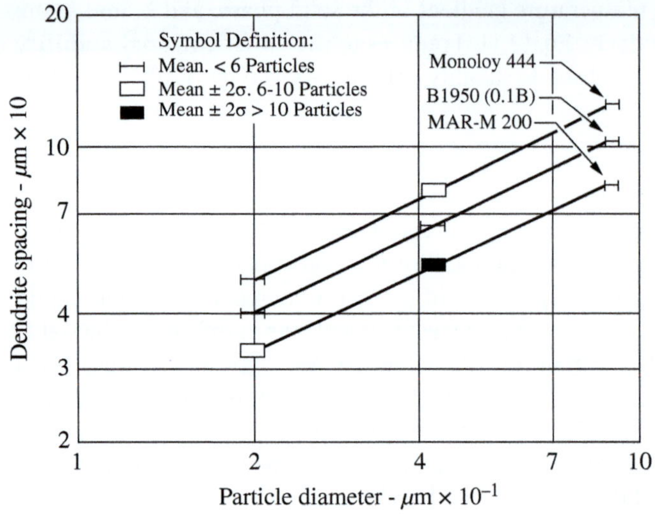

Figure 3.42 Powder size dependence of SDAS in some Ni-based superalloys. Source: Patterson et al. [189].

Although the power law relationship reported between SDAS and powder size is universally applicable, the values of the experimental constants k_p and m critically depend on atomization conditions and alloy composition. The results shown in Figure 3.42 clearly reveal the influence of chemical composition on SDAS. The three alloys shown in Figure 3.42 exhibit similar composition, except for the carbon content. Patterson et al. [189] concluded that SDAS is inversely proportional to carbon content, since for the three alloys shown in Figure 3.42 the carbon content decreases in the order of MAR-M 200, B1950, and Monoloy 444. In contrast, the results of Samuel [34] suggest that the addition of Zr to Al—Li alloys leads to an increase in SDAS corresponding to each powder size studied. Another example has been shown in Figure 3.33, for SDAS measurements in binary Al—Si alloys. The experimental constants k_p and m in Eq. (3.120), therefore, differ from material to material.

The reported relationship between **dendrite arm spacing** and powder size demonstrates the importance of powder size on segregation; however, such a relationship offers little insight into the mechanisms of microstructure refinement. For example, it is difficult to explain why for the same powder size range there are significant differences in SDAS for different alloy compositions. Therefore, investigations on microstructure refinement often seek to relate SDAS to available solidification variables [57, 120, 182, 188]. The three most frequently used variables are average **cooling rate**, solidification time, and interfacial velocity. A frequently used approach is to plot SDAS as a function of the average **cooling rate**. This approach is derived from the study of the relationships between microstructure and **cooling rate** under conventional **solidification conditions**. The determination of such an SDAS–**cooling rate** relationship requires reliable experimental measurement of the **cooling rate**. Figure 3.43 shows the dependence of SDAS on the **cooling rate** for Al–3.4–3.6 wt%Cu as summarized by Jones [182]. From the data shown in this figure, SDAS clearly decreases with an increasing **cooling rate**. The dependence of SDAS on the **cooling rate** can be described by a power law relationship. A general form of this empirical relationship is

$$\mathrm{SDAS} = k_c \dot{T}^{-n} \tag{3.121}$$

where k_c and n are experimental constants. Similar power law relationships have been reported by various authors for different materials, with differences in the magnitude of the experimental constants [57].

The correlation of SDAS with the average **cooling rate** using a power law equation does not provide any physical insight into the process of rapid solidification that generally occurs during atomization. Under constrained growth conditions, such as those that are present in directional solidification, dendrites adjust their spacing to minimize the **constitutional undercooling** ahead of the solid–liquid interface. If the spacing between the dendrites is too large to satisfy the **constitutional undercooling** condition, new dendrites will branch out to reduce the **dendrite arm spacing**. If the dendrite arm spacing is too small, however, some of the slow-growing dendrites will be engulfed by their neighbors. This self-adjusting tendency of dendrites has been confirmed by direct experimental observations in transparent materials [192]. From a **constitutional undercooling** standing point it is deduced that SDAS or PDAS

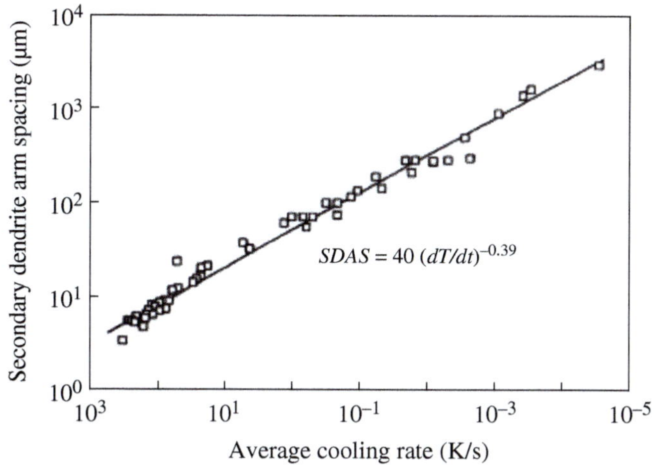

Figure 3.43 Relationships between average cooling rate and secondary dendrite arm spacing. Source: Reproduced by permission of Claitor's Publishing Division [182].

should be a function of the product of the thermal gradient and the interfacial velocity, G_1V_{s1} [81]. This product is the average cooling rate because

$$\dot{T} = \frac{\partial T}{\partial x}\frac{\partial x}{\partial t} = G_1V_{s1} \tag{3.122}$$

where x is the coordinate parallel to the interface growth direction.

Alternatively, SDAS is sometimes expressed as a function of local solidification time at the solid–liquid interface. The local solidification time is related to the average cooling rate by [57]

$$t_f = \frac{\Delta T_0}{\dot{T}} \tag{3.123}$$

Combining Eq. (3.123) with (3.121) yields the relationship between SDAS and local solidification time:

$$\text{SDAS} = k_s t_f^n \tag{3.124}$$

where k_s is an experimentally determined constant, n has the same meaning as that used in Eq. (3.121), and t_f is the local solidification time. While the physical meaning of the relationship between SDAS and solidification time is not obvious, it is indicative of the self-adjustment of dendrites to **constitutional undercooling** and of the effect of dendrite arm coarsening in the mushy zone (semiliquid, semisolid).

Experimentally, SDAS is found to be strongly coupled with the solid–liquid interfacial velocity. To that effect, there are numerous studies available, particularly on eutectic systems. In eutectic systems, the eutectic spacing (ES), i.e. the spacing between alternative phases, is taken as a parameter that is indicative of the scale of the microstructure. Figure 3.44 shows the eutectic spacing of Al–33 wt%Cu from various investigations as summarized by Jones [182]. The empirical relationship deduced from these data suggests [182]

$$(\text{ES})V_{s1}^{1/2} = k_e \tag{3.125}$$

where ES is the eutectic spacing, and k_e is an experimentally determined constant. The value of k_e as determined by Jones for Al–33 wt%Cu is 9.4 μm$^{3/2}$ s$^{1/2}$ [182]. A similar dependence of eutectic spacing on interfacial velocity was also observed in an Al—Al_6 Mn eutectic system where the constant k_e was empirically determined to have a value of 34.0 μm$^{3/2}$ s$^{1/2}$ [193]. Accordingly, the relationship between ES and solid–liquid interfacial velocity expressed by Eq. (3.125) appears to be universally applicable to eutectic systems [193–195].

Several theoretical models have been developed in an effort to correlate the scale of the microstructure with the interfacial velocity. The microstructural scale in such models is generally described as the dendrite tip radius [135, 168, 187, 196, 197]. For dendrite growth, Langer and Küller-Krumbhaar [198, 199] proposed that dendrite tips can grow closer to a marginally stable state. Applying the Mullins–Sekerka stability analysis to dendrite tips, Langer and Küller-Krumbhar [198] obtained a criterion for the dendrite tip radius as

$$R_d = \lambda_s \tag{3.126}$$

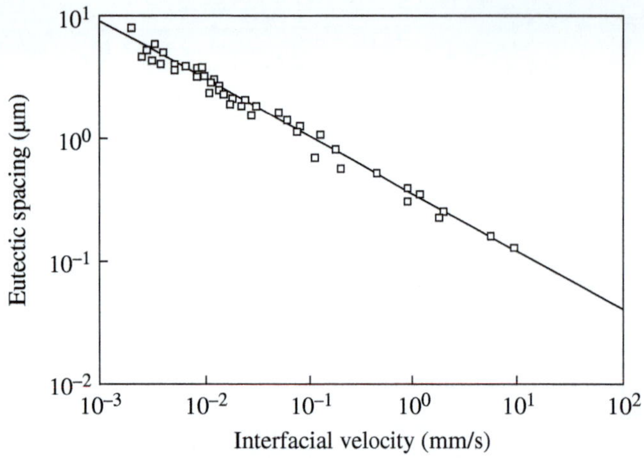

Figure 3.44 Eutectic spacing, ES, as a function of solid–liquid interfacial velocity for Al–33 wt%Cu. The solid line represents (ES) $V_{s1}^{1/2} = k_e$ with $k_e = 9.4\ \mu m^{3/2}\ s^{1/2}$. Source: Reproduced by permission of Claitor's Publishing Division [182].

where R_d is the radius of dendrites growing under steady-state conditions, and λ_s is the critical wavelength of the solid–liquid interface at the limit of stability. Using the above criterion and the Mullins–Sekerka analysis, Kurz et al. [135, 198] obtained a general expression for the radius of a dendrite as

$$\Xi V_{s1}^2 - BV_{s1} + G_1 = 0 \tag{3.127}$$

where

$$\Xi = \frac{\pi^2 \Gamma}{P_c^2 D_1^2} \tag{3.128}$$

and

$$B = \frac{m_1 C_o (1-k)\xi_c}{D_1[1-(1-k)Iv(P_c)]} \tag{3.129}$$

Kurz et al. [135] examined the change in $R_d^2 V_{s1}$ as a function of interfacial velocity by plotting $R_d^2 V_{s1}$ versus V_{s1}. The results of this analysis are shown in Figure 3.45 [135]. It is evident that for low Péclet numbers, i.e. at relatively low

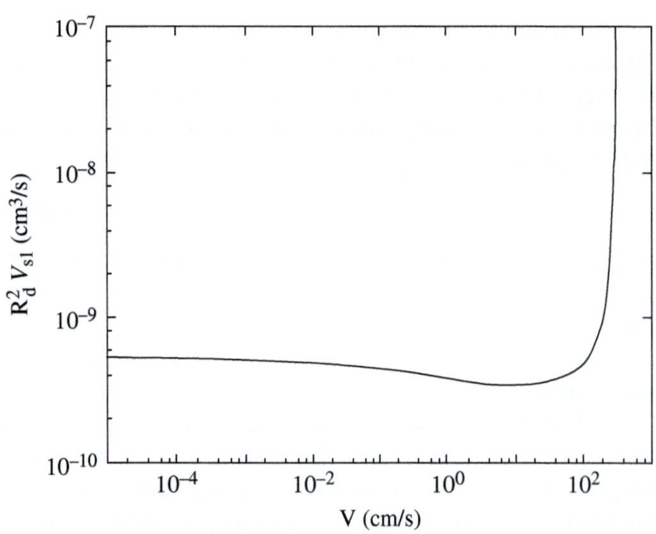

Figure 3.45 Magnitude of $R_d^2 V_{s1}$ from Eq. (3.140) as a function of interfacial velocity for Ag–5 wt%Cu alloy with zero temperature gradient [135]. Source: Kurz et al. [135].

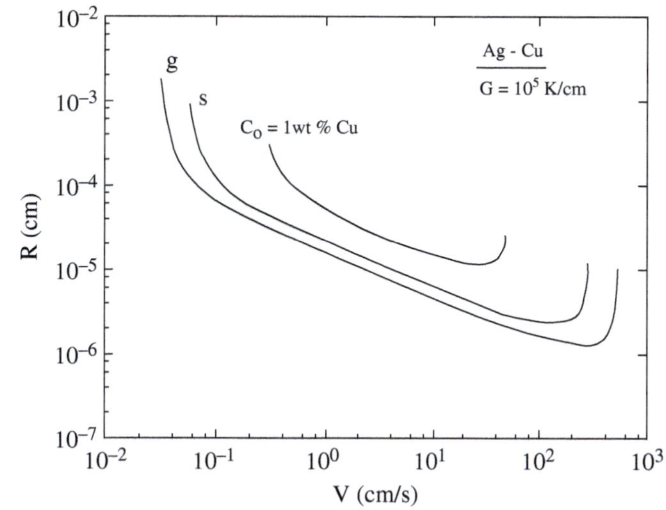

Figure 3.46 Dendrite tip radius, R_d, as function of growth rate, V_{sl}, and initial composition, C_o, with constant k_o and m for several Ag–Cu alloys at $G_1 = 10^5$ Kcm^{-1}. Source: Kurz et al. [135].

interfacial velocity conditions ($V_{sl} < 10^2$ cm s^{-1}), $R_d^2 V_{s1}$ remains essentially constant. Analytically, under the condition of $G_1 = 0$, Eqs. (3.127)–(3.129) may be simplified into [135]

$$R_d^2 V_{s1} = \frac{4\pi^2 D_1 \Gamma[1-(1-k)Iv(P_c)]}{C_o m_1 (k-1)\xi_c} \tag{3.130}$$

Moreover, for small Péclet number conditions, $\xi_c \to 1$ and $Iv(P_c) \to 0$, Eq. (3.130) may be simplified as

$$R_d^2 V_{s1} = k_d \tag{3.131}$$

where k_d is a constant. This relationship obtained by Kurz et al. [135] is similar to the empirical relationships determined between **dendrite arm spacing** and interfacial velocity. Figure 3.46 shows the dendrite tip radius for Ag—Cu alloys as a function of interfacial velocity as predicted based on Eqs. (3.127)–(3.129) [135]. A power law relationship is present between dendrite radius and interfacial velocity only at intermediate interfacial velocities. At very low and very high interfacial velocities, the dendrite radius approaches infinity, corresponding to the stability limits of a planar interface. It should be noted, however, that the relationship expressed by Eq. (3.131) is between dendrite radius and interfacial velocity. Moreover, the experimentally determined **dendrite arm spacings**, although intimately linked to the dendrite radius, are significantly affected by coarsening during solidification and during post-solidification cooling.

3.6.2.2 Grain Size

The presence of rapid solidification inevitably leads to considerable grain refinement. In splat cooling, for example, grain sizes on the order of 0.01 μm have been reported [200]. The grain size that is formed in rapidly solidified materials is generally reported to be primarily dependent on **cooling rate** and undercooling. Vogt and Frommeyer [201] and Cantor [202], to cite two examples, studied the effect of **cooling rate** during splat quenching on grain size refinement in Fe–6.3 wt%Si, Ni–3 ~ 8 wt%Al and 316 stainless steel. Figure 3.47 shows the dependence of grain size on solid-state cooling rate as reported by Vogt and Frommeyer [201]. In this study, solid-state cooling rate was used to characterize the thermal conditions that are present during solidification. The results clearly show that a grain size refinement was achieved by increasing the solid-state cooling rate. A similar trend of grain refinement with the increasing **cooling rate** was also observed by Cochrane et al. [120] for Ni-based alloys solidified using the drop tube technique.

The dependence of the grain size on undercooling is generally studied using the drop tube or levitation methods in which undercooling may be readily quantified [120, 138]. Experimental studies of the influence of undercooling on grain size generally reveal an abrupt grain refinement when the degree of undercooling experienced by the liquid exceeds a certain critical value [120, 138, 203–205]. Accordingly, at undercoolings below this critical value, the microstructure is characterized by a coarse-grained dendritic morphology. Moreover, it is consistently reported that

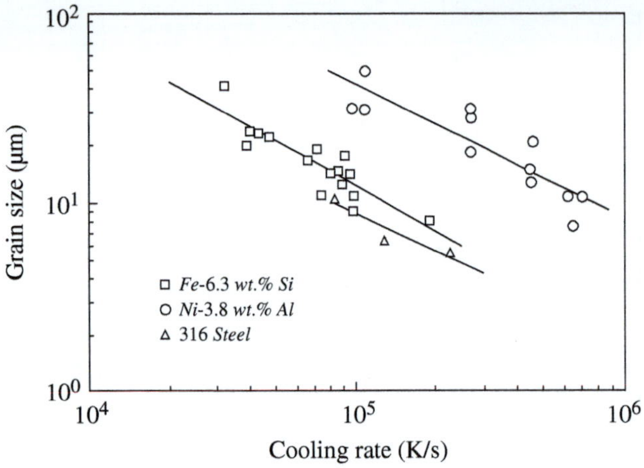

Figure 3.47 Measured grain size as a function of cooling rate. Source: Vogt and Frommeyer [201].

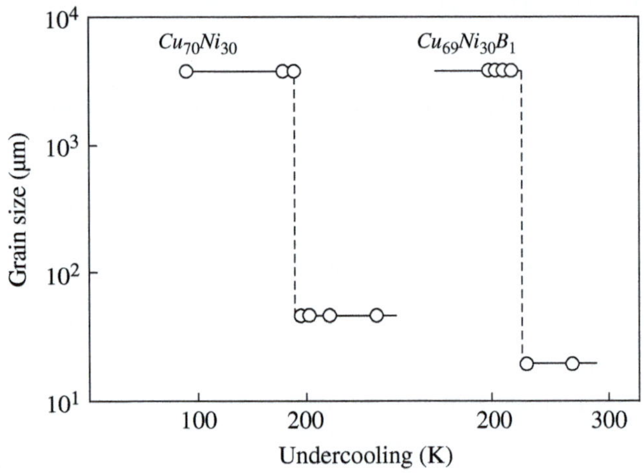

Figure 3.48 Grain size in $Cu_{70}Ni_{30}$ and $Cu_{69}Ni_{30}B_1$ powders as a function of undercooling. Source: Reproduced by permission of the American Institute of Physics [138].

the grain size decreases with increasing undercooling [205]. Under conditions of **critical undercooling**, there is a sudden decrease in grain size of up to two orders of magnitude [120, 138, 203, 204]. Such a sudden change in grain size with undercooling is illustrated in Figure 3.48 [138]. Beyond the **critical undercooling**, grain size decreases with a further increase in undercooling; the refinement, however, is gradual.

Several mechanisms have been proposed to explain the observed grain size refinement with increasing undercooling [120, 138, 203–205]. These mechanisms involve

(a) **copious nucleation**,
(b) **grain growth**,
(c) **dendrite arm fragmentation**, and
(d) **solid-state coarsening**.

In the **copious nucleation** mechanism, grain size refinement is the result of a sudden increase in **nucleation rate** due to the presence of homogeneous nucleation. This condition occurs only when the latent heat released by the solid–liquid front does not hinder **homogeneous nucleation** ahead of the growth front. Accordingly, the **copious nucleation** mechanism becomes possible when the rate of heat extraction exceeds that of the energy released during **recalescence**. When **recalescence** is insignificant, solidification occurs isothermally in the undercooled liquid. In this case, the resulting grain size will depend on the competition between **nucleation** and **grain growth** kinetics. A model for grain size evolution under these conditions has been proposed and may be expressed as [206]

$$d_G = \left(\frac{V_{s1}}{I}\right)^{1/4} \tag{3.132}$$

where d_G is grain size. When solidification occurs against a surface, such as in splat quenching, the rate of heterogeneous nucleation on the quenching surface may dictate the final grain size. Under this condition, and when **recalescence** is ignored, the grain size may be estimated using a similar formulation as that expressed by Eq. (3.132) [207] as

$$d_G = \left(\frac{8V_{s1}}{\pi I_s}\right)^{1/3} \tag{3.133}$$

where I_s the surface heterogeneous nucleation rate with the units of $m^{-2}\ s^{-1}$.

Dendrite arm fragmentation and coarsening (recrystallization) may be responsible for the final grain size. In related work, Willnecker et al. [138] studied grain refinement behavior in undercooled $Cu_{70}\ Ni_{30}$ and $Cu_{69}\ Ni_{30}\ B_1$ alloys. The results of Willnecker et al. demonstrate that at low undercoolings the relationship between interfacial velocity and undercooling can be described using a power law relationship. Above a **critical undercooling**, the interfacial velocity deviates significantly from a power law relationship. Interestingly, the abrupt grain refinement that is often observed in undercooled alloys occurs at the same temperature at which the **solid–liquid interfacial velocity** deviates from a power law relationship. Willnecker et al. proposed that the deviation of interfacial velocity from a power law relationship is accompanied by an increase in solute trapping by the growing dendrites. In this case, solute enrichment at the dendrite root facilitates dendrite remelting during **recalescence**. Hence, the dendrite arms that break free from the solid–liquid interface provide heterogeneous nucleation centers and contribute to a decrease in grain size [138]. Experimental support to this mechanism is provided by the grain refinement observed by Cochrane et al. [120]. The results of Cochrane et al., obtained with undercooled $Fe_{65}Ni_{35}$, revealed a similar abrupt transition in grain refinement with undercooling. The fine grains observed in this study exhibited a distinct directionality as they were observed to radiate from a nucleation center (see Figure 3.22a). Since dendrite arm fragments provide nuclei ahead of the growing front, such a directionality is expected. On the contrary, if a **copious nucleation** mechanism operates, the microstructure should exhibit a random orientation.

The importance of recrystallization in determining grain size was suggested by Cochrane et al. [120] and Greer [207]. Greer proposed that when the rate of crystal growth controls, the grain size should be inversely proportional to the cooling rate $1/\dot{T}$. Such a relationship is given by

$$d_G \propto \dot{T}^{-1} \tag{3.134}$$

Cochrane et al. [120] plotted the experimentally determined grain size in Q170 Ni_{30} as a function of the cooling rate. As shown in Figure 3.49, Cochrane et al. found that when plotted using a double logarithmic scale, the square grain size versus inverse cooling rate curve exhibits two distinct linear regimes. The results in this figure reveal a deviation from a linear behavior at a cooling rate of approximately $\dot{T} = 10^4 K\ s^{-1}$. Cochrane et al. concluded that for $\dot{T} > 10^4 K\ s^{-1}$ recrystallization (coarsening) is suppressed, and hence the linear relationship between the grain size and cooling rate is retained. When $\dot{T} < 10^4 Ks^{-1}$, however, the final grain size is controlled by recrystallization. Theoretically, if the final grain size is governed by post-solidification recrystallization, grain size should be proportional to the square root of the cooling rate as [208]

$$d_G \propto \dot{T}^{-0.5} \tag{3.135}$$

Cochrane et al. obtained a similar empirical relationship to that expressed by Eq. (3.135) when $\dot{T} < 10^4 K\ s^{-1}$ [120]

$$d_G \propto \dot{T}^{-0.59} \tag{3.136}$$

Therefore, the available results suggest that for low cooling rate conditions, coarsening during post-solidification recrystallization is perhaps the primary mechanism that dictates the final grain size.

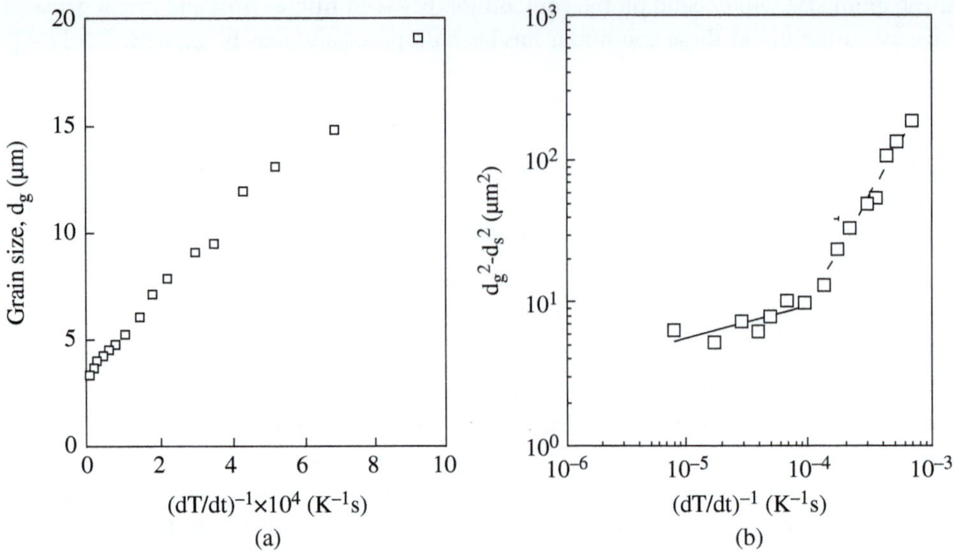

Figure 3.49 The variation of (a) d_g and (b) $d_g^2 - d_s^2$ as a function of $(dT/dt)^{-1}$ for droplets of $Cu_{70}Ni_{30}$. Source: Cochrane et al. [120].

3.6.3 Phase Selection

The microstructure that is present following the completion of solidification is the result of various competitive kinetic processes. As result, the **phases** present may be equilibrium or non-equilibrium depending on solidification kinetics and the operative mechanisms of phase selection [19, 35–40, 49, 98, 99, 131, 133, 136]. The presence of various phases and their relative proportion are dependent on two primary processes: competition during nucleation and competition during growth. Nucleation kinetics determine the possibility of the occurrence of a potential phase, whereas growth kinetics primarily govern the relative proportion of the various phases. The following microstructural features are often used to characterize phase selection during solidification:

(a) phase composition [37, 131, 141, 142, 209, 210],
(b) phase fraction [37, 98, 133, 141], and
(c) **phase morphology** [35, 37, 38, 49].

In the present context, phase composition denotes the various phases that comprise the solidification microstructures.

Phase composition is a result of the competition between the relevant stable and **metastable phases**. For example, in the microstructure of atomized powders, amorphous, stable, and metastable crystalline phases are frequently present simultaneously. In crystalline materials there are many examples of phase competition. Some examples of crystalline systems which involve phase competition include Al—Mn [211, 212], Al—Fe [35, 37], Fe—Ni [131], and Ti—Al [38, 141, 209]. Graves et al. [141], for example, studied phase selection in TiAl droplets produced during rotating electrode atomization. The composition of the TiAl alloy used in this study was Ti–50 at%Al which falls in the single-phase TiAl intermetallic regime in the equilibrium phase diagram. Graves et al. however, observed two major phases in the atomized microstructure: the TiAl equilibrium phase and the metastable $\alpha_2(Ti_3Al)$ phase. The proportion of phases present in the microstructure was noted to depend on the powder size (or cooling rate). Whereas the equilibrium TiAl phase was predominant in coarse powders, the metastable $\alpha_2(Ti_3Al)$ was predominant in fine droplets (e.g. less than 44 μm). The fraction of metastable $\alpha_2(Ti_3Al)$ phase was noted to increase with increasing cooling rate or decreasing droplet size.

Thoma and Perepezko [131] studied the microstructural transitions that occur in undercooled Fe—Ni droplets using a drop tube technique. In this alloy system the b.c.c. **metastable phase** (δ) competes with the f.c.c. equilibrium phase (γ) in the presence of undercooling. In their study, Thoma and Perepezko [131] deliberately controlled the environmental conditions by filling the drop tube chamber with either He or a He–2 wt%H_2 gas mixture. Figure 3.50 summarizes the fraction of metastable δ phase present as a function of droplet size. It is evident that the formation

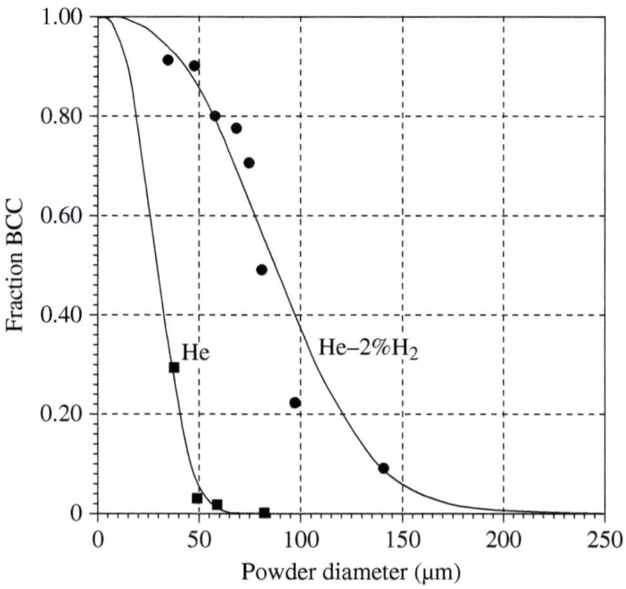

Figure 3.50 Fraction of the metastable b.c.c. phase in Fe–15 wt%Ni powders as a function of powder size. Powders were processed in He and He–2 vol%H_2 gases. Source: Thoma et al. [131] / Reproduced with permission from Springer Nature.

of the δ phase is favored in fine droplets or alternatively under high cooling rates or high undercooling conditions. A significant difference in the fraction of δ phase was present in droplets solidified under a pure He environment relative to that present in droplets solidified under a He–2 wt%H_2 gas mixture was noted. Accordingly, the fraction of the δ phase is much higher in droplets solidified under a He–2 wt%H_2 environment relative to that present in droplets solidified under He. Since there is no significant difference in cooling conditions between He and He–2 wt%H_2, the observed difference in the fraction of the δ phase was attributed to changes in surface nucleation kinetics. To that effect, Thoma and Perepezko [131] argued that the presence of H_2 in the He–2 wt%H_2 gas mixture increased the degree of undercooling experienced by the droplets by reducing any oxide phases present on the droplet's surface. Accordingly, nucleation kinetics were thought to be responsible for the observed phase selection.

The influence of nucleation on phase selection may be rationalized by invoking thermodynamic and kinetic principles. This may be illustrated using simple free energy and kinetic diagrams. As an example, Figure 3.51 shows

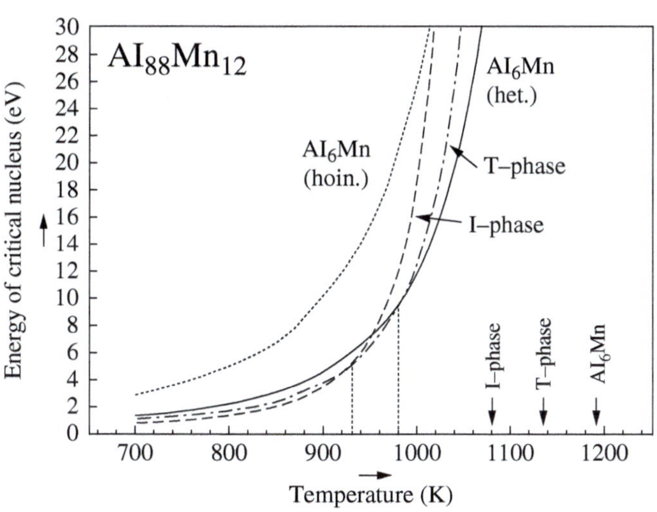

Figure 3.51 Critical free energy of nucleation for Al_6Mn, T-phase, and I-phase as a function of temperature in an $Al_{88}Mn_{12}$ alloy melt. Source: Reproduced by permission of Elsevier Science from F. Gillessen and D. M. Herlach (1991) *Mater. Sci. Eng.*, **A134**, 1220–1224 [211].

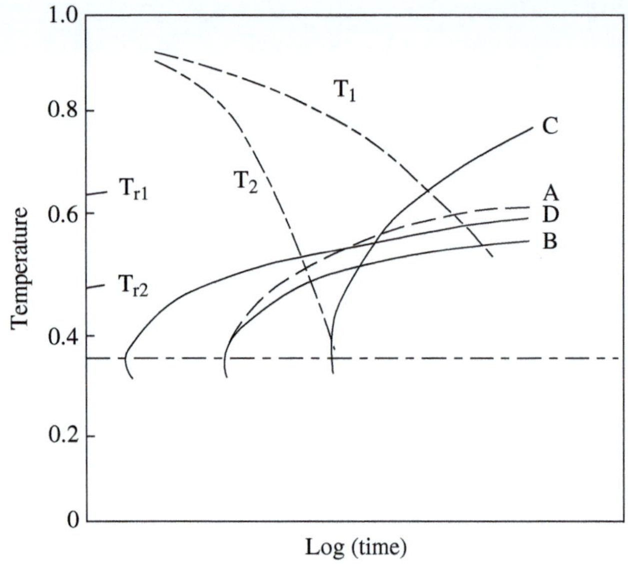

Figure 3.52 Time–temperature transition diagram representing different nucleation kinetics that may occur during continuous cooling of undercooled liquids; A–D represent hypothetical phases. Source: Boettinger and Perepezko [36] / Reproduced with permission of The American Institute of Mining, Metallurgical and Petroleum Engineers.

the critical free energy of nucleation, ΔG^*, for several phases, i.e. equilibrium Al_6Mn, metastable T-phase, I-phase, and undercooled liquid in an Al–12 at% Mn alloy [211]. In the case of Al_6Mn, the magnitude of the critical free energy was calculated assuming both homogeneous and heterogeneous [$f(\theta) = 0.45$] nucleation. According to the data shown in this figure, **homogeneous nucleation** of the T-phase and I-phase are energetically favorable at high undercoolings, whereas heterogeneous nucleation of Al_6Mn is favorable at low undercoolings, i.e. in the temperature range of $T_1 > T > 980$ K. This thermodynamic condition, however, does not guarantee the nucleation of these **metastable phases**. Kinetically, to **nucleate** a low-temperature phase, the liquid has to be quenched to the proper undercooling temperature, thereby passing through a regime where other high-temperature phases are stable. Accordingly, if the cooling rate is not sufficiently high to avoid nucleation of the high-temperature phases, the high-temperature phases will **nucleate** first. The kinetics associated with phase selection can be illustrated by using a temperature–time transition diagram (TTT diagram) for nucleation. One example of such a diagram is shown in Figure 3.52 [36, 213]. In this diagram, the nucleation kinetics of hypothetical phases A, B, C, and D are presented in terms of reduced nucleation temperature T_r versus the logarithm of time. Although the nucleation temperature of phase C is higher, the nucleation kinetics associated with this phase are relatively sluggish. Hence, the nucleation of this phase can be bypassed if the cooling rate is higher than a certain critical value as indicated by the cooling curve $\dot{T}_2$. Moreover, under certain cooling conditions, as defined by the cooling curve $\dot{T}_1$ several phases may **nucleate** simultaneously in the liquid.

Cochrane et al. [37] studied phase competition kinetics in a more complex Al–8 wt%Fe system. The droplet size range studied varied between 50 and 2000 μm. In this study, a systematic change in phase composition and fraction droplet size was observed.

The pie graphs shown in Figure 3.53 summarize the phase composition that was observed in each droplet size range studied. The fraction of each phase is indicated by the area that the phase occupies in the pie graph. The phases that are formed in atomized Al–8 wt%Fe include

(a) primary $Al_{13}Fe_4$,
(b) $Al - Al_{13}Fe_4$ lamella eutectic structure,
(c) $Al - Al_6Fe$ metastable cellular and rod-like eutectic, and
(d) coarse and microcellular $\alpha - Al$.

It is clear from Figure 3.53 that each phase predominates a particular droplet size or cooling rate range. For example, the equilibrium $Al_{13}Fe_4$ and the metastable Al–Al_6Fe eutectic phases appear primarily in coarse

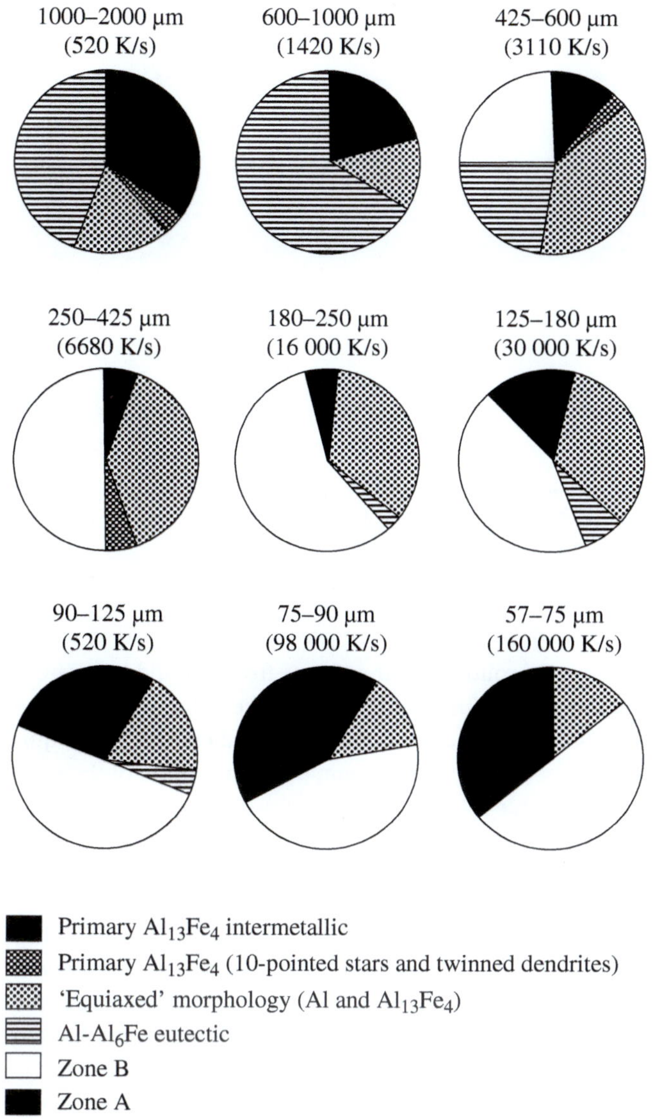

Figure 3.53 Fraction of droplet population in each size range which commenced solidification with a particular structure. The data shown in the brackets are the mean cooling rate calculated between the liquidus and the eutectic temperature. Source: Reproduced by permission of Elsevier Science from R. F. Cochrane, P. V. Evans, and A. L. Greer (1991) *Mater. Sci. Eng.*, **133A**, 803–806 [37].

droplets. The extended solid solution of α-Al is preferentially observed in fine droplets. Based on phase selection kinetics in the Al–8 wt%Fe system, Cochrane et al. [37] proposed that phase competition during growth was responsible for the observed phase selection. In their analysis, Cochrane et al. assumed that all the phases observed in the microstructure may **nucleate** and that the fastest-growing phases will overwhelm the solidification microstructure. They calculated the growth kinetics of the various phases observed in Al–8 wt%Fe droplets using the dendrite and eutectic growth models developed by Trivedi et al. [196, 214]. The results of their calculations are shown in Figure 3.54 in terms of the growth velocity of the various phases as a function of liquid temperature (or undercooling) [37]. The results of Cochrane et al. demonstrate that at high temperatures the equilibrium $Al_{13}Fe_4$ grows at the fastest rate. Hence, the phase growing with the fastest velocity was noted to change with decreasing temperature in the following sequence: Al–$Al_{13}Fe_4$ eutectic phase, Al–Al_6Fe eutectic phase, and α-Al phase. This sequence is identical to the observed transition of phase composition with decreasing droplet size [37].

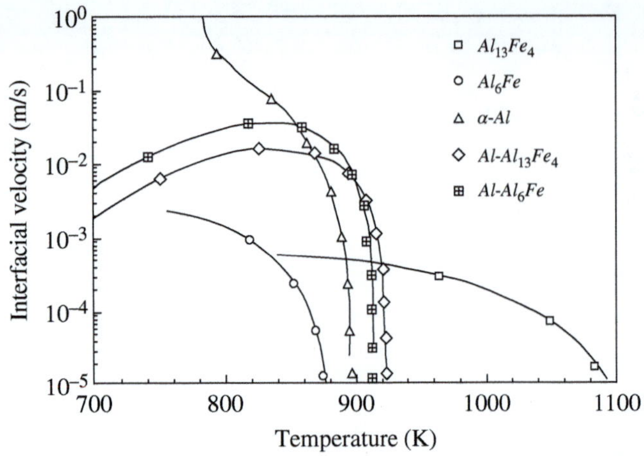

Figure 3.54 The calculated growth velocities of various possible structures in undercooled liquid Al–8 wt%Fe as a function of bulk liquid temperature. Source: Reproduced by permission of Elsevier Science from R. F. Cochrane, P. V Evans and A. L. Greer (1991) *Mater. Sci. Eng.*, **133A**, 803–806 [37].

3.6.4 Solute Redistribution

Under equilibrium **solidification conditions**, the composition of a solid phase that forms from a liquid is different from the liquid composition. This difference is the result of the equilibrium partitioning of the **solute** between the solid and the liquid phases. The relative composition of the solid phase and the liquid phase is dictated by the equilibrium phase diagram. When equilibrium is always maintained in a solidifying system, the composition of the solid phase that forms in the various stages of **solidification** obeys a simple relationship:

$$C_s = \frac{k_0 C_0}{f_s(k_0 - 1) + 1} \tag{3.137}$$

where f_s is the fraction of the solid, and C_s and C_0 are the compositions of the solid and that of the original alloy. Complete diffusion in the liquid and solid phases is required to maintain this equilibrium condition. It is evident that at the beginning of **solidification**, i.e. when $f_s = 0$, the composition of the solid is $C_s = k_0 C_0$ whereas, at the completion of solidification, i.e. $f_s = 0$ the composition of the solid is C_0. Although such equilibrium solidification should result in no solute segregation, this condition is rarely achieved under practical solidification conditions.

Under normal solidification conditions, diffusion in the liquid and solid phases is often limited, and as a result, equilibrium is maintained only at the local solid–liquid interface. In this case, partitioning of the solute leads to solute enrichment in the material which solidifies last. When the segregation distance is significant, for example, in ingot cast materials, a post-solidification homogenization treatment is necessary to homogenize the distribution of the alloying elements. Solute redistribution during conventional solidification conditions has been the subject of numerous investigations [215–217]. As a result, there are a variety of formulations available that may be used to calculate solute partitioning among phases during solidification. In related studies, Brody and Flemings [217] analyzed the solute distribution during conventional solidification by assuming interfacial equilibrium, limited solute diffusion in the solid phase, and complete diffusion in the liquid. A mass balance formulation under these assumptions yields [217]

$$\left(C_1 - C_s^i\right) L \; d\, f_s = L(1 - f_s)dC_1 + \frac{D_s}{V_{s1}} dC_s^i \tag{3.138}$$

where C_1 is the composition of the liquid phase; C_s^i is the composition of the solid at the solid–liquid interface; and L is the diffusion path length. The interfacial velocity can be approximated by the ratio between diffusion path length and local solidification time, i.e. $V_{sl} = L/t_f$. For example, in dendritic solidification, if the **dendrite arm spacing** is assumed to be twice the diffusion path length, i.e. $L = 1/2\text{DAS}$, then Eq. (3.138) reduces to [217]

$$C_s^i \left(\frac{1 - k_0}{k_0}\right) d\, f_s = \left(\frac{1 - f_s}{k}\right) dC_s^i + \delta dC_s^i \tag{3.139a}$$

with

$$\delta = \frac{4D_s t_f}{\text{DAS}^2} \tag{3.139b}$$

Integration of Eq. (3.139a,b) leads to a formulation of solute distribution in the solid phase as [217]

$$C_s^i = kC_0\left(1 - \frac{f_s}{1 + \delta k_0}\right)^{k-1} \tag{3.140}$$

The non-equilibrium solidification conditions that are present during atomization often preclude the assumption of local solid–liquid equilibrium at the interface. In this case, local solute partitioning at the solid–liquid interface is not solely governed by thermodynamic conditions. Instead, solute partitioning may be limited by diffusion and interfacial growth kinetics. The partition coefficient is, therefore, expected to be greater than its equilibrium value, k_0. Under this condition and when solute diffusion in the solid is limited, the solid solubility of elements may be extended over their equilibrium values as determined by the equilibrium phase diagram. The phenomenon of solid solubility extension has been studied extensively throughout rapid solidification literature [19, 39, 40, 49, 99, 139, 142, 218, 219]. Based on theoretical and experimental studies, three mechanisms have been proposed in an attempt to rationalize the reported solubility extension, including

(a) featureless growth above the absolute stability criteria,
(b) **diffusionless solidification** (or alternatively, **massive solidification**), and
(c) solute trapping.

The first two mechanisms, i.e. **featureless solidification** and **diffusionless solidification**, are responsible for producing segregation-free microstructures. When conditions are such that the **solid–liquid interfacial velocity** is above that predicted by the absolute stability criterion, even lateral segregation is suppressed. Accordingly, the microstructure that forms is segregation free. The condition of absolute stability is defined based on Eqs. (3.117)–(3.119). Experimentally, such solidification microstructures have been frequently observed in ultrafine droplets [19, 39, 49] and in other materials solidified under extremely high cooling rates [39, 83, 220]. In **droplets**, segregation-free microstructures may be present throughout entire ultrafine droplets [49] or locally near the position where solidification is presumed to have been initiated [19, 39]. In related work, Boettinger and Coriell [83] studied solute partitioning behavior in Al–1 wt%Cu and Al–5 wt%Cu alloys quenched using an electron beam surface melting technique. They observed a segregation-free planar microstructure in both alloys. In Al–1 wt%Cu the cellular-to-planar transition occurred in the interfacial velocity range of 0.1–0.2 ms^{-1}, whereas, in Al–5 wt%Cu a similar transition occurred at higher interfacial velocities, i.e. between 0.5–0.6 ms^{-1}. The results obtained from a statistical analysis of the transition velocities between cellular and segregation-free microstructures are shown in Figure 3.55 and are compared with those of theoretical predictions [83]. According to Figure 3.55, the theoretically calculated interfacial velocities corresponding to absolute stability appears to be greater than the experimentally observed values.

The **diffusionless solidification** (or **massive solidification**) mechanism is thought to be present when the interfacial velocity is much greater than the diffusion length, i.e. when

$$V_{1s} \geq V_t = \frac{D}{a_0} \tag{3.141}$$

where V_t is the critical velocity for **massive solidification**. Since both featureless and **diffusionless solidification** may be responsible for segregation-free solidification, the critical interfacial velocity for segregation-free microstructures is determined by a competition between these two processes. Hence, the process that requires a lower critical interfacial velocity dictates the formation of a segregation-free microstructure. As an example of the competition between featureless and **diffusionless solidification**, Figure 3.56 shows the critical velocities calculated by Hehmann et al. [139] for Mg-based alloys. Hehmann et al. calculated the critical velocities for featureless and diffusionless solidification using the absolute stability criterion of Eq. (3.117) and the diffusionless condition of Eq. (3.141). The results of these calculations are presented as a function of alloy composition for the Mg—Al and Mg—Ca systems (Figure 3.56). It is evident that featureless solidification requires a lower interfacial velocity than that required for diffusionless solidification in dilute alloys [139]. In the elevated alloy concentration regime, however, **diffusionless solidification** requires a lower interfacial velocity. Therefore, in dilute alloys, segregation-free solidification is stability controlled, whereas in the concentrated alloys, it is **massive solidification** controlled [139].

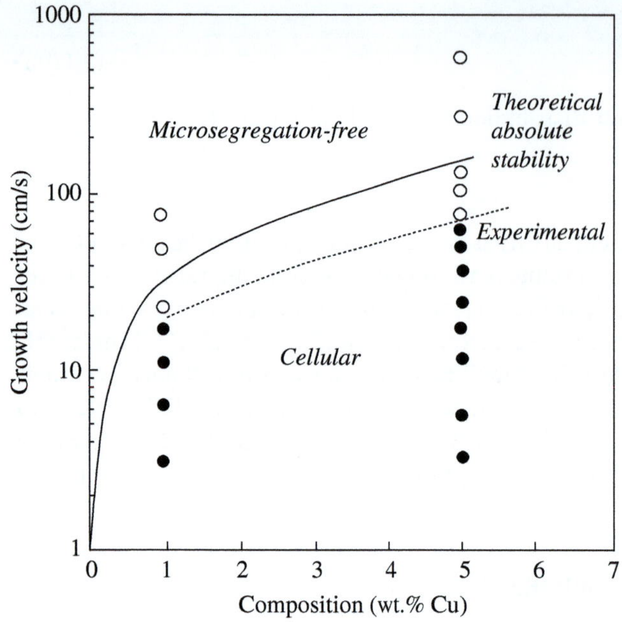

Figure 3.55 Summary of microstructure obtained by electron beam surface melting and resolidification of Ag–1 wt%Cu and Ag–5 wt%Cu alloys at various growth rates: open circles represent microsegregation-free microstructures; closed circles represent cellular or dendritic microstructures. Source: Reproduced by permission of Elsevier Science from W. J. Boettinger and S. R. Coriell (1984) *Mater. Sci. Eng.*, **65**, 27–36 [83].

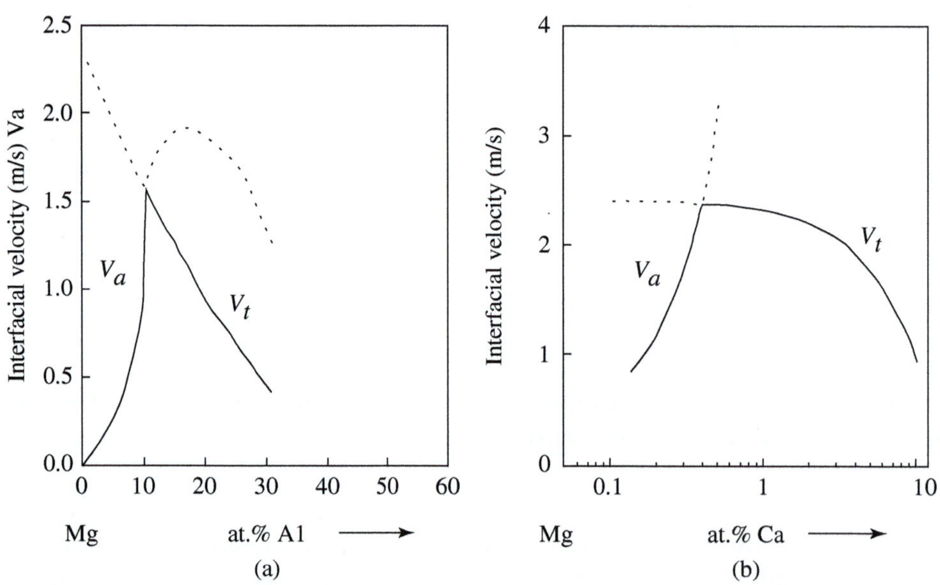

Figure 3.56 Calculated solidification front velocity required for the formation of a segregation-free solid in (a) Mg–Al and (b) Mg–Ca systems. Source: Reproduced by permission of Elsevier Science from F. Hehmann, F. Sommer, and B. Predel (1990) *Mater. Sci. Eng.*, **A125**, 249–265 [139].

Solute trapping results from the competition between diffusion and interfacial growth kinetics. The partition coefficient in this case is greater than the equilibrium value but solidification may not be segregation free. Experimentally, the extension of solid solubility has been noted in non-planar microstructures formed under rapid solidification conditions. For solute trapping, there are several theoretical models that are available for the dependence of partition

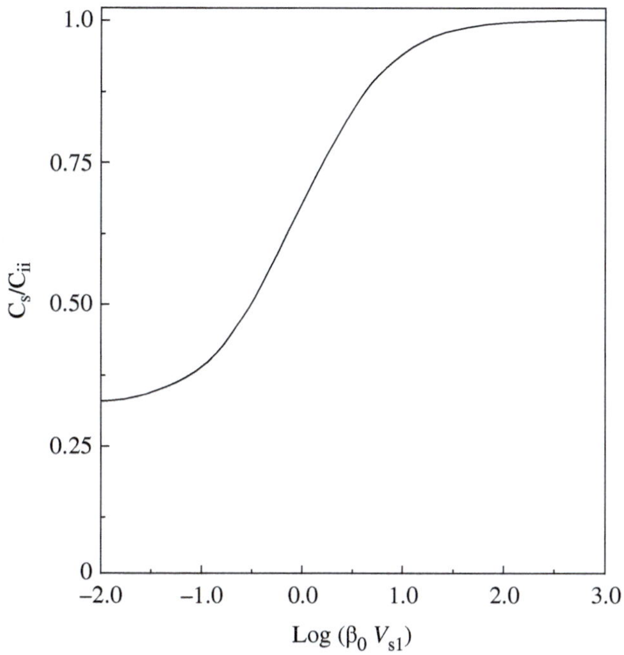

Figure 3.57 Arsenic (k_o = 0.3) segregation in Si calculated using stepwise growth model. Source: Adapted from Aziz [221].

coefficient on interfacial velocity [221–225]. The solute trapping model developed by Aziz [221, 222] for continuous growth predicts

$$k(V_{s1}) = \frac{k_0 + \beta_0 V_{s1}}{1 + \beta_0 V_{s1}} \tag{3.142}$$

where $k(V_{sl})$ is the partition coefficient at an interfacial velocity V_{s1}, and $\beta_0 = a_0/D_1$. Figure 3.57 shows the interfacial partition coefficient as a function of dimensionless velocity, $V_{sl}\beta_0 = V_{SL}a_0/D_1$. [221]. It is worth noting that at very low velocities Eq. (3.142) reduces to the equilibrium partition coefficient. At very high interfacial velocities, however, the partition coefficient approaches unity.

3.7 Summary

In this chapter, we reviewed the transfer of thermal energy which occurs as soon as the molten metal leaves the heated region inside of an atomizer. The transfer of thermal energy during atomization critically influences the evolution of the microstructure and hence plays an important role in determining the structure of a solidified droplet. This chapter discussed the heat transfer and solidification phenomena that occur during droplet flight and the associated microstructural development that occurs inside of an atomized droplet. The thermal conditions in the metal prior to atomization influence the microstructure in a process that is described as superheating. As the heat is transferred from the rapidly cooling droplet into the cooler gas environment, there are several mechanisms that influence the rate of thermal energy dissipation, such as conduction, convection, and radiation. The information reviewed in this chapter reveals that there are several process parameters that affect the transfer of thermal energy from a **droplet**. Perhaps one of the most important process parameters is the velocity of the gas being used to atomize the metal. The magnitude of the convective heat transfer coefficient is critically influenced by the velocity differential between the droplet and the atomization gas. Gas selection also influences the transfer of thermal energy from the droplet to the environment. The chapter also describes the formation of the first solid–solid nuclei inside of a droplet, a process known as nucleation, and discusses the various conditions that lead to the formation and development of the solid phase. Overall, the solidification process experienced by atomized droplets is complex and depends upon the physical parameters of the metal being atomized, as well as the physical values of the gas being used to disintegrate the metal. There is a wealth of published papers describing changes in solidification morphology as a function of solidification rate and our knowledge in this area has been useful in designing atomizers as well as in selecting optimal parameters that will yield desirable microstructures. Nevertheless, despite the vast amount of fundamental knowledge in this area, there still remains room for productive research, particularly when atomizing novel metals and alloys.

Nomenclature

Variable	Description	Units
a	Interatomic distance	m
Bi	Biot number	—
C_p	Constant pressure heat capacity	J mol^{-1} K^{-1}
$C_\mathrm{p,g}$	Constant pressure heat capacity of atomization gas	J mol^{-1} K^{-1}
$C_\mathrm{p,l}$	Constant pressure heat capacity of liquid	J mol^{-1} K^{-1}
C_l	Solute composition in liquid	—
C_s	Solute composition in solid	—
C_drag	Fluid drag coefficient	—
d	Droplet diameter	m
d_G	Grain size	m
d_0	Capillarity length	m
D_l	Self-diffusion coefficient of liquid	m^2 s^{-1}
DAS	Dendrite arm spacing	m
E_a	Activation energy for atomic transport	J mol^{-1}
ES	Eutectic spacing	m
f_n	Fraction of nucleant-free droplets	—
f_s	Fraction of solid	—
F_g	Gravitational force	N
F_d	Fluid drag force	N
F_b	Buoyant force	N
g	Gravitational acceleration	m s^{-2}
G_l	Temperature gradient in liquid	K m^{-1}
G_lc	Constitutional temperature gradient	K m^{-1}
G_s	Temperature gradient in solid	K m^{-1}
ΔG_v	Volume free energy	J m^{-3}
ΔH_f	Heat of fusion	J mol^{-1}
h	Heat transfer coefficient	W m^{-2} K^{-1}
h_c	Convective heat transfer coefficient	W m^{-2} K^{-1}
h_cs	Static heat transfer coefficient	W m^{-2} K^{-1}
h_r	Radiative heat transfer coefficient	W m^{-2} K^{-1}
H	Enthalpy	J mol^{-1}
$I(T)$	Nucleation rate	m^{-3} s^{-1}
J_c	Convective thermal energy flux	W m^{-2}
J_r	Radiative thermal energy flux	W m^{-2}
k	Thermal conductivity	W m^{-1} K^{-1}
k_g	Thermal conductivity of atomization gas	W m^{-1} K^{-1}
k_o	Equilibrium partition coefficient	—
K_m	Interface kinetic coefficient	m s^{-1} K^{-1}
m_d	Mass of droplet	kg
n_1	Number of liquid atoms per unit volume	m^{-3}
n_s	Number of catalytic sites per unit volume	m^{-3}
N_Pr	Prandtl number	—

Variable	Description	Units
N_v	Number density of volume nucleants	m^{-3}
N_s	Number density of surface nucleants	m^{-2}
Nu	Nusselt number	—
P_c	Solutal Péclet number	—
P_t	Thermal Péclet number	—
r_d	Droplet radius	m
r_e	Radius of embryo	m
r_e^*	Critical radius of embryo	m
Re	Reynolds number	—
R_d	Dendrite radius	m
R_n	Radius of nucleant	m
SDAS	Secondary dendrite arm spacing	m
S_e	Surface area of embryo	m^2
ΔS_f	Molar entropy of fusion	$J\,mol^{-1}\,K^{-1}$
Ste	Stefan number	—
t_f	Solidification time	s
T	Temperature	K
T_g	Temperature of atomization gas	K
T_l	Liquidus temperature	K
T_n	Nucleation temperature	K
ΔT	Undercooling	K
ΔT_i	Interfacial undercooling	K
ΔT_{cr}	Critical undercooling	K
ΔT_o	Freezing temperature range	K
ΔT_S	Superheat	K
$\Delta \overline{T}$	Dimensionless bulk undercooling	—
$\Delta \overline{T}_c$	Dimensionless solutal undercooling	—
$\Delta \overline{T}_i$	Dimensionless interfacial undercooling	—
$\Delta \overline{T}_r$	Dimensionless curvature undercooling	—
$\Delta \overline{T}_t$	Dimensionless thermal undercooling	—
$\dot{T}$	Cooling rate	$K\,s^{-1}$
V_a	Critical interfacial velocity for absolute stability	$m\,s^{-1}$
V_c	Critical interfacial velocity for constitutional stability	$m\,s^{-1}$
V_d	Velocity of droplet	$m\,s^{-1}$
V_e	Volume of embryo	m^3
V_g	Velocity of atomization gas	$m\,s^{-1}$
V_{sl}	Interfacial velocity	$m\,s^{-1}$
V_{step}	Step velocity	$m\,s^{-1}$
V_t	Critical velocity for massive solidification	$m\,s^{-1}$
α	Thermal diffusivity	$m^2\,s^{-1}$
ε_r	Surface emissivity	—
Γ	Gibbs-Thompson factor	m K
η_g	Dynamic viscosity of atomization gas	Pa s
η_l	Dynamic viscosity of liquid	Pa s
λ	Step height	m
λ_v	Decay coefficient	m
ν	Jump frequency	s^{-1}
ν_g	Kinematic viscosity of atomization gas	$m^2\,s^{-1}$

Variable	Description	Units
θ_w	Wetting angle	degree
θ	Dimensionless temperature	—
Θ	Dimensionless cooling rate	—
ρ	Density	$kg\,m^{-3}$
ρ_d	Droplet density	$kg\,m^{-3}$
σ_r	Stefan-Boltzmann constant	$Wm^{-2}\,k^{-4}$
σ_{lc}	Interfacial energy between liquid phase and nucleant	$J\,m^{-2}$
σ_{sc}	Interfacial energy between solid phase and nucleant	$J\,m^{-2}$
σ_{sl}	Solid–liquid interfacial free energy	$J\,m^{-2}$
v_d	Volume of droplet	m^3
Ω	Molar volume	$m^3\,mol^{-1}$
$\overline{\Omega}$	Average molar volume	$m^3\,mol^{-1}$
ψ	Dimensionless enthalpy	—

References

1 Lavernia, E.J., Ayers, J.D., and Srivatsan, T.S. (1992). *Int. Mater. Rev.* 37 (1): 1.
2 Lavernia, E.J. (1989). *Int. J. Rapid Solid.* 5: 47.
3 Annavarapu, S., Apelian, D., and Lawley, A. (1988). *Metall. Trans. A.* 19A: 3077.
4 Gupta, M., Mohamed, F.A., and Lavernia, E.J. (1991). *Mater. Sci. Eng.* A144: 99.
5 Lavernia, E.J., Baram, J., and Gutierrez, E. (1991). *Mater. Sci. Eng.* A132: 119.
6 Liang, X., Kim, H.K., Earthman, J.C., and Lavernia, E.J. (1992). *Mater. Sci. Eng.* A153: 646.
7 Annavarapu, S. and Doherty, R. (1993). *Int. J. Powder Metall.* 29: 331.
8 Morris, D.G. and Morris, M.A. (1991). *J. Mater. Res.* 6: 361.
9 Brooks, R.G., Leatham, A.G., Coombs, J.S., and Moore, C. (1977). *Metall. Metal Form.* 44: 157.
10 Fiedler, H.C., Sawyer, T.F., Kopp, R.W., and Leatham, A.G. (1987). *JOM* 39 (8): 28.
11 Lavernia, E.J., Rai, G., and Grant, N.J. (1986). *Mater. Sci. Eng. A* A79: 211.
12 Zhou, J. and Duszczyk, J. (1990). *J. Mater. Sci.* 25: 4541.
13 Savage, S.J. and Froes, F.H. (1984). *J. Meteorol.* 36 (4): 20.
14 Backmark, U., Bäckström, N., and Arnberg, L. (1986). *Powder Metall. Int.* 18 (338): 442.
15 Lawley, A. (1977). *Powder Metall Powder Tech.* 13: 168.
16 Matyja, H., Geissen, B.C., and Grant, N.J. (1968). *J. Inst. Met.* 96: 30.
17 German, R.M. (1984). *Powder Metallurgy Science*. Princeton, New Jersey: MPIF.
18 Gessinger, G.H. (1984). *Powder Metallurgy of Superalloys*. London: Butterworths.
19 Salas, O. and Levi, C.G. (1988). *Int. J. Rap. Solid.* 4: 1.
20 Perei, J., Mahoney, J.F., Duwez, P., and Kalensher, B.E. (1980). *Rapid Solidification Processing, Principles and Technologies II* (ed. R. Mehrabian, B.H. Kear, and M. Cohen), 287. Baton Rouge, Louisiana: Claitor's Publishing Division.
21 Levi, C.G. and Mehrabian, R. (1983). *Rapid Solidification Technology: Source Book* (ed. R.L. Ashbrook), 403. Ohio: ASM, Metals Park.
22 Steinberg, J., Lord, A.E. Jr., Lacy, L.L., and Johnson, J. (1981). *Appl. Phys. Lett.* 38: 135.
23 Hofmeister, W.H., Evans, N.D., Bayuzick, R.J., and Robinson, M.B. (1986). *Metall. Trans. A.* 17A: 973.
24 Weber, J.K.R., Nordine, P.C., Goretta, K.C., and Poeppel, R.B. (1994). *J. Mater. Res.* 9: 1657.
25 Bertero, G.A., Hofmeister, W.H., Robinson, M.B., and Bayuzick, R.J. (1991). *Metall. Trans. A.* 22A: 2723.
26 Wang, C.C. and Smith, C.S. (1950). *Trans. Metall. Soc. AIME* 188: 136.
27 Zhang, D.L. and Cantor, B. (1992). *Acta Metall. Mater.* 40: 2951.
28 Turnbull, D. and Cech, R.E. (1950). *J. Appl. Phys.* 21: 411.
29 Perepezko, J.H., Graves, J.A., and Mueller, B.A. (1987). *Processing of Structural Materials by Rapid Solidification* (ed. F.H. Froes and S.J. Savage), 13. Ohio: ASM International, Materials Park.

30 Orme, M. and Muntz, E.P. (1990). *Phys. Fluid. J. A* 2: 1124.
31 Orme, M. (1991). *Phys. Fluid. J. A* 3: 2936.
32 Fussell, R.S., Kirchner, H.O.K., Prinz, F.B., and Weiss, L.E. (1994). *J. Therm. Spray Tech.* 3: 148.
33 Sampath, S. and Herman, H. (1993). *JOM* 46 (3): 42.
34 Samuel, F.H. (1986). *Metall. Trans. A.* 17A: 73.
35 Boettinger, W.J., Bendersky, L., and Early, J.G. (1986). *Metall. Trans. A.* 17A: 781.
36 Boettinger, W.J. and Perepezko, J.H. (1985). *Rapidly Solidified Crystalline Alloys* (ed. S.K. Das, B.H. Kear, and C.M. Adam), 21. Warrendale, Pennsylvania: AIME.
37 Cochrane, R.F., Evans, P.V., and Greer, A.L. (1991). *Mater. Sci. Eng.* 133A: 803.
38 McCullough, C., Valencia, J.J., Levi, C.G., and Mehrabian, R. (1990). *Mater. Sci. Eng.* 124A: 83.
39 Boettinger, W.J. (1988). *Mater. Sci. Eng.* 98: 123.
40 Kelly, T.F., Cohen, M., and Vander Sande, J.B. (1984). *Metall. Trans. A.* 15A: 819.
41 Lavernia, E.J., Baram, J., and Gutierrez, E. (1991). *Mater. Sci. Eng.* 132A: 119.
42 Molnárová, O., Málek, P., Lukáč, F., and Chráska, T. (2016). Spark Plasma Sintering of a Gas Atomized Al7075 Alloy: Microstructure and Properties. *Materials.* 9 (12): 1004. https://doi.org/10.3390/ma9121004.
43 Cohen, M., Kear, B.H., and Mehrabian, R. (1980). *Rapid Solidification Processing: Principle and Technologies, II* (ed. R. Mehrabian, B.H. Kear, and M. Cohen), 1. Louisiana: Claitor's Publishing Division, Baton Rouge.
44 Perepezko, J.H., Rasmussen, D.H., Anderson, I.E., and Loper, C.R. Jr., (1979). *Solidification and Castings of Metals*, 169. London: The Metal Society.
45 Hsu, S.C., Chakravory, S., and Mehrabian, R. (1978). *Metall. Trans. B.* 9B: 221.
46 Flemings, M.C. (1976). *Report of the ARPA Materials Research Council.* Ann Arbor, USA: University of Michigan.
47 Hirth, J.P. (1978). *Metall. Trans. A.* 9A: 401.
48 Levi, C.G. (1988). *Metall. Trans. A.* 19A: 699.
49 Levi, C.G. and Mehrabian, R. (1982). *Metall. Trans. A.* 13A: 13.
50 Mehrabian, R. (1982). *Int. Met. Rev.* 27 (4): 185.
51 Predecki, P., Mullendore, A.W., and Grant, N.J. (1969). *Trans. Metall. Soc. AIME* 233: 1581.
52 Harbur, D.R., Anderson, J.W., and Maraman, W.J. (1969). *Trans. Metall. Soc. AIME* 245: 1055.
53 Ramati, S.D.E., Abbaschian, G.J., Backman, D.G., and Mehrabain, R. (1978). *Metall. Trans. B.* 9B: 279.
54 Hong, S., Backman, D.G., and Mehrabian, R. (1979). *Metall. Trans. B.* 10B: 229.
55 Gillen, A.G. and Cantor, B. (1985). *Acta Metall.* 33: 1813.
56 Banan, M., Gray, R.T., and Wilcox, W.R. (1991). *J. Cryst. Growth* 113: 557.
57 Mehrabian, R. (1978). *Rapid Solidification Processing: Principles and Technologies* (ed. R. Mehrabian, B.H. Kear, and M. Cohen), 9. Baton Rouge, Louisiana: Claitor's Publishing Division.
58 Mehrabian, R., Hsu, S.C., Levi, C.G., and Kou, S. (1981). *Advances in Metal Processing* (ed. J.J. Burke, R. Mehrabian, and V. Weiss), 13. New York: Plenum Press.
59 Lee, E. and Ahn, S. (1994). *Acta Metall. Mater.* 47: 3231.
60 Gutierrez-Miravete, E., Lavernia, E.J., Trapaga, G.M. et al. (1989). *Metall. Trans. A.* 20A: 71.
61 Gupta, M., Mohamed, F.A., and Lavernia, E.J. (1992). *Metall. Trans. A.* 23A: 831.
62 Liu, H., Rangel, R.H., and Lavernia, E.J. (1994). *Acta Metall. Mater.* 42: 3277.
63 Grant, P.S., Cantor, B., and Katgerman, L. (1993). *Acta Metall. Mater.* 41: 3097.
64 Bewlay, B.P. and Cantor, B. (1990). *Metall. Trans. B.* 21B: 899.
65 Ranz, W.E. and Marshall, W.R. (1952). *Chem. Eng. Prog.* 48: 173.
66 Mathur, P., Apelian, D., and Lawley, A. (1989). *Acta Metall.* 37: 429.
67 Bewlay, B.P. and Cantor, B. (1989). *Mater. Sci. Eng.* 118: 207.
68 Gutierrez-Miravete, E. and Lavernia, E.J. (1989). *Physical Chemistry of Powder Metals, Production and Processing* (ed. W.M. Small), 175. Warrendale, Pennsylvania: TMS.
69 Grant, P.S., Cantor, B., and Katgerman, L. (1993). *Acta Metall. Mater.* 41: 3109.
70 Grant, P.S. and Cantor, B. (1991). *Cast Metals.* 4: 140.
71 Liu, J., Arnberg, L., Bäckström, N. et al. (1988). *Mater. Sci. Eng.* 98: 43.
72 Connelly, S., Coombs, J.S., and Medwell, J.O. (1986). *Metal Powder Rep.* 40: 653.
73 Lavernia, E.J., Gutierrez-Miravete, E., Szekely, J., and Grant, N.J. (1988). *Int. J. Rapid Solid.* 4: 89.
74 Chen, C.P., Shang, H.M., and Jiang, Y. (1992). *Int. J. Numer. Methods Fluids* 15: 233.

75 Szekely, J. (1979). *Fluid Flow Phenomena in Metals Processing*. New York: Academic Press.

76 Kurten, H., Raasch, J., and Rumpf, H. (1966). *Chemie-Intenieur-Technik* 38: 941.

77 McDonell, V.G., Lavernia, E.J., and Samuelsen, G.S. (1989). *Synthesis and Analysis in Materials Processing: Advances in Characterization and Diagnosis of Ceramic and Metal Particulate Processing* (ed. E.J. Lavernia, H. Henein, and I. Anderson), 29. Warrendale Pennsylvania: TMS.

78 Turnbull, D. (1956). *Solid State Physics*, vol. 3 (ed. F. Seitz and D. Turnbull), 225. New York: Academic Press.

79 Uhlmann, D.R. and Chalmers, B. (1965). *Ind. Eng. Chem.* 57: 19.

80 Uhlmann, D.R., Hays, J.F., and Turnbull, D. (1966). *Phys. Chem. Glasses.* 7: 159.

81 Flemings, M.C. (1974). *Solidification Processing*. New York: McGraw-Hill.

82 Turnbull, D. (1950). *J. Appl. Phys.* 21: 1022.

83 Boettinger, W.J. and Coriell, S.R. (1984). *Mater. Sci. Eng.* 65: 27.

84 Turnbull, D. (1952). *J. Chem. Phys.* 20: 411.

85 Miyazawa, Y. and Pound, G.M. (1974). *J. Cryst. Growth.* 23: 45.

86 Chen, H.S. and Turnbull, D. (1967). *Acta Metall.* 15: 369.

87 Perepezko, J.H. and Rasmussen, D.H. (1978). *Metall. Trans. A.* 9A: 1490.

88 Dubey, K.S. and Ramachandrarao, P. (1984). *Acta Metall.* 32: 91.

89 Hoffman, J.D. (1958). *J. Chem. Phys.* 29: 1192.

90 Jones, D.R.H. and Chadwick, G.A. (1971). *Philos. Mag.* 24: 995.

91 Thompson, C.V. and Spaepen, F. (1979). *Acta Metall.* 27: 1855.

92 Frankel, J. (1955). *Kinetic Theory of Liquids*. New York: Dover.

93 Hirai, N. and Eyring, H. (1958). *J. Appl. Phys.* 29: 810.

94 Hirai, N. and Eyring, H. (1959). *Polymer Sci.* 37: 51.

95 Flory, P.J. (1953). *Principles of Polymer Chemistry*. Ithaca, New York: Cornell University Press.

96 Sanchez, I.C. (1974). *J. Appl. Phys.* 45: 4204.

97 Christian, J.W. (1975). *The Theory of Transformation in Metals and Alloys*, 2e. Oxford, United Kingdom: Pergamon Press.

98 Kelton, K.F., Greer, A.L., and Thompson, C.V. (1983). *J. Chem. Phys.* 79: 6261.

99 Spaepen, F. and Trunbull, D. (1976). *Rapidly Quenched Metals II* (ed. N.J. Grant and B.C. Giessen), 205. Cambridge, Massachusetts: MIT Press.

100 Turnbull, D. (1969). *Contemp. Phys.* 10: 473.

101 Pan, L.M., Saunders, N., and Tsakiropoulos, P. (1989). *Mater. Sci. Technol.* 5: 609.

102 Beyer, R.T. and Ring, E.M. (1972). *Liquid Metals, Chemistry and Physics* (ed. S.Z. Beer), 450. New York: Marcel Dekker.

103 Pound, G.M. (1958). *Liquid Metals and Solidification*, 87. Metals Park: ASM, Ohio.

104 Rasmussen, D.H. and Loper, C.R. Jr., (1976). *Acta Metall.* 24: 117.

105 Perepezko, J.H., Mueller, B.A., and Ohsaka, K. (1987). *Undercooled Alloy Phases* (ed. E.W. Colling and C.C. Koch), 289. Warrendale, Pennsylvania: TMS-AIME.

106 Clyne, T.W. (1984). *Metall. Trans. B* 15B: 369.

107 Rasmussen, D.H. and Mackenzie, A.P. (1973). *J. Chem. Phys.* 59: 5003.

108 Rasmussen, D.H. and Loper, C.R. Jr., (1975). *Acta Metall.* 23: 1215.

109 Thompson, C.V. and Spaepen, F. (1983). *Acta Metall.* 31: 2021.

110 Willnecker, R., Herlach, D.M., and Feuerbacher, B. (1988). *Mater. Sci. Eng.* 98: 85.

111 Cech, R.E. and Turnbull, D. (1951). *J. Meteorol.* 191: 242.

112 Thompson, C.V. and Spaepen, F. (1982). *Materials Research Society Symposium Proceedings*, vol. 9 (ed. G.E. Rindone), 603. Amsterdam: North-Holland.

113 Spaepen, F. (1975). *Acta Metall.* 23: 729.

114 Perepezko, J.H. (1980). *Proceedings of the 2nd International Conference on Rapid Solidification Processing* (ed. V.A. Reston, R. Mehrabian, B.H. Kear, and M. Cohen), 56. Baton Rouge, Louisiana: Claitor's Publishing Division.

115 Cooper, K.P., Anderson, I.E., and Perepezko, J.H. (1982). *Proceedings of the 4th International Conference on Rapidly Quenched Metals* (ed. T. Masumoto and K. Suzuki), 107. Japanesse Institute of Metals.

116 Turnbull, D. and Cech, R.E. (1950). *J. Appl. Phys.* 21: 804.

117 Turnbull, D. (1956). *Solid State Phys.* 3: 226.
118 Pauli, J., Ryall, J., and Delamore, G.W. (1976). *Seri. Metall.* 10: 845.
119 Turnbull, D. (1950). *Trans. AIME* 188: 1140.
120 Cochrane, R.F., Herlach, D.M., and Feuerbacher, B. (1991). *Mater. Sci. Eng.* A133: 706.
121 Allen, W.P. and Perepezko, J.H. (1991). *Metall. Trans. A.* 22A: 765.
122 Gupta, M., Lane, C., and Lavernia, E.J. (1992). *Seri. Metall. Mater.* 26: 825.
123 Wu, Y. and Lavernia, E.J. (1992). *Metall. Trans. A.* 23A: 2923.
124 Fletcher, N.H. (1958). *J. Chem. Phys.* 29: 572.
125 Fletcher, N.H. (1963). *J. Chem. Phys.* 38: 237.
126 Turnbull, D. and Fisher, J.C. (1949). *J. Chem. Phys.* 17: 71.
127 Cech, R.E. (1956). *Trans. AIME* 206: 585.
128 Kelly, T.F. and Vander Sande, J.B. (1983). *Chemistry and Physics of Rapidly Solidified Materials: Proceedings* (ed. B.J. Berkowitz and R.O. Scattergood), 35. Warrendale, Pennsylvania: TMS.
129 Kim, Y.W., Lin, H.M., and Kelly, T.F. (1987). *Undercooled Alloy Phases* (ed. E.W. Collings and C.C. Koch), 109. Warrendale, Pennsylvania: TMS.
130 Kim, Y.W., Lin, H.M., and Kelly, T.F. (1985). *Proceedings of the Electron Microscopy Society of America* (ed. G.W. Bailey), 46. San Francisco, California: San Francisco Press.
131 Thoma, D.J. and Perepezko, J.H. (1992). *Metall. Trans. A.* 23A: 1347.
132 Adkins, N.J.E. and Tsakiropoulos, P. (1991). *Mater. Sci. Eng.* A133: 767.
133 Drehman, A.J. and Turnbull, D. (1981). *Seri. Metall.* 15: 543.
134 Chattopadhyay, K. and Raviprasad, K. (1989). *Key Eng. Mater.* 38–39: 83.
135 Kurz, W., Giovanola, B., and Trivedi, R. (1986). *Acta Metall.* 34: 823.
136 Munitz, A., Elder-Randall, S.P., and Abbaschian, R. (1992). *Metall. Trans. A.* 23A: 1817.
137 Sekhar, J.A. (1991). *J. Cryst. Growth* 109: 113.
138 Willnecker, R., Herlach, D.M., and Feuerbacher, B. (1990). *Appl. Phys. Lett.* 56: 324.
139 Hehmann, F., Sommer, F., and Predel, B. (1990). *Mater. Sci. Eng.* A125: 249.
140 Bassler, B.T., Hofmeister, W.H., Carro, C., and Bayuzick, R.J. (1994). *Metall. Mater. Trans. A* 25A: 1301.
141 Graves, J.A., Perepezko, J.H., Ward, C.H., and Froes, F.H. (1987). *Seri. Metall.* 21: 567.
142 Allen, W.P. and Perepezko, J.H. (1991). *Metall. Trans. A.* 22A: 753.
143 Levi, C.G. and Mehrabian, R. (1982). *Metall. Trans. A.* 13A: 221.
144 Gill, W.N., Jang, J., Mollendorf, C., and Adam, C.M. (1984). *J. Cryst. Growth* 66: 351.
145 Rains, R.K. and Kadlec, R.H. (1970). *Metall. Trans. A.* 1A: 1501.
146 Himmelblau, D.M. and Bischoff, K.B. (1968). *Process Analysis and Simulation; Deterministic Systems*, 43. New York: Wiley.
147 Webster, D., Wald, G., and Crements, W.S. (1981). *Metall. Trans. A.* 12A: 1495.
148 Grant, N.J., King, S., and Wang, W. (1980). *Mater. Sci. Eng.* 44: 213.
149 Wu, Y. and Lavernia, E. J. (1992). Presentation at 1992 TMS Spring Meeting, March, 1992, San Diego, California.
150 Cahn, J.W. (1960). *Acta Metall.* 8: 554.
151 Cahn, J.W., Hilling, W.B., and Sears, G.W. (1964). *Acta Metall.* 12: 1421.
152 Chalmers, B. (1958). *Growth and Perfection of Crystals* (ed. R.H. Doremus, D. Turnbull, and B.W. Roberts), 291. New York: Wiley.
153 Turnbull, D. (1949). *Thermodynamics in Metallurgy*. Metals Park, Ohio: ASM.
154 Spaepen, F. and Turnbull, D. (1982). *Laser Annealing of Semiconductors* (ed. J.M. Poate and J.W. Mayer), 15. New York: Academic Press.
155 Coriell, S.R. and Turnbull, D. (1982). *Acta Metall.* 30: 2135.
156 Frank, F.C. (1949). *Disc. Faraday Soc.* 5: 48.
157 Volmer, M. (1935). *Kinetik de Phasenbildung*. Dresden and Leipzig: Steinkopff.
158 Becker, R. and Doring, W. (1935). *Ann. Phys.* 24: 719.
159 Hilling, W.B. (1959). *Kinetics of High Temperature Processes* (ed. W.D. Kingery), 127. Cambridge, Massachusetts, MIT: Technology Press.
160 Colligan, G.A. and Bayles, B.J. (1962). *Acta Metall.* 10: 895.
161 Walker, J.L. (1964). *Principles of Solidification* (ed. B. Chalmers), 114. New York: Wiley.

162 Kramer, J.J. and Tiller, W.A. (1965). *J. Chem. Phys.* 42: 257.

163 Rigney, D.A. and Blakely, J.M. (1966). *Acta Metall.* 14: 1375.

164 Jackson, K.A. (1958). *Liquid Metals and Solidification*, 174. Metals Park: ASM, Ohio.

165 Hofmeister, W.H., Bayuzick, R.J., and Robinson, M.B. (1991). *Rev. Sci. Instrum.* 61: 2220.

166 Schleip, E., Willnecker, R., Herlach, D.M., and Gorier, G.P. (1988). *Mater. Sci. Eng.* A98: 39.

167 Lipton, J., Glicksman, M.E., and Kurz, W. (1984). *Mater. Sci. Eng.* 65: 57.

168 Lipton, J., Kurz, W., and Trivedi, R. (1987). *Acta Metall. Mater.* 35: 957.

169 Lipton, J., Glicksman, M.E., and Kurz, W. (1987). *Metall. Trans. A.* 18A: 341.

170 Mullins, W.W. and Sekerka, R.F. (1964). *J. Appl. Phys.* 35: 444.

171 Trivedi, R. and Kurz, W. (1986). *Acta Metall.* 34: 1663.

172 Abbaschian, G.J. and Flemings, M.C. (1983). *Metall. Trans. A.* 14A: 1147.

173 Kattamis, T.Z. and Flemings, M.C. (1966). *Trans. TMS-AIME* 236: 1523.

174 Southin, R.T. and Weston, G.M. (1974). *J. Australian Inst. Met.* 19: 93.

175 Hogan, L.M. (1979). *Physics of Materials* (ed. D.W. Borland, L.M. Clarebrough, and A.J.M. Moore), 111. Australia: University of Melbourne.

176 Mondolfo, L.F. (1982). *Grain Refinement in Castings and Welds* (ed. G.J. Abbaschian and S.A. Davis), 3. Warrendale, Pennsylvania: TMS-AIME.

177 Kattamis, T.Z. and Flemings, M.C. (1967). *Trans. Am. Foundrymen's Soc.* 75: 191.

178 Munitz, A. and Abbaschian, G.J. (1987). *Undercooled Alloy Phases* (ed. C.C. Koch and E.W. Collings), 23. Warrendale, Pennsylvania: TMS-AIME.

179 McDevitt, D.D. and Abbaschian, G.J. (1982). *Chemistry and Physics of Rapidly Solidified Materials* (ed. B.J. Berkowitz and R.O. Scattergood), 49. Warrendale, Pennsylvania: TMS-AIME.

180 Boettinger, W.J., Coriell, S.R., and Sekerka, R.F. (1984). *Mater. Sci. Eng.* 65: 27.

181 Domingue, J.A., Kissinger, R.D., Boesch, W.J., and Tien, J.K. (1982). *Rapid Solidification Processing: Principle and Technologies, III*. In: *National Bureau of Standards* (ed. R. Mehrabian), 259. Gaithersburg, Maryland.

182 Jones, H. (1978). *Rapid Solidification Processing: Principle and Technologies* (ed. R. Mehrabian, B.H. Kear, and M. Cohen), 28. Baton Rouge, Louisiana: Claitor's Publishing Division.

183 Joly, P.A. and Mehrabian, R. (1974). *J. Mater. Sci.* 9: 1446.

184 Tiller, W.A., Jackson, K.A., Rutter, J.W., and Chalmers, B. (1953). *Acta Metall.* 1: 428.

185 Chalmers, B. (1982). *Principles of Solidification* (ed. E. Robert), 150. Malabar, Florida: Krieger Publ. Co.

186 Coriell, S.R. and Sekerka, R.F. (1980). *Rapid Solidification Processing: Principle and Technologies II* (ed. R. Mehrabian, B.H. Kear, and M. Cohen), 9. Baton Rouge, Louisiana: Claitor's Publishing Division.

187 Trivedi, R., Sekhar, J.A., and Seetharaman, V. (1989). *Metall. Trans. A.* 20A: 769.

188 Grant, N.J. (1978). *Rapid Solidification Processing: Principle and Technologies* (ed. R. Mehrabian, B.H. Kear, and M. Cohen), 230. Baton Rouge, Louisiana: Claitor's Publishing Division.

189 Patterson, R.J. II, Cox, A.R., and van Reuth, E.C. (1983). *Rapid Solidification Technology Source Book* (ed. R.L. Ashbrook), 414. Materials Park, Ohio: ASM.

190 Miracle, D.B., Williams, R.A., and Lipsitt, H.A. (1982). *Rapid Solidification Processing: Principle and Technologies, III* (ed. R. Mehrabian), 234. Gaithersburg, Maryland: National Bureau of Standards.

191 Cebulak, W.S. (1982). *Rapid Solidification Processing: Principle and Technologies, III* (ed. R. Mehrabian), 324. Gaithersburg, Maryland: National Bureau of Standards.

192 Jackson, K.A., Hunt, J.D., Uhlmann, D.R., and Seward, T.P. III, (1966). *Trans. AIME* 236: 149.

193 Juarez-Islas, J.A. and Jones, H. (1987). *Acta Metall.* 35: 499.

194 Beghi, G., Piatti, G., and Street, K.N. (1971). *J. Mater. Sci.* 6: 118.

195 Beghi, G. and Pellegrini, G. (1980). *J. Mater. Sci.* 15: 2403.

196 Trivedi, R., Lipton, J., and Kurz, W. (1987). *Acta Metall.* 35: 965.

197 Kurz, W. and Fisher, D.J. (1980). *Acta Metall.* 29: 11.

198 Langer, J.S. and Küller-Krumbhaar, H. (1977). *J. Cryst. Growth* 42: 11.

199 Langer, J.S. and Küller-Krumbhaar, H. (1978). *Acta Metall.* 26: 1681.

200 Jones, H. (1977). *Vacancies '76* (ed. R.E. Smallman and J.E. Harris), 175. United Kingdom: Materials Society London.

201 Vogt, E. and Frommeyer, G. (1985). *Rapidly Solidified Materials* (ed. P.W. Lee and R.S. Carbonara), 291. Metal Park, Ohio: ASM.
202 Cantor B. (1985). *Proceedings of NATO Advanced Research Workshop*, March 1985, Theuern, Bavaria, Martinus Nijhoff Publishers, The Netherlands.
203 Devaud, G. and Turnbull, D. (1987). *Acta Metall.* 35: 765.
204 Kattamis, T.Z. (1976). *J. Cryst. Growth* 34: 215.
205 Kattamis, T.Z. and Mehrabian, R. (1974). *J. Vac. Sci. Tech.* 11: 1118.
206 Johnson, W.F. and Mehl, R.F. (1939). *Trans. AIME* 135: 416.
207 Greer, A.L. (1991). *Mater. Sci. Eng.* A133: 16.
208 Reed-Hill, R.E. (1973). *Physical Metallurgy Principles* (ed. D. van Nostrands), 305. New York.
209 Hyman, M.E., McCullough, C., Levi, C.G., and Mehrabian, R. (1991). *Metall. Trans. A.* 22A: 1647.
210 McCullough, C., Miller, D.S., and Levi, C.G. (1993). *Metall. Trans. A.* 24A: 1481.
211 Gillessen, F. and Herlach, D.M. (1991). *Mater. Sci. Eng.* A134: 1220.
212 Shechtman, D., Blech, I., Gratias, D., and Cahn, J.W. (1984). *Phys. Rev. Lett.* 53: 1951.
213 Perepezko, J.H. (1984). *Mater. Sci. Eng.* 65: 125.
214 Trivedi, R., Magnin, P., and Kurz, W. (1987). *Acta Metall.* 35: 971.
215 Scheil, E. (1942). *Z. Metallkd.* 34: 70.
216 Pfann, W.G. (1952). *Trans. AIME* 194: 747.
217 Brody, H.D. and Flemings, M.C. (1966). *Trans. TMS-AIME* 236: 615.
218 Evans, P.V. and Greer, A.L. (1988). *Mater. Sci. Eng.* 98: 357.
219 Evans, P.V., Vitta, S., Hamerton, R.G. et al. (1990). *Acta Metall.* 38: 233.
220 Boettinger, W.J., Shechtman, D., Schaefer, R.J., and Biancaniello, F.S. (1984). *Metall. Trans. A.* 15A: 55.
221 Aziz, M.J. (1982). *J. Appl. Phys.* 53: 1158.
222 Aziz, M.J. (1983). *Appl. Phys. Lett.* 43: 552.
223 Jackson, K.A., Gilmer, G.H., and Leamy, H.J. (1980). *Laser and Electron Beam Processing of Materials, Proceedings of Symposium of Materials Research Society* (ed. C.W. White and P.S. Peercy), 104. New York: Academic Press.
224 Wood, R.F. (1982). *Phys. Rev. B* 25: 2786.
225 Cahn, J.W., Coriell, S.R., and Boettinger, W.J. (1980). *Laser and Electron Beam Processing of Materials, Proceedings of Symposium of Materials Research Society* (ed. C.W. White and P.S. Peercy), 89. New York: Academic Press.

4

Composite Powders for Additive Manufacturing

4.1 Introduction

Additive manufacturing (AM) of **metal matrix composite**s and **intermetallic matrix composite**s (MMCs and IMCs) has become increasingly active area of research [1–11]. AM offers an attractive alternative to available techniques such as **liquid phase processing** [12–23], **solid phase processing** [24–26], and **rheocasting** [27, 28] of MMCs, as it becomes possible to manufacture net-shape or **near-net shape components** in one step. MMCs and IMCs are typically comprised of a metal or intermetallic matrix reinforced with **ceramic** phases. MMCs and IMCs combine the properties of the matrix, that is, high toughness, strength, ductility, and machinability, with those of the **ceramic** reinforcements, that is, high strength, stiffness, and thermal stability, yielding high specific strength, modulus, wear resistance and thermal stability [29–32]. In many stiffness, strength, thermal stability, and wear-critical applications, composite materials potentially offer a better performance relative to that of their unreinforced counterparts. For example, a 50% increase in elastic modulus may be achieved by substituting unreinforced Al alloys with Al alloy-based discontinuously reinforced MMCs. Moreover, such a substitution may potentially produce a 10% reduction in structural weight [33].

The matrix materials that are commonly used in MMCs and IMCs include **Al-based alloys**, **Ti-based alloys**, **Ni-based alloys**, **steels**, and **intermetallic compounds**, to name a few. The reinforcement materials include **carbides**, such as SiC, **TiC**, and **B_4C**; oxides, such as **Al_2O_3**; **nitrides**, such as Si_3N_4 and AlN and **BN**, and; elemental materials, such as **graphite** and Si. Most of these reinforcement phases exhibit a high wetting angle ($\theta > 90°$) with molten matrix materials. This **nonwetting behavior** between reinforcements and matrices renders the incorporation of reinforcements a challenging task. Accordingly, various processing techniques have emerged because of the effort to incorporate **ceramic** reinforcements into metal and intermetallic matrices. These techniques may be classified into three general categories: **liquid-phase processes**, solid-phase processes, and two-phase processes (i.e. solid and liquid, such as **rheocasting**). In addition, reinforcements, when placed in intimate contact with matrices at elevated temperatures, will often develop undesirable interfacial reactions. The products of such chemical reactions are often brittle interfacial phases that are generally detrimental to the mechanical properties.

The above processing methods have been applied, with varying degrees of success to the synthesis of composites. In the **powder metallurgy** approach, for example, matrix powders and reinforcing phases, such as particles and **whiskers**, are mixed by **mechanical blending (solid phase process)**. The mixture is subsequently pressed to a packing density of up to 75% of the theoretical density, degassed, and consolidated by hot pressing, rolling, extrusion, solid phase sintering, or **super liquidus sintering** [26, 29–33]. The primary advantage associated with this method is the potential to attain a uniform reinforcement distribution and a rapidly solidified matrix microstructure. This technology has been developed by various commercial manufactures such as Alcoa (Pennsylvania, PA, USA), Ceracon Inc. (Sacramento, CA, USA), DWA Composite Specialties Inc. (Chatsworth, CA, USA), Naval Surface Weapons Center, White Oak Laboratory (Silver Spring, MD, USA), and Advanced Composite Materials Corp. (Greer, SC, USA). The primary drawbacks associated with this methodology are twofold: first, the need to handle fine, highly reactive powders, and second, the problems that are associated with the **clustering** of reinforcements [26, 29–33]. In addition, when **super liquidus sintering** is employed for consolidation, extensive **interfacial interactions** between reinforcements and metallic matrices may occur. The liquid phase methods include primarily the Dural process and liquid infiltration. In the Dural process, reinforcements are premixed with a liquid matrix into a **composite slurry** by mechanical agitation. The **slurry** is then poured into a shaped mold to produce a bulk solid

Metallic Powders for Additive Manufacturing: Science and Applications, First Edition.
Enrique J. Lavernia, Kaka Ma, Julie M. Schoenung, James F. Shackelford, and Baolong Zheng.

composite [12, 13, 27, 28, 32–35]. In the **liquid infiltration method**, reinforcements, continuous or discontinuous, are prepared into a porous preform. Molten metal is then pressurized into the preform to produce a composite material [16–23]. The problems that are associated with liquid phase methods are reinforcement **clustering** in discontinuously reinforced composites; interfacial reactions, reinforcement damage, and **reinforcement–reinforcement contact** reinforced composites. The non-uniform distribution of ceramic reinforcements in discontinuously reinforced MMCs can be partially solved using **rheocasting (two-phase process)**. In **rheocasting**, the ceramic particles are introduced into the matrix at temperatures that fall within the solid–liquid two-phase regime. The mixture is vigorously agitated to form a low-viscosity slurry which may be readily poured into a shaped mold [27, 28]. Due to the semiliquid characteristics of the matrix, ceramic particles are mechanically entrapped and are prevented from segregating and agglomerating. Although the rheocasting process is carried out under a relatively low temperature as compared to liquid phase processes, prolonged exposure of reinforcements to the molten matrix is not completely avoided.

Additive manufacturing of composite materials offers several advantages, such as limited exposure of the ceramic particles to elevated temperatures; the ability to selectively add reinforcing particles during deposition, as a function of time and spatial location, thereby providing the ability to fabricate graded or layered composite materials, and; the potential to combine metal-ceramic combinations that would otherwise be incompatible due to the formation of undesirable interfacial products. There are several approaches that have been successfully implemented to incorporate ceramic particles during additive manufacturing. First, one may co-inject particles into metal powders during **atomization**, thereby generating composite powders that can be used as feedstock during AM. Second, ceramic particles can be mechanically milled together with ceramic particles to generate **composite powder aggregates** which can be used as the source material for AM. Third, ceramic particles and metal powders can be co-injected simultaneously into the melt pool during AM.

In the sections that follow, we begin with a discussion of the fundamental factors that govern the interactions of ceramic particles and metal droplets during **co-injection**. To address properly the underlying phenomena the present chapter is divided into five sections. Section 4.2 discusses the various approaches that have been explored in the synthesis of composites using atomization: **ceramic co-injection**, atomization of **premixed composites**, and **reactive spray deposition**. Section 4.3 discusses the fundamental interactions that occur between ceramic particles and metallic droplets during atomization, and factors that influence reinforcement incorporation. Section 4.4 discusses the interaction mechanisms that occur between ceramic particles and the solid–liquid interface during **solidification**, such as particle engulfment behavior and particle entrapment behavior. Finally, Section 4.5 addresses **mechanical milling** (MM) and **direct ceramic co-injection** during additive manufacturing.

4.2 Fabrication Methods

Atomization and **co-injection** studies can be traced back to the direct atomization and deposition (also known as spray deposition) of MMCs, first reported during the mid-1980s and early 1990s, that involved the simultaneous spraying of metals and ceramic particles, a process that is referred to as co-deposition [29, 36–44]. Most of this early work involved using pure Al and commercial Al alloys as metal matrices [29, 36–44]. Since then, a wide variety of matrix-reinforcement combinations has been successfully processed using this approach. The matrix materials that have been used include **Al-based alloys** such as Al–Cu [45–52], Al–Fe [53], Al–Li [54–57], Al–Mg–Si [44, 50, 51, 57–62], Al–Si [63–69], Al–Ti [70] and Al–Zn–Mg–Cu [51, 71–73]; Cu alloys [73–75]; Fe-based alloys [76–78]; **Ti-based alloys** [79–81], and; **intermetallic compounds**, such as CoSi [82], $MoSi_2$ [83, 84], and **Ni_3Al** [85–88]. The reinforcements that are frequently used include **Al_2O_3, graphite, SiC**, TiB_2, TiC, and **steels** in both powder and fiber forms.

4.2.1 Atomization and Co-injection

Atomization and co-injection have been successfully used for the synthesis of composite powder materials [29, 36–44, 48, 49, 54–67]. The term co-injection is used to describe the simultaneous **injection** of particles into atomized droplets. Experimentally, co-injection is generally accomplished by entraining the reinforcements using flowing gas and aiming the gas–reinforcement mixture at the dispersion of atomized droplets. In principle, this method is suitable for the synthesis of discontinuously reinforced composite powders, including fibers and

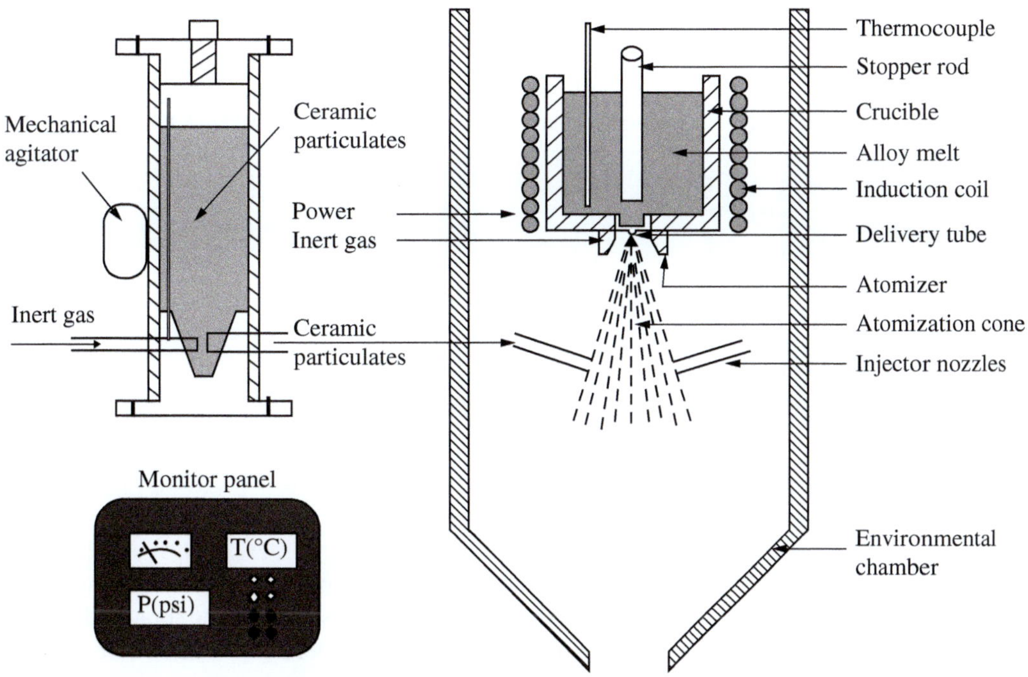

Figure 4.1 Schematic diagram illustrating atomization and co-injection. Source: Wu et al. [63] / Springer Nature.

whiskers. Inspection of the available scientific literature, however, reveals that co-injection has been primarily used for the fabrication of particle-reinforced composite powders presumably, because of the availability and low costs that are associated with the particle type of reinforcements. A schematic diagram of the equipment typically used for the atomization and co-injection of composites is shown in Figure 4.1. Atomization and co-injection involve three consecutive steps: atomization, ceramic particle co-injection, and **solidification**. During atomization, the matrix material is first disintegrated into a fine dispersion of droplets using a high-velocity gas. Although atomization is accomplished in the same manner as that used for unreinforced metals and alloys (see Chapters 2 and 3), it is sometimes necessary to modify the relevant process parameters such as gas pressure and gas-to-metal flow rate ratio. Figure 4.2 shows a simplified diagram illustrating the atomization and co-injection [63]. Following atomization, the droplets travel under the combined forces of fluid drag and gravity and experience **solidification**. Prior to complete

Figure 4.2 The atomization and co-injection process.

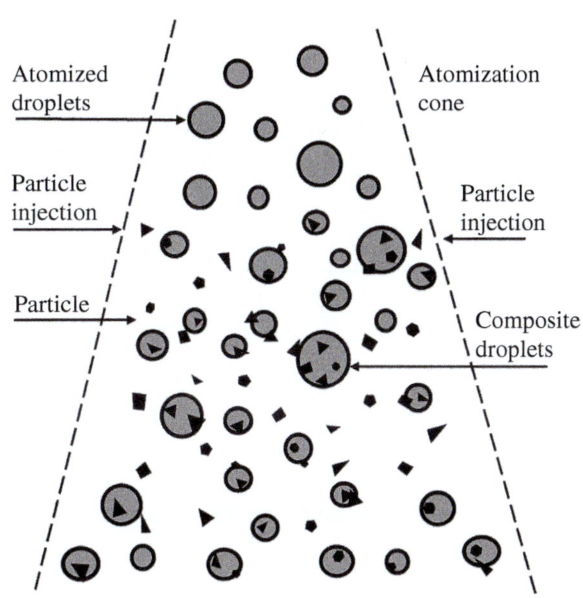

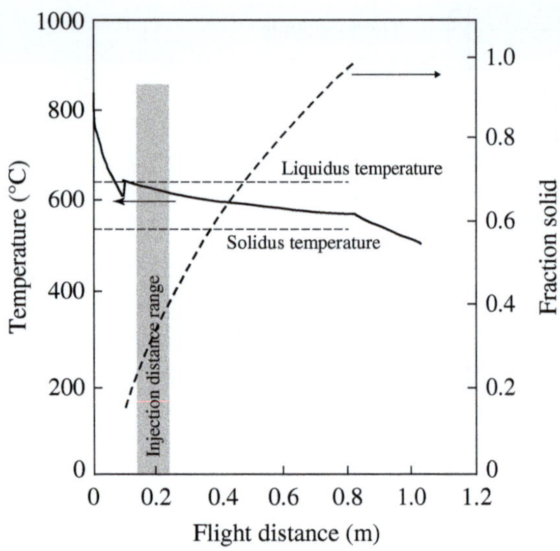

Figure 4.3 Calculated temperature and fraction of solid in atomized Al–Li (average size) droplet as a function of flight distance. Source: Adapted from JOM (formerly Journal of Metals) Vol. 45, No. 7, 1993, pp 424–49, a publication of The Minerals, Metals & Materials Society, Warrendale, PA 15086 [56].

solidification of the droplets, ceramic particles are injected into the atomized spray using one or several gas injectors. This injection procedure allows ceramic particles to interact with the atomized droplets. Finally, the spray of droplets interdispersed with ceramic particles is collected in a cyclone separator. The distance between the atomizer and the injection position is defined as the **injection.**

To understand the influence of droplet solidification during **ceramic co-injection** Gupta et al. [56] calculated the solidification history of an average-sized droplet as a function of flight distance based on a **simplified enthalpy model** as described in detail in Chapter 3. Figure 4.3 shows the calculated fraction of solid phase and droplet temperature as a function of flight distance for Al–Li droplets with an average size of 84 μm [56]. The corresponding injection distances are also indicated in this figure. As shown in Figure 4.3, the average-sized droplets are in the semiliquid state during co-injection. Typically, droplets will contain less than 40 vol% of solid phases at the injection distance [56]. Proper selection of an injection distance must involve several considerations. First, the selected distance must facilitate proper dispersion of ceramic particles to achieve the desirable reinforcement volume fraction and distribution. For example, selection of an injection distance that is too close to the atomizer may lead to nozzle freezing. Alternatively, selection of an injection distance that is far from the atomization region will result in a low yield of injected particles. Second, the injection distance must be selected to minimize or avoid considerable **interfacial interactions** that may evolve as particles are placed in intimate contact with liquid and partially liquid droplets.

In atomization and co-injection, ceramic particles are typically accelerated and directed to the droplet spray by means of gas injectors. Generally, two or more injectors may be used in a single experiment depending on the desired volume fraction of reinforcements. Although a variety of designs may be effectively used to entrain ceramic particles with appropriate gas flow controls, there are two basic types of injectors that have been used for this purpose, namely, a **fluidized bed injector** and a **coaxial injector**. Injectors are generally located outside of the environmental chamber at a location where gas injection is facilitated directly into the chamber. Figure 4.4 shows a typical fluidized bed

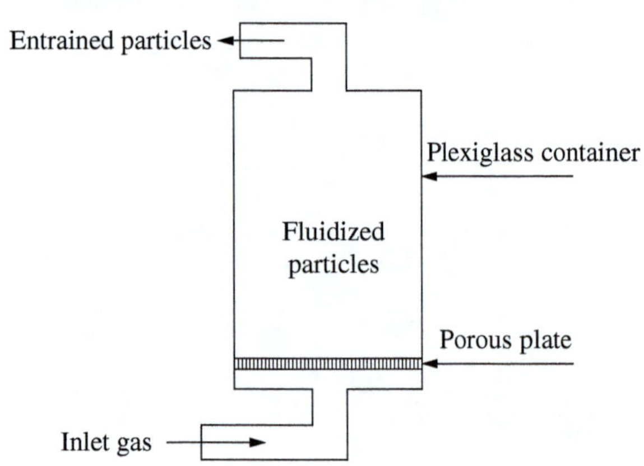

Figure 4.4 A fluidized bed injector. Source: Reproduced by permission of TMS and ASM International [70].

injector [70]. In a fluidized bed injector, ceramic particles are placed inside of a tube containing a porous material (supporter) in a configuration which is identical to that of a static fluidized bed. During injection, a pressurized gas (typically inert) passes through the fluidized bed and entrains the ceramic particles. As the gas pressure increases the fluidized bed begins to expand, accompanied by a decrease in bulk density. When some critical pressurizing condition is achieved, the **drag force** that the fluid exerts on the particles equals the **gravitational force** which holds the particles inside the injector container. This condition is known as the minimum fluidization condition. Under the minimum fluidization condition, particles are free to move relative to each other in the container. The expansion of the fluidized bed can be expressed using a void fraction function [70, 89]:

$$f_{\mathrm{void}} = 1 - \frac{\rho_{\mathrm{p}}}{\rho_{\mathrm{t}}} \tag{4.1}$$

where f_{void} is the volume fraction of void, ρ_{p} is the packing density of particles, and ρ_{t} is the theoretical density of the particle material. The void fraction approaches unity with continued fluidization when the injection gas pressure satisfies the minimum fluidization condition. The velocity that the particles achieve under the minimum fluidization condition is defined as the **terminal velocity** which is given by [70, 89]

$$V_{\mathrm{t}} = \frac{d_{\mathrm{p}}^2(\rho_{\mathrm{t}} - \rho_{\mathrm{g}})g}{18\eta_{\mathrm{g}}} \tag{4.2}$$

where V_{t} is the **terminal velocity**; d_{p} is the particle size; g is the gravitational acceleration, and; ρ_{g} and η_{g} are the density and dynamic viscosity of the injection gas, respectively. If one assumes that the velocity of the injection gas inside of the fluidized injector container is v_{g}, then the mass flow rate of the injection gas can be calculated from

$$\dot{m}_{\mathrm{g}} = \frac{\pi D_{\mathrm{in}}^2}{4} \rho_{\mathrm{g}} v_{\mathrm{g}} \tag{4.3}$$

where $\dot{m}_{\mathrm{g}}$ is the mass flow rate of the injection gas, and D_{in} is the inner diameter of the injector container. The mass flow rate of the reinforcement particles flowing out of a fluid bed injector is therefore proportional to the gas flow rate. Accordingly

$$\dot{m}_{\mathrm{p}} = K_{\mathrm{in}} \dot{m}_{\mathrm{g}} \tag{4.4}$$

where $\dot{m}_{\mathrm{p}}$ is mass flow rate of the reinforcement particles, and K_{in} is an injector constant which is related to injector design and must be parametrically established from fluidization experiments. Gupta et al. [56], for example, determined the fluidized bed constants, K_{in}, for two fluidized bed injectors. The fluidized bed constants, as summarized in Table 4.1, depend on fluidized bed volume and injection gas type. Corresponding values of K_{in} are 1.2 and 0.58 for N_2 and Ar for a fluidized bed volume of 4.73×10^{-4} m^3, and; 1.2 for both N_2 and Ar for a fluidized bed volume of 1.74×10^{-3} m^3. The disadvantage associated with a fluidized bed injector is the relatively low mass flow rate of particles. As an example, Figure 4.5 illustrates the mass flow rates of reinforcement particles and the associated volume fraction of reinforcements as computed for a metal flow rate of 40 g s^{-1} and for fluidized bed volumes of 4.73×10^{-4} m^3 and 1.74×10^{-3} m^3, respectively. It is evident that the volume fraction of reinforcements is generally less than 10% [56].

The geometry of a **coaxial injector** is illustrated schematically in Figure 4.6 [70]. This type of injector is comprised of two coaxial tubes: an injector tube and a nozzle tube which are displaced by a separation distance. During injection,

Table 4.1 Typical operating parameters for fluid bed injectors.

Parameters	Fluidized bed #1	Fluidized bed #2
Volume (m^3)	4.73×10^{-4}	1.74×10^{-3}
Height of fluid bed (m)	0.18	0.18
Fluidizing pressure (MPa)	0.69	0.69
Injector constant (N_2)	1.2	1.2
Injector constant (Ar)	0.58	1.2

Source: Adapted from JOM (formerly Journal of Metals) Vol. 45, No. 7, 1993, pp 424–49, a publication of The Minerals, Metals & Materials Society, Warrendale, PA 15086 [56].

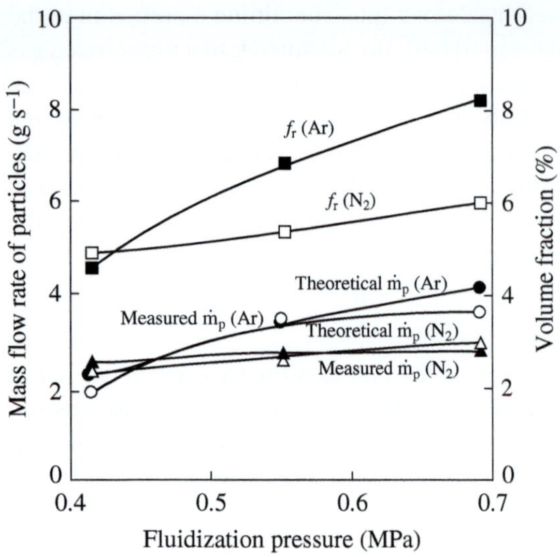

Figure 4.5 Measured and computed mass flow rate, $\dot{m}_p$, and volume fraction, f_r, of **SiC** particles as a function of injection gas type and fluidization pressure for a fluidized bed volume of $1.74 \times 10^{-3}\ m^3$. Source: Adapted from JOM (formerly Journal of Metals) Vol. 45, No. 7, 1993, pp 424–49, a publication of The Minerals, Metals & Materials Society, Warrendale, PA 15086 [56].

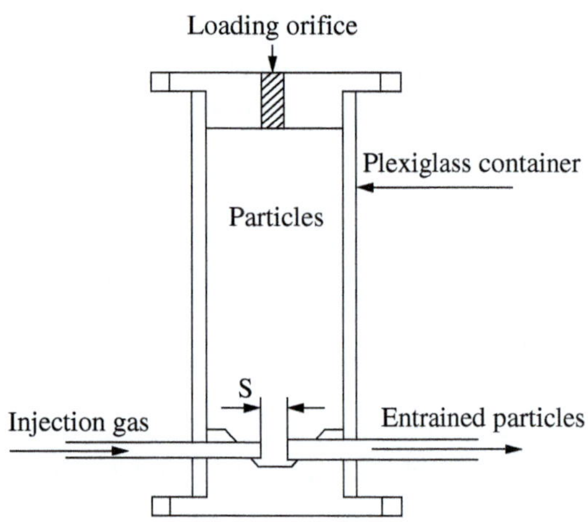

Figure 4.6 A coaxial injector where *S* is the separation distance between injector tube and nozzle tube. Source: Adapted from Gupta et al. [70].

pressurized gas passes through the gas injector tube into the nozzle tube. The injector tube feeds particles into the nozzle tube by generating an aspiration pressure at the separation. Particles are entrained by the injection gas and accelerated in the nozzle tube. Reliable theoretical models for mass flow rate are currently unavailable, because of the complexity associated with particle entrainment in coaxial injectors. Experimentally, however, the mass flow rate may be determined by the variation in mass of particles in the injector as a function of time:

$$\dot{m}_p = \frac{\Delta M_p}{\Delta t} \tag{4.5}$$

where t is the injection time and M_p is the mass of particles in the injector. Zeng et al. [90] suggested that in a coaxial injector, the mass flow of particles is a function of the physical properties of injection gas and particles, injection pressure differential, Δp_{in}, injector tube diameter, d_{in}, separation distance, S, and nozzle tube diameter, d_{noz}. Consequently, for given injection gases and particles the mass flow rate is a function of the following:

$$\dot{m}_p = \dot{m}_p(\Delta p_{in}, d_{in}, S, d_{noz}) \tag{4.6}$$

Zeng et al. [90] studied the flow characteristics in a coaxial injector as illustrated in Figure 4.6 using **SiC** particles with an average size of 15 μm. Figure 4.7 illustrates that the mass flow rate of particles increases with increasing injection pressure in the 0.14–0.21 MPa range. Thereafter, a quasi-steady state is reached where the mass flow rate varies marginally with increasing injection pressure. Figure 4.8 shows that the mass flow rate of particles increases with increasing separation between injector tube and nozzle tube. The injector operating conditions used in obtaining the results in Figure 4.8 are summarized in Table 4.2 [90]. It is evident that a quasi-steady state flow rate is attained at a separation distance greater than 0.76 cm. It is worth noting that the mass flow rates obtained using the coaxial

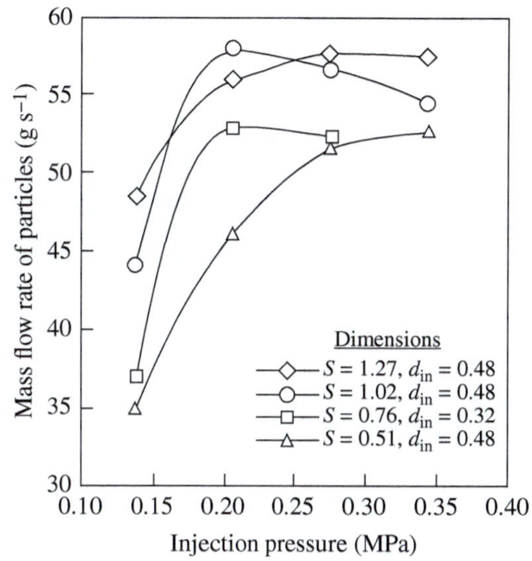

Figure 4.7 Mass flow rate of particles as a function of injection pressure measured for a coaxial injector where S is the separation distance and d_{in} is the diameter of the injector tube (the diameter of the nozzle tube used is 1.27 cm). Source: Adapted from Zeng et al. [90].

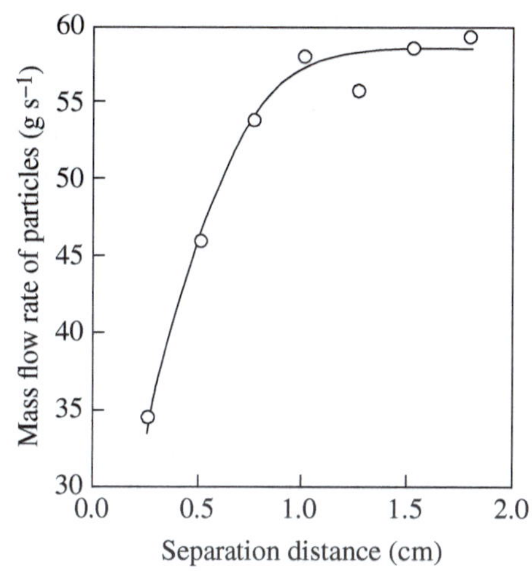

Figure 4.8 Mass flow rate of particles as a function of separation distance measured for a coaxial injector for the following conditions: injector tube diameter = 0.48 cm; nozzle tube diameter = 1.27 cm and injection pressure = 0.21 MPa. Source: Adapted from Zeng et al. [90].

Table 4.2 Operating parameters for a typical coaxial injector.

Parameter	Symbol	Value	Units
Volume of injector	V_{in}	Variable	cm^3
Separation distance	S	Variable	cm
Injection pressure	ΔP_{in}	0.21	MPa
Injection gas	–	N_2	—
Injector tube diameter	d_{in}	0.48	cm
Nozzle tube diameter	d_{noz}	1.25	cm

Source: Adapted from Zeng et al. [90].

type of injector are in the range of 30–60 g s^{-1} for injection pressures in the 0.1–0.5 MPa range which are much greater than those associated with fluidized bed injectors (e.g. <10 g s^{-1} for injection pressures in the 0.4–0.7 MPa range) (see Figures 4.5, 4.7, and 4.8).

It is evident from the previous discussion that in atomization and co-injection experiments ceramic particles are incorporated into the matrix while the latter is in a semiliquid state. Several advantages associated with this particular method include reduced contact temperature between ceramic reinforcements and matrix, minimized **interfacial interactions**, rapidly solidified microstructures, and limited reinforcement segregation [29, 53, 70, 91, 92]. In atomization and co-injection, ceramic particles are incorporated into the matrix by means of one or several gas injectors. Hence, in principle, it is feasible to synthesize MMC droplets reinforced with more than one type of reinforcement. Such materials reinforced with a mixture of reinforcements are referred to as hybrid composites. The fabrication of hybrid MMCs may also be realized by injecting mixtures of different types of ceramic particles.

4.2.2 Atomization of Premixed MMCs

In the atomization of **premixed MMCs**, composites are utilized as starting materials for atomization experiments [46, 47, 54, 68]. Therefore, the reinforcement phases may be introduced by using one of the two methods illustrated in Figure 4.9. The first method involves in situ formation of reinforcement phases. In this approach, the matrix material is first melted, and reinforcements are formed in the crucible by **chemical reactions** between the molten matrix and gaseous or solid phases. The **composite slurry** is then atomized, and composite powders are produced. In the second approach either a solid MMC is used as the starting material for melting and atomization or, alternatively, particles and molten matrix are mixed into a **slurry**. The **slurry** is then atomized into composite droplets.

There are, however, several disadvantages that are associated with the atomization of **premixed MMCs**. First, when reinforcing phases are introduced by remelting **premixed MMCs**, reinforcing phases are exposed to the liquid matrix materials for a relatively long period of time. In this case, thermal exposure and the associated interfacial reactions, depending on the relevant reaction kinetics, may not be avoided. Interfacial reactions limit the combinations of matrix and reinforcements that may be successfully synthesized using this approach. Second, the **fluidity** of a composite slurry decreases with increasing particle volume fraction. Consequently, at elevated reinforcement volume fractions, it may be difficult to deliver the **composite slurry** into a gas atomizer. Viscosity, therefore, presents a limit to the volume fraction of reinforcements that can be successfully incorporated using this method.

4.2.3 Reactive Atomization

In **reactive atomization**, reinforcement phases are introduced into the metal or intermetallic matrix by promoting **chemical reactions** during atomization and during deposition to produce thermodynamically stable dispersoids [76–78, 85, 87, 93]. Based on the conditions of the reacting phases, **reactive atomization** may be classified into one of the three categories: **gas–liquid interactions**; liquid–liquid interactions, and; **liquid–solid interactions** as illustrated in Figures 4.9–4.11 [87].

4.2.3.1 Gas–Liquid Interactions

Gas–liquid interactions represent, perhaps, the most studied area in reactive atomization. This approach combines atomization and reaction into a single-step process to synthesize composite materials as illustrated in Figure 4.10 [85]. It offers an opportunity for in situ, continuous control over alloy composition and **chemical reactions** between atomized droplets and reactive atomization gas. By carefully selecting alloying additive and reactive gas combinations

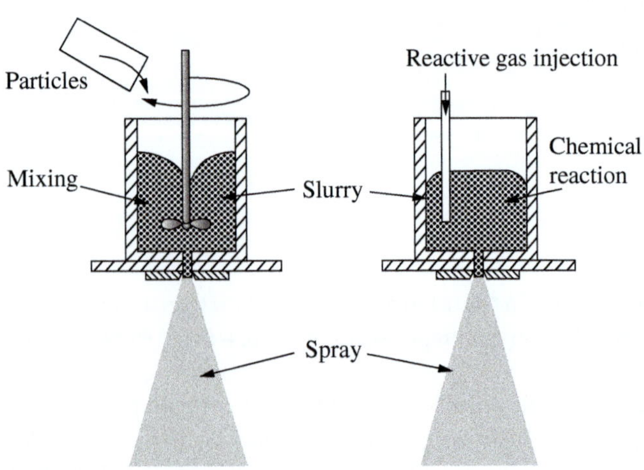

Figure 4.9 The gas atomization synthesis of premixed MMC powders.

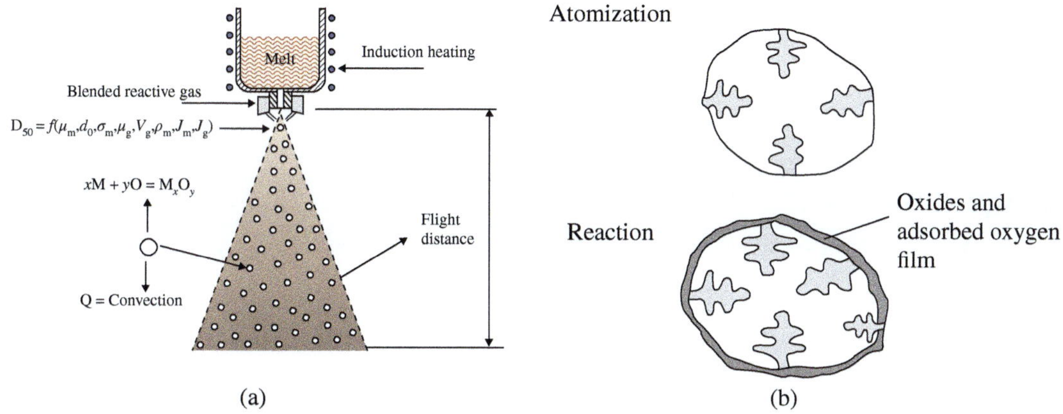

Figure 4.10 The **reactive spray deposition** process involving liquid–gas reactions: (a) reactive atomization and (b) formation of surface oxides.

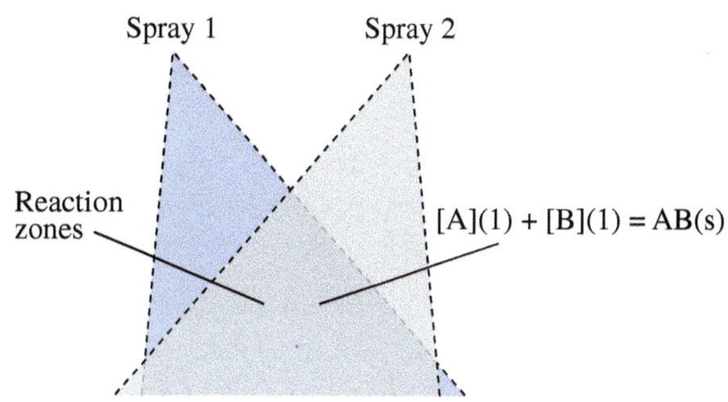

Figure 4.11 Reactive atomization utilizing liquid–liquid reactions.

based on **thermodynamic considerations**, it is possible to synthesize materials containing in situ dispersoids, such as **carbides**, **nitrides**, and **oxides** using this approach. During reactive atomization, a molten alloy is atomized using a reactive gas mixture. Chemical reactions occur between the matrix material and the reactive gas during atomization. In some cases, the reactively atomized powders are captured in flight, a process known as reactive atomization and deposition. Some of these experimental results are described next.

Zeng et al. [85] conducted a series of experiments on the reactive atomization of $\mathbf{Ni_3Al}$ intermetallics using O_2–N_2 mixtures. Table 4.3 summarizes some of the important processing parameters used in this study [85]. The **oxygen** content in the gas mixtures used was in the 5–35 wt% range. In this study, rare earth element Y was added into $\mathbf{Ni_3Al}$ to promote the following oxidation reactions:

$$\mathrm{NiAl[Al](l)} + \mathrm{O_2(g)} \rightarrow \mathrm{NiAl(l)} + \mathrm{Al_2O_3(s)} \tag{4.7}$$

Table 4.3 Typical experimental variables used in reactive atomization of $\mathbf{Ni_3Al}$.

Material	**$Ni_3Al + 2\%$ Y (wt%)**
Atomization gas	$N_2 + O_2^*$
Atomization pressure	2.41 MPa (350 psi)
Pouring temperature	1600 °C
Melt flow rate	$21\,g\,s^{-1}$
Gas flow rate	$9\,g\,s^{-1}$ (estimated)
Flight distance	27.9 cm

a) Containing 5, 15, and 35 wt% O_2, respectively.
Source: Adapted from ASM International and The Minerals, Metals & Materials Society [85].

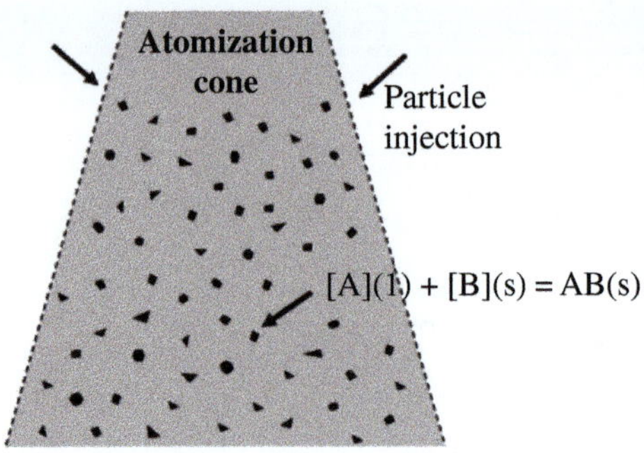

Figure 4.12 TEM micrographs showing the presence of Y_2O_3 in reactively spray-deposited Ni_3Al-Y atomized with an $N_2 + 5$ vol% O_2 gas mixture. Source: Zeng et al. [85]/Reproduced with permission from Springer Nature.

Table 4.4 Oxygen and yttrium content in reactively spray-deposited $\mathbf{Ni_3Al}$.[a)]

Experiment number	Atomization gas	Yttrium (wt%)	Oxygen (wt%)
1	N_2	0.3	0.03
2	$N_2 + 5$ wt% O_2	0.29	0.20
3	$N_2 + 15$ wt% O_2	0.07	0.15
4	$N_2 + 35$ wt% O_2	0.58	2.38

a) Chemical analysis data.
Source: Adapted from ASM International and The Minerals, Metals & Materials Society [85].

and

$$NiAl[Y](l) + O_2(g) \rightarrow NiAl(l) + Y_2O_3(s) \tag{4.8}$$

where "g," "1," and "s" denote gaseous, liquid, and solid phases, respectively. The results of this study demonstrated that metal oxides were formed by reactive spraying with O_2-containing gases [85]. As shown in Figure 4.12, the **oxides** were identified to be primarily Y_2O_3 [85]. Such **oxides** generally exhibit a spheroidal morphology with sizes in the range of a few nanometers to approximately one-tenth of a micrometer. Table 4.4 summarizes the oxygen content in reactively sprayed $\mathbf{Ni_3Al}$ determined using chemical analysis [85]. It is evident that a high O_2 content in the atomization gas and a high Y content in the matrix material promote the formation of **oxides**. For example, the material atomized with an $N_2 + 5$ wt% O_2 gas mixture contains 0.2 wt% of oxygen, whereas that atomized with an $N_2 + 35$ wt% O_2 gas mixture contains up to 2.38 wt% of oxygen.

To introduce $\mathbf{Al_2O_3}$ dispersoids in the Cu matrix Perez and Morris [75] conducted reactive atomization experiments using a Cu–1 wt% Al alloy. In this study, $N_2 + O_2$ gas mixtures were used for atomization to promote the following oxidation reaction:

$$Cu[Al](l) + O_2(g) \rightarrow Cu[1] + Al_2O_3(s) \tag{4.9}$$

The results of this study reveal that the oxygen and Al contents in as-spray-deposited preforms depend on the O_2 concentration in the atomization gas (see Table 4.4) [75]. When the gas mixture contained less than 2 vol% of O_2, the oxygen content in the spray-deposited preform was not detectable. Oxygen contents increased and Al content decreased in the spray-deposited preforms with increasing O_2 concentration in the atomization gas. For example, for a gas mixture containing 8 vol% of O_2, the volume fraction of $\mathbf{Al_2O_3}$ calculated from the oxygen content was 1%, whereas for a gas mixture containing 15 vol% of O_2, the volume fraction of $\mathbf{Al_2O_3}$ decreased to 0.04% because of the significant loss of Al during spray deposition. In this latter case, the extra oxygen present in the spray-deposited preforms was found to be in the form of Cu oxides, that is, CuO and Cu_2O [75]. The feasibility of forming TiC in a steel matrix using CH_4 (methane) as the reactive atomization gas was also investigated by Mathur et al. [73].

In related studies, Unigame et al. [76, 77] and Ranganathan et al. [78] investigated the reactive spray atomization and deposition behavior of **steels** to promote the formation of Al_2O_3 particles as dispersoids through the following oxidation reactions:

$$Fe[Al](l) + O_2(g) \rightarrow Fe(l) + Al_2O_3(s) \tag{4.10}$$

and

$$Fe[Al](l) + [Fe](l) + O_2(g) \rightarrow Fe(l) + FeAl_2O_4(s) \tag{4.11}$$

To provide insight into the mechanisms that govern the evolution of dispersoids from gas–liquid reactions, Zeng et al. [85, 94] analyzed the kinetics involved in the formation of $\mathbf{Al_2O_3}$ in a Ni intermetallic. They proposed that there are five possible steps that may be rate-limiting: (1) mass transfer of oxygen in the gas boundary layer surrounding droplets, (2) adsorption of oxygen atoms at the surface of droplets, (3) mass transfer of oxygen through the oxide film, (4) selective chemical reaction of oxygen with reactive elements, (5) mass transfer of oxygen in droplets (see Figure 4.13). Zeng et al. [85] compared the mass transfer of oxygen in the boundary layer (step 1) with internal diffusion in the droplet (steps 3 and 5), using the Biot number for mass transfer:

$$B_i = \frac{k_d d}{D_i} \tag{4.12}$$

where d is the droplet diameter; D_i is the oxygen diffusivity with "i" being "l" for liquid and "ox" for oxide film, and, k_d is the convective mass transfer coefficient of oxygen. The parameter k_d can be estimated using the Sherwood number, *Sh*, for a spherical droplet:

$$Sh = \frac{k_d d}{D_g} = 2.0 + 0.60\left(\frac{V_d}{\nu_g}\right)^{1/2}\left(\frac{\nu_g}{D_g}\right)^{1/3} \tag{4.13a}$$

for

$$2 \leq Re \leq 800 \tag{4.13b}$$

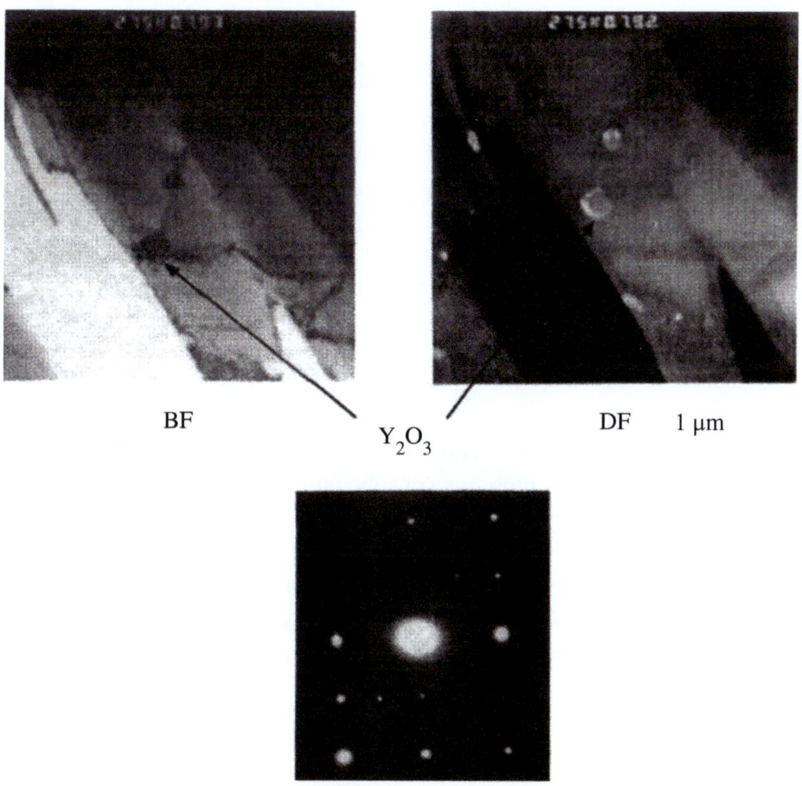

Figure 4.13 Possible **rate-limiting steps** to the reaction between droplet and oxygen where $C_{O2\infty}$ is the oxygen content in the atomization gas mixture. The numbers in parentheses refer to the various reaction steps. Source: Adapted from ASM International and The Minerals, Metals & Materials Society [85].

where D_g is the diffusivity of oxygen in the **gaseous phase**; V is the relative velocity; ν_g is the kinematic viscosity of the gas, and; Re is Reynolds number. From the average droplet diameter of 60 µm and the values of gas viscosity, oxygen diffusivity, and relative velocity of 0.132 cm^2 s^{-1}, 0.178 cm^2 s^{-1}, and 133 ms^{-1}. Zeng et al. [85] estimated that the values of Sh and $k_d d$ were 15.35 and 2.73 cm^2 s^{-1}, respectively. The Biot numbers were calculated to be 3.42×10^6 and 5.25×10^{19}, using $D_l = 8 \times 10^{-7}$ cm^2 s^{-1} [95] for internal diffusion, and $D_{ox} = 5.2 \times 10^{-20}$ cm^2 s^{-1} [96] for diffusion through the **oxide** film, respectively. Consequently, mass transfer through the boundary layer should be much faster than diffusion inside the droplet, and diffusion through the **oxide** film. Zeng et al. [85] further noted that the diffusivity of oxygen through $\mathbf{Al_2O_3}$ is 5.2×10^{-20} cm^2 s^{-1}, whereas that of Al is 3.3×10^{-20} cm^2 s^{-1} at 1200 °C [96]. By simply comparing these data with the diffusivity of oxygen in $\mathbf{Ni_3Al}$ (8×10^{-7} cm^2 s^{-1}), it was concluded that the mass transfer through the oxide film is slower than the mass transfer inside the droplet. Hence, oxides may behave as a diffusion barrier at the surface of the droplet. Zeng et al. [85] further proposed that during the initial stages of atomization, the oxide film produced may be discontinuous, so that the rate of chemical reaction, instead of diffusion, becomes the **rate-limiting step**.

Based on the assumption that diffusion within the oxide film is the **rate-limiting step**, Liu et al. [97] formulated a mathematical model to simulate the oxidation kinetics that are present during reactive atomization. In this study, the temperature of a droplet was calculated from the Fourier equation:

$$\frac{\partial T_i}{\partial t} = \alpha_i \frac{1}{r^2} \frac{\partial}{\partial r}\left(r^2 \frac{\partial T_i}{\partial r}\right) \quad 0 < r < \frac{d}{2}, 0 < t \tag{4.14}$$

where r is the spherical coordinate, and T_i and α_i are the temperature and thermal diffusivity, respectively, with "i" being "s" for solid, "l" for liquid, and "ox" for oxide. At the droplet surface, forced convective and radiative heat exchanges contribute to rapid heat extraction. Liu et al. [97] argued that the frictional heat produced by the interaction between gas and droplet reduces the heat extraction rate. This effect becomes important in a supersonic atomization gas where the compressibility of gas is significant [98]. Hence, Liu et al. [97] adopted a modified Newton's law [98] to describe the cooling of a droplet in high-speed atomization gas:

$$k_i \frac{\partial T_i}{\partial r}\bigg|_{r=d/2} = h_c \left[\left(T_g + r_e \frac{V^2}{2gJc_{p,g}}\right) - T_i\big|_{r=d/2}\right] - \sigma_r \varepsilon_r T_i^4\big|_{r=d/2} \tag{4.15}$$

where k_i and h_c are the conductivity of droplet material and the convective heat transfer coefficient, respectively; r_e, g, and J are the recovery factor, gravitational acceleration, and mechanical equivalent of heat, respectively; $c_{p,g}$ and T_g are the heat capacity and temperature of the atomization gas, respectively, and; σ_r and ε_r are the Stefan–Boltzmann constant and surface emissivity, respectively. The term, $r_e[V^2/(2gJc_{p,g})]$, in Eq. (4.15) represents the effect of frictional heat on the surface cooling of the droplet. When solidification occurs from a concentric growth center, an energy balance condition must be satisfied at the solid–liquid interface, that is, the release rate of the latent heat of fusion equals the net thermal energy conduction rate across the interface:

$$\rho_s \Delta H_f \frac{dR_s}{dt} = k_s \frac{\partial T_s}{\partial r}\bigg|_{r=R_s^+} - k_l \frac{\partial T_l}{\partial r}\bigg|_{r=R_s^-} \tag{4.16}$$

where ρ_s and ΔH_f are the density of solid phase and the heat of fusion, respectively; T_s and T_l are the temperatures of solid and liquid phases, respectively; k_s and k_l are the thermal conductivities of solid and liquid phases, respectively, and; R_s is the instant diameter of the solid phase. The solid–liquid interface velocity, dR_s/dt, depends not only on the conduction heat flow rate but also on the solidification kinetics.

Liu et al. [97] considered that when an oxide film forms at the droplet surface, the thermal energy released raises the local temperature in the vicinity of the oxide/liquid (or oxide/solid) interface. An energy balance condition must then be satisfied at the oxide/matrix interface, that is, the release rate of the oxidation reaction heat must equal the net heat conduction rate across the interface:

$$\frac{1}{2}\frac{\rho_{Al}}{M_{Al}} \Delta H_r \frac{dR_r}{dt} = k_{ox} \frac{\partial T_{ox}}{\partial r}\bigg|_{r=R_r^+} - k_i \frac{\partial T_i}{\partial r}\bigg|_{r=R_r^-} \tag{4.17}$$

where ρ_{Al} and M_{Al} are the density and molar weight of Al, respectively; ΔH_r is the heat of reaction; R_r is the radial position of the oxide/matrix interface, and; k_{ox} and T_{ox} are the thermal conductivity and temperature of the oxide film, respectively. Since it was assumed that the diffusion of oxygen in the oxide film is the **rate-limiting step**, the position and velocity of the oxide/matrix interface can be calculated directly using the parabolic law for high-temperature oxidation [99]:

$$R_r = \frac{d}{2} - (2D_{ox}c_{\infty}t)^{1/2} \tag{4.18}$$

and

$$\frac{dR_r}{dt} = -\frac{1}{2}\left(\frac{2D_{ox}c_{\infty}}{t}\right)^{1/2} \tag{4.19}$$

where D_{ox} is the diffusivity of oxygen in the oxide film and c_{∞} is oxygen concentration in the oxidation atmosphere.

Liu et al. [97] calculated the thickness of $\mathbf{Al_2O_3}$ film during atomization of $\mathbf{Ni_3Al}$. Figure 4.14 illustrates the variation in oxide film thickness as a function of flight distance and droplet size. The results indicate that the oxide film forms within a very short flight time, less than 1.2 μs, on the droplet surface, and then grows into the interior. Once an oxide film forms, its growth will slow down drastically because of the low diffusivity of oxygen through the oxide film. It is evident that oxide thickness increases with increasing droplet size. For example, the steady-state thickness values of $\mathbf{Al_2O_3}$ film corresponding to droplet diameters of 30, 60, and 108 μm are 1.8, 2.0, and 2.3 nm, respectively. This behavior is a result of the high temperature associated with large droplets. Figure 4.15 shows the effect of the oxygen concentration in the atomization gas on $\mathbf{Al_2O_3}$ film thickness on a 60 μm droplet [97]. It is evident that the formation and growth of the oxide film follow similar trends when the oxygen concentration increases from 5 to 35 wt%. The oxide film thickness increases proportionally with increasing oxygen concentration.

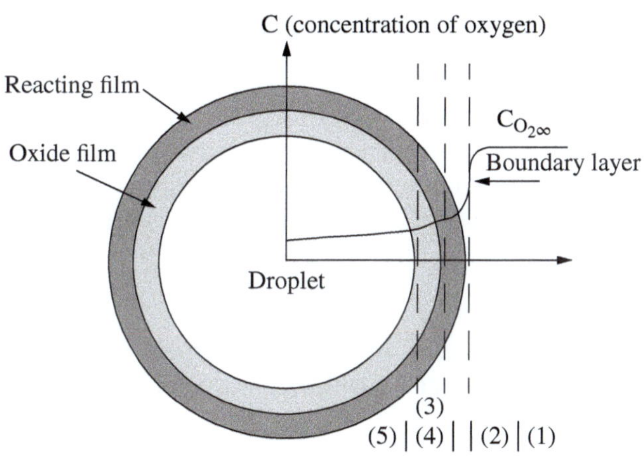

Figure 4.14 Effects of droplet size on the formation and growth of oxide film for three droplet sizes. Source: Reproduced from Acta Metall Mater., 42, H. Liu, R.H. Rangel and E.J. Lavernia, modeling of reactive atomization and deposition processing of **Ni3Al**, p. 3277, copyright (1994), with kind permission from Elsevier Science Ltd, The Boulevard, Landford Lane, Kidlington OX5 1GB, UK [97]/with permission from Elsevier.

Figure 4.15 Effects of oxygen concentration in the atomization gas on the formation and growth of oxide film. Source: Reproduced from Acta Metall Mater., 42, H. Liu, R.H. Rangel and E.J. Lavernia, modeling of reactive atomization and deposition processing of Ni3Al, p. 3277, copyright (1994), with kind permission from Elsevier Science Ltd, The Boulevard, Landford Lane, Kidlington OX5 1GB, UK [97] / with permission from Elsevier.

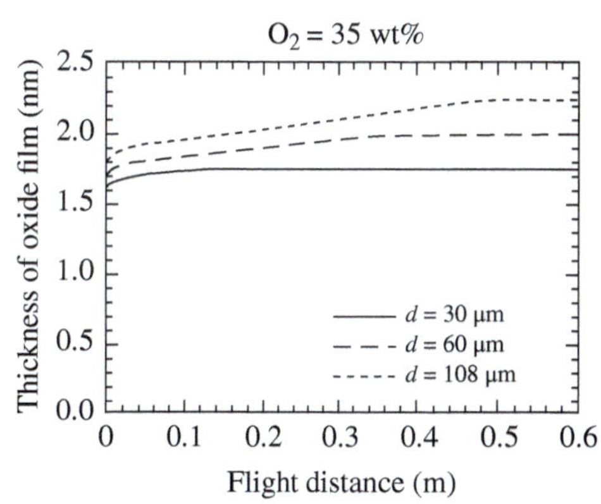

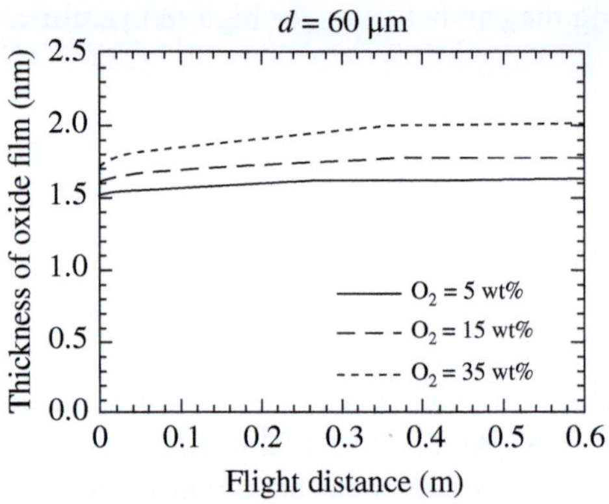

Figure 4.16 Reactive atomization utilizing liquid–solid reactions.

The steady-state thickness values of the $\mathbf{Al_2O_3}$ film corresponding to oxygen concentrations of 5, 15, and 35 wt% which Liu et al. [97] calculated were 1.6, 1.8, and 2.0 nm, respectively (see Figure 4.16). The overall volume fraction of the surface oxides in atomized powders estimated based on the calculated oxide film thickness was 0.02%. In view of the low volume fraction of Al_2O_3 calculated for atomization, Zeng et al. [94] proposed that the residence time of the droplets in the atomization stage is significantly shorter relative to the time that they are exposed to reactive gas during deposition. The volume fraction of oxides should therefore be governed by deposition. Experimental support to this suggestion is provided by the results obtained by Ranganathan et al. [78] who noted that the oversprayed Fe–Al powders produced during **reactive spray deposition** using N_2–O_2 mixtures were devoid of traces of oxides. However, detailed results regarding the reaction kinetics during the deposition stage are not currently available.

4.2.3.2 Liquid–Liquid Interactions

Reactive atomization using liquid–liquid interactions involves the mixing of two molten metals either prior to or following atomization, as illustrated in Figure 4.11. This methodology has not been explored in detail. It has been suggested that, in principle, reactions between two liquid streams may produce refractory particle dispersions [87]. The size of the resultant dispersoids may be controlled by the cooling rate of the droplets [87]. One example of **liquid–liquid reactions** is provided by Cu alloys as

$$\mathrm{Cu[Ti](l) + Cu[B](l) \rightarrow Cu(l) + TiB_2(\,s)} \tag{4.20}$$

in which alloying elements Ti and B from two different sprays of Cu react leading to the formation of TiB_2 dispersoids. Preliminary results obtained using this approach to produce Cu alloys with extremely fine dispersions of TiB_2 particles are encouraging [100].

4.2.3.3 Liquid–Solid Interactions

Reactive atomization and deposition using **liquid–solid interactions** are illustrated in Figure 4.12. In this process, solid particles of one of the reactants are injected into the atomized spray of a matrix. The particles react with the matrix during atomization and subsequently during deposition to produce reinforcement phases. This process has been used to produce stainless steels containing TiN particle dispersions [87, 101]. In this case, stainless steels containing 2 wt% Ti were spray processed with Cr_3N_2 or Cr_xN co-injection. During interactions, the chromium nitride particles react with the matrix to produce TiN particles according to the following reaction:

$$\mathrm{[Ti](l) + Cr_x\,N(\,s) \rightarrow Ti[Cr] + TiN(s)} \tag{4.21}$$

Available experimental results demonstrate that stainless steels containing up to 2 vol% of TiN particles may be produced using this method [87, 101].

Liquid–solid reactions have also been used to produce TiC particles in Fe–5 wt% Ti alloy [74]. Majagi et al. [74] co-injected cast iron (Fe–2.56C) particles (d_{50} <45 μm) into a Fe–5 wt% Ti alloy during atomization. Their theoretical and experimental results demonstrated that Fe–C particles completely penetrated atomized Fe–Ti droplets during co-injection (see Figure 4.17). Carbon in cast iron particles is expected to react with Ti in Fe–Ti solution during atomization and during deposition to form TiC particles according to

$$\mathrm{Fe[Ti](l) + Fe[C](s) \rightarrow Fe + TiC(s)} \tag{4.22}$$

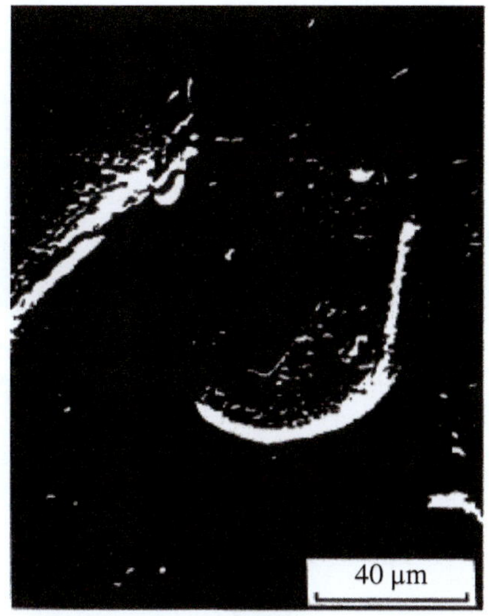

Figure 4.17 Evidence of penetration of Fe–C particles into Fe–Ti droplets during atomization and co-injection. Source: Majagi et al. [74]/Reproduced with permission from Minerals, Metals, & Materials Society.

Some other **liquid–solid interactions** have also been suggested for this approach. These interactions include

$$Ti(l) + SiC(s) \rightarrow Ti[Si] + TiC(s) \tag{4.23}$$

$$Cu[Al](l) + CuO(s) \rightarrow Cu + Al_2O_3(s) \tag{4.24}$$

and

$$Fe[Ti](l) + Fe[B](s) \rightarrow Fe + TiB_2(s) \tag{4.25}$$

4.3 Incorporation of Reinforcements During Co-injection

As discussed in Section 4.2, there are several approaches that have been successfully utilized to incorporate reinforcing particles during spray deposition. Inspection of the relevant literature reveals that the most widely used approach, however, involves the co-injection of a stream of inert gas with entrained particles into the atomized droplet spray (see Figure 4.2). The problem posed by the incorporation of a distribution of ceramic particles into a metallic matrix during co-injection may be best analyzed in the context of a generalized gas-to-metal transition. A review of the available scientific literature demonstrates that the gas-to-metal transfer problem is not limited to atomization. In **liquid-phase processing** of MMCs, for example, gas injection and/or mechanical stirring are often applied to incorporate ceramic particles into molten matrices [102, 103]. The gas-to-metal transition of ceramic particles is also encountered in novel surface modification technologies in which the surface of a metallic material is partially melted using high-energy sources such as laser or electron beams [104–106]. Simultaneously, ceramic particles are incorporated into the molten metal pools by means of gas injection to produce a layer of abrasive-resistant MMC. Ayers et al. [104, 105], for example, studied the surface modification of Al alloys using ceramic injection. In this study, the surface of a commercial 6061 Al alloy was scanned and melted using a CW–CO_2 laser beam. Simultaneously, TiC particles were injected into the pool of molten metal by means of gas injection. The results reported by Ayers et al. [104, 105], shown in Figure 4.18, reveal that a composite layer of approximately 1 mm in thickness was produced on the surface of 6061 Al [104, 105].

To address the problem of gas-to-metal transition in the context of co-injection of ceramic particles into atomized droplets, the discussion that follows is divided into three sections. The first section describes the physical mechanisms that are thought to be responsible for the incorporation of ceramic particles into atomized droplets. The second section discusses the various models that are formulated to elucidate the process of particle penetration into droplets. The third section addresses particle penetration into **solid droplets**.

4.3.1 Incorporation Behavior of Reinforcements

Available experimental evidence suggests that there are several interaction mechanisms that govern the incorporation of ceramic particles into matrices during atomization and co-injection illustrated in Figure 4.2. These mechanisms are

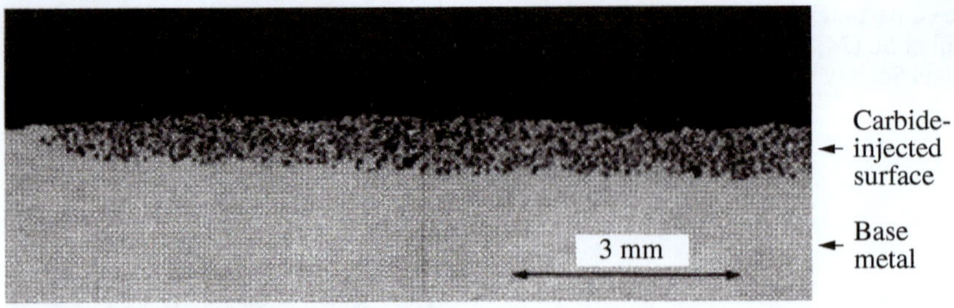

Figure 4.18 TiC particles incorporated into the surface layer of a 6061 Al matrix by laser surface melting and particle co-injection. Source: Ayers [104]/from Elsevier.

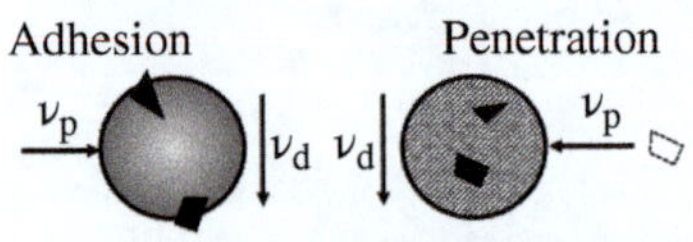

Figure 4.19 Incorporation of ceramic particles during atomization and co-injection.

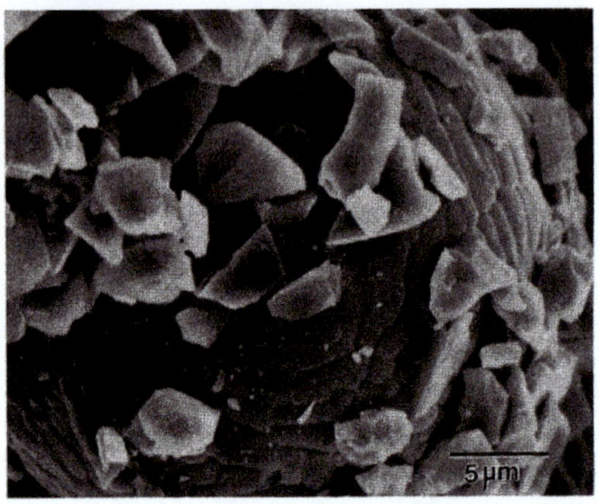

Figure 4.20 The presence of **SiC** particles on the surface of Al–Li/**SiC** powders [46]. Source: Lavernia et al. [55]/Reproduced with permission from Springer Nature.

illustrated in Figure 4.19 and include the following: penetration of ceramic particles into droplets and the interaction of the particles with the **solidification front** [55]. In the case when the injected particles lack sufficient kinetic energy to penetrate droplets they adhere to droplet surfaces (Figure 4.20). Ceramic particles may penetrate atomized droplets when the particles possess sufficient kinetic energy relative to that of the droplets [55].

Penetration of particles into atomized droplets as illustrated in Figure 4.19 has been confirmed by numerous investigators by examining the microstructure of powders collected from atomization experiments [63, 64, 73, 74, 86, 107]. Wu and Lavernia [64], for example, reported the penetration of **SiC** and TiB_2 reinforcements into Al–Si droplets. In related work, Liang and Lavernia [86] studied the atomization and co-injection of Ni_3Al-based IMCs by co-injecting TiB_2 particles into the atomized spray. They found that TiB_2 particles penetrated the Ni_3Al droplets during co-injection. An example of the Ni_3Al/TiB_2 powders produced during atomization and co-injection in Figure 4.21 [86]. Penetration of reinforcement particles into droplets during atomization has also been observed in other materials. Mathur et al. [73], for example, co-injected W particles into a spray of Cu droplets to synthesize Cu-based MMCs. The results of this study showed that W particles penetrated the Cu droplets during co-injection. As a result, the W particles were uniformly embedded in the oversprayed Cu powders.

The interactions that occur between co-injected ceramic particles and atomized droplets are dependent on factors such as atomization and co-injection parameters, droplet solidification conditions, physical properties of the particles and those of the droplets, as well as the wetting behavior between the particles and the droplets. With reference to penetration behavior, Wu and Lavernia [64] conducted a systematic study on the Al–Si/SiC system. To that effect, they atomized an Al–4 wt% Si alloy into droplets and co-injected SiC particles during atomization. To provide insight into the factors that affect the penetration behavior of ceramic particles, Wu and Lavernia studied the distribution and volume fraction of SiC particles in atomized droplets as a function of droplet size. They observed that the distribution of ceramic particles in atomized droplets depends critically on droplet size. As shown in Figure 4.22 in droplets that are smaller than 98 μm, SiC particles are relatively uniformly distributed, whereas in large droplets (>180 μm) SiC

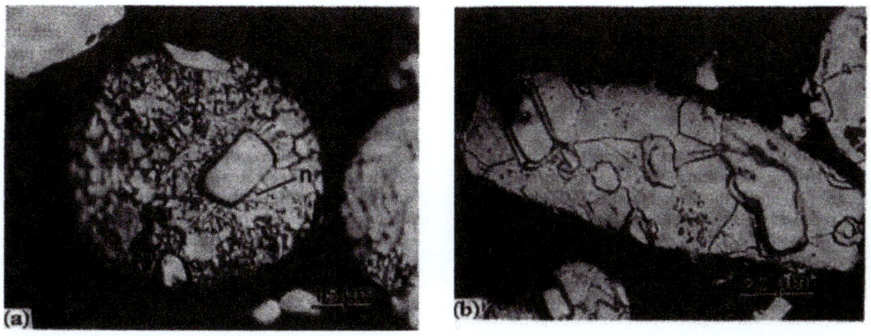

Figure 4.21 The microstructures of Ni_3Al/TiB_2 powders showing the penetration of TiB_2 particles into Ni_3Al droplets (powder size: 63–90 μm). Source: Liang et al. [86] / from Elsevier. (a) Spherical powder and (b) elongated powder.

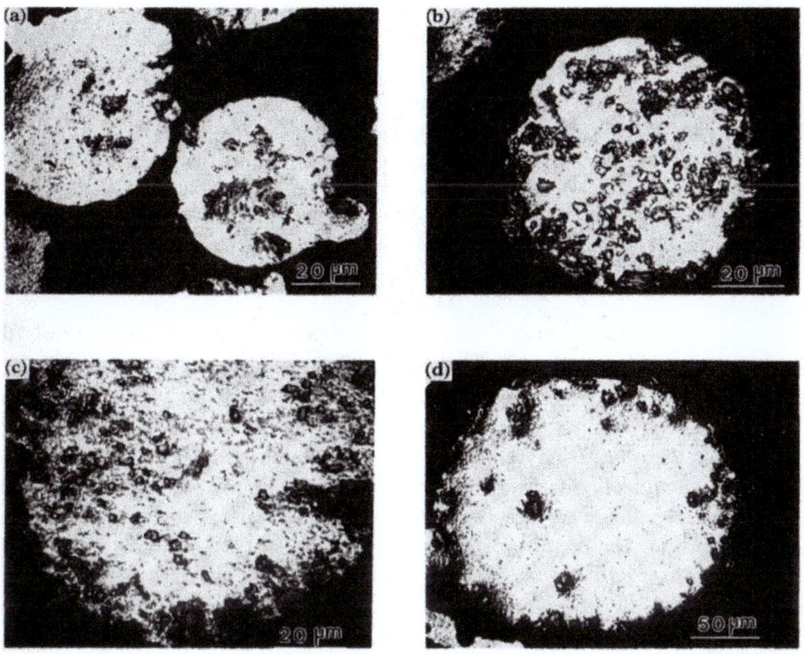

Figure 4.22 Optical micrographs showing the distribution of SiC particles in Al–4 wt% Si droplets of various diameters *d*: (a) $d = 43$ μm; (b) $d = 83$ μm; (c) $d = 116$ μm, and; (d) $d = 230$ μm. Source: Zhang et al. [64]/from Elsevier.

particles are segregated to droplet surface regions [64]. For example, in the 230 μm droplets shown in Figure 4.22, SiC particles are present primarily at droplet surfaces. Wu and Lavernia further determined the volume fraction of the SiC particles that are present in the atomized droplets. The results of their study, as shown in Figure 4.23, reveal that the volume fraction as a function of droplet size exhibits a maximum value at intermediate droplet size (e.g. 98 μm for Al–4 wt% Si/SiC system). Accordingly, both increasing and decreasing droplet size lead to a decrease in the amount of ceramic particles that are incorporated into the droplets during co-injection.

The penetration behavior of co-injected particles, as illustrated in Figures 4.22 and 4.23, may be rationalized in part by considering droplet solidification behavior. To that effect, a simple approach is to compare the droplet temperature with the liquidus temperature at the injection distance and assuming no solidification. Accordingly, if the temperature of a droplet is greater than the liquidus temperature the droplet may be safely assumed to be completely molten. If the temperature falls below the liquidus temperature the droplet may be in one of three states: undercooled liquid, partially solid, and completely solid. Such a comparison was made by Wu and Lavernia [64] by calculating the temperature of an Al–4 wt% Si droplet under Newtonian conditions and assuming a thermal energy balance:

$$\rho_l c_{p,l} V_d \frac{dT}{dt} = -hA_d(T - T_g) \tag{4.26}$$

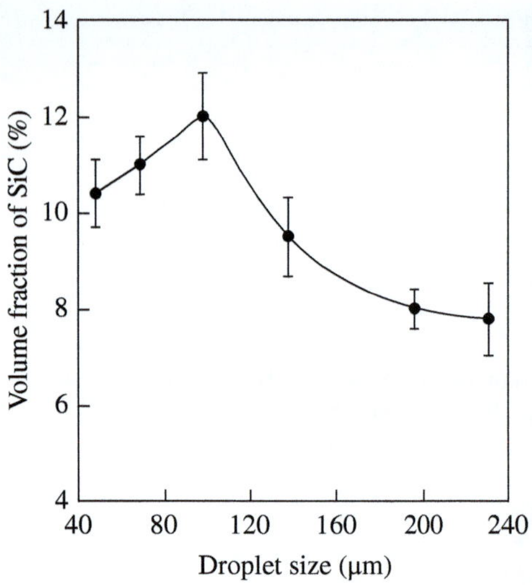

Figure 4.23 Volume fraction of SiC particles in Al–4 wt% Si droplets as a function of droplet size. Source: Adapted from TMS and ASM International [64].

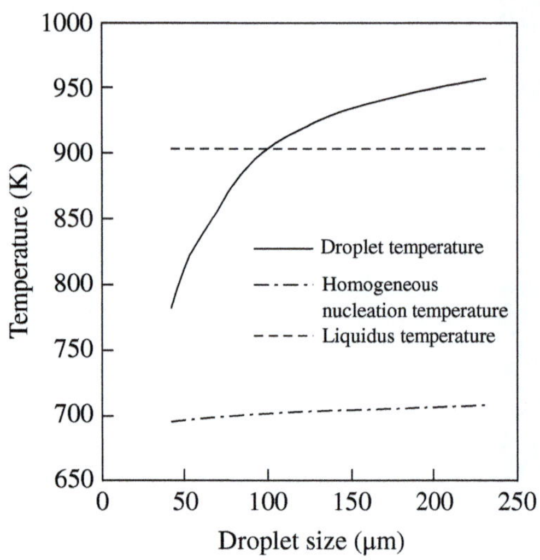

Figure 4.24 Comparison of droplet temperature (assuming no solidification) with liquidus temperature and homogeneous nucleation temperature for Al–4 wt% Si droplets. Source: Adapted from TMS and ASM International [64].

where V_d is the droplet volume; A_d is the droplet surface area; ρ_l and $c_{p,l}$ are the density and heat capacity of the liquid; T is the droplet temperature, and; T_g is the atomization gas temperature. The results of these calculations are summarized in Figure 4.24 for Al–4 wt% Si droplets as a function of droplet size and are compared with the liquidus temperature [64]. As expected, the temperature of liquid droplets decreases rapidly with decreasing droplet diameter. Accordingly, at the injection distance, 230 μm droplets are still 51 K above the liquidus temperature, T_L, whereas 42 μm droplets are already 121 K below T_L. A comparison of the droplet temperatures and the liquidus temperature, as shown in Figure 4.24, yields the following findings for droplets of different sizes during co-injection. First, when the droplet diameters are greater than 98 μm, they are above the liquidus temperature, and consequently, are completely liquid. Second, when droplets are less than 98 μm, their temperatures fall below the liquidus temperature. Consequently, such droplets may be in one of three possible states: liquid, semiliquid, or solid. From an analysis of the solidification characteristics of droplets at the injection distance, Wu and Lavernia proposed that there are two mechanisms that are responsible for the observed distribution and the volume fraction of SiC particles in droplets: penetration of particles into droplets and segregation of particles during penetration [64]. The first mechanism will be addressed in Section 4.3.2. The second mechanism in which it is proposed that particles may segregate to the surface of droplets during solidification implies that to attain a homogeneous distribution of particles penetration alone is not sufficient. Accordingly, the interactions of the particles with the **solidification front** at the droplet interior must be taken into consideration; this phenomenon is discussed in detail in Section 4.4.

The observed decrease in reinforcement volume fraction, as illustrated in Figure 4.23, may be the result of increasing penetration resistance with decreasing droplet size. Penetration resistance is greater in small droplets because of

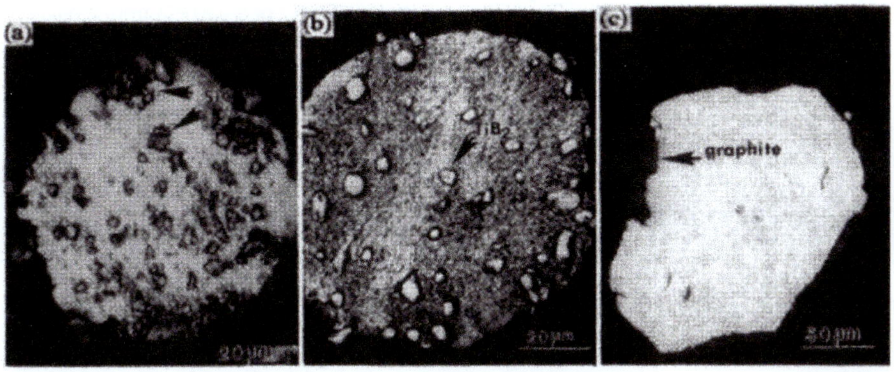

Figure 4.25 Micrographs showing the presence of ceramic particles in Al alloy droplets produced by atomization and co-injection: (a) Al–4 wt% Si/SiC; (b) Al–4 wt% Si/TiB_2, and; (c) 6061 Al/graphite. Source: Flemings et al. [63] / from Springer Nature.

the elevated fraction of solid. The influence of solid fraction on penetration behavior will be discussed in detail in Sections 4.3.2 and 4.3.3. On the contrary, the decrease in reinforcement volume fraction with increasing droplet size observed in the large droplet size regime (e.g. >98 μm as indicated by Figure 4.23) is the result of reinforcement redistribution during solidification. As the droplet size increases, the probability of particle segregation increases. The factors that influence reinforcement redistribution during solidification will be discussed in detail in Section 4.4.

Regarding the effect of particle type on incorporation behavior during atomization and co-injection, available experimental results demonstrate that penetration is dependent critically on the physical and chemical characteristics of both particles and droplets. Figure 4.25 shows the microstructure of Al–Si/SiC, Al–Si/TiB_2 and 6061 Al/**graphite** droplets produced by atomization and co-injection [63]. This figure reveals that penetration is extensive in the Al–Si/SiC and Al–Si/TiB_2 systems. However, penetration of **graphite** into 6061 droplets appears to be difficult. For example, as shown in Figure 4.25, a relatively large number of ceramic particles are present throughout the cross-section of Al–Si/SiC and Al–Si/TiB_2 droplets. However, in the 6061 Al/**graphite** system, **graphite** particles are present primarily at droplet surfaces. Wu et al. [63] attributed these experimental observations to the differences in density and wetting behavior among the various types of particles. Accordingly, a large particle density and a low wetting angle with the liquid will facilitate penetration [63]. The differences in penetration behavior associated with various reinforcement particles have also been noted by other investigators working on the atomization and co-deposition synthesis of composites. Mathur et al. [73], for example, co-injected W and Al_2O_3 particles into a Cu droplet spray to synthesize Cu-based composites. They compared the microstructure of oversprayed Cu/W and Cu/Al_2O_3 powders and reported significant differences in penetration behavior. When co-injected with W, a large fraction of W particles was observed in the Cu droplets. In contrast, extensive penetration was not observed in the Cu/Al_2O_3 system. In this latter case, Al_2O_3 particles were noted to reside primarily at the surface of Cu powders. Interestingly, Mathur et al. noted that W particles were uniformly distributed in the spray-deposited microstructure of Cu/W MMCs. In this case, W particles were observed both intragranularly and transgranularly. In contrast, in the spray-deposited Cu/Al_2O_3 MMCs, extensive segregation of Al_2O_3 was observed. The observed differences in penetration behavior of W and Al_2O_3 were attributed to the large differences in density between these two types of particles as well as to their large difference in wettability with molten Cu. Mathur et al. suggested that since W exhibits a high density and a good wettability (i.e. low wetting angle) with Cu its extent of penetration into Cu droplets exceeds that of Al_2O_3 [73].

4.3.2 Penetration of Semiliquid Droplets

To enhance our understanding of the fundamental mechanisms that govern **droplet penetration** during co-injection researchers working in this area have proposed several physical models [63, 64, 74, 108]. The penetration of ceramic particles may occur either during the collision with atomized droplets or subsequently during interactions with the spray dispersion. In both cases, particles are at a relative velocity with the droplets. An analogous problem of particle penetration into a matrix material has also been addressed by researchers working in the areas of gas injection and centrifugal incorporation of ceramic reinforcements into molten alloys [102, 103]. Injection of ceramic particles into

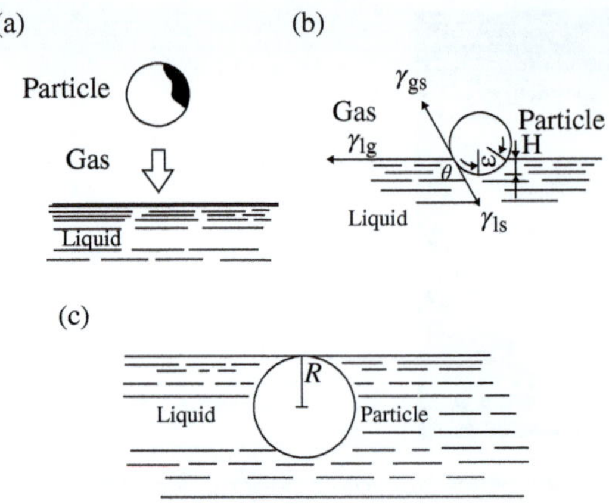

Figure 4.26 Steps involved in the transfer of a particle from a gas into a liquid: (a) before penetration; (b) particle on liquid surface, and; (c) after penetration. R is particle radius; θ is **wetting angle**; H is penetration distance, and; γ_{gs}, γ_{lg}, γ_{ls} are the gas–solid, liquid–gas, and liquid–solid surface energies, respectively. Source: Reproduced by permission of the Institute of Materials [12].

a molten alloy is illustrated schematically in Figure 4.26 [12]. In this process, ceramic particles are accelerated using a **carrier gas** and are directed onto the surface of a molten alloy. When the injection velocity is sufficiently high, the ceramic particles penetrate the liquid surface leading to incorporation. The particle penetration behavior that is encountered during gas injection onto a liquid surface is analogous in that both processes involve the dynamic transfer of reinforcement particles from a gaseous phase into a matrix material. The only significant difference is that in atomization, the matrix material is in the semiliquid, liquid, or solid states. The formulation of a physical model to describe the process of particle penetration during co-injection typically involves an **energy balance**, a **force balance**, or combinations thereof [63, 64, 74, 108]. Such models are discussed in detail below.

4.3.2.1 Energy Balance

An energy balance formulation considers the changes in energy that are associated with the particle penetration process. For the penetration phenomena illustrated in Figure 4.26, a particle possesses a certain amount of kinetic energy which is a result of the relative motion between the particle and the matrix material (droplet or bulk liquid). This kinetic energy provides the driving force for penetration. In the specific case when the matrix is molten, penetration of a particle causes an increase in the free energy of the particle/matrix system, when the particle exhibits a large **wetting angle** with the liquid, that is, greater than 90°. Accordingly, a simple energy balance suggests that when the increase in free energy in the particle/matrix system resulting from the particle incorporation is offset by the kinetic energy of the particle, penetration will occur. The kinetic energy of a particle is well established and is given by

$$E_k = \frac{1}{2} m_p V^2 \tag{4.27}$$

where E_k is the kinetic energy of the particle relative to the matrix material; m_p is the mass of the particle, and V is the relative velocity between the particle and the matrix. In the case of gas injection incorporation, as illustrated in Figure 4.26, the liquid metal is stationary and V becomes the particle velocity. In the case of collisions between a droplet and a particle, as encountered during atomization and co-deposition, both the particle and the droplet are in motion, and V is then the relative velocity between them. The increase in free energy in the particle/matrix system due to penetration (e.g. gas-to-matrix transfer) is a result of the replacement of a particle/gas interface by a particle/matrix interface. The interfacial free energy prior to penetration is the **interfacial energy** between the particle and the environmental gas:

$$\Delta E_{pg} = A_p \gamma_{pg} \tag{4.28}$$

where ΔE_{pg} is the total **interfacial energy** between the particle and the gas; A_p is the particle surface area, and; $\gamma_{pg}$is the **interfacial energy** between the particle and the gas. When the particle is completely immersed in the liquid the total **interfacial energy** becomes

$$\Delta E_{pl} = A_p \gamma_{pl} \tag{4.29}$$

where ΔE_{pl} is the total **interfacial energy** between the particle and the liquid, and; γ_{pl} is the **interfacial energy** between the particle and the liquid. The net increase in interfacial free energy is therefore the difference between Eqs. (4.28) and (4.29). Accordingly

$$\Delta E_s = A_p(\gamma_{pl} - \gamma_{pg}) \tag{4.30}$$

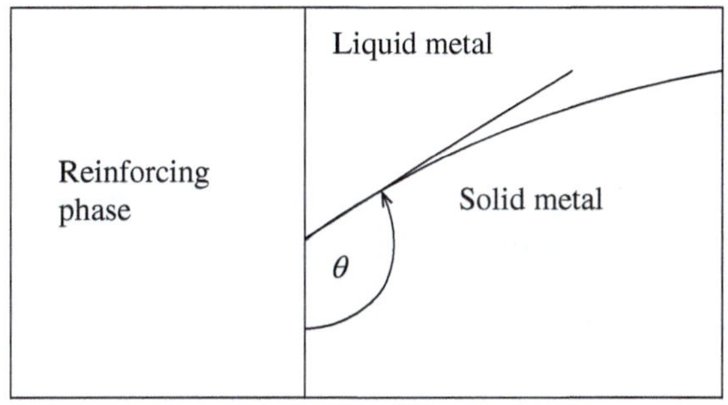

Figure 4.27 Definition of the **wetting angle**, θ, between solid metal and the reinforcement phase. Source: Reproduced by permission of the Institute of Materials and ASM International [23].

where ΔE_s is the increase in interfacial free energy due to penetration. Considering the schematic illustration shown in Figure 4.27 [23], the particle/gas interfacial energy, γ_{pg}, and the particle/liquid interfacial energy, γ_{pl}, are related by Young's equation [109]. From a surface tension force balance consideration, Young's equation is given by

$$\gamma_{pg} = \gamma_{pl} + \gamma_{lg} \cos\theta \tag{4.31}$$

where γ_{lg} is the **interfacial energy** between the liquid and the environmental gas, and; θ is the **wetting angle** between the liquid and the particle. The values of these two parameters γ_{lg} and θ are readily available in the literature, although the accuracy of many published results must be scrutinized due to variability in the experimental techniques utilized. Some of the published interfacial energy and **wetting angle** values are summarized in Table 4.5 [110, 111] and Table 4.6 [112–117, 119, 120]. Combining Eqs. (4.30) and (4.31) yields an expression of the increase in interfacial

Table 4.5 Surface energies of pure liquid metals at the melting temperatures.

Metal	Surface energy (mJ m^{-2})	Temperature (K)	References
Al	866	933	[110]
Cu	1300	1356	[110]
Fe	1880	1809	[111]
Mg	560	922	[111]
Ni	1780	1728	[111]
Ti	1650	1940	[111]

Table 4.6 **Wetting angle**, θ, between liquid metals and ceramics.

Metal	Ceramic	Temperature (K)	θ (degree)	References
Al	Al_2O_3	NR	110	[102]
	Graphite	1273	155	[112]
	SiC	933	130	[113]
	TiB_2	NR	70	[114]
	ZrO_2	1173	145	[115]
Cu	Al_2O_3	1423	100–155	[116]
	SiC	1408	106–167	[117]
	ZrO_2	1373	126	[115]
Ni	Al_2O_3	—	108–111	[118]
Si	SiC	1723	33	[119]

NR, not reported.

free energy in terms of the liquid–gas interfacial energy and **wetting angle** as

$$\Delta E_{\rm s} = -A_{\rm p}\gamma_{\rm lg}\cos\theta \tag{4.32}$$

Accordingly, in the simple energy balance approach, a particle may penetrate a liquid (droplet or bulk) when its kinetic energy is greater than the free energy increase caused by penetration. The mathematical expression of this criterion is therefore

$$\Delta E_{\rm k} \geq -A_{\rm p}\gamma_{\rm lg}\cos\theta \tag{4.33}$$

Alternatively, Eq. (4.33) may be expressed in terms of the relative velocity as

$$V^2 \geq \frac{-2A_{\rm p}\gamma_{\rm lg}\cos\theta}{m_{\rm p}} \tag{4.34}$$

If one further assumes that the particle is spherical, Eq. (4.34) may be simplified in terms of readily measurable parameters, that is, particle diameter and particle density:

$$V^2 \geq \frac{-12\gamma_{\rm lg}\cos\theta}{\pi\rho_{\rm p}d_{\rm p}} \tag{4.35}$$

where $\rho_{\rm p}$ and $d_{\rm p}$ are the density and diameter of the particle.

4.3.2.2 Force Balance

The force balance approach is effectively used when considering the penetration of a ceramic particle into a liquid as present during centrifugal incorporation. Centrifugal processes are fundamentally different from gas injection in that a centrifugal force is imparted on the particle because of centrifugal rotation. At low rotational speeds, the centrifugal force is balanced by a resistance force that is generally attributable to the surface tension of the liquid. Accordingly, no penetration occurs at low rotational speeds [102]. However, when the rotational speed is sufficiently high, the centrifugal force acting on the particle will be able to overcome the resistance forces, and incorporation of particles may occur. Kacar et al. [102] studied the penetration behavior of ceramic particles into liquid metals during centrifugal rotation. They defined that a force is positive if it is pointing from the particle into the liquid. In this study, Kacar et al. considered the effect of three independent forces on penetration behavior: inertial force, surface energy force, and buoyant force. Accordingly, the total force acting on a particle during centrifugal rotation is given by

$$\Sigma F = ma\left(1 - \frac{\rho_{\rm l}}{\rho_{\rm p}}\right) + K_{\rm s}\left(\frac{m_{\rm p}}{\rho_{\rm p}}\right)^{1/3}\gamma_{\rm lg}\cos\theta \tag{4.36}$$

where $\rho_{\rm l}$ is the density of the liquid; $m_{\rm p}$ is the mass of the particle and $K_{\rm s}$ is a shape factor. The condition for particle penetration (incorporation) is defined as

$$\Sigma F > 0 \tag{4.37}$$

Kacar et al. further considered two general cases: (i) $\rho_{\rm p} > \rho_{\rm l}$ and non-wetting ($\theta \geq 90°$) and (ii) $\rho_{\rm p} < \rho_{\rm l}$ and wetting ($\theta < 90°$). In both cases, the critical acceleration required for incorporation may be deduced from Eq. (4.36) as

$$a_{\rm cr} = \frac{K_{\rm s}\gamma_{\rm lg}\cos\theta}{\rho_{\rm l} - \rho_{\rm p}}\left(\frac{\rho_{\rm p}}{m_{\rm p}}\right)^{2/3} \tag{4.38}$$

where $a_{\rm cr}$ is the critical acceleration for penetration which can be converted to the force required for penetration. Kacar et al. [102] studied the penetration behavior of Al_2O_3, ZrO_2 and SiC particles into liquid Al during centrifugal rotation. They observed that for each material there is a critical acceleration, above which penetration of particles occurs. They further observed that the critical acceleration or force decreases with increasing particle size. For example, for the Al/ZrO_2 system, the magnitudes of the critical acceleration corresponding to particle sizes of 2.8 and 5.6 μm are 236 and 79 $\rm ms^{-2}$, respectively.

Majagi et al. [74], formulated a force balance model by considering the impingement of a particle onto a two-dimensional liquid surface along the gravity direction as illustrated in Figure 4.28. It was considered that the

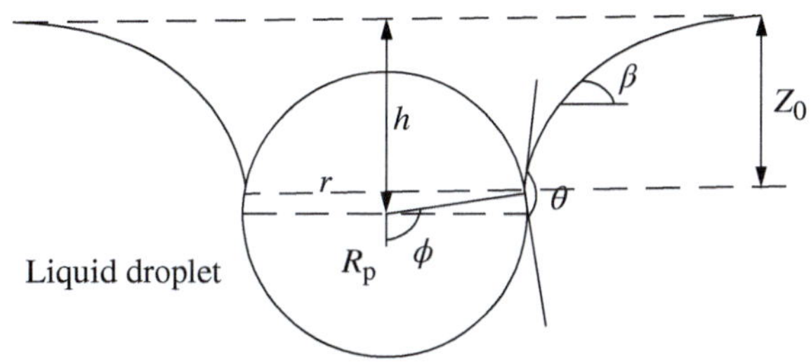

Figure 4.28 The penetration process of a spherical particle into a liquid where R_p, r, h, and Z_0 are the radius of the particle, the radius of the meniscus, the depth of the particle center below the original liquid surface, and the maximum vertical distance of a point on meniscus above the three-phase contact, respectively, and; β, θ, and ϕ are the inclination of meniscus to the horizontal plane, the **wetting angle** and the semiepical angle, respectively. Source: Adapted from TMS and ASM International [74].

resistance forces to penetration arise from surface tension and hydrostatic pressure. For the configuration illustrated in Figure 4.28, the surface tension force is:

$$F_s = -\pi d_p \sigma_1 \sin\phi \sin(\phi + \theta) \tag{4.39}$$

where F_s is the surface tension force; d_p is the particle diameter, and; ϕ is the semi-epical angle as defined in Figure 4.28. The hydrostatic pressure on the immersed part, F_b, (buoyancy) is given by [74]

$$F_b = -\frac{\pi}{2\Delta} d_p^3 \rho_1 g (1 - \cos\phi)^2 (2 + \cos\phi) \tag{4.40}$$

The hydrostatic pressure directly above the contact area, F_h, is [74]:

$$F_h = \frac{\pi}{4} d_p^2 g \sin^2\phi \rho_p Z_0 \tag{4.41}$$

where ρ_p is the density of the particle and Z_0 is defined in Figure 4.28. The **gravitational force**, F_g, is calculated from [74]:

$$F_g = \frac{\pi}{6} d_p^3 \rho_p g \tag{4.42}$$

The force associated with particle velocity change, F_v, is calculated from [74]:

$$F_v = \frac{\pi}{12P_d} d_p^3 \rho_p \left(V_0^2 - V_1^2\right) \tag{4.43}$$

where V_0 and V_1 are velocities of particles prior to and following penetration, respectively, and; P_d is **penetration depth**. Majagi et al. [74] considered that if the net force acting on a particle in its impinging direction is greater than zero, penetration will occur.

Majagi et al. applied the force balance model described by Eqs. (4.39)–(4.43) to the penetration problem of ceramic particles into metallic droplets during atomization and co-injection. They reported that the most important factors that govern the penetration behavior are particle density, **wetting angle**, and particle size. The results of these calculations are summarized in Figure 4.29. It is evident that the velocity required for penetration increases with increasing wetting angle. The primary driving force for penetration is the kinetic energy of the particle. For a given initial velocity, a greater particle density or a larger particle size represents a greater kinetic energy. Accordingly, an increase in particle density and size would lead to a decrease in the minimum velocity required for penetration. Such a tendency is shown in Figure 4.26b,c [74]. A similar trend of decreasing minimum penetration velocity with decreasing wetting angle, and with increasing particle size and density was also noted by other researchers [63, 108].

4.3.2.3 Combined Energy and Force Balance

The two energy and force balance approaches are inherently simple and offer considerable insight into the particle penetration problem. The simplicity, however, arises from the assumption that fluid drag is less important than surface tension in resisting particle penetration. Such an assumption is justifiable when considering the incorporation of particles into a low-viscosity liquid matrix using gas injection or centrifugal rotation or when considering the penetration behavior of reinforcement particles into metallic droplets which are in the liquid state. However, as has been shown in Chapter 3 during atomization a considerable proportion of droplets are in the semiliquid or solid states at the injection distances. Since the presence of solid phases in the semiliquid droplets will significantly increase the viscosity [121], the effect of viscous drag on particle penetration must be taken into consideration. As an example, Figure 4.30 shows the effect of solid fraction in a semiliquid slurry on the apparent viscosity of an Al–4.5Cu–1.5Mg (wt%) alloy as obtained experimentally by Kattamis and Piccone [121]. Figure 4.30 reveals a rapid increase in apparent viscosity with increasing fractions of solid even at relatively small solid fractions.

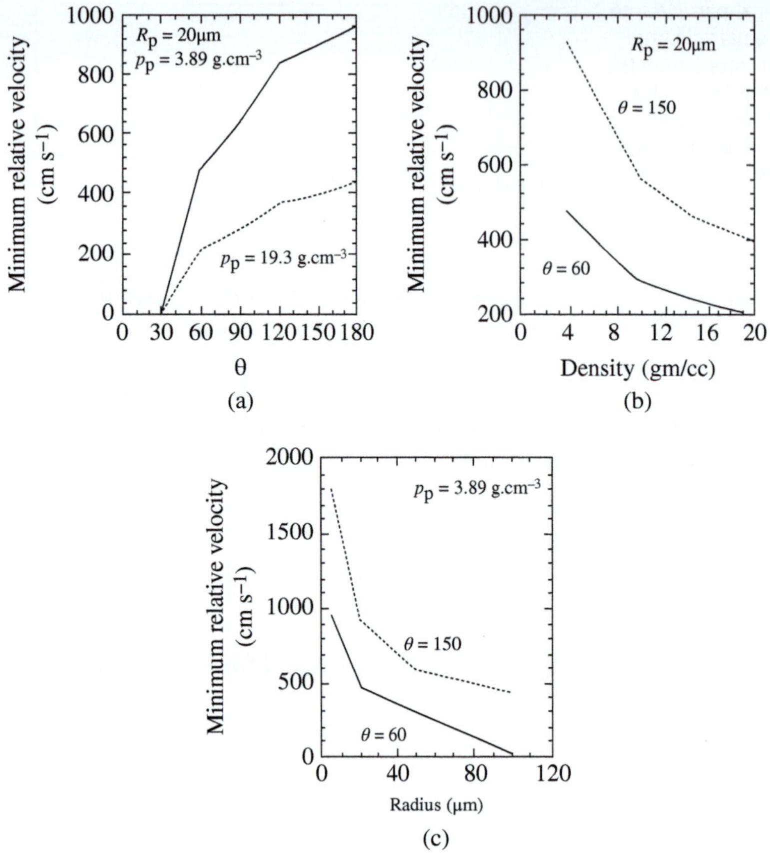

Figure 4.29 Effects of the wetting angle (a), particle density (b), and particle size (c) on the minimum penetration velocity of particles into droplets during atomization and co-injection. Source: Adapted from TMS and ASM International [74].

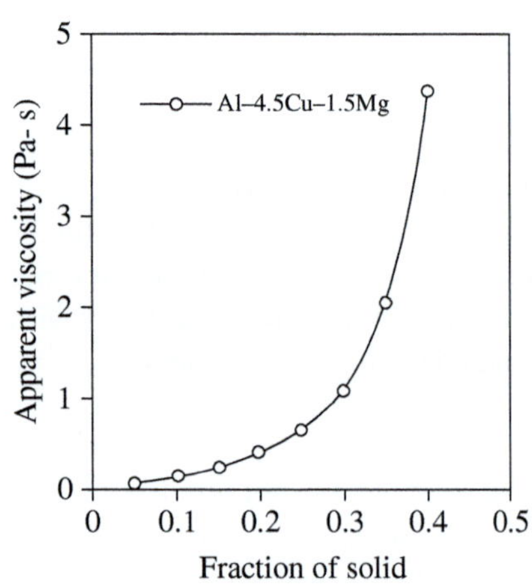

Figure 4.30 Apparent viscosity of Al–Cu–Mg alloy as a function of solid fraction. Source: Reproduced from Materials Science and Engineering A 131 T.Z. Kattamis and T.J. Picone Rheology of Semisolid A–4.5% Cu 1.5% Mg Alloy, p 265 (1991) with kind permission from Elsevier Science SA, Lausanne, Switzerland [121]/with permission from Elsevier.

In related studies, We et al. [63] and Zhang et al. [108] calculated the penetration behavior of ceramic particles such as Al_2O_3, **graphite**, SiC, and $T1B_2$ during atomization. These studies addressed the penetration of droplets by co-injected ceramic particles during atomization. Since during collision, a particle is at a relative velocity with a droplet, in these studies the kinetic energy, as given by Eq. (4.27), is the driving force for penetration. Instead of using a simple energy balance or a force balance, Wu et al. and Zhang et al. considered the changes in the driving force and the resistance forces that occur during penetration. As an example, Figure 4.31 illustrates the relative position between particles and droplets during penetration [63]. Wu et al. further considered that the resistance forces to

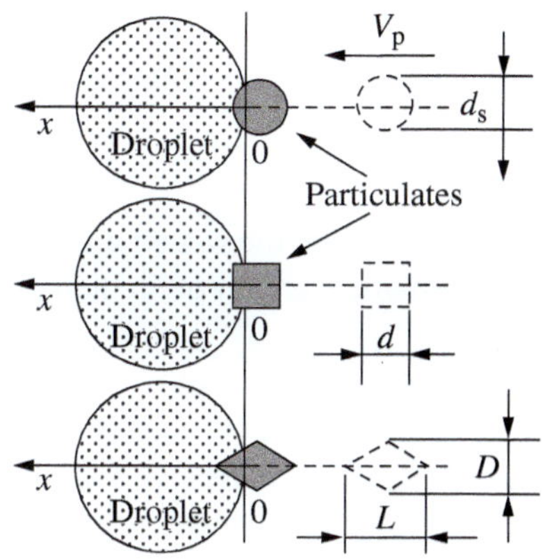

Figure 4.31 The relative positions between particles with different geometries and droplets during penetration; d_s, d, L, and D are the particle dimensions for all three geometries. Source: Adapted from TMS and ASM International [63].

particle penetration arise from two separate sources, namely, surface tension and fluid drag. Accordingly, the total resistance force acting on a particle, F_r, is given by [63]:

$$F_r = F_s + F_d \tag{4.44}$$

where F_s and F_d are the forces due to surface energy changes during penetration and fluid drag due to the motion of a particle inside of a droplet, respectively.

In the process of penetration, the surface tension force is represented as the derivative of the surface energy as a function of the particle penetration distance, x:

$$F_s = -\frac{d\Delta E_s}{dx} \tag{4.45}$$

where F_s is the surface tension force. The associated surface energy term, ΔE_s in Eq. (4.45) is comprised of three contributions during penetration: particle/gas interfacial energy, particle/droplet interfacial energy, and droplet/gas interfacial energy. Accordingly, the surface energy in the droplet–particle–gas system at any time is given by the following summation:

$$\Delta E_s = S_{pl}\gamma_{pl} + S_{pg}\gamma_{pg} + S_{lg}\gamma_{lg} \tag{4.46}$$

where S_{pl}, S_{pg}, and S_{lg} are the contact areas between particle and liquid, particle and gas, and liquid and gas, respectively. Therefore, the surface tension resistance becomes

$$F_s = -\left[\gamma_{pl}\frac{dS_{pl}}{dx} + \gamma_{pg}\frac{dS_{pg}}{dx} + \gamma_{lg}\frac{dS_{lg}}{dx}\right] \tag{4.47}$$

Furthermore, using Young's equation (4.31) and considering that $dS_{pg} = -dS_{pl}$, Eq. (4.47) may be simplified into

$$F_s = \frac{dS_{pl}}{dx}\gamma_{lg}\cos\theta - \frac{dS_{lg}}{dx}\gamma_{lg} \tag{4.48}$$

The fluid drag force is a result of the motion of a particle in intimate contact with a liquid phase. In the case of a particle immersed in a fluid, the **drag force** arises from two sources: the pressure that the fluid exerts on the particle, and the frictional drag at the fluid/moving particle interface. In fluid mechanics, however, the total drag force which the fluid exerts on a particle is generally computed utilizing a single equation in the form of [89, 122]:

$$F_d = -S_p K f \tag{4.49}$$

where S_p is the projected area of the particle in the traveling direction, and K is a kinetic factor and is proportional to the square of particle velocity. Therefore, the fluid drag acting on a moving particle may be approximated by [89]

$$F_d = -\frac{1}{2}S_p\rho_l V^2 f \tag{4.50}$$

where ρ_l is the density of the fluid and V is the velocity of the particle. The frictional factor, f, in Eq. (4.34) is a function of the dimensionless Reynolds number which is given by [89]

$$f = \begin{cases} 24/Re & 0 < Re \leq 1 \\ 24/Re^{0.646} & 1 < Re \leq 400 \\ 0.5 & 400 < Re \leq 3\times 10^5 \end{cases} \tag{4.51}$$

The dimensionless Reynolds number, *Re,* is related to the relative velocity between the liquid and the particle dimension in the liquid as [89]

$$Re = \frac{d_w V}{v_l} \tag{4.52}$$

where d_w may be the wetted diameter of the particle in the liquid; V is the relative velocity of the particle, and; v_l is the kinematic viscosity of the liquid.

4.3.2.4 Penetration Depth

With Eqs. (4.44), (4.48), and (4.50) the force acting on a particle by a droplet may be calculated at any time. In the case of a particle at a relative velocity, V_0, that collides with a droplet, its instant velocity may then be calculated from an integration of the forces acting on it [64]

$$V = V_0 + \int_0^t \frac{F_r}{\rho_p V_p} dt \tag{4.53}$$

where ρ_p and V_p are the density and volume of the particle, respectively, and; t is the elapsed time from the beginning of impact. The corresponding **penetration depth**, P_d, of the particle at time t may be calculated by integrating the velocity from $t = 0$, corresponding to the initial velocity of V_0, to a time t:

$$P_d = \int_0^t V(t) dt \tag{4.54}$$

4.3.2.5 Particle Type, Morphology, and Solid Fraction

To provide insight into the penetration mechanisms of particles in practical MMC systems Wu et al. [63] and Zhang et al. [108] formulated physical models based on Eqs. (4.44)–(4.54). There are two aspects of these calculations that are worth mentioning. First, to evaluate the penetration behavior of various types of particles and to compare the calculated results with experimental observations, these studies selected more practical particle/droplet combinations. Accordingly, the penetration behavior of Al_2O_3, **graphite**, SiC, and $T1B_2$ into Al droplets was considered. Second, to make calculations under more realistic conditions the droplets in these studies were assumed to be in a semiliquid state containing various fractions of solid.

In considering the effect of particle morphology, Wu et al. [63] calculated the penetration behavior by assuming three distinct particle morphologies, including spherical, cubic, and duplex cones as illustrated in Figure 4.31 [63]. Figure 4.32 compares the effect of particle morphology on the penetration behavior of ceramic particles into liquid droplets [63]. The minimum velocity required for penetration in this figure is presented as a function of the effective particle size for cubic, spherical, and duplex-cone particles. The effective size of a particle is defined herein as the diameter of a sphere of identical volume as a random-shaped particle. It is evident from the results shown in Figure 4.32 that the minimum, penetration velocities are similar for particles which exhibit the same effective size

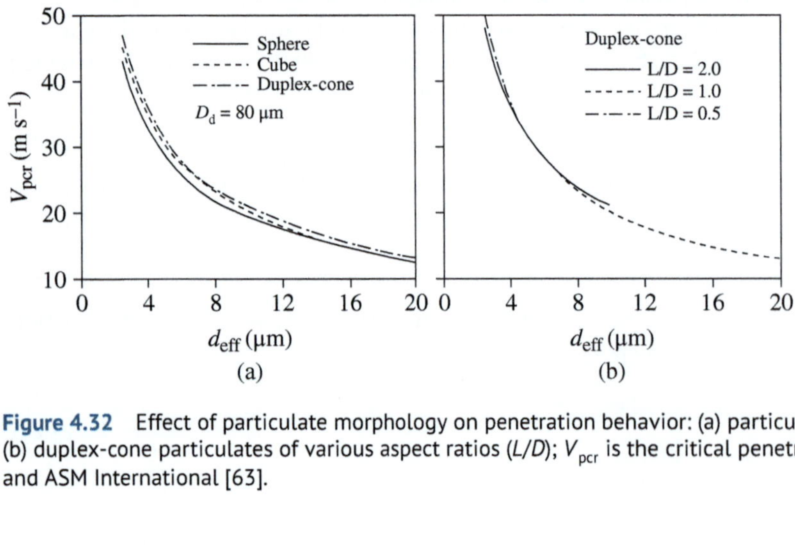

Figure 4.32 Effect of particulate morphology on penetration behavior: (a) particulates of various morphologies and (b) duplex-cone particulates of various aspect ratios (*L/D*); V_{pcr} is the critical penetration velocity. Source: Adapted from TMS and ASM International [63].

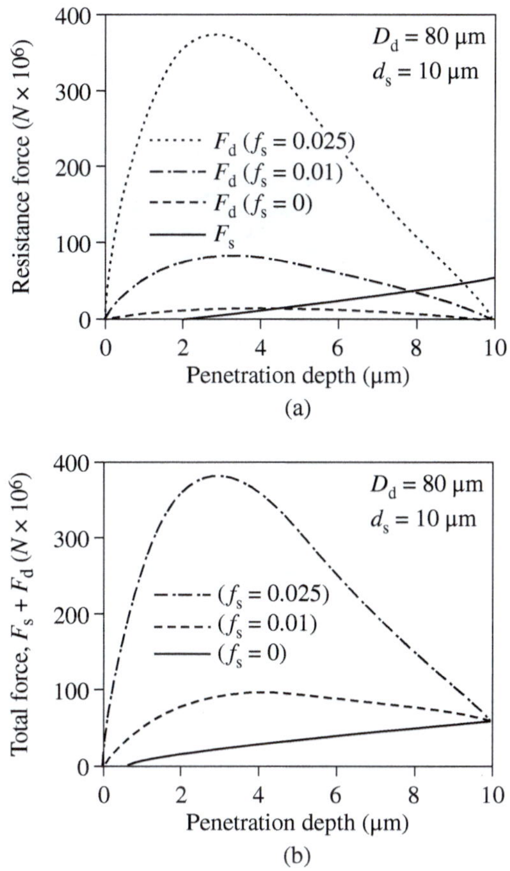

Figure 4.33 Comparison of the penetration resistance of SiC particles in Al droplets; (a) surface tension resistance and fluid drag and (b) total resistance force where f_s is the solid fraction; F_d is the fluid drag force; F_s is the surface tension force; D_d is the droplet size, and; d_s is the particle size. Source: Adapted from TMS and ASM International [63].

but different morphologies. For example, as shown in Figure 4.32b, for each effective particle size considered, there are only marginal differences between the minimum penetration velocities of duplex-cone particles with various aspect ratios (e.g. L/D values).

Wu et al. [63] calculated the relative magnitude of the surface tension force and the fluid drag force as a function of the penetration distance (i.e. the distance that a particle enters a droplet) and the fraction of solid. The results of these calculations are summarized in Figure 4.33 for both surface tension and fluid drag forces that an 80 μm diameter liquid droplet exerts on a spherical SiC particle of 10 μm in diameter. As demonstrated by the results shown in Figure 4.33, when a droplet is in the liquid state, fluid drag is greater than surface tension resistance at the initial stages of penetration. However, surface tension resistance increases with increasing penetration distance. When the penetration depth is greater than the radius of the particle, surface tension resistance becomes much greater than fluid drag. Regarding the effect of solid fraction, the results shown in Figure 4.33 indicate that fluid drag becomes increasingly important with increasing fraction of solid. At solid fractions greater than 0.025, the **drag force** becomes much greater than that of the surface tension force. Therefore, fluid drag is an important factor affecting the penetration behavior of particles into droplets that are in the semiliquid state.

To illustrate the effect of the fraction of solid on penetration, Wu et al. [63] and Zhang et al. [108] calculated the minimum penetration velocity for spherical particles. The results of these calculations are summarized in Figure 4.34 as a function of particle size and solid fraction contained in the semiliquid droplets [63]. According to the results shown in Figure 4.34, the critical penetration velocity increases rapidly with an increasing fraction of solid. This observation is consistent with the penetration resistance illustrated in Figure 4.33 which shows that when the fraction of solid is greater than 0.025, fluid drag dominates the penetration behavior. Wu et al. [63] further compared the minimum penetration velocity of different ceramic particles. Their results, summarized in Figure 4.35, indicate that for identical particle size the minimum velocity required for penetration increases in the sequence: TiB_2, Al_2O_3, SiC, and **graphite**. Consequently, the penetration ability is greatest for TiB_2 and decreases in the order: TiB_2, Al_2O_3, SiC, and **graphite**. This sequence is in the same order as the decrease in particle density and increase in wetting angle. The wetting angles of ceramic particles commonly encountered in atomization

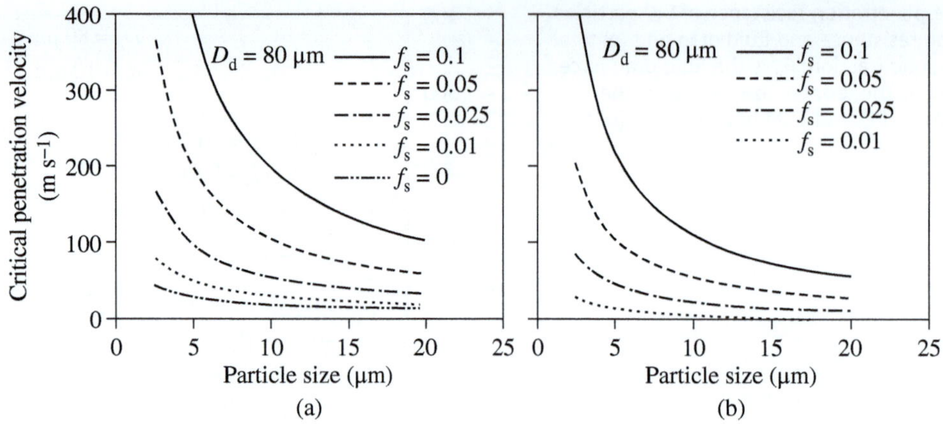

Figure 4.34 Critical penetration velocity as a function of particle size and the fraction of solid for a spherical particle into an Al droplet (80 μm in diameter): (a) SiC; (b) TiB_2. Source: Adapted from TMS and ASM International [63].

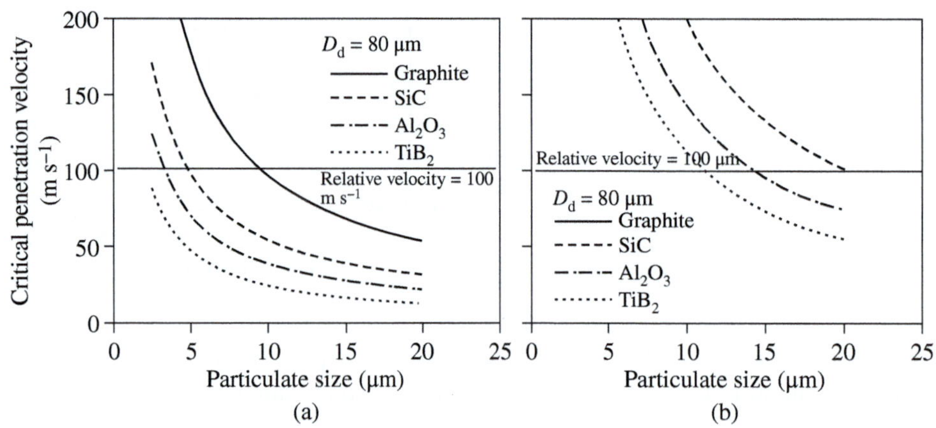

Figure 4.35 Critical penetration velocity as a function of particulate size and the fraction of solid for various particle chemistries into an Al droplet (80 μm in diameter): (a) fraction of solid, f_s = 0.025 and (b) f_s = 0.10. Source: Adapted from TMS and ASM International [63].

and co-injection are summarized in Table 4.6 [25, 102, 112, 119]. Wu et al. [63] and Zhang et al. [108], therefore, concluded that except for the fraction of solid, particle density and wetting angle are the critical factors which govern particle penetration behavior during atomization and co-deposition. Since in the liquid state, only surface tension resistance is important, both wetting angle and particle density govern penetration. In the semiliquid state, however, fluid drag is the most important resistance. In this case, solid fraction and particle density govern penetration.

4.3.3 Penetration of Solid Droplets

For a particle to penetrate a **solid droplet**, impingement of the particle must lead to plastic deformation on the droplet surface. The problem concerning the penetration behavior of a hard ceramic particle into a relatively soft metallic surface has been addressed by Tiwari and Herman [123]. The dynamic pressure, P, exerted by a ceramic particle on a **solid droplet** can be calculated by a simplified equation [123], since the elastic moduli of ceramics are much higher than that of metallic materials:

$$P = 0.558\left[\rho_p\left(\frac{E}{1-v_p^2}\right)^4\right]^{0.2} V^{0.4} \tag{4.55}$$

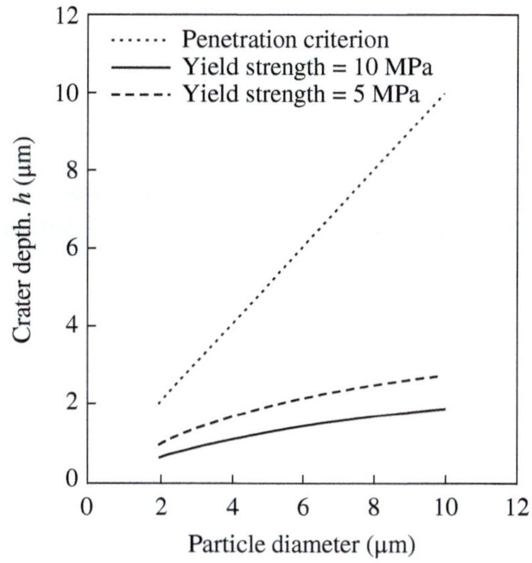

Figure 4.36 Depth of the crater, *h*, formed on the surface of **solid droplets** by SiC particles as a function of particle diameter, d_p. The dotted line represents crater depth equivalent to the SiC particle diameter, and the dashed line and solid line represent the calculated depth when the yield strength of the alloy is assumed to be 5.0 and 10.0 MPa, respectively. Source: Adapted from ASM International and The Minerals,Metals & Materials Society [64].

where E and ν_p are the elastic modulus and Poisson's ratio of the droplet, respectively; ρ_p is the particle density, and; V is the relative velocity between the particle and the droplet. The depth of crater, d_c, caused by the impingement can be expressed as [123]

$$d_c = \left[\frac{0.14 m_p V^2}{\pi (d_p/2 - d_c/3) \sigma_{ym}} \right]^{0.5} \tag{4.56}$$

where σ_{ym} is the yield strength of the droplet material and m_p is the mass of the particle. Based on this formation, it has been proposed [64, 123] that when the crater depth is greater than the size of the particle, penetration is possible. Otherwise, the particle will adhere to the droplet surface. Figure 4.36 summarizes the calculated crater depth using Eq. (4.56) for the impingement of a SiC particle on a solid Al–Si droplet. The dotted line in this figure represents the penetration criterion; the dashed line and solid line represent the crater depths formed on droplets with yield strengths of 5.0 and 10.0 MPa, respectively. In both cases, the crater depths are less than the critical value for penetration. Consequently, it was concluded that penetration may not occur in **solid droplets** for the conditions studied.

4.4 Particle Behavior During Solidification

The mechanisms that govern particle incorporation during atomization and co-injection, discussed in Section 4.3, illustrate that particle penetration occurs when droplets are in a completely molten or partially molten state. Accordingly, the particles may be able to travel in the molten material following penetration, and the final distribution of particles in the matrix becomes critically dependent on the interactions that occur between particles and solid–liquid interfaces. The problem of particle redistribution during solidification is not limited to atomization and co-injection. In liquid phase processes, such as ingot casting, there is considerable redistribution of particles during solidification [22, 124–126]. When such particles reside under the action of a **gravitational force** or are rejected by **solidification fronts**, extensive reinforcement segregation will be present in the final microstructure. In atomization and co-injection, there are three possible interaction mechanisms that may influence the final distribution of ceramic particles following penetration. These include the engulfment of particles by the **solidification front**, **mechanical entrapment** of particles by the solidification front, and particle-induced nucleation. These mechanisms are discussed in more detail subsequently.

4.4.1 Engulfment of Reinforcements by Solid–Liquid Interface

When an advancing solid–liquid front encounters an extraneous particle, it may **engulf** the particle or reject it. Engulfment is a process which involves the liquid-to-solid transition of a particle or particles. Whether a solid–liquid interface engulfs or rejects a ceramic particle critically depends on the kinetic and thermodynamic conditions that are present. The engulfment/rejection of a particle is also governed by the physical properties of the particle and those of the matrix [124, 125, 127–131]. When ceramic particles are suspended in a matrix, as in liquid or semiliquid processing, they will tend to be rejected by the approaching solid–liquid interface resulting in the segregation of reinforcements into the last zone to solidify [22, 124, 126]. In certain metal–ceramic systems, however, the interfacial energies between the **solid phases** and the ceramics are relatively low. As a result, the engulfment of the ceramic particles becomes energetically favorable. Under such conditions, the engulfment of ceramic particles by the solid–liquid interface occurs during solidification [12, 132]. However, in most ceramic/metal systems the interfacial energy between the ceramic and the solid metal is relatively high. Consequently, ceramic reinforcements tend to be rejected by the solid–liquid interface in such systems.

The engulfment/rejection behavior of particles by advancing solid–liquid interfaces has been confirmed by direct experimental observations in the solidification of organic matrix composites which are optically transparent [127, 129, 133]. In non-transparent metal or IMCs, the engulfment/rejection behavior of particles has been studied using indirect methods in which the solid–liquid interfacial velocity may be carefully controlled. Examples in this area are the Bridgman technique [128] and the wedge mold solidification technique [134]. In both techniques, the interfacial velocity may be determined with some degree of accuracy and hence the engulfment of particles may be studied as a function of interfacial velocity. Although in these studies various combinations of matrices and particles have been used, the experimentally observed particle engulfment/rejection phenomena are qualitatively similar. At relatively low interfacial velocities, particles are rejected by the advancing solid–liquid interface [127, 129, 133]. When the solid–liquid interfacial velocity approaches some critical value, engulfment occurs. Moreover, the critical engulfment velocity generally increases with decreasing particle size. An empirical relationship relating the critical velocity with particle size may be expressed in the general form of [127–129, 133]:

$$V_{\mathrm{cr}} r_{\mathrm{p}}^{n} = C \tag{4.57}$$

where V_{cr} is the **critical interfacial velocity** at which the transition between rejection and engulfment of particles occurs; r_{p} is the particle radius; n is an exponential factor, and; C is an experimental constant. The exponent, n, is in the range of 0.28–0.90 for most ceramic Al alloy systems [128].

The interaction behavior between ceramic particles and the solid–liquid interface during MMC solidification has attracted considerable interest. Attempts have been made by various authors to correlate the observed particle engulfment/rejection behavior with solidification conditions, the physical properties of the matrix materials, and those of ceramic particles [124, 127, 128, 135, 136]. Available theoretical models typically involve the calculation of a **critical interfacial velocity** for engulfment. The parameters most frequently used in these models are the particle size and interfacial velocity. Consequently, the engulfment/rejection phenomena in various matrix–particle combinations have been evaluated as a function of interfacial velocity and particle size. Moreover, attempts have been made to develop expressions with a form that is like the empirical relationship described by Eq. (4.57) [124, 127–131, 135]. Physically, the engulfment/rejection behavior is complicated by the various phenomena that are present during the solidification of composite materials such as **diffusion**, fluid flow, heat transfer, and phase transformations [124, 127–131, 135]. Accordingly, the formulations associated with the various models differ significantly, depending on the physical process that is being emphasized in each one of these models. The models that have been developed include those based on thermodynamic equilibrium [129], those based on a force balance on the particles [128, 131, 133], and those based on a material balance during transport in the matrices [127].

The simplest model of particle engulfment was developed by Omenyi and Neumann based on purely **thermodynamic considerations** [129]. They proposed that the engulfment of extraneous particles by an advancing solid–liquid interface can be predicted from simple thermodynamic properties. When the net free energy change due to the liquid-to-solid transition of a particle is less than zero, the engulfment of the particle by a solid–liquid interface will be energetically favorable. This argument led to the criterion of engulfment as [129]:

$$\Delta G_{\mathrm{NET}} \leq 0 \tag{4.58}$$

where ΔG_{NET} is the net change in free energy because of engulfment. For the complete transition of a spherical particle from a liquid phase into a **solid phase**, the net free energy change involved is [129]

$$\Delta G_{\mathrm{NET}} = 4\pi r_{\mathrm{p}}^{2}(\gamma_{\mathrm{ps}} - \gamma_{\mathrm{pl}}) \tag{4.59}$$

where γ_{ps} and γ_{pl} are the interfacial energies between particle and solid and between particle and liquid, respectively. The primary advantage associated with Eq. (4.59) lies in its simplicity. However, the engulfment of a particle by a solid–liquid interface is a kinetic process, and hence the thermodynamic approach can only predict the possibility of engulfment. Consequently, Eq. (4.59) fails to predict a critical engulfment velocity which is normally noted from experimental observations. Since the engulfment behavior is determined by the sign of the net free energy change, Eq. (4.59) provides no useful information regarding the particle size dependence of engulfment behavior [129]. Moreover, because engulfment is a kinetic process engulfment of ceramic particles is sometimes observed, even when engulfment is energetically unfavorable under equilibrium solidification conditions.

4.4.1.1 Mass Balance

Uhlmann et al. [127] developed a kinetic model that may be used to predict particle engulfment by considering a mass balance in the gap created between a particle and a solid–liquid front. It is considered that if a particle is rejected by a solid–liquid front, matrix material must be supplied continuously into the gap region between the particle and the interface so that the separation between them can be maintained during solidification. Furthermore, Uhlmann argued that the interfacial free-energy-driven self-diffusion should be responsible for the transport of material into the gap region. Figure 4.37 illustrates schematically the configuration between a solid–liquid interface and a particle during solidification [124]. Uhlmann et al. further assumed that the variation in interfacial energy with the separation distance between the particle and the interface is described by a power law relationship [127]:

$$\Delta\sigma = \Delta\sigma_o \left(\frac{d_o}{d_g}\right)^n \tag{4.60}$$

where d_o and d_g are the minimum possible separation and the actual distance between the particle and the interface, respectively; $\Delta\sigma$ is the interfacial energy change, and $\Delta\sigma_o$ is its value corresponding to a separation distance of d_o. When rejection of a particle occurs, the matrix material consumed by the advancing solid–liquid interface must be balanced by the material transported into the gap region by self-diffusion. When the interfacial velocity exceeds a certain critical value, the rate of material consumption by solidification may exceed that of mass transport. In this case, the particle will be engulfed. Such mass conservation considerations were used to determine a critical velocity for rejection/engulfment transition [127]

$$V_{cr} = \frac{1}{2}(n+1)\left(\frac{a_o \Delta H_f V_0 D_l}{kTr_p^2}\right) \tag{4.61}$$

where n is a constant ($n = 5$), a_o is the interatomic distance, k is the Boltzmann's constant, V_0 is the atomic volume; T is the temperature; ΔH_f is the volumetric heat of fusion, and D_l is the self-diffusion coefficient in the liquid.

4.4.1.2 Force Balance

Direct experimental observations in organic matrix–ceramic particle systems indicate that there is no observable gap under the detection limit of an optical microscope between the interface and the particles under steady-state pushing conditions. Therefore, the actual gap that separates the interface and the particle is relatively small by many investigators, that is, on the order of 10^{-7} to 10^{-8} m [127–129, 131, 133]. Chernov et al. [131] argued that when the separation distance between the particle and the interface becomes very small, van de Waals forces will be active.

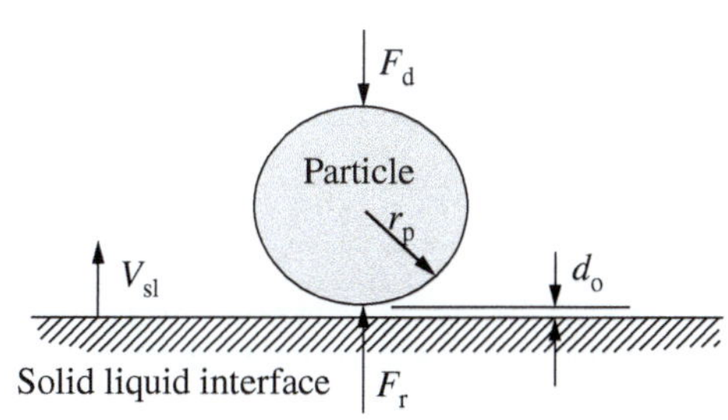

Figure 4.37 Forces acting on a particle pushed by a solid–liquid interface moving at a velocity V_{sl} where F_d and F_r are the fluid drag and resistance forces, respectively, and d_o is the minimum separation. Source: Adapted from TMS and ASM International [124].

As a result, there will be a disjoining pressure acting on the particle. If one considers the force that results in particle motion in the direction of interfacial progress as positive, then for the gap between two flat surfaces the force resulting from this disjoining pressure is [131]

$$F_r = \frac{\pi B r_p}{d_g^n} \tag{4.62}$$

where B is a constant; d_g has its original meaning, and the exponent n is 2 or 4 depending on matrix properties; for a metal matrix n equals 2 [131].

When a particle is rejected by an advancing solid–liquid interface, it moves at a relative velocity with the liquid phase. The hydrodynamic pressure generated on the front surface of the particle due to this relative motion tends to push the particle back toward the interface [131]. This hydrodynamic force is proportional to the particle velocity. Under steady-state pushing conditions, the particle velocity equals the solid–liquid interfacial velocity. Accordingly, the hydrodynamic force acting on a particle is given by [131]

$$F_h = -\frac{6\pi\eta_l V_{sl} r_p^2}{d_o} \tag{4.63}$$

where η_l is the dynamic viscosity of the liquid, d_o is the minimum possible separation distance between particle and interface, and V_{sl} is the interfacial velocity. Chernov et al. [131] argued that the particle will be rejected into the liquid if the hydrodynamic force is less than the repulsive force. Hence, particle engulfment occurs when the hydrodynamic force exceeds the repulsive force. A force balance then leads to the formulation of the **critical interfacial velocity** for rejection/engulfment transition

$$V_{cr} = \frac{0.14B}{\eta_l r_p}\left(\frac{\alpha}{B r_p}\right)^{1/3} \tag{4.64}$$

where α is a constant and is on the order of 10^2 erg cm^{-2} [131]

4.4.1.3 Thermal Field

In the models, the engulfment/rejection behavior of a particle ahead of a solid–liquid interface is addressed without considering the possible effects of the particle on the **interfacial growth behavior**. In fact, the presence of particles ahead of a solid–liquid interface will alter the thermal and solidification fields and the concomitant interfacial morphology [133, 137, 138]. This effect will, in turn, influence the particle engulfment/rejection behavior. Chernov and Mel'Nikova [137] and Surappa and Rohatgi [138] analyzed the influence of particles on the thermal field ahead of a solid–liquid interface and proposed that the thermal properties of the particle and those of the liquid may govern the particle/interface interactions. The influence that a particle exerts on the growth of the solid–liquid interface is critically dependent on the relative magnitudes of the thermal conductivity of the particle and the liquid. As shown in Figure 4.38, if the thermal conductivity of the particle is greater than that of the liquid a depression will form underneath the particle [133]. In this case, engulfment is mechanically facilitated. If, on the other hand, the thermal conductivity of the particle is less than that of the liquid a protuberance will form on the interface behind the particle due to the reduced heat flow to the interface. Consequently, the particle may roll over and fall into the adjacent liquid.

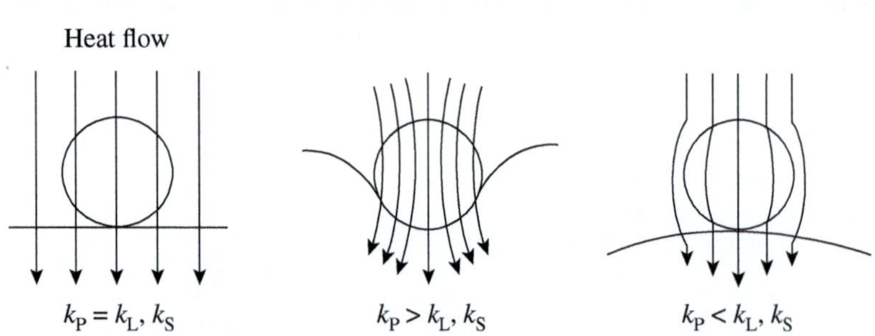

Figure 4.38 Modification in solid–liquid interface shape expected from the difference in heat conductivity between the foreign particle, k_p, and the liquid phase, k_L. Source: Reproduced by permission of Elsevier Science from G. F. Bolling and J. Cissé (1971) *J. Cryst. Growth*, **10**, 56 [133].

Such a process will facilitate rejection. A thermal conductivity criterion was proposed accordingly by Chernov and Mel'Nikova [137] for engulfment as

$$\frac{k_p}{k_l} > 1 \tag{4.65}$$

Surappa and Rohatgi [138] further modified this model and proposed that it is the relative heat diffusivity that influences the interfacial morphology and the engulfment behavior. In this case, the criterion for engulfment becomes

$$\left(\frac{k_p c_{p,p} \rho_p}{k_l c_{p,l} \rho_l}\right)^{1/2} > 1 \tag{4.66}$$

where $c_{p,\,p}$ and $c_{p,\,l}$ are the constant pressure heat capacities of the particle and liquid, respectively. Equations (4.65) and (4.66) have been used to predict particle engulfment with various degrees of success. One of the major drawbacks of these models is that they cannot predict a critical velocity for the rejection/engulfment transition.

4.4.1.4 Thermal Field and Force Balance

Stefanescu et al. [128] also used a force balance approach to formulate a critical velocity for the rejection/engulfment transition. In their approach, the repulsive force acting on a particle is considered to result from an elevation in interfacial energy when the separation distance between a particle and a solid–liquid interface falls in the range of 10^{-7}–10^{-8} m. The variation in interfacial energy with separation distance is assumed to be of an identical form to that described by Eq. (4.60). The force resulting from this variation in free energy tends to push the particle away from the solid–liquid interface. This separation force may be calculated from [128]

$$F_r = -\frac{\pi r_p \Delta\sigma}{n-1} \tag{4.67}$$

Stefanescu et al. further considered that the same hydrodynamic force as that given by Eq. (4.63) is present when a particle moves at a relative velocity with the surrounding liquid. Hence, the hydrodynamic force increases with increasing interfacial velocity. At the critical velocity corresponding to a rejection/engulfment transition, a force balance yields the expression:

$$V_{cr} = \frac{\Delta\sigma d_o}{6(n-1)\eta_l r_p} \tag{4.68}$$

Stefanescu et al. further incorporated the effect of the thermal conductivity difference between the particle and the liquid in front of the solid–liquid interface on particle engulfment. In their model, it is postulated that the presence of ceramic particles in front of a solid–liquid interface will alter the thermal field ahead of the interface. When the thermal conductivity effect is introduced, the critical velocity formulation of Eq. (4.68) may be modified into

$$V_{cr} = \frac{\Delta\sigma d_o}{6(n-1)\eta_l r_p}\left(2 - \frac{k_p}{k_l}\right) \tag{4.69}$$

where k_p and k_l are the thermal conductivities of the particle and the liquid, respectively.

The models have been developed based on the various physical processes that are present during the interaction between an extraneous particle and a solid–liquid interface. The critical velocities for rejection/engulfment transitions that are predicted using these models, however, differ from each other by several orders of magnitude. The applicability of the critical velocity formulations derived from each model may depend on the kinetic and solidification conditions that are imposed during experimentation. For the thermal and kinetic conditions that are normally present during MMC solidification, the applicability of these various models awaits further verification.

4.4.1.5 Engulfment During Droplet Solidification

The engulfment behavior of particles during droplet solidification is not a well-understood phenomenon. The experimentally observed decrease in particle volume fraction with increasing droplet diameter in the large droplet size range (i.e. >98 μm), as shown in Figure 4.23, suggests that particle engulfment is preferred for small droplet sizes, whereas rejection becomes increasingly important with increasing droplet size. Accordingly, this experimental observation may be simply rationalized by noting that in large droplets the solid–liquid interfacial velocity is relatively low, and therefore particles are rejected to the surface of the droplets during solidification. Wu and Lavernia [64] addressed

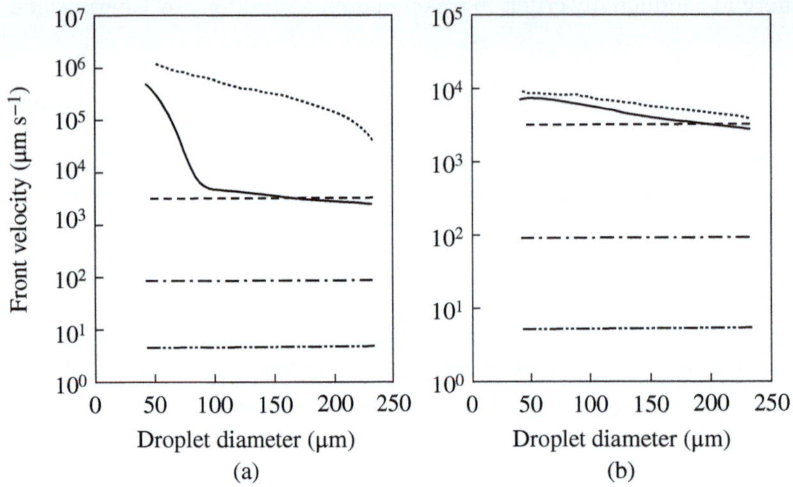

Figure 4.39 Comparison of the solidification front velocities of Al–4 wt% Si droplets calculated for (a) homogeneous and (b) heterogeneous nucleation undercooling ($\Delta T = 40$ K) for two solid fractions, $f_s = 0.2$, and $f_s = 0.3$ [64] with the critical front velocity, V_{cr}, necessary for SiC particulate engulfment according to the models proposed by Stefanescu et al. [128], Bolling and Cissé [133], and Chernov et al. [131]. Source: Adapted from TMS and ASM International [64].

the effect of droplet size on the particle engulfment behavior by comparing the critical engulfment velocities with the solid–liquid interfacial velocities that are present in droplets. In particular, the models developed by Stefanescu et al. [128], Bolling and Cissé [133] and Chernov et al. [131] were used to calculate the critical engulfment velocities. In this study, the solidification front velocity (i.e. solid–liquid interfacial velocity), V_{sl}, was calculated for Al droplets by considering both dynamic and equilibrium solidification. The results shown in Figure 4.39a were calculated by assuming that solidification starts with homogeneous nucleation undercooling (i.e. in the undercooling range of 195–208 K, for droplets in the size range of 42–230 μm) [64]. The results shown in Figure 4.39b were calculated by assuming that solidification starts with an initial heterogeneous undercooling of $\Delta T = 40$ K. The factors that influence solidification front velocity have been discussed in detail in Chapter 5, and hence only the critical equations will be presented here. In this study the interfacial velocity in undercooled liquid was calculated by using the kinetic relationship between solidification front velocity (V_{sl}), undercooling (ΔT), and kinetic growth coefficient (K_m) [139]:

$$V_{sl} = K_m \Delta T \tag{4.70}$$

The solidification front velocity under equilibrium solidification conditions was calculated from an energy balance consideration. When Newtonian conditions are applicable and in the absence of significant undercooling, the thermal energy released during the progress of the solid–liquid interface must be balanced by the thermal energy that is dissipated from the surface of a droplet. For a single concentric solidification event, under these limiting conditions the interfacial velocity is given by:

$$V_{sl} = \frac{r_d^2 h(T_l - T_g)\Omega}{r^2 \left[(c_{p,s} - c_{p,l})T_l + \Delta H_f\right]} \tag{4.71}$$

where r and r_d are the instantaneous interfacial position and the radius of the droplet, respectively; T_l and T_g are the liquidus temperature of the droplet and the temperature of the atomization gas, respectively; $c_{p,s}$ and $c_{p,l}$ are the constant pressure molar heat capacities of the solid and that of the liquid, respectively; h is the heat transfer coefficient, and; Ω and ΔH_f are the molar volume and heat of fusion of the droplet material, respectively. The results of Wu and Lavernia [64], as summarized in Figure 4.39, reveal that for the range of undercooling considered the solid–liquid interfacial velocity decreases with increasing droplet size and increasing fraction of solid. Also shown in this figure are the critical interfacial velocities that are required for the engulfment of a 10 μm SiC particle. As shown in Figure 4.39, the critical velocities predicted by the models of Bolling and Cissé [133] and Chernov et al. [131] are well below the typical interfacial front velocities present in the rapidly solidified droplets. The behavior predicted based on the model of Stefanescu et al. [128] (Figure 4.39) appears to provide the best results. This model predicts a critical velocity that is lower than the solidification front velocity for small droplets and a critical velocity that is higher than the solidification front velocity for large droplets. Consequently, engulfment is preferred in small droplets and rejection is preferred in large droplets. Such a prediction is in general agreement with available experimental observations on the segregation behavior of SiC particles in Al–Si droplets, as shown in Figure 4.22.

4.4.2 Mechanical Entrapment of Reinforcements by Solidification Fronts

As a result of the non-transparent nature of most MMC materials, the engulfment phenomenon is generally studied by examining the segregation of particles. In such studies, the observation of the entrapment of particles is affected, not only by the physical engulfment as discussed in Section 4.4 but also by mechanical entrapment. Moreover, all the theoretical models consider the ideal case of a single particle in front of a moving solid–liquid interface. When there is a dispersion of particles present in front of a solid–liquid interface and when there are multiple solidification events, the effects of particle–particle interactions and particle–interface interactions must be taken into consideration. Figure 4.40 illustrates schematically two possible mechanisms of mechanical entrapment of particles by solidification fronts as proposed based on experimental observations of particle distribution in droplets [64]. As shown in Figure 4.40, a particle may be immobilized in between grains or primary dendrites when solidification fronts from several directions mechanically lock the particle [64]. In this case, the particle is rejected by a solidification front, until several solidification fronts, originating at different nuclei, converge. A particle may also be mechanically entrapped in between secondary dendrite arms. For such a mechanism to be viable, the secondary dendrite arm spacing must be comparable with the size of the particles. Such a mechanism is illustrated schematically in Figure 4.40a [64].

The mechanical entrapment of particles by dendrites has been confirmed experimentally by examining the solidification microstructure of MMCs [23, 132]. One extreme example of this phenomenon is provided by the segregation of reinforcement particles to the last solidification region, as commonly observed in the microstructure of MMCs processed under slow solidification conditions. Jin and Lloyd [132], for example, studied the solidification microstructure of Al–7 wt% Si reinforced with 15 vol% of SiC particles. In this study, SiC particles were mixed with a molten Al–7 wt% Si alloy. The **composite slurry** was then cast into an iron mold. Jin and Lloyd observed that at a low cooling rate, that is, 7 Ks^{-1}, the SiC particles segregated to dendrite boundaries and formed an interdendritic network as shown in Figure 4.41 [23, 132]. In this type of microstructure, the extent of particle–particle contact is considerable. It is worth noting that under slow cooling rate conditions, the scale of the dendrite arms is relatively large, leading to long-distance segregation of particles. When the dendrite arm spacing is small, as in the case of droplet solidification

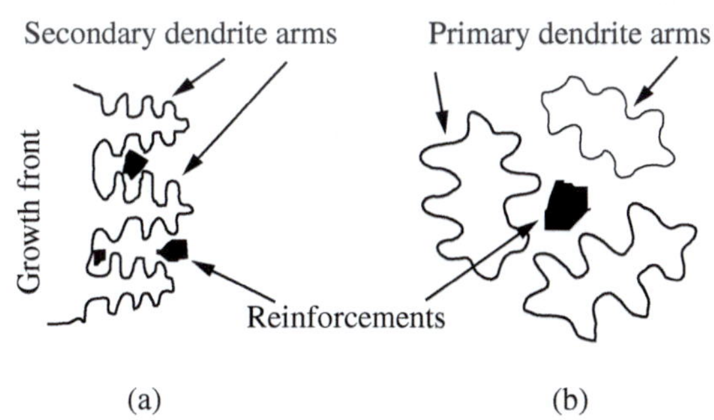

Figure 4.40 Schematic diagram showing the mechanical entrapment of ceramic particles by multiple solidification fronts: (a) by secondary dendrite arms and (b) by primary dendrite arms. Source: Adapted from TMS and ASM International [64].

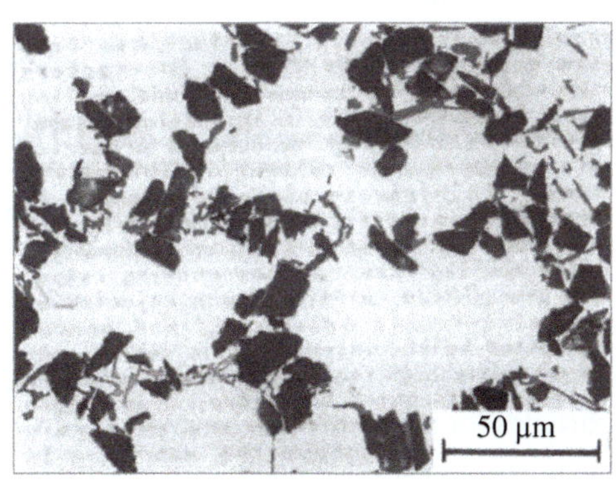

Figure 4.41 Microstructure of Al–7 wt% Si–15 vol% of SiC composite solidified at a cooling rate of 7 Ks^{-1} showing the segregation of SiC particles to dendrite boundaries. Source: Mortensen and Jin [23].

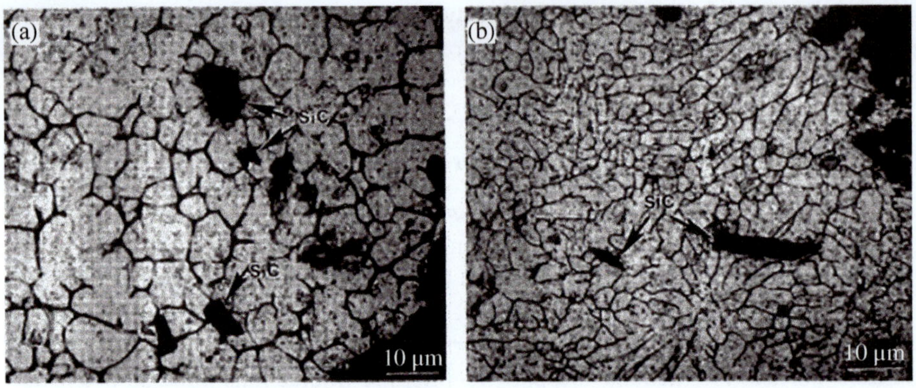

Figure 4.42 Microstructure of spray-atomized and co-injected Al−4 wt% Si/SiC powders showing the presence of SiC particles at (a) secondary dendrite boundaries; (b) primary dendrite boundaries (droplet size = 230 μm). Source: Zhang et al. [64]/from Elsevier.

or splat quenching, the particles are rejected only a short distance before becoming mechanically entrapped by multiple solidification fronts. Consequently, the macroscopic appearance of the MMC reveals a homogeneous reinforcement distribution. In a related study, Wu and Lavernia [64] examined the location of SiC particles in Al–4 wt% Si/SiC droplets produced by atomization and co-injection. The results shown in Figure 4.42 illustrate the mechanical entrapment of SiC particles by secondary dendrite arms (Figure 4.42a) and by primary dendrite arms (Figure 4.42b). Evidence of particle entrapment has also been reported in as-spray-deposited MMCs. One example in this area is provided by the results reported by Gupta et al. [140] on Al–Li/SiC MMCs. In this work, the distribution of SiC particles was analyzed by examining their relative location with respect to grain boundaries. Gupta et al. observed that a considerable amount of SiC particles were present within the grains. Since the grain boundary is likely to be the last position of a solid–liquid interface during solidification, entrapment of SiC particles must have occurred. Such entrapment may result from one or a combination of the following mechanisms: engulfment by the solidification front during atomization, or mechanical entrapment by dendrite arms during atomization.

4.4.3 Reinforcement-Induced Nucleation

4.4.3.1 Free Energy Effects

In Sections 4.4.1 and 4.4.2, two important mechanisms of particle entrapment during MMC solidification were discussed, namely, engulfment and mechanical entrapment. A third mechanism that may lead to particle entrapment is particle-induced nucleation. There are extensive experimental results which support this premise, particularly in ingot cast Al–Si MMCs. Cissé et al. [135], for example, added Al_2O_3 particles directly into hypereutectic Al–Si alloy melts. The **composite slurry** was then allowed to solidify under ingot casting conditions. They observed that Al_2O_3 particles can effectively nucleate primary Si phase crystals. Interestingly, nucleation of the Al phase was, however, not enhanced by the presence of the Al_2O_3 particles. One example of the nucleation of primary Si on the surface of Al_2O_3 particles is illustrated in Figure 4.43 [12, 135].

Similar observations of Si nucleation on Al_2O_3 and other ceramic particles have also been documented by other researchers working on the solidification processing of MMCs [12, 132, 136, 141, 142]. To that effect, the nucleation of Si and eutectic Al–Si phase on SiC and **graphite** particles are frequently reported throughout the literature [132, 136, 142]. Lloyd and Jin [136], for example, reported nucleation of an Al–Si eutectic phase on SiC particles in the last regime to solidify in hypoeutectic Al–Si alloys. In a comprehensive review on the solidification processing of MMCs, Mortensen and Jin [23] also reported the engulfment of SiC particles by solidification fronts of primary Si crystals in a hypereutectic Al–16 wt% Si alloy reinforced with 15 vol% SiC particles (see Figure 4.44). The cooling rate experienced by the material shown in Figure 4.44 is 7 Ks^{-1}. These experimental results suggest that ceramic particles which exhibit low interfacial energy with solid Si may promote the nucleation of Si crystals and thereby become entrapped in the matrix. The nucleation of Si on Al_2O_3, SiC, and **graphite** particles under slow cooling rate conditions may be partially attributed to the lower interfacial energy between solid Si and particles relative to that between liquid phase and particles (or alternatively, the low wetting angle between solid Si and particles). Consequently, nucleation of Si on such particles is energetically favored [23].

Figure 4.43 Micrograph showing nucleation of primary Si crystals on the surface of porous Al_2O_3 particles in an ingot cast Al–16 wt% Si composite. Source: Rohatgi et al. [12]/from Taylor & Francis Group.

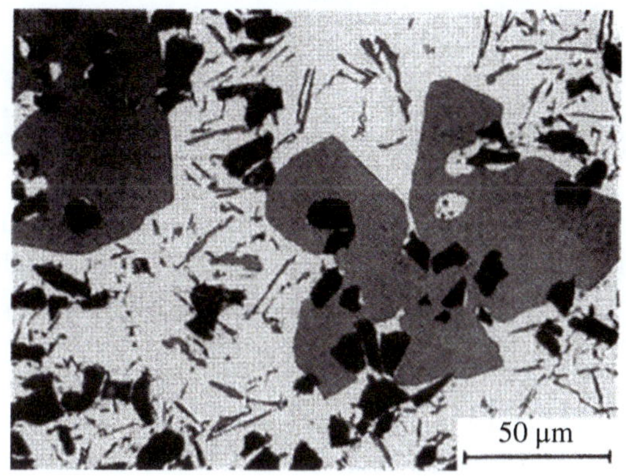

Figure 4.44 Microstructure of Al–16 wt% Si/15 vol% SiC composite (cooling rate = 7 Ks^{-1}) showing the engulfment of SiC particles by Si crystals. Source: Mortensen and Jin [23]/from Taylor & Francis Group.

4.4.3.2 Thermal Effects

The entrapment of particles because of particle-induced nucleation may also be rationalized based on thermal effects. Thermal effects are thought to be important in cases when the particles are not sufficiently wetted by the matrix, and hence the nucleation of **solid phases** on particle surfaces becomes energetically favorable only when the particles are at a lower temperature than the surrounding liquid phase which is the typical condition during atomization and co-injection. Wu and Lavernia [64], for example, reported nucleation of α-Al on SiC and $T1B_2$ particles during atomization and co-injection. As shown in this figure, the α-Al phase appears to originate from SiC particles and grows radially from a nucleation center. An important observation of SiC-induced nucleation in Al–Si/SiC composite droplets is that such nucleation occurs primarily in powders that are less than 98 μm [64]. Such droplets are typically below the liquidus temperature during particle co-injection and hence are undercooled [64]. It was suggested that the nucleation of α-Al on SiC particles observed in atomized droplets differs from the nucleation of Si on SiC as observed under conventional solidification conditions [64]. In the latter, the low interfacial energy between solid Si and SiC enhances the probability of surface nucleation. In the case of α-Al, however, available experimental evidence suggests that SiC exhibits a high wetting angle (or high interfacial energy) with solid α-Al. Mortensen et al. [126], for example, studied the remelting and solidification behavior of a fiber-reinforced Al–4.5Cu/SiC composite using the Bridgman method (directional solidification). In this study, a SiC-fiber-reinforced Al–4.5Cu MMC was prepared using an infiltration method. During directional solidification, the fiber-reinforced sample was placed in a Bridgman furnace, remelted, and directionally solidified. To study the solidification behavior, Mortensen et al. [126] quenched the sample prior to the completion of solidification. Figure 4.45 shows that in the as-quenched Bridgman sample, the fibers are surrounded by a refined microstructure which is formed presumably during quenching. Since in directional solidification, fibers, and liquid alloy are at the same temperature, the refined microstructure must be completely molten prior to quenching. Consequently, the fibers are surrounded by a solute-rich liquid during solidification which separates them from dendrites. These results indicate that when the growth of α-Al occurs at a temperature that is identical to that of SiC fibers, α-Al tends to avoid the fibers during solidification. Consequently, in this case, nucleation of α-Al on SiC is not energetically favorable.

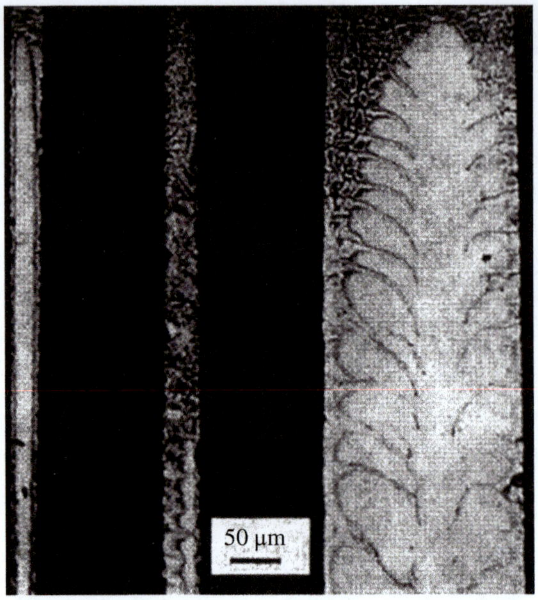

Figure 4.45 Longitudinal section of a directionally solidified fiber-reinforced Al/SiC MMC showing the dendrite tip region. Source: Mortensen et al. [126]/from Elsevier.

In rationalizing the ceramic-particle-induced nucleation behavior, Zhang [143] suggested that during atomization and co-injection, the ceramic particles are generally injected at a much lower temperature (e.g. 300 K) relative to that of the molten droplets. Consequently, the thermal gradient induced by such particles in the droplets may promote the nucleation of **solid phases** even in the case when the interfacial energies between the particle and the solid phases are relatively high. Zhang [143] calculated the temperature distribution inside of an atomized Al droplet in the presence of a cold ceramic particle based on simple heat transfer considerations. In these calculations, it was assumed that the ceramic particle becomes completely immersed in a molten droplet during co-injection. Following penetration, the droplet containing a ceramic particle travels at a relative velocity with the surrounding gas. To simplify the problem formulation, it was assumed that a ceramic particle at an initial temperature, T_{po}, is present concentrically within a metallic droplet at the beginning of the heat transfer calculations. This simplified condition is illustrated schematically in Figure 4.46. Consequently, the droplet loses its thermal energy to the cold particle and the surrounding gas environment. In this case, the temperature distribution in the droplet and the particle may be calculated from Eq. (4.14). In these calculations, the initial temperature of the particle, T_{po}, was assumed to be identical to the temperature prior to penetration. Moreover, the initial temperature of the droplet was assumed to be the temperature of a droplet at a distance corresponding to the point of **ceramic co-injection** and was also calculated from heat transfer considerations. The corresponding boundary conditions are described as follows. At the droplet gas interface, that is, when $r = R$, the conduction heat flux must be balanced by the heat flux across the droplet surface:

$$-k_l \frac{\partial T}{\partial r}\bigg|_{r=R} = h(T - T_g) \tag{4.72}$$

where k_l is the thermal conductivity of the liquid, h is the heat transfer coefficient across the droplet surface and T_g is the temperature of the surrounding gas environment. Two heat transfer conditions were assumed at

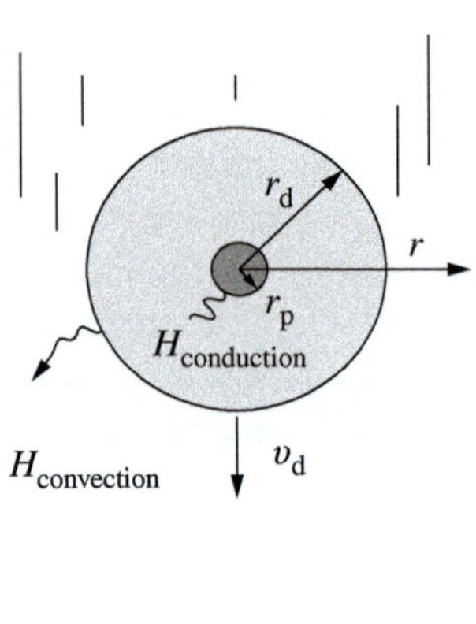

Figure 4.46 The relative positions of a particle and a droplet at the beginning of the heat transfer calculations where r_d and r_p are the radius of droplet and particle, respectively; $H_{conduction}$ and $H_{convection}$ are the enthalpies of conduction and convection, respectively and v_d is droplet velocity. Source: Zhang et al. [143]/with permission from Elsevier.

the particle/liquid interface: conduction controlled, and convection controlled. When interfacial heat transfer is conduction controlled, the boundary condition at the particle/liquid interface is given by

$$k_{\mathrm{l}} \left.\frac{\partial T}{\partial r}\right|_{r=r_{\mathrm{p}}} = k_{\mathrm{p}} \left.\frac{\partial T}{\partial r}\right|_{r=r_{\mathrm{p}}}$$
$$T_{\mathrm{ip}} = T_{\mathrm{il}} \tag{4.73a}$$

where T_{ip} and T_{il} are temperatures at the particle/liquid interface on the particle and the liquid side, respectively, and; r_{p} is particle radius. When interfacial heat transfer is convection controlled, the boundary at the particle/liquid interface is given by

$$k_{\mathrm{l}} \left.\frac{\partial T}{\partial r}\right|_{r=R_{\mathrm{p}}} = k_{\mathrm{p}} \left.\frac{\partial T}{\partial r}\right|_{r=R_{\mathrm{p}}} = -h_{\mathrm{cpl}}(T - T_{\mathrm{l}}) \tag{4.73b}$$

where h_{cpl} is the heat transfer coefficient at the particle/liquid interface. At the common center of the particle and the droplet, spherical symmetry of the model requires that

$$\left.\frac{\partial T}{\partial r}\right|_{r=0} = 0 \tag{4.74}$$

Zhang [143] calculated the temperature distribution inside of an Al droplet containing a ceramic particle, by solving the governing equations using a numerical approach. The calculated transient temperature profile is summarized in Figure 4.47 for a 10 μm Si C particle embedded within a 100 μm Al droplet. As shown in this figure, a large thermal gradient is present only at the droplet/particle interface. In addition, such a temperature gradient is present only for a short period of time of less than 10 μs. In the areas away from the particle/droplet interface, the temperature profile changes only slightly with the distance from the droplet center. Accordingly, the temperature profile in regions of the droplet that are away from the particle remains largely unaffected by the presence of the particle.

To estimate an upper limit of undercooling, Zhang [143] calculated the homogeneous nucleation rate within the bulk body in the droplet and the heterogeneous nucleation rate at the particle surface by assuming steady-state nucleation. Under steady-state nucleation conditions, the homogeneous nucleation rate is given by

$$I_{\mathrm{homo}} = \frac{k_{\mathrm{v}}}{\eta_{\mathrm{l}}} \exp\left[-\frac{\Delta G^*}{kT}\right] \tag{4.75}$$

where k is the Boltzmann's constant, ΔG^* is the free energy of nucleation, η_{l} is the liquid viscosity and k_{v} is a constant which is relatively insensitive to temperature. The magnitude of k_{v} is on the order of IQ^{30} [144]. The corresponding heterogeneous nucleation rate is given by

$$I_{\mathrm{heter}} = \frac{n_{\mathrm{s}}}{n_{\mathrm{l}}} \frac{k_{\mathrm{v}}}{\eta_{\mathrm{l}}} \exp\left[-\frac{\Delta G^*_{\mathrm{het}}}{kT}\right] \tag{4.76}$$

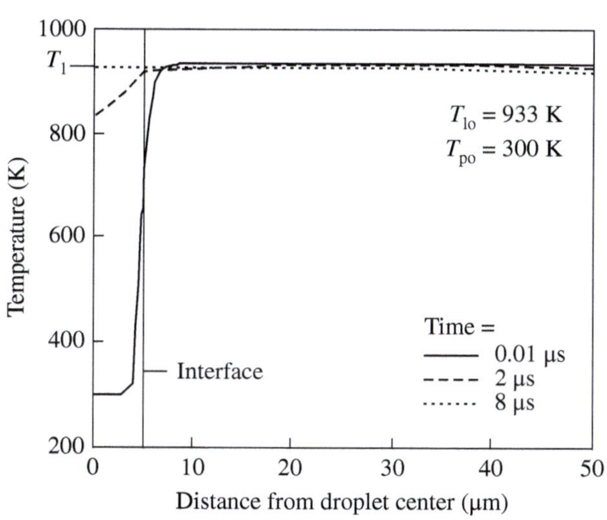

Figure 4.47 The temperature distribution at the interior of an Al droplet containing a SiC particle in the center where T_{lo} = initial droplet temperature and T_{po} = initial particle temperature [143].

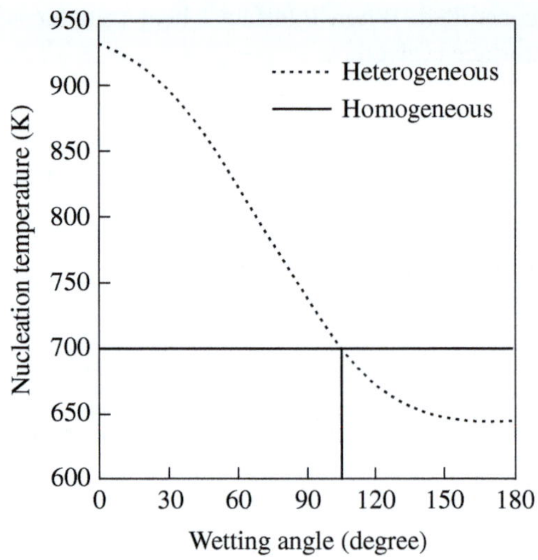

Figure 4.48 Comparison of the heterogeneous and homogeneous nucleation temperatures when droplet and particle are at identical initial temperatures. Source: Zhang et al. [143]/with permission from Elsevier.

where I_{heter} is the heterogeneous nucleation rate; n_s is the number of catalytic sites per unit volume of liquid and n_l is the number of liquid atoms per unit volume, and; $f(\theta)$ is a function of the wetting angle as given by Eq. (4.56). In this study, the extent of undercooling and the corresponding nucleation temperatures were calculated from the homogeneous and heterogeneous nucleation rates. Figure 4.48 shows the relative importance of homogeneous and heterogeneous nucleation as a function of the wetting angle between the solid phase and the particle for a 100 μm Al droplet containing a 10 μm particle. In calculating the nucleation temperatures shown in Figure 4.48, it was assumed that when penetration occurs the particle and droplet are at an identical initial temperature of T_l (i.e. the melting temperature of Al). As shown in Figure 4.48, for low wetting angles, nucleation is limited by heterogeneous nucleation on the surface of the particle. The heterogeneous nucleation temperature, however, decreases rapidly with increasing wetting angle. At wetting angles that exceed 106°, the heterogeneous nucleation temperature is less than the homogeneous nucleation temperature. Hence, when the wetting angle between the particle and solid phase is greater than 106°, the particle will not promote heterogeneous nucleation when the particle is at the same initial temperature as the droplet. This is the condition encountered in commonly utilized Al/ceramic systems such as Al/Al_2O_3, Al/**graphite**, and Al/SiC.

As shown in Figure 4.47 when the particle is at a lower initial temperature relative to that of the droplet, the interfacial temperature between the particle and the droplet is much lower than the bulk temperature of the droplet during an initial transient period. Considering this effect, Zhang [143] defined a bulk nucleation temperature which is the average temperature of a droplet when nucleation occurs at the particle/droplet interface. Figure 4.49 illustrates the calculated bulk nucleation temperature in a droplet when a particle at an initial temperature of 300 K is present

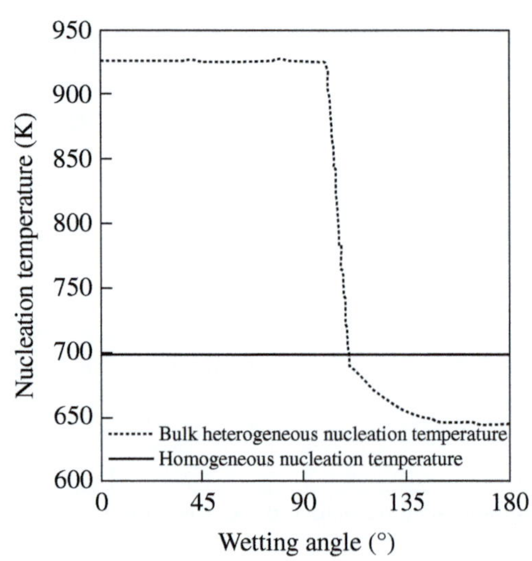

Figure 4.49 Influence of ceramic particle on the bulk undercooling in an Al droplet (initial particle temperature = 300 K, initial droplet temperature = 933 K). Source: Zhang et al. [143]/with permission from Elsevier.

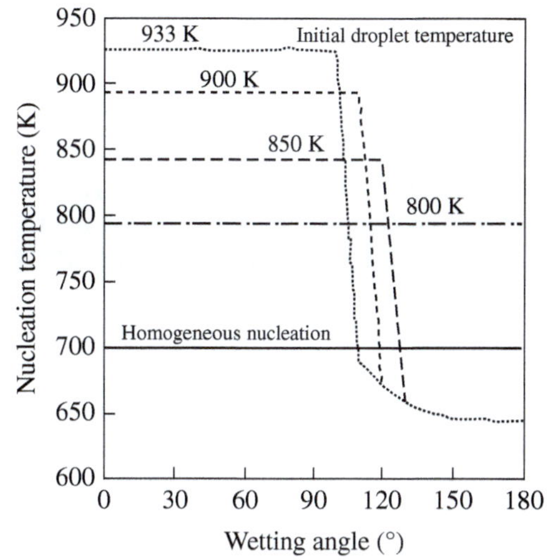

Figure 4.50 Effect of initial undercooling on the nucleation temperature of a droplet containing a ceramic particle at an initial temperature of 300 K. Source: Zhang et al. [143]/with permission from Elsevier.

in the center of a droplet which is at the liquid's temperature. According to the results presented by Zhang [143], at low wetting angles heterogeneous nucleation is induced at the particle surface partly because of the large thermal gradient present which leads to a local temperature depression, and partly because of the low interfacial energy between the solid phase and the particle. In this case, when nucleation starts at the particle surface, the liquid at the interface is already highly undercooled, whereas the bulk liquid in the droplet is still at a relatively low undercooling. However, the thermal shock effect is limited to low wetting angles, that is, smaller than 110°. At high wetting angles, homogeneous nucleation prevails (see Figure 4.49). Zhang [143] further studied the effects of thermal shock when a droplet is initially undercooled. The results of these calculations are summarized in Figure 4.50 for various initial undercoolings for a particle temperature of 300 K. The results indicate that the effect of thermal shock increases with increasing initial droplet undercooling. Moreover, when the initial undercooling is greater than 130 K (e.g. an initial droplet temperature = 800 K) heterogeneous nucleation occurs at the particle surface regardless of wetting angle. According to these results, ceramic particles that typically exhibit large wetting angles with solid phases may also induce heterogeneous nucleation. In this case, however, heterogeneous nucleation is dependent on thermal shock, that is, the initial temperature difference between the particle and the droplet as well as on the initial undercooling of the droplets. This latter observation agrees with results which suggests that heterogeneous nucleation of α-Al on SiC particles occurs preferentially in small droplets. It should be noted, however, that the estimation of undercooling assumes of steady state nucleation. The validity of such steady-state nucleation condition in the case of thermal shock induced by particle injection awaits verification.

4.5 Other Methods for Fabricating MMC Powders

Several production methods have been used to fabricate MMC powders for AM, but each method produces vastly different sizes, morphologies, and compositions of powder, and these characteristics influence the properties of AM components. Due to its ability to produce spherical powders that achieve the desired flowability, injectability, and apparent density in AM, gas atomization (GA) is typically chosen for AM metallic powder production. In the case of GA MMC powders, the reinforcement ceramic particles need to be well distributed inside of the powders. As discussed in Section 4.3.3, in this chapter, one approach involves the co-injection of ceramic particles into the atomization spray field. With careful control of the co-injection conditions, it is possible to achieve a relatively homogenous dispersion of ceramic particles in the atomized powders, although interactions with the solidification front can lead to segregation of the ceramic particles, as described in Section 4.4.2. However, obtaining a homogeneous dispersion of the reinforcement phase in the case of nanoscale ceramic particles, carbon nanotubes (CNTs), or ceramic whiskers and fibers, presents some challenging conditions because of the highly irregular surface morphologies. Such types of reinforcement particles are better suited for mechanical dispersion methods such as those associated with MM, **surface coating**, and **reaction synthesis**. In this section, we will discuss the benefits and limitations of some of these methods to produce composite powders that can be used as feedstock for AM processes.

4.5.1 Mechanical Milling and Cryomilling

To address the reinforcement homogeneity issues posed by other powder production methods, MM and alloying have been used to disperse reinforcement particles within metal matrix powders. In this approach, matrix and reinforcement powders are mixed and placed together with hardened metal balls in a sealed milling vial or attritor. The vial is then agitated in a **planetary**, linear, or cyclical motion, or attritor ball mills to repeatedly induce particle fracture and cold welding of the metal matrix. Parameters such as **ball-to-powder ratio**, **milling speed**, **milling duration**, and **process control agents (PCAs)** may be used to tailor the milled powder size, microstructure, and reinforcement distribution [145, 146]. MM under a liquid nitrogen environment, a process termed **cryomilling**, is sometimes used to limit oxygen exposure, accelerate plastic deformation mechanisms, and promote desirable reactions with the nitrogen environment [147]. Both milling and **cryomilling** tend to produce agglomerate sizes that are typically smaller than gas-atomized powder. These techniques are discussed in more detail below.

MM has been successfully used to fabricate Al MMC powders for AM processes, since alloys such as $AlSi_{10}Mg$, are ductile and hence facilitate homogeneous encapsulation of the reinforcing particles during milling. The effect of **milling duration** on producing 95.0 wt% $AlSi_{10}Mg$ and 5.0 wt% nano-TiC composites via planetary milling were reported [148], and results show that both particle and grain sizes were refined after 15 hours of milling. The composite powders exhibited a uniform nano-TiC dispersion after milling and after AM laser beam melting (LBM), resulting in a 32% increase in hardness and a 20% increase in tensile strength with no loss in ductility as compared to the properties of unreinforced $AlSi_{10}Mg$ components. The authors of this study argued that the use of a high energy density (laser power) during AM effectively increased the Marangoni convection effect in the melt pool thereby leading to enhanced recirculation of the ball-milled and surface-coated MMC powders. Results from this and other studies show that there is a close interrelation between the reinforcement distribution in the powder MMC feedstock, the AM power and scanning speed, and the final microstructure and mechanical properties of AM-produced components.

Porosity is an important issue with AM components, especially when produced using non-spherical powders such as those generated using MM. In related studies, $AlSi_{10}Mg$-based composites with CNT reinforcements were fabricated via ball milling and AM, and the resultant densification behavior and microstructure were studied [149]. In other work, planetary milling was utilized to fabricate Al–Al_2O_3 composites specifically designed for powder bed fusion (PBF) [150]. In this study, powders milled up to 20 hours experienced a transformation from flat, irregular particles to rounded particles with preferable flow properties. It was reported that the predominance of particle cold welding early in the milling process resulted in an increase in particle size from 15 to 50 μm after 10 hours of milling. Further milling decreased the particle size to 35 μm and created semi-spherical particle morphology. A reduction in grain size from 71 nm after 8 hours of milling to 48 nm after 20 hours of milling was noted. Subsequent consolidation of the 20-hour milled powder with AM (specifically PBF) at various laser power and scanning speeds demonstrated that flowable, PBF-feasible aluminum matrix composite powder can be achieved using MM.

During MM, small amounts (2.0–5.0 wt%) of a PCA are frequently used [145, 151]. This additive, typically stearic acid, acts as a lubricant to prevent excessive inter-particle welding, as well as to minimize the adhesion of the material being processed to the milling balls and equipment. The use of PCAs is particularly important when grinding ductile materials such as Al and Al alloys. Unfortunately, PCAs are often incorporated into the milled material and are difficult to remove after milling, sometimes requiring degassing at relatively elevated temperatures (i.e. >375 °C). When such milled powders are used in AM, the PCA degrades under high temperatures and risks forming hydrocarbon defects in the AM-deposited materials. These defects can lead to premature failure of components, and hence MM of ductile metals without the use of PCAs requires additional research.

Fe-based alloys are the most popular materials used for AM processing. The production of Fe-based nanocomposites using MM has gained some interest given the potential of producing alloys and composites with improved properties. Inspection of the published literature shows that Fe-based composite powders include oxides, **carbides**, and intermetallic phases. For example, MM Fe-based oxide-dispersion-strengthened (ODS) alloy composites combine the high-temperature strength and thermal stability of the oxide dispersion with excellent resistance to oxidation, carburization, and hot corrosion. Table 4.7 summarizes some MM processing parameters that have been used for the synthesis of Fe-based nanocomposites [146].

MM at cryogenic temperatures, **cryomilling**, is another MMC production method that allows for dispersoids to be fully encapsulated within matrix particles and provides several benefits. Typical cryomilling involves milling

Table 4.7 MM processing parameters for the synthesis of Fe-based nanocomposites [146].

		Process parameters							
Matrix	**Reinforcement**	**Mill**	**Milling vial**	**Milling medium (h)**	**Speed (rpm)**	**BPR**	**PCA**	**Atmosphere**	**Milling time (h)**
Fe	Fe_3O_4	SPEX 8000	Hardened steel	Hardened steel		10 : 1		Argon	20
Fe	Fe_3O_4	SPEX 8000	Hardened steel	Hardened steel		10 : 1		Argon	30
Fe	Fe_yO	Fritsch P6	Hardened steel	Hardened steel	500	9 : 1		Air	20
Fe	Fe_3C	Fritsch P7						Argon	
Fe	SiO_2	Fritsch P5	Hardened SS	Hardened SS	200	10 : 1		Air	80
Fe	SiO_2	High-energy mill							60
Fe	SiO_2		Cr-steel	Cr-steel				Argon	60
α-Fe	$Nd_2Fe_{14}b$	SPEX 8000	Hardened steel	Hardened steel				Argon	48
α-Fe	$Nd_2Fe_{14}b$	High-energy ball mill						Argon	5
Fe	Al_2O_3	SPEX 8000	Carbon steel	Carbon steel		13 : 1		Air	48
Fe	Al_2O_3	SPEX 8000	Tool steel	Tool steel		10 : 1		Argon	24
Fe(Al)-Fe_3Al	Al_2O_3	Planetary ball mill			360	7 : 1		Argon	50
Fe	Cu	SPEX 8000	Hardened steel	Hardened steel		4 : 1		Argon	20
Fe	Bi	AGO-2 planetary ball						Argon	0.667
Fe	(TM)-carbides	Planetary ball mill			150			Argon	180
Fe	NbC	SPEX 8000						Argon	40
Fe_3Al	TiC	Planetary mill (QM-1SP2)			360			Argon	50
$(Fe,Ti)_3Al$	Al_2O_3	Planetary ball mill	Hardened Cr steel	Hardened Cr steel	500	10 : 1		Argon	100
Fe(Al)		High-energy GN-2	Hardened steel	Hardened steel	700	10 : 1	$CH_3(CH_2)_5CH_3$	Argon	70
Fe–Cr–Al	TiB_2	Planetary mill (PM200)	Tool steel	Tool steel	500	20 : 1		Argon	48

Figure 4.51 SEM images showing (a) the morphology of 5083 Al/B_4C composite powders cryomilled for eight hours, and (b) the B_4C particle distribution in the cross sections of 5083 Al/B_4C composite powder milled for six hours. Source: Ye et al. [153]/with permission from Elsevier.

constituent powders and liquid nitrogen or argon slurry in a Szegvari attritor. Cryomilling achieves higher microstructural control when synthesizing Al-based nanocomposites as compared to materials processed via traditional mechanical alloying [146, 147]. For example, cryomilling can achieve the desired nanocrystalline microstructure more quickly than room temperature MM [147, 152]. Moreover, during cryomilling, there is minimal welding to the milling equipment, thus eliminating the necessity for PCAs as in room temperature milling. Homogeneously distributed in situ and ex situ aluminum nanocomposite powders have been fabricated using this method [153, 154].

In related studies, cryomilling was successfully employed by Ye et al. to fabricate particulate B_4C-reinforced Al matrix nanocomposite powders [153]. The composite powders were milled for different times to investigate underlying microstructural evolution and helped establish the formation mechanism for the nanocomposite during cryomilling. In this study, commercial 5083 Al alloy powder (90 wt%) and particulate B_4C (10 wt%) were blended and then cryomilled under a liquid nitrogen atmosphere. Cryomilling was carried out at a temperature of −180 °C using a Szegvari attritor with a **ball-to-powder ratio** of 32 : 1 and a rotation speed of 180 rpm. Stearic acid (0.2 wt%) was added into the milling chamber as a PCA to prevent the severe adhesion of the Al onto the chamber and milling balls. Figure 4.51 shows the morphology of 5083 Al/B_4C composite powders cryomilled for eight hours, and the B_4C particle distribution in the cross sections of 5083 Al/B_4C composite powder milled for six hours. During cryomilling, the particles are flattened, adhere together, and develop a flakelike morphology. Gradually, the flake-like particles become fine equiaxed particles, and the surfaces of these particles are highly irregular. After six hours of cryomilling, further milling results in very little change in the powder morphology. The particles underwent repeated cold welding and fracturing during cryomilling. In the early stage of cryomilling, the particle size became larger, and the size distribution became broader. After six hours of cryomilling, the particle size approaches a constant value and the size distribution is narrow, as shown in Figure 4.52a. As a result, both large and small composite particles were driven to an intermediate size with a narrower size distribution. The size of the B_4C particles did not change significantly and remained in the micron size range, due to the extreme hardness and impact resistance of B_4C, as shown in Figure 4.52b. The coarse-grained 5083 Al was subdivided into microbands very quickly at the early stage of cryomilling and was gradually refined to equiaxed grains with a minimum grain size of 23 nm. After cryomilling for six hours, a steady state had been achieved and the B_4C particles were uniformly distributed in the nanocrystalline 5083 Al matrix with randomly oriented, equiaxed grains, as shown in Figure 4.53. The authors concluded that composite powders with uniformly distributed B_4C particles and strong Al/B_4C interfaces were successfully fabricated.

Cryomilling can promote **mechanochemical reactions** which may be used to tailor MMC powder composition. Direct contact between the liquid nitrogen and Al during slurry cryomilling can lead to the in situ formation of **nanoscale aluminum nitride (AlN)** and aluminum oxynitride (Al(O,N)) phases [154, 155]. When homogeneously distributed, these phases also act as nucleation sites, effectively increasing grain refinement and strengthening. Other **ceramic** particles such as oxides and carbides may also be added in the milling process to encourage in situ reinforcement synthesis during cryomilling [146, 147].

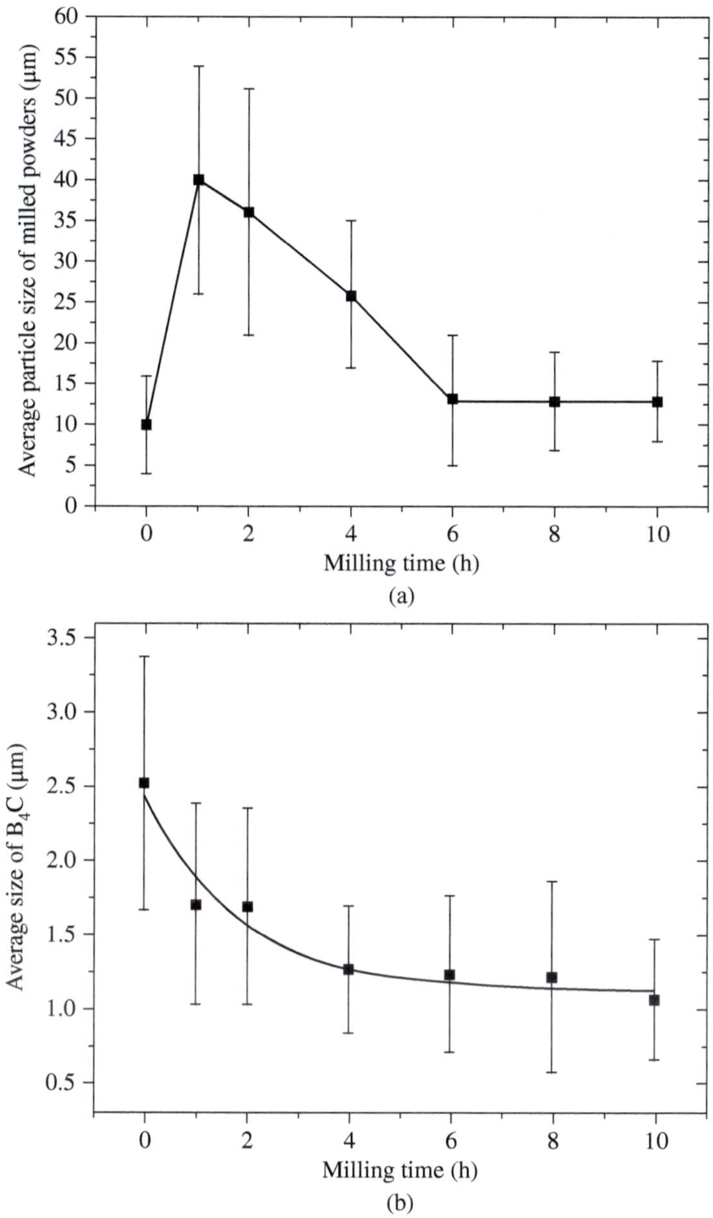

Figure 4.52 Variation in (a) the average particle size for the 5083 Al/B_4C composite powder during cryomilling, and (b) the average B4C particle size during cryomilling of the 5083 Al/B_4C composite powder. Source: Ye et al. [153]/with permission from Elsevier.

If oxidation and nitridation reactions are to be avoided, for example for certain metals where oxide and **nitride** formation may adversely affect properties, cryomilling may be performed under conditions such that the liquid nitrogen does not come in direct contact with the milling media. To that effect, a vibratory cryomilling vial with liquid nitrogen cooling channels was designed to fabricate high-purity metallic nanoparticles [156]. Spex SamplePrep (Metuchen, NJ, USA) also produces a line of Freezer/Mill that submerge milling vials in liquid nitrogen, avoiding nitrogen and oxygen contamination. Different from traditional rotary attritor and spherical impactors which induce shear fracture in the liquid nitrogen slurry, a Spex cryomill utilizes linear pulverization from a single cylindrical impactor inside a milling vial submerged in liquid nitrogen. The kinetics of this process are similar to those of the high-energy ball milling of the Spex 8000, but they occur at cryogenic temperatures [145]. Compared to attritor cryomilling, the Spex cryomill avoids nitridation and the development of **AlN** dispersoids, because the milling material does not come in direct contact with liquid nitrogen. The impactor-to-powder mass ratio, milling time,

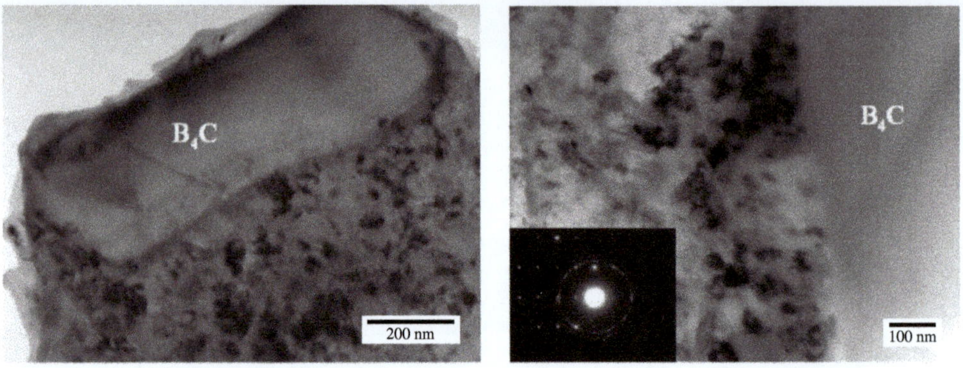

(a) (b)

Figure 4.53 Bright-field TEM images for the 5083 Al/B_4C composite powder cryomilled for eight hours: (a) lower magnification and (b) higher magnification with the inset as the SAD pattern from the interface. Source: Ye et al. [153]/with permission from Elsevier.

PCAs, and grinding impactor geometry are the key parameters for this process. This type of mill has been used to fabricate an Al/CuO thermite composite powder but remains unexplored for AM feedstock powder [157].

Cryomilling techniques have been utilized to fabricate nanostructured Al MMC powder, but their customizable potential for AM processes remains to be explored. To render MMC powders as attractive feedstock material for use in AM processing, the specific powder morphology and size distribution required for AM must be met. The studies reviewed in this section suggest that although some progress has been achieved, much work remains to be done to accomplish the goal of establishing milling/cryomilling as mainstream methods to produce AM feedstock powder [148, 150, 153].

4.5.2 Surface Coating

Various methods have been successfully developed to produce MMC powders suitable for AM via **surface coating** techniques. To that effect, mechanical mixing or blending has been used extensively but these approaches have only achieved limited success. In related work, $AlSi_{10}Mg$ composites containing nano- and micro-TiB_2 reinforcements were produced using a ball mill but without the grinding media [158]. In this work, the authors reported that the density of the AM-deposited $AlSi_{10}Mg$/nano-TiB_2 composites critically depends on the energy density used during AM, and moreover they noted the absence of any cracking in the fabricated material. In other studies, Nano-TiC was distributed onto spherical Ti powders through mechanical mixing in a planetary mill without any grinding media [159], as shown in Figure 4.54. While the same correlation between LBM energy density and resulting densification was found, these composites exhibited undesirable grain coarsening as linear energy density increased from 250 to 1000 J m^{-1}. Several studies have made the same grain coarsening observation when using high energy densities in laser AM processing of $AlSi_{10}Mg$- and Ni-based, surface-coated composites [159, 160]. These results indicate that there

Figure 4.54 Mechanically mixed Ti/nano-TiC composites. Source: Gu et al. [159]/from Elsevier/CC BY-NC-ND 3.0.

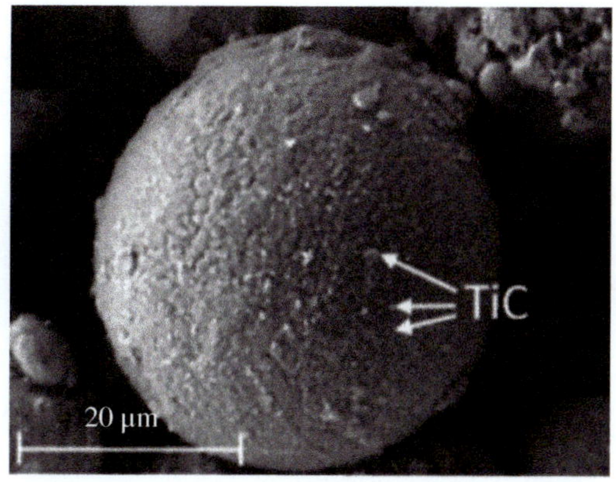

Figure 4.55 Al7075/nano-TiC composite particle prepared by surface inoculation. Source: Tasche et al. [161]/from Springer Nature.

is a critical range of AM energy densities that must be identified in order to obtain homogeneous microstructures with minimal grain growth.

Surface inoculation has also been used to synthesize MMC and MMNC powders for AM. In one study, TiC nanoparticles were ultrasonically dispersed on gas atomized Al7075 particles in an electrolytic solution [161] as shown in Figure 4.55. The TiC nanoparticles were subsequently dry etched onto the matrix oxide surface leading to a homogeneous TiC dispersion. It was reported that preferable epitaxial grain growth occurs during AM LBM fabrication.

A method of nano-functionalization has also been explored involving the use of an electrostatic assembly technique to distribute tungsten carbide (WC) nanoparticles on gas-atomized $AlSi_{10}Mg$ powder [162]. Nanostructured functionalized surfaces are materials that contain nanoscale features produced in a well-controlled manner and involve nanocoating, film deposition, or in situ surface modifications at the nanoscale level, to improve material performance [163]. The results of this work show that a uniform reinforcement distribution in AM PBF processed MMCs leads to an increase in strength and wear resistance as compared to those of unreinforced $AlSi_{10}Mg$ structures. Residual porosity was found to exist within deposited materials with both functionalized and non-functionalized powders and attributed to an increase in volumetric energy absorption that results from the increased reinforcement volume. However, few details were provided as to the applicability and scalability of such an electrostatic assembly method.

The direct coating of matrix or reinforcement particle surfaces offers the benefit of incorporating multiple phases while retaining the spherical morphology of the atomized powder, or in the case of irregular particles, improving their flowability. Moreover, surface reactions between powder and coating can increase mechanical bonding between matrix and reinforcement phase thereby leading to enhanced dispersion during AM. In related work, Ti6Al4V-based MMCs using Ti6Al4V powder with Ni-coated TiC particles as the feedstock materials were processed by AM directed energy deposition (DED) [1]. In this study, Ni-coated TiC particles (33 wt% Ni) were produced via a chemical vapor deposition (CVD) Fast Fluidized Bed process, with a size range of 10–80 μm, a mean size of 36 μm, and a median size of 34 μm. Figure 4.56a shows the scanning electron microscopy (SEM) (backscattered electron imaging [BSE]) micrographs of Ni-coated TiC particles. The morphology of the coated TiC particles is more spherical than that of uncoated TiC particles. Spherical particles are beneficial for the delivery systems that are typically used in AM DED processing. As shown in Figure 4.56b, the TiC particles are completely encapsulated by the Ni coating with a minimum thickness of about 2 μm. There was no evidence indicating the presence of micro-scale defects, such as voids and cracks, at the interface between Ni coating and the TiC ceramic core. It was also observed that some fine TiC particles (1–5 μm), which presumably originate from inter-particle collisions in the CVD reactor, were embedded in the Ni coating layer.

IN625 Ni-based MMCs components were also deposited using AM DED process with Ni-coated and un-coated TiC reinforcement particles to provide insight into the influence of interfaces on MMCs [2]. The flowability of mixed powder, interface between the matrix and the TiC particles, the interaction between laser beam and TiC particles, and mechanical properties of the DED deposited MMCs were effectively improved by using Ni-coated TiC particles. The TEM studies showed that the Ni coating was comprised of single-phase, nearly equiaxed grains with a size range from 10 to 100 nm [2], indicating that the synthesized Ni coating is nanocrystalline. The presence of a nanostructure in the coating can be rationalized based on a rapid quenching mechanism during cooling from the liquid state.

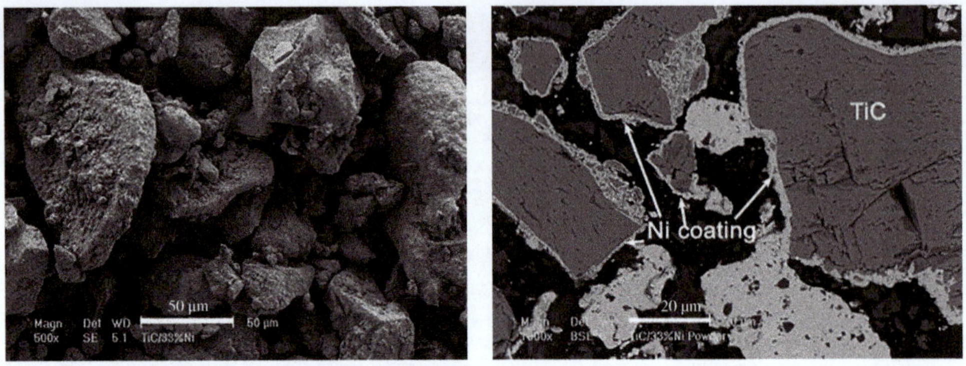

(a) (b)

Figure 4.56 SEM morphology of Ni-coated TiC particles, (a) surface, and (b) cross-section. Source: Zheng et al. [1]/from Springer Nature/CC BY-NC 2.0.

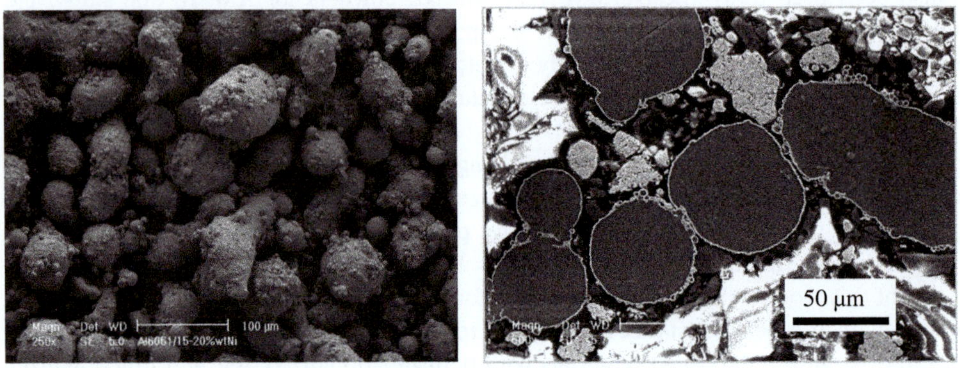

(a) (b)

Figure 4.57 (a) SEM morphology of Ni coated Al6061 particles, showing the mix of spherical and elongated particles which have fine satellite particles on the surface, and (b) SEM micrograph of a polished cross-section showing uniformity of 1 μm thick Ni coating. Source: Zheng et al. [166]/from Taylor & Francis Group.

As such, it has been argued that rapid displacement of the liquid/solid interface will lead to an equiaxed structure with nanoscale dimensions, due to limited growth time during rapid solidification [164, 165]. No apparent reaction and defects, such as pores, cracks, or dislocations, were observed at the interface between the nano-crystalline Ni coating and TiC particles, and the interface is relatively smooth and continuous. The TiC core and the Ni coating appeared to be well bonded.

A hybrid, Al/Al_3Ni metallic foam was in situ synthesized via AM DED of Ni-coated 6061 Al powder in the absence of a foaming agent by Zheng et al. [166]. The powders used were Al6061 particles coated with 15–20 wt% Ni via a CVD fast-fluidized bed process [167], as shown in Figure 4.57. The powders had a mix of 45–120 μm spherical and elongated particles, having fine 1 μm satellite particles attached to their surface. During DED processing, the Ni coating reacted with the Al matrix, resulting in the simultaneous formation of a fine dispersion of Al_3Ni, and a high-volume fraction of porosity. As a reinforcement phase, the **intermetallic compound** formed particles with a size range of 1–5 μm and a volume fraction of 63% in the solid cell wall area, with accompanying 35–300 μm pores with a 60% volume fraction.

4.5.3 Reaction Synthesis

The concept of forming reinforcing particles in situ has been widely applied to traditional manufacturing routes, such as castings, and is generally described as reaction synthesis. The chemical reactions that lead to the formation of these particles can be controlled by manipulating heat input and other solidification parameters. This concept has also been applied to AM, as illustrated by the formation of nanoscale intermetallic and oxide reinforcement particles in LBM using mixtures of Al, AlMg1SiCu, and $AlSi_{10}Mg$ with Fe_2O_3 [168]. This work showed that a homogeneous dispersion of particulates, coupled with the high cooling rate that is associated with LBM led to heterogeneous nucleation and a resultant increase in hardness for the alloy compositions studied. The behavior of the Al/SiC chemistry and associated interfacial reactions during LBM has also been studied in terms of resultant SiC particle size [169]. Results showed that coarse ($D_{50} = 50\,\mu m$) SiC particles blended with $AlSi_{10}Mg$ powders were less prone to react and form a tertiary Al_4SiC_4 phase than fine ($D_{50} = 5\,\mu m$) SiC particles during LBM. In fact, not all in situ reactions are preferable, such as it was observed degradation of SiC in an aluminum matrix during LBM [170]. Interfacial formation of aluminum carbide (Al_4C_3) plates led to poor wetting, degradation of mechanical properties, and crack initiation.

Reaction synthesis can also lead to challenges, as illustrated by the difficulties encountered while trying to control phase formation while attempting to generate high-strength γ-TiAl intermetallic compounds from Ti alloys via electron beam melting (EBM) [171]. Similarly, work on MM and reacting of Ti and Al powders showed the formation of the detrimental $TiAl_3$ phase. Reaction synthesis of pre-alloyed Ti–Al–Nb–Cr–Fe powder generated net shapes with α_2-Ti_3Al and γ-TiAl phases, but vaporization of aluminum decreased the amount of the desirable γ-TiAl phase [172].

Reaction synthesis has been demonstrated to be a viable method for producing MMCs during AM; however, consistent and reproducible MMCs via AM will require more research. The effects of energy density and feedstock powder parameters on the resultant porosity, microstructure, and mechanical properties of AM-synthesized MMCs need to be addressed in further work.

MMC powders are becoming an increasingly important material feedstock for the AM of MMCs. The potential for enhanced mechanical performance due to grain refinement and dislocation strengthening in AM MMCs is high; however, AM process parameters need to be optimized to avoid the formation of undesirable defects, such as porosity, residual stresses, and hot cracking.

Mechanically milled MMCs exhibit similar solidification issues as surface-coated MMCs when subjected to AM PBF processing. To tailor MMC energy absorptivity, a critical energy density is needed to minimize porosity, retain uniform reinforcement distribution, and maximize the MMC strengthening mechanisms. It is interesting to note that morphology has not been identified as a limiting factor for the use of **mechanically milled** powders in AM PBF. However, powder flowability is a critical factor for achieving full-density AM components, especially for PBF. Mechanically milled and cryomilled powders are known to have irregular morphologies that are expected to hinder flowability and thereby increase porosity. However, no study has examined flowability and AM part densification as a function of the mode of MMC powder production.

Each MMC powder production method can achieve various dispersion states of the reinforcement phase. For example, **surface coating** methods provide a reinforcement distribution that relies on Marangoni convection during AM to achieve uniformity, whereas MM and cryomilling produce uniform reinforcement distribution prior to AM. There are also few studies that quantify and systematically compare various MMC powder production methods to the resultant reinforcement distribution, microstructure, and mechanical properties of AM-produced components. More studies should examine the microstructural and mechanical effects of AM MMC components with various MMC powder feedstock production methods.

4.6 Summary

In this chapter, we described the various processing methods that have been used to manufacture composite powders in an effort to take advantage of the combination of a metal and a **ceramic** and the benefits that one can gain from properly designing a material that contains both soft and hard phases. Composite powders can be manufactured in a number of different ways and in this chapter, we first described the generation of composite powders using atomization. The chapter starts with a discussion of the fundamental factors that govern the interactions of **ceramic** particles and metal droplets during co-injection. The co-injection of **ceramic** particles into an atomized dispersion of droplets can be used as an effective method to prepare composite powders which in turn can be used directly for additive manufacturing. The underlying phenomena that govern these interactions are complex and involve a combination of thermal, fluid, and mass transfer phenomena. This chapter also discussed the process of reacting a metal droplet with a gas phase to generate an oxide phase in situ. This process can be effectively used to prepare composite powders containing both a metal and an oxide phase. The interaction of the **ceramic** particles with the solidifying metal is complex and ultimately governs the physical location of the **ceramic** particles in a metal droplet. Other methods beyond atomization, including MM and cryomilling, **surface coating**, and reaction synthesis, for fabricating MMC powders were also briefly described.

Nomenclature

Variable	Description	Units
a	acceleration	$M\,s^{-2}$
a_{cr}	critical acceleration	$M\,s^{-2}$
A_d	surface area of droplet	m^2
Bi	Biot number	—
$c_{p,l}$	constant pressure heat capacity of liquid phase	$J\,kg^{-1}\,K^{-1}$
$c_{p,g}$	constant pressure heat capacity of gas	$J\,kg^{-1}\,K^{-1}$
$c_{p,p}$	constant pressure heat capacity of particle	$J\,kg^{-1}\,K^{-1}$
$c_{p,r}$	constant pressure heat capacity of reinforcement	$J\,kg^{-1}\,K^{-1}$
D	droplet diameter	m
d_c	depth of crater	m
d_g	separation between particle and solid–liquid interface	m
d_{in}	injector tube diameter	m
d_{noz}	nozzle tube diameter	m
d_o	minimum separation between particle and solid–liquid interface	m
d_p	particle size	m
D	diffusivity	$m^2\,s^{-1}$
D_g	diffusivity of oxygen in gaseous phase	$m^2\,s^{-1}$
D_{in}	inner diameter of the injector container	m
D_l	self-diffusion coefficient in liquid	$m^2\,s^{-4}$
E	elastic modulus	GPa
f_m	volume fraction of matrix	—
f_r	volume fraction of reinforcements	—
f_s	fraction of solid phase	—
f_{void}	volume fraction of void	—
$f_{w,r}$	weight fraction of reinforcements	—
g	gravitational acceleration	ms^{-2}
h	heat transfer coefficient	$Wm^{-2}\,K^{-1}$
h_c	convective heat transfer coefficient	$Wm^{-2}\,K^{-1}$
ΔH_f	heat of fusion	$I\,kg^{-1}$
ΔH_r	heat of reaction	$J\,kg^{-1}$
k	Boltzmann's constant	$J\,K^{-1}$
k_i	thermal conductivity	$Wm^{-1}\,K^{-1}$
k_l	thermal conductivity of liquid phase	$Wm^{-1}\,K^{-1}$
k_{ox}	thermal conductivity of the oxide film	$Wm^{-1}\,K^{-1}$
k_p	thermal conductivity of particle	$Wm^{-1}\,K^{-1}$
k_s	thermal conductivity of solid phase	$Wm^{-1}\,K^{-1}$
K_{in}	injector constant	—
K_m	kinetic growth coefficient	$ms^{-1}\,K^{-1}$

Variable	Description	Units
m_c	mass of composite material	kg
$\dot{m}_g$	mass flow rate of gas	$kg\,s^{-1}$
$\dot{m}_p$	mass flow rate of particles	$kg\,s^{-1}$
m_r	mass of reinforcement	kg
n_s	number of catalytic sites per unit volume	m^{-3}
n_l	number of liquid atoms per unit volume	m^{-3}
p	percentage of porosity	-
p	dynamic pressure	MPa
Δp_{in}	injection pressure	MPa
r_d	radius of droplet	m
r_p	radius of particle	m
Re	Reynolds number	—
S	separation distance	m
Sh	Sherwood number	—
S_{pl}	contact area between particle and liquid	m^2
S_{pg}	contact area between particle and gas	m^2
S_{lg}	contact area between liquid and gas	m^2
T	temperature	K
T_e	reference temperature	K
T_g	temperature of gas	K
T_l	liquidus temperature	K
T_s	solidus temperature	K
ΔT	undercooling	K
V	relative velocity	ms^{-1}
V_{cr}	critical interfacial velocity of engulfment	ms^{-1}
V_d	volume of droplet	m^3
V_{sl}	solid–liquid interfacial velocity	ms^{-1}
V_t	terminal velocity	ms^{-1}
α_l	thermal diffusivity of liquid	$cm^2\,s^{-1}$
ε_l	surface emissivity	—
ϕ	semiepical angle	degree
γ_{lg}	interfacial energy between liquid and gas	$J\,m^{-2}$
γ_{pg}	interfacial energy between particle and gas	$J\,m^{-2}$
γ_{pl}	interfacial energy between particle and liquid	$J\,m^{-2}$
γ_{ps}	interfacial energy between particle and solid	$J\,m^{-2}$
η_g	dynamic viscosity of gas	Pa s
η_l	dynamic viscosity of liquid	Pa s
θ	wetting angle	degree
ρ_c	density of composite	$kg\,m^{-3}$
ρ_g	density of gas	$kg\,m^{-3}$
ρ_l	density of liquid phase	$kg\,m^{-3}$
ρ_m	density of matrix	$kg\,m^{-3}$
ρ_p	packing density of particles	$kg\,m^{-3}$
ρ_r	packing density of reinforcement	$kg\,m^{-3}$

Variable	Description	Units
ρ_s	density of solid phase	kg m^{-3}
ρ_t	theoretical density of particles	kg m^{-3}
Ω	molar volume	m^3 mol^{-1}
$\Delta\sigma$	change in interfacial energy	J m^{-2}
σ_r	Stefan–Boltzmann constant	Wm^{-2} K^{-4}
v_g	velocity of the injection gas	ms^{-1}
ν_g	kinematic viscosity of atomization gas	m^2 s^{-1}
ν_l	kinematic viscosity of liquid	m^2 s^{-1}
ν_P	Poisson's ratio	—

References

1 Zheng, B., Smugeresky, J.E., Zhou, Y. et al. (2008). *Metall. Mater. Trans. A* 39A: 1196–1205.
2 Zheng, B., Topping, T., Smugeresky, J.E. et al. (2010). *Metall. Mater. Trans. A* 41A: 569–573.
3 Li, L., Wang, J., Lin, P., and Liu, H. (2017). *Ceram. Int.* 43: 16638–16651.
4 Behera, M.P., Dougherty, T., and Singamneni, S. (2019). *Procedia Manuf.* 30: 159–166.
5 Li, N., Liu, W., Wang, Y. et al. (2021). *Chin. J. Mech. Eng.* 34: 38.
6 Dadkhah, M., Mosallanejad, M.H., Iuliano, L., and Saboori, A. (2021). *Acta Metall. Sinica* 34: 1173–1200.
7 Promakhov, V., Matveev, A., Klimova-Korsmik, O. et al. (2022). *Metals* 12: 525.
8 Tang, S., Ummethala, R., Suryanarayana, C. et al. (2021). *Adv. Eng. Mater.* 23: 2100053.
9 Seetharaman, S. and Gupta, M. (2021). Additive manufacturing of metal matrix composites. In: *Encyclopedia of Materials: Composites* (ed. D. Brabazon), 209–229. Oxford: Elsevier.
10 Raj Mohan, R., Venkatraman, R., Raghuraman, S. et al. (2022). *Scanning* 2022: 5610333.
11 Vály, L., Grech, D., Neubauer, E. et al. (2017). *Key Eng. Mater.* 742: 129–136.
12 Rohatgi, P.K., Asthana, R., and Das, S. (1986). *Int. Met. Rev.* 31: 115.
13 Mehrabian, R. (1988). *Mater. Res. Soc. Symp.* 120: 3.
14 Dural Aluminum Composites Corporation. Duralcan Metal Matrix Composites May-June Data Report Package, Dural Aluminum Composites Corporation, San Diego, CA, USA.
15 Clyne, T.W., Bader, M.G., Cappieman, G.R., and Hubert, P.A. (1985). *J. Mater. Sci.* 20: 85.
16 Mortensen, A., Masur, L.J., Cornie, J.A., and Flemings, M.C. (1989). *Metall. Trans. A.* 20A: 2535.
17 Masur, L.J., Mortensen, A., Cornie, J.A., and Flemings, M.C. (1989). *Metall. Trans. A.* 20A: 2549.
18 Mortensen, A. and Wong, T. (1990). *Metall. Trans. A* 21A: 2257.
19 Calhoun, R.B. and Mortensen, A. (1992). *Metall. Trans. A* 23A: 2291.
20 Mortensen, A. and Michaud, V. (1990). *Metall. Trans. A* 21A: 2059.
21 Michaud, V. and Mortensen, A. (1992). *Metall. Trans. A* 23A: 2263.
22 Mortensen, A., Michaud, V.J., and Flemings, M.C. (1993). *JOM* 45 (1): 36.
23 Mortensen, A. and Jin, I. (1992). *Int. Mater. Rev.* 37: 101.
24 Krishnamurthy, S., Kim, Y.W., Das, G., and Froes, F.H. (1990). Metal and Ceramic Matrix Composites. In: (eds. R.B. Bhagat, A.H. Clauer, P. Kumar, and A.M. Ritter), 145. Warrendale, PA: TMS.
25 Jian, M.K., Bhanuprasad, V.V., Kamatt, S.V. et al. (1993). *Int. J. Powder Metall.* 29: 267.
26 Rack, H.J. (1988). Light-Weight, High-Performance Metal Matrix Composites. *Processing and Properties of Powder Metallurgy Composites* (ed. K. Vedula, A.A. Ritter, and P. Kumar), 155. Warrendale, PA: TMS.
27 Fascetta, E.F., Riek, R.G., Mehrabian, R., and Flemings, M.C. (1973). *Trans. AFS* 81: 81.
28 Mehrabian, R., Riek, R.G., and Flemings, M.C. (1974). *Metall. Trans. A* 5A: 1899.
29 Ibrahim, I.A., Mohamed, F.A., and Lavernia, E.J. (1991). *J. Mater. Sci.* 26: 1137.
30 McKimpson, M.G., Pohlenz, E.L., and Thompson, S.R. (1993). *JOM* 45 (1): 26.

31 Hunt, J.W.H., Osman, T.M., and Lowandowski, J.J. (1993). *JOM* 45 (1): 30.

32 Mortensen, A. (1990). *Fabrication of Particle Reinforced Metal Composites, Proceedings of an International Conference*, 217. Materials Park, OH: ASM International.

33 Geiger, A.L. and Jackson, M. (1989). *Adv. Mater. Process* 7: 23.

34 Lloyd, D.J. (1991). *Acta Metall. Mater.* 39: 59.

35 Mummery, P.M. and Derby, B. (1991). *12th International Riso Symposium on Materials Science*, 535. Roskilde, Denmark: Riso National Laboratory.

36 Singer, A.R.E. (1983). *Ann. CIRP* 32: 145.

37 Willis, J. (1988). *Met. Mater.* 4: 485.

38 Buhrmaster, C.L., Clark, D.E., and Smart, H.O. (1988). *J. Met.* 40: 44.

39 Kojima, K.A., Lewis, R.E., and Kaufman, M.J. (1989). *Aluminum-Lithium Alloys V* (ed. E.A. Starke and J.T.H. Sanders), 85. Birmingham: MCE Publications.

40 White, J., Palmer, I.G., Hughes, I.R., and Court, S.A. (1989). *Aluminum-Lithium Alloys V* (ed. E.A. Starke and J.T.H. Sanders), 1635. Birmingham: MCE Publications.

41 Llorca, J., Ruiz, J., Healy, J.C. et al. (1994). *Mater. Sci. Eng. A.* A185: 1.

42 Gupta, M., Mohamed, F.A., and Lavernia, E.J. (1989). *Proceedings of the 17th International Symposium on Advances in Processing and Characterization of Ceramic Metal Matrix Composites* (ed. H. Mostaphaci), 236. Oxford, UK: Pergamon Press.

43 Gupta, M., Mohamed, F.A., and Lavernia, E.J. (1990). *Mater. Manuf. Processes* 5: 165.

44 Wu, Y. and Lavernia, E.J. (1991). *JOM* 43 (8): 16.

45 Zhang, J., Perez, R.J., Gupta, M., and Lavernia, E.J. (1993). *Seri. Metall. Mater.* 28: 91.

46 Wu, Y. (1994). Interaction Mechanisms between Ceramic Particulates and Metallic Matrices and Mechanical Behavior fo Al-Si-X (X=SiC, TiB_2) MMC, Synthesized using Spray Atomization and Co-Injection. PhD Thesis, University of California at Irvine, Irvine, CA, USA.

47 Wu, M., Srivatsan, T.S., Pickens, J.R., and Lavernia, E.J. (1992). *Seri. Metall. Mater.* 27: 761.

48 Gupta, M., Srivatsan, T.S., Mohamed, F.A., and Lavernia, E.J. (1993). *J. Mater. Sci.* 28: 2245.

49 LLorca, J., Martin, A., Ruiz, J., and Elices, M. (1993). *Metall. Trans. A* 24A: 1575.

50 Kahl, W. and Peupp, J. (1990). *Swiss Mater.* 2 (4): 17.

51 Eliasson, J. and Sandstrom, R. Application of Aluminium Matrix Composites. *Key Engineering Materials*, vol. 104, 105, 3–36. Switzerland: Trans Tech Publications, Ltd ISSN: 162-9795.

52 Carvalho, M.H., Marcelo, T., and Carvalhinhos, H. (1992). *J. Mater. Sci.* 27: 2101.

53 Kim, N.J., Park, W.J., Ahn, S., and Elias, L. (1994). *High Performance Metal and Ceramic Composites, in Light Weight Alloys for Aerospace Applications III* (ed. E.W. Lee and N.J. Kim), p. 137. Warrendale, PA: TMS.

54 Gupta, M., Bowo, K., Lavernia, E.J., and Earthman, J.C. (1993). *Seri. Metall. Mater* 28: 1053.

55 Gupta, M., Ibrahim, I.A., Mohamed, F.A., and Lavernia, E.J. (1991). *J. Mater. Sci.* 26: 6673.

56 Gupta, M., Mohamed, F.A., and Lavernia, E.J. (1991). *Int. J. Rapid. Solid.* 6: 247.

57 Gupta, M., Mohamed, F.A., and Lavernia, E.J. (1992). *Metall. Trans. A* 23A: 831.

58 Perez, R.J., Zhang, J., and Lavernia, E.J. (1992). *Seri. Metall. Mater.* 27: 1111.

59 Zeng, X. and Lavernia, E.J. (1992). *Int. J. Rapid. Solid.* 7: 219.

60 Zhang, J., Perez, R.J., and Lavernia, E.J. (1994). *Acta Metall. Mater.* 42: 395.

61 Zhang, J., Gungor, M.N., and Lavernia, E.J. (1993). *J. Mater. Sci.* 28: 1515.

62 Perez, R.J., Zhang, J., Gungor, M.N., and Lavernia, E.J. (1993). *Metall. Trans. A* 24A: 701.

63 Wu, Y., Zhang, J., and Lavernia, E.J. (1994). *Metall. Mater. Trans. B* 25B: 135.

64 Wu, Y. and Lavernia, E.J. (1992). *Metall. Trans. A* 23A: 2923.

65 Wu, Y. and Lavernia, E.J. (1993). *Proceedings of International Conference on Advanced Synthesis of Engineered Structural Materials, 31 August–2 September 1992* (ed. E.J. Lavernia, F.H. Froes, and J.J. Moore), 253. Materials Park, OH, San Fransisco, CA, USA: ASM International.

66 Wu, Y. and Lavernia, E.J. (1994). *Processing and Fabrication of Advanced Materials III* (ed. T.S. Srivatsan, J.J. Moore, and V.A. Ravi), 501. Warrendale, PA: TMS.

67 Srivatsan, T.S. and Lavernia, E.J. (1994). *Composite Eng.* 4 (4): 459.

68 Gupta, M., Lane, C., and Lavernia, E.J. (1992). *Seri. Metall. Mater.* 26: 825.

69 McLelland, A.R.A., Atkinson, H.V., Kapranos, P., and Kirkwood, D.H. (1991). *Mater. Lett.* 11 (1, 2): 26.

70 Gupta, M., Juarez-Islas, J., Frazier, W.E. et al. (1992). *Metall. Trans. B* 23B: 719.

71 Armanie, K.P. and Zaidi, M.A. (1987). Metallographic Examination of Osprey MMC Sample. Technical Report, Alcoa Technical Center, Pittsburgh, PA, USA, p. 1.

72 Maher, P.P., Cantor, B., and Katgerman, L. (1990). *International Conference on Advanced Aluminum and Magnisium Alloys* (ed. T. Khan and G. Effenberg), 659. Materials Park, OH: ASM International.

73 Mathur, P., Kim, M.H., Lawley, A., and Apelian, D. (1990). *Powder Metallurgy: Key to Advanced Materials Technology*, 55. Materials Park, OH: ASM International.

74 Majagi, S.I., Ranganathan, K., Lawley, A., and Apelian, D. (1992). Spray Forming of Metal Matrix Composites. *Microstructural Design by Solidification Processing* (ed. E.J. Lavernia and M.N. Gungor), 139. Warrendale, PA: TMS.

75 Perez, J.F. and Morris, D.G. (1994). *Seri. Metall. Mater.* 31: 231.

76 Unigame ,Y. (1989). An analysis of oxide dispersion strengthening of ferrous alloys during spray casting. MS Thesis, Drexel University, Philadelphia, PA, USA.

77 Unigame,Y., Lawley, A., Apelian, D. (1993). In situ spray casting of dispersion strengthened alloys — I: thermodynamics and reaction kinetics. Powder Metallurgy World Congress, Paper No. 14E-T6-6. Kyoto, Japan: Japan Powder Metallurgy Association.

78 Ranganathan, K., Lawley, A., Apelian, D. (1993). In situ spray casting of dispersion strengthened alloys — II: experimental studies. -6 1993 Powder Metallurgy World Congress , Paper No. 14E-T6. Kyoto, Japan: Japan Powder Metallurgy Association.

79 Zhao, Y.Y. and Grant, P.S. (1993). *B. Cantor* 169 (2): 263.

80 Zhao, Y.Y., Grant, P.S., and Cantor, B. (1993). *J. Physique* 3: 1685.

81 Grant, P.S., Zhao, Y.Y., Li, J.H. et al. (1995). *Science and Technology of Rapid Solidification and Processing* (ed. M.A. Otooni), 109. Dordrecht, The Netherlands: Kluwer.

82 Baskin, D., Wolfenstine, J., and Lavernia, E.J. (1994). *J. Mater. Res.* 9: 362.

83 Jeng, Y.L., Wolfenstine, J., and Lavernia, E.J. (1993). *Seri. Metall. Mater.* 28: 453.

84 Jeng, Y.L., Lavernia, E.J., and Wolfenstine, J. (1993). *Seri. Metall. Mater.* 29: 107.

85 Zeng, X., Liu, H., Chu, M.G., and Lavernia, E.J. (1992). *Metall. Trans. A* 23A: 3394.

86 Liang, X. and Lavernia, E.J. (1992). *Mater. Sci. Eng. A* A153: 654.

87 Lawley, A. and Apelian, D. (1994). *Powder Metall.* 37: 123.

88 Zeng, X. and Lavernia, E.J. (1995). *Seri. Metall. Mater.* 32: 991.

89 Geiger, G.H. and Poirier, D.R. (1973). *Transport Phenomena in Metallurgy*. Reading, Massachusetts: Addison-Wesley.

90 Zeng, X. (1995). Dissertation, Micro-structure and Properties of Ni3Al and Si3N4 Processed Using Reactive Atomization, Department of Chemical Engineering and Materials Science. University of California, Irvine.

91 Ibrahim, I.A., Mohamed, F.A., and Lavernia, E.J. (1990). *International Conference on Advanced Aluminum and Magnesium Alloys* (ed. T. Khan and G. Effenberg), 745. Materials Park, OH: ASM International.

92 Lavernia, E.J. (1991). *SAMPE Q.* 22: 2.

93 Liu, H., Zeng, X., and Lavernia, E.J. (1993). *Scri. Metall.* 29: 1341.

94 Zeng, X., Nutt, S.R., and Lavernia, E.J. (1995). *Metall. Mater. Trans. A* 26: 817.

95 Brandes, E.A. (1983). *Smithells Metals Reference Book*, 6e. London, United Kingdom: Butterworths.

96 Oishi, Y. and Kingery, W.D. (1970). *J. Chem. Phys.* 33: 480.

97 Liu, H., Rangel, R.H., and Lavernia, E.J. (1994). *Acta Metall. Mater.* 42: 3277.

98 Johnson, H.A. and Rubesin, M.W. (1949). *Trans. ASME* 71: 447.

99 Kubaschewski, O. and Hopkins, B.E. (1962). *Oxidation of Metals and Alloys*, 2e. New York: Academic Press.

100 Lee, A.K., Sanchez-Caldera, L.E., Chun, J.H., and Suh, N.P. (1989). *Multicomponent ultrafine microstructures, MRS Conference Series*, vol. 132 (ed. L.E. McCandish et al.), 87. Pittsburgh, PA, USA: Metals Research Society.

101 Leatham, A.G., Ogilvy, A., and Elias, L. (1993). The osprey process: current status and future possibilities. In: *P/M in Aerospace, Defense and Demanding Applications Conference*. San Diego, CA, USA.

102 Kacar, A.S., Rana, F., and Stefanescu, D.M. (1991). *Mater. Sci. Eng., A* A135: 95.

103 Rohatgi, P.K., Asthana, R., Yadav, R.N., and Ray, S. (1990). *Metall. Trans. A* 21A: 2073.

104 Ayers, J.D. (1984). *Wear* 97: 249.

105 Ayers, J.D., Schaefer, R.J., and Robey, W.P. (1981). *J. Met.* 33 (8): 19.

106 Ayers, J.D. and Bolster, R.N. (1984). *Wear* 93: 193.
107 Liang, X. and Lavernia, E.J. (1993). *JOM* 45 (7): 50.
108 Zhang, J., Wu, Y., and Lavernia, E.J. (1994). *Acta Metall. Mater.* 42: 2955.
109 Young, T. (1805). *Trans. Roy. Soc.* 95: 65.
110 Murr, L.E. (1975). *Interfacial Phenomena in Metals and Alloys*. London, UK: Addison-Wesley.
111 Allen, B.C. (1972). *Liquid Metals: Chemistry and Physics* (ed. S.Z. Beer). New York: Marcel Dekker.
112 Nakae, H., Yamamoto, K., and Sato, K. (1991). *Mater. Sci. JIM* 32: 531.
113 Laurent, V., Chatain, D., and Eustathopoulos, N. (1987). *J. Mater. Sci.* 22: 244.
114 Liao, X.A. and Liu, Y.X. (1990). *Proceedings of Light Metals 1990*, vol. 1990, 409. TMS, Warrendale, PA; Anaheim, CA, USA.
115 Ueki, M., Naka, M., and Okamoto, I. (1986). *J. Mater. Sci. Lett.* 5: 1261.
116 Chaklader, A.C.D. (1989). *Processing of Ceramic and Metal Matrix Composites* (ed. H. Mostaghaci), 257. Elmsford, NY: Pergamon Press.
117 Nogi, K. and Ogino, K. (1988). *Trans. Jpn. Inst. Met.* 29: 805.
118 Levitt, A.P. and Wolf, S.M. (1970). *Whisker Technology* (ed. A.P. Levitt), 245. New York: Wiley Interscience.
119 Bevolo, A.J., Campisi, G.J., Shanks, H.R., and Schmidt, F.A. (1980). *J. Appl. Phys.* 51: 5390.
120 Kanetkar, C.S., Kacar, A.S., and Stefanescu, D.M. (1988). *Metall. Trans. A* 19A: 1833.
121 Kattamis, T.Z. and Piccone, T.J. (1991). *Mater. Sci. Eng. A* A131: 265.
122 Poirier, D.R. and Geiger, G.H. (1994). *Transport Phenomena in Materials Processing*. Warrendale, PA: TMS.
123 Tiwari, R. and Herman, H. (1991). *Seri. Metall. Mater.* 25: 1103.
124 Stefanescu, D.M., Moitra, A., Kacar, A.S., and Dhindaw, B.K. (1990). *Metall. Trans. A* 21A: 231.
125 Russell, K.C., Cornie, J.A., and Oh, S.Y. (1986). Particulate wetting and particle: Solid interface phenomena in casting metal matrix composites. *Interfaces in Metal Matrix Composites* (ed. A.K. Dhingra and S.G. Fishman), 61. Warrendale, PA: TMS/AIME.
126 Mortensen, A., Cornie, J.A., and Flemings, M.C. (1988). *Metall. Trans. A* 19A: 709.
127 Uhlmann, D.R., Chalmers, B., and Jackson, K.A. (1964). *J. Appl. Phys.* 35: 2986.
128 Stefanescu, D.M., Dhindaw, B.K., Kacar, S.A., and Moitra, A. (1988). *Metall. Trans. A* 19A: 2847.
129 Omenyi, S.N. and Neumann, A.W. (1976). *J. Appl. Phys.* 47: 3956.
130 Sasikumar, R. and Kumar, M. (1991). *Acta Metall. Mater.* 39: 2503.
131 Chernov, A.A., Temkin, D.E., and Mel'nikova, A.M. (1976). *Sov. Phys. Crystall.* 21: 369.
132 Jin, I. and Lloyd, D.J. (1990). *Fabrication of Particle Reinforced Composites* (ed. J. Masounave and F.G. Hamel), 47. Materials Park, OH: ASM International.
133 Bolling, G.F. and Cisse, J. (1971). *J. Cryst. Growth* 10: 56.
134 Wu, Y., Liu, H., and Lavernia, E.J. (1994). *Acta Metall. Mater.* 42: 825.
135 Cissé, J., Bolling, G.F., and Kear, H.W. (1975). *Metall. Trans. B* 6B: 195.
136 Lloyd, D.J. and Jin, I. (1988). *Metall. Trans. A* 19A: 3107.
137 Chernov, A.A. and Mel'Nikova, A.M. (1965). *Sov. Phys. Crystal.* 10: 672.
138 Surappa, M.K. and Rohatgi, P.K. (1981). *J. Mater. Sci. Lett.* 16: 562.
139 Cahn, J.W., Hilling, W.B., and Sears, G.W. (1964). *Acta Metall.* 12: 1421.
140 Gupta, M., Mohamed, F.A., and Lavernia, E.J. (1991). *Mater. Sci. Eng.* A144: 99.
141 Sugishita, J., Fujiyoshi, S., Imura, T., and Ishi, M. (1982). *Wear* 81 (2): 181.
142 Krishnan, B.P. and Rohatgi, P.K. (1984). *Met. Tech.* 11: 41.
143 Zhang, Q., Rangel, R.H., and Lavernia, E.J. (1996). *Acta Materialia* 44: 3693–3703.
144 Boettinger, W.J. and Perepezko, J.H. (1985). *Rapidly Solidified Crystalline Alloys* (ed. B.H. Kear, C.M. Adam, and S.K. Das), 21. Warrendale, PA: AIME.
145 Suryanarayana, C. (2001). *Progress in Materials Science* 46: 1–184.
146 Suryanarayana, C. and Al-Aqeeli, N. (2013). *Prog. Mater. Sci.* 58: 383–502.
147 Witkin, D.B. and Lavernia, E.J. (2006). *Prog. Mater. Sci.* 51: 1–60.
148 Wang, H. and Gu, D. (2014). *J. Compos. Mater.* 49: 1639–1651.
149 Wang, L.-Z., Chen, T., and Wang, S. (2017). *Optik* 143: 173–179.
150 Han, Q., Setchi, R., and Evans, S.L. (2017). *Int. J. Adv. Manuf. Technol.* 88: 1429–1438.
151 J.R. Groza, E.J. Lavernia, J.F. Shackelford, M.T. Powers (Eds.) *Materials Processing Handbook*, CRC Press, 2007.

152 Zhu, X., Zhang, X., Wang, H. et al. (2003). *Scr. Mater.* 49: 429–433.
153 Ye, J., Lee, Z., Ahn, B. et al. (2006). *Metall. Mater. Trans. A* 37: 3111–3117.
154 Li, Y., Liu, W., Ortalan, V. et al. (2010). *Acta Mater.* 58: 1732–1740.
155 Ma, K. and Schoenung, J. (2010). *Philos. Mag. Lett.* 90: 739–751.
156 Kumar, N. and Biswas, K. (2015). *Rev. Sci. Instr.* 86: 083903.
157 Badiola, C., Schoenitz, M., Zhu, X., and Dreizin, E.L. (2009). *J. Alloys Compounds* 488: 386–391.
158 Aversa, A., Marchese, G., Lorusso, M. et al. (2017). *Adv. Eng. Mater.* 19: 1700180.
159 Gu, D., Wang, H., Chang, F. et al. (2014). *Phys. Procedia* 56: 108–116.
160 Marchese, G., Aversa, A., Lorusso, M. et al. (2018). *Metals* 8: 175.
161 Tasche, L., Hoyer, K.-P., Zhuravlev, E. et al. *Surface Inoculation of Aluminium Powders for Additive Manufacturing Guided by Differential Fast Scanning Calorimetry, TMS 2019 148th Annual Meeting & Exhibition Supplemental Proceedings*, vol. 2019, 485–493. Cham: Springer International Publishing.
162 Martin, J.H., Yahata, B.D., Clough, E.C. et al. (2018). *MRS Commun.* 8: 297–302.
163 Lunello, L., Potrich, C., Pasquardini, L., and Pederzolli, C. (2016). Nanostructured functionalized surfaces. In: *Encyclopedia of Nanotechnology* (ed. B. Bhushan), 2727–2734. Dordrecht, Netherlands: Springer.
164 Lavernia, E.J., Ayers, J.D., and Srivatsan, T.S. (1992). *Int. Mater. Rev.* 37: 1–44.
165 Herlach, D.M. (1994). *Mater. Sci. Eng.* R12 (4, 5): 177–272.
166 Zheng, B., Li, Y., Smugeresky, J.E. et al. (2011). *Phil. Mag.* 91: 3473–3497.
167 He, J., Zhou, Y., Baker, D. et al. (2003). *Ceramic Powder Encapsulated by Nanocrystalline Metallic Coating*, vol. 2003, 167–172. Chicago, IL: MS&T.
168 Dadbakhsh, S. and Hao, L. (2012). *J. Alloys Compounds* 541: 328–334.
169 Chang, F., Gu, D., Dai, D., and Yuan, P. (2015). *Surf. Coat. Technol.* 272: 15–24.
170 Ocelík, V., Vreeling, J.A., and De Hosson, J.T.M. (2001). *J. Mater. Sci.* 36: 4845–4849.
171 Cormier, D., Harrysson, O., Tushar, M., and West, H. (2008). *Res. Lett. Mater. Sci.* 2007: 1–4.
172 Gussone, J., Hagedorn, Y.-C., Gherekhloo, H. et al. (2015). *Intermetallics* 66: 133–140.

5

Diagnostic and Characterization Techniques

5.1 Introduction

The successful commercial exploitation of **spray atomization** and deposition technologies will depend on the implementation of *in situ* diagnostic techniques to monitor and control critical processing variables, such as the size and velocity of the powders as well as the surface temperatures of the preform. However, **atomization** and deposition processes are generally difficult to monitor as a result of the large number of interrelated variables that are involved [1]. Some of the latest developments in this field have occurred primarily in four areas: laser light diffraction techniques, **high-speed cinematography**, thermal imaging, and surface ionization. All these techniques have their own advantages and shortcomings as related to monitoring online spray parameters. A review of process monitoring techniques and an assessment of current sensor technology, as it is related to spray atomization, has been reported recently in the literature [2–5]. The purpose of this chapter is to present a review of the **diagnostic technologies** that are available for the *in situ* monitoring of melt atomization and deposition: Section 5.2 describes **flow visualization** techniques; Section 5.3 introduces **particle image velocimetry (PIV)**; Section 5.4 discusses the **particle counting, sizing, and velocity probe (PCSV-P)**; Section 5.5 addresses the **high-speed cinematography** techniques; Section 5.6 describes high-speed off-axis holographic cinematography; Section 5.7 covers the mechanisms of **infrared thermal imaging**; Section 5.8 summarizes **phase Doppler particle analysis (PDPA)**; Section 5.9 reviews the principle of **surface ionization for monitoring particles (SIMPs)**; and finally Section 5.10 addresses **intelligent sensors**.

5.2 Flow Visualization Techniques

A comprehensive understanding of **atomization** requires detailed information on related phenomena, such as breakup of the molten metal, thickness of the metal liquid film at the exit of the delivery tube, and interactions of the atomization gas with the metal film being atomized. Inspection of the published literature shows that combinations of **optical visualization** techniques and **high-speed cinematography** have widely been used to study single-phase flow (i.e. just gas flow during atomization), especially for the case of **close-coupled gas atomization (CCGA)**. Other optical methods, such as PIV and **pulse laser imaging (PLI)** techniques, are more useful for studies involving two-phase flow such as when liquid/gas interactions are being studied. The PIV technique uses a tracking method involving seeding the flow with particles and tracking their movement via consecutive image frames; the motion of particles inside the flow is then used to calculate the velocity and direction of the flow. The PLI imaging technique operates on a principle that involves producing a double-pulse laser beam, which can be used to create two consecutive images split by a very short time delay. Because of a short exposure time (~15 μs) between each laser pulse and using a high-resolution imaging technique, liquid/gas interactions can be captured for very short time intervals.

Ünal published an early study (1980) using the high-speed imaging technique with a high exposure flash to investigate the atomization of Al alloy [6]. He also combined an **optical visualization** technique (Schlieren) with a high-speed imaging system to study flow separation and the issue of metal backflow during CCGA in 1989. In this study, combination of the optical Schlieren imaging system with the high-speed imaging technique showed that the back-stream flow problem occurred as the result of flow separation of the gas from the external wall of the melt nozzle as a result of the formation of a negative pressure gradient associated with shockwaves [6]. Mates and Settles [7, 8]

Metallic Powders for Additive Manufacturing: Science and Applications, First Edition.
Enrique J. Lavernia, Kaka Ma, Julie M. Schoenung, James F. Shackelford, and Baolong Zheng.

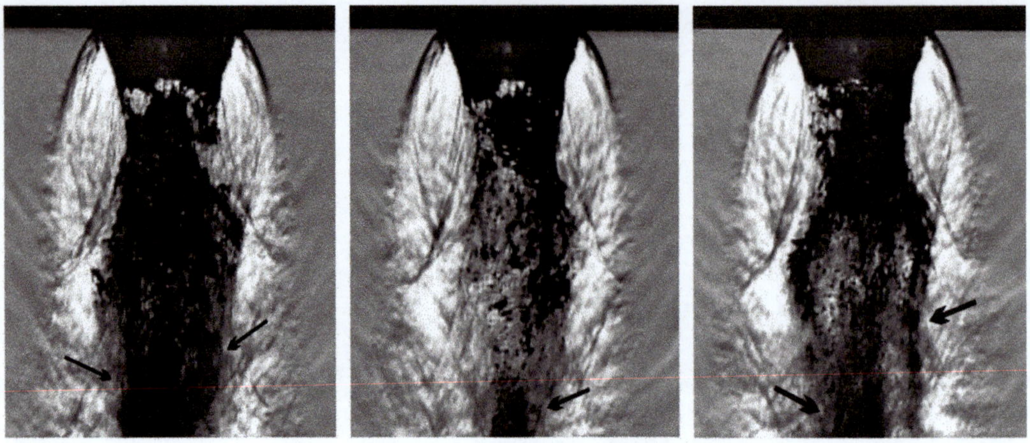

Figure 5.1 Random still images of breakup phenomena at the gas atomization of Sn with high-pressure gas atomization gas atomizer system (arrows show the fine droplets). Source: Mates and Settles [7].

extended the early studies of Ünal using Schlieren and a CCD camera capable of imaging 1/30th of a second and strobe flashlight source with duration of 1.2 μs. They used the Schlieren technique to study the single-phase flow of different gas atomizer configurations and combined Schlieren and high-speed imaging technique to visualize the breakup of Tin (Sn) melt during two-phase flow; some of their results are shown in Figure 5.1 [7]. These individual images show that the primary breakup of the molten tin is highly unsteady. From one frame to the next, both the amount and spatial distribution of liquid in the **flow field** varies significantly. Their results also show that different **atomizer designs** will influence the gas flow pattern in the vicinity of the melt delivery nozzle. In addition, they used two-phase flow analyses to reveal the chaotic nature of the gas and melt interactions during the CCGA process [7].

The history using Schlieren technique can be traced back to the seventeenth century and the work by Robert Hooker in 1665 [8]. Since then, various researchers have developed this technique over the past centuries. Early in the twentieth century, Albert Töpler worked on this technique (1906) [8]. Physicist Robert Wood [8] (1868–1955) used this method for visualizing shock wave formation in a supersonic air tunnel during World War II in Germany. Subsequently, this method became an important tool in fluid dynamic laboratories for designing military aircraft and missiles.

Schlieren imaging and shadowgraphy are visualization techniques that can be effectively used to reveal phenomena in fluids and solids using a high-speed camera. The basic principle is to translate invisible light refraction into changes in light intensity, which can be perceived as light and dark by human eyes. **Refraction** is defined as phase speed changes of light as it passes through a transparent medium. The theory of the Schlieren technique is based on **refractive index** gradients in transparent media. These cause light rays to bend (refract) in the direction of increasing **refractive index**, which is due to light velocity reduction in a higher-refractive-index material. Refraction is defined as phase speed changes of light as it passes through a transparent medium. The refractive index of gas, according to the Gladstone–Dale relation, is related to gas density [8]:

$$n = K\rho + 1 \tag{5.1}$$

where K is the Gladstone–Dale constant, ~0.23 $cm^3\,g^{-1}$ for air at standard conditions, and ρ is the gas density. The basic Schlieren system arrangement is relatively simple, and it involves using two convex lenses or geometric optics and a point of light source. There are different ways for optically arranging the appearance of the Schlieren in an image of the field of interest. In the **Töpler Schlieren method**, there is a source of light, two convex lenses with specific focal length and a knife-edge commonly a razor blade blocking the straight light but passing the distorted light to the screen or camera lens, which is needed for capturing the Schlieren images. The selection of the light source depends on the test or objects to be viewed and can be a laser, a tungsten-halogen, or an LED light. A schematic diagram of a typical Töpler Schlieren arrangement as applied to gas atomization is shown in Figure 5.2. The light beams may be refracted when encountering an area with different temperature and density within the same material. A variation in the index of light reflection in air causes different light distortion around the object [8]. The relation between density

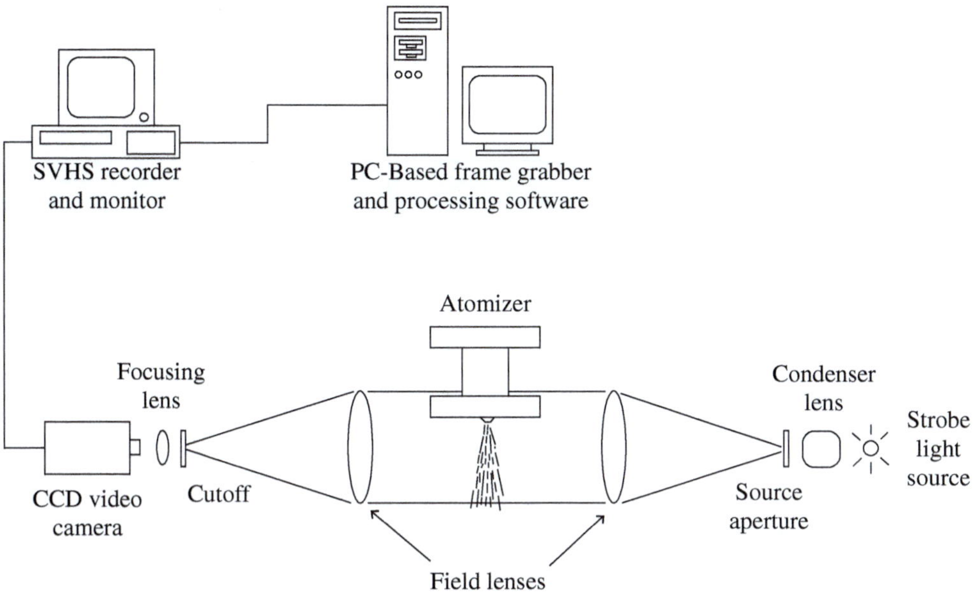

Figure 5.2 Schematic arrangement of Schlieren equipment for the atomization process. Source: Mates and Settles [7].

gradient and refractive index variations is given by the Gladstone–Dale equation:

$$G(\lambda) - \frac{n-1}{\rho A} \tag{5.2}$$

where $G(\lambda)$ is defined as:

$$G(\lambda)\,2.2244\times 10^{-4}\times\left[1+\left(\frac{6.7132\times 10^{-n}}{\lambda}\right)2\right] \tag{5.3}$$

where λ is the light wavelength (m), n is the light reflective index, and $G(\lambda)$ is the Gladstone–Dale number [8].

The production of Schlieren images requires using perfectly parallel light beams. By positioning the light source at the focal point of a large convex lens, the diverging rays strike the lens and refracted to form wide parallel beams. These beams pass through the air and as an incident to the second converging lens. Some of the light becomes remain parallel as they pass through the air between the spaces of these two lenses and bring to the focus at the focal point of the second lens [8]. These beams continue and form an inverted image on the screen. However, to produce a Schlieren image on the screen, an obstacle, or a knife-edge, which is commonly a razorblade, must be placed at the focal point of the second lens, which blocks the straight light coming from the second lens at the focal point and allow the refracted light to pass to the screen [8].

The same phenomena may occur with two parabolic mirrors with special arrangement known as Z-type arrangement (Figure 5.3) [7, 8]. Choosing the right Schlieren arrangement between Töpler and Z-type depends on different factors such as setup location and cost of mirrors; generally, mirrors are more expensive than lenses with the same diameter.

In the gas atomization process, as the high-velocity gas is first injected into the atomization region and then propagates into the atomization chamber, the compressible flow creates a temperature and pressure gradient, i.e. a gradient in refractive index in the air. In the visualization system, a light ray passing orthogonally through this heterogeneous medium will experience curvatures, and their integral – the ray deflection – can be visualized by Schlieren imaging and shadowgraphy. In the study of the gas flow behavior in the gas atomization process, using both Schlieren techniques is one of the best approaches to provide insight into the gas flow field around the melt delivery nozzle. For instance, Settles et al. [8] used Töpler arrangement to study the gas flow pattern for two **atomizer designs** and Ünal [6] used a Z-type mirror arrangement to investigate the back-stream flow problem during CCGA.

The mirror system can be single-mirror or two-mirror, but the latter one is more common and compelling when two mirrors are available, because the parallel beam between mirrors provides the least ambiguous image and is essential for any quantitative work. The Z-type two-mirror Schlieren system is the most popular method, consisting

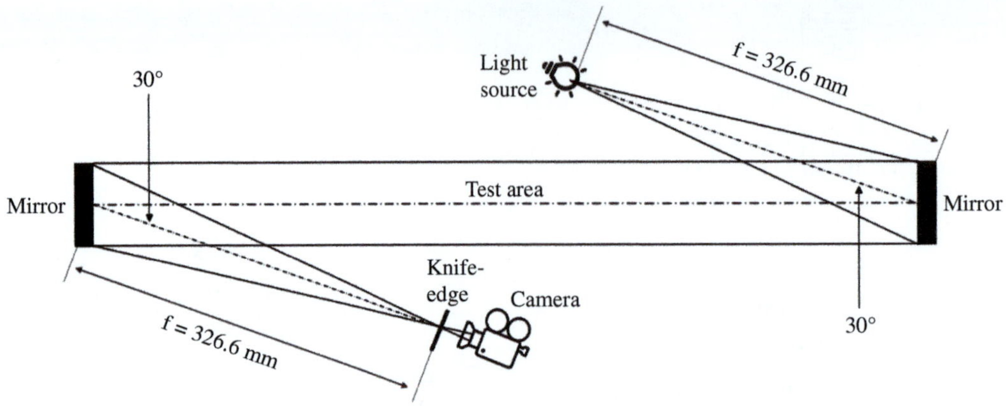

Figure 5.3 Z-type two-mirror Schlieren system. Source: Settles et al. [8]/Springer Nature.

of a diverging illuminator beam, a converging analyzer beam, and a parallel beam between two mirrors. The system captures the jet flow information within the test area (parallel beam) [8].

The **shadowgraph technique** differs from the Schlieren imaging technique mainly in the following aspects: (i) Schlieren setup includes a knife-edge (cutoff) while Shadowgraph does not. The orientation of the cutoff affects the image obtained. A simple cutoff only detects ray refractions with components perpendicular to it, e.g. a horizontal cutoff detects vertical refractive index gradient [8]; and (ii) the percentage of cutoff used determines the Schlieren sensitivity, while sensitivity of the Shadowgraph depends on the optical path length. In general, the Schlieren technique is more sensitive than the **Shadowgraph technique**. When the cutoff percentage is too low, the Schlieren effect is very slight, and it resembles a shadowgraph. When the cutoff percentage is too high, it results in over-ranging [9], i.e. the brightness is too low and local flow details cannot be observed; (iii) The Schlieren illumination level responds to the first derivative of the refractive index, while that of Shadowgraph responds to the second derivative of the refractive index. Although Schlieren has much higher sensitivity in general, Shadowgraph is more sensitive and produces clearer lines in case gas flows with shockwaves or turbulence [8].

Biancaniello and Ridder studied the **flow fields** in gas and liquid streams using **flow visualization** techniques (Schlieren, shadowgraph and 0.5/as flash photography) [10]. From their work, typical Schlieren photographs of the gas-only flow for atomizer plenum pressures of 5.7, 6.5, and 8.5 MPa are shown in Figure 5.4. Figure 5.4b is a Schlieren photograph (knife-edge parallel to flow axis) of the gas-only flow with an atomizer plenum pressure of 6.5 MPa, which is the optimum aspiration condition. The wave pattern in Figure 5.4b is of higher contrast, with a well-developed "diamond"-type wave pattern. In their work, optical imaging of density waves via Schlieren and shadowgraph photography was used to study jet dynamics in an inert gas atomization atomizer. The density wave images showed that a characteristic wave structure predominates at the maximum aspiration operating condition. This structure was not

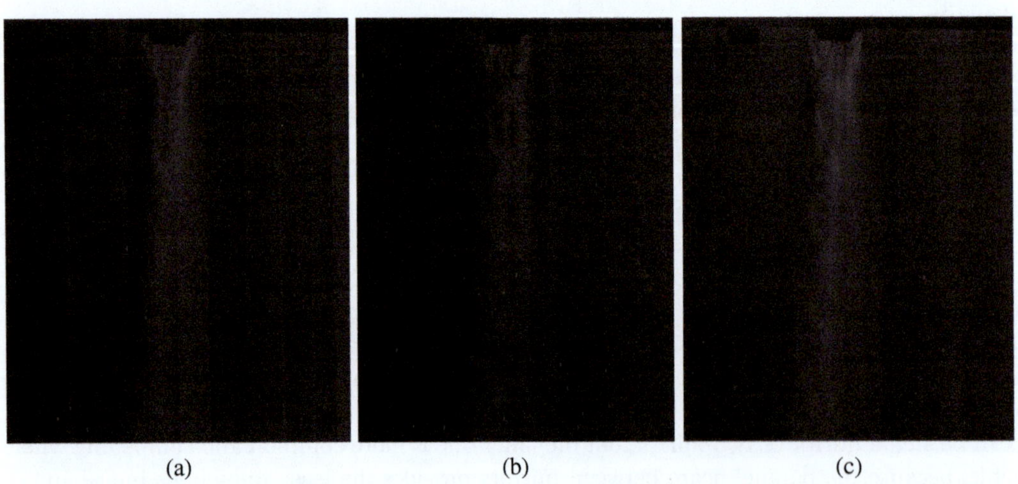

Figure 5.4 Schlieren photographs of gas jet flow: (a) atomizer pressure of 5.7 MPa, (b) atomizer pressure of 6.5 MPa (ideal), and (c) atomizer pressure of 8.5 MPa. Source: Biancaniello et al. [10]/from Elsevier.

present under any other operating condition. Flash photography of the liquid exiting the delivery nozzle showed not only the importance of aspiration pressure but also the importance of the shape of the inner nozzle bore in producing a sheet of liquid. When applied to the atomization of metals, the results will allow the process to achieve desired particle sizes at higher metal flow rates.

High-speed video imaging has been used to observe and analyze the melt flow breakup during gas atomization. Markus et al. investigated jet breakup in liquid metals during twin fluid atomization using high-speed video images [11]. They reported on the influence of different material properties on the primary disintegration process of a metal melt in a free-fall atomizer configuration and applied a linear instability analysis of the high-speed video images. The linear stability analysis can be used to predict the quality of surface instabilities in melt disintegration in a free-fall atomizer. Anderson et al. [7] used visualization of the atomization gas flow and melt breakup to study various atomization nozzle designs [12]. They used **high-speed cinematography** with self-illumination from the molten Fe-based alloys, using a Phantom 7.1 high-speed video camera from Vision Research with a Nikon 85 mm f/1.8D AF Nikkor lens set to f/16. The resolution was set to maximize the active pixels at framing rates, between 2000 and 8000 frames per second (fps), that were appropriate to capture the motion of the molten metal, starting with the onset of atomization, i.e. after a full stream had been established and the first evidence of attack by the atomization gas became apparent. The information presented in the single frames included (i) the onset of gas flow acting on the free-falling stream, (ii) the start of melt transport across the flat base of the tip in response to the force of the gas recirculation, (iii) further melt transport across the tip, (iv) nearly complete melt tip transport and brief truncation of the stream, (v) complete melt filming at the tip periphery and the start of the full atomization process, and (vi) a later spray image of minor stream/film wandering away from the front of the pour tube periphery.

The above discussed optical testing methods can all be classified as noncontact measurement techniques because they cause no disturbances to the **flow field** being tested. Progress in these and other noncontact measurement technologies can be attributed to the rapid development of high-speed optical technologies in combination with the availability of increasing computer power. With the rapid development of optical technologies, experimental studies of two-phase flows have continued to increase.

5.3 Particle Image Velocimetry (PIV)

The atomization process is influenced, not only by numerous process parameters, such as **gas pressure, atomizer design, superheat temperature**, and the velocity of the gas jets, but also by the physical and chemical properties of the gas and molten metal; including viscosity, reactivity, and density, to name a few. In this section, we address the study of metal breakup during atomization, as well as the spatial velocity distribution of the droplets in the melt spray using the method of PIV. PIV is an experimental method, based on image cross-correlation techniques, that can be used for the determination of flow velocity fields. The use of PIV for the measurement of fluid velocities first appeared in the literature in the 1980s [13, 14]. The significant scientific and technical progress that have been achieved in the past decade or so in the fields of lasers, image recording, data analysis techniques, and computer technology have led to further development of the PIV technique. Presently, PIV has become a powerful tool for studying the behavior of particle laden flow fields. The scientific fundamentals and applications of PIV have been reviewed in some excellent papers [15–19] and in a comprehensive book by Raffel et al. [20]. Therefore, in this section, we introduce the PIV technologies focusing on the applications in melt atomization.

As a nonintrusive optical method, PIV is used to determine the velocity of particles over a two-dimensional field of view (FOV) of a flow. The basic principle behind this method is the illumination of the FOV of a flow with a light sheet. This light sheet is produced by a pulse laser followed by a cylindrical lens. The pulse laser emits two pulses consecutively within a short time interval. A camera records synchronous to the laser two images of the scattered light from small moving particles and is located perpendicular to the light sheet. Due to the temporal separation of the laser pulses, the local positions of the mapped particles are shifted. The images are evaluated applying a two-dimensional discrete cross-correlation of small interrogation areas from each image. The coordinates of the cross-correlation peak are the mean spatial displacement in the respective interrogation area. Subpixel accuracy is achieved by a Gaussian fit to the area around the correlation peak. This measure takes much more information into account, thus giving a more precise displacement vector. This vector divided by the temporal separation of the two laser pulses is the mean velocity vector for the interrogation area.

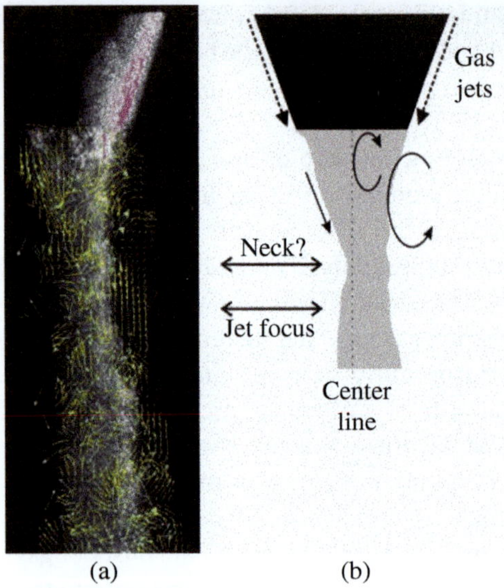

Figure 5.5 (a) Image of atomization with superimposed velocity vectors calculated using the PIV technique from an image pair with 5 μs separation, and (b) the approximate trajectory of the gas jets is shown by the dashed lines. The focus of the jets and an apparent neck in the melt plume are also indicated. Source: Mullis et al. [23]/from Elsevier.

Inspection of the scientific literature reveals a few studies that describe implementing the PIV method for velocity measurements of FOV in melt atomization processes. Most measurements of velocities have been performed by Laser Doppler Anemometry (LDA) or Phase Doppler Anemometry (PDA) but at large distances from the nozzle [21, 22]. Mullis et al. investigated the flow during CCGA using high-speed imaging [23]. Notably, they demonstrated the possibility of obtaining velocity data from two images by applying PIV evaluation software; however, quantitative information of the velocity field was not provided in this study. They used a defocused, pulsed Nd:YAG laser to produce pairs of still images during CCGA with an effective shutter speed of 6 ns, and these images were subsequently used to reconstruct the velocity field within the atomization plume using the PIV technique. Further detailed analysis of the still image pairs produced by the pulsed laser system was obtained by applying PIV mapping software. A typical image, with the superimposed velocity field, obtained using this procedure is shown in Figure 5.5a [23]. The most obvious features from the image are also shown in sketch form to the right of the PIV image, restricted to the primary region close to the nozzle. Two dashed arrows shown in Figure 5.5b [23] represent the approximate alignment of the **gas jets**, estimated from the atomizer configuration. As the gas is in choked flow, it will expand freely upon exiting the jet and the indicated focus is for reference only. The information obtained was used to supplement the interpretation of the previous studies into CCGA using high frame rate (18 000 fps) video imaging coupled with quantitative Fourier analysis for both a research scale metal atomizer and an analog (water) atomizer [24, 25].

Measurements of the velocity distribution of copper droplets in the De Laval atomization process have been reported by Planche et al. [26]. In this study, they implemented the PIV technique in an atomization chamber and measured the droplet velocities at different axial distances from the nozzle exit. As a result, they found a close relationship between velocity data and droplet size: with increasing pressure of the atomization gas, the relative velocity between gas and moving liquid droplets also increases and the higher shear disintegration yields finer powder. Kirmse and Chaves measured the two-dimensional velocity distribution of ligaments and drops during melt atomization [27]. They described the setup and the application of the PIV method for measurements of the droplet velocity distribution in a melt atomization spray of a vacuum inert gas atomization facility in detail. The data evaluation of the velocity measurements shows only small differences between different metals. The melt jet breaks up into a hollow cone spray at approximately 60 mm from the nozzle in both investigated cases. The velocity is reported to fall in the range between 120 and 210 m s^{-1}, in good quantitative agreement with those obtained via numerical simulations by Zeoli et al. [28].

The basic conditions of PIV measurements in flows, like water or air, differ from those of the velocity measurements in melt atomization sprays [27]. In the first case, small particles within the flow radiate scattered light which is then received by the camera. If the number of particles is small, additional particles can be seeded to achieve enough imaged centers of scattered light over the entire FOV. In the second case, the particles are ligaments and drops of molten steel. It is not possible to increase the number of radiating particles by seeding the spray. Therefore, melt atomization sprays are sparse sprays compared to dense atomization sprays such as fuel sprays. The conditions that are present in **high-velocity oxy-fuel** (HVOF), for example, lead to the formation of a sparse spray [29].

In the region near the **atomization** nozzle, large ligaments are present because **atomization** begins with large structures. These structures scatter light with large light intensities which lead to blooming effects on the image sensor. The light spots expose more pixels of the sensor as the real spot size would do. This results in inaccurate velocity data and in reduced spatial resolution after the evaluation using PIV software very near to the nozzle.

In Kirmse and Chaves' work, the camera used in the PIV equipment is a high dynamic 14-bit-cooled CCD camera "PCO 1600" (PCO AG, Kelheim, Germany) [27]. The resolution of the image sensor is $Hp = 1600$ pixels X $Wp = 1200$ pixels with a square pixel size of 7.4 mm. A typical value of the maximum displacement of scattering light centers within the two images is $\Delta\text{pix} \leq 10$ pixels. Then, the temporal separation τ is calculated by $\tau = \Delta\text{pix}\ H/(Hp\ |u|)$. The values of $\tau_1 = 2$ ms and $\tau_2 = 3$ ms were applied during our measurements. The PIV laser is a 200 mJ Nd:YLF-pulsed laser "Nano L200-15 PIV" (Litron Lasers Ltd., Rugby, UK). The wavelength of the laser radiation is 532 nm, and its maximum repetition rate per laser head is 15 Hz. The pulse width is between 6 and 9 ns. The two cavities can be controlled individually. Therefore, precise trigger pulses from external digital delays were used in this study. The deviation of the temporal separation τ was found to be 1 ns measured with a digital storage oscilloscope. The light sheet was generated by passing of the laser radiation through a cylindrical lens with a focal length of 5 mm. In the range of the FOV, the light sheet was thinner than 1 mm.

The PIV software package "Fluere" by Kyle P. Lynch was used for the data evaluation [27, 30]. This software is highly optimized for parallel processing and runs fast on multiple platforms. The paired images were cross-correlated using four runs of the multi-pass processing: the height and width of the interrogation area decreased from 64×64 pixel^2 to 16×16 pixel^2 during the runs. The displacement of the interrogation windows was always 50% of its height and width, and the adjacent interrogation areas overlapped.

Figure 5.6 illustrates the velocity measurements of molten steel particles, performed during metal atomization in a vacuum inert gas atomization facility (ALD Vacuum Technologies GmbH, Hanau, Germany) [27]. This facility

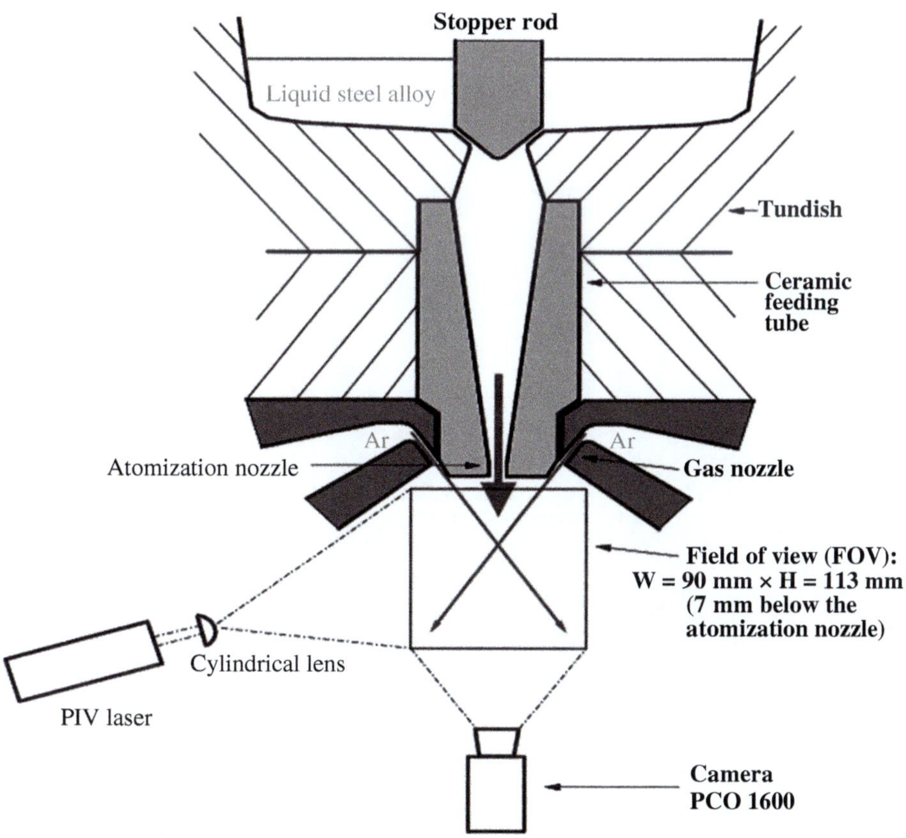

Figure 5.6 Schematic illustration of the vacuum inert gas atomization facility with the field of view of the PIV measurements, the locations of the PIV laser with the cylindrical lens producing the light sheet and the camera PCO 1600. Source: Kirmse et al. [27]/John Wiley & Sons.

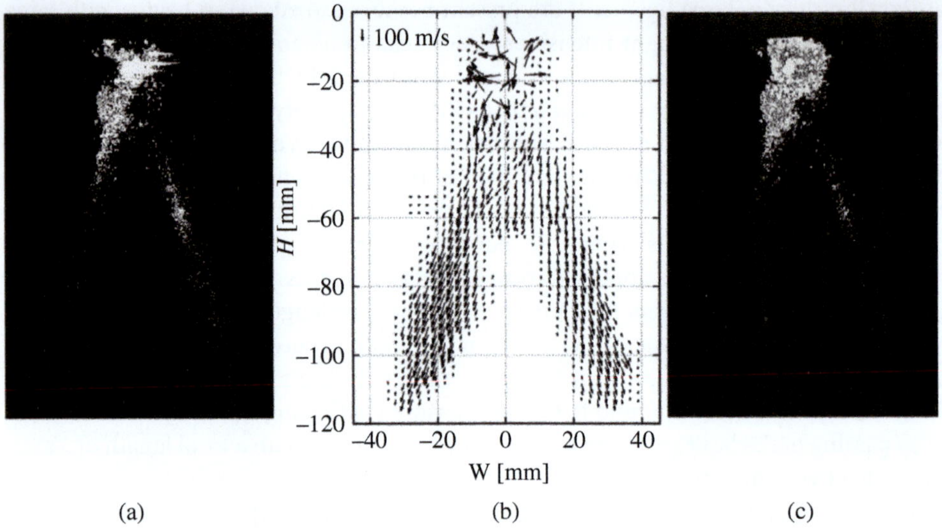

Figure 5.7 The measurement results of the left (a) and the right (c) images are the original pictures of the double-exposure shot at the time of 0.1 s during the melt atomization of X5CrNi18-18 steel. (b) The vector plot depicts the result of the evaluation of the two images using the algorithm of the PIV software. Source: Korobeinikov et al. [27]/from Springer Nature/CC BY 4.0.

comprises the vacuum induction melting furnace and an inert gas atomizer. The FOV of the velocity measurements had a width of $W = 90$ mm and a height of $H = 120$ mm and was located through the jet axis with 7 mm below the atomization nozzle. This distance was determined by the location of the observation windows at the facility. The measurement results are shown as a vector plot in Figure 5.7 and as contour maps in Figure 5.8. The length of the vertical vector in the upper left corner illustrates a velocity value of $100\ \mathrm{m\,s^{-1}}$. For Figure 5.7, the distribution of the particle velocities shows a cross section through a hollow cone below this height. Only every second velocity vector is displayed here to enhance the visibility. The temporal development of the velocity magnitude of the particle velocity is

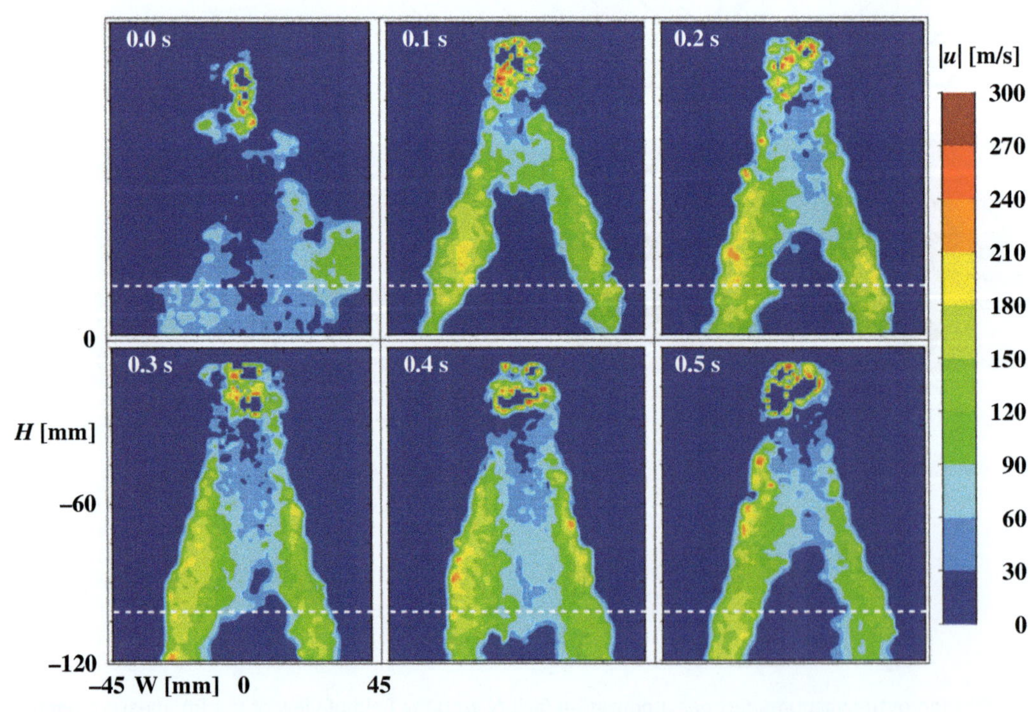

Figure 5.8 Temporal evolution of the velocity magnitude of the particle velocity during the melt atomization of X5CrNi18-10 steel. Source: Korobeinikov et al. [27]/from Springer Nature/CC BY 4.0.

shown in Figure 5.8 from the beginning of the **atomization** at the time of $t = 0.0$ s to $t = 0.5$ s. The white-dashed lines are added at a height of −100 mm, near which the temperature of the particles reaches the liquidus temperature of the metal.

The contour maps of the magnitude of the instantaneous velocity distribution represent a cross section through the center of the atomized spray with the width of the light sheet thinner than 1 mm. The interaction of the argon gas jets with the melt begins at $t = 0.0$ second. The disintegration is clearly observable from $t = 0.1$ second and later, as shown in Figure 5.8. The spray cone is stabilized, and the velocity distributions become approximately similar. The peaks of high velocity are detected shortly below the nozzle in the case of X5CrNi18-10 steel. The melt was broken up into a hollow cone spray at about 60 mm from the nozzle. The atomized melt attains the shape of a hollow cone below that distance. The development of a hollow cone spray is due to rotational flow in the crucible. With further increase of the distance to the nozzle, the hollow cone disintegrates into melt film ligaments and droplets with similar velocities as at the breakup location.

PIV is a powerful experimental fluidics technique relying on image pairs to gather a **flow field** of targeted samples. The PIV technique can achieve quantitative measurement of instantaneous turbulent velocity fields of melt film ligaments and droplets of melt atomization as an extension of qualitative **flow visualization**.

5.4 Particle Counting, Sizing, and Velocity Probe (PCSV-P)

The PCSV-P can be used to obtain online measurements of particle size, velocity, and concentration [2, 31]. The system uses several pieces of hardware including a computer and an electronic probe. It can be used for large-scale industrial systems as well as small laboratory systems. The PCSV-P uses a fiber-optic-based system within a water-cooled probe to allow remote measurements of the above-mentioned parameters.

This system uses well-understood light scattering techniques to obtain information from the atomized powder. A low power laser is utilized to illuminate particles as they randomly pass through a sample volume (see Figure 5.9). The sample volume is determined by the intersection of the laser beam and the receiver optics. A detector picks up the scattered light from the laser and transfers the signal to a photomultiplier, which converts the signal to a voltage pulse that can be interpreted by the signal processor. The PCSV-P technique utilizes an intensity deconvolution algorithm which relates the scattered light signals to particle size and concentration. Since the technique uses a very small sample volume size, statistically there will be only one particle in that volume whose combination of diffraction intensity and duration can be correlated to give its size and velocity.

During measurement, the probe is placed inside a spray chamber at a fixed distance below the nozzle. The exact placement is unique for each system, and optimum placement has to be determined experimentally. The PCSV-P can be calibrated using a rotating disk (14 mm in diameter). The disk consists of monodispersed clear holes which have been photoetched into an opaque substrate. These holes provide diffraction scattering centers which in turn provide data for calibration of the instrument. The calibration procedure is basic and rather simple, making the probe easy to set up and use.

During operation, the probe provided reasonably accurate data on particle size and velocity [32]. Figure 5.10 shows the PCSV-P probe size measurements for two different atomization pressures (690 and 860 kPa). The resulting curves

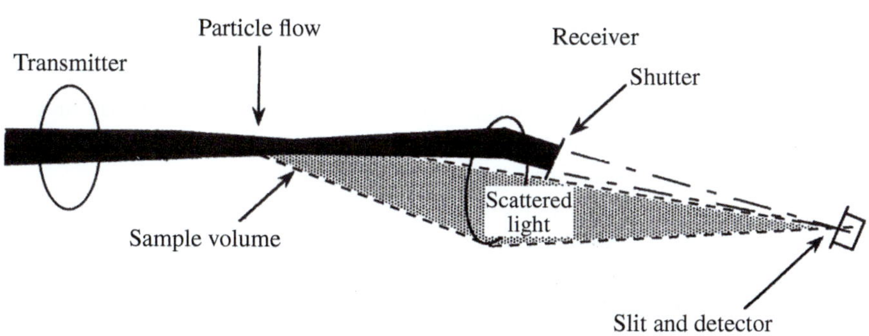

Figure 5.9 PCSV-P optics. Source: Henein et al. [2]/Pascal and Francis/CC BY 4.0.

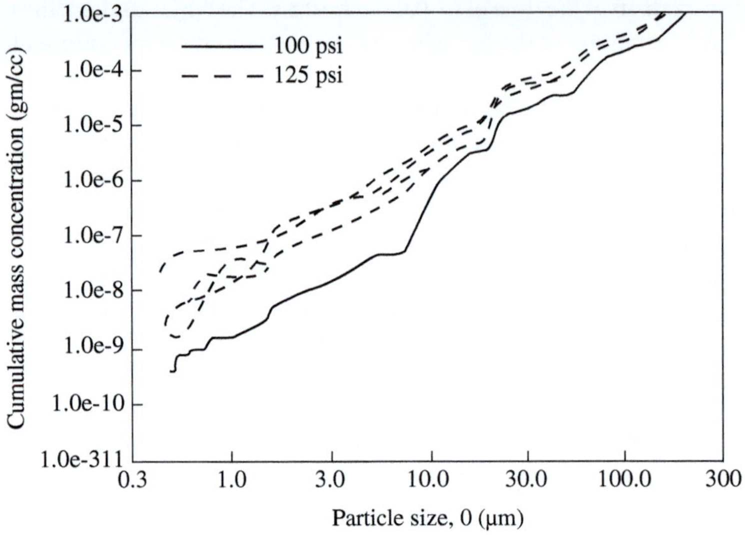

Figure 5.10 PCSV-P results for 689 kPa (100 psi) and 862 kPa (125 psi) showing reproducibility of measurements. Source: Henein et al. [2]/Pascal and Francis/CC BY 4.0.

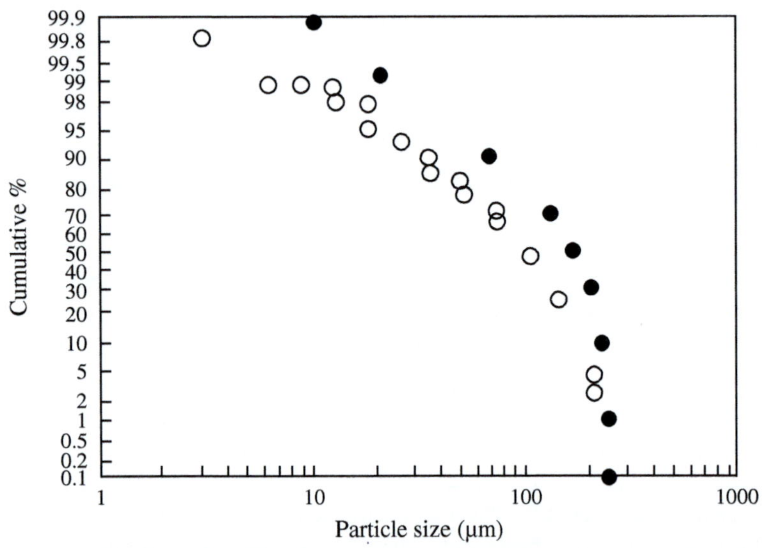

Figure 5.11 Log probability plot results comparing Micro-Trac (°) with PCSV-P (•). Source: Henein et al. [2]/Pascal and Francis/CC BY 4.0.

correlate well with the theoretical values. In **atomization**, it is expected that higher gas nozzle pressure will yield a finer powder and the values of the curves in Figure 5.10 substantiate this premise. Velocity measurements have been extrapolated from the sampling rate and time frequency of the measurements, but no specific data were provided to check the associated accuracy.

Figure 5.11 compares the results of particle size measurements made with the PCSV-P probe and those made with Micro-Trac instruments for the same collection of atomized powders. It is evident that both techniques show general agreement in that the average particle size decreases with increasing atomization pressure. This trend is illustrated by the results in Figures 5.12 and 5.13. Moreover, the PCSV-P probe results predict a larger particle size than that measured with the Micro-Trac instrument. The differences between the two methods have been attributed to several reasons. First, larger particles (<600 μm) were sieved from the collected powder prior to Micro-Trac measurements to remove coarse metal fragments which inevitably accumulate during start-up and shutdown phases of the atomization process. This experimental bias would cause the results to be weighted toward the smaller particle sizes. Also, the Micro-Trac can measure particles in a larger size range (0.5–300 μm) than PCSV-P, and therefore it is slightly more accurate.

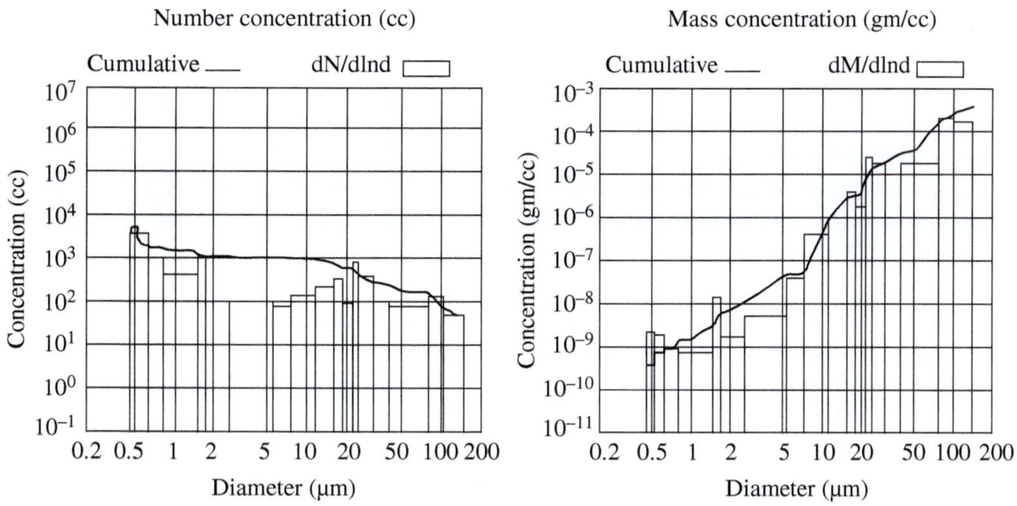

Figure 5.12 PCSV-P results of particle size distribution curves for atomization pressure of 689 kPa. Source: Henein et al. [2]/ Pascal and Francis/CC BY 4.0.

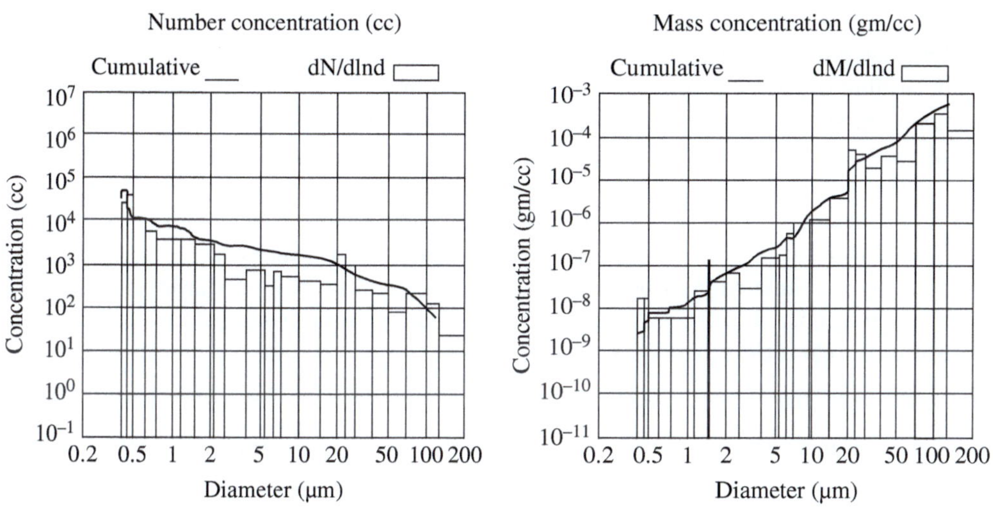

Figure 5.13 PCSV-P results of particle size distribution curves for atomization pressure of 862 kPa. Source: Henein et al. [2]/ Pascal and Francis/CC BY 4.0.

There are several factors regarding the limitations of PCSV-P that should be noted. First, for large particle sizes (above 100 μm), the shape of particles normally becomes more and more irregular with increasing particle size. Consequently, the accuracy of the size measurements decreased rapidly with increasing particle size. Second, the PCSV-P size measurements are dependent to a significant extent on the absorption and diffraction characteristics of the material being measured. These two factors account for some of the discrepancies associated with this monitoring method.

Overall, this technique provided valuable and reliable information within certain experimental limits as previously mentioned. The instrument is well suited to perform reproducible and simultaneous measurements of velocity, size, and concentration of particles. It is also very capable of producing data at high temperatures, which makes it useful when atomizing high-temperature alloys. In addition, the equipment associated with PCSV-P is flexible, allowing forward and backward scattering geometries; this renders the technique adaptable to numerous spray chamber configurations. Furthermore, rapid signal processing (500 kHz) allows real-time information to be available on the quality of the atomized alloy. As a result, when the PCSV-P is properly implemented in a research or industrial setting, it can serve to provide real-time monitoring of several of the above-mentioned operational parameters during **spray atomization**.

5.5 High-Speed Cinematography/Video

In an effort to reach the goal of instantaneous particle measurement, investigators have attempted various configurations using conventional photography/cinematography techniques [33]. In most of these studies, the underlying optics are based on **Fraunhofer diffraction** theory, which is used to model the diffraction of waves when plane waves are incident on a diffracting object, and the diffraction pattern is viewed at a distance satisfying the **Fraunhofer condition** and when it is viewed at the **focal plane** of an imaging lens [34, 35]. In contrast, the diffraction pattern created near a diffracting object is given by the Fresnel diffraction equation. The **Fraunhofer diffraction** equation was named in honor of Joseph von Fraunhofer and is widely used in particle holography [36]. **Fraunhofer diffraction** is also widely used as an optical diagnostic tool for characterizing the size distribution of particle ensembles that are large in comparison with the wavelength. There are a number of excellent papers on this topic [37–39], and commercial instruments for sizing particles based on **Fraunhofer diffraction** are readily available. The typical **Fraunhofer diffraction** instrument configuration is illustrated in Figure 1, where the light diffracted by particles along the line of sight of the probe laser beam is collected by a lens and directed to the back focal (Fourier transform) plane. Unscattered light is focused on-axis at the back **focal plane** of the transform lens, and scattered light is directed toward off-axis positions on the transform plane. A schematic of an array of semicircular annular photodetectors is also shown at the detection plane of Figure 1. The transform lens effectively translates the far-field angular distribution of the diffracted light into a radial r spatial distribution at the back **focal plane** via the transformation equation $r = f\theta$. Each particle size produces a distinctive diffraction pattern signature, which is mixed with diffraction patterns from the other sizes at the detector plane. The essence of the particle sizing problem is to disentangle these signatures by carefully selected sampling of the diffraction pattern followed by a mathematical inversion to determine the particle size spectrum. In the context of adaptive control of the system, the goal is to extract the maximum amount of useful size distribution information with a minimum number of measurements. The maximum information obtainable is dependent on the noise levels in the measurements, the sampling or detection scheme, the stability or robustness of the inversion method, the actual size distribution, and the allowable level of uncertainty. Ideally, a laser diffraction system would have variable detector geometry and size resolution, which could both be adjusted at runtime depending on the measurement conditions.

Consider particles illuminated by a collimated laser beam as shown in Figure 5.14. A monodispersed ensemble of spherical particles that are large compared to the wavelength would produce the characteristic Airy diffraction pattern as described by **Fraunhofer diffraction** theory [38]:

$$i(\theta) = I_{inc} / k^2 \cdot J_1{}^2(\alpha\theta) / \theta^2 \cdot \alpha^2 N_{tot}(\alpha) \tag{5.4}$$

where $i(\theta)$ is scattered intensity at the angle θ measured from the laser beam axis, I_{inc} is the irradiance of the incident beam, J_1 is the first-order Bessel function of the first kind, and $\alpha = \pi D/\lambda$ is the particle size parameter, where D is the

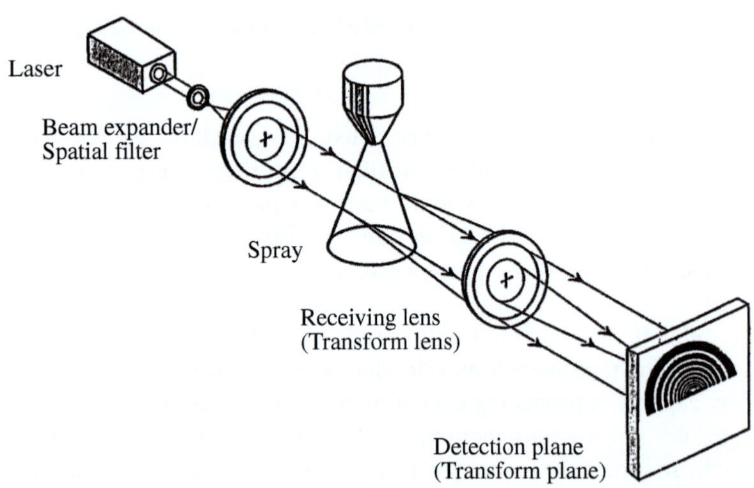

Figure 5.14 Schematic of a conventional laser diffraction particle sizing instrument. The spatial filter/beam expander optics produce a collimated laser beam of typically several millimeters diameter, which is passed through the particle field (a spray shown here). Source: Hirleman et al. [39]/Optica Publishing Group.

particle diameter and λ is the laser wavelength. The wavenumber k is $2\pi/\lambda$ and N_{tot} (α) is the total number of particles of size α in the laser beam. Equation (5.4) is valid if the following conditions are met: near-forward small-angle scattering ($\sin\theta \approx \theta$); anomalous diffraction is negligible (refractive scatter does not contribute); N_{tot} (a) is large, and the particles are randomly positioned and uniformly illuminated; vignetting is negligible; and finally multiple scattering is negligible.

In most systems of engineering interest, a distribution of particle sizes, or a polydispersion, is generally encountered. The composite scattered intensity profile is a linear combination of the characteristic profiles of each size with a weighting coefficient equal to the number of particles of that size in the sample volume. The basic equation of **Fraunhofer diffraction** particle sizing is a Fredholm integral equation of first order and first kind involving an

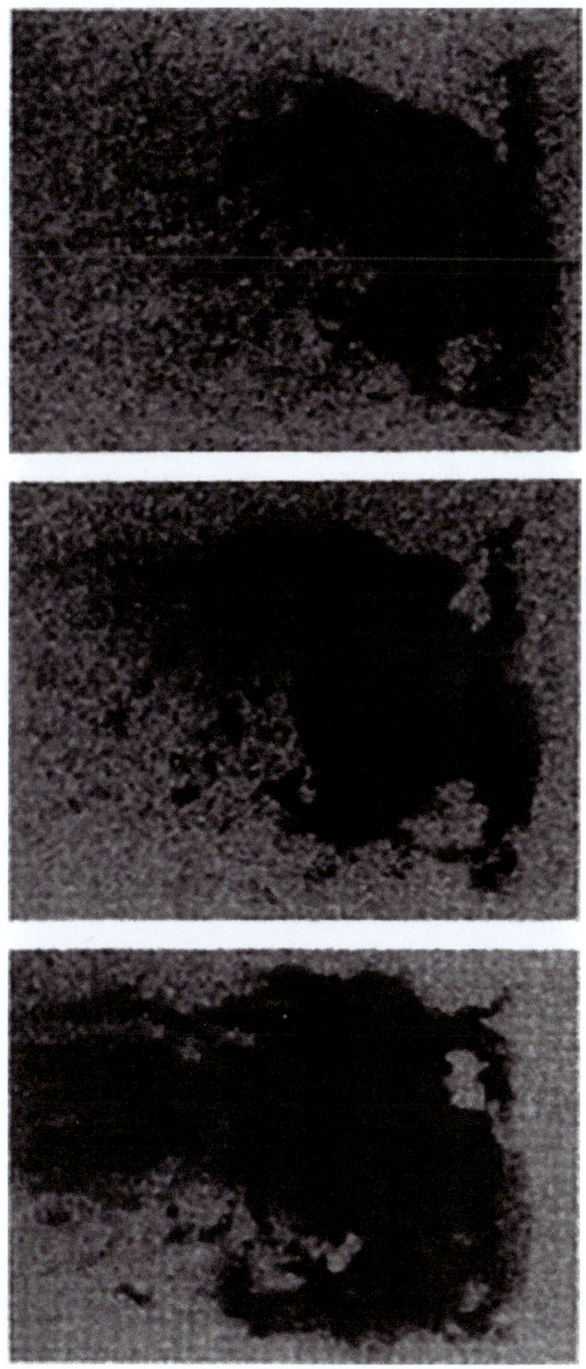

Figure 5.15 Images taken from a l,0000 fps movie. Source: Reprinted with permission. National Institute of Standards and Technology, 1992, rights reserved.

integral over all particle sizes [38, 40]:

$$i(0) = I_{inc}/k^2 \int_0^\infty J_1^2(\alpha\theta)/\theta^2 \cdot \alpha^2 n(\alpha) d\alpha \tag{5.5}$$

where $n(\alpha)$ is an unnormalized probability density function representing the number-based particle frequency spectrum (i.e. $n(\alpha)d\alpha$ is the number of particles in the laser beam with sizes between α and $\alpha + d\alpha$). The inverse problem requires measurement of $i(\theta)$ or more precisely measurement of $i(\theta)$ integrated over some finite receiving aperture at a finite number of scattering angles. Finally, $n(\alpha)$ is determined by solving (inverting) Eq. (5.5).

High-speed cinematography involves the use of a camera with the capabilities for high shutter speed and high-speed film [1]. It is usually set up at one of the windows of the spray chamber which allows a clear view of the spray cone. The procedure can provide particle distribution, nozzle phenomena information, and velocity information from photographs. Consecutive frames similar to those shown in Figure 5.15 are used to track a single particle [41]. By measuring the distance that the particle travels between frames, the velocity of the particle can be calculated based on the time interval between frames. Distribution of the particles and nozzle phenomena can also be directly observed from consecutive frames. High-speed cinematography, however, lacks fine detail due to the coarseness of high-speed film (see Figure 5.15), and it also requires time to develop the film to obtain the results. Recently, this process has been taken a step further by incorporating laser holography [3].

The high-speed video system operates on a similar principle as high-speed cinematography. In a high-speed video system, however, events of interest are recorded on video tapes [5, 42]. Such a system is capable of recording at a speed that exceeds 1000 full frames per second. The hardware required for such video systems includes an illumination light source (i.e. a high-intensity floodlight or a stroboscope synchronized to the camera), a high-speed video camera, and an image analysis subsystem. The light source is not required when the material is self-illumining. In principle, high-speed video systems can be used to record liquid breakup, as well as droplet formation, solidification, and deformation. One example in this area is provided by the work of Trapaga et al. [42]. They used Spin-Physics Ektapo 1000 high-speed video processor to study the deformation behavior of Cu droplets impinging onto a substrate. Figure 5.16 illustrates schematically the experimental setup used in this study. During measurement, a droplet is heated and melted using a magnetic levitator. The droplet is then allowed to free fall onto a flat substrate. For a transparent substrate, three high-speed cameras are used to record the deformation process of the droplet with their locations shown in Figure 5.16. The cameras used by Trapaga et al. are capable of recording at a speed of 1000 frames per second at

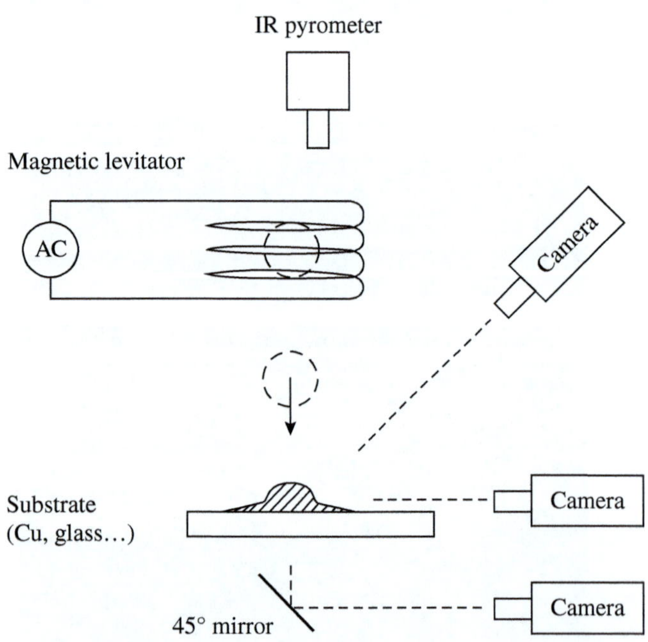

Figure 5.16 The experimental setup for measuring droplet deformation. Source: Reprinted with permission from, Advances in Powder Metallurgy & Particulate Materials– 2011 (Figure 9 Page 02-25), Metal Powder Industries Federation, 105 College RoadEast, Princeton, New Jersey, USA.

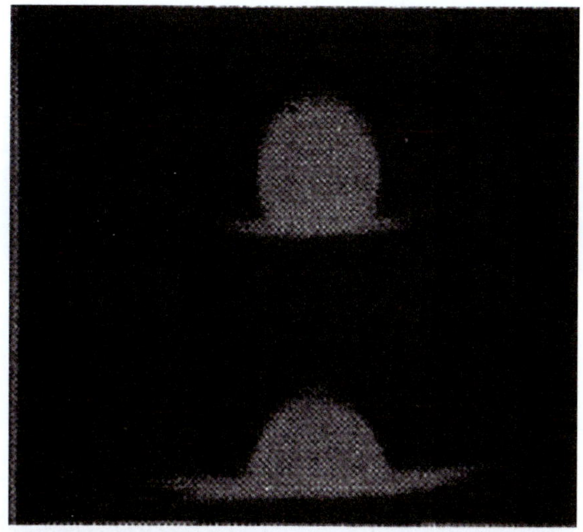

Figure 5.17 Photographs obtained from high-speed video recordings illustrating the spreading of a Cu droplet on a Cu substrate. Source: Harding et al. [42]/from Springer Nature.

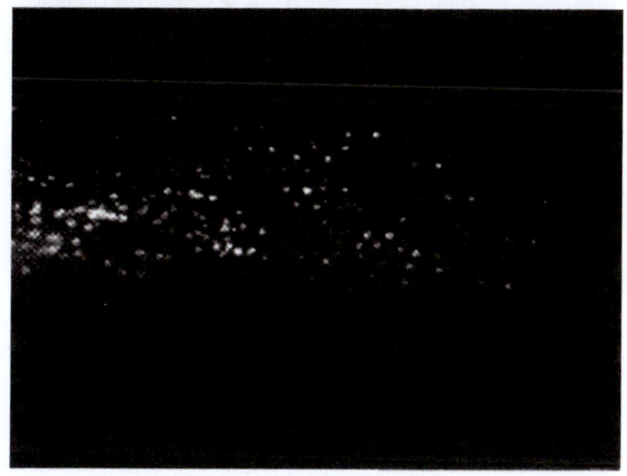

Figure 5.18 Double-exposure image of a direct current plasma spray. Source: Gleiter [5]/from Springer Nature.

a resolution of 240 × 190 pixels per frame. Figure 5.17 illustrates the photographs obtained directly from high-speed recordings which reveal the spreading of a droplet during impingement.

High-speed video with intensive laser strobe illumination is especially useful for imaging in multi-droplet systems. Typical processes that may be analyzed include arc welding, gas atomization, thermal spray, plasma spray, and twin wire spray [5]. When a double-shuttering system is used, this high-speed video system becomes a very powerful tool for measuring droplet velocity in multi-droplet systems. In this case, a single video frame is exposed twice by illuminating a droplet with two consecutive laser pulses separated by a certain length of time. Consequently, each particle appears as a pair of bright spots in a double-exposed image. One example of such an image is shown in Figure 5.18 taken during direct current plasma spraying [5]. Knowing the separation time between the laser pulses, and measuring the separation distance between the spots, the velocity of each particle in the image can be accurately measured.

5.6 High-Speed Off-Axis Holographic Cinematography

Another technique which improves on existing technology and can be used to extract real-time information from **atomization** is **high-speed off-axis holographic cinematography**. This technique, originally developed by Lauterborn, Judt, and Schmitz [3], has been utilized to access information on turbulent cavitation bubble clouds in fluid flow, but the experimental setup may be readily adopted to **spray atomization**. The basic premise of this method relies on high-speed strobe laser imaging which is capable of freezing atomization phenomena.

The system consists of a copper vapor laser which emits a quasi-infinite series of short (~35 ns) light pulses with an energy of ~1 MJ. The light comprises two spectral lines (yellow $\lambda = 578.2$ nm and green $\lambda = 510.6$ nm). The yellow light

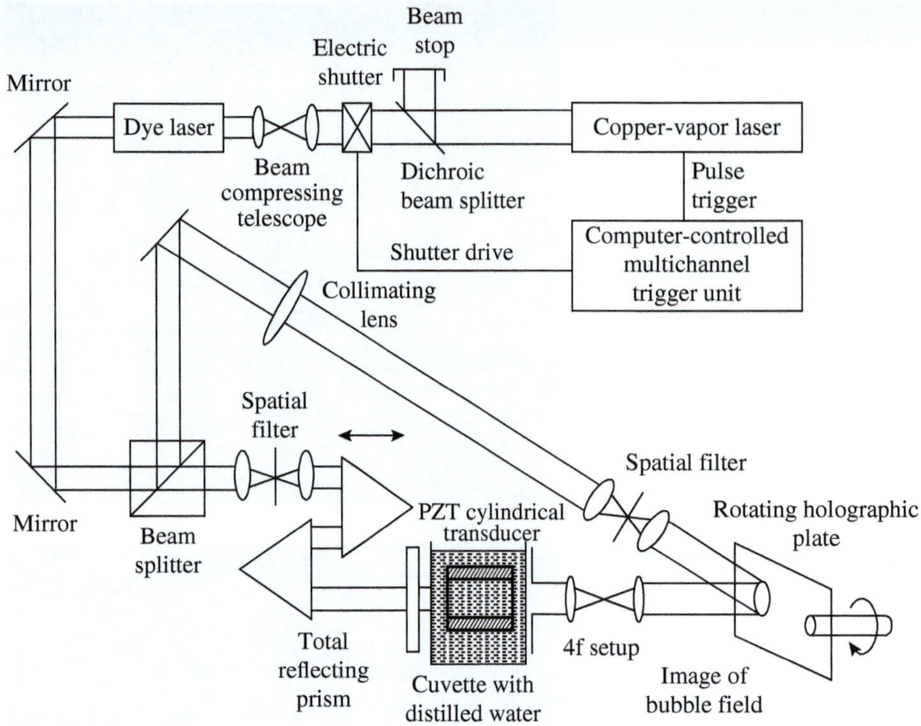

Figure 5.19 High-speed off-axis holographic cinematography. Source: Reproduced by permission of the Optical Society of America [3].

is used to expose the holographic images, and the green light is used to pump a dye laser to improve the coherence and wavelength tuneability of the copper vapor laser. This is done because the coherence of the copper-vapor laser pulse is very low, making direct holographic recording severely limited. A schematic diagram of the setup can be seen in Figure 5.19. A computer-controlled shutter and laser emit a fixed number of pulses which produce an equivalent number of holograms. The pulses leaving the dye laser are deflected by mirrors onto a beam splitter to provide two beams for off-axis holography: a reference beam and an object beam. The object beam traverses a delay line of prisms so that its path length to the holographic plate can be adjusted to approximately that of the reference beam. The light at this point serves to illuminate a given volume of the chaotic phenomena of interest (atomization) [5.1, 5.7]. The scene is then imaged using two lenses in front of the holographic plate. The plate itself is placed on a rotating axle which permits it to rotate at ~250 rpm. While it is rotating, the lenses focus each pulse and form a hologram roughly 1 cm^2 on a separate part of the plate (see Figure 5.20). Each of the holograms is a successive image of a spatial event a fraction of a second apart (216 μs between frames). The information from the separate holographic images can be

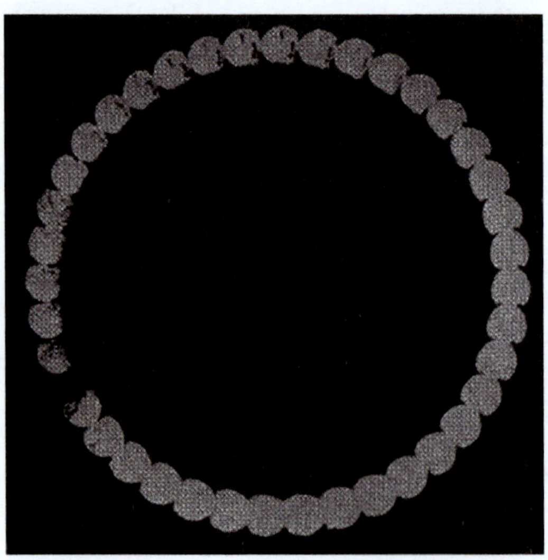

Figure 5.20 Example of a series of holograms on the rotating holographic plate. Source: Lauterborn et al. [3]/from Optica Publishing Group.

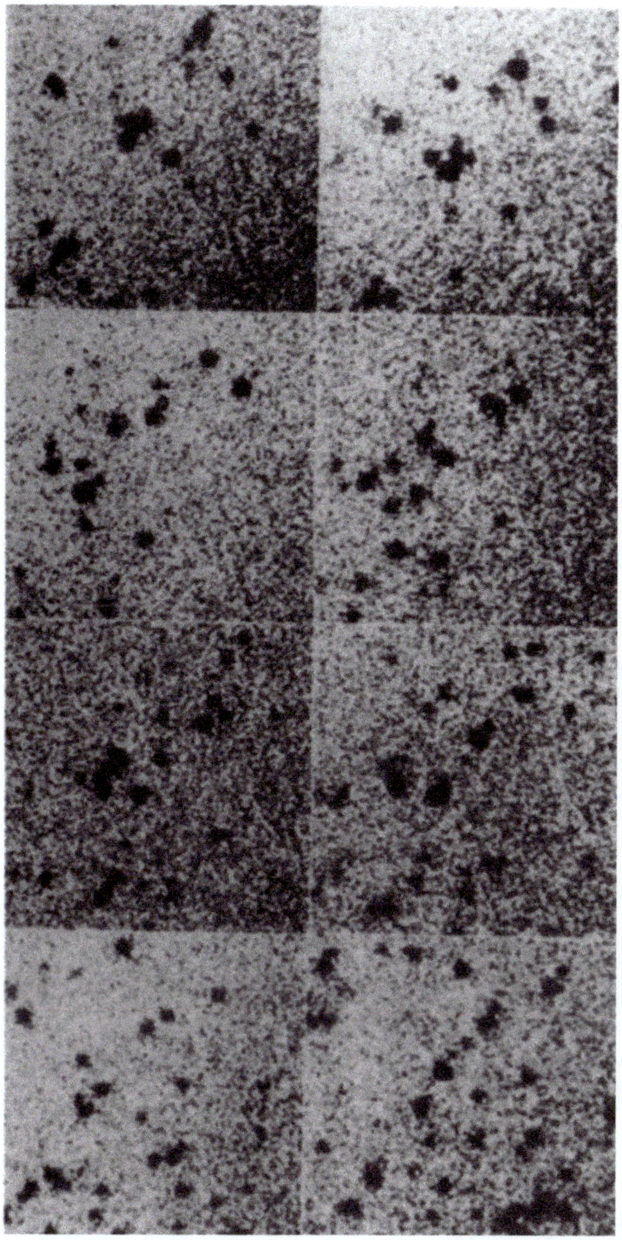

Figure 5.21 Series of images from successive holograms. The time between frames is 216 μs. Time proceeds from left to right and from top to bottom. Source: Lauterborn et al. [3]/from Optica Publishing Group.

used to obtain information such as particle size and size distribution about **atomization**. Figure 5.21 shows a series of images from successive holograms revealing individual particles moving in space. The consecutive frames can also be used to track a single particle and measure its velocity based on movement through consecutive frames and the time between the frames.

In summary, the off-axis holographic cinematography can be successfully used to obtain diagnostic information about spray processing. The system has the advantage of providing three-dimensional information about particle characteristics and offers basically an infinite field of depth in selecting a sample volume. The very short pulse and high light energy of the laser system coupled with the pumped dye are capable of freezing atomization events. In addition, the system can be combined with a digital image processing unit which could provide slow motion three-dimensional pictures of the **atomization** process. The shortcomings of this system include the limited resolution which is possible when dealing with very small particles and holographic plates. Also, limited velocity information can be gathered from the consecutive frames. However, continuous improvements in this technology will make this process an important tool for monitoring atomization events.

5.7 Infrared Thermal Imaging

An **infrared thermal imaging** system consists of a thermal camera positioned at a spray chamber window and a computer for image analysis [43, 44]. The system operates on the basis that the intensity of infrared radiation falling on the detector (camera) is proportional to the temperature to the fourth power. It is inherently assumed that radiation emitted by the surrounding atmosphere and other background sources are taken into account and compensated for. In addition, it is necessary to calibrate each individual camera with each specific chamber in an effort to compensate for variations in background absorption and emitting characteristics of each unique experimental apparatus.

Once the equipment is properly calibrated and in place, it can provide interesting insight of atomization phenomena as well as monitor the process. A series of infrared thermographs are taken at various time intervals.

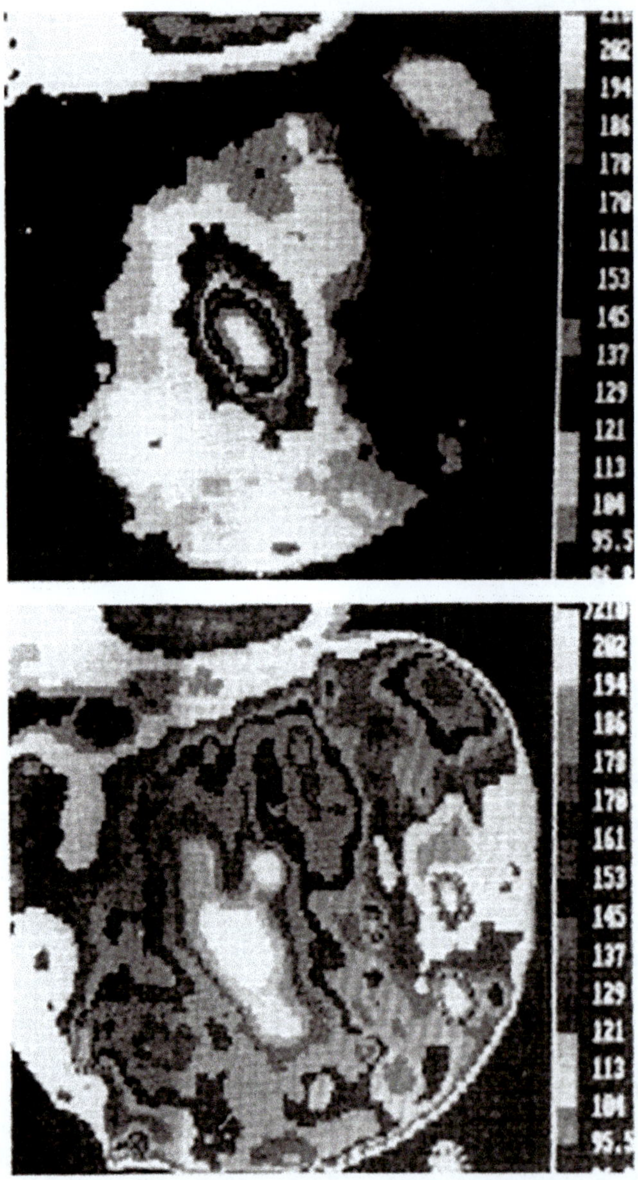

Figure 5.22 Thermographs at various time intervals during spray deposition of a Sn-38 wt%Pb alloy; numbers on the graphs refer to temperatures. Source: Grant et al. [43]/from Elsevier.

The thermographs are then examined with the aid of a computer to obtain specific information about the atomized spray and the deposited material, such as temperature distribution across the preform and the thermal history. The instant feedback available on video screens is a desirable online diagnostic feature of this system. The temperature resolution of existing systems is on the order of 0.1 °C·shade^{-1}. As one example in this area, Cantor and Bewlay [44] applied the infrared technology to the determination of the temperature history of spray-deposited performs during deposition. The measurements resulted in the thermographs as shown in Figure 5.22. The thermographs offer information about the instant temperature distribution on the deposition surface. The temperature history of particular points of interest can also be determined from consecutive graphs. This observation supports the premise that droplets deposited at the onset of spraying, directly onto the substrate, are more efficiently quenched by the substrate than droplets deposited later onto a warmer material. The thermographs can also reveal information about the distribution of particles and cooling characteristics of the spray. The shape and color of the spray cone can be correlated to the size distribution of the droplets. It can also be deduced from the thermal data that larger droplets within the spray tend to reside in central regions of the gas flow because of their larger inertia, whereas smaller droplets are more likely to be swept toward outer regions of the diverging spray cone. Information such as this can be used to better understand the events occurring during **atomization**.

The aforementioned infrared technique has demonstrated that useful information about atomization parameters can be obtained using thermal imaging. There are, however, still several problems associated with this approach. First, **infrared thermal imaging** systems tend to be expensive and cumbersome due to the need for cryogenic operation of the camera. Second, sophisticated computer algorithms are required to decipher the shades of the thermal maps for transient systems such as **atomization**. Third, the presence of both liquid and solid phases within the impinging droplets makes accurate calibration of the system difficult. More recently, developments in advanced pyroelectric ceramics for thermal infrared detection and imaging by What more [31] have decreased the price of the equipment and raised the maximum temperature of efficient operation. These developments will allow greater resolution and increased ease of operation. The General Electric Corporation (Lynn, MA, USA) also has made progress on intelligent algorithms for deciphering the thermal maps in transient systems [45]. Together, these two advancements may bring this technology to the forefront of diagnostic techniques for atomization and deposition processes.

5.8 Phase Doppler Particle Analysis (PDPA)

PDPA is a relatively well-established technique that has been implemented in many existing diagnostic systems for atomization studies [46]. PDPA is basically a single particle counting technique that can provide information such as particle size, velocity vector, and particle concentration in a small given volume. The hardware consists of a transmitter, receiver, and a signal processor. A schematic of the phase Doppler interferometer can be seen in Figure 5.23. The entire system operates on the simple principle of measuring the phase shift of light diffracted from particles in the path of the optics [47]. The optics are comprised of a continuous gas laser such as helium–neon (He–Ne) or argon ion used for illumination of the particles. A beam splitter is utilized to divide the highly collimated light into two parallel beams of equal power. A high-quality lens then focuses the two beams to the same point. This region where the two beams converge is the sample volume space where particle measurements are made. In this volume, interference between the two beams results in spatial modulation of the light intensity which can be thought of as a grating of light and dark stripes on a macroscopic scale. A particle passing through this region scatters the light whose temporal intensity pattern corresponds to the gratings spatial intensity pattern. The correlation or differences of the patterns is related to the particle velocity. The scattered light from the two beams is collected by the receiver lens and focused through an aperture onto photodetectors. The scattered light waves produce a doppler shift signal which can be analyzed and correlated to the particle velocity. The magnitude of the doppler shift on the detector can be related to the size of the particle. It should be noted that the system assumes a spherical particle when it produces size information. In practical atomization conditions, droplets or particles may not always be spherical. For the most accurate results, three detectors should be used to record the Doppler shift. Then by knowing the relationship between phase shift and powder size (see Figure 5.24), the size and velocity of the particle can be quantified.

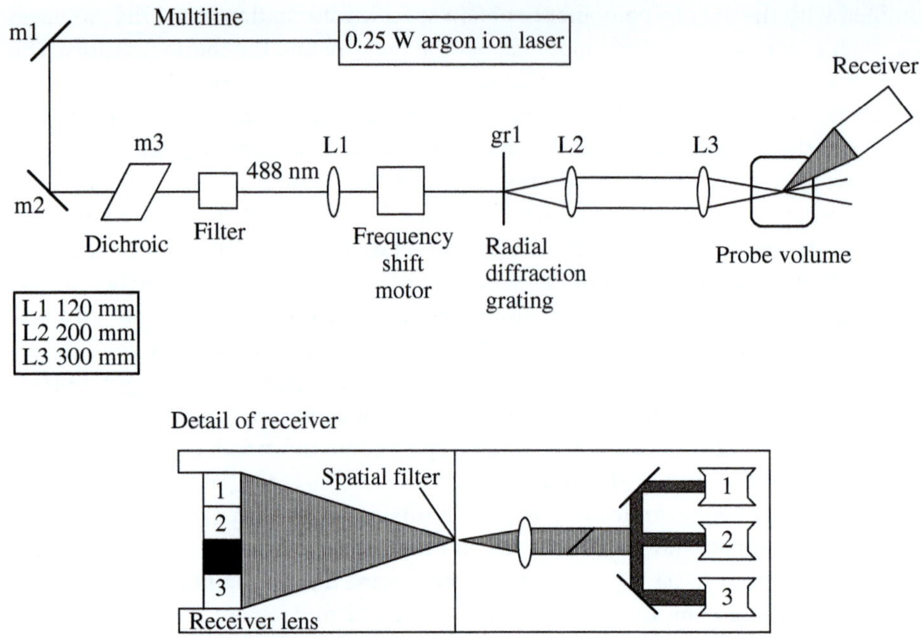

Figure 5.23 The phase Doppler interferometer. Source: Reproduced by permission of TMS and ASM International [47].

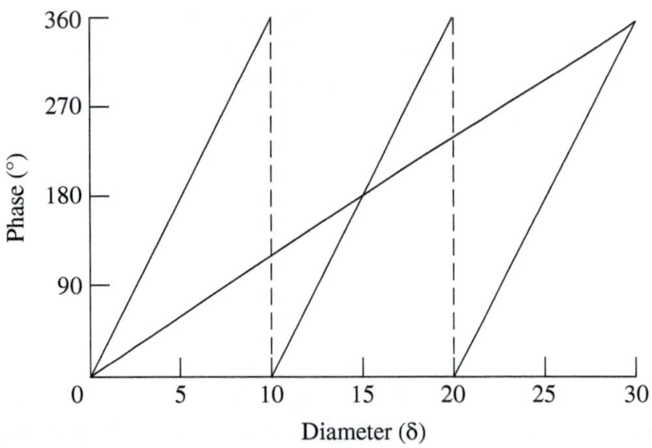

Figure 5.24 Relationship between phase shift and particle size. Source: Reproduced by permission of the *International Journal of Powder Metallurgy*, APMI International, Princeton, NJ [48].

The PDPA system can be calibrated from velocity and turbulence data which can be inferred from seed particle measurements. Calibration is not necessary for velocity measurements; however, in order to withdraw any other relevant information, the system should be normalized for all unique particles and atomization conditions. Once the system is fully set up and calibrated, it is capable of measuring particle sizes from 1 to 3000 μm. This technique offers several advantages over some of the other methods already reviewed. This method relies on the measurement of frequency modulations and time intervals which make it independent of signal attenuations and environmental effects. The linear relationship between the Doppler shift and particle size allows for accurate size measurements. The major limitation that is associated with PDPA is that the shape and optical absorption characteristics of the powders can skew the accuracy of the measurements. New versions of this technique are concentrating on overcoming these obstacles, and future systems should provide even more accurate data. Overall, PDPA is a well-established monitoring technique which is currently being utilized by many facilities employing atomization units. With continued improvements, this technique has the potential of becoming the dominant monitoring tool for diagnostic systems.

5.9 Surface Ionization For Monitoring Particles (SIMP)

The diagnosis techniques that are discussed in the previous sections are all based on optical principles. Consequently, such techniques are subjected to complex alignment of optics and measurement error associated with lens contamination by the atomized materials. Another type of method that exists for obtaining information from atomization is based on an electric principle which avoids the problems associated with lens contamination.

A relatively new instrument called the SIMP relies on interpretation of the electronic information that is discharged by atomized particles [44, 48, 49]. When an atom or molecule whose ionization potential is comparable with the work function of a metal, encounters a heated metal surface, the atom or molecule may leave the surface as a positive ion. The SIMP instrument collects ions produced during this process using a special electrode. The burst of ions accompanying the impact of a particle containing surface-ionizable impurities produces a sizable electrical pulse. Many atoms such as Na, K, Ba, Ca, Th, and U have this characteristic. The instrument utilizes a resistance-heated tungsten wire placed in the path of the falling droplets. Current is run through the wire, and due to IR heating from its intrinsic resistance, it is maintained at a constant temperature. The potential difference between the filament and the collection electrode is monitored. The electrical signals that are produced can be correlated with particle size. The technique relies on the detection of the impurity ions which are common in most alloys and can detect particles as small as 100 μm. The diameter of the wire is small making it a nonintrusive technique. The system only responds to detecting alkali and other surface ionizable constituents [1, 50]. Other particles such as combustion products, water mist, and photochemical smog do not affect the detection capabilities of the system.

This method of monitoring particle size is effective, simple, and inexpensive, but it does hold some disadvantages. The relationship between electrical pulse height and particle diameter is nonlinear and depends on material characteristics. This makes it rather difficult to calibrate the instrument. Also, the range of particle size detection is limited, making the system inapplicable to some atomization units. However, it still offers an economical alternative to other optical diagnostic units for certain specific applications.

5.10 Intelligent Sensors

The final diagnostic tool which will be introduced in this section encompasses several of the methods already mentioned but adds an intelligent process control system. In fact, significant progress has been made toward the goal of automated process control of melt atomization [51–56]. However, the dynamic nature of the atomization process makes it difficult to achieve reproducible results, and hence complete control over the properties (powder size, distribution, and morphology) and microstructure of the final power remains elusive. Understanding and quantifying some of the critical atomization parameters, such as nozzle size, **gas pressure**, and metal flow rate, will allow melt atomization to become a more predictable fabricating methodology.

Typifying some of the pioneering work being done on **intelligent sensors**, Ridder et al. [41] was able to correlate nozzle geometry to spray cone shape and particle size [41]. Then by using this information as a database and utilizing other diagnostic tools such as infrared and laser diffraction sensors, they have been able to set up a primitive intelligent control system for atomization [51, 52]. For their study, Ridder et al. [7] utilized the supersonic inert gas metal atomization (SIGMA) system located at the National Institute of Standards and Technology (NIST).

Using **high-speed cinematography** and laser holography, Ridder et al. were able to isolate the turbulent phenomenon that is typically present at the nozzle [41]. The atomizing nozzle assembly used by SIGMA is shown schematically in Figure 5.25. The actual assembly is comprised of 18 gas nozzles arranged in a circle around a liquid delivery tube. The specific dimensions and identification of the nozzle features are provided in the Nomenclature, and three of the features shown in Figure 5.25 were addressed. P_0, **gas jet** stagnation pressure, P_r, the reservoir pressure, and P_{dt}, liquid delivery tube pressure are the three quantities which ultimately dictate the characteristics of the gas flow field and consequently the properties of the atomized droplets. By modeling of these parameters, together with Schlieren photography and laser-pulsed holography, Ridder et al. proposed that the wave patterns of the gases at the nozzle resembled those in Figure 5.26. **Gas pressure** measurements for three different gases (nitrogen, argon, and helium) were made at the important locations of the nozzle (P_0, P_r, and P_{dt}). The results were compiled and plotted in Figures 5.27 and 5.28. With computer analysis and modeling, Ridder et al. were able to correlate the peaks of the aspiration curves to specific points of the wave pattern in Figure 5.26. Finally, using the information derived from the

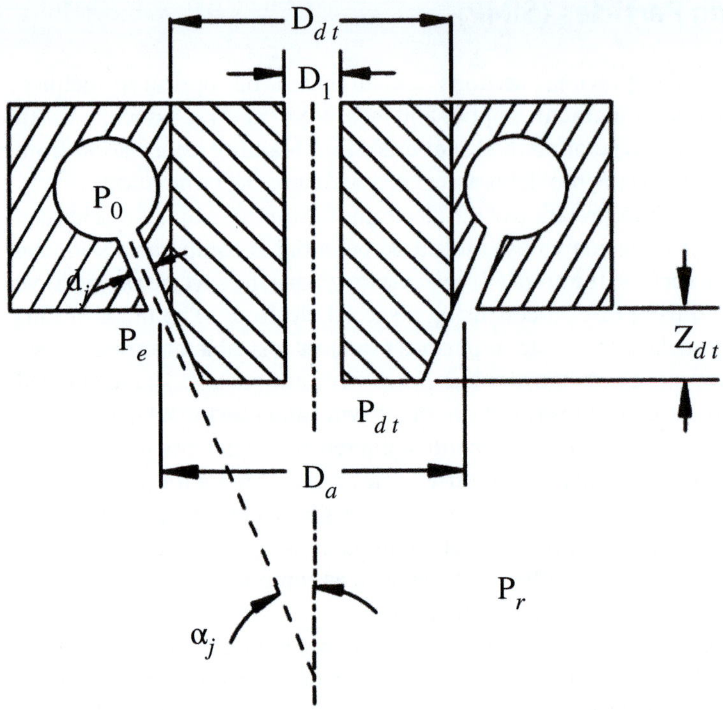

Figure 5.25 Schematic cross section of the nozzle. Source: Ridder et al. [41]/Springer Nature.

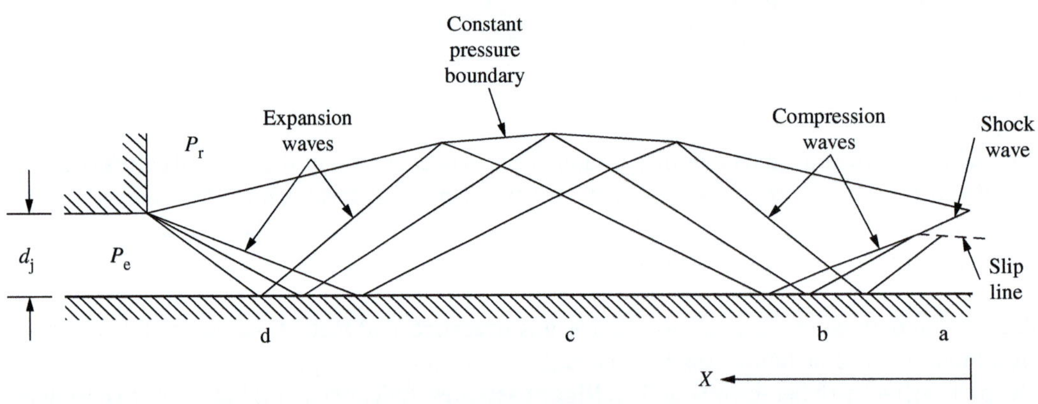

Figure 5.26 Schematic illustration of the wave pattern at the nozzle. (a)–(d) Positions of the liquid delivery tube relative to the jet at conditions a, b, c, and d in Figure 5.28. Source: Ridder et al. [41]/Springer Nature.

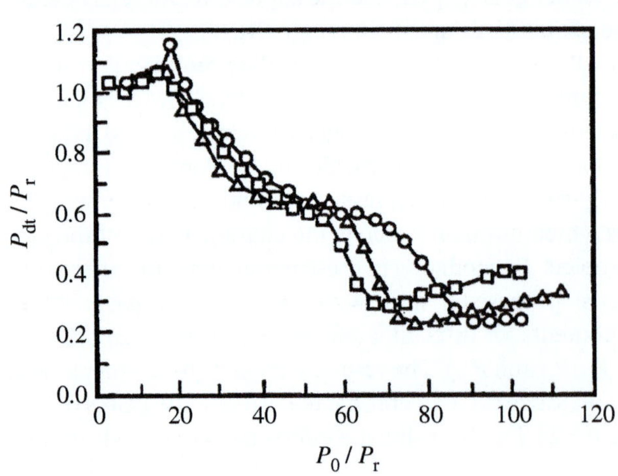

Figure 5.27 Aspiration curves for (Δ) nitrogen, (□) argon, and (○) helium. Source: Osella et al. [52].

Figure 5.28 Generalized aspiration curve showing the predominant features of the curves in Figure 5.27. (a)–(d) The nozzle locations shown in Figure 5.27. Source: Osella et al. [52].

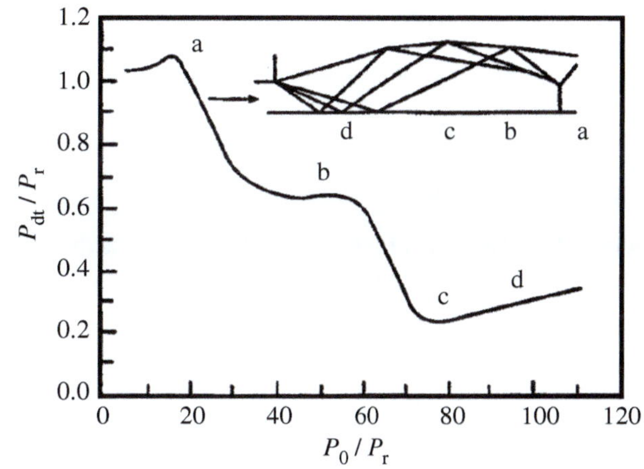

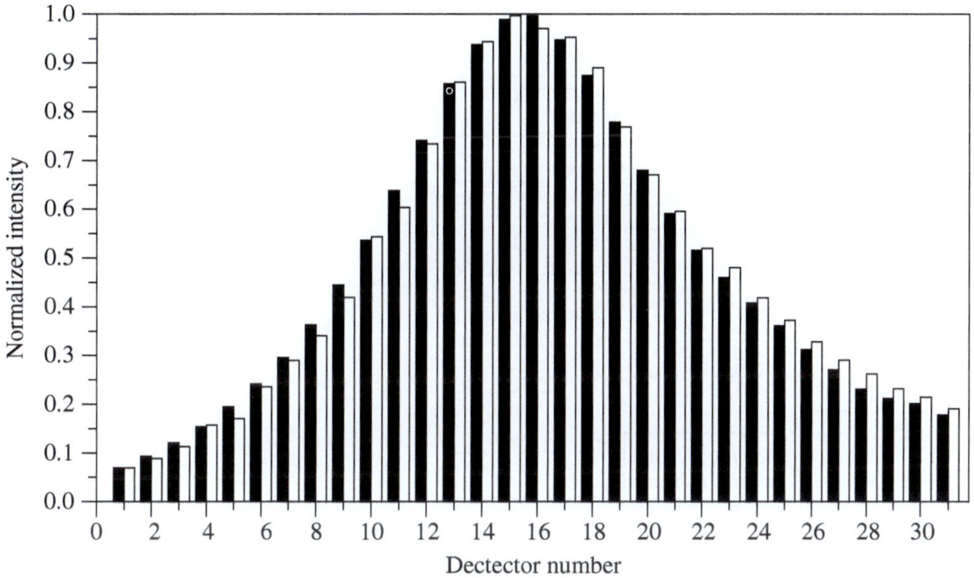

Figure 5.29 Comparison of (■) modeled and (□) measured light intensity. Source: Ridder et al. [41]/Springer Nature.

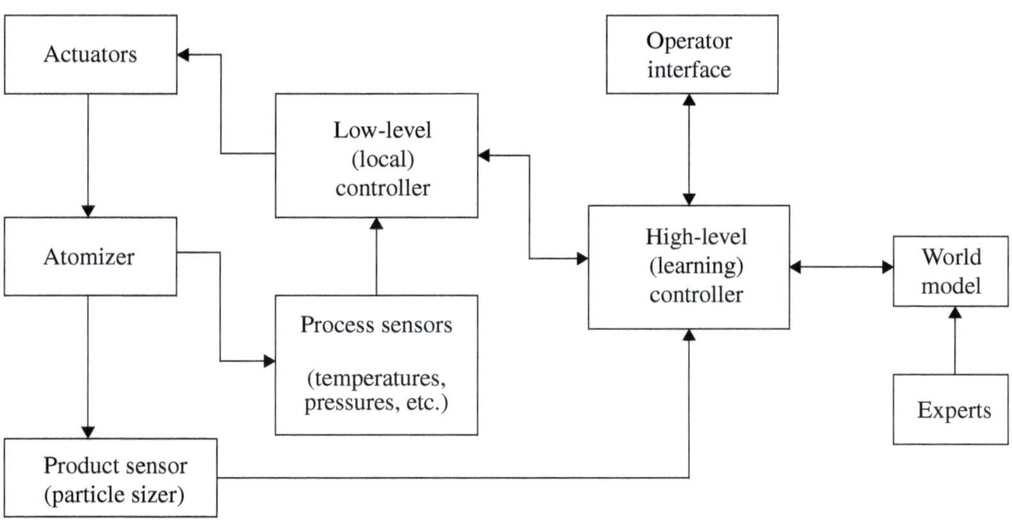

Figure 5.30 Logic loop for **intelligent sensors**. Source: Ridder et al. [41]/Springer Nature.

nozzle analysis, Ridder et al. [41] were able to formulate primitive mathematical equations to model the behavior of the gases. In turn, these equations were used to formulate input–output relationships for the computer control of the dynamic systems such as melt atomization. The form of these equations is as follows:

$$X_{x+1} = f(X_k, U_k) \tag{5.6}$$

$$Y_k = g(X_k, U_k) \tag{5.7}$$

where X_k is the state of the system at time k, U_k is the control input at time k, Y_k is the measurable output of the system at time k, X_{x+1} is the state of the system at time $k+1$, and the functions f and g are the next-state and output functions,

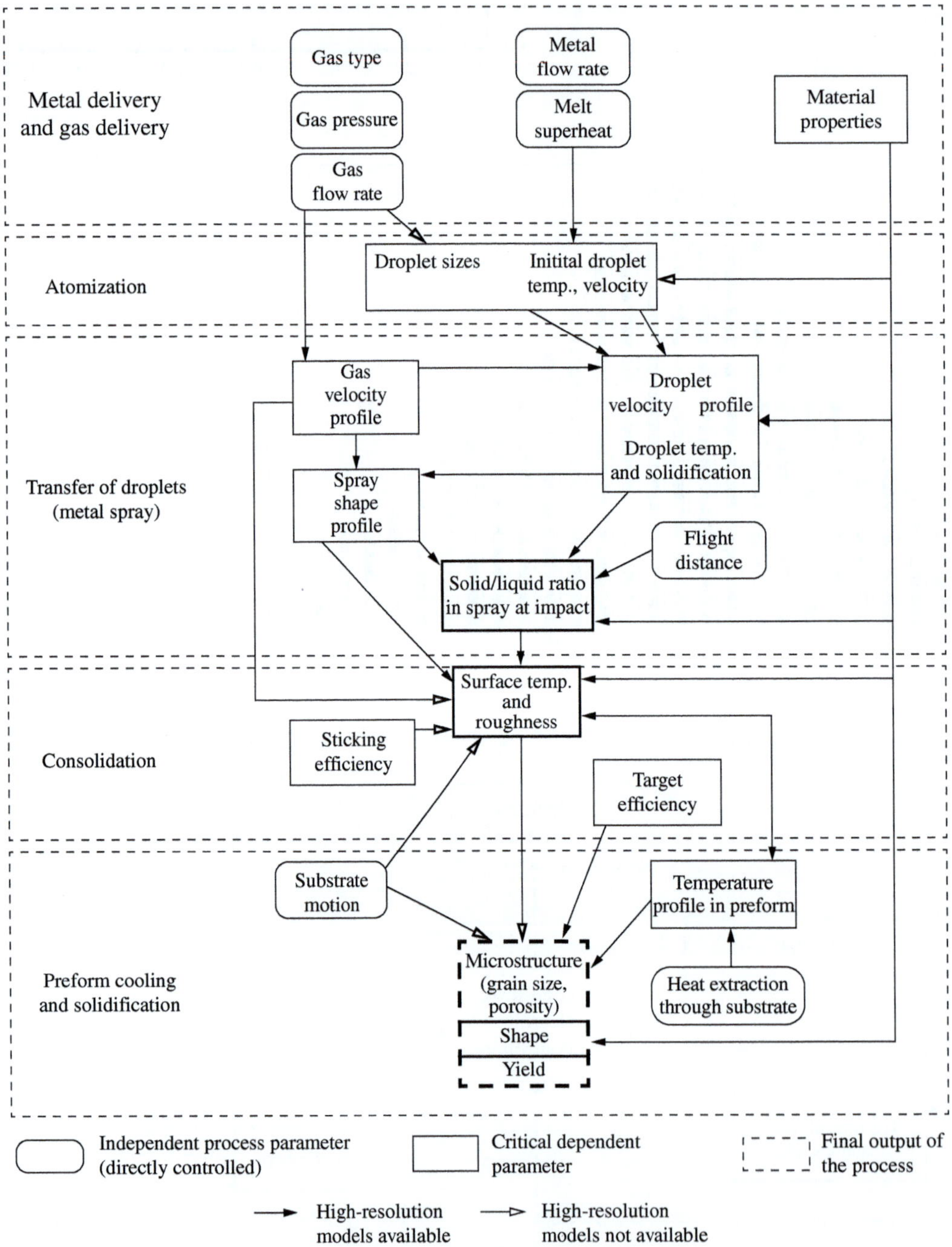

Figure 5.31 Flow chart depicting the interlinking of independent and dependent process parameters to the various stages of Osprey processing. Source: Mathur et al. [54].

Figure 5.32 Schematic diagram showing the three stages of gas atomization. Source: Biancaniello et al. [51]/with permission from Elsevier.

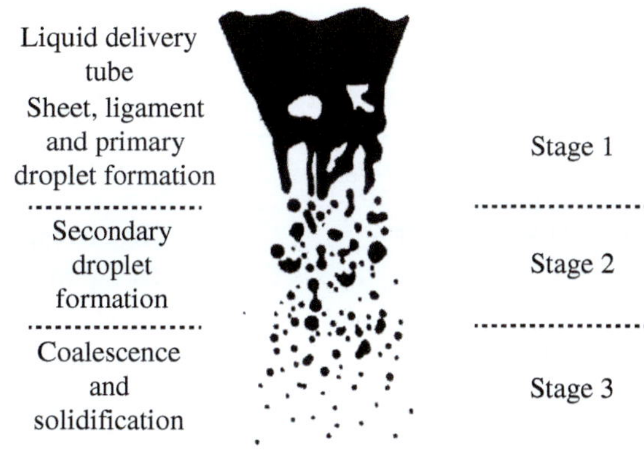

respectively [41, 51]. They were then able to use those equations to write basic if-then algorithms to predict particle size and distribution. By using laser diffraction techniques, they were able to correlate their predicted and experimental results as shown in Figure 5.29. As is evident from Figure 5.29, their initial results looked very promising.

Eventually, Ridder et al. envision a system where, by monitoring gas pressures, nozzle geometry, viscosity of atomization liquid, gas type, chamber pressure, etc., they would be able accurately to control particle size and distribution which will, in turn, dictate the materials performance [53]. They had a system which can monitor some of these parameters and react based on if-then statements. Their system follows the logic of Figure 5.30. The high-level controllers' decision-making strategy was based on the principle of Hierarchical Control [57]. In this approach, the strategy was goal-directed. For the controller's highest-level goal, the objective was decomposed into several more elementary subgoals. All the subgoals must be achieved to accomplish the objective. All the preliminary work and models done to support this method appear very encouraging [41, 52].

Mathur et al. had also made some very significant contributions to **intelligent sensors** [54]. His group, working on optimizing the Osprey™ process, had developed a computer-controlled spray-casting apparatus. Mathur et al. had focused their research on optimizing control of eight independent process parameters such as super heat, gas flow rate, and substrate to nozzle distance (see Figure 5.31), which would enable them in turn to control the dependent parameters such as solid/liquid ratio in spray at impact, surface temperature, grain size, etc. (see Figure 5.31). By trial and error, Mathur et al. had compiled a database of optimum independent process parameters settings. This information, together with mathematical equations which related all the parameters to final material properties, had allowed them to program a computer-controlled spray-casting apparatus which could monitor itself and producing desired end products with specified material characteristics [54]. The flow chart of Figure 5.31 depicted the interlinking of the process parameters. Coupling the optimum independent process parameter settings with appropriate sensor and control technology had enabled Mathur et al. to achieve remarkable process control and near automation of melt atomization [54].

The information discussed in this section suggests that intelligent control of atomization is not yet a reality. The atomization phenomenon is not yet fully described by mathematical representations. The complex stages of atomization ranging from sheet, ligament, and primary droplet formation to secondary droplet formation to coalescence and solidification are extremely difficult to quantify (see Figure 5.32). However, more of the gas flow and nozzle events can now be represented by equations [41]. This, together with some of the other diagnostic techniques, is bringing the industry one step closer to full diagnostic control.

5.11 Summary

All the sensors and diagnostic techniques reviewed herein are aimed at controlling atomization events which will in turn allow control of the particle size and distribution. Most of the methods mentioned are fundamentally unique. Some utilize optical measurements to obtain powder characterization information, while others use electrical signals or geometrical features. Each of these methods offers its own advantages for each particular application. In addition, each sensor also exhibits some type of shortcoming inherent to its methodology (see Table 5.1 for a comparison of all the reviewed methods). Evaluation of each method and its potential diagnostic application will yield the sensor which is best suited to a particular users need. There is no single best or worst diagnostic tool for atomization,

Table 5.1 Summary of diagnostic methods.

Method	Advantages	Disadvantages
Particle counting, sizing, and velocity probe (PCSV-P) fiber-optic-based system	Simultaneous measurements of velocity, size, and concentration High-temperature operation Backward and forward	Inaccurate for particles larger than 100 μm in diameter Measurements sensitive to material characteristics
High-speed cinematography	Simple Inexpensive	Poor resolution Takes time to develop film
High-speed off-axis holographic cinematography	Provides 3-D information about particle characterization Infinite field of depth Provides particle size and distribution information	Complex setup Limited resolution Limited velocity information
Infrared thermal imaging (intensity proportional to temperature to the fourth power)	Provides droplet cooling data and size distribution information instant feedback	Needs complicated computer algorithm to decipher thermographs Complex calibration
Phase Doppler particle analyzing (PDPA)	Measures size, velocity, and concentration Unaffected by signal attenuations or environmental effects	Based on spherical geometry of droplets Optical absorption of droplets may affect results
SIMP monitor electrical signal correlated with particle size	Simple nonintrusive inexpensive	Dependent on material impurities Limited particle size detection Complex calibration
Intelligent sensors Correlate operational parameters to properties of final material	Automated operation Reproducible results Full diagnostic control	Limited mathematical models currently available Complicated computer algorithms

but rather a best-suited application. Continued research and integration of the available diagnostic methods will advance atomization science and help yield more reproducible results with custom powder characteristics.

This successful commercial application of **spray atomization** technologies will depend on the implementation of diagnostic techniques that can be used to monitor and control process variables. In this chapter, we discuss the latest developments that have occurred in an effort to monitor atomization processes, including laser light diffraction techniques, **high-speed cinematography**, thermal imaging, and surface ionization. The chapter describes the benefits and shortcomings of these techniques and presents results from the literature reporting on the use of these techniques for the atomization of metals. It is evident that it is challenging not only to collect the relevant data but also to interpret the meaning of this data. The very high density of droplets inside of the atomized spray renders it difficult to accurately distinguish the behavior and location of individual droplets. Moreover, the thermal energy that radiates from elevated temperature droplets also makes it difficult to accurately assess their precise temperature as they travel from the atomization region into the collecting chamber. Nevertheless, the material reviewed in this chapter reveals that progress has been made in both monitoring and controlling the process of atomization, and that as work in this area continues to move forward it will be possible to optimize the generation of particular powder sizes and associated distributions more efficiently.

References

1 Siegel, M.W. and Henein, H. (1989). *Synthesis and Analysis in Materials Processing: Characterization and Diagnostics of Ceramic and Metal Particulate Processing* (ed. E.J. Lavernia, H. Henein, and I. Anderson), 1. Warrendale, Pennsylvania: TMS.

2 Henein, H., Meyer, P.L., Holve, D.J., and Kuhni, M.A. (1992). *Int. J. Powder Metall.* 28: 149.

3 Lauterborn, W., Judt, A., and Schmitz, E. (1993). *Opt. Lett.* 18: 4.
4 Miller, C.E. (1991). *Adv. Imaging* 8: A38.
5 Hoffman, T. (1991). *Adv. Mater. Proc.* 140: 37.
6 Ünal, A. (1989). *Metall. Trans. B* 20: 613.
7 Mates, S.P. and Settles, G.S. (1995). A flow visualization study of the gas dynamic of liquid metal atomization nozzles. In: *Advances in Powder Metallurgy and Particulate Materials, Part 1* (ed. M. Phillips and J. Porter), 15. Princeton, NJ: Metal Powder Industries Federation.
8 Settles, G.S. (2001). *Schlieren and Shadowgraph Techniques: Visualizing Phenomena in Transparent Media*. Springer Science & Business Media.
9 Jonassen, D.R., Settles, G.S., and Tronosky, M.D. (2006). *Opt. Lasers Eng.* 44: 190.
10 Biancaniello, F.S., Espina, P.I., Mattingly, G.E., and Ridder, S.D. (1989). *Mater. Sci. Eng., A* 119: 161.
11 Markus, S., Fritsching, U., and Bauckhage, K. (2002). *Mater. Sci. Eng., A* 326: 122.
12 Anderson, I.E., Rieken, J.R., Meyer, J. et al. (2011). *Visualization of Atomization Gas Flow and Melt Break-up Effects in Response to Nozzle Design Description*. Ames, IA: Ames Laboratory (AMES).
13 Lauterborn, W. and Vogel, A. (1984). *Annu. Rev. Fluid Mech.* 16: 223.
14 Dudderar, T.D., Meynart, R., and Simpkins, P.G. (1988). *Opt. Lasers Eng.* 9: 163.
15 Stanislas, M., Okamoto, K., and Kähler, C. (2003). *Meas. Sci. Technol.* 14: R63.
16 Stanislas, M., Okamoto, K., Kähler, C.J., and Westerweel, J. (2005). *Exp. Fluids* 39: 170.
17 Stanislas, M., Okamoto, K., Kähler, C.J. et al. (2008). *Exp. Fluids* 45: 27.
18 Dabiri, D. (2009). *Exp. Fluids* 46: 191.
19 Adrian, R.J. (1991). *Annu. Rev. Fluid Mech.* 23: 261.
20 Raffel, M., Willert, C.E., Wereley, S.T., and Kompenhans, J. (2007). *Particle Image Velocimetry: A Practical Guide*. Springer.
21 Bewlay, B.P. and Cantor, B. (1989). *Mater. Sci. Eng., A* A118: 207.
22 Ziesenis, J. and Bauckhage, K. (2003). *Part. Part. Syst. Char.* 20: 290.
23 Mullis, A.M., McCarthy, I.N., and Cochrane, R.F. (2011). *J. Mater. Process. Technol.* 211: 1471.
24 Mullis, A., Adkins, N., Aslam, Z. et al. (2008). *Int. J. Powder Metall* 44: 55.
25 McCarthy, I., Adkins, N., Aslam, Z. et al. (2009). *Powder Metall.* 52: 205.
26 Planche, M.P., Khatim, O., Dembinski, L. et al. (2012). *Powder Technol.* 229: 191.
27 Kirmse, C. and Chaves, H. (2016). *Steel Res. Int.* 87: 1295.
28 Zeoli, N., Tabbara, H., and Gu, S. (2012). *Appl. Phys. A* 108: 783.
29 Chaves, H., Herbst, S., Skupsch, C., and Therm, J. (2012). *Spray Technol.* 21: 882.
30 Antipas, G.S.E. (2013). *Powder Metall.* 56: 317.
31 Whatmore, R.W. (1991). *Ferroelectrics* 118: 241.
32 Ridder, S.D., Biancaniello, F.S., Espina, P.I., and Miller, S.A. (1989). *Synthesis and Analysis in Materials Processing: Characterization and Diagnostics of Ceramic and Metal Particulate Processing* (ed. E.J. Lavernia, H. Henein, and I. Anderson), 73. Warrendale, Pennsylvania: TMS.
33 Meng, H. and Hussain, F. (1991). *Fluid Dyn. Res.* 8: 33.
34 Born, M. and Wolf, E. (1999). *Principles of Optics: Electromagnetic Theory of Propagation, Interference, and Diffraction of Light*. London: Cambridge University Press.
35 Jenkins, F.A. and White, H.E. (1957). *Fundamentals of Optics*. New York, NY, USA: McGraw-Hill.
36 Tyler, G.A. and Thompson, B.J. (1976). *Optica Acta* 23: 688.
37 Hirleman, E.D. (1984). Particle sizing by optical, nonimaging techniques. In: *Liquid Particle Size Measurement Techniques* (ed. J.M. Tishkoff, R.D. Ingebo, and J.B. Kennedy), 35. Philadelphia: American Society of Testing Materials.
38 Bayvel, L.P. and Jones, A.R. (1981). *Electromagnetic Scattering and Its Applications*. London: Applied Science Publishers.
39 Hirleman, E.D. and Dellenback, P.A. (1989). *Appl. Opt.* 28: 4870.
40 Hirleman, E.D. (1988). *Part. Charact.* 4: 128.
41 Ridder, S.D., Osella, S.A., Espina, P.I., and Biancaniello, F.S. (1992). *J. Powder Metall.* 28: 133.
42 Trapaga, G., Mathys, E.F., Valencia, J.J., and Szekely, J. (1992). *Metall. Trans. B* 23B: 701.
43 Cantor, B., Kim, W.T., Bewlay, B.P., and Grant, P.S. (1989). *Seri. Metall. Mater.* 23: 1651.

44 Cantor, B. and Bewlay, B.P. (1990). *Metall. Trans. B* 21B: 899.

45 Gataska, F. (1990). *Personal Communication.* Lynn, Massachusetts: General Electric Corp.

46 Perez, R., Zhang, J., Gungor, M., and Lavernia, E.J. (1993). *Metall. Trans. A* 24A: 701.

47 Mcdonell, V.G., Lavernia, E.J., and Samuelsen, G.S. (1989). *Synthesis and Analysis in Materials Processing: Characterization and Diagnostics of Ceramic and Metal Particulate Processing* (ed. E.J. Lavernia, H. Henein, and I. Anderson), 13. Warrendale, Pennsylvania: TMS.

48 Jiang, G., Henein, H., and Siegel, M.W. (1990). *J. Powder Metall.* 26: 253.

49 Kumar, M., Balasubramanian, T.V., and Misra, R.D.K. (1990). *Mater. Sci. Lett.* 9: 147.

50 Salgado, L., Filho, F.A., Neves, M.D.M.D., and De Souza, O.C. (1991). *Mater. Sci. Eng., A* A133: 692.

51 Biancaniello, F.S., Conway, J.J., Espina, P.I. et al. (1990). *Mater. Sci. Eng., A* A124: 9.

52 Osella, S.A., Ridder, S.D., Biancaniello, F.S., and Espina, P.I. (1991). *J. Mater. Eng. Perform.* 43: 18.

53 Biancaniello, F.S., Espina, P.I., Mattingly, G.E., and Ridder, S.D. (1989). *Mater. Sci. Eng., A* 9: 161.

54 Mathur, P., Annavarapu, S., Apelian, D., and Lawley, A. (1989). *J. Mater. Eng. Perform.* 41: 23.

55 Annavarapu, S., Apelian, D., and Lawley, A. (1990). *Metall. Trans. A* 21A: 3237.

56 Lawrynowicz, D.E. and Lavernia, E.J. (1995). *J. Mater. Sci.* 30: 1125.

57 Wilson, I.D. (1979). *Int. J. Control* 29: 899.

6

Atomization Improvements for Additive Manufacturing

6.1 Introduction

The critical powder characteristics that are most relevant to **additive manufacturing (AM)** processes include **average size**, size distribution, **morphology, satellite particles, porosity, yield, chemistry**, and of course **cost**. Not surprisingly, efforts to improve the performance of atomized powder for AM processes often involve parametric studies of the influence of process parameters on the resultant powder characteristics. Despite progress in the formulation of theoretical frameworks that can be used to calculate powder characteristics from fundamental principles, accurate predictions have been hindered by the large number of concurrent variables affecting the atomization process, as described in Chapter 2. In most cases, the usefulness of an equation is limited to experimental parameter space and property ranges, such as **fluid type**, pressure range, **atomizer type**, and **alloy** type. Even theoretical studies attempting to establish universal curves for **drop-sized** prediction [1, 2] await confirmation by means of more extensive experimental studies. Nevertheless, these studies have helped establish an understanding of the key parameters that affect the atomization process. Furthermore, some consensus has been achieved regarding the effect that some of the variables have on atomization, while strong disagreement continues to exist regarding the role of other variables.

6.2 Gas and Metal Flow Rates

It has been consistently found and documented in several independent studies that as the pressure of the atomizing gas increases, with other variables being kept constant, the mass median powder diameter decreases [3–9]. This has been attributed primarily to an increase in the mass flow rate of the gas with a concomitant increase in the gas/metal ratio. An examination of all the empirical equations discussed confirms the importance of mass flow rate in determining the distribution of powder particle size.

Gas pressure is an important variable that has a complex effect on the **gas atomization (GA)** process. Gas pressure can affect the length of the **supersonic core** in the gas jets [6], and spreading of the gas jets can lead to an increase in powder size [10]. Gas pressure during atomization has also been reported to increase **metal flow rate** by creating a **low-pressure zone** at the exit of the **metal delivery tube**, which effectively aspirates the metal [11–14]. Thompson [15] reported an increase in average powder size with an increase in gas pressure due precisely to this effect. Couper and Singer [12] documented a series of **"flow pressure maps"** to demonstrate how metal flow rate could be increased by aspiration (negative pressure) or retarded by bubbling (positive pressure). This phenomenon is a result of two competing aspects: (i) gas pressure and (ii) the placement of the metal delivery tube.

The magnitude of the under pressure (aspiration) or overpressure (pressurization) at the tip of the metal delivery tube also depends on the atomization pressure [11–14]. Accordingly, care must be exercised when selecting the position of the metal delivery tube with respect to the existing atomization gas jet, primarily because variations in the position of the metal delivery tube have a profound effect on atomization efficiency.

It has been proposed that the pressure effects result from, or are affected by, an occurrence of vortices, which form due to an **acoustic feedback phenomenon** [11]. The acoustic feedback effect is active whenever space is left between the metal delivery tube and the **supersonic core** of the atomization gas stream. Vortices are known to be present in the mixing region of a supersonic flow at distances downstream from the nozzle orifice [16]. These vortices

Metallic Powders for Additive Manufacturing: Science and Applications, First Edition.
Enrique J. Lavernia, Kaka Ma, Julie M. Schoenung, James F. Shackelford, and Baolong Zheng.

result from downstream-traveling disturbances in the jet, generated by upstream-propagating sound waves emitted by the supersonic jet itself [17]. The disturbance train develops into a long vortex, whose axis is a helix, enclosing the flow. An interaction of sound waves traveling upstream with the eddies traveling downstream results in the formation of a standing pattern. This is the acoustic feedback effect on supersonic jets [16, 17]. Any sound-reflecting material in the vicinity of the **gas jet** can alter or reduce the acoustic feedback, which is also influenced by the inlet pressure and/or the gas spread angle. The acoustic feedback effect must be accounted for to understand the observed pressure variations at the metal delivery tube orifice, in addition to **turbulent mixing, shock wave formation** and **reflection**, and other related **gas dynamics phenomena**.

The relevant factors related to gas and metal flow rates that influence powder quality for AM are then summarized as follows:

- The optimal values of the gas and metal flow rates that yield powder sizes should be optimized for a particular AM technology in terms of size, distribution, and **solidification microstructure**.
- The metal flow rate will affect the production capacity, thereby affecting the **cost** of the powder.
- The **gas flow rate** will affect the powder cooling rate, and hence the solidification microstructure, so it needs to be tailored to the **alloy** being atomized.

6.3 Gas Velocity

The dependence of **gas velocity** on pressure is a well-documented phenomenon that has been studied thoroughly in fluid mechanics [10, 18].

A mathematical model of high-speed gas flowing over the surface of a liquid layer was developed by Bradley [1, 2] and discussed in Chapter 2 (Section 2.7.2). The results of Bradley's empirical analysis suggest that a decrease in average powder particle size occurs when the relative velocity between gas and metal is increased. Other investigations [19–21] present experimental evidence supporting this suggestion. This behavior has been rationalized not only in terms of an increase in mass flow rate that accompanies the velocity increase but also in terms of **secondary droplet disintegration mechanisms**.

A droplet formed by the **primary disintegration** of a bulk liquid may be subjected to additional pressure forces and undergo further disintegration of dynamic pressure due to the gas stream velocity exceeds the restoring force due to **surface tension**. This condition, termed the maximum stability criterion, is given by [22]

$$\frac{1}{2}\rho_{\mathrm{g}}(U_{\mathrm{rel}})^2 = C\frac{\sigma_{\mathrm{l}}}{d_{\mathrm{D}}} \tag{6.1}$$

where U_{rel} is the relative velocity, ρ_{g} is gas density, σ_{l} is the surface tension of liquid, and d_{D} is the droplet diameter.

The behavior of nonferrous metals during gas atomization using conical nozzle jets was studied by Naida et al. [23, 24]. Using calculated **gas velocity** contours of various forms, depending on system parameters, they determined the atomization variables necessary to obtain a particular yield of powder.

The relationship between gas velocity and atomization variables is complex. In atomization, the gas leaves the nozzle at high speeds but slows down during its travel to the metal contact point [7, 25, 26]. Rai et al. [10] showed that, for a given gas pressure, the powder size is strongly dependent on the $\mathrm{d}t/R$ ratio, where $\mathrm{d}t$ is the throat diameter of each **gas jet** and R is the over which the **gas jet** spreading occurs. Mottern [27] used numerical solutions to the Navier–Stokes equations to calculate the maximum orifice length that precludes the **velocity relaxation** process from interfering with the formation of a stabilizing film in the gas jets.

The important role of **gas dynamics** in the determination of atomizing characteristics has resulted in complex nozzle designs. The ultrasonic gas atomizer (USGA) is an example of such a design [10, 12–14, 28–32]. The USGA principle uses high pressure coupled with a flow interference mechanism, similar to that in a conventional whistle (Hartmann whistle), to generate high-energy supersonic gas flow velocities. Exit gas velocities greater than Mach 2 have been reported for USGA atomization [33].

The exit gas velocity depends on the design of the atomizer being used as well as on **gas chemistry** (i.e. Ar, N_2, or He). Therefore, when producing powders for AM, the following factors should be accounted for:

- Exit gas velocity should be tailored to match the desired average powder size and **size distribution**.
- Exit gas velocity should be optimized to minimize the generation of satellite powders.
- Exit gas velocity should be matched to minimize **gas recirculation** inside of the atomization chamber.

6.4 Physical Characteristics of the Gas and Melt

In gas atomization of metals, the role of melt properties such as viscosity, **surface tension**, and properties of the gas such as density and viscosity are complex. Lewis et al. [34] report a 60% decrease in average powder particle size resulting from the same corresponding percentage decrease in gas viscosity, other variables being kept constant. However, when gas density was reduced to one-seventh of its original value for a constant gas viscosity and ratio of liquid to gas, the average powder particle size increased by a factor of 2, despite an increase in gas velocity.

Putimtsev [35] atomized a number of different metals and **alloys** and found both **surface tension** and **liquid metal viscosity** to influence the shape of the atomized powders. In this study, it was observed that the tendency to form spherical particles was strongly favored by low melt viscosity and high melt surface tension. Putimtsev [36] also evaluated the effects of viscosity on powder size of a number of iron alloys (Fe, Fe–Al, and Fe–Si) and concluded that as the melt viscosity is increased, it becomes progressively more difficult to form finer powders. While results from this and other investigations [7, 37] suggest that viscosity is more important than surface tension in controlling powder size distribution, Klar and Shafer [38] concluded otherwise. Results from their work on the atomization of copper alloys revealed that a decrease in average powder particle size results from a decrease in surface tension. The decrease in surface tension of the melt was obtained by increasing the **oxygen** content, and the effect of melt viscosity on powder distribution was found to be negligible.

The extent to which a molten metal is heated above its melting point **(superheat)** has an influence on the resulting melt atomization characteristics. There are limited theoretical analyses on the effect that melt superheat has on the powder size distribution. There exists, however, ample experimental evidence suggesting that an increase in melt super-heat promotes a finer powder size distribution [5, 6, 15, 18, 36, 39]. Thompson [15] suggests that this phenomenon (e.g. increased melt superheat leads to a finer powder size distribution) is due to the dependence of the melt density on temperature. As the temperature of the metal is increased, the actual volume of metal passing through the pouring tube is not changed, but the resultant mass is less than that obtained at lower temperatures. This results in an increase in the gas/metal mass flow, which in turn promotes a finer powder size. Thompson's rationalization, however, fails to account for the variation of other properties with temperature, such as melt viscosity and surface tension.

Inspection of the scientific literature reviewed in Chapter 2 demonstrates that most of these investigations were conducted with **water, wax/polyethylene mixtures, oils**, and **molten waxes**. Two notable exceptions were the work of Tallmadge [7], which involved nitrogen atomization of cast iron, and the study of Lubanska [8], which led to the development of an equation applicable to the gas atomization of liquid metals. The lack of experimental atomization data for metals and alloys may be attributed to the difficulties associated with an atomization experiment involving molten metals. The extreme reactivity, akin to that of molten metals, for example, severely limits the types of crucible materials that may be used to contain the metal during heating prior to atomization. Other significant factors associated with the atomization of molten metals include atmosphere during melting, chamber geometry, and atomizer design, to cite a few examples.

The extensive work reviewed in prior chapters demonstrates that the **physical characteristics** of the gas and melt critically influence the resultant powder size distribution. It is also evident, however, that it is difficult to predict the relative importance that some of these physical parameters, such as surface tension and viscosity, have on the atomization of metals and alloys. These inherent characteristics of molten metals lead to major discrepancies between experimental results and numerical predictions, and not surprisingly, only a few of these correlations, such as the one proposed by Lubanska (see Chapter 2) are commonly used in investigations involving molten metals. Therefore, as far as powders for AM are concerned, the following factors should be considered when selecting the **physical characteristics** of the gas and melt.

- Avoid possible reactions between molten alloy and atomization gas.
- Select melt superheat considering the need to balance excessive vaporization of alloying elements (melt superheat too high) while minimizing the possibility of premature melt solidification (melt superheat too low).
- Select **gas chemistry** considering the need to balance high cooling rate (e.g. He) with **cost** (e.g. He is more expensive than Ar or N).

6.5 Powder Size Distribution and Other Variables

In powder-based AM, there are particle size requirements that generally depend on the type of AM process. For example, in the case of **powder bed fusion (PBF)**, the preferred powder size range is generally smaller than 45 μm, whereas in the case of **directed energy deposition (DED)**, the size range commonly used ranges between 25 and 150 μm. Published results suggest that large, non-spherical, and fused particles can disrupt the deposition process and cause the formation of defects in AM components. Given that gas atomization is the most common method of metal powder production, and the optimization of the gas atomization process is of particular interest to both powder metallurgy and metal AM research. The details on the theory of powder size distribution can be referred to the Section 2.5 powder size distribution in Chapter 2. The optimization of gas atomization for improving the powder size distribution involves both experimental and numerical simulation approaches [40, 41].

The first investigation on gas atomization for metal powder production and particle size is attributed to S. Thompson in 1948 [15]. He used a close-coupled gas atomizer to study the effects of gas pressure, melt temperature, and melt flow rate on powder size distribution. In another work, Ayers and Anderson [42] studied the impact on stagnation pressure on ***powder size*** and their results suggest that the best powder yield is obtained when the static pressure at the melt inlet is minimized. Ünal [43] studied the atomization process of an aluminum alloy for different atomizing gases. He used **helium**, nitrogen, and **argon** as the atomizing gas and studied various combinations of stagnation pressures, gas-to-melt flow rate ratios, and melt superheat temperatures. He observed a slight variation in powder size distribution when the melt temperature is increased above 1100 K (for Al alloy). This is due to the temperature dependence of melt viscosity and surface tension. Helium provided the finest powder distribution, and the powder sizes increased with increasing gas density (**argon** provided the coarsest powder distribution). He correlated the mean diameter of the powder distribution to be directly proportional to the square root of the melt flow rate. In a subsequent study, Ünal [44] utilized **Schlieren** images to study the supersonic flow characteristics in gas-only flow in a close-coupled gas atomizer. Miller et al. [45] studied the influence of axisymmetric and non-axisymmetric gas nozzle geometries and found that non-axisymmetric nozzles provide a better yield of finer powder. Strauss [46] used preheated gas to increase the gas momentum without raising the gas pressure and managed to reduce the mean diameter of the powder distribution. However, the lower limit of the diameter range remained unchanged irrespective of the gas temperature. Preheating the atomizing gas also reduces the operation cost as it lowers the gas consumption. In a subsequent study [47], he introduced a new parameter, the normalized gas energy rate, which correlates well with the mean diameter at a wide range of operating conditions. The normalized gas energy rate is defined as the ratio between gas kinetic energy and melt mass flow rate.

The formation of an open or **closed wake** condition during atomization is another operating condition that has been studied extensively. A **closed wake** occurs when the flow circulation region that develops below the melt-tip is independent of the surrounding flow structures. It is due to a normal shock that can appear in the vicinity of this flow region. This normal shock, also called a Mach disk, acts as a shield isolating this region from the surrounding environment. Ting et al. [48] studied this phenomenon to investigate its effect on powder yield, and observed the development of a wake closure pressure for their specific atomizer geometry. Their results show that a **closed wake** condition can lead to a finer powder yield as the interaction between the Mach disk, and the melt stream creates pulsating characteristics in the melt stream.

Generally, an increase in melt flow rate is directly correlated with larger powder particles. It was reported that the particle size increased proportionally to the square root of the melt flow rate [49]. Ünal also studied the effects of melt temperature on the atomization process. At all temperatures used in this study, 775–925 °C, atomization proceeded successfully, and results indicated that melt temperatures above 825 °C led to minor powder refinement. Powders atomized at temperatures around 775 °C were slightly coarser with a wider dispersion in particle size distribution and this was attributed to lower surface tension and viscosity of the melt.

In terms of the influence of gas selection, a study by Janowski et al. [50] found no significant variations in powder characteristics as a function of backfill gas or atomizing gas. In this study, stainless steel 304L rods were gas atomized using **Argon** in five different atomization runs. Results were obtained by analyzing powder production at the same gas/metal flow rate ratios and indicated that higher gas pressures, around 2.12 MPa, did not produce finer powders relative to those powders produced with lower gas pressures, around 1.56 MPa [50]. Another experimental and numerical study by Lavernia and coworkers showed that the measured mean powder size is almost unchanged when gas pressure changes between 5.52 and 2.62 MPa because of aspiration metal effects [51, 52]. These results demonstrate that gas pressure is not an independently influential parameter and does not have significant effects on the powder yield or particle size distribution. This study also examined the mechanical properties of nitrogenated stainless steel powders after **hot isostatic pressing (HIP)** consolidation. The authors noted varying amounts of internal porosities in powders were a result of differences in chemical behavior of the gas/metal system, and atomizing gas pressure. This observation was attributed to solubility of the atomizing gas in the liquid and solid metal. These results are similar to those found for the atomization of Rene-95 alloy by VanStone et al. [53].

The droplet size distribution for gas-atomized various alloys has been reported to follow a lognormal distribution [40, 54–57]. The mass probability density function, $p(d)$, of the droplet-size distribution can be expressed by [58–60]:

$$p(d) = \frac{1}{\sqrt{2\pi}\ln\sigma_g}\exp\left[-\frac{(\ln d - \ln d_{50})^2}{2(\ln\sigma_g)^2}\right] \tag{6.2}$$

where d is the droplet size. Generally, the powder size distribution is represented in terms of a cumulative frequency, $f(d_i)$, which is defined as the fraction of powders that fall in the size range that is smaller than d_i. Another often used parameter for describing atomized power size is the mass median diameter d_{50}, which is defined as the droplet size that corresponds to the 50% cumulative frequency. d_{50} can be well predicted for gas atomizing Al alloys powder by using a correlation developed by Lubanska [56]:

$$d_{50} = K_d D_n\left[\frac{\eta_m}{\eta_g W}\left(1 + \frac{\dot{M}}{\dot{G}}\right)\right]^{\frac{1}{2}} \tag{6.3}$$

where K_d is a constant, D_n is the **melt stream diameter** (i.e. the nozzle diameter), η_m ($m^2\ s^{-1}$) and η_g ($m^2\ s^{-1}$) are the kinematic viscosity of the melt and gas, respectively, $\dot{M}$ ($kg\,s^{-1}$) and $\dot{G}$ ($kg\,s^{-1}$) are the melt and gas flow rates, respectively, W is the Weber number, $W = v^2\rho_m D/\gamma_m$, where ρ_m ($kg\,m^{-3}$) and γ_m ($J\,m^{-2}$) are the density and surface tension of the melt, respectively. σ_g is the geometric standard deviation characterizing the spread of the droplet size distribution centered around d_{50}, and can be estimated by the following empirical equation [40, 61]:

$$\sigma_g = q d_{50}{}^j \tag{6.4}$$

where q and j are constants, and the unit of d_{50} is micrometer herein.

Figure 6.1 shows the effect of gas pressure and composition on droplet mass median size [51] of Al alloy. The mass median diameter, d_{50}, decreases with increasing gas pressure, and evidently decreases with increasing fraction of He when used He + Ar gas mixture as atomizing gas. The effect of gas pressure on decreasing of droplet size is also apparent when gas pressure is under 2.76 MPa.

Numerical analysis provides useful insight into the thermal and momentum behavior of the atomized droplet dispersion, even if the size distribution of droplets is deduced from empirical data. For example, numerical modeling provided useful information regarding the complex inter-relationships between the independent atomization parameters, such as atomization gas composition, gas pressure, melt superheat temperature, and gas/melt mass flow ratio, and their influence on the microstructure of solidification. Regarding the effects of He fraction in the atomizing gas, the results show that an increase in He content generally causes a decrease in powder size due to its low density which results in an increase in gas velocity, as illustrated in Figure 6.1. However, the cooling effects of He gas on the melt viscosity also need to be considered for high-viscosity materials. The high cooling capacity of He gas increases melt viscosity faster, as compared to N_2, Ar, and other gas mixtures. A rapid increase in viscosity might have an adverse effect on the primary break-up of melt, as well as on the secondary break-up mechanisms. This might lead to an increase in the average powder size. In addition, an increase in the atomization gas pressure can generally cause a decrease in the powder size distribution. However, the presence of aspiration phenomena at the point of atomization can affect the melt flow rate and shift the droplet distribution toward a large size [62]. Such complex interactions, quite common in GA, make it difficult to accurately predict the exact particle size distribution and overall cooling behavior of droplets.

The discussion of powder sizes is generally centered around the average powder diameter. The spread of powder size distribution is an important property that will determine the yield of powders that lie within a specified size range. Although the powder sizes obtained during atomization may be accurately described by using a variety of distributions, normal, log-normal, etc., atomized metals typically exhibit a log-normal distribution of powder sizes.

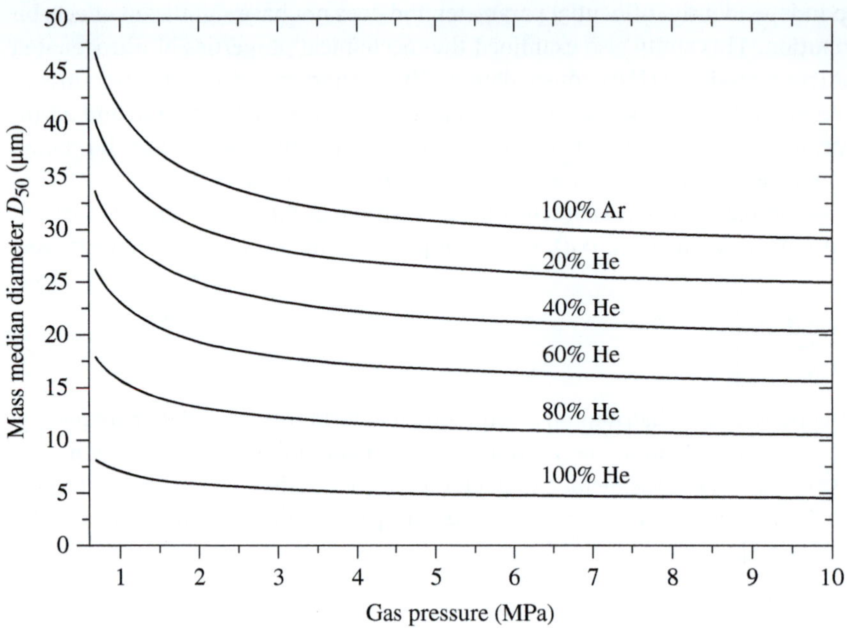

Figure 6.1 Effect of gas pressure and composition on particle size. Source: Zheng et al. [51]/Springer Nature.

For such cases, one way to express the spread of the distribution is by using the geometric standard deviation [63]. The geometric standard deviation, σ_g, is defined as

$$\sigma_g = \frac{X_{84}}{X_{50}} \tag{6.5}$$

where X_{84} is the cumulative 84 weight percent undersize, and X_{50} is the cumulative 50 weight percent undersize; both numbers are obtained from a log probability plot.

Several authors have shown if d_m (mass median diameter) is plotted vs. σ_g (geometric standard deviation) on a double logarithmic scale, a straight-line results; that is, $\sigma_g = ad_m{}^b$ (a, b = constants) [8, 64, 65]. Studies on the atomization characteristics of a high-strength aluminum alloy [66] report values of 0.527 and 0.377 for a and b, respectively. Hence, any variations in mass median diameter resulting from variations in the atomization parameters, discussed above, will affect the spread of the distribution.

Other variables also have effects on the atomization of metals and alloys, although their precise effects are less clearly understood. The ambient pressure of the medium in which the atomization process occurs has been reported to influence both the powder size and distribution [67, 68], while another independent investigation found no correlation between the two variables [19]. The effect of the angle selected between the molten metal stream and the impinging gas stream has been reported to affect the resultant powder distribution [7, 23, 69]. There is, however, difficulty in interpreting these results, due to intrinsic variations in atomizing conditions parameters coupled with variations in atomizer designs.

6.6 Powder Morphology

The **particle morphology** is an important characteristic of the powder that influences AM processing in terms of powder flowability and packing density. This characteristic is dependent on the process used to manufacture the powders as described in Chapter 1. In general, gas atomization (GA) produces mostly spherical-shaped particles due to the relatively lower solidification rate as compared with that of **water atomization (WA)**, for example, leading to increased flowability and apparent density. These characteristics render GA powders ideal for most currently used AM systems. Results show that gas-atomized powders are preferred for AM applications where powder flow is essential. In comparison to gas atomization, powders produced by plasma atomization or plasma rotation electrode techniques are sometimes preferred because of their highly spherical morphology and smooth surfaces with fewer shape irregularities. Bulk particle irregularities (morphology) are more commonly seen for water-atomized powder, while surface irregularities (satellites) are more frequently associated with GA powder [41, 70]. The characteristics and formation of satellites in GA powder will be described in detail in Section 6.7. In general, high sphericity means good **flowability** of the powder. Spherical particles have a smaller contact surface area and therefore friction and interlocking between particles are lower than that of the particles with irregular morphology.

As described in Chapter 2, the powder production process via melt atomization can be divided into the following steps: primary breakup into ligaments, ligament breakup and secondary breakup, and/or spheroidization, while simultaneously cooling and solidification take place. Apart from the thermodynamic conditions of the atomization gas, melt properties such as viscosity, density, surface tension, heat capacity, and thermal conductivity will influence the mechanisms that govern spheroidization. Melt viscosity is a quantity that describes the resistance of the melt to shear forces. The higher the viscosity, the stronger the forces required to deform the melt will be. In the case of the spheroidization process, more energy (from surface tension forces and inertia) is dissipated during deformation due to higher viscosity. Therefore, a decrease in the time required for full spheroidization with increasing melt viscosity is generally observed. If the viscosity is too high, the resistance to spheroidization exceeds the other forces involved (surface tension and inertia forces) and spheroidization becomes very difficult. Surface tension is the driving force for the spheroidization process. Surface tension does not affect spheroidization time, but solely influences the frequency of the oscillations. Melt density proportionally affects inertia forces, which keep the oscillating melt in motion and increasing melt density leads to a longer spheroidization process.

A lower viscosity and a higher surface tension of the melt generally result in a higher level of sphericity of the atomized powder. During gas atomization, the molten material cools down rapidly and solidifies as it contacts the atomization gas due to the thermal difference between gas and melt. However, when using a preheated atomizing gas, the lower temperature differential between gas and molten metal results in a lower heat transfer, and consequently the atomized droplets solidify at a rate that is comparatively slower than when using low-temperature atomization gas [71]. This lower solidification rate tends to favor the formation of spherical droplets, as the residence time in the liquid phase is increased [72]. Atomization results show that for constant atomization pressure, the mean diameter of the powder particles decreases with increasing atomization gas temperature and that for a given powder size, sphericity increases with increasing gas-to-melt flow ratio (G/M). Similarly, an increase in melt superheat also leads to an increase in sphericity of the atomized powders.

In related studies, Schwenck et al. [73] investigated the influence of atomization gas pressure and melt mass flow on the characteristics of CuSn10 powder. They found that sphericity depends on the particle size so that larger particles show lower sphericity. Gao et al. [74] studied the influence of atomizing gas pressure, delivery tube diameter, and melt superheat on particle size and morphology distribution of AlSi10Mg powders produced by double-nozzle gas atomization using different parameters. They reported that an increase in gas pressure and melt superheat or a decrease in the delivery tube diameter both led to a reduction in powder size, as a result of differences in melt mass flow rates and gas-to-metal ratios. Vlachos and Chang [75] researched the relationship between **flowability** and powder physical properties (e.g. particle size, size distribution, and morphology). These authors observed that, despite research indicating that reduced powder size is associated with increased powder cohesion and, hence, poor **flowability**, adding controlled amounts of fine powders to a powder mixture does not necessarily reduce the **flowability** of the mixture. In addition, a change in flowability can be noticeable if particle size is reduced to a size that is at least ten times smaller than the largest sizes. They also reported that elongated and irregular powders have poor flowability [75].

The morphological properties of powders depend on their size as well as on the material being atomized [76]. According to Özbilen [76] gas-atomized powder particle shape and surface texture also depend on the type of atomizing gas, the **oxygen** level, and the operating pressure. The coarse powder becomes more irregular in shape and rougher on the surface when atomization is carried out under oxidizing conditions. This occurs due to the presence of an inhomogeneous surface oxide film thickness.

Powder morphology is generally studied via scanning electron microscope (SEM). As an example, for gas-atomized Al-17Si powder, it was reported that the SEM morphology of powder particles reveals that smaller particles possess spherical morphology, whereas large-size particles typically have an irregular and elongated morphology [77]. Powders in the intermediate size range exhibit a mixture of spherical and irregular morphologies. It is worth noting that even though smaller powders tend to exhibit a spherical morphology, their flowability is sometimes poor as a result of their higher surface area and the presence of van der Waals' forces.

Various shape parameters, such as **circularity**, convexity, solidity, and aspect ratio, have been employed by image analysis systems (e.g. Morphologi G3) to characterize the powder morphology [78]. It is unlikely that a single shape parameter can perfectly characterize particles for all applications. However, **circularity** would not be appropriate in an application where both spikey and elliptical particles are present. Since the gas-atomized particles are generally spherical in spherical the **circularity** of gas-atomized powders is commonly used [79]. **Circularity** is the ratio of the circumference of the projected/equivalent area of a particle to the actual perimeter of the particle according to the definition by Morphologi G3 [78]. The **circularity** of powder particles can be calculated based on the equation suggested by the ISO 9276-6 standard [80]:

$$\text{Circularity} = \frac{4\pi A}{P^2} \tag{6.6}$$

where P and A are the measured circumference and area covered by a particle projection, respectively. **Circularity** can quantify how close the shape is to a perfect spherical morphology. Given the fact that the **circularity** of an ideal sphere is unity, deviation from the ideal spherical shape results in lower sphericities. A perfect spherical morphology has **circularity** of 1.0, while a very narrow, elongated particle has **circularity** close to 0. It was reported that the shape parameter **circularity** of the gas atomized 316L SS powder particles (size 20–63 μm) is in a range of 0.93–0.97 [79], with a higher circularity for the smaller particles. For the large particles with a size of 200 μm, the **circularity** is about 0.90. Small particles experience rates of cooling and solidification that are higher than those of large powders, thereby decreasing the chance of inter-powder collisions and other events (i.e. satellite formation) that can lead to irregular morphologies.

The sphericity of powder powders can be measured using **circularity** and aspect ratio parameters. These are important parameters that can be used to quantify the morphology of a group of powder particles. As an example, in related work, powder morphology was analyzed using a Malvern G3 optical measurement system [81]. Since this is a particle imaging system, it is possible to quantify the **circularity** and aspect ratio of particles. The values of **circularity** and aspect ratio of gas-atomized powders with high gas atomization pressure (labeled as PH atomizer) and low atomization pressure (labeled as PS atomizer) for each size range are summarized in Figure 6.2 [81]. The PH curves show a constant **circularity** and aspect ratio with increasing particle size range. On the other hand, the PS curves show a decrease in these characteristics with increasing particle size range and these results agreed with SEM images.

In fact, describing the shape of powders either qualitatively or quantitatively can be extremely challenging, especially when the particles are highly irregular and show no resemblance to primitive shapes. Nevertheless, qualitative

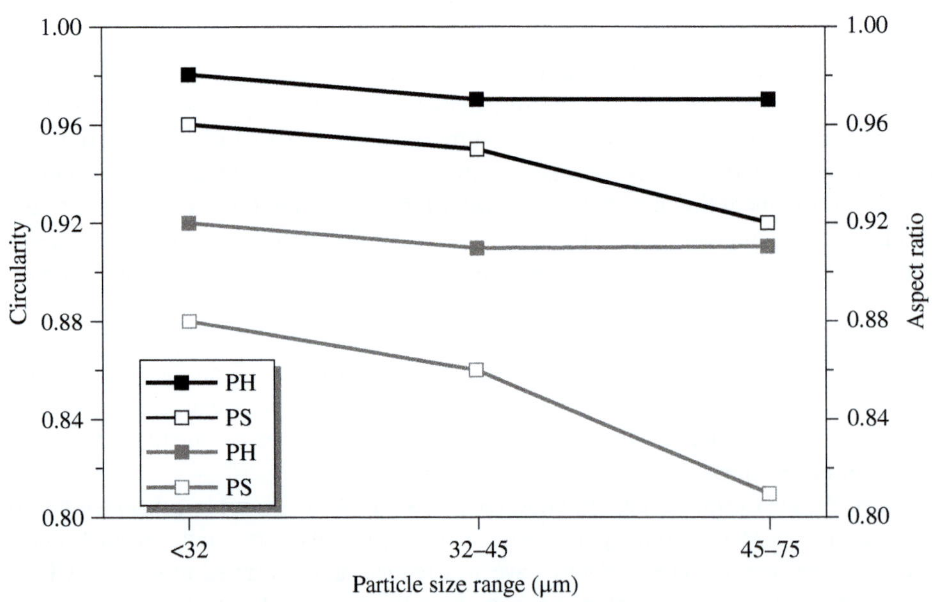

Figure 6.2 Circularity and aspect ratio of the gas atomized powders with high gas atomization pressure (PH) and low atomization pressure (PS). Source: [81]/ABM/Licensed under CC BY 4.0.

Table 6.1 Various shape factors used for particle shape characterization.

Name of shape factor	Equation	References
Circularity	$4\pi A/P^2$	[80, 85]
Aspect ratio	l/L	[86]
Elongation	$\text{Log}_2(a/b)$	[87]
Dispersion	$\text{Log}_2(\pi ab)$	[87]
Roundness	r_1/R	[88]
Flatness	r_2/R	[88]
Perimeter-to-area ratio	P^2/A	[89]

A – The projected area of the particle; P – Perimeter of the particle; l – Minor axis length perpendicular to the major axis; L – Major axis length connecting the two most distant points on the projection area A; a – Major axis length of Legendre ellipse of particle; b – Minor axis length of Legendre ellipse of particle perpendicular to a; r_1 – Radius of curvature of sharpest developed edge of particle; r_2 – Radius of curvature of the most convex direction of particle; R – Mean radius of the particle.

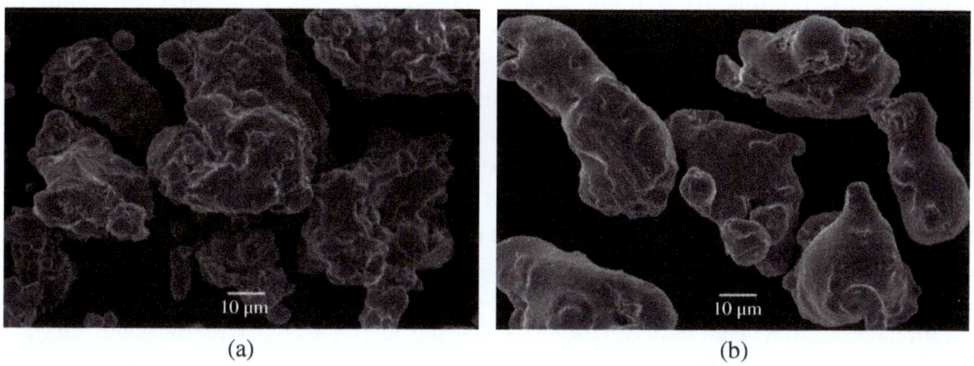

Figure 6.3 Different water-atomized 316L stainless steel powder morphology with different apparent densities (a) 3.09 g cm^{-3}, and (b) 3.56 g cm^{-3}. Source: Hoeges et al. [90]/with permission of Elsevier.

descriptions of powder particles have been standardized [82] and are used to describe the general shape of powder particles. Despite these efforts, quantification of particle shape remains challenging given the large number of factors involved. This problem has been approached by formulating equations that yield dimensionless ratios, more commonly known as shape factors [83, 84], some of which are listed in Table 6.1.

Gas-atomized powder is assumed to be the powder most suited for AM due to its morphology and flow qualities. As an alternative, powder particles produced with WA have high irregular morphology, low apparent density, and a slightly higher concentration of **oxygen**. This is attributed to the faster cooling and solidification rates during the melt atomization and the use of **water** to disintegrate the metal flow into powder [90, 91]. Figure 6.3 shows the typical morphology of water-atomized 316L SS powder and related apparent density.

The resulting **particle morphology** of WA powder can be modified by tuning the atomization process, altering the production route, or including a post-treatment procedure in the process chain. In related work, Boisvert et al. [92] conducted a study in which a magnesium treatment of the melt prior to the atomization was performed to improve the shape of WA steel powder. The technique creates conditions that promote the formation of more sphere-like particles. The shape improvement is caused by an increase in the surface tension of the melt due to the presence of magnesium. The improved sphericity of the Mg-treated particles is caused by a larger solidification time and a smaller spheroidization time of the droplets. The creation of a continuously renewed insulating Mg gas layer during the solidification of the droplets results in a larger solidification time. The smaller spheroidization time is a result of a larger surface tension of the melt from the reaction of Mg with dissolved sulfur. The larger surface tension of the melt also caused a decrease in the amount and size of internal pores. Schade et al. [93] described a process of manufacturing

WA powders specifically for AM. The process enables the production of fine WA powders by using high-pressure water atomization (HPWA), followed by the removal of irregular particles from the obtained powder. Although the described methods confirm the possibility of altering the shape of WA powders, achieving a more spherical powder leads to an increase in manufacturing costs.

6.7 Powder Satellites

The production of high-quality metallic powders is critical for enabling AM technology to attain its full potential. However, commonly used powders frequently contain a high proportion of "satellites" which degrade flowability during processing and can negatively affect component quality. Satellites are the small powders that sometimes appear on the surface of larger powders as spherical irregularities. These are formed as a result of the interactions that occur when small fully solidified powders impact liquid or semi-liquid droplets, due to their differences in velocity and temperature. The existence of these satellite particles negatively affects the metal AM process due to mechanical interlocking that occurs between surface irregularities during powder feeding. Improvements to gas atomization to suppress powder satellite formation can contribute an important powder quality enhancement for powder utilization in AM. Gas-atomized powders with minimal satellite concentration will have smoother spherical surfaces that improve flowability for AM DED as well as the layer spreading characteristics for AM PBF processing.

Generally, a wide range of particle sizes is found in gas-atomized powders [41, 76, 94, 95], while satellite formation has been generally observed only in the form of fine particles (typically <5 μm) that attached to large ones but not found between powders of comparable size. This suggests that the different acceleration profiles of gas-atomized droplets or particles with different size are an important cause of satellite formation because fine particles flow at higher acceleration rates and higher speeds than those of large ones and welding/sticking of particles can occur during impact even when the large particle is fully solidified. Adhesion will also occur when the large particle is in a fully liquid state (droplets) but undercooled, resulting in an embedded satellite. When the large particle is a liquid droplet, the satellite may either dissolve into the liquid or simply remain embedded on the surface, depending on the actual momentum and thermal fields. For satellite formation, there must be a relatively large particle, on which fine particles collide and remain attached.

There are different types of interactions that occur between atomized fine particles and coarse particles during gas atomization due to differences in their corresponding momentum and thermal conditions. Most of the types of particle/particle interactions are schematically illustrated in Figure 6.4, which provides a succinct summary of the variety of satellite particle attachment processes that can occur during the interaction of a solidified fine-sized powder particle and a larger particle [96]. The interactions that can lead to satellite formation occur immediately after droplet formation in the atomization plume as small solid particles interact with liquid or semi-liquid powders. The interactions between particles that occur at the highest temperatures will result in the coalescence of droplets, and at the lowest temperatures will result in elastic collisions between particles. Both cases do not necessarily generate satellite projections. Inspection of SEM micrograph of a typical gas-atomized NiAl powders produced in our laboratories at UCI [97], as shown in Figure 6.5, reveals examples of both **welded satellite** projections and sintered or **"stuck on" satellites.**

Gas atomization research has led to multiple hypotheses to rationalize the formation of excessive satellite decoration on the resulting powders [41, 61, 95], but typically agree that this problem is related to the production of a broad-size distribution of particles in an atomized spray. Generally, there are two separate causes that have been proposed to rationalize the formation of satellites or projections on GA powders, which prevent smooth flowability and lead to challenging packing behavior in AM. In one scenario that depends primarily on particle size [3], satellite attachment is attributed to the unavoidable "rendezvous" of fine powders with coarser powders during their flight downstream in the spray of an atomizer. If droplets are in a central position in the spray cone, the fine droplets accelerate faster in this high-velocity gas flow, and cool and solidify before the larger droplets in the droplet size distribution of the atomized spray. The fines could eventually impinge or weld onto the larger fully or partially molten droplets, attaching to larger particles that move slower in the atomization gas flow during free-fall and cooling in the spray chamber. During atomization, large particles need more time to solidify and move more slowly than fine ones during the early breakup stages of atomization. Finer powder particles, solidifying and moving faster than large particles, can catch and stick to the large particles that are still liquid or in a mushy state during atomization. This relative

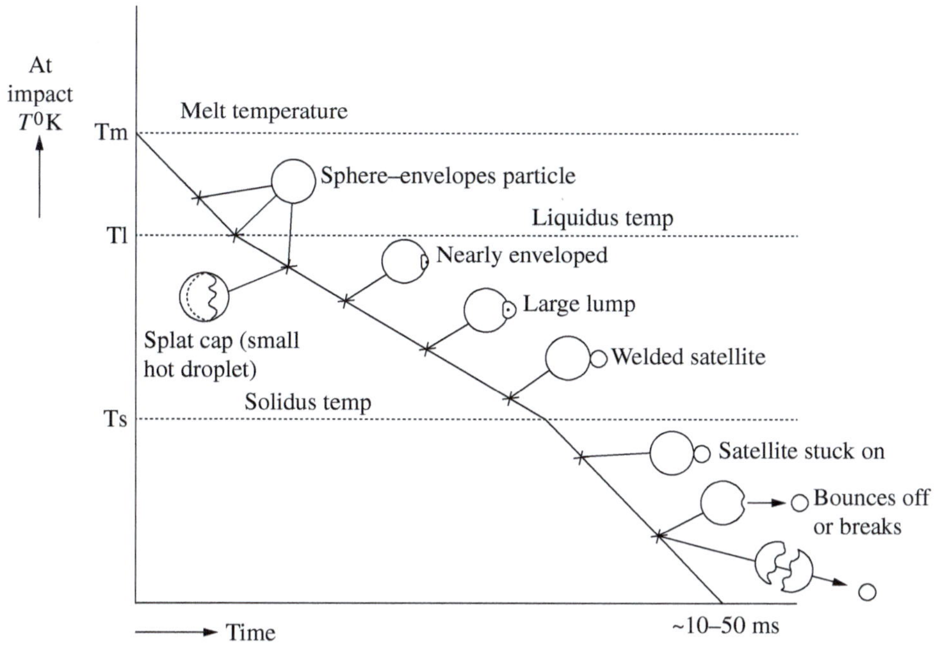

Figure 6.4 Effect of temperature of cooling large droplet on shape after impact by cold particles. Source: Dunkley et al. [96]/Metal Powder Industries Federation.

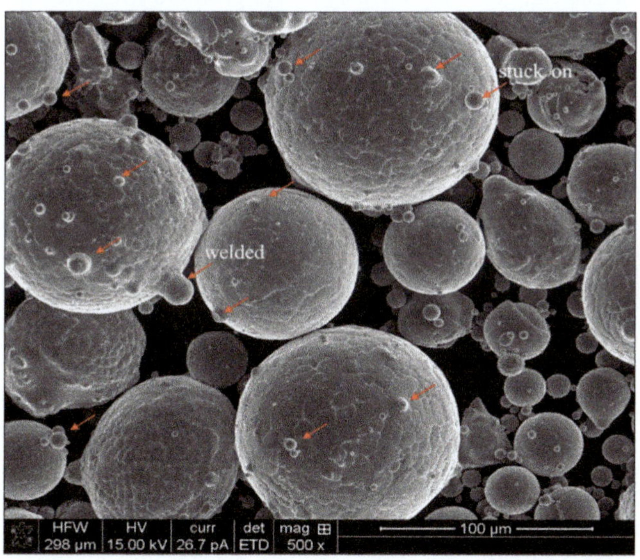

Figure 6.5 SEM micrograph of a typical gas-atomized Ni-10 wt% Al powder used for AM DED processing [97].

velocity difference causes collisions between fine and large particles, leading to sintering or welding, and satellite formation.

Another explanation considers spray chamber designs that are typically used for the industrial production of metal powders and offers alternative design options that can reduce the formation of satellites. In this case, the underlying mechanism [98] relates to the entrainment of "clouds" of fine solidified particles into the exterior region of the atomization spray **"cone,"** where they weld themselves onto larger droplets. This suggestion is supported by experimental observations made during atomization in chambers fitted with observation ports which show that just after a few seconds of operation, the view of the atomization plume is totally obscured by a dense **cloud** of dust circulating in the atomizing chamber [96]. Fine droplets are located on the outer region of the spray cone and will cool faster and solidify into powders that can be propelled up the spray chamber walls by a major external recirculation effect [98], that continues to feed fines into this recirculating cloud. Once the fine solidified powders reach the top of the chamber, they will be attracted to the atomization spray and can become entrained in the freshly generated droplet spray and can be welded onto larger droplets in the spray.

In terms of the influence of gas atomization process parameters, it has been reported that atomizing gas selection can influence satellite formation [76]. Generally, metal powders are formed by a three-step breakup process in conventional subsonic gas atomization, in which fine powder is formed and is removed from the impingement point of the gas on the liquid metal stream (see Chapter 3). Under turbulent conditions that are typical of subsonic gas atomization processes, satellites are produced by a process of collisions or impingement of fine solidified powder particles onto the coarser molten or semi-molten particles during various solidification stages. Because the gas heat transfer coefficient that is associated with He is much higher than that with Ar, for example, the powders atomized with He solidify more rapidly, providing less occasion for satellite formation, as reported in Özbilen's study [76]. In contrast, the relatively slow solidification rates that are associated with Ar-atomized powders facilitate satellite formation in subsonic gas atomization. The occurrence of satellites is reduced if the cooling rate is increased because coarse particles will have solidified before the collisions have an opportunity to occur. The higher cooling rate that is associated with He atomizing gas reduces the time available for powder welding and satellite formation in atomized powders, resulting in low satellite concentration in gas-atomized powders.

The sticking of small particles onto the large ones also depends on the morphology and surface condition of the large particles, hence affecting satellite formation [76]. When atomization is carried out under oxidizing conditions, the coarse powder becomes more irregular in shape and rougher on the surface, which provides conditions that facilitate fine powder particles attaching to coarse ones during in-flight collisions by welding, and forming more satellites, as observed in published studies [99, 100]. The **oxidation** characteristics of the liquid metals being atomized also play an important role in determining the morphology and surface roughness of powder particles. For example, metals that do not readily oxidize during atomization, such as Cu and Sn, generally form spherical powders with smooth surfaces that decrease the probability of satellite formation during interparticle impact [76].

The feeding problems that are associated with **powder satellites** during AM processes have motivated research into approaches to effectively eliminate satellite formation on gas-atomized powders. In related studies, Mostaghimi et al. [101] examined the influence of shape and the satellite content on the flowability of FeMnAlSi powders. They found a substantial improvement in flowability when powders were produced using a satellite-minimizing system [101], in which the number of satellites can be reduced by using an additional gas flow that limits recirculation in the atomization chamber.

In other studies aimed at reducing the satellite formation, Dunkley and Telford [96] reported that they developed an anti-satellite system, which withdraws gas from the atomizing chamber, removes the powder in it, and introduces the cleaned gas around the atomization plume. The designed system could fit into a conventional vertical inert gas atomizer to suppress satellite formation. They studied a range of metals, including non-ferrous and ferrous alloys, and showed a marked reduction in satellites, which could lead to a 15% increase in bulk apparent density and much-improved powder flow properties.

In terms of suppressing the satellite formation due to **gas recirculation** and the formation of a fine powder cloud, Anderson et al. [102] experimentally suppressed much of the fine powder cloud by using a narrow spray chamber (30 cm inner diameter) that limits the external recirculation flow effects based on numerical simulations. Their results showed a significant improvement in the sphericity and surface smoothness of the resulting powder [98], although some satellite formation was still observed. There are considerable challenges to this method of satellite suppression, including the avoidance of "splatted" particulates from premature collision with the spray chamber wall, especially during atomization conditions that produce a broad spray cone [103].

Xin-yu et al. [104] reported the use of two gas-flow-regulation strategies to prevent fine particle entrainment in gas recirculation zone, thus restricting the satellite formation. One strategy proposed involves the introduction of ancillary gas flow, and another one involves using an improved design of a step-shape atomization chamber. In this research, the evolution of macroscopic flow patterns and particle trajectories were studied based on numerical simulations using the ANSYS Fluent code. The results show that the fine particle entrainment in the gas recirculation zone can be effectively prevented when the ancillary gas flow is introduced at the position of $R/2$ (R is the radius of atomization chamber) away from the chamber center with the ratio of ancillary flow rate to atomization flow rate larger than 0.8. The optimized step-shaped atomization chamber is with a step width of 300 mm and a step height of around 575–600 mm, respectively, as shown in Figure 6.6b [104]. To demonstrate this concept, Ti alloy powder was

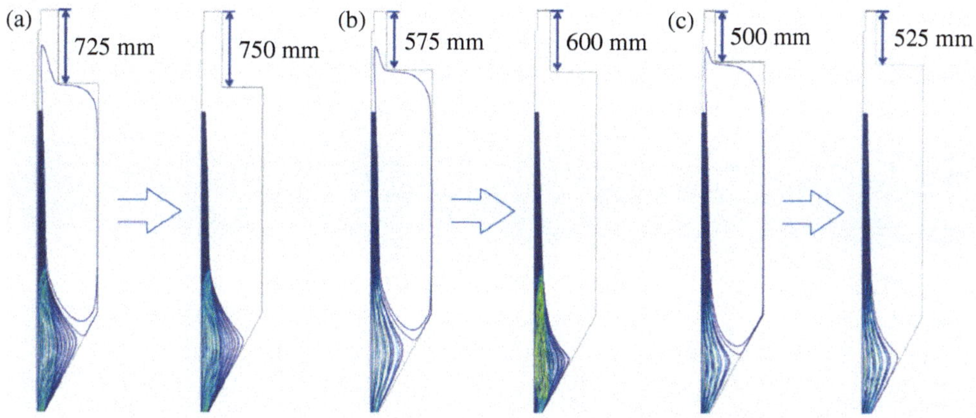

Figure 6.6 Step dimensions that effectively restrict the fine particle entrainment: (a) $D = 400$ mm, $\Delta H = 725-750$ mm; (b) $D = 300$ mm, $\Delta H = 575-600$ mm; (c) $D = 200$ mm, $\Delta H = 500-525$ mm. Source: [104].

atomized by applying the above gas-flow-regulation strategies while their size distribution, sphericity, the outgrowth rate (outgrowth rate was used for evaluating the satellite rate, $\eta = 1 - [1/(n_{\text{outgrowth}} + 1)]$, where $n_{\text{outgrowth}}$ is the number of outgrowth detected on the particle by using the particle shape analyzer [74]) was characterized, respectively. The results show that the outgrowth rate of the powders has been reduced by about 45%, compared with those prepared without the gas flow-regulation strategies.

To effectively suppress the satellite formation in gas-atomized powder, more numerical simulation, and modeling on the influence of gas atomization chamber design and experiments to verify the results should be performed, to improve the satellite suppression effects and prevent or reduce the formation of satellite particles that seem promising.

6.8 Powder Porosity

The presence of pores in atomized powders can negatively affect microstructure evolution during AM. Pores can form during atomization when the atomization gas becomes "trapped" during solidification and hence it is not released into the environment. Pores can form either in the interior of powders, or at powder surfaces and negatively affect the performance of AM-deposited structures. The formation of pores during AM is linked to the quality of the starting powder. The work performed by Susan et al. [105] suggests a direct relationship between AM deposit porosity and gas-atomized powders used for deposition. The **gas entrapped** in gas-atomized powder can be retained in AM-deposited components in the form of gas porosity. The **powder porosity** is greatly influenced by the selected manufacturing method. Powders produced by gas atomization (GA) generally exhibit a have a higher percentage of pores, as compared to equivalent powder manufactured by plasma rotating electrode process (PREP), for example. Surface porosity, which is sometimes observed in GA powders, is typically non-existent in PREP powder. **Gas entrapment** during GA generally occurs as the atomized metal experiences simultaneous deformation and solidification in the spray field.

A straightforward approach to mitigate the presence of gas pores in AM materials is to ensure that the starting powder is free from any internal pores or other surface defects. As discussed in Chapter 2, GA techniques make use of an inert gas, typically Ar or He, to protect from oxidation and to promote convective cooling during solidification. During the formation of the droplets and particles, the inert gas interacts with the liquid metal, leading to the entrainment of significant levels of gas (up to 28%) within the particles, irrespective of the alloy or the production technique [106]. The pores entrapped in gas-atomized powder are nominally present in different sizes. Large-sized internal porosity (pore dia. >10–90% of powder dia.) from trapped atomization gas is most prevalent in coarser size gas-atomized powders (dia. >70 μm) [102, 106]. Small-sized pores (pore dia. <5% of powder dia.) from trapped inter-dendritic solidification as sub-micro porosity are related to alloy "mushy" zone (liquid + solid) range. These pores are also more apparent in coarser powders due to a slower solidification rate [107]. The internal porosity in GA powder can be characterized using optical microscopy (OM) or SEM on powder samples with a polished cross-section [108]. Micro-computed

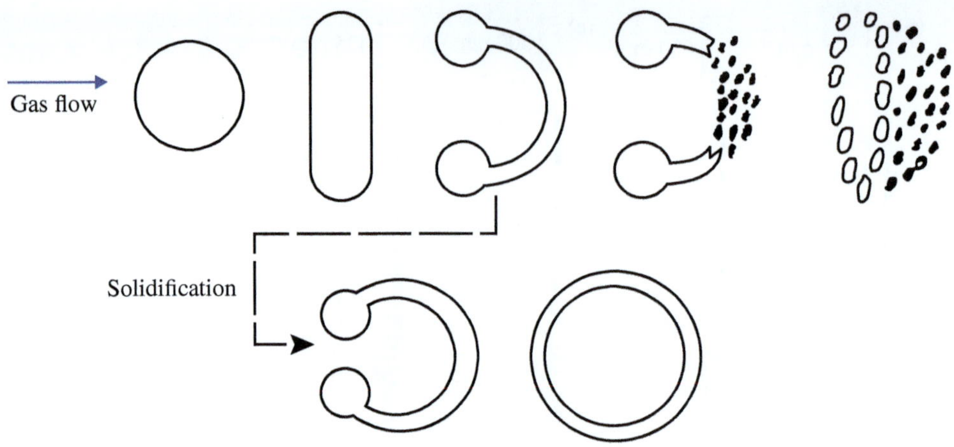

Figure 6.7 Schematic of the bag breakup and collapse mechanism proposed by Rabin et al. [106]. Source: Rabin et al. [106]/ Powder Metallurgy Technology.

X-ray tomography (μ-CT) can be used as a non-destructive imaging technique for studying internal porosities. μ-CT scans the sample using X-ray to detect any pores in GA powder [109].

It is necessary to examine the droplet formation mechanisms during gas atomization to understand how gas entrapment and other defects form. In terms of the mechanisms that govern gas entrapment during atomization, there exists extensive literature [3, 41, 106] concerning the breakup of a melt stream into droplets in the presence of a relatively high **gas velocity** based on the energetics of the atomization gas interactions with the molten metal. There are independent droplet breakup mechanisms identified to operate over different regimes, depending on the value of the Weber number [106], which can be written as:

$$We = dv^2 D_0/\gamma \tag{6.7}$$

where d is the flow field density, v is the relative velocity, D_0 is the initial particle diameter and γ is the surface tension. Among these mechanisms, one is termed as bag breakup process and illustrated in Figure 6.7, which schematically shows two options for the bag breakup mechanism of liquid droplet formation at high gas velocities ($12 < We < 50$). During solidification the bag may either shatter or, may close in on itself with sufficient viscosity rise, entrapping an atomization **gas bubble**. This process is like the blowing of a bubble except that, instead of forming a stable bubble, the bag bursts giving rise to many smaller droplets. The breakup of the melt into droplets can occur in a dynamic sequence with droplet cooling and solidification. When bag breakup is stimulated at high gas velocities, a melt fragment or large droplet becomes shaped into a bag-like sheet that spreads in a direction normal to the gas flow. The bag would shed and shatter into small droplets from its periphery. Alternatively, if the melt viscosity rises sufficiently, the sheet would collapse on itself to form a large hollow sphere drop with a trapped atomization gas inside. Therefore, the energetics of the breakup process should be reduced to suppress the generation of hollow spheres. However, it is generally a challenge to achieve this without precise control of the atomization process. The typical examples of trapped gas pores from commercial gas-atomized powders are shown in Figure 6.8, with the powder containing porosity to

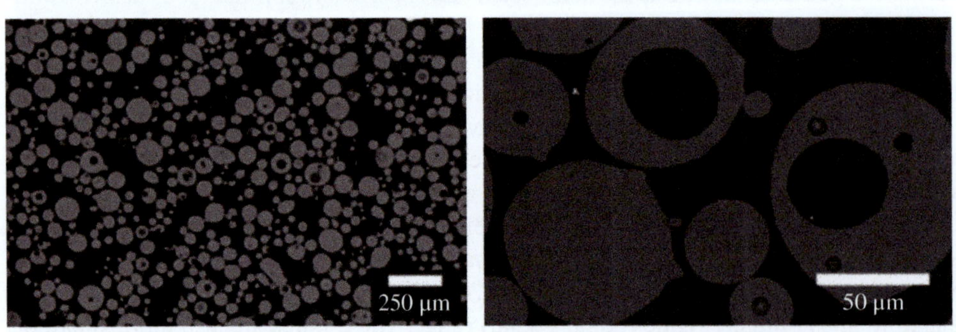

Figure 6.8 SEM micrographs of inert gas atomized MAR-M-247 powder in the size range of 45–106 μm, shown in polished cross-sections at low and high magnification, respectively. Source: Anderson et al. [98]/with permission of Elsevier.

be up to 28% (183/653) of the powders from one vendor, and 21% (121/575) of the powders provided by another vendor [102].

As described in previous publications [3, 110–112], there are several types of ligament and direct droplet formation mechanisms that lead to instabilities and droplet pinch-off without any apparent opportunity to trap atomization gas. Suppressing powder porosity formation requires the use of an alternate atomization mechanism with lower energy than that corresponding to the "bag breakup and collapse" mechanism described above [106]. However, it remains a challenge to develop gas atomization configurations and associated parameters that promote only these lower energy droplet formation mechanisms, while avoiding bag breakup and collapse. Some recent results from a gas-atomization experiment [98] appear to make progress toward the suppression of internal porosity by using atomization gas with reduced kinetic energy, where only 4% of the powders (13/304) had apparent trapped gas.

Porosity was found to consistently decrease with decreasing particle size which is in agreement with results from previous studies [113, 114]. The relationship between porosity and particle size is highly influenced by liquid breakup behavior [113], where larger ligaments tend to entrap larger amounts of gas, a phenomena which can be described by the bag breakup mechanism [106]. In one batch of gas-atomized powder, gas concentration and porosity in powder particles both tend to increase in a non-linear fashion with increasing particle size. There is a large increase in gas content and porosity near a particle size of about 75 μm [113], after which the quantities remain relatively constant, irrespective of alloy systems and gas atomization technique used. In addition to gas entrapment, a secondary contribution from fine inter-dendritic solidification shrinkage may also explain some finer voids [115–117].

The formation of pores in gas-atomized powders can be attributed to several factors [118]: (i) **Diffusion** of the dissolved gas from the molten metal to the atmosphere during cooling. (ii) **Cavitation** induced by supersonic or subsonic atomization conditions. (iii) Formation of shrinkage cavities during solidification (excluding amorphous materials). (iv) Gas entrapment during the formation and recombination of droplets. The combination of these factors induces the formation and growth of pores, which depends on the alloys, temperature of the molten metals, and atomization conditions.

Based on "bag breakup and collapse" mechanism [106] described above, the use of a low-pressure/energy gas during spraying could reduce the gas entrapment during the breakup of the molten-metal stream and subsequent quenching during droplet flight. However, a decrease in the gas pressure causes a decrease in the breakup force, which results in larger atomized powder particles and a higher proportion of irregularly shaped particles.

Gas bubbles in a melt can be removed via **diffusion**, given enough time and thermal energy. Similarly, the gas bubbles inside of molten metal droplets, including those originating from gas entrapment or gas discharge from the molten metal, can be removed with enough time available prior to solidification. The presence of an oxide film on the particle surface may hinder the **diffusion** and removal of the internal gas [118]. This is particularly challenging as cooling and solidification occur rapidly after droplet formation. Full solidification prior to complete gas removal will result in the formation of internal pores. Minimizing the formation of surface oxide layers could remove the gas from the droplet, resulting in a lower total volume of internal pores.

In related studies, Yodoshi et al. [118] investigated the phenomenon of pore formation in gas-atomized amorphous $Fe_{76}Si_9B_{10}P_5$ powder particles under different atomization. The powder porosity was measured using synchrotron radiation X-ray computed tomography, which allowed the detection of small pores with diameters <10 μm. The quantification of the porosity, i.e. the pore diameter and volume ratio, under different gas atomization conditions were determined. They found the volume ratio of the pores with the use of low-pressure/energy Ar as the atomization gas was lower than that with the use of high-pressure Ar. The use of a low-pressure gas during atomizing results in the increase of atomized particle size. When using Ar-3 or 7% H_2 mixtures as the atomizing gas, the number, and volume of pores were also decreased without affecting the particle size and cooling rate. The presence of H_2 as a reducing gas can suppress the surface oxidation of the droplets during gas atomization, which can allow the trapped gas bubbles to be efficiently removed before solidification. Both studies demonstrated that the pore fraction in gas-atomized powder can be decreased using low-pressure and H_2-containing atomizing gas.

Figure 6.9 illustrates the mechanism of pore formation inside of a powder induced by gas entrapment [118]. Gas atomization involves the primary breakup of a molten metal stream under a high-pressure gas, and the subsequent stripping breakup of large droplets into relatively small droplets [40, 110]. When small droplets collide with large droplets, they can coalesce into larger ones, which may result in gas entrapment. The trapped gas forms a spherical bubble and could be discharged to the exterior of droplets. The droplet surface can be oxidized due to the presence of O_2 in the atomizing gas, resulting in a hard and slaggy (highly viscous) oxide film. The oxide film can inhibit the

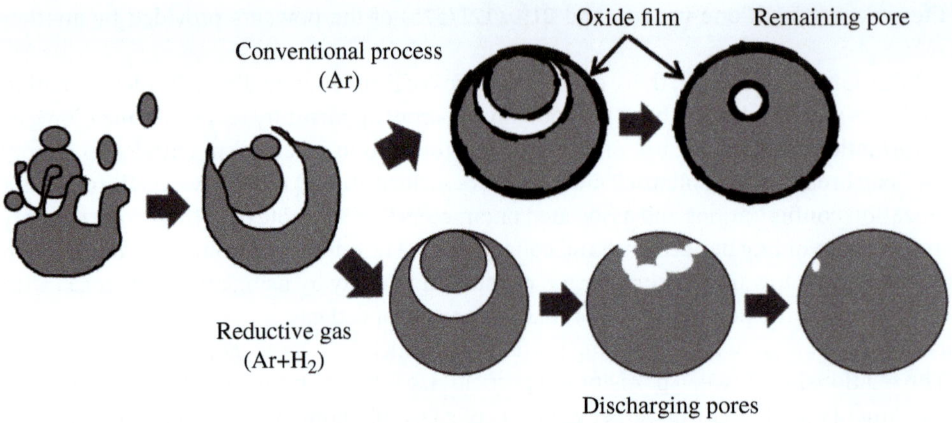

Figure 6.9 Schematic illustration of the pore formation/elimination mechanism in the atomized powders. Source: Yodoshi et al. [118]/Japan Society of Powder and Powder Metallurgy.

formation of spherical particles and the removal of trapped gas during the collision and coalescence process. When the atomization gas contains a reducing gas, such as H_2, the formation of surface oxides could be minimized, and the droplet surface remains relatively clean during cooling. This allows readily coalescence of the droplets and removal of gas to occur before the particle is solidified due to the decreased surface tension. Because gas atomization involves rapid cooling in an extremely short time, the slight difference in the deformation resistance of melt provided by the surface oxidation could have significant effects on the final quantity of pores in the atomized powder [118].

Suppressing **powder porosity** formation in gas-atomized powders remains a significant challenge that impedes the development and processing of novel materials and applications in AM processing. In fact, many questions remain regarding the precise mechanisms that govern the formation of internal pores during GA, and accordingly, reliable, and reproducible approaches that can be implemented to eliminate pores from GA have yet to be established.

6.9 Summary

There is a need to understand and assess our knowledge about atomized powder characteristics and their impact on the AM process. The powder characteristics, such as morphology, particle size and distribution, **flowability**, satellites, and porosity, are important to ensure powder performance in AM processing and the built components. When producing atomized powders for AM, the following factors of gas atomization should be optimized:

- Process parameters, such as gas-to-metal mass flow ratios, gas pressure and **chemistry**, atomization environment, atomizer, and chamber should be optimized in terms of both powder quality (i.e. solidification microstructure, contaminants, etc.), **powder size** and distribution, morphology, **powder satellites**, and porosity.
- The selection of the above factors needs to consider costs and thereby balance the need for optimal powder characteristics at minimum cost.

Nomenclature

Variable	Description	Units
A	The projected area of particle	μm^2
a	Major axis length of Legendre ellipse of particle	μm
b	Minor axis length of Legendre ellipse of particle perpendicular to a	μm
D_n	Melt stream diameter	mm
d	Droplet size	μm
d_D	Droplet diameter	μm

Variable	Description	Units
$\dot{G}$	gas flow rates	$kg\,s^{-1}$
g	Gravitational acceleration	$m\,s^{-2}$
L	Major axis length connecting the two most distant points on the projection area A	μm
l	Minor axis length perpendicular to the major axis	μm
$\dot{M}$	melt flow rates	$kg\,s^{-1}$
P	Perimeter of particle	μm
$p(d)$	Mass probability density function	
R	The mean radius of the particle	μm
r_1	The radius of curvature of the sharpest developed edge of a particle	μm
r_2	The radius of curvature of the most convex direction of a particle	μm
X_{84}	Cumulative 84 weight percent undersize	
X_{50}	Cumulative 50 weight percent undersize	
U_{rel}	Relative velocity	$m\,s^{-1}$
σ_l	The surface tension of the liquid	MPa
η_g	Kinematic viscosity of the gas	$m^2\,s^{-1}$
η_m	Kinematic viscosity of the melt	$m^2\,s^{-1}$
ρ_g	Gas density	$kg\,m^{-3}$

References

1 Bradley, D. (1973). *J. Phys. D: Appl. Phys.* 6: 1724.
2 Bradley, D. (1973). *J. Phys. D: Appl. Phys.* 6: 2267–2272.
3 Klar, E. and Fesko, J.W. (1984). *Atomization, Metals Handbook-Ninth Edition*, vol. 7, 25–51. Metals Park, OH: American Society for Metals.
4 Hirschhorn, J.S. (1969). *Introduction to Powder Metallurgy*. New York: American Powder Metallurgy Institute.
5 Gummeson, P.U. (1972). *Powder Metall. Int.* 15: 67–94.
6 Singer, A.R.E., Coombs, J.S., and Leatham, A.G. (1974). *Modern Developments in Powder Metallurgy*, vol. 8 (ed. H.H. Housner and W.E. Smith), 263. New York: Plenum.
7 Tallmadge, J.A. (1978). *Powder Metallurgy Processing, New Techniques and Analyses* (ed. H. Kuhn and A. Lawley), 1. New York: Academic Press.
8 Lubanska, H. (1970). *J. Metals* 22: 45–49.
9 Kim, K.Y. and Marshall, W.R. (1971). *AlChe* 17: 575.
10 Rai, G., Lavernia, E., and Grant, N.J. (1985). *J. Metals* 37: 17–29.
11 Baram, J. (1988). *Mater. Sci. Eng.* 98: 65–69.
12 Couper, M.J. and Singer, R.F. (1984). Rapidly solidified aluminum alloy powder produced by optimization of the gas atomization technique. In: *Proceedings of 5th International Conference on Rapidly Quenched Metals* (ed. S. Steeb and H. Warlimont). North Holland, Amsterdam.
13 Veistinen, M., Lavernia, E.J., Baram, J., and Grant, N.J. (1989). *Int. J. Powder Metall.* 25 (2): 89–92.
14 Baram, J., Veistinen, M., Lavernia, E.J. et al. (1988). *J. Mater. Sci.* 23: 2457–2463.
15 Thompson, J.S. (1948). *J. Inst. Metal* 74: 101.
16 Davies, M.G. and Olderfield, D.E.S. (1962). *Acustica* 12: 257.
17 Glass, D.R. (1986). *A/AA J.* 6: 1890.
18 Szekely, J. and Themelis, N.J. (1971). *Rate Phenomena in Process Metallurgy*. New York: Wiley-Interscience.
19 Weiss, M.A. and Worsham, C.H. (1959). *ARS J.* 4: 252–259.
20 Nukiyama, S. and Tanasawa, Y. (1938). *J. Trans. Soc. Mech. Eng. (Japan)* 4: 86–93.

21 Kim, M.H. and Jones, H. (1981). Effect of process variables in gas-jet atomization and production of multilayer deposits. In: *Proceedings of 4th International Conference on Rapidly Quenched Metals*. Sendai, Japan.
22 See, J.B. and Johnston, G.H. (1978). *Powder Technol.* 21: 119–133.
23 Naida, Y.I., Nichiporenko, O.S., and Fen'Ko, M.A. (1975). *Sov. Powders Metall. Met. Ceram., translated from Prorshkovaya Metall.* 11 (442): 94–97.
24 Naida, Y.I. (1971). *Sov. Powder Metall. Met. Ceram., translated from Poroshkovaya Metall.* 3 (21): 187–190.
25 Lawley, A. (1977). *Int. J. Powder Metall. Powder Technol.* 13: 169.
26 Anderson, A.R. and Johns, F.R. (1955). *Jet Propul.* 25: 13–15.
27 Mottern, J.W. (1978). *AlChE Symp. Ser.* 24: 75–81.
28 Anderson, I.E., Figliola, R.S., and Morton, H. (1991). *Mater. Sci. Eng.* 148: 101–114.
29 Amberg, L. (1984). *Scand. J. Metall.* 13: 391–392.
30 Mizutani, Y., Uga, Y., and Nishimoto, T. (1972). *Bull. Lap. Soc. Mech. Eng.* 15: 620–627.
31 Domalavage, P., Ashdown, C., O'Handley, R.C., and Grant, N.J. (1983). *Mater. Sci. Eng. Lett.* 57: Ll–L2.
32 Anand, V., Kaufman, A.J., and Grant, N.J. (1980). *Rapid Solidification Processing Principles and Technologies* (ed. R. Mehrabian, B. Kear, and M. Cohen), 273–286. Baton Rouge, LA: Claitors.
33 Backmark, U., Backstrom, N., Amberg, L. (1985). Production of Metal Powder by Ultrasonic Gas Atomization. Swedish Institute for Metals Research Report no. IM 2051.
34 Lewis, H.C., Edwards, D.G., Goglia, M.J. et al. (1948). *Ind. Eng. Chem.* 40: 67–74.
35 Putimtsev, B.N. (1968). *Sov. Powder Metall. Met. Ceram.* 11: 839–844.
36 Putimtsev, B.N. (1972). *Sov. Powder Metall. Met. Ceram.* 11: 171–175.
37 Hall, E.J. (1924). U.S. Patents 1,501,449, and (1925) 1,545,253.
38 Klar, E. and Shafer, W.M. (1975). *Int. J. Powder Metall. Powder Technol.* 11: 241–245.
39 Small, S. and Bruce, T.J. (1968). *Int. J. Powder Metall.* 4: 7–17.
40 Lavernia, E.J., Srivatsan, T.S., and Rangel, R.H. (1992). *J. Atomiz. Spr.* 2: 253–274.
41 Lavernia, E.J. and Wu, Y. (1996). *Spray Atomization and Deposition*. John Wiley & Sons, Inc.
42 Ayers, J.D. and Anderson, I.E. (1985). *JOM* 37: 16–21.
43 Ünal, A. (1989). *Metall. Trans. B* 20: 613–622.
44 Ünal, A. (1989). *Metall. Trans. B* 20: 61–69.
45 Miller, S.A., Miller, R.S., Mourer, D.P., and Christensen, R.W. (1997). *Int. J. Powder Metall.* 33: 37.
46 Strauss, J.T. (1999). *Met. Powder Rep.* 54: 24–28.
47 Strauss, J. (2000). Improvements in close-coupled atomization: an Empirical Approach. TMS Annual Meeting in Nashville, Tennessee, p. 83.
48 Ting, J., Peretti, M.W., and Eisen, W.B. (2002). *Mater. Sci. Eng., A* 326: 110–121.
49 Ünal, A. (1987). *Mater. Sci. Technol.* 3: 1029–1039.
50 Janowski, G.M., Biancaniello, F.S., and Ridder, S.D. (1992). *Metall. Trans. A* 23: 3263–3272.
51 Zheng, B., Lin, Y., Zhou, Y., and Lavernia, E.J. (2009). *Metal. Mater. Trans. B* 40B: 768–778.
52 Zheng, B., Lin, Y., Zhou, Y., and Lavernia, E.J. (2009). *Metal. Mater. Trans. B* 40B: 995–1004.
53 VanStone, R., Rizzo, F., and Radavich, J. (1980). *Proceedings of the 2nd International Conference on Rapid Solidification Processing, Principles and Applications*, 260. Reston, VA: Claitor's Publishing Division.
54 Liang, X. and Lavernia, E.J. (1993). *Mater. Sci. Eng., A* 161: 221–235.
55 Grant, P.S., Cantor, B., and Katgerman, L. (1993). *Acta Metall. Mater.* 41: 3097–3108.
56 Lubanska, K. (1970). *JOM* 22: 45–49.
57 Unal, A. (1989). *Metall. Trans. B* 20B: 61–69.
58 Crow, E.L. and Shimizu, K. (1988). *Lognormal Distributions: Theory and Applications*, 2. New York, NY: Marcel Dekker.
59 Smith, J.E. and Jordan, M.L. (1964). *J. Colloid. Sci.* 19: 549.
60 Lefebvre, A.H. (1989). *Atomizations and Spray*, 85. Bristol, PA: Taylor & Francis.
61 Lawley, A. (1992). *MPIF*, 74. NJ: Princeton.
62 Thompson, J.S. (1948). *J. Inst. Met.* 74: 101–132.
63 Allen, T. (1981). *Particle Size Measurement*, 3e. London: Chapman Hall.
64 Klar, E. and Shafer, W.M. (1972). *Powder Metallurgy for High Performance Applications* (ed. J.J. Burke and B. Weiss), 57–68. Syracuse, NY: Syracuse University Press.

65 Rao, P. (1973). Shape and other properties of atomized metal powders. Ph.D. thesis, Drexel University, Philadelphia.

66 Marinkovich, J., Mohamed, F.A., Pickens, J.R., and Lavemia, E.J. (1989). *J. Metals* 41: 36–34.

67 Dombrowski, N. and Hooper, P.C. (1962). *Chem. Eng. Sci.* 17: 291–305.

68 Grant, N.J. (1972). *Powder Metallurgy for High Performance Applications* (ed. J.J. Burke and B. Weiss), 88–97. Syracuse, NY: Syracuse University Press.

69 Fraser, R.P., Dombrowski, N., and Routley, J.H. (1963). *Chem. Eng. Sci.* 18: 339–353.

70 Ahsan, M.N., Pinkerton, A.J., Moat, R.J., and Shackleton, J. (2011). *Mater. Sci. Eng., A* 528: 7648–7657.

71 Lohner, H., Fritsching, U., Bauckhage, K.A. (2002). *Melt Atomization by Heated Gases*. 'ILASS-Europe 2002, Zaragoza 9–11 Sep. 2002'

72 Li, X.-g., Zhu, Q., Shu, S. et al. (2019). *Powder Technol.* 356: 759–768.

73 Schwenck, D., Ellendt, N., Fischer-Bühner, J. et al. (2013). Effect of process parameters on powder quality. In: *SDMA. 5th International Conference on Spray Deposition and Melt Atomization*, 12. Bremen, Germany; Bremen: SDMA.

74 Gao, C.-f., Xiao, Z.-y., Zou, H.-p. et al. (2019). *Trans. Nonferrous Met. Soc. China* 29: 374–384.

75 Vlachos, N. and Chang, I.T.H. (2011). *Powder Technol.* 205: 71–80.

76 Özbilen, S. (1999). *Powder Metall.* 42: 70–78.

77 Goudar, D.M., Srivastava, V., and Rudrakshi, G. (2017). *Eng. J.* 21: 155–168.

78 M.I. Ltd. (2013). Morphologi G3 user manual. Enigma Business Park, Malvern, UK.

79 Cui, C., Uhlenwinkel, V., Schulz, A., and Zoch, H.-W. (2020). *Metals* 10: 61.

80 I. ISO. (2008). ISO. Geneva, p. 23.

81 Araújo, A.P.M.d., Micheloti, L., Kiminami, C.S. et al. (2021). *Tecnol. Metal. Mater. Min.*, vol. 18: e2316.

82 ASTM B243-13. (2014). Standard Terminology of Powder Metallurgy.

83 Barrett, P.J. (1980). *Sedimentology* 27: 291–303.

84 Sutton, A., Kriewall, C.S., Leu, M.C., and Newkirk, J.W. (2016). *The Solid Freeform Fabrication 2016, Proceedings of the 26th Annual International Solid Freeform Fabrication Symposium – An Additive Manufacturing Conference.* University of Texas at Austin.

85 Cox, E.P. (1927). *J. Paleontol.* 1: 179–183.

86 Schneiderhöhn, P. (1954). *Heidelberger Beiträge Zur Mineral. Und Petrogr.* 4: 172–191.

87 Mikli Valdek, E., Kaerdi, K.H., and Pritt, K. (2001). Characterization of powder particle morhpology. *Proc. Estonian Acad. Sci. Eng.* 7: 22–34.

88 Wentworth, C. (1922). *U.S. Geol. Surv. Bull.* 39: 91–114.

89 Janoo, V.C. (1998). Quantification of shape, angularity, and surface texture of base course materials. *Special Report - Corps Eng. U.S. Army, Cold Regions Research and Engineering Laboratory* 98: 22.

90 Hoeges, S., Zwiren, A., and Schade, C. (2017). *Met. Powder Rep.* 72: 111–117.

91 Fedina, T., Sundqvist, J., Powell, J., and Kaplan, A.F.H. (2020). *Addit. Manuf.* 36: 101675.

92 Boisvert, M., Christopherson, D., Beaulieu, P., and L'Espérance, G. (2017). *Mater. Des.* 116: 644–655.

93 Schade, C.T., Murphy, T.F., and Walton, C. (2014, Metal Powder Industries Federation, Orlando, Florida). Development of atomized powders for additive manufacturing, World Congress. *Powder Metall. Part. Mater.* 2014: 215–226.

94 Lawley, A. (1992). *Atomization, the Production of Metal Powder*. Princeton, New Jersey: MPIF.

95 Dunkley, J.J. (1998). *Atomization, ASM Handbook, Powder Metal Technologies and Applications*, vol. 7. Materials Park, OH: ASM International.

96 Dunkley, J.J. and Telford, B. (2002, Metal Powder Industries Federation, Princeton, NJ). Control of "Satellite" particles in gas atomization, PM2TEC 2002 World Congress. *Adv. Powder Metall. Part. Mater.* 2002: 3-103–103-110.

97 Zheng, B., MacDonald, B.E., Belcher, C.H. et al. (2023). Microstructure evolution and mechanical behavior of Ni-NiAl functionally integrated materials (FIMs) processed via directed energy deposition (DED), TMS 2023. San Diego, CA USA.

98 Anderson, I.E., White, E.M.H., Tiarks, J.A. et al. (2017). Fundamental progress toward increased powder yields from gas atomization for additive manufacturing. In: *Advances in Powder Metallurgy and Particulate Materials-2017* (ed. R. Goto and J.T. Strauss), 136–146. Princeton, NJ: Metal Powder Industries Federation.

99 Özbilen, S., Ünal, A., and Sheppard, T. (1991). *Powder Metall.* 34: 53–61.

100 Ünal, A. (1989). *Mater. Sci. Technol.* 5: 1027–1033.

101 Mostaghimi, F., Fischer-Bühner, J., Heemann, L. et al. (2018). Anti-satellite system for the improvement of powder quality in additive manufacturing using a FeMnAlSi alloy. In: *Proceedings of the Euro PM2018*. European Powder Metallurgy Association, Bilbao, Spain (Bilbao: EPMA).

102 Anderson, I.E., White, E.M.H., and Dehoff, R. (2018). *Curr. Opin. Solid State Mater. Sci.* 22: 8–15.

103 Byrd, D.J., Tiarks, J., Riedemann, T. et al. (2017). Verification of gas atomization results with in-situ particle size analysis. In: *Advances in Powder Metallurgy and Particulate Materials-2017* (ed. R. Goto and J.T. Strauss), 115–126. Princeton, NJ: Metal Powder Industries Federation.

104 He, X.-y., Xing-gang, L.I., Huang, Y.-h., and Zhu, Q. (2022). *Powder Metall. Technol.* 40 (4): 302–317.

105 Susan, D.F., Puskar, J.D., Brooks, J.A., and Robino, C.V. (2006). *Mater. Charact.* 57: 36–43.

106 Rabin, B.H., Smolik, G.R., and Korth, G.E. (1990). *Mater. Sci. Eng., A* 124: 1–7.

107 Boettinger, W.J. and Perepezko, J.H. (1985). Fundamentals of rapid solidification. In: *Rapidly Solidified Crystalline Alloys* (ed. S.K. Das, B.H. Kear, and C.M. Adam), 21–58. Warrendale, PA: The Metallurgical Society, Inc.

108 Zhong, C., Chen, J., Linnenbrink, S. et al. (2016). *Mater. Des.* 107: 386–392.

109 Ahsan, M.N., Bradley, R., and Pinkerton, A.J. (2011). *J. Laser Appl.* 23: 022009.

110 Yule, A.J. and Dunkley, J.J. (1994). *Atomization of Melts: for Powder Production and Spray Deposition*. USA: Oxford University Press.

111 Mates, S.P., Ridder, S.D., and Biancaniello, F.S. (2000). *Comparison of the supersonic length and dynamic pressure characteristics of discrete-jet and annular close-coupled nozzles used to produce fine metal powders*. TMS Annual Meeting in Nashville, Tennessee, 71–81.

112 Dombrowski, N. and Johns, W.R. (1963). *Chem. Eng. Sci.* 18: 203–214.

113 Chen, G., Zhou, Q., Zhao, S.Y. et al. (2018). *Powder Technol.* 330: 425–430.

114 Cheng, M., Wu, J., Lu, Z.-G. et al. (2021). *Acta Metall. Sinica (English Letters)* 34: 1386–1394.

115 Tokarz, T. and Radavich, J.F. (1982). *Microstruct. Sci.* 10: 51–56.

116 Eaton, H.E. and Bornstein, N.S. (1978). *Metall. Trans. A* 9: 1341–1342.

117 Backmark, U., Backstrom, N., and Arnberg, L. (1986). *Powder Metall. Int.* 18: 422–424.

118 Yodoshi, N., Endo, T., and Masahashi, N. (2021). *Mater. Trans.* 62: 1549–1555.

7

Atomization of Alloys

7.1 Introduction

In a typical commercial aircraft, **aluminum 2xxx** (Al–Cu) and **aluminum 7xxx** (Al–Zn–Mg) series comprise about 80% of the structural weight [1]. Not surprisingly, novel processing techniques such as additive manufacturing (AM) are being aggressively researched for applications in the aerospace and transportation industries. There is a broad variety of metals and alloys that are being considered as candidate materials for AM. Rather than attempting to provide a comprehensive overview of every single family of metal powders that is being used in AM processes, we will concentrate on the most widely used systems, namely, alloys based on Al, Fe, Ni, Ti, Co, and the new family of **high-entropy alloys (HEAs)**. The application of AM to a particular alloy family depends on the intended application, but the overall goal is typically to fabricate net-or near-net-shape components in a single step, thereby avoiding the complexities and costs associated with typical casting, plastic deforming, and machining operations. In some cases, it has been demonstrated that the performance of AM components can exceed those of the conventionally processed counterpart materials, an observation that is generally attributed to the relatively high rates of solidification that can be present during AM. What is interesting is the fact that most of the original work done to develop AM techniques was done using traditional **stainless steel** powders, which stands in stark contrast to the wide variety of alloys being considered today. In Sections 7.2–7.7 that follow, we describe published results obtained for atomized powders. In some cases, we provide detailed information on the performance of powders consolidated via various techniques, such as **spray deposition**, **extrusion**, etc., to illustrate the performance metrics that can be attained with these approaches, and hence the likely comparison goals of AM components. Moreover, the cooling rates that are attainable during AM ($\sim 10^2$–10^4 K s^{-1}) are similar to those present during spray deposition ($\sim 10^3$ K s^{-1}), which provides an interesting comparison between these two approaches.

7.2 Aluminum-Based Alloys and Powders

Aluminum (Al), being the second most abundant metallic element, possesses numerous favorable properties such as its appearance, lightweight, mechanical properties, corrosion resistance, and fabricability, which make its alloys a cost-effective choice for many applications. In the manufacturing industry, Al alloys have become increasingly popular over the past few decades due to their favorable properties. To obtain a more comprehensive understanding of their structural performance and to develop accurate and reliable designs, many research projects have been carried out.

The family of Al alloys can be categorized according to their form and main alloying elements, which include Si, Cu, Mg, Mn, Sn, Ni, and Zn. The two primary classifications of Al alloys are casting alloys and wrought alloys, which are further subdivided into the categories of heat-treatable and nonheat-treatable. Although Al alloys and powders have gained significant attention in the AM industry, their low laser absorption, high thermal conductivity, and reduced powder flowability have presented many challenges in the AM process. Currently, mainly Al–Si alloys are processed using AM technology. However, in recent years, numerous efforts have been made to develop new compositions and approaches, specifically designed for laser-based AM processes, to address the aforementioned problems and explore the potential application in advanced componentry.

Metallic Powders for Additive Manufacturing: Science and Applications, First Edition.
Enrique J. Lavernia, Kaka Ma, Julie M. Schoenung, James F. Shackelford, and Baolong Zheng.

7.2.1 Al-Based Alloy Powders

Despite the fact that a number of methods have been used to produce Al and Al alloy powders, including gas and **centrifugal atomization**, **ultrasonic** or **pulsed atomization**, **melt spinning** with attrition, and **mechanical alloying**, the majority of Al and Al alloy powders that are commercially available are produced by air or inert gas atomization (GA) [2].

The inert gas-atomized Al alloy powders can be produced with a particle morphology that is spherical, spheroidal, or nodular by varying the atomizing conditions and modifying the atomization process details. Generally, uniformity of composition and a fine microstructure are characteristics of powder particles as a result of the presence of a high cooling rate of 10^4–10^6 K s^{-1} [3]. The gases that are primarily used in inert gas atomization are Ar, He, and Ni. The properties of Al alloy powders change with variations in the atomization process parameters. In general, powders with a spherical morphology are better suited for AM, given their excellent flowability. However, spherical and occasionally nodular powders will also work, depending on the AM technology used, the powder flow properties, and other characteristics of the powder.

Commercial manufacturing of aluminum and aluminum alloy powders involves two types of atomization: atomization using air and atomization using inert gases such as Ar, N, or He (IGA). There are various types of atomization configurations that can be used for commercial manufacturing of powders, and these are described in Chapter 2; these can include, for example, horizontal or vertical configurations. The particular configuration depends often on whether production rate or particle size needs to be optimized [4]. In general, horizontal atomization systems can lead to faster rates of production but with coarser particle size distributions (PSDs).

In a typical atomization production run involving the spray chamber is first purged with dry air gas to minimize undesirable contaminants, such as water vapor. The molten metal is prepared in a holding furnace and transferred to the atomization melt chamber. The liquid metal is typically superheated to a level of 100–150 °C above the liquidus to avoid the possibility of the molten Al freezing in the pour tube before the melt exits into the spray chamber. Gas nozzles encircle the metal delivery tube and are arranged such that compressed dry air impinges upon the molten stream, disintegrating the liquid metal into droplets. Immediately following the melt disintegration, cooling air from the ambient atmosphere enters the spray chamber to cool the molten droplets and passivate the surface of the droplets with an oxide film [4]. Because fine Al powder forms an explosive mixture with air over a wide range of metal-to-air ratios, Al powder plants typically have explosion hazards associated with them. The intent of the passivation process is to purposely oxidize the Al powders to minimize powder explosions during postatomization powder handling, which are critical during commercial production when large amounts of powders are involved. The droplets are quenched by the gaseous atmosphere in the chamber and subsequently solidify as powder particles. These powder particles, together with a substantial volume of cooling air, are then drawn through a chiller chamber into the collection system consisting of either one or multiple sets of cyclones. After the cyclones, the powder is transported in an atmosphere of inert gas to screens and stored under inert gas into drums and bins or onto transportation systems, such as trucks or rail cars.

The process described above for air atomization is similar to that used during IGA of Al powders, but in this case, the spray chamber is isolated from the ambient air. Prior to atomization, the spray chamber of an IGA process is purged with an inert gas to remove residual gases. Sometimes, when passivation of the nascent Al metal surfaces is desirable, the purge gas will not completely remove oxygen and gaseous water from the spray chamber.

The various mechanisms that disrupt the stream of molten metal during atomization are described in Chapter 2, and their influence can appear as peaks in frequency distribution plots [5]. The PSD of atomized powders can be tuned by modifying process parameters and screen sizes to produce a wide range of products from Al granules (typically +200 mesh) to fine atomized powders (up to 99%, −325 mesh). Control of particle size during atomization can be achieved by the selection of the rate of atomization (metal delivery rate). In general, decreasing the metal flow rate leads to an increase in the average particle size. The median diameter is approximately proportional to the square root of the rate of atomization. As metal flow rate is increased and the powder becomes coarser for any given atomizing conditions, in general, the spread in powder sizes (as measured by σ_g) also increases.

The **morphology** of atomized Al powders is critically influenced by the amount of oxygen in the gas phase during atomization. Particles atomized in an inert gas show spherical features, whereas air-atomized particles are of irregular shape (also known as "nodular" powders). This effect of oxygen on morphology is related to the pinning effect of the oxide nuclei that form on the droplets before they have solidified [6]. A high density of such nuclei forms on a droplet's surface when there is a high level of oxygen in the gas phase leading to effective pinning at those points. Whereas at low oxygen levels or in inert gas, the rate of oxidation is slow and will not cause pinning, in which case surface tension forces can maintain the spherical shape of the droplet until solidification. Results obtained using gases with varying oxygen content have shown that the change in morphology takes place rather sharply at an oxygen level between 2% and 4%. Powders made in a gaseous atmosphere containing 2% oxygen or less show the same degree of **sphericity** as inert gas-atomized powders. When oxygen exceeds 4%, the deviation from **sphericity** approaches those levels observed in air-atomized powders. The dependence of particle morphology on oxygen content results from the distortion of the droplets by **oxide** formation on their surfaces.

Particle size and **superheat** also influence the degree of **sphericity** of powder particles. Smaller particles tend to be closer to spherical than larger particles manufactured under the same conditions. Another feature of atomized Al powders, essential in milling and other applications, is the surface quality of the particles. Finer powders generally have excellent surface quality. **Satellites** form by warm/cold welding of smaller particles onto large particles during atomization. If the collision occurs before the larger particle has solidified, the **satellite** particle will be embedded.

One of the significant challenges in the processing of Al powder is the surface oxide coating on the powder. Al reacts readily with moisture or free oxygen in the air during manufacture to form an oxide coating on the powder surface. The amount of surface oxide formed during atomization strongly depends on the oxygen level of the atomizing atmosphere and the particle size of the powder produced. IGA spherical powders contain substantially less oxygen than air-atomized powders. The surface oxide coating of Al powder is typically around 4 nm or higher in thickness and is a significant obstruction in liquid phase sintering [7]. The oxygen content of fine powders atomized in the air can approach 1% level by weight. For coarser powders, 0.4% oxygen is more typical. A mean oxide thickness can be calculated from the oxygen content of the powder if the specific surface area has also been determined. It is then found that the thickness of oxide is, in fact, greater for coarser powders despite the lower overall oxygen content. The finest powders, under 10 μm in size, show a mean oxide thickness of 4–5 nm, whereas, with coarser powders, the thickness can reach 10 nm. In P/M and metal injection molding (MIM) technologies, the presence of magnesium in small percentages helps to break up the oxide layer and aids the achievement of a better sintering process. Most of the research work on Al AM has involved the use of low melting point, cast Al alloy compositions. However, several other studies have also been conducted, using wrought alloys such as 6061 and 7075. The 6000 series alloy powders have outstanding properties including high ductility, machinability, thermal conductivity, electrical conductivity, and corrosion resistance. The 7000 series alloy powders, with a high zinc content, are known for their excellent mechanical properties and are heat treatable to achieve even higher strength.

7.2.2 Al–Si Alloys

Al–Si alloys exhibit a low coefficient of **thermal expansion**, excellent **wear resistance**, and relatively high **elastic modulus** owing to the presence of a relatively high-volume fraction of Si phases and represent perhaps the most commonly used Al alloys for AM, partly because of their low tendency to crack during processing. Improvement in such properties may be achieved through increasing Si content, especially to the **hypereutectic** composition regime. Unfortunately, **ingot** metallurgy (IM) aluminum alloys in the **hypereutectic** Si regime exhibit poor mechanical properties. The slow cooling rate conditions typical of the IM process give rise to coarse primary Si crystals in combination with a high-volume fraction of the **interdendritic** eutectic phase rendering inherent brittleness.

In view of the important drawbacks of IM, considerable research efforts have been directed toward the synthesis of **hypereutectic** Al–Si alloys through rapid solidification with noticeable refinement in microstructure [8–25]. Improvement in properties in rapidly solidified Al–Si alloys over those of IM materials is not only a result of microstructural refinement but also a result of the improvement in alloying ability. Al–Si alloys synthesized using rapid solidification are heavily alloyed. The typical compositions that have been investigated are summarized in Table 7.1 [8, 13, 14, 18–22]. The primary alloying elements include Si, Cu, Mg, Fe, and Ni. Si is primarily responsible for the low coefficient of **thermal expansion** and high wear resistance. Cu and Mg promote strengthening through solution and precipitation hardening. The precipitation hardening phases formed by Cu and Mg with Al, such as Al_2Cu, Mg_2Si, β-Al_2CuMg and $Cu_2Mg_8Si_5Al_4$, are relatively stable up to 150 °C beyond which extensive coarsening occurs. Elements such as Fe and Ni improve elevated temperature mechanical properties through precipitation of intermetallic phases, such as δ-Al_4FeSi_2 and β-Al_5FeSi that are relatively stable at temperatures up to 300 °C [18, 19].

Table 7.1 Chemical compositions of hypereutectic Al–Si alloys.

Si	Cu	Mg	Fe	Ni	References
16.0	—	—	—		[16]
20.0	—	—	—	—	[16]
24.0	—	—	—	—	[16]
20.0	3.0	1.0	—	—	[20]
20.0	3.1	1.3	0.3	—	[21]
20.0	—	—	5.0	—	[14]
20.0	—	—	5.0	5.0	[13]
20.0	3.0	1.0	5.0	—	[18]
20.0	3.0	1.0	—	7.5	[19]
17.0	4.5	0.6	—	—	[8]
17.5–19.0	4.2–4.5	1.0–1.2	—	1.4–1.6	[22]

Al–Si is, perhaps, one of the alloy systems that best illustrate the potential of **spray deposition** processing for microstructural modification and mechanical property improvement. Direct comparison of the microstructures of IM and spray-deposited 390 alloys revealed that the coarse Si blocks and eutectic Al–Si phase that are generally associated with IM materials were suppressed in the spray-deposited microstructure [23]. A similar modification in Si-containing phases is generally reported for spray-deposited hypereutectic Al–Si alloys [8, 14, 21, 22, 24, 25]. The size of the Si particles in spray-deposited materials falls between those associated with IM and P/M materials. The results of Duszczyk et al. [21] on extruded Al–Si alloys, for example, demonstrated that the sizes of Si particles in IM, spray-deposited, and P/M materials were 70, 5–11, and 2–7 μm, respectively (see Table 7.2). Considerable improvements in strength and ductility through **spray deposition** and powder metallurgy processing relative to **ingot** casting have been reported for both conventional alloy compositions and for novel experimental alloys (see Table 7.3) [8, 14, 16–22]. For example, as summarized in Table 7.3, the strength and ductility of the as spray-deposited 390 Al alloy are equivalent to those of the IM 390 Al alloys of identical chemical composition despite the presence of porosity in the former material (<2.4%). In the as-extruded condition, the strength of the spray-deposited 390 Al is lower than that of the RS-P/M Al–Si alloys of similar composition. The ductility of the spray-deposited alloy is,

Table 7.2 Size of Si crystals in Al–Si alloys.

Alloy (wt%)	Processing	Phase constitution	Size of Si (μm)	References
Al–17Si–4.5Cu–0.6Mg (390 Al)	IM	Primary Si	80	
		Eutectic Si	—	[8]
	SD	Primary Si	5.0	[8]
Al–20Si–5Fe	SD	Primary Si	2.0	[18]
Al–20Si–5Fe–3Cu–lMg	RS-P/M	Primary Si	5.0	
		Eutectic Si	0.3	[18]
Al–20Si–3Cu–lMg	RS-P/M	Primary Si	5.0	
		Eutectic Si	1.0	[20]
Al–20Si–3Cu–lMg	RS-P/M	Si	2–7	[21]
	IM	Si	70	[21]
390 Al	SD	Si	5–11	[21]
A1–12Si	PSD	Primary Si	<0.5	[16]

IM, ingot metallurgy; RS-P/M, rapid solidification powder metallurgy; SD, spray deposition; PSD, plasma spray deposition.

Table 7.3 Room temperature mechanical properties of Al–Si alloys.

Alloy	Temper	σ_{ys} (MPa)	σ_{UTS} (MPa)	El (%)	References
Al–20Si (RS-P/M)	Extruded	112	206	10.0	[16]
Al–20Si (PSD)	As sprayed	—	250	0.5	[16]
Al–20Si–5Fe (SD)	Extruded	280	—	1.6	[14]
Al–20Si–3Cu–lMg (RS-P/M)	Extruded	262	360	3.9	[20]
	Extruded + T6	321	430	3.7	[20]
Al–20Si–3Cu–lMg–5Fe (RS-P/M)	Extruded	—	510	—	[12]
390 Al (IM, Acurad)	As cast	240	280	<1.0	[26]
	As cast + T6	365	365	<1.0	[26]
A390 Al (IM, permanent mold)	As cast	200	200	<1.0	[26]
	As cast + T6	310	310	<1.0	[26]
A390 Al (IM)	Extruded + T6	374	387	1.3	[23]
A390 Al (SD)	As sprayed	218	276	0.9	[8]
	Extruded	233	328	7.2	[8]
	Extruded + T6	506	532	3.0	[23]

IM, ingot metallurgy; RS-P/M, rapid solidification powder metallurgy, SD, spray deposition; PSD, plasma spray deposition.

however, significantly improved over that of both RS-P/M alloys and the IM alloys. For example, the elongation of the as-extruded Al–17Si–4.5Cu–0.6Mg listed in Table 7.3 is 7.2%, whereas that for extruded RS-P/M Al–20Si–3Cu–l Mg is 3.9%. In the T6 temper, the strength and ductility of spray-deposited 390 Al are significantly improved over that of the IM material. For example, the results in Table 7.3 indicate that the yield strength and ductility of spray-deposited 390 are 506 MPa and 3.0%, whereas those of the IM 390 are 374% and 1.3%. These improvements were rationalized on the basis of the microstructural refinement and homogeneous distribution of alloying elements that are associated with the spray-deposited materials [23].

Recently, developed alloy compositions have also been the subject of study using **spray deposition**. Some notable examples in this area are Al–20Si–X (X = Fe, Ni), Al–20Si–3.1Cu–1.3Mg, and N403 (Al–17.5–19Si–4.2–4.5Cu–1.4–1.6Ni–1.0–1.2Mg–1.0Fe (Mn, Cr, Ti, Zr_5 V) wt%) [14, 15, 21, 22, 24, 25]. Alloy N403 is designed to exhibit high tensile and **fatigue** strength at both room and elevated temperatures [22, 24, 25]. The elevated temperature strength is acquired through the precipitation of **transition metal element**-containing phases. Available results on these experimental alloys suggest significant improvements in microstructure, elevated temperature strength, fatigue strength,

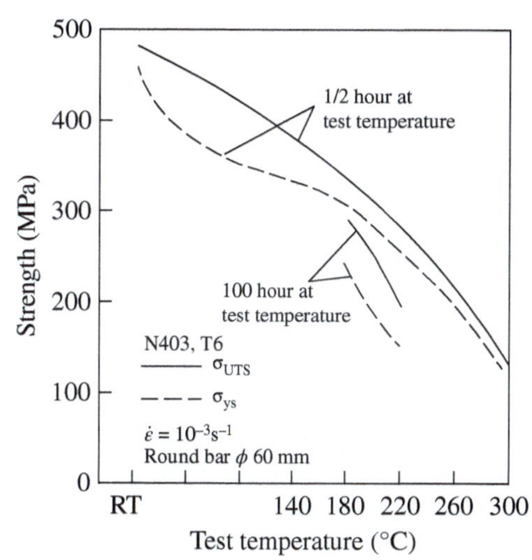

Figure 7.1 Elevated temperature strength of spray-deposited N403. Source: Reproduced by permission of Alusuisse-Lonza Services AG [22].

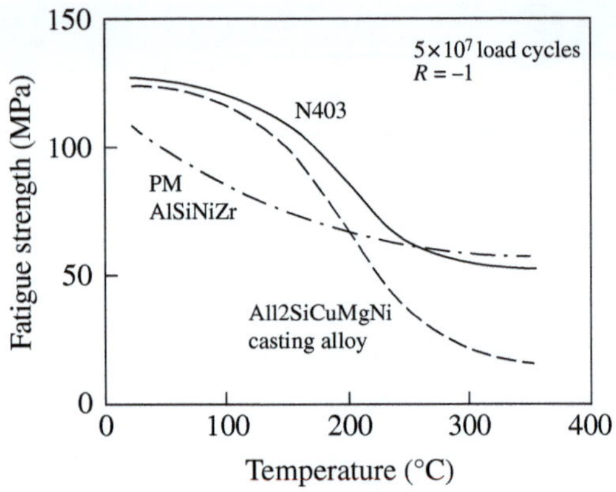

Figure 7.2 Fatigue behavior of N403 (a spray-deposited alloy) as compared with that of IM and P/M Al–Si alloys as a function of testing temperature. Source: Reproduced by permission of Alusuisse-Lonza Services AG [22].

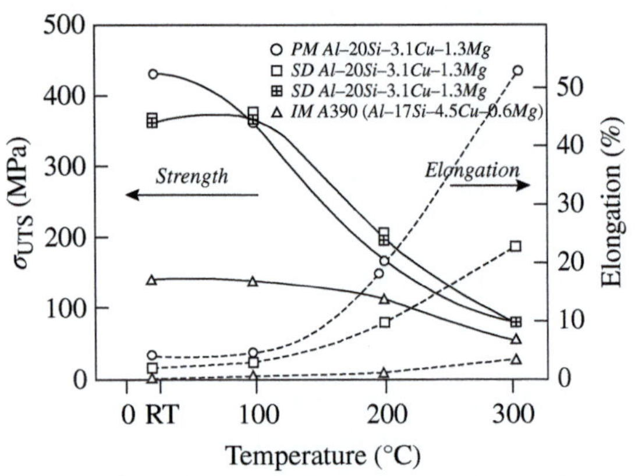

Figure 7.3 Tensile properties of Al–Si alloys; IM, ingot metallurgy; P/M, powder metallurgy (extruded); and SD, spray deposition (extruded). Source: Reproduced by permission from Duszczyk et al. [21].

and wear resistance through rapid solidification processing, i.e. **spray deposition** and RS-P/M, relative to ingot casting [22, 24, 25]. The spray-deposited N403 alloy, for example, exhibits an excellent combination of tensile and fatigue strength. As summarized in Figure 7.1, the yield strengths of this alloy at room temperature and 180 °C are 440 and 310 MPa, respectively [22]. Moreover, the 5×10^7 cycle fatigue strength is significantly higher than that of a P/M AlSiNiZr alloy at temperatures less than 200 °C (see Figure 7.2). It is also evident that the fatigue strength of spray-deposited N403 is much higher than that of an optimized AlSiCuMgNi casting alloy [22]. In related studies, Duszczyk et al. [21] also noted significant improvement in the tensile properties of spray-deposited Al–Si alloys relative to those of IM materials (see Figure 7.3). The results from the same study show that spray-deposited Al–Si alloys exhibit the best wear resistance. The wear loss of spray-deposited materials is much lower than those of P/M and IM materials of similar or identical compositions [21]. In view of the fact that the size of the Si particles in the spray-deposited Al–Si was larger than that in the RS-P/M material, the higher wear resistance of the spray-deposited material relative to that of the RS-P/M material was rationalized using the argument that a slightly larger particle size may contribute to resist fretting wear and that the presence of oxide stringers in the RS-P/M material may be detrimental to wear properties [21].

7.2.3 Al–Cu Alloys

Commercial **aluminum 2024** contains copper, iron, manganese, chromium, silicon, and other elements from trace to significant amounts, which precipitate as coarse (20–60 μm) brittle constituent phases during slow **ingot** solidification. Such impurity-related phases are normally present in the last zone to solidify and are detrimental to mechanical properties. **Rapid quenching**, however, suppresses the segregation of elements and formation of coarse intermetallics, leading to improved mechanical behavior [27]. Table 7.4 summarizes the tensile data for a commercial Al alloy 2024 ingot product as compared with those of rapidly solidified splat-quenched and spray-deposited materials consolidated by hot extrusion [26–29]. The mechanical properties of the rapidly solidified splat alloy were improved over those of the ingot product, and the properties of the alloy obtained by spray deposition showed further improvements over the other two. The improved mechanical properties of the rapidly solidified materials resulted from the

Table 7.4 Comparison of mechanical properties of extruded Al 2024 alloys produced using various methods.[a)]

Method	σ_{ys} (MPa)	σ_{UTS} (MPa)	El (%)	RA (%)	K_{IC} (MPa m$^{1/2}$)	References
SD + Ext.	417	586	16	23	31	[27]
SD + Ext.	402	562	15	22	31	[27]
SQ + Ext.	351	501	24	24	—	[28]
IM + Ext.	310	441	20	—	—	[29]
IM + Ext.	329	469	19	—	31	[26]

a) Properties determined in the T4 condition.
SD, spray-deposited; SQ, splat quenched; IM, ingot cast; Ext, extruded; σ_{ys}, yield stress; σ_{UTS}, ultimate tensile strength; EI, elongation index; RA, reduction in area; and K_{IC}, plane-strain fracture toughness.

substitution of coarse segregated phases in the conventional material by Al_3Fe, Al_6Mn, and Al_7Cr, all present as dispersion strengtheners in the rapidly solidified materials [26–29]. The improvements realized in the spray-deposited material as compared with the splat quenched material were attributed to minimization of **oxide** stringers and unbroken oxide films in the former material [26–29].

The microstructural refinement (i.e. grain sizes <50 μm and refined intermetallic phases) and enhanced alloying flexibility that are characteristic of **atomized materials** have led to the development of novel aluminum alloy compositions containing elevated amounts of iron, nickel, cerium, zirconium, chromium, cobalt, and manganese [30, 31]. Examples in this area are the N202 alloy (Al–6.1–6.5Cu–1.7–1.9Mn–0.4–0.5Mg–0.9–1.1 Zr(V, Ti) wt%) and the family of ES2x alloys (Al–5–6Cu–0.5–0.8Mn–0.5Mg–0.5Ag–1.0Zr (V, Ti) wt%) [32, 33]. Transition elements are introduced to take advantage of the refinement in precipitate size and the enhancement in metastable solid solubility. The N202 alloy, for example, was designed as a heat-treatable AlCuMn alloy, which exhibits high strength under both low and high cycle fatigue conditions.

Moreover, the addition of transition metal elements such as Zr, Ti, and V leads to high strength at both room and elevated temperatures up to 210 °C [32]. The ES2x alloys were developed based on N202 to produce 2xxx alloys with improved room and elevated temperature strength.

7.2.4 Al–Transition Metal Alloys

Al–transition metal (Al–TM) alloys contain **transition metal elements**, such as Fe, V, and Cr, and sometimes **rare earth** elements, such as Ce, as major alloying elements. These alloying elements form dispersion strengthening phases with A1 during solidification. The dispersion strengthening phases are thermally stable because of the low solubility and diffusivity of transition and **rare earth** elements in Al. Consequently, the temperature capability of such alloys may be extended beyond the limit of conventional Al alloys. Owing to the low solubility of transition elements, however, such alloys may be synthesized only using rapid solidification methods [26, 30–41]. Currently, there are two **rapid solidification powder metallurgy (RS-P/M)** technologies that have been successfully applied in this alloy system. In the first method, the starting alloys are rapidly solidified by **planar flow casting (PFC)** into ribbons. The ribbons are then comminuted into powders and densified into a bulk material by powder metallurgical methods such as **extrusion**, rolling, or forging. In the second method, rapidly solidified powders are produced by gas atomization, followed by densification procedures that are similar to that used for ribbons. Certain Al–TM alloys such as Al-Fe–V–Si and Al–Fe–Ce have also been produced using spray deposition to reduce the number of processing steps and to mitigate the oxygen contamination that is commonly associated with RS-P/M techniques [42–44]. Although spray-deposited Al–TM alloys exhibit high ductility, their yield and ultimate tensile strengths have been noted to be significantly lower than those of the materials produced using RS-P/M techniques (see Table 7.5) [42–44].

7.2.5 Al–Li Alloys

The density of aluminum is reduced by about 3% for each weight percent addition of **lithium**, while **Young's modulus** is increased by about 6% as a result of the presence of **lithium**, both in solid solution and in the coherent δ'

Table 7.5 Room temperature mechanical properties of Al–TM alloys.

Alloy (wt%)	Condition	Test temp. (°C)	σ_{ys} (MPa)	σ_{UTS} (MPa)	El (%)	K_{IC} (MPa $m^{-3}/^2$)	References
Al–6Fe–1V–2Si	SD + Ext.	20	217	—	21.5	—	[42]
	SD + Ext.	400	65	—	23.5	—	[42]
Al–8Fe–4Ce	SD + Rolling	RT	182	224	9.0	—	[43]
	RS-P/M sheet	RT	490	565	9.0	—	[43]
Al–8.5Fe–1.2V–1.7Si	As-SD	25	251	281	9.5	18.5	[44]
	SD + Ext.	25	271	313	18.0	21.0	[44]
	RS-P/M	25	353	398	7.2	—	[44]
8009	PFC[a)]	25	385	—	6.1	—	[44]
8009	As-SD	420	85	91	24.0	—	[44]
Al–8.5Fe–1.2V–1.7Si	SD + Ext.	420	103	112	24.0	—	[44]
	RS-P/M	420	213	222	6.8	—	[44]
8009	PFC[a)]	420	118	—	7.9	—	[44]

a) Data determined from graphs; RS-P/M, rapid solidification powder metallurgy.
SD, spray deposition; Ext, extruded; PFC, planar flow casting; RT, room temperature, σ_{ys}, yield stress; σ_{UTS}, ultimate tensile strength; EI, elongation index; and K_{IC}, plane-strain fracture toughness.

(Al_3U) phase. In principle, the weight savings in aircraft structural parts manufactured with lithium-containing alloys could reach 15% [45]. Unfortunately, Al–Li alloys exhibit poor deformation and fracture behavior because of strain localization, which results from planar slip associated with shearable precipitates and because of the formation of coarse, brittle intermetallic phases during slow solidification. Some of the approaches taken for improving toughness in Al–Li–X alloys involve (i) encouraging dislocation cross-slip or precipitate bypassing by increasing mismatch in the Al–Al_3U system; (ii) introducing secondary precipitation systems; and (iii) utilizing additional dispersion-hardening systems. Other approaches taken include **grain refinement** through additions of manganese, zirconium, chromium, and cobalt, and minimization of tramp elements (potassium, sodium, and sulfur) through alloy control [45–55].

Many of the problems associated with ingot cast Al–Li alloys may be alleviated to a certain extent by RS-P/M processing. The high solidification rates associated with RS-P/M processing help to minimize segregation of lithium, refine grain size and size of secondary phases. Although significant success has indeed been achieved, the RS-P/M approach introduced another problem, namely, oxygen contamination. The large surface-to-volume ratio associated with powders, together with the reactivity of Al and Li with oxygen, makes it impossible to eliminate oxides from P/M materials [55]. The oxide contamination problem is made worse by the presence of Li and Mg. Once it became clear that, to process Al–Li alloys, it is desirable to promote rapid quenching conditions while minimizing oxide contamination, atomization processing became an attractive selection.

In studies on high-strength Al alloys with lithium additions, improvements in room temperature tensile ductility at equivalent strength levels when compared with those of RS-P/M material have been reported (Table 7.6) [27, 46, 47]. The crack growth rates for the alloys obtained by spray deposition were found to be significantly lower than those for ingot cast (IM) 2020 up to a stress intensity level, ΔK, of 17 MPa $m^{1/2}$ (Figure 7.4). The improvement in **fatigue crack growth** rate for the spray-deposited X2020 alloy was attributed to a decrease in oxide volume fraction [47].

Spray deposition has also been applied to several commercial Al–Li alloys including 2090, 8090 and Weldalite™ 049 [35, 36, 46, 48, 49, 55]. Extensive work has been conducted at Cospray Advanced Aluminum Materials, Banbury, UK, and Alcan International, Banbury, UK, on the spray deposition of 8090. The mechanical properties of spray-deposited 8090 are summarized in Table 7.6. The room temperature tensile strengths of spray-deposited 8090 are generally higher than those of the IM materials of identical chemical compositions. These improvements have been proposed to be a result of the refinement in the T2 (Al_6CuLi_3) phase [46, 48, 49]. The results obtained by Alcan International, Banbury, UK, and Lockheed, Palo Alto, CA, USA have demonstrated that the size of the T2 phase may be reduced by an order of magnitude through spray deposition processing relative to that in materials processed through direct chill (DC) casting [54]. Similar results of improved strength in spray-deposited materials over that of IM materials

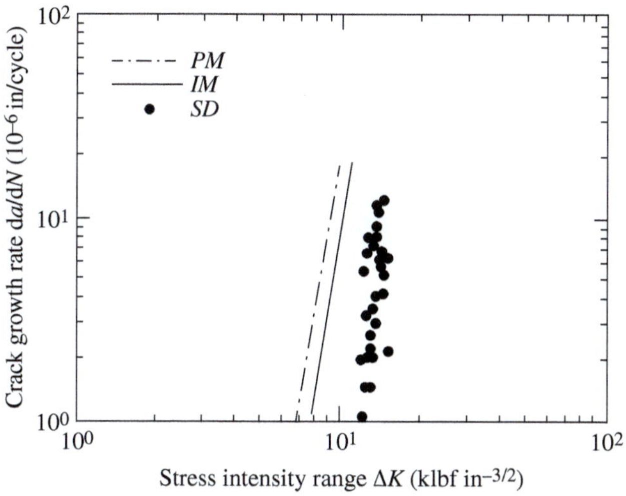

Figure 7.4 Constant load amplitude **fatigue crack growth** behavior of spray-deposited X2020 (SD) compared with that of IM 2024 and P/M X2020. Source: Adapted from Lavernia et al. [47].

Table 7.6 Room temperature mechanical properties of extruded lithium-containing Al alloys.

Alloy	Processing	σ_{ys} (MPa)	σ_{UTS} (MPa)	El (%)	References
2024-1.0Li (T4)	RS TR	388	524	21	[46]
2024-3.0Li (T6)	RS TR	571	583	5	[46]
2024-1.6Li (T4)	RS P/M	382	486	8	[46]
2024-1.6Li (T4)	SD	362	513	16	[46]
X2020 (T6)	IM	531	570	3	[27]
X2020 (T6)[a)]	RS P/M	622	649	5	[27]
X2020 (T7)[b)]	RS P/M	623	650	5	[27]
X2020 (T6)[b)]	SD	648	666	7	[27]
X2020 (T7)[c)]	SD	587	621	7	[27]
8090 (T6)	IM	480	520	5	[48]
8090 (T6)	SD	501	541	5	[49]
Weldalite™ 049 (T6)	IM	500	540	16	[55]
Weldalite™ 049 (T6)	SD	638	665	12	[55]

a) 1.55Li, extrusion ratio 30 : 1, T6 single age, T7 double age;
b) 1.74Li, extrusion ratio 28 : 1;
c) 1.74Li extrusion ratio 14 : 1.
RS TR, rapid solidified twin rolled; RS P/M, rapid solidified powder metallurgy; SD, spray deposition; IM, ingot metallurgy; σ_{ys}, yield stress; σ_{UTS}, ultimate tensile strength; EI, elongation index.

were also obtained in a Weldalite™ 049 alloy [55]. As summarized in Table 7.6, at equivalent ductility levels, the yield strengths of spray-deposited Weldalite™ 049 are more than 100 MPa higher than those of its IM counterpart. The improvements in room temperature strength have been attributed to the refinement in grain size and the size of precipitates associated with spray deposition processing.

The application of spray deposition in experimental high-Li aluminum alloys has also yielded promising results. Meschter et al. [56], for example, reported a significant improvement in the notched tensile behavior of a spray-deposited Al–4wt%Li–lwt% Cu–0.2wt%Zr alloy over that shown by the same material but in the rapidly solidified P/M form. Spray deposition processing is attractive for Al–Li alloys especially in high-Li concentrations of greater than 2.5wt%Li when IM processing is limited by severe Li segregation during slow solidification.

Table 7.7 Room temperature mechanical properties of Al–Li alloys.

Alloy (wt%)	Processing	σ_{ys} (MPa)	σ_{UTS} (MPa)	El (%)
Al-3.51Li-1.08Cu-0.63Mg-0.37Zr	SD + rolling (T6: 190 °C, 16 h)	414	519	5.8
	SD + rolling (T6: 170 °C, 24 h)	388	471	6.4
Al-3.51Li-1.08Cu-0.63Mg-0.37Zr	SD + extrusion (T6: 190 °C, 16 h)	558	616	4.5
	SD + extrusion (T6: 170 °C, 24 h)	557	584	3.6
Al-2.24Li-1.48Cu-1.20Mg-0.26Zr	SD + rolling (T6: 190 °C, 16 h)	460	516	12.2
	P/M (T6: 190 °C, 16 h)	473	536	6.4
Al-2.24Li-1.48Cu-1.20Mg-0.26Zr	SD + rolling (T6: 170 °C, 24 h)	464	497	10.5
	P/M (T6: 190 °C, 16 h)	507	543	5.6

P/M, powder metallurgy; and SD, spray deposition.
Source: Adapted from Zhang et al. [57].

Table 7.7 summarizes the mechanical properties of two spray-deposited Al–Li alloys [57]. The results obtained on the high-Li aluminum alloy (Al–3.51Li–1.08Cu–0.63 Mg–0.37Zr wt%) demonstrate that a relatively high strength and ductility can be achieved by direct rolling following spray deposition. The results in Table 7.7 also indicate that the strength of spray-deposited high-Li aluminum alloy may be improved further by extrusion following spray deposition without significant decrease in ductility. This improvement demonstrates the important role of microstructural refinement on the mechanical behavior of Al–Li alloys. The results for the low-Li aluminum alloy (Al–2.24Li–1.48Cu–1.20 Mg–0.26Zr wt%), as summarized in Table 7.7, reveal that spray-deposited materials exhibit a lower strength than that of identical materials produced using the RS-P/M route. This behavior is generally attributed to the high-volume fraction of oxides that are present in RS-P/M materials [46, 57]. The principal drawback with the RS-P/M process appears, at present, to be the negative effects of the high oxygen content on ductility. For RS-P/M processing, at least one dimension of the starting material, i.e. powders or flakes, must be small in order to meet the heat transfer requirements for rapid solidification. Therefore, further consolidation is always necessary to form bulk, three-dimensional products. As a result of the relatively high surface-to-volume ratio characteristic of powders and flakes, the alloys produced usually contain a relatively high oxygen content as compared with spray-deposited or IM materials. As examples, the oxygen contents in some of the alloys are summarized in Table 7.8 [46]. The oxide films resist welding between powder particles during consolidation and form oxide stringers following forging, extrusion and rolling, which are detrimental to the mechanical properties of high-strength Al alloys. Both unwelded prior particle boundaries (PPBs) and oxide stringers provide easy locations for crack formation and propagation during loading.

The properties of RS-P/M Al alloys are strongly affected by the presence of oxides in the final product. In particular, the fracture characteristics of rapidly solidified Al–Li alloys are enormously degraded because of the presence of oxide stringers in the fully densified hot-worked and heat-treated alloys [58–60]. Kang and Grant [60] studied a rapidly solidified P/M X2020 alloy, which achieved significant improvements in room temperature tensile properties and an 11–15% increase in the **specific modulus**, E/ρ, when compared with the corresponding values for the IM material. The fracture toughness values, however, fell in some cases, to less than 50% of those for ingot-based Al alloys. **Fatigue crack growth** behavior is also affected by the presence of oxides in the microstructure of RS-P/M Al alloys. Comparison between RS-P/M X2020-T6 and IM X2020-T6 shows that rapidly solidified alloys are superior in the low ΔK region but that the **fatigue crack growth** rates of P/M alloys are higher than those of IM alloys in the intermediate and high ΔK regions [58–60]. This behavior has been associated with the interaction between a growing sharp crack and high-volume fraction of oxide stringers.

7.2.6 Al–Zn–Mg–Cu Alloys

High-strength Al–Zn–Mg–Cu alloys (**7xxx**) are heavily alloyed with Zn, Mg, and Cu that form the major strengthening phases such as intermediate η' and equilibrium η [$Mg(Zn,Al,Cu)_2$]. In principle, further improvements in mechanical properties in this alloy system may be achieved through grain refinement. Accordingly, spray deposition processing

Table 7.8 Oxygen content in Al alloys.

Alloy (wt%)	Process	Oxygen content (wt%)
7075 Al	H_2O atomization	0.457[a)]
	Inert gas atomization	0.209[a)]
	Rotating cup process	0.0066[b)]
7075 Al + Ni + Zr	Spray deposition + extrusion	0.011[b)]
X2020-B (Al–3.6Cu–3.03Li–0.49Mn)	Spray deposition + extrusion	0.010[b)]
Al–3.6Cu–3.0Li–0.5Mn	Spray deposition + HIP	0.0080[b)]
Al–3Li–l.5Cu–lMg–0.2Zr	Atomization + extrusion	0.051[c)]
	Ingot cast	0.005[c)]
Al–3.2Li–2.1Cu–0.83C	Mechanical alloying + extrusion	0.65
7090 Al	Atomization + extrusion	0.2–0.5
Al–3.7Ni–l.5Fe	Air atomization	0.386[c)]
	Splat flakes	0.139[c)]
	Centrifugal atomization	0.020[c)]

a) Oxygen content determined from a measured amount of Al_2O_3, $MgAl_2O_4$, and MgO using neutron activation analysis;
b) oxygen content determined using the inductively coupled plasma technique;
c) oxygen content determined using fast neutron activation analysis.
Source: Reprinted from *Mater. Sci. Eng.*, A98, E. J. Lavemia and N. J. Grant, spray deposition of metals: a review, 381–394, 1988 with kind permission from Elsevier Science – NL, Sara Burgerhartstraat 25, 1055 KV Amsterdam, The Netherlands [46].

has been used to achieve rapidly solidified microstructures with low oxygen content [61–70]. An important effort in the spray deposition of high-strength Al 7xxx alloys is the modification of alloy compositions through the addition of transition metal elements such as Ni and Zr. For example, available experimental results on a modified 7075 alloy (7075 + 1wt%Ni and 0.8wt%Zr) suggest that the fine dispersions of Al_3Ni and Al_3Zr that are produced during spray deposition and subsequent thermomechanical treatment helped to improve the mechanical properties (see Table 7.9) [61, 63, 68, 70, 71]. The spray-deposited and hot-extruded material in the T4 + T6 condition exhibited a tensile strength, yield strength, and elongation of 817, 741 MPa, and 9%, respectively, as compared to those of 600%, 583%, and 9% for IM 7075 Al. The fatigue strength of spray-deposited 7075 modified with Ni and Zr was also considerably greater than that of IM 7075 (see Figure 7.5) [62]. Although the IM 7075 alloy modified with 1wt%Ni and 0.8wt%Zr also exhibited a high strength, its ductility was relatively low, being less than 2%. The combination of high strength and ductility associated with the spray-deposited 7075 + 1.0Ni + 0.8Zr alloy was a result of grain refinement and the presence of a fine dispersion of Ni- and Zr-containing phases. Transmission electron microscopy and scanning electron microscopy studies of the hot-extruded spray-deposited alloy in the T6 condition revealed that the microstructure consisted of fine grains 2–7 μm in diameter with extensive intragranular precipitates which included, among others, the Al_3Ni phase [72].

In principle, the strength of the 7xxx alloys may be improved by increasing the Zn content close to the theoretical limit associated with phase equilibria while maintaining proper ratios with Cu and Mg. In conventional IM, the practical limit of Zn content is attributable to the extensive **macro-segregation** and cracking that accompany slow solidification. Accordingly, new experimental alloys with enhanced alloying have been studied extensively using RS-P/M and spray deposition processing [61, 67, 70]. Such alloys typically exhibit higher Zn and transition metal element contents relative to those of conventional 7075 Al. For example, alloy N707 [Al–10.8–11.4Zn–2.2–2.5Mg–1.0–1.2Cu–0.25–0.32Zr–<0.2Si, Fe (wt%)] developed by Alusuisse-Lonza Service Ltd., Switzerland was designed to incorporate high levels of Zn and Zr that are suitable for spray deposition processing. The high Zn and Zr contents lead to high peak age strength without a significant loss in ductility. As shown in Table 7.9, this alloy exhibits a superior combination of strength and ductility relative to those of IM 7075 alloy. Moreover, it also shows superior fatigue properties relative to those of conventional 7075 (Figure 7.6) [68, 69].

Table 7.9 Room temperature mechanical properties of 7xxx alloys.

Alloy (wt%)	Temper	Orientation	σ_{ys} (MPa)	σ_{UTS} (MPa)	El (%)	K_{IC} (MPa m$^{-3/2}$)	References
7075 + 1.0Ni + 0.8 Zr (SD + Ext.)	T6	L	717	751	7	—	[63]
7075 + l.0Ni-f-0.8 Zr (SD + Ext.)	T4 + T6	L	741	817	9	—	[63]
7075 + 1.0Ni + 0.8 Zr (RS-P/M + Ext.)	T6	L	672	682	10	—	[71]
7075 + 1.0Ni + 0.8 Zr (IM + Ext.)	T6	L	750	762	2	—	[71]
7075 (IM)	T6	L	583	600	9	—	[71]
Eural (low enthalpy SD + Ext.)	T6	L	790	810	4.9	—	[61]
Eural (high enthalpy SD + Ext.)	T6	L	762	798	2.2	—	[61]
	T7X	L	713	728	5.4	—	[61]
Eura2 (low enthalpy SD + Ext.)	T6	L	807	819	7.1	—	[61]
Eural (RS-P/M + Ext.)	T6	L	716	735	1.9	—	[61]
Eural (RS-P/M 4–Ext.)	T7X	L	660	691	6.7	—	[61]
N707 (SD)	T6	L	760	775	8.0	—	[68]
	T7	L	690	695	8.0	—	[68]
Al–11Zn–2Mg–lCu–0.3Zr (SD + Ext.)	T6	L	705	719	29.0	—	[70]
	T6	L-T	620	634	8.2	38.2	[70]
	T6	T-L	—	—	—	17.3	[70]
	T7	L	503	536	17.9	—	[70]
	T7	L-T	482	521	11.5	75.5 (J_{1C})	[70]
	T7	T-L	—	—	—	37.8	[70]

IM, ingot metallurgy; RS-P/M, rapid solidification powder metallurgy; SD, spray deposition; ext., extruded; Eural, a 7xxx alloy containing 10–13 Zn; Eura2, Eural composition with reduced Cr and Mn contents; N707 is Al–10.8–11.4Zn–2.2–2.5Mg–1.0–1.2Cu–0.25–0.32Zr–<0.2Si, Fe (wt%).

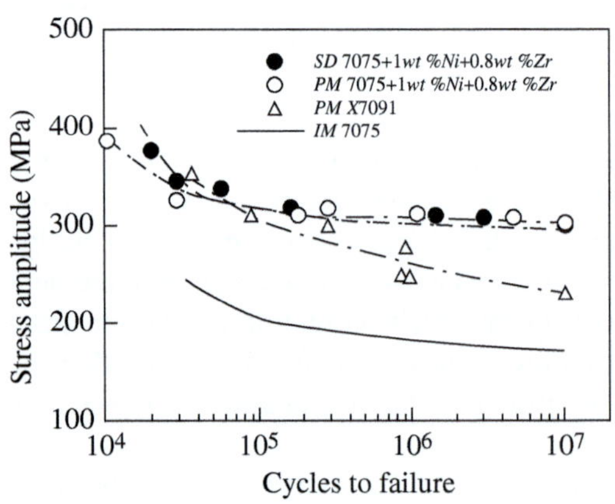

Figure 7.5 Smooth bar fatigue behavior of a spray-deposited and hot-extruded 7075 + l.0 Ni + 0.8 Zr alloy as compared with those of other high-strength Al alloys. Source: Adapted from Lavernia et al. [62].

The high fatigue resistance associated with the N707 alloy was considered to be primarily a result of the fine, homogeneous microstructure of the spray-deposited material [68, 69]. Results on spray-deposited alloys with identical or similar compositions to that of N707 have also demonstrated improvements in tensile properties as compared with those of conventional and RS-P/M 7xxx alloys [25, 61, 67, 70]. The results of Mächler et al. [70], summarized in Table 7.9, show that the yield strength of a spray-deposited Al–11Zn–2Mg–lCu–0.3Zr in the T6 temper is 709 MPa with a tensile elongation of 29%. As summarized in Table 7.9, this alloy also exhibits relatively high strength and ductility in the L-T (Long transversal) direction. De Sanctis [61] compared the microstructure and mechanical properties of EURA1 (a 7xxx alloy with 10–13 wt%Zn) and EURA2 (EURA1 composition with reduced Cr and Mn contents) synthesized using RS-P/M processing and spray deposition. The results of this study,

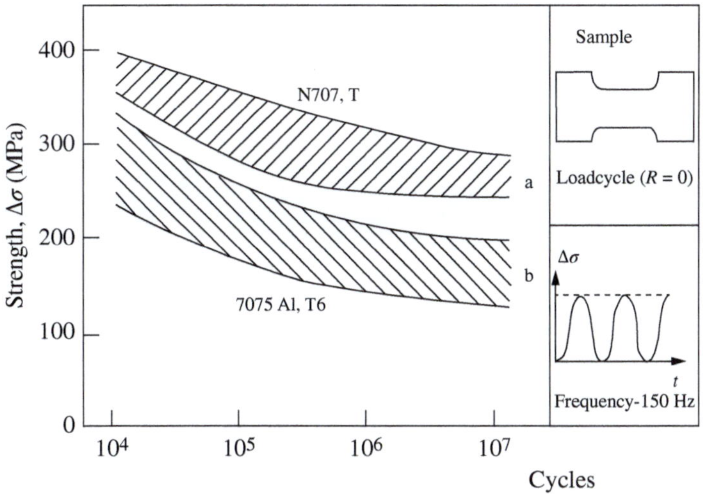

Figure 7.6 Fatigue strength profiles of N707 under various processing conditions as compared with data measured for 7075 Al in the form of plates, sheets, and forgings. Source: Reproduced by permission of Swiss Materials, Verlag Dr. Felix Wüst AG [69].

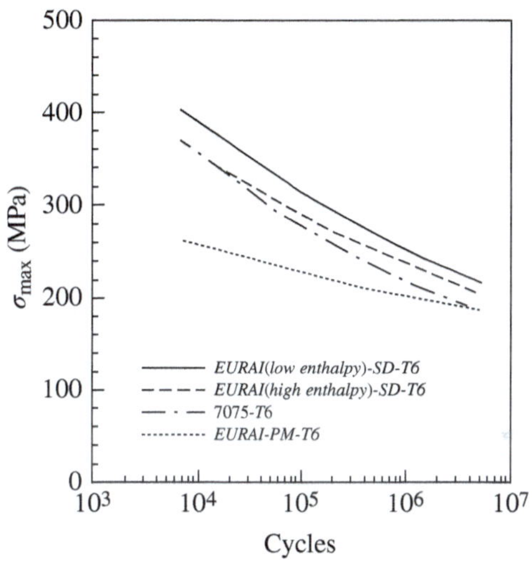

Figure 7.7 Fatigue behavior of spray-deposited and RS-P/M EURA extruded alloys. Source: Reproduced by permission of Elsevier Science from M. de Sanctis [61].

also summarized in Table 7.9, reveal that the spray-deposited alloy generally exhibits higher strength and ductility than the RS-P/M alloy in the T6 temper. The smooth bar fatigue strength of the EURA alloys is the highest among those of the IM 7075 A1 and RS-P/M EURA alloys. The results summarized in Figure 7.7 reveal that the strength of spray-deposited EURA1-T6 alloy for a fatigue life of 10^6 cycles is approximately 15% higher than that of the IM 7075 [61]. It is also evident that the RS-P/M EURA1 alloy exhibits a much lower fatigue strength than the spray-deposited EURA1 alloy. Regarding **fatigue crack growth** rate, only limited results are available in the range greater than 10^{-6} mm cycle^{-1}. **Fatigue crack growth rate** of EURA alloys appears to be similar to that of the IM 7075 Al but is much lower than that of the RS-P/M 7090. The relatively low fatigue strength associated with the RS-P/M EURA alloys was attributed to their high oxygen and hydrogen contents [61]. The results indicated that the oxygen content in RS-P/M EURA alloys was in the 1000–3000 ppm range, whereas that in spray-deposited alloys was in the 50–200 ppm range. The hydrogen content measured for the RS-P/M and spray-deposited EURA alloys was in the range of 1–4 and 0.1–0.2 ppm, respectively. The fracture toughness of spray-deposited high-Zn 7xxx aluminum alloys has been noted to be higher as compared with that of the conventional 7xxx alloys and high-strength RS-P/M alloys [67, 68, 70]. As summarized in Figure 7.8, the spray-deposited Al–11Zn–2Mg–lCu–0.3Zr alloy in L-T orientation exhibits superior strength over fracture toughness ratios [70]. In the T-L orientation, the fracture toughness is relatively low but is still comparable with those of high-strength RS-P/M alloys (see Table 7.9).

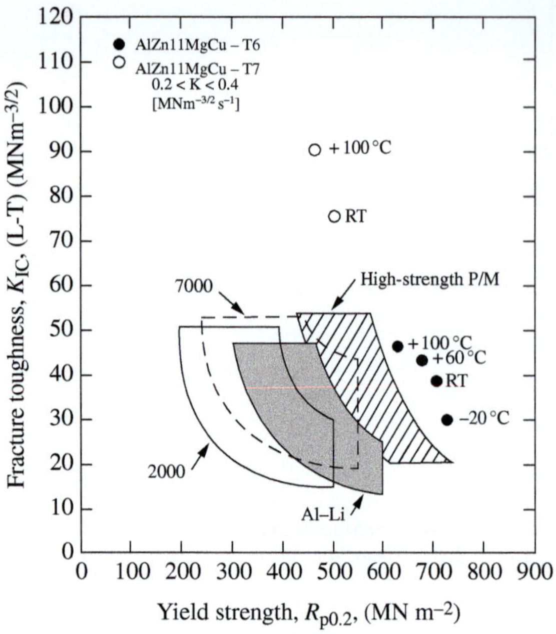

Figure 7.8 Fracture toughness and strength of spray-deposited and extruded Al–11Zn–2Mg–lCu–0.3Zr (wt%) as compared with other high-strength Al alloys. Source: Mächler et al. [70]/John Wiley & Sons.

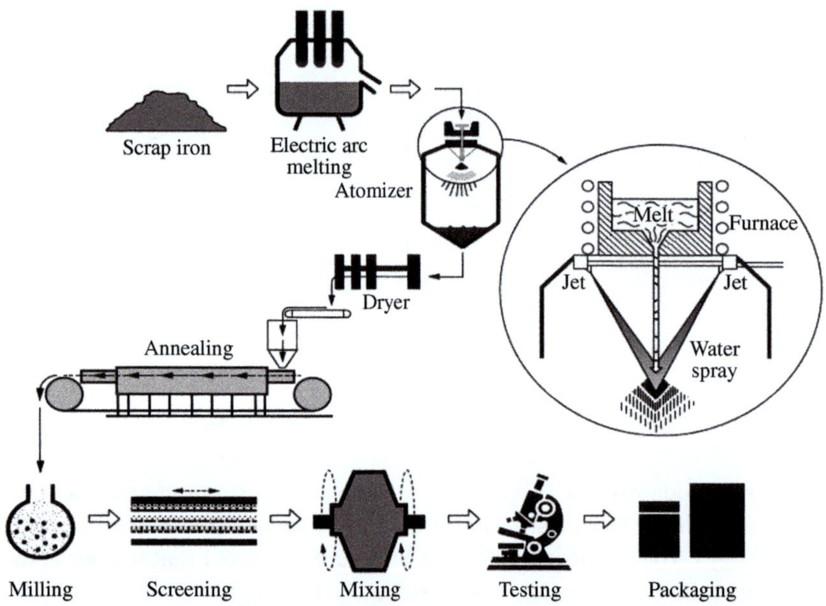

Figure 7.9 Typical water atomization process. Source: German [96]/Metal Powder Industries Federation.

7.3 Iron-Based Alloys and Powders

Iron-based alloys are produced as both atomized powers as well as-spray-deposited materials. Most of the oxide phases commonly found in iron-based rapidly solidified alloys are easily reduced; consequently, the primary motivation behind the numerous studies on the spray deposition of these alloys has been the reduction in the number of processing steps as compared to the IM and powder metallurgy (P/M) methods, and the improvement in the properties through rapid solidification and microstructural modification [73–87], while **stainless steel** powders are the most popular materials currently used in metal-based AM.

7.3.1 Fe-Based Alloy Powders

Fe-based steel powders represent the most significant tonnage of raw materials used in powder metallurgy (P/M) and AM because they are inexpensive to manufacture compared to nonferrous metals. Fe-based steel powders generally possess superior properties, particularly strength, than other metals and alloys, and they can be alloyed and heat treated [88]. Atomization processes are widely used to make iron and steel powders. Atomization consists of the disintegration of a molten metal stream into ligaments and, eventually, droplets, rapidly cooled by conductive and convective heat transfer.

In the gas atomization of Fe and Fe alloys, melting occurs under a protective atmosphere or under a vacuum, to protect reactive constituents from becoming oxidized. Such systems are designed for the dry collection of the atomized powder, and the atomization tanks range from 6 to 10 m (20 to 33 ft) [89] to ensure the solidification of the powder particles before they reach the bottom of the atomizing tank. Similar to nonferrous alloys, horizontal gas atomization is used sometimes because it is more economical. The absence of oxidation during inert gas atomization leads to oxygen contents that are low, typically less than 200 ppm.

Oxide dispersion strengthened (ODS) ferritic stainless steels are candidate materials for structural components in nuclear energy fusion reactors. There is a molten metal processing technique for the simplified production of precursor powder for ODS ferritic **stainless steel** alloys [90, 91]. These ODS alloys are candidates for high-temperature applications within future-generation thermal power reactors. This novel process is known as **gas atomization reaction synthesis (GARS)** as a simplified approach to generate precursor oxide dispersion forming powders directly, thus eliminating the time-intensive **mechanical alloying** (MA) step from the ODS fabrication sequence [91, 92]. During GARS, a reactive atomization gas (i.e. Ar-O_2) is used to oxidize the surfaces of nascent ferritic **stainless steel** alloy droplets during primary break up and rapid solidification of the atomized powders. The fast reaction kinetics associated with GARS promotes the formation of a nano-metric metastable Cr-enriched surface oxide layer. This metastable oxide phase can transport a prescribed amount of O into the consolidated microstructure, which resides along PPBs. Elevated temperature heat treatments were then used to drive solid-state internal O exchange reactions between this metastable oxide phase and Y-enriched intermetallic precipitates, forming highly stable nano-metric mixed oxide (i.e. Y–(Hf or Ti)–O) dispersoids throughout the alloy microstructure [93, 94]. These alloys are termed chemical reservoir (CR) alloys due to the unusual O exchange reaction between the strategically distributed metastable Cr-enriched oxide phase (i.e. O reservoir) and Y-enriched intermetallic precipitates (i.e. Y reservoir). Another similar route to produce ODS ferritic stainless steels with composition Fe–14Cr–2W–0.3Ti–0.3Y2O3 (wt%) involves surface treatment of gas-atomized powder followed by reactive synthesis (STARS) [95]. In this study, starting powders containing Ti and Y are obtained by gas atomization. Then, a metastable Cr- and Fe-rich oxide layer forms on the surface of the powder particles. During consolidation by hot isostatic pressing (HIP), the metastable oxide layer at PPBs dissociates, the oxygen diffuses toward saturated solutions or metallic Ti- and Y-rich particles, and Y–Ti–O nano-oxides (mainly Y_2TiO_5) precipitate in the ferritic matrix.

Water atomization (WA) of molten iron and steel is used commercially for the high-volume production of low-carbon iron and alloy steel powders. Factories have an annual output ranging from 20 000 to more than 200 000 metric tons [88]. Most manufacturing facilities follow the process flow depicted in Figure 7.9 [96]. The water atomizing process typically starts with a source of raw material, usually **scrap steel**, which is melted in an **electric arc furnace** and refined. Some manufacturers have an additional melting/refining step that uses a ladle refining furnace to control alloying additions and improve the control of nonmetallic inclusions more precisely. Sometimes, a manufacturer can use high-purity molten Fe-3%C from an on-site smelter. The refined melt is atomized using high-pressure water jets to disintegrate the liquid metal stream. After atomizing, steel powder is dewatered and dried. The melting process uses molten steel containing 0.1–0.3% C, giving the as-atomized powder a complex, martensitic microstructure.

Additionally, oxygen from the water and air aspirated into the atomizer creates an oxide shell on the iron powder. Oxygen content is typically in the range of 0.1–0.2%, depending on the alloy. Excess carbon and oxide are undesirable for compaction and are removed during an annealing process performed at a temperature between 800 and 900 °C in a hydrogen atmosphere. The annealing process produces a partially sintered cake, which is crushed and milled. The finished iron powder is packaged through a final screening process [96–98].

Iron and steel powders made by water atomization have irregular shapes. While the surface is rough, the powders are nearly fully dense and have little internal porosity. The powder is screened to a 60 mesh (<250 mm) size and typically has a mean particle size, *d*50, between 100 and 150 μm. The irregular surface of the iron particles promotes interlocking and reasonable green strength in die compaction. The level of residual elements is relatively low for most commercially available water-atomized plain iron powders. The number of residual elements such as carbon, chromium, nickel, molybdenum, sulfur, phosphorus, and silicon are controlled to deficient levels, which is essential to ensure consistent pressing and sintering behavior. The manganese level can be maintained slightly higher, ranging from 0.08% to 0.20%, imparting some primary alloy strengthening to the final P/M steel [99].

Because alloying elements are homogeneously distributed throughout the powder particles, the overall compressibility is usually lower than plain iron powders due to the increased strength of the powder particles. This is a drawback when powders are pressed at room temperature, as in the traditional P/M process, but this characteristic becomes irrelevant for hot powder forging applications. That is why fully prealloyed powders are commonly used in powder forging applications. The most common prealloyed elements are molybdenum and nickel because they are easily reduced under conventional sintering conditions. Manganese and chromium additions have traditionally been avoided because they result in higher powder oxygen content, which is difficult to remove during sintering and causes inconsistency in final P/M parts. However, prealloyed powders containing chromium and manganese have been commercialized.

In the case of stainless steel powders mainly intended for conventional press and sinter processing, or the so-called compacting-grade powders, it is highly essential that green compacts with adequate green strength can be produced using commercially available compacting presses and tooling. For this reason, water atomization is the preferred powder manufacturing process because it renders the powder particles highly irregular. It is also more economical than gas atomization. Stainless steel powders used for consolidation through AM, MIM, **extrusion**, and HIP are usually made by gas atomization, and these powder metallurgy (P/M) techniques typically produce fully dense parts.

The cooling rates in water atomization are higher than in gas atomization, although both are sufficient to produce powders free of **macro-segregation** after processing and yield homogenous microstructures. Some typical parameters for both types of atomization are listed in Table 7.10 [89].

The water atomization of stainless steel also typically consists of a power unit, a melting furnace, an atomization tank, and a collection vessel. For stainless steels, high-frequency induction furnaces are commonly used because they permit, through induction stirring, efficient alloying of the various constituents of steel.

The melting of stainless steel for water atomization differs from the melting practice used in wrought and cast stainless steel processing. Because little refining occurs during melting in an induction furnace, the melting charge must be protected from excessive oxidation during melting. Furthermore, decarburization is not an option because there can be no oxygen/oxygen–argon injection. Hence, much of the carbon introduced with raw materials are transferred to the product. This mandates the use of low-carbon iron and low-carbon ferroalloys. Similarly, no manganese is removed in the process; hence the raw material selected must also be low in manganese content. Both requirements restrict the

Table 7.10 Typical industrial process parameters for water and gas atomization of stainless steels.

Parameter	Water atomization	Gas atomization
Type of furnace	Induction	Induction
Furnace capacity	Up to 4500 kg (10 000 lb)	Up to 5000 kg (11 000 lb)
Melting	Open air	Open air
Atomizing medium	Water	Argon (nitrogen)
Water pressure	11–18 MPa (1500–2500 psi)	—
Gas pressure	—	0.76–2.6 MPa (110–380 psi)
Water flow rate (single-orifice nozzle)	200–400 l min^{-1} (53–106 gal min^{-1})	—
Gas flow rate (single-orifice nozzle)	—	20–40 m^3 min^{-1} (800–1600 scfm)
Metal flow rate	45–90 kg min^{-1} (100–200 lb min^{-1})	45–90 kg min^{-1} (100–200 lb min^{-1})

Source: Warzel [89]/ASM International.

amount and type of scrap that can be used for the manufacture of compacting-grade stainless steel powder. After the melting charge reaches its atomization temperature, typically 50–75 °C above the melting point of the alloy [89], a thin layer of slag is removed. After this, the induction furnace is gradually tilted to allow the liquid metal to flow, via a runner, into the tundish. From the bottom of the tundish, the molten metal emerges as a well-defined stream within the atomization chamber, disintegrating into a powder. Oxidation of the powder decreases significantly with decreasing pouring or atomizing temperature [100]. Low oxygen content powders can be produced by partly filling the atomization tank with water and purging it with nitrogen. Nitrogen purging of the atomization chamber is also essential to prevent the risk of explosion due to the reaction of nascent hydrogen gas released during atomization with the oxygen-rich atmosphere.

The control of physical powder characteristics is vital to producing stainless steel powders that exhibit sufficient properties to allow component manufacture. The powder's PSD and chemical composition affect numerous properties critical to component manufacture and application performance. As in gas atomization, particle size in water atomization is controlled by the pressure of the atomizing medium. Particle size is primarily controlled by water pressure in water atomization. Higher pressures will produce finer powders. Water pressures near 14 MPa can disintegrate the molten stainless steel stream into a predominantly –100 mesh (<150 mm) powder used for most structural parts. V-jet nozzles are widely used for directing high-pressure water onto liquid metal. Various water jet configurations are possible through different arrangements of such nozzles [101], which in turn give rise to various PSDs. When atomized at constant pressure and constant flow rates of metal and atomizing medium, PSDs form straight lines when plotted on log-normal paper. It is easy to manipulate particle size data because only two numbers, dm, the mean mass diameter, and Φg, the geometric standard deviation, a measure of the breadth or width of a distribution, define the entire PSD. The value of Φg is obtained as the ratio of the particle size diameters taken at 84.1% and 50.0% of the cumulative undersized weight plot, respectively [102].

Several important variables must be considered to produce a narrow PSD, and among these, atomizer design is critical because it determines the interaction of the liquid metal with the high-pressure water. With a sound water atomization system, it is possible to produce powders with a Φg of approximately 2.0. This value corresponds to a powder where two-thirds of its PSD (by volume) have an upper-to-lower size ratio of 4 to 1.

The particle shape of a water-atomized powder significantly influences the apparent density, flow properties, green strength, and compressibility of the powder. It also affects AM and sintered properties, including dimensional change and mechanical properties. In this case, the fundamental property of importance is probably the particle coordination number because irregularly shaped particles form more contact points with neighboring particles during pressing. Although statistical methods for determining particle shape have been developed, the apparent density or, better yet, the tap density of a narrow size fraction of a powder is a convenient measure of particle shape that can be quickly and easily determined and correlated with other properties. Quite a broad range of particle shapes can be produced by water atomization. Some shape control is possible by modifying chemical composition, but standard control is through the atomizing head design. Improved atomization efficiency is attributed to the so-called two-step atomization technique, which produces finer powders than those possible with either gas or water atomization alone. Independent control of the two closely linked types of atomization allows for broad control of particle shape. Physical powder characteristics that increase the apparent density of a powder, for example, a more spherical particle shape or a PSD that provides improved packing improves compressibility at the expense of green strength.

The control of powder chemistry is of equal importance to that of the physical characteristics of a powder. For water-atomized grades of 304L and 316L that have nickel contents close to their upper ranges (13% and 14%, respectively), it is advisable to keep their carbon contents to ≤0.02%. While achieving such low-carbon contents with an induction furnace and typical commercial sintering is possible, it is safer to forgo the benefits of high nickel content and seek compressibility improvements through other means. The potential for intergranular corrosion represents an ever-present challenge because of the slow cooling rates employed in most industrial sintering operations.

As mentioned earlier, a significant difference between water and gas atomization is the substantial oxidation occurring with the former. After its deoxidation with ferro-silicon, the liquid stainless steel has a low oxygen content of less than 200 ppm. However, it reacts accordingly when it encounters water, like most other metals. A primary goal in the water atomization of stainless steels is to produce powders with oxygen contents as low as possible yet sufficient to render the particle shape irregular for adequate green strength.

7.3.2 Stainless Steels

The microstructure and mechanical properties of a number of spray-deposited **stainless steels** have been investigated, including 12Cr, En. 58J [73], Sanicro 28 [79, 80], 9Cr-Mo [87], and 316L [74, 81, 82]. The reported mechanical properties of these materials, summarized in Table 7.11, illustrate that spray-deposited stainless steels generally exhibit mechanical properties that are equivalent to those of the conventional IM materials [74, 81, 86, 103, 104]. Brooks et al. [73] compared the microstructure and mechanical properties of 12Cr stainless steels produced by **spray-forging** and those produced by conventional forging (ingot cast plus forging). They reported that the spray-deposited 12 wt%Cr following extrusion had a similar equiaxed microstructure as that was present in the as-spray-deposited condition. As shown in Table 7.11, the room temperature tensile strengths of the spray-forged 12Cr stainless steels are equivalent to those of the conventionally forged material in the longitudinal direction. Moreover, the mechanical properties of such spray-forged 12Cr stainless steels were isotropic. When compared with the transverse properties of the conventionally forged material, however, the spray-forged material showed a superior elongation (19% vs. 7%) and a better reduction in area (57% vs. 17%). In the same study, however, Brooks et al. reported that the mechanical properties of an 18/10/3 (EN.58J)-type stainless steel were equivalent in both spray-forged and conventionally forged conditions.

Austenitic stainless steels such as 316L are nonage hardenable. The strength of such alloys, however, may still be improved considerably through grain size refinement because of the well-known **Hall–Petch effect** as

$$\sigma = \sigma_i + kd_g^{-n} \tag{7.1}$$

where σ is strength; σ_i is frictional stress; d_g is grain size; and k and n are experimental constants (positive values). The advantages of spray deposition processing in such materials are the direct grain refinement achievable in the as-spray-deposited conditions and indirect grain refinement through improved workability [81, 82]. Üçok et al. [81] and Ibrahim et al. [82] studied the microstructure and mechanical properties of spray-deposited 316L stainless steel in various tempers. In this study, it was observed that the as-spray-deposited condition, 316L, exhibited an equiaxed grain structure with an average grain size of 10–40 μm and a density corresponding to 93–99% of the theoretical value. Üçok et al. [81, 82] reported that the material can be completely densified by a 50% hot rolling pass. It is evident from

Table 7.11 Room temperature tensile properties of spray-deposited stainless steels.

Material and condition	σ_{ys} (MPa)	σ_{UTS} (MPa)	El (%)	RA (%)	References
12 Cr					
Spray forged (isotropic properties)	608	745	19	57	[73]
Conventional forging (longitudinal)	590	739	23	61	[73]
Conventional forging (transverse)	614	757	7	17	[73]
En. 58J					
Spray forged (isotropic properties)	220	552	65	65	[75]
Conventional forging	215	524	55	65	[75]
316L					
Spray-deposited wrought alloy	545	661	40	—	[74]
Conventional wrought alloy	519	664	40	—	[74]
Spray-deposited + 50% HR at 1273 K	580	770	46	68	[81]
Spray-deposited + 50% HR at 1273 K + 80% CR at 298 K	1290	1450	6	45	[81]
Spray-deposited + 90% CR at 298 K	1290	1380	3	55	[81]
Spray-deposited + 80% CR at 77 K	1670	1920	3	22	[81]
Spray-deposited + 80% CR at 77 K + 873 K 10 min	1280	1610	3	49	[81]
Spray-deposited + 80% CR at 77 K + 973 K 10 min	1190	1400	5	51	[81]
Ingot grade 316L + 1323 anneal	250	580	50	60	[103]

AC, air cooling; HR, hot rolling; CR, cold rolling.

the results shown in Table 7.11 that the spray-deposited and rolled 316L stainless steel exhibited an excellent combination of mechanical properties. For example, the 50% hot-rolled material has a yield strength of YS = 580 MPa, an ultimate tensile strength of UTS = 770 MPa and an elongation of El = 46%. These values compared favorably with those of the commercially supplied 316L sheets (YS = 250 MPa, UTS = 580 MPa, El = 50%) and those of the wrought 316L as reported by Rickinson et al. [74] (YS = 519 MPa, UTS = 664 MPa, El = 40%). The results summarized in Table 7.11 also illustrate the improvements in strength that may be achieved by hot and cold working associated with proper annealing heat treatment. For example, the results of Üçok et al. [81] and Ibrahim et al. [82] demonstrated that a cryogenic rolling of 80% at 77 K led to the complete transition from an austenite structure into a martensite structure. The yield and ultimate strengths of such cryogenically rolled materials were 1670 and 1920 MPa, respectively. Üçok et al., Ibrahim et al. [81, 82] reported that reversion treatments by annealing at 873 and 973 K for 10 minutes led to an ultra-fine austenitic grain structure and a high strength. For example, the grain sizes corresponding to these two annealing temperatures were 0.1 and 0.3 μm, respectively. The yield strength values of the cryogenically rolled materials were 1280 and 1190 MPa corresponding to 873 and 973 K reversion treatment, respectively.

7.3.3 Tool Steels

High-speed steels (HSSs) normally contain elevated concentration of elements such as C, Co, Cr, Mo, V, and W. Such materials acquire high bend strength, hardness, and wear resistance through the formation of metal carbides in the form of MC, M_2C, $M_{23}C_6$, and M_6C. When HSSs are processed utilizing conventional ingot casting techniques, coarse carbides, and eutectic phases form during slow solidification. Moreover, the distribution of carbides in materials processed under slow solidification conditions is generally nonuniform. Both P/M and spray deposition methods are suitable for producing HSS microstructures with a uniform distribution of fine carbides [74–77]. Rickinson et al. [74] studied the microstructures and properties of HSSs (M2, M15) produced utilizing spray deposition and observed considerable microstructural refinement as compared with IM microstructures. The secondary dendrite arm spacings determined in the as-spray-deposited materials were equivalent to those produced at effective cooling rates in the 10^3–10^4 K s^{-1} range. The grain size observed in the spray-deposited HSSs was also refined relative to that present in the IM materials and was in the grade range of Snyder Graff 20 as compared with that of Snyder Graff 8–12 in IM materials. Moreover, Rickinson et al. observed that in spray-deposited M2 steel preforms the feathery M_6C-MC type eutectic phases typical of slow solidification conditions were completely suppressed. No change in chemical composition was reported during the atomization-splatting cycle, and the oxygen and nitrogen levels were reported to be similar to those for air-melted ingots. Rickinson et al. noted that the primary carbide precipitates that were sometimes suppressed in splat-quenched materials were not eliminated in the spray-deposited materials. The average size of the carbides was, however, reduced to the 2–3 μm range. These refined microstructures in spray-deposited HSSs were consistently observed in other studies [75–77]. Itami et al. [76], for example, compared the average carbide sizes observed in IM, RS-P/M and spray-deposited HSSs. They observed that the carbides in spray-deposited preforms were in a globular morphology and were uniformly distributed. The sizes of carbides in spray-deposited, RS-P/M and IM materials were <6, <2 and 10–20 μm, respectively. Khare and Laporte [77], conducted a systematic study on the carbide size in spray-deposited M15 HSS. They noted that the morphology of carbides in the spray-deposited materials was approximately spherical. The size of the largest carbides in spray-deposited materials was less than 12.7 μm with the average size in the range of 2.1–5.1 μm. The spray-deposited M2 HSS reportedly developed a better response to heat treatment because of the microstructural refinement. The austenitizing time required to achieve optimum combination of mechanical properties was reduced for spray-deposited materials relative to that for ingot materials [74].

With reference to the mechanical properties and performance of the high-speed tool steels, available results show that spray-deposited materials are generally equivalent to RS-P/M materials. The mechanical properties and performance of spray-deposited tool steels are superior relative to those of the IM materials of equivalent chemical compositions [74, 76, 78, 105]. Nama et al. [105], for example, compared the three-point bending strength of a spray-deposited high-speed tool steel (Fe–2.3C–18Cr–2 Mo–1V in wt%) with that of the RS-P/M materials of identical chemical composition. In this study, the bending strengths of the spray-deposited materials were determined under two different conditions: in the as-spray-deposited condition, and in the spray-deposited and forged condition. Prior to bending tests, the materials were subjected to quenching at 1303 K followed by annealing at 773 K to achieve the desired hardness of HRC 62. Nama et al. observed that the bending strength of the HSS was low in the as-spray-deposited condition,

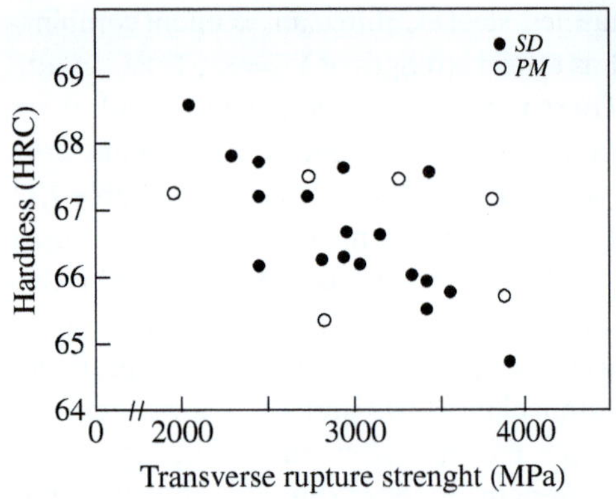

Figure 7.10 Transverse rapture strength and hardness of spray-deposited (SD) and forged, and powder metallurgy (P/M) high-speed steels. Source: Reproduced by permission from Itami et al. [76].

being approximately 45% of that of the P/M material. This phenomenon was attributed to the presence of porosity in the as-spray-deposited condition. The bending strength of spray-deposited materials was, however, greatly improved by forging. The results of Nama et al. revealed that the strength of the spray-deposited materials was increased to 3800 MPa by forging to an area reduction of 4 : 1, equivalent to that of the RS-P/M material. This observation agrees with the results of Ikawa et al. [78], which show that the bending strength of spray-deposited HSSs was improved by approximately 100% following densification by HIP.

In related studies, Itami et al. [76] compared the hardness, wear resistance, and cutting performance of HSS endmills (Fe–1.19 C–0.41 Mn–0.24 Si–4.12 Cr–5.31 Mo–6.21 W–3.1 V–8.59 Co in wt%) produced utilizing IM, RS-P/M and spray deposition methods. For wear resistance tests, the materials were heat treated to a hardness of HRC 67 by quenching from 1100 to 1200 °C followed by tempering between 480 and 600 °C. The results of these studies, summarized in Figure 7.10, show that the spray-deposited materials exhibited transverse rupture strength that was equivalent to that of the RS-P/M materials at each of the hardness levels studied [76]. However, the spray-deposited materials showed a better wear resistance than the RS-P/M materials. For example, under identical testing conditions, the wear rate of the spray-deposited materials is 25% lower than that of the RS-P/M materials. Itami et al. further compared the cutting performance of the IM, RS-P/M and spray-deposited materials. The results shown in Figure 7.11 indicate that the cutting performance of the HSS produced using different methods depends on the workpiece materials used [76]. When SUS304 stainless steel was used as a working piece, the cutting performance of the RS-P/M and spray-deposited materials was similar, and both were superior to that of the IM material. When SKD11 tool steel was used as the workpiece, the cutting distances (the cutting distance before failure) of the IM and RS-P/M materials were equivalent. The cutting distance of the spray-deposited material was, however, much longer than both the IM and the RS-P/M materials [76]. Although the precise reason of the high wear resistance and good cutting performance observed for spray-deposited materials is not clear, Itami et al. postulated that the carbide size characteristic

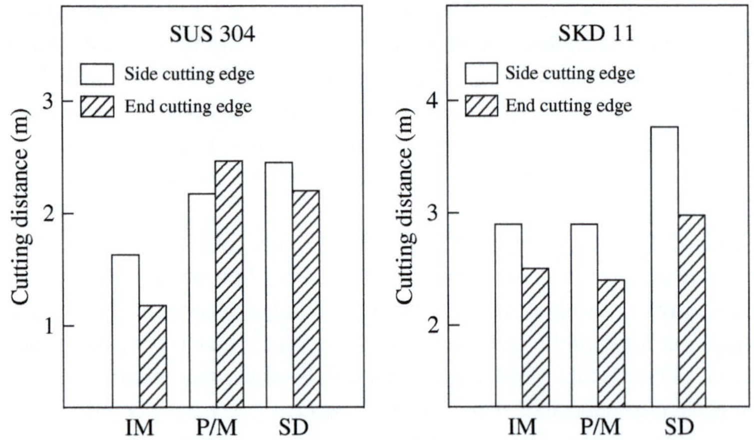

Figure 7.11 Cutting performance (cutting distance) of IM, P/M, and spray-deposited and forged high-speed steel endmills for different workpieces: SUS 304 and SKD 11. Source: Reproduced by permission from Itami et al. [76].

Table 7.12 Room temperature tensile properties of spray-deposited stainless steels.

Material and condition	σ_{ys} (MPa)	σ_{UTS} (MPa)	El (%)	Charpy (J)	References
9Cr-Mo					
Spray-deposited + 50% hot rolling at 1033 K	622	800	18	—	[87]
Ingot + 1313 K, 1 h/AC + 1033 K, 1 h/AC	771	885	11	—	[87]
En.8					
Spray-deposited wrought alloy	357	567	25	9.8	[86]
Conventional wrought alloy	366	543	30	11.6	[86]
En.24					
Spray-deposited wrought alloy	772	869	19	28.9	[86]
Conventional wrought alloy (longitudinal)	772	896	25	38.1	[86]
Conventional wrought alloy (transverse)	717	855	16	20.8	[86]

AC, air cooling.

of spray-deposited materials might be more suitable for producing a good wear resistance. Ikawa et al. studied the microstructure and mechanical properties of high-speed tool steels. In this work, HSSs were spray-deposited into mill rolls. The results of this study show that the wear rate of the spray-deposited materials was reduced to approximately 1/2–1/6 of that of the IM materials of identical temper (HRC 63–64). The roll test results conducted in this study indicated that the roll lives of the spray-deposited rolls were two to three times longer than those of the IM rolls [78].

7.3.4 Other Iron-Based Materials

Megusar et al. [87] studied the room temperature mechanical properties of spray-deposited 9 wt%Cr–1 wt%Mo air-hardening ferric steel (nuclear reactor candidate material). The material produced in this study was 95% dense and was rolled directly at the tempering temperature (1033 K) used for conventional wrought material to combine the rolling and tempering treatment into a single operation. Megusar et al. [87] found that the room temperature tensile strengths of a hot-rolled 9 wt%Cr–1 wt%Mo obtained by spray deposition were lower than those of the ingot material (see Table 7.12). The ductility of the spray deposition material, however, was higher (18% vs. 11%). The microstructure and properties of spray-deposited low-carbon steels, including En.3A, En.8, En.16, En.19, En.24, En.31, En.33, and SAE1020, have also been extensively studied [82, 86]. A comparison of the mechanical properties of spray-forged and conventionally forged low-carbon steels was made by Brooks et al. [86]. The mechanical properties summarized in Table 7.12 for spray-forged and conventionally forged En.8 and En.24 illustrate the similarities between spray-forged low-carbon steels and conventionally forged materials. Brooks et al., however, reported that the mechanical properties of the spray-deposited steels are more isotropic than those of conventionally forged steels.

7.4 Nickel-Based Alloys and Powders

Most Ni-based **superalloys** contain significant to large amounts of alloying elements that are present to attain high-temperature strength and thermal stability. Concurrent with the increase in alloying elements are the increase in **macro-segregation** and decrease in hot workability. Accordingly, research efforts have involved the synthesis of various superalloys utilizing one of the three following processes: RS-P/M [106–108], **plasma spray deposition** [109–111], and spray atomization and deposition [106, 107, 112–115]. One of the advantages associated with the RS-P/M methods is the potential of producing rapidly solidified microstructures that are essentially free of macro-segregation with reduced micro-segregation. The RS-P/M processes are, however, susceptible to the problems associated with oxygen pickup and ceramic contamination [107]. Plasma spray deposition offers similar advantages to the RS-P/M methods in that macro-segregation-free microstructures are readily produced. Moreover, this method provides the possibility of near-net-shape forming which may effectively eliminate the need of cold or hot working.

Plasma spray deposition has been utilized to synthesize superalloy preforms with excellent structural integrity and is being developed to produce near-net-shape components [109–111]. Since preatomized powders are used as the feeding materials in plasma deposition, extensive oxygen pickup is not avoided. Moreover, the principal drawbacks are the low production rates and high costs. Accordingly, among the aforementioned production methods, spray deposition is being considered as an approach to minimize oxide contamination, manufacture near-net-shapes, and reduce the need for intermediate working steps. Spray deposition has been used for the synthesis of a wide variety of superalloy compositions. At the Howmet Corporation (Whitehall, MI, USA) and the General Electric Company (Schenectady, NY, USA), for example, spray deposition is presently used to produce aerospace-quality nickel-based superalloys including IN100, IN713, IN718, René 41, René 80, René 88DT, René 95, René 125, Merl 76, Waspaloy, and AF2-1DA-6 [107, 112–116]. At the David Taylor Research Center (Annapolis, ML, USA), the spray deposition method has been tested to produce IN625 for marine pipes [106, 117]. Other alloys produced using spray deposition include Nimonic 115 [118], Astroloy [119], and U720 [115]. The results obtained with these materials are described in Sections 7.4.1–7.4.4.

7.4.1 Ni-Based Alloy Powders

High-quality superalloy powders are a prerequisite for high-performance P/M Ni-based superalloys and AM. A large proportion of all powder used to produce P/M superalloys is prealloyed. Chapters 1 and 2 of this book describe several methods for producing metal powders, including inert gas atomization [119]. In addition, to reduce the inclusions and refine the particle size of the powder as much as possible, novel metal powder preparation methods have been developed, such as the plasma rotating electrode process (PREP), the spark plasma discharge spheroidization (SPDS) process [120], and **electrode induction gas atomization (EIGA)** [121, 122].

The principal commercial powder-making processes are gas atomization and **vacuum atomization**. In gas atomization of superalloys, high-purity feed raw materials are vacuum induction melted and subsequently atomized using inert gas. Typically, the equipment consists of a top-melt chamber where high-purity raw materials are vacuum induction melted. The metal is delivered to a nozzle to form a small metal stream atomized by high-pressure gas using either a continuous annular-type or discrete individual nozzle. The liquid metal is atomized by high-pressure gas jets. Generally, the gas-atomized powder is solidified at a rate from 10^3 to 10^5 K s^{-1}. The resulting powder, which is highly spherical, is collected in a bottom chamber or cyclone collector. The PSD resulting from this process depends mainly on the surface tension, g, viscosity, Z, and density, r, of the melt and the velocity, v, of the atomizing gas and the atomizer design. Oxygen content is of the order of 100 ppm, depending on particle size. Argon gas is commonly used to atomize superalloys. Nitrogen gas atomization can also be used to reduce costs and eliminate entrapped argon. The use of nitrogen usually requires slight modifications in composition, such as lower carbon content, to accommodate the nitrogen increase associated with the atomization process.

Vacuum atomization is a commercial practice based on the rapid evolution of gas; the principle that when molten metal is supersaturated with gas under pressure and suddenly exposed to a lower pressure, the gas expands, comes out of solution, and causes the liquid metal to be atomized. The equipment arrangement consists of two vertical chambers connected by a transfer tube. The metal is vacuum induction melted in the lower chamber. The molten metal is forced upward through the transfer tube when the lower chamber is pressurized with gas. As the metal rises into the upper chamber, it is atomized by escaping gases. The resulting powder falls to the bottom of the upper chamber and through a powder drain into a collection container. This method is considered equivalent to vacuum melt gas atomization. The atomized powder particles, which are solidified at a rate of about 103 K s^{-1}, fall to the bottom of the upper chamber. The cooled powder is collected under a vacuum in another chamber, sealed, and backfilled with an inert gas. This technique can atomize up to 1000 kg of superalloy spherical powder in one cycle. This technology has been employed for LC Astroloy, MERL 76, and IN-100 alloys [123].

Gas atomization is the most common process for commercial quantities of Ni-based powder. A stream of liquid metal is hit by gas jets with high velocity, thus breaking the stream with kinetic energy that is transferred to the metal and scattering the droplets. The droplets then solidify rapidly, becoming powder particles with a particle size usually below 150 μm. These are then collected at the bottom of the atomization tank. Compared to water atomization, gas atomization often yields more spherical particles. However, when producing finer powders by gas atomization, it can be observed that smaller particles have attached themselves to the larger particles during the atomization process. These smaller particles are called **"satellites"** and can affect the flow properties of the metal powder.

By partially filling an atomization tank with water, it is possible to achieve different results when atomizing an alloy. Without water, the cooling rate of the powder particles is lower, and the oxygen content in the powder is low. When using water cooling a higher cooling rate can be achieved, but the powder will have slightly higher oxygen content. An important parameter when controlling particle size during gas atomization is the gas-to-metal ratio. The median particle size is inversely related to the square root of this ratio, meaning a higher gas flow rate for a specific metal flow rate will result in finer powder particles. Other parameters to consider during gas atomization are cooling rate, which depends on tapping temperature, atomization pressure, free fall distance, metal flow rate, and temperature in the atomization tank. The metal flow rate depends on the size of the tundish nozzle, the liquid metal level in the tundish, and the pressure in the atomization tank [124].

After production, powders are screened to remove oversized particles and blended to obtain a uniform distribution. All powder handling is performed to minimize the possibility of introducing foreign material into the powder, which requires the use of specially designed stainless steel containers, valves, and inert and air handling of powder in clean rooms. Powder is customarily screened to minimize the inclusion size in the final part. Depending on the application, powder sizes ranging from 60 to 325 mesh (45–250 μm) are typically used. Prior to being loaded into containers for consolidation, superalloy powders are evaluated for cleanliness using techniques such as water elutriation, which separates nonmetallic inclusions from the powder for counting, sizing, and identification. A consolidated powder sample is also evaluated by conducting large bar (e.g. 13 mm diameter by 50 mm gage length, or 0.5 in. by 2 in.) tensile and fatigue tests. Cleanliness and fatigue life are evaluated based on the inclusion population found on the fracture surface. In the early 1990s, a laboratory test known as heavy liquid separation (HLS) was developed, which used the density differences between ceramics and the base alloy to separate and analyze inclusions [125]. The frequency and character of inclusions found using high-sensitivity ultrasonic inspection are also used to characterize and determine an acceptable inclusion population.

Stainless or mild steel containers are used for powder consolidation and require thorough cleaning before loading. A "final filter" is commonly used to rescreen the powder as it enters the container and ensure that no oversized particles are in the consolidated part. The powder is heated under a vacuum to remove air and adsorbed moisture. Different manufacturers use various combinations or modifications of these outgassing techniques.

For specific gas atomization and application of Ni-based alloy powder, high-temperature Ni-base superalloy materials were developed as porous coatings for application onto the near-surface channeled superalloy substrates for advanced airfoil cooling applications [126]. High-pressure gas atomization (HPGA) has been used to produce two Ni-based superalloy compositions. A model ternary Ni-based superalloy composition (Ni–13Al–10 at%Cr) was initially used to optimize the HPGA process to target the more extensive size range of powders from 100 to 200 μm [126]. The ternary alloy atomization was performed at 300 psi with Ar gas and a pour temperature of 1700 °C. After the first successful powder-making trial, similar atomization parameters were used to produce powder of alloy composition targeted for high-temperature oxidation resistance, Ni–13Al–10Cr–1Re–2Ta2W–2Ru–2.5Pt–0.1at%Hf. The atomization parameters included a lower Ar gas pressure (130 psi) and a pour temperature of 1750 °C. The resulting powder composition was determined using inductively coupled plasma mass spectrometry (ICPMS) analysis.

The microstructure, transformation temperatures, **magnetic properties**, and the influence of heat treatment on Ni–Mn–Ga powders (Ni49.0 Mn26.5 Ga25.5) were studied in detail [127]. The diameter of the metal delivery tube was 1.5 mm, and the atomization gas was argon. The alloy was atomized at 1353–1581 K using a melt chamber pressure of 0.25–0.3 bars and an atomization pressure of 40–50 bars. The powders' phase and magnetic transformation temperatures were considerably lower than similar conventionally ball-milled alloys. Nevertheless, heat treatment of the atomized powder was found to recover the magnetic properties of the materials, probably by releasing the residual strains generated by the rapid solidification during the gas atomization process. The sintering of powders started at a significantly lower temperature when compared to the powder prepared by ball milling of bulk Ni–Mn–Ga-alloys.

Superalloy powder is also produced using the PREP and the **centrifugal atomization** process. In PREP, a plasma arc is used to melt the end of a rapidly rotating prealloyed bar. The high-speed rotation of the bar causes liquid droplets to spin from the end of the bar and solidify in flight through a helium atmosphere. In **centrifugal atomization**, a molten stream of prealloyed metal is poured onto a spinning disk and atomized by centrifugal force. The emitted liquid particles are further broken up and subsequently cooled by high-pressure helium gas. One example of this technique is the rotating electrode process (REP) used in the early production of the IN-100 and Rene95 powders. In this process, a bar of the specified composition, 15–75 mm in diameter, acts as a consumable electrode. A direct current electric arc between the consumable electrode and a stationary tungsten negative electrode melts the end of this rotating positive electrode. The process is carried out in helium gas. Centrifugal force causes spherical molten droplets to come off the high-speed rotating electrode. These droplets rapidly solidify and are collected at the bottom of the chamber. The plasma rotating electrode (PREP) modifies the REP process and differs from the former in that the plasma is used to melt the consumable electrode. Cooling rates amount to 105 K s^{-1} for IN-100 [128]. The rotating electrode technique is restricted because of the problem in rapidly solidified superalloy consumable electrode manufacturing [30]. Currently, this method is in production for rotating aircraft disks, shafts, and blade retainers.

The powders are screened to remove oversized particles and produce a uniform distribution. All the powder handling processes are designed to eliminate the possibility of introducing foreign material into the powder. This involves using specially designed stainless steel containers, valves, and inert gas handling or air handling of powder in clean rooms.

To solve the problem of ceramic inclusions, **EIGA** technology has been designed and developed by the ALD Vacuum Technologies Company (Hanau, Germany). It can be seen from Figure 1.10 in Chapter 1 of this book that a superalloy electrode is immersed in a conical induction coil in this method. The molten metal falls directly into the atomization zone, and thus using a ceramic crucible can be avoided. Therefore, EIGA prevents the introduction of ceramics from the crucibles. The electrode bottom is immersed in an induction coil, where it melts and produces dripping or streaming liquid metal, which directly breaks down and is atomized by the high-pressure inert gas from the nozzles. Finally, the molten droplets cool, and particles are obtained. In the past, EIGA was mainly used to produce powders of Ti, Nb, Zr, and other low melting point alloys [36]. It also produces super-clean and super-spherical superalloy powders through EIGA by process parameter and nozzle optimization [122].

EIGA has successfully prepared high-purity FGH4096 Ni-based superalloy powder. The powder particles have high sphericity and smooth surface. The solidification microstructure of the particles ranging between 0 and 50 μm exhibits cellular structure. In particles ranging between 50 and 100 μm, cellular structure, and dendritic microstructure occur. In particular, the dendritic microstructure increases rapidly when particle size exceeds 80 μm, and the cellular morphology will decrease simultaneously. When the particle size is larger than 100 μm, the solidification microstructure of particles is dendritic. The reason why the surface solidification microstructure transforms from cellular to dendritic is probably a result of different cooling rates since the fine particles cool faster than the coarse ones. In addition, no hollow particles range between 0 and 150 μm, while few hollow particles are present when the particle size exceeds 150 μm. This indicates that EIGA can effectively prevent thermally induced porosity caused by hollow particles [120].

7.4.2 Inconel Alloys

Spray deposition processing of Inconel alloys has been reported in a number of studies [106, 107, 113, 116, 129–131]. The microstructure of these alloys is generally characterized by fine, equiaxed grains with a grain size of less than 30 μm in the as-spray-deposited condition [113]. More importantly, macro-segregation and γ' eutectic phases were not observed in the spray-deposited materials [116]. Using optimum spray deposition conditions, the porosity content may be reduced to less than 1% [113, 116]. Fiedler et al. [113, 129] evaluated a spray-deposited IN718 and compared its properties with those of materials processed utilizing alternative casting and powder metallurgy methods. They noted that the tensile strength, ductility, and stress rupture life of the IN718 in the as-spray-formed condition were similar to those of wrought ingot alloy and were superior to those of the cast alloy (see Figure 7.12) [129]. The room and high-temperature mechanical properties of the spray-deposited IN718 were further improved by forging or by HIP prior to heat treatment (Figure 7.12 and Table 7.13). Compared to the wrought cast material, the spray-deposited IN718, forged or hot isostatically pressed, exhibits higher strength and tensile elongation [129, 132, 133]. The 649 °C/724 MPa stress rapture life of the spray-deposited IN718 was significantly improved relative to those of IM wrought and powder hot isostatic pressed materials. For example, the spray-deposited IN718 exhibited a stress rapture life in the range of 201–327 hours, whereas those materials produced using the IM wrought and powder HIP methods were in the range of 100–193 hours [129, 132, 133]. Improved strength and stress rapture life of spray-deposited IN718 relative to its wrought and powder metallurgy counterparts have also been reported by other investigators [115].

Nell et al. [131] spray-deposited an IN100 alloy onto a rotating, nitrogen-cooled cylinder to produce a seamless tube. The low strain rate forgeability of the spray-deposited IN100 tube was reported to be exceptional, considering the fact that the as-deposited material had an oxygen content of 250 weight ppm and a carbon content of 0.24 wt%. The high oxygen content of the spray-deposited material noted in this study resulted from poor control of the atmosphere prior

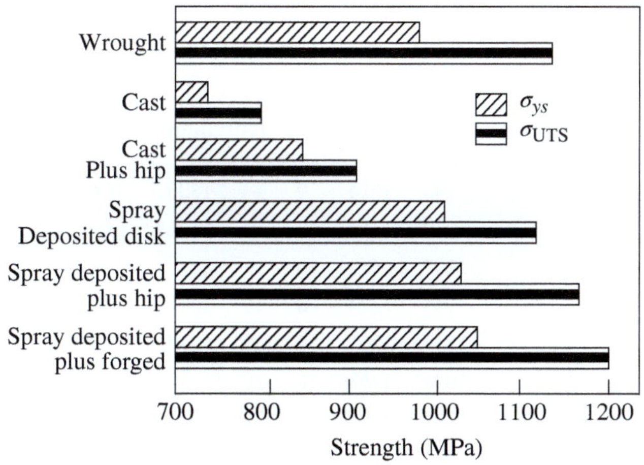

Figure 7.12 Tensile strength of IN718 alloys with different processing histories. The spray-deposited alloy was produced using N_2 as an atomization gas [129].

Table 7.13 Tensile and stress rupture properties of IN718 alloys [129].

	Tensile properties (room temperature)				Tensile properties (649 °C)				Stress rapture (649 °C)	
Condition	σ_{ys} **(MPa)**	σ_{UTS} **(MPa)**	**El (%)**	**RA (%)**	σ_{ys} **(MPa)**	σ_{UTS} **(MPa)**	**El (%)**	**RA (%)**	**Stress (MPa)**	**Life (h)**
Wrought	1186	1358	21	37	1000	1158	21	31	724	193
Cast[a)]	917	1089	11	—	—	—	—	—	586	100
Cast + HIP[b)]	965	1144	16	—	—	—	—	—	—	—
SD	1213	1351	9	17	1013	1117	14	25	724	201
SD + HIP	1262	1393	12	24	1027	1165	16	30	724	327
SD + forging	1310	1455	11	20	1048	1200	14	28	724	243

a) from Ref. [132];
b) from Ref. [133].
SD, spray-deposited; HIP, hot isostatically pressed;
Source: Reproduced with permission.

to spray deposition. The room temperature and 650 °C tensile properties of the spray-deposited materials densified using forging and HIP were found to be within the range of values reported for other RS-P/M methods (Tables 7.14 and 7.15) [74, 106, 130–137]. The spray-deposited materials in the hot isostatically pressed and isothermally forged conditions displayed stress rupture strength levels at 650 °C that were comparable to those of both gas-atomized P/M IN100 and hot isostatically pressed MERL 76, but the material obtained by spray deposition displayed a more stable microstructure than that of the atomized alloy.

IN625 alloy exhibits a strong tendency of segregation, a high flow stress and a high degree of strain hardening. As a result, the cost associated with IM processing of this alloy is high. Accordingly, this alloy has been considered for processing using powder metallurgy methods (P/M) or spray deposition processing in order to produce tubular-shaped products for marine pipe applications [106]. In the spray deposition approach, tubes were produced by directly depositing atomized droplets on a rotating cylindrical substrate. The microstructure of such spray-deposited tubes exhibited a fine grain size in the range of 30–50 μm. The grain size following HIP was reported to be in the 22–32 μm range. Ta and Nb carbide precipitates were observed in the spray-deposited and annealed conditions, primarily at triple points. In related studies, Moran and Palko [106] compared the strength and ductility of as-spray-deposited and P/M IN625 to determine their feasibility for **extrusion**.

They observed that both strength and ductility of the as-spray-deposited IN625 were lower than those of the P/M material. This phenomenon was attributed to the relatively small amount of hot working applied to the spray-deposited material. Table 7.14 summarizes the tensile test results obtained from IN625 alloys produced using

Table 7.14 Room temperature tensile properties of superalloys.

Condition	Processing	Heat treatment	σ_{ys} (MPa)	σ_{UTS} (MPa)	El (%)	RA (%)	Grain size (μm)	References
IN 100	As cast	—	848	1014	9	—	1500	[132]
IN 100	Subsonic atomized (Ar) + extruded	a)	1069	1517	14	14	11–12	[134]
IN 100	Subsonic atomized (Ar) + extruded	—	1207	1682	20	16	8	[135]
IN 100	Subsonic atomized (Ar) + hot isostatically pressed	—	945	1124	8	10	20–250	[135]
IN 100	Spray-deposited (tube) + hot isostatically pressed	b)	1117	1200	4	8	10–20	[131]
IN 100	Spray-deposited (tube) + isoforged	b)	1048	1462	15	10	5–10	[131]
IN 100	Atomized + extruded + isoforged (gatorized)	c)	1138	1600	22	22	6–11	[136]
Meri 76	Subsonic atomized (Ar) + hot isostatically pressed	b)	1007	1538	24	25	8–16	[137]
Merl 76[d)]	Subsonic atomized (Ar) + hot isostatically pressed	b)	965	1482	15	15	8–16	[137]
Merl 76	Spray-deposited (disc preform) + hot isostatically pressed	b)	1151	1503	20	—	20–40	[137]
Merl 76	Spray-deposited (mounded preform) + isoforged	b)	1158	1400	14	—	20–40	[137]
IN 625	IM + roll extrusion (35%) + 982 °C annealing	—	496	993	40	—	—	[106]
IN 625	P/M + roll extrusion (35%) + 982 °C annealing	—	510	965	47	—	—	[106]
IN 625	SD + roll extrusion (41%) + 1038 °C annealing	—	483	1007	39	—	—	[106]

a) 1177 °C, 4 h, oil quenched +649 °C, 22–24 h, air cooled +750 °C, 8 h, air cooled;
b) 11 163 °C, 2 h, oil quenched +871 °C, 40 min, air cooled +982 °C, 45 min +649 °C, 24 h, air cooled +760 °C, 16 h, air cooled;
c) 1121 °C, 2 h, oil quenched +871 °C, 40 min, air cooled +982 °C, 45 min +649 °C, 24 h, air cooled +760 °C, 4 h, air cooled.
d) Minimum acceptable by Pratt and Whitney.
Source: E.J. Lavernia.

Table 7.15 650 °C stress rupture properties of superalloys.

Condition	Processing	Heat treatment	Stress (MPa)	Rupture life (h)	El (%)	RA (%)	Grain size (μm)	References
IN 100	Spray-deposited (tube) + hot isostatically pressed	a)	965	26	1	4	10–20	[131]
IN 100	Spray-deposited (tube) + hot isostatically pressed	a)	690	> 316	—	—	10–20	[131]
IN 100	Spray-deposited (tube) + isoforged	a)	1034	10	—	—	5–10	[131]
IN 100	Spray-deposited (tube) + isoforged	a)	965	72	2	10	5–10	[131]
IN 100	Atomized + extruded + iso forged (gatorized)	b)	1034	60	—	—	6–11	[131]
Merl 76	Subsonic atomized (Ar) + hot isostatically pressed	a)	965	169	11	8	8–16	[137]

a) 1163 °C, 2 h, oil quenched +871 °C, 40 min, air cooled +982 °C, 45 min + 649 °C, 24 h, air cooled +760 °C, 16 h, air cooled.
b) 1121 °C, 2 h, oil quenched +871 °C, 40 min, air cooled +982 °C, 45 min + 649 °C, 24 h, air cooled +760 °C, 4 h, air cooled.
Source: Reproduced by permission of The Metal Powder Industries Federation, 105 College Road, East Princeton, NJ 0854066 [131].

different methods. In the cold-rolled and annealed condition, the materials produced using ingot casting, P/M, and spray deposition exhibit similar mechanical properties. Moreover, the tensile elongations of the spray-deposited materials are inferior to that of the P/M materials.

7.4.3 René Alloys

Alloys such as René 41 (no W), René 80, and René 95 typically contain elevated amounts of Cr, Co, Mo, W, and other elements. The principal advantages of rapid solidification in such alloys are reduced macro-segregation and refined grain size. Grain refinement may lead to excellent hot workability and low-temperature strength. In terms of applications that require a high creep resistance, however, grain refinement may not be an advantage. In a study of the spray deposition of nickel-based superalloys, Bricknell [107] compared the microstructure, chemistry, and mechanical properties of a René 80 alloy, produced using powder atomization and consolidation (RS-P/M); plasma deposition; and spray atomization and deposition. In this study, the René 80 alloy was spray-deposited separately using Ar and N_2 as atomization gases. When utilizing the optimum spray deposition conditions, the René 80 maintained a fine, equiaxed grained microstructure with a grain size in the range of 6–12 μm. In addition, no profound mid-temperature ductility loss was found for the low oxygen spray-deposited material. Mid-temperature brittleness was typically associated with RS-P/M materials because of the high oxygen contents. The as-spray-deposited grain size, as reported for René 95, was on the order of 30 μm [113]. It was reported that under the optimum spray deposition conditions, the relative density of the as-deposited materials may reach 99.4% for Ar atomized and 99.9% for N_2-atomized materials.

Bricknell [107] compared the oxygen pickup in the three typical processes, namely, RS-P/M, **plasma spray deposition**, and spray atomization and deposition. He noted that only the material produced utilizing spray deposition exhibited a low oxygen pickup. In contrast, both RS-P/M and plasma deposition require the use of preatomized powders. Consequently, oxygen content in such materials was noted to be significantly higher than in spray-deposited materials. For example, for the same starting alloy, the oxygen content observed in spray-deposited René 80 (Ar atomized) was 42 ppm, that in plasma-spray-deposited material was 458 ppm, and that in RS-P/M René 80 was 264 ppm. It was suggested that, in René 80, a high oxygen content may lead to mid-temperature embrittlement by oxygen diffusion along grain boundaries. Hence, Bricknell [107] further compared the temperature dependence of the mechanical properties of René 80 produced utilizing the three aforementioned methods. The results of this comparison, shown in Figure 7.13, indicate that both RS-P/M and plasma-spray-deposited René 80 exhibited significant mid-temperature brittleness [107]. This behavior was evidenced by the low ductility in the temperature range of 800–1000 °C. Moreover, no significant mid-temperature ductility loss was observed for the low oxygen content René 80 produced utilizing spray deposition.

Table 7.16 summarizes the tensile properties of spray-deposited René alloys where they are compared with alloys of identical chemical compositions but produced utilizing alternative methods [107, 112, 115, 118, 138, 139]. Under identical temper ($1250\,°C\,1\,h^{-1}ST + 1095\,°C\,4\,h^{-1} + 1050\,°C\,4\,h^{-1} + 845\,°C\,16\,h^{-1}$), the strengths of the spray-deposited materials are generally higher than those of the cast alloy. For example, for testing temperatures of up to 800 °C, the strengths of the spray-deposited materials are 10–20% greater than those of the cast alloys. The tensile elongation of the spray-deposited materials is, however, lower relative to that of the cast alloys. Fiedler et al. [112, 113] compared the strength and stress rapture properties of spray-deposited and hot isostatic pressed René 95 with those of the materials produced utilizing other consolidation and thermal processing methods (see Table 7.16). The René 95 alloy in the spray-deposited and heat-treated condition (1106–1110 °C solutioning + 760 °C aging) exhibited reasonably high strength ($\sigma_{ys} = 1097$ MPa, $\sigma_{UTS} = 1470$ MPa) and tensile elongation (EI = 11.8%). The strengths of the spray-deposited material changed slightly following forging; the tensile ductility was, however, improved from 11.8% to 21.9% [112, 113]. When compared to the materials processed utilizing other methods as shown in Table 7.16, the spray-deposited and forged René 95 exhibits an excellent combination of mechanical properties. The strengths of spray-deposited and forged René 95 are comparable to those produced utilizing casting and powder metallurgy techniques. The tensile ductility and stress rupture life are, however, much greater than those of the materials produced utilizing nonspray techniques. When compared with the wrought alloy, the spray-deposited and ring-rolled René 41

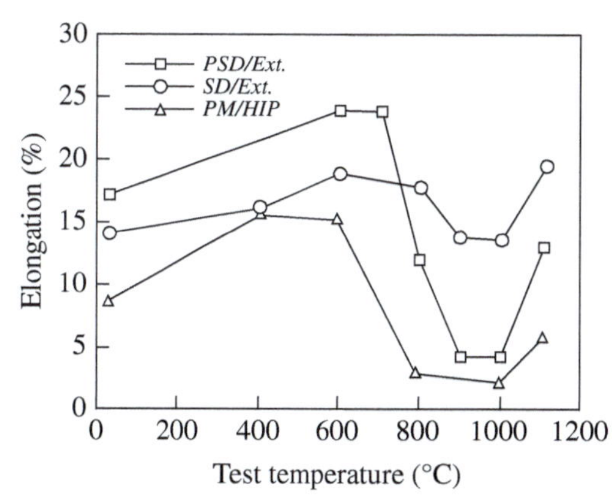

Figure 7.13 Tensile elongation of heat-treated René 80 as a function of test temperature (heat treatment: solutioning at 1250 °C for one hour followed by 1095 °C/4 h + 1050 °C/4 h + 845 °C/16 h): SD, spray-deposited; PM, powder metallurgy; PSD, plasma spray-deposited; Ext, extruded; and HIP, hot isostatically pressed. Source: Reproduced by permission of ASM International and The Minerals, Metals & Materials Society [107].

Table 7.16 Tensile properties of René alloys produced utilizing various methods.

Processing	Test temp. (°C)	σ_{ys} (MPa)	σ_{UTS} (MPa)	El (%)	RA (%)	t_f (h)	References
René 80							
Spray formed (Ar)	25	954	1333	—	—	—	[107]
Spray formed (N_2)	25	906	1181	—	—	—	[107]
Cast	25	853	1026	—	—	—	[107]
René 95							
Cast + wrought	650	1180	1460	8.8	—	44	[138]
	650	929	1250	9.0	13.0	—	[118]
Powder + HIP	650	1140	1500	9.3	—	—	[138]
	650	1100	1500	13.8	13.4	—	[118]
Powder + EF	650	1140	1500	12.4	16.2	—	[118]
	650	1097	1449	13.0	15.0	—	[139]
Spray formed	650	1097	1470	11.8	11.8	-	[112]
Spray formed and forged	650	1154	1465	21.9	32.9	179	[112]
René 41							
Spray + RR	25	60	1398	21.0	26.5	244	[115]
Cast + wrought	25	848	1124	14.0	17.5	—	[115]
Spray + RR	649	896	1311	32.0	40.0	344	[115]
Cast + wrought	649	896	1286	14.0	16.0	—	[115]

RA, reduction in area; t_f, stress rapture time, and RR, ring rolled.

shows higher strength and ductility presumably because of the elimination of macro-segregation and the refinement in grain size.

7.4.4 Other Superalloys

Prichard and Dalal [115, 116] evaluated the room and elevated temperature mechanical properties of spray-deposited Waspaloy. In this study, the mechanical properties of spray-deposited hot isostatically pressed Waspaloy rings were determined in the longitudinal, tangential, and radial directions. The results of these studies, summarized in Figure 7.14, indicate that such spray-deposited and hot isostatically pressed materials do not exhibit any significant

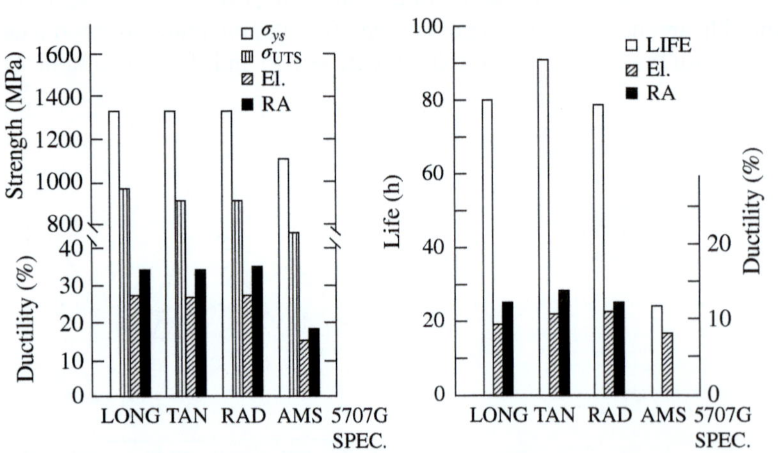

Figure 7.14 Tensile and stress rupture properties of spray-deposited, hot isostatically pressed Waspaloy rings compared to wrought specification AMS-5707G Source: Reproduced by permission of TMS and ASM International [116].

directionality [116]. Moreover, the results shown in Figure 7.14 show that the spray-deposited and hot isostatically pressed Waspaloy meets all of the property requirements as designed by the specification AMS 5707G. Related studies show that the strength and stress rupture properties of the spray-deposited and ring-rolled Waspaloy were superior to those of its wrought counterpart [115]. Ogata et al. [130] showed that the room temperature tensile properties of spray-deposited MERL 76 upset forged (thickness reduction of 6 to 1 at 1393 K) or hot isostatically pressed (at 1436 K and 19 $kg\,m^{-2}$) were comparable with those produced by other RS-P/M methods (Table 7.14) [130, 137]. In other studies, Evans et al. [118] reported that the combination of fine equiaxed grains, low oxide content, and minimal macro-segregation enables the normally nonforgeable cast MAR-M 200 to be hot worked after spray deposition. In the same study, the rupture life of spray-deposited Nimonic 115 alloy at 116 MPa and 980 °C was found to be between 100 and 123 hours, compared with the wrought ingot product life of 75 hours.

7.5 Titanium-Based Alloy and Powders

7.5.1 Ti-Based Alloys

The extraordinary combination of properties exhibited by titanium (Ti) and its alloys has made them highly desirable for various industries. Ti was first discovered in 1791 by William Gregor, a Cornish Clergyman and amateur Mineralogist [140]. The shiny gray, malleable, and ductile metal exhibits properties between **aluminum** and iron, and with a density of 4.5 $g\,cm^{-3}$, it is heavier than the first but lighter than the latter. Its melting point is high at 1668 °C and can reach an **elastic modulus** of up to 107 GPa. An abundant element that has transformed the world is mainly used in its alloyed form, by which extraordinary properties, including light weight-to-strength ratio, high corrosion, and temperature resistance, are exploited. Ti alloys are excellent for aerospace, biomedical, and automotive applications. Furthermore, as a compound, it is used as a pigment, a smoke screen for skywriting, and an abrasive. However, the cost of making the alloy is high due to the nonconventional processing necessary to adapt the material to the environment it will be utilized in. Nonetheless, the alloys' versatility due to their properties justifies the cost.

Ti alloys have a wide range of mechanical and physical properties. Its allotropy in the solid state primarily drives these properties. Two primary crystallographic ***allotropes*** exist in Ti alloys: the hexagonal close-packed (HCP) α-phase and the higher-temperature body-centered cubic (BCC) β-phase. BCC's atomic packing factor is 0.68, while HCP's is 0.74. HCP is, thus, more densely packed than BCC crystals. The tensile strength of these alloys lies between 200 and 1400 MPa. The crystallographic metallurgy of Ti is altered predominately when in the pure metals at 882 °C. It was reported [141] that below this transformation temperature, the alloy has an HCP structure known as alpha (α), while when above it, the structure exhibits a BCC known as beta (β). Ti is an allotropic material with a phase transformation from the α phase, an HCP crystal structure, to the β phase, a BCC structure. The Ti transition from $\alpha \rightarrow \beta$ occurs at elevated temperatures occurring at 882 °C for commercially pure Ti and 995 °C for Ti–6Al–4V. This temperature is known as the β-transus.

Ti alloys contain different alloying elements that are usually classified into α- or β-stabilizer, depending on whether they increase or decrease the α–β transformation temperature of 882 °C reported for pure Ti. Alloying elements such as oxygen, nitrogen, **aluminum**, and vanadium are some stabilizers added to Ti alloy. The addition of these elements governs and influences the β-transus temperature and thereby influences microstructure solidification transformation. These stabilizing elements can be grouped into two main categories: (i) α-stabilizers: elements that form with less than four bounds with atoms to dissolve in the α phase, thus an increase β-transus temperature. The most common elements added to Ti alloy are, e.g. O and Al. It has been reported that an increase of more than >8% Al can result in brittle formation in Ti alloys [142, 143]; (ii) β-stabilizers: these elements, such as vanadium and chromium, decrease the transformation temperature and are known to exhibit lower α phase solubility. These elements can further be subcategorized into either beta-isomorphous or beta-eutectoid elements. Elements are also added to increase the solubility of different materials. Such stabilizers also lower the α-type characteristics within a Ti alloy. Apart from this, the electronic resistance of Ti alloys is also reduced because of the addition of such β stabilizers, which could be deformed conveniently. Dai et al. [144] studied Ti–6Al–4V and found that only Al and Sn elements can increase the ω phase in comparison to other the β, α′, α″, and ω microstructure.

Ti alloys can be categorized into four main groups based on their atomic crystal structure, namely, α phase (HCP) alloys, near-α phase alloys, β phase (BCC) alloys, and α + β phase (mixed) alloys [145, 146]. The alloying elements

Table 7.17 Alloying elements and their effects on Ti alloys [144, 147].

Effects	Alloying Elements
α – stabilizer	Al, Ga; Interstitial: N, O, C
β – isomorphous stabilizer	Mo, V, W, Ta
β – eutectoid stabilizer	Cu, Mn, Cr, Fe, Ni, Co, Si; Interstitial: H
Strengthening elements	Sn, Zr (solid solubility in α and β)

can be categorized according to their effect on the stability of the α- and β-phases. Table 7.17 shows the alloying elements (stabilizers) that affect microstructure development [144, 147]. Those that raise the transformation temperature are known as α-stabilizers (Al, O, N, and Ga) and those that decrease the transformation temperature are called β-stabilizers (Mo, V, W, and Ta). Cu, Mn, Fe, Ni, Co, Si, and H are also β-stabilizers but are not typically used since they lead to the formation of a eutectoid [140, 146, 148]. Al is a very effective α-strengthening element at ambient and elevated temperatures up to 550 °C and its low density is an additional advantage. The residual elements (or impurities) such as C, N, Si, and Fe raise the strength and lower the ductility of Ti alloys. The group α-Ti alloy represents the commercially pure Ti alloy grade with O and Fe as the primary alloying elements. α-Ti alloys generally have creep resistance superior to β-alloys and are preferred for high-temperature applications. The absence of a ductile-to-brittle transition, a feature of β-alloys, makes α-alloys suitable for the temperature of cryogenic applications [141, 145]. These group alloys are characterized by satisfactory strength, toughness, and weldability, but poorer forgeability than β-alloys.

Near-α Ti alloys contain a large percentage of α-stabilizers and small amount of β-stabilizers like Mo and V, represented by Ti–3Al–2.5V (Ti–3–2.5), Ti–8Al–1Mo–1V (Ti–8–1–1), and Ti–6Al–2Sn–4Zr–2Mo (Ti–6–2–4–2S). These group alloys are mainly used at operating temperatures between 400 and 520 °C [149].

β-Ti alloys are referred to high hardenability Ti alloys due to the β-stabilizers. The typical alloys include Ti–15V–3Cr–3Al–3Sn (Ti–15–3), TIMETAL 21S (Ti–15Mo–2.7Nb–3Al–0.2Si), Ti–3Al–8V–6Cr–4Mo–4Zr, and Ti–10V–2Fe–3Al alloys [143]. The transition elements, such as Mo, V, and Nb, present in these alloys tend to decrease the temperature of the α- to β-phase transition and promote development of the BCC β-phase. They have excellent forgeability over a wider range of forging temperatures than α-alloys [150]. β-Ti alloys also exhibit high stress corrosion resistance, can be weldable, heat-treated, and offer fabrication advantages, producing sheets owing to their cold rolling capabilities. β-Ti alloys are widely used because they possess the right balance of mechanical properties, including plasticity, castability, weldability, and thermal conductivity [151, 152].

Two-phase α + β Ti alloys, a mixture of α- and β-phases, may contain between 10% and 50% β-phases at room temperature. The common α + β alloys are Ti–6Al–4V (Ti64 or IMI 318), Ti–6Al–2Sn–4Zr–6Mo (Ti–6–2–4–6), and Ti–6Al–2Sn alloys. The properties of these alloys can be controlled through heat treatment, which is used to adjust the amounts and types of β-phase element. A good combination of their properties serves to ensure better performance within the temperature range of 315–400 °C [150]. Three different microstructures can form depending on the thermal–mechanical process used: a lamellar structure, an equiaxed structure, and a duplex microstructure. The combination of α and β stabilizers at different ratios includes typical alloys of Ti–6Al–4V, Ti–6Al–2Sn–4Zr–6Mo, and Ti–6Al–2Sn, which are among the most commonly used commercial Ti alloys [145], which are typically used in the range of 315–400 °C [142]. The addition of Al strengthens the α phase while also increasing $\alpha + \beta \leftrightarrow \beta$ transition and can result in a reduction of the density. However, adding V, a β–stabilizer, reduces transformation temperature [145, 147].

Ti readily reacts with oxygen at 1200 °C in air and 610 °C in pure oxygen, forming Ti oxide [153, 154]. Oxygen is a potent solution strengthener for Ti alloys, but an excess will compromise ductility and fracture toughness [155, 156]. To meet the oxygen requirement of industrial standards for final manufactured components, which is less than 0.2 wt%, [153, 154], most applications require the oxygen content in Ti powder to be less than 0.15 wt%.

The most used Ti alloy in industrial applications is Ti–6Al–4V. It contains 90%Ti, 6%Al, and 4%V, as evident from its name, although the material can contain small amounts of other components, such as oxygen, hydrogen, nitrogen, and iron [157]. Ti–6Al–4V has been extensively studied due to its exceptional properties and is commonly used in AM applications due to its lightweight, high strength, good weldability, corrosion resistance, biocompatibility, and cost savings associated with the efficient use of the material in AM [158, 159]. The alloy does have poor thermal conductivity, which is known to cause complications when machining parts and has implications on the printing process because it induces high thermal gradients. All these studies, and others, published over the last three decades, have helped manufacturers to enhance their qualities [141, 148].

Ti–6Al–4V accounts for over 50% of all Ti alloy usage, mainly in the medical and aerospace industries, making it a well-understood system in conventional forged and cast forms. Ti–6Al–4V is a two-phase alloy consisting primarily of an HCP α phase and a small amount of BCC β. Microstructures are most commonly lamellar, though equiaxed and bimodal structures are possible. The mechanical properties strongly depend on the microstructures formed with these two phases; understanding how these phases form through the printing process is critical for predicting performance in an application. It should be noted that there are some disadvantages of using Ti alloys. It is been reported that the energy required for sponge extraction for Ti is 16 times more than steel, hence the high cost [148]. The material has been found to lose its strength when operating at elevated temperatures above 350–400 °C.

Conventional manufacturing of Ti–6Al–4V parts includes but is not limited to ingot casting or forging, as well as the involvement of multi-axis processes that produce up to 85% of material waste and high cost and time [160]. Previously successful work has shown results are favorable for AM fabricated Ti–6Al–4V parts demonstrating compatibility and, in some cases, superiority to cast parts. However, microstructure and mechanical property anisotropy has also been reported, along with various methods to alleviate it, such as modification of process control, part orientation, and thermal monitoring.

Ti–6Al–4V is an α–β alloy, which means that thermomechanical processing can be used to form several microstructures. This way, a highly customized alloy can be produced for specific applications. For instance, to produce an alloy with a homogeneous microstructure, the solidification of Ti–6Al–4V should be accomplished at a high solidification rate. The microstructure of this alloy is highly dependent on the heat treatment and processing history; therefore, these factors also play an essential role in defining the mechanical properties of this alloy. High-temperature X-ray diffraction can be used to monitor these phase transformations and to analyze the kinetics of this process [161].

Numerous heat treatments have been developed to alter the morphology and volume fraction of $\alpha + \beta$ in Ti–6Al–4V. For example, when heating Ti–6Al–4V above the β-transus temperature, the entire microstructure enters the BCC β phase regime, and then upon cooling back below the β-transus, the grains transform to an acicular structure of transformed α platelets and retained β platelets between them. These platelets are commonly termed laths.

7.5.2 Ti-Based Alloy Powders

High-quality and spherical Ti alloy powder is an essential raw material for net-shape fabrication via powder metallurgy (P/M), MIM, and AM. The cost and quality of Ti powder, including spherical Ti alloy, have been a significant hurdle that prevented P/M Ti from being adopted for a wide range of applications. The demand for spherical and high-purity Ti powder has brought wide attention to the properties and cost, as well as to powder-producing processes. The performance of Ti components manufactured from powder has a strong dependence on the quality of the powder. Understanding Ti alloy powder's quality, properties, and production methods is crucial.

With the advance of AM technologies in the recent decade, the manufacturing of Ti components using **selective laser melting (SLM)**, directed energy deposition (DED), and electron beam melting (EBM) techniques emerged as one of the most critical areas of Ti manufacturing [162, 163]. One challenge for developing these manufacturing technologies is to have high-quality and low-cost spherical Ti alloy powder. Other advanced net-shape manufacturing methods, including MIM and HIP, use spherical Ti or Ti alloy powders to make bulk materials and components. The critical characteristics of spherical Ti powder include particle size and size distributions, flowability, and chemical compositions, especially oxygen content. The requirements of the PSD vary with applications: <45 μm for MIM, 10–45 μm for SLM, 10–45 μm for cold spraying, 45–106 μm for EBM, and 30–200 μm for DED [162].

Commercial spherical Ti powder production methods include gas atomization (GA), plasma atomization (PA), PREP, and **hydride–dehydride (HDH)**. Table 7.18 shows some of the current powder atomization processes used to manufacture powder for AM processes. The PREP powder is widely known to have high purity and a near-perfect spherical shape. Nevertheless, the particle size of PREP powder is typically coarse (e.g. 50–350 μm) [164], as shown in Figure 7.15, which is coarser than the ideal size desired for SLM, EBM, or MIM applications. The fine spherical powder can be produced via GA and PA methods. Typical particle sizes of GA and PA Ti alloy powders range from

Table 7.18 Summary of powder characteristics by manufacturing process [142].

Manufacturing process	Particle size (μm)	Advantages	Disadvantages	Common uses
Water atomization	0–500	High throughput Range of particle sizes Only requires feedstock in ingot form	Postprocessing required to remove water Irregular particle morphology Satellites present Wide PSD Low yield of powder between 20 and 150 μm	Nonreactive
Gas atomization (inc. EIGA)	0–500	Wide range of alloys available Suitable for reactive alloys Only requires feedstock in ingot form High throughput Range of particle sizes Use of EIGA allows for reactive powders to be processed Spherical particles	Satellites present Wide PSD Low yield of powder between 20 and 150 μm	Ni, Co, Fe, Ti (EIGA), Al
Plasma atomization	0–200	Extremely spherical particles	Requires feedstock to either be in wire form or powder form High cost	Ti (Ti64 most common)
Plasma rotating electrode process	0–100	High-purity powders Highly spherical powder	Low productivity High cost	Ti Exotics
Centrifugal atomization	0–600	Wide range of particle sizes with very narrow PSD	Difficult to make extremely fine powder unless very high speed can be achieved	Solder pastes, Zinc of alkaline batteries, Ti and steel shot
Hydride–dehydride process	45–500	Low cost option	Irregular particle morphology High interstitial content (H, O)	Ti6/4 Limited to metals which forms a brittle hydride

10 to 300 μm [165, 167]. The atomized powder can be classified to produce desired size ranges, but classification reduces the yield of a usable size range, increasing the cost of the material. The low yield of fine powder produced by the current commercial methods is the leading technical reason for the high powder cost used for the advanced net shaping processes, especially for AM. Therefore, the recent R&D efforts have primarily focused on improving the yield of fine powder (<45 μm) with an acceptable increase in operating and feedstock cost, mainly driven by the boost in the market of the powder-bed-based AM.

Ti powder's elemental chemical compositions and physical properties' requirements for AM application include high purity, sphericity, flowability, and no trapped gas porosity. One of the critical challenges for producing Ti powder is the control of oxygen content in the powder, especially fine powder. In general, the oxygen content of Ti powder is inversely proportional to the particle sizes [169]. In other words, the smaller the particle size, the higher the oxygen content [167]. The powder flowability can be affected by a few factors, including powder shape and size, interparticle friction, type of materials, and environmental factors [170, 171]. Generally, powder with good flowability should have a spherical shape, and the particle sizes should be reasonably large. The flowability of the powder decreases with decreasing particle size. Furthermore, the powders must have good apparent density and tap density, affecting the density and uniformity of manufactured parts. Spherical Ti alloy powder with low oxygen and good flowability is in high demand. High-quality and spherical Ti alloy powders that meet these requirements, especially powders for AM Ti alloys, are all very costly and in short supply, which hinders the development of Ti for broad applications using AM

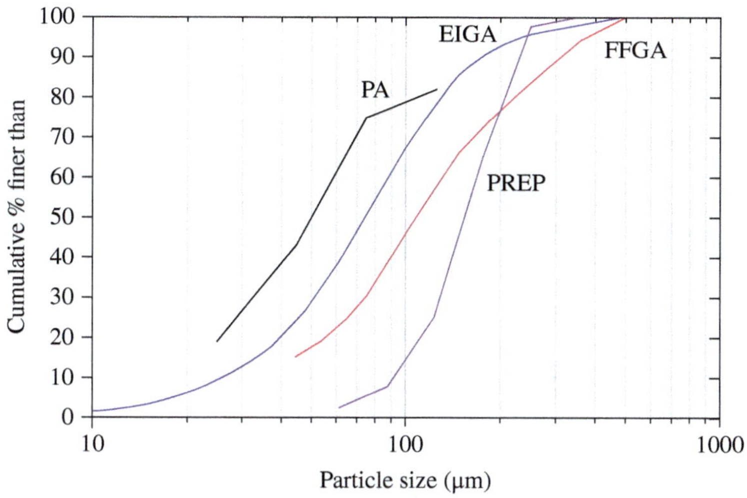

Figure 7.15 Typical particle size distribution of Ti alloy powder produced by free fall gas atomization (FFGA) [165], **electrode induction gas atomization (EIGA)** [166], plasma atomization (PA) [167], and plasma rotating electrode process (PREP) [164]. *Note*: The powder size distribution curves of EIGA, PA, and PREP are for Ti–6Al–4V. Moll [165] did not specify which Ti alloy. Reported PSDs are quoted from the corresponding references. Not meant to be the limit of the methods. Source: Sun et al. [168]/Springer Nature.

and other manufacturing techniques. Therefore, it is strongly necessary to develop new methods to produce low-cost and high-quality Ti alloy powders that meet chemical composition requirements and physical properties.

Ti alloy powders are primarily manufactured with chemical electrolytic or atomization methodologies. Most current atomization processes use energy sources, such as high-pressure inert gas, plasma torches, and induction coils with a nozzle to melt metals, forming spherical droplets and solidifying into powder particles [3, 142, 168]. The quality of the powder is directly related to the production technique. A variety of techniques are used: GA, water atomization (WA), rotary atomization (RA), PREP, PA, and others. Some atomization techniques yield irregular shapes, like WA and RA, others have many satellites (like GA), and some are highly spherical and smooth (like PREP and PA). Figure 7.16 shows the powder quality of Inconel 718 powder produced by different powder atomization processes. These results reveal various powder surface morphologies and different amounts and geometries of internal pores. Porosity in the powder feedstock is typical for certain production techniques, like gas atomization, that entrap inert gas during atomizing droplets from the melt. This trapped gas can be transferred to the AM part due to rapid solidification, resulting in powder-induced porosity in the fabricated materials [173]. These pores are typically spherical, resulting from the vapor pressure of the entrapped gas. Higher-quality powders produced via the PREP do not contain such pores and have been used to eliminate powder-induced porosity in DED and PBF systems [162].

Most commercial spherical Ti powder in the current market is produced by atomization or plasma spheroidization (PS) [142]. Among the atomization techniques, GA, PA, and PREP are used commercially to produce spherical Ti alloy powders. All atomization processes consist of three main integrated steps: melting, atomization, and solidification. Melting can be accomplished by vacuum induction, plasma arc melting, induction drip melting, or direct plasma heating. Atomization involves breaking a melt into droplets using high-energy gases, which solidify during the flight in a cooling chamber under inert gas protection, as described in Chapter 2. The related atomization methods, such as "free fall," "confined," or "closed" nozzles, have been developed to improve the quality of powder manufacturing. Droplets can also be formed by spinning a liquid stream off of a disc, causing molten droplets to form and undergo centrifugal acceleration away from the center of spin. The droplet will subsequently solidify during flight. The atomizing Ti alloy processes are usually conducted in high-purity inert gas, Ar or He, as the atomizing media to minimize oxygen pick up, reducing metal oxidation, which can usually be controlled within the range of 100–500 ppm depending on the specific process and size of the powder. The low heat capacity of gas as the atomizing media means that the metal droplets have a relatively long solidification time resulting in spherical powder. The PSD can range from 10 to 150 μm, which is influenced principally by the gas-to-melt flow rate ratio during atomization. Each specific atomization technique varies in detail from others, but they all share the same three main steps as described earlier.

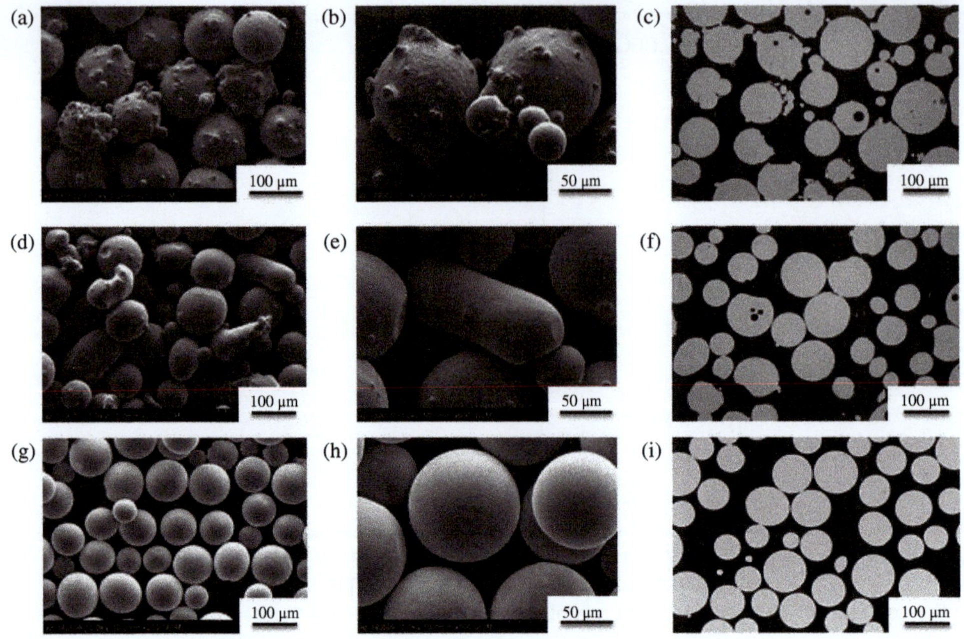

Figure 7.16 SEM of AM powder manufactured by (a) gas atomization (GA), (b) Satellite particles on GA powder, (c) gas pores from GA powder, (d) rotary atomization (RA) processed (e) elongated particle by RA, (f) smaller gas pores than GA powder, (g) Plasma Rotating Electrode Process (PREP), (h) PREP spherical particle, (i) no gas pores. Source: Sames et al. [162, 172]/John Wiley & Sons, Inc.

Crucible Materials Corporation developed the gas atomization of Ti alloys in the 1980s [174, 175]. In this process, elemental raw materials or prealloyed Ti alloy ingots were molten with induction skull melting in a water-cooled copper crucible vessel under a vacuum or controlled atmosphere. The typical process proceeds as follows. First, the Ti alloy of interest is homogenized by holding the material in the molten state for a specific time. Second, the melt is poured into a refractory metal nozzle located in a tundish and then atomized using high-pressure inlet gas, which is typically Ar rather than He, primarily based on economic considerations.

Another variation of GA is EIGA, whereby a metal bar feedstock is rotated and melted by an induction coil. This method is mainly used when processing reactive alloys such as Ti alloys as molten Ti is very reactive to most common metals and ceramics. EIGA, as shown in Figure 1.10, was developed by ALD Vacuum Technologies Company (Hanau, Germany) [176] to produce "ceramic-free" powder, in which the melt is not in contact with any ceramic components that might introduce contamination to Ti alloy powder [177]. A prealloyed rod (25–70 mm) is rotated at a very slow speed and melted in a conical induction coil [178]; then, the melt falls into a gas nozzle to be broken into fine droplets. The diameter of the electrode rod can be increased up to 90 mm [166] or 120 mm [177] to increase productivity. To minimize the presence of contaminants during atomization, a gas atomization apparatus with a Ti coating on the inner wall of the atomization chamber and other components in the flow path was designed and developed [179]. Compared to PA, the cost of powder is lower due to the use of gas, but less spherical and has more satellites than PA powder.

The principle of PA is shown in Figure 7.17a. PA is the process of melting a wire spool metal feedstock with a plasma torch and cooling it in an inert tower [180]. PA was developed to produce fine and spherical powder in 1996 [181, 182]. In the PA process, a prealloyed wire, e.g. 1/16 in or 1/8 in., is fed into a hot zone, around 10 000 K, heated by plasma torches. The wire is melted and broken into droplets that can be cooled rapidly with a typical cooling rate of 10–1000 °C s^{-1}. The plasma-atomized Ti powder has high purity because the melt does not contact any solid materials that may contaminate the powder before solidification.

The yield of fine powder and the energetics of PA can be controlled by varying the diameter and the feed rate of the wire, the inlet gas pressure (or gas-to-metal ratio), the angle of attack between the wire and plasma jets, and the distance between the wire and the plasma outlet [183]. Published data show that the yield of fine Ti–6Al–4V powder can be improved from 39.9% to 59.6% by increasing the G/M from 8.7 to 12.9 and shortening the distance from the

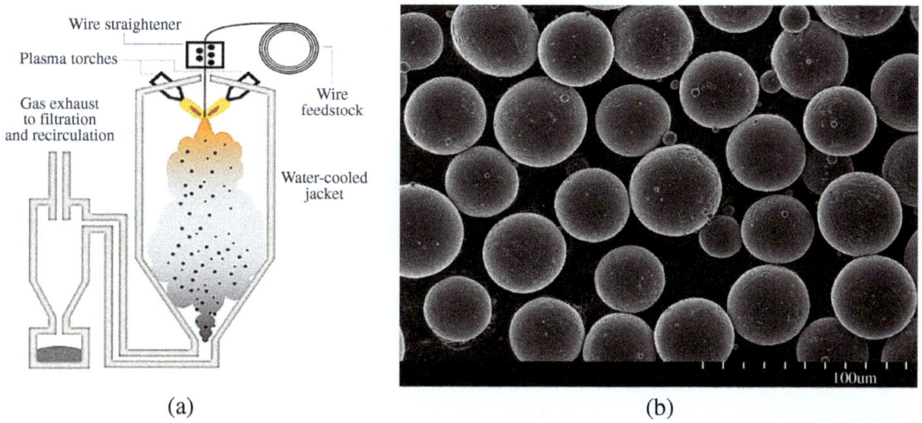

Figure 7.17 AP&C GE Additive (a) advanced plasma atomization process, and (b) produced SEM image of PA Ti-6Al-4V (45–106 μm). Source: Wheeler [180].

wire and the plasma outlet from 25 to 19 mm [184]. Furthermore, the production rate can be significantly improved by adding an induction coil to preheat the wire before being fed into the plasma [183].

Spherical particles with a small amount of relatively small satellites are shown in Figure 7.17b above. The atomization process environment operates in a vacuum to minimize metal oxidation levels and reduce gas entrapment. Although the process is limited to alloys available in a wire spool batch, this allows the traceability of the powder batches. According to powder manufacturer AP&C GE Additive, PA powder exhibits the following characteristics: spherical particle morphology with minimal satellites or internal pores. The level of metal oxidation is reasonably high, with particle sizes up to 200 μm. It also possesses good flow and packing properties required for AM process.

The plasma-atomized Ti alloy powder has near-perfect sphericity and fewer satellite particles than the gas-atomized powder. The issues of internal porosity resulting from trapped gas during atomization and some satellite particles are still concerns for PA Ti alloy powder. Unlike EIGA powder, the PA process has much better spherical particles than EIGA. The main drawback of this process is that the feedstock has to be in the form of wires. In addition to the high cost of Ti wires, this technique can only be used to produce an alloy powder if the wire form of the alloy is available. However, the cost of powder can be higher compared to EIGA powder. The PREP atomization is known to produce better spherical powder but comes at a higher cost [142].

The PREP was initially developed as a refinement of the production method called the REP that Nuclear Metals, Inc. developed in the 1960s [185]. In PREP, the arc from a tungsten-tipped cathode melts the metal electrode rod in the REP. The rod (usually with diameters of 89 mm or 63.5 mm) spins at speeds of 3000–15 000 rpm [186], so the liquid melt is spun off from the electrode surface to form droplets because of centrifugal forces. The ejected droplets solidify to form solid spherical particles during flight. S. Abkowitz reported the production of spherical Ti Alloy, Ti-7Al-2Nb-1Ta, powder using REP in 1966, and the particle size was approximately 150 μm [187]. It is worth noting, however, that discrete tungsten particles were found in HIPed Ti-6Al-4V fabricated using REP powder, which is detrimental to fatigue properties [188]. Later, the heat source was replaced with a transferred arc plasma torch to avoid the introduction of tungsten inclusions [164]. In REP, Helium is preferred because of its improved heat transfer properties and electric arc characteristics [164].

PREP is widely used to fabricate spherical Ti alloy powders, in part, because of the purity that it can achieve. During PREP processing, the melt does not come into contact with other metals or ceramics before solidification. PREP Ti powder has very few pores because the metal droplets are produced by centrifugal forces rather than by high-pressure gas. PREP powder has fewer satellite particles compared with the other production methods. As discussed earlier, the satellite particles formed during gas atomization are likely a result of the backflow of very fine particles to the spray plume. In PREP, the droplets fly radially away from the metal surface with a centrifugal force. Because the droplets moved in order, the chance of collisions of droplets and particles to form satellites is very low [186].

The challenges of producing high-purity, low-cost Ti, and Ti alloy powders have fueled continuing research and development efforts to optimize existing technologies and develop novel approaches.

For example, the **HDH** process can produce titanium, zirconium, vanadium, and tantalum powders. The process consists of performing a reversible reaction on the material by hydrating Ti alloy at elevated temperatures

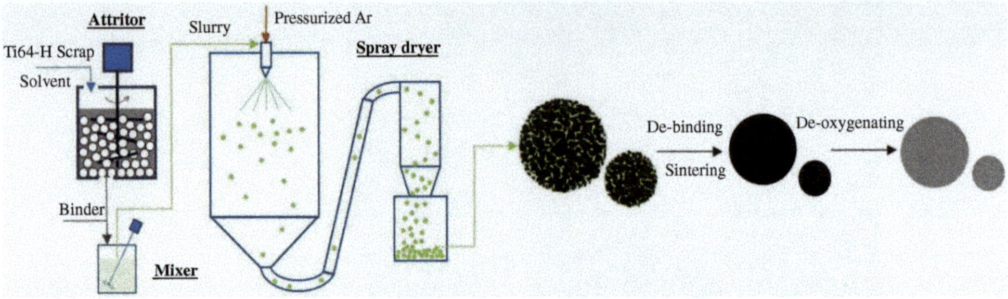

Figure 7.18 Flowchart of **granulation-sintering-deoxygenation (GSD)** method. Source: Sun et al. [198]/with permission of Elsevier.

(~650 °C) for long periods, allowing the brittle phase of the Ti alloys to be crushed and milled for finer powder size. Powder is then reheated under a high vacuum at ~350 °C to remove the hydrogen [169, 189]. The result is an angular-shaped (nonspherical) powder that can be used successfully for fabrication in traditional powder production methods. HDH powder is considered a low-cost and lower-grade metal powder, as it is irregular. HDH powders are produced via mechanical processes and have relatively high amounts of oxides and other contaminants, including chlorine. HDH-produced particle sizes can range up to 500 μm due to their irregular morphology, and flowability and packing density are generally lower than those of spherical powder. Therefore, HDH powders are usually not used for AM but are more suited to traditional methods, such as pressing sintering, MIM, cold isostatic pressing, and HIP [189, 190].

Efforts to use lower cost and lower quality powder produced by the HDH process in the EBM process have demonstrated that this powder type can lead to issues with porosity [190]. Consequently, HIPing and other special double-melting techniques have been used as approaches to reduce porosity in both HDH and HDH-blended powders. These results underscore the importance of powder quality to microstructure and the potential to use both processing and postprocessing to overcome feedstock issues.

PS is widely used to make irregular powder spherical. In PS, irregular powder, such as that produced by HDH, is heated in a plasma gas stream to melt the particles. The energy from the plasma torch melts the powders allowing them to solidify into near-perfect spheres as they travel inside an environmental chamber [191]. The unique characteristic of PS is that the particle sizes do not change during plasma processing. Plasma-spheroidized particles typically have a nearly perfect spherical morphology compared to other atomized powders [192]. Feedstock materials for PS can be HDH powder or any irregularly shaped Ti powders [193–196]. The challenge for PS is to produce fine spherical Ti alloy powder with low oxygen at a low cost. It was reported that using Ti hydride powder as feedstock helps to improve the yield of fine powder of PS [197]. The feed powder largely determines the impurity level of the plasma-spheroidized Ti powder. The availability of low oxygen and low-cost fine Ti powder is currently limited and costly. Another potential issue of the PS process for making Ti alloy powder is at risk of losing the low melting point element (e.g. Al) due to evaporation at the plasma temperatures.

Recently, a new approach, **granulation-sintering deoxygenation** (GSD), was developed for making spherical Ti powder [198, 199], as shown in Figure 7.18, illustrating the critical steps of this process. GSD process includes three main stages: (i) Granulation, Ti alloy hydride or Ti hydride with master alloy, e.g. hydrogenated from Ti sponge or Ti alloy scrap, is milled to fine particles and then granulated to spherical granules in the desired size range using spray-drying; (ii) Sintering, the spherical granules are sintered to obtain dense Ti particles. (iii) Deoxygenation, the densified spherical Ti powder with high oxygen content, is deoxygenated with Mg to desired purity. The spherical fine Ti alloy powder produced by the GSD method is low cost as it has much lower waste and can use low-cost starting powder. The yield from the GSD process is near 100%, with a very narrow powder size distribution without much yield loss, as all the powder that is either under or oversized particles can be recirculated through the process.

In addition to approaches involving spheroidization spheroidizing or producing particles in a molten state, research has also demonstrated that it is possible to modify particle morphology in the solid state via mechanical means [200, 201]. For example, the flowability of irregularly shaped particles can be improved by removing sharp angles on the particles through high-speed blending or high sheer milling.

The powder quality is characterized in accordance with ASTM standards. To determine the safe use of powder in process or to handle in the subsequent P/M and AM processing, a minimum ignition energy (MIE) test can be carried out following standard BS EN 13821:2002 to ensure and minimize any fire ignition of powder due to metal reaction with other metal during handling [202].

In summary, the dominant processes for producing Ti alloy powders are gas atomization, PA, and PREP, each of which has advantages and disadvantages. Significant progress in new production techniques has been made in improving the yield of high-quality spherical fine Ti alloy powder and reducing the cost to make the fine powder affordable for a broader range of end-use applications, including AM.

7.6 Cobalt-Based Alloys and Powder

7.6.1 Co-Based Alloys

Cobalt (Co) can only be found in a chemically combined form in the earth's crust, except for small deposits in natural meteoric iron alloys [203]. Co–Cr compositions, also known as **"Stellites,"** were first patented by Elwood Haynes in 1907 and have been in use ever since. Commercial Co-based alloys are often derived from the Co–Cr–W and Co–Cr–Mo ternaries. These alloys possess excellent magnetic properties, corrosion resistance, **wear resistance**, and high-temperature strength. The properties of Co-based alloys arise from the crystallographic nature of Co, the solid-solution-strengthening effects of Cr, W, and Mo, the formation of metal carbides, and the corrosion resistance imparted by Cr. Harder Co alloys are used for resistance to wear, while tougher Co compositions are employed for high-temperature applications such as gas-turbine vanes and buckets. Due to their excellent biocompatibility, Co-based alloys are also used to produce artificial joints.

Cobalt alloys typically exhibit an face-centered cubic (FCC) crystalline structure and share similar alloying elements like Cr, Ni, Mo, and N with stainless steels. The formation of a passive layer of chromium oxide is responsible for their corrosion resistance, similar to stainless steels. However, cobalt alloys tend to have a higher carbon content than implantable stainless steels, resulting in the distribution of carbides throughout the material and promoting increased adhesive wear resistance. Furthermore, cobalt alloys exhibit excellent corrosion resistance compared to stainless steel due to their high wear resistance.

High-temperature alloys possess a unique set of properties, including high strength and toughness, resistance to degradation under thermomechanical cycling, and good stability in severe environments such as oxidation and corrosion [204, 205]. Although conventional Co-based superalloys are less strong than Ni-based superalloys, they maintain their strength at higher temperatures [206]. Table 7.19 provides the nominal compositions of representative Co-based alloys that have been extensively studied.

It should be noted that Co-based superalloys possess a coherent γ–γ' (Co_3(Al,W)) microstructure, which is almost identical to that of Ni-based superalloys. This feature enables Co–Al–W superalloys to exhibit exceptional high-temperature properties. The metastable Co_3Al γ' is stabilized by alloying with W, giving these Co-based alloys a melting temperature that is 100–150 °C higher than current Ni-based superalloys [207].

The solvus temperature of γ' precipitates is a critical factor that influences the high-temperature behavior of Co-base superalloys. Table 7.20 displays the γ' solvus temperature for some Co-based alloys [208]. Other Co–Al–W-based alloys that have been studied include Co–9Al–7W–2Ta–0.02B, Co–9Al–7W–2Ta–0.02B–2Hf, and Co–9Al–7W–2Ta–0.02B–2Nb, as well as Co–9Al–(9–x)W–2Ta–0.02B–xMo [212].

Cobalt-based superalloys have a similar FCC structure to Ni-based alloys at high temperatures. In 1971 [213], it was first reported that two-phase alloys within the ternary Co alloy system consist of an FCC (γ) matrix strengthened by coherent L12-ordered precipitates (γ'), analogous to Ni-based alloys. An intermetallic γ'-(L12)-ordered FCC phase in the Co–Al–W system was reported in 2006 [214], resulting in a new class of high-temperature alloys that may increase the maximum operating temperatures of gas-turbine engines. Notably, other intermetallic results were reported for the Co system, such as Co_3Ti [215] or Co_3Ta [216], but none of these alloys was capable of undergoing heat treatment at high temperatures. However, there are some matters needing to be addressed, such as a lower γ' solvus temperature compared to Ni-based alloys [214, 217], increased mass density due to the large amounts of W needed to stabilize the γ'-phase, and reduced strength and yield stress anomaly compared to Ni_3(Al, Ti). Nonetheless, the existence of this new class of precipitation-strengthened Cobalt-based alloys opens the door to the possibility of obtaining a dual phase γ/γ' with satisfactory properties at high temperatures, like nickel alloys.

The advantages of Co-based alloys include higher melting temperatures and flatter stress rupture curves, which provide a useful stress capability to a higher absolute temperature than polycrystalline Ni- or Fe-based superalloys.

Table 7.19 Compositions of representative Co-based alloys.

		Nominal composition (wt%)															
Alloy	**Description**	**Co**	**Ni**	**Al**	**W**	**Ta**	**Ti**	**Cr**	**Mo**	**Fe**	**Mn**	**Si**	**C**	**B**	**Zr**	**Hf**	**Other**
Co-based γ + γ′ superalloys																	
Co–9Al–9W	Low γ′ fraction	Bal.		3.7	24.6												
Co–9Al–11W	High γ′ fraction	Bal.		3.7	28.4												
Co–12Al–11W	Near γ′ single phase	Bal.		4.7	29.4												
Co–9Al–10W–2Ta	Stable γ′	Bal.		3.4	25.4	5.1											
Co–8Al–8W–1.5Ta–4.5 Cr	Stable γ′, oxidation	Bal.		3.1	21.2	4		3.5									
Co–7Al–8W–6Ti	Stable γ′	Bal.		2.7	22.5		4.5										
Co–10Ni–9Al–9W–0.06C–0.01 B	Co–Ni-based	Bal.	8.7	3.7	24.6								0.001	0.002			
Co–29Ni–10Al–6W–6Cr–2Ta	Co–Ni-based	Bal.	26	4	17.5	6.6		5									
Conventional Co-based alloys																	
Haynes188	Wrought alloy		22		14			22		<3	<1.5	0.35	0.1	<0.015			0.03La
L605	Wrought alloy		10		15			20		<3	1.5	<0.4	0.1				
FSX414	Cast alloy		10		7			29		<2	<1	<1	0.25				

Source: Suzuki et al. [206]/Annual Reviews.

Table 7.20 γ′ solvus temperature in the current and reported Co-based alloys (at%).

Alloy	**γ′ solvus temperature (°C)**	**References**
Co–7Al–7W–4Ti–2Ta	1157	[208]
Co–7Al–6W–4Ti–2Ta–1Mo	1143	[208]
Co–9Al–10W	1029	[209]
Co–9Al–10W–2Mo	1018	[209]
Co–9Al–7W	965	[210]
Co–9Al–7W–2Mo	977	[210]
Co–9Al–8W–2Ta	1087	[211]
Co–9Al–8W–2Ta–2Mo	1087	[211]

Source: Zhou et al. [208]/with permission of Elsevier.

These alloys also exhibit superior hot corrosion resistance, high concentration of Co, Ni, Cr, and W providing good resistance against lead oxides, sulfur oxides, and other corrosive compounds in the combustion gas turbine. In addition, Co-based alloys exhibit superior thermal fatigue resistance and weldability compared to Ni-based alloys.

Co–Cr-based alloys (Stellite) have been widely used in various applications such as automotive, aerospace, marine, petrochemical, and medical due to their desirable properties such as high strength, wear/corrosion resistance, hardness, and biocompatibility [218–220]. Chromium, tungsten, and molybdenum alloying elements contribute to strengthening cobalt-based alloys with the formation of a solid solution matrix, and with the presence of carbon, carbide precipitates can contribute to the strengthening mechanism [221, 222]. Co–Cr alloys are commonly used in biomedical applications such as dental restoration, e.g. dental bridge, dental crown, denture, etc. Traditional processing methods for the part production of these alloys include casting, powder metallurgy, and metal forming [223, 224].

Metal AM has become a popular method for producing parts in industries such as aerospace and biomedical, where customization and low production volumes are often necessary. Co-based alloys are particularly suitable for AM due to the convenience it offers compared to traditional manufacturing techniques.

7.6.2 Co-Based Alloy Powders

Co and Co alloy powders are used in various applications, including the production of superalloys, permanent magnets, and hard-facing alloys [203]. Co-based alloys are also used in oral and orthopedic surgery due to their excellent properties [203]. In the powder metallurgy industry, Co is a significant material for improving the high-temperature properties of superalloys, the cutting capability of sintered carbide and high-speed tool steels, and the toughness of high-strength steel [203]. Co alloy powders are produced using various techniques, including atomization, the hydro-metallurgical technique, reduction of oxides, carbonyl processing, the electrolytic method, and the amalgam technique [203]. This section focuses on atomization, while other production methods can be found in related literature [203, 225].

Atomized Co-based alloy powders are widely used in high-temperature applications, such as hard-facing, and gas and water atomization are the commonly used fabrication methods for producing hard-facing powders [226]. Commercial Co-based hard-facing alloys can be classified into two types, those containing the **Laves phase** and carbide-containing Co alloys. The Laves phase is a kind of intermetallic compound that has a hexagonal structure similar to M_7C_3 carbides, but with a lower hardness value, typically 1000–1200 HB [226]. Laves phase materials retain mechanical properties over a wide range of temperatures and are mainly used for plasma spray hard-facing operations, such as plasma spray coating of gas-turbine blades. Two Laves phase-containing cobalt-based alloys available for hard-facing applications are Co–28Mo–8Cr–2Si and Co–28Mo–17Cr–3Si, which contain at least 50 vol% of Laves phase [226]. A typical carbide-containing Co alloy, with a composition of Co–28Cr–4W–1.1C, is a eutectic alloy containing 16–17 vol% M_7C_3-type carbides in a Co-base matrix alloyed with Cr and W [226].

Carbide-containing Co alloys have a unique microstructure consisting of a matrix and a dark phase called M_7C_3. The matrix is the lightest colored phase and has a FCC crystal structure. On the other hand, M_7C_3 has a HCP crystal structure with a composition of $(Cr_{0.85}Co_{0.14}W_{0.01})_7C_3$. The presence of carbides imparts resistance to abrasive wear, while the matrix provides resistance to corrosion and/or elevated temperature hardness retention and defines galling wear properties.

Gas atomization is the widely used technique for producing Co alloy powders, as it minimizes oxygen content and contamination. Some producers prefer vacuum melting and inert-gas atomization using nitrogen or argon. The oxygen content in these atomized powders is significantly below 1000 ppm, while the nitrogen level varies from 600 to 2000 ppm [123].

In the plasma-transferred arc hard-facing process, the powder is introduced into a combined arc/plasma stream to form a molten pool on the workpiece. The alloy is carried from a powder feeder to the plasma torch in a stream of argon gas, which forms the plasma and is directed from the torch into the arc effluent where it is melted, and the fusion is bonded to the base metal. The powder used for this process ranges from about 150–45 μm in size. Coarser powders can also be used because the transferred arc easily melts such powder. Deposits ranging from 0.64 to 3.2 mm thick can be produced rapidly in a single pass [227].

The Co–Cr–W–C system alloys retain high hardness at elevated temperatures, mostly from carbide and solid-solution strengthening rather than precipitation hardening. The alloys RCoCrC and RCoCrB are higher hardness modifications of alloy RCoCrA because of higher carbon and tungsten contents. Alloy RCoCrC contains a primary M7C3, a eutectic M_7C_3 plus matrix, and a secondary M6C. Alloy RCoCrB is similar to alloy RCoCrA, but it has more M_6C and, therefore, contains inter-dendritic eutectic carbide (chromium and tungsten rich) [227].

Co-based hard-facing powders are ideal for use in applications that require wear resistance, corrosion resistance, and high-temperature properties. Duflos and Stohr [228] compared the microstructures of gas-atomized powders and melt-spun ribbons of Satellite 6 (27.7Cr–4.0W–1.43Si–1.22C-Bal.Co) and investigated the influence of quench rates

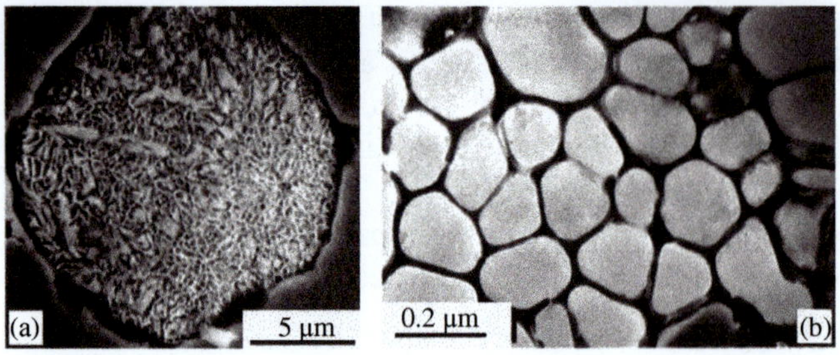

Figure 7.19 Cellular structure observed in a 15 μm diameter gas-atomized Co alloy powder particle: (a) scanning electron micrograph (SEM) on a polished section; (b) transmission electron micrograph (TEM) on a thin foil. Source: Duflos et al. [228]/Springer Nature.

on resulting microstructures. Gas atomization with nitrogen was used for Satellite 6, while cobalt-based alloys were spun to ribbon by **melt spinning** with a copper roller. As shown in Figure 7.19, the resulting powders exhibited a cellular structure, with a cell diameter of 0.1–0.2 μm in a 15 mm diameter particle of the gas-atomized powder [228]. Sandvik Osprey Company utilizes gas atomization technology to produce Co alloy fine powders, with commercial powders containing 20–29.0 wt% Cr, 1.0–10 wt% Ni, 2.0–15.0 wt% W, 0.75–3.0 wt% Fe, and 0.02–2.0 wt% C [229]. The powders have a predominately spherical shape and a mean particle size ranging from 6.7 to 12.0 μm, depending on the grade (Co 212, Co 502, and Co 90).

Gas atomization has been used to atomize Co-based metallic glass or amorphous powders [230]. **Amorphous alloys** are a specific group of alloys where the atoms are arranged without repeating the long-range order [230]. To obtain an amorphous structure, molten alloys must be cooled at a rate higher than the critical cooling rate. High cooling rates block dynamic atom diffusion, enabling the development of an amorphous solid-state structure of metal alloys [231].

Gas atomization is an effective method for fabricating metallic glass powder due to its ability to obtain spherical and chemically homogeneous powder particles, which is essential to maintain high **glass-forming ability (GFA)** of the alloys [231–235]. In contrast, in-situ alloying of blended powder by mechanical ball milling can be cost-effective, but it often requires heat treatment to homogenize the chemical composition, which is unfavorable for fabricating metallic glasses [236].

Experimental results showed that fully vitrification of powders depends on the size of atomized particles [231–235, 237] and GFA of the processed alloy [238]. The pressure of the gas used for gas atomization has been found to have a correlation with PSD [239]. Fully amorphous powders were obtained from gas atomization for particle size <45 μm.

The production of bulk metallic glasses through AM has been successfully achieved using metallic glass powders obtained via gas atomization [240]. Ti-based and Fe-based metallic glass samples have been built through SLM using gas-atomized metallic glass powders with varying particle sizes [241, 242]. Although the Fe-based metallic glass powder used in SLM was not in amorphous state and contained a volume of crystalline phase of approximately 10%, the built parts were nearly amorphous.

Metallic glasses offer new possibilities for obtaining materials that are harder, more durable, and have better anticorrosion and magnetic properties. Cobalt-based **amorphous alloys** are known for their exceptionally high hardness [243, 244]. For instance, the microhardness of (CoFe)YB bulk metallic glasses is approximately 1336 HV, while that of CoNiTaB is 1410 HV [243, 244]. Co-based metallic glasses with ultra-high strength and high hardness are ideal surface-coating materials that can enhance the hardness and wear resistance of substrate materials [31]. A Co-based amorphous coating was fabricated from atomized powder by laser cladding, after preparing Co-based amorphous powder with high GFA through gas atomization technique [232]. Before gas atomization, a master alloy of $Co_{55}Ta_{10}B_{35}$ (at%) was synthesized using pure Co, Ta, and B. The resulting amorphous powders below 300 mesh were then sieved for laser experiments to ensure consistent flowability.

Co is a natural ferromagnetic material with a magnetic moment of 1.7–1.75 μB. CoFeSiB bulk metallic glass soft magnetic alloys are excellent candidates for small-sized inductor or transformer applications due to their high electrical resistivity [245]. The most popular soft magnetic Co-based alloys, such as Vitrovac and its modifications, are characterized by relatively low coercivity, high saturation magnetization, low eddy current, and high magnetic permeability [245]. Moreover, the supercooled liquid region (ΔT_x) for some Co-based alloys can reach $111 \pm 5\,°C$, enabling the development of samples with various shapes [246].

Metallic glass powders with high GFA in the form of powder were atomized to develop Bulk Metallic Glasses (BMGs) via AM. Co-based metallic glass powders of $Co_{47.6}B_{21.9}Fe_{20.4}Si_{5.1}Nb_5$ (at%) and $Co_{42}B_{26.5}Fe_{20}Ta_{5.5}Si_5Cu_1$ (at%) were developed through gas atomization [247]. The alloying process was carried out in an induction vacuum furnace and maintained at high temperature for one hour to ensure that all elements were melted and mixed uniformly. The cast bars were remelted under vacuum in a crucible and then gas atomized using high-pressure Ar. The resulting atomized powders were annealed at *Tg* and *Tx*, and their properties were analyzed using Mössbauer and magnetic measurements. The obtained powders were spherical and chemically homogeneous, indicating that gas atomization was an efficient method for producing Co-based metallic glass powder. Annealing resulted in an increase in mechanical properties such as hardness and **elastic modulus** of $Co_{47.6}B_{21.9}Fe_{20.4}Si_{5.1}Nb_5$ and $Co_{42}B_{26.5}Fe_{20}Ta_{5.5}Si_5Cu_1$, which was attributed to the crystallization of the amorphous structure. The magnetic study revealed that $Co_{47.6}B_{21.9}Fe_{20.4}Si_{5.1}Nb_5$ and $Co_{42}B_{26.5}Fe_{20}Ta_{5.5}Si_5Cu_1$ were soft magnetic and semi-hard magnetic materials, respectively.

7.7 High-Entropy Alloys and Powders

7.7.1 High-Entropy Alloys

In the realm of materials science, alloying has been a longstanding technique employed to impart desired properties to a material. It typically involves adding small amounts of secondary elements to a primary element. In recent times, there has been a surge of interest in a novel alloying approach known as HEAs. This technique involves the incorporation of multiple principal elements in high concentrations to create new materials with unique properties [248].

HEAs are designed based on a multi-principal element approach to engineering a unique **random solid solution (RSS)** crystalline structure. In this structure, each constituent element has an equal probability of occupying a given lattice site. The presence of RSS phases in certain HEAs is responsible for the exceptional chemical, physical, mechanical, and functional properties that have been reported in published literature.

HEAs have revolutionized the world of alloys by breaking free from the limitations of phase diagrams and opening a vast central region that offers a multitude of possibilities for creating new alloys with novel properties. Over the past two decades, HEAs have evolved into **complex concentrated alloys** (CCAs), which possess even greater levels of compositional and microstructural complexities that can meet the requirements of alloys for extreme service conditions. The unique advantage of HEAs lies in their ability to explore a practically limitless multi-dimensional compositional space. Despite their discovery in 2004, only limited regions of this compositional space have been investigated so far. Given the extraordinary properties exhibited by HEAs, it is not surprising that there has been a significant increase in research interest in them in recent years, and this trend is expected to continue.

The earliest reports of HEAs date back to 2004 when two research groups, Prof. B. Cantor and Prof. J.-W. Yeh simultaneously published papers describing the development of crystalline multicomponent alloys [249, 250]. Cantor's initial work can be traced back to his undergraduate thesis in the late 1970s. His research was later adapted into another project in 1998. In 2004, Yeh published several studies, including the one where he first introduced the term "HEAs"[251–253]. These studies described the concept of synthesizing metallic alloys with near-equi-atomic concentrations of five or more constituent elements. Upon characterizing these alloys, it was found that far fewer phases were formed than initially predicted based on the Gibbs phase rule [249].

Furthermore, some compositions exhibit a single crystalline structure where each constituent element has an equal probability of occupying a given lattice site. These uncommon observations piqued the interest of the materials science community. The number of publications on HEAs has grown exponentially from 2004 to 2020. Interestingly, this growth is in stark contrast to the linear increase of publications on single-phase HEAs [248, 254]. The publications on HEAs primarily focus on three aspects: understanding the mechanisms behind the exceptional properties exhibited by these alloys, gaining a fundamental understanding of their unique structure, and developing the ability to predict HEA compositions.

In 2006, to explain the observations of simple RSS phases in early HEA studies, Yeh proposed four core effects of multi-principal element composition [255]. Yeh proposed four core effects of multi-principal element composition to explain the observations of simple RSS phases in early HEA studies. These effects are as follows:

(1) The high-entropy effect, which explains the formation of an RSS phase by thermodynamic stabilization due to a maximization of mixing entropy.
(2) Severe lattice distortion inherent to the RSS, which has a significant effect on the mechanical and physical properties.
(3) Sluggish diffusion kinetics that potentially cause the nanocrystalline stability and corrosion resistance observed in HEAs.
(4) The cocktail effect, which treats HEAs as a composite material of elements selected for desired properties.

As more experimental results have been published on a variety of HEA compositions, the validity of these four core effects has been studied, debated, and, in some cases, confirmed. The community's current understanding of each effect is discussed below.

Several HEAs exhibit exceptional properties, including new deformation mechanisms that surpass those of conventional alloys. Variants of the equimolar Cantor alloy have been developed with fewer alloying elements or nonequimolar compositions. One of the reasons for this development is to reduce the stacking-fault energy to promote twinning-induced plasticity (TWIP) [256, 257], and/or transformation-induced plasticity (TRIP) [254, 258, 259] and, thus, to realize materials with superior strain hardening ability. In a set of experiments on Cantor-related HEAs with composition Fe80−xMnxCo10Cr10, where the Mn content was varied between $x = 30$ at% and 45 at%, it was observed that the Mn-rich alloy (45 at%) deforms by developing complex dislocation patterns. On the other hand, the 40 at% Mn variant was found to be a TWIP alloy, while the 35 at% Mn variant was a TRIP alloy, reflecting the influence of the reduction in stacking-fault energy with decreasing Mn content. This observation is illustrated in Figure 7.20 [260]. The X-ray diffraction spectra in the above microstructure images indicate the dominant deformation mechanisms of the respective alloys. The peaks in the spectra correspond to the phase fractions present in millimeter-sized sample regions. The images also show the associated deformation mechanisms, with the Fe80−xMnxCo10Cr10 alloys displaying different deformation mechanisms based on their Mn content. Specifically, the Mn-rich alloy (45 at%) deforms through the development of complex dislocation patterns, while the 40 at% Mn variant exhibits TWIP and the 35 at% Mn variant undergoes TRIP [260].

HEAs continue to garner significant interest in the material science community due to their exceptional mechanical and functional properties. Specifically, single-phase FCC HEAs are primarily composed of near-equi-atomic combinations of 3D transition metals. Among these, the CrMnFeCoNi alloy has been extensively studied and has

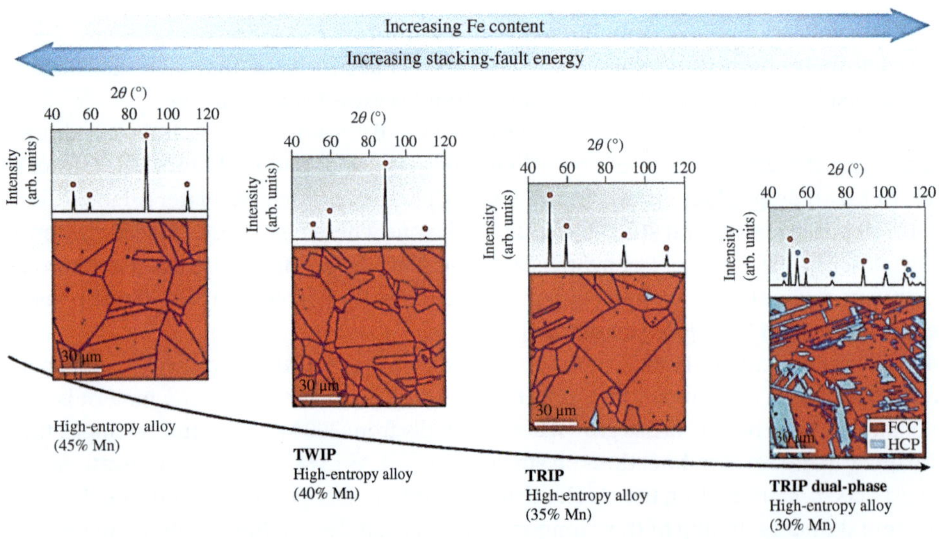

Figure 7.20 Tuning the stacking-fault energy and the phases in a set of nonequimolar derivatives of the Cantor alloy. The phase fractions (red, FCC; blue, HCP) for the alloy system Fe80−xMnxCo10Cr10 change as a function of Mn content. Source: Adapted from Li et al. [260]

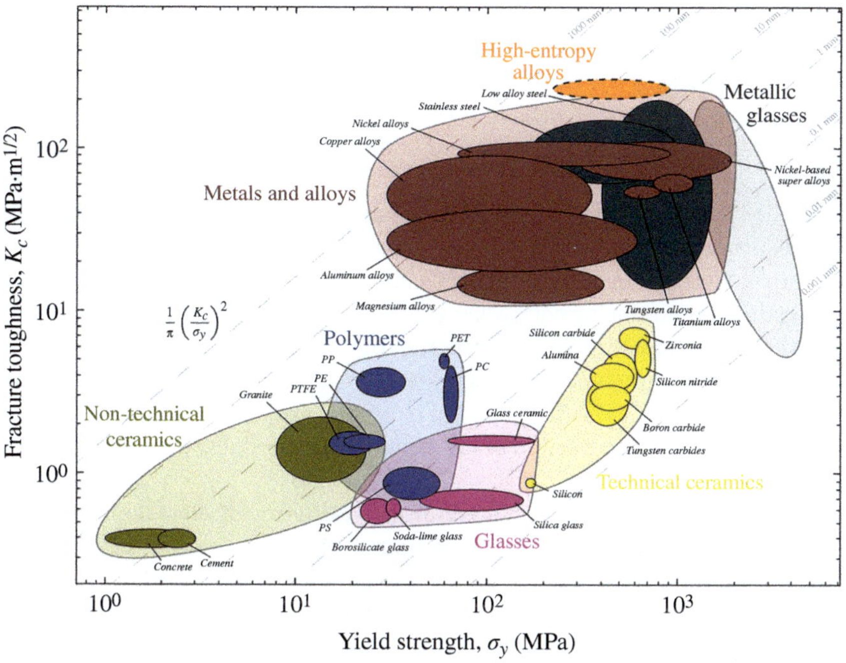

Figure 7.21 Ashby map presenting fracture toughness as a function of tensile yield strength including HEAs and many other material systems. Source: Gludovatz et al. [261]/American Association for the Advancement of Science.

demonstrated remarkable mechanical properties. Gludovatz and colleagues have reported on the high strength, high fracture toughness, and high work hardening rate of this alloy, which is attributed to the enhanced solid solution strengthening, deformation twinning, and microstructure refinement induced by the high-entropy effect [261] synthesized the equi-atomic CrMnFeCoNi composition via arc melting followed by cold forging, a rolling process, and annealing to achieve a sheet sample with a refined equiaxed grain structure. The fracture toughness and tensile yield strength of this material are compared to those of common material systems in Figure 7.21 [261]. The fracture toughness of CrMnFeCoNi, a single-phase FCC HEA composed of 3D transition metals in near-equi-atomic proportions, is exceptionally high. Although its yield strength does not exceed that of all metals and alloys, it exhibits noteworthy mechanical properties. This is attributed, in part, to the formation of planar glide dislocations that generate cell structures, promoting strain hardening during plastic deformation. Otto et al. [262] conducted an experiment on the CrMnFeCoNi alloy by arc melting and cold rolling it to study its mechanical behavior at cryogenic temperatures. The results showed that the alloy exhibited increased yield strength at low temperatures while maintaining its ductility. This was attributed to the addition of deformation-induced nanoscale twinning that further enhanced strain hardening, thus preventing the onset of necking in each sample. These observations, coupled with the alloy's exceptional fracture toughness, make it a potential candidate for cryogenic applications. However, further research is needed to evaluate its feasibility for implementation [254].

The field of HEAs can be classified into three categories: single-phase, multiphase, and complex-engineered HEAs. These classifications are helpful in accurately interpreting the exceptional results that continue to emerge in this field. To discover more promising HEAs, researchers are relying on the development of empirical rules, thermodynamic parameters, and computational modeling. The remarkable momentum of this field is expected to continue, as HEA research advances toward the application of these materials for a variety of engineering designs.

7.7.2 High-Entropy Alloy Powders

The production of advanced materials through AM has led to significant interest in the fabrication of high-quality metallic powders. The effectiveness and quality of AM products primarily rely on the quality of the initial powder [263]. The characterization of prealloyed powders is important for the fabrication of high-quality metallic powders used in AM. This section discusses the fabrication of HEA powders through the atomization process and the investigation of their powder characteristics.

The processing methods of HEAs have a significant impact on their microstructure and mechanical properties [264]. In the past, most HEAs were fabricated using conventional arc melting, casting methods, and **mechanical alloying** [265, 266]. Producing HEAs with complex geometries for structural applications using the traditional cast technique imposes significant constraints due to the many structural defects present in the resulting alloys. Additional thermal–mechanical procedures are required to eliminate these defects and homogenize the microstructure, which can be costly and inefficient.

Compared to conventional methods, powder metallurgy-associated AM technology has emerged as a new transformative technology that has attracted increasing attention due to its many advantages. These advantages include short processing time, net-shape fabrication, unrivaled design freedom, process flexibility, and less material wastage. This indicates that it has the potential to be a method for producing complex metallic parts [267, 268]. The quality of parent metal powders strongly influences the structural and mechanical properties of final AM components. Therefore, the fabrication of high-quality feedstock powders has garnered significant attention in AM technology. To meet the stringent specifications of AM processing, alloy powders must fulfill several requirements, such as powder flowability, narrow PSD, and high bulk density. This ensures the production of high-quality AM products with high density, dimensional precision, and low surface roughness.

Traditionally, HEA powders have been synthesized rapidly by **mechanical alloying** of elemental powders with equi-atomic ratios [266]. The prolonged high-energy milling (30–60 hours) of different elemental powders of equi-atomic ratio for the synthesis of HEA powders leads to the easy contamination of the powder by milling materials. While mechanical alloying is a well-established technique for synthesizing nanocrystalline materials with uniform microstructures and compositional homogeneity, the irregular shape of the resulting powders makes them unsuitable for use in AM processes. Water atomization is a cost-effective method for producing metal powders, but it yields powders with rough, irregular shapes due to the high impact energy of high-water pressures and rapid cooling rates. Such irregularly shaped powders are less desirable for AM processes because they significantly affect powder flowability, powder-layering capability, and packing density [269].

The gas atomization process for rapid solidification is a popular and effective method for producing powders that offer good compositional homogeneity, a low risk of oxidation, and the potential for mass production. In comparison to mechanically alloyed powder, gas-atomized powder has higher purity and greater homogeneity in both composition and morphology [270]. Recently, significant efforts have been directed toward producing complex HEA parts using prealloyed gas-atomized powders through AM technology. The spherical shape and narrow size distribution of the gas-atomized powders are considered essential factors that contribute to improved powder flowability during material processing. As a result, gas-atomized powders are preferred for AM applications [271, 272]. A study conducted by Wang et al. utilized gas-atomized HEA powders to fabricate CoCrFeMnNi HEA via AM technology. The gas-atomized powders were thoroughly characterized to ensure their suitability for AM processes [273]. Li et al. investigated the microstructure and mechanical properties of CoCrFeMnNi HEAs produced through the SLM technique. They found that the mechanical properties of the SLM-HEAs were improved compared to the as cast CoCrFeMnNi HEA due to the pile-up of dislocations and lattice distortion, which were the main strengthening mechanisms of the AM-HEAs [274]. Eißmann et al. utilized gas-atomized prealloy powders to fabricate a single-phase CoCrFeMnNi HEA via EBM technology. However, the effect of powder quality on the resulting EBM-HEA product has yet to be explored [275].

To evaluate the appropriateness of powders for AM purposes, it is essential to examine their characteristics such as uniformity, compressibility, shape, size distribution, surface oxides, impurity concentrations, and tap density. A study was conducted on the Al0.5CoCrFeMnNi HEA powders that were atomized to investigate the impact of their PSD on their potential use in AM [276]. The desired Al0.5CoCrFeMnNi HEA was synthesized by weighing high-purity metallic components of Co, Cr, Fe, Mn, Ni, and Al according to stoichiometric ratio. The mixture was melted in a high-frequency induction furnace and the melt stream was atomized in high-purity argon atmosphere through a boron nitride melt delivery nozzle with an orifice diameter of 5 mm. The rapidly solidified atomized HEA powders were collected from the bottom chamber and classified into four different size distributions (< 20, 20–63, 63–106, and 106–150 μm). The shape of the obtained alloy powders was spherical with smooth surfaces and fewer satellite powders. The PSD was analyzed using a laser diffraction particle size analyzer. The powder characteristics such as tap density (4.7 $g\,cm^{-3}$), compressibility (8.39%), angle of repose (AOR) (23.57°), and flat plate angle (25.33°) were measured and indicated that the powder flowability can be improved. The scanning electron micrograph (SEM)–EDS results showed that the alloy powders exhibited a homogeneous chemical composition regardless of particle size.

The secondary dendrite arm spacing decreased with decreasing particle size and increasing cooling rate, leading to an increase in the hardness of HEA powders.

The CrFeNiMn alloy is an equi-atomic Co-free HEAs that has been previously categorized as a multiphase alloy, and is considered a promising option for high strength and radiation-resistant applications [277], can be fabricated into a single or near single-phase alloy by using gas atomization. Similar research has been conducted with alloys consisting of Cr, Fe, Ni, and Mn elements, but always in nonequi-atomic concentrations, i.e. $Fe_{40}Mn_{28}Ni_{32-x}Cr_x$ ($x=4$, 12, 18, 24 (at%)) [278] and $FeNiMnCr_{18}$ [277, 279]. The particle size of gas-atomized powders determines their final grain size and phase structure due to variations in cooling rate.

The phase structures and nano-hardness of gas-atomized equi-atomic CrFeNiMn HEA powder were investigated in detail [280]. The equi-atomic CrFeNiMn powder was produced from commercial purity raw materials of Cr, Fe, Ni (>99.6 wt%) and Mn (98.9 wt%) using close-coupled gas atomization in this study. The raw elements were melted in an Al_2O_3 crucible under Ar atmosphere with a melt temperature of 1920 K, and atomized at a pressure of 1.2 MPa using a close-coupled atomizer with an outlet diameter of 2.5 mm, resulting in a melt mass flow rate of 252 kg h^{-1}. An anti-satellite was also employed to improve powder flowability. The melt droplets solidified with the help of Ar, which acted as an inert and cooling gas. The cooling rate of the melt droplets varied from 102 to 105 K s^{-1} depending on the droplet diameter and gas-to-melt mass flow ratio (GMR) [281, 282]. The microstructure of the gas-atomized powder was investigated by sieving the powder into four different particle sizes: <20, 20–45, 45–90, and 90–150 μm. The findings suggest that the size distribution of the powder has an impact on its composition and structure. Gas atomization of the duplex-phase-structured CrFeNiMn alloy powder led to a nearly single-phase FCC structure, with a small amount of BCC phase embedded in the primary FCC phase. The weight percentage of the finely dispersed BCC phase decreased with increasing particle size (i.e. with decreasing cooling rate). The gas-atomized CrFeNiMn powder exhibited a nearly perfect spherical shape with a fine surface pattern. The grain size within the powder particles varied from 2.5 ± 0.4 to 6.0 ± 0.6 μm, depending on the particle size. The average nano-hardness of the particle sizes <20 and 90–150 μm, determined from nanoindentation, was 4.25 ± 0.73 and 3.96 ± 0.60 GPa, respectively.

The fabrication and characterization of CoCrFeNiMn HEA powder processed by gas atomization were also investigated [283]. Spherical entropy alloy powder of CoCrFeNiMn was produced using a precisely controlled large-scale gas atomization process in this study. The alloy's high ductility and mechanical properties at low temperatures were also studied. The gas atomization process was performed using a high-capacity gas atomizer, the Dongyang Induction Melting Furnace, which employed Ar gas for producing high-purity spherical HEA powder. Co, Cr, Fe, Ni, and Mn with purities of 99.99%, 99.5%, 99.9%, 99.9%, and 99.5%, respectively, were utilized as raw materials. A vacuum induction melting furnace was used to first prepare the master alloy, which was subsequently heated to around 1700 °C in an Ar atmosphere using a magnesia crucible [283]. The gas atomization process was used in this study to produce spherical CoCrFeNiMn HEA powder by atomizing the molten metal, which was delivered through a 4.5 mm diameter orifice, with 25 bar of Ar gas. The fabricated powder particles were automatically separated by particle size. Powders with a size under 53 μm had almost perfect spherical shapes and smooth surfaces without satellites or other surface defects. All of the powder was confirmed to form a single FCC phase, and the five elements were uniformly distributed within the spherical powder. It was confirmed that gas atomization process produced high-purity, equi-atomic, and single-phase spherical HEA powder on a large scale.

As previously mentioned, numerous HEAs exhibit promising mechanical properties and resistance to oxidation [284, 285]. AlxCoCrFeNi is one of the most extensively studied HEA systems due to its low-cost and abundant constituent elements, along with excellent mechanical properties. The microstructure of AlxCoCrFeNi can be modified by increasing the Al mole ratio, resulting in a single FCC solid solution, duplex FCC plus BCC solid solutions, or a single BCC solid solution. Using gas atomization process, C.-C. Yang et al. produced HEA powders of $Al_{0.5}CoCrCuFeNi$, $Al_{0.5}CoCrCuFeNiSi_{1.2}$, and $Al_{0.5}CoCrCuFeNiSi_{2.0}$ and compared their hardness to those of the respective cast alloys [286].

A mixture of pure elements of Al, Co, Cr, Cu, Fe, Ni, and Si with 99.9 wt% purity was induction melted in a copper crucible under a nitrogen atmosphere prior to gas atomization. Using the high-pressure nitrogen gas atomization method, $Al_{0.5}CoCrCuFeNi$, $Al_{0.5}CoCrCuFeNiSi_{1.2}$, and $Al_{0.5}CoCrCuFeNiSi_{2.0}$ HEA powders were prepared. The master ingot was heated to the operating temperature of $T_m = 1700$ K under vacuum conditions (1×10^{-2} torr). Subsequently, the molten alloy was ejected through a nozzle with a N_2 pressure of 2.2 MPa. The gas-atomized powders were sieved and divided into different size ranges from 40 (350 mesh) to 149 (100 mesh) μm. All alloy powders had spherical shapes with smooth surfaces. The median PSD, *d*50, for different alloy compositions was about

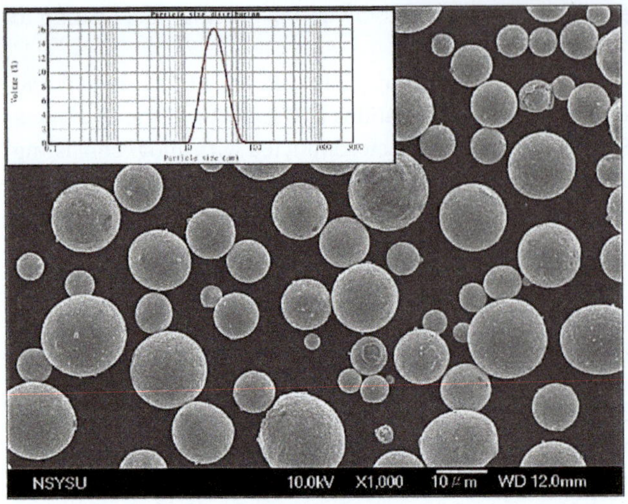

Figure 7.22 SEM micrograph and particle size distribution (inset) of $Al_{0.5}$CoCrCuFeNi alloy powders. Source: Yang et al. [286]/with permission of Elsevier.

27 μm, as shown in Figure 7.22 [286]. By adding silicon, the gas atomization process can decrease the occurrence of segregation at a faster cooling rate. Additionally, the solid solution strengthening effect can enhance the powder's hardness from 200.5 to 751.8 HV without compromising the alloy's crystallinity. The author's group employed a combination of nonequilibrium processes, such as GAs, cryomilling (CM), and spark plasma sintering (SPS), to fabricate a nanocrystalline (NC) multiphase AlCoCrCuFeNi HEAs [287]. The study explored the nonequilibrium microstructural changes in nanocrystalline multiphase AlCoCrCuFeNi HEAs fabricated using a combination of GAs, cryomilling (CM), and SPS. The investigation uncovered nanoscale deformation structures such as dislocation tangles and deformation bands. The study also examined the impact of the nonequilibrium nature of the GA, CM, and SPS processes on **grain refinement** and their effects on the mechanical properties.

The AlCoCrCuFeNi HEA alloy melt was gas-atomized using a confined nozzle atomizer. Prior to melting and atomization, both the melting and cooling chambers were evacuated with a mechanical rotary vacuum pump and then filled with nitrogen gas. The HEA master alloy was induction heated in an inert gas atmosphere for 30 minutes to ensure complete dissolution of all elements. The alloy's temperature was monitored using a thermocouple located inside a stopper rod to control the melting temperature and delivery of the alloy into the atomization zone.

The gas atomization process was carried out at a temperature of 1550 °C and a gas pressure of 300 psi. The melt was atomized by 18 concentric gas jets in the nozzle, and the resulting powder was cooled down to room temperature in the inert gas atmosphere of the atomizer. Finally, the powders were collected in air.

Figure 7.23a displays the SEM morphology of the AlCoCrCuFeNi HEA alloy powders produced through atomization. The **morphology** of atomized powders can be affected by surface tension and liquid metal viscosity. In this case, the gas atomization process resulted in spherical morphology for the AlCoCrCuFeNi HEA powders. Figure 7.23b

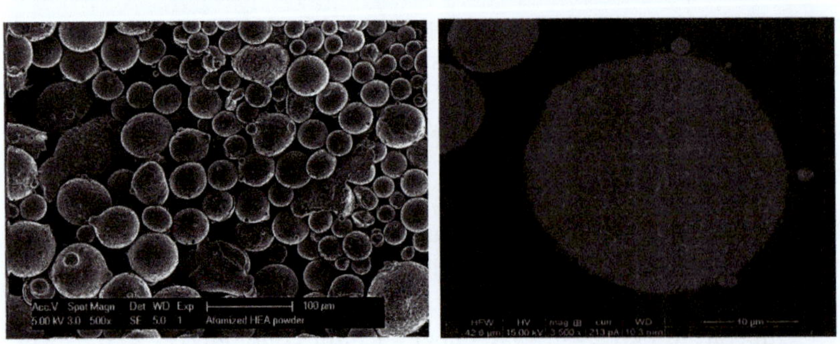

(a) (b)

Figure 7.23 SEM micrographs of (a) morphology and (b) cross-section of gas-atomized AlCoCrCuFeNi HEA powder. Source: Zheng et al. [287] with permission of Elsevier.

shows the SEM cross-section micrographs of the AlCoCrCuFeNi HEA powder. The gas atomization process led to much less phase separation in the HEA powder than in cast HEA due to the high cooling rate that depresses phase separation.

The studies discussed in this section have shown that HEA powders produced through gas atomization have favorable morphology, chemical composition, and flowability. Such powders have the potential to be utilized in AM applications.

7.8 Summary

This chapter provides a comprehensive summary of the atomization process used for the production of metallic alloy powders, which are widely used in net shaping fabrication through powder metallurgy and AM. GAs is identified as the dominant technology for producing high-quality spherical metallic alloy powders, due to their spherical morphology, high flowability, and homogeneous chemical composition. With the increasing demand for high-quality metallic alloy powders in AM technologies, there is a need to understand the various powder preparation processes, which allow precise control of the powder's physical characteristics and chemical composition, and to obtain prescribed attributes for different applications.

The chapter provides an overview of the processing, microstructure, and mechanical properties of various Al-, Fe-, Ni-, Ti-, and Co-based alloys, as well as the new family of HEA, which are mostly developed related to AM processing. The basic techniques of commercial and emerging atomization methods, such as GA, PA, and PREP, are discussed in detail, with specific information related to GA processing parameters and resulting powder properties of different alloys reviewed.

Despite the maturity of GA technology, two main issues remain with GA-prepared powder particles, namely, satellite particles and gas pores within the metal particles. However, the desired powder morphology and particle size can be achieved through the identification of optimal atomization parameters. Powder qualities are known to affect powder flowability, which can improve the injection-ability and spread-ability of powder particles for the AM process. Given that AM technology is different from other metallic powder-based part production processes, additional processing improvements and analysis of powder morphology, powder size distribution (PSD), physical characteristic, chemistry, and powder recycling are needed to control and improve the powders used in AM.

In summary, this chapter highlights the importance of the atomization process to produce metallic alloy powders for AM.

Nomenclature

Variable	Description	Units
AOR	Angle of repose	—
CCA	Complex concentrated alloys	—
c_o	Crack dimension	m
CM	Cryomilling	—
d_g	Grain size	m
d_p	Reinforcement particle size	m
DED	Directed energy deposition	—
E	Elastic modulus	GPa
E_c	Elastic modulus of composite	GPa
E_m	Elastic modulus of matrix	GPa
E_r	Elastic modulus of reinforcement	GPa
E'	Storage modulus	GPa

Variable	Description	Units
E''	Loss modulus	GPa
EBM	Electron beam melting	—
EIGA	Electrode induction gas atomization	—
El	Elongation	%
f_r	Resonant frequency	s^{-1}
GARS	Gas atomization reaction synthesis	—
GFA	Glass-forming ability	—
GMR	Mass flow ratio	—
GSD	Granulation-sintering-deoxygenation	—
H	Hardness	MPa
HDH	Hydride–dehydride	
HEA	High-entropy alloy	—
HIP	Hot isostatic pressing	—
HLS	Heavy liquid separation	—
HPGA	High-pressure gas atomization	—
HSS	High-speed steels	—
ICPMS	Inductively coupled plasma mass spectrometry	—
IGA	Inert gas atomization	—
IM	Ingot metallurgy	—
K_c	Fracture toughness	$MPa\,m^{1/2}$
MA	Mechanical alloying	—
MIE	Minimum ignition energy	—
MIM	Metal injection molding	—
$2N_f$	Number of fatigue reversals to failure	—
p	Percentage of porosity	—
ODS	Oxide dispersion strengthened	—
P	Peak load	N
PA	Plasma atomization	—
P/M	Powder metallurgy	—
PREP	Plasma rotating electrode process	—
PSD	Particle size distribution	—
PS	Plasma spheroidization	—
Q	Activation energy	$J\,mol^{-1}$
Q^{-1}	Inverse quality factor	—
R	Gas constant	$JK^{-1}\,mol^{-1}$
RA	Rotary atomization	—
REP	Rotating electrode process	—
RSS	Random solid solution	—
SD	Spray-deposited	
SLM	Selective laser melting	—
SPS	Spark plasma sintering	—
SPDS	Spark plasma discharge spheroidization	—
t	Time	s
T	Temperature	K

Variable	Description	Units
TRIP	Transformation-induced plasticity	
TWIP	Twinning-induced plasticity	—
$V_{\mathrm{f,e}}$	Effective volume fraction of the reinforcement	—
$V_{\mathrm{f,r}}$	Volume fraction of the reinforcement	—
WA	Water atomization	—
δ	Logarithmic decrement	—
ε_0	Strain amplitude	—
ε_{a}	Anelastic strain	—
ε_{e}	Elastic strain	—
$\Delta\varepsilon_{\mathrm{e}}/2$	Elastic strain amplitude	—
$\varepsilon'_{\mathrm{f}}$	Fatigue ductility coefficient	—
$\Delta\varepsilon_{\mathrm{p}}/2$	Plastic strain amplitude	—
$\dot{\varepsilon}$	Strain rate	s^{-1}
φ	Loss angle	degree
η	Loss factor	—
η_{c}	Loss factor of composite	—
η_{m}	Loss factor of matrix	—
η_{r}	Loss factor of reinforcement	—
ρ	Density	$\mathrm{g\,cm^{-3}}$
ρ_{t}	Theoretical density	$\mathrm{g\,cm^{-3}}$
σ	Strength	MPa
σ_0	Stress amplitude	MPa
σ_{i}	Lattice frictional stress	MPa
σ_{ys}	Yield strength	MPa
σ_{UTS}	Ultimate tensile strength	MPa
σ'_{f}/E	Fatigue strength coefficient	—
ψ	Specific damping capacity	—

References

1 Lewis, R.E., Palmer, I.G., Eckvall, J.C. et al. (1983). *Proceedings of the 3rd International Conference on Rapid Solidification Processing* (ed. R. Mehrabian), 613. Gaithersburg, MD: National Bureau of Standards.

2 Samal, P. and Newkirk, J. (ed.) (2015). Production of aluminum and aluminum-alloy powder. In: *Powder Metallurgy*, vol. 7, 0. ASM International.

3 Dunkley, J.J. (1998). Atomization. In: *ASM Handbook: Powder Metal Technologies and Applications*, vol. 7 (ed. P.W. Lee, Y. Trudel, R. Lacocca, and R.M. German), 35–52. ASM International Publishers.

4 Williams, J.E. (1984). *Production of Aluminum Powder*, ASM Metals Handbook, 9e, vol. 7, 125–130. Metals Park, OH: American Society for Metals.

5 Ünal, A. (1990). *Powder Metall.* 33: 53–64.

6 Özbilen, S., Ünal, A., and Sheppard, T. (1989). Influence of oxygen on morphology and oxide content of gas atomized aluminum powders. In: *Physical Chemistry of Powder Metals Production and Processing* (ed. W.M. Small), 489–505. TMS.

7 German, R. (1994). Liquid phase sintering. In: *Powder Metallurgy Science*, 2e, 374–375. Princeton: Metal Powder Industry Federation.

8 Wu, Y. (1994). Interaction mechanisms between ceramic particulates and metallic matrices, and mechanical behavior of Al-Si-X(x=SiC, TiB_2) MMCs synthesized using spray atomization, co-injection and deposition. PhD Dissertation, University of California at Irvine, California.

9 Liu, Q.Y., Li, Q.C., and Liu, Q.F. (1989). *Acta Metall. Mater.* 39: 2497.

10 Fathalla, N. (1989). *J. Mater. Sci.* 24: 2488.

11 Hafiz, M.F., Fathalla, N., and Moshref, S.B. (1990). *Z. Metallkd.* 81: 70.

12 Zhou, J., Duszczyk, J., and Korevaa, B.M. (1990). *J. Mater. Sci.* 26: 3041.

13 Hehmann, F., Bienvenu, Y., Durand, M. et al. (1990). *PM'90, International Conference on Powder Metallurgy*, vol. 1, 216. London: Institute of Metals.

14 Zhou, J. and Duszczyk, J. (1991). *Hot Deformation of Aluminum Alloys* (ed. T.G. Langdon, H.D. Merchant, J.G. Morris, and M.A. Zaidi), 473. Warrendale, PA: TMS.

15 Estrada, J.L. and Duszczyk, J. (1990). *J. Mater. Sci.* 25: 1381.

16 Murakami, K., Takuno, N., Okamoto, T., and Yoshinari (1992). *Mater. Sci. Eng. A* 154A: 93.

17 Ojha, S.N., Jha, J.N., and Singh, S.N. (1991). *Scr. Metall. Mater.* 25: 443.

18 Zhou, J., Duszczyk, J., and Korevaar, B.M. (1991). *J. Mater. Sci.* 26: 824.

19 Zhou, J., Duszczyk, J., and Korevaar, B.M. (1992). *J. Mater. Sci.* 27: 4247.

20 Zhou, J. and Duszczyk, J. (1990). *J. Mater. Sci.* 25: 4541.

21 Duszczyk, J., Estrada, J.L., Korevaar, B.M. et al. (1987). *P/M Aerospace Materials Conference*, 26. Shrewsbury: MPR Publishing Service Ltd.

22 (1990). *Spray-Deposited High Performance Materials: N403.* Neuhausen, Switzerland: Alusuisse-Lonza Services Ltd.

23 Wu, Y., Cassada, W.A., and Lavernia, E.J. (1995). *Metall. Mater. Trans. A* 26: 1235.

24 Kahl, W. and Peupp, J. (1990). *Met. Powder Rep.* 45: 4.

25 Zheng, J.H. and Kahl, W. (1990). *Advanced Aluminum and Magnesium Alloys* (ed. T. Khan and G. Effenberg), 317. Materials Park, OH: ASM.

26 Baker, H. (ed.) (1979). *Metals Handbook and Properties and Selection, Nonferrous Alloys and Pure Metals*, 2. Metals Park, OH: ASM.

27 Lavernia, E.J., McKewan, G.W., and Grant, N.J. (1986). *Prog. Powder Metall.* 42: 457.

28 Sankaran, K.K. and Grant, N.J. (1975). *Mater. Sci. Eng., A* A44: 213.

29 Speide, M.O. (1975). *Metall. Trans. A* 6A: 631.

30 Lavernia, E.J., Rai, G., and Grant, N.J. (1985). *Mater. Sci. Eng. A* 79A: 211.

31 Savage, S.J. and Frose, F.H. (1984). *JOM* 36 (4): 20.

32 (1990). *Spray-deposited High Performance Materials: N202.* Neuhausen, Switzerland: Alusuisse-Lonza Services Ltd.

33 White, J., Mingard, K., Hughes, I.R., and Palmer, I.G. (1994). *Powder Metall.* 37: 129.

34 Llorca, J., Ruiz, J., Elices, M., and Beevers, C.J. (1994). *Mater. Sci. Eng., A* 185A: 1.

35 Willis, T.C., White, J., Jordan, R.M., and Hughes, I.R. (1987). The microstructure and properties of Al–SiCp composite materials produced by spray deposition. *Solidification Processing Conference*, September 1987, Sheffield University, UK.

36 White, J., Willis, T.C., Hughes, I.R., and Jordan, R.M. (1988). *TMS Annual Meeting*. Phoenix, AZ, USA: The Minerals, Metals & Materials Society (TMS).

37 Ruhr, M. and Baram, J. (1991). *Metall. Trans. A* 22A: 2503.

38 Baram, J. (1991). *Metall. Trans. A* 22A: 2515.

39 W. M. Griffith, R.E. Sanders and, G.J. Hildeman, *High Strength Powder Metallurgy Aluminum Alloys* (ed. M.J. Kear, G.J. Hildeman), TMS-AIME, Warrendale, PA, 1982, p. 209.

40 Paris, H.G., Bilman, F.R., Cebulak, W.S., and Petit, J.I. (1980). *Rapid Solidification Processing, Principles, and Technologies* (ed. R. Mehrabian, B.H. Kear, and M. Cohen), 331. Baton Rouge, LA: Claitor's Publishing Division.

41 Adam, C.M. and Bourdeau, R.G. (1980). *Rapid Solidification Processing, Principles, and Technologies* (ed. R. Mehrabian, B.H. Kear, and M. Cohen), 246. Baton Rouge, LA: Claitor's Publishing Division.

42 Kim, N.J., Park, W.J., Ahn, S., and Elias, L. (1994). *High Performance Metal and Ceramic Composites* (ed. K. Upadhya), 137. Warrendale, PA: TMS.

43 Chu, M.G., Denzer, D.K., Chakrabarti, A.K., and Billman, F.R. (1988). *Mater. Sci. Eng.* 98: 227.

44 Hariprasad, S., Sastry, S.M.L., Jerina, K.L., and Lederich, R.J. (1993). *Metall. Trans. A* 24A: 865.
45 Palmer, I.G., Lewis, R.E., and Crooks, D.D. (1981). *Aluminum-Lithium Alloys* (ed. J.T.H. Sanders and J.E.A. Starke), 241. Warrendale, PA: AIME.
46 Lavernia, E.J. and Grant, N.J. (1988). *Mater. Sci. Eng., A* 98A: 381.
47 E. J. Lavernia, T. Ando, N.J. Grant (1986). Structures and Properties of Lithium Containing X2020 Aluminum Alloys Produced by Liquid Dynamic Compaction. (eds. P. Lee and R. Carbonara), 1st International Conference on Rapidly Solidified Materials, 29. Metals Park, OH: ASM.
48 White, J., Palmer, I.G., Hughes, I.R., and Court, S.A. (1989). *Aluminum-Lithium Alloys V* (ed. J.T.H. Sanders and J.E.A. Starke), 1635. Birmingham: MCE Publ.
49 Brooks, P.J.S. and White, J. The production of particulate reinforced MMCs and other advanced aluminum materials by the Cospray process.
50 Saunders, N. (1990). *Calphad* 14: 61.
51 Saikawa, S., Nakai, K., Sugiura, Y., and Kamio, A. (1995). *J. Jpn. Inst. Light Met.* 45 (7): 385.
52 Shashidhar, S.R., Kumar, A.M., and Hirth, J.P. (1995). *Metall. Mater. Trans. A* 26A: 2269.
53 Lee, S.H., Lee, J.S., and Woo, K.D. (1995). *J. Korean Inst. Met. Mater.* 33: 252.
54 Elias, L.G., Ogilvy, A.J., and Leatham, A.G. (1990). *Advanced Aluminum and Magnesium Alloys* (ed. T. Khan and G. Effenberg), 325. Materials Park, OH: ASM.
55 Wu, M., Srivatsan, T.S., Pickens, J.R., and Lavernia, E.J. (1992). *Scr. Metall. Mater.* 27: 761.
56 Meschter, P.J., Lederich, R.J., O'Neal, J.E. et al. (1986). *International Conference on Aluminum Alloys*, Charlottesville, VA, USA.
57 Zhang, Y.C., Wu, Y., Hao, Y.Y. et al. (1989). *Aluminum-Lithium Alloys V* (ed. J.T.H. Sanders and J.E.A. Starke), 95. Birmingham: MCE Publ.
58 Meschter, P.J., Lederich, R.J., and O'Neal, J.E. (1981). *Aluminum-Lithium Alloys II* (ed. E.A. Starke and T.H. Sanders), 419. London: Institute of Metals.
59 Gysler, A., Cooks, R., and Starke, J.E.A. (1981). *Aluminum-Lithium Alloys II* (ed. J.E.A. Starke and J.T.H. Sanders), 263. London: Institute of Metals.
60 Kang, S. and Grant, N.J. (1981). *Aluminum-Lithium Alloys II* (ed. J.E.A. Starke and J.T.H. Sanders), 419. London: Institute of Metals.
61 DeSanctis, M. (1991). *Mater. Sci. Eng., A* 141A: 103.
62 Lavernia, E.J. and Grant, N.J. (1986). *Int. J. Rapid Solid.* 2: 93.
63 Lavernia, E.J., Rai, G., and Grant, N.J. (1986). *Int. J. Powder Metall.* 22 (1): 9.
64 Duan, X., Hao, Y., and Yoshida, M. (1993). *Int. J. Powder Metall.* 29 (2): 149.
65 Plies, J.B. and Grant, N.J. (1993). *Int. J. Powder Metall.* 29 (2): 149.
66 Fitzpatrick, J.M., Lewis, R.E., and Crooks, D.D. (1985). Direct Processing of Aluminum Powder Metallurgy Alloys. Technical Report, Final Report: *AFWAL-TR-84-4176*. Palo Alto, CA: Lockheed Missiles & Space Company, Inc., 1.
67 Juárez-Islas, J.A., Perez, R., Lengsfeld, P., and Lavernia, E.J. (1994). *Mater. Sci. Eng., A* 179/180A: 614.
68 (1990). *Report Spray-deposited High Performance Materials: N707*. Neuhausen, Switzerland: Alusuisse-Lonza Services Ltd.
69 Kahl, W. and Leupp, J. (1990). *Swiss Mater.* 2 (4): 17.
70 Mächler, R., Uggowitzer, P.J., Solenthaler, C. et al. (1991). *Mater. Sci. Technol.* (ed. R.W. Cahn et al.) 7: 447.
71 Domalavage, P.K., Grant, N.J., and Gefen, Y. (1975). *Metal. Trans., A* 14A: 1599.
72 Ando, T., Lavernia, E.J., and Grant, N.J. (1986). *Mater. Lett.* 5 (1): 17.
73 Brooks, R.G., Moore, C., Leatham, A.G., and Coombs, J.S. (1977). *Powder Metall.* 2: 100.
74 Rickinson, B.A., Kirk, F.A., and Davies, D.R.G. (1981). *Powder Metall.* 25 (1): 1.
75 Leatham, A.G. and Brooks, R.G. (1984). *Mod. Dev. Powder Metall.* 15: 157.
76 Itami, T., Ikawa, Y., and Kumagai, K. (1991). *Proceedings of the 2nd Japan International SAMPE Symposium.*
77 Khare, A.K. and Laporte, A.J. (1985). *Characterization of Spray Formed Test Pieces Produced by Osprey.* Neuhausen, Switzerland.
78 Ikawa, Y., Itami, T., Kumagai, K. et al. (1990). *ISIJ Int.* 30: 756.
79 Leatham, A.G. and Lawley, A. (1993). *Int. J. Powder Metall.* 29 (4): 321.

80 Leatham, A.G., Ogilvy, A., and Elias, L. (1991). The Osprey Process Current Status and Future Possibilities. *P/M in Aerospace, Defence and Demanding Applications Conference*. Princeton, NJ, San Diego, CA USA: MPIF.

81 Üçok, I., Ando, T., and Grant, N.J. (1991). *Int. J. Powder Metall.* 27 (3): 237.

82 Ibrahim, I.A., Ando, T., and Grant, N.J. (1992). *Int. J. Rapid Solid.* 7: 35.

83 Ebalard, S., Hacquin, A., and Bienvenu, Y. (1994). *LaRevue de Métallurgé – CIT/Science et Génie des Materaux* 91: 721.

84 Igharo, M. and Wood, J.V. (1989). *Powder Metall.* 32: 124.

85 Ebalard, S. and Cohen, M. (1991). *Mater. Sci. Eng., A* 133A: 297.

86 Brooks, R.G., Leatham, A.G., Coombs, J.S., and Moore, C. (1977). *Metall. Metal Form.* 44: 157.

87 Megusar, J., Lavernia, E.J., Domalavage, P. et al. (1984). *J. Nucl. Mater.* 122/123A: 789.

88 Marucci, M.L. and Catanese, J.A. (2015). Powder metallurgy carbon and low-alloy steels. In: *ASM Handbook: Powder Metallurgy*, vol. 7 (ed. E.P. Samal and J. Newkirk), 311. ASM International.

89 Warzel, R.T. (2015). Manufacture of stainless steel powders. In: *ASM Handbook: Powder Metallurgy*, vol. 7 (ed. P. Samal and J. Newkirk), 421. ASM International.

90 Rieken, J., Anderson, I., Kramer, M. et al. (2010). Microstructure evolution of gas atomized iron-base ODS alloys. *Int. J. Powder Metall.* 46: 17–32.

91 Rieken, J.R., Anderson, I.E., Kramer, M.J. et al. (2012). *J. Nucl. Mater.* 428: 65–75.

92 Anderson, I.E., Lograsso, B.K., and Ellis, T.W. (1994). Gas atomization synthesis of refractory or intermetallic compounds and supersaturated solid solutions. US Patent No. 5,368,657.

93 Rieken, J.R., Anderson, I.E., and Kramer, M.J. (2010). *Int. J. Powder Metall. (Princeton, NJ)* 46: 17–31.

94 Rieken, J.R., Anderson, I.E., and Kramer, M.J. (2010). *Int. J. Powder Metall.* 46: 9–12.

95 Pazos, D., Cintins, A., de Castro, V. et al. (2018). *Nucl. Mater. Energy* 17: 1–8.

96 German, R. (2005). *Powder Metallurgy and Particulate Materials Processing*. Metal Powder Industries Federation.

97 Lawley, A. (1992). *Atomization – The Production of Metal Powders*. Metal Powder Industries Federation.

98 Beiss, P., Dalal, K., and Peters, R. (2002). *International Atlas of Powder Metallurgical Microstructures*. Metal Powder Industries Federation.

99 Metal Powder Industries Federation (2012). *Materials Standards for PM Structural Parts MPIF Standard 35*. Metal Powder Industries Federation.

100 Dautzenberg, N. and Gesell, H. (1976). *Powder Metall. Int.* 8 (1): 14–17.

101 Gummeson, P.U. (1972). *Powder Metall.* 15 (29): 67–94.

102 Allen, T. (1981). *Particle Size Measurement*, 3e. Chapman and Hall.

103 Hucek, H.J. and Ward, J.S. (1987). *Structural Alloys Handbook, Metals and Ceramics Information Center*. Columbus, OH: Battelle Columbus Division.

104 Singer, A.R.E. (1991). *Mater. Sci. Eng., A* 135A: 13.

105 Nama, Y., Saito, M., Matsushita, T., and Takigawa, H. Properties of tool steel by spray forming process, (Preprint).

106 Moran, A.L. and Palko, W.A. (1988). *Powder Metallurgy*, 1. Princeton, NJ: APMI.

107 Bricknell, R.H. (1986). *Metall. Trans. A* 17A: 583.

108 Sims, C.T., Stoloff, N.S., and Hagel, W.C. (1987). *Superalloys II*. New York: Wiley.

109 Smith, R.W., Paliwal, M., and Apelian, D. (1983). *3rd International Conference on Rapid Solidification Processing* (ed. R. Mehrabian), 105. Gaithersburg, MD: National Bureau of Standards.

110 Smith, R.W., Rigney, D.V., Smith, J.S., and Rairden, J.R. (1983). *Proceedings of the 3rd International Conference on Rapid Solidification Processing* (ed. R. Mehrabian), 469. Gaithersburg, MD: National Bureau of Standards.

111 Jackson, M.R., Rairden, J.R., Smith, J.S., and Smith, R.W. (1981). *J. Met.* 33 (11): 23.

112 Fiedler, H.C., Sawyer, T.F., and Kopp, R.W. (1987). Spray Formed René 95. Technical Report, GE Infor. Ser. *87 CRD034*. Schenectady, New York: General Electric.

113 Fiedler, H.C., Sawyer, T.F., and Kopp, R.W. (1987). *J. Met.* 39 (8): 28.

114 Leatham, A.G., Ogilvy, A., and Elias, L. (1992). *Spray Forming – Commercialisations and Applications, Spray Forming: Science*, Technology and Applications Seminar. San Francisco, CA, Princeton, NJ: APMI.

115 Dalal, R.P., Prichard, P.D., and Tom, T. (1993). *International Gas Turbine and Aeroengine Congress and Exposition*, 93-GT–424. Cincinnati, OH: ASME.

116 Prichard, P.D. and Dalal, R.P. (1992). *Superalloys 1992* (ed. S.D. Antolovich, R.W. Antolovich, R.W. Stusrud, et al.), 205. Warrendale, PA: TMS.

117 Moran, A.L. and Palko, W.A. (1987). *Prog. Powder Metall.* 43: 711.

118 Evans, R.W., Leatham, A.G., and Brooks, R.G. (1985). *Powder Metall.* 28: 13.

119 Gessinger, G.H. and Bomford, M.J. (1974). *Int. Metall. Rev.* 19: 51–76.

120 Jia, C.L., Ge, C.C., and Yan, Q.Z. (2016). *Chin. Phys. B* 25: 320.

121 Hohmann, M. and Pleier, S. (2005). *Acta Metall. Sin. (English Lett.)* 18: 15.

122 Gerling, R., Hohmann, M., and Schimansky, F.P. (2007). *Mater. Sci. Forum.* 539: 2693.

123 Neikov, O.D., Naboychenko, S.S., and Murashova, I.B. (2019). Production of cobalt and cobalt-alloy powders. In: *Handbook of Non-Ferrous Metal Powders* (ed. O.D. Neikov, S.S. Naboychenko, and N.A. Yefimov), 670. Elsevier.

124 Neikov, O.D. (2007). Gas atomized powders. In: *Handbook of Non-Ferrous Metal Powders: Technologies and Applications* (ed. O.D. Neikov, S. Naboychenko, I.B. Mourachova, et al.), 123. Cambridge, MA: Elsevier.

125 Roth, P.G., Murray, J.C., Morra, J.E., and Hyzak, J.M. (1994). *Powder Metall. Part. Mater.* 2: 1–12.

126 White, E.M.H., Heidloff, A., Byrd, D. et al. (2016). *Powder Metall.* 59: 181–187.

127 Nilsén, F., Aaltio, I., Ge, Y. et al. (2015). *Mater. Today: Proc.* 2: S879–S882.

128 Coyne, J., Couts, W., Chen, C., and Roehm, R. (1980). Powder Metallurgy Superalloys. Aerospace Materials for the 1980's 1, 1980.

129 Fiedler, H.C., Sawyer, T.F., and Kopp, R.W. (1986). Spray Forming – An Evaluation Using IN 718. Technical Report, GE Infor. Ser. *86 CRD113*. Schenectady, New York: General Electric.

130 Ogata, K., Lavernia, E.J., Rai, G., and Grant, N.J. (1986). *Int. J. Rapid Solid.* 2 (1): 21.

131 Nell, J., Rai, G., Lavernia, E.J., and Grant, N.J. (1986). Microstructures and Properties of AN IN-100 Tube Produced by Liquid Dynamic Compaction. *Proceedings of the 1986 Annual Powder Metallurgy Conference and Exhibition*, Princeton, NJ: APMI.

132 (1977). *High Temperature High Strength Nickel Based Superalloys*, 3e. Hungtington, West Virginia: International Nickel Company.

133 Bailey, P.G. and Schweikert, N.H. (1980). *US Army Mantech J.* 5: 28.

134 Larson, J.M. (1973). *Mod. Dev. Powder Metall.* 8: 537.

135 Moskowitz, L.N., Pelloux, R.M., and Grant, N.J. (1972). *Superalloys-Processing*. Warrendale, PA: AIME.

136 Annis, C.G., Wallace, R.M., and Sims, D.L. (1976). An Interpolative Model for Elevated Temperature Fatigue Crack Propagation. Technical Report, *AFML-TR- 76–176*. Dayton, OH: US Air Force Materials Laboratory.

137 Eng, R.D. and Evans, D.J. (1982). Hot Isostatically Pressed Manufacture of High Strength MERL 76 Disk and Seal Shapes. Technical Report, *NTIS Rep. N82-26439*. Gaithersburg, MD: National Technical Information Services, US Department of Commerce, 65.

138 Bartos, J.L. (1978). *Review of Superalloy Powder Metallurgy Processing for Aircraft Gas Turbine Applications*, ASTM Special Technical Publication 672, 564.

139 Chang, D.R., Krueger, D.D., and Sprague, R.A. (1984). *Superalloy 1984*, 245. Warrendale, PA: AIME.

140 Williams, J.C. and Belov, A.F. (2013). *Titanium and Titanium Alloys: Scientific and Technological Aspects*, vol. 3. Springer US.

141 Donachie, M.J. (2000). *Titanium: A Technical Guide*, vol. II. The Materials Information Society.

142 Dawes, J., Bowerman, R., and Trepleton, R. (2015). *Johnson Matthey Technol. Rev.* 59: 243–256.

143 Froes, F.H. and Bomberger, H.B. (1985). *JOM* 37: 28.

144 Dai, J.H., Wu, X., Song, Y., and Yang, R. (2012). *J. Appl. Phys.* 112: 123718.

145 Abraham, M. (2017). *Encyclopedia of Sustainable Technologies*. Cambridge: Elsevier.

146 Ezugwu, E.O., Da Silva, R.B., Sales, W.F., and Machado, A.R. (2017). Overview of the machining of titanium alloys. In: *Encyclopedia of Sustainable Technologies* (ed. M.A. Abraham), 487–506. Oxford: Elsevier.

147 Ding, R. and Guo, Z.X. (2004). *Mater. Sci. Eng., A* 365: 172–179.

148 Boyer, R., Collings, E.W., and Welsch, G. (1994). *Materials Properties Handbook: Titanium Alloys*. ASM International.

149 Ezugwu, E.O. and Wang, Z.M. (1997). *J. Mater. Process. Technol.* 68: 262–274.

150 Campbell, F.C. (2008). Titanium. In: *Elements of Metallurgy and Engineering Alloys* (ed. F.C. Campbell), 533–544. ASM International.

151 Abbaschian, R. and Reed-Hill, R.E. (2008). *Physical Metallurgy Principles*. Stamford, CT: Cengage Learning.
152 Al-bermani, S.S., Blackmore, M.L., Zhang, W., and Todd, I. (2010). *Metall. Mater. Trans. A* 41A: 3422–3434.
153 ASTM F1580 (2012). *Standard Specification for Titanium and Titanium-6 Aluminum-4 Vanadium Alloy Powders for Coatings of Surgical Implants*. West Conshohocken, PA: ASTM International.
154 ASTM F2924 (2014). *Standard Specification for Additive Manufacturing Titanium-6 Aluminum-4 Vanadium with Powder Bed Fusion*. West Conshohocken, PA: ASTM International.
155 Liu, Z. and Welsch, G. (1988). *Metall. Trans. A* 19: 527.
156 Horiya, T. and Kishi, T. (1994). *Nippon Steel Tech. Rep.* 62: 85.
157 Campbell, G.H., LaGrange, T.B., King, W.E. et al. (2005). *Solid-Solid Phase Transformations in Inorganic Materials 2005*. Pheonix, AZ: TMS and ASM.
158 Tshephe, T.S., Akinwamide, S.O., Olevsky, E., and Olubambi, P.A. (2022). *Heliyon* 8: e09041.
159 Zhang, X., Liu, S., Liu, Y. et al. (2023). *Metals* 13: 462.
160 Gaytan, S.M., Murr, L.E., Medina, F. et al. (2009). *Mater. Technol.* 24: 180–190.
161 Pederson, R. (2002). *Microstructure and Phase Transformation of Ti-6 Al-4 V*. Robert Pederson.
162 Sames, W.J., List, F.A., Pannala, S. et al. (2016). *Int. Mater. Rev.* 61: 315–360.
163 Gu, D.D., Meiners, W., Wissenbach, K., and Poprawe, R. (2012). *Int. Mater. Rev.* 57: 133–164.
164 Roberts, P.R. (1989). *PM Conference: Advances in Powder Metallurgy*, 427–438. San Diego, CA: Metal Powder Industries Federation.
165 Moll, J.H. (2000). *JOM* 52: 32.
166 Garcia, W. (2015). *Titanium Europe Conference 2015*. Birmingham, UK: International Titanium Association.
167 Smagorinski, M.E. and Tsantrizos, G. (2002). *Proceedings of the 2002 World Congress on Powder Metallurgy & Particulate Materials*. Orlando, FL: Metal Powder Industries Federation.
168 Sun, P., Fang, Z.Z., Zhang, Y., and Xia, Y. (2017). *JOM* 69: 1853–1860.
169 McCracken, C.G., Motchenbacher, C., and Barbis, D.P. (2010). *Int. J. Powder Metall.* 46: 19.
170 Spierings, A.B., Voegtlin, M., Bauer, T., and Wegener, K. (2016). *Prog. Addit. Manuf.* 1: 9.
171 Chan, L.C.Y. and Page, N.W. (1997). *Powder Technol.* 90: 259.
172 Sames, W.J., Medina, F., Peter, W.H. et al. (2014). Effect of process control and powder quality on Inconel 718 produced using electron beam melting, Superalloy 718 and derivatives. 8th International Symposium on Superalloy 718 and Derivatives, TMS (The Minerals, Metals & Materials Society), Pittsburgh, PA.
173 Susan, D.F., Puskar, J.D., Brooks, J.A., and Robino, C.V. (2006). *Mater. Charact.* 57: 36–43.
174 Yolton, C.F. (1989). *PM in Aerospace and Defense Technologies* (ed. F.H. Froes), 123–131. Seattle, WA: Metal Powder Industries Federation.
175 Yolton, C.F. (1992). Method for producing titanium particles. US Patent 5,084,091.
176 Hohman, M. and Ludwig, N. (1994). Method for producing semi-finished products from NiTi shape memory alloys. US Patent 5,284,329 A.
177 Franz, H., Plochl, L., and Schimansky, F.P. (2008). Recent advances of titanium alloy powder production by ceramic-free inert gas atomization. *24th Annual ITA Conference Titanium 2008*. Las Vegas, NV, USA: International Titanium Association.
178 Pleier, S., Goy, W., Schaub, B. et al. (2004). EIGA—An innovative production method for metal powder from reactive and refractory alloys. *International Conference on Powder Metallurgy & Particulate Materials*. Princeton, NJ: Metal Powder Industries Federation, 2–49.
179 Hanusiak, W.M. and McBride, D.R. (2013). Titanium powder production apparatus and method. Patent Application 13/414,769.
180 Wheeler, A. (2015). *Third Plasma Atomization Reactor Put Into Service*. AP&C.
181 Entezarian, M., Allaire, F., Tsantrizos, P., and Drew, R.A.L. (1996). *JOM* 48: 53.
182 Tsantrizos, P.G., Allaire, F., and Entezarian, M. (1998). Method of production of metal and ceramic powders by plasma atomization. US Patent 5,707,419.
183 Dion, C.A.D., Kreklewetz, W., and Carabin, P. (2016). Plasma apparatus for the production of high quality spherical powders at high capacity. Publication number WO2016191854 A1.
184 Larouche, F., Balmayer, M., and Trudeau-lalonde, F. (2017). Plasma atomization metal powder manufacturing processes and systems therefore. Publication number WO2017011900 A1.
185 Kaufmann, A.R. (1963). Method and apparatus for making powder. US Patent 3,099,041.

186 Miller, S.A. and Roberts, P.R. (1990). *ASM Handbook: Powder Metal Technologies and Applications*, vol. 7, 97–101. Materials Park, OH: ASM International.

187 Abkowitz, S. (1966). *JOM* 18: 458.

188 Vaughan, R.F., Blenkinsop, P.A., and Morton, P.H. (1982). Comparative evaluation of forged Ti-6A1-4V bar made from shot produced by rep and CSC processes. In: *Titanium and Titanium Alloys: Scientific and Technological Aspects*, vol. 3 (ed. J.C. Williams and A.F. Belov), 2377–2388. Boston, MA: Springer.

189 McCracken, C.G., Barbis, D.P., and Deeter, R.C. (2011). *Powder Metall.* 54: 180–183.

190 Medina, F. (2013). Reducing metal alloy powder costs for use in powder bed fusion additive manufacturing: improving the economics for production. Materials Science and Engineering, vol PhD. El Paso, TX: University of Texas.

191 Boulos, M. (2004). *Metall. Powder Rep.* 59: 16.

192 Barbis, D.P., Gasior, R.M., Walker, G.P. et al. (2015). Titanium powders from the hydride-dehydride process. In: *Titanium Powder Metallurgy: Science, Technology and Applications* (ed. M. Qian and F.H. Froes), 101–105. Oxford: Butterworth-Heinemann.

193 Araci, K., Mangabhai, D., and Akhtar, K. (2015). Production of titanium by the Armstrong process. In: *Titanium Powder Metallurgy: Science, Technology and Applications* (ed. M. Qian and F.H. Froes), 149–162. Oxford: Butterworth-Heinemann.

194 Mellor, I., Grainger, L., Rao, K. et al. (2015). Titanium powder production via the Metalysis process. In: *Titanium Powder Metallurgy: Science, Technology and Applications* (ed. M. Qian and F.H. Froes), 51–67. Waltham, MA: Elsevier.

195 Zhang, Y., Fang, Z.Z., Xia, Y. et al. (2016). *Chem. Eng. J.* 286: 517.

196 Zhang, Y., Fang, Z.Z., Xia, Y. et al. (2017). *Chem. Eng. J.* 308: 299.

197 Liu, C.C., Lu, X., Zhang, L. et al. (2016). *Powder Metall.* 59: 229.

198 Sun, P., Fang, Z.Z., Xia, Y. et al. (2016). *Powder Technol.* 301: 331.

199 Fang, Z.Z., Xia, Y., Sun, P., and Zhang, Y. (2016). Production of substantially spherical metal powders. US Patent 9,421,612.

200 Zhang, Y., Fang, Z.Z., Sun, P. et al. (2016). *J. Am. Chem. Soc.* 138: 6916.

201 Sun, Y.Y., Gulizia, S., Oh, C.H. et al. (2015). *JOM* 67: 564.

202 Yang, L., Hsu, K., Baughman, B. et al. (2017). *Additive Manufacturing of Metals: The technology, Materials, Design and Production*. Springer.

203 Neikov, O.D., Naboychenko, S.S., and Murashova, I.B. (2018). Production of cobalt and cobalt-alloy powders. In: *Handbook of Non-Ferrous Metal Powders, Technologies and Applications*, 2e (ed. O.D. Neikov, S.S. Naboychenko, and N.A. Yefimov), 670. Elsevier Science.

204 Long, H., Mao, S., Liu, Y. et al. (2018). *J. Alloys Comp.* 743: 203–220.

205 Pollock, T.M. and Tin, S. (2006). *J. Propul. Power* 22: 361–374.

206 Suzuki, A., Inui, H., and Pollock, T.M. (2015). *Annu. Rev. Mater. Res.* 45: 345–368.

207 Pollock, T.M. (2016). *Nat. Mater.* 15: 809–815.

208 Zhou, H.J., Xue, F., Chang, H., and Feng, Q. (2018). *J. Mater. Sci. Technol.* 34: 799–805.

209 Xue, F., Wang, M., and Feng, Q. (2012). *Superalloys 2012*, 813–821. Warrendale, PA: TMS.

210 Ooshima, M., Tanaka, K., Okamoto, N.L. et al. (2010). *J. Alloys Comp.* 508: 71–78.

211 Bauer, A., Neumeier, S., Pyczak, F., and Göken, M. (2010). *Scr. Mater.* 63: 1197–1200.

212 Zhong, F., Fan, F., Li, S., and Sha, J. (2016). *Prog. Nat. Sci.: Mater. Int.* 26: 600–612.

213 Lee, C.S. (1971). *Precipitation-Hardening Characteristics of Ternary Cobalt – Aluminum – X Alloys*. The University of Arizona.

214 Sato, J., Omori, T., Oikawa, K. et al. (2006). *Science* 90: 312.

215 Viatour, D.C.P. and Drapier, J.M. (1973). *Cobalt* 3: 67.

216 Drapier, D.C.J.M. (1968). *Cobalt* 39: 63.

217 Pollock, T.M., Dibbern, J., Tsunekane, M. et al. (2010). *JOM* 58: 62.

218 Rodriguez, P., Mostafaei, A., and Chmielus, M. (2019). *J. Undergrad. Res.* 4: 63–67.

219 Kuznetsov, V.V., Filatova, E.A., Telezhkina, A.V., and Kruglikov, S.S. (2018). *J. Solid State Electrochem.* 22: 2267–2276.

220 Rosalbino, F. and Scavino, G. (2013). *Electrochim. Acta* 111: 656–662.

221 Khoddamzadeh, A., Liu, R., Liang, M., and Yang, Q. (2012). *Compos. Part A Appl. Sci. Manuf.* 43: 344–352.

222 Ahmed, R., Ashraf, A., Elameen, M. et al. (2014). *Wear* 312: 70–82.

223 Giacchi, J.V., Morando, C.N., Fornaro, O., and Palacio, H.A. (2011). *Mater. Charact.* 62: 53–61.

224 Gülsoy, H.Ö., Özgün, Ö., and Bilketay, S. (2016). *Mater. Sci. Eng., A* 651: 914–924.

225 Warlimont, H. and Martienssen, W. (2018). *Handbook of Metal Powders*. Springer.

226 Poster, A.R. (1966). *Handbook of Metal Powders*. Reinhold Publishing Corporation.

227 Madan, D.S. (1988). Metal and metal alloy powders for welding, hardfacing, brazing, and soldering. In: *ASM Handbook*, vol. 7, 1065–1082. ASM International Publishers.

228 Duflos, F. and Stohr, J.F. (1982). *J. Mater. Sci.* 17 (12): 3641–3652.

229 Sandvik (2018). *Cobalt Alloy Metal Powder*. Sandvik Osprey Company.

230 Stoica, M. (2017). *Fe-Based Bulk Metallic Glasses: Understanding the Influence of Impurities on Glass Formation*. Berlin/Heidelberg: Springer-Verlag.

231 Żrodowski, Ł., Wróblewski, R., Choma, T. et al. (2021). *Materials* 14: 2541.

232 Liu, X., Bi, J., Meng, Z. et al. (2021). *Tribol. Int.* 162: 107142.

233 Lin, T.-J., Sheu, H.-H., Lee, C.-Y., and Lee, H.-B. (2021). *J. Alloys Comp.* 867: 159132.

234 Zheng, B., Lin, Y., Zhou, Y., and Lavernia, E.J. (2009). *Metall. Mater. Trans. B* 40B: 768–778.

235 Zheng, B., Lin, Y., Zhou, Y., and Lavernia, E.J. (2009). *Metall. Mater. Trans. B* 40B: 995–1004.

236 Katz-Demyanetz, A., Koptyug, A., and Popov, V.V.I. (2020). n-Situ alloying as a novel methodology in additive manufacturing. *2020 IEEE 10th International Conference on Nanomaterials: Applications Properties (NAP)*, Sumy, Ukraine, pp. 02SAMA05-01–02SAMA05-04.

237 Ciftci, N., Yodoshi, N., Armstrong, S. et al. (2020). *J. Mater. Sci. Technol.* 59: 26–36.

238 Zhao, Y.F., Si, J.J., Song, J.G. et al. (2014). *Mater. Sci. Eng., B* 181: 46–55.

239 Shi, Y., Lu, W., Sun, W. et al. (2022). *J. Mater. Sci. Technol.* 105: 203–213.

240 Zheng, B., Zhou, Y., Smugeresky, J.E., and Lavernia, E.J. (2009). *Metall. Trans. A* 40A: 1235–1245.

241 Deng, L., Wang, S., Wang, P. et al. (2018). *Mater. Lett.* 212: 346–349.

242 Nong, X.D., Zhou, X.L., and Ren, Y.X. (2019). *Opt. Laser Technol.* 109: 20–26.

243 Liang, X., Li, Y., Bao, F. et al. (2021). *Intermetallics* 132: 107135.

244 Zhang, G., Zhang, H., Yue, S. et al. (2019). *Intermetallics* 107: 47–52.

245 Ackland, K., Masood, A., Kulkarni, S., and Stamenov, P. (2018). *AIP Adv.* 8: 056129.

246 Kim, J.T., Hong, S.H., Kim, Y.S. et al. (2019). *J. Non-Cryst. Solids* 505: 310–319.

247 Kuś, A., Pilarczyk, W., Małachowska, A. et al. (2021). *Materials* 14: 7357.

248 MacDonald, B.E., Fu, Z., Zheng, B. et al. (2017). *JOM* 69: 2024–2031.

249 Cantor, B., Chang, I.T.H., Knight, P., and Vincent, A.J.B. (2004). *Mater. Sci. Eng., A* 213: 375–377.

250 Yeh, J.W., Chen, S.K., Lin, S.J. et al. (2004). *Adv. Eng. Mater.* 6: 299–303.

251 Hsu, C.-Y., Yeh, J.-W., Chen, S.-K., and Shun, T.-T. (2004). *Metall. Mater. Trans. A* 35: 1465.

252 Yeh, J.-W., Lin, S.-J., Chin, T.-S. et al. (2004). *Metall. Mater. Trans. A* 35: 2533.

253 Huang, P.-K., Yeh, J.-W., Shun, T.-T., and Chen, S.-K. (2004). *Adv. Eng. Mater.* 6: 74.

254 George, E.P., Raabe, D., and Ritchie, R.O. (2019). *Nat. Rev. Mater.* 4: 515–534.

255 Yeh, J.W. (2006). *Eur. J. Control* 31: 633.

256 Deng, Y., Tasan, C.C., Pradeep, K.G. et al. (2015). *Acta Mater.* 94: 124–133.

257 Li, Z. and Raabe, D. (2017). *JOM* 69: 2099–2106.

258 Li, Z., Pradeep, K.G., Deng, Y. et al. (2016). *Nature* 534: 227–230.

259 Li, Z., Körmann, F., Grabowski, B. et al. (2017). *Acta Mater.* 136: 262–270.

260 Li, Z., Tasan, C.C., Springer, H. et al. (2017). *Sci. Rep.* 7: 1–7.

261 Gludovatz, B., Hohenwarter, A., Catoor, D. et al. (2014). *Science* 345: 1153.

262 Otto, F., Dlouhý, A., Somsen, C. et al. (2013). *Acta Mater.* 61: 5743.

263 Sutton, A.T., Kriewall, C.S., Leu, M.C., and Newkirk, J.W. (2017). *Virtual Phys. Prototyp.* 12: 3–29.

264 Alshataif, Y.A., Sivasankaran, S., Al-Mufadi, F.A. et al. (2020). *Met. Mater. Int.* 26: 1099–1133.

265 Jiang, L., Lu, Y., Dong, Y. et al. (2014). *Intermetallics* 44: 37–43.

266 Kumar, D., Maulik, O., Bagri, A.S. et al. (2016). *Mater. Today: Proc.* 3: 2926–2933.

267 Seifi, M., Salem, A., Beuth, J. et al. (2016). *JOM* 68: 747–764.

268 Kok, Y., Tan, X.P., Wang, P. et al. (2018). *Mater. Des.* 139: 565–586.

269 Manakari, V., Parande, G., and Gupta, M. (2016). *Metals* 7: 2.
270 Mullis, A., Bigg, T., and Adkins, N. (2015). *J. Alloys Comp.* 648: 139–148.
271 Brif, Y., Thomas, M., and Todd, I. (2015). *Scr. Mater.* 99: 93–96.
272 Ocelík, V., Janssen, N., Smith, S., and De Hosson, J.T.M. (2016). *JOM* 68: 1810–1818.
273 Wang, P., Huang, P., Ng, F.L. et al. (2019). *Mater. Des.* 168: 107576.
274 Li, R., Niu, P., Yuan, T. et al. (2018). *J. Alloys Comp.* 746: 125–134.
275 Eißmann, N., Klöden, B., Weißgärber, T., and Kieback, B. (2017). *Powder Metall.* 60: 184–197.
276 Lee, Y., Nagarjuna, C., Song, J.-W. et al. (2021). *Powder Metall.* 64: 219–227.
277 Wu, Z., Bei, H., Otto, F. et al. (2014). *Intermetallics* 46: 131–140.
278 Stepanov, N., Shaysultanov, D., Tikhonovsky, M., and Salishchev, G. (2015). *Mater. Des.* 87: 60–65.
279 Kumar, N.K., Li, C., Leonard, K. et al. (2016). *Acta Mater.* 113: 230–244.
280 Lehtonen, J., Ge, Y., Ciftci, N. et al. (2020). *J. Alloys Comp.* 827: 154142.
281 Ciftci, N., Ellendt, N., Von Bargen, R. et al. (2014). *J. Non-Cryst. Solids* 394: 36–42.
282 Ciftci, N., Ellendt, N., Coulthard, G. et al. (2019). *Metall. Mater. Trans. B* 50: 666–677.
283 Park, T.G., Lee, S.H., Lee, B. et al. (2018). *Arch. Metall. Mater.* 63: 1055–1059.
284 Wang, Y., Li, B., Ren, M. et al. (2008). *Mater. Sci. Eng., A* 491: 154–158.
285 Li, B., Wang, Y., Ren, M. et al. (2008). *Mater. Sci. Eng., A* 498: 482–486.
286 Yang, C.-C., Hang Chau, J.L., Weng, C.-J. et al. (2017). *Mater. Chem. Phys.* 202: 151–158.
287 Zheng, B., MacDonald, B., Fu, Z. et al. (2020). *Microstructural Evolution and Mechanical Behavior of an AlCoCrCuFeNi High Entropy Alloy during Non-equilibrium Powder Metallurgical Processing TMS2020.* San Diego, CA: TMS.

Part II

Powders in Additive Manufacturing

8

Overview of Metal Additive Manufacturing Technologies

Additive manufacturing (AM) generally refers to the technologies that make products by building materials up layer by layer, that is, additively, rather than subtracting material from a large block of preexisting material as in traditional manufacturing and machining. In AM, **three-dimensional (3D)** physical objects are created or joined as specified by **computer-aided design (CAD)** model data [1]. AM is advantageous in its capability of making complex structures with less material waste and reducing the number of parts required for an application. At present, AM technologies are used for a broad range of applications in the engineering industry, aerospace, defense, architecture, medicine, toys, and entertainment. In May 2022, the White House announced that the Biden Administration would support the effort of AM Forward, a voluntary compact among large manufacturers to help smaller U.S.-based suppliers increase the use of AM [2]. In this chapter, we provide an overview of metal AM technologies, starting with the first section on their history, followed by the introduction to several specific metal AM methods and the corresponding processing principles.

8.1 History of Metal Additive Manufacturing Techniques

Early concept development of machines for metal AM and relevant experimental set-ups date back to the 1970s of the twentieth century and emerged from efforts originally aimed at the **3D printing** of polymers. The market and community of metal AM have been growing and changing tremendously since Deckard et al. at the University of Texas – Austin initiated the development of what is yet arguably the first machine in the history of metal AM, in which a laser was used to selectively melt layers of polymer and, later, metal [3, 4]. A variety of processes have been developed in the past decades to additively manufacture 3D solid parts using feedstock that can be powder, wires, or sheets. Wohlers and Gornet provided a detailed report of the history of AM from the initial commercialization of stereolithography in 1987 to April 2015 [5]. Based on the information from Wohlers' report, plus the progress in recent years, Figure 8.1 displays the timeline for the development of AM technologies. The early research in the 1980s focused on powder-bed-based laser sintering, which was patented and copyrighted as **selective laser sintering (SLS)** in 1986 [3, 4]. DTM Corp., which has been acquired by 3D Systems, developed and marketed the first industrial SLS machine in 1992 [6]. Sachs and Cima at Massachusetts Institute of Technology (MIT) patented a process called "3D printing" that used inkjet printing to deposit binder [7]. Nowadays, the use of 3D printing has expanded to describe all forms of AM, whereas the original inkjet binder deposition method developed at MIT is known as **binder jetting (BJT)**, which is also used for AM of polymers and ceramics in addition to metals [8]. Metal BJT stood in the shadow of **powder bed fusion (PBF)** technologies for nearly 20 years until Hewlet-Packard (HP) and Desktop Metal announced their commitment to enter the AM market. In 1993, researchers in Sweden developed another powder bed technique that used an **electron beam (EB)** instead of a laser as the energy source, resulting in the patent of **electron beam melting (EBM)**, which was later licensed by Arcam AB (GE bought Arcam AB in 2016) [9, 10]. In 1994, EOS launched **direct metal laser sintering (DMLS)** [11]. As opposed to using powder bed as the basis, another class of AM techniques was developed to deposit feedstock directly into a **melt pool** in the 1990s, currently referred to as **directed energy deposition (DED)** techniques. Early DED machines using wires or filaments as feedstock traced their history to metal welding. In 1995, researchers at Sandia National Laboratories developed an approach to directly feed metal powder into a melt pool created by a laser heat source. This technology was first commercialized and trademarked as **Laser Engineered Net Shaping (LENS™)** [12]. One of the first commercial

Metallic Powders for Additive Manufacturing: Science and Applications, First Edition.
Enrique J. Lavernia, Kaka Ma, Julie M. Schoenung, James F. Shackelford, and Baolong Zheng.

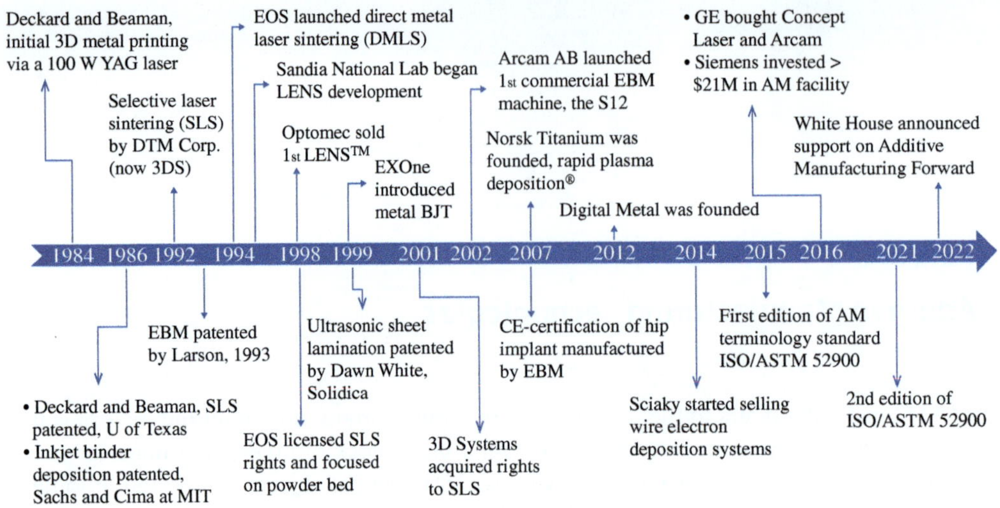

Figure 8.1 History of metal AM technologies development – the timeline.

LENS systems was sold by Optomec in 1998. In 2007, the conventional **plasma arc welding** process was adopted by Norsk Titanium as an AM technology that focused on aircraft and space applications due to the use of a large amount of titanium wires [13]. With the same aim of the application of large-scale titanium blanks, Sciaky, Inc. introduced its wire EB DED system in 2014 [14]. The fourth category of metal AM processes uses sheets as the feedstock and welds the metal sheets together to form bulk parts, named **sheet lamination (SHL)**. An SHL process that combined ultrasonic welding and computer numerical control (CNC), named ultrasonic object consolidation, was originally developed and patented in 1999 [15].

Starting in the twenty-first century, the development of commercial AM systems and market growth entered a fast-paced stage. Meanwhile, standardization of terminology and process control emerged as a critical milestone to facilitate communication between people involved in the AM field but with different focused areas. The ASTM International Committee F42 on AM Technologies released its survey on AM design rules in July 2010, which led to the ballot and approval of Specification for the Data Exchange Format for AM standard in May 2011 as the first non-terminology standard [16]. A few years later, the first edition of standard terminology for AM was issued under the fixed designation ISO/ASTM 52900 in 2015, which was recently replaced by the second edition in 2021 [1]. Following this standard, Figure 8.2 provides an overview of the categories of AM processes for metals and the relevant feedstock types. By its name "AM," the basic and common principle for the various AM processes is to bring together and join the units of feedstock material by successive addition, most commonly layer by layer, to build physical 3D parts with CAD geometries. Variations in process principle do exist, depending on how the feedstock material

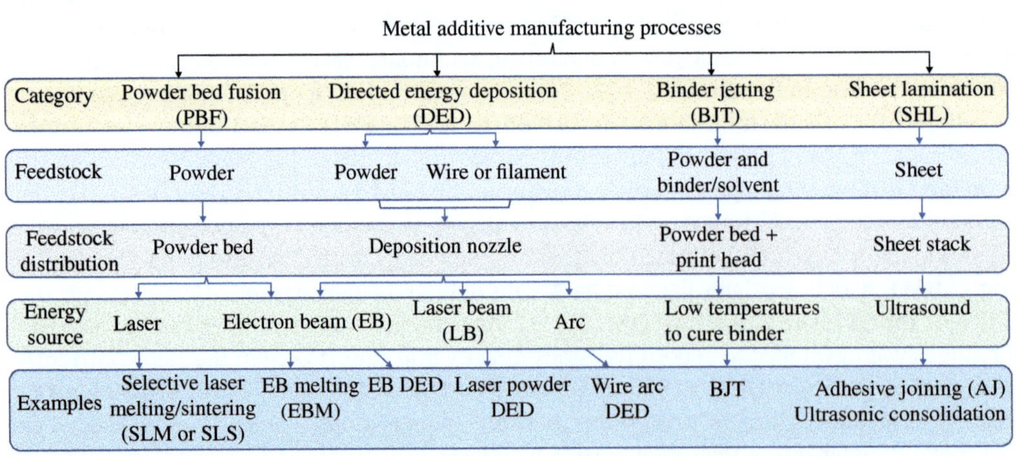

Figure 8.2 Overview of the categories of metal AM processes. Terminologies are based on [1].

is delivered to the system and how it is distributed to the build surface to be added to the part, as well as the energy (heat) sources (e.g. laser, EB, arc plasma, etc.). In metal AM, the feedstock material can typically come in the form of powder, filament/wire, or sheet. Powder feedstock can be distributed layer by layer in a powder bed or deposited by a nozzle in DED. In Sections 8.2–8.5, the process principles of specific AM processes are discussed, with more depth given to PBF and DED as the scope of this book focuses on the role of metal powder in AM, whereas in BJT, the binder, typically a liquid-state polymer solvent, is key to the process; and in SHL, the feedstock material is in the form of sheets rather than powder. Also, PBF and DED are more widely adopted for AM of metals compared to BJT and SHL.

8.2 Powder Bed Fusion (PBF)

According to ISO/ATSM 52900, PBF is generally defined as an AM process in which thermal energy selectively fuses regions of a powder bed [1]. The source of the thermal energy can be **laser beam (LB)**, **EB**, or **infrared light** (**IrL**). Common terms for PBF processes that have been used in the literature include SLS, **selective laser melting (SLM)**, **laser metal fusion (LMF)**, **laser beam melting (LBM)**, EBM, and so forth. The use of various terms to describe the PBF process resulted from multiple mechanisms that are possible for the fusion to occur: solid-state sintering, liquid-phase sintering, chemically induced binding, and full melting. At the fundamental level, regardless of the fusion mechanism, energy absorption and heat transfer govern the processing principles.

8.2.1 PBF Processing Principles

Figure 8.3 schematically illustrates the typical equipment setup of PBF processes featuring (a) SLM or SLS [17] and (b) EBM [18]. A build platform, also called a build plate, is required to hold the powder, and build upon it. In the SLM or SLS process, the LB passes through a system of lenses and is reflected by a mirror onto the surface of the build platform. The movement of the LB along the planar directions (x and y) is controlled by the mirrors to follow the paths defined in the CAD model. After a layer of powder is selectively melted, the platform moves downward (z-direction), and a recoating blade or roller pushes another layer of fresh powder from the powder feed reservoir and spreads it across the top of the previous layer. Then the laser scanning process and powder spreading are repeated until the part is finished. The part is surrounded by powder throughout the process and the excess powder not sintered in each layer serves as a support structure for overhanging features. The powder is fused together primarily by solid-state bonding in SLS, in contrast to melting in SLM. In PBF-LB, the building process takes place inside an enclosed chamber filled with an inert gas, typically argon, to minimize oxidation and degradation of the powdered material. In comparison,

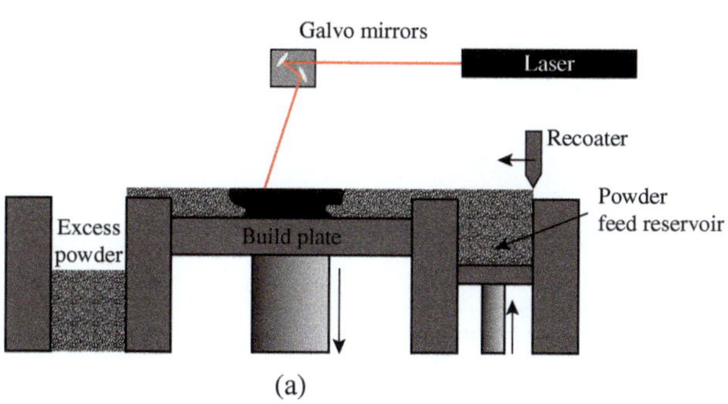

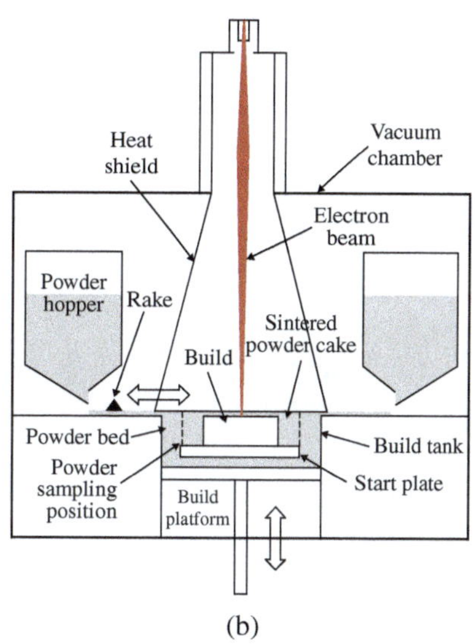

Figure 8.3 Schematic diagrams of the typical setup of PBF processes (a) SLM or SLS. Source: Spears and Gold [17] / Springer Nature / Public Domain CC BY 4.0 and (b) EBM. Source: Gruber et al. [18] / Springer Nature / Public Domain CC BY 4.0.

the EBM process is enclosed in a high vacuum chamber made up of a building tank with a vertically adjustable process platform, two powder hoppers, and a rake system for powder delivery, as shown in Figure 8.3b. In the EBM process, electrons are emitted from a heated filament cathode made from tungsten (W) for some older models of EBM machines, lanthanum hexaboride (LaB_6), or cerium hexaboride (CeB_6) in more recent models of EBM equipment. The mechanism is similar to those used in electron microscopy but uses much higher currents (1–50 mA). These electrons are then accelerated through the anode to a velocity between 0.1 and 0.4 times the speed of light, producing a high-energy EB. Since the EB gun is stationary and mechanical parts are not needed to deflect the beam, high scanning speeds (up to 1000 m s^{-1}) and high positioning accuracy are achieved in EBM. The magnetic field of lenses brings the e-beam into focus. The building process starts by filling the build tank with powder, forming a bed on which the start plate can lie. A computer-controlled EB scans the powder surface based on the desired shape as specified by a CAD model. When the e-beam hits the metal powder, the high kinetic energy is transformed into heat to melt the powder in the powder bed. After one layer of the powder is fused to the build, the build platform moves downward, and then the new powder is fed from the powder hopper by gravity and is raked into layers across the table. The built specimen is lowered onto the built table with each layer melt cycle until the completion of the build. Advantages of EBM technology include the ability to process reflective metal powders such as Al alloys, to create complex parts with overhangs, and to generate biocompatible components. The disadvantages of the EBM process include high surface roughness, frequent failure of the filament, and the formation of an additional "lip" along the edges of the fabricated parts.

It is important to understand the solidification process during PBF as it determines the initial microstructure of the parts that are fused or deposited, including but not limited to the phase distribution and grain morphology. The layer-by-layer solidification process is determined by energy absorption and heat transfer, which is essentially an energy-powder-fusion zone interaction. When the energy source is a laser, for example, SLM and SLS, the absorption of the energy in metals is physically through the excitation of free electrons in the conduction band by the monochromatic laser photons. Figure 8.4a shows the heat transfer paths in melt pool formation and solidification in an SLM process [17]. The heat transfer modes primarily consist of radiative loss, convection to the fill gas, conduction to the build part or substrate, and conduction to the powder bed surrounding the build part. It should be noted that the heat conduction may be inhibited by the powder bed due to its green packing density, that is, air gaps that remain between powder particles. The surface **laser power density** $I_0(\mathrm{W \cdot m^{-2}})$ is commonly used to describe the state of the melt pool in PBF-LB processes:

$$I_0 = P_{abs}/S_0 \tag{8.1}$$

where P_{abs} is the absorbed laser power and S_0 is the interaction surface area.

Another laser power parameter that is more practical is the **volumetric energy density** $E_v(\mathrm{J \cdot m^{-3}})$, which is defined in Eq. (8.2) using I_0 and the laser scan speed V_0.

$$E_v = I_0(\mathrm{W \cdot m^{-2}})/V_0(\mathrm{m \cdot s^{-1}}) \tag{8.2}$$

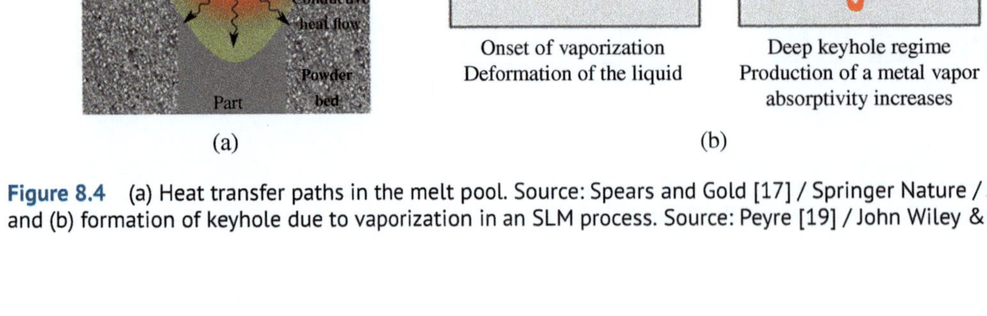

Figure 8.4 (a) Heat transfer paths in the melt pool. Source: Spears and Gold [17] / Springer Nature / Public Domain CC BY 4.0 and (b) formation of keyhole due to vaporization in an SLM process. Source: Peyre [19] / John Wiley & Sons.

The amount of laser radiation I ($W \cdot m^{-2}$) depends on the absorption coefficient α (m^{-1}), following Eq. (8.3):

$$I = I_0 \exp(-\alpha z) \tag{8.3}$$

where I_0 is the incident laser power intensity ($W \cdot m^{-2}$), also called surface power density, and z is the interaction depth (m). The absorption of radiation controls the overall laser-powder interaction efficiency while the heat conduction defines the contours of fusion isotherms.

Some metals, like Al, Cu, and Mg, have very high electrical and thermal conductivity, leading to a high reflectivity R (>90%), in contrast to transition metals such as Fe, Ti, Ni, and so forth, which have $50\% < R < 75\%$ for a near-infrared laser. It should be noted that the transition from a solid state (powder) to a liquid state **(melt pool)** can cause an increase in absorptivity A ($A = 1 - R$): from +2% for Ti to +49% for Ni [20]. The I_0 in PBF is often $\sim 10^{10}\ W \cdot m^{-2}$, leading to the interaction regime exceeding the vaporization threshold. Consequently, a vapor capillary is formed, also named a keyhole, as shown in Figure 8.4b. **Keyhole** formation is a well-known phenomenon in laser welding. Therefore, a conduction welding regime is commonly used to analyze the heat transfer, and the cross-section of the fusion zones is typically regarded as quasi-elliptical. High-speed microscopic imaging was developed by researchers at Lawrence Livermore National Laboratory to monitor the **melt pool** flow dynamics and keyhole pore defects during PBF-LB [21–23]. For a thorough description of the physics of heat transfer and the associated equations, the readers are recommended to refer to a book chapter by Peyre in [19].

Heat transfer modes in EBM are similar to those in SLM described previously, except that the convective heat transfer is significantly reduced due to the vacuum environment. As a result, radiative heat loss from the surface of the part and conductive heat loss to the powder surrounding the part are the dominant modes of heat transfer. Also, EBM operates at elevated temperatures (400–1000 °C) due to the need for high temperatures (>2000 °C) to heat the cathode filament to trigger electron emission. Thus, it is necessary to consider the thermal history of the part built via EBM with respect to solidification and the hold temperature [24]. The fused part that solidifies from the **melt pool** is kept at an elevated temperature until all layers are finished. Another primary difference between laser PBF and PBF-EB arises from the absorption of the beam. PBF-EB leads to a greater penetration thickness, 10–50 μm, in contrast to only a few nanometers in PBF-LB. The melt pool geometry is primarily determined by the correlation of beam velocity to beam power. Increasing beam diameter is a way to decrease the thermal gradient of the melt pool and slow down solidification.

8.2.2 Feedstock Powder for PBF

The **spreadability** of the powder is key to the layer uniformity of the build part in PBF processes. To manufacture parts with satisfying structural and/or functional properties, it is important to achieve homogeneous spreading of powder to control the layer thickness and roughness. Therefore, to ensure the high quality of the build parts, the first step is to understand and choose feedstock powder with appropriate characteristics, such as shape, size, and surface roughness, according to different PBF processes. Impurities (e.g. the presence of a passivating oxide layer on the powder particle surface) and moisture should be also considered. In general, the built parts made from a powder bed with a higher packing density have lower internal stresses, smaller part distortion, lower porosity, and smaller surface roughness [25]. Typically, spherical particle morphology is desirable as spherical particles exhibit better flowability and achieve a higher packing density in the powder bed compared to non-spherical particles. Non-spherical powder particles tend to mechanically interlock and entangle with one another, and thus, obstruct the flow of the powder. Consequently, heterogeneous layers of powder particles with varying packing densities may form on the top of the previous layer that has turned into a solid surface of the build part, leading to the formation of defects like pores and incomplete melting [25]. In situ X-ray imaging and high-speed photography techniques are used to probe the powder flow behavior, laser-powder interactions in and near the **melt pool**, as well as the thermal history and the consequent microstructure [22, 26–28]. *In situ* high-speed transmission X-ray imaging enabled characterization of laser-metal interactions at nanosecond and microsecond timescales [28]. Matthews et al. used high-speed imaging to reveal a denudation phenomenon of titanium alloy and steel alloy powders under varying laser conditions and ambient gas pressures in PBF-LB [29]. A recent study by Mussatto et al. showed that the powder particle sphericity and surface texture dictated how the spreader velocity and the layer thickness would affect the quality of powder bed topography in spread layers [30]. It was found that powder flowability alone did not guarantee the uniform quality of layers. Rather, the spreader velocity played a substantial role – low spreading velocities ($\leq 8\ cm\ s^{-1}$) gave rise to better uniformity. It was also reported that powders with a wide particle size distribution and a high amount of fine particles (<25 μm) led to uniform and dense powder beds. However, such fine particles tend to be subjected to segregation. Another work on PBF-LB of AlSi10Mg also suggested that the dimensional accuracy of the build parts decreased when fine particles (D_{50} = 9 μm) were used, compared to the situation of using coarse powder (D_{50} = 40 μm) [31]. To sum up, a high relative particle packing density with high powder flowability is desired with properties related to particle size, surface topology, and shape distribution. It is commonly accepted that powders for PBF-LB processes have an average particle size (25–45 μm), which is smaller than that for EBM (40–105 μm). A more specific and in-depth discussion on the effect of powder on the metal AM processes is provided in Chapters 10 and 12.

8.2.3 Post-processing After PBF

It is very rare that the parts in the as-built condition right after the PBF processes can be directly used in practice. Post-processing treatments are frequently required. For instance, excess powder, support structures, and substrate material must be removed. PBF processes require powder to either be vacuumed out of the part for laser melting, or blasted off using loose powder for EBM, referred to as a depowdering process. Support structures, for mechanical and thermal support, are frequently used in PBF and must be mechanically removed by application of force or cutting. The build plate is typically adhered to the finished part and must be cut off. Subsequently, various heat or thermo-mechanical treatments are commonly applied to relieve residual stress (annealing), further densify the parts (hot isostatic pressing), and/or improve mechanical performance (solutionizing and aging). Secondary machining is also sometimes applied to improve surface finish and/or to achieve desired tolerances.

8.3 Directed Energy Deposition (DED)

DED refers to AM processes in which focused energy generates a **melt pool** into which the feedstock is deposited from nozzles. The origin of DED is traced to welding technology. One of the most studied commercial DED machines is the LENS process originally developed at Sandia National Laboratories and commercialized by Optomec Company, Albuquerque, NM, as mentioned in Section 8.1. The energy sources for DED can be a focused LB, EB, or plasma arc. The feedstock material for DED can be in the form of powder or wire (Figure 8.2). As the scope of this book is on metal powders for AM, the discussion in the following sections focuses on the powder-based DED-LB.

8.3.1 DED Processing Principles

In DED, metal powders are delivered from powder hoppers through a coaxial powder nozzle (the most common case), for example, four nozzles equally spaced around a deposition head in LENS, using a pressurized inert carrier gas. Figure 8.5a illustrates the schematic setup of a laser powder DED system [32]. The powder is injected into the focal

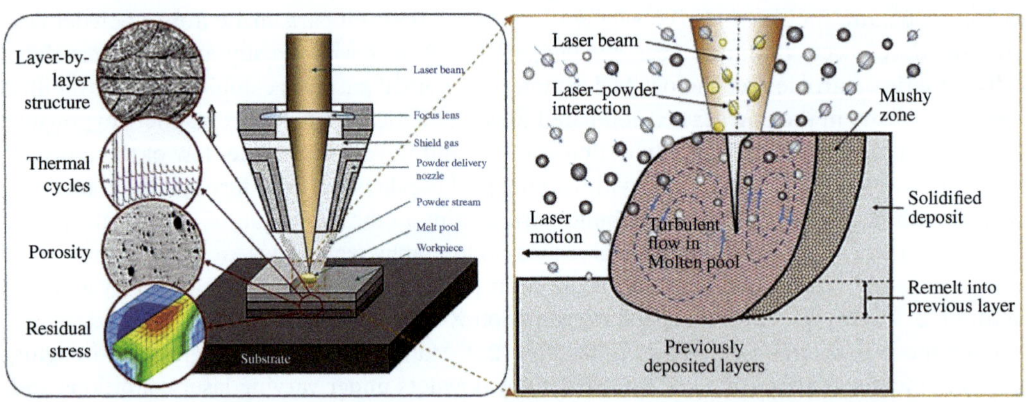

Figure 8.5 Schematic diagram illustrating the attributes and scientific issues related to laser powder DED processing: (a) microstructure, multiple interfaces, thermocycles, defects, and residual stress, and (b) interaction among injected powder, laser beam, and melt pool. Source: Zheng et al. [32] / with permission of Elsevier.

point of the LB, where a **melt pool** is formed by the focused LB. In other words, the laser first interacts with the powder stream projected over an interaction distance before being absorbed by the substrate or the previously deposited layers. The **melt pool** increases in size as the distance to the substrate decreases, resulting from the injection and melting of the powder particles. The substrate, on which the powder is being deposited, can move in a plane in the x and y directions, and as the laser power is switched on and off according to a CAD model, the targeted cross-section of the component is fabricated. The focused LB and deposition head are moved up in the z-direction as each layer is completed, and a fully dense metal component can be deposited. As opposed to PBF, there is no pre-placed powder for support in the DED platform, so the build stands on its own on the substrate. The deposition process is performed in a glove box filled with inert gas (e.g. Ar), where the inert gas is used to shield the **melt pool** from atmospheric oxygen for better control of properties, and to promote layer-to-layer adhesion by providing better surface wetting.

DED is a complex thermal process that involves rapid heating, melting, rapid solidification of the molten powder into a build part, and potential vaporization [33, 34]. The far-from-equilibrium process such as high cooling rates ($\sim 10^2$–10^4 °C s^{-1}) in DED results in the formation of metastable non-equilibrium microstructures [35]. Although DED processes are often compared to welding technologies, there are significant differences, including remarkable thermal gradients within a small melt pool, a fast-moving heat source, multiple scan layers with different directions and hatch patterns, as well as exposure of the material to multiple thermal cycles. These factors all contribute to the melt pool dynamics (Figure 8.5b), solidification behavior, and the resultant microstructure. In the melt pool, feedstock powder already starts to interact with the LB when it is in-flight, leading to turbulent flow, as opposed to the static powder bed in PBF. The common scheme between PBF and DED processes is that both involve simultaneous melting of the surface layer of powder and the remelting or reheating of the underlying layers. In other words, thermal cycling occurs during the deposition process as the material cools and reheats. The shield gas in DED has a flow rate more than 15 times higher than that of the fill gas in PBF-LB, resulting in a larger convective heat transfer in DED. On the other hand, DED is less energetic ($I_0 = 10^8 \sim 10^9$ W · m^{-2}) than PBF, and thus, vaporization of the metal melt (keyhole regime) does not often occur, except for elements with a low vaporization point, for instance, Al in Ti-6Al-4V.

Compared to the PBF technology, DED has the following major advantages: (i) refined microstructures in the build part because of small heat-affected zone and high cooling rates, (ii) feasibility of gradient deposition of multiple materials within one part, and (iii) capability of adding materials and features to existing structures and components, which enables DED to be a repair method. Laser powder DED is a highly targeted solution for repairing high-performance metal components. This technique can be used to precisely add material to worn or damaged areas with minimal heat effect, even for the most sensitive thin wall components such as those found in gas turbine engines [36]. DED can also be used to do *in situ* alloying or grading of the material composition with independently controlled powder delivery hoppers and/or by controlling deposition parameters, producing location-specific properties, that is, varying properties at different locations of the part [37–39]. This is very important for AM of functionally graded materials, which is difficult to be realized using other metal AM technologies [40]. However, it is also important to note that DED is inferior to PBF in terms of surface roughness, geometric accuracy and resolution, and making overhanging features.

8.3.2 Feedstock Powder for DED

Just like the PBF processes, the DED process and the quality of the build parts are also significantly dependent on the distinct properties of the powder feedstock material, including chemical composition, melting temperature, thermal conductivity, reflectivity, specific heat capacity, melt viscosity, melt surface tension, spectral emissivity, and so forth [33]. As mentioned in Section 8.3.1, in DED, feedstock powder is delivered from hoppers, which utilize a perforated disc mechanism rotating at a controlled speed to meter powder into a stream of inert gas. The inert gas stream carries the powder through the nozzles of the deposition head onto the melt pool. The powder streams diverge to some extent when they are injected from the nozzles, forming conical streams. These powder cones intersect in the vicinity of the melt pool, making inter-particle collisions inevitable. The relatively long journey the feedstock powder travels through, in contrast to the powder bed in PBF, makes flowability a critical powder property in the DED processes, compared to the term **spreadability** used in PBF [41–43]. The shape and particle size distribution of the feedstock powder are the key elements that control the powder flow behavior, as well as the dimensional accuracy and properties of the near-net shape components. The surface finish of the deposited parts is a strong function of the powder particle size, with a small powder particle size giving a better surface finish [36]. Spherical particles are ideal for the DED processes, due to the good flowability provided by the sphericity, as discussed in Section 8.2.2. But several studies show that powder particles with irregular shapes, such as ball-milled powder, are also feasible in DED-LB [33, 44]. The suitable particle size for DED-LB is in the range of 45–150 μm, while for DED-EB, the powder particle size can be 40–80 μm [33, 45]. If the powder particles are too small, agglomeration will occur, associated decrease in flowability, and clog the delivery path from the powder hopper to the nozzle. If the powder particle size is too large, it becomes unsuitable for a gas-carried-particle process or it can clog the nozzle. The stability of the melt pool requires a narrow particle size distribution due to the turbulent dynamic flow in the interaction volume. Given that the thermal history during the DED process involves the small size of the melt pool, large thermal gradients, and displacement of the solid/liquid interface, a detailed study of the powder particle-pool interactions remains a formidable challenge. If the feedstock powder is a ball-milled powder that typically exhibits nanocrystalline structure, when the powder experiences a high temperature but no melting prior to impingement onto the melt pool, the maintenance of a nanocrystalline or formation of ultrafine-grained microstructure will depend on the thermal stability of the feedstock powder. A high-speed camera can be used to characterize the particle-droplet melting and particle-pool interaction behavior during the DED process. Also, high-speed imaging enables the characterization of the morphology and structure of the feedstock powder in different stages before entering the melt pool [42, 46–50], as shown in Figure 8.6 [50]. Figures 8.6b,e,f reveal notable details as the powder particles travel and impinge on the melt pool. These particles remained on the surface for ~0–600 μs before being absorbed into the melt. In some cases, particles bounced away from the surface after interacting with particles already settled on the surface.

The other major type of feedstock for DED is wires or filaments. Handling metal wires is safer than metal powders. Using wires as the feedstock is also advantageous in cost of raw materials [51], deposition speed (greater than that of powder deposition) [52], as well as eliminating the need for the process of powder recycling [53]. One key point to consider for using metal wires in DED is how to make the wire into lengths suitable for layer-by-layer deposition. The diameter of the wires used in DED commonly varies between 0.8 and 1.6 mm [52]. Since the scope of this book is powder, here we briefly discuss the influence of wires on the processes using the example of **wire arc additive manufacturing (WAAM)** [53]. WAAM involves the use of arc welding equipment that allows for high deposition rates [54]. Hence, the wires should first exhibit good arc stability and good feeding features such as surface texture, which can be improved by applying proper drawing speeds and specific lubricants on the wire surface. Winding of the wires during the deposition process is of concern to achieve good quality and properties of the final builds. To

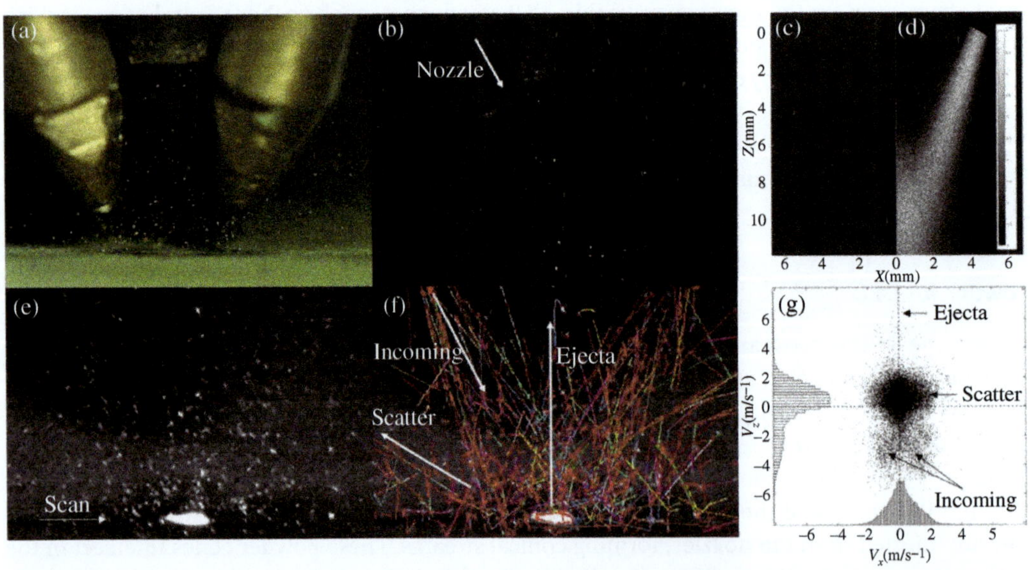

Figure 8.6 Characterization of powder spray pattern with high-speed imaging. (a) Straight particle trajectories are diffusely reflected by a substrate. (b) A single nozzle spray is observed. (c) Lower-speed video (10 kHz) provides a one million particle dataset to identify the spatial concentration of the spray (d). In (e) a single track of material is deposited. (f) Three particle flight modes – incoming particles sprayed from the nozzle, diffusely scattered particles, and particles ejected rapidly upward by metal vaporizing from the melt pool surface. (g) The plot of velocity components. Source: Reproduced with permission from Haley et al. [50] / Elsevier.

optimize the remanence values, it is important to observe the tight tolerance and the straightness of the wires [52]. High-quality production in WAAM is yet limited, due to several materials processing challenges and defects caused by elevated levels of heat inputs [53, 55]. Therefore, wire-fed DED processes are considered complementary to, rather than competing with, the powder-fed DED processes.

8.3.3 Post-processing After DED

Similar to the PBF technology, post-processing treatment is typically needed for the build parts made by DED. First, substrates are used to support the deposition process and the build parts are bonded to the substrate after the DED process. Hence, the substrate needs to be removed before any other post-processing. DED processes produce near-net-shape parts, implying that it is necessary to apply secondary processing, such as machining or polishing, to reach the required tolerances and surface quality. DED processes often result in high surface roughness of build parts. The primary causes of high surface roughness include: (i) unmelted or partially melted powder particles adhered to the surface due to low heat input and large powder particles [56, 57] and balling due to Rayleigh instability at high laser scanning speeds that break the melt pool into small islands that are then dragged to the outer edges of the melt pool [58], (ii) stair stepping effect that restricts all layered manufacturing processes, particularly when inclined or curved surfaces are formed [59], and (iii) splashing of molten material [60]. In addition, annealing is often required to relieve the high residual stress resulting from the high cooling rates and thermal cycling in DED processes.

8.4 Metal Binder Jetting

BJT is often regarded as a powder bed-based AM process. However, it is distinct from PBF as the process mechanism of BJT is a sequence of multiple steps: depositing binder on metal powder, curing the binder to hold the metal powder together, sintering the powder, and removing the binder material (debinding), with an optional step of infiltration with a second metal, whereas PBF directly melts and/or sinters the material as discussed in Section 8.2. BJT is also used for AM of ceramics nowadays. ExOne is currently the main manufacturer of BJT printers.

8.4.1 BJT Processing Principles

Figure 8.7 shows the schematic diagram of the working principle of a BJT process [61]. The process starts by spreading a thin layer of powder into the powder bed, then the printhead selectively deposits binder droplets into the powder bed. The binder droplets are absorbed by the powder bed to fill the empty gaps between the powder particles, acting like glue to join the powder particles together and to hold the cross section of each layer together while creating a bond between each layer and its immediate neighbor layers on the top and at the bottom. The printhead contains hundreds of small orifices, each ejecting mono-dispersed droplets with volumes in the order of pico-liter. The printing plate (piston) is then lowered to allow another layer of powder to be spread. The powder delivery system works similar

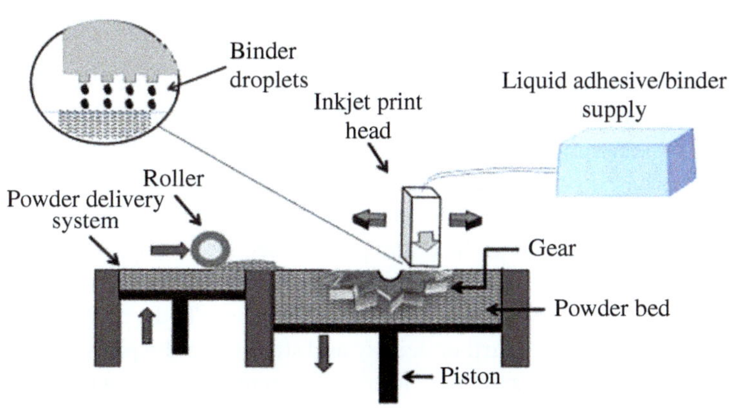

Figure 8.7 Schematic diagram of the working principle of BJT. Source: Gupta et al. [61] / with permission of Elsevier.

to that in the PBF, with the piston raising the position of the powder feeder and the roller spreading a new layer of powder into the powder bed. The process repeats until the build part is completed, with unused powder (around 95%) recycled. After the part is printed, it will be exposed to a relatively elevated temperature to cure the binder to create a green body. In contrast to PBF and DED using focused LB or EB, BJT does not employ high-power heat sources. Therefore, the working principle of BJT is not relevant to laser-powder or electron-powder interaction. Rather, the physics of binder droplet formation, the droplet–surface, and droplet–powder bed interaction dominate the process principle of BJT. The binder materials are typically polymers in solvent or aqueous solution, for which viscosity, density, and surface tension directly affect the droplet formation process. Toyserkani et al. discussed the physics and governing equations for the binder droplet in BJT [62]. As the scope of this book focuses on metal powders, readers are recommended to refer to [62] for detailed theory and physics of binder materials.

8.4.2 Feedstock Powder for BJT

The characteristics of the feedstock powder strongly affect many aspects of the BJT operations, part design requirements, and economics. As the BJT processes share the same mechanism in the powder delivery system, the general discussion relevant to the feedstock powder in PBF in Section 8.2.2 is also applicable here. For instance, the spherical shape of the powder is desirable to achieve good flow and **spreadability**. One of the primary concerns in BJT is the packing characteristics of the powder layer differently compared to PBF. That is because the densification during printing in BJT is negligible, whereas laser or e-beam sinter the powder bed during building in PBF. The powder packing/spread density and pore uniformity are incorporated into the green part during printing via BJT, which significantly impacts the final properties and dimensional accuracy of the final part. The green part density determines the shrinkage if the part is to be sintered for full density. For monosized spherical powders, the theoretical packing density is approximately 60% of the true density [63].

In addition, particle size distribution plays a key role in the post-processing sintering stage of green parts for BJT. Several studies have revealed that multimodal particle size distributions by mixing powders of different particle sizes are beneficial to increase the green density [64–67]. Higher green density reduces the shrinkage required for densification. However, it is also important to note that an increase in green density from multimodal powders does not necessarily mean the final sintered density will be increased. A comparison of the green part and sintered densities for different powder mixtures (monomodal, bimodal, and trimodal) of metal BJT parts suggested that multimodal powders achieved densities >90% in the final sintered part only when sintering aids were used [68]. It still remains a prime obstacle to spreading fine particles well in layer-wise AM processes, for both PBF and BJT [63]. Powders with fine particle sizes have more surface area for a given volume, making them sensitive to environmental conditions. In addition, powder size can alter slumping significantly. Taking blended powder solid-super solidus liquid phase sintering as an example, when a blend of coarse–fine powders contains coarse powder with a particle size ≥5 times larger than the size of the fine fraction sintered below its solidus point, the mixture can be sintered over wider temperature ranges with no slumping [69]. This method was demonstrated in the densification of Inconel 718 and is stated to be dependent on powder characteristics, fine particle fraction, and binder chemistry [70].

8.4.3 Post-processing After BJT

When the printing step is finished, it is common that a curing step is required to dry the binder to give the "joined" powder its green strength. This process is called curing and depowdering. Figure 8.8 illustrates the multiple steps of post-printing treatment for BJT [71]. The powder bed is removed from the BJT printer and heated in an oven for curing at 180–200 °C for a prescribed number of hours based on the build box volume and binder characteristics (Figure 8.8d). Then depowdering is employed by removing the loose powder in the build box via a vacuum and careful, manual brushing. In some cases, the depowdering step is skipped as the entire powder bed is cured followed by slight machining. Then the green parts are moved directly to the densification step. Metal green parts made with BJT after the curing step often remains in a fragile green-body state (relative density ~60% or lower). Thus, additional post-processing, such as sintering and/or infiltration, is mandatory to further densify and strengthen the parts. Various factors need to be considered in determining the proper post-processing cycle and parameters, including but not limited to material composition, powder particle size, sintering atmosphere, temperature, holding time, and so forth. Certain design factors also must be considered for the densification stage. As shrinkage occurs during sintering, the

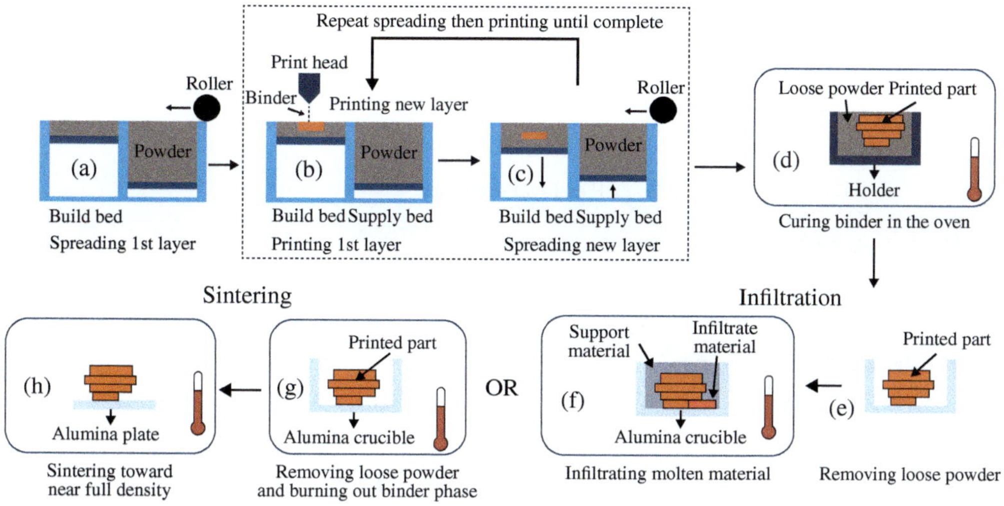

Figure 8.8 Schematic flow chart of the post-processing after the green body of parts is built via BJT. Source: Do et al. [71] / with permission of Elsevier. (a) build bed spreading 1st layer, (b) and (c) repeat spreading and printing until the part is completed; (d) curing binder in the oven; (e) removing loose powder to obtain green body of the part; (f) infiltrating molten material; (g) and (h) sintering to get near full density.

initial design of the part in BJT must be able to meet the required tolerances and dimensions after undergoing densification. Sintering of single alloys to full density tends to result in highly warped geometries whereas infiltration often produces accurate features. It is significant to consider design factors such as section size to ensure the binder can be effectively removed. The overall post-process is multistep and more complex than those needed for PBF or DED. However, as no high temperature or power is involved during the BJT process, it is anticipated that fewer issues associated with residual stresses and distortion are of concern. In addition, BJT can print large and complex components without the need for support structures. BJT also generally outperforms PBF or DED processes in forming small features, overhang structures, struts, and microchannels with high accuracy.

8.5 Sheet Lamination (SHL)

In metal SHL, metal foils or sheets are stacked and laminated using welding, adhesive, heat, or pressure. SHL is distinct from the other three categories of metal AM discussed above as the feedstock is sheets rather than powders. As a result, the working mechanism is the stacking of precision-cut metal sheets into 2D part slices created from a 3D object. Figure 8.9 is a schematic diagram showing the process of SHL [72]. The material sheet is rolled on the building platform with a heated roller. Successive sheets are fed onto the platform where they are cut precisely using a mechanical cutter or a laser cutter. A key feature of SHL is the order in which sheets are applied and cut/machined. The stacked sheets may be either cut to the specified geometry prior to adhesion or machined post-adhesion (form-then-bond) or vice versa (bond-then-form). Other bonding mechanisms can be adhesively joined or metallurgically bonded using brazing, diffusion bonding, resistance welding, or ultrasonic consolidation [15, 73–75]. The excess materials after cutting are left as support, and after completion of the process can be removed and recycled. Compared with most powder-based AM technologies like PBF and laser powder DED, objects created via SHL do not require support structures, as the unused material sheets that surround the build part hold everything in place. The ultrasonic consolidation process is one of the most common technologies for metal SHL. The ultrasonic SHL process consists of the following steps: (i) the substrate is milled to achieve a flat deposition surface; (ii) tailings are blown off the substrate; (iii) Material is deposited for a given layer in metal tapes through ultrasonic welding, (iv) the edges of the tape are trimmed from the given layer to match the desired part geometry, (v) layers are iterated until a part is finished, and (vi) milling operations are engaged as required to produce channels, holes, or other features. The optional milling or machining operation enables the incorporation of cooling channels and embedded structures. Thus, ultrasonic SHL is regarded as a hybrid AM method. A comprehensive review of this type of metal AM technology is provided by Hehr and Norfolk [76].

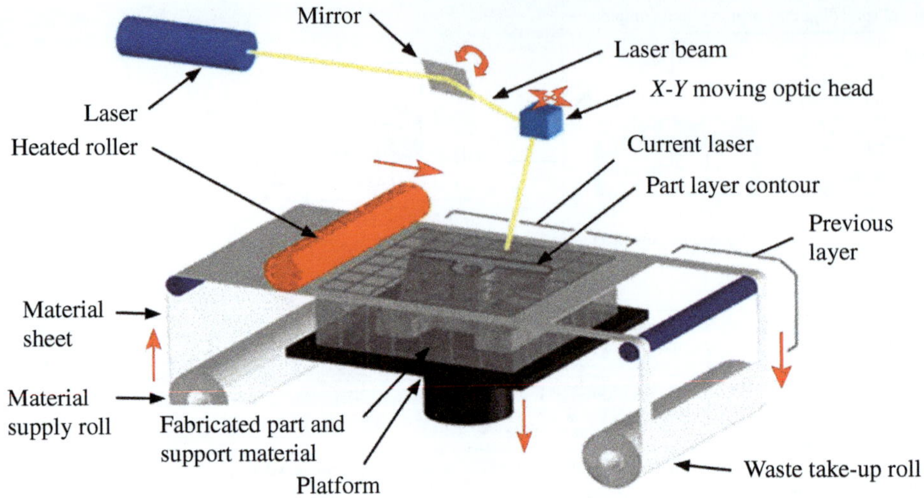

Figure 8.9 Schematic diagram of the sheet lamination process. Source: Mercado Rivera et al. [72] / Springer Nature.

SHL is advantageous in reducing geometric distortion as the original metal sheets retain their properties, relatively easy manufacturing of large-scale parts and structures, good surface finish, and economical cost. However, without post-processing, parts made via SHL have a lower surface quality and lower dimensional accuracy compared to PBF processes [77]. Adhesively joined parts may not perform well in shear and tensile loading conditions. And anisotropic properties are prevalent in the builds made via SHL due to the type of joining process. In addition, it is time-consuming to remove the excess parts of laminates after the formation of the target object compared to DED and PBF methods.

8.6 Summary

This chapter provided an overview of several metal AM technologies, with an emphasis on the working principles. The corresponding requirements of feedstock powder and post-processing treatments were briefly discussed. Although a detailed discussion of computational efforts in the AM field is not included in this chapter, **integrated computational materials engineering** (**ICME**) plays an important role in advancing metal AM technologies, including but not limited to, optimal design of shapes, proper selection of processing parameters, development of new alloy compositions, and microstructure prediction [78–80]. Common modeling and simulation approaches include Calphad, finite element modeling and finite element analysis, phase field model, and so forth [79–82]. Discussion on computational work will be provided in Chapter 10. While every individual category of AM technologies has its own unique capabilities and advantages, it is common that the feedstock powder characteristics are key to the process control and the quality of the final parts. Therefore, the next few chapters will discuss the effects of powder in more detail, starting with the powder-laser-melt pool interactions in AM in Chapter 9.

Acronym/Nomenclature

3D	three dimensional
AM	additive manufacturing
BJT	binder jetting
CAD	computer-aided design
CNC	computer numerical control
DED	directed energy deposition
DED-EB	electron beam-directed energy deposition
DED-LB	laser beam-directed energy deposition

DMLS	direct metal laser sintering
EBM	electron beam melting
LENS™	Laser Engineered Net Shaping
PBF	powder bed fusion
PBF-EB	electron beam powder bed fusion
PBF-LB	laser beam powder bed fusion
SHL	sheet lamination
SLS	selective laser sintering
SLM	selective laser melting
WAAM	wire arc additive manufacturing
I_0	surface laser power density ($W \cdot m^{-2}$)
P_{abs}	the absorbed laser power
S_0	the interaction surface area
E_v	volume energy density ($J \cdot m^{-2}$)
V_0	laser scanning speed ($m \cdot s^{-1}$)
α	absorption coefficient (m^{-1})
z	interaction depth (m)

References

1 (2021). ISO/ASTM 52900:2021(en) Additive manufacturing — General principles — Fundamentals and vocabulary. ISO/TC 261 and ASTM Committee F42. https://www.iso.org/obp/ui/#iso:std:iso-astm:52900:ed-2:v1:en.

2 Using Additive Manufacturing to Improve Supply Chain Resilience and Bolster Small and Mid-Size Firms. *The White House*. (2022). https://www.whitehouse.gov/cea/written-materials/2022/05/09/using-additive-manufacturing-to-improve-supply-chain-resilience-and-bolster-small-and-mid-size-firms/ (accessed 3 February 2022).

3 Deckard, C.R. (1986). *Part Generation by Layer-Wise Selective Laser Sintering*. Austin, TX: University of Texas – Austin.

4 Deckard, C.R. Method and apparatus for producing parts by selective sintering. US4863538A, 05 September 1989. https://patents.google.com/patent/US4863538A/en (accessed 16 September 2022).

5 Wohlers, T. and Gornet, T. (2016). *History of additive manufacturing*. Wohlers Associate, Inc. https://wohlersassociates.com/wp-content/uploads/2022/08/history2016.pdf.

6 Gibson, D.V. and McClure, P.F. (1992). The Commercialization of New Technologies: the Case of DTM Corp. p. 3.

7 Sachs, E.M., Haggerty, J.S., Cima, M.J. et al. Three-dimensional printing techniques. US5204055A, 20 April 1993. https://patents.google.com/patent/US5204055A/en (accessed 16 September 2022).

8 Mostafaei, A., Elliott, A.M., Barnes, J.E. et al. (2021). Binder jet 3D printing – process parameters, materials, properties, modeling, and challenges. *Prog. Mater. Sci.* 119: 100707. https://doi.org/10.1016/j.pmatsci.2020.100707.

9 About GE Additive | GE Additive. https://www.ge.com/additive/who-we-are (accessed 16 September 2022).

10 Larson, R. (1998). Method and device for producing three-dimensional bodies. US5786562A, 28 July 1998. https://patents.google.com/patent/US5786562A/en (accessed 16 September 2022).

11 The Story of EOS GmbH and Industrial 3D Printing. https://www.eos.info/en/topnavi/about-us/history (accessed 16 September 2022).

12 News Release – Laser Engineered Net Shaping (LENS). https://www.sandia.gov/media/lens.htm (accessed 16 September 2022).

13 Norsk Titanium | Company. https://www.norsktitanium.com/company#history (accessed 16 September 2022).

14 Metal Additive Manufacturing with EBAM® Technology | Sciaky. https://www.sciaky.com/additive-manufacturing/electron-beam-additive-manufacturing-technology (accessed 16 September 2022).

15 White, D. (2003). Ultrasonic consolidation of aluminum tooling. *Adv. Mater. Processes* 161: 64–65.

16 Kawalkar, R., Dubey, H.K., and Lokhande, S.P. (2022). A review for advancements in standardization for additive manufacturing. *Mater. Today: Proc.* 50: 1983–1990. https://doi.org/10.1016/j.matpr.2021.09.333.

17 Spears, T.G. and Gold, S.A. (2016). In-process sensing in selective laser melting (SLM) additive manufacturing. *Integr. Mater. Manuf. Innov.* 5 (1): 16–40. https://doi.org/10.1186/s40192-016-0045-4.

18 Gruber, H., Henriksson, M., Hryha, E. et al. (2019). Effect of powder recycling in electron beam melting on the surface chemistry of alloy 718 powder. *Metall. Mater. Trans. A* 50 (9): 4410–4422. https://doi.org/10.1007/s11661-019-05333-7.

19 Peyre, P. (2022). The physics of metal additive manufacturing processes. In: *Additive Manufacturing of Metal Alloys 1* (ed. P. Peyre and É. Charkaluk), 151–199. John Wiley & Sons, Ltd https://doi.org/10.1002/9781394163380.ch3.

20 Bretonnet, J.-L. (2005). Conductivité électrique des métaux liquides. *Techniques de l'ingénieur Étude et propriétés des métaux*, vol. base documentaire: TIP551WEB, no. ref. article: m69. Editions T.I. https://doi.org/10.51257/a-v1-m69.

21 Matthews, M.J., Rubenchik, A., Guss, G. et al. (2015). Characterization of melt-flow dynamics in selective laser melting (SLM) processes. In: *CLEO: 2015*, ATh3A.5. San Jose, California: OSA Technical Digest (online) (Optica Publishing Group) https://doi.org/10.1364/CLEO_AT.2015.ATh3A.5.

22 Calta, N.P., Thampy, V., Lee, D.R.C. et al. (2020). Cooling dynamics of two titanium alloys during laser powder bed fusion probed with in situ X-ray imaging and diffraction. *Mat. Des.* 195: 108987. https://doi.org/10.1016/j.matdes.2020.108987.

23 Forien, J.-B., Calta, N.P., DePond, P.J. et al. (2020). Detecting keyhole pore defects and monitoring process signatures during laser powder bed fusion: a correlation between in situ pyrometry and ex situ X-ray radiography. *Addit. Manuf.* 35: 101336. https://doi.org/10.1016/j.addma.2020.101336.

24 Sames, W.J., List, F.A., Pannala, S. et al. (2016). The metallurgy and processing science of metal additive manufacturing. *Int. Mater. Rev.* 61 (5): 315–360. https://doi.org/10.1080/09506608.2015.1116649.

25 Vock, S., Klöden, B., Kirchner, A. et al. (2019). Powders for powder bed fusion: a review. *Progr. Addit. Manuf.* 4 (4): 383–397. https://doi.org/10.1007/s40964-019-00078-6.

26 Scipioni Bertoli, U., Guss, G., Wu, S. et al. (2017). In-situ characterization of laser-powder interaction and cooling rates through high-speed imaging of powder bed fusion additive manufacturing. *Mat. Des.* 135: 385–396. https://doi.org/10.1016/j.matdes.2017.09.044.

27 Ashby, A., Guss, G., Ganeriwala, R.K. et al. (2022). Thermal history and high-speed optical imaging of overhang structures during laser powder bed fusion: a computational and experimental analysis. *Addit. Manuf.* 53: 102669. https://doi.org/10.1016/j.addma.2022.102669.

28 Martin, A.A., Calta, N.P., Hammons, J.A. et al. (2019). Ultrafast dynamics of laser-metal interactions in additive manufacturing alloys captured by in situ X-ray imaging. *Mater. Today Adv.* 1: 100002. https://doi.org/10.1016/j.mtadv.2019.01.001.

29 Matthews, M.J., Guss, G., Khairallah, S.A. et al. (2016). Denudation of metal powder layers in laser powder bed fusion processes. *Acta Mater.* 114: 33–42. https://doi.org/10.1016/j.actamat.2016.05.017.

30 Mussatto, A., Groarke, R., O'Neill, A. et al. (2021). Influences of powder morphology and spreading parameters on the powder bed topography uniformity in powder bed fusion metal additive manufacturing. *Addit. Manuf.* 38: 101807. https://doi.org/10.1016/j.addma.2020.101807.

31 Balbaa, M.A., Ghasemi, A., Fereiduni, E. et al. (2021). Role of powder particle size on laser powder bed fusion processability of AlSi10mg alloy. *Addit. Manuf.* 37: 101630. https://doi.org/10.1016/j.addma.2020.101630.

32 Zheng, B., Haley, J.C., Yang, N. et al. (2019). On the evolution of microstructure and defect control in 316L SS components fabricated via directed energy deposition. *Mater. Sci. Eng., A* 764: 138243. https://doi.org/10.1016/j.msea.2019.138243.

33 Svetlizky, D., Das, M., Zheng, B. et al. (2021). Directed energy deposition (DED) additive manufacturing: physical characteristics, defects, challenges and applications. *Mater. Today* 49: 271–295. https://doi.org/10.1016/j.mattod.2021.03.020.

34 Zheng, B., Xiong, Y., Nguyen, J. et al. (2009). Powder additive processing with laser engineered net shaping (LENS) in powder metallurgy research trends. In: *Powder Metallurgy Research Trends* (ed. L.J. Smit and J.H. Van Dijk), 125–190. New York: Nova Science Publishers, Inc.

35 Zheng, B., Zhou, Y., Smugeresky, J.E. et al. (2008). Thermal behavior and microstructure evolution during laser deposition with laser-engineered net shaping: part II. Experimental investigation and discussion. *Metall. Mater. Trans. A* 39 (9): 2237–2245. https://doi.org/10.1007/s11661-008-9566-6.

36 Atwood, C.L., Ensz, M.T., Greene, D.L. et al. (1999). Laser Spray Fabrication for Net-Shape Rapid Product Realization LDRD. United States, April 1999. SAND99-0739. https://doi.org/10.2172/5657.

37 Kim, S.H., Lee, H., Yeon, S.M. et al. (2021). Selective compositional range exclusion via directed energy deposition to produce a defect-free Inconel 718/SS 316L functionally graded material. *Addit. Manuf.* 47: 102288. https://doi.org/10.1016/j.addma.2021.102288.

38 Kelly, J.P., Elmer, J.W., Ryerson, F.J. et al. (2021). Directed energy deposition additive manufacturing of functionally graded Al-W composites. *Addit. Manuf.* 39: 101845. https://doi.org/10.1016/j.addma.2021.101845.

39 Yan, L., Chen, Y., and Liou, F. (2020). Additive manufacturing of functionally graded metallic materials using laser metal deposition. *Addit. Manuf.* 31: 100901. https://doi.org/10.1016/j.addma.2019.100901.

40 Ansari, M., Jabari, E., and Toyserkani, E. (2021). Opportunities and challenges in additive manufacturing of functionally graded metallic materials via powder-fed laser directed energy deposition: a review. *J. Mater. Process. Technol.* 294: 117117. https://doi.org/10.1016/j.jmatprotec.2021.117117.

41 Kiani, P., Scipioni Bertoli, U., Dupuy, A.D. et al. (2020). A statistical analysis of powder flowability in metal additive manufacturing. *Adv. Eng. Mater.* 22: 2000022. https://doi.org/10.1002/adem.202000022.

42 Jiang, S., Zheng, B., and Schoenung, J.M. (2022). Directed energy deposition of metal matrix composites: computational and experimental comparison of powder particle flow behavior. *J. Mater. Res. Technol.* 16: 516–529. https://doi.org/10.1016/j.jmrt.2021.12.009.

43 Iams, A.D., Gao, M.Z., Shetty, A. et al. (2022). Influence of particle size on powder rheology and effects on mass flow during directed energy deposition additive manufacturing. *Powder Technol.* 396: 316–326. https://doi.org/10.1016/j.powtec.2021.10.059.

44 Fullenwider, B., Kiani, P., Schoenung, J.M. et al. (2019). Two-stage ball milling of recycled machining chips to create an alternative feedstock powder for metal additive manufacturing. *Powder Technol.* 342: 562–571. https://doi.org/10.1016/j.powtec.2018.10.023.

45 Hofmeister, W. and Griffith, M. (2001). Solidification in direct metal deposition by LENS processing. *JOM* 53 (9): 30–34. https://doi.org/10.1007/s11837-001-0066-z.

46 Wolff, S.J., Webster, S., Parab, N.D. et al. (2021). In-situ observations of directed energy deposition additive manufacturing using high-speed X-ray imaging. *JOM* 73 (1): 189–200. https://doi.org/10.1007/s11837-020-04469-x.

47 Wang, H., Gould, B., Haddad, M. et al. (2022). In situ high-speed synchrotron X-ray imaging of laser-based directed energy deposition of the alloying process with dissimilar powders. *J. Manuf. Processes* 75: 1003–1011. https://doi.org/10.1016/j.jmapro.2022.01.055.

48 Webster, S., Wolff, S., Bennett, J. et al. (2019). Porosity formation and meltpool geometry analysis using high-speed, in situ imaging of directed energy deposition. *Microsc. Microanal.* 25 (S2): 2556–2557. https://doi.org/10.1017/S1431927619013515.

49 Wolff, S.J., Wu, H., Parab, N. et al. (2019). In-situ high-speed X-ray imaging of piezo-driven directed energy deposition additive manufacturing. *Sci. Rep.* 9 (1): 962. https://doi.org/10.1038/s41598-018-36678-5.

50 Haley, J.C., Schoenung, J.M., and Lavernia, E.J. (2018). Observations of particle-melt pool impact events in directed energy deposition. *Addit. Manuf.* 22: 368–374. https://doi.org/10.1016/j.addma.2018.04.028.

51 Cunningham, C.R., Flynn, J.M., Shokrani, A. et al. (2018). Invited review article: strategies and processes for high quality wire arc additive manufacturing. *Addit. Manuf.* 22: 672–686. https://doi.org/10.1016/j.addma.2018.06.020.

52 Thomas, M. (2022). Raw Materials: Metal Powders and Wires. In: *Additive Manufacturing of Metal Alloys 1* (ed. P. Peyre and É. Charkaluk), 97–149. Wiley https://doi.org/10.1002/9781394163380.ch2.

53 Tomar, B., Shiva, S., and Nath, T. (2022). A review on wire arc additive manufacturing: processing parameters, defects, quality improvement and recent advances. *Mater. Today Commun.* 31: 103739. https://doi.org/10.1016/j.mtcomm.2022.103739.

54 Wu, B., Pan, Z., Ding, D. et al. (2018). A review of the wire arc additive manufacturing of metals: properties, defects and quality improvement. *J. Manuf. Processes* 35: 127–139. https://doi.org/10.1016/j.jmapro.2018.08.001.

55 Tawfik, M.M., Nemat-Alla, M.M., and Dewidar, M.M. (2021). Enhancing the properties of aluminum alloys fabricated using wire + arc additive manufacturing technique – a review. *J. Mater. Res. Technol* 13: 754–768. https://doi.org/10.1016/j.jmrt.2021.04.076.

56 DebRoy, T., Wei, H.L., Zuback, J.S. et al. (2018). Additive manufacturing of metallic components – process, structure and properties. *Prog. Mater Sci.* 92: 112–224. https://doi.org/10.1016/j.pmatsci.2017.10.001.

57 Brückner, F. and Leyens, C. (2017). 3 – Hybrid laser manufacturing. In: *Laser Additive Manufacturing* (ed. M. Brandt), 79–97. Woodhead Publishing https://doi.org/10.1016/B978-0-08-100433-3.00003-8.

58 Wei, H.L., Mukherjee, T., Zhang, W. et al. (2021). Mechanistic models for additive manufacturing of metallic components. *Prog. Mater Sci.* 116: 100703. https://doi.org/10.1016/j.pmatsci.2020.100703.

59 Yasa, E., Poyraz, O., Solakoglu, E.U. et al. (2016). A study on the stair stepping effect in direct metal laser sintering of a Nickel-based superalloy. *Procedia CIRP* 45: 175–178. https://doi.org/10.1016/j.procir.2016.02.068.

60 Qiu, C., Panwisawas, C., Ward, M. et al. (2015). On the role of melt flow into the surface structure and porosity development during selective laser melting. *Acta Mater.* 96: 72–79. https://doi.org/10.1016/j.actamat.2015.06.004.

61 Gupta, K., Jain, N.K., and Laubscher, R. (2017). Chapter 4 – Advances in Gear Manufacturing. In: *Advanced Gear Manufacturing and Finishing* (ed. K. Gupta, N.K. Jain, and R. Laubscher), 67–125. Academic Press https://doi.org/10.1016/B978-0-12-804460-5.00004-3.

62 Toyserkani, E., Sarker, D., Ibhadode, O.O. et al. (2021). Binder jetting and material jetting, physics and modeling. In: *Metal Additive Manufacturing* (ed. E. Toyserkani, D. Sarker, O.O. Ibhadode, et al.). Wiley https://doi.org/10.1002/9781119210801.ch6.

63 Baker, P.R. (1997). Three dimensional printing with fine metal powders. Thesis, Massachusetts Institute of Technology. https://dspace.mit.edu/handle/1721.1/46287 (accessed 16 September 2022).

64 Ye, X., Li, Y., Ai, Y. et al. (2018). Novel powder packing theory with bimodal particle size distribution-application in superalloy. *Adv. Powder Technol.* 29 (9): 2280–2287. https://doi.org/10.1016/j.apt.2018.06.012.

65 Wu, H., Cheng, Y., Liu, W. et al. (2016). Effect of the particle size and the debinding process on the density of alumina ceramics fabricated by 3D printing based on stereolithography. *Ceram. Int.* 42 (15): 17290–17294. https://doi.org/10.1016/j.ceramint.2016.08.024.

66 Bai, Y., Wagner, G., and Williams, C.B. (2017). Effect of bimodal powder mixture on powder packing density and sintered density in binder jetting of metals. *ASME. J. Manuf. Sci. Eng.* 139 (8): 081019. https://doi.org/10.1115/1.4036640.

67 Gregorski, S.J. (1996). High green density metal parts by vibrational compaction of dry powder in three dimensional printing process. Thesis, Massachusetts Institute of Technology. https://dspace.mit.edu/handle/1721.1/8179 (accessed 16 September 2022).

68 Ziaee, M. and Crane, N.B. (2019). Binder jetting: a review of process, materials, and methods. *Addit. Manuf.* 28: 781–801. https://doi.org/10.1016/j.addma.2019.05.031.

69 Liu, J. and Ryneson, M.L. (2004). Blended powder solid-supersolidus liquid phase sintering. US6746506B2, 08 June 2004. https://patents.google.com/patent/US6746506B2/en (accessed 16 September 2022).

70 Nandwana, P., Elliott, A.M., Siddel, D. et al. (2017). Powder bed binder jet 3D printing of Inconel 718: densification, microstructural evolution and challenges. *Curr. Opin. Solid State Mater. Sci.* 21 (4): 207–218. https://doi.org/10.1016/j.cossms.2016.12.002.

71 Do, T., Kwon, P., and Shin, C.S. (2017). Process development toward full-density stainless steel parts with binder jetting printing. *Int. J. Mach. Tools Manuf* 121: 50–60. https://doi.org/10.1016/j.ijmachtools.2017.04.006.

72 Mercado Rivera, F.J. and Rojas Arciniegas, A.J. (2020). Additive manufacturing methods: techniques, materials, and closed-loop control applications. *Int. J. Adv. Manuf. Technol.* 109 (1): 17–31. https://doi.org/10.1007/s00170-020-05663-6.

73 Yi, S., Liu, F., Zhang, J. et al. (2004). Study of the key technologies of LOM for functional metal parts. *J. Mater. Process. Technol.* 150 (1): 175–181. https://doi.org/10.1016/j.jmatprotec.2004.01.035.

74 Nowotny, S., Himmer, T., Techel, A. et al. (2002). Recent developments in laminated tooling by multiple laser processing. *ICALEO* 2002 (1): 166023. https://doi.org/10.2351/1.5066179.

75 Xu, B., Wu, X., Lei, J. et al. (2013). Research on micro-electric resistance slip welding of copper electrode during the fabrication of 3D metal micro-mold. *J. Mater. Process. Technol.* 213 (12): 2174–2183. https://doi.org/10.1016/j.jmatprotec.2013.06.009.

76 Hehr, A. and Norfolk, M. (2019). A comprehensive review of ultrasonic additive manufacturing. *Rapid Prototyping J.* 26 (3): 445–458. https://doi.org/10.1108/RPJ-03-2019-0056.

77 Ahn, D., Kweon, J.-H., Choi, J. et al. (2012). Quantification of surface roughness of parts processed by laminated object manufacturing. *J. Mater. Process. Technol.* 212 (2): 339–346. https://doi.org/10.1016/j.jmatprotec.2011.08.013.

78 Motaman, S.A.H., Kies, F., Köhnen, P. et al. (2020). Optimal design for metal additive manufacturing: an integrated computational materials engineering (ICME) approach. *JOM* 72 (3): 1092–1104. https://doi.org/10.1007/s11837-020-04028-4.

79 Smith, J., Xiong, W., Yan, W. et al. (2016). Linking process, structure, property, and performance for metal-based additive manufacturing: computational approaches with experimental support. *Comput. Mech.* 57 (4): 583–610. https://doi.org/10.1007/s00466-015-1240-4.

80 Smith, J., Xiong, W., Cao, J. et al. (2016). Thermodynamically consistent microstructure prediction of additively manufactured materials. *Comput. Mech.* 57 (3): 359–370. https://doi.org/10.1007/s00466-015-1243-1.

81 Mozaffar, M., Ndip-Agbor, E., Lin, S. et al. (2019). Acceleration strategies for explicit finite element analysis of metal powder-based additive manufacturing processes using graphical processing units. *Comput. Mech.* 64 (3): 879–894. https://doi.org/10.1007/s00466-019-01685-4.

82 Cao, Y., Lin, X., Kang, N. et al. (2021). A novel high-efficient finite element analysis method of powder bed fusion additive manufacturing. *Addit. Manuf.* 46: 102187. https://doi.org/10.1016/j.addma.2021.102187.

9

Powder–Laser–Melt Pool Interactions

9.1 Introduction

It is crucial to comprehend the physical interactions that take place between **powder**, **laser beam**, and **melt pool** in the vicinity of the melt pool in order to effectively regulate the **additive manufacturing** (AM) process and achieve the desired microstructure. To achieve accurate control of complex AM systems, it is necessary to have a comprehensive understanding of the process dynamics, which can cover a wide range of parameters including powder properties, melt pool characteristics, laser beam parameters, and geometric features. In terms of control of process parameters, there are various real-time **closed-loop** control techniques that have been demonstrated to enhance machine capabilities and geometric control, understanding the underlying physical phenomena is still essential for achieving optimal performance [1–4]. Although "active" systems can assist in controlling the process, they do not entirely eliminate the need to comprehend the underlying process. For example, deliberately modifying a single process parameter to regulate geometry in **selective laser melting** (SLM) 316L stainless steel (SS) can cause unanticipated changes to the microstructure [5]. In contrast to the aforementioned approach, certain studies have focused on investigating and improving "passive" stability phenomena, where an open-loop system can keep constant set of process conditions, such as the distance between the powder delivery nozzle and the surface of the component being manufactured (this is sometimes described as **working distance**, WD) for **directed energy deposition** (DED) processing [6], by implementing a self-regulating feedback loop [7]. Some studies have reported success in achieving passive control of working distance drift by exploiting the fact that alterations in the working distance can impact laser focus, resulting in a modification of laser spot size and potential modulation of mass capture efficiency [8, 9]. An essential question that arises is whether there are any other feedback loops, whether existing or potential, that could either stabilize or destabilize the build due to the complex physics involved.

Various tools are being examined to tackle the issue of how process conditions may lead to a stabilizing feedback loop. On the experimental front, one commonly used and simple method is to conduct parametric experimental sweeps on a simplified geometry and then perform metallurgical analysis to isolate machine conditions [10–12] and develop operational heuristics [13]. Furthermore, the use of in situ monitoring can facilitate the tracking of various parameters such as melt pool geometries, temperatures, and DED powder spray patterns, which can be correlated with the resulting deposit properties. On the theoretical front, certain components of DED systems can be modeled analytically with suitable assumptions. One widely used example is the Rosenthal thermal model, which involves a moving point heat source across a semi-infinite plane [14]. Rosenthal model was originally developed for welding but has been used to great effect in DED processing [15–17]. In work by Pinkerton, it is proposed that there are three aspects that need to be taken into consideration [18]: the powder flow streams [19]; the melting, flow, and solidification occurring at the laser–substrate interaction zone, as well as the thermal history of the part as a whole during the building process [19–22].

The various AM technologies discussed in this book are rapidly growing and are the focus of intense technical attention due to their potential to produce high-quality components with location-specific composition and extensive microstructural control. Despite the numerous experimental and modeling efforts devoted to the study of AM, very few studies directly observe the interactions between individual powder particles and the liquid metal pool with a high enough temporal frequency to accurately describe these discrete contact events. As an AM component is produced by accumulating layers deposited with small powder particles, it is essential to understand all the phenomena that occur during the laser–powder–melt pool interactions. Notably, since the frequency and nature of

Metallic Powders for Additive Manufacturing: Science and Applications, First Edition.
Enrique J. Lavernia, Kaka Ma, Julie M. Schoenung, James F. Shackelford, and Baolong Zheng.

these powder-pool impingements significantly influence the overall process behavior and remain a poorly quantified fundamental building block, the goal of this chapter is to review the available information. This chapter is organized as follows: first, the laser and laser-material interaction are briefly introduced, followed by detailed discussions on the laser-material and melt pool interaction during DED and **powder bed fusion** (PBF), respectively.

9.2 Laser and Laser-Material Interactions

The foundation for the laser was established by Einstein in 1917 when he introduced the phenomenon of **stimulated emission** [23]. Maiman experimentally demonstrated the first laser by using a ruby crystal in 1960 [24]. The term **LASER** stands for "Light Amplification by Stimulated Emission of Radiation." Different types of lasers exist, each with a specific wavelength. The principle behind lasers involves pumping electrons from the ground state to a higher energy level, creating an inversion in population between the two states. This sets the stage for stimulated emission, where incoming radiation causes the emission of coherent radiation in phase with the incident radiation. By employing a resonant cavity, the radiation with the specific wavelength can be amplified, producing a beam of highly monochromatic, coherent, and directional light. The beam can be transported from the cavity through the use of semi-transparent mirrors [25, 26].

Lasers are classified into five main types based on their gain medium: gas lasers (such as CO_2), solid-state lasers (like Nd:YAG), fiber lasers, liquid lasers (dye lasers), and semiconductor lasers (laser diodes). Each type has a characteristic wavelength. These types can be further divided into subcategories based on their mode of operation, namely, continuous-wave lasers and pulsed lasers. In laser welding and AM, both power density and total power are required to be high, which limits the types of lasers that can be used [26].

High-power lasers are essential for many industrial applications such as cutting, welding, and marking. Currently, there are two main types of high-power lasers: CO_2 and Nd:YAG lasers [26]. CO_2 lasers use a gas mixture of N_2, He, and CO_2 to generate a wavelength of 10.6 μm through vibration transitions in the CO_2 molecule. The maximum power output of continuous-wave CO_2 lasers is typically in the kilowatt range. Nd:YAG lasers, on the other hand, use a Nd-doped $Y_3Al_5O_{12}$ crystal as the active medium, producing a wavelength of 1.06 μm. This makes it possible to transport the beam through optical fibers and achieve high-power densities [27]. Nd:YAG lasers are commonly used in three-dimensional applications due to their flexible fiber optic beam delivery, while their high metal absorption coefficient makes them well-suited for metalworking applications. Fiber lasers are a special type of solid-state laser that uses an optical fiber mixed with a rare-earth element as the gain medium. Fiber lasers have many advantages, including a straighter and smaller beam that provides more precise results, a small footprint, good electrical efficiency, low maintenance, and low operating costs. They are commonly used for metal cutting, welding, marking, and surface treatment applications due to their high electro-optic conversion efficiency and high beam quality. In recent years, there has been a trend to replace CO_2 lasers with newer generation Nd:YAG and fiber lasers due to their superior performance and cost-effectiveness [4, 26, 27].

9.2.1 Laser–Matter Interactions

When a laser beam strikes a surface, the radiation is either reflected, absorbed, or transmitted [27].

The sum of the fractions of reflected, absorbed, and transmitted energy is equal to one. The total energy that interacts with the target can be expressed by the following equation:

$$E = R \cdot E + A \cdot E + T \cdot E \tag{9.1}$$

Here, the coefficients R, A, and T represent the **reflectivity**, **absorptivity**, and **transmissivity** of the material, respectively. The absorption of laser energy by metals is typically low, around 10–20% of the total energy, due to the high reflectivity of metals, particularly at longer laser wavelengths like those used in CO_2 and Nd:YAG lasers. The amount of absorbed laser energy can be increased by redirecting the reflected energy back onto the target [28].

The intensity of a laser propagating into a material decreases exponentially with increasing depth, following the Beer–Lambert's expression [29]:

$$I(z,t) = I_0(t)(1 - R - T)e^{-Az} \tag{9.2}$$

where A is absorptivity, which depends on the medium, laser wavelength, and intensity, as well as the inverse of absorption length; I_0 (=AE) is the initial incident intensity at the surface of the material; and t is time. The beam intensity I at a depth z is quantified by this equation. The penetration depth of laser energy for a given material is defined as $p = 1/A$, where approximately 63% of the initial energy has been absorbed [27].

If the power applied is enough, lasers can melt, boil, or create **plasma**. When the **energy density** per unit area is sufficient to create plasma, a vapor cavity or keyhole forms around the laser beam in the material. The motion of the liquid metal in the melt pool becomes highly turbulent in this state [26]. Most laser processes are thermally driven, and they offer precise spatial and temporal control of the heat input, which is crucial for melting and vaporization. However, the amount of energy absorbed by the material is an important factor that influences microstructure. Laser power density is determined by dividing the laser power by the spot area and the scan speed, as shown in Eq. (9.3):

$$I = P/(vd) \tag{9.3}$$

where P is laser power, v is laser scan speed, and d is laser spot size in diameter. Increasing the spot size and scan speed for a given laser power will result in a decrease in power density.

Delivery of the laser beam is also crucial in order to attain optimal deposition conditions. The beam must be transported from the source to the target and appropriately focused to achieve the desired energy density and spot size. Reflective-transmissive and fiber optic beam delivery systems are commonly used for this purpose. Reflective-transmissive systems use a series of reflective mirrors to direct the beam to the target, while fiber optic systems use a fiber optic cable to transmit the beam. The reflective-transmissive method has been commonly used in DED technology, while fiber optic systems are commonly used with Nd:YAG lasers.

9.2.2 Laser-Material Processing

Lasers have gained wide acceptance and are considered reliable tools in the materials' processing industry. They find diverse applications in various fields. Laser-material processing can be broadly categorized into four main categories based on their applications. There are several categories encompassing various manufacturing processes. These categories comprise deposition forming, which focuses on creating near net-shape or fully finished products. Additionally, there is AM, which involves innovative techniques such as 3D printing. Another category is joining, which encompasses processes like welding, brazing, and other similar methods. Furthermore, machining falls under this classification, involving processes like cutting, drilling, and similar operations. Lastly, there is surface engineering, which includes processes such as cladding, heat treatment, and other similar techniques. Figure 9.1 illustrates the three main classes of laser processes based on the desired degree of heating and phase transformation: heating without melting/vaporizing, melting without vaporizing, and vaporizing [27], which shows the domain for different laser-material processing techniques as a function of laser power and interaction time. Surface heating

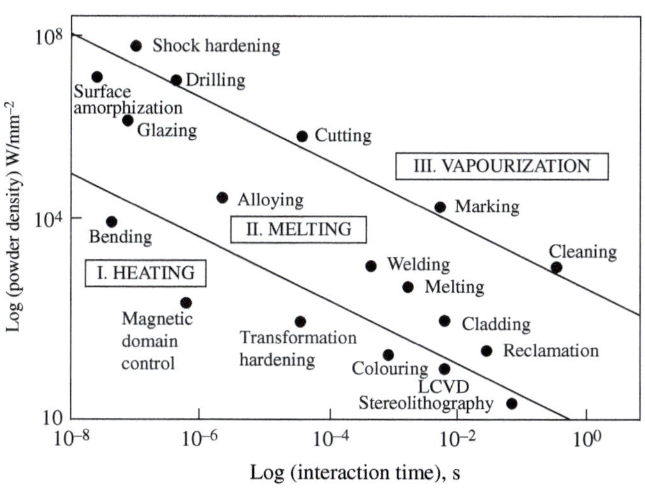

Figure 9.1 Schematic of process map in terms of laser power density as a function of interaction time for different laser-material processes. Source: Adapted from Steen et al. [27].

without surface melting, which is required for transformation hardening, bending, and magnetic domain control, relies on low power density. In contrast, high-power density is necessary for surface melting, glazing, cladding, welding, and cutting. For cutting, drilling, and machining, a very short interaction/pulse time requires delivery of a substantially high-power density.

Laser-material processing has the advantage of flexibility. By utilizing a multi-axis workpiece positioning system or robot, a single laser beam can perform drilling, cutting, welding, deposition, and heat-treating processes. As laser-material processing is a noncontact process, energy transfer occurs through radiation. The aim of material processing is to produce products with the correct shape, geometry, and dimensions. With laser-material processing, it is possible to manufacture finished products directly from raw powder materials without the need for any intermediate operations [4, 27, 30–33]. Fabrication in a single step is a significant advancement and is highly desirable due to the savings in time, cost, material, and labor when compared to traditional fabrication methods.

9.3 Laser-Material Interactions During DED Processing

In DED, an inert gas stream delivers powder at a controlled rate through nozzles focused on the melt pool. The powder streams diverge upon exiting the **nozzles**, forming conical powder streams [4, 34]. The powder **cones** generated by the gas stream intersect near the melt pool, resulting in interparticle collisions and interactions between the laser beam, powder, and melt pool. Additionally, the DED process is thermal and involves the heating, melting, possible vaporization, and re-solidification of the deposited materials. The resulting deposited materials have a layered structure and undergo multiple **thermal cycles**, typically leading to the presence of pores and residual stresses, as illustrated in Figure 8.5a [35]. In the vicinity of melt pool, the injected powder particles interact with both the laser beam and the melt pool, as depicted in Figure 8.5b. The formation of a "**keyhole**," a metal vapor created in addition to the metal melt, can occur under certain deposition conditions and can be influenced by the **convective** and **turbulent** flow associated with the melt pool. The occurrence of these phenomena depends on the thermal and physical properties of the deposited material, as well as on various process parameters such as laser power intensity profile, powder flow rate, velocity and trajectory of sprayed powder particles, scan sequence, and timing of laser passes. Thus, a fundamental understanding of the underlying mechanisms governing DED is necessary to achieve optimal properties through proper microstructure control. The deposited materials typically have a layered structure, undergo multiple thermal cycles, and contain pores and residual stresses near the melt pool, as shown in Figure 8.5a.

9.3.1 Inflight Particle Heating

In the process of DED deposition, powder is dispensed from containers using a perforated disc mechanism that rotates at a specific speed to meter the powder into a stream of inert gas. This gas stream then carries the powder through nozzles that are aimed at the melt pool. As the powder streams exit the nozzles, they diverge slightly, creating conical patterns. These cones intersect near the melt pool, resulting in interparticle collisions.

The powder particles that are ejected from the nozzle interact with the laser beam, which is dependent on the process conditions and the local density of the powder. The powder particles absorb thermal energy and attenuate the laser beam. As a result, the powder particles that reach the surface of the substrate located below the nozzle are heated or melted depending on the laser power and flow conditions [34, 36]. Figure 9.2a depicts a thermal image of the powder particles heated by the laser beam, while Figure 9.2b shows the geometric relations. The amount of heat absorbed by the particles during flight is determined by factors such as particle density, thermophysical properties, particle morphology and size distribution, free flight duration through the beam, and gas velocity [22].

The partitioning of energy during DED deposition of 316L SS using a Nd:YAG laser was investigated in a study by Pinkerton and Li [37]. Numerical and experimental analyses showed that the substrate reflected approximately 54% of the laser power, while 30% was absorbed by the substrate, 11% was reflected by the powder, 4% was lost due to dispersed powder, and only 1% could be attributed to the deposited powder. Another investigation examined the effect of laser power, powder flow rate, and particle size on energy absorption and studied the retention time and trajectories of powder particles within the beam [38].

The distribution of laser energy during DED deposition is affected by the **working distance**, which is defined as the distance between the nozzle bottom and the top surface of the deposited material. The powder flow field

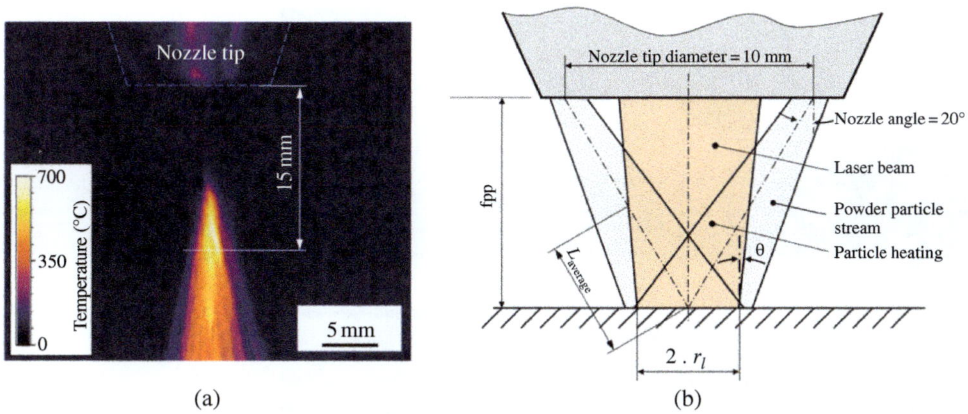

Figure 9.2 (a) Thermal image of the heated powder particles at the nozzle exit for a 1000 W laser power and (b) geometric relations. Source: Arrizubieta et al. [36]/with permission of Elsevier.

causes the WD to reach a stable value, which is heavily influenced by the accumulation of thermal energy in the part and the dispersion of powder flow. The energy absorbed by the powder can either be transferred to the melt pool when the powder enters the pool or is dissipated into the chamber if the powder particle fails to integrate into the part, as depicted in Figure 9.3 [34]. Particles that do not rebound and travel through the laser beam have a specific threshold z height. If particles are below this threshold, they will integrate into the pool, but if they are above it, they will have a noncaptured trajectory [34].

Developing a comprehensive understanding of the laser beam–powder–melt pool interactions during DED is a challenging task due to the unique thermal history involved. This is because the small size of the melt pool, the presence of large thermal gradients, and the fast movement of the solid/liquid interface present obstacles. The localized deformation during impact, temperature variation, and microstructure can vary depending on the position in the melt pool when the feedstock powder experiences a high temperature prior to impingement onto the melt pool. To study the attenuation effects of particles to the laser beam, particle melting, and particle–melt pool interaction behavior during DED, high-speed photography, and thermal imaging are valuable tools. Additionally, high-speed thermal imaging can provide detailed information about the thermal behavior, including temperature gradient and cooling rate, in the vicinity of the melt pool, as shown in Figure 9.4 by Haley et al. [39].

The presence of a **residence time** for particles floating on the surface of the melt pool accounts for some of the observed differences in morphology in particles attached to free surfaces, which can lead to nonuniformity in

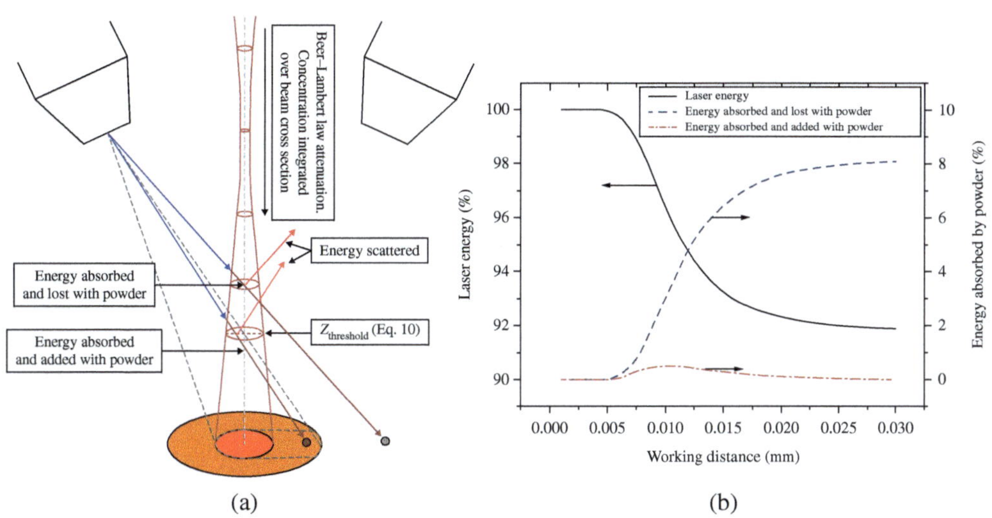

Figure 9.3 Energy partitioning as the laser is absorbed and scattered by powder in flight, (a) geometric relations between the powder flow, laser beam, and melt pool, and (b) the variation of laser energy and energy absorbed by powder with the working distance. Source: Haley et al. [34]/with permission of Elsevier.

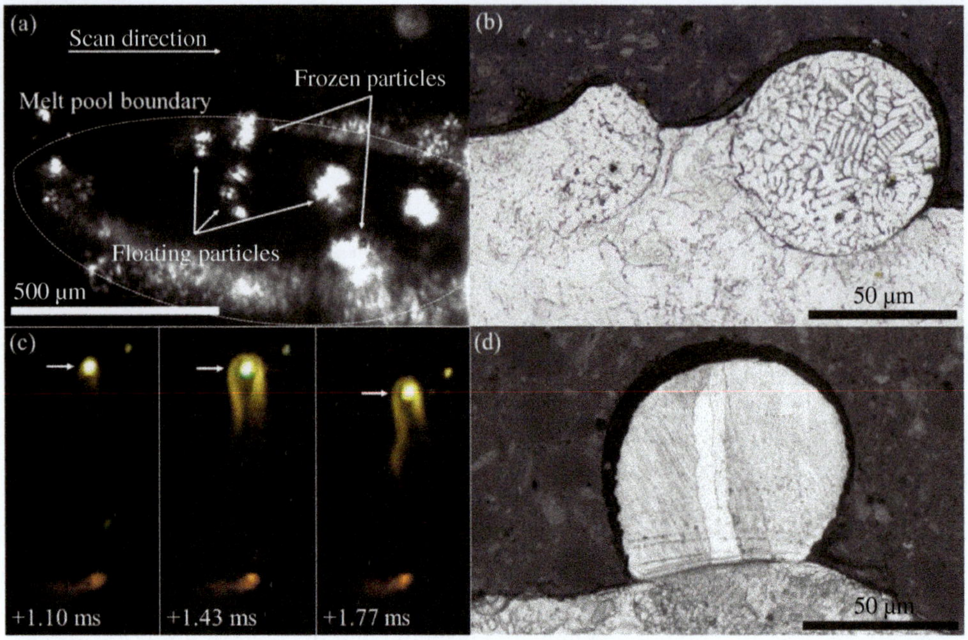

Figure 9.4 Two distinct formation origins for particles attached to external surfaces. In (a), particles can float on the surface of the melt pool or freeze to the bottom when nearer the melt pool periphery; this can result in particles never fully melting as the melt pool solidifies around them, as shown in (b). In (c), a slow-moving particle passes through the path of the heating laser, melting and partially vaporizing it. These particles can land and solidify on available surfaces, as seen in (d). Source: Haley et al. [39]/with permission of Elsevier.

surface finish of DED components. This issue can introduce dimensional inaccuracies and potentially cause particle detachment during service, which may contaminate the surrounding media. Two mechanisms for particle attachment to the substrate are identified in Figure 9.4a,c [39], along with associated cross-sectional micrographs. In (a), particles floating on the surface of the melt pool can cause the melt pool to solidify around the unmolten particle, resulting in the microstructure shown in Figure 9.4b with a clear boundary and spherical shape. In (c), a particle passing through the path of the laser can melt and vaporize, generating a molten droplet that can fall back to the build and refreeze on the surface. The microstructure in Figure 9.4d exhibits a fine cellular solidification pattern growing from the contact surface and conforming to the original track surface. High-speed photography and thermal imaging can provide useful information about the particle–laser beam–melt pool interactions and can help characterize the particle attenuation effects on the laser beam, particle melting, and particle–melt pool interaction behavior during DED. Additionally, high-speed thermal imaging can provide detailed information about thermal behavior, such as temperature gradients and cooling rates, in the vicinity of the melt pool. Figure 9.4 from Haley et al. demonstrates the usefulness of these techniques. However, due to the small size of the melt pool, the presence of large thermal gradients, and the fast movement of the solid/liquid interface, establishing a comprehensive understanding of the laser beam–powder–melt pool interactions remains a formidable challenge in DED.

The quality of the surface finish of the deposited parts is significantly influenced by the size of the powder particles, with smaller particles generally resulting in a better surface finish [40]. According to [41], the surface roughness of DED deposited IN690 alloy is approximately 12 μm. The suitable particle size for DED deposition is in the range of 30–150 μm [42]. Small particles can result in agglomeration and reduced flowability, while larger particles are not appropriate for gas-carried particle processes. Additionally, a narrow particle size distribution is necessary for maintaining the stability of the laser-generated melt pool [40].

9.3.2 Thermal Behavior of Melt Pool

During the DED process, the laser beam causes a small and fast-moving melt pool on the deposit surface, with any remaining beam energy impinging on the surface. To elucidate the mechanisms governing microstructure evolution during DED, it is essential to understand the interactions between the laser beam and the melt pool, along with the

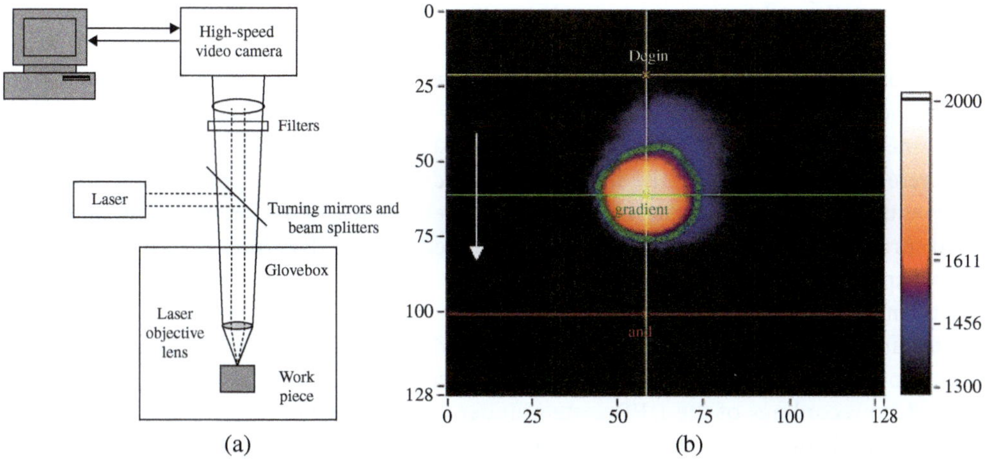

Figure 9.5 (a) Experimental setup showing thermal measurement system positioned above the DED system, (b) A thermal image of a WC–Co sample looking from the top when layer 5 was being deposited. Source: Xiong et al. [47]/with permission of Elsevier.

local thermal conditions in the pool. Noncontact thermal imaging techniques, including visible and near-infrared (IR) radiation pyrometry, provide a useful tool to establish the thermal profile of the melt pool and the associated cooling rate [42–47]. In a related study, a single-wavelength high-speed digital charge-coupled device (CCD) video **camera** was used to monitor the thermal images of a 316L SS line build during the DED process [46]. The experimental setup utilized a broadband pass filter centered at 650 nm and a telephoto lens in the image path. By analyzing the solidification interface temperature of 316L SS at 1650 K, it was determined that the melt pool size increases as the laser power increases up to 275 W. The findings indicate that the cooling rate is approximately 103 K s^{-1}, and the highest cooling rate can be achieved with low laser power and high scan speed. Additionally, another related study utilized a high-speed digital CCD video camera positioned from above to observe the melt pool [42, 47]. Compared to the side setup reported in [46], the camera used in this experimental setup remains stationary and has the same focal point as the laser. This allows the camera to stay in focus and capture the melt pool from any X, Y, and Z positions. The thermal behavior of WC–Co cermet during DED was investigated using in-situ top view **high-speed thermal imaging** (shown in Figure 9.5) in combination with **finite element modeling** (FEM) [47], in order to provide fundamental insight into the factors that influence microstructure evolution. The temperature in degrees K is represented by the colorized image, with the size of the image displayed in pixels on the X- and Y-axes. The white arrow indicates the direction in which the laser beam is moving. By using in-situ high-speed thermal imaging, it was possible to characterize temperature gradients and cooling rates in the vicinity of the melt pool, while the 3D FEM allowed for simulation of the entire sample.

The **two-wavelength pyrometer** is a thermal imaging system that has gained significant interest due to its reliance on the ratio of the relative intensities of radiation at two different wavelengths. One advantage of this approach is its independence of emissivity, which provides a more precise temperature measurement [45], with a margin of error between $\pm 6\,°C$. The imaging pyrometer was used to investigate the thermal behavior of DED-processed 316L SS, with a dynamic range of 1500–2500 K. The temperature profile provided information on the temperature gradient distribution and cooling rate in the melt pool and surrounding area. Results showed that the temperature gradient from the center of the pool was around 10^2–10^3 K mm^{-1}, and the cooling rate was around 10^2–10^4 K s^{-1} in the DED-processed zone. However, one limitation of existing thermal imaging methods is their inability to capture the entire thermal history of deposited components, especially the temperature variations in the solidified materials.

9.3.3 Interactions Between Particles and Melt Pool

Real-time monitoring can offer valuable insights into how process parameters impact powder flow, interactions between the laser and particles or melt pool, dynamics of the melt pool, and formation of porosity. A recent study was carried out to characterize the interactions between particles in flight and with the melt pool using a high-speed camera [39], which generated helpful information for the analysis and understanding of particle melting and particle–pool interactions during the DED process. Several frames shown in Figure 9.6 depict intriguing details of the powder particles as they move and collide with the melt pool [39]. The impact of individual powder particles on the melt was observed to cause rippling and floating. These particles would then sink beneath the surface after remaining there for up to 600 μs. The duration of their stay on the surface influenced their interaction with impinging particles, which could lead to the rejection of some particles. To obtain velocity distribution statistics under DED conditions, the trajectories of powder particles were traced after exiting the feed nozzles. Additionally, a three-phase computational fluid dynamics (CFD) model was used to study the impact of various factors on particle behavior, including wettability effects. This model applied surface tension and was coupled with observed melting dynamics of individual 316L SS particles [48]. The impact of particle size, impact velocity, melt pool and particle temperature, surface tension, and thermophysical properties of the material on residence time sensitivity was examined and discussed.

The spatial and velocity distribution of the powder flow cloud was analyzed by examining select frames from video recordings of the DED system captured under various conditions, as depicted in Figure 8.6 [39]. This figure depicts several frames captured from video recordings of the DED system under various conditions. The first image, (a), clearly shows the motion of individual particles, indicating that the chosen particle size and gas flow parameters lead to straight trajectories after leaving the spray nozzles and diffuse reflection from the rough substrate surface. Image (b) presents a higher magnification of a single nozzle, revealing a low number of particle–particle interactions at the nozzle exit, but a high concentration of particles with fast-spinning satellites that are indiscernible at the selected frame rate. Image (c) captures a larger field of view of the entire spray pattern, recorded at a lower frame rate (10 000 frames per second), which allowed the identification of over one million particles and the determination of the powder cloud's particle concentration. Finally, image (e) shows the quantification of the velocity of the powder cloud during the deposition of a single track of material by tracking particle trajectories using the open software MOSAIC [49]. In (f), trajectories are visually emphasized using color. Analysis of the velocity components revealed the presence of three distinct particle groups. The first group comprises incoming particles that move at a speed of 2–5 $m\,s^{-1}$ immediately after being sprayed. The second group consists of scattered and rebounding particles, which move more slowly in all directions. Lastly, a small group of particles can be observed moving rapidly upward at speeds of up to 6–8 $m\,s^{-1}$, as they are caught and ejected by the plume of metal vaporizing off the surface of the melt pool. The concentration distributions and trajectories obtained from this analysis provide a fundamental dataset for further process estimation calculations.

With higher magnification and sampling rates, it is possible to observe the consumption of individual powder particles by the melt pool. Although some particles pass directly beneath the surface without significant change in speed, most particles impact the surface, causing a ripple, and then float along the surface before sinking below it. This behavior was observed by Haley et al. and is shown in Figure 9.6 [39]. The surface residence time of a small sample of 36 particles over 21 ms was found to have an average of 370 μs. However, a significant scatter in residence times was observed, with the shortest recorded at 32 μs and the longest at 1210 μs. It is worth noting that the available data may exclude residence times shorter than one frame (<16 μs) or longer than the maximum video length (>1600 μs) due to technological limitations of high-speed videography. Nonetheless, these observations indicate the existence and approximate time scale of the phenomenon.

The observed scatter in surface residence times could be partly attributed to the dispersion in particle sizes. If the melt's surface tension is strong enough to resist the particle's initial penetration, the particle would need to melt before being absorbed into the pool. The time required for melting would depend heavily on various factors, including the particle's mass. Heat accumulates in the particle through conductive/convective heat transfer from the melt pool and radiative heat transfer from the laser. Both heat transfer pathways depend on the particle's exposed surface area, which is a function of r^2, while the particle's mass is a function of r^3, suggesting that particle size could influence the residence time.

Besides floating on the surface of the melt pool, particles can also solidify at the bottom of the pool, as seen in Figure 9.4, where particle motion suddenly stops instead of drifting along the surface. This is more likely to happen at the edges of the pool, where the pool is shallower and at a lower temperature with higher viscosity. When a solid-state thermal connection is established between the particle and the bottom of the melt pool, it provides a pathway for incoming heat dissipation, which can extend the time before the particle fully melts. As a result, these frozen particles are observed to persist beyond the length of the captured video (>1600 μs) more frequently.

Various factors influence the behavior of particles upon impacting the surface of the melt pool, including the wettability and surface properties of both the particle and liquid, as well as the viscosity of the melt, which is affected

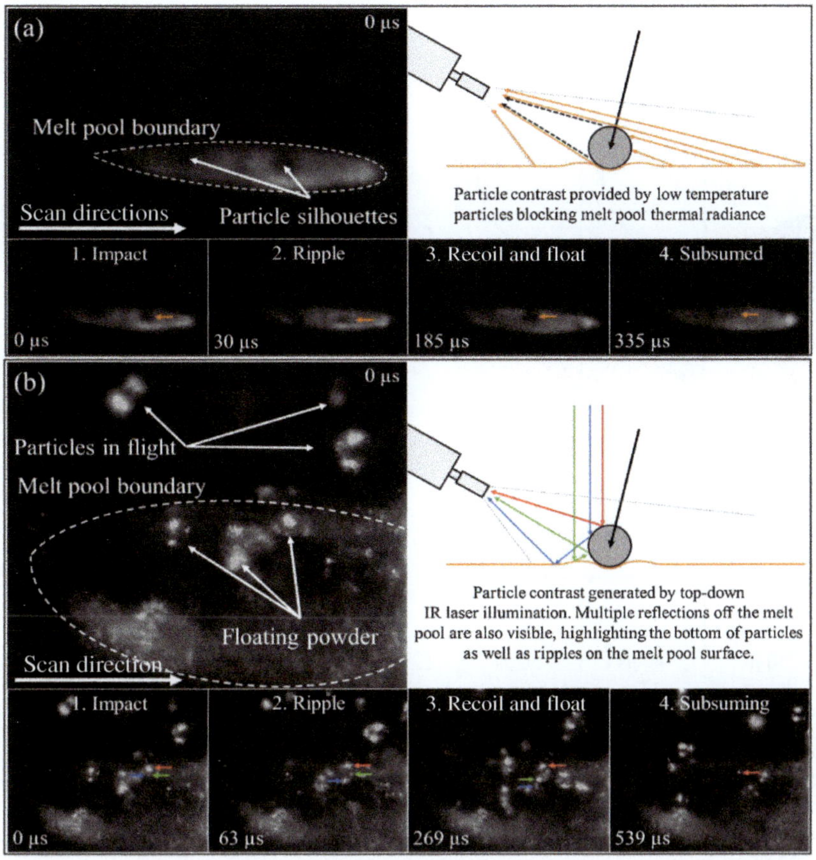

Figure 9.6 High-speed videos demonstrating powder–melt pool interactions. Video is available in supplementary material. Two illumination conditions were used, particles seen in silhouette from thermal radiance collected from the melt pool (a) as well as from unfiltered reflected illumination from the primary laser (b). Particles do not penetrate the surface, but instead impact (1), displace molten material forming a ripple (2), float for a brief residence period (3), before subsuming (4). Color-coded arrows highlight generated contrast points tracking the particle through these characteristic stages. Source: Haley et al. [39]/with permission of Elsevier.

by the melt temperature. A high-temperature gradient in the melt pool results in a corresponding viscosity gradient. The particles can exhibit different behaviors such as being trapped by surface tension and oscillating, rebounding off the surface, or penetrating the liquid if their velocity exceeds a certain critical value [50], leaving a vapor cavity behind them which collapses in various ways [51]. The medium fidelity model can simulate the different scenarios that arise when particles impact the surface of the melt pool. This is accomplished by adjusting the surface energies to modify the contact angle, as shown in Figure 9.7. The outcome is determined by the interaction between the particle's initial kinetic energy and the energy required to submerge and wet the particle. [48].

Laser-based metal AM provides a novel pathway to produce net-shaped, lightweight metal matrix composite (MMC) components [4, 52]. For DED of MMCs, the starting powder feedstock involves a mixture of metal and ceramic particles with a range of sizes, morphologies, and densities. During DED processing, powders should be efficiently injected into the melt pool for good final built quality. The metallic powder used in DED systems is usually spherical or near-spherical shaped with diameters ranging from 25 to 150 μs [4, 52]. Injecting ceramic powder into a melt pool presents a challenge due to the typical characteristics of reinforcement ceramic particles, such as small particle sizes, irregular morphologies, low density, and poor wettability to the melt [52]. Such a heterogeneous mixture of physical characteristics makes it difficult to establish optimum deposition conditions, and hence extensive parametric experiments are often needed to optimize the microstructure and performance of a final component. To provide insight into the behavior of feeding metal and ceramic powders and interaction among them with melt pool during DED, Jiang et al. reported on a study that uses two complementary approaches [53]: The first approach involved creating a three-dimensional COMSOL numerical model to simulate the movement of particles considering

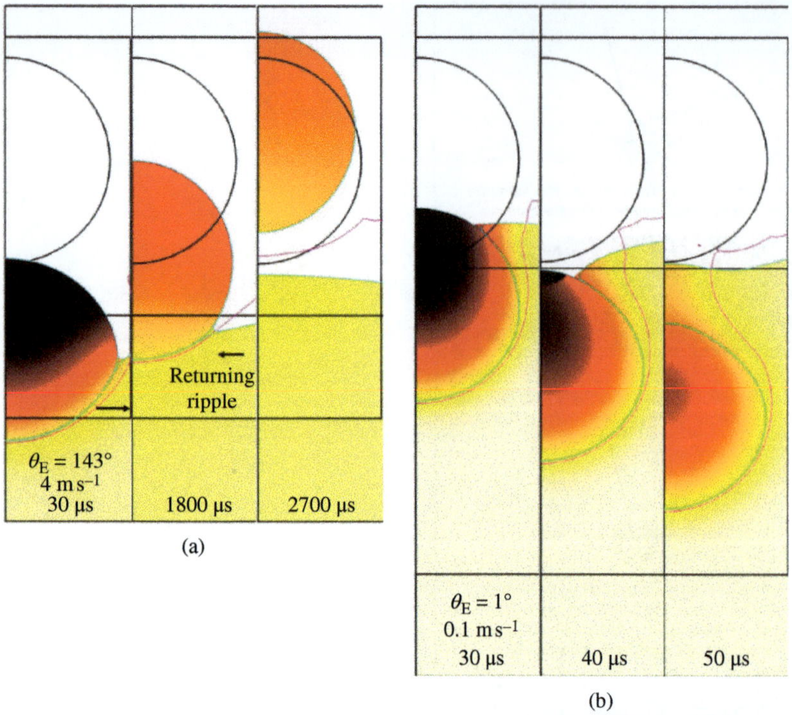

Figure 9.7 Low and high wettability particles that demonstrate different behavioral regimes; in (a), the particle is not energetically and strongly bound to the melt pool by surface tension, so a ripple returning from the perimeter from the simulation domain is able to cast it upward and off the melt pool. In (b), the particle wets so well that the energy threshold barrier to penetrating the melt has disappeared; even with a very low initial velocity of 0.1 m s^{-1} it is drawn into the melt rapidly. Source: Haley et al. [48]/with permission of Elsevier.

gas flow, interactions between particles and gas, as well as collisions with walls. In the second approach, high-speed photography was utilized to observe and compare the flow behavior of metallic particles (Inconel 718) and ceramic particles (TiC). The findings demonstrate noticeable disparities in particle velocity and spatial distribution between Inconel 718 and TiC particles. These variations can be attributed to differences in particle sizes, morphologies, material densities, and collision behavior [53]. By understanding these differences, the targeted compositions and homogeneous reinforcement distribution can be attained in DED components through proper material and process parameter selection, and the custom design of tailored microstructures for heterogeneous functionality can also be achieved.

High-temperature gradient and related vaporization of the metal matrix from melt pool can cause turbulent flow of the TiC reinforcement particles above the surface of the melt pool and may result in reduced particle capture efficiency. The partial pressure of the resultant metal vapor could reach as high as several atmospheres [54]. In their study, Sen et al. employed high-speed photography to investigate the interaction between metal and ceramic particles injected into a melt pool during DED processing. Figure 9.8 presents snapshots extracted from the recorded high-speed video, depicting the behavior of pure Inconel 718 and Inconel 718 + TiC composite materials during the DED deposition process [53]. The Inconel 718 particles are captured by the melt pool more easily than the TiC particles due to the size and density of the Inconel 718 particles being larger than the TiC particles. TiC particles (bright in the image), on the other hand, are shown to be significantly influenced by the vaporization torrent from the melt pool. TiC particles tend to be blown upward by the vapor and experience turbulent flow due to their fine particle size and low density. This may result in the TiC particles having difficulty being captured by the melt pool surface, and thus may lead to decreased TiC content in the final deposited MMC part [55–57], and consequently influence on the microstructure compositions and properties of the deposited MMCs.

Significant efforts have been devoted to the problem of injection, interfaces, and distribution of ceramic particles in metallic matrix of MMCs during AM processing [52, 58–60]. In recent research, a metallic coating has been applied to the surface of the reinforcement material. This coating serves to enhance the injection of fine ceramic particles into

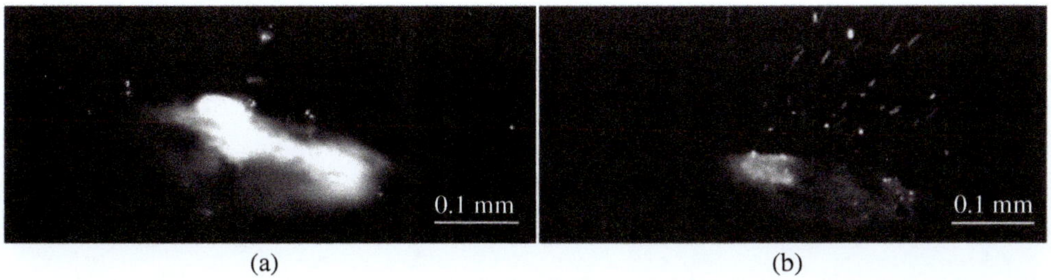

Figure 9.8 Snapshots from the high-speed videos showing the melt pool for: (a) the pure Inconel 718 and (b) the blended Inconel 718 + TiC. Videos were recorded at a frame rate of 20 000 frames per second. Source: Jiang et al. [53]/with permission of Elsevier.

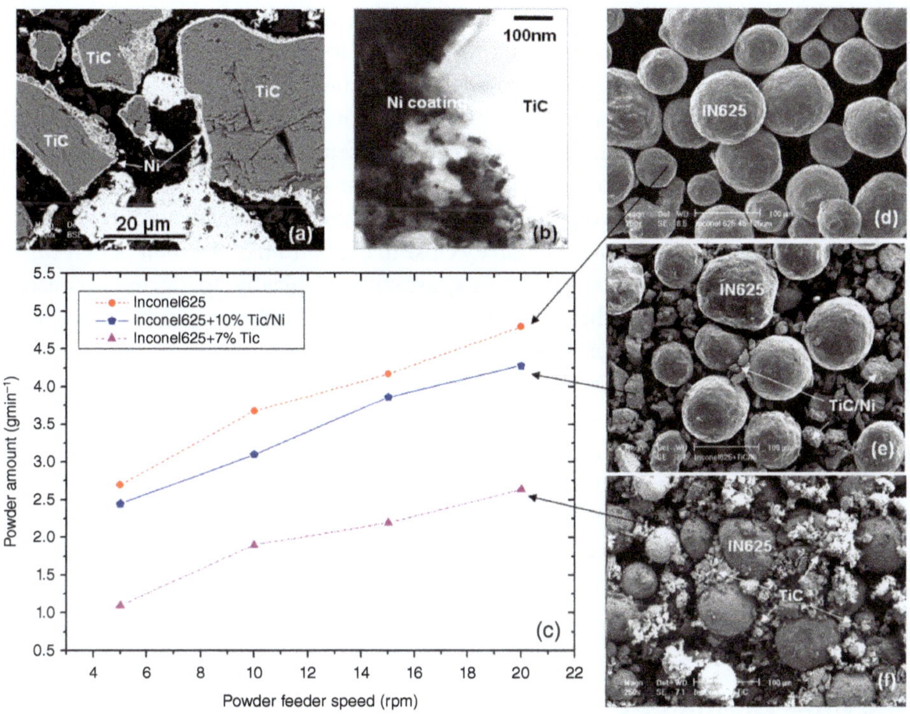

Figure 9.9 Variation of powder flowability with different mixed powders: (a) SEM and (b) TEM morphology of Ni-coated TiC particles, (c) Curve of powder delivery amount with feeder speed, and feeding powder of (d) gas-atomized IN625, (e) mixed powder of IN625 + 10%TiC/Ni, and (f) mixed powder of IN625 + 7%TiC. Source: Zhen et al. [58]/Springer Nature.

the melt pool and modify the interfacial structure, ultimately promoting improved wettability during the processing stage [52, 58]. Figure 9.9a exhibits scanning electron microscopy (SEM) backscattered electron (BSE) micrographs of TiC particles coated with nickel (TiC/Ni) via chemical vapor deposition (CVD). The micrographs indicate that the irregularly shaped TiC particles are entirely encapsulated by a 2 μm thick Ni coating. The TEM studies, as given in Figure 9.9b, showed that the coating is comprised of single-phase, nearly equiaxed grains with a size range from 10 to 100 nm, indicating that the Ni coating is nanocrystalline. No apparent reaction or defects, such as pores, cracks, or dislocations, were observed at the interface between the nanocrystalline Ni coating and TiC particles. The TiC core and the Ni coating appear to be well bonded, at least from a physical standpoint. Compared to the uncoated TiC particles, the morphology of the coated TiC particles appears more spherical and smoother. The spherical shape of the particles is advantageous for the delivery system utilized in DED processing, as it enhances flowability. The researchers investigated the variation in powder flowability by collecting the delivered powder within a two-minute timeframe at different feeder rotational speeds, as depicted in Figure 9.9c–f. The flowability of the mixed powder, which includes the coated particles, is approximately 12% lower than that of gas-atomized IN625 powder. However, it is significantly higher, around 108%, than the flowability of the mixture containing uncoated ceramic powder [58].

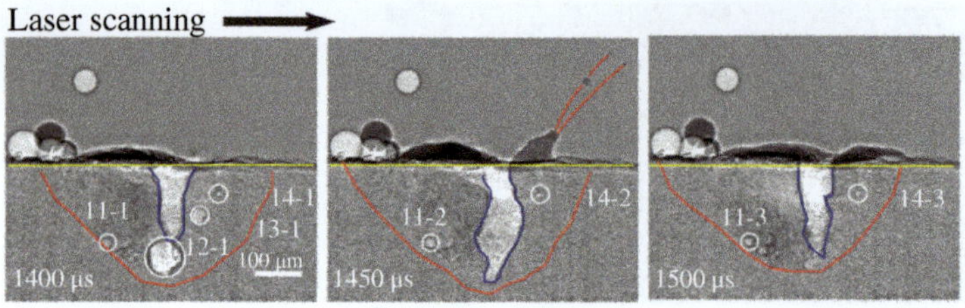

Figure 9.10 Evolution of cavity, melt pool, porosity, and spattering in a DED simulation experiment with a laser power of 250 W and a scan speed of 100 mm s^{-1}. Source: Wolff et al. [61]/Springer Nature.

The interaction between laser and ceramic powder is different from that of metals, and ceramics are much harder to process by laser than metals [26]. The use of metal-coated ceramic particles has shown improvements in laser power absorption compared to uncoated ceramic particles. Incorporating Ni-coated TiC particles has effectively enhanced the flowability of the mixed powder, the interface between the matrix and TiC particles, the interaction between the laser beam and TiC particles, as well as the mechanical properties of the DED deposited MMCs [58].

A different research project utilized high-speed **X-ray** imaging to examine the powder flow and laser–melt pool interactions in DED processing [61]. The study referenced utilized a custom DED setup to capture high-resolution images of Ti--6Al--4V powder deposition and examine how laser–melt pool interactions affect the formation of porosity [61]. Figure 9.10 presents intriguing insights into the development of gas pores and the transformation of pores from a keyhole cavity. The diagram illustrates that the implosion of a cavity close to the bottom of the melt pool results in the expulsion of a particle from the surface, a phenomenon known as **spattering**. Typically, spattering is caused by high-pressure gradients or recoil pressure generated by the vapor-plasma plume and helps to stabilize the melt pool during DED processing, but it can also cause surface blemishes or increased roughness.

In summary, the use of visualization techniques, including high-speed imaging, has enabled a better understanding of laser-materials' interaction and microstructural evolution during DED materials processing. As these techniques continue to improve, they will play a crucial role in answering fundamental scientific questions related to porosity and residual stresses. However, as DED is increasingly used to manufacture complex material systems, including MMCs and functionally gradient materials, the sophistication of novel visualization methods must continue to adapt to ensure precise control of the many interrelated process parameters involved. Overall, the ongoing development and implementation of advanced visualization techniques are critical to advancing the capabilities and applications of DED in materials science and engineering.

9.4 Laser-Material Interactions During PBF Processing

PBF or SLM is a rapidly developing AM technology capable of producing high-precision parts [62, 63]. During the PBF process, a spreader (blade or roller) is used to initially distribute thin layers of metallic powder. Then, one or multiple focused high-energy beams (laser or E-beam) with high-energy density are applied to melt the powder. The melted regions subsequently solidify and fuse together, forming a layer of the built component. This process of layering powder, melting, and solidification is repeated until the final part is achieved.

PBF technology has been broadly investigated and widely applied, however, it still faces challenges in controlling the deposited materials and part quality, which is highly affected by the defect formation, e.g. balling effects, lack of fusion and pores, residual stresses, etc. [64]. The detection and prevention of internal defects while ensuring nondestructive quality assurance poses a significant challenge. Mitigating internal defects requires a thorough understanding of their formation mechanism and an improvement in PBF process conditions. These defects are closely associated with thermophysical phenomena occurring in the powder bed, including melting of powder, droplet spattering, and evaporation, which result from the interaction between the laser, powder, and melt pool. A comprehensive understanding of the underlying thermal and physical phenomena, as well as the defect formation mechanisms in PBF, is crucial for material and production control. While some research work has been carried out in recent years, **X-ray imaging** has been used to directly capture laser–powder interactions, but it is limited by spatial resolution [65]. **Numerical simulation** has been reported in published literature as an alternative method for obtaining detailed information on the flow dynamics of the melt pool and the thermal behavior during powder melting and solidification [66, 67]. The aim of this section is to provide insight into the behavior of powder and melt pool during interactions with the laser beam in the context of laser–powder–melt pool interactions.

9.4.1 Powder Layer Characteristics and Spreading

In this section, we focus on specific aspects related to the characteristics of layering powder and powder distribution in PBF, which are crucial for the successful AM printing of components. The powder feedstock is characterized by its chemical composition, morphology, particle size distribution, flowability, and apparent density. The characteristics of the powder layer are determined by several factors, including particle shape, size distribution, surface chemistry, packing density, rheology, and thermal properties such as **absorptivity** and **conductivity**. All these factors influence the behavior of the powder and the subsequent laser processing. [68–70]. Ensuring sufficient powder flowability and high particle packing density are also critical factors in achieving successful material spreading and part densification during PBF processing [71].

As discussed in Chapters 1 and 2 on atomization, metallic powders generated through different techniques can exhibit varying morphologies and size distributions, which have a direct impact on the flowability and packing density of the feedstock powder during PBF processing. Generally, powders with near-spherical shapes and narrow Gaussian-like particle size distributions are preferred for PBF processing as they lead to better quality powder layers. In theory, powders with a wide size distribution and a sufficient amount of fine particles can achieve higher packing densities. Some studies have been carried out [68, 72–74] to establish the relationship between powder packing and particle size distributions. A primary finding has been that the packing efficiency of a coarse-sized powder packing system can be enhanced by introducing fine particles with sizes comparable to the voids or vacancies present between the particles, as illustrated in Figure 9.11 [68].

The powder layer packing height can also be affected by the powder's packing arrangement. Wischeropp et al. [76] reported a study on the measurement of an actual powder layer height and packing density in a single layer in PBF. The authors drew the conclusion that the height of the actual powder layer in a single layer was 4–5.5 times greater than the height achieved by a powder spreader. This was attributed to the presence of spatter in the powder and the denudation effect, both of which affected the quality of the powder layer.

As discussed in Section 8.2, **powder spreading** is the first and a key step in the PBF process, which dictates the quality of a powder bed and consequently influences the quality of the AM printing outcome. There are many parameters affecting the quality of final components in powder-based AM processing. The powder morphology and particle size distribution have a direct influence on powder flowability and tap/packing density, which in turn affect the powder spreading process [77, 78]. The mechanisms of powder deposition may be different depending on the PBF system, but generally, a rake or a roller is employed to distribute powder over a built platform. In PBF systems, a slider with various brush types can be utilized to spread the powder.

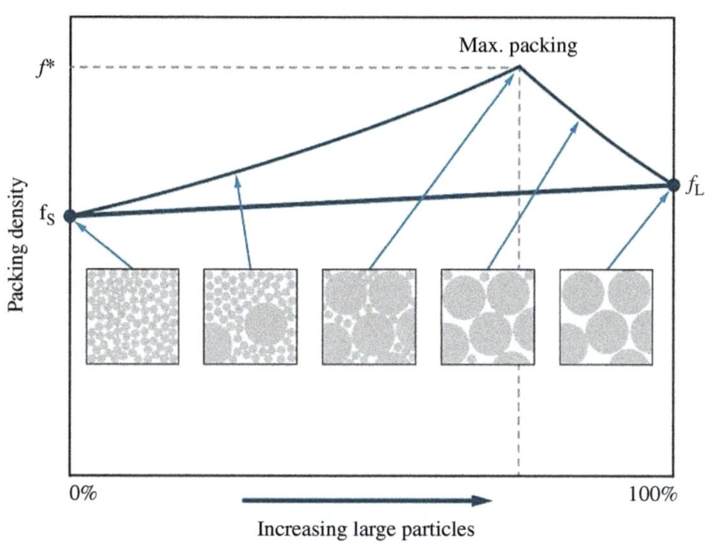

Figure 9.11 Schematic of particle size variations vs. packing densities [68, 75].

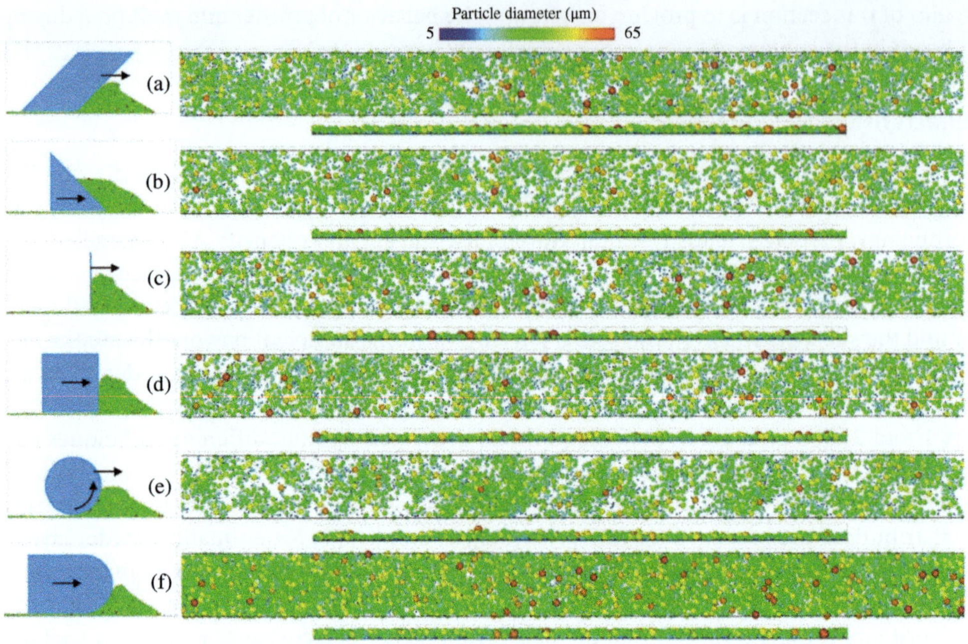

Figure 9.12 Top and side views of powder layers spread by different spreaders: (a) inclined blade, (b) declined blade, (c) vertical blade, (d) wide blade, (e) roller, and (f) round blade. Source: Wang et al. [79]/with permission from Elsevier.

The effects of spreader geometry and dimension on powder spreading were examined using the discrete element method [79, 80]. The study found that the round blade system had the largest deposited particle volume and the largest contact force exerted on the underlying particle layer, resulting in less segregation in the powder layer. The study showed that a compact region in the powder pile exists, and round and inclined surfaces of blade spreaders allow more particles in the compact region to be layered compared with vertical blades. The formed powder layer was denser due to larger forces exerted on the underlying part with the round and inclined blades. The inhomogeneity of powder layers was caused by particle burst phenomenon resulting from particle motion conflict in the compact region rather than the large forces. The study also found that roller systems had the largest particle motion conflict, resulting in sparse and inhomogeneous powder layers with small layer gaps, whereas the size segregation in blade systems was not as severe as the roller systems[79]. Figure 9.12 illustrates the effects of various spreader geometries on powder spreading.

Several factors should be considered when spreading metallic powder, including interparticle interactions, the interaction between the powder and the walls of the process chamber and substrate, friction between particles, particle size, and blade velocity [69]. Escano et al. [81] used high-speed X-ray imaging to study the particle dynamics (the evolution of the repose angle) of 316L SS powder during the spreading process. The authors conducted an experiment to compare the effect of two different average particle size distributions (D_{50} = 67 μm and D_{50} = 23 μm) on powder spreading under the same process conditions (a constant spreading speed of 11.5 mm s^{-1}). The results show that the finer particle size had a higher angle of repose, indicating reduced flowability and stronger interparticle forces within the powder system. Using discrete element method, Chen et al. investigated powder spreading mechanisms in PBF and their influence on the packing density of a powder layer [82]. This study identified three mechanisms that affect the packing density of powder deposition: cohesion effect, wall effect, and percolation effect. The cohesion effect is caused by the Van der Waals forces that affect particle agglomeration. The wall effect is defined by the impact of particle size and powder layer thickness on the occurrence of voids in a powder layer. The percolation effect occurs in bimodal powder particle systems and leads to particle segregation, which is an unfavorable phenomenon in laser PBF. These effects are illustrated in Figure 9.13 [82].

Numerous studies have been conducted to explore the interaction between spherical particles or a particle and a wall surface [83–86]. The electrostatic interactions between powder particles, or between a particle and a surface, result in Van der Waals forces, which vary in strength depending on the size and shape of the particles, and the contact area between their surfaces. This phenomenon has been extensively studied in the literature [87, 88].

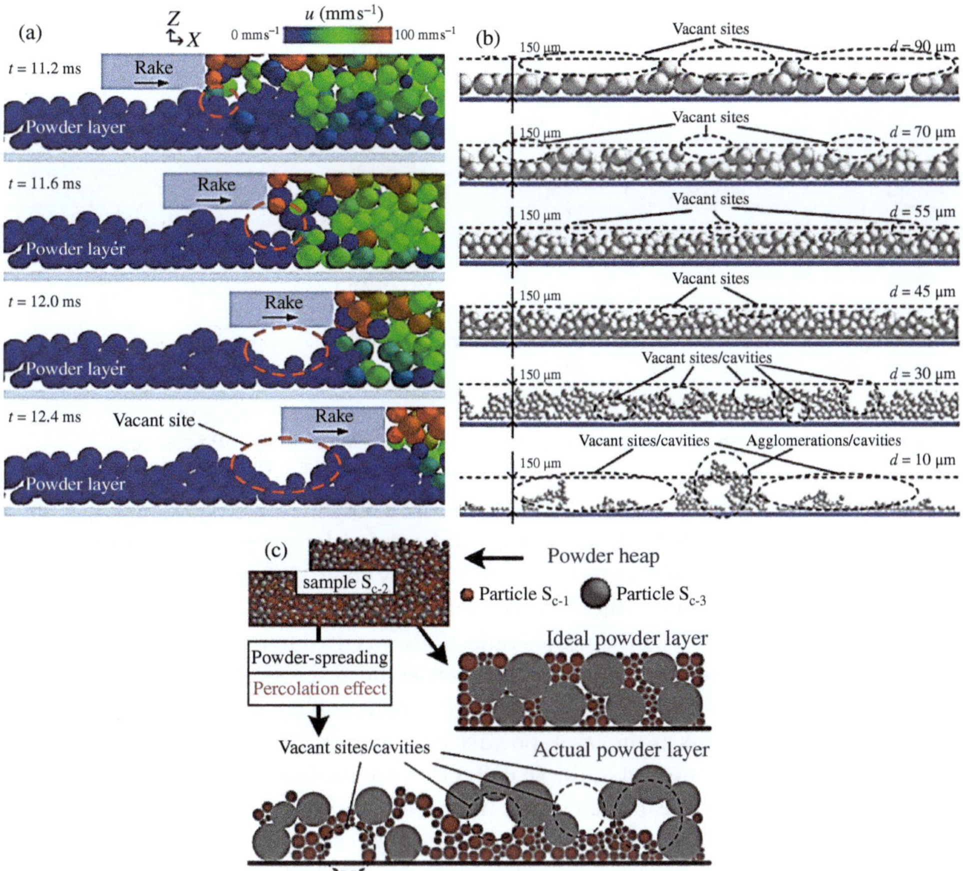

Figure 9.13 Discrete element method simulation results for three types of powder spreading mechanisms: (a) dynamic wall effect, the formation of voids due to the dynamic wall effect, (b) the cohesion effect, deposited powder layer with particle size, and (c) the percolation effect, powder layer resulted from the bimodal sample, according to [82]. Source: Chen et al. [82]/with permission of Elsevier.

The Van der Waals forces also play a critical role in the powder spreading process in PBF as it governs the formation of powder–powder contacts and hinders their relative motion within the powder bed [70, 89].

9.4.2 Laser Beam–Powder Interactions

PBF is a promising AM technology for producing metal components. To understand the phenomena that occur during the laser–powder–melt pool interactions for each single track, it is crucial to comprehend the entire process. During the laser–matter interactions, the powder absorbs, transmits, or reflects some of the incoming laser beam. When the powder absorbs sufficient laser energy, it melts and forms a melt pool, which dissipates some of its heat energy to sinter the surrounding particles. The sintered powder layer is commonly found along the edges of a build feature, which increases the surface roughness of the AM part. If the input laser energy is high enough, it can vaporize the elements from the melt pool, and with further increase in input laser energy, the laser beam ionizes the metal vapor and forms plasma.

The temperature at the melt pool center is much higher than that on the edges of the melt pool, resulting in a very high-temperature gradient across the melt pool surface. This gradient causes the fluid inside the melt pool to flow from a region with high surface tension (the edges) to the bottom of the melt pool, and then from the bottom of the melt pool to a region with low surface tension (the center), recirculating again. This phenomenon is known as **Marangoni convection** [90], which is induced by a surface tension gradient by temperature, concentration, or both gradients at the interface. As the melt pool cools down rapidly, it solidifies.

In order to gain insight into the thermal and physical phenomena that occur during the laser–powder interactions in the PBF process, researchers have utilized the discrete element method to study a randomly packed powder.

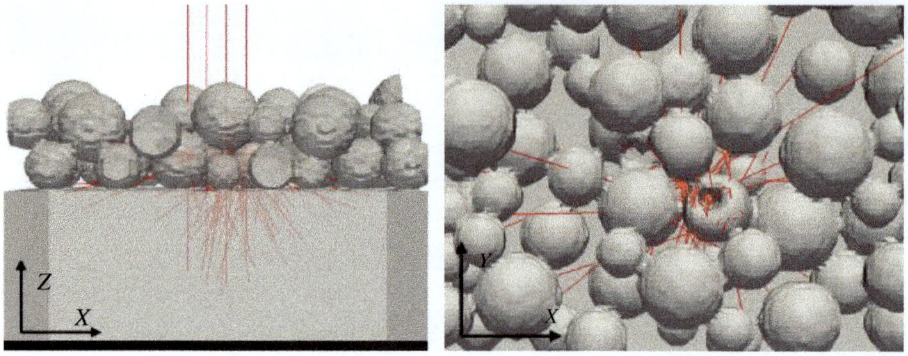

(a) (b)

Figure 9.14 The travel path of laser rays during PBF processing: (a) penetration of rays in the X–Z plane; and (b) top view of multiple reflection in the X–Y plane. The laser rays are indicated by the red lines. Source: Li et al. [91]/EDP Sciences.

They then employed CFD coupled with **volume of fluid** to simulate the laser–powder interactions [91]. A model that incorporates a three-phase approach using CFD coupled with volume of fluid was developed to capture the melt pool dimensions and dynamics. The model considers crucial factors such as the mushy zone, surface tension, Marangoni effect, and recoil pressure [91, 92]. To accurately analyze the thermal and fluid dynamics during the melting and solidification process in PBF, a laser ray tracing model with multiple reflections was used to represent the laser rays. Additionally, a **discrete transfer radiation model** was employed to describe the absorption and transmission of the laser rays with the powder. This approach allowed for the capture of melt pool flow dynamics and thermal behavior during the process, facilitating a detailed analysis of the laser–powder interactions. The travel path of the laser rays in the PBF system was traced, and the penetration and multiple **reflections** of the laser rays are illustrated and shown in Figure 9.14 [91]. The simulation successfully captured various thermal and physical phenomena, including full and partial melting of the powder. Nonuniform melting was observed due to variations in spatial layer thickness within the randomly distributed powder bed. The temperature distribution in the simulation domain illustrated the laser travel path and its interaction with the media, including the gas, powder, and substrate.

Figure 9.14 can be used to visualize the path of laser rays during laser–powder interaction for better comprehension of the process [91]. Initially, four collimated laser beams are directed toward the PBF system, passing through the gas media as shown in Figure 9.14a, until they hit the solid powder bed surface. As depicted in Figure 9.14b, the solid surface becomes concave due to the mass transfer to liquid, indicating that only the region beneath the initial laser–powder interaction is melted. Multiple reflections are generated based on the randomly distributed powder and the spherical morphology of particles, causing the laser beams to scatter into the surrounding powder and gas domain. These multiple transfer paths of laser rays can enhance laser energy absorption. [93]. In Figure 9.14a, it is shown that the laser rays can penetrate deeply into the powder and even reach the substrate. This can result in the melting of the substrate, or build plate, which is necessary to create a bond between the newly deposited layer and the previous layers.

Gunenthiram et al. employed an instrumented PBF setup equipped with a high-speed camera (>10 000 images s^{-1}) to perform in-situ observation of single tracks fabricated during PBF processing for real-time visualization of laser–powder interactions. They analyzed the results both during and postdeposition [94]. The study examined the physical causes behind powder denudation and the behavior of the melt pool. The denudation of powder was observed not only on the final powder beds but also in the high-speed videos, which revealed a lateral flow of powder toward the melt pool. This flow was found to be limited to the hotter zone of the melt pool, where a vertical vapor flow was also observed [95]. High-speed videos revealed a lateral powder flow toward the melt pool and a vertical vapor flow, which oscillated between a normal direction and a 30° rear inclination (Figure 9.15) [94]. The inward motion of powder particles toward the melt pool was assumed to be caused by the vapor flow. The denudation zone width and the attraction of particles fluctuated due to these oscillations. Capillary-driven particle flow toward the melt pool edge was also observed in a thin width close to the melt pool edge [94].

The PBF process involves complex interactions among laser, powder, and melt pool, which can cause the movement of powder particles within the melt pool due to fluid drag. However, explaining the dynamics and complexity of particle movement behavior during the process remains a challenge. Most studies have focused on the movement behavior

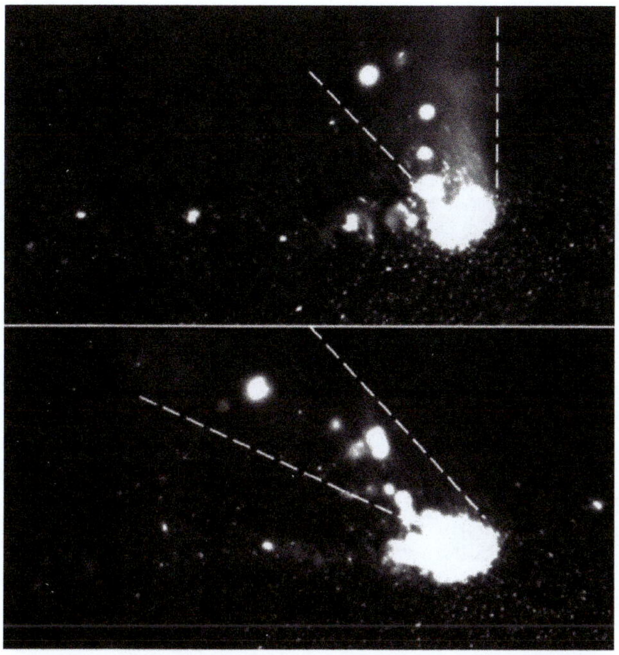

Figure 9.15 High-speed imaging of the vapor flow fluctuations at 0.33 m s^{-1} – 320 W during laser–powder interactions in PBF processing. Source: Gunenthiram et.al. [94]/American Institute of Physics.

of near-spherical particles produced by gas atomization and their interactions with the laser beam within the powder bed system [96–101]. The production of high-quality components in AM is facilitated by using gas-atomized powders, which possess excellent flowability and high sphericity. These properties enable an even distribution of the powder, resulting in superior product quality [68]. Chen and Yan developed a computational multiphase flow model to simulate the interactions that occur between the powder particles and the gas phase during the PBF process [102]. The study calculated the drag force exerted on near-spherical powder particles in various regions of the powder bed. They found that the metal vapor flow, along with the vortex flow of the surrounding gas, was the primary cause of powder spattering and denudation [102], which are addressed in the next section in this chapter. Haeri et al. [103] examined how the behavior of powder is affected by laser radiation pressure (the force exerted by a laser beam on a surface due to the transfer of momentum from photons to the surface) and found that this force can have a comparable impact to aerodynamic forces.

Moreover, they argued that, in the vapor jet region, the drag force may dominate the radiation pressure. Volpp [104] used high-speed imaging to observe particle movement outside of the process area and concluded that the temperature and pressure increase around the laser-material interaction zone can initiate the particle transfer toward the melt pool. It was observed that the transfer of particles toward the melt pool in the area outside the processing zone and concluded that in addition to the gas flow induced by vapor, the increase in temperature and pressure around the processing area can also cause particle movement toward the melt pool. In related studies, a computational flow model was used to simulate the interaction between the gas phase and powder particles in PBF [102]. The research aimed to compute the drag force acting on powder particles in different regions within the powder bed. Additionally, to comprehend the behavior of particle movement in a powder system, it is crucial to analyze the interaction between particle pairs.

The movement behavior of gas-atomized and water-atomized powder particles in PBF was investigated by Fedina et al. using high-speed imaging [70]. The study measured **particle velocities**, **entrainment distances**, and **denudation zones** for both gas- and water-atomized powders. The sequence of images illustrated the motion of gas-atomized and water-atomized powder particles in the particle entrainment region in front of the melt pool, as illustrated in Figure 9.16. The entrainment of gas-atomized powder particles toward melt pool region was initiated 1.6 mm from the edge of the melt pool, whereas the distance was 0.6–0.8 mm for the water-atomized powder. The differences in observed behavior were related to the variations in particle shape of the two types of powder. The processing of water-atomized powder resulted in a 16% narrower denudation zone for a low **volumetric energy density**

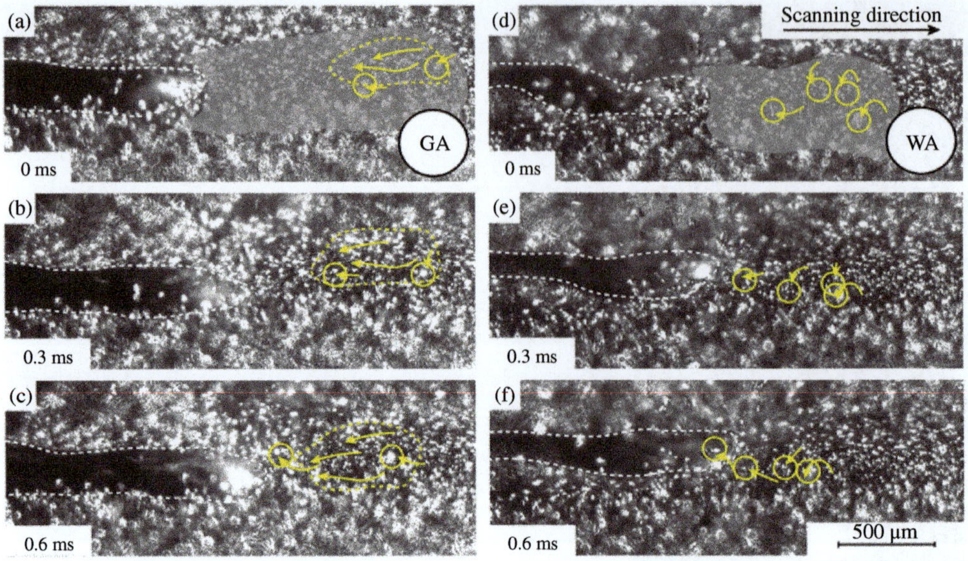

Figure 9.16 A sequence of images from high-speed imaging, highlighting the motion of gas-atomized (GA) (a, b, c) and water-atomized (WA) (d, e, f) powder particles in the particle entrainment region in front of the melt pool. Source: Fedina et al. [70]/with permission from Elsevier.

compared to gas-atomized powder. The denudation width difference decreased with increasing volumetric energy density, most likely due to a steeper pressure gradient near the melt pool, which diminishes the impact of powder shape. X-ray computed microtomography was also utilized to estimate the **drag force** acting on the powder particles of various morphologies. The results showed that the radial drag force exerted on gas-atomized powder was 64% greater than when using water-atomized powder. Moreover, if the water-atomized powder particles were of elongated shape, the drag force decreased by almost an order of magnitude, demonstrating the importance of the particle morphology in the process dynamics. Nonspherical powders tend to establish contacts with neighboring particles, thus significantly affecting the powder movement, entrainment, and denudation behaviors. In addition to that, the impact of drag force on the particle dynamics varies, depending on the particle shape and its position in the powder bed.

9.4.3 Spatter and Denudation Formation

During PBF processing, **spatter ejections** and **powder denudation** phenomena are commonly observed alongside variations in melt pool dimensions and instabilities. The formation of spatter is generally considered detrimental to the PBF process and leads to a decrease in deposited material and component quality. Spattering refers to the ejection of small metal droplets or powder particles from the interaction zone between the laser, powder, and melt pool. There are two main mechanisms that cause spatter formation: recoil pressure coupled with the Marangoni effect [105] and vapor-driven particle entrainment by an induced gas flow [106]. Spatter is formed when the liquid gains enough kinetic energy to surpass the capillary pressure of the melt. The movement of the liquid is closely tied to the intensity of the vapor recoil force, which is determined by the amount of energy applied. [99]. When the temperature of the melt pool surpasses the vaporization point, it triggers the creation of a vapor jet, leading to the expulsion of molten material. The upward jet then encounters the vapor plume present above the melt pool and disintegrates into small droplets, which are akin to spatters, while traveling through the laser beam. The occurrence of spattering is a widespread event during PBF processing [107–109]. Two types of spatters have been identified in PBF processing. The first type, known as droplet spatter or vapor-driven spatter, occurs when molten material is torn and ejected directly from the melt pool. This happens when the temperature of the melt pool exceeds the vaporization temperature, causing melt expulsion. The second type of spatter, known as powder spatter or entrainment-driven spatter, occurs when nonmolten powder particles surrounding the melt pool are blown away by the ambient gas flow.

A monitoring system was developed by Ikeshoji et al. [110], which combines fringe projection for surface morphology measurement and thermal field measurement with a high-speed camera. The results of the monitoring show that the melt pool was asymmetrical and had a spindle shape, and that spatter was mainly emitted from the built part

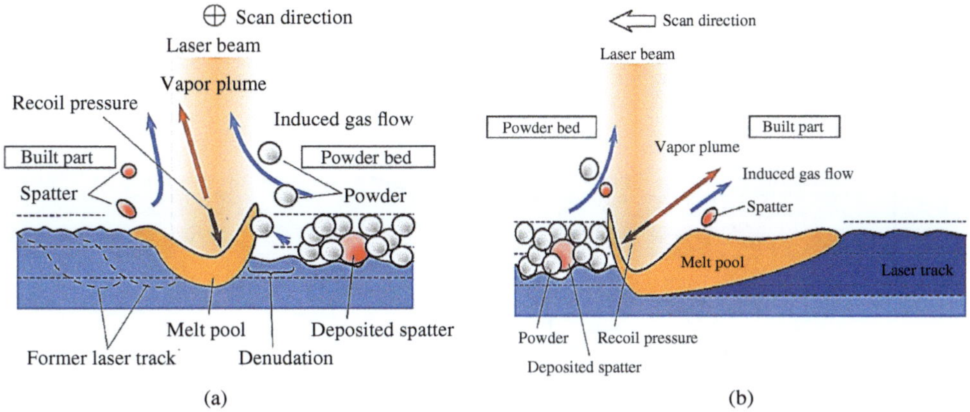

Figure 9.17 Schematic of the melt pool behavior, spatter, and denudation zone formation in PBF, (a) the transverse section of the laser scan direction and (b) longitudinal section; melt pool length is shortened because of space limitations. Source: Ikeshoji et al. [110]/CC BY 4.0/Springer Nature.

solidified side of the melt pool. Additionally, it was observed that the surface morphology of the built part under the powder layer influenced the stability of the melt pool. Consequently, a graphical representation for heterogeneous surfaces was proposed and is depicted in Figure 9.17 [110].

When a powder bed is melted using a focused laser beam to form a melt pool, it can result in the emission of metal vapor. This vapor may escape from the melt, and the resulting vapor pressure can push the surface of the melt downwards, creating a cavity known as a keyhole [97, 111–114]. The vapor plume above the melt surface generates a jet flow, which in turn induces an upward flow of the surrounding gas. This induced gas flow leads to a region denuded of powder particles located next to the laser tracks. The powder particles that are entrained by this process are then introduced into the melt pool, where they can be ejected as spatter [115]. Metal vapor jets induce the **Bernoulli effect** under high atmospheric pressure. However, under low atmospheric pressure, where atmospheric gas transport can be seen as a molecular flow, the metal vapor expands locally at the laser beam spot. This expansion leads to an outward flow that pushes the surrounding powder particles resulting in denudation, as explained by Matthews et al. [54]. When the atmospheric pressure is low, the recoil pressure is also reduced, leading to lower spatter emission. The circulation of atmospheric gas is insufficient to decrease the pressure enough to cause the vapor jet to transform into an expansion flow. Nevertheless, the atmospheric gas flow can still impact the direction in which high-flying spatter is expelled.

The primary cause of metal particle ejection during PBF is widely acknowledged to be the recoil pressure mechanism [97, 107, 116]. However, Ly et al. reported that vapor-driven particle entrainment by an induced gas flow could be the primary mechanism responsible for generating ejecta [99]. Spatter formation, which is commonly observed during laser welding, is the process of molten droplet ejection from the melt pool [117]. In the case of PBF, ejecta generation involves mostly droplet ejection from a surrounding powder bed. There are several driving forces for ejecta generation, which can be categorized as follows: spatters resulting from melt pool surface instability, as well as hot and cold ejecta [118]. As previously mentioned, spatters are formed by the recoil pressure mechanism, and they are directly ejected from the melt pool. Due to the similarity in the melting and cooling processes as in gas atomization of powder droplets, spatters usually have a spherical shape. According to Ly et al., the ejection velocity of spatters can reach up to 3–8 $m\,s^{-1}$ [99]. Hot ejecta refers to the powder particles that pass through the laser beam and get entrained by the vapor jet. The velocity of hot particle ejecta depends on their size; smaller particles can reach higher velocities (ranging from 6 to 20 $m\,s^{-1}$). Powder particles that do not appear to melt when passing behind the laser field are referred to as cold ejecta [118, 119]. The ejection velocities of this type range from 2 to 4 $m\,s^{-1}$ [99]. The degree of spattering in PBF is significantly influenced by the laser power and scan speed. Environmental pressure also plays a critical role in the spattering behavior by regulating the divergence angle of the vapor jet (the angle of vapor jet measured from the surface of melt pool, as illustrated θ_1, θ_2, and θ_3 in Figure 9.18 [97]) and the inert gas flow [97, 106]. When a higher energy density is used during the PBF process, there is a greater occurrence of spattering because of the elevated energy input and the formation of a keyhole [106, 116]. High laser power and low scan speed can lead to overheating, whereas high scan speed and low laser power promote the **balling effect** due to underheating [120].

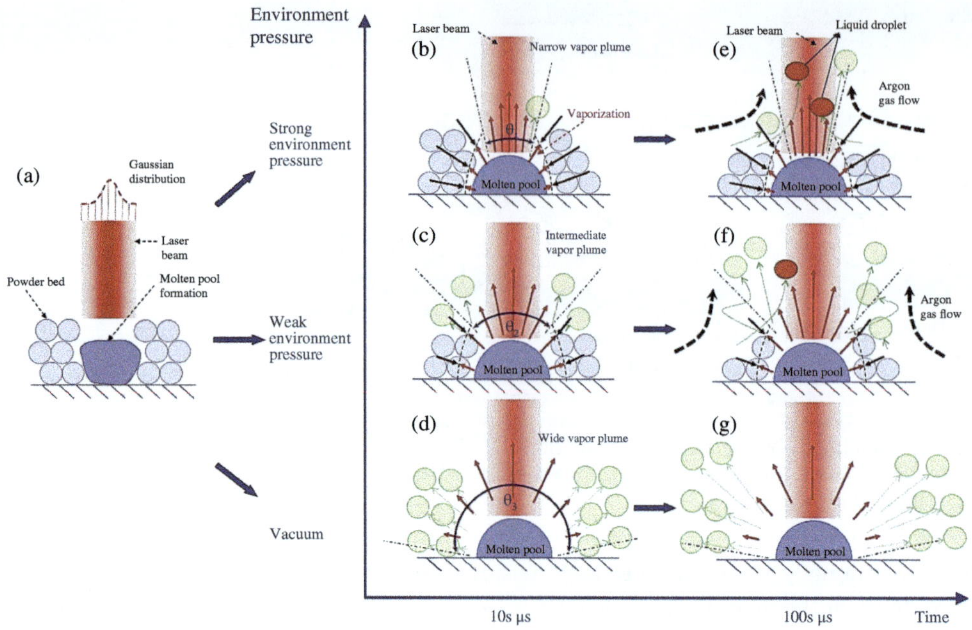

Figure 9.18 Influence of environmental pressure in a PBF system on powder spatter behavior . The diagram illustrates the behavior of powder spattering over time on the horizontal axis, while the effect of environmental pressure is depicted on the vertical axis. (a) A Gaussian laser beam melts the powder. After a brief period (~10 μs in this study), vaporization initiates from the molten pool, and vapor escapes as a narrow vapor jet under high environmental pressure (b), an intermediate vapor jet under low environmental pressure (c), or a wide vapor plume in a vacuum (d). The diverging angles of the vapor jet, represented by $\theta_1 < \theta_2 < \theta_3$, increase as the environmental pressure decreases. The red arrows and black arrows represent the influence of vapor pressure and environmental pressure, respectively, with the length of the arrows indicating the relative pressure intensity. The trajectories of the moving powder particles are depicted by green arrows. After the formation of the vapor jet for a certain duration (~100 μs), the vapor jet induces an argon gas flow under high environmental pressure (e) and low environmental pressure (f). This induced argon gas flow generates and carries adjacent powder particles (shown in blue) toward the region of the molten pool. If the powder particles enter the laser beam area, they will be heated and melted, resulting in hot spatters (depicted in red) (g). The powder ejection driven by the expanding vapor plume in a free manner (shown in green). The vapor plume continuously expands and ejects all the powder particles in its path. Source: Guo et al. [97]/with permission of Elsevier.

The balling phenomenon occurs when the viscosity of the melt pool is high due to a relatively low temperature, which causes poor flowability in the melt pool [121]. Mutke et al. reported that the use of high laser scan speeds can lead to the balling effect caused by the **Plateau-Rayleigh instability**, which is characterized by surface tension-driven break-up of the melt pool [122]. Insufficient wetting of the substrate can also intensify the balling effect by increasing the surface free energy.

Guo et al. [97] investigated the importance of the environmental pressure on spatter behavior. The results from this study show that when the melt pool surface reaches the boiling point, intense vaporization occurs, resulting in metal vapor and droplet ejection from the melt pool area. Environmental pressure has a significant impact on the vapor formation and subsequent outcomes. The intensity of chamber pressure affects the divergence angle of the vapor jet and gas flow behavior. In vacuum conditions, the metal vapor expands freely, resulting in a large divergence angle. The absence of argon gas in the environment does not generate gas flow or entrain particles toward the melt pool. High environmental pressure attracts the gas flow and powder particles toward the vapor field above the melt pool, promoting particle entrainment. However, with the increase of environmental pressure, the total amount of spatter and powder ejecta decreases, forming a narrower vapor jet than observed under low pressure or vacuum conditions, as depicted in Figure 9.18 [97]. Hence, increasing the chamber pressure in PBF can alleviate spattering, leading to an enhancement in the surface quality of the end products.

Studies conducted on powder spattering have revealed that a significant portion of ejected powder undergoes oxidation, resulting in a change in their particle size distribution. This, in turn, introduces oxygen to the powder bed and causes disruptions in the powder layer. The accumulation of spatter on the powder bed can increase surface roughness during the melting of subsequent layers, lead to porosity, and act as a potential source of defects such as **lack-of-fusion** pores in the final product [99, 123].

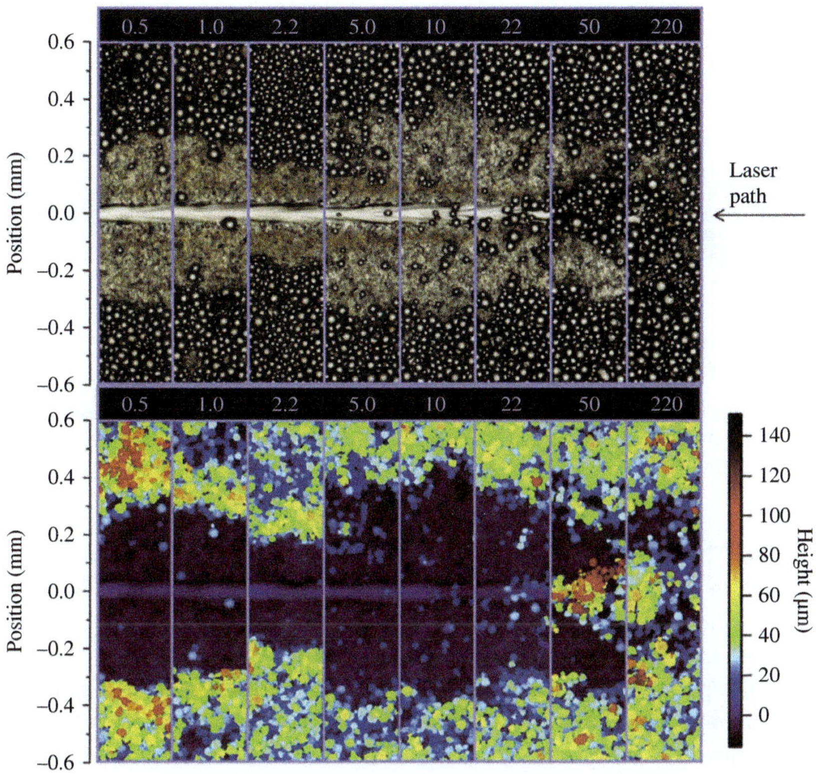

Figure 9.19 Denudation zone formation around a solidified track [54, 124]. Montage of 1.2 × 0.25 mm optical micrographs (top) and height maps (bottom) of the solidified melt track within a powder layer following scanning laser exposure at 225 W and 1.4 m s^{-1} as a function of ambient Ar pressure (shown above image slices in Torr).

The interaction between the laser beam and the powder bed in PBF can also result in the denudation phenomenon and particle movement. Denudation refers to the depletion of particles in an area near the laser scan path, as depicted in Figure 9.19 [54, 124]. Three distinct regions can be identified near the laser path center, namely, track accumulation zone, the denuded zone, and the background powder zone. According to [99], the main factor responsible for denudation of powder near a solidified track is the gas flow induced by the intensive evaporation of the melt pool during laser processing. This gas flow is caused by the pressure drop inside the vapor plume due to the Bernoulli effect [124]. The width of the denuded area around the laser scan path, also known as denudation width, is influenced by various process parameters and ambient gas pressure. The degree of denudation is affected by the induced gas flow, which in turn affects the quality of the final PBF components [124]. The formation of pores or voids due to denudation is a common outcome. These defects can negatively impact the quality of the final product and lead to a deterioration of material properties [99]. The shape and structure of powder particles can also affect the magnitude of denudation since they significantly influence the movement of particles in a powder bed. In PBF, particle movement is primarily driven by particle entrainment induced by gas flow. Previous studies have investigated the behavior of individual particles in a powder bed and their interaction with the laser beam [96–101].

9.4.4 Powder Degradation

Several studies have demonstrated that the degradation of powder after PBF processing is influenced by a range of factors, such as the processing conditions, the environment, and the feedstock material [125–128]. Particle size, powder agglomeration degree, and chemical purity are important factors that affect the quality of powder. Wang et al. conducted a study on the impact of laser energy on spattering, its formation principle, appearance, and composition [106]. Increasing laser energy input from 0.32×10^5 W cm^{-3} to 1.30×10^5 W cm^{-3} resulted in an increase in spattering intensity and quantity, with metal liquid reaching a height of 11 cm. The major sources of spattering were recoil pressure, Marangoni effect, and heat effect in the melt pool, resulting in three different types of spattering morphology as shown in Figure 9.20 [106]. This study addressed the impact of laser energy on spattering, spatter formation

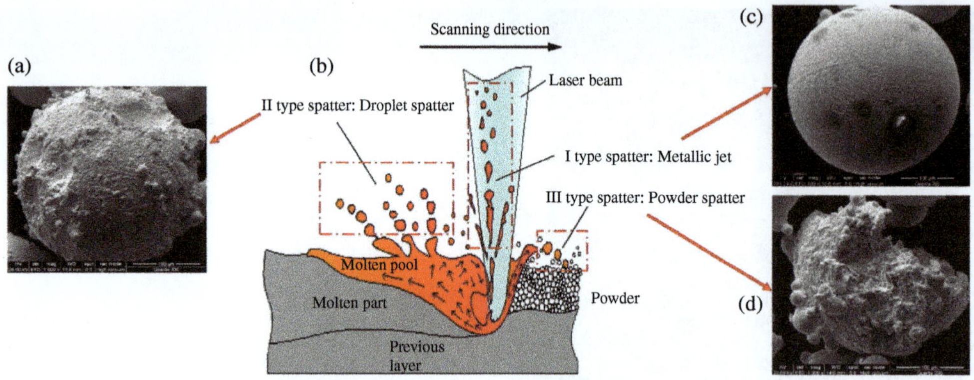

Figure 9.20 Formation mechanisms of different types of spatter: (a) morphology of spherical splashing (type-I splashing); (b) schematic illustration of type spatters during laser melt pool interaction; (c) morphology of coarse spherical morphology (type-II splashing); (d) morphology of irregular splashing (type-III splashing). Source: Wang et al. [106]/with permission of Elsevier.

principles, spatter appearance, and composition. The solidified spatter particles had an average size of around 162 μm, which is much larger than the original powder size of 32 μm and had various appearances. The composition of spatter powders was almost the same as the original powders, but the contents of O, Si, and C increased significantly. Spatter particles were embedded in both the surface and interior of SLM-fabricated parts.

After the powder is reused multiple times, the particle size distribution tends to slightly increase due to powder agglomeration and the introduction of ejecta to the processed powder. However, the loss of fine particles can result in less absorption of laser energy, leading to a decrease in heat input [125]. An elevation in the oxygen content within the powder may alter the behavior of the melt pool, influence laser absorption, and facilitate the formation of oxides on the surfaces of the particles [87, 97, 129, 130]. Satellites present on particle surfaces should be considered when handling the powder as they increase the surface area of particles and result in a more irregular powder shape. Additionally, they promote agglomeration and particle–particle interaction, which can lead to the development of metal–metal contacts.

Defects caused by powder spattering can lead to various issues, including lack-of-fusion pores in the final components [131]. The spatters generated from the melt pool during PBF tend to be larger in size compared to those due to powder particles. Therefore, when they fall onto the powder bed, they have the potential to adversely affect the mechanical properties and increase the surface roughness of the final parts [99]. In addition, regardless of their source, all ejecta have a higher oxygen content compared to virgin powder. This results in an overall increase in oxygen content in the recycled powder and inside the build. Figure 9.21 illustrates the potential areas where ejecta can

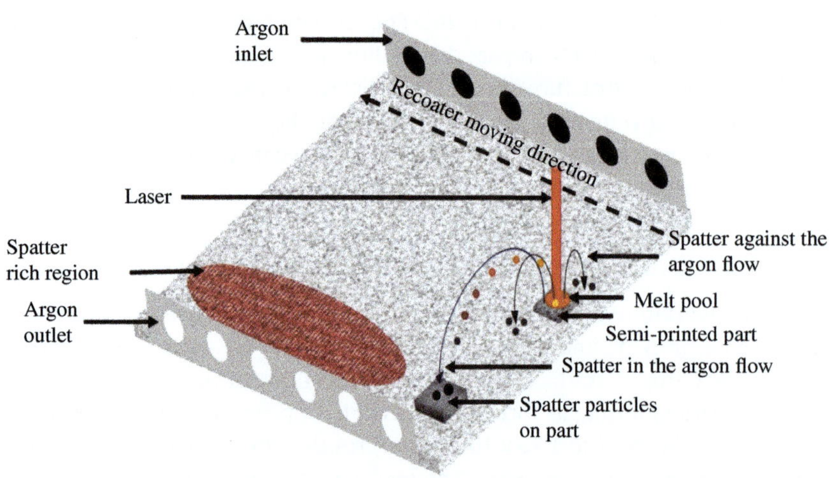

Figure 9.21 Schematic representation of spatter formation, types, and effects of spatial distribution of ejecta on the build plate. Source: Esmaeilizadeh et al. [118]/with permission of Elsevier.

land on a powder bed during the PBF process [118]. In addition to spatter formation, PBF may also produce some byproducts in the form of condensate or nanoparticles. Reports indicate that soot and fully oxidized powder particles can form [99, 125].

9.5 Summary

This chapter discussed the laser–powder–melt pool interactions in both DED and PBF AM processes. The current technological challenges such as porosity and residual stresses could be overcome by better understanding these interactions. However, the underlying formation mechanisms of defects and the melt pool dynamics remain unclear. Most of the published studies and theories on laser–powder–melt pool interaction are related to PBF processes, using both ex-situ and in-situ characterization techniques. While these techniques can partially elucidate the evolution of laser–powder–melt pool interactions, they are experimentally challenging. Advanced techniques such as X-ray radiography could provide further understanding of complex laser–matter interactions. Future studies should aim to validate existing theories and numerical simulations on melt pool behavior.

Nomenclature

Variable	Description	Units
A	Absorptivity of material	
AM	Additive manufacturing	—
CCD	Charge-coupled device	—
CFD	Computational fluid dynamics	—
d	Laser spot size in diameter	mm
D_{50}	Average particle size	μm
DED	Directed energy deposition	—
DEM	Discrete element method	
DTRM	Discrete transfer radiation model	
E	Laser energy	W
FEM	Finite element modeling	
GA	Gas atomization	
I	Intensity of a laser propagating into a material	
MP	Melt pool	
EBM	Electron beam melting	—
PBF	Powder bed fusion	—
R	Reflectivity of material	
P	Laser power	
PSD	Particle size distribution	
SLM	Selective laser melting	—
SS	Stainless steel	—
T	Transmissivity of material	
V	Laser scan speed	
VOF	Volume of fluid	
WA	Water atomization	—
WD	Working distance	—

References

1 Everton, S.K., Hirsch, M., Stravroulakis, P. et al. (2016). *Mater. Des.* 95: 431–445.

2 Tapia, G. and Elwany, A. (2014). *J. Manuf. Sci. Eng.* 136: 060801–060810.

3 Smugeresky, J.E., Zheng, B., Ajdelsztajn, L. et al. (2005). *134th TMS Annual Meeting & Exhibition* (13–17 February 2005). San Francisco, CA.

4 Zheng, B., Xiong, Y., Nguyen, J. et al. (2009). Powder additive processing with laser engineered net shaping (LENS). In: *Powder Metallurgy Research Trends* (ed. L.J. Smit and J.H.V. Dijk), 125–190. Hauppauge, NY, New York: Nova Science Publishers, Inc.

5 Wang, D., Song, C., Yang, Y., and Bai, Y. (2016). *Mater. Des.* 100: 291–299.

6 Xiong, Y., Smugeresky, J.E., and Schoenung, J.M. (2009). *J. Mater. Process. Technol.* 209: 4935–4941.

7 Turichin, G.A., Klimova, O.G., Zemlyakov, E.V. et al. (2015). *Phys. Procedia* 78: 397–406.

8 Tang, Z.-J., Liu, W.-W., Wang, Y.-w. et al. (2020). *Int. J. Adv. Manuf. Technol.* 108: 3437–3463.

9 Knapp, C., Lienert, T., Carpenter, J., and Kovar, D. (2016). In situ monitoring of directed energy deposition. Presented at the MS&T, Salt Lake City, Utah.

10 Balu, P., Leggett, P., Hamid, S., and Kovacevic, R. (2013). *Mater. Manuf. Processes* 28: 173–182.

11 Kumar, S., Sharma, V., Choudhary, A.K.S. et al. (2013). *Int. J. Adv. Manuf. Technol.* 67: 2681–2687.

12 Sun, Y. and Hao, M. (2012). *Opt. Lasers Eng.* 50: 985–995.

13 de Oliveira, U., Ocelík, V., and De Hosson, J.T.M. (2005). *Surf. Coat. Technol.* 197: 127–136.

14 Rosenthal, D. (1946). *The Theory of Moving Sources of Heat and Its Application to Metal Treatments*. Cambridge: ASME.

15 Sammons, P.M., Bristow, D.A., and Landers, R.G. (2013). *J. Manuf. Sci. Eng.* 135: 054501–054507.

16 Beuth, J. and Klingbeil, N. (2001). *JOM* 53: 36–39.

17 Gockel, J., Klingbeil, N., and Bontha, S. (2016). *Metall. Mater. Trans. B* 47: 1400–1408.

18 Pinkerton, A.J. (2015). *J. Laser Appl.* 27: S15001.

19 Balu, P., Leggett, P., and Kovacevic, R. (2012). *J. Mater. Process. Technol.* 212: 1598–1610.

20 Michaleris, P. (2014). *Finite Elem. Anal. Des.* 86: 51–60.

21 Zheng, B., Zhou, Y., Smugeresky, J.E. et al. (2008). Part I numerical calculations. *Metall. Mater. Trans. B* 39A: 2228–2236.

22 Manvatkar, V., De, A., and DebRoy, T. (2014). *J. Appl. Phys.* 116: 124905.

23 Einstein, A. (1917). *Phys. Z.* 18: 121.

24 Maiman, T.H. (1960). *Nature* 187: 493.

25 Smith, W.V. and Sorokin, P.P. (1966). *The Laser*. McGraw-Hill.

26 Ready, J.F. and Farson, D.F. (2001). *LIA Handbook of Laser Materials Processing*. Laser Institute America: Magnolia Publishing, Orlando, FL.

27 Steen, W.M. (2003). *Laser Material Processing*. London: Springer-Verlag.

28 Weerasinghe, V.M. and Steen, W.M. (1987). *Applied Laser Tooling*, 183. Dordrecht: Martinus Nijhoff Publishers.

29 Mayerhöfer, T.G., Pahlow, S., and Popp, J. (2020). *ChemPhysChem* 21: 2029–2046.

30 Duley, W.W. (1986). *Laser Surface Treatment of Metals: NATO-ASI Series (E), No.115* (ed. C.W. Draper and P. Mazzoldi), 3. Boston: Martinus Nijhoff.

31 Laeng, J., Stewart, J.G., and Liou, F.W. (2000). *Int. J. Prod. Res.* 38: 3973.

32 Mordike, B.L. (1993). *Materials Science and Technology* (ed. R.W. Cahn et al.). Weinheim, VCH.

33 Draper, C.W., Preece, C.M., Jacobson, D.C. et al. (1980). Laser alloying of deposited metal films on nickel. In: *Laser and Electron Beam Processing of Materials* (ed. C.W. White and P.S. Peercy), 721. NewYork: Academic Press.

34 Haley, J.C., Zheng, B., Bertoli, U.S. et al. (2019). *Mater. Des.* 161: 86–94.

35 Zheng, B., Haley, J.C., Yang, N. et al. (2019). *Mater. Sci. Eng., A* 764: 138243.

36 Arrizubieta, J.I., Lamikiz, A., Klocke, F. et al. (2017). *Int. J. Heat Mass Transfer* 115: 80–91.

37 Pinkerton, A.J. and Li, L. (2004). *Proc. Inst. Mech. Eng. Part B: J. Eng. Manuf.* 218: 363–374.

38 Pinkerton, A.J. (2007). *J. Phys. D: Appl. Phys.* 40: 7323–7334.

39 Haley, J.C., Schoenung, J.M., and Lavernia, E.J. (2018). *Addit. Manuf.* 22: 368–374.

40 Atwood, C.L., Griffith, M.L., Harwell, L.D. et al. (1999). *Sandia Report SAND99-0739*, 1–30. Sandia National Laboratories.

41 Lewisa, G.K. and Schlienger, E. (2000). *Mater. Des.* 21: 417–423.

42 Hofmeister, W., Griffith, M., Ensz, M., and Smugeresky, J. (2001). *JOM* 53: 30–34.

43 Griffith, M.L., Ensz, M.T., Puskar, J.D. et al. (2000). *Materials Research Society, V625, Symposium Y Proceeding April*.

44 Griffith, M.L., Schlienger, M.E., Harwell, L.D. et al. (1999). *Mater. Des.* 20: 107–113.

45 Wei, W., Zhou, Y., Ye, R. et al. (2002). Investigation of the thermal behavior during the LENSTM process. In: *Proceedings of the International Conference on Metal Powder Deposition for Rapid Manufacturing*. San Antonio, TX, United States: Metal Powder Industries Federation, pp. 128–135.

46 Hofmeister, W., Wert, M., Smugeresky, J.E. et al. (1999). *JOM-e* (http://www.tms.org/pubs/journals/JOM/9907/Hofmeister/Hofmeister-9907.html) 51.

47 Xiong, Y., Hofmeister, W.H., Cheng, Z. et al. (2009). *Acta Mater.* 57: 5419–5429.

48 Haley, J.C., Schoenung, J.M., and Lavernia, E.J. (2019). *Mater. Sci. Eng., A* 761: 138052.

49 Sbalzarini, I.F. and Koumoutsakos, P. (2005). *J. Struct. Biol.* 151: 182–195.

50 Wang, A., Song, Q., Ji, B., and Yao, Q. (2017). *Powder Technol.* 311: 408–415.

51 Aristoff, J.M. and Bush, J.W.M. (2009). *J. Fluid Mech.* 619: 45–78.

52 Zheng, B., Smugeresky, J.E., Zhou, Y. et al. (2008). *Metall. Mater. Trans. A* 39A: 1196–1205.

53 Jiang, S., Zheng, B., and Schoenung, J.M. (2022). *J. Mater. Res. Technol.* 16: 516–529.

54 Matthews, M.J., Guss, G., Khairallah, S.A. et al. (2016). *Acta Mater.* 114: 33–42.

55 Hegab, H.A. (2016). *Manuf. Rev.* 3: 1–17.

56 Collins, P.C., Brice, D.A., Samimi, P. et al. (2016). *Annu. Rev. Mater. Res.* 46: 63–91.

57 Clyne, T.W. and Withers, P.J. (1993). *An Introduction to Metal Matrix Composites*. New York: Cambridge University Press.

58 Zheng, B., Topping, T., Smugeresky, J.E. et al. (2010). *Metall. Mater. Trans. A* 41A: 569–573.

59 Li, N., Liu, W., Wang, Y. et al. (2021). *Chin. J. Mech. Eng.* 34: 38.

60 Tang, S., Ummethala, R., Suryanarayana, C. et al. (2021). *Adv. Eng. Mater.* 23: 2100053.

61 Wolff, S.J., Wu, H., Parab, N. et al. (2019). *Sci. Rep.* 9: 962.

62 Lu, Z.L., Cao, J.W., Jing, H. et al. (2013). *Virtual Phys. Prototyp.* 8: 87–95.

63 Biemond, J.E., Hannink, G., Verdonschot, N. et al. (2013). *Sci. Mater. Med.* 24: 745–753.

64 Zhang, B., Li, Y., and Bai, Q. (2017). *Chin. J. Mech. Eng.* 30: 515–527.

65 Leung, C.L.A., Marussi, S., Atwood, R.C. et al. (2018). *Nat. Commun.* 9: 1–9.

66 Cook, P.S. and Murphy, A.B. (2020). *Addit. Manuf.* 31: 100909.

67 Wei, H.L., Mukherjee, T., Zhang, W. et al. (2021). *Prog. Mater Sci.* 116: 100703.

68 Tan, J.H., Wong, W.L.E., and Dalgarno, K.W. (2017). *Addit. Manuf.* 18: 228–255.

69 Meier, C., Weissbach, R., Weinberg, J. et al. (2019). *J. Mater. Process. Technol.* 266: 484–501.

70 Fedina, T., Sundqvist, J., and Kaplan, A.F.H. (2022). *Powder Technol.* 395: 720–731.

71 Gu, D.D., Meiners, W., Wissenbach, K., and Poprawe, R. (2012). *Int. Mater. Rev.* 57: 133–164.

72 Funk, J.E. and Dinger, D. (1994). *Predictive Process Control of Crowded Particulate Suspensions*, 1e. US: Springer.

73 Furnas, C.C. (1931). *Ind. Eng. Chem.* 23: 1052–1058.

74 Westman, A.E.R. (1936). *J. Am. Ceram. Soc.* 19: 127–129.

75 Bennett, J.P. and Smith, J.D. (2012). *Fundamentals of Refractory Technology*. John Wiley & Sons.

76 Wischeropp, T.M., Emmelmann, C., Brandt, M., and Pateras, A. (2019). *Addit. Manuf.* 28: 176–183.

77 Vock, S., Klöden, B., Kirchner, A. et al. (2019). *Progr. Addit. Manuf.* 4: 383–397.

78 Mussatto, A., Groarke, R., O'Neill, A. et al. (2021). *Addit. Manuf.* 38: 101807.

79 Wang, L., Yu, A., Li, E. et al. (2021). *Powder Technol.* 384: 211–222.

80 Bicanic, N. (2004). Discrete element methods. In: *Encyclopedia of Computational Mechanics*, vol. 1 (ed. E. Stein, D. Borst, and T.J.R. Hughes), 311–337. Wiley.

81 Escano, L.I., Parab, N.D., Xiong, L. et al. (2018). *Sci. Rep.* 8: 15079.

82 Chen, H., Wei, Q., Zhang, Y. et al. (2019). *Acta Mater.* 179: 158–171.

83 Jones, R., Pollock, H.M., Geldart, D., and Verlinden-Luts, A. (2004). *Ultramicroscopy* 100: 59–78.

84 Tsuji, Y., Tanaka, T., and Ishida, T. (1992). *Powder Technol.* 71: 239–250.

85 Jordam Caserta, A., Navarro, H.A., and Cabezas-Gómez, L. (2016). *Powder Technol.* 302: 462–479.
86 Thornton, C., Cummins, S.J., and Cleary, P.W. (2013). *Powder Technol.* 233: 30–46.
87 Thummler, F. and Oberacker, R. (1991). *Introduction to Powder Metallurgy*. London: The Institute of Materials.
88 Hamaker, H.C. (1937). *Physical Metallurgy*, 3e, vol. 4, 1058–1072. Elsevier, BV.
89 Chen, H., Wei, Q., Wen, S. et al. (2017). *Int. J. Mach. Tools Manuf.* 123: 146–159.
90 Khairallah, S.A., Anderson, A.T., Rubenchik, A., and King, W.E. (2016). *Acta Mater.* 108: 36–45.
91 Li, E., Wang, L., Zou, R. et al. (2021). Investigation of laser-powder interaction in laser powder bed fusion process in additive manufacturing, EPJ Web of Conferences. *Powders Grains* 249 (2021): 12002.
92 Li, E.L., Wang, L., Yu, A.B., and Zhou, Z.Y. (2021). *Powder Technol.* 381: 298–312.
93 Wang, L., Li, E.L., Shen, H. et al. (2020). *Powder Technol.* 363: 602–610.
94 Gunenthiram, V., Peyre, P., Schneider, M. et al. (2017). *J. Laser Appl.* 29.
95 Masmoudi, A., Bolot, R., and Coddet, C. (2015). *J. Mater. Process. Technol.* 225: 122–132.
96 Bertoli, U.S., Guss, G., Wu, S. et al. (2017). *Mater. Des.* 135: 385–396.
97 Guo, Q., Zhao, C., Escano, L.I. et al. (2018). *Acta Mater.* 151: 169–180.
98 Bidare, P., Bitharas, I., Ward, R.M. et al. (2018). *Acta Mater.* 142: 107–120.
99 Ly, S., Rubenchik, A.M., Khairallah, S.A. et al. (2017). *Sci. Rep.* 7: 4085.
100 Martin, A.A., Calta, N.P., Hammons, J.A. et al. (2019). *Mater. Today Adv.* 1: 100002.
101 Tran, H.-C. and Lo, Y.-L. (2018). *J. Mater. Process. Technol.* 255: 411–425.
102 Chen, H. and Yan, W. (2020). *Acta Mater.* 196: 154–167.
103 Haeri, S., Haeri, S., Hanson, J., and Lotfian, S. (2020). *Powder Technol.* 368: 125–129.
104 Volpp, J. (2019). *Procedia Manuf.* 36: 26–32.
105 Scriven, L. and Sternling, C. (1960). *Nature* 187: 186–188.
106 Wang, D., Wu, S., Fu, F. et al. (2017). *Mater. Des.* 117: 121–130.
107 Liu, Y., Yang, Y., Mai, S. et al. (2015). *Mater. Des.* 87: 797–806.
108 King, W.E., Anderson, A.T., Ferencz, R.M. et al. (2015). *Appl. Phys. Rev.* 2: 041304.
109 Chivel, Y. (2014). Ablation phenomena and instabilities under laser melting of powder layers. In: *8th International Conference on Photonic Technologies LANE*, pp. 1–7.
110 Ikeshoji, T.-T., Yonehara, M., Kato, C. et al. (2022). *Sci. Rep.* 12: 20384.
111 Zhao, C., Fezzaa, K., Cunningham, R. et al. (2017). Real-time monitoring of laser powder bed fusion process using high-speed X-ray imaging and diffraction. *Sci. Rep.* 7 (3602): 2017.
112 Cunningham, R., Zhao, C., Parab, N. et al. (2019). *Science* 363: 849–852.
113 Young, Z.A., Guo, Q., Parab, N.D. et al. (2020). *Addit. Manuf.* 36: 101438.
114 Guo, Q., Zhao, C., Qu, M. et al. (2019). *Addit. Manuf.* 28: 600–609.
115 Bidare, P., Bitharas, I., Ward, R. et al. (2018). *Int. J. Adv. Manuf. Technol.* 99: 543–555.
116 Taheri Andani, M., Dehghani, R., Karamooz-Ravari, M.R. et al. (2018). *Addit. Manuf.* 20: 33–43.
117 Nagel, F., Brömme, L., and Bergmann, J.P. (2019). *J. Laser Appl.* 31: 022005–022008.
118 Esmaeilizadeh, R., Ali, U., Keshavarzkermani, A. et al. (2019). *J. Manuf. Processes* 37: 11–20.
119 Gasper, A.N.D., Szost, B., Wang, X. et al. (2018). *Addit. Manuf.* 24: 446–456.
120 Ye, D., Zhu, K., Fuh, J.Y.H. et al. (2019). *Opt. Laser Technol.* 111: 395–406.
121 Ye, D., Hsi Fuh, J.Y., Zhang, Y. et al. (2018). *ISA Trans.* 81: 96–104.
122 Mutke, C., Geenen, K., Röttger, A., and Theisen, W. (2018). *Mater. Charact.* 145: 337–346.
123 Fedina, T., Sundqvist, J., and Kaplan, A.F.H. (2021). *Mater. Today Commun.* 27: 102241.
124 Matthews, M.J., Guss, G., Khairallah, S.A. et al. (2017). Denudation of metal powder layers in laser powder-bed fusion processes. In: *Additive Manufacturing Handbook, Product Development for the Defense Industry*, 1e (ed. A.B. Badiru, V.V. Valencia, and D. Liu), 677–692. CRC Press.
125 Heiden, M.J., Deibler, L.A., Rodelas, J.M. et al. (2019). *Addit. Manuf.* 25: 84–103.
126 Asgari, H., Baxter, C., Hosseinkhani, K., and Mohammadi, M. (2017). *Mater. Sci. Eng., A* 707: 148–158.
127 Ardila, L.C., Garciandia, F., González-Díaz, J.B. et al. (2014). *Phys. Procedia* 56: 99–107.
128 Hadadzadeh, A., Baxter, C., Amirkhiz, B.S., and Mohammadi, M. (2018). *Addit. Manuf.* 23: 108–120.
129 Lutter-Günther, M., Bröker, M., Mayer, T. et al. (2018). *Proc. CIRP* 74: 33–38.
130 Simonelli, M., Tuck, C., Aboulkhair, N.T. et al. (2015). *Metall. Mater. Trans. A* 46: 3842–3851.
131 Gasper, A.N.D., Hickman, D., Ashcroft, I. et al. (2019). *Powder Technol.* 354: 333–337.

10

Influence of Powder Chemistry on Additive Manufacturing

10.1 Introduction

The characteristics of the **feedstock powder** play a critical role in the metal additive manufacturing (AM) processes and the microstructure and properties of the final parts. A large fraction of the research in the metal AM community has focused on **chemical compositions** that are available in cast and wrought alloys, with the aim of achieving or improving the properties of original alloys. However, most of the commercially available powder alloys were not designed specifically for AM processes. Rather, these powders were designed for a given set of **thermo-mechanical processing (TMP)** routes typically to conventional powder metallurgical manufacturing processes. The nature of layer-by-layer thermal cycling associated with the build cycles in AM leads to a complex thermal history; hence, the use of existing alloys may result in sub-optimal microstructures in the as-printed conditions. Developing new alloy compositions, especially ones that are specifically tailored for AM, is a key challenge to be addressed in the current AM field, as powder chemistry is the first dominant factor that determines the printability of the alloys in AM. Therefore, the scope of this chapter focuses on the influence of powder chemistry on the AM processes and the quality of the additively manufactured parts.

10.2 Alloy Compositions

Despite growing interest and tremendous research efforts, only a limited number of commercial alloys have achieved a relatively developed stage in AM so far, such as stainless steels, pure titanium (Ti) and Ti-6Al-4V alloys, nickel (Ni)-based superalloys, aluminum (Al)-silicon (Si) alloys, and cobalt (Co)-chromium (Cr) alloys [1]. Table 10.1 provides a summary of common alloys currently available or being developed for AM. It is important to note that not all powders are compatible with all specific AM platforms. Also, Table 10.1 is not an exhaustive list as the powder manufacturing industry continues developing new custom-designed alloys. Proper or optimal processing windows vary significantly for different base alloys in a select AM technology. Even for nominally the same base alloy, e.g. Ti–6Al–4V (by weight percent, wt%), the processing parameters reported in the literature differ substantially, despite using the same AM technology [2–4]. The **chemical composition** of the **feedstock powder** affects the microstructure formation by changing **solidification thermodynamics** and/or **kinetics**, or by solid-state **phase transformations**, partly due to different compositions having varied thermal properties and different absorption of, the reflectance of, and/or reactions with the AM energy source. For example, parts made from commercial Inconel 718 (a Ni–Fe alloy) powder via electron beam powder bed fusion (PBF-EB) underwent undesirable solid **phase transformations** during the building process, as the optimal powder bed temperature for crack-free deposition is coincidentally the same temperature at which metallurgically undesirable phase transformations occur [5, 6].

Modifying alloy compositions to take advantage of AM **solidification** variables can improve alloy properties. The interaction of alloying elements is recognized in promoting desirable microstructural phases and solid-solution effects that will improve printability or enhance properties. Montero-Sistiaga et al. reported that the addition of 3 and 4 wt% of silicon to the aluminum alloy (AA) AA7075 base powder by **mechanical mixing** improved its processability by **selective laser melting (SLM)** [7]. The modification of the powder chemistry not only increased the density of the parts manufactured by SLM but also reduced microcracks due to **grain refinement**. Martin et al. added zirconium (Zr) nanoparticles into AA7075 and AA6061 to control solidification during the laser beam

Metallic Powders for Additive Manufacturing: Science and Applications, First Edition.
Enrique J. Lavernia, Kaka Ma, Julie M. Schoenung, James F. Shackelford, and Baolong Zheng.

Table 10.1 Common alloys currently available for AM, including commercial and R&D.

Fe-based	Ni-based	Cu-based	Ti-based	Al-based	Co-based	Refractory and misc.
SS 316L SS 304L SS 347 SS 420 SS 431 SS 17-4PH SS 15-5PH H13 LS 4140 LS 4340 Invar 36 JBK-75	IN 625 IN 718 IN738 Haynes 214 Haynes 230 Haynes 282 Hastelloy-X C-276 CM247LC René 80 Monel K-500 Waspaloy	Pure Cu GlidCop GRCop42 GRCop84 CuSn10 C-18150 C-18200 CU110	Ti6Al4V (different grades) γ-TiAl Ti 4822 Ti 6242	AlSi10Mg A205 F357 AA 6061 AA 4047 AA 7050 AA 2024 AA 2319	CoCr CoCrMo CoCrW Stellite® 6 Stellite® 21 Haynes 188	W W-25Re Mo Mo-41Re Mo-47.5Re Ta C-103 Noble metals: Au, Ag, Pt Bimetallic: GRCop-84/IN625, C18150/IN625
Steel-MMC NASA HR-1	Ni-MMC Rene N5 CMSX-4		β-Ti alloys Ti-Nb	Al-MMC Scalmalloy (Al-Mg-Sc) Al-Ce-X	SB-CoNi-10	Multi-principal element alloys

Note: this table is not a completely inclusive list. Acronyms: SS – stainless steel; PH – precipitation hardenable; LS – low alloy steels; MMC – metal matrix composites; IN – Inconel.
Source: Data were obtained from [1, 11].

powder bed fusion (PBF-LB) process for better processability [8]. Crack-free and fully dense ($99.0 \pm 0.1\%$) parts were obtained in this study for the alloy composition containing 1.5 wt% Zr. The AA7075-1.5Zr part exhibited fine-grained, equiaxed microstructures that can accommodate strains more easily due to the thermally induced shrinkage. Powder additives considerably influence the thermo-kinetics of the **melt pool**. Wimmer et al. found that the cooling rate of the molten track decreased with adding the aluminum alloy AlSi10Mg to stainless steel 316L powder in PBF-LB [9]. They also reported that the microstructure of the parts contained an increased formation of the body-centered cubic (BCC) phase as the fraction of AlSi10Mg increased. In a more recent study, Wimmer et al. also reported that the addition of AlSi10Mg powder (up to 20 wt%) to the stainless steel 316L powder led to a 48% lower variation of the melt pool depth compared to using 316L powder without additives in PBF-LB [10].

Changes in powder chemistry also result in varying optical and thermal properties that are of importance for the metal AM processes. Specifically, in laser-based AM processes, since the powder particles are molten or sintered by the laser, the absorption and the reflection of the laser at a given wavelength, as well as the surface topography of powder, significantly influence the selection of process parameters and the stability of the melt pool. Powder chemical compositions that have a higher reflectance and lower absorption (optical properties) require higher energy input in AM processes. For instance, comparing Ti alloys and Al alloys, Al alloys require a higher **laser power density** input than Ti alloys in PBF-LB, despite that the melting temperature for Al (~660 °C) is distinctively lower than that of Ti (1668 °C) [12]. This is because Al has a higher reflectance of 0.9, absorption coefficient on the order of 10^{10}, and more than 10 times higher thermal conductivity ($200\ \text{W m K}^{-1}$) than Ti (reflectance of 0.5–0.6, absorption coefficient on the order of 10^5, and thermal conductivity $17\ \text{W m K}^{-1}$). Those optical data correspond to a wavelength of 1064 nm (e.g. Nd:YAG laser). Figure 10.1 displays the minimum laser power needed to melt different metals using a pulsed Nd:YAG laser and the absorptivity vs. wavelength of different metals for selected lasers.

Variations in the powder composition that result in different thermal properties will also influence the **melt pool** and the **heat transfer** to the surrounding powder particles in a powder bed and subsequently affect cooling rates, residual stress, distortion, and crack formation [12]. Even for the same nominal base alloy, the powder chemistry varies depending on production processes and powder vendors [12, 14]. Taking Ti–6Al–4V as an example, different grades are defined based on the concentrations of the alloying elements. ASTM B265 defines the limits of the concentrations of the elements in Ti–6Al–4V Grade 5 are Al (5.5–6.5 wt%), O (Max. 0.2 wt%), N (Max. 0.05 wt%), and H (Max. 125 ppm) [15]. Spherical powder of Ti–6Al–4V can be produced by gas atomization (GA), plasma atomization (PA), or plasma rotating electrode process [16]. In the work by Meier et al., four batches of powder provided by three

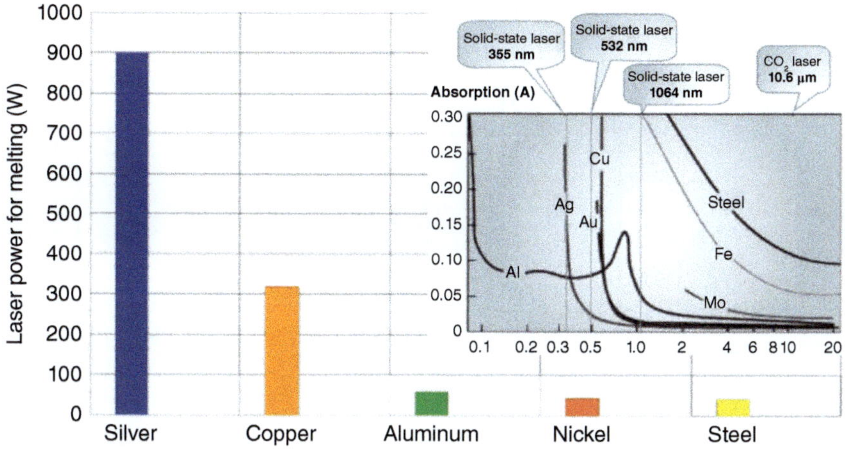

Figure 10.1 Minimum laser power needed to melt different metals with a pulsed Nd:YAG laser and absorptivity versus wavelength of different metals for selected lasers [13].

different powder producers were investigated for PBF-LB [12]. While they found that the chemical compositions of three batches of the powder (nominally produced by GA and PA, respectively) were within the limits according to ASTM B265, especially the crucial oxygen, nitrogen, and hydrogen contents. Two out of the three batches reached oxygen levels close to the value for Grade 23 (0.14 wt% O), whereas the fourth batch of powder has 0.23 wt% O, which exceeded the oxygen concentration limit of Grade 5 by 0.03 wt%. In addition, two batches of the PA powder showed increasing Al levels from the powder surface to the interior with corresponding decreasing levels of Ti and V. The concentrations of O, H, and N strongly affect mechanical properties. Meier et al. reported that the tensile properties of the AM parts were significantly affected by the **chemical composition** of the powder, particularly dissolved oxygen, and polluting satellites. The presence of TiO_2 and Al_2O_3 oxide layer on the powder surface affected energy intake [12].

In addition, powder surface chemistry can affect the spreading behavior, **flowability**, and the evolution of the layer characteristics during AM [17, 18]. The powder surface chemical state evolves due to interaction forces between powder particles, and the cohesive forces between the particles may reduce the powder packing density, smoothness, and layer uniformity [19, 20]. Wu et al. investigated the effect of surface moisture level (adsorbed water molecules) of Ti–6Al–4V powder on **spreadability** and **flowability** via discrete element numerical simulations combined with powder characterization experiments [18]. Three types of powder with different surface moist levels were studied: as-received powder without any treatment, moistened powder with surface H_2O molecules, and exsiccated powder with grown nano-scaled oxides film. Their results, summarized in Figure 10.2, illustrate how the surface H_2O molecules can facilitate the formation of the meniscus or liquid capillary neck between two adhered particles, and thus promote the capillary effect between particles. Consequently, the resistance during particle motion was increased, leading to the decrease of powder flowability and spreadability, and the degradation of powder layer quality. They concluded that the exsiccated powder, which grew a surface film, exhibited deteriorated flowability but preferable spreadability compared to the as-received powder. When the surface roughness of the exsiccated powder caused by the growth of the surface oxide film increased, the powder layer showed enlarged pore volume, uneven pore distribution, and finally a lower packing fraction than that of the as-received powder. Therefore, long-term pre-treatment that yields powder with a thick surface oxide film is not desirable for powder bed-based AM, and the chemical reactions on powder that cause large film roughness should be prevented in powder storage and treatment processes. Likewise, Cordova et al. measured the spreadability of pre-treated (air drying and vacuum drying) and moisturized powders for four alloys used in PBF-LB, including Inconel 718, Ti–6Al–4V, AlSi10Mg, and Scalmalloy [17]. They reported that AlSi10Mg was the most susceptible alloy to moisture and oxygen pick-up, significantly decreasing the spreadability and relative density on the build platform. In contrast, the moisture and oxygen content of Inconel 718, Ti–6Al–4V, and Scalmalloy remained at the same level, lower than that of AlSi10Mg, even after moisturizing, which was attributed to the **powder morphology** – Inconel 718 and Ti–6Al–4V powders in this study had both a smoother surface and a spherical shape, and Scalmalloy presented a better corrosion behavior and a more regular powder shape compared to AlSi10Mg.

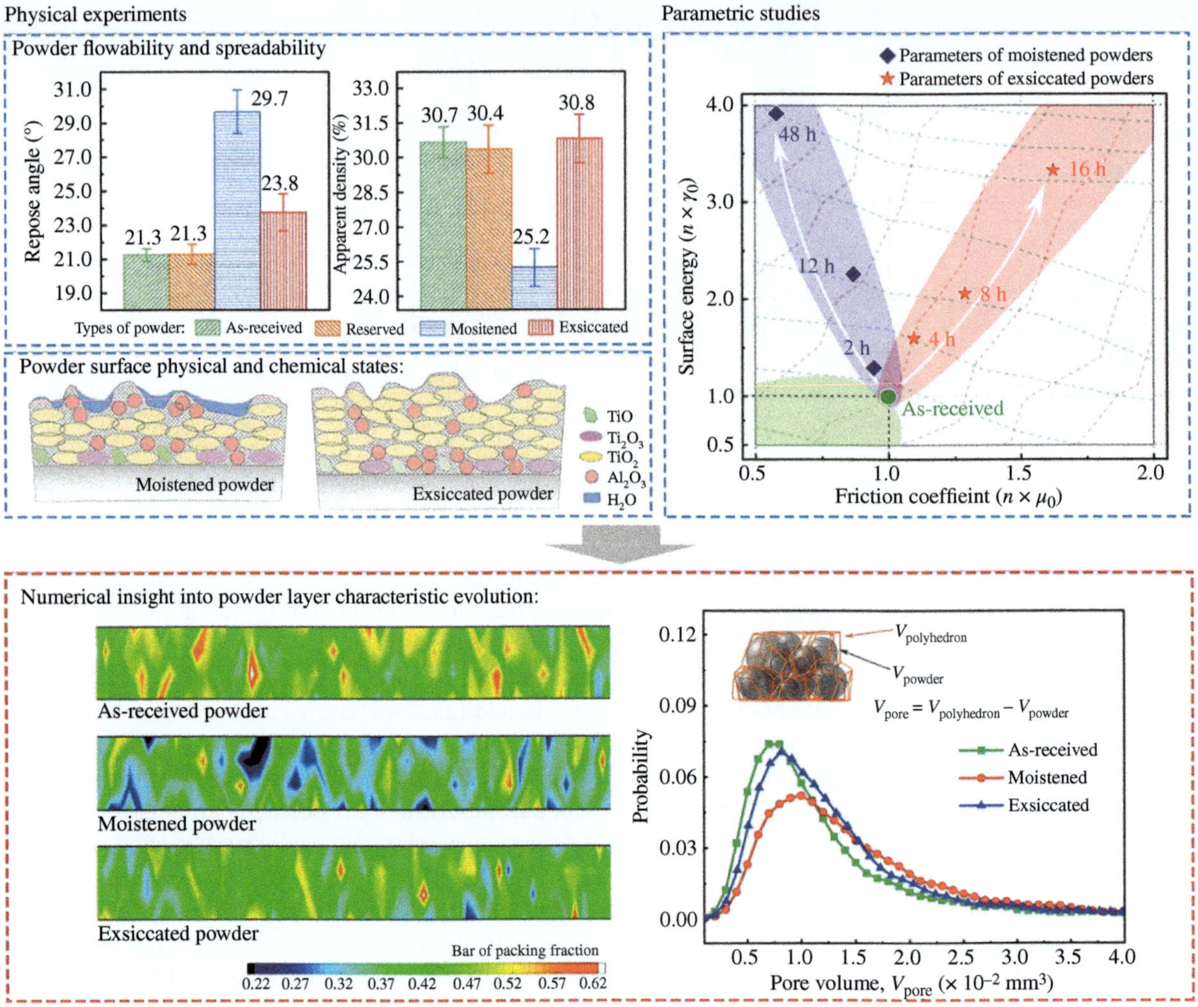

Figure 10.2 Effect of powder surface moisture on powder flowability, spreadability, and powder layer characteristics [18].

These aforementioned studies suggest there are opportunities for the design of alloy compositions that are tailored to the specific thermo-chemical, thermo-mechanical, and thermo-kinetic aspects of the AM process. As trial-and-error methods are not an efficient approach for the design and/or selection of alloy compositions for AM, researchers have been highly motivated to develop an **integrated computational materials engineering (ICME)** approach for the design of alloy powders for AM, particularly for control of **solidification** and/or solid-state transformations that are very likely sensitive to AM thermal characteristics. To design alloys that can achieve targeted solidification microstructure and solid-state transformations after AM, a typical methodology couples solidification interface response function theories with multicomponent thermodynamic calculations via CALPHADs (Calculation of Phase Diagrams). Pollock et al. conducted a case study to design Ni-base superalloys that can be printed with predominantly directional (columnar) grain growth with a minimal tendency for stray grains and targeted fractions of γ' phase within a build manufactured by PBF-EB [1]. Through the optimization calculations by writing an advanced programming interface with ThermoCalc® software [21], they designed two Ni alloys (i) Ni6666: Ni-6Cr-6Al-6Ta (by wt%) and (ii) Ni106: Ni-10Cr-6Al (by wt%) specifically for the PBF-EB process and the building geometry in their work. Frederick et al. claimed that subtle variations in geometry with reference to processing conditions close to the boundaries might usher in **equiaxed grain** growth [22]. In another case study, Pollock et al. utilized a suite of computational tools, including density functional theory (DFT), cluster-assisted statistical mechanics (CASM), phase field dislocation dynamics, and CALPHAD calculations, combined with experimental work, to explore a large multicomponent compositional space for AM of Co–Ni superalloys via PBF-LB and PBF-EB [1, 23]. One of the key findings was the identification of a final composition giving rise to a desirable two-phase microstructure, designated SB-CoNi-10, having a nominal composition of Co-36.5Ni-13.2Al-6.0Cr-3.5Ta-1.0W (at%) [1, 24]. Wang et al. applied the CALPHAD-based ICME framework to design the composition of copper-bearing high strength low alloy steels with the aim of achieving the optimal amount of strengthening precipitates such as Ni_3Ti and Fe_2SiTi [25]. To design printable alloys for various AM processes, combinations of predictive tools are important given a high dimensional alloy design space and the complex physics in metal AM.

10.3 Impurities and Segregation

Impurities are typically present in AM metallic components resulting from either intentional addition or unintentional contamination from powder production processes or the AM process. Elemental **segregation** plays a vital role in the solidification structure and mechanical behavior of the AM components. Here several select studies are discussed to address the influence of impurities and segregation.

Song's team used a multi-scale model to quantitatively predict the segregation features of Nb in Ni-based superalloys made by laser beam-**directed energy deposition** (DED-LB) [26]. Their simulation results suggested that the Nb concentration was dependent on the molten pool site-specific solidification conditions: the average Nb concentration in the interdendritic region decreased gradually with an increasing cooling rate from the bottom to the top of the molten pool. The Nb concentration in the γ-Ni matrix was positively correlated to the temperature gradient, solidification rate, and cooling rate during non-equilibrium solidification. It was also observed that the long-chain Nb distribution in the interdendritic region would further enrich into droplets as the solidification proceeded, due to the surface tension of the melt. The droplet-like Nb segregation in the interdendritic region finally transformed to (γ + Laves) eutectic microstructure in the solid state. It was suggested that the Nb segregation and the volume fraction of the brittle Laves phase could be reduced by increasing the cooling rate and changing the solidification conditions of the **melt pool** via laser modulation and water cooling. Bermingham and Dargusch explored the formation of segregation defects, known as β-flecks, in the β-Ti alloy (Ti-3Al-8V-6Cr-4Mo-4Zr) and proposed that AM could be a promising method to produce high-quality metastable β-titanium alloys free of segregation defects [27]. β-flecks form when β-stabilizing solutes (e.g. Cr) segregate, leading to localized phase instabilities, which are detrimental to the aging response and mechanical properties of the alloy. In their work, it was observed that β-flecks initially formed during the deposition of the top layer; however, each new subsequent layer and thermal cycle promoted solid-state diffusion of alloy elements, which effectively healed the segregation defect.

On the other hand, segregation, when added deliberately, can also be utilized to alter the residual stress states at the interdendritic and grain boundary regions, and consequently reduce **hot tearing**. In a study of SLM of an Al_xCoCrFeNi **high entropy alloy (HEA)** by Sun et al., it was reported that the hot crack density could be dramatically decreased in a certain composition range [28]. The mechanism was attributed to Al segregation to the interdendritic regions and grain boundaries during the solidification of the specific alloy composition. First, Al was ejected from the primary dendritic face-centered cubic (FCC) phase and then segregated into the interdendritic regions, followed by spinodal decomposition in these Al-enriched regions to finally form the ordered B2 NiAl and disordered BCC Cr phases. The B2/BCC phases have a higher molar volume and lower homologous temperatures, which accommodate the inherent residual strain. The B2/BCC grains, together, serve as an effective mechanical work absorber and reduce the overall tensile residual contents. Based on these results, it was claimed that grain boundary segregation could provide a new pathway to systematically counteract the hot tearing problem in AM of metals and alloys.

Wu et al. investigated the effects of Ti **impurity** on the crystal structure and mechanical properties of additive manufactured U-14 at% Nb alloy [29]. The key finding from this study is that the formation of γ'-tetragonal phase was promoted with an increasing Nb equivalent (Nb_{eq}) composition that represents the Ti **impurity** level, at the expense of α''-monoclinic phase. This structural change resulted in a yield strength double that of water quenched α'' and a strain-induced transformation to α'' with superelastic strains to 4.5%. In another study of SLM of pure copper, it was found that the addition of carbon nanoparticles improved the optical absorption and flowability of gas-atomized Cu powder [30]. The addition of the carbon impurities into the Cu powder via **mechanical mixing** not only reduced the required laser energy density to obtain dense parts but also promoted *in-situ* de-oxidization, and thus, purified the Cu part during the SLM process as compared to the virgin powder feedstock. However, the fabricated parts exhibited modest tensile and electrical properties. The microstructural analysis demonstrated that the carbon **impurity** in the powder resulted in the presence of a segregated phase, which was rich in phosphorus, oxygen, and carbon, and

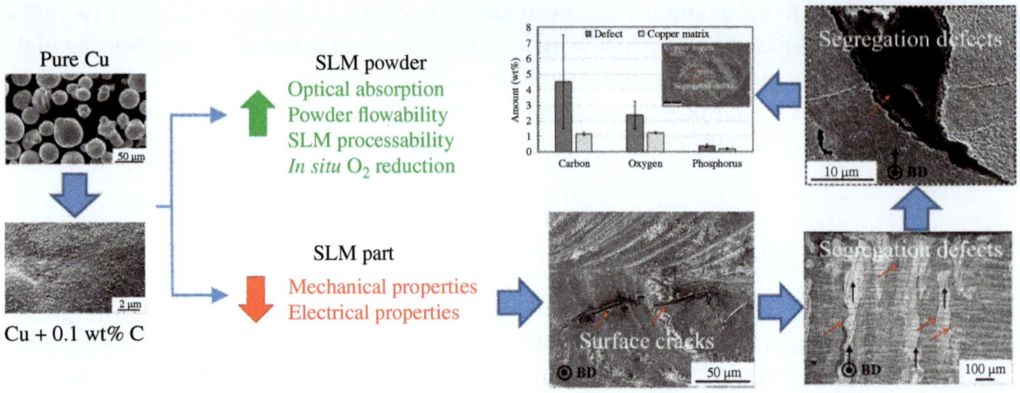

Figure 10.3 Effect of carbon on segregation defects and the crack formation in Cu parts made by SLM [30].

changed the orientation of the segregation phase, as shown in Figure 10.3. The Cu + 0.1 wt% C part showed a low electrical conductivity, mainly due to the presence of a phosphorus **impurity**, forming a solid solution with copper. The lower tensile properties were linked to the segregation of carbon, oxygen, and phosphorus-based impurities, as well as their orientation along the grain boundaries of the grains with epitaxial growth.

10.4 High Entropy Alloys (Multi-Principal Element Alloys)

HEAs, also called **multi-principal element alloys (MPEAs)**, represent complicated chemistries that generally contain three to five principal metallic elements in equiatomic ratios, in contrast to traditional alloys in which only one or two elements are dominant. Increasing the number of the principal elements tends to increase the entropy of the alloy system, which was the origin of the name HEA when it was first reported independently by Yeh et al. [31] and Cantor et al. [32] in 2004. Due to the controversy on the truth of high mixing entropy for different chemical compositions, nowadays it is recommended to use the term MPEAs. MPEAs are attractive for various industrial applications, particularly high-temperature applications, such as turbine blades, bond coatings in thermal barrier coatings, and radiation-resistant materials, because of their superior mechanical performance and high thermal stability. MPEAs are characteristic of their microstructure consisting of simple solid solutions, commonly FCC or BCC structure, with severe lattice distortion [33]. Several studies reported that MPEAs were likely to exhibit composition segregation or to form intermetallic compounds in casting and forging due to the presence of multiple principal elements and severe lattice distortion [33–36]. In contrast, the large **undercooling** and the ultrahigh cooling rate of AM technologies can mitigate the chemical composition segregation issue by accelerating the movement of the solid–liquid interface [37]. In the layer-by-layer AM process, the diffusion of the multiple elements in MPEAs is suppressed by solidification to form a solid solution. Increasing cooling rates also lead to the change in microstructure from dendrites to chemically homogeneous cellular structures [37–39]. Furthermore, high nucleation rates are induced by the large undercooling and fast-moving solid–liquid interface in the AM process, resulting in fine grain structures [40, 41]. Those phenomena and advantages are the main driving force for AM of MPEAs, in addition to the general capability of making complex geometries and near-net-shaped components.

Three AM processes have been primarily used to make MPEA components, namely directed laser deposition (DLD), SLM, and selective electron beam melting (SEBM) [42]. The latter two belong to the PBF category while DLD is in the DED category. A review paper by Ostovari Moghaddam et al. provided an overview of the mechanical properties achieved in various MPEA parts fabricated by AM, as shown in Figure 10.4 [42]. As different MPEAs have distinct and complex chemical compositions and subsequently different physical characteristics (interaction with or responses to laser beam or e-beam), there is not yet a clear correlation between the feedstock chemistry and the processing parameters for optimizing the microstructure and properties of MPEAs in AM. It is generally agreed that combinations and integration of physics-based computational modeling, carefully designed experiments, and data-driving machine learning approaches are in demand to identify suitable process windows for specific MPEAs. Most current literature investigated how various processing parameters affected the defect formation and microstructure in MPEAs, rather

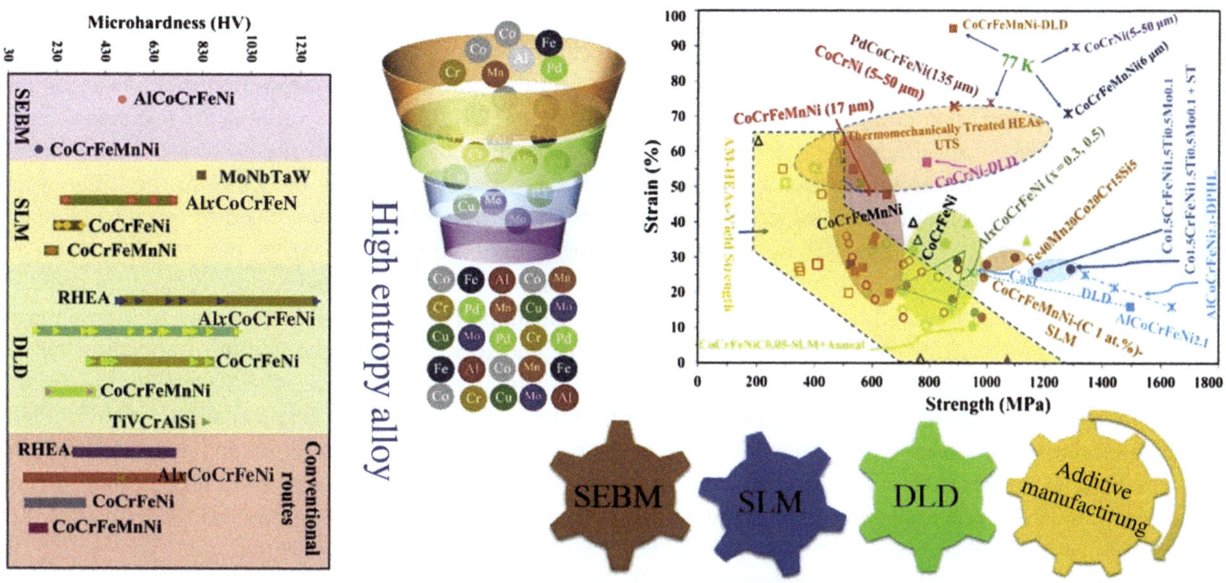

Figure 10.4 Overview of the mechanical properties achieved in HEAs made by AM [42].

than how the chemistry of MPEAs influences the selection of the process. For instance, the work by Dada et al. indicated that a laser power of 1200–1600 W combined with a scanning speed of 8–12 mm s^{-1} gave rise to reduced processing defects, compared to other parameters, when AlTiCrFeCoNi and AlCoCrFeNiCu MPEAs were fabricated by DED-LB [43]. Tong et al. uncovered that the solidification microstructure of CoCrNiFeMn MPEAs via DED-LB was dendritic **columnar grains** in the build, which became coarser with increasing laser power [44]. The horizontal sections of the **columnar grains** were equiaxed with increasing grain size as the laser power was increased.

The chemistry of the MPEAs first determines the printability in AM. Not all MPEAs are feasible for AM. For instance, intrinsic properties such as high laser reflectivity, high concentration of volatile elements, high thermal conductivity, and/or high melting points, make it challenging or even impossible to find optimal processing parameters to produce fully dense or crack-free MPEAs [45]. The complexity in the chemistry of MPEAs makes the AM process optimization complicated and further exerts a challenge on the tailoring of microstructure and properties for these alloys. The earliest MPEA investigated by a PBF method was FeCoCrNi, which confirmed the feasibility of AM of MPEAs [46]. Brif et al. stated that an FeCoCrNi MPEA fabricated by PBF-LB exhibited only FCC solid solution, which was consistent with the microstructure of the FeCoCrNi MPEA made by induction and arc melting [46]. Despite that few studies have investigated how the **chemical composition** of the **feedstock powder** of MPEAs directly changes the selection of AM processing parameters, it has been reported that variations in powder chemistry affect phase formation and phase transition/evolution in AM of MPEAs. Taking Al-containing MPEAs as an example, as-built AlFeCoCrNi via PBF-LB processing exhibited a disordered BCC (A_2) + ordered BCC (B_2) structure [47], as opposed to the simple FCC solid solution in FeCoCrNi MPEA (containing no Al) reported by Brif et al. [46]. Niu et al. reported that the B_2 phase was first formed at the **melt pool** boundary where a higher cooling rate occurred. The amount of the A_2 phase decreased and the amount of the B_2 phase increased with increasing laser volumetric energy density. Sun et al. suggested that for Al_xCoCrFeNi produced with PBF-LB, the phase transformed from FCC to BCC/B_2 + FCC and then to BCC/B_2 as the Al content increased from 0% to 1%, and all BCC/B_2 phases emerged at the interdendritic and grain boundaries [28]. Likewise, a **phase transformation** from BCC to FCC was observed in AlCrCuFeNi_x made by PBF-LB. As the Ni ratio (x) increased from 1.0 to 3.5, a lamellar FCC phase formed. while the relative fraction of basketweave and B_2 matrix decreased and changed to a spherical morphology [48]. All these studies suggest that when the chemical composition of the **feedstock powder** changes, the AM processing parameters need to be adjusted. Otherwise, the final microstructure of the MPEA builds will be changed. DED has been also applied to produce MPEAs and is particularly popular to fabricate MPEA coatings. So far, successful deposition of MPEAs via DED includes 3D transition metal MPEA, Al-added MPEA, and refractory MPEA coatings to give improved wear resistance, corrosion resistance, and/or strength.

The **feedstock powders** for AM of MPEAs can be prepared by GA of a MPEA melt or mechanical alloying that involves mixing pure metal or different alloy composition powders via high-energy ball milling, in which severe plastic deformation, repeated fracture, and rewelding occur until a uniform composition is achieved in the powder [49]. Compared to other alloys, it is more difficult to produce MPEA powders suitable for AM technology. As discussed in Chapter 8, different AM technologies have stringent requirements on the powder characteristics, including particle size, flowability, oxygen content, and powder surface roughness. In addition, the differences in solubility, mixing entropy, and melting point among the multiple principal elements need to be considered to avoid liquid phase or intermetallic compound formation during the production of MEPA powders [50, 51]. For refractory MPEAs, their poor machinability and high melting temperatures make it even more difficult in powder production [52]. Atomization methods are promising in the fabrication of MPEA powders that have high sphericity (good flowability and packing density), uniform composition, and low oxygen content. But atomization has a relatively high cost and low efficiency. In contrast, mechanical alloying may produce MPEA powders nominally with any ratios of metallic elements and high solid solubility, but the flowability and/or spreadability of ball-milled powders are relatively poor due to the irregular shape of the particles. Plus, the ball milling media tend to bring in contaminations [53]. Hence, producing pre-alloyed powders for MPEAs with high solid solubility, homogeneous composition, and high sphericity remains the major challenge in the development of AM MPEAs. Second, the complex chemistry, i.e. three to five principal elements, of MPEAs complicates the selection and optimization of AM process parameters, which will further pose a challenge to microstructure control and tailoring of properties and performance, sometimes even preventing successful manufacturing of any high-quality MPEA parts. Therefore, significant research efforts are in demand to continue the progress of AM of MPEAs.

10.5 Metal Matrix Composites

In addition to the **MPEAs**, adding ceramics to a metal or alloy matrix to form composites represents another level of complex chemistry in AM. **Metal matrix composites (MMCs)** have had a well-established history in metallurgy since the 1960s, for which several articles and books have provided comprehensive reviews [54–58]. MMCs generally take advantage of the ductility and toughness of the metal matrix and the high hardness and high modulus of the reinforcement to achieve a combination of properties that make MMCs favorable materials in severe working conditions, such as high mechanical loads, high-wear/friction, and high temperatures, in the advanced aerospace and automobile industries [59]. This section focuses on the effect of the reinforcement chemical composition in the context of producing MMCs via AM. AM technologies like PBF and DED have been successfully utilized to produce fully dense Al-, Fe- Ti-, TiAl-, and Ni-based MMCs [60]. Common ceramic reinforcements in MMCs include Al_2O_3, B_4C, BeO, SiC, TiB_2, WC, TaC, NbC, and TiC, with SiC being most commonly used commercially, followed by Al_2O_3 and TiC by a significant margin [60]. While it is well known that the addition of reinforcements in MMCs can refine the grain size of the matrix, the understanding of how the ceramic reinforcements influence the AM processes, particularly the **melt pool** dynamics, is relatively limited.

As discussed in Section 10.2, the absorption ability of materials and reflectivity are important aspects affecting the laser-powder interaction. The absorptivity also significantly depends on the wavelengths of the energy source and temperature [13]. Most metals exhibit low absorption of the laser when the wavelength is larger than 1 μm (Figure 10.1). Al has a high reflectivity of laser energy while Si has a very high absorptivity (~70%). Due to the limited solubility of Si in Al, pre-alloyed Al–Si powders typically contain some pure Si particles, which readily absorb the laser energy and then dissipate heat to the other alloying elements (e.g. Al) in the powders. This may explain the successful AM of Al–Si-based alloys [61]. Likewise, the addition of ceramic reinforcements may reduce the laser reflectivity of the metal matrix, which is essential for laser-based AM processes to improve energy efficiency. Gao et al. reported that a low laser power (~100 W) could enable good formability of a TiN-reinforced AlSi10Mg nanocomposite via SLM because the laser reflectivity of the composite powder was reduced from 62% to 25% due to the addition and uniform distribution of TiN nanoparticles [62]. It was also predicted that the temperature of unreinforced Al–12Si alloy can reach approximately 1500 °C, while that of SiC reinforced Al composite powders may increase further since the laser absorptivity of SiC is about 78%, in contrast to that of Al being as low as 7% [63]. Gu et al. investigated the effect of SiC and TiB_2 reinforcement in Al alloy matrix composites by calculating laser absorptivity percentages along the laser beam path during SLM [64]. They found that the interaction between the laser beam and MMC powder particles was enhanced by the reinforcement, as illustrated by the increased laser ray track spots on the powder particle surfaces. With the laser beam moving, a randomly packed SiC-reinforced AlSi10Mg powder bed and a TiB_2-reinforced AlSi10Mg powder bed exhibited an absorptivity that fluctuated from 0.37 to 0.59 and 0.34 to 0.49, respectively, significantly higher than that of AlSi10Mg powder (ranged from 0.19 to 0.32). Moreover, the results indicated that SiC reinforcement was more effective in enhancing laser absorption behaviors than TiB_2 [64]. On the other hand, the incorporation of ceramic reinforcement particles may increase the viscosity of the melt and thus, decrease the flowability, subsequently resulting in porosity in the parts [60]. In addition, ceramic reinforcing particles in the MMC powders can induce local modifications in the **melt pool**, depending on their thermodynamic and physical properties [65]. The surface tension properties of the metal matrix, which affect the wettability of the ceramic reinforcement, also need to be considered as a key factor that influences the shape and stability of the melt pool due to Marangoni convection [66]. A detailed investigation into the influence of the addition of reinforcing particles on melt pool behavior requires the use of specialized observation techniques, such as thermal imaging, high-speed camera, and synchrotron X-ray diffraction for the process zone [67–71].

Similar to MPEAs, the production of MMC **feedstock powders** for AM primarily relies on two methods – atomization and **mechanical milling**, as shown in Figure 10.5. Due to the distinct characteristics between ceramics and metals resulting from totally different atomic bonding types, many challenges are caused by the poor compatibility with the reinforcement-matrix interface, contamination of the powder particle surfaces, irregular shape of ball-milled powders, and so forth. MMC powders produced by **mechanical milling** retard the full advantage of the reinforcements and thus restrict the property improvement [60]. *In-situ* chemical reactions between metal matrix and reinforcement materials to produce reinforced MMC powders may effectively solve the issue of poor compatibility and achieve the comprehensive performance of the composites [57, 72]. *In-situ* reaction method is further discussed in Section 10.6. Other difficulties in AM of MMCs include the spreading of the heterogeneous powder, distributing small ceramic particles that also typically have lower densities than metals, and forming a strong bond between the metal and the ceramic [73]. The following influencing factors need to be considered when choosing a ceramic reinforcement for a specific metal matrix: melt pool stability, resulting microstructure, temperature profile, and affinity for certain defects.

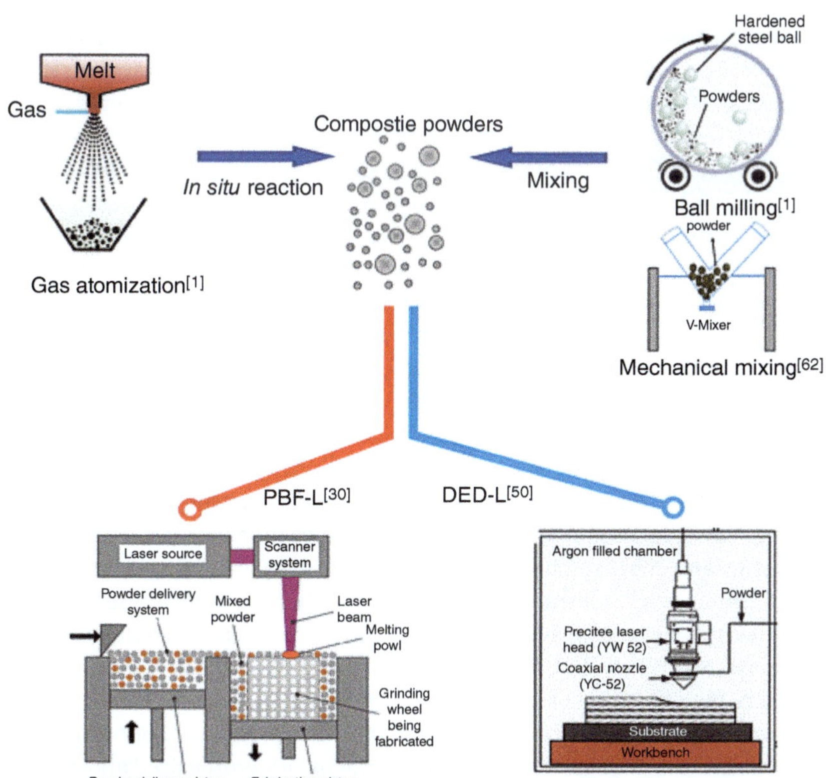

Figure 10.5 Illustration of production of MMC powders that are then used in AM to make bulk samples [74].

10.6 *In-Situ* Alloying (In-Process Alloying)

AM of MPEAs and MMCs, as discussed in Sections 10.4 and 10.5, can use pre-alloyed powders as feedstock in PBF and DED. However, the composition range for pre-alloyed powder is narrow. The limited availability and high costs of pre-alloyed powders make AM of new alloys inflexible and expensive [75]. Alternatively, ***in-situ* alloying**, also referred to as **in-process alloying**, has been explored to enable the fabrication of complex material compositions during certain manufacturing stages, through reactions between powders of different chemical compositions. This strategy can enhance the flexibility of the AM methods by benefiting from the high laser local energy, and thus, ***in-situ* alloying** is a potential way of moving towards an on-demand alloy design framework in response to the growing needs of the industry. A general fundamental mechanism is that the high cooling rates and thermal cycling associated with the layer-by-layer nature of AM enable far-from-equilibrium conditions that subsequently enable chemical compositions that are not achievable via traditional alloying methods. It is noteworthy that **in-process alloying** is usually possible in DED platforms by using different powder hoppers, while PBF technologies rely more on pre-alloyed or pre-blended mixtures of powders to enable **in-process alloying**. Based on the review article on ***in-situ* alloying** in laser-based AM processes by Mosallanejad et al. [75], ***in-situ* alloying** procedures started with the DED process in the early 2000s while PBF-LB activities on this topic started about 10 years later. Pioneering work was done via DED rather than PBF due to the larger laser spot size and the higher laser power in DED, which allowed for improved mixing of the alloying elements. While various alloys have been explored for ***in-situ* alloying** in AM, the top two most-studied alloys are Ti-based and Ni-Ti alloys, mainly related to their excellent combination of specific mechanical properties and outstanding corrosion resistance, which makes them very suitable for biomedical and aerospace applications.

As *in-situ* alloying in AM involves blending and reaction between two or more pure elemental or commercially available alloy powders, the physical properties, such as melting point, specific heat capacity, enthalpy of fusion, and absorptivity, of each powder need to be carefully considered. The enthalpy of mixing is particularly of interest, as it influences the melt pool temperature and subsequently the synthesis of new alloy compositions via the *in-situ* alloying approach. For instance, alloys that have exothermic mixing reactions with negative heat of mixing (e.g. Fe–Cr–Ni [76], Ni–Al [77]), can lead to a temperature gradient in the melt pool, which subsequently affects the mass transfer by inducing the **Marangoni effect** (surface tension driven fluid flow). In contrast, Liu et al. claimed that the temperature increase resulting from the enthalpy of mixing would not affect the melt pool temperature; rather, it was only limited to the reaction zone in the adiabatic condition [77]. Despite the minor debate on the effect on the melt pool, several papers have unanimously suggested that the required laser energy could be lowered, and thus, the energy efficiency could be increased, attributed to the exothermic reaction of *in-situ* alloying in both PBF and DED techniques [78–80]. Specifically, in a study of PBF-LB processing of a Ti–B mixture, thermal models were developed to treat both the laser beam energy and the exothermic reaction energy between Ti and B powders as the energy source [78]. The modeling results, also verified by experiments, revealed that the exothermic reaction energy was an important energy source for printing the TiB parts and reduced the energy input needed from the laser beam. These studies support that the powder chemistry plays a vital role in *in-situ* alloying as the chemical compositions of the **feedstock powder** determine the alloying reaction energy, and hence, could be utilized to facilitate the AM process. On the other hand, powders with significantly different melting points due to different chemistry are likely to experience unmelted particles if the particles of the element with higher melting temperature are too large [81, 82].

Another perspective of feedstock powder chemical composition effect for *in-situ* alloying lies in mitigating **texture** in the microstructure. It is well known that the intrinsic characteristics of AM techniques, i.e. rapid cooling-heating cycles during the layer-by-layer building and directional **thermal gradient**, lead to a severely textured microstructure that may not be favorable in numerous applications. Some studies preheated the substrates to reduce the texture. An alternative solution to counter the **thermal gradient** effect is adding alloying elements (i.e. changing chemistry), which can act as either nucleating agents or columnar-equiaxed-transition promoters to alter the solidification mode. For example, Tedman-Jones et al. developed a theory based on phase diagrams to predict how some alloying elements, namely, W, Nb, and Mo, affect **grain refinement** in Ti alloys [83]. They used phase diagrams, as they are indicative of the constituents' atomic properties, to predict the efficiency of the alloying element as a nucleant. While many researchers aim to discover the process-properties correlations for *in-situ* alloying in AM and dedicated efforts to the effects of the process parameters on the microstructure and transformation behavior of *in-situ* synthesized alloys, the role of the initial chemistry of the powders, no matter if it is pre-mixed or in-process blended during the reaction, merits more in-depth investigation to develop a broader range of printable alloys.

10.7 Summary

The chemical composition of the feedstock powder is the first factor that determines the printability using various AM technologies and the quality of the final builds, before physical characteristics, such as particle size distribution, apparent density, and morphology, are considered. This is because the thermophysical properties, including melting point, specific heat capacity, enthalpy of fusion, and reflectivity, are dependent on the chemical composition of the powder. Alloy powders consisting of elements with vastly different laser absorption properties tend to have varying local energy absorption, which is further influenced by chemical macro-segregation in the powders. Consequently, the melting and solidification mode during the AM process is affected. High-quality powders, especially with low levels of impurities, are typically high cost, a significant challenge of AM, which also poses a major barrier to the early stages of alloy design for AM. This motivates the ICME approach for the design of alloy powders for AM. Promising pathways for alloy design for AM include: (i) tailoring of powder surfaces to achieve desired microstructure in the complex melt pool environment, (ii) the design of alloys that possess solidification characteristics and solid-state transformations that may be insensitive to AM thermal signatures, thereby reducing the uncertainty brought about by geometrical variations and processing effects, and (iii) fully considering the potential effects of the powder chemistry on the selection and optimization of AM process parameters. Research on AM of more complex chemistries, such as MPEA and MMCs, is in progress and calls for more careful characterization and control of feedstock powders. *In-situ* alloying during AM processes shows immense potential in developing new alloys from small quantities to large amounts.

Nomenclature

BCC	body-centered cubic
CALPHAD	calculation of phase diagrams
CASM	cluster-assisted statistical mechanics
DED	directed energy deposition
DED-LB	laser beam-directed energy deposition
DFT	density functional theory
DLD	directed laser deposition
FCC	face-centered cubic
GA	gas atomization
HEAs	high entropy alloys
ICME	integrated computational materials engineering
MMCs	metal matrix composites
MPEAs	multi-principal element alloys
PA	plasma atomization
PBF	powder bed fusion
PBF-EB	electron beam powder bed fusion
PBF-LB	laser beam powder bed fusion
SEBM	selective electron beam melting
SLM	selective laser melting
TMP	thermo-mechanical processing

References

1 Pollock, T.M., Clarke, A.J., and Babu, S.S. (2020). Design and tailoring of alloys for additive manufacturing. *Metall. Mater. Trans. A* 51 (12): 6000–6019.

2 Onuike, B. and Bandyopadhyay, A. (2018). Additive manufacturing of Inconel 718 – Ti6Al4V bimetallic structures. *Addit. Manuf.* 22: 844–851.

3 Sterling, A.J., Torries, B., Shamsaei, N. et al. (2016). Fatigue behavior and failure mechanisms of direct laser deposited Ti–6Al–4V. *Mater. Sci. Eng., A* 655: 100–112.

4 Liu, S. and Shin, Y.C. (2019). Additive manufacturing of Ti6Al4V alloy: a review. *Mater. Des.* 164: 107552.

5 Polonsky, A.T., Echlin, M.P., Lenthe, W.C. et al. (2018). Defects and 3D structural inhomogeneity in electron beam additively manufactured Inconel 718. *Mater. Charact.* 143: 171–181.

6 Shassere, B., Greeley, D., Okello, A. et al. (2018). Correlation of microstructure to creep response of hot isostatically pressed and aged electron beam melted Inconel 718. *Metall. Mater. Trans. A* 49 (10): 5107–5117.

7 Montero-Sistiaga, M.L., Mertens, R., Vrancken, B. et al. (2016). Changing the alloy composition of Al7075 for better processability by selective laser melting. *J. Mater. Process. Technol.* 238: 437–445.

8 Martin, A., Vilanova, M., Gil, E. et al. (2022). Influence of the Zr content on the processability of a high strength Al-Zn-Mg-Cu-Zr alloy by laser powder bed fusion. *Mater. Charact.* 183: 111650.

9 Wimmer, A., Kolb, C.G., Assi, M. et al. (2020). Investigations on the influence of adapted metal-based alloys on the process of laser beam melting. *J. Laser Appl.* 32 (2): 022029.

10 Wimmer, A., Hofstaetter, F., Jugert, C. et al. (2022). In situ alloying: investigation of the melt pool stability during powder bed fusion of metals using a laser beam in a novel experimental set-up. *Prog. Addit. Manuf.* 7 (2): 351–359.

11 Gradl, P., Mireles, O., and Andrews, N. (2018). Intro to additive manufacturing for propulsion systems. AIAA Propulsion and Energy 2019.

12 Meier, B., Warchomicka, F., Petrusa, J. et al. (2022). Influence of powder production process and properties on material properties of Ti6Al4V manufactured by L-PBF. *Int. J. Adv. Manuf. Technol.* 123 (5): 1577–1588.

13 Naeem, M. (2013). Laser processing of reflective materials. *Laser Tech. J.* 10 (1): 18–20.

14 Sun, Y., Aindow, M., and Hebert, R.J. (2018). Comparison of virgin Ti-6Al-4V powders for additive manufacturing. *Addit. Manuf.* 21: 544–555.

15 Subcommittee: B10.01 (2020). ASTM B265-20a: standard specification for titanium and titanium alloy strip, sheet, and plate. https://doi.org/10.1520/B0265-20A

16 Sun, P., Fang, Z.Z., Zhang, Y., and Xia, Y. (2017). Review of the methods for production of spherical Ti and Ti alloy powder. *JOM* 69 (10): 1853–1860.

17 Cordova, L., Bor, T., de Smit, M. et al. (2020). Measuring the spreadability of pre-treated and moisturized powders for laser powder bed fusion. *Addit. Manuf.* 32: 101082.

18 Wu, Q., Qiao, C., Fan, W. et al. (2022). Insights into surface chemistry induced powder layer characteristic evolutions in additive manufacturing. *Powder Technol.* 407: 117635.

19 Meier, C., Weissbach, R., Weinberg, J. et al. (2019). Critical influences of particle size and adhesion on the powder layer uniformity in metal additive manufacturing. *J. Mater. Process. Technol.* 266: 484–501.

20 Wang, L., Li, E.L., Shen, H. et al. (2020). Adhesion effects on spreading of metal powders in selective laser melting. *Powder Technol.* 363: 602–610.

21 Andersson, J.-O., Helander, T., Höglund, L. et al. (2002). Thermo-Calc & DICTRA, computational tools for materials science. *Calphad* 26 (2): 273–312.

22 Frederick, C.L., Plotkowski, A., Kirka, M.M. et al. (2018). Geometry-induced spatial variation of microstructure evolution during selective electron beam melting of Rene-N5. *Metall. Mater. Trans. A* 49 (10): 5080–5096.

23 Murray, S.P., Pusch, K.M., Polonsky, A.T. et al. (2020). A defect-resistant Co–Ni superalloy for 3D printing. *Nat. Commun.* 11 (1): 4975.

24 Stewart, C.A., Murray, S.P., Suzuki, A. et al. (2020). Accelerated discovery of oxidation resistant CoNi-base γ/γ' alloys with high L12 solvus and low density. *Mater. Des.* 189: 108445.

25 Wang, X., Sridar, S., and Xiong, W. (2020). Thermodynamic investigation of new high-strength low-alloy steels with heusler phase strengthening for welding and additive manufacturing: high-throughput CALPHAD calculations and key experiments for database verification. *J. Phase Equilib. Diffus.* 41 (6): 804–818.

26 Xiao, W., Xu, Y., Xiao, H. et al. (2021). Investigation of the Nb element segregation for laser additive manufacturing of nickel-based superalloys. *Int. J. Heat Mass Transfer* 180: 121800.

27 Ng, C.H., Bermingham, M.J., and Dargusch, M.S. (2021). Eliminating segregation defects during additive manufacturing of high strength β-titanium alloys. *Addit. Manuf.* 39: 101855.

28 Sun, Z., Tan, X., Wang, C. et al. (2021). Reducing hot tearing by grain boundary segregation engineering in additive manufacturing: example of an Al_xCoCrFeNi high-entropy alloy. *Acta Mater.* 204: 116505.

29 Wu, A.S., Brown, D.W., Clausen, B., and Elmer, J.W. (2017). The influence of impurities on the crystal structure and mechanical properties of additive manufactured U–14at% Nb. *Scr. Mater.* 130: 59–63.

30 Jadhav, S.D., Dadbakhsh, S., Vleugels, J. et al. (2019). Influence of carbon nanoparticle addition (and impurities) on selective laser melting of pure copper. *Materials* 12 (15): 2469.

31 Yeh, J.-W., Chen, S.-K., Lin, S.-J. et al. (2004). Nanostructured high-entropy alloys with multiple principal elements: novel alloy design concepts and outcomes. *Adv. Eng. Mater.* 6 (5): 299–303.

32 Cantor, B., Chang, I.T.H., Knight, P., and Vincent, A.J.B. (2004). Microstructural development in equiatomic multicomponent alloys. *Mater. Sci. Eng., A* 375–377: 213–218.

33 Zhang, Y., Zuo, T.T., Tang, Z. et al. (2014). Microstructures and properties of high-entropy alloys. *Prog. Mater Sci.* 61: 1–93.

34 Jafary-Zadeh, M., Khoo, K.H., Laskowski, R. et al. (2019). Applying a machine learning interatomic potential to unravel the effects of local lattice distortion on the elastic properties of multi-principal element alloys. *J. Alloys Compd.* 803: 1054–1062.

35 Zhou, Y.J., Zhang, Y., Kim, T.N., and Chen, G.L. (2008). Microstructure characterizations and strengthening mechanism of multi-principal component AlCoCrFeNiTi0.5 solid solution alloy with excellent mechanical properties. *Mater. Lett.* 62 (17): 2673–2676.

36 Wang, X.F., Zhang, Y., Qiao, Y., and Chen, G.L. (2007). Novel microstructure and properties of multicomponent $CoCrCuFeNiTi_x$ alloys. *Intermetallics* 15 (3): 357–362.

37 Li, H., Huang, Y., Jiang, S. et al. (2021). Columnar to equiaxed transition in additively manufactured CoCrFeMnNi high entropy alloy. *Mater. Des.* 197: 109262.

38 Joseph, J., Jarvis, T., Wu, X. et al. (2015). Comparative study of the microstructures and mechanical properties of direct laser fabricated and arc-melted $Al_xCoCrFeNi$ high entropy alloys. *Mater. Sci. Eng., A* 633: 184–193.

39 Ishimoto, T., Ozasa, R., Nakano, K. et al. (2021). Development of TiNbTaZrMo bio-high entropy alloy (BioHEA) super-solid solution by selective laser melting, and its improved mechanical property and biocompatibility. *Scr. Mater.* 194: 113658.

40 Vikram, R.J., Murty, B.S., Fabijanic, D., and Suwas, S. (2020). Insights into micro-mechanical response and texture of the additively manufactured eutectic high entropy alloy AlCoCrFeNi2.1. *J. Alloys Compd.* 827: 154034.

41 Xie, Y., Liang, J., Zhang, D. et al. (2020). Sustaining strength–ductility synergy of CoCrFeNiMn high entropy alloy by a multilevel heterogeneity associated with nanoparticles. *Scr. Mater.* 187: 390–394.

42 Ostovari Moghaddam, A., Shaburova, N.A., Samodurova, M.N. et al. (2021). Additive manufacturing of high entropy alloys: a practical review. *J. Mater. Sci. Technol.* 77: 131–162.

43 Dada, M., Popoola, P., Mathe, N. et al. (2020). Process optimization of high entropy alloys by laser additive manufacturing. *Eng. Rep.* 2 (10): e12252.

44 Tong, Z., Ren, X., Jiao, J. et al. (2019). Laser additive manufacturing of FeCrCoMnNi high-entropy alloy: effect of heat treatment on microstructure, residual stress and mechanical property. *J. Alloys Compd.* 785: 1144–1159.

45 Li, C., Ferry, M., Kruzic, J.J., and Li, X. (2022). Review: multi-principal element alloys by additive manufacturing. *J. Mater. Sci.* 57 (21): 9903–9935.

46 Brif, Y., Thomas, M., and Todd, I. (2015). The use of high-entropy alloys in additive manufacturing. *Scr. Mater.* 99: 93–96.

47 Niu, P.D., Li, R.D., Yuan, T.C. et al. (2019). Microstructures and properties of an equimolar AlCoCrFeNi high entropy alloy printed by selective laser melting. *Intermetallics* 104: 24–32.

48 Luo, S., Su, Y., and Wang, Z. (2021). Microstructural evolution and mechanisms in additively manufactured AlCrCuFeNix complex concentrated alloys via selective laser melting. *J. Alloys Compd.* 870: 159443.

49 Vaidya, M., Muralikrishna, G.M., and Murty, B.S. (2019). High-entropy alloys by mechanical alloying: a review. *J. Mater. Res.* 34 (5): 664–686.

50 Popov, V.V., Katz-Demyanetz, A., Koptyug, A., and Bamberger, M. (2019). Selective electron beam melting of Al0.5CrMoNbTa0.5 high entropy alloys using elemental powder blend. *Heliyon* 5 (2): e01188.

51 Wang, P., Huang, P., Ng, F.L. et al. (2019). Additively manufactured CoCrFeNiMn high-entropy alloy via pre-alloyed powder. *Mater. Des.* 168: 107576.

52 Kang, B., Lee, J., Ryu, H.J., and Hong, S.H. (2018). Ultra-high strength WNbMoTaV high-entropy alloys with fine grain structure fabricated by powder metallurgical process. *Mater. Sci. Eng., A* 712: 616–624.

53 Joo, S.-H., Kato, H., Jang, M.J. et al. (2017). Structure and properties of ultrafine-grained CoCrFeMnNi high-entropy alloys produced by mechanical alloying and spark plasma sintering. *J. Alloys Compd.* 698: 591–604.

54 Sharma, D.K., Mahant, D., and Upadhyay, G. (2020). Manufacturing of metal matrix composites: a state of review. *Mater. Today Proc.* 26: 506–519.

55 Chawla, K.K. (2006). Metal matrix composites. In: *Materials Science and Technology* (ed. R.W. Cahn, P. Haasen, and E.J. Kramer), 122–179. Wiley-VCH Verlag GmbH & Co. KGaA.

56 Ibrahim, I.A., Mohamed, F.A., and Lavernia, E.J. (1991). Particulate reinforced metal matrix composites – a review. *J. Mater. Sci.* 26 (5): 1137–1156.

57 Tjong, S.C. and Ma, Z.Y. (2000). Microstructural and mechanical characteristics of in situ metal matrix composites. *Mater. Sci. Eng., R* 29 (3, 4): 49–113.

58 Ma, K., Lavernia, E.J., and Schoenung, J.M. (2017). Particulate reinforced aluminum alloy matrix composites: effect of microconstituents. *Rev. Adv. Mater. Sci* 48: 91–104.

59 Miracle, D.B. (2005). Metal matrix composites – from science to technological significance. *Compos. Sci. Technol.* 65 (15): 2526–2540.

60 Li, N., Liu, W., Wang, Y. et al. (2021). Laser additive manufacturing on metal matrix composites: a review. *Chin. J. Mech. Eng.* 34 (1): 38.

61 Sercombe, T.B. and Li, X. (2016). Selective laser melting of aluminium and aluminium metal matrix composites: review. *Mater. Technol.* 31 (2): 77–85.

62 Gao, C., Xiao, Z., Liu, Z. et al. (2019). Selective laser melting of nano-TiN modified AlSi10Mg composite powder with low laser reflectivity. *Mater. Lett.* 236: 362–365.

63 Li, X.P., Wang, X.J., Saunders, M. et al. (2015). A selective laser melting and solution heat treatment refined Al–12Si alloy with a controllable ultrafine eutectic microstructure and 25% tensile ductility. *Acta Mater.* 95: 74–82.

64 Gu, D., Yang, Y., Xi, L. et al. (2019). Laser absorption behavior of randomly packed powder-bed during selective laser melting of SiC and TiB_2 reinforced Al matrix composites. *Opt. Laser Technol.* 119: 105600.

65 Xiong, Y., Hofmeister, W.H., Smugeresky, J.E. et al. (2012). Investigation of atypical molten pool dynamics in tungsten carbide-cobalt during laser deposition using in-situ thermal imaging. *Appl. Phys. Lett.* 100 (3): 034101.

66 Khairallah, S.A., Anderson, A.T., Rubenchik, A., and King, W.E. (2016). Laser powder-bed fusion additive manufacturing: physics of complex melt flow and formation mechanisms of pores, spatter, and denudation zones. *Acta Mater.* 108: 36–45.

67 Xiong, Y., Hofmeister, W.H., Cheng, Z. et al. (2009). In situ thermal imaging and three-dimensional finite element modeling of tungsten carbide–cobalt during laser deposition. *Acta Mater.* 57 (18): 5419–5429.

68 An, N., Shuai, S., Hu, T. et al. (2022). Application of synchrotron X-ray imaging and diffraction in additive manufacturing: a review. *Acta Metall. Sinica (English Letters)* 35 (1): 25–48.

69 Wang, H., Gould, B., Haddad, M. et al. (2022). In situ high-speed synchrotron X-ray imaging of laser-based directed energy deposition of the alloying process with dissimilar powders. *J. Manuf. Processes* 75: 1003–1011.

70 Wolff, S.J., Webster, S., Parab, N.D. et al. (2021). In-situ observations of directed energy deposition additive manufacturing using high-speed X-ray imaging. *JOM* 73 (1): 189–200.

71 Haley, J.C., Schoenung, J.M., and Lavernia, E.J. (2018). Observations of particle-melt pool impact events in directed energy deposition. *Addit. Manuf.* 22: 368–374.

72 Wang, H., Li, G., Zhao, Y., and Chen, G. (2010). In situ fabrication and microstructure of Al_2O_3 particles reinforced aluminum matrix composites. *Mater. Sci. Eng., A* 527 (12): 2881–2885.

73 Jiang, S., Zheng, B., and Schoenung, J.M. (2022). Directed energy deposition of metal matrix composites: computational and experimental comparison of powder particle flow behavior. *J. Mater. Res. Technol.* 16: 516–529.

74 Tang, S., Ummethala, R., Suryanarayana, C. et al. (2021). Additive manufacturing of aluminum-based metal matrix composites – a review. *Adv. Eng. Mater.* 23 (7): 2100053.

75 Mosallanejad, M.H., Niroumand, B., Aversa, A., and Saboori, A. (2021). In-situ alloying in laser-based additive manufacturing processes: a critical review. *J. Alloys Compd.* 872: 159567.

76 Li, W., Chen, X., Yan, L. et al. (2018). Additive manufacturing of a new Fe-Cr-Ni alloy with gradually changing compositions with elemental powder mixes and thermodynamic calculation. *Int. J. Adv. Manuf. Technol.* 95 (1): 1013–1023.

77 Liu, W. and Dupont, J.N. (2003). In-situ reactive processing of nickel aluminides by laser-engineered net shaping. *Metall. Mater. Trans. A* 34 (11): 2633–2641.

78 Hu, Y. and Li, J. (2017). Selective laser alloying of elemental titanium and boron powder: thermal models and experiment verification. *J. Mater. Process. Technol.* 249: 426–432.

79 Zhang, B., Chen, J., and Coddet, C. (2013). Microstructure and transformation behavior of in-situ shape memory alloys by selective laser melting Ti–Ni mixed powder. *J. Mater. Sci. Technol.* 29 (9): 863–867.

80 Samuel, S., Nag, S., Scharf, T.W., and Banerjee, R. (2008). Wear resistance of laser-deposited boride reinforced Ti-Nb–Zr–Ta alloy composites for orthopedic implants. *Mater. Sci. Eng., C* 28 (3): 414–420.

81 Yadroitsev, I., Krakhmalev, P., and Yadroitsava, I. (2017). Titanium alloys manufactured by in situ alloying during laser powder bed fusion. *JOM* 69 (12): 2725–2730.

82 Kang, N., Li, Y., Lin, X. et al. (2019). Microstructure and tensile properties of Ti-Mo alloys manufactured via using laser powder bed fusion. *J. Alloys Compd.* 771: 877–884.

83 Tedman-Jones, S.N., McDonald, S.D., Bermingham, M.J. et al. (2019). A new approach to nuclei identification and grain refinement in titanium alloys. *J. Alloys Compd.* 794: 268–284.

11

Physical Powder Characteristics and Additive Manufacturing

11.1 Introduction

The **physical characteristics** of powder affect the quality, consistency, and properties of additively manufactured components. These physical characteristics are often determined by the **powder production method**. In this chapter, the physical characteristics of powders and how they are characterized are described, followed by a review of select powder production methods including **powder reuse** and **powder recycling**. The relationship between the powder production methods and **physical powder characteristics** is also discussed.

11.2 Characterization of Physical Powder Properties

The most commonly studied and characterized physical powder properties are **particle size**, **particle size distribution**, and **particle morphology**. These physical characteristics can be measured and quantified for individual particles and/or batches of powder. These characteristics are important in and of themselves for additive manufacturing, especially since some of the additive manufacturing techniques have particle size constraints that must be accounted for, such as **nozzle diameter** constraints in directed energy deposition (DED) or **layer thickness** in powder bed fusion (PBF). These physical characteristics are also important, however, because they also affect the processability of the powder in various additive manufacturing techniques. The **packing density** of powder, which is ultimately determined by the characteristics of particle size, particle size distribution, and morphology, is particularly important for PBF techniques. Likewise, powder **flowability**, which is also determined by these characteristics, is particularly important for DED techniques. For some of these physical powder characteristics, there are well-established characterization methods and **standards**. For others, a range of methods are currently used in practice, yet standardization does not yet exist [1]. A variety of methods used to characterize the physical powder properties are described and summarized below. Prior to characterization, a representative sample of powder is required. Thus, **powder sampling** methods and standards are described first.

11.2.1 Powder Sampling

Due to the large sizes of a typical powder batch, reliable powder sampling is a critical step to ensure that the powder used in the process meets the required quality and consistency standards. The main objective of powder sampling is to collect a small amount of sample powder that accurately represents the properties of the bulk quantity.

When dealing with large quantities of powder, generating a small amount that is representative of the bulk can be extremely challenging. The recommended practice for bulk sampling is to collect samples from a **moving stream** of powder [2]. However, when this is not feasible, **static sampling** procedures can also be employed [1]. A commonly used device to extract samples from powder loads at rest is a **sample thief** [1]. A sample thief is a long tube with a sampling chamber and a sharp end that can be inserted into a container of powder. To extract powder samples, a sample thief is inserted into the desired location, the chamber is opened and filled with powder, closed, and then removed. It is crucial to sample from different sites within the container to detect any variations in powder properties. The selection of sampling sites should be informed by an understanding of factors such as possible flow and segregation of powder during transport.

Metallic Powders for Additive Manufacturing: Science and Applications, First Edition.
Enrique J. Lavernia, Kaka Ma, Julie M. Schoenung, James F. Shackelford, and Baolong Zheng.

ASTM B215-20 standard practices [3] describe procedures for obtaining samples of metal powders for testing powders being transferred from blenders or storage tanks, as well as powders already packaged in a container. This standard also includes instructions for collecting metal powder from a moving stream. For sampling a single container, a rectangular vessel should be passed at a constant speed when the container is 1/4, 1/2, and 3/4 full. For multiple containers, three portions of the sample should be taken when the first container is 1/2 full, in the middle of the run, and near the end of the run.

After obtaining samples from the bulk lot, the materials need to be further subdivided into manageable quantities. The common sampling devices used for subdividing bulk samples into laboratory samples include **cone and quartering**, **table sampling**, **chute splitting**, **spin riffling**, and **scoop sampling** [1, 4]. Cone and quartering involve creating a cone-shaped heap of powder, flattening it with a spatula, and then dividing it into four equal sections. One of these sections is then selected, and the process is repeated until the desired sample size is obtained, which is typically 1/16th the size of the original sample [1]. The process can be repeated until the desired sample size is obtained. In addition to this approach, Figure 11.1 shows illustrations of a few sampling devices, including a table sampler, a chute splitter, and a spin riffler. Table sampling divides the bulk sample of powder by allowing it to flow along an inclined plane with obstacles and holes. Chute splitting involves the use of a series of chutes to divide the sample into two separate collection trays. Spin riffling utilizes a **hopper** to pour the bulk sample onto a vibratory chute that leads to a rotating ring of sample containers, providing an efficient way to obtain a representative sample without the occurrence of segregation.

Allen and Khan [5] compared the reliability of different sampling techniques by examining binary mixtures of coarse and fine sand. For each sampling technique, 16 samples were examined to obtain the standard deviations. The results are presented in Table 11.1, suggesting that the spinning riffler is the most reliable over all other sampling techniques investigated, while cone and quartering is the least. A summary of the pros and cons of each sampling device is presented in Table 11.2 [2].

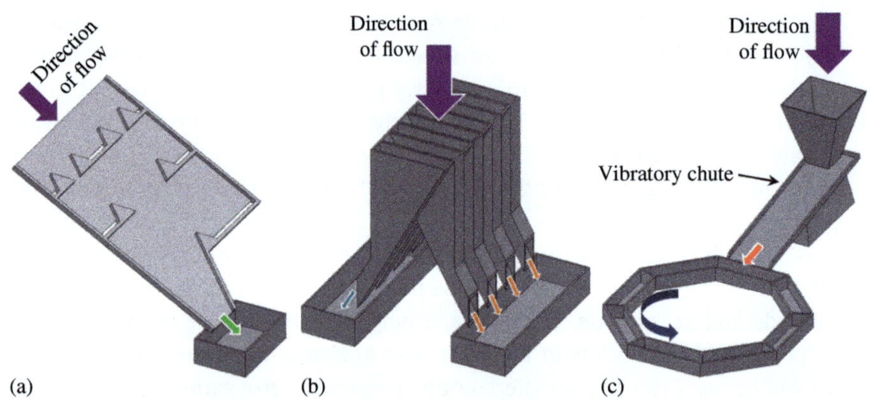

Figure 11.1 Illustrations of (a) a table sampler, (b) a chute splitter, and (c) a spin riffler. Source: Cooke et al. [1]/U.S Department of Commerce/Public domain.

Table 11.1 Comparison of reliability of selected sampling methods using coarse-fine sand mixtures.

Sampling technique	Percentage of standard deviation (%)
Cone and quartering	6.81
Scoop sampling	5.14
Table sampling	2.09
Chute splitting	1.01
Spinning riffling	0.146

Source: Allen [4]/with permission of Elsevier.

Table 11.2 Comparison of advantages and disadvantages of sampling techniques.

Sampling technique	Advantages	Disadvantages
Cone and quartering	Good for powders with poor flow characteristics	Operator-dependent
Scoop sampling	Reliable for homogeneous and nonflowing powder	Particle segregation
Table sampling	Ability to separate large quantities of material	Dependent on the initial feed
Chute splitting	Ability to reduce powder samples in half after one pass	Operator bias
Spinning riffling	Reliable for free-flowing powder	Inability to do large quantities of powder efficiently

Source: Jillavenkatesa et al. [2]/U.S Department of Commerce/Public domain.

11.2.2 Particle Size and Particle Size Distribution

Particle size is one of the most important physical characteristics of powder particles. In additive manufacturing, particle size limits, for instance, the minimum layer thickness. Particle size distributions affect packing density as well as **spreadability** and flowability. The **particle diameter** is commonly used to define the particle size of a spherical particle, while for a nonspherical particle, it is common to calculate an "**equivalent spherical diameter**" based on either equivalent volume or equivalent surface area [4].

To represent the size distribution of a population of powders, one method is to create a histogram that plots the frequency of equally spaced size categories, $g(D_i)$, against the powder size. Another way is to use a cumulative frequency, $G(D_i)$, which represents the fraction of powders falling in a size range smaller than D_i. As illustrated in Figure 11.2, the three most important parameters that are used to characterize a particle size distribution are the **modal**, the **mean**, and the **median diameter** [2]. The modal diameter is defined as the powder size that corresponds to the maximum frequency and, thus, in a graphical representation, is the size that is represented by the peak of the differential size distribution curve. The median diameter is defined as the particle size that corresponds to the 50% cumulative frequency. Finally, the mean diameter is defined as the weighted average size of the particles. Many statistical parameters can be derived from the particle size distribution. For example, D10, D50, and D90 are usually reported, which are used to indicate the particle size where 10%, 50%, and 90% of the particles have a smaller diameter than that value, respectively. It is important to note that distribution functions, such as those shown in Figure 11.2, may be constructed based on different powder characteristics such as mass, number, volume, and surface area.

Various methods have been used in the literature to determine the particle size and particle size distribution, including **sieving**, microscopy-based techniques, **gravitational sedimentation**, and **laser light diffraction**.

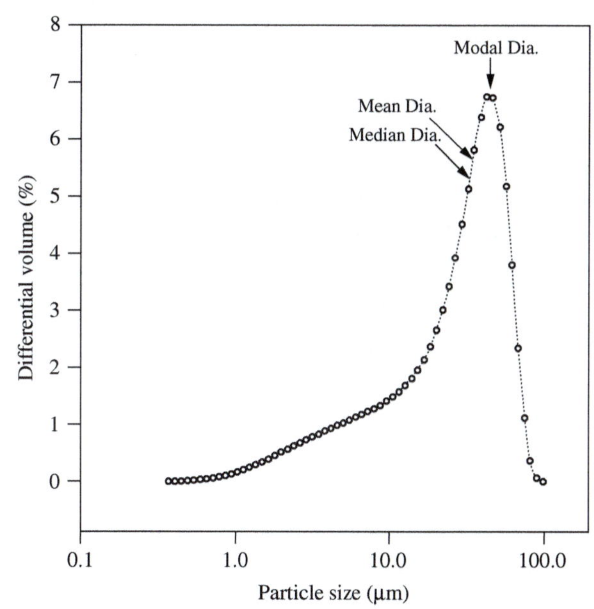

Figure 11.2 Particle size distribution histogram. Source: Jillavenkatesa et al. [2]/U.S Department of Commerce/Public domain.

Sieve analysis is commonly used due to its simplicity, low cost, and ease of interpretation. Sieving involves passing a powder through a series of stacked sieves with decreasing **mesh sizes**. The particle diameter determined by sieving is defined by the length of the minimum square aperture through which the particles can pass. **ASTM B214-22** [6] provides detailed instructions for the dry sieve process of metal powders or mixed powders. According to this standard, a series of sieves should be arranged consecutively based on the size of their openings, with the largest sieve on top and a solid pan beneath the lowest one. The powder is placed in the top sieve, covered, and the entire apparatus is secured into a **sieve shaker**. Sieve mesh sizes can range from 45 to 850 μm. ASTM B214-22 also advises that the sieve shaker be set to a rotation speed within the range of 270–300 rotations per minute and a tapping frequency of 140–160 taps per minute. After the sieving procedure is completed, one needs to carefully separate each sieve and measure the weight of the powder left on each screen. Particle size distributions can be plotted as a histogram representing **percentage mass retained** on each screen as a function of aperture opening size, or **cumulative percentage of mass** that is finer or coarser as a function of aperture opening size. The total mass of the powder retained after sieving should be within ±2% difference from the mass of the original charge to avoid significant error in the particle size distribution results.

Microscopy-based techniques involve the use of **optical light microscopes**, **scanning electron microscopes** (SEMs), or **transmission electron microscopes** (TEMs) to visually identify the size information of individual particles. Typically, microscopy-based measurement of particle size is based on the projected area that is perpendicular to the viewing orientation. Microscopy-based techniques offer a significant advantage over other particle size analysis methods by allowing the determination of particle shape and morphology besides enabling direct measurement of particle size. These techniques can be applied to study a wide range of materials with a broad distribution of particle sizes, ranging from the nanometer to the millimeter scale. However, improper sampling and sample preparation can lead to erroneous results when using microscopy-based techniques, particularly due to the small sample sizes that can be analyzed. Therefore, great caution should be exercised during sampling and sample preparation when utilizing this method. To generate meaningful data, it is necessary to well disperse the particles, prepare a specimen that is representative of the sample, and measure an adequate number of particles [2]. Due to the large number of particles being observed, microscopy-based size analysis would be greatly aided using automated or semi-automated computer systems and algorithms. The number of particles that need to be analyzed to have results with some degree of confidence is larger for powder with a wider particle size distribution [2]. **Powder dispersion** is often achieved by dispersing the powders in a solution, with exceptions when the nature of the powder being studied requires that the sample cannot be exposed to such a solution. Ultrasonication and ball mills can be used to break up **agglomerates**, while appropriate operational parameters need to be determined to ensure no generation of contamination, excessive heat, and fragmentation. The solution is then transferred to a slide, a stub, or a grid for microscopy analysis. Greater precautions are necessary when dealing with powders in a dry state. A uniform layer of particles can be spread out on a glass slide and then picked up using adhesive tape [2]. If the particles are too fine or have a large aspect ratio, achieving a uniform spread may be difficult, which may lead to errors [2].

Gravitational sedimentation is another commonly used technique for particle size and particle size distribution characterization. This technique utilizes **X-ray absorption** or light scattering to determine the particle concentration at different depths within a monitored container when particles settle down under gravity in a fluid medium or liquid. **ASTM B761-17** [7] describes a standard test method to determine particle size distribution by X-ray monitoring of gravity sedimentation. By monitoring the concentration of the particles at various depths by X-ray, the settling velocities of the particles can be determined, and the particle diameter corresponding to that velocity can be calculated using **Stokes' equation** [2]:

$$v = \frac{D^2(\rho_p - \rho_L)g}{18\eta_L} \tag{11.1}$$

where v is the settling velocity of the particles, D is the particle diameter, ρ_p and ρ_L are the particle and fluid density, respectively, g is the gravitational acceleration, and η_L is the viscosity of the fluid. A basic requirement for properly applying this method is that all the particles are of uniform density and composition. Therefore, the settling velocity of different particles in the fluid is influenced only by the particle size. The test method can be applied for particles with a particle size range of 0.1–100 μm. Moreover, the size and settling velocity relationship used in this testing approach necessitates that the particles possess a **Reynolds number** of 0.3 or lower, ensuring that the sediment stays within the laminar flow regime. In order to judge the quality of obtained results, the Reynolds number (*Re*) of the largest particle present should be calculated using the following equation [7]:

$$Re = \frac{D^3(\rho_p - \rho_L)\rho_L g}{18\eta_L^2} \tag{11.2}$$

where D is the diameter of the particle, ρ_p is the particle density, ρ_L is the suspending liquid density, g is the acceleration of gravity, and η_L is the suspending liquid viscosity.

Laser light diffraction involves dispersing a sample of the material in a liquid or aspirating a dry sample in a carrier gas, then circulating it through a light beam [2]. The light beam is scattered by all the particles passing through it, and photodetector arrays collect the scattered light and convert it into electrical signals for analysis. Using theories such as Fraunhofer Diffraction, Mie Scattering, or a combination of both, the analyzed signal can be converted into a particle size distribution. The Fraunhofer model is not suitable for particles smaller than 2–4 μm. In these cases, it is more appropriate to use instruments that utilize the Mie optical model, which can analyze particles as small as 0.1 μm. Light diffraction-based size characterization offers rapid analysis of particle size, relatively simple sample preparation and is one of the inexpensive techniques. However, there are still many limitations when applying this technique to nonspherical particles.

11.2.3 Particle Morphology

The morphology of powder particles significantly influences how the particles spread, flow, and pack together. In terms of laser-based PBF, for example, powder morphology affects the packing density of the powder bed, and, therefore, the density, surface roughness, and accuracy of the build parts. The sampling methods described in Section 11.2.1 can be followed to ensure a representative sample of powder is taken before characterizing particle morphology.

Microscopy techniques are commonly used to characterize powder morphology. Optical microscopy, electron microscopy, or scanning probe microscopy can be used to observe the morphology of the powder particles. The **X-ray computed tomography** technique allows for visualization of the powder morphology in three dimensions. The output of this technique typically consists of hundreds or thousands of cross-sectional images that can be assembled into a three-dimensional representation of the sample [8]. Contrast is generated by variations in the X-ray attenuation properties of different materials within the sample, and mathematical algorithms are employed to reconstruct the internal structure based on this information. The choice of technique depends on the desired resolution, sensitivity, and sample preparation requirements.

To promote consistency in the description of powder particle morphology, **ASTM B243-22** [9] defines several particle morphology descriptors, such as **acicular powder** (needle-shaped particles), **flake powder** (flat or scale-like particles), **granular powder** (particles with approximately equidimensional, nonspherical shapes), **needles** (elongated rod-like particles), **nodular powder** (irregular particles with knotted, rounded, or similar shapes), **platelet powder** (powder composed of flat particles with considerable thickness), and **spherical powder** (globular-shaped particles). However, these terms are qualitative in nature and lack scientific rigor. Alternative methods result in more quantitative metrics. For instance, Figure 11.3 illustrates an image analysis approach to quantify particle morphology using the **Legendre ellipse**. The Legendre ellipse is an ellipse with the center point at the object's centroid and having the same geometrical moments up to the second order with the original particle area [11], where a and b are the major and secondary half-axes of the Legendre ellipse, respectively. Different shape parameters are defined to quantitatively characterize particle morphology. Aspect ratio is often used to describe the ellipticity of a particle

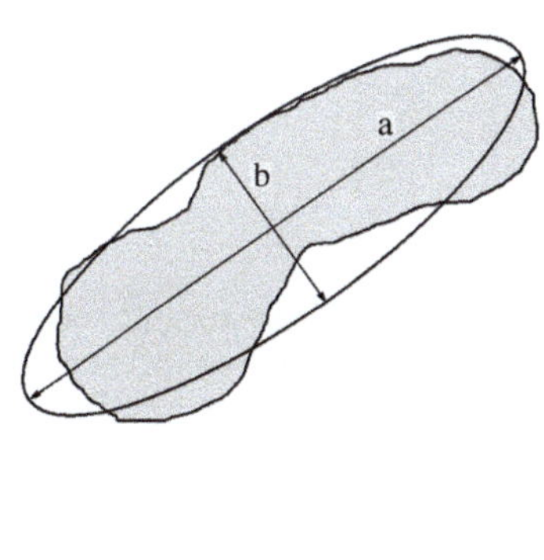

Figure 11.3 Particle morphology fitted to a Legendre ellipse. Source: Mikli et al. [10]/Estonian Academy Publishers.

and is defined by:

$$\text{Aspect ratio} = a/b \tag{11.3}$$

Elongation, which is defined as:

$$\text{Elongation} = \log_2(a/b) \tag{11.4}$$

increases for ellipses with a larger aspect ratio. The elongation of a circle is 0, and of an ellipse with the ratio of axes 1 : 2 is 1.

A common parameter to characterize particle irregularity is **roundness**. Roundness [12], sometimes called **circularity** [13–15], is defined as,

$$\text{Roundness or Circularity} = (4\pi A)/P^2 \tag{11.5}$$

While in some other studies, roundness is defined as [10],

$$\text{Roundness} = P^2/(4\pi A) \tag{11.6}$$

where P is the perimeter, and A is the area of the object. Similarly, a **roundness factor** is defined as:

$$\text{Roundness factor} = P/d_A \tag{11.7}$$

where P is the perimeter, and d_A is the diameter of the circle with the same area as the particle.

Another important characteristic of powder particles is the **angularity** [10], which refers to the degree of sharpness of edges and corners. Angularity affects the flow properties of powder particles, such as their packing density and flowability. It has also been assumed that the angularity is directly related to particle abrasiveness. A numerical parameter called **spike parameter** can be calculated based on image analysis of the particle morphology [16]. Several methods have been suggested for calculating this parameter. One such method is the linear fit approach, which involves tracing the particle boundary in fixed steps using a set of triangles with height h and apex angle θ, as illustrated in Figure 11.4a. A **spike value** is first defined as [10]:

$$\text{Spike value} = h\cos\frac{\theta}{2} \tag{11.8}$$

and the spike parameter [10]:

$$\text{Spike parameter} = \frac{1}{n}\sum\frac{1}{m}\sum\frac{\text{SV}_{\max}}{h_{\max}} \tag{11.9}$$

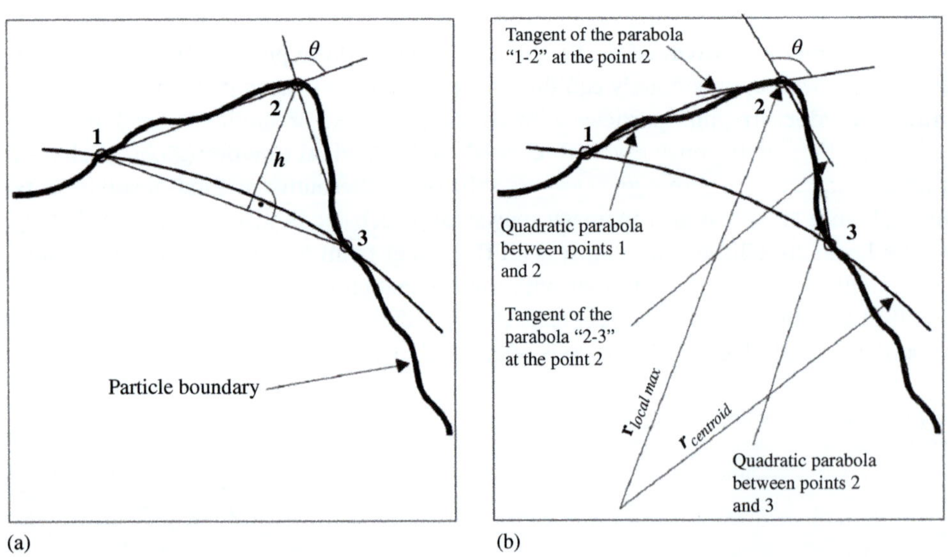

Figure 11.4 Schematic illustration of the methods used to measure the spike parameter by: (a) constructing triangles around the particle boundary and (b) based on fitting quadratic segments to protruding sections of the particle boundary. Source: Mikli et al. [10]/Estonian Academy Publishers.

where m is the number of valid spike values (SV) for a given step size, and n is the number of different step sizes used. A modified angularity parameter based on fitting quadratic segments has also been developed [16]. This method counts spikes outside of a fitted circle. For each spike outside of the circle, the local maximum radius is identified, which corresponds to the apex of the spike, as shown in Figure 11.4b [10]. Differentiating the polynomials at the apex yields the apex angle θ and the spike value-quadratic fit can be calculated as:

$$\text{Spike value} = \cos\frac{\theta}{2} \tag{11.10}$$

The spike parameter determined with this modified quadratic fit method is calculated by averaging the SV.

11.2.4 Powder Flow Characteristics

Flowability affects the feeding of powder through the hopper or printing nozzle during DED-type additive manufacturing. A good flowability is also vital to achieving a homogeneous and compact powder bed in PBF techniques, which is critical to providing good layer quality and avoiding defects in the printed objects. In general, the flow performance is affected by the interparticle friction between the powder particles, such as mechanical friction, the mechanical interlocking of particles, Van der Waals forces, capillary forces, electrostatic forces, and magnetic forces [17]. The powder characteristics that influence flowability include particle size distribution, density of metal or alloy, morphology of particles, roughness of particle surfaces, and moisture content. Moreover, the flowability of metal powders is not an inherent property; it depends not only on the physical properties of the powder but also on the stress state, equipment used, and handling method [18, 19]. Figure 11.5 illustrates different flowability testing techniques. Due to the complexity of the powder flow, agreement is still lacking on which technique is the most suitable for any specific additive manufacturing process.

Flow rate measurements are typically carried out using a flowmeter **funnel**, as seen in Figure 11.6, which allows the powder to flow out under the influence of gravity. The **Hall flowmeter funnel** (Figure 11.6a) has a smaller orifice of 2.5 mm, while the **Carney funnel** (Figure 11.6b) has a larger orifice of 5.0 mm. The **Gustavsson funnel** (Figure 11.6c) is another setup that has an orifice with a 2.5 mm orifice exit diameter. But the **cone angle** of the funnel is 30° instead of 60°, as used in Hall and Carney funnels. The Carney or Gustavsson funnels are often used for the powder that cannot flow through the Hall flowmeter (fine powders and poorly lubricated powders). The flow rate can be determined using either a static or dynamic flow method. In the static flow method, the user blocks the funnel opening with a dry finger and pours a predetermined mass or volume of powder. After loading all the powder without any tapping, vibration, or movement of the funnel, the timing of the flow starts upon removing the finger and stops when the last of the powder has flowed through the opening. The dynamic flow method does not require the user to block the orifice. The timing begins as soon as the powder starts flowing out of the funnel and stops when

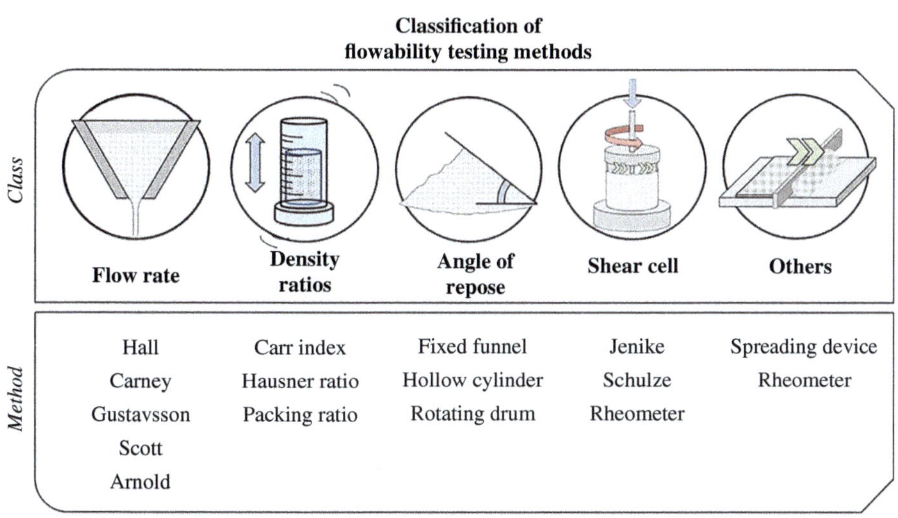

Figure 11.5 Flowability measurement methods. For each class of measurement, commonly used methods are listed. Source: Figure adapted from Marchetti et al. [17].

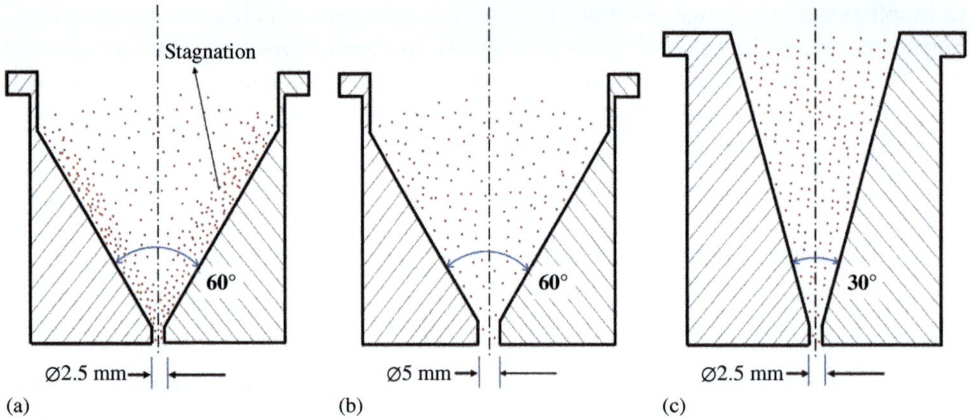

Figure 11.6 (a) Hall flowmeter funnel, (b) Carney funnel, and (c) Gustavsson funnel. Source: Muthuswamy et al. [20]/Springer Nature.

all the powder has finished flowing through the funnel. The flow rate value in seconds per gram can be measured according to **ASTM B213-20** [21] using the Hall flowmeter funnel. The time taken for the entire powder sample (50 g, as per ASTM B213-20) to exit the funnel is measured and then divided by the mass of the powder sample. Measuring the flow by volume instead of by mass eliminates the variable of powder density. **ASTM B855-22** [22] describes the volumetric flow rate measurement using the **Arnold meter** and Hall flowmeter funnel. The acquisition of the correct volume of powder (20 cm^3) requires the use of an Arnold meter. The flow rate value is determined by dividing the measured time taken for all the powder to exit the funnel by the volume of the powder sample, and it is reported in units of seconds per cm^3. Since the flow rate, either by mass or by volume, is defined as a time length, a higher value means a lower flowability.

The **density ratio**, which is a measure of the ability of powder particles to pack, can also impact the flowability. The packing of a powder is quantified through its bulk density. A distinction should be noted between the bulk density of powder particles in a loose or a dense packing condition. For example, as outlined in **ASTM B703-21** [23], the **apparent density** of powder particles in a loose packing condition can be measured using the Arnold meter by slowly sliding the test powder over the cavity in a die block. The most common example to promote a denser packing is achieved after the powder is gently shaken or tapped and allowed to settle to a constant volume, as described in **ASTM B527-23** [24]. When comparing different powders, the ratio between high and low packing densities, or **Hausner Ratio**, R_H is commonly used. This value is expressed as the ratio between the **tap density** and the apparent density. A high value of the Hausner ratio indicates a high variation between the starting loose packing density and the final dense packing density, which suggests poor flow of the powder particles.

The **angle of repose** is an important measure of powder flowability and can be defined as the steepest angle measured from the horizontal plane at which a pile of powder comes to a stop from flowing. It is complementary to the **avalanche angle** (AA), which is the angle at which a stationary powder begins to flow. Flowability can be classified based on the angle of repose, according to Carr (Table 11.3) [19]. Various methods can be employed to measure the angle of repose, including the **fixed funnel method**, where the powder is allowed to fall through a funnel onto a flat surface to form a pile; the **hollow cylinder method**, which involves filling a hollow cylinder with powder and allowing it to settle into a pile while carefully lifting the cylinder, and the **tilting box method**, where the powder runs along a tilted planar surface until it comes to a stop, and the angle of tilt is measured. The **revolving cylinder/drum method** is used to determine the **dynamic angle of repose**. This method involves placing powder particles inside a cylinder that has a transparent side. The cylinder is then rotated at a fixed speed, causing the materials to move and rotate within it. As the rotation continues, the materials reach a maximum angle, known as the dynamic angle of repose.

The assessment of powder flow properties can also be conducted through the method of **powder rheometry**. The FT4 Powder Rheometer, depicted in Figure 11.7, is a commonly used instrument for powder flow measurements [17, 19, 25]. Its working principle involves a twisted blade being compelled to traverse a powder bed along a predefined helical trajectory. The motion of the blade generates the forces that cause the powder to deform and flow, which are

Table 11.3 Carr's classification of flow properties by the angle of repose.

Angle of repose	Flow properties
$\alpha < 30°$	Very free flowing
$30° < \alpha < 38°$	Free flowing
$38° < \alpha < 45°$	Passable to fair flow
$45° < \alpha < 55°$	Cohesive
$55° < \alpha$	Very cohesive

Source: Zegzulka et al. [19]/Springer Nature/CCBY 4.0/Public domain.

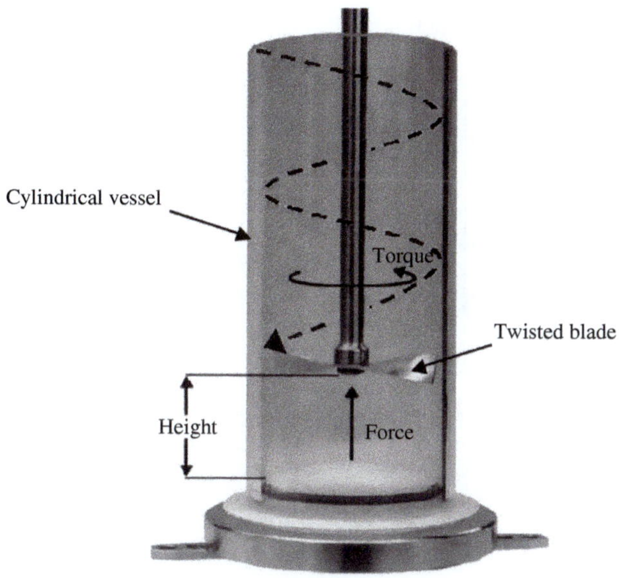

Figure 11.7 Schematic diagram of the FT4 Powder Rheometer. Source: Leturia et al. [25]/Reproduced with permission from Elsevier.

continuously quantified to determine the flow energy (E) required to enable powder flow. Different flow patterns can be attained by varying the axial and rotational velocities of the blade, as well as its direction of rotation (either clockwise or anti-clockwise).

Despite the various techniques to measure powder flowability, a single flow parameter is often unreliable, as they cannot accurately represent actual powder behavior in additive manufacturing [26, 27]. To gain insight into how powder characteristics affect flow behavior in metal additive manufacturing processes, Kiani et al. [27] conducted a systematic study using three distinct techniques for measuring powder flowability: an FT4 powder rheometer, a Revolution powder analyzer, and a modified funnel method. Gas- or water-atomized 316L stainless steel and AlSi10Mg powders from six different vendors were investigated. The **Pearson correlation coefficient** between the analyzed powder characteristics was visualized using Corrplot, as shown in Figure 11.8.

The study found that powders that can pack more densely (higher **apparent density increase**, ADI) tend to stick together more (higher shear stress, τ) and resist flow initiation, leading to poor flow (higher **maximum break angle**, MBA). Spherical particles tend to pack more efficiently than irregularly shaped ones, which have low **aspect ratios** (AR_{50}). Fine particles (smaller particle size, D50) tend to agglomerate and experience interparticle friction, making flow difficult. The study also examined **powder flow rate** (FR, in units of g/s), AA, and **specific energy** (SE) to quantify continuous flowability. Finally, the study concluded that understanding the behavior of particles during hopper discharge is important for additive manufacturing systems, and powders that resist the initiation of flow will lead to inconsistencies and porosity in final parts.

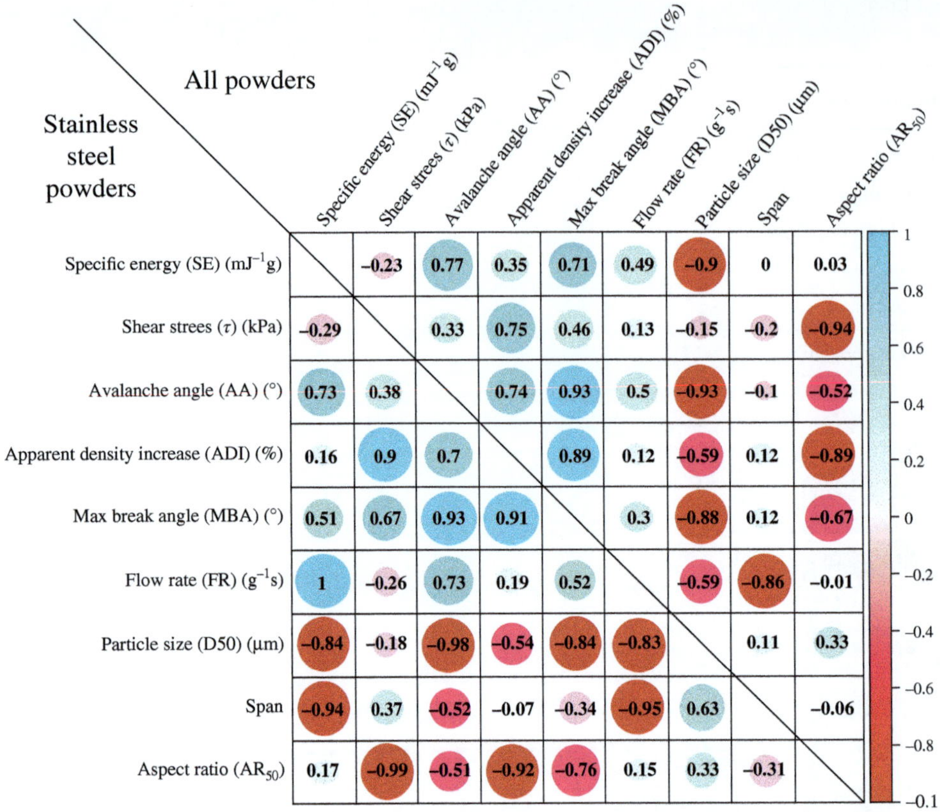

	Specific energy (SE)	Shear strees (τ)	Avalanche angle (AA)	Apparent density increase (ADI)	Max break angle (MBA)	Flow rate (FR)	Particle size (D50)	Span	Aspect ratio (AR$_{50}$)
Specific energy (SE) (mJ^{-1}g)		−0.23	0.77	0.35	0.71	0.49	−0.9	0	0.03
Shear strees (τ) (kPa)	−0.29		0.33	0.75	0.46	0.13	−0.15	−0.2	−0.94
Avalanche angle (AA) (°)	0.73	0.38		0.74	0.93	0.5	−0.93	−0.1	−0.52
Apparent density increase (ADI) (%)	0.16	0.9	0.7		0.89	0.12	−0.59	0.12	−0.89
Max break angle (MBA) (°)	0.51	0.67	0.93	0.91		0.3	−0.88	0.12	−0.67
Flow rate (FR) (g^{-1}s)	1	−0.26	0.73	0.19	0.52		−0.59	−0.86	−0.01
Particle size (D50) (μm)	−0.84	−0.18	−0.98	−0.54	−0.84	−0.83		0.11	0.33
Span	−0.94	0.37	−0.52	−0.07	−0.34	−0.95	0.63		−0.06
Aspect ratio (AR$_{50}$)	0.17	−0.99	−0.51	−0.92	−0.76	0.15	0.33	−0.31	

Figure 11.8 The Pearson correlation coefficient between powder characteristics for stainless steel powders and all powders (both stainless steel and AlSi10Mg). Source: Kiani et al. [27]/John Wiley & Sons.

11.3 Powder Production Methods

The list of metal powder production methods is extensive, as summarized in Figure 11.9 [28] and described in detail in Chapter 7. Atomization processes can produce fine metal powders with uniform particle size and consistent physical and chemical properties and, therefore, are currently the most commonly used methods to produce metal powders that are suitable for additive manufacturing. Atomization may be used, at least in principle, for any metal or alloy system that can be melted. It involves breaking up a molten metal stream into small droplets, which then solidify into metal powders. Depending on the atomization media, atomization processes can be grouped into gas, water, plasma, and centrifugal atomization. **Gas atomization** (GA) uses a stream of high-pressure gas to atomize the molten metal; **water atomization** uses a stream of high-pressure water to atomize the molten metal; **plasma atomization** uses a plasma torch to atomize the molten metal; and **centrifugal atomization** uses a rapidly rotating disk to atomize the molten metal. The mean particle size, particle size distribution, and particle morphology of metal powders produced by atomization are strongly affected by the processing variables, such as flow rate of melt, melt temperature, gas velocity (for GA), or disc rotation peripheral velocity (for centrifugal atomization), and physical properties of the melt.

Other techniques, such as **mechanical milling**, **melt spinning**, **plasma rotating electrode process** (PREP), and **chemical processes**, are also used to produce metal powders for additive manufacturing. Chemical processes, for example, include the reduction of metal-oxide ores by hydrogen, carbon monoxide, and ammonia gas to produce metal powders, such as iron and cobalt. By milling the oxide ore and carefully controlling the reduction atmosphere and processing time, powders with well-controlled particle sizes can be produced using this method. In addition to these established powder production methods listed in Figure 11.9, this field is constantly evolving with new methods such as hybrid gas-water and plasma atomization, wire induction heating GA, and dual-stage ball milling [29]. The size and shape of the powder particles produced by each method can vary, depending on the specific process conditions and the material used. The method for powder production is chosen based on a variety of factors, such as cost,

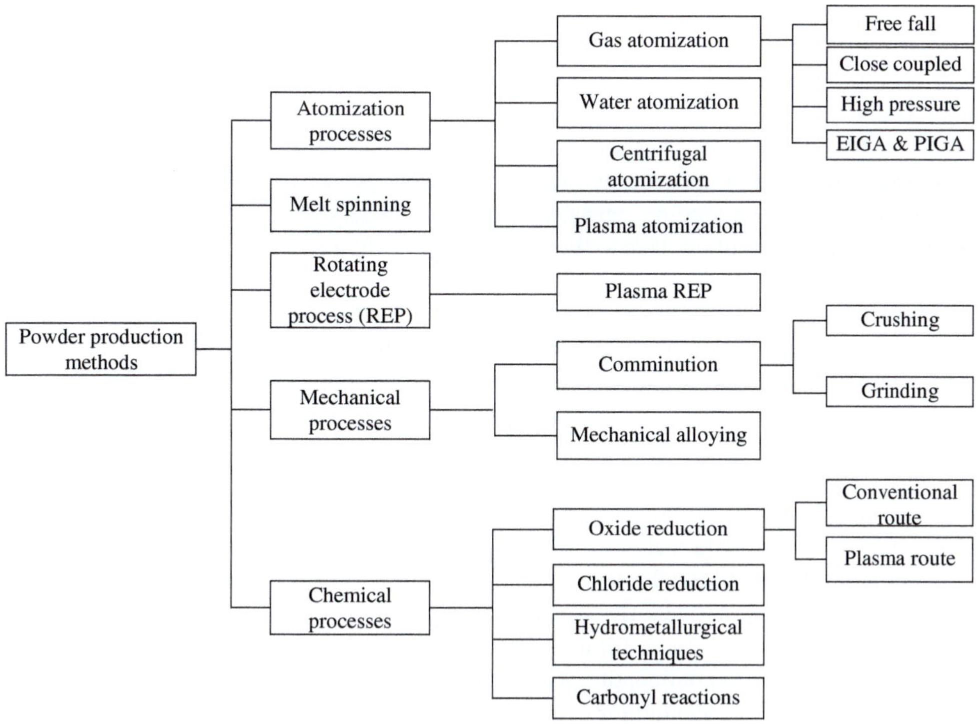

Figure 11.9 Categorization of established metal powder production methods. Source: Antony et al. [28]/Springer Nature.

production scale, the physical and chemical properties of the powders to be produced, and the desired morphology of the powders.

11.3.1 Gas Atomization

During GA, the bulk liquid metal is broken up into small particles by high-velocity gas jets (argon, nitrogen, or air), solidifying into spherical powder particles. This process is shown schematically in Figure 1.8 [30]. The metals are melted and superheated in an inert gas atmosphere and held at temperature for homogenization. The molten metal is then delivered to the nozzle and atomized by jets of gas into fine droplets. The kinetic energy of the gas is transferred to the molten metal mainly in the form of surface energy, providing the driving force for atomization. Finally, the solidified powder particles are allowed to cool down in the inert gas atmosphere.

Among all metal powder production methods, GA is the most widely used method to produce metal powders for additive manufacturing, especially for prealloyed powders, due to its advantages, including:

(a) Flexibility: GA can produce a wide range of alloy compositions with extensive control over resulting powder characteristics and properties.
(b) High quality: The powders produced by GA are clean and have a fine and uniform microstructure, which leads to high-quality parts.
(c) Cost-effectiveness: GA is a relatively low-cost method for producing metal powders, especially when compared to alternative methods such as mechanical milling.
(d) Scalability: The process is easily scalable and can produce tonnage quantities of metal powders, making it suitable for industrial-scale production.

11.3.2 Water Atomization

Water atomization employs jets of water as the atomization media to disintegrate the molten metal into fine droplets. A representative water atomization process is shown in Figure 1.16 [31]. As-received materials are melted and superheated to a temperature suitable for atomization. The molten metal is delivered to the nozzle and atomized as it is hit by high-pressure water jets. The pressure of the jet and the angle relative to the melt can be varied to modify the particle size ranges and morphology. The molten metal is broken up into fine droplets, which are rapidly solidified by the surrounding water. The cooling rates are typically one to two orders of magnitude larger than that for GA [32]. After collecting the resulting powders, postprocessing filtering and drying are needed to remove water and residual moisture.

The lower operating cost of water atomization makes water-atomized powder attractive to some additive manufacturing applications. This powder production method typically results in powders with higher oxygen content, irregular shape, and a wider particle size distribution. Due to these reasons, it is often reported that water-atomized powder, when used as the feedstock for additive manufacturing processes, leads to low packing density, poor powder flowability, and low bulk density of the build component [33–37]. The differences between using water and gas-atomized powders can be reduced by optimizing the additive manufacturing process parameters for water-atomized powder, such as laser power, hatch distance, laser scan speed, and spreading speed [34].

11.3.3 Mechanical Milling

Mechanical milling is one of the most widely used powder production approaches in powder metallurgy, which involves repeated deformation, fracturing, and cold welding. For different applications, mechanical milling equipment comes in various forms, including shaker mills, vibratory mills, tumbler mills, horizontal and vertical attritors, and planetary ball mills. Mechanical milling is particularly suitable for the synthesis of powders of complex compounds, intermetallics, metal-ceramic composites, supersaturated solid solutions, and amorphous alloys. The process is relatively low-cost and allows for the production of fine, uniform powders with controlled particle size and shape. A major drawback of mechanical milling is the potential for contamination from the milling media and the atmosphere. The selection of proper milling media and milling energy is critical for successful powder production.

To address the increasing need for sustainable feedstock production for metal additive manufacturing, increased research attention has focused on applying the mechanical milling approach to produce powders from **recycled waste products**, including shavings, chips, and flakes. In the study by Canakci and Varol [38], segments of AA7075 aluminum alloy were produced by machining a rod billet on a lathe. The chips were then crushed using a double roller crusher and milled in high-purity Ar-purged tungsten carbide vials using a high-energy planetary ball mill. The milling took place at 400 rpm in an Ar atmosphere, and methanol was used to prevent cold welding. A ball milling model showed that ball-chip-ball collisions dominated at milling times of 0.5–5 hours, while ball-powder-ball collisions dominated at times over 5 hours. The particle size decreased with increasing milling time. In another study by Fullenwider et al. [39], serrated 304L stainless steel chips of length 5–20 mm were utilized. The results indicated that using 20 mm balls in milling decreased the size of the particles in the final powder while using 6 mm balls altered the particle morphology to spherical or near-spherical. These findings provided the rationale for using a dual-stage ball milling process to produce spherical or near-spherical particles with diameters ranging from 38 to 150 μm.

11.4 Powder Reuse, Recycling, and Recovery

The production of powder feedstock that is suitable for additive manufacturing is expensive, with a typical unit cost of \$40–500/kg [29]. Although additive manufacturing processes are often described as cleaner production methods with no or limited waste, powder-based metal additive manufacturing still suffers from low material efficiency [40]. In PBF and binder jetting, regardless of the size of the component, a large volume of powder is required depending on the build chamber size. About 6.7 wt% of the powder introduced into the laser powder bed machine was consumed, including only 2–3 wt% solidified into the components, according to Sartin et al. [41]. In DED, only a small fraction of powder feedstock ejected from the nozzles reaches the melt pool, leading to a powder efficiency (the ratio of powder that was solidified in the component to the total amount of fed powder) as low as 5% [40, 42] on average, and values of 3–33% [43] and 50% [44, 45] have been reported. In addition, safe disposal of powder is not only costly but also has an excessive environmental impact. Therefore, powder recycling and reuse are beneficial from both an economic and a green engineering standpoint, although concerns remain regarding the feasibility of the direct use of material that has been exposed to previous use in additive manufacturing. Some effects of repeated powder recycling on powder morphology, chemistry, and microstructure are discussed in Section 11.6.

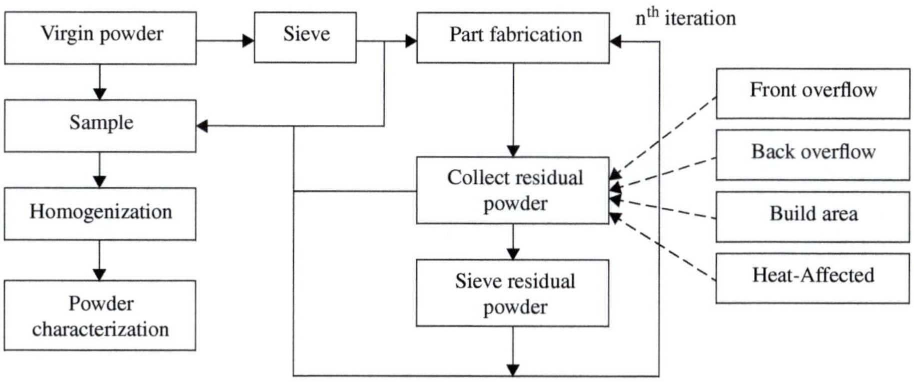

Figure 11.10 Flow diagram for powder reuse procedure in powder bed fusion. Source: Sutton et al. [46]/with permission of Elsevier.

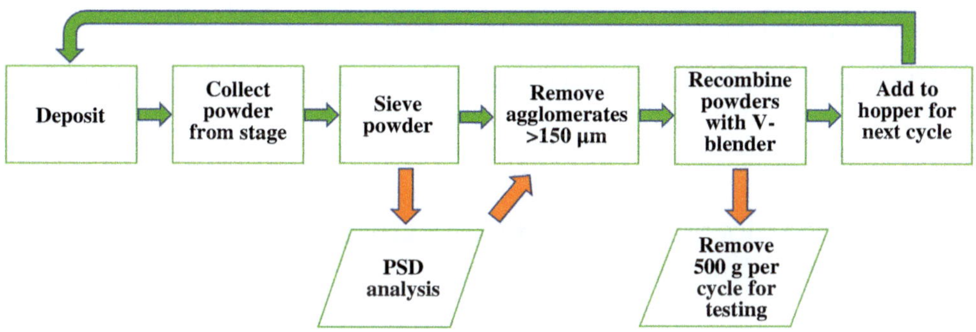

Figure 11.11 Flow diagram for powder reuse methodology in directed energy deposition. Source: Terrassa et al. [43]/with permission of Elsevier.

Because of the cost and waste of powder during the deposition process, many studies and industrial applications now include recycled powder in the feedstock. Figures 11.10 and 11.11 show examples of the reuse and sampling procedures for stainless steel powder in laser PBF and DED, respectively. The methodology mainly includes an iterative process of three main steps: (i) depositing the component after the powder is loaded into the machine, (ii) collecting the residual powder from the machine, and (iii) powder recovery, typically by (consecutive) sieving to remove sintered agglomerated particles and fines, i.e. particles smaller than a specified value.

Strategies for powder reuse are divided into two categories in general, depending on whether the as-received powder is added to rejuvenate the recycled powder for the next build cycle [29]. In the first strategy, powder from a single batch or with the same condition is used for each build cycle, as illustrated in Figure 11.12. With this strategy, the powder will degrade collectively with less variations in powder properties. In the second strategy, the as-received powder is continuously added to the recycled powder after each or a certain number of build cycle(s) [48–51]. Compared with the first one, the second strategy increases powder longevity and reduces material waste but leads to high inhomogeneity and poor traceability of powder properties. The small fractions of powder that reach very high cycle numbers can potentially be detrimental to part quality. Due to these disadvantages, this strategy is not desirable for industries where reliability is critical, such as aerospace, nuclear, and biomedical [47].

Plasma spheroidization as a newly developed method for powder recovery. This technique melts the surface layers of the particles and transforms the used feedstock into dense and spherical powder, while the particle surface becomes smoother and the oxygen, nitrogen, and hydrogen contamination is decreased [52–54]. This technique can also improve flowability and packing density and eliminate particle porosity of the degraded powder [55]. Examining the energy consumption associated with this process and utilizing the powder after spheroidization treatment in the additive manufacturing process to create components are essential next steps to ensure feasibility.

In summary, a wide range of recycling, recovery, and reuse approaches are used in powder additive manufacturing. These strategies have been developed on a case-by-case basis, with a limited number of replication studies and no consensus on the best practices for recycling feedstock powders. Moreover, the powder recycling strategy applied in some

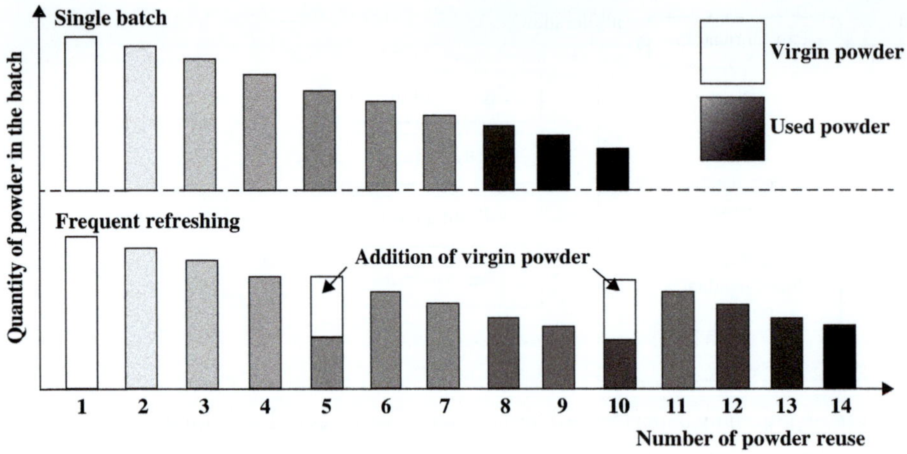

Figure 11.12 Schematic illustration of two different powder reuse strategies. Darker bars represent powder that has experienced more reuse cycles. Source: Moghimian et al. [47]/with permission of Elsevier.

studies is not clearly described [56–60]. For future applications, it is necessary to develop standards and best practice recommendations on recycling strategies, including key process parameters, powder sampling and quantification strategies, and powder characteristics that should be controlled during recycling and reuse.

11.5 Influence of Powder Production Methods and Parameters On Powder Properties and Additive Manufacturing

Different production methods and process parameters can lead to variations in particle size distribution, morphology, surface roughness, porosity, chemical purity, and flowability. Therefore, the choice of powder production methods is critical in additive manufacturing, as it affects the performance of the powder in the subsequent processing steps.

Gao et al. [15] investigated the effect of using specific gases during the atomization of metal powder feedstock. The researchers characterized two nitrogen-atomized and one argon-atomized 316L stainless steel powders using traditional powder flow measurement tools, including Hall flow rate, angle of repose, apparent density, tap density, and rotating drum and annular shear rheological tools. The study found that traditional flowability characterization tools and particle size measurements were not able to differentiate the powders, but rheological testing using a Revolution Powder Analyzer and an FT4 Powder Rheometer allowed for differences in the powders to be identified. The argon-atomized powders exhibited the lowest AAs when tested in the rotating drum and improved flow, which was attributed to the higher sphericity of the powder. The selected powder characteristics and flowability properties are presented in Tables 11.4 and 11.5. The research highlights the importance of choosing appropriate characterization tools to accurately assess powder flow properties.

The research conducted by Sun et al. [61] compared the characteristics of gas-atomized Ti–6Al–4V powders from six different vendors. The authors followed the **ASTM F3049** standard [64] to examine the differences in microstructures, particle size distributions, chemistries, surface appearances, flow behavior, and packing densities of the powders. While the study provided detailed descriptions of the similarities and differences among the various powders, as listed in Tables 11.4 and 11.5, the authors emphasized that further research is necessary to understand how these discrepancies in powder properties impact the properties of additively manufactured parts.

Cui et al. studied the effects of PREP parameters on the size distribution of Ti–6Al–4V powder. They investigated the influence of rotation speed, electrode diameter, plasma current, additional gas, and gas flow rate on powder formation during PREP [14, 65]. The results revealed that increasing the rotation speed or diameter of the electrode resulted in a decrease in mean powder size, while an increase in plasma current led to an increase in the mean powder diameter. Furthermore, the study investigated the effects of additional gas flow, and it was found that the use of a He atmosphere resulted in smaller powder compared to the Ar atmosphere. Increasing the flow rates at low gas flow rates decreases the mean particle size, which was attributed to the disturbance effect, while at high gas flow rates, the cooling effect dominates, resulting in an increase in the mean particle size.

Table 11.4 Influence of powder production method and process parameters on primary powder characteristics.

		Particle size distribution															
		Light/Laser scattering			Image analysis			Sieving			μ-CT			Sphericity			
Material	Powder production method	D10 (μm)	D50 (μm)	D90 (μm)	D10 (μm)	D50 (μm)	D90 (μm)	D10 (μm)	D50 (μm)	D90 (μm)	D10 (μm)	D50 (μm)	D90 (μm)	D10	D50	D90	Reference
Austenite stainless steel	GA, nitrogen A	23	35	53	12	23	38							0.70	0.91	0.98	[15]
	GA, nitrogen B	20	30	44	10	21	32							0.72	0.91	0.98	
	GA, argon	20	32	51	11	20	32							0.74	0.94	0.99	
Ti–6Al–4V	GA, AP&C				49	64	93										[61]
	GA, Hoeganaes				51	68	95										
	GA, Puris				38.1	65	102										
	GA, TIMET				52.5	79	113										
	GA, ATI				54.1	101	146										
	GA, Praxair				57.8	70	95										
	GA										25	36	46	0.46	0.79	0.91	[62]
	PA										26	37	50	0.64	0.93	0.97	
	PA										20	33	44	0.55	0.84	0.93	
	GA				43	70	98										[63]
	PREP				60	105	120										
	PA				42	60	92										
Ni-based alloy 625	GA	21	32	45	20	28	44	19	30	42	17	25	43	0.96			[35]
	WA	18	33	54	20	42	63	14	29	44	7	16	39	0.76			

GA: gas atomization, WA: water atomization, PA: plasma atomization, PREP: plasma rotating electrode process.

Compared to Ti–6Al–4V, there are far fewer studies on producing Al alloy powders. Özbilen [66] conducted a study to investigate how the particle characteristics of commercially pure Al powders were affected by changes in atomizing gas and gas pressure. The atomizing gas pressure varied between 1.06–1.85 MPa for argon runs and 1.02–1.85 MPa for helium runs during Al powder production. Particle size was analyzed using dry sieving and wet sieving. The size distributions revealed a bimodal distribution of all powder batches. Argon gas led to a coarser mean particle size compared to helium gas, which was attributed to the formation of fewer fine particles (sub-10 μm). The mean particle size of Al decreased as the gas pressure was reduced. Argon with low pressure caused the production of generally globular or spherical powder particles. However, at high argon pressure, the fine size range of the particles was spherical, while the coarse particles had oblong and cylindrical shapes, as well as more satellites attached to the surface. When helium gas was used, many agglomerates of very fine particles, as well as dumbbell or cylindrical-shaped coarse and fine particles, were observed.

Powder characteristics are also highly dependent on the different powder production methods. Mostafaei et al. systematically analyzed the characteristics of Ni-based alloy 625 powder particles that were atomized using GA or water atomization methods [35]. The morphology and size distribution of the water and argon-atomized powders were examined using various methods such as sieving, optical image analysis, scanning electron microscopy, laser particle size analysis, and micro-computed X-ray tomography (μ-CT). As expected, the water-atomized powder particles had irregular morphology, whereas the gas-atomized particles were more spherical. SEM micrographs of water

Table 11.5 Influence of powder production method and process parameters on powder flow properties.

Material	Powder production method	Apparent density, ρ_{App} (g cm^{-3})	Tap density, ρ_{Tap} (g cm^{-3})	Hausner ratio, P_{Tap}/ρ_{App}	Hall flow rate (s/50 g)	Basic flow energy (mJ)	Specific energy (mJ g^{-1})	Reference
Austenite stainless steel	GA, nitrogen A	4.38 ± 0.01	4.53 ± 0.01	1.14 ± 0.01	13.3 ± 0.2	867 ± 27	2.82 ± 0.11	[15]
	GA, nitrogen B	4.35 ± 0.01	5.07 ± 0.01	1.16 ± 0.01	14.3 ± 0.2	873 ± 12	2.97 ± 0.09	
	GA, argon	4.53 ± 0.01	5.19 ± 0.02	1.15 ± 0.01	13.1 ± 0.1	886 ± 10	3.00 ± 0.47	
Ti–6Al–4V	GA, AP&C	2.52 ± 0.005	2.76	1.09	26.6 ± 0.1			[61]
	GA, Hoeganaes	2.44 ± 0.010	2.78	1.14	26.9 ± 0.1			
	GA, Puris	2.47 ± 0.009	2.86	1.16	26.2 ± 0.4			
	GA, TIMET	2.64 ± 0.004	2.78	1.05	26.8 ± 0.1			
	GA, ATI	2.60 ± 0.004	2.94	1.13	24.5 ± 0.1			
	GA, Praxair	2.45 ± 0.015	2.79 ± 0.03	1.14	26.5 ± 0.03			
	GA	2.39 ± 0.012	2.55 ± 0.022	1.07	35 ± 0.4	294 ± 14	2.85 ± 0.03	[62]
	PA	2.61 ± 0.015	2.72 ± 0.017	1.04	25 ± 0.07	265 ± 4	1.61 ± 0.01	
	PA	2.58 ± 0.016	2.70 ± 0.00	1.05	26 ± 0.05	280 ± 14	1.73 ± 0.06	
	GA		2.38		33.5			[63]
	PREP		2.59		29.6			
	PA		2.59		31.8			

GA: gas atomization, WA: water atomization, PA: plasma atomization, PREP: plasma rotating electrode process.

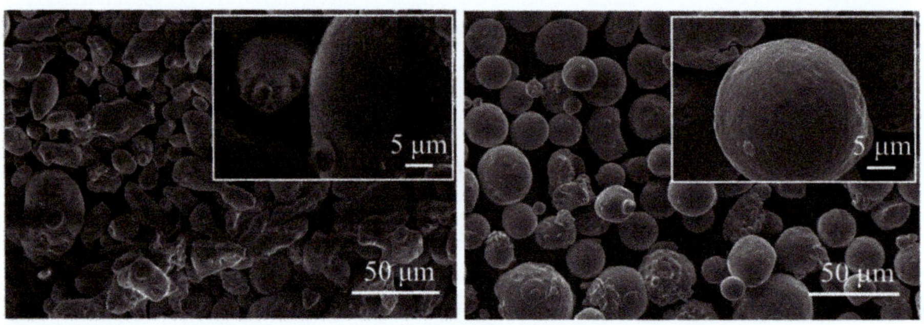

(a) (b)

Figure 11.13 SEM micrographs of Ni-based alloy 625 powders produced by (a) water atomization and (b) gas atomization. Insets show micrographs at higher magnification. Source: Mostafaei et al. [35]/Reproduced with permission from Elsevier.

and gas-atomized Ni-based alloy 625 powders are shown in Figure 11.13. Some powder characterization results are presented in Table 11.4. The study not only compared different powders but also evaluated the strengths and weaknesses of various characterization techniques. The authors concluded that μ-CT with higher resolution provides more accurate size and morphology measurements of a large amount of powder particles than other powder characterization methods, such as sieving, optical microscopy, and laser diffraction particle analysis. Additionally, μ-CT offers the possibility to nondestructively visualize pores in a large volume of powder.

Ti–6Al–4V has been extensively researched as a material for additive manufacturing, and as such, Ti–6Al–4V powder particles produced through different methods have been studied and compared. In the study by Brika et al. [62], three powder batches were produced, with one being gas-atomized (powder 1) and two plasma atomized. One of the plasma-atomized batches (powder 2) had a similar particle size distribution to the gas-atomized batch, while the other contained a greater amount of fine particles (powder 3). The powder particles were analyzed using μ-CT to examine their morphology, size, and density. The flowability was evaluated using Hall and Gustavsson flowmeters, as well as an FT4 powder rheometer. They found that the plasma atomized powder batch (powder 2) with higher sphericity displays better rheological behavior and more efficient particle packing, favoring the formation of a highly dense powder bed. The presence of a greater number of fine particles in powder 3 negatively affects the flow properties due to higher interparticle friction. The authors further introduced a performance metric based on established rheology-based criteria to determine the overall suitability of the powder for the laser PBF process.

Gas-atomized and plasma-atomized Ti–6Al–4V powders that have been recycled were investigated by combining μ-CT with mathematical analysis by Garboczi and Hrabe [67]. Each scanned particle was analyzed in 3D and various characteristics of the particles, including volume, surface area, length, width, thickness, curvature, moment of inertia, and porosity, were generated. The results show that the recycled plasma-atomized powder particles are more spherical than the recycled gas-atomized powder particles. According to their findings, the aspect ratios of particles determined in 2D differ from those obtained in 3D, questioning the validity of using 2D methods to predict particle size accurately.

Karlsson et al. [68] examined Ti–6Al–4V powders produced by gas-atomized and plasma-rotating electrode processes with size ranges of 45–100 and 25–45 μm, respectively. The particle surface structure and size distributions of the powders were investigated using scanning electron microcopy. They showed powder particles produced by GA had a rougher surface, larger particle size, and broader size distribution than those produced by PREP. Guo et al. [69] investigated Ti–6Al–4V powders that were synthesized using two different methods: **electrode induction melting gas atomization** (EIGA) and PREP. Figure 11.14 displays SEM micrographs of the powders, indicating that the EIGA powder is mostly spherical, with small satellites attached to the surface of the larger particles (Figure 11.14a,c). On the other hand, the PREP powder is spherical in shape, with a smooth surface free from satellites (Figure 11.14b,d). Laser scattering was used to analyze the particle size distribution, as presented in Figure 11.14e. The EIGA powder exhibited an average particle size (D50) of 93 μm, along with a broader size distribution, while PREP had an average particle size (D50) of 135 μm with a narrower size distribution.

Three types of Ti–6Al–4V powders, produced through GA, plasma atomization, and PREP, were examined using various techniques such as SEM, mass spectrometer gas analyzer, and synchrotron X-ray computed tomography to compare their microstructure, porosity, and argon content [63]. The powders were sieved to achieve identical particle sizes below 150 μm, although they had varying particle size distributions (Table 11.4). The flowability of the powders (Table 11.5) was found to be similar across all three types. The study found that the PREP powder exhibited considerably better surface smoothness and sphericity than the gas-atomized and plasma-atomized powders. Furthermore, the results demonstrated that the PREP powder had the lowest porosity and argon content, whereas the gas-atomized powder had the highest values.

Overall, the available literature on powder production and characterization for additive manufacturing highlights the crucial role of process parameters control and powder production method selection in achieving desired powder properties for subsequent processing. However, a lack of consistency in the analysis techniques used to evaluate powder characteristics poses a challenge in comparing results across different studies. Standardizing powder characterization procedures can be helpful in addressing this issue. Additionally, a significant proportion of the existing studies have focused on a limited number of materials, such as Ti–6Al–4V, Ni-based alloys, and stainless steel powders, despite the expanding range of materials that can be printed using additive manufacturing. Therefore, exploring the production methods for additional powders is necessary to expand the scope of additive manufacturing applications.

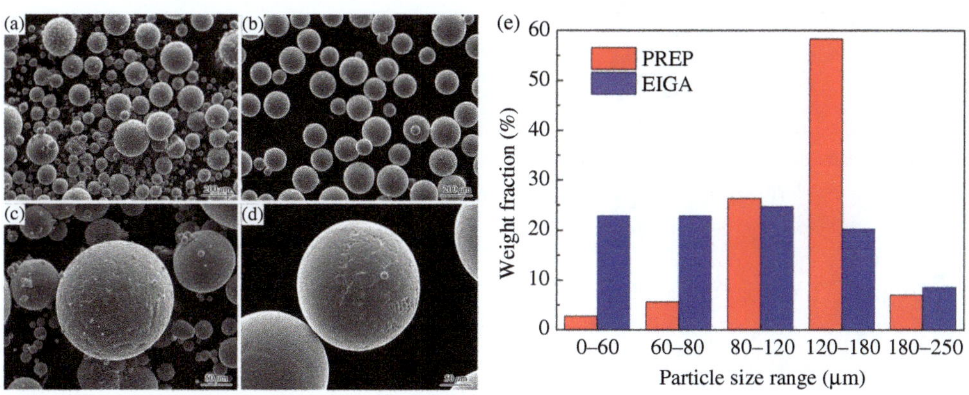

Figure 11.14 Microstructure of Ti–6Al–4V powder processed by (a, c) EIGA and (b, d) PREP. (e) Particle size distributions for electrode induction melting gas-atomized and plasma rotating electrode processed Ti–6Al–4V powders. The powders were sieved by 60 standard mesh (powder particle size ≤250 μm). Source: Guo et al. [69]/Reproduced from Springer Nature.

11.6 Influence of Powder Reuse, Recycling, and Recovery on Powder Characteristics and Additive Manufacturing

Adopting and reusing metal powder in additive manufacturing processes has gained significance due to its potential to reduce material waste, cost, and environmental impact. However, a crucial concern remains: will repeated exposure to the additive manufacturing environment and recycling procedures lead to changes in the physical and chemical properties of the powder, ultimately affecting the quality of the final products? In addition to the impact of the number of powder reuse cycles, the additive manufacturing condition and recycling procedures are also important factors that may affect the characteristics of the powder. To fully embrace the benefits of powder recycling, it is essential to investigate the effects of powder reuse on the properties of metal powder particles. By understanding how the characteristics of the powder evolve over many reuse cycles, manufacturers can optimize their recycling strategies and ensure high-quality end products.

Several alloys have been studied to determine the influence of powder reuse for additive manufacturing, including Ti–6Al–4V [48–51, 56, 57, 70–74], Ni-based alloys [50, 59, 60, 72, 75–78], stainless steels [46, 79–83], and Al-based alloys [82, 84–87].

As-received and recycled Ti–6Al–4V powder (Figure 11.15a, b) for electron beam melting (EBM) and a Ni-alloy powder (Figure 11.15c, d) for selective laser melting (SLM) have been investigated by Strondl et al. [56]. The recycled EBM powder was treated by blasting to break bonds between particles, while the recycled SLM powder was sieved (<63 μm) after the machine use. They found that powder reuse has a marginal influence on particle size distribution and aspect ratio (Figure 11.15). Blasting can cause mechanical impact marks on the powder surface, as seen in Figure 11.15a,b. Possibly due to the impact marks on the powder particle surfaces, together with a slightly increased fraction of small particles, the recycled EBM powder showed lower flowability than the new powder. The opposite outcome was found for the powder intended for SLM. The reason for this could be a decreased amount of finer particles compared with the new powder owing to the powder layering process during SLM.

The influence of reuse cycles (up to 21 cycles) on **extra-low interstitials** (ELIs) Ti–6Al–4V powder for EBM has been investigated in terms of powder composition, particle size distribution, particle morphology, and flowability [88]. With an increase in reuse cycles, the particle size distribution became narrower, although there was very little change in D50 (vol.) values (Table 11.6). The powder particles became less spherical and rougher after 11 reuse cycles. After 21 cycles of reuse, flowability improved by a modest amount, as indicated by the decreased powder flow rate and tap density (apparent density remained essentially unchanged). Please note that the powder flow rate values in Table 11.6 are in units of s/50 g; thus, a lower flow rate value indicates better flowability.

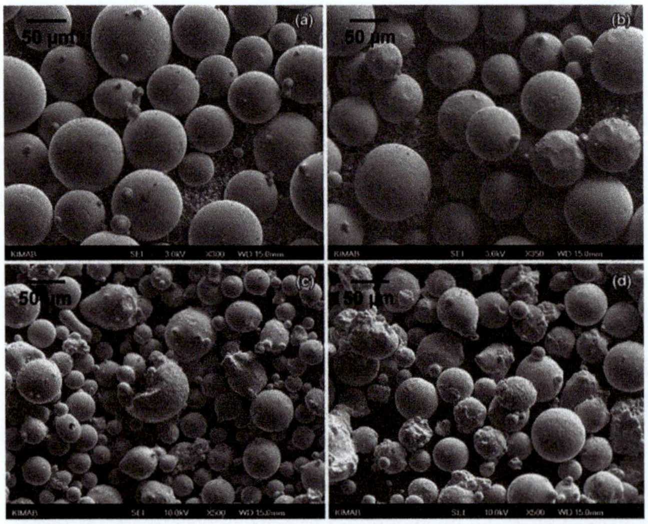

Figure 11.15 (a) As-received and (b) recycled Ti–6Al–4V powder for electron beam melting. (c) As-received and (d) recycled Ni-alloy powder for SLM. Source: Strondl et al. [56]/Reproduced from Springer Nature.

Table 11.6 Effect of powder reuse on powder characteristics.

Material	Condition	Flow rate (s/50 g)	Avalanche angle	Avalanche energy (mJ kg^{-1})	Apparent density, ρ_{App} (g cm^{-3})	Tap density, ρ_{Tap} (g cm^{-3})	Particle size (μm) D10	D50	D90	Reference
Ti–6Al–4V	As-received					2.84				[56]
	EBM recycled					2.83				
Ti–6Al–4V	As-received	32.5			2.56	2.96				[88]
	10 cycles	29.5			2.58	2.94		72.7[a)]		
	21 cycles	28.3			2.57	2.88		73.2[b)]		
Ti–6Al–4V	As-received	1.6					24.6	38.6	59.9	[71]
	100 cycles	3.4					19.7	30.6	47.3	
Inconel 718	As-received		51.8	22.1	4.50	5.26	14.6	27.1	49.0	[60]
	50 cycles	16.2	44.3	11.9	4.35	5.12	19.2	30.8	49.6	
ATI 718	As-received	11.5					26.5	31.5	49.7	[78]
	5 cycles	12.3					26.5	40.1	72.3	
	9 cycles	13.5					24.3	40.2	84.6	
Stainless steel	As-received		34.8	8.4	4.35	4.72				[46]
	4 cycles									
	7 cycles		33.3	7.2	4.45	4.80				
AISI 316L stainless steel	As-received							22.0		[82]
	50 cycles							29.0		
AISI 316L stainless steel	As-received						7.0	16.9	26.9	[58]
	50 cycles						8.2	15.4	27.5	
AlSi10Mg	As-received							11.0		[82]
	50 cycles							13.0		

a) After 2 reuse cycles.
b) After 16 reuse cycles.
Source: Adapted from [29].

In addition to the studies highlighted in Table 11.6, Popov et al. [57] used SEM to investigate the morphological defects of gas-atomized Ti–6Al–4V powder after being recycled up to 69 times in EBM. The findings are illustrated in Figure 11.16. Morphological defects observed in the recycled powder can be categorized into several types: elongated particles, agglomerated satellites, and particles with molten satellites. These defects occurred due to overheating, melting, and mechanical impacts during the additive manufacturing and recycling processes, including sieving.

Quintana et al. evaluated the impact of ELI Ti–6Al–4V powder reuse in SLM on powder characteristics [70]. The reused powder was analyzed using two sampling methods: (i) before printing, 200 g of powder were collected from the dispenser, and (ii) during the SLM process, double cone-shaped trap-capsules, which contain approximately 60 g of powder, were vertically printed along with other production parts. The study found that as the number of reuse cycles increased (from 1 to 31), there was a slight decrease in the D85 particle size, as measured by both sampling methods. Although the recycled powder maintained a mostly spherical shape, it had higher surface roughness and fewer satellites. The flowability of the powder increased with reuse cycles, while the tap density showed a slight decline, which might be due to the decrease in satellites and the sintering of fine particles during the SLM process.

Denti et al. [71] conducted a study to investigate how powder recycling affects the laser-beam powder bed fusion (PBF-LB) of Ti–6Al–4V. Their findings demonstrated that the original Ti–6Al–4V powder particles exhibited a predominantly normal Gaussian-type particle size distribution and were nearly spherical, with only a few particles being slightly elongated or broken. To recycle the powder, all unfused particles, including loose powder remaining in the build chamber and any overflow, were extracted from the PBF-LB machine once the printing job was completed. The powder was then thoroughly mixed and sifted through an 80 μm sieve under an inert gas. Despite sieving used during recycling, satellite particles were found attached to the powder surface, and a few aggregates were present in

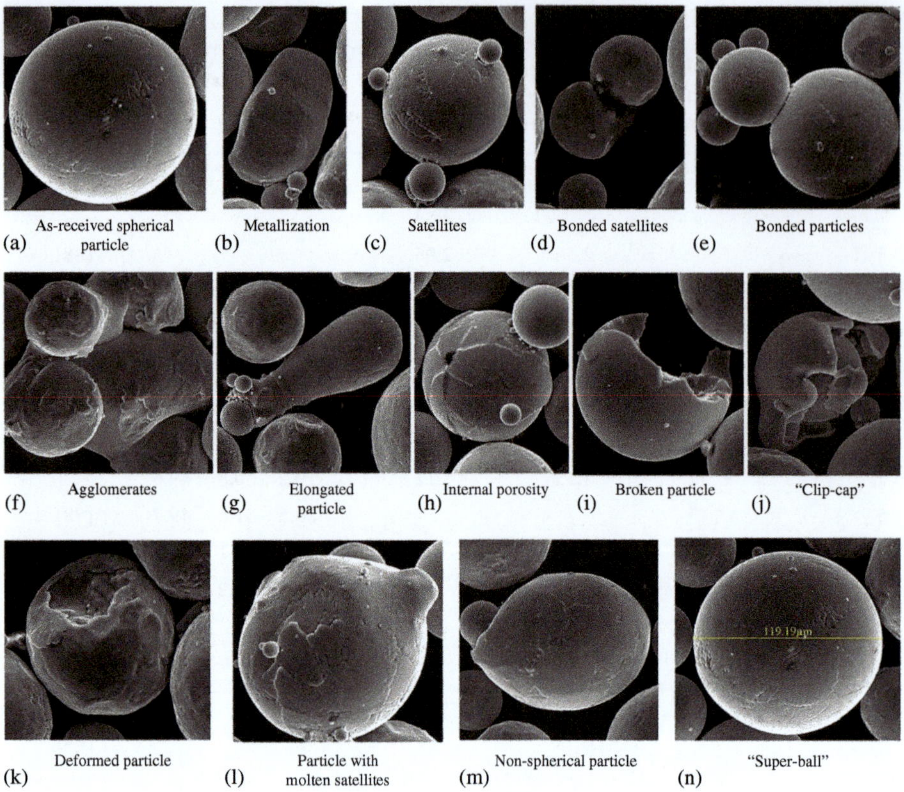

Figure 11.16 Classification of Ti-6Al-4V powder defects in EBM: (a) as-received spherical particle, (b–n) recycled powder showing a variety of defects. Source: Popov et al. [57]/Reproduced with permission from Elsevier.

the powder after 100 reuse cycles. Powder reuse led to a slight narrowing of the particle size distribution, a decrease in D50 values, and a large drop in flowability (increase in powder flow rate value, in units of s/50 g), as seen in Table 11.6.

The utilization of GA-produced Inconel 718 powder for up to 50 cycles resulted in several modifications in powder characteristics [60]. There was a noticeable change in the particle size distribution, with increased D10, D50, and D90 values (as shown in Table 11.6) and decreased circularity. The amplified particle sizes were attributed to the presence of **spatters** and satellites in the reused powder (see Chapter 9 for more discussion on spattering.) Furthermore, both the apparent density and tap density of the reused powder experienced small decreases, leading to improved flowability (as shown in Table 11.6). The authors suggested that the increased flowability might be due to a decrease in the concentration of small particles in the powder or particle oxidation.

When gas-atomized Ni superalloy ATI 718 powder was reused for up to nine cycles, it underwent changes in particle size and morphology [78]. There was a noticeable increase in the number of large-sized particles, many of which had nonspherical agglomerated or fused morphologies. The particle size distribution became flattened. Furthermore, the D50 value for the powder increased, and flowability decreased (as indicated in Table 11.6).

Heiden et al. [58] conducted a comparison of various properties of as-received (atomized) and reused (up to 30 cycles) 316L stainless steel powders in PBF-LB. The reused powder showed small changes in the particle size distribution (as shown in Table 11.6), a minor decrease in circularity, and a slight reduction in aspect ratio. Both as-received and reused powders exhibited spherical, angular/irregular, and rod-like particle shapes, with satellites (ranging in size from nm to μm) present in both types of powders. However, with reuse, there was a slight increase in the proportion of irregular-shaped particles. The as-received powder particles exhibited distinct grain boundaries on their surfaces, whereas reused powder particles were slightly deformed and did not exhibit visible grain boundaries on the surfaces. The particles that did not contain satellites had decreased surface roughness with reuse, hypothesized to be due to the loss of dendritic exteriors.

According to a study on gas-atomized AISI 304L stainless steel powder [46], recycling the PBF-LB powder resulted in a more spherical morphology than the as-received powder. Additionally, slight increases in both apparent and tap densities led to increased flowability (as shown in Table 11.6). As the reuse cycles increased, the particles in the recycled powder became slightly larger and more elongated. Regarding particle morphology, the recycled powder differed significantly from the as-received powder after the seventh reuse cycle. Specifically, the recycled powder contained large-sized particles, some highly spherical, while others were irregular.

As-received and reused powders (50 cycles) of 316L stainless steel and AlSi10Mg used for the PBF-LB process were compared in terms of morphology, particle size distribution, and circularity [82]. The particle size distribution of each powder slightly increased with reuse, as shown in Table 11.6. The as-received and reused 316L stainless steel powders were primarily spherical, with satellites only present in the as-received powder. The as-received and reused AlSi10Mg powders had similar morphologies, but satellites were present in the reused powder. The mean circularity of 316L stainless steel powder was relatively consistent with reuse, while the mean circularity of AlSi10Mg powder decreased.

Upon reviewing the studies on Ti–6Al–4V powder, it is evident that there is no consistent trend regarding the D50 values with reuse. Certain studies indicate that D50 values either decreased or remained unchanged with reuse (after 15–100 cycles [49, 51, 70, 71]), while others observed an increase (after 12 cycles [89]). In the case of Ni-based alloys, a consistent trend can be observed across multiple studies: particle size increased with powder reuse, regardless of the specific parameter examined, such as D50 [59, 60, 75, 78] or "weighted average" [76], and across a range of up to nine cycles and 50 cycles. However, for AlSi10Mg alloy powder, few replication studies focused on the same powder property (such as D50), making it challenging to identify consistent trends [84–87]. Overall, there is little coherence in the trends of the properties and powder characteristics that were commonly assessed, as summarized in this section.

11.7 Postproduction Methods for Treating Powders

Metal powders suitable for additive manufacturing processes in the current market are mainly produced by GA or PREP, both of which have a high operating cost. As a result, the costs of powders and finished parts are high for most additive manufacturing operations and applications. Cost-effective spherical metal powders are suitable for additive manufacturing and are needed for the broader application of additive manufacturing. Postproduction methods that modify powder characteristics, in combination with powder recycling or low-cost powder production methods, have been explored in this regard. In this section, various postproduction methods for treating powders are discussed.

Inductively-coupled plasma process (ICPP) is a spheroidization technique developed by Tekna Plasma Systems, Inc. that employs a high-temperature plasma generated by high-frequency induction coils. The plasma torch used in the spheroidization system includes a plasma torch, as illustrated in Figure 11.17a [90]. The process uses high-frequency induction coils to generate a high-temperature plasma within a confinement tube. Gas flows through the tube, and the resulting plasma heats and partially or fully melts the powder particles, causing them to become spherical droplets under surface tension. ICPP has been used to treat powders of a variety of materials, such as commercially pure Ti [92], Ti–6Al–4V alloy powder [91, 93], 316L stainless steel [94], AlSi10Mg [95], and refractory high-entropy alloy NiTaMoW [96] to improve their suitability for additive manufacturing.

SEM micrographs of Ti–6Al–4V powders before and after ICPP are shown in Figure 11.17b,c [91]. The as-received powder particles exhibit irregular morphologies with rough edges and sharp corners. The spheroidized powders had almost perfect sphericity, smooth surfaces, good dispersion, narrower particle size distribution, and improved flow properties compared to the as-received powders [91]. This approach has been investigated for producing cost-effective titanium spherical powders using low-cost hydrogenation–dehydrogenation treated titanium powder [92].

316L stainless steel powder underwent ICPP to enhance its use in powder 3D printing [94]. The as-received powder shows a unimodal particle size distribution in the 10–100 μm range. ICPP powder showed a bimodal size distribution, generating fine particles in the 0.1–10 μm range. The dynamic flowability was found to be lower for the spheroidized powder. The density of 3D-printed SS 316L pieces with spheroidized powder was comparable to that with commercial powder.

High-temperature fluidization is when a fluid (gas or liquid) is passed through a bed of powder at a high velocity, causing the powder to become suspended and exhibit fluidic behavior. To create a cost-effective method for preparing powders suitable for additive manufacturing, researchers have investigated the effects of fluidization treatment on the morphology, particle size distribution, and flowability of hydrogenated-dehydrogenated Ti powder [97]. The powder was processed in a fluidized bed reactor under a steady flow of high-purity Ar gas at varying temperatures for

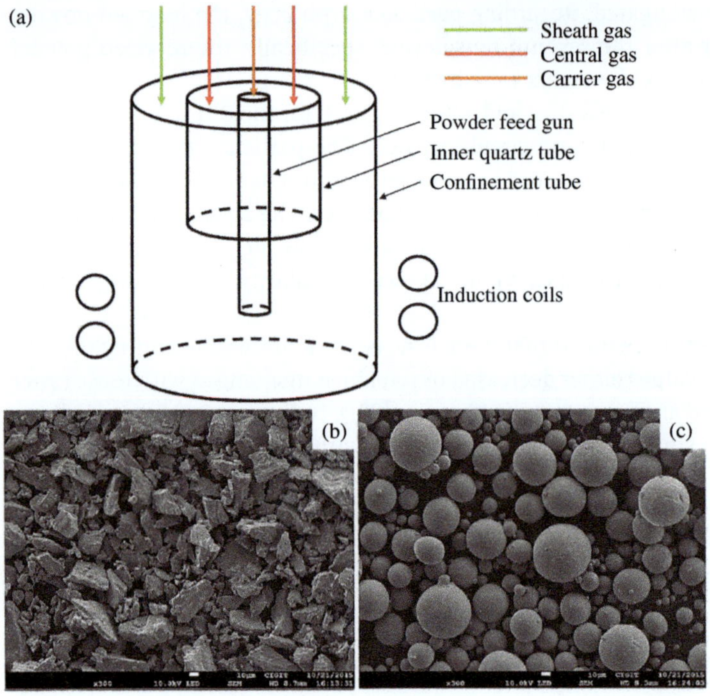

Figure 11.17 (a) Schematic of torch's components in the plasma spheroidization system [90]. (b and c) SEM micrographs of the as-received and the plasma spheroidized Ti–6Al–4V powders, respectively. Source: Wei et al. [91]/Elsevier.

10 minutes. The treated powder had smoother edges and a narrower particle size distribution than the as-received powder, with some smaller particles partially embedded in or adhered to larger-sized particles. The flowability of the powder increased slightly between 450 and 500 °C but decreased between 500 and 600 °C due to moderate sintering.

Fluidization treatment was employed to treat a mechanically milled WMoTaTi refractory high-entropy alloy [98]. Fluidization effectively improves particle sphericity and thus makes its spreading performance suitable for SLM. The micro-morphology, microstructure, and spreadability of treated powder were studied. Powders, after fluidizing, exhibited a higher sphericity and narrower particle size distribution, as seen in Figure 11.18. The Hall flow rate of the fluidized powders was 35.3 ± 0.2 s/50 g, a significant improvement over the ball-milled powders. The fluidized refractory high-entropy alloy powder also demonstrated improved packing and spreadability.

Mechanical ball milling has also been investigated as a promising technique for treating powders and modifying their characteristics for additive manufacturing. Ball milling led to marked changes in various physical properties of hydrogenated–dehydrogenated prepared Ti powder [99, 100]. Compared to the as-received powder, the ball-milled powder exhibited modified morphology (grinding off the edges of the powder led to a change from nonspherical to near-spherical morphology); narrower particle size distribution; and decreased D10, D50, and D90 values, which were attributed to collisions and friction that occurred during the milling. Additionally, multiple methods determined powders' flowability/rheological properties before and after ball milling. The ball-milled powder showed a lower Hausner ratio, resulting in an increase in flowability index, lower angle of repose, lower degree of compression (defined as the difference between the tap density and the apparent density divided by the tap density), and higher specific surface area. The results confirmed that the ball-milled titanium powder achieved good flowability despite the poor flowability of the as-received powder.

The characteristics of 17-4 stainless steel powder produced by **cryomilling**, i.e. attrition ball milling of powder at cryogenic temperatures or cryogenic liquid (e.g. argon or nitrogen), have been examined [101]. Degassing was often performed afterward to remove stearic acid blended with the powder before cryomilling. Gas-atomized, as-cryomilled, and cryomilled-degassed-sieved powders were compared. For **degassing**, the powder was heated under flowing argon to 325 °C at 10 °C/min, held for six hours, and then cooled to room temperature. GA-produced powder particles were generally spherical and contained pores and satellites. The cryomilled powder particles were plate-like and displayed a noticeable increase in particle size. The cryomilled-degassed-sieved powder had more

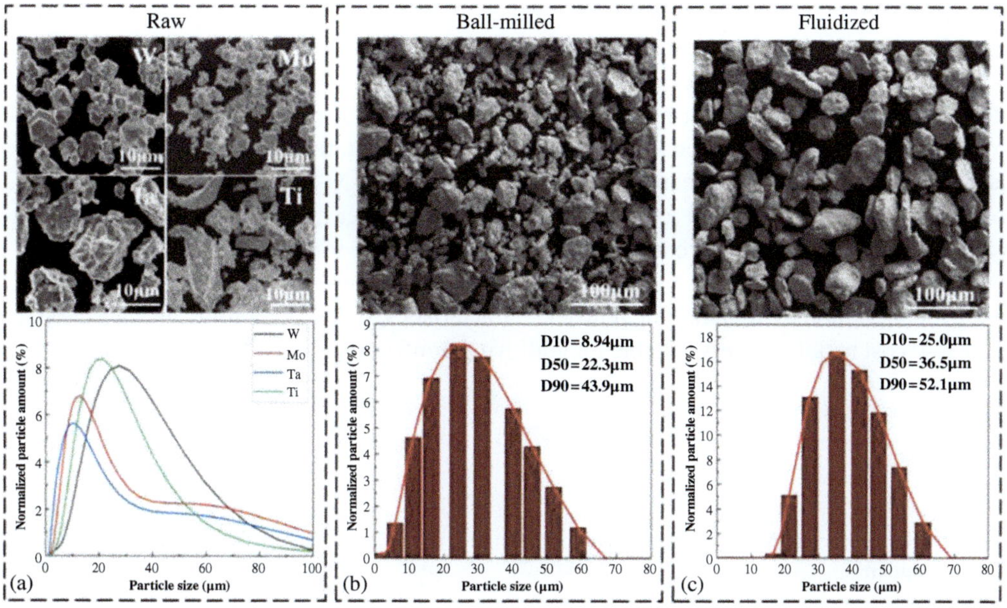

Figure 11.18 WMoTaTi refractory high-entropy alloy powder morphologies and the particle size distributions measured by laser diffraction of the (a) constituent, (b) ball-milled, and (c) fluidization treated powders. Source: Liu et al. [98]/Reproduced from Taylor and Francis group.

remarkable flowability than the GA-produced powder, as measured by a Brookfield powder flow tester, possibly due to the reduction in fine particles due to the cryomilling.

To ensure optimal performance during additive manufacturing, powders must be moisture-free, as moisture can cause clumping, inconsistent flow, and poor surface finish. Therefore, various postproduction **drying methods** are employed to remove moisture from powders. In a study by Cordova et al., the effects of **vacuum drying** (<1 Pa, at 85 °C, for 12 hours) and **air-drying** (at 150 °C, for 20 minutes) on the microstructure and flowability of four standard metal powders (Inconel 718, Ti–6Al–4V, AlSi10Mg, and Scalmalloy) were investigated, along with their as-received and moisturized (50 °C and 80% humidity) states [102]. Microstructure analysis using SEM showed no differences in powder morphology between the powders in their as-received, wet, and dried states. However, the flowability of the vacuum-dried powders, as measured by the Hall flowmeter, increased for all four materials, indicating that vacuum drying is an effective drying method. Two novel applicator tools were specifically designed in the study to assess the spreadability in the thin layer application of PBF-LB powders. Nevertheless, the results obtained from these applicator tools indicate that the variations in drying treatments did not significantly improve the spreadability in thin layer application.

11.8 Summary

This chapter gives an overview of the physical characteristics of metal powders that are relevant for their use in additive manufacturing, including the powder characterization approaches, powder production methods along with powder reuse and powder recycling, the influences of powder production methods and process variables on the physical powder characteristics, and finally the effects of postproduction powder treatment methods.

The physical properties of metal powders used in additive manufacturing, particularly particle size and size distribution, particle morphology, and powder flowability, play a crucial role in the additive manufacturing process and the quality of the final products. Various characterization approaches have been developed to measure and quantify these physical powder properties. However, standardizing powder characterization procedures is essential, yet lacking, for consistent analysis and comparison of results across different studies. Powder sampling standards are also critical to obtaining representative samples for characterization.

Atomization processes, such as gas, water, plasma, and centrifugal atomization, are the most commonly used methods to produce metal powders suitable for additive manufacturing. Other techniques like mechanical milling, plasma rotating electrodes, and chemical processes are also being explored to produce metal powders. The production methods and process variables affect the mean particle size, particle size distribution, morphology, and flowability of powder particles.

Powder recycling and reuse in additive manufacturing have gained importance, driven by the demand for cost and waste reduction. However, there needs to be more consensus on best practices and standards for recycling strategies. Current literature focuses on a limited number of materials, including Ti–6Al–4V, Ni-based alloys, stainless steels, and Al-based alloys. Thus, there is a need to explore recycling methods for a broader range of powders. Factors like the number of reuse cycles, additive manufacturing conditions, and recycling procedures can affect powder characteristics. Investigating the effects of powder reuse can optimize recycling strategies and ensure high-quality products. Postproduction methods are being explored for modifying powder characteristics, such as spheroidicity, particle size, and flowability, to produce low-cost metal powders suitable for additive manufacturing.

Nomenclature

Variable	Description	Units
A	Area	mm^2
ADI	Apparent density increase	%
AA	Avalanche angle	degree
AM	Additive manufacturing	—
AR_{50}	Aspect ratio	—
D	Particle diameter	m
d_A	Diameter of circle	m
DED	Directed energy deposition	—
E	Flow energy	mJ
EBM	Electron beam melting	—
EIGA	Electrode induction melting gas atomization	—
ELI	Extra-low interstitials	—
FR	Flow rate	$g\,s^{-1}$; or s/50 g
g	Gravitational acceleration	$m\,s^{-2}$
GA	Gas atomization	—
h	Height	m
ICPP	Inductively-coupled plasma process	—
m	Number of valid spike values	—
MBA	Maximum break angle	degree
n	Number of step sizes	—
P	Perimeter	m
PA	Plasma atomization	—
PBF	Powder bed fusion	—
PBF-LB	Powder bed fusion – laser beam	—
PREP	Plasma rotating electrode process	—
Re	Reynolds number	—
R_H	Hausner Ratio	—
SE	Specific energy	$mJ\,g^{-1}$
SEM	Scanning electron microscopy	—
SLM	Selective laser melting	—

Variable	Description	Units
SV	Spike values	—
v	Settling velocity of the particles	$m\,s^{-1}$
WA	Water atomization	—
α	Angle of repose	degree
η_L	Viscosity of the fluid	Pa s
ρ_{App}	Apparent density	$g\,cm^{-3}$
ρ_{Tap}	Tap density	$g\,cm^{-3}$
ρ_p	Particle density	$kg\,m^{-3}$
ρ_L	Fluid density	$kg\,m^{-3}$
θ	Apex angle	degree
τ	Shear stress	kPa
μ-CT	Micro-computed X-ray tomography	—

References

1 Cooke, A.L. and Slotwinski, J.A. (2012). *Properties of Metal Powders for Additive Manufacturing: A Review of the State of the Art of Metal Powder Property Testing*. NIST.

2 Jillavenkatesa, A., Lum, L.-S.H., and Dapkunas, S. (2001). *NIST Recommended Practice Guide: Particle Size Characterization*. NIST.

3 ASTM B215-20 (2020). *Standard Practices for Sampling Metal Powders*. West Conshohocken: ASTM International.

4 Allen, T. (2003). *Powder Sampling and Particle Size Determination*. Elsevier.

5 Allen, T. and Khan, A.A. (1970). Critical evaluation of powder sampling procedures. *Chem. Eng.* 238: 108–112.

6 ASTM B214-22 (2022). *Standard Test Method for Sieve Analysis of Metal Powders*. West Conshohocken: ASTM International.

7 ASTM B761-17 (2021). *Standard Test Method for Particle Size Distribution of Metal Powders and Related Compounds by X-Ray Monitoring of Gravity Sedimentation*. West Conshohocken: ASTM International.

8 Slotwinski, J., Garboczi, E., Stutzman, P. et al. (2014). Characterization of metal powders used for additive manufacturing. *J. Res. Natl. Inst. Stand. Technol.* 119: 460–493.

9 ASTM B243-22 (2022). *Standard Terminology of Powder Metallurgy*. West Conshohocken: ASTM International.

10 Mikli, V., Kaerdi, H., Kulu, P., and Besterci, M. (2001). Characterization of powder particle morphology. *Proc. Est. Acad. Sci.* 7 (1): 22–34.

11 Kulu, P., Tumanok, A., Mikli, V. et al. (1998). Possibilities of evaluation of powder particle granulometry and morphology by image analysis. *Proc. Est. Acad. Sci.* 4 (1): 3–17.

12 Wu, J.-S., Ho, H.-O., and Sheu, M.-T. (2001). A statistical design to evaluate the influence of manufacturing factors on the material properties and functionalities of microcrystalline cellulose. *Eur. J. Pharm. Sci.* 12 (4): 417–425.

13 Hellén, L. and Yliruusi, J. (1993). Process variables of instant granulator and spheroniser: III. Shape and shape distributions of pellets. *Int. J. Pharm.* 96 (1): 217–223.

14 Cui, Y., Zhao, Y., Numata, H. et al. (2021). Effects of process parameters and cooling gas on powder formation during the plasma rotating electrode process. *Powder Technol.* 393: 301–311.

15 Gao, M.Z., Ludwig, B., and Palmer, T.A. (2021). Impact of atomization gas on characteristics of austenitic stainless steel powder feedstocks for additive manufacturing. *Powder Technol.* 383: 30–42.

16 Stachowiak, G.W. (1998). Numerical characterization of wear particles morphology and angularity of particles and surfaces. *Tribol. Int.* 31 (1): 139–157.

17 Marchetti, L. and Hulme-Smith, C. (2021). Flowability of steel and tool steel powders: a comparison between testing methods. *Powder Technol.* 384: 402–413.

18 Vasilenko, A., Glasser, B.J., and Muzzio, F.J. (2011). Shear and flow behavior of pharmaceutical blends – method comparison study. *Powder Technol.* 208 (3): 628–636.

19 Zegzulka, J., Gelnar, D., Jezerska, L. et al. (2020). Characterization and flowability methods for metal powders. *Sci. Rep.* 10 (1): 21004.

20 Muthuswamy, P. (2022). Influence of powder characteristics on properties of parts manufactured by metal additive manufacturing. *Lasers Manuf. Mater. Process.* 9 (3): 312–337.

21 ASTM B213-20 (2020). *Standard Test Methods for Flow Rate of Metal Powders Using the Hall Flowmeter Funnel.* West Conshohocken: ASTM International.

22 ASTM B855-22 (2022). *Standard Test Method for Volumetric Flow Rate of Metal Powders Using the Arnold Meter and Hall Flowmeter Funnel.* West Conshohocken: ASTM International.

23 ASTM B703-21 (2021). *Standard Test Method for Apparent Density of Metal Powders and Related Compounds Using the Arnold Meter*. West Conshohocken: ASTM International.

24 ASTM B527-23 (2023). *Standard Test Method for Tap Density of Metal Powders and Compounds.* West Conshohocken: ASTM International.

25 Leturia, M., Benali, M., Lagarde, S. et al. (2014). Characterization of flow properties of cohesive powders: a comparative study of traditional and new testing methods. *Powder Technol.* 253: 406–423.

26 Clayton, J., Millington-Smith, D., and Armstrong, B. (2015). The application of powder rheology in additive manufacturing. *JOM* 67 (3): 544–548.

27 Kiani, P., Scipioni Bertoli, U., Dupuy, A.D. et al. (2020). A statistical analysis of powder flowability in metal additive manufacturing. *Adv. Eng. Mater.* 22 (10): 2000022.

28 Antony, L.V.M. and Reddy, R.G. (2003). Processes for production of high-purity metal powders. *JOM* 55 (3): 14–18.

29 Lewis, G. (2022). Aspects of the powder in metal additive manufacturing: a review. *World J. Eng. Technol.* 10 (2): 363–409.

30 Zheng, B., Lin, Y., Zhou, Y., and Lavernia, E.J. (2009). Gas atomization of amorphous aluminum powder: Part II. Experimental investigation. *Metall. Mater. Trans. B* 40 (6): 995–1004.

31 Asgarian, A., Tang, Z., Bussmann, M., and Chattopadhyay, K. (2020). Water atomisation of metal powders: effect of water spray configuration. *Powder Metall.* 63 (4): 288–299.

32 Zheng, B. and Lavernia, E.J. (2011). Melt atomization. In: *Handbook of Atomization and Sprays: Theory and Applications* (ed. N. Ashgriz), 837–848. Boston, MA: Springer US.

33 Li, R., Shi, Y., Wang, Z. et al. (2010). Densification behavior of gas and water atomized 316L stainless steel powder during selective laser melting. *Appl. Surf. Sci.* 256 (13): 4350–4356.

34 Hoeges, S., Zwiren, A., and Schade, C. (2017). Additive manufacturing using water atomized steel powders. *Met. Powder Rep.* 72 (2): 111–117.

35 Mostafaei, A., Hilla, C., Stevens, E.L. et al. (2018). Comparison of characterization methods for differently atomized nickel-based alloy 625 powders. *Powder Technol.* 333: 180–192.

36 Fedina, T., Sundqvist, J., Powell, J., and Kaplan, A.F.H. (2020). A comparative study of water and gas atomized low alloy steel powders for additive manufacturing. *Addit. Manuf.* 36: 101675.

37 Abdelwahed, M., Bengtsson, S., Casati, R. et al. (2021). Effect of water atomization on properties of type 4130 steel processed by L-PBF. *Mater. Des.* 210: 110085.

38 Canakci, A. and Varol, T. (2015). A novel method for the production of metal powders without conventional atomization process. *J. Cleaner Prod.* 99: 312–319.

39 Fullenwider, B., Kiani, P., Schoenung, J.M., and Ma, K. (2019). Two-stage ball milling of recycled machining chips to create an alternative feedstock powder for metal additive manufacturing, *Powder Technology*, 342: 562–571.

40 Ma, K., Smith, T., Lavernia, E.J., and Schoenung, J.M. (2017). Environmental sustainability of laser metal deposition: the role of feedstock powder and feedstock utilization factor. *Proc. Manuf.* 7: 198–204.

41 Sartin, B., Pond, T., Griffith, B. et al. (2017). 316L powder reuse for metal additive manufacturing. In: *2017 International Solid Freeform Fabrication Symposium*, 2430–2446. University of Texas at Austin.

42 Carroll, P.A., Pinkerton, A.J., Allen, J. et al. (2006). The effect of powder recycling in direct metal laser deposition on powder and manufactured part characteristics. In: *Proceedings of AVT-139 Specialists Meeting on Cost Effective Manufacture via Net Shape Processing*, 18 - 1–18-10. NATO Research and Technology Organisation.

43 Terrassa, K.L., Haley, J.C., MacDonald, B.E., and Schoenung, J.M. (2018). Reuse of powder feedstock for directed energy deposition. *Powder Technol.* 338: 819–829.

44 Koike, R., Takemura, S., Kakinuma, Y., and Kondo, M. (2018). Enhancement of powder supply efficiency in directed energy deposition based on gas-solid multiphase-flow simulation. *Proc. CIRP* 78: 133–137.

45 Takemura, S., Koike, R., Kakinuma, Y. et al. (2019). Design of powder nozzle for high resource efficiency in directed energy deposition based on computational fluid dynamics simulation. *Int. J. Adv. Manuf. Technol.* 105 (10): 4107–4121.

46 Sutton, A.T., Kriewall, C.S., Karnati, S. et al. (2020). Characterization of AISI 304L stainless steel powder recycled in the laser powder-bed fusion process. *Addit. Manuf.* 32: 100981.

47 Moghimian, P., Poirié, T., Habibnejad-Korayem, M. et al. (2021). Metal powders in additive manufacturing: a review on reusability and recyclability of common titanium, nickel and aluminum alloys. *Addit. Manuf.* 43: 102017.

48 Wei, C., Ma, X., Yang, X. et al. (2018). Microstructural and property evolution of Ti6Al4V powders with the number of usage in additive manufacturing by electron beam melting. *Mater. Lett.* 221: 111–114.

49 Carrion, P.E., Soltani-Tehrani, A., Phan, N., and Shamsaei, N. (2019). Powder recycling effects on the tensile and fatigue behavior of additively manufactured Ti-6Al-4V parts. *JOM* 71 (3): 963–973.

50 Cordova, L., Campos, M., and Tinga, T. (2019). Revealing the effects of powder reuse for selective laser melting by powder characterization. *JOM* 71 (3): 1062–1072.

51 Ghods, S., Schultz, E., Wisdom, C. et al. (2020). Electron beam additive manufacturing of Ti6Al4V: evolution of powder morphology and part microstructure with powder reuse. *Materialia* 9: 100631.

52 Boulos, M. (2004). Plasma power can make better powders. *Met. Powder Rep.* 59 (5): 16–21.

53 Zhang, Y., Fang, Z.Z., Sun, P. et al. (2016). Thermodynamic destabilization of Ti-O solid solution by H_2 and deoxygenation of Ti Using Mg. *J. Am. Chem. Soc.* 138 (22): 6916–6919.

54 Fang, Z.Z., Paramore, J.D., Sun, P. et al. (2018). Powder metallurgy of titanium – past, present, and future. *Int. Mater. Rev.* 63 (7): 407–459.

55 Boulos, M.I. (2012). New frontiers in thermal plasmas from space to nanomaterials. *Nucl. Eng. Technol.* 44 (1): 1–8.

56 Strondl, A., Lyckfeldt, O., Brodin, H., and Ackelid, U. (2015). Characterization and control of powder properties for additive manufacturing. *JOM* 67 (3): 549–554.

57 Popov, V.V., Katz-Demyanetz, A., Garkun, A., and Bamberger, M. (2018). The effect of powder recycling on the mechanical properties and microstructure of electron beam melted Ti-6Al-4V specimens. *Addit. Manuf.* 22: 834–843.

58 Heiden, M.J., Deibler, L.A., Rodelas, J.M. et al. (2019). Evolution of 316L stainless steel feedstock due to laser powder bed fusion process. *Addit. Manuf.* 25: 84–103.

59 Yi, F., Zhou, Q., Wang, C. et al. (2021). Effect of powder reuse on powder characteristics and properties of Inconel 718 parts produced by selective laser melting. *J. Mat. Res. Technol.* 13: 524–533.

60 Paccou, E., Mokhtari, M., Keller, C. et al. (2021). Investigations of powder reusing on microstructure and mechanical properties of Inconel 718 obtained by additive manufacturing. *Mater. Sci. Eng. A* 828: 142113.

61 Sun, Y., Aindow, M., and Hebert, R.J. (2018). Comparison of virgin Ti-6Al-4V powders for additive manufacturing. *Addit. Manuf.* 21: 544–555.

62 Brika, S.E., Letenneur, M., Dion, C.A., and Brailovski, V. (2020). Influence of particle morphology and size distribution on the powder flowability and laser powder bed fusion manufacturability of Ti-6Al-4V alloy. *Addit. Manuf.* 31: 100929.

63 Chen, G., Zhao, S.Y., Tan, P. et al. (2018). A comparative study of Ti-6Al-4V powders for additive manufacturing by gas atomization, plasma rotating electrode process and plasma atomization. *Powder Technol.* 333: 38–46.

64 ASTM F3049-14 (2021). Standard guide for characterizing properties of metal powders used for additive manufacturing processes. West Conshohocken: ASTM International.

65 Cui, Y., Zhao, Y., Numata, H. et al. (2020). Effects of plasma rotating electrode process parameters on the particle size distribution and microstructure of Ti-6Al-4V alloy powder. *Powder Technol.* 376: 363–372.

66 Özbilen, S. (1999). Influence of atomising gas pressure on particle shape of Al and Mg powders. *Powder Technol.* 102 (2): 109–119.

67 Garboczi, E.J. and Hrabe, N. (2020). Particle shape and size analysis for metal powders used for additive manufacturing: technique description and application to two gas-atomized and plasma-atomized Ti64 powders. *Addit. Manuf.* 31: 100965.

68 Karlsson, J., Snis, A., Engqvist, H., and Lausmaa, J. (2013). Characterization and comparison of materials produced by electron beam melting (EBM) of two different Ti–6Al–4V powder fractions. *J. Mat. Process. Technol.* 213 (12): 2109–2118.

69 Guo, R.-P., Xu, L., Zong, B.Y.-P., and Yang, R. (2017). Characterization of prealloyed Ti–6Al–4V powders from EIGA and PREP process and mechanical properties of HIPed powder compacts. *Acta Metall. Sin. (Engl. Lett.)* 30 (8): 735–744.

70 Quintana, O.A., Alvarez, J., Mcmillan, R. et al. (2018). Effects of reusing Ti-6Al-4V powder in a selective laser melting additive system operated in an industrial setting. *JOM* 70 (9): 1863–1869.

71 Denti, L., Sola, A., Defanti, S. et al. (2019). Effect of powder recycling in laser-based powder bed fusion of Ti-6Al-4V. *Manuf. Technol.* 19 (2): 190–196.

72 Chandrasekar, S., Coble, J.B., Yoder, S. et al. (2020). Investigating the effect of metal powder recycling in electron beam powder bed fusion using process log data. *Addit. Manuf.* 32: 100994.

73 Cao, Y., Delin, M., Kullenberg, F., and Nyborg, L. (2020). Surface modification of Ti-6Al-4V powder during recycling in EBM process. *Surf. Interface Anal.* 52 (12): 1066–1070.

74 Shanbhag, G. and Vlasea, M. (2021). Powder reuse cycles in electron beam powder bed fusion—variation of powder characteristics. *Materials (Basel)* 14 (16): 4602.

75 Ardila, L.C., Garciandia, F., González-Díaz, J.B. et al. (2014). Effect of IN718 recycled powder reuse on properties of parts manufactured by means of selective laser melting. *Phys. Proc.* 56: 99–107.

76 Renderos, M., Girot, F., Lamikiz, A. et al. (2016). Ni based powder reconditioning and reuse for LMD process. *Phys. Proc.* 83: 769–777.

77 Gruber, H., Henriksson, M., Hryha, E., and Nyborg, L. (2019). Effect of powder recycling in electron beam melting on the surface chemistry of alloy 718 powder. *Metall. Mater. Trans. A* 50 (9): 4410–4422.

78 Rock, C., Ledford, C., Garcia-Avila, M. et al. (2021). The influence of powder reuse on the properties of nickel super alloy ATI 718^{TM} in laser powder bed fusion additive manufacturing. *Metall. Mater. Trans. B* 52 (2): 676–688.

79 Gorji, N.E., Saxena, P., Corfield, M. et al. (2020). A new method for assessing the recyclability of powders within Powder Bed Fusion process. *Mater. Char.* 161: 110167.

80 Soltani-Tehrani, A., Pegues, J., and Shamsaei, N. (2020). Fatigue behavior of additively manufactured 17-4 PH stainless steel: the effects of part location and powder re-use. *Addit. Manuf.* 36: 101398.

81 Delacroix, T., Lomello, F., Schuster, F. et al. (2022). Influence of powder recycling on 316L stainless steel feedstocks and printed parts in laser powder bed fusion. *Addit. Manuf.* 50: 102553.

82 Mohd Yusuf, S., Choo, E., and Gao, N. (2020). Comparison between virgin and recycled 316L SS and AlSi10Mg powders used for laser powder bed fusion additive manufacturing. *Metals* 10 (12): 1625.

83 Zhang, J., Hu, B., Zhang, Y. et al. (2018). Comparison of virgin and reused 15-5 PH stainless steel powders for laser powder bed fusion process. *Prog. Addit. Manuf.* 3 (1): 11–14.

84 Asgari, H., Baxter, C., Hosseinkhani, K., and Mohammadi, M. (2017). On microstructure and mechanical properties of additively manufactured AlSi10Mg_200C using recycled powder. *Mater. Sci. Eng. A* 707: 148–158.

85 Del Re, F., Contaldi, V., Astarita, A. et al. (2018). Statistical approach for assessing the effect of powder reuse on the final quality of AlSi10Mg parts produced by laser powder bed fusion additive manufacturing. *Int. J. Adv. Manuf. Technol.* 97 (5): 2231–2240.

86 Maamoun, A.H., Elbestawi, M., Dosbaeva, G.K., and Veldhuis, S.C. (2018). Thermal post-processing of AlSi10Mg parts produced by selective laser melting using recycled powder. *Addit. Man.* 21: 234–247.

87 Raza, A., Fiegl, T., Hanif, I. et al. (2021). Degradation of AlSi10Mg powder during laser based powder bed fusion processing. *Mat. Des.* 198: 109358.

88 Tang, H.P., Qian, M., Liu, N. et al. (2015). Effect of powder reuse times on additive manufacturing of Ti-6Al-4V by selective electron beam melting. *JOM* 67 (3): 555–563.

89 Seyda, V., Kaufmann, N., and Emmelmann, C. (2012). Investigation of aging processes of Ti-6Al-4 V powder material in laser melting. *Phys. Proc.* 39: 425–431.

90 Sehhat, M.H., Chandler, J., and Yates, Z. (2022). A review on ICP powder plasma spheroidization process parameters. *Int. J. Ref. Met. Hard Mater.* 103: 105764.

91 Wei, W.-H., Wang, L.-Z., Chen, T. et al. (2017). Study on the flow properties of Ti-6Al-4V powders prepared by radio-frequency plasma spheroidization. *Adv. Powder Technol.* 28 (9): 2431–2437.

92 Liu, M., Huang, L., Ren, C. et al. (2022). Effect of radiofrequency plasma spheroidization treatment on the laser directed energy deposited properties of low-cost hydrogenated-dehydrogenated titanium powder. *Materials (Basel)* 15 (4): 1548.

93 Liu, C.C., Lu, X., Zhang, L. et al. (2016). Fabrication of micro-fine spherical Ti–6Al–4V alloy powders based on hydrogen decrepitation and plasma spheroidisation. *Powder Metall.* 59 (4): 229–235.

94 Ji, L., Wang, C., Wu, W. et al. (2017). Spheroidization by plasma processing and characterization of stainless steel powder for 3D printing. *Metall. Mater. Trans. A* 48 (10): 4831–4841.

95 Wang, L., Liu, Y., and Chang, S. (2016). Fabrication of spherical AlSi10Mg powders by radio frequency plasma spheroidization. *Metall. Mater. Trans. A* 47 (5): 2444–2453.

96 Liu, B., Duan, H., Li, L. et al. (2021). Microstructure and mechanical properties of ultra-hard spherical refractory high-entropy alloy powders fabricated by plasma spheroidization. *Powder Technol.* 382: 550–555.

97 Ding, W., Chen, G., Qin, M. et al. (2019). Low-cost Ti powders for additive manufacturing treated by fluidized bed. *Powder Technol.* 350: 117–122.

98 Liu, C., Zhu, K., Ding, W. et al. (2022). Additive manufacturing of WMoTaTi refractory high-entropy alloy by employing fluidised powders. *Powder Metall.* 65 (5): 413–425.

99 Dong, Y.P., Li, Y.L., Zhou, S.Y. et al. (2021). Cost-affordable Ti-6Al-4V for additive manufacturing: powder modification, compositional modulation and laser in-situ alloying. *Additive Manuf.* 37: 101699.

100 Wang, D.W., Zhou, Y.H., Yao, X.Y. et al. (2022). Inheritance of microstructure and mechanical properties in laser powder bed fusion additive manufacturing: A feedstock perspective. *Mater. Sci. Eng. A* 832: 142311.

101 Kellogg, F., Kudzal, A., Mock, C. et al. (2020). *Feasibility of Cryomilled 17-4 Stainless Steel Powder as Feedstock for Additive Manufacturing*. CCDC Army Research Laboratory.

102 Cordova, L., Bor, T., de Smit, M. et al. (2020). Measuring the spreadability of pre-treated and moisturized powders for laser powder bed fusion. *Addit. Manuf.* 32: 101082.

12

Microstructure Evolution and Powder Effects

12.1 Introduction

Metal **additive manufacturing (AM)** is a **rapid solidification** process involving repeated **thermal cycling** while building a part layer-by-layer [1]. Recurring heating and cooling cause an intrinsic heat treatment effect in the previously deposited layers, and thus, the local microstructure of the AM metallic part experiences a complicated evolution, leading to the development of multiscale **hierarchical microstructures** – ranging from microscale **columnar grains** [2], submicron **dislocation cells** and networks [3], to nanoscale **precipitates** in some cases [4, 5], as displayed in Figure 12.1 [1]. Coupled with the complex **fluid dynamics** that occur in the **melt pool** and during the fusion of the melt pool, micro-, and submicron pores are also commonly observed in the metal AM parts due to **lack of fusion** or **gas entrapment** [6, 7]. Given a broad range of metal AM methods and systems, as discussed in Chapter 8, the microstructures of additively manufactured metallic components are influenced by a variety of process parameters, among which the power of the **heat source**, **beam spot size**, **scan speed**, **hatch spacing**, and **layer thickness** are the most commonly investigated parameters in the literature [6, 8]. While postprocess treatments, such as **annealing** and **tempering**, lead to additional **microstructural evolution**, the scope of this chapter is to discuss the microstructure formation and evolution during the AM process and in the as-printed condition, along with the mechanisms that control the microstructure, i.e. **solidification kinetics** and **solid-state sintering**. The effects of powder on microstructure evolution are particularly addressed.

12.2 Grain Structure and Phase Composition

12.2.1 Columnar-to-Equiaxed Transition (CET)

As described in Sections 9.3 and 9.4, the laser-powder interaction in metal AM is complicated due to the small size of the melt pool, large **thermal gradients**, and the fast movement of the **liquid–solid interface**. These characteristics of metal AM processes lead to complex transient material consolidation processes that directly affect the resulting microstructural features such as **grain size** and **crystal texture**, which in turn influence the mechanical properties of the deposited materials, such as strength, hardness, and residual stress. One critical feature distinguishing metal AM from other manufacturing methods is their tendency to maintain **epitaxy** during the layer-by-layer deposition process [2]. **Epitaxial solidification** is characteristic of **heterogeneous nucleation** by which a solid grain or nucleus is nucleated from a liquid in contact with a substrate. In general, the thermal conditions of metal AM promote epitaxial growth of **columnar grains** [2], representing one challenge in AM as **columnar grain** structures often give rise to undesired **anisotropic** properties. The growth of **columnar grains** is geometrically related to the energy source's scan speed and the melt pool's shape. **Feedstock powder** plays a role from the perspective of material properties such as **thermal diffusivity**. High scan speed, high power, and powder with low thermal diffusivity promote a comet shape for the melt pool, whereas low scan speed, low power, and powder with high thermal diffusivity result in the formation of a near-spherical melt pool. The growth of the columnar grains follows the direction of the largest thermal gradient in the melt pool [9]. At a specific position along the melt pool boundary, the columnar grains grow parallel to the melt pool's prevalent thermal gradient. Therefore, straight and broad columnar grains are produced when the melt pool is comet-shaped, as the direction of the maximum thermal gradient in a comet-shaped

Metallic Powders for Additive Manufacturing: Science and Applications, First Edition.
Enrique J. Lavernia, Kaka Ma, Julie M. Schoenung, James F. Shackelford, and Baolong Zheng.

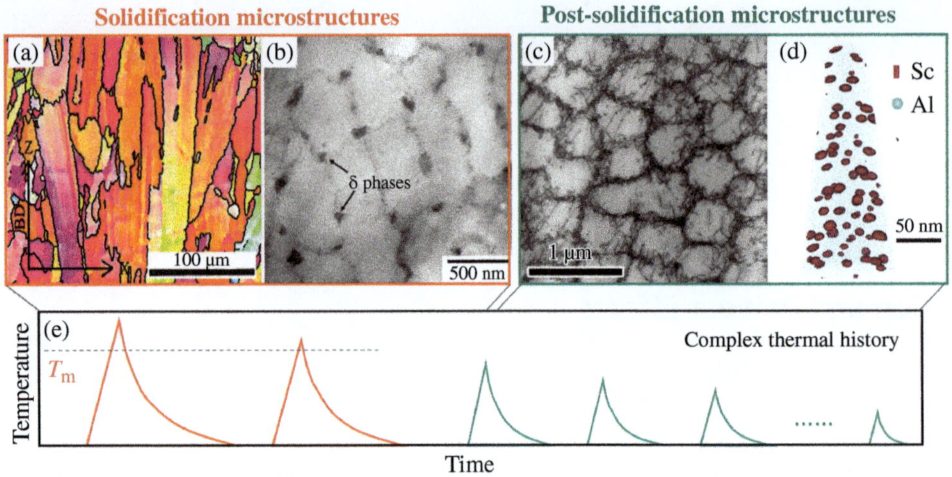

Figure 12.1 Microstructure evolution of metals fabricated by AM. (a) Typical columnar grain structure, (b) secondary phases distributed along grain boundaries, (c) dislocation cells in postsolidification structures, and (d) nanoprecipitates in the matrix. (e) Cyclic thermal history leading to the formation of multistage microstructures during AM. Source: Liu et al. [1]/with permission from Elsevier.

melt pool does not change significantly during solidification. In contrast, a spherical or near-spherical melt pool tends to produce curved and tapered columnar grains due to a shift in the direction of the maximum **temperature gradient** in the liquid from the melt pool boundary to the liquid-free surface [2]. It is worth noting that other factors may also affect the shape of the melt pool such as convection, surface tension gradients, and electromagnetic forces.

Since a columnar grain structure is linked to undesirable anisotropy, the **columnar-to-equiaxed transition (CET)** is a phenomenon that has been of interest in conventional solidification for a long time [10–13]. Several studies on CET have revealed that **undercooling**, ΔT, ahead of the columnar grains plays a crucial role in nucleation and growth of **equiaxed grains** [14, 15]. A few numerical studies regarding the CET phenomenon in AM processes have confirmed that the following factors are critical in the formation of columnar grains and the subsequent transition to equiaxed grains: a critical undercooling for **grain nucleation** ΔT_n [16], **constitutional supercooling** ΔT_{CS} [17], thermal gradient G [18, 19], and the **inoculant** density [16, 20]. First, to understand the grain morphology in the metallic parts made by AM, let us define several geometrical factors and solidification parameters. The angle between the energy source scan direction and the **grain growth** direction is taken as θ, and the steady-state nominal **grain growth** rate is denoted by R_N, and v is used for energy source scan speed. Then we have:

$$R_N = v \cdot \cos\theta \tag{12.1}$$

The major **dendritic growth** direction for metals with a cubic crystal structure is <100>. In other words, there are preferred **grain growth** directions, which makes the local grain growth rate of the crystalline grains, defined as R_L, always higher than R_N. If the angle between the <100> crystallographic orientation and the normal to the melt pool boundary is denoted by φ, the relationship between R_L and R_N can be expressed as Eq. (12.2):

$$R_L = R_N / \cos\varphi \tag{12.2}$$

It is apparent that R_L increases with increasing **misorientation** between the grain growth direction and the direction of the maximum thermal gradient G in the melt pool. As the melt pool moves along the energy source scan direction, the extent of the constitutional undercooling ahead of the advancing interface controls the stability of the solidification front, which is influenced by R_L, the thermal gradient in the melt pool, and the presence of alloying or impurity elements in the feedstock powder. When the ratio of G/R_L is high, a **planar solidification** front is promoted. When the ratio is low, the crystallized grain morphology changes to **cellular**, cellular-dendritic, or dendritic depending on the ΔT_{CS} ahead of the advancing liquid–solid interface front. The melt pool is undercooled by extensive **segregation** of the **solute atoms**, increasing the temperature of the crystalized solid with respect to the liquid. The CET starts to occur when the G/R_L ratio reaches a critical low value for a given material system containing a fixed number of **heterogeneous nucleation** sites [2].

The thermal gradient (also called **temperature gradient**), G, typically has the unit of $\mathrm{K\,m^{-1}}$ and is defined as follows when three dimensions are considered [19]:

$$G = \sqrt{G_x^{\,2} + G_y^{\,2} + G_z^{\,2}} \tag{12.3}$$

where G_x, G_y, and G_z are temperature gradients at the liquid–solid interface along the x, y, and z directions, respectively; i.e. $G_x = \frac{\mathrm{d}T}{\mathrm{d}x}$; $G_y = \frac{\mathrm{d}T}{\mathrm{d}y}$; *and* $G_z = \frac{\mathrm{d}T}{\mathrm{d}z}$. Then the liquid–solid interface advancing velocity, also often named growth or **solidification velocity**, R ($\mathrm{m\,s^{-1}}$), can be calculated using cooling rate $\left|\frac{\mathrm{d}T}{\mathrm{d}t}\right|$ and **temperature gradient** as formulated below [19]:

$$R = \frac{\left|\frac{\mathrm{d}T}{\mathrm{d}t}\right|}{\sqrt{G_x^{\,2} + G_y^{\,2} + G_z^{\,2}}} \tag{12.4}$$

Gäumann et al. used the Kurz–Giovanola–Trivedi model for rapid solidification processes by neglecting the nucleation undercooling (ΔT_n) at high thermal gradients ($G \sim 10^6\ \mathrm{K\,m^{-1}}$) and simplified the underlying thermodynamics and kinetics of the CET to the following equation [9]:

$$\frac{G^n}{R} = a\left\{\left(\frac{-4\pi N_0}{3\ln(1-\Phi)}\right)^{1/3} \cdot \left(\frac{1}{n+1}\right)\right\}^n \tag{12.5}$$

where N_0 is the nucleation site density, Φ is the volume fraction of equiaxed grains, n and a are the alloy constants. N_0 depends on the composition of the alloy and undercooling. Higher values of N_0 correspond to decreased probability of the formation of equiaxed grains during solidification. By rearranging Eq. (12.5), the ratio $\frac{G^n}{R}$ can be directly correlated to the volume fraction of equiaxed grains (Φ) formed during the solidification of the melt pool. Theoretically, Φ can be increased by decreasing the temperature gradient G and increasing the solidification rate R. That is why equiaxed grains are expected to primarily form in the top layer of a single melt pool [21]. However, during the AM of multilayer metallic parts, the top layer of the melt pool in the previous layer is partially or entirely remelted during the deposition of the next layer, which results in the loss of equiaxed grains and production of a fully columnar grain morphology in the final metallic component.

Using a nickel (Ni)-base superalloy, Inconel 718 (IN718), previous studies by Raghavan et al. have shown that plotting the spatial and temporal variations of G and R in the solidification map could effectively distinguish the G–R ranges for columnar grains, mixed grains, and equiaxed grains, respectively [19]. It is reported that the grain structure is columnar as the melt pool begins to solidify, and then the grain morphology moves toward the mixed region during solidification, which finally turns into an equiaxed region at the end of solidification. Raghavan et al. first proposed that a spot melting strategy could effectively acquire the CET or fully equiaxed grain structures in IN718 during **electron beam (EB)** based AM process, according to their numerical modeling of heat transfer [19]. Karimi et al. reported that CET could be promoted via site-specific melting in **powder bed fusion (PBF)** – EB using line order number function, which affected the cooling rate and temperature gradient, and thus, controlled grain morphology and texture [22]. Figure 12.2 illustrates the mechanism of how line scan offset changed solidification conditions in the melt pool and induced *in-situ* recrystallization, finally yielding an equiaxed grain structure. Liu et al. also investigated the mechanism of CET in IN718 during PBF-EB, via a combination of numerical modeling and experimental verification [17]. They constructed the G–R plot on a map of solidified grain structures with superimposed simulated results and experimental points for spot and line melting from studies by Raghavan et al. [19], Dehoff et al. [23], Babu and coworkers [24], Promoppatum, et al. [25], as shown in Figure 12.3a [17]. Representative grain structures in the melt pool are illustrated by the **inverse pole figure (IPF)** map obtained via **electron backscatter diffraction (EBSD)** in Figure 12.3b–d for spot melting, and in Figure 12.3e,f for line melting, respectively, to demonstrate the CET achieved by applying these novel melting strategies. The upper-left region, i.e. high G values and low R values (high G/R ratio) produces columnar grains, while equiaxed grains are promoted in the right-lower region corresponding to low G/R ratios. Researchers have discovered that by increasing the scan speed (indicated by the yellow arrow) and/or raising the preheating temperature (indicated by the green arrow), it is possible to decrease the temperature gradient and enhance the rate of solidification. As a result, the G–R map moves from the columnar grains toward the equiaxed grain patterns. In a more recent study, Raghavan et al. investigated the correlation between spot-melt parameters, build geometry, and solidification microstructure evolution in IN718 during the PBF-EB process, the results of which

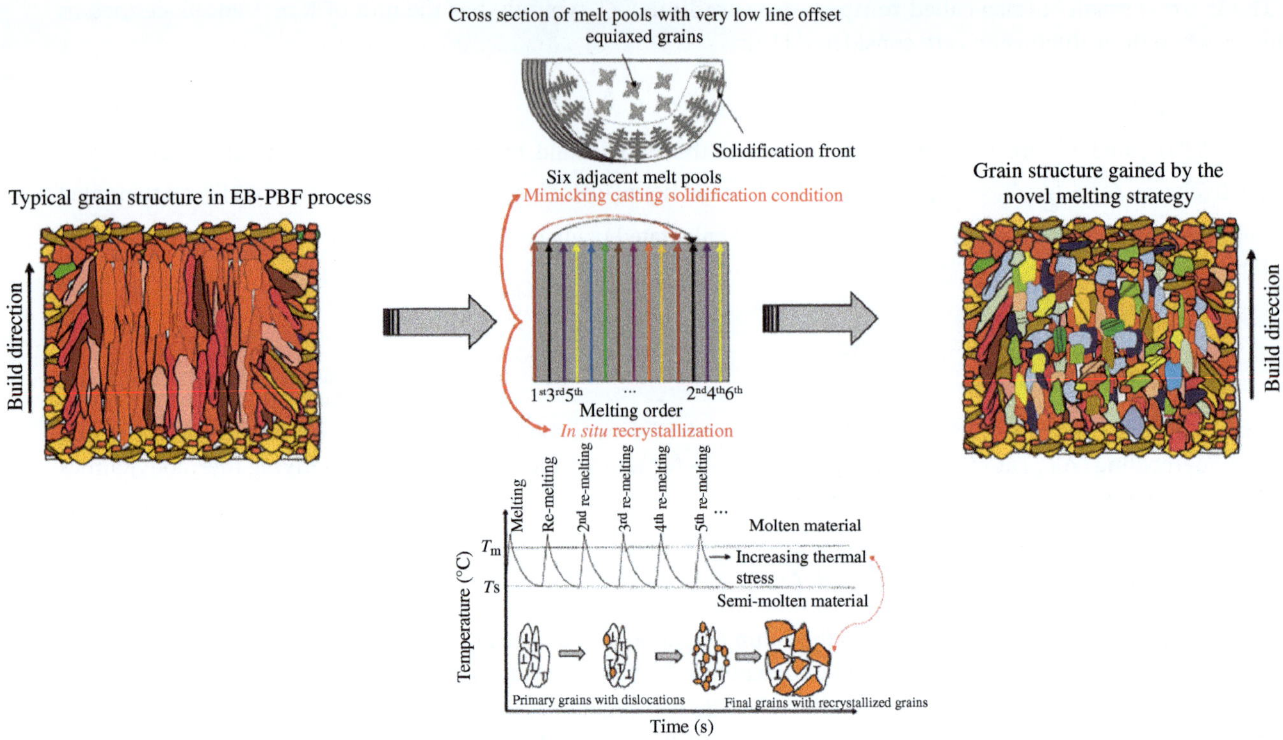

Figure 12.2 Columnar-to-equiaxed grain transition via site-specific melting strategy in PBF [22].

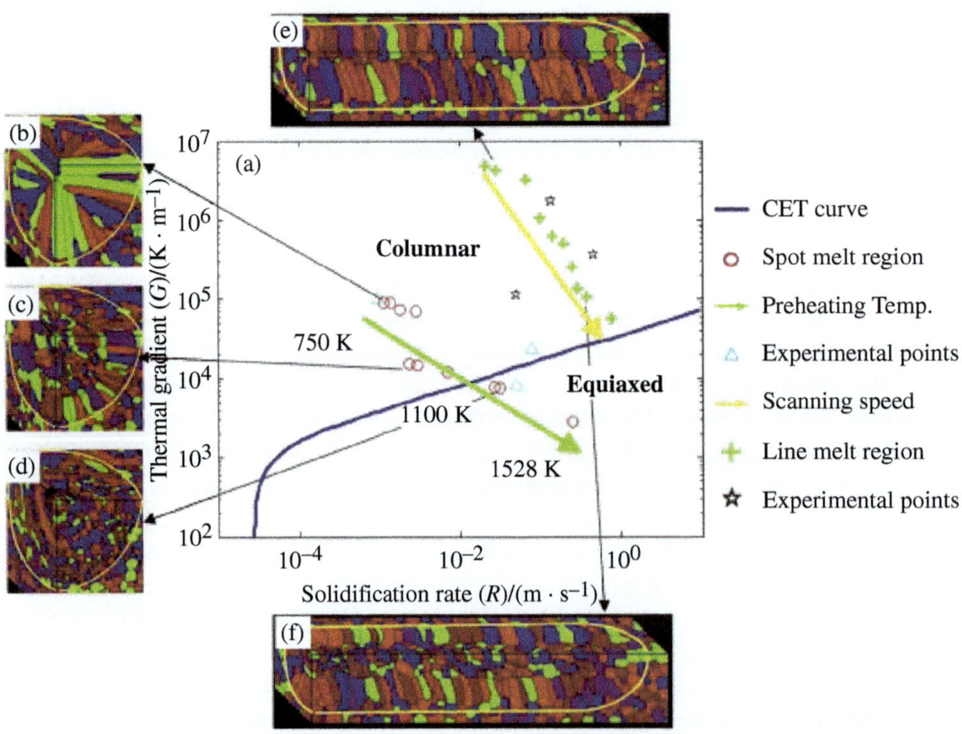

Figure 12.3 (a) The temperature gradient, G, vs. solidification rate, R, plotted on a solidification map for IN718. (b)–(d) are the representative grain morphologies in the melt pool for the spot melting strategy under variational preheating temperatures; (e) and (f) are the representative grain morphologies in the melt pool for the line melting strategy with increasing scan speeds [17].

suggest that the area energy density per layer during the AM process needs to scale with the cross-sectional area of the layer being melted to achieve and sustain an equiaxed solidification microstructure [26].

To reduce the anisotropy in mechanical properties resulting from the primary columnar grains with fairly strong textures in the AM parts, it is desirable to promote the CET. Based on the work discussed above, it seems that the CET is predominantly controlled by varying AM processing parameters, e.g. the scan speed, substrate preheating temperatures, and melting strategies. However, it is important to address that the **feedstock powder** prevalently affects the CET via its thermo-physical properties, particularly thermal diffusivity, thermal conductivity, **specific heat capacity**, and **melting point**, which are critical input material parameters to calculate thermal gradient G and solidification rate R. The powder particle size and morphology also influence the heat transfer pathway, as the premelting particle–particle contact area depends on the powder morphology. Thus, the cooling rate and the undercooling in front of the liquid–solid interface are greatly influenced by the feedstock powder characteristics. The chemistry of the powder is the dominant factor, as discussed in Chapter 10, that influences the microstructure of the printed parts. The powder effect can be utilized to promote the CET by adding inoculants, **alloying**, or **dilute solute** additions, all of which involve modification of the chemical composition.

For instance, the study by Martin et al. suggests that a general approach of adding nanoparticles to the feedstock powder of aluminum (Al) alloys is effective in promoting fine equiaxed microstructures by nucleation of new grains during **selective laser melting (SLM)**, a PBF AM method [27]. ZrH_2 was selected as the nanoparticles because of its stability in air and ability to decompose at the melting temperature to form a favorable Al_3Zr nucleant phase that has more than 20 crystallographic interfaces matching with the primary **face centered cubic (FCC)** Al phase in the Al 7075 and Al 6061 base alloys. The formation of those lattice-matched Al_3Zr nanoparticles at the instant of melting provides a high density of low-energy-barrier heterogeneous nucleation sites ahead of the solidification front, which decreases the critical amount of undercooling needed to induce equiaxed growth. Figure 12.4 demonstrates

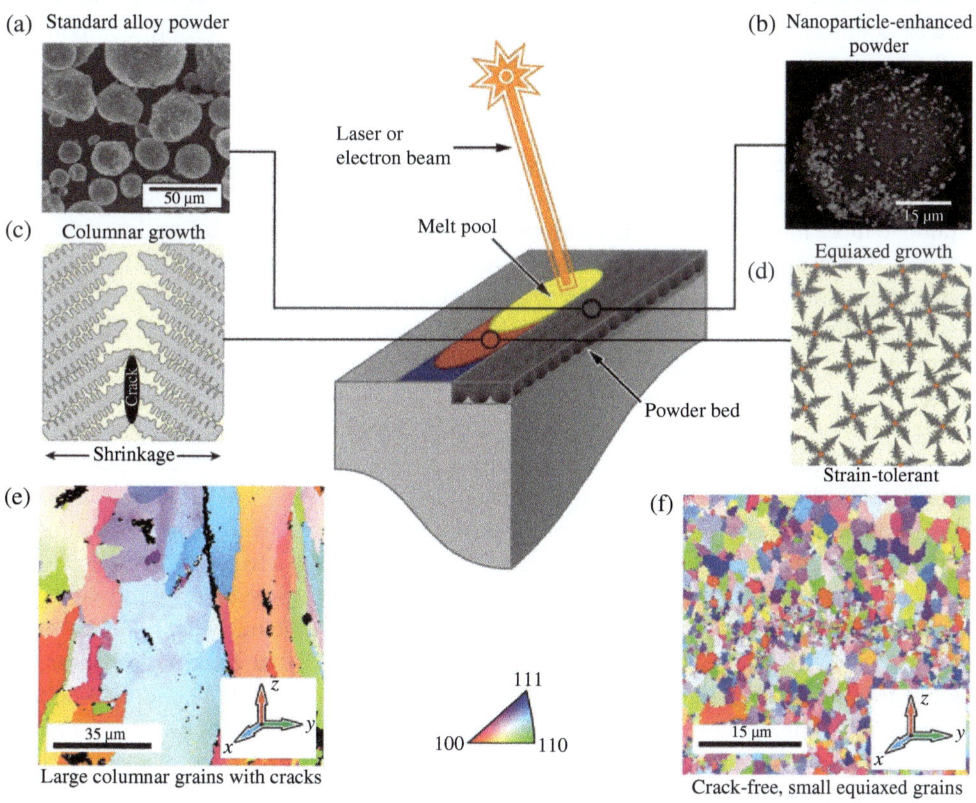

Figure 12.4 Modification of feedstock powder with nanoparticles promotes the CET. (a) Conventional Al 7075 powder feedstock; (b) Al 7075 powder functionalized with nanoparticles. (c) Typical columnar growth of dendrites in solidification microstructure in AM of many alloys, including Al 7075; (d) Suitable nanoparticles can induce heterogeneous nucleation and facilitate equiaxed grain growth; (e) Intolerable microstructure with large columnar grains and periodic cracks when 3D-printed using unmodified feedstock powder, as illustrated by the IPF. (f) Powder functionalized with nanoparticles produces crack-free, fine equiaxed grains. Source: Martin et al. [27]/Springer Nature.

how incorporating nanoparticles into the feedstock powder affects the solidification microstructure and finally leads to a crack-free equiaxed grain structure, as opposed to the large columnar grains with cracks when standard alloy powder is used in SLM [27]. It is necessary to have both an energetically favorable condition and many nucleation sites for nucleation and growth of new grains ahead of the solidification front. Despite that undercooling is present in the unmodified material (Figure 12.4a,c) to provide an energetically favorable condition, a substantially higher energy barrier needs to be overcome to induce homogeneous nucleation; thus, columnar growth is preferred, as shown in Figure 12.4e. In contrast, the nanoparticle-enhanced powder (Figure 12.4b) provides a high density of low-energy-barrier heterogeneous nucleation sites ahead of the solidification front, leading to the equiaxed grain growth in solidification microstructure (Figure 12.4d). Eventually, a fine equiaxed structure (Figure 12.4f) can be achieved under the same processing conditions as those used with the unmodified powder.

Another method to promote CET via the powder effect is the addition of dilute solutes. Recent studies discovered that additions of solute elements to titanium (Ti) and some Ti-based alloys could produce equiaxed microstructures when processing these materials using AM [28, 29]. Two physical mechanisms were proposed to explain the effect of the solute addition: (i) chemical grain refinement by controlling the **growth restriction factor**, Q, and (ii) increased **freezing range** due to the alloying addition coupled with the rapid cooling rates associated with AM techniques [28]. The **growth restriction factor** Q describes the effect of solute elements on the growth of solid–liquid interface of the new grains during the solidification of alloys; and Q controls the rate of development of a **constitutional supercooling** zone [30, 31]. Q is calculated based on the slope of the liquidus line on the phase diagram, m_l, the solute concentration in the corresponding alloy c_0, and the equilibrium partition coefficient, k, of the added solute. For a binary alloy, Q follows Eq. (12.6) below [30, 31]:

$$Q = m_l \cdot c_0 \cdot (k - 1) \tag{12.6}$$

where $k = \frac{c_s}{c_l}$; c_s and c_l are the solute concentrations at the interface for the solid and liquid, respectively. The units for m_l, c_0 and k are K/wt%, wt%, and unitless, respectively, which give the unit for Q is K. For a multi-component alloy, the total Q is estimated by Eq. (12.7) given below [31]:

$$Q_{\text{total}} = \sum [m_{li} \cdot c_i \cdot (k_i - 1)] \tag{12.7}$$

where i corresponds to different solute elements. Note that Eq. (12.7) is valid for dilute alloying systems or alloys with solutes having limited or no interactions among the constitutional elements of the alloy.

Higher values of Q correspond to more nucleation and the resultant microstructure moves toward equiaxed grains. Since it is recognized that Al and V provide limited constitutional undercooling in Ti–6Al–4V alloys [32], extensive effort has focused on modifying the chemistry of the alloy powder to control the solidification of the high-temperature β phase during AM, specifically, the identification of the solute elements with high Q in Ti, building upon the growth restriction theory developed for casting [33]. In the study by Zhang et al., it is demonstrated that the addition of copper (Cu) to Ti not only led to a full transition of columnar grains to equiaxed grains but also refined the **grain size** of the prior β phase by two orders of magnitude [34]. The Cu solutes have a high capacity to form a large constitutional supercooling ΔT_{CS} zone in front of the liquid–solid interface, contributing to a high value of Q, which can override the negative effect of a high thermal gradient in the melt pool during AM. Likewise, Simonelli et al. revealed that adding a small amount of iron (Fe) powder, up to 3 wt% (weight percent), into Ti-6Al-4V powder gave rise to a mixed columnar and equiaxed grain morphology in the parts made by PBF-Laser Beam (LB) [35]. Although a truly randomly oriented equiaxed grain structure was not achieved, the use of the dilute solute Fe effectively suppressed the formation of columnar prior-β grains. Other promising solutes for refining prior-β grains in Ti alloys include silicon (Si) [36–38], beryllium (Be) [39], and chromium (Cr) [40, 41]. Nevertheless, it is important to note that certain disadvantages are associated with the use of those solute elements. For instance, contrasting evidence exists regarding whether intermetallic silicides may form during solidification and then embrittle the Ti alloy [36, 37]. The utilization of Be presents a challenge due to the potential health risks involved [39]. For Cr, it has been discovered that $TiCr_2$ Laves precipitates form at grain boundaries in the parts due to AM's complex cyclic thermal history, which deteriorates the properties of the Ti alloy [40, 41].

In related studies, the selection of solute elements, also known as alloying additions, was based on their ability to expand the freezing range of the specific alloy through dilution [32, 42–44]. This strategy takes into account the fast-cooling rates associated with AM processes. Consequently, when the alloy's freezing range is significantly widened, achieving a substantial degree of undercooling becomes possible, which initiates the solidification process.

As a result, a considerably high rate of solid nucleation occurs within the melt pool, leading to a greater number density of solid nuclei. This phenomenon proves advantageous for the CET, ultimately forming equiaxed grains. Welk et al. investigated the effect of adding Fe, Ni, and Molybdenum (Mo) to commercially pure (CP) Ti and Ti–6Al–4V by using computational thermodynamics (CALPHAD) [42]. The results predicted that alloying with Fe and Ni would increase the CP-Ti and Ti alloys' freezing range; thus, the CET was promoted with rapid cooling rates during AM processing. In contrast, the addition of Mo was reported to have limited influence on the freezing range, which was not sufficient to produce a CET during AM processing. The effects of the Fe, Ni, and Mo on the Ti alloys were confirmed experimentally by using **laser beam-directed energy deposition (DED-LB)** [42]. Welk et al. also introduced a method to determine the minimum concentration of the solute required to induce a CET during DED-LB, which involves the strategy of producing gradient builds with increasing contents of Ni or Fe, which suggested that the minimum concentration needed to be 3 wt% for each solute element. Controversy regarding the effect of Mo addition was reported in [32, 43, 44]. These studies demonstrated that Mo could expand the freezing range of Ti–6Al–4V to destabilize the planar growth of Ti–6Al–4V and finally refine the prior-β grains.

Some alloying elements can be added as solutes to Ti alloys to utilize the coupled action of both high Q and heterogeneous nuclei for CET. Several studies showed that boron (B), tungsten (W), and several rare-earth elements or oxides (La, Y, etc.) could produce a significant constitutional undercooling and an increase in the nuclei number density to promote equiaxed grain formation in CP-Ti and Ti-based alloys [29, 45–48]. Out of those solute elements, B has a deleterious effect on ductility due to the precipitation of titanium borides. The use of rare-earth oxides and W resulted in a composite microstructure of heterogeneous particles distributed in Ti alloys' typical $\alpha + \beta$ microstructure after laser AM, indicating poor nucleation efficiency. It is still a matter of ongoing research to elucidate the effect of these secondary phases distributed in the microstructure on the mechanical properties of the alloy, but it appears the influence is deleterious [35].

In addition to modifying the chemistry of the powder, an increase in the powder feed rate is reported to be effective in producing a moderate thermal condition, which promotes the formation of equiaxed grains in a single cladding layer of a β-Ti alloy (Ti5Al2Sn2Zr4Mo4Cr) [49]. With the further increase in the powder feed rate, the continuity of the epitaxial growth of the cellular structure in the overlap zone is broken off, and a fully equiaxed microstructure was obtained. Generally, the preservation of equiaxed β grains in the overlap zone during depositing of a new cladding layer is critical for grain morphology control.

12.2.2 Phase Composition

While Section 12.2.1 focuses on the microscale grain structure, the phase composition in a metallic part made by AM should also consider submicron and nanoscale secondary phases and/or precipitates. The solidification process in the melt pool dominates the formation and morphology of the primary phase grains. In contrast, the formation of secondary phases and the size, morphology, and distribution of these secondary phases are controlled during the final stage of the solidification or the postsolidification process. The evolution of secondary phases and precipitates is affected by the intrinsic cyclic heat treatment due to the track-by-track, layer-by-layer building process of AM. As discussed in Chapter 10, the chemical composition of the feedstock powder is the first dominant factor determining the phase composition and microstructure of the metallic AM parts. Therefore, it is not easy to provide a generic discussion on the phase composition for different metals and alloys. One common scheme, however, is the formation of an interdendritic secondary phase during the final stage of solidification in AM [50]. Fine precipitates form in the regions between the primary columnar dendritic grains due to the segregation of **solute atoms** as the melted metal solidifies, as illustrated schematically in Figure 12.5a [1]. As the solid–liquid interface front moves rapidly during solidification in AM, the solute atoms do not get sufficient time to completely diffuse out of the solvent matrix in the small melt pool, leading to a supersaturated solid solution in the primary columnar phase. Exceptions occur for some alloys containing elements with small partition coefficients, which allow the solute atoms to be rejected into the interdendritic liquid more readily. Consequently, secondary phases form in the intercolumnar regions. Examples include Al–Si alloys [51], Ni superalloys containing Nb [52], a **multi-principal element alloy (MPEA)** $Al_xCoCrFeNi$ [53], and Ag–Cu–Ge alloys [54]. Experimental results showing the formation of interdendritic secondary phases in these alloys are presented in Figure 12.5b–e. Those alloys contain elements that have different solubility in the solid phase from that in liquid, and thus, these elements (e.g. Si in Al alloys, Nb in Ni superalloys, Al in a $Al_xCoCrFeNi$ MPEA, and Cu in Ag–Cu–Ge) tend to segregate to the liquid surrounding the solidified dendritic arm. Subsequently, the

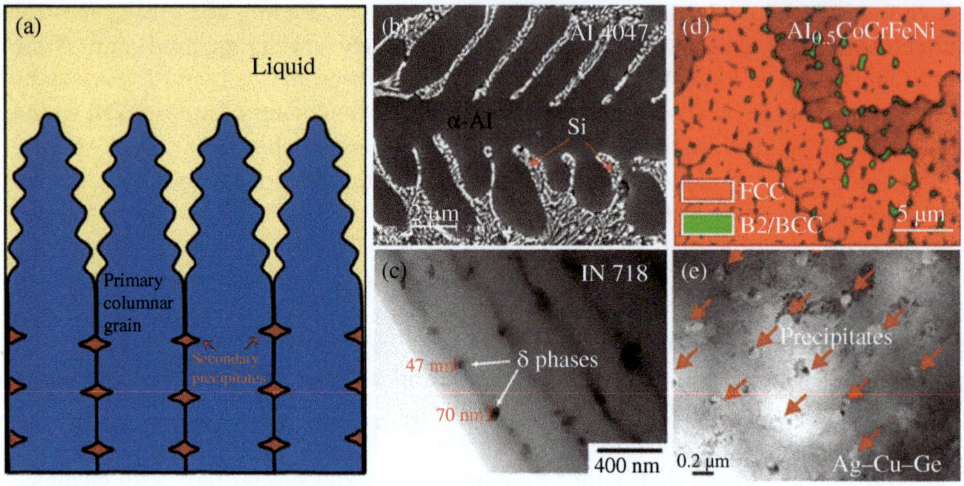

Figure 12.5 (a) Schematic diagram showing the interdendritic secondary phases that form at the final stage of solidification during AM [1]; (b) Scanning electron microscopy (SEM) image of the interdendritic Si phase in an Al 4047 alloy fabricated by DED. Source: Dinda et al. [51]/Springer Nature. (c) Interdendritic δ phase that formed due to segregation of Nb in IN718 fabricated by DED. Source: Kuo et al. [52]/with permission from Elsevier. (d) Electron backscatter diffraction phase map showing the interdendritic body centered cubic (BCC) phase that formed due to Al atom segregation in an SLM printed Al0.5CoCrFeNi high-entropy alloy. Source: Sun et al [53]/with permission from Elsevier. (e) Discontinuous Cu-rich precipitates with periodic arrangements in an Ag-Cu-Ge alloy. Source: Wang et al. [54]/Springer Nature/CC BY 4.0.

local thermodynamics of the alloy is altered by the segregation, providing the driving force for the formation of the secondary phase. Despite the fact that these interdendritic secondary phases act as a strengthening factor, this type of microstructure is disadvantageous for ductility since segregation at grain boundaries results in a preferable crack path in fracture. Particularly, brittle secondary phases result in hot cracking during the fabrication of some alloys; hence, it is important to eliminate these brittle secondary phases in AM owing to the high residual stress accumulation [50]. In general, to reduce the number of secondary phases, a higher cooling rate is preferrable so that less time is provided for the solute atoms to diffuse and segregate in the interdendritic regions.

After a single melt pool is solidified, a cyclic heating effect is exerted on the previously solidified regions, leading to further microstructural evolution. When the further microstructural evolution only involves temperatures below the melting point, it can be referred to as the postsolidification microstructure of AM, which should be differentiated from postprocessing microstructure [1]. The complex, cyclic thermal history from this point, combined with high residual stresses accumulated in the built material [55, 56], results in the formation of a high density of dislocations, which may further evolve to form **dislocation cells** [3, 57], as shown in Figure 12.1c,e. In addition, a precipitation reaction may occur in the previously solidified regions as it is subjected to an intrinsic heat treatment effect resulting from the successive thermal cycling, as shown by the **atom probe tomography (APT)** in Figure 12.1d [5]. Both formation processes of the **dislocation cells** and precipitates are regarded as postsolidification microstructures in AM, as the processes happen at temperatures below the melting point, i.e. through solid-state diffusion rather than solidification. Since defects and residual stress in AM parts are discussed in depth in Chapters 13 and 14, this section focuses on the precipitates that can form due to thermal exposure postsolidification.

In traditional heat treatment for precipitation, peak temperature and holding time (i.e. aging conditions) are two critical factors that control microstructure development after the material is solution heat treated to achieve a supersaturated solid solution. One beneficial feature of AM is its capability to produce uniform supersaturation of alloying elements in the solidification microstructure because of its high cooling rate and rapid solidification. In other words, the solidification microstructure in AM is easily induced to form precipitates upon aging heat treatment. Aging hold time is generally on the order of tens of hours to achieve peak hardening in conventional postprocessing heat treatment for precipitation-hardened alloys, making the AM interval time appear too short to induce significant precipitates. For instance, the inter-track time is typically less than one second in PBF and DED due to the fast scan speeds; even considering the interlayer time that includes the printing time, dwell time, relocating time, and recoating time (only in PBF), the time scale is still in the range of tens of seconds, depending on the sample size and scan strategy [58, 59]. Hence, although the solidification microstructure in AM parts provides a supersaturated solid solution, the precipitation reaction in the solid-state is often suppressed in AM due to the extremely short interval time, i.e. the aging holding time [5, 60]. Nevertheless, some studies have used the strategy of artificially extending the interlayer time to induce the formation of nanoprecipitates during AM. For example, Kürnsteiner et al. tailor – designed a Fe19Ni5Ti (wt%) alloy by setting a 120-second pause time after the deposition of every four layers during DED-LB [61]. This intentional extension of interlayer dwell time gave rise to a sandwich structure consisting of alternating soft and hard layers in the alloy. During the pause, the temperature decreased to a point below the martensitic transformation temperature of the steel. In addition, the subsequent temperature spikes triggered the precipitation of a nanoscale Ni–Ti η-phase in the heat-affected zone.

In addition to the processing parameters of AM, it was already discussed in Section 12.2.1 how the powder influences the solidification microstructure by the various thermo-physical properties of different chemical compositions, including liquidus temperature, solidus temperature, density, viscosity, thermal conductivity, and specific heat. Likewise, the powder also influences the formation of secondary phases and their characteristics in the AM parts. For example, when Mo was added to a Ti–2Al alloy, it was observed that as the Mo content increased, the amount of secondary α laths is increased, and the size of both primary and secondary α laths are reduced [62]. In Chapter 11, several types of powder were described in terms of their distinct features in morphology, size distribution, sources, and fabrication methods. Specifically, ball milling (or mechanical milling) can generate powders with customized alloy compositions but with nonspherical shapes as opposed to spherical gas-atomized powder. Khademzadeh et al. [63–65] used mechanically milled elemental blends of nonspherical Ti and Ni powders as starting material for the micro-DED of Ni–Ti alloys. According to their findings, mechanical alloying could facilitate the formation of the Ni–Ti phase. In other words, the mechanical milling process provided the energy required for the formation of the melt pool so that the laser scan rate could be increased three times and, consequently, increased the tendency for the formation of the equiatomic Ni–Ti phase instead of undesired intermetallic phases. DED-LB has been explored to examine and compare the effect of using gas-atomized (GA) Ti–6Al–4V powder and plasma-rotating electrode process (PREP) powders [66]. The outcomes indicate that it is advantageous to use PREP powder over GA powder, as PREP powder exhibits three times less porosity. Furthermore, while Ti–6Al–4V parts deposited from both types of powder show a distinctive, epitaxial prior β grains microstructure that transits to α-lathes and retained β during cooling, the parts deposited using PREP powder exhibit lower surface roughness, reduced intralayer porosity, higher deposition rate, and longer columnar prior β-grains with a fully lamellar $\alpha + \beta$ microstructure. Meier et al. also investigated the influence of powder production process and the resultant powder properties on the quality of Ti–6Al–4V parts manufactured by PBF-LB [67]. They reported that the density of the build parts was primarily affected by rheology and chemical pollution on the powder surface, particularly surface oxide layers that formed during the different powder production processes. Plasma-atomized powder exhibits better sphericity but thicker surface oxide layers than GA powder [67]. The parts made from the plasma atomized powder exhibit brittle dynamic fracture behavior, suggested by Charpy impact testing results, which is attributed to the oxides on the powder surface.

12.3 Solidification Kinetics

In general, solidification refers to the process by which a material in molten status cools and transforms to solid during the fabrication of a part. As discussed in Chapters 8 and 9, during metal AM, a high-energy source, laser or EB, is used to selectively melt and fuse metal powders or wires layer-by-layer. The melted metal rapidly cools and solidifies to form the desired shape. Solidification kinetics play a vital role in determining the microstructural features, such as **grain size**, grain morphology, and the presence of defects like pores or cracks, which ultimately influence the mechanical properties of the final part. The classical solidification theory has been introduced in Section 12.2, where CET was discussed, which generally suggests that as the G/R ratio is decreased, the order of the dominant grain morphology is expected to be as follows – the planar, cellular, columnar dendritic, and equiaxed dendritic structures during the solidification processes of alloys, as shown in Figure 12.6 [68].

To start with, solidification consists of nucleation and grain growth. A brief review of heterogeneous nucleation is provided here, as heterogeneous nucleation is prevalent in the AM process. The shape and critical size of the solid nucleus are governed by the balance among several energy terms, i.e. interfacial free energies between the liquid (the metal melt) and the solid nucleus γ_{SL}, between the metal substrate and the liquid γ_{ML}, and between the solid

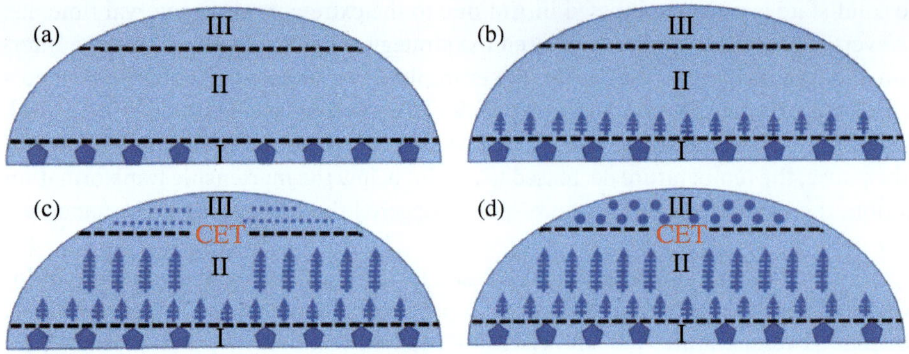

Figure 12.6 Schematic illustration of the CET behavior: (a) the formation of planar and cellular grains, (b) the formation of columnar grains, (c) the continuous columnar grain growth and the nucleation of equiaxed grains, and (d) the formation of CET behavior [68].

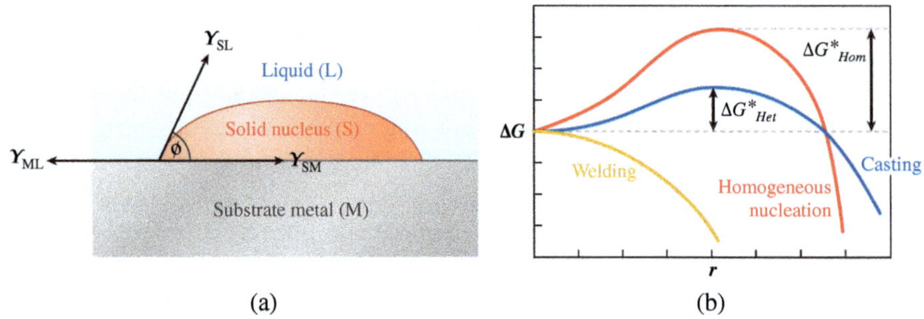

Figure 12.7 (a) Solid nucleus in contact with substrate metal and liquid. (b) Schematic of free energy change associated with heterogeneous nucleation in casting and welding, along with homogeneous nucleation [69].

nucleus and the metal substrate γ_{SM}, following the formula Eq. (12.8) below at equilibrium. The schematic diagram is provided in Figure 12.7a [69]:

$$\gamma_{ML} = \gamma_{SM} + \gamma_{SL} \cos \phi \tag{12.8}$$

In most cases in AM, the chemical composition in the solid nucleus is almost the same as the nominal chemical composition of the substrate material, which makes the interfacial energy between the solid nucleus and the metal substrate negligible, i.e. $\gamma_{SM} \approx 0$; for the same reason, the interfacial energies between the solid nucleus and the liquid and that between the metal substrate and the liquid are approximately the same, i.e. $\gamma_{SL} \approx \gamma_{ML}$. Therefore, $\cos \phi \approx 1$, which gives that ϕ is almost zero, finally leading to epitaxial columnar solidification. From the energy perspective, as shown in Figure 12.7b, the critical nucleus radius r^* is derived by setting the first derivative of the total free energy change ΔG with respect to r equal to zero.

The energy barrier to be overcome for heterogeneous nucleation to proceed, ΔG^*_{HET}, is given as Eq. (12.9) [69]:

$$\Delta G^*_{HET} = \frac{16\pi}{3}\left(\frac{\gamma_{SL}}{\Delta G_v}\right)^3 S(\phi) \tag{12.9}$$

where ΔG_v is the free energy change per unit volume associated with the formation of the nucleus; $S(\phi)$ is a shape factor dependent on the contact angle ϕ given by

$$S(\phi) = \frac{(2 + \cos \phi)[1 - \cos \phi]^2}{4} \tag{12.10}$$

Apparently, $S(\phi) \rightarrow 0$ in epitaxial solidification since $\phi \rightarrow 0$, implying that the energy barrier for heterogeneous nucleation to proceed during epitaxial solidification is negligible. In stark contrast, as the mold material and the cast materials are dissimilar in conventional casting processes, ϕ is finite, and hence, $S(\phi)$ and ΔG^*_{HET} have finite values [69].

Now that the basic thermodynamics of solid nucleation in AM is provided, we can move on to kinetics. Since the energy barrier for heterogeneous nucleation is negligible, the final microstructure is primarily determined by the grain growth following nucleation and the associated kinetics. Solidification kinetics in metal AM can be described by several mathematical equations that relate to the cooling and solidification process. First, the cooling rate, $\dot{T}$, can be determined using Newton's law of cooling [70], which relates the rate of heat transfer to the temperature difference between the molten metal and its surrounding environment:

$$\dot{T} = \frac{d[T(t)]}{dt} = -k_1(T_l - T_{env}) \qquad (12.11)$$

where k_1 is a material-dependent cooling coefficient, noting that the subscript 1 is added here to distinguish this cooling coefficient from other constants such as the Rosenthal constant to be introduced later. T_l is the temperature of the molten metal, and T_{env} is the temperature of the environment surrounding the melt pool. The cooling rate is one key factor determining the solidification kinetics in AM and the microstructure of the parts. As discussed in Section 12.2.1, high cooling rates lead to fine grains and enhance the strength of the parts, while slow cooling rates result in coarse columnar grains and decreased mechanical properties. Process parameters, such as laser power, scan speed, and part geometry, influence the cooling rate.

The solidification time (t_s) represents the time required for the molten metal to completely solidify. It can be calculated using the Rosenthal equation [71]:

$$t_s = K\,(\rho\, C_p)/(k\alpha^2) \qquad (12.12)$$

where K is the Rosenthal constant, which is a function of the geometry and process conditions; ρ is the density of the material (kg m^{-3}); C_p is the **specific heat capacity** of the material ($Jkg^{-1}K^{-1}$); and k is the thermal conductivity of the material ($Wm^{-1}K^{-1}$), while α is the **thermal diffusivity** of the material ($m^2\ s^{-1}$). Another important equation for the solidification range is ΔT_{solid}, which represents the temperature range over which solidification occurs. It can be estimated using the Scheil–Gulliver equation [72]:

$$\Delta T_{solid} = (T_m - T_s)[1 - \exp(-\beta)] \qquad (12.13)$$

where T_m is the alloy's melting temperature; T_s is the solidus temperature of the alloy; and β is Scheil–Gulliver parameter that is material-dependent.

It is necessary to point out that Eqs. (12.11)–(12.13) provide simplified representations of the complex solidification process in AM. The accuracy of these equations may vary depending on the specific metal, alloy, and process conditions. Overall, the basic fundamental physics of metal AM is identical to that of multi-pass welding whereby a **heat source** locally melts the material as it passes, and the molten metal solidifies before being partially melted again on the next pass [73]. However, the key distinguishing feature between welding and metal AM is the complex thermal conditions that arise from the cyclic heat treatment and the numerous process parameters that influence local thermal conditions in a small melt pool. In PBF-based AM, once the high-energy beam bombards it, the local temperature of the metallic powder instantaneously increases above the melting point, forming a microscale melt pool in a short period. Subsequently, the heat source scans quickly to the next site, and the melt pool solidifies rapidly under the influence of previously deposited layers serving like a cooling substrate [74, 75]. Consequently, an extremely high cooling rate of $\sim 10^7$ K/s and a steep temperature gradient of up to $\sim 10^7$ K/m can be achieved during the metal AM process [76]. In DED-based AM, the cooling rates are not as high as that in PBF; but it has been measured to be on the order of 10^3 K s^{-1} [77]. High thermal gradients can lead to the formation of residual stresses, which subsequently affect the dimensional accuracy and induce distortion or cracking in the printed parts. As described in Section 12.2.1, the kinetics of grain growth is determined by the G/R ratio (ratio of the temperature gradient to the solidification growth rate), and it is significantly affected by the solute elements in the alloys being printed. During grain growth, solute atoms are rejected by the solidified phase and move into the metal melt, creating a solute accumulation layer ahead of the solid–liquid interface that expands the freezing range. Consequently, constitutional supercooling arises because of micro-segregation. The constitutional supercooling and morphological stability of the solid–liquid interface during crystal growth work together to greatly influence the final fast-solidification microstructure. Owing to the large cooling rates in AM of metallic objects, there exists a high liquid–solid interface velocity [76]. The nonequilibrium conditions governing the solidification of the melt pool in AM are accompanied by solute trapping phenomena when the liquid–solid boundary moves faster than the diffusivity of the solute constituent in the liquid phase [78]. Therefore, the chemical compositions of the feedstock powder significantly influence the solidification kinetics via the solute effect.

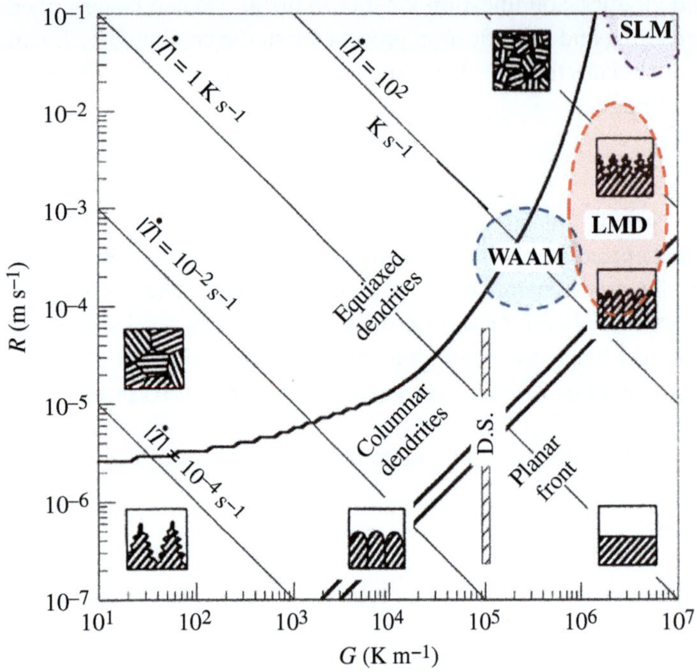

Figure 12.8 Schematic diagram showing the nucleation and grain growth, including the CET, for AM techniques with different thermal gradients, cooling rates, and solidification velocities. The image is adapted from reference [80]. To be consistent with the symbols used in this chapter, the original symbol V in [80] has been changed to R.

Significant resources have been dedicated to the development of increasingly sophisticated models and simulations, which encompass additional factors and considerations. The aim is to enhance our overall comprehension of solidification kinetics in metal AM. In related work, Ou et al. developed a three-dimensional heat transfer and fluid flow model of **wire arc additive manufacturing (WAAM)** to calculate temperature and velocity fields, deposit shape and size, cooling rates, and solidification parameters. They verified the model via experimental results for an H13 tool steel [79]. They revealed that convection is the primary heat transfer mechanism inside the molten pool. Higher scan speed increased the cooling rate but reduced the G/R ratio, indicating reduced stability of planar front solidification of components. Higher deposition rates could be achieved by increasing the power input, using thicker wires, and higher wire feed rates. Figure 12.8 presents a schematic diagram of how the different solidification kinetics, which is a combination of cooling rate $|\dot{T}|$, **solidification velocity** R, and **thermal gradient** G, give rise to various **grain nucleation** and grain growth morphologies [80]. The effect of time is implicitly included in the cooling rate $|\dot{T}|$ and **solidification velocity** R. The cooling rates in various metal AM technologies are as follows, from high to relatively low, SLM (on the order of 10^6 K s^{-1}), DED-LB such as **laser engineered net shaping (LENS^{TM})** $\sim 10^5$ K s^{-1}, PBF-EB such as **electron beam melting (EBM)** $\sim 10^4$ K s^{-1}, and WAAM $\sim 10^2$ K s^{-1}. Hence, SLM produces finer and more equiaxed grain structures, relative to the other AM methods.

In addition, surface tension is also a key factor in solidification kinetics. Surface tension-driven fluid flow, referred to as the **Marangoni effect**, accounts for most of the fluid in the melt pool during the AM procedure [81]. **Marangoni convection** plays a critical role in the mass transfer phenomenon in the melt pool and enhances elemental diffusion [82]. The melt pool is locally heated by the small spot of the energy source, resulting in a large **thermal gradient** that has an order of magnitude of $\sim 10^7$ K m^{-1} [76]. Consequently, there would be a surface tension gradient on the top surface, promoting local melt flow [83]. In the absence of active surface elements affecting the surface tension, Marangoni forces drive the melt from the middle to the periphery of the pool, forming a convection pattern, leading to the bottom part of the melt being lifted upward with the surface material transferring downward [6]. Besides improving the diffusion process, Marangoni forces contribute to the complete melting of the particles by dragging them inside the melt pool, contributing to the chemical homogeneity of *in-situ* synthesized AM products [84]. When a powder mixture is used in metal AM, local energy absorption is significantly influenced by the macro-segregation of the constituent powders with elements with vastly different laser absorption properties, subsequently changing the melting and solidification kinetics [85]. Likewise, for powders having constituents with significantly different melting points, it is highly possible to experience unmelted particles, particularly when the particles of the element with higher melting temperatures are large [43, 86].

All these studies mentioned above suggest the powder effect on the solidification kinetics in metal AM. Understanding solidification kinetics is crucial for achieving desired material properties, controlling microstructure, and ensuring the integrity of the manufactured components. To optimize the solidification kinetics in metal AM, process parameters, such as laser power, scan speed, and **layer thickness**, can be adjusted to control the cooling rate and ensure proper solidification. Additionally, advanced process monitoring and control techniques, such as in situ thermal imaging or real-time feedback systems, can be employed to understand and manage solidification kinetics during the AM process. Above all, the powder effect needs to be further investigated.

12.4 Solid-State AM

Most of the metal AM technologies rely on melting and solidification, i.e. liquid-to-solid phase transformation. However, as discussed in Sections 12.2 and 12.3, challenges and limitations exist associated with repeated local melting and thermal complexity that result in complicated and undesired microstructures. Metallic AM parts made by repeated melting and solidification are prone to suffer from solidification-related defects such as residual stress, distortion, elemental segregation, volatile element loss, hot cracking, and porosity [6, 87]. The spatially variable thermal cycles and large **temperature gradients** result in inhomogeneous microstructure and properties. Therefore, efforts and research have been devoted to solid-state AM methods as an alternative [88]. Tuncer and Bose reviewed solid-state metal AM, dividing it into two broad categories – sintering-based and **mechanical deformation**-based, primarily depending on the **metallurgical bonding** mechanisms [88]. As the bonding mechanism and temperature profile in solid-state AM methods dictates the consequential microstructure in the parts, it is important to differentiate the heating mechanism between the two categories of solid-state AM. Examples of sintering-based solid-state AM include **binder jetting (BJT)** [89], **metal extrusion** [90, 91], and **screen printing**, **stencil printing** [92]. These sintering-based solid-state AM technologies employ uniform external heating to consolidate a 3D-printed powder compact at a high fraction of the melting temperature (>0.75 T_m) of the manufactured material. The bonding mechanism is primarily **atomic diffusion** between the powder particles, which is generally facilitated by the thermal reduction of the oxide layer on powder surfaces. The driving force is the reduction in the surface energy so as to reduce the total system energy by reducing the surface area through material bonding. Sintering-based AM methods can utilize and leverage the relatively mature theory and knowledge of powder metallurgy and sintering technology [93, 94]. It also offers the opportunity to achieve equiaxed grains in all directions in the final parts, improving microstructure homogeneity, with some exceptions. When the temperature profile is not designed appropriately, grain coarsening will happen to the microstructure during sintering. Typically, external stress is absent in sintering-based AM prior to or during sintering; thus, the microstructure of the printed part is dictated dominantly by the starting powder characteristics, chemistry, and sintering conditions. The microstructure is prone to contain large voids caused by print defects that sintering cannot eliminate. Typical microstructures of parts made by sintering-based AM comprise equiaxed, uniform grains free of deformation-related texture. However, different microstructures can emerge in cases of alloys that tend to form elongated grains due to the chemical composition, such as the β phase grains in some Ti alloys. Some other alloys are subjected to partial liquid-phase sintering, formation of low-melting secondary phase at grain boundaries, precipitation, or abnormal grain growth that may occur during sintering at high temperatures, such as when using the BJT process [95, 96]. BJT has been discussed in Chapter 8, Section 8.4. Mostafaei et al. reported that when the processing parameters in BJT, such as **layer thickness**, binder saturation, printhead binder droplet size, and drying time, are similar, the particle size distribution of the starting powder plays a determining role for supersolidus liquid-phase sintering, pore removal and final microstructure [95]. Figure 12.9 shows the effect of powder particle size distribution on the microstructure of builds made from Inconel 625 (a nickel-based superalloy). A wider powder particle size distribution gives rise to denser microstructures, while higher sintering temperature results in liquid-phase formation at the grain boundaries. It is also important to note the isotropic equiaxed grains that are present through the microstructure. Fine powders yield nonuniform, anisotropic linear shrinkage during sintering, which is unfavorable for designing complex structures. In comparison, **metal extrusion** AM is competitive in terms of its capability to build parts with

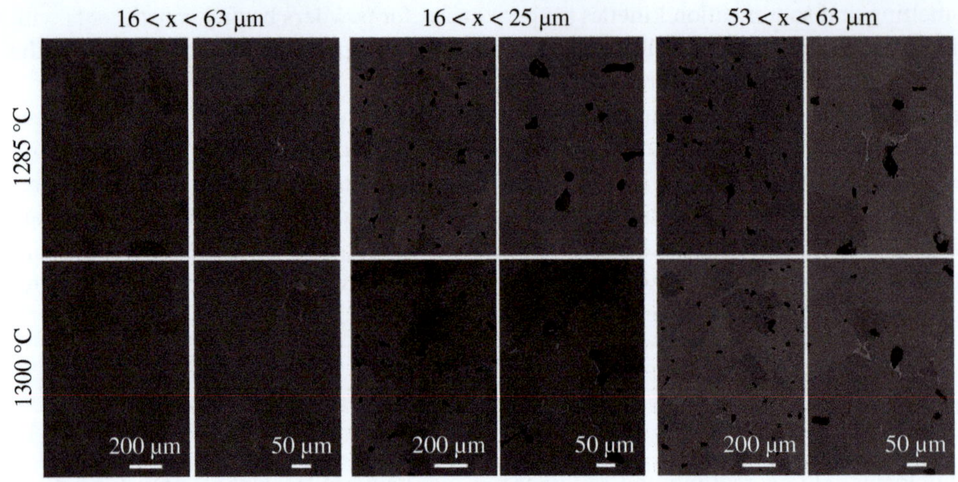

Figure 12.9 Microstructures of Inconel 625 alloy made by BJT as a function of particle size distribution and sintering temperature. Source: Mostafaei et al. [95]/with permission from Elsevier.

many features in a single process, including overhangs, enclosed cavities, and cooling channels. These characteristic features are achieved by separating the part and the support by depositing an inert layer that does not sinter or react with the build material [95].

In contrast to providing thermal energy as the driving force for bonding mechanisms in sintering-based AM, **mechanical deformation**-based solid-state AM techniques utilize kinetic energy to facilitate bonding. The source of kinetic energy can come from high-frequency ultrasonic waves or rotational friction accompanied by a normal force or supersonic impact. In deformation-based solid-state AM, heat is a byproduct of the process rather than an external energy source. The amount of local temperature rise depends on the physical and thermal properties of the material being built and the substrate material. Typical microstructures of the components produced by deformation-based processes consist of recrystallized fine grains in regions where severe plastic deformation occurs. The temperature rise is often local at the interfaces, where material deposition or joining occurs due to both friction and plastic deformation. Examples of this category of solid-state AM include ultrasonic AM like the **sheet lamination (SHL)** described in Chapter 8 Section 8.5, **additive friction stir deposition (AFSD)** [97, 98], and **cold spray additive manufacturing (CSAM)** [99, 100]. In ultrasonic AM, the feedstock is in the form of sheets or foils, for which the bonding mechanism relies on mechanical disruption of the oxide layers between the asperities of the foils. The ultrasonic vibrations induce scrubbing motion, followed by dynamic shearing to facilitate metallic bonding between layers. Severe plastic deformation with high-strain rate causes adiabatic heating, but the maximum temperature achieved by ultrasonic vibrations is up to a fraction of 0.2–0.5 T_m [88]. Localized strain can result in void formation, causing brittle cracking at the sheet interfaces [101]. Due to the limited temperature involved in ultrasonic SHL, the microstructure of the feedstock is highly likely to be retained in the final parts. Specifically, elongated grains with a certain rolling texture are typically present in the feed material, which is also observed in the built parts. Pores are prone to form at the joining interfaces in the case of insufficient metal plastic flow and shear, leading to discontinuous metallurgical bonding. Relative to the high extent of uniformity in microstructure obtained by sintering-based solid-state AM, the parts made by SHL exhibit a gradient from refined equiaxed grains at the interface to elongated grains with a rolling texture that is present in the starting material – sheets or foils, as shown in Figure 12.10a [88, 102]. Despite the texture (anisotropy) carried on from the feedstock, SHL provides the ease of manufacturing laminar composites by joining dissimilar materials without melting at the interface.

Rather than using high-frequency ultrasonic vibrations, vigorous friction by rubbing or rotation can also be employed to consolidate metals by bonding, like AFSD, a relatively new thermomechanical additive process inspired by the SHL AM method [97, 98]. The feed material for AFSD is in either rod or powder form. The material flow in the friction zone is key to the bonding, microstructure, and mechanical properties of the final part. In AFSD, the heating at the interface between the feedstock material and the base plate is promoted by rotational motion in conjunction with a normal force due to friction, which "softens" and severely deforms the feedstock material, leading to metallurgical bonding [98]. The maximum temperature induced by friction in AFSD is up to 0.6–0.9 times

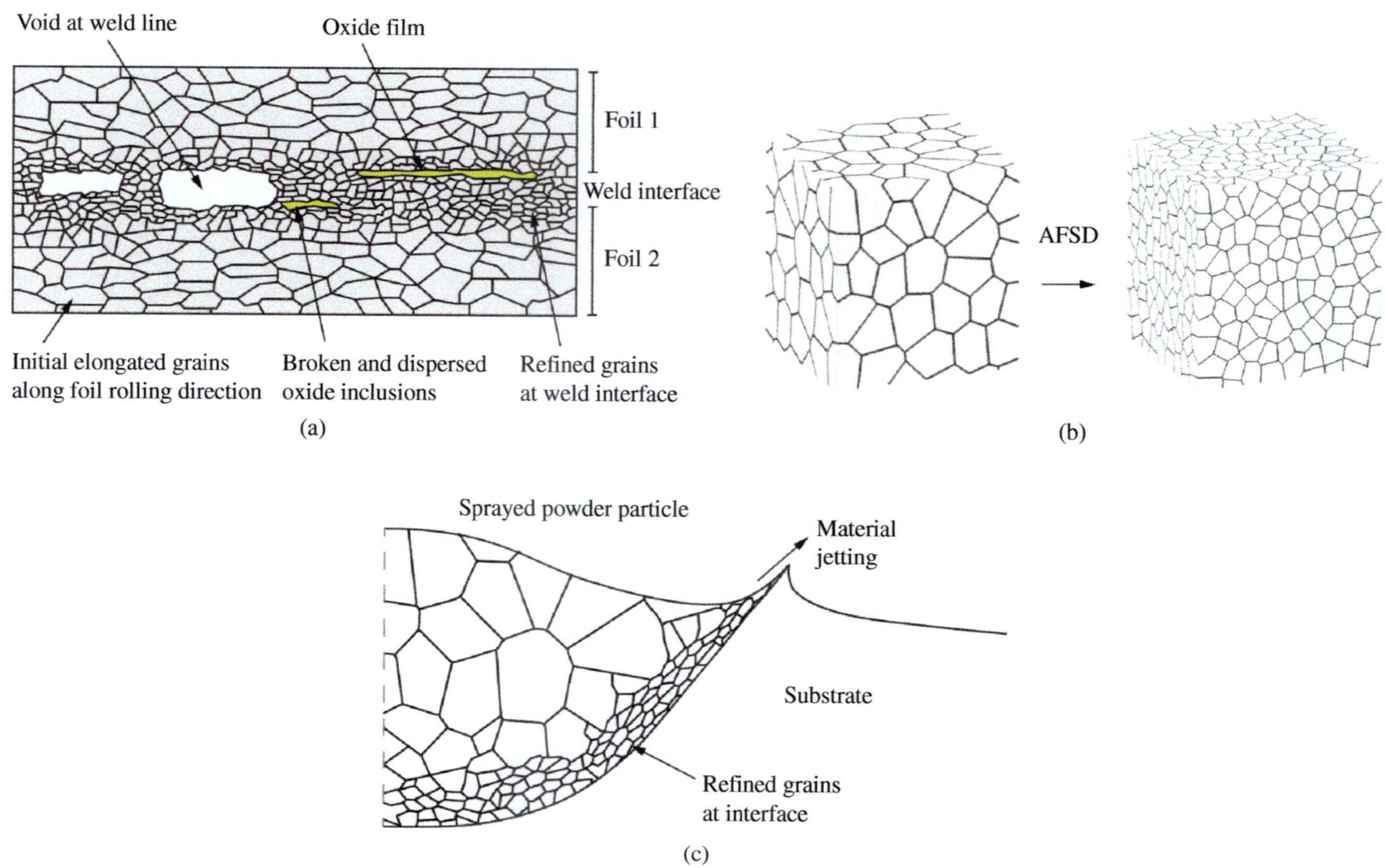

Figure 12.10 Schematic illustration of microstructures of the parts made by mechanical deformation-based solid-state AM techniques: (a) ultrasonic SHL, (b) AFSD, and (c) CSAM [88, 102].

the melting point of the material being printed, T_m [103]. So far, the following materials have been reported to be printable via AFSD: Al and Mg using powder feedstock [104, 105], and Inconel 625 and AISI 310 stainless steel using solid rods as feedstock [106, 107]. Since AFSD is a relatively newly developed solid-state AM method, there is not yet a systematic investigation into the processing – microstructure – property correlation. It is agreed that critical parameters that control the material flow and heat generation in the "stir zone" include the geometry of the tool, rotational speed, traverse velocity, feed rate of feedstock, substrate properties, and cooling media [88]. Typical microstructures in AFSD components are comprised of fine equiaxed grains randomly distributed in all directions, with some variations such as dynamic recrystallization and grain refinement, which are schematically illustrated in Figure 12.10b [88].

CSAM, as another major deformation-based solid-state AM technique, can retain the original properties of the feedstock to produce oxide-free deposits and not adversely influence underlying substrate materials during manufacture [99]. In cold spray, the powder is injected out of a nozzle and then impacted onto the substrate at a super high velocity, i.e. with high kinetic energy, and the fusion of the powder into the final part relies greatly on the kinetic energy of the powder particles rather than thermal energy [100]. Due to severe plastic deformation, dynamic recrystallization and grain refinement occur at the level of a single particle in CSAM parts. The powder particles are accelerated to a critical impact velocity by a propulsive gas with the design of suitable spraying parameters, including the type of gas, gas temperature, and pressure. Using a gas that has a lower molecular weight (e.g. helium) with higher gas temperature and pressure, higher particle velocities can be achieved, yielding higher deposition efficiency and stronger bonding. Due to the limitation of solid-state bonding, solid-state AM is mostly limited to ductile metals [88]. To establish strong bonding and achieve high-efficiency deposition, the speed of the powder particles must exceed a critical impact velocity. The critical velocity for a material depends on the material properties of the powder and substrate (e.g. density, particle size, ductility, strength, thermal properties, and preheating temperature) and the impact parameters (e.g. impact angle and substrate and particle initial temperatures). The degree of plastic deformation decreases from the impact surface to the center of the particle, leading to a gradient in **grain size** through the particle cross-section, as shown in Figure 12.10c [102]. The presence of surface oxides on the feedstock powder is a key aspect that affects the deposit quality and microstructure of the CSAM part. Powders with higher oxygen content deteriorate the deposition efficiency. Common powders used in CSAM are spherical with particle sizes between 20 and 50 μm [102]; however, it also has been reported that low-cost sponge-like irregular Ti powders effectively facilitate densification by collapsing at the moment of impact on the substrate [108]. Since the oxides on the feedstock surface are disrupted mechanically during cold spray, fragments of broken oxides are observed in CSAM builds. Other parameters that affect the deposit quality in CSAM parts include the powder feed rate, nozzle traverse speed, standoff distance, scanning step, and nozzle trajectory.

12.5 Summary

In a broad sense, the microstructure of metallic parts produced by AM is determined by two fundamental factors – the process parameters and alloy compositions (powder effect). These two factors are also interactive or interdependent, as discussed in Chapter 10. Development and improvement of AM technologies necessitate a thorough understanding of the underlying physics behind melting, solidification, and microstructure evolution coupled with multiple mass and energy transport modes. Therefore, this chapter provides an explanation of the prevalent columnar grain structure in the parts that melt-solidification-based AM manufactures, and the mechanisms of CET are discussed based on the solidification kinetics. Microstructural features are also discussed at a smaller scale than the primary grains, such as interdendritic secondary phases, nanoprecipitates, and dislocation cells. The powder effect on **microstructural evolution** is addressed, particularly the chemistry of the powder via the solute effect. In comparison, solid-state AM techniques utilize different consolidation and fusion mechanisms, exhibiting the advantageous tendency to form uniform equiaxed grain structures.

Nomenclature

AFSD	additive friction stir deposition
APT	atom probe tomography

BJT	binder jetting
CET	columnar-to-equiaxed transition
CSAM	cold spray additive manufacturing
CP	commercially pure
DED	directed energy deposition
EB	electron beam
EBM	electron beam melting
EBSD	electron backscatter diffraction
FCC	ace centered cubic
GA	gas atomized
IPF	inverse pole figure
LB	laser beam
LENS™	laser engineered net shaping
LMD	laser metal deposition
MPEA	multi-principal element alloy
PBF	powder bed fusion
PREP	plasma-rotating electrode process
SHL	sheet lamination
SLM	selective laser melting
WAAM	wire arc additive manufacturing
ΔT	undercooling (K)
ΔT_n	critical undercooling for grain nucleation (K)
ΔT_{CS}	constitutional supercooling (K)
ΔT_{solid}	the solidification temperature range, also called freezing range
$\dot{T}$	cooling rate (K/s)
G	thermal gradient (K/m)
Q	growth restriction factor
R	solidification velocity (m/s)
r	solid nuclei radius
T	temperature (K)
T_{m}	melt temperature (K)
t	time (s)
t_{s}	solidification time
θ	the angle between the energy source scan direction and the grain growth direction
R_{L}	local growth rate of the crystalline grains
R_{N}	the stead-state nominal grain growth rate
v	energy source scanning speed
φ	misorientation between the grain growth and the direction of the maximum thermal gradient
$S(\phi)$	shape factor
N_0	nucleation site density
$\emptyset$	the volume fraction of equiaxed grains
m_{l}	the slope of the liquidus line on the phase diagram
c_0	the solute concentration in the initial alloy powder
c_{s}	the solute concentrations in the solid at the solid–liquid interface
c_{l}	the solute concentrations in the liquid at the solid–liquid interface
k	the equilibrium partition coefficient of the added solute element in alloy powder
γ_{SL}	interfacial free energy between the liquid (the metal melt)- and the solid nucleus
γ_{ML}	interfacial free energy between the metal substrate and the liquid
γ_{SM}	interfacial free energy between the solid nucleus and the metal substrate
ΔG^*_{HET}	critical energy for heterogeneous nucleation to proceed
ΔG_v	the free energy change per unit volume associated with formation of the solid nucleus
ρ	density of the material (kg m^{-3})

C_p	specific heat capacity of the material ($Jkg^{-1}\,K^{-1}$)
k	thermal conductivity of the material ($W m^{-1}\,K^{-1}$)
α	the thermal diffusivity of the material ($m^2\,s^{-1}$)
K	the Rosenthal constant

References

1 Liu, Z., Zhao, D., Wang, P. et al. (2022). Additive manufacturing of metals: microstructure evolution and multistage control. *J. Mater. Sci. Technol.* 100: 224–236.

2 Basak, A. and Das, S. (2016). Epitaxy and microstructure evolution in metal additive manufacturing. *Annu. Rev. Mater. Sci.* 46: 125–149.

3 Liu, L., Ding, Q., Zhong, Y. et al. (2018). Dislocation network in additive manufactured steel breaks strength–ductility trade-off. *Mater. Today* 21 (4): 354–361.

4 Damon, J., Koch, R., Kaiser, D. et al. (2019). Process development and impact of intrinsic heat treatment on the mechanical performance of selective laser melted AISI 4140. *Addit. Manuf.* 28: 275–284.

5 Jägle, E.A., Sheng, Z., Wu, L. et al. (2016). Precipitation reactions in age-hardenable alloys during laser additive manufacturing. *JOM* 68 (3): 943–949.

6 DebRoy, T., Wei, H.L., Zuback, J.S. et al. (2018). Additive manufacturing of metallic components – process, structure and properties. *Prog. Mater Sci.* 92: 112–224.

7 Reijonen, J., Revuelta, A., Riipinen, T. et al. (2020). On the effect of shielding gas flow on porosity and melt pool geometry in laser powder bed fusion additive manufacturing. *Addit. Manuf.* 32: 101030.

8 Sames, W.J., List, F.A., Pannala, S. et al. (2016). The metallurgy and processing science of metal additive manufacturing. *Int. Mater. Rev.* 61 (5): 315–360.

9 Gäumann, M., Henry, S., Cléton, F. et al. (1999). Epitaxial laser metal forming: analysis of microstructure formation. *Mater. Sci. Eng., A* 271 (1): 232–241.

10 Hunt, J.D. (1984). Steady state columnar and equiaxed growth of dendrites and eutectic. *Mater. Sci. Eng.* 65 (1): 75–83.

11 Gandin, C.-A. (2000). From constrained to unconstrained growth during directional solidification. *Acta Mater.* 48 (10): 2483–2501.

12 Badillo, A. and Beckermann, C. (2006). Phase-field simulation of the columnar-to-equiaxed transition in alloy solidification. *Acta Mater.* 54 (8): 2015–2026.

13 Ares, A.E. and Schvezov, C.E. (2007). Influence of solidification thermal parameters on the columnar-to-equiaxed transition of aluminum-zinc and zinc-aluminum alloys. *Metall. Mater. Trans. A* 38 (7): 1485–1499.

14 Dong, H.B. and Lee, P.D. (2005). Simulation of the columnar-to-equiaxed transition in directionally solidified Al–Cu alloys. *Acta Mater.* 53 (3): 659–668.

15 Martorano, M.A., Beckermann, C., and Gandin, C.-A. (2003). A solutal interaction mechanism for the columnar-to-equiaxed transition in alloy solidification. *Metall. Mater. Trans. A* 34 (8): 1657–1674.

16 Li, X. and Tan, W. (2018). Numerical investigation of effects of nucleation mechanisms on grain structure in metal additive manufacturing. *Comput. Mater. Sci.* 153: 159–169.

17 Liu, P., Wang, Z., Xiao, Y. et al. (2019). Insight into the mechanisms of columnar to equiaxed grain transition during metallic additive manufacturing. *Addit. Manuf.* 26: 22–29.

18 Plotkowski, A., Kirka, M.M., and Babu, S.S. (2017). Verification and validation of a rapid heat transfer calculation methodology for transient melt pool solidification conditions in powder bed metal additive manufacturing. *Addit. Manuf.* 18: 256–268.

19 Raghavan, N., Dehoff, R., Pannala, S. et al. (2016). Numerical modeling of heat-transfer and the influence of process parameters on tailoring the grain morphology of IN718 in electron beam additive manufacturing. *Acta Mater.* 112: 303–314.

20 Haines, M., Plotkowski, A., Frederick, C.L. et al. (2018). A sensitivity analysis of the columnar-to-equiaxed transition for Ni-based superalloys in electron beam additive manufacturing. *Comput. Mater. Sci.* 155: 340–349.

21 Liu, S., Zhu, H., Peng, G. et al. (2018). Microstructure prediction of selective laser melting AlSi10Mg using finite element analysis. *Mater. Des.* 142: 319–328.

22 Karimi, P., Sadeghi, E., Ålgårdh, J. et al. (2021). Columnar-to-equiaxed grain transition in powder bed fusion via mimicking casting solidification and promoting in situ recrystallization. *Addit. Manuf.* 46: 102086.

23 Dehoff, R.R., Kirka, M.M., Sames, W.J. et al. (2015). Site specific control of crystallographic grain orientation through electron beam additive manufacturing. *Mater. Sci. Technol.* 31 (8): 931–938.

24 Raplee, J., Plotkowski, A., Kirka, M.M. et al. (2017). Thermographic microstructure monitoring in electron beam additive manufacturing. *Sci. Rep.* 7 (1): 43554.

25 Promoppatum, P., Yao, S.-C., Pistorius, P.C., and Rollett, A.D. (2017). A comprehensive comparison of the analytical and numerical prediction of the thermal history and solidification microstructure of Inconel 718 products made by laser powder-bed fusion. *Engineering* 3 (5): 685–694.

26 Raghavan, N., Stump, B.C., Fernandez-Zelaia, P. et al. (2021). Influence of geometry on columnar to equiaxed transition during electron beam powder bed fusion of IN718. *Addit. Manuf.* 47: 102209.

27 Martin, J.H., Yahata, B.D., Hundley, J.M. et al. (2017). 3D printing of high-strength aluminium alloys. *Nature* 549 (7672): 365–369.

28 Nartu, M.S.K.K.Y., Welk, B.A., Mantri, S.A. et al. (2023). Underlying factors determining grain morphologies in high-strength titanium alloys processed by additive manufacturing. *Nat. Commun.* 14 (1): 3288.

29 Mendoza, M.Y., Samimi, P., Brice, D.A. et al. (2017). Microstructures and grain refinement of additive-manufactured Ti-xW alloys. *Metall. Mater. Trans. A* 48 (7): 3594–3605.

30 Kozlov, A. and Schmid-Fetzer, R. (2012). Growth restriction factor in Al-Si-Mg-Cu alloys. *IOP Conf. Ser. Mater. Sci. Eng.* 27 (1): 012001.

31 Mitrašinović, A.M. and Robles Hernández, F.C. (2012). Determination of the growth restriction factor and grain size for aluminum alloys by a quasi-binary equivalent method. *Mater. Sci. Eng., A* 540: 63–69.

32 Vrancken, B., Thijs, L., Kruth, J.-P., and Van Humbeeck, J. (2014). Microstructure and mechanical properties of a novel β titanium metallic composite by selective laser melting. *Acta Mater.* 68: 150–158.

33 Tedman-Jones, S.N., McDonald, S.D., Bermingham, M.J. et al. (2019). Investigating the morphological effects of solute on the β-phase in as-cast titanium alloys. *J. Alloys Compd.* 778: 204–214.

34 Zhang, D., Qiu, D., Gibson, M.A. et al. (2019). Additive manufacturing of ultrafine-grained high-strength titanium alloys. *Nature* 576 (7785): 91–95.

35 Simonelli, M., McCartney, D.G., Barriobero-Vila, P. et al. (2020). The influence of iron in minimizing the microstructural anisotropy of Ti-6Al-4V produced by laser powder-bed fusion. *Metall. Mater. Trans. A* 51 (5): 2444–2459.

36 Bermingham, M.J., McDonald, S.D., Dargusch, M.S., and StJohn, D.H. (2008). The mechanism of grain refinement of titanium by silicon. *Scr. Mater.* 58 (12): 1050–1053.

37 IBRAHIM, K.M., HUSSEIN, A.H., and ABDELKAWY, M. (2013). Effect of Si-addition as a grain refiner on microstructure and properties of Ti-6Al-4V Alloy. *Trans. Nonferrous Met. Soc. China* 23 (7): 1863–1874.

38 Mereddy, S., Bermingham, M.J., StJohn, D.H., and Dargusch, M.S. (2017). Grain refinement of wire arc additively manufactured titanium by the addition of silicon. *J. Alloys Compd.* 695: 2097–2103.

39 Bermingham, M.J., McDonald, S.D., StJohn, D.H., and Dargusch, M.S. (2009). Beryllium as a grain refiner in titanium alloys. *J. Alloys Compd.* 481 (1): L20–L23.

40 Banerjee, R., Collins, P.C., and Fraser, H.L. (2002). Phase evolution in laser-deposited titanium-chromium alloys. *Metall. Mater. Trans. A* 33 (7): 2129–2138.

41 Zhang, Y.Z., Meacock, C., and Vilar, R. (2010). Laser powder micro-deposition of compositional gradient Ti–Cr alloy. *Mater. Des.* 31 (8): 3891–3895.

42 Welk, B.A., Taylor, N., Kloenne, Z. et al. (2021). Use of alloying to effect an equiaxed microstructure in additive manufacturing and subsequent heat treatment of high-strength titanium alloys. *Metall. Mater. Trans. A* 52 (12): 5367–5380.

43 Yadroitsev, I., Krakhmalev, P., and Yadroitsava, I. (2017). Titanium alloys manufactured by in situ alloying during laser powder bed fusion. *JOM* 69 (12): 2725–2730.

44 Collins, P.C., Banerjee, R., Banerjee, S., and Fraser, H.L. (2003). Laser deposition of compositionally graded titanium–vanadium and titanium–molybdenum alloys. *Mater. Sci. Eng., A* 352 (1): 118–128.

45 Bermingham, M.J., McDonald, S.D., and Dargusch, M.S. (2018). Effect of trace lanthanum hexaboride and boron additions on microstructure, tensile properties and anisotropy of Ti-6Al-4V produced by additive manufacturing. *Mater. Sci. Eng., A* 719: 1–11.

46 Bermingham, M.J., StJohn, D.H., Krynen, J. et al. (2019). Promoting the columnar to equiaxed transition and grain refinement of titanium alloys during additive manufacturing. *Acta Mater.* 168: 261–274.

47 Barriobero-Vila, P., Gussone, J., Stark, A. et al. (2018). Peritectic titanium alloys for 3D printing. *Nat. Commun.* 9 (1): 3426.

48 Ueda, K., Nakaoka, S., and Narushima, T. (2013). β-grain refinement of α+β-type Ti–4.5Al–6Nb–2Fe–2Mo alloy by using rare-earth-oxide precipitates. *Mater. Trans.* 54 (2): 161–168.

49 Zhang, Q., Zhang, S., Zheng, M. et al. (2020). Effects of powder feed rate on formation of fully equiaxed β grains in titanium alloys fabricated by directed energy deposition. *Metals* 10 (4): 521.

50 Aboulkhair, N.T., Simonelli, M., Parry, L. et al. (2019). 3D printing of aluminium alloys: additive manufacturing of aluminium alloys using selective laser melting. *Prog. Mater Sci.* 106: 100578.

51 Dinda, G.P., Dasgupta, A.K., Bhattacharya, S. et al. (2013). Microstructural characterization of laser-deposited Al 4047 alloy. *Metall. Mater. Trans. A* 44 (5): 2233–2242.

52 Kuo, Y.-L., Horikawa, S., and Kakehi, K. (2017). The effect of interdendritic δ phase on the mechanical properties of Alloy 718 built up by additive manufacturing. *Mater. Des.* 116: 411–418.

53 Sun, Z., Tan, X., Wang, C. et al. (2021). Reducing hot tearing by grain boundary segregation engineering in additive manufacturing: example of an AlxCoCrFeNi high-entropy alloy. *Acta Mater.* 204: 116505.

54 Wang, Z., Xie, M., Li, Y. et al. (2020). Premature failure of an additively manufactured material. *NPG Asia Mater.* 12 (1): 30.

55 Chen, W., Voisin, T., Zhang, Y. et al. (2019). Microscale residual stresses in additively manufactured stainless steel. *Nat. Commun.* 10 (1): 4338.

56 Hu, D., Grilli, N., Wang, L. et al. (2022). Microscale residual stresses in additively manufactured stainless steel: computational simulation. *J. Mech. Phys. Solids* 161: 104822.

57 Mughrabi, H. (1983). Dislocation wall and cell structures and long-range internal stresses in deformed metal crystals. *Acta Metall.* 31 (9): 1367–1379.

58 Xu, W., Lui, E.W., Pateras, A. et al. (2017). In situ tailoring microstructure in additively manufactured Ti-6Al-4V for superior mechanical performance. *Acta Mater.* 125: 390–400.

59 Mohr, G., Altenburg, S.J., and Hilgenberg, K. (2020). Effects of inter layer time and build height on resulting properties of 316L stainless steel processed by laser powder bed fusion. *Addit. Manuf.* 32: 101080.

60 Kürnsteiner, P., Wilms, M.B., Weisheit, A. et al. (2017). Massive nanoprecipitation in an Fe-19Ni-xAl maraging steel triggered by the intrinsic heat treatment during laser metal deposition. *Acta Mater.* 129: 52–60.

61 Kürnsteiner, P., Wilms, M.B., Weisheit, A. et al. (2020). High-strength Damascus steel by additive manufacturing. *Nature* 582 (7813): 515–519.

62 Zhang, F., Yang, M., Clare, A.T. et al. (2017). Microstructure and mechanical properties of Ti-2Al alloyed with Mo formed in laser additive manufacture. *J. Alloys Compd.* 727: 821–831.

63 Khademzadeh, S., Parvin, N., Bariani, P.F., and Mazzucato, F. (2015). Effects of micro laser sintering process parameters on quality of nickel-titanium single tracks and thin walls. *Met. Mater. Int.* 21 (6): 1081–1090.

64 Khademzadeh, S., Parvin, N., and Bariani, P.F. (2015). Production of NiTi alloy by direct metal deposition of mechanically alloyed powder mixtures. *Int. J. Precis. Eng. Manuf.* 16 (11): 2333–2338.

65 Khademzadeh, S., Zanini, F., Bariani, P.F., and Carmignato, S. (2018). Precision additive manufacturing of NiTi parts using micro direct metal deposition. *Int. J. Adv. Manuf. Technol.* 96 (9): 3729–3736.

66 Ahsan, M.N., Pinkerton, A.J., Moat, R.J., and Shackleton, J. (2011). A comparative study of laser direct metal deposition characteristics using gas and plasma-atomized Ti–6Al–4V powders. *Mater. Sci. Eng., A* 528 (25): 7648–7657.

67 Meier, B., Warchomicka, F., Petrusa, J. et al. (2022). Influence of powder production process and properties on material properties of Ti6Al4V manufactured by L-PBF. *Int. J. Adv. Manuf. Technol.* 123 (5): 1577–1588.

68 Li, H., Huang, Y., Jiang, S. et al. (2021). Columnar to equiaxed transition in additively manufactured CoCrFeMnNi high entropy alloy. *Mater. Des.* 197: 109262.

69 Das, S. (2003). Physical aspects of process control in selective laser sintering of metals. *Adv. Eng. Mater.* 5 (10): 701–711.

70 Davidzon, M.I. (2012). Newton's law of cooling and its interpretation. *Int. J. Heat Mass Transfer* 55 (21): 5397–5402.

71 Rosenthal, D. (1941). Mathematical theory of heat distribution during welding and cutting. *Weld. J.* 20 (5): 220s–234s.

72 Schaffnit, P., Stallybrass, C., Konrad, J. et al. (2015). A Scheil–Gulliver model dedicated to the solidification of steel. *Calphad* 48: 184–188.

73 Andersson, J. and Sjöberg, G.P. (2012). Repair welding of wrought superalloys: Alloy 718, Allvac 718Plus and Waspaloy. *Sci. Technol. Weld. Joining* 17 (1): 49–59.

74 Guo, Q., Zhao, C., Qu, M. et al. (2019). In-situ characterization and quantification of melt pool variation under constant input energy density in laser powder bed fusion additive manufacturing process. *Addit. Manuf.* 28: 600–609.

75 King, W.E., Anderson, A.T., Ferencz, R.M. et al. (2015). Laser powder bed fusion additive manufacturing of metals; physics, computational, and materials challenges. *Appl. Phys. Rev.* 2 (4): 041304.

76 Hooper, P.A. (2018). Melt pool temperature and cooling rates in laser powder bed fusion. *Addit. Manuf.* 22: 548–559.

77 Jeong, J., Webster, S., Liao, S. et al. (2022). Cooling rate measurement in directed energy deposition using photodiode-based planck thermometry (PDPT). *Addit. Manuf. Lett.* 3: 100101.

78 Ahmad, N.A., Wheeler, A.A., Boettinger, W.J., and McFadden, G.B. (1998). Solute trapping and solute drag in a phase-field model of rapid solidification. *Phys. Rev. E* 58 (3): 3436–3450.

79 Ou, W., Mukherjee, T., Knapp, G.L. et al. (2018). Fusion zone geometries, cooling rates and solidification parameters during wire arc additive manufacturing. *Int. J. Heat Mass Transfer* 127: 1084–1094.

80 Bermingham, M., StJohn, D., Easton, M. et al. (2020). Revealing the mechanisms of grain nucleation and formation during additive manufacturing. *JOM* 72 (3): 1065–1073.

81 Fuhrich, T., Berger, P., and Hügel, H. (2001). Marangoni effect in laser deep penetration welding of steel. *J. Laser Appl.* 13 (5): 178–186.

82 Le, T.-N. and Lo, Y.-L. (2019). Effects of sulfur concentration and Marangoni convection on melt-pool formation in transition mode of selective laser melting process. *Mater. Des.* 179: 107866.

83 Aversa, A., Moshiri, M., Librera, E. et al. (2018). Single scan track analyses on aluminium based powders. *J. Mater. Process. Technol.* 255: 17–25.

84 Mosallanejad, M.H., Niroumand, B., Aversa, A., and Saboori, A. (2021). In-situ alloying in laser-based additive manufacturing processes: a critical review. *J. Alloys Compd.* 872: 159567.

85 Simonelli, M., Aboulkhair, N.T., Cohen, P. et al. (2018). A comparison of Ti-6Al-4V in-situ alloying in selective laser melting using simply-mixed and satellited powder blend feedstocks. *Mater. Charact.* 143: 118–126.

86 Kang, N., Li, Y., Lin, X. et al. (2019). Microstructure and tensile properties of Ti-Mo alloys manufactured via using laser powder bed fusion. *J. Alloys Compd.* 771: 877–884.

87 Mukherjee, T., Zuback, J.S., De, A., and DebRoy, T. (2016). Printability of alloys for additive manufacturing. *Sci. Rep.* 6: 19717.

88 Tuncer, N. and Bose, A. (2020). Solid-state metal additive manufacturing: a review. *JOM* 72 (9): 3090–3111.

89 Ziaee, M. and Crane, N.B. (2019). Binder jetting: a review of process, materials, and methods. *Addit. Manuf.* 28: 781–801.

90 German, R.M. (1997). Injection molding of metals and ceramics. *Powder Metall.* 42: 157–160.

91 Thompson, Y., Gonzalez-Gutierrez, J., Kukla, C., and Felfer, P. (2019). Fused filament fabrication, debinding and sintering as a low cost additive manufacturing method of 316L stainless steel. *Addit. Manuf.* 30: 100861.

92 Pan, J., Tonkay, G.L., Storer, R.H. et al. (2004). Critical variables of solder paste stencil printing for micro-BGA and fine-pitch QFP. *IEEE Trans. Electron. Packag. Manuf.* 27 (2): 125–132.

93 German, R. (2014). *Sintering: from Empirical Observations to Scientific Principles.* Butterworth-Heinemann.

94 Anselmi-Tamburini, U., Gennari, S., Garay, J.E., and Munir, Z.A. (2005). Fundamental investigations on the spark plasma sintering/synthesis process: II. Modeling of current and temperature distributions. *Mater. Sci. Eng., A* 394 (1): 139–148.

95 Mostafaei, A., Rodriguez De Vecchis, P., Nettleship, I., and Chmielus, M. (2019). Effect of powder size distribution on densification and microstructural evolution of binder-jet 3D-printed alloy 625. *Mater. Des.* 162: 375–383.

96 Mostafaei, A., Elliott, A.M., Barnes, J.E. et al. (2021). Binder jet 3D printing—process parameters, materials, properties, modeling, and challenges. *Prog. Mater Sci.* 119: 100707.

97 Palanivel, S. and Mishra, R. (2017). Building without melting: a short review of friction-based additive manufacturing techniques. *Int. J. Addit. Subtr. Mater. Manuf.* 1: 82.

98 Mishra, R.S., Haridas, R.S., and Agrawal, P. (2022). Friction stir-based additive manufacturing. *Sci. Technol. Weld. Joining* 27 (3): 141–165.

99 Yin, S., Cavaliere, P., Aldwell, B. et al. (2018). Cold spray additive manufacturing and repair: Fundamentals and applications. *Addit. Manuf.* 21: 628–650.

100 Li, W., Yang, K., Yin, S. et al. (2018). Solid-state additive manufacturing and repairing by cold spraying: a review. *J. Mater. Sci. Technol.* 34 (3): 440–457.

101 Sridharan, N., Gussev, M., Seibert, R. et al. (2016). Rationalization of anisotropic mechanical properties of Al-6061 fabricated using ultrasonic additive manufacturing. *Acta Mater.* 117: 228–237.

102 Lee, C. and Kim, J. (2015). Microstructure of kinetic spray coatings: a review. *J. Therm. Spray Technol.* 24 (4): 592–610.

103 Yu, H.Z., Jones, M.E., Brady, G.W. et al. (2018). Non-beam-based metal additive manufacturing enabled by additive friction stir deposition. *Scr. Mater.* 153: 122–130.

104 Mukhopadhyay, A. and Saha, P. (2020). Mechanical and microstructural characterization of aluminium powder deposit made by friction stir based additive manufacturing. *J. Mater. Process. Technol.* 281: 116648.

105 Calvert, J.R. (2015) Microstructure and mechanical properties of WE43 alloy produced via additive friction stir technology. Master thesis. Virginia Tech. http://hdl.handle.net/10919/55816

106 Rivera, O.G., Allison, P.G., Jordon, J.B. et al. (2017). Microstructures and mechanical behavior of Inconel 625 fabricated by solid-state additive manufacturing. *Mater. Sci. Eng., A* 694: 1–9.

107 Dilip, J.J.S., Janaki Ram, G.D., and Stucker, B.E. (2012). Additive manufacturing with friction welding and friction deposition processes. *Int. J. Rapid Manuf.* 3 (1): 56–69.

108 MacDonald, D., Fernández, R., Delloro, F., and Jodoin, B. (2017). Cold spraying of armstrong process titanium powder for additive manufacturing. *J. Therm. Spray Technol.* 26 (4): 598–609.

13

Defect Formation and Powder Effects

13.1 Introduction

While the benefits of **additive manufacturing (AM)** have made this technology an attractive alternative to conventional methods, the resulting parts produced by AM will be acceptable only if **defects** produced during manufacturing are not at a significantly greater level than produced by those conventional methods. In this chapter, we focus on various defects that can occur during AM and the relation of powders to the production of those defects. Specifically, we explore the degree of **porosity**, **cracking** and **delamination**, **interfacial structure** and **grain size**, chemical **segregation**, and **surface roughness**. We also acknowledge that some applications, such as biomedical, require controlled levels of porosity.

13.2 Porosity

The types of porosity in AM are summarized in Table 13.1 [1, 2]. Gas entrapment can originate from the feedstock, elemental evaporation during the melting of the alloy, or entrapment of the inert shielding gas in the molten pool. Keyholes arise from a high energy density during deposition causing localized vaporization and gas entrapment. Insufficient energy input can lead to a **lack of fusion** (LoF). The shapes of these resulting porosities vary widely. **Gas pores** are generally the smallest and most spherical in nature. **Keyholes** are larger and less spherical, sometimes elongated in the build direction. LoF pores are large and similar in size to the melt pool and irregular in shape and contain unmelted powder particles. These various pores can arise in both **powder bed fusion (PBF)** and **directed energy deposition (DED)** types of AM. The mitigation of porosity-related defects in DED using a laser beam (DED-LB) is done by process optimization, feedstock control, and **hot isostatic pressing (HIP)** as a post-treatment [2]. As with DED-LB, HIP is especially useful for closing pores in PBF using laser beam (PBF-LB) manufactured parts and has led to the widespread use of PBF-LB in the aerospace industry [3, 4]. A dramatic example of the control of porosity through process optimization is shown in Figure 13.1 [4] for Ti6Al4V samples produced by PBF-LB as a function of laser power. The corresponding porosity morphologies are shown in Figure 13.2 with images produced by **X-ray computed tomography** (CT), a characterization tool discussed below [5].

For PBF-LB, various factors lead to a LoF of powders and keyhole porosity [1]. Despite the stochastic nature of PBF-LB, pores can be carried over to the finished parts due to gas entrapment during the gas atomization process. Moisture can contribute to pore formation in reactive powders such as Al-based alloys. Pretreatment of the feedstock to remove surface moisture can help to achieve density values greater than 99.5%. Also, if the energy deposited per unit volume of powder is not sufficient, defects can grow due to low fusion rates and poor consolidation of the powder. If the energy is excessive, overheating can lead to keyhole pore formation.

In the AM of thin-walled tubular specimens of Ti6Al4V and 17-4 PH by PBF-LB, defect characterizations were done both by 2D digital microscopy and 3D computed tomography [6]. The 3D analysis showed a lack of significant variability in defect distributions along the building height but noticeable defect density variations across the wall thickness, with more critical defects closer to the inner and outer surfaces. Optimization of the process parameters, surface machining, and HIP reduced the overall volume fraction of defects leading to a more uniform and random distribution of defects. The 2D analyses showed larger and a higher density of defects in thicker specimens apparently

Metallic Powders for Additive Manufacturing: Science and Applications, First Edition.
Enrique J. Lavernia, Kaka Ma, Julie M. Schoenung, James F. Shackelford, and Baolong Zheng.

Table 13.1 Types of porosity in additive manufacturing [1, 2].

Porosity type	Source of porosity	Characteristics
Entrapped gas	Entrapment in feedstock powder; shielding gas entrapment; selective evaporation	Nearly spherical and generally less than 100 μm
Keyholes	Excessive heating leading to evaporation	Larger than 10–20 μm and less spherical; may be elongated in the build direction
Lack of fusion (LoF)	Inadequate heating	Generally irregular and larger than 100 μm; entrapped particles

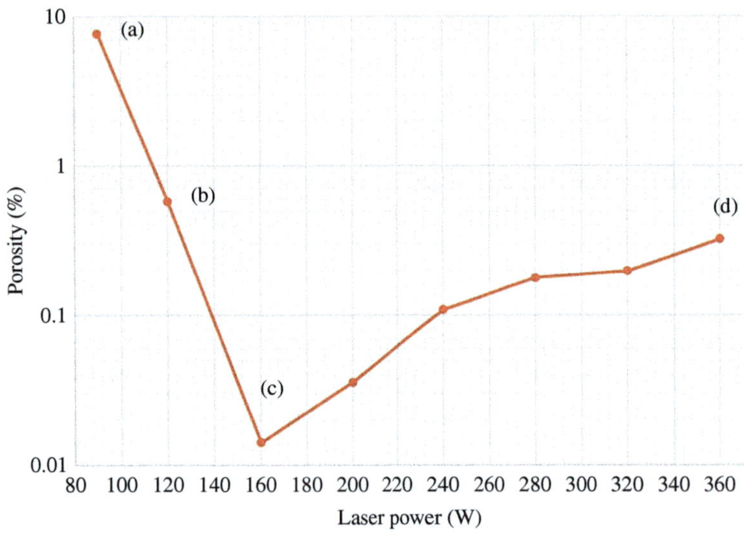

Figure 13.1 The porosity in Ti6Al4V samples produced by powder bed fusion using a laser beam (PBF-LB) as a function of laser power. Point (a) represents an excessive lack of fusion (LoF), point (b) represents a normal LoF, point (c) represents an optimal power level, and point (d) represents excessive heating that leads to keyhole porosity. Source: du Plessis et al. [4]/with permission of Elsevier.

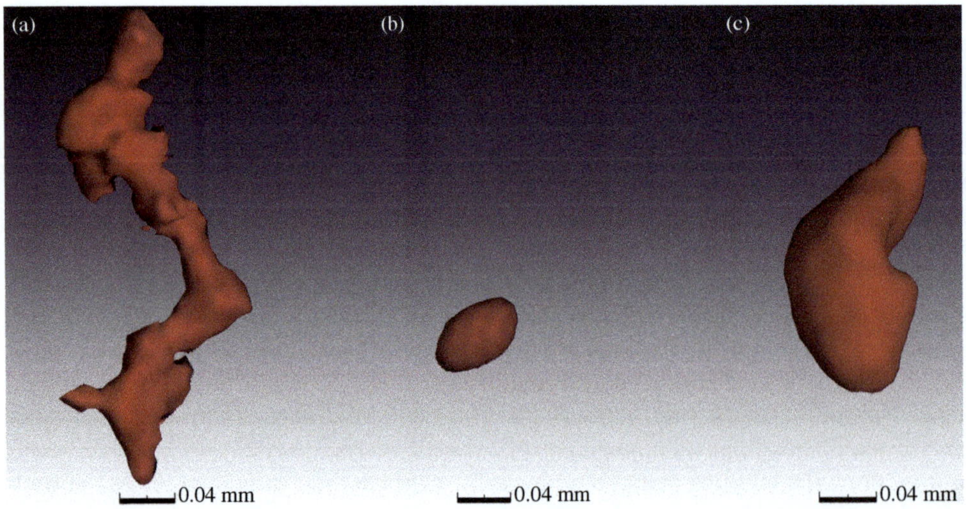

Figure 13.2 A summary of the three types of pore morphologies in specimens of Ti6Al4V produced by powder bed fusion using a laser beam (PBF-LB): (a) lack of fusion (LoF), (b) gas porosity, and (c) keyhole porosity. The build direction is vertical in all images. Source: du Plessis [5]/with permission of Elsevier.

correlating with differences in cooling rates and/or beam-material interactions. Both 3D and 2D analyses showed that larger defects are more critical for fatigue performance due to decreasing aspect ratio and circularity/sphericity. The 2D and 3D analyses showed consistent results for the spatial distribution of defects, but 3D analyses provided more detailed images of the morphology of interconnected, large LoF defects. In the characterization of porosity formation in cubic specimens of Ti6Al4V produced by PBF-LB, a visual summary of the three types of pore morphologies was given as shown in Figure 13.2. This study indicated that keyhole pore formation is minimized at larger power settings in combination with higher scan speeds [5]. Understanding the detailed production of porosity defects is greatly aided by various computational methods including the **finite element method (FEM)** and related mesoscopic simulations [7]. Porosity in as-fabricated specimens of Ti6Al4V produced by a PBF using an electron beam (PBF-EB) system was found to be less than 50 μm in diameter and resulted from argon entrapment. Their removal by HIP produced a significant improvement in fatigue strength (from 407–441 MPa at 10^7 cycles to 538–607 MPa at 10^7 cycles) [8, 9].

X-ray computed tomography (CT) has been very useful for characterizing porosity and related defects such as cracks. This nondestructive inspection technique has the benefit of monitoring such features over time including crack formation and allowing estimates of life expectancy. The technique is less effective for large parts or for alloys with high atomic number elements, which can be X-ray absorbing. In those cases, smaller coupon samples are recommended for process monitoring [5, 10, 11]. CT has not only proven to be beneficial for identifying structural defects such as porosity in AM but it is also being used to improve predictive model-based qualification processes and calibrate in-process monitoring [12]. In addition to X-ray CT, **ultrasonic testing (UT)**, another common **nondestructive testing (NDT)** technology, is useful for characterizing AM products [13]. In studying parts produced by PBF-LB, elongated pores caused by incomplete fusion between layers were found in a wide range of sizes between 50 and 500 μm. Such features can be observed nondestructively by 150 MHz fundamental frequency plane wave ultrasonic transducers. On the other hand, smaller gas pores are generally below the limit of detection for current commercially available ultrasonic transducers [14]. Nonetheless, ultrasonic measurements of changes in the effective bulk modulus and the effective density can be sensitive indicators of the presence of porosity. This has shown promise for mapping porosity in specific regions of complex structures produced by AM. Figure 13.3 illustrates the use of **effective bulk modulus elastography (EBME)** for this purpose [13]. Such ultrasonic techniques also monitor interatomic dimensions as related to elastic behavior leading to the ability to provide a detailed characterization of **residual stresses** [15].

The **flowability** of powder is of fundamental importance to ensure that defects such as porosity are to be avoided in AM [16]. Flowability is defined by ISO/ASTM 52907 as the "capability of a powder to displace powder particles with respect to each other or relative to another surface, under the effect of one or more motive forces." A standard device for measuring flowability by specifically measuring powder cohesiveness involves monitoring flow patterns in a rotating drum. This cohesiveness in turn provides an assessment of the spreadability of the powder in the PBF form of AM.

The characterization of the **particle size distribution (PSD)** of the powder feedstock is important in that PSD plays a critical role in powder flowability. In addition, particle morphology including **surface roughness**

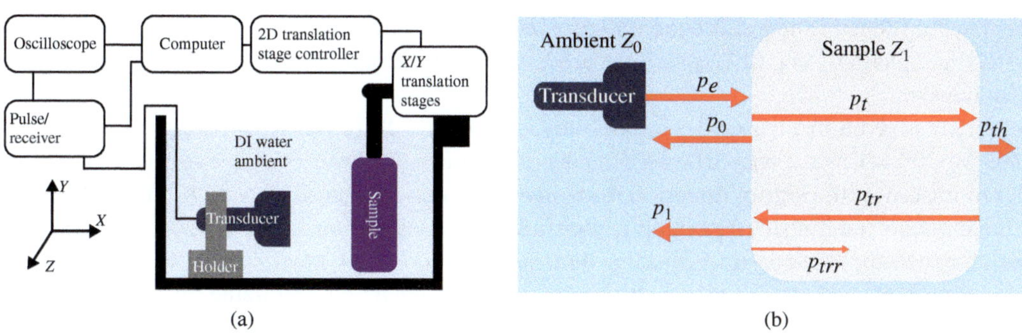

Figure 13.3 (a) The experimental set-up for the effective bulk modulus elastography (EBME) technique. (b) Schematic showing the principle of EBME, viz. using the acoustic impedance and speed of sound to calculate the effective bulk modulus and effective density. Source: Yang et al. [13]/MDPI/Licensed under CC BY 4.0.

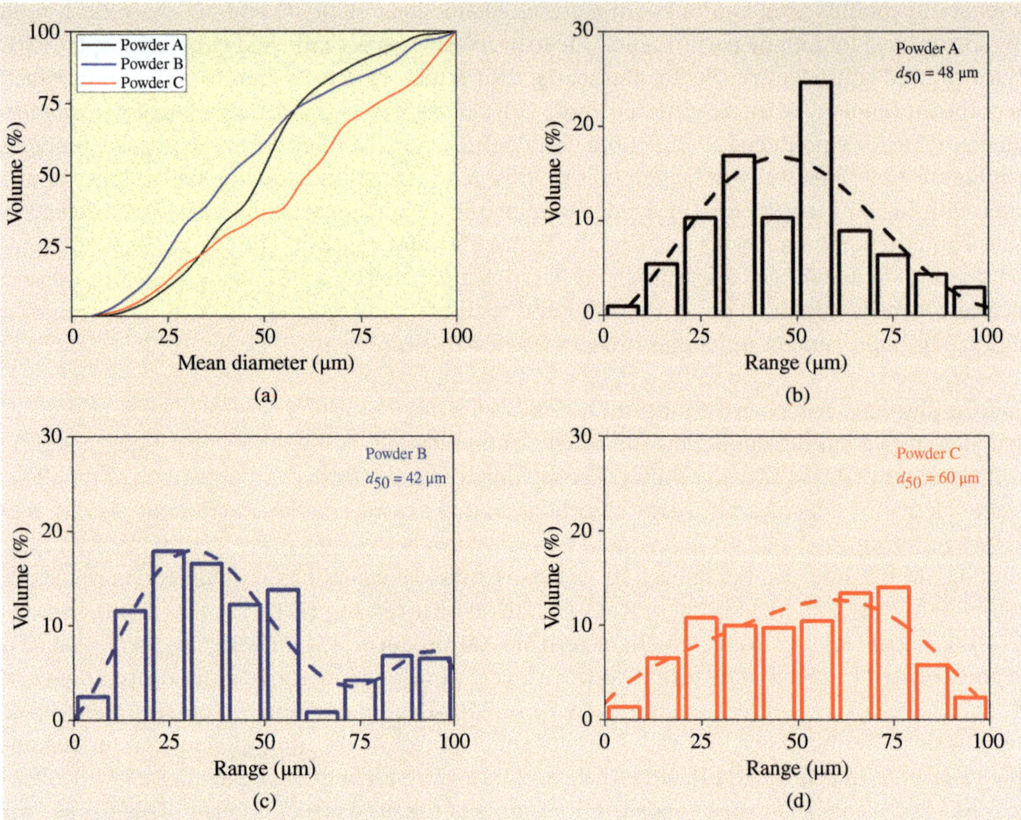

Figure 13.4 (a) A comparison of the particle size distributions (PSD) of three AlSi12 alloy powders for powder bed fusion manufacturing using a laser beam, along with PSD bar charts for (b) powder A, (c) powder B, and (d) powder C. Source: Baitimerov et al. [17]/MDPI/Licensed under CC BY 4.0.

influences interparticle friction and, in turn, flowability. For example, the quality of PBF-LB-produced AlSi12 alloys has been investigated as a function of PSDs [17]. Three powders with distinctively different distributions are shown in Figure 13.4. The PSD of powder A was associated with the highest density product (99.4%, compared to approximately 95% for powders B and C), with that powder having the lowest span defined as

$$\text{span} = (d_{90} - d_{10})/d_{50} \tag{13.1}$$

where d_{10} is the diameter representing the smallest 10% of the distribution, d_{50} is the mean diameter, and d_{90} represents the diameter for 90% of the distribution. In this case, the flowability of the powder is greater for smaller values of the span. We shall see in Section 15.2 that for some other systems, a wider distribution (larger values of the span) provides a higher packing density and a subsequent higher part density. Other powder characteristics are also important such as the relative sphericity of particles, various process parameters, and, in this study, slight variations in the Si content of each powder.

Another example of a PSD is given in Figure 13.5 for an alloy steel 12CrNi2Y (0.745 w/o Cr, 1.85 w/o Ni, 1.71 w/o Y, along with 0.605 Mn, 0.144 C, <0.3 w/o Si) used for fabricating nuclear power equipment by the DED-LB technique. This PSD is reminiscent of the **lognormal distribution** of powders introduced in Section 2.5. The powder was produced by vacuum induction melting inert gas atomization [18]. This alloy demonstrates superior performance to ordinary carbon steels for such applications and is discussed further in Section 13.4. Maximizing the density of the powder feedstock is also important, as that packing density has a significant effect on minimizing the porosity in the final part. The span of the distribution ensures that smaller particles can fill interstices among the packing of larger particles [19].

While we have focused on the general effort to minimize porosity in order to produce high-density parts, there are applications that require controlled levels of porosity. Biomedical systems such as implants requiring bone growth into porous surfaces of prostheses are important examples. Metallic **biomaterials** are widely used for medical devices

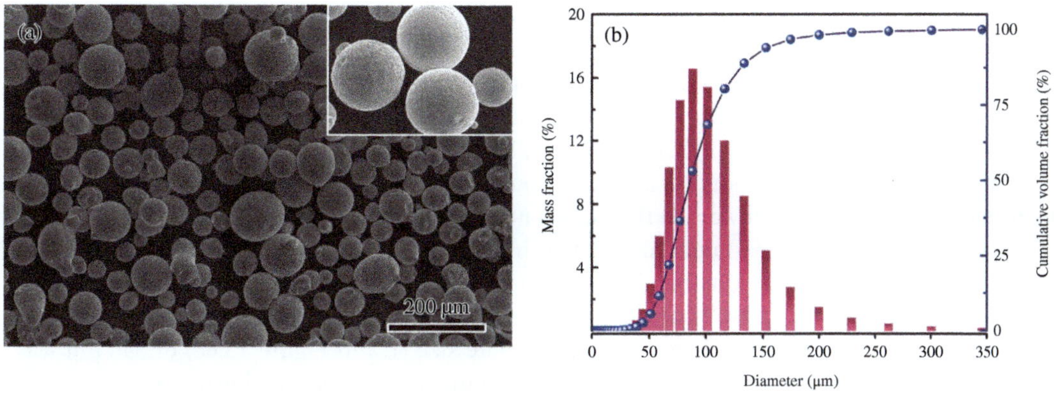

Figure 13.5 (a) The SEM morphology of a spherical powder of an alloy steel 12CrNi2Y used for fabricating nuclear power equipment by the directed energy deposition using a laser beam (DED-LB) technique corresponds to (b) its particle size distribution. Source: Guan et al. [18]/with permission of Elsevier.

for humans, and, since the 1960s, metallic prostheses have been used in total hip replacement (THR) surgery, one of the most successful and cost-effective operations in all of medicine. Since the 1970s, porous surfaces have been used in many THR designs for fixation by bone ingrowth [20].

The ability to produce controlled porosity by AM techniques has been studied in conjunction with biomedical applications of the type just described. For example, a porous β-type Ti-40Nb alloy has been developed for biomedical applications by PBF-LB technology [21]. Cylindrical samples were produced by hot pressing as well as PBF-LB. The hot-pressed samples were comprised of $\alpha + \beta$ phase mixtures and a total porosity of $3 \pm 0.4\%$ was present as interstices among the powder particles. The PBF-LB samples had small amounts of α and α'' as precipitates within the β phase and a total porosity of $17 \pm 1\%$. In both cases, the porosity was not interconnected but was homogeneously distributed within the cylinder volume. The porosity in the case of PBF-LB was produced by the contamination of the powder with O, C, and N during mechanical alloying, a topic covered in Section 12.2. The high level of porosity was thought to be the result of the particles having a more pronounced oxide layer on their surface leading to a thermal barrier to the local melting process in PBF-LB. The compressive elastic modulus of the PBF-LB cylinders was 33 ± 2 GPa, comparable to the value for healthy cancellous bone. Both the PBF-LB and hot-pressed samples demonstrated good *in vitro* biocompatibility.

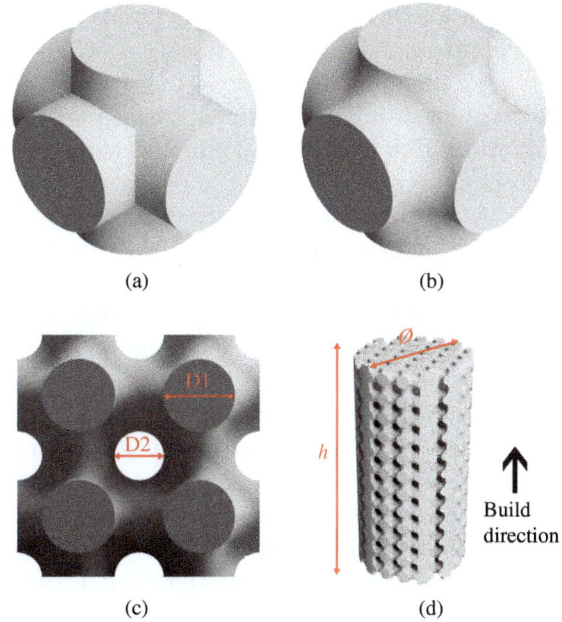

Figure 13.6 (a) Intersecting cylinders define a triply periodic minimal surface (TPMS) with (b) the Schwarz primitive (Schwarz P) unit cell being the TPMS with maximum fluid permeability that can be repetitively stacked as in (c), with (d) representing the overall cylindrical shape of Ti6Al4V samples produced by PBF-LB for use as biomedical scaffolds. Source: Soro et al. [22]/With permission of Elsevier.

In another study, Ti6Al4V alloy, one of the most widely used engineered materials in orthopedic surgery, was processed by PBF-LB to produce highly ordered, interconnected porosity [22]. The goal was to create biomedical scaffolds for the purpose of matching the mechanical properties of bone and providing a geometry conducive to bone ingrowth. The geometry of the interconnected porosity was based on a **Schwarz primitive** (Schwarz P) unit cell, an example of a triply periodic minimal surface (TPMS) as shown in Figure 13.6. This experimental study evaluated the Schwarz P-based porosity, which numerical studies indicated could be ideal for both mechanical and biological performance. While developed by Hermann Schwarz in the nineteenth century as an exercise in differential geometry to demonstrate minimal periodic surfaces, biomedical scientists in the early part of the twenty-first century have noted that the Schwarz P porous medium exhibits maximum fluid permeability and is then potentially ideal for osseointegration. Samples were produced with 64% porosity resulting in a stiffness of 22.3 GPa and a yield strength of 160 MPa, values comparable to that of cortical bone. A minimal pore size (D2 in Figure 13.6c) of 596 μm was produced which is suitable for osseointegration. Comparison between numerical models and experimental results indicated that **surface roughness** caused by powder adhesion did not have a significant effect on static mechanical properties.

13.3 Cracking and Delamination

Cracking and delamination are common problems in layered manufacturing given that rapid heating and cooling are inherent components of the process [2]. The thermal stresses associated with DED are especially high and make these defects a special concern in the form of AM [23]. As a practical matter, cracks can be considered a special form of porosity [1].

Delamination can be defined as a separation between two consecutive layers or between the first deposited layers and the baseplate. This defect occurs when the interlayer residual stress is higher than the material's yield strength [24]. Delamination typically occurs when the powder is unmelted or partially melted or if there is insufficient remelting of layers beneath the melt pool [25]. High-stress concentrations at the interface between the build and the baseplate are common sources of delamination.

Overall, **cracking** and delamination can serve as significant barriers to the adaption of AM more broadly in manufacturing applications. Those alloys that are susceptible to **cracking** in fusion welding are likely to exhibit this defect in AM. There are three major types of **cracking** in AM parts [2]. The first is **solidification cracking** (or hot cracking) along grain boundaries; the result of the greater contraction of the top (hotter) layers relative to the lower (cooler) layers. This can sometimes be the result of excessive energy applied in the process. The second is **liquation cracking** in the partially melted (or mushy) zone. This type results from the melting of grain boundary precipitate due to rapid heating to a temperature below the liquidus followed by both solidification and contraction upon cooling producing tensile stresses. Liquation cracking can occur in alloys such as Ni-based superalloys that have large differences between solidus and liquidus temperatures, in alloys such as Ti6Al4V that exhibit large solidification shrinkage, and in alloys such as Al-based alloys that exhibit large thermal contraction. The third is **ductility dip cracking**, a form of solid-state intergranular cracking exhibited by some fcc alloys at elevated temperatures.

The use of **ultrasonic vibration** during the powder deposition process in DED technology has been shown to reduce or eliminate microcracking, along with refining the **grain size** and general improvements in the levels of porosity and surface roughness [26]. Figure 13.7 shows the significant reduction in both porosity and microcracking produced by the addition of ultrasonic vibration in producing samples using the **laser-engineered net shaping (LENS^{TM})** technique. In this study, thin-walled samples of AISI 630, a precipitation-hardening stainless steel also known as 17-4 alloy, were fabricated with and without the addition of a 41 kHz ultrasonic signal provided by a piezoelectric ceramic transducer. The authors suggested that the ultrasonic vibration could break up gas-evolved pores and homogenize the molten pool with a lower level of residual stress.

Intermetallic compounds such as iron aluminides have attractive properties but can be challenging to produce by conventional methods such as casting, powder metallurgy, or hot extrusion. Such processes are costly and can result in brittle parts with coarse microstructures. Some promise has been shown using DED-LB for Fe-(28 at%Al), although substantial cracking was observed. Preheating the substrate to 200 °C, however, suppressed the cracking upon rapid cooling [27]. Also, several factors make AM by DED for aluminum alloys challenging, due to high surface

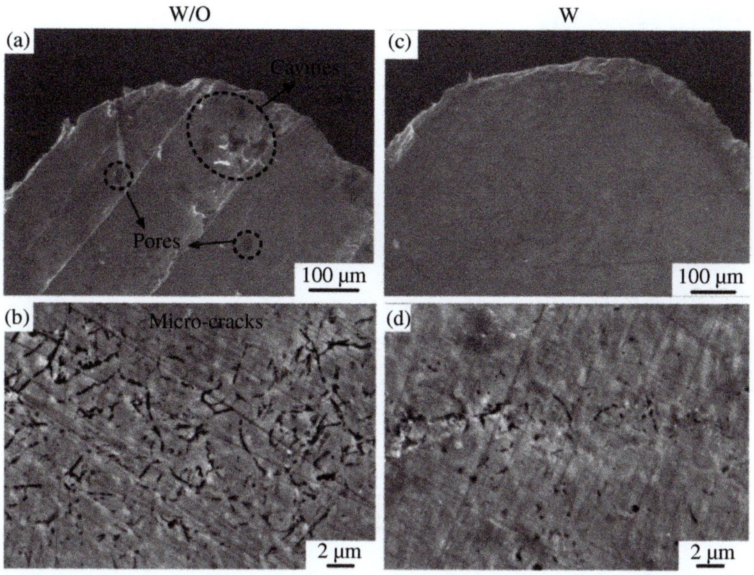

Figure 13.7 Comparisons of porosity, cavities, and microcracks in AISI 630 thin walls fabricated by the laser-engineered net shaping (LENS™) technique without (a and b) and with (c and d) ultrasonic vibration. Source: Cong et al. [26]/with permission of Elsevier.

reflectivity and thermal conductivity combined with high cooling rates. Pore formation and hot cracking can be resulting defects. Avoidance of these defects has been found in a limited number of aluminum alloys, mainly AlSi and AlSi10Mg, due to their improved melt pool wettability and a lower melting temperature provided by the high Si content [28, 29]. While cracking is a major concern in AM due to the potential for such defects because of rapid heating/cooling cycles and related residual stresses, AM also allows the use of new and custom alloys that potentially are more resistant to cracking. Examples include custom aluminum alloys, such as AlSi10Mg mentioned above, along with 7A77, 6061-RAM2, and Scalmalloy [3].

Inconel 718 is a nickel-based superalloy that is attractive for aerospace applications due to its high-temperature strength, as well as high creep and corrosion resistance. Samples of Inconel 718 produced by DED-LB exhibited a comparable fatigue strength to wrought samples at elevated temperatures, but the fatigue resistance at room temperature was slightly lower [30]. On the other hand, the fatigue resistance of the DED-LB samples in the low cycle fatigue (LCF) regime, showed slightly lower fatigue resistance at elevated temperatures compared to their room temperature performance. This was explained by the formation of brittle needle-like δ and Laves phases on grain boundaries at the elevated temperatures that increased the fatigue crack growth rate. In the high cycle fatigue (HCF) regime, the DED-LB samples showed similar fatigue resistance between room temperature and elevated temperatures, likely due to surface oxidation at the higher temperature retarding crack growth. Overall, crack initiation in the DED-LB samples in high-temperature fatigue testing was found to initiate from the surface, indicating no effect of pores or LoF on the fatigue behavior. In addition, DED technology has been shown to produce high-quality mechanical properties for impeller and airfoil components compared to conventional manufacturing techniques. For example, in repairing turbine blades and a fuel nozzle swirler, Inconel 718 was deposited without causing cracks that are associated with such repairs using typical welding or laser deposition processes [31].

Functionally graded materials (FGMs) are being explored for their potential as parts to be produced by DED AM [32]. As with conventionally produced FGM, brittle intermetallic phases and hot cracking are problems. It has been shown, however, that a NiCr intermediate transition layer applied using the LENS technique was effective in preventing the production of brittle intermetallic phases such as FeTi and Fe_2Ti when joining Ti6Al4V to 410 SS [33]. The NiCr layer reduced the cracking and delamination due to brittle intermetallics, as well as the associated residual stresses. Similarly, vanadium has been used as a transition material along the build direction in functionally graded Ti6Al4V to 304L SS [34]. They found that a gradual change in the concentration of V did not prevent the formation of brittle intermetallics such as FeTi. In fact, a few transition layers of 100% V were needed to produce a sufficient barrier between the Ti6Al4V and 304L SS alloys to prevent such intermetallics and associated cracking in the build.

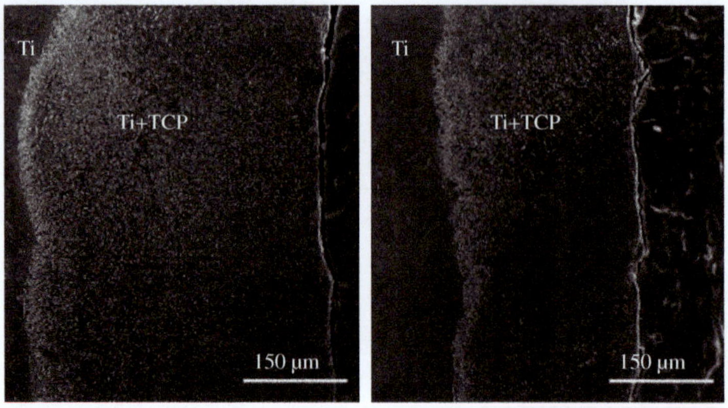

(a) (b)

Figure 13.8 Scanning electron micrographs show tricalcium phosphate (TCP) coatings produced on Ti by the LENS™ technique provide a crack-free interface for improving an orthopedic implant's osseointegration *in vivo*. The coatings were produced with a laser power of (a) 500 W and (b) 400 W. Source: Roy et al. [35]/with permission of Elsevier.

While cracking and delamination have been frequent concerns in AM as well as conventional metal processing, using the LENS technique to coat tricalcium phosphate (TCP) onto Ti was found to be crack-free and has provided a competitive technology to such traditional ceramic coatings that are used to improve an orthopedic implant's osseointegration *in vivo* [35]. Figure 13.8 shows typical cross-sectional micrographs indicating that the coating thickness increased from 250 to 400 μm as the laser power was increased from 400 to 500 W.

13.4 Interfacial Structure and Grain Size

The microstructures associated with AM were discussed in Chapter 12. In this section, we will discuss the interfacial structure and grain size as microstructural features that can be considered defects. In doing so, we will see that these "defects" are often associated with the other defect categories in this chapter in complex ways. In fact, the complex interrelationships of different types of defects have been seen in previous sections and will continue to rise throughout the remainder of the chapter.

For example, it should also be noted that examples of cracking in Section 13.3 are often initiated at the grain and interfacial boundaries [2]. Also in Section 13.3, the difficulty of processing conventional aluminum alloys by AM

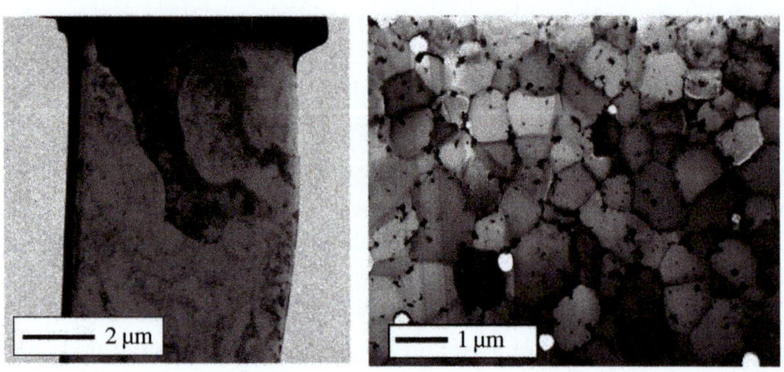

(a) (b)

Figure 13.9 Bright-field transmission electron micrographs show (a) columnar grains in an additively manufactured 6061 aluminum alloy in contrast to (b) finer, equiaxed grains resulting from using an yttria stabilized zirconia (YSZ) additive. The less-than-optimal columnar microstructure can be considered the defect. Source: Opprecht et al. [36]/with permission of Elsevier.

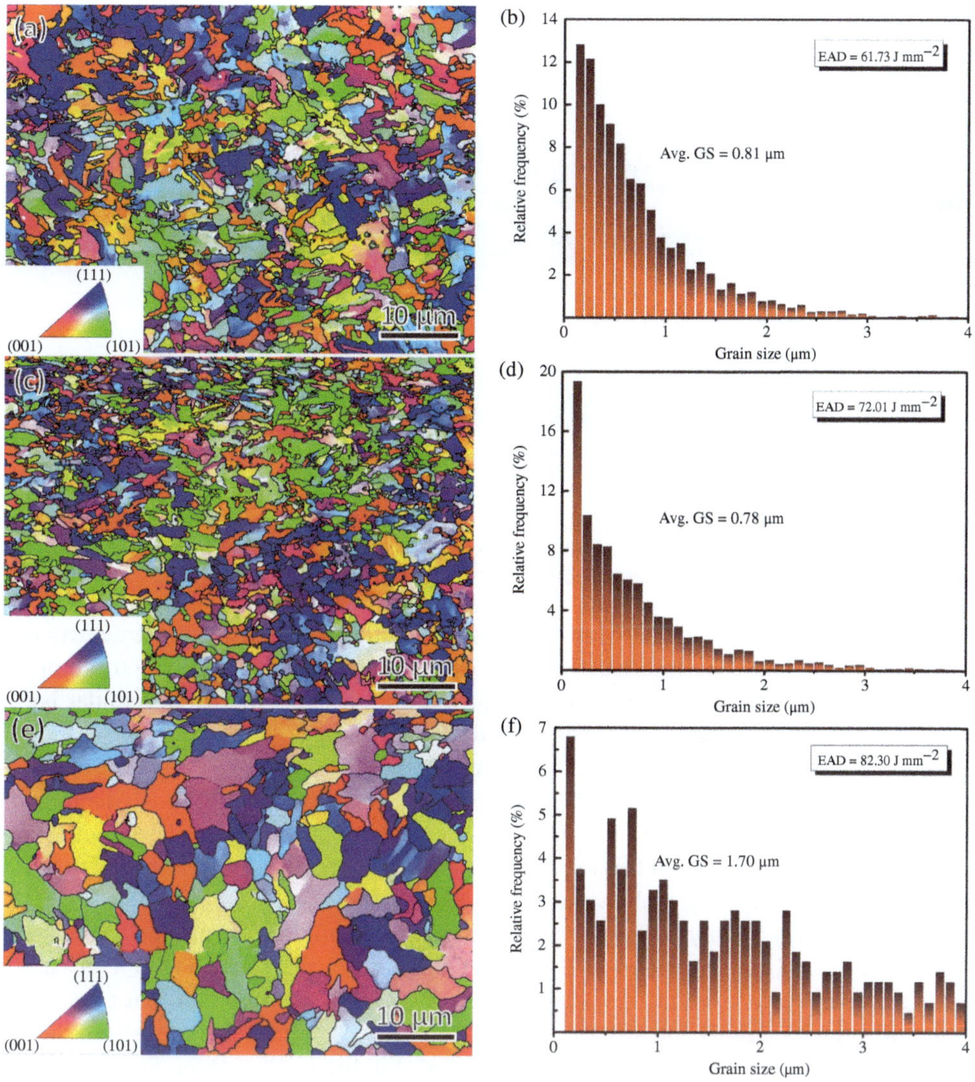

Figure 13.10 While the particle size distribution for a 12CrNi2Y alloy steel used in the nuclear industry was introduced in Section 13.2 (Figure 13.5), the resulting microstructures vary for different laser energy area densities (EAD) in directed energy deposition using a laser beam (DED-LB) additive manufacturing. Inverse pole figure (IPF) maps and grain size distributions are shown for three different EADs. (a) and (b) show the IPF map and grain size distribution for an EAD of 61.73 J/mm^2 while (c) and (d) show the same for an EAD of 72.01 J/mm^2 and Figure (e) and (f) represent an EAD of 82.30 J/mm^2. Source: Guan et al. [18]/ with permission of Elsevier.

was discussed, along with the need to focus on custom aluminum alloys such as AlSi10Mg for successful AM processing. It is worth noting that cracking that can accompany the AM of aluminum alloys has, in some cases, been avoided by additives such as zirconium oxide and some rare earth oxides. This positive result was attributed to finer, equiaxed grains as a form of grain refinement to avoid the development of large columnar structures [36]. In this case, the less-than-optimal microstructure can be considered the defect. Figure 13.9 contrasts the columnar structure (a) with the equiaxed structure (b). Furthermore, the refined microstructure shows both intra- and intergranular Zr precipitates arising from the zirconium oxide additive. The overall result was a crack-free sample of the widely used 6061 aluminum alloy successfully processed by the PBF-LB technique by using the controlled mixing of 6061 and yttria-stabilized zirconia (YSZ) powders.

The PSD for alloy steel used in the nuclear industry was introduced in Section 13.2 (Figure 13.5). The resulting microstructures for three different laser energy area densities (EAD) in DED-LB-produced samples are shown in

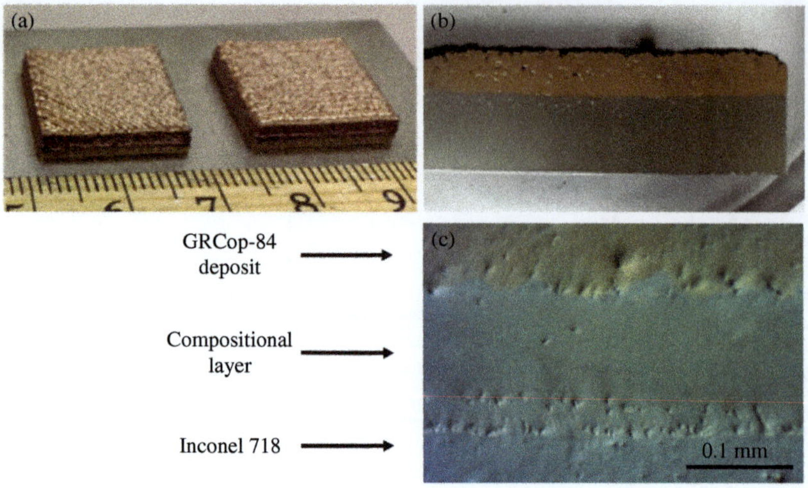

Figure 13.11 The successful interfacial bonding of a copper alloy, GRCop-84, on Inconel 718 is achieved using the LENS™ technique with a compositional gradation approach: (a) the uncut sample geometry, (b) a sample cross section, and (c) an optical image of the interface. Source: Onuike et al. [37]/with permission of Elsevier.

Figure 13.10. The rapid heating/cooling cycles associated with AM produce relatively complex microstructures of bainite, ferrite, and carbide. The optimal density in this study (98.95%) was produced by the EAD of 72.01 J mm^{-2} (Figure 13.10c and d). With increasing EAD, the microstructures transform from polygonal ferrite to granular bainite [18]. Again, the less-than-optimal microstructures can be considered defects.

As DED is a melt-cast process, a diffuse interface produces good metallurgical bonding [2]. As an example, Figure 13.11 illustrates the successful interfacial bonding of a high-temperature copper alloy, GRCop-84, deposited on Inconel 718 using the LENS technique with a compositional gradation approach. This copper alloy layer produced a more than 300% increase in the thermal conductivity of this materials system for applications such as rocket nozzle walls. The concentration gradient across the interface is shown in Figure 13.12 [37].

Figure 13.13 shows the effect of laser energy density (LED) on grain size and grain morphology in producing Inconel 718 parts by the LENS technique. In general, and as shown in Figure 13.13a, the grain size goes up with increasing energy density. Nonetheless, the average grain size during AM processing is significantly less than that produced by wrought processing (4–8 μm over the LED range for AM versus approximately 30 μm for the wrought product). The smaller grain size is due to the much higher cooling rate (1000–1500 K s^{-1}) in the LENS system. Figure 13.13b shows the morphological feature of **dendrite arm spacing (DAS)** as a function of LED. Again, the DAS feature goes up with increasing energy density. Also, the range of DAS values between 2 and 6 μm for the AM products is considerably smaller than the typical DAS in the wrought Inconel 718 material (10–40 μm). Figure 13.14 shows the grain morphology for a range of LED values. In this system, optimal mechanical performance is associated with smaller grain sizes leading to a recommendation of an LED value less than 137.5 J mm^{-2} [38]. One should also note that a detailed discussion of DAS is given in Section 3.6.2.

As shown in Figure 8.5 in Chapter 8, multiple interfaces that can be produced by the layer-by-layer build-up in the LENS version of DED AM can be viewed as both an advantage and a challenge. For example, in the production of 316L SS components, a columnar grain structure due to epitaxial growth from the previous solid interface was observed in each layer [39]. As a result, the LENS process allows local control of the microstructure and texture by manipulating the solidification front. Various images of these multiple interfaces in 316L SS samples are summarized in Figure 13.15. Electron microscopy revealed high dislocation densities at both the boundaries and the interiors of growth cells. Nonetheless, good quality structures were generally produced, although defects such as gas porosity and lack-of-fusion porosity were found in specimens with improper laser focus.

Process modeling is showing promise for guiding processing parameter selection in producing high (99%) density parts with minimal defects [40]. Figure 13.16 shows a useful processing parameter S for DED processing of Ti6Al4V alloys where S is defined as

$$S = P/(vG) \tag{13.2}$$

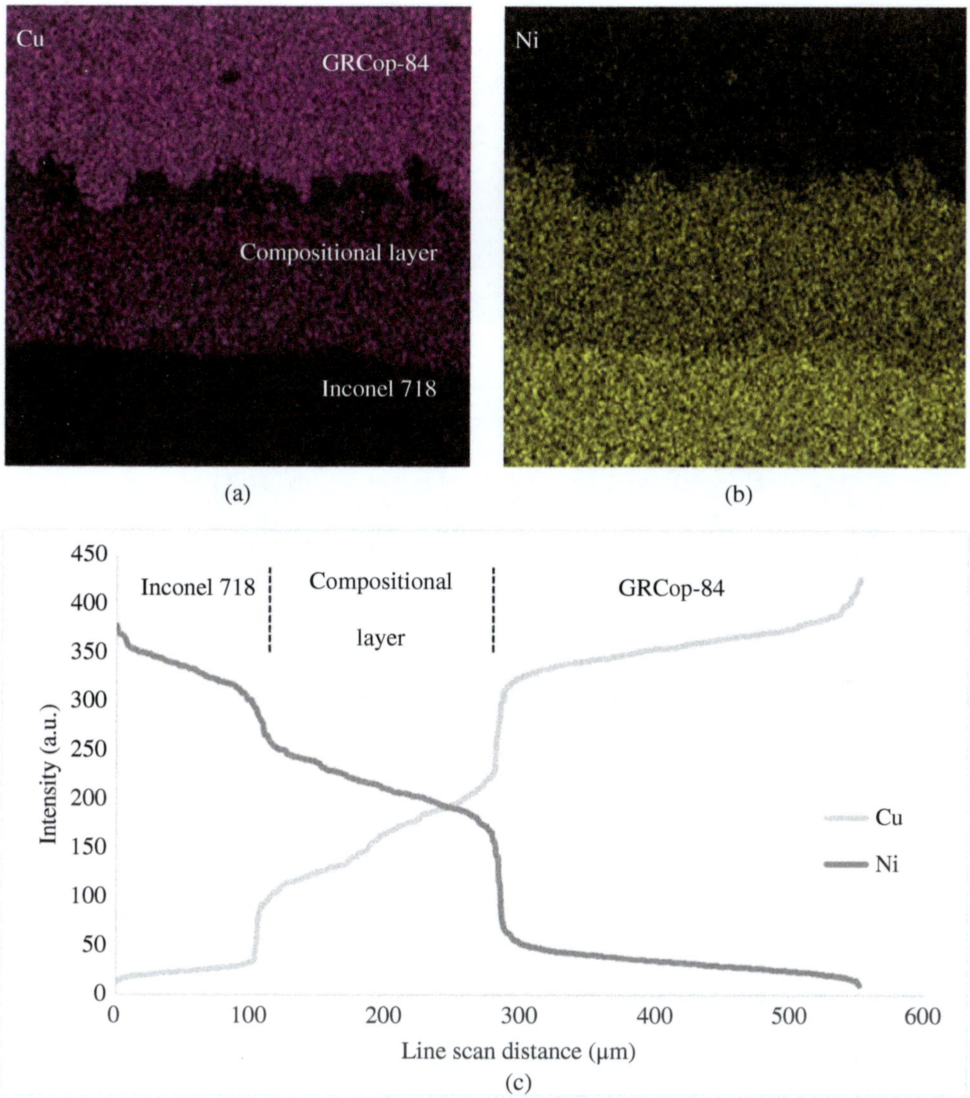

Figure 13.12 Energy dispersive spectroscopy (EDS) mapping across the interface shown in Figure 13.11: (a) the Cu map, (b) the Ni map, and (c) the compositional gradation plot. Source: Onuike et al. [37]/with permission of Elsevier.

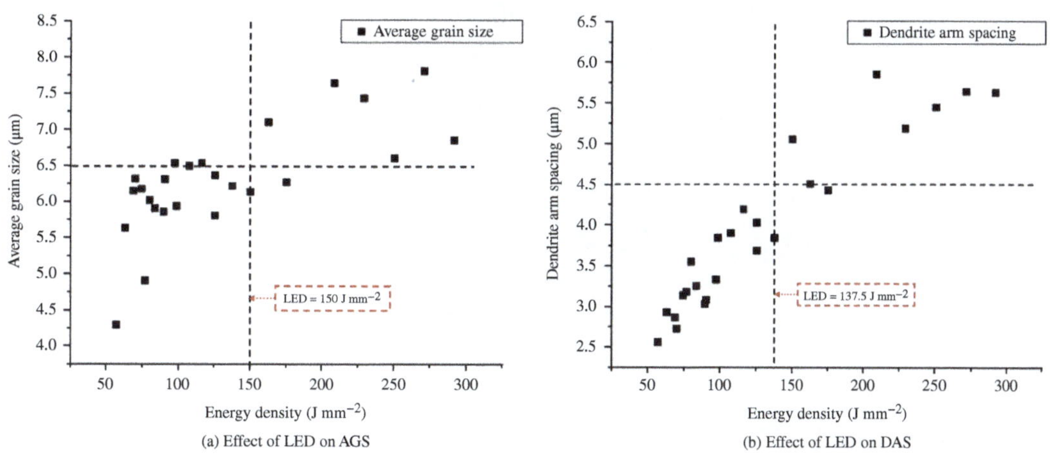

Figure 13.13 The effect of laser energy density (LED) on grain size and grain morphology in producing Inconel 718 parts by the LENS^{TM} technique: (a) the grain size goes up with increasing energy density, and (b) the morphological feature of dendrite arm spacing (DAS) also goes up with increasing energy density. Source: Liu et al. [38]/with permission of Elsevier.

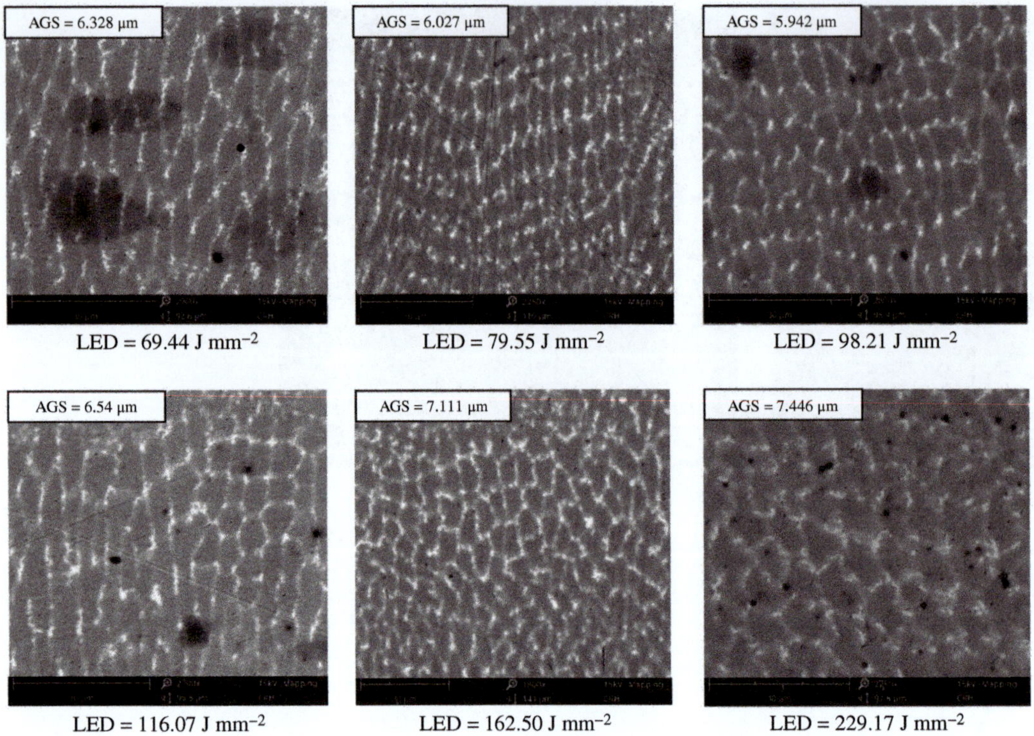

Figure 13.14 The grain morphology of the Inconel 718 system shown in Figure 13.13 for a range of laser energy density (LED) values. Source: Liu et al. [38]/with permission of Elsevier.

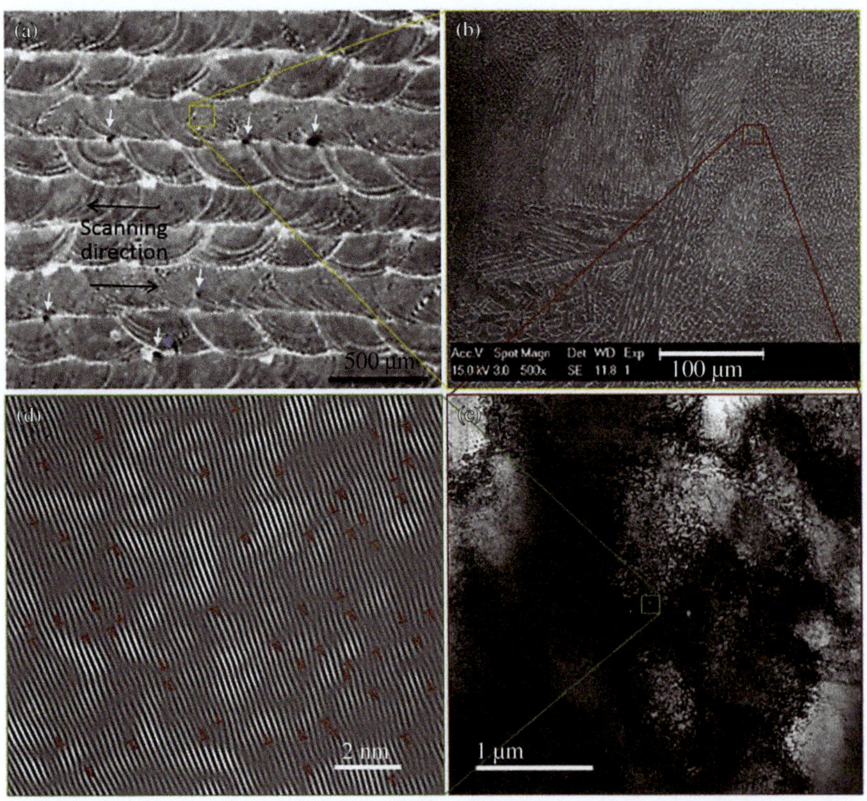

Figure 13.15 Multiple interfaces that occur in the production of 316L SS components by the LENS™ technique are shown in (a) an optical micrograph of the chemically etched patterns, (b) a scanning electron micrograph of multiple columnar colonies growing in different directions, (c) a bright field transmission electron micrograph of the interface between intercellular structure, revealing a high density of tangled dislocations, and (d) a high-resolution transmission electron micrograph showing a high density of individual dislocations at intercellular interfaces. Source: Zheng et al. [39]/with permission of Elsevier.

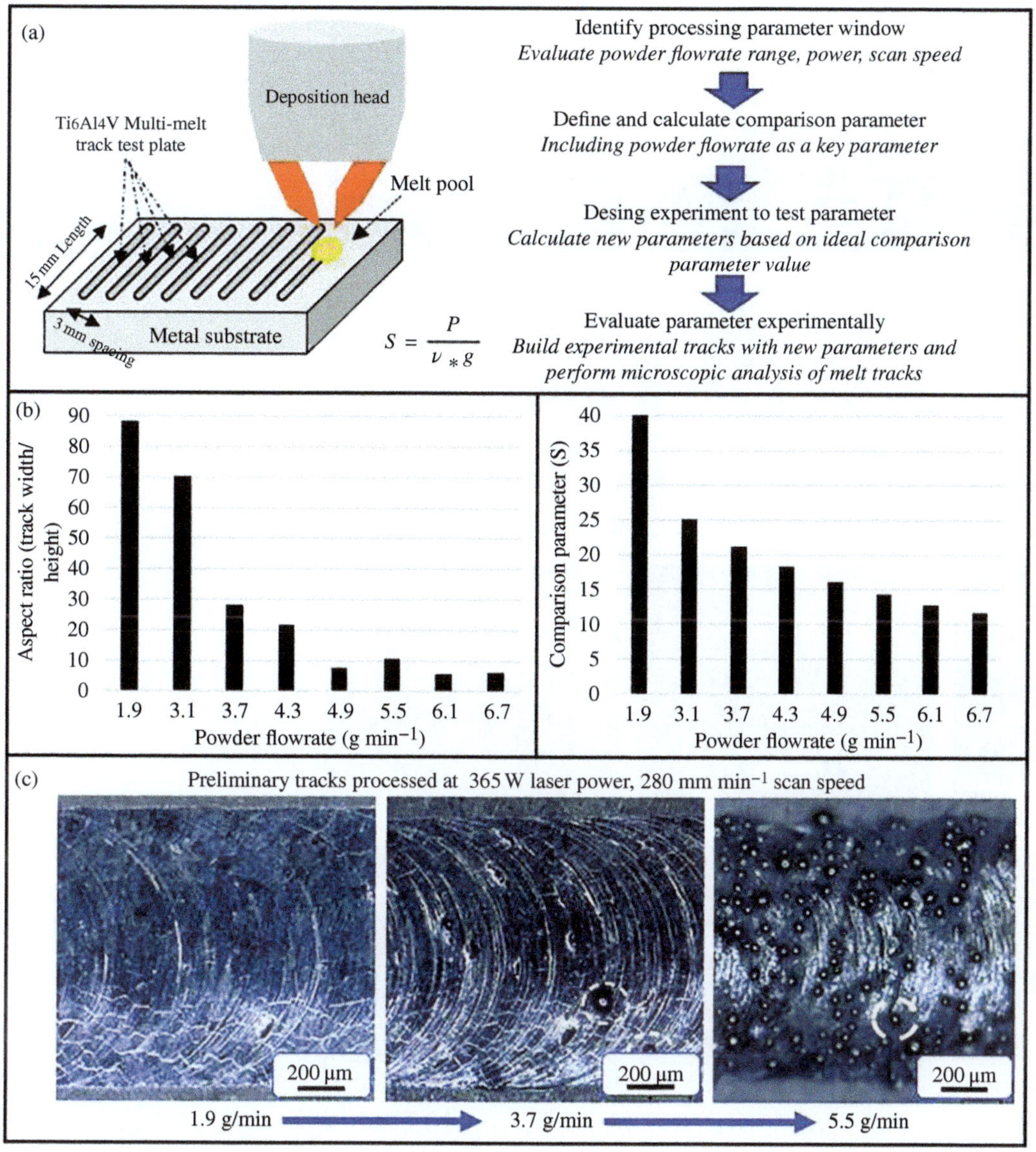

Figure 13.16 (a) LENS™ schematic, (b) initial deposition track characteristics as a function of flow rate, and (c) images of builds at various flow rates. Source: Traxel et al. [40]/with permission of Elsevier.

where P is the laser power (in $J\,s^{-1}$), v is the laser scan speed (in $mm\,s^{-1}$), and G is the powder mass flow rate (in $g\,s^{-1}$). In this study, previous modeling for PBF technology was adapted to the LENS form of DED processing with the goal of an experimental determination of optimal processing parameters. In these experiments, it was found that increasing the powder flow rate in S decreases the overall energy input as more mass is delivered to the melt pool. They found that sufficient energy must be supplied to the melt pool to ensure complete melting of the particles, the main factor for a quality build that avoids significant regions of unmelted particles.

Interfacial bonding is of fundamental importance in composite systems. The DED form of AM has potential benefits for producing **metal matrix composites (MMCs)** due to its controlled multi-feed powder hopper. This system allows the deposition of particle reinforcements within the MMCs with a controlled distribution. For example, the low solubility of carbon in both the α and β phases of titanium makes TiC a suitable reinforcing particulate for Ti-based MMCs. Figure 13.17 shows a range of interfacial structures produced within Ti6Al4V-based MMCs for TiC levels

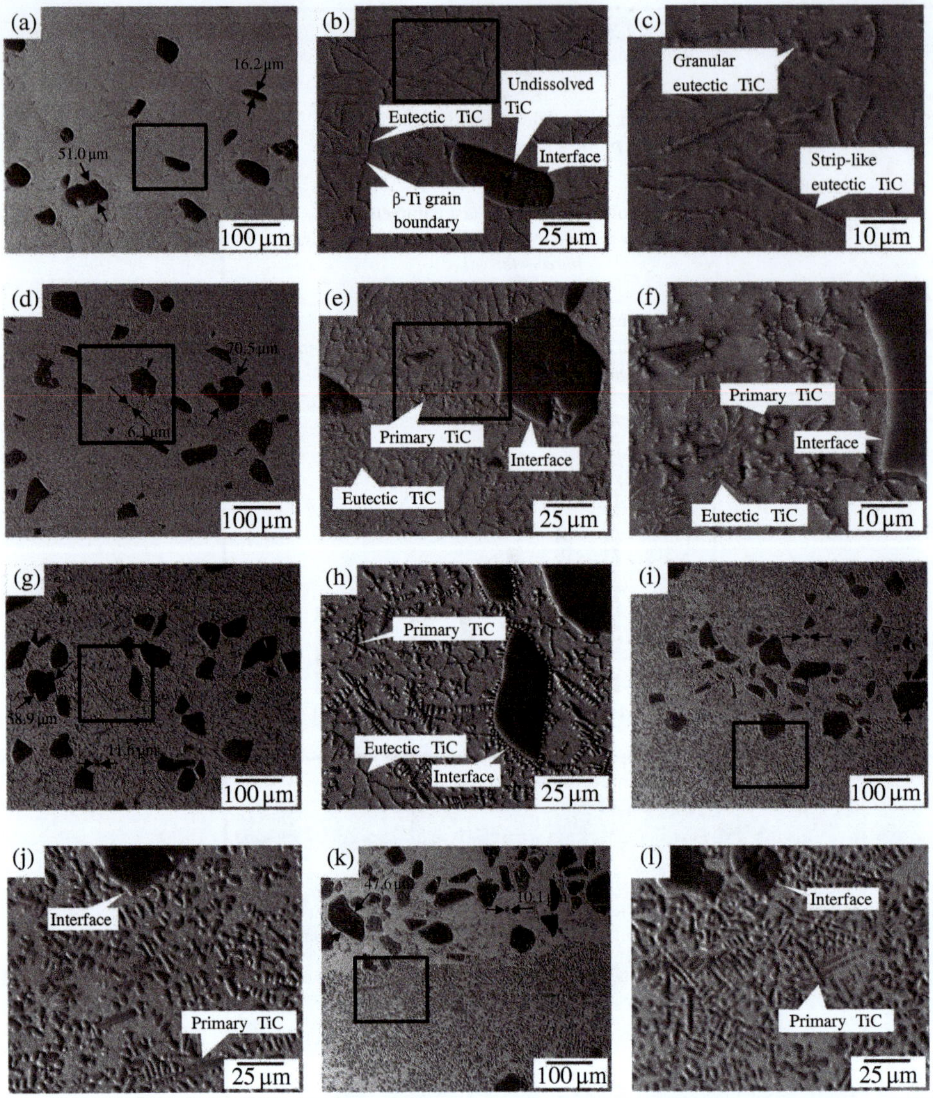

Figure 13.17 A range of interfacial structures are found in microstructures of Ti6Al4V-based metal matrix composites (MMC) for various TiC levels: (a), (b), and (c) 5 vol%, (d), (e), and (f) 10 vol%, (g) and (h) 15 vol%, (i) and (j) 20 vol%, and (k) and (l) 30 vol%. Source: [41]/with permission of Elsevier.

between 5 and 30 vol% [41]. The images show combinations of the α and β phases of Ti along with primary and eutectic phases of resolidified TiC and some unmelted TiC particles. These TiC/Ti6Al4V MMCs have become popular in the aerospace and automotive industries due to improved mechanical behavior compared to the Ti6Al4V alloy. The flexibility of producing parts by DED AM is proving to be more efficient than conventional methods such as powder metallurgy and *in situ* casting, resulting in greatly reduced production times. Figure 13.18 shows the tensile test results for the range of TiC levels shown in Figure 13.17. The 5 vol% TiC addition produces a significant improvement in the ultimate tensile strength (UTC) compared to the matrix alloy (1226 MPa compared to a value of 1091 MPa for Ti6Al4V). A 10 vol% TiC addition gives a UTC comparable to the Ti6Al4V value, but further TiC additions show a steady decrease compared to the matrix alloy. This decline was associated with dendritic TiC phases and undissolved TiC particles in microstructures for TiC additions above 10 vol%. Furthermore, Figure 13.19 shows that heat treatment of 950 °C for 10 hours diminishes the UTC for the 5 vol% TiC MMC only slightly to 1202 MPa while increasing

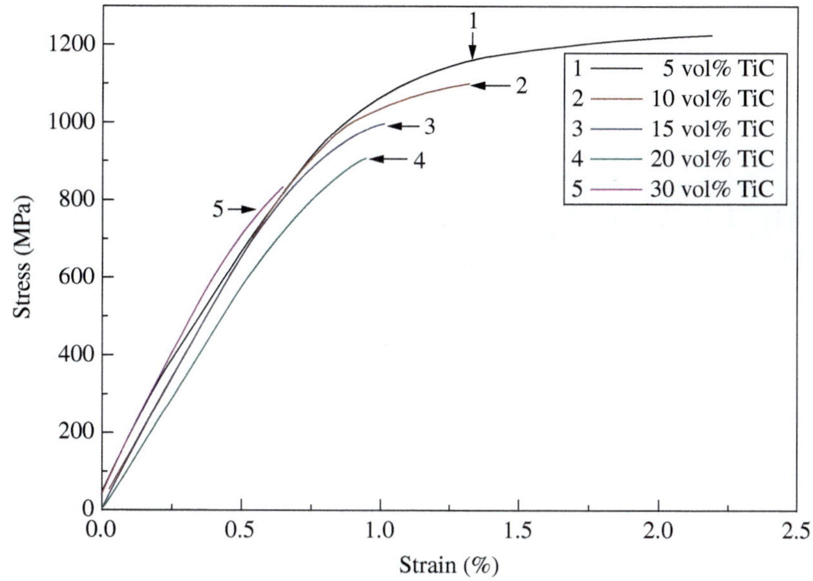

Figure 13.18 Tensile test results for the TiC/Ti6Al4V metal matrix composites shown in Figure 13.17. Source: Wang et al. [41]/with permission of Elsevier.

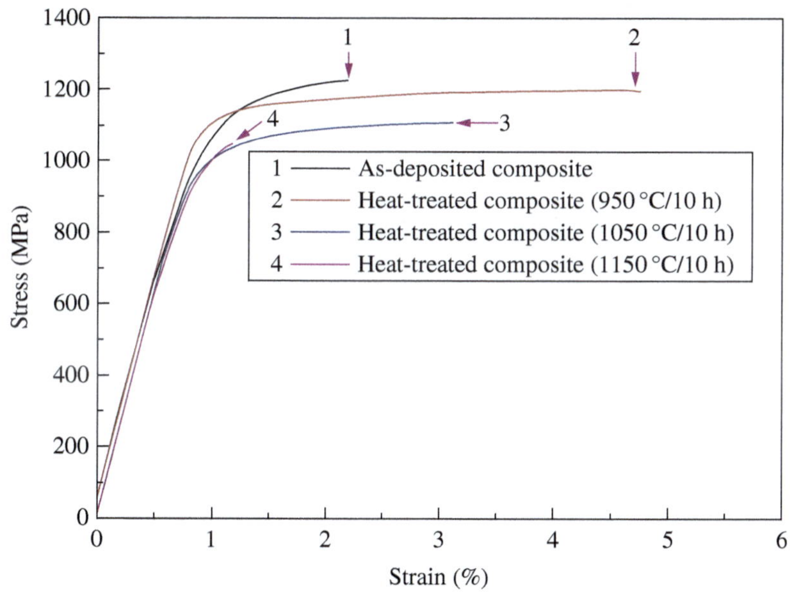

Figure 13.19 Tensile test results for the 5 vol% TiC/Ti6Al4V composite shown in Figure 13.17 for the as-deposited system and with various heat treatments. Source: Wang et al. [41]/with permission of Elsevier.

the elongation at failure from 1.31% to 3.95%. In summary, the 5 vol% TiC/Ti6Al4V MMC exhibited optimal tensile properties compared to the Ti6Al4V base alloy, an indication of strong interfacial bonding between matrix and reinforcing particulates.

Another useful application of DED AM is the ability to repair high-value parts. The relationship of grain size in the original part to the build layer has been studied in model systems [42]. Figure 13.20 gives the detailed microstructural images for a build layer on top of a hot rolled 316L SS sheet shown as electron backscatter diffraction (EBSD) maps created within a scanning electron microscope. Clearly, the substrate grains are smaller and more equiaxed than the grains in the repaired part. The magnifying box within the normal plan shows the epitaxial

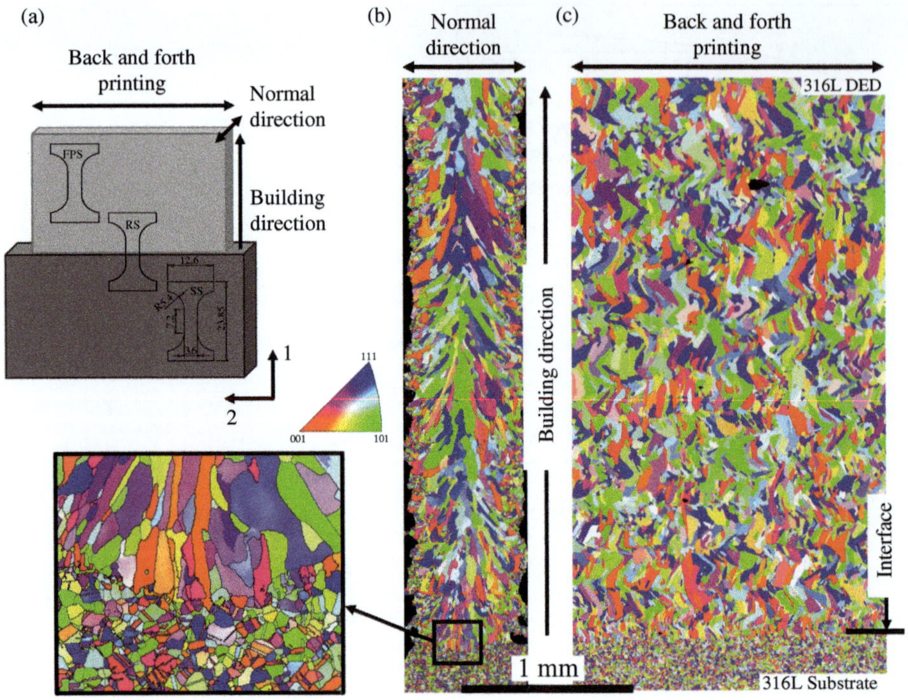

Figure 13.20 (a) Sample system in which a surface layer is repaired on top of a hot rolled 316L SS sheet, along with an electron backscatter diffraction (EBSD) map of (b) normal and (c) printing plans. The magnifying box within the (b) normal plan shows the epitaxial growth of grains at the interface indicating a solid metallurgical bond. Source: Balit et al. [42]/with permission of Elsevier.

growth of grains at the interface indicating a solid metallurgical bond. As illustrated in Figure 13.20a, three dog bone-shaped test specimens were produced. Details of the tensile testing results on these parts are given in Section 15.4.

13.5 Segregation

Chemical segregation is another microstructural feature, along with the interfacial structure and grain size topics from Section 13.4, that is frequently identified as a defect in AM processing. For example, in analyzing the precipitation kinetics and phase compositions during the PBF-LB of Inconel 625, elemental segregation was found to be the root cause of unusual precipitation behavior [43]. Heavy elements such as Nb and Mo were found to be locally segregated in inter-dendritic regions. The segregation was ascribed to localized departures from the nominal Nb and Mo levels in the powder feedstock. An initial **post-processing** heat treatment at 870 °C for stress relief led to platelet-like δ phase precipitates with deleterious effects on other mechanical properties, but a higher temperature heat treatment at 1150 °C for one hour removed those δ phase precipitates providing an overall homogenization of the alloy.

The challenges in processing aluminum alloys by AM were introduced in Section 13.3. The widely used Al 7xxx series alloys are good examples. Although conventionally produced Al 7xxx alloys display outstanding mechanical properties, the low boiling point elements such as Zn, Cu, and Mg can be depleted in AM processing leading to porosity and LoF in the as-deposited alloy. To address that problem, Ni-coated Al 7050 powder has been used, allowing improved laser energy absorbance. This, in turn, has allowed the reduction of the applied laser energy in the process [44]. Although the Ni coating helped to reduce porosity, the segregation of Ni in the inter-dendritic boundaries led to the formation of a brittle intermetallic phase and a degradation in mechanical properties.

Shape memory alloys undergo solid-to-solid phase transformations induced by certain temperature or stress changes, allowing the recovery of permanent strains [45]. The most common example is NiTi (Nitinol) which has been used in a wide range of applications and is especially well known for use in various medical devices. AM is considered an attractive alternative to conventional processing, especially considering the frequent requirement

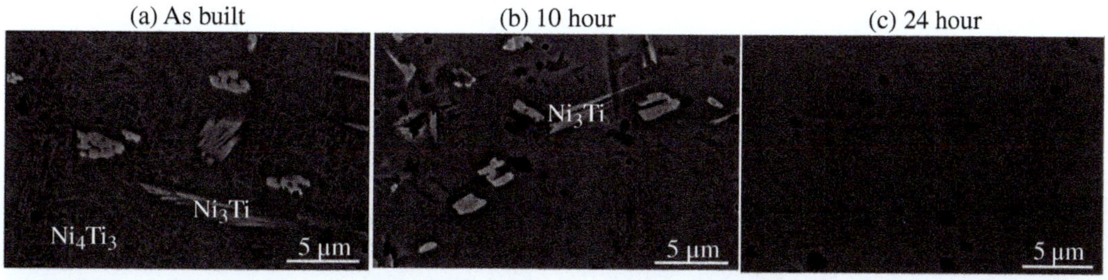

Figure 13.21 Backscatter electron images show how an homogenization heat treatment at 950 °C can remove the Ni_4Ti_3 precipitates and Ni_3Ti second phase particles present in (a) an as-built Ni53Ti47 (at%) shape memory alloy. (b) After 10 hours, these phases are diminished, and (c) after 24 hours, they are fully removed. Source: Hamilton et al. [46]/with permission of Elsevier.

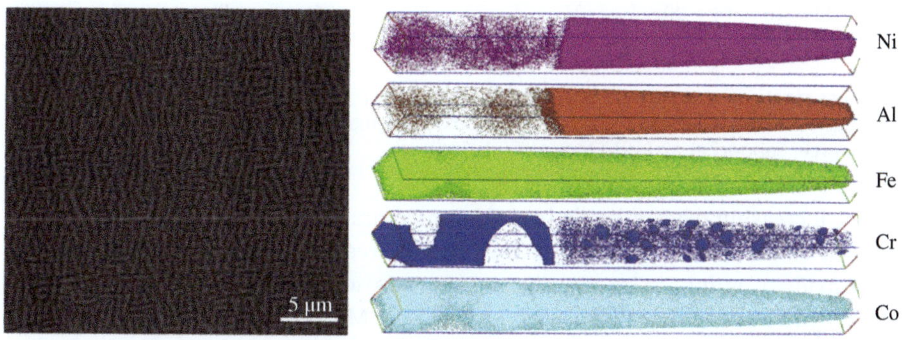

Figure 13.22 Atom probe tomography indicates that, in an Al_xCoCrFeNi alloy aged at 600 °C for 168 hours, Fe and Co were homogeneously dispersed, while Ni, Al, and Cr exhibited severe segregation across two conjugate B2 phases. Source: Wang et al. [47]/with permission of Elsevier.

of complex part shapes that can be produced to near net shape by AM. DED, in particular, is promising, as it can allow processing by either a mixture of pure Ni and Ti powders or a pre-alloyed NiTi feedstock [32]. Elemental segregation, however, has been observed to produce chemical composition anisotropy along the build direction in the AM processing of Ni53Ti47 (at%) [46]. Nonetheless, Figure 13.21 shows how Ni_4Ti_3 precipitates and Ni_3Ti second phase particles in the as-built part are removed by a 24-hour homogenization heat treatment at 950 °C.

High entropy alloys (HEAs) have shown attractive properties such as improved strength and fracture toughness compared to conventional alloy chemistries. A number of these HE alloys have proven to be amenable to AM [32]. One system that has been extensively studied as a candidate for DED AM is Al_xCoCrFeNi due not only to its unique properties but also to the low cost of its elemental constituents [47]. In fact, the microhardness of the alloy has been reported to increase by 13% compared to its as-cast counterpart due to the increased melt pool cooling rate and the increased temperature that produced a refined grain size [48]. It is worth noting however while Fe and Co were homogeneously dispersed, Ni, Al, and Cr exhibited severe segregation across two conjugate B2 (BCC) phases (formed by spinodal decomposition) as measured by atom probe tomography and shown in Figure 13.22 [47].

Closely related to the segregation of impurities to interfacial boundaries is the fact that many alloys being processed by AM have a high affinity for surface oxidation in the presence of atmospheric oxygen. This point was raised in conjunction with the production of high porosity β-type Ti-40Nb alloy for biomedical applications by PBF-LB technology as discussed in Section 13.2. The significant level of porosity was found to be the result of the contamination of the powder with O, C, and N during mechanical alloying. More specifically, the high level of porosity was thought to be the result of the particles having a more pronounced oxide layer on their surface leading to a thermal barrier to the local melting process in PBF-LB. Also, in Section 13.3, we saw that crack growth in samples of Inconel 718 produced by DED-LB was retarded in the HCF regime likely due to surface oxidation. While oxidation proved beneficial in those two previous cases, oxidation is often a challenge in AM processing. In the case of AM processing of stainless steels, the high Cr content has a high oxygen affinity that can lead to a degradation of mechanical properties [32]. For example, Figure 13.23 shows the reduction in ductility in 316L stainless

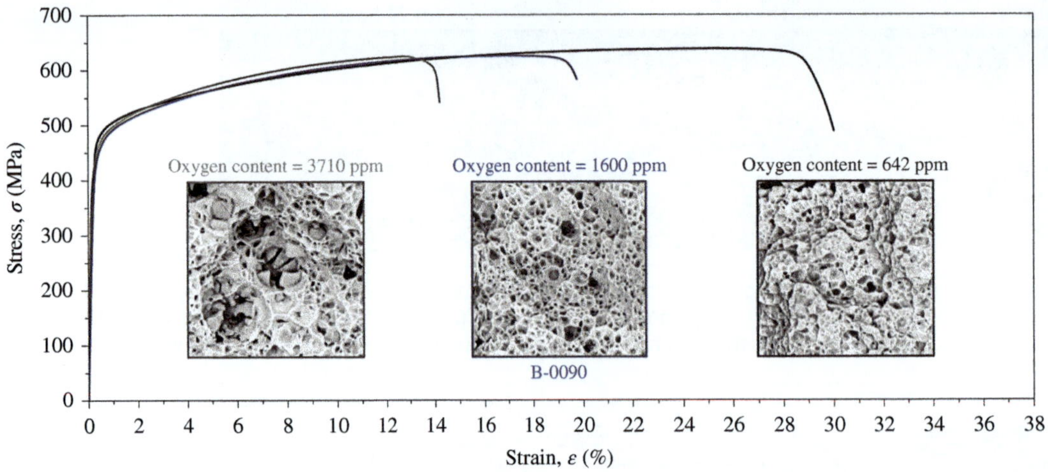

Figure 13.23 The ductility in 316L stainless steel samples produced by DED is seen to drop significantly with increasing oxygen levels. The micrograph insets indicate that the fracture surface contained oxide inclusions and that their size increased markedly with increasing oxygen levels. Source: Saboori et al. [49]/Elsevier/CC BY 4.0.

steel samples produced by DED under increasing oxygen levels [49]. The micrograph insets in Figure 13.23 indicate that the fracture surface contained oxide inclusions and that their size increased markedly with increasing oxygen levels. Chemical analysis of the oxides was done by energy dispersive spectroscopy (EDS) using field emission scanning electron microscopy (FESEM) and showed that they were rich in Si and Mn, minor elements in the 316L composition.

13.6 Surface Roughness

Surface roughness is a significant concern for AM. Although the various AM technologies are near-net-shape processes, post-processing is generally necessary to meet required tolerances and surface quality. Within the suite of AM technologies, the DED process tends to have a higher degree of roughness than PBF-LB due to generally larger feedstock powder size and larger layer thickness [2]. On the other hand, PBF-EB also tends to have a higher surface roughness than PBF-LB due to larger feedstock powder sizes as well [3]. Part geometries are also, in general, factors in the degree of surface roughness found on the part. For example, build angle constraints limit overhangs to a maximum of 45° relative to the vertical, and surface roughness depends on the overhang angle. Down-facing surfaces often have rougher surfaces with excess material and irregular surface morphology. In turn, this roughness can act as stress concentrator and crack initiation locations [3]. For DED AM, the sources of surface roughness include partially melted powder particles, Raleigh instability in the molten pool, a "stair-stepping" effect on inclined and curved surfaces, and the splashing of molten material [2]. The adverse effects of surface roughness (diminished dimensional tolerances and reduced fatigue strength) can generally be mitigated by increased heat input, smaller layer thickness, finer powder particles, and post-processing by HIP and chemical/electrochemical processing [2].

For example, **spattering** is a term used to describe the ejection of a particle from the surface of a part due to the collapse of a cavity near the bottom of the melt pool. Such spattering occurs in the presence of large pressure gradients induced by the vapor plasma plume. The evolution of such a cavity within the melt pool and subsequent spattering is shown in Figure 13.24, which was produced by an *in situ* high-speed X-ray imaging system at the Advanced Photon Source at the Argonne National Laboratory [50]. In these images of DED processing of Ti6Al4V alloy done with a laser power of 250 W and a scan speed of 100 mm s^{-1}, the horizontal red line represents the substrate surface, and the first frame at 1400 μs shows an initial set of pores prior to spattering. Pores 11, 13, and 14 are entrapped gas porosity, while 12 is a keyhole pore associated with the collapse of the cavity. Spattering is seen in the second image at 1450 μs, as pores 12 and 13 are engulfed by the cavity and the remaining pores grow larger. The third image at 1500 μs, after spattering, shows the shrinking of the pores and cavity associated with the cooling of the smaller melt

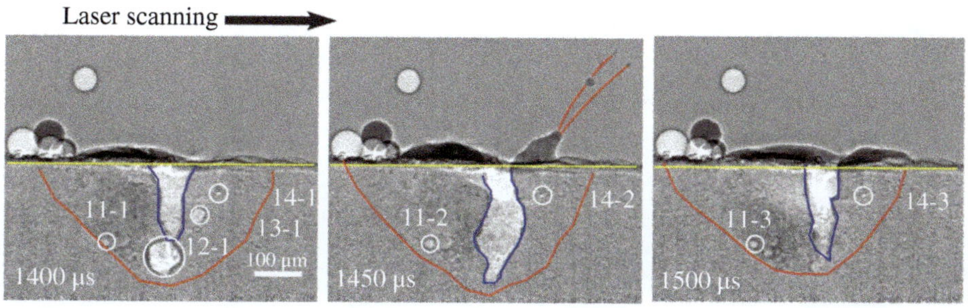

Figure 13.24 In these *in situ* high-speed X-ray images of the DED processing of Ti6Al4V alloy, the horizontal red line represents the substrate surface, and spattering is seen in the second image at 1450 μs, as pores 12 and 13 are engulfed by the cavity and the remaining pores grow larger. The material ejected during spattering can contribute to surface roughness when eventually coming to rest on the surface. Source: Wolff et al. [50]/Springer Nature/CC BY 4.0.

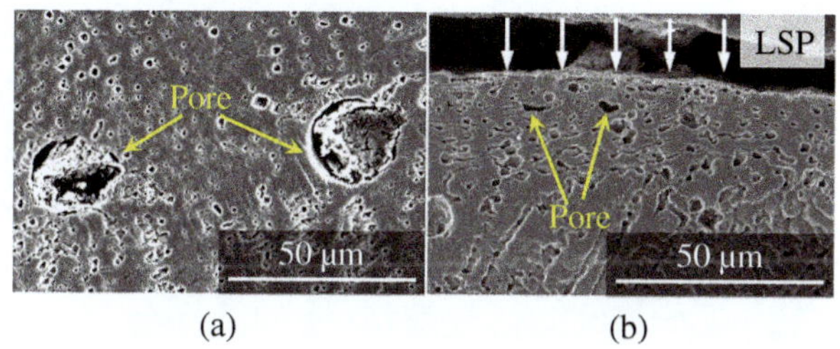

Figure 13.25 (a) Morphology of pores in the surface layer of an as-built specimen of high entropy CrMnFeCoNi alloy. (b) As an alternative to hot isostatic pressing (HIP) as a post-treatment for AM-produced parts, laser shock peening (LSP) generated sufficient plastic deformation on the surface to effectively close the pores shown in (a). Source: Tong et al. [51]/with permission of Elsevier.

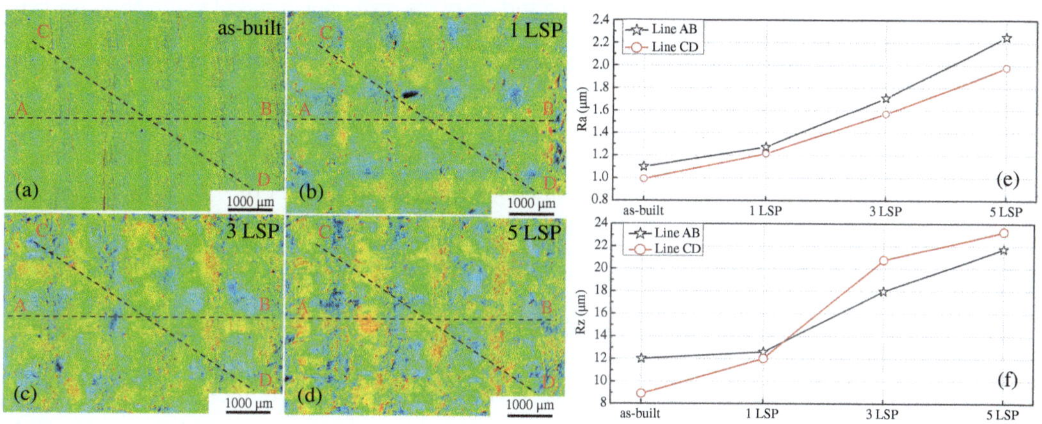

Figure 13.26 (a) As-built morphology of the high entropy CrMnFeCoNi alloy shown in Figure 13.25, along with (b) 1 LSP treatment, (c) 3 LSP treatments, and (d) 5 LSP treatments. (e) Surface roughness Ra and (f) maximum average height of the roughness profiles Rz along lines shown in (a) to (d). Source: Tong et al. [51]/with permission of Elsevier.

pool. The material ejected during spattering can contribute to surface roughness when eventually coming to rest on the surface.

While HIP has been a common post-treatment for AM-produced parts, **laser shock peening (LSP)** has also been explored as a means to improve the tensile properties of a DED-produced high entropy CrMnFeCoNi alloy. While the shock peening generated sufficient plastic deformation on the surface to effectively close pores as shown in Figure 13.25, the surface roughness increased with the shock peening cycles, roughly doubling in magnitude after 5 LSP cycles, as shown in Figure 13.26 [51]. The effect of LSP on residual stress is discussed in Section 14.6.

Along with porosity and residual stress, surface roughness is a significant concern for HCF and LCF. Such fatigue performance was initially inferior in most cases for AM-produced components compared to that for conventionally produced parts [3]. Considerable progress has been made, however, in mitigating these issues, and now defects have been sufficiently minimized such that mechanical performance, including fatigue performance, is often comparable to traditionally manufactured parts [52]. A study of the AM of dental materials has provided a relationship between fatigue life and surface roughness in titanium alloys [53]. Specifically, Ti6Al4V was fabricated using conventional casting, which was subsequently rolled, as well as by PBF-LB and PBF-EB. The surface roughness was characterized

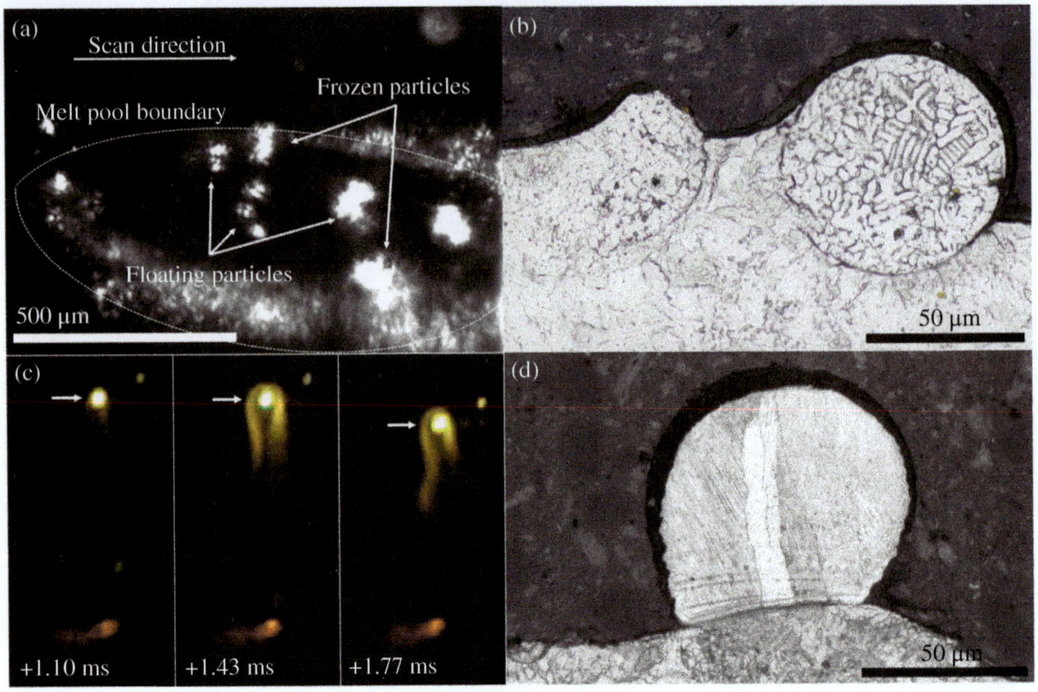

Figure 13.27 (a) and (b) Images showing how particles can float on the surface of the melt pool or freeze to the bottom when near the melt pool periphery, resulting in unmelted particles and surface roughness. (c) and (d) Images showing a slow-moving particle passing through the path of the heating laser and ultimately landing on the surface, again contributing to surface roughness. Source: Haley et al. [55]/with permission of Elsevier.

by scanning electron microscopy and profilometry. The fatigue life was measured by the three-point bending of beam specimens. The sequence of increasing surface roughness was rolled < as-cast < PBF-LB < PBF-EB, with fatigue life decreasing in the same sequence. Their data can be represented by the simple relationship [8]:

$$\ln(\text{fatigue life}) = -0.34\ln(\text{surface roughness}) + \text{constant} \tag{13.3}$$

indicating that increasing the surface roughness from 3 to 1000 Ra roughly decreases the fatigue life by an order of magnitude (from 10^5 to 10^4 cycles). On the other hand, varying the processing parameters in another study of Ti6Al4V produced by PBF-LB showed that reducing the surface roughness was done at the expense of increasing the residual stress, an indication of the challenges of reducing defects and the need for process optimization in the trade-off of these various defects [54].

The relationship of surface roughness to the dynamics of melting of individual powder particles has been illustrated by imaging the melt pool at a rate as high as 200 000 frames per second with a pixel resolution of up to 3.6 μm [55]. This study indicated that particles often impact and float on the surface of the melt pool for several hundred microseconds before melting into the pool. The authors concluded that the particles that attach to the surface during this residence time constitute a primary source for surface roughness. They also found that some incoming particles rebound from these attached particles, reducing the efficiency of powder delivery in the DED-LB process and possibly contributing to surface roughness. Furthermore, these attached particles can detach during service producing wear debris. Two distinct particle formation origins are shown in Figure 13.27.

In the discussion of porosity detection in AM products produced by PBF-LB, ultrasonics was seen to be a useful tool. Ultrasonics has also been used to inspect for unfused powder particles, a potential source of surface roughness. The unfused powder sizes were generally in the range of 100–150 μm. Such features in Ti6Al4V printed parts can be observed nondestructively by high-frequency plane wave ultrasonic transducers, although this is not possible in other commonly printed materials due to the higher speed of sound in those materials [13]. It is worth noting that the typical degree of surface roughness on AM-produced parts can reduce the sensitivity of some NDT techniques such as UT. On the other hand, surface roughness has little effect on the performance of **X-ray computed tomography (CT)** [3].

13.7 Summary

Parts produced by AM will be acceptable alternatives to conventionally produced ones only if defects produced during manufacturing are not at a significantly greater level. In focusing on various defects that can occur during AM and the relation of powders to the production of those defects, we explored the degree of porosity, cracking and delamination, interfacial structure and grain size, chemical segregation, and surface roughness. We also acknowledged that some applications, such as biomedical, require controlled levels of porosity. A wide variety of pore sizes and shapes can arise in both PBF and DED types of AM. The mitigation of porosity can be done by process optimization, feedstock control, and post-treatments such as HIP. Controlled levels of porosity for biomedical applications have been produced by carefully controlled powder chemistries that prevent complete densification as well as incorporating scaffold geometries into build patterns. Cracking and delamination are common problems in layered manufacturing given rapid and repetitive heating and cooling. The thermal stresses associated with DED are especially high. Delamination can occur as a separation between two consecutive layers or between the first deposited layers and the baseplate. Delamination occurs when the interlayer residual stress is higher than the material's yield strength due to unmelted or partially melted powder or insufficient remelting of layers beneath the melt pool. Preheating the substrate, avoiding the formation of brittle intermetallic phases, and ultrasonic vibration of powders have proven effective in preventing cracking and delamination.

Interfacial structure and grain size are microstructural "defects" that are often associated with other defect categories in complex ways. For example, cracking is often initiated at the grain and interfacial boundaries and can sometimes be avoided by selective additives that produce grain refinement. Optimizing the PSD for efficient powder density is helpful for minimizing the risk of cracking and delamination, along with optimizing laser EAD. Chemical segregation is another microstructural feature that can be a defect in AM processing. For example, elemental segregation can cause unusual precipitation behavior that requires post-processing heat treatment optimization to produce overall homogenization of the alloy. Closely related to interfacial segregation is the affinity for surface oxidation in the presence of atmospheric oxygen which can be both a benefit and a challenge. Finally, surface roughness is a significant concern for AM frequently requiring post-processing to meet required tolerances and surface quality. The DED process tends to have a higher degree of roughness than PBF-LB due to generally larger feedstock powder size and layer thickness. Similarly, PBF-EB tends to have a higher surface roughness than PBF-LB. The sources of surface roughness can include partially melted powder particles, Raleigh instability in the molten pool, a "stair-stepping" effect on inclined and curved surfaces, and the splashing of molten material. The adverse effects of surface roughness can generally be mitigated by increased heat input, smaller layer thickness, finer powder particles, post-processing by HIP, LSP, and chemical/electrochemical processing.

Nomenclature

AM	additive manufacturing
DAS	dendrite arm spacing
d_{50}	diameter of the PSD mean
d_{10}	diameter of the smallest 10% of the PSD
d_{90}	diameter representing 90% of the PSD
DED	directed energy deposition
DED-LB	directed energy deposition using a laser beam
EBSD	electron backscatter diffraction
EBME	effective bulk modulus elastography
EAD	energy area density
EDS	energy dispersive spectroscopy
FESEM	field emission scanning electron microscopy
FEM	finite element method
FGM	functionally graded materials
HCF	high cycle fatigue
HEA	high entropy alloys

HIP	hot isostatic pressing
LoF	lack of fusion
LED	laser energy density
LENS™	laser engineered net shaping
P	laser power
v	laser scan speed
LSP	laser shock peening
LCF	low cycle fatigue
MMC	metal matrix composite
NDT	nondestructive testing
PSD	particle size distribution
PBF	powder bed fusion
PBF-LB	powder bed fusion using a laser beam
PBF-EB	powder bed fusion using an electron beam
G	powder mass flow rate
Schwarz P	Schwarz primitive
THR	total hip replacement
TCP	tricalcium phosphate
TPMS	triply periodic minimal surface
UTC	ultimate tensile strength
UT	ultrasonic testing
CT	X-ray computed tomography
YSZ	yttria stabilized zirconia

References

1 Sola, A. and Nouri, A. (2019). *J. Adv. Manuf. Process.* 1: 1. https://doi.org/10.1016/j.mattod.2021.03.020.
2 Svetlisky, D., Das, M., Zheng, B. et al. (2021). *Mater. Today* 49: 271. https://doi.org/10.1016/j.mattod.2021.03.020.
3 Blakey-Milner, B., Gradl, P., Snedden, G. et al. (2021). *Mater. Des.* 209: 110008. https://doi.org/10.1016/j.matdes.2021.110008.
4 du Plessis, A. and Macdonald, E. (2020). *Addit. Manuf.* 34: 101191. https://doi.org/10.1016/j.addma.2020.101191.
5 du Plessis, A. (2019). *Addit. Manuf.* 30: 100871. https://doi.org/10.1016/j.addma.2019.100871.
6 Sanaei, N., Fatemi, A., and Phan, N. (2019). *Mater. Des.* 182: 108091. https://doi.org/10.1016/j.matdes.2019.108091.
7 Khairallah, S.A., Anderson, A.T., Rubenchik, A. et al. (2016). *Acta Mater.* 108: 36. https://doi.org/10.1016/j.actamat.2016.02.014.
8 Frazier, W.E. (2014). *J. Mater. Eng. Perform.* 23: 1917. https://doi.org/10.1007/s11665-014-0958-z.
9 Svensson, M. (2009). Ti6Al4V manufactured with electron beam melting (EBM): Mechanical and chemical properties. In: *Aeromat 2009*. Dayton, OH.
10 du Plessis, A., Yadroitsev, I., Yadroitsava, I. et al. (2018). *3D Print. Addit. Manuf.* 5: 227. https://doi.org/10.1089/3dp.2018.0060.
11 du Plessis, A., Waller, J., le Roux, S. et al. (2020). X-ray computed tomography inspection in metal additive manufacturing: The role of witness specimens. In: *Structural Integrity of Additive Manufactured Materials and Parts*, 139. Conchohoken, PA: ASTM International https://doi.org/10.1520/STP163120190109.
12 Serafino, N. and Steiner, D. (2021). *Metal AM* 7: 201.
13 Yang, T., Mazumder, S., Jin, Y. et al. (2021). *Materials* 14: 4929. https://doi.org/10.3390/ma14174929.
14 Everton, S.K., Hirsch, M., Stravroulakis, P. et al. (2016). *Mater. Des.* 95: 431. http://dx.doi.org/10.1016/j.matdes.2016.01.099.
15 Pantawane, M.V., Yang, T., Jin, Y. et al. (2021). *Mater. Sci. Eng., A* 811: 140990. https://doi.org/10.1016/j.msea.2021.140990.
16 Neveu, A. (2021). *Metal AM* 7: 175.

17 Baitimerov, R., Lykov, P., Zherebtsav, D. et al. (2018). *Materials* 11: 742. https://doi.org/10.3390/ma11050742.

18 Guan, T., Chen, S., Chen, X. et al. (2019). *J. Mater. Sci. Technol.* 35: 395. https://doi.org/10.1016/j.jmst.2018.10.024.

19 Liu, A. and Khalid, R. (2022). *Metal AM* 8: 203.

20 Eliaz, N. (2019). *Materials* 12: 407. https://doi.org/10.3390/ma12030407.

21 Zhuravleva, K., Bonisch, M., Prashanth, K.G. et al. (2013). *Materials* 6: 5700. https://doi.org/10.3390/ma6125700.

22 Soro, N., Attar, H., Wu, X. et al. (2019). *Mater. Sci. Eng., A* 745: 195. https://doi.org/10.1016/j.msea.2018.12.104.

23 Alimardani, M., Toyserkani, E., Huissoon, J.P. et al. (2009). *Opt. Lasers Eng.* 47: 1160. https://doi.org/10.1016/j.optlaseng.2009.06.010.

24 DebRoy, T., Wei, H.L., Zuback, J.S. et al. (2018). *Prog. Mater. Sci.* 92: 112. https://doi.org/10.1016/j.pmatsci.2017.10.001.

25 Sames, W.J., List, F.A., Pannala, S. et al. (2016). *Int. Mater. Rev.* 61: 315. https://doi.org/10.1080/09506608.2015.1116649.

26 Cong, W. and Ning, F. (2017). *Int. J. Mach. Tool Manuf.* 121: 61. https://doi.org/10.1016/j.ijmachtools.2017.04.008.

27 Rolink, G., Vogt, S., Sencekova, L. et al. (2014). *J. Mater. Res.* 29: 2036. https://doi.org/10.1557/jmr.2014.131.

28 Ding, Y., Muniz-Lerma, J.A., Trask, M. et al. (2016). *MRS Bull.* 41: 745. https://doi.org/10.1557/mrs.2016.214.

29 Ye, H. (2003). *J. Mater. Eng. Perform.* 12: 288. https://doi.org/10.1361/105994903770343132.

30 Nezhadfar, P.D., Johnson, A.S., and Shamsaei, N. (2020). *Int. J. Fatigue* 136: 105598. https://doi.org/10.1016/j.ijfatigue.2020.105598.

31 Xue, L., Li, Y., Chen, J. et al. (2015). Proceed. *ASME Turbo Expo* 6: GT2015-43971. https://doi.org/10.1115/GT2015-43971.

32 Svetlizky, D., Zheng, B., Vyatskikh, A. et al. (2022). *Mater. Sci. Eng., A* 840: 142967. https://doi.org/10.1016/j.msea.2022.142967.

33 Sahasrabudhe, H., Harrison, R., Carpenter, C. et al. (2015). *Addit. Manuf.* 5: 1. https://doi.org/10.1016/j.addma.2014.10.002.

34 Reichhardt, A., Dillon, R.P., Borgonia, J.P. et al. (2016). *Mater. Des.* 104: 404. https://doi.org/10.1016/j.matdes.2016.05.016.

35 Roy, M., Krishna, B.V., Bandyopadhyay, A. et al. (2008). *Acta Biomater.* 4: 324. https://doi.org/10.1016/j.actbio.2007.09.008.

36 Opprecht, M., Garandet, J.-P., Roux, G. et al. (2020). *Acta Mater.* 197: 40. https://doi.org/10.1016/j.actamat.2020.07.015.

37 Onuike, B., Heer, B., and Bandyopadhyay, A. (2018). *Addit. Manuf.* 21: 133. https://doi.org/10.1016/j.addma.2018.02.007.

38 Liu, Z., Kim, H., Liu, W. et al. (2019). *J. Manuf. Process.* 42: 96. https://doi.org/10.1016/j.jmapro.2019.04.020.

39 Zheng, B., Haley, J.C., Yang, N. et al. (2019). *Mater. Sci. Eng., A* 764: 138243. https://doi.org/10.1016/j.msea.2019.138243.

40 Traxel, K.D., Malihi, D., Starkey, K. et al. (2020). *Manuf. Lett.* 25: 88. https://doi.org/10.1016/j.mfglet.2020.08.005.

41 Wang, J., Li, L., Tan, C. et al. (2018). *J. Mater. Process. Technol.* 252: 524. http://dx.doi.org/10.1016/j.jmatprotec.2017.10.005.

42 Balit, Y., Guevenoux, C., Tanguy, A. et al. (2020). *Mater. Lett.* 270: 127632. https://doi.org/10.1016/j.matlet.2020.127632.

43 Zhang, F., Levine, L.E., Allen, A.J. et al. (2018). *Acta Mater.* 152: 200. https://doi.org/10.1016/j.actamat.2018.03.017.

44 Singh, A., Ramakrishnan, A., Baker, D. et al. (2017). *J. Alloys Compd.* 719: 151. https://doi.org/10.1016/j.jallcom.2017.05.171.

45 Hamilton, R.F., Bimber, B.A., Taheri Andani, M. et al. (2017). *J. Mater. Process. Technol.* 250: 55. https://doi.org/10.1016/j.jmatprotec.2017.06.027.

46 Hamilton, R.F., Bimber, B.A., and Palmer, T.A. (2018). *J. Alloys Compd.* 739: 712. https://doi.org/10.1016/j.jallcom.2017.12.270.

47 Wang, R., Zhang, K., Davies, C. et al. (2017). *J. Alloys Compd.* 694: 971. https://doi.org/10.1016/j.jallcom.2016.10.138.

48 Kunce, I., Polanski, M., Karczewski, K. et al. (2015). *J. Alloys Compd.* 648: 751. https://doi.org/10.1016/j.jallcom.2015.05.144.

49 Saboori, A., Piscopo, G., Lai, M. et al. (2020). *Mater. Sci. Eng., A* 780: 139179. https://doi.org/10.1016/j.msea.2020.139179.

50 Wolff, S.J., Wu, H., Parab, N. et al. (2019). *Sci. Rep.* 9: 962. https://doi.org/10.1038/s41598-018-36678-5.

51 Tong, Z., Liu, H., Jiao, J. et al. (2020). *Addit. Manuf.* 35: 101417. https://doi.org/10.1016/j.addma.2020.101417.

52 Yadroitsev, I., Yadroitsava, I., du Plessis, A. et al. (ed.) (2021). *Fundamentals of Laser Bed Fusion of Metals*, First edition. Elsevier https://www.elsevier.com/books/fundamentals-of-laser-powder-bed-fusion-ofmetals/yadroitsev/978-0-12-824090-8.

53 Chan, K.S., Koike, M., Mason, R.L. et al. (2013). *Metall. Mater. Trans. A* 44A: 1010. https://doi.org/10.1007/s11661-012-1470-4.

54 Artzt, K., Mishurova, T., Bauer, P.P. et al. (2020). *Materials* 13: 3348. https://doi.org/10.3390/ma13153348.

55 Haley, J.C., Schoenung, J.M., and Lavernia, E.J. (2018). *Addit. Manuf.* 22: 368. https://doi.org/10.1016/j.addma.2018.04.028.

14

Residual Stress and Powder Effects

14.1 Introduction

Although any thermomechanical manufacturing process can result in the formation of **residual stresses**, the successful application of **3D printing** technology to the production of structural metal parts has, from the outset, raised a concern that the rapid thermal cycling of each volume element in the structure could lead to greater levels of **residual stress** than are produced in conventional methods, including traditional powder metallurgical techniques. Despite that concern, using **additive manufacturing (AM)** to repair parts has been shown to produce less residual stress than doing such repairs with the conventional technique of welding [1]. In that case, AM techniques have the benefit of lower heat input during the repair process. An example will be illustrated in Section 14.5.

Residual stress is a term that has already appeared in Part II of this book and frequently so in Chapter 13 on defects in the AM process. In this chapter, we begin with an overview of the methods of measuring residual stress and then focus on the roles of powder characteristics (size and morphology, as well as physical and chemical properties) and process parameters, along with the **pre-processing** heat treatment and **post-processing** treatments that can help minimize the level of residual stress in the final part.

14.2 Measuring Residual Stress

In Chapter 13, we saw that the rapid heating–cooling cycles during layer-by-layer deposition in AM lead to various defects. The same is true to produce residual stress [2]. Residual stress can be defined as the stresses remaining within a material in the absence of external forces or thermal gradients after processing [3]. Three types of residual stress can be defined:

Type I: Macro residual stresses that are on a scale larger than the grain size of the material,
Type II: Micro residual stresses that vary on a scale of a single grain, and
Type III: Micro residual stresses that exist within a grain because of dislocations and other crystalline defects [3].

A wide portfolio of techniques is available for the measurement of residual stress. These techniques can be classified as destructive or nondestructive in nature. **Destructive testing** involves some degree of mechanical damage that might affect the reuse of the test piece. One of the more destructive techniques is the sectioning method [3]. Sectioning involves cutting out sections of a part and measuring the axial and/or bending deformation that results from the release of the stress caused by the cutting. Successful measurements are possible when care is taken to avoid excessive plasticity or heat in the cutting process [4]. The strains released by the cutting process are generally measured by strain gauges. The strips may exhibit both axial (membrane) and bending deformations.

The **contour method** is a variation of the sectioning method in which a high-precision measurement of surface curvature is made after a single cut in the vicinity of residual stress [5]. The cut is made by an **electric discharge machine (EDM)**, and the resulting contour produced by residual stress redistribution is measured by a **coordinate measuring machine (CMM)**. As the peak-to-valley magnitude of the contour will be on the order of 10–100 μm, precision metrology is required. The measured displacement data are analyzed with a finite element model of the specimen. A very good correlation has been demonstrated between the contour method and the nondestructive

Metallic Powders for Additive Manufacturing: Science and Applications, First Edition.
Enrique J. Lavernia, Kaka Ma, Julie M. Schoenung, James F. Shackelford, and Baolong Zheng.

neutron diffraction technique (described below) in characterizing the residual stress in the vicinity of a linear friction weld [6].

A popular and less destructive method is hole drilling [3]. It is a simple and quick method that is both accurate and reliable. The mechanical damage to the part is confined to the relatively small (about 1.8 mm diameter and up to 2.0 mm deep) drilled hole and, in some cases, does not prevent its reuse. The hole drilling releases some of the residual stress in the immediate area, and the corresponding strains are measured by nearby strain gauges [7]. Standard equations allow the residual stress to be calculated from the strain values measured by the adjacent strain gauges.

A variety of **nondestructive testing (NDT)** technologies are available for residual stress measurement. **X-ray diffraction (XRD)** is a well-established method in which the measurement of the interplanar spacing within individual crystalline grains of material serves as an atomic-scale strain gauge [8, 9]. In this method, conventional Bragg angle measurements of interplanar spacing are made at two or more angles relative to the surface normal of the test piece. Differences in the measured lattice spacings for the same crystallographic planes (Miller indices, *hkl*) at the different surface orientations can be related to the residual stress in the surface layer. For multiple angles of inspection (ψ) relative to the surface normal, the residual stress (σ) at an angle (ϕ) relative to the principal stress direction is given by

$$\sigma_\phi = \frac{E}{(1+\nu)\sin^2\psi}\left(\frac{\sin\theta_n}{\sin\theta_i} - 1\right) \tag{14.1}$$

where E is the elastic modulus, ν is the Poisson's ratio, and θ_n and θ_i are the Bragg angles for diffraction experiments made relative to the surface normal ($\psi = 0°$) and the inclined angle, respectively [9].

A significant limitation of this method is the shallow penetration depth of typical X-ray sources; for example, stress information in typical steels will be known only to a depth of 0.013 mm with a laboratory-scale X-ray diffractometer. The variation in stress as a function of depth can be done by applying the destructive technique of etching away surface layers. Substantial improvement in X-ray penetration is possible with synchrotron X-ray sources [3, 10]. These high-energy and highly penetrating beams are, however, limited to large user facilities.

Neutron sources are generally even more penetrating than synchrotron sources, allowing residual stress values to be mapped through relatively large parts with a spatial resolution of approximately 1 mm^3 [8–11]. A limitation in carrying out **neutron diffraction** is that neutron sources capable of such measurements are, like synchrotron beams, generally in large user facilities as opposed to individual laboratories. Figure 14.1 illustrates the configuration of the spectrometer for materials research at temperature and stress (SMARTS) facility at the Lujan Neutron Scattering Center at the Los Alamos National Laboratory.

Ultrasonic testing (UT) is a popular NDT technique, and ultrasonics has been applied to the measurement of residual stress. The central principle is that the velocity of ultrasound varies approximately linearly with the level of stress in each material [8]. A typical measurement involves a transducer producing an ultrasonic pulse with a frequency of several megahertz into a test piece, and the time-of-flight of the pulse over a known distance provides the velocity. While this ultrasonic technique provides the promise of three-dimensional nondestructive measurements, there are numerous challenges. Several metal characteristics can also affect the velocity, including crystallographic

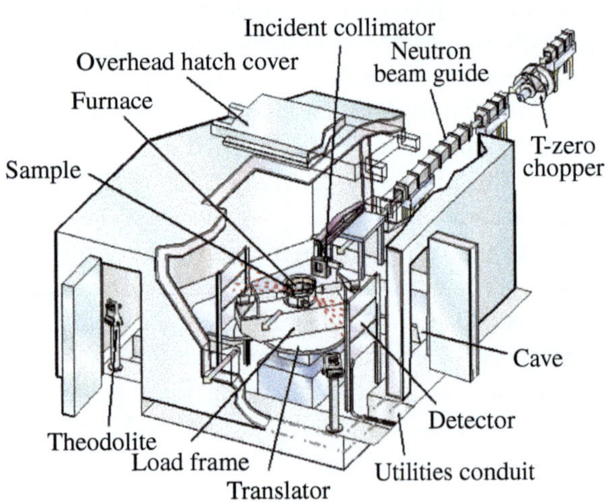

Figure 14.1 The spectrometer for materials research at the temperature and stress (SMARTS) facility at the Lujan Neutron Scattering Center at the Los Alamos National Laboratory. Source: Courtesy of the Los Alamos National Laboratory.

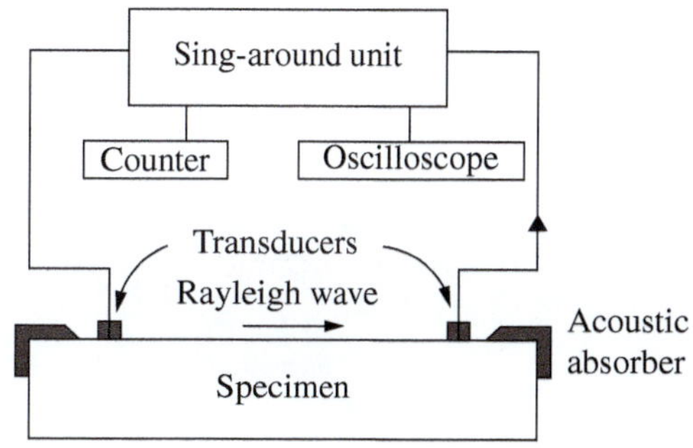

Figure 14.2 The "sing around" technique of measuring surface residual stress from the velocity of Rayleigh surface waves. Source: Chakrapani et al. [14]/ASM International.

texture, multiple phases, precipitates, and dislocation structures. Sorting out these competing effects can be difficult. Nonetheless, considerable effort has been applied to developing ultrasonic velocity measurements, and the technique is increasingly being used for residual stress mapping [3]. Recent advances in UT applications for characterizing AM products have been reviewed [12]. The **effective bulk modulus elastography (EBME)** technique illustrated in Figure 13.3 as a means of monitoring porosity in AM parts can also monitor interatomic dimensions leading to the ability to provide a detailed characterization of residual stresses [13]. An example will be given in Section 14.5. In addition to bulk ultrasonic waves, surface waves can be used to measure residual stress. For example, **Rayleigh waves** were identified by Lord Rayleigh in the late nineteenth century as a form of surface acoustic waves (SAW) and are now an established form of NDT for stress measurement [14]. A schematic of a so-called "sing around" technique is illustrated in Figure 14.2, in which a new Rayleigh wave is triggered each time a previous wave is detected. It is worth noting that the Rayleigh wave technique can also be used to monitor surface roughness, the focus of Section 13.6.

The **Barkhausen noise analysis (BNA)** technique measures the number and magnitude of reorientations produced by the expansion and contraction of magnetic domains in a ferromagnetic material under an external magnetic field [8]. The "noise" is the somewhat random set of pulses that are created by the reorientations and detected by a sensing coil. In a well-calibrated materials system, the BNA signal can be correlated to the level of residual stress, although such a calibration is not a trivial task. The depth of measurement of BNA is relatively small (up to 0.2 mm for surface-hardened components) but substantially greater than that of the typical XRD method [3].

Along with the experimental measurement of residual stress in AM parts, the **computational modeling** of residual stress production has been a useful, complementary approach [2]. The modeling is based on the numerical solution of thermal-stress equilibrium equations. Following a large body of literature describing fusion welding, these models can provide the evolution of stresses and displacements in the 3D geometry of a part as a function of time, while experimental data are typically limited to a few, discrete points. Nonetheless, the complexity of the modeling requires that the modeling results be validated by high-quality experimental data. An additional challenge is the computational cost of models with sufficient spatial and temporal resolution [15].

14.3 Powder Characteristics

In Section 2.5, the distribution of metallic powder particle sizes was introduced in detail, and, in Section 13.2, the role of the powder **particle size distribution (PSD)** as well as the particle morphology was seen to play a central role in the **spreadability** and **flowability** of powders in AM. The spreadability and flowability, in turn, were seen to be significant factors in the ultimate quality of the AM parts, including the minimization of product defects such as porosity. In the same way, particle size and morphology play a central role in ensuring minimal levels of residual stress in the final AM parts [16]. In fact, optimal processing and minimal residual stress levels require an overall understanding of the laser beam-powder-melt pool interaction, as illustrated in Figure 14.3 for the **powder bed fusion** using a laser beam (PBF-LB) technique [16, 17].

In Section 13.2, a specific case study of aluminum alloys was discussed regarding the role of powders in minimizing porosity defects. Residual stress makes the AM of aluminum alloys a significant challenge. For these alloys, the physical and chemical properties of aluminum are important factors, along with the size and morphology of the powder particles. The high **coefficient of thermal expansion (CTE)** of Al makes its alloys highly susceptible to shrinkage, cracking, and deformation during the severe thermal cycles involved in AM [18]. The repetitive thermal cycling corresponds to a nonhomogeneous heat distribution leading to thermal expansion and contraction of the solidified structure and high residual stress zones around the molten pool. In addition, the low density of aluminum leads to

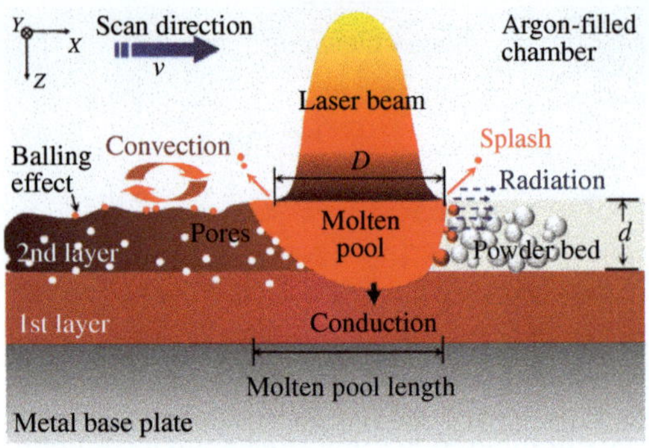

Figure 14.3 Schematic overview of the laser beam-powder-melt pool interaction for the powder bed fusion using a laser beam (PBF-LB) technique. Source: Shi et al. [17]/with permission of Elsevier.

poor powder flowability which affects the quality of the build [19]. Aluminum also has high moisture absorption and surface oxidation which are detrimental to laser-based deposition [18, 20]. As a result, PBF-LB and **directed energy deposition (DED)** technology have generally focused on various AlSi alloys such as the case study in Section 13.2 that involved PBF-LB. These alloys are good candidates for AM as they exhibit excellent laser absorption, high corrosion resistance, good weldability, and good mechanical performance [19, 21].

14.4 Pre-processing Heat Treatment

The **preheating** of the substrate as well as the building chamber and printed part during deposition are popular and effective techniques for mitigating residual stress [2, 15]. The residual stress reduction is the result of reducing the overall thermal gradients in the printed part during printing. Nonetheless, the pre-processing heat treatment does not eliminate residual stress entirely, and additional post-processing might be required. Post-processing treatments are the focus of Section 14.6.

Numerous examples are available to illustrate the effectiveness of the preheating step. For example, in the powder bed fusion using an electron beam (PBF-EB) technique, the electron beam can be scanned rapidly across the surface of the powder prior to the actual melting step. This preheating serves to minimize residual stress [1]. As noted in Section 13.3, preheating the substrate when producing the **intermetallic** compound Fe-(28 at%Al) by directed energy deposition using a laser beam (DED-LB) suppressed cracking associated with residual stress [22]. Substrate preheating is also one of the most significant methods for residual stress reduction in wire arc additive manufacturing (WAAM) [23]. Figure 14.4 indicates the basis for this reduction with the dramatic lowering of the thermal gradients in the first 10 layers of a build of 1.2 mm diameter HO8Mn2Si deposition were associated with increasing levels of preheating [24].

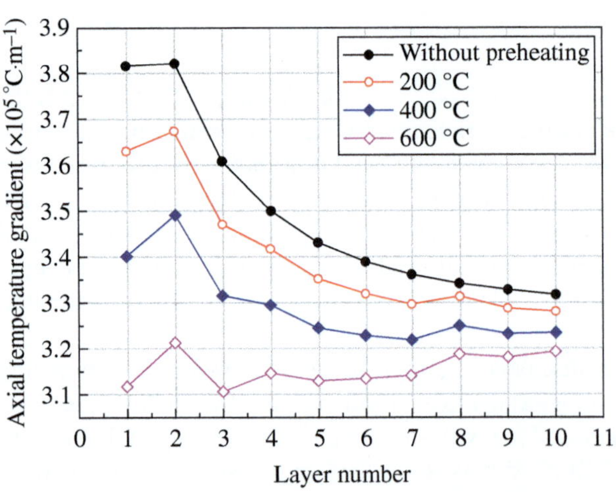

Figure 14.4 The lowering of the thermal gradients in the first 10 layers of a build of 1.2 mm diameter HO8Mn2Si deposition wire by wire arc additive manufacturing (WAAM) illustrates the benefits of substrate preheating for reducing residual stress. Source: Xiong et al. [24]/with permission of Elsevier.

14.5 Process Parameters

From the outset of this chapter, we have noted that residual stress is a direct consequence of the rapid thermal cycling of each volume element in the structure during the various AM technologies. In this section, we focus on the specifics of the AM process that can produce a part with an acceptable level of residual stress, which can be defined as comparable to that present in conventionally produced manufacturing technologies.

Considerable benefit has been found in optimizing the scan strategy during deposition by DED-LB, including the laser scan speed and the scan pattern [2]. Effective strategies include shorter deposition length, scanning in smaller regions, "spiraling-in" patterns as opposed to "spiraling-out," and a decreased layer thickness below the melt pool [15, 25]. Also, the track width and hatch spacing should be set to allow bead overlap [25].

Figure 14.5 shows a set of laser deposition patterns that were compared in terms of the resulting mechanical and thermal properties in DED-LB manufacturing [26]. The patterns were (a) a simple raster, (b) one in which the beam spirals from the outer edge of the area toward the interior (i.e. spiral in), (c) one in which the beam spirals from the interior toward the outer edge (spiral out), and (d) a fractal known as a Hilbert curve. Figure 14.6 shows the resulting

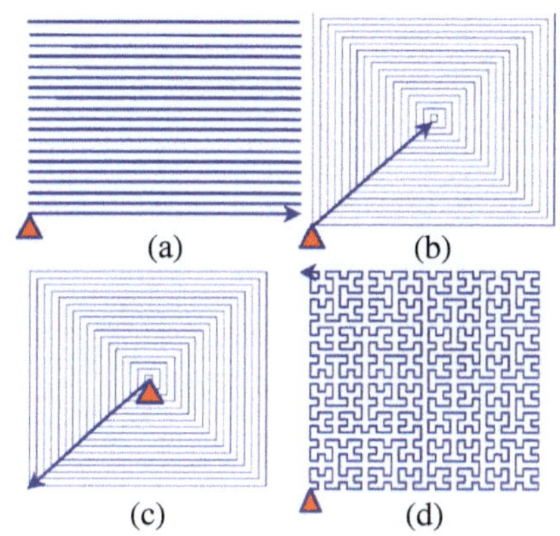

Figure 14.5 Laser deposition patterns in directed energy deposition using a laser beam (DED-LB) manufacturing: (a) a simple raster, (b) spiral inward, (c) spiral outward, and (d) a fractal known as a Hilbert curve. Source: Yu et al. [26]/with permission of Elsevier.

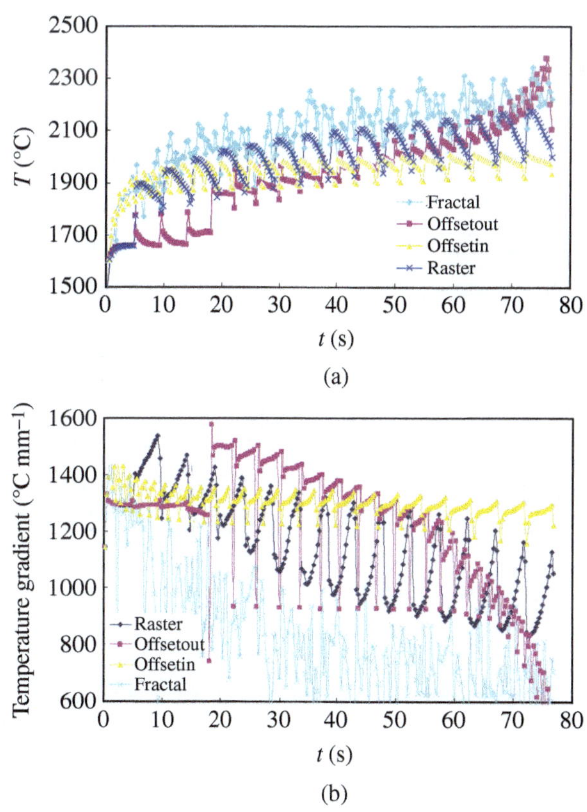

Figure 14.6 The resulting transient (a) maximal temperatures and (b) maximal temperature gradients with time for each of the deposition patterns in Figure 14.5. Part (a) shows that the fractal pattern produced the highest pool temperature resulting in a larger pool size. Part (b) shows that the fractal and spiral patterns produced the lowest temperature gradients, consistent with lower residual stresses. Source: Yu et al. [26]/with permission of Elsevier.

transient (a) maximal temperatures and (b) maximal temperature gradients with time for each of the deposition patterns. Part (a) shows the temperature oscillation for the melt pool corresponding to the periodic heating/cooling cycles, along with a steady heat accumulation with deposition time. The temperature data also show that the fractal pattern produced the highest pool temperature resulting in a larger pool size. Part (b) shows that the fractal and spiral in patterns produced the lowest temperature gradients, consistent with lower degrees of deformation and corresponding residual stress. It should be noted that the "spiral in" build strategy just mentioned has also been shown to be effective for reducing residual stresses in WAAM processing [23]. Another processing step for the WAAM technique that has been shown to reduce residual stress is the cold-work process in which a load of up to 100 kN is imposed on a roller over a deposited layer to produce plastic deformation of the surface and recrystallize the grain in the following deposition. This cold-work step has shown promise in reducing the residual stress in aluminum and titanium alloys but has generally been confined to simple geometry parts [23].

Process parameter adjustments are, however, material sensitive [27]. For example, increasing the interlayer dwell time from 0 to 40 seconds during the deposition of Inconel 625 provided additional cooling and a reduction in residual stress from ~710 to ~566 MPa. On the other hand, the same increase in dwell time in the deposition of Ti6Al4V led

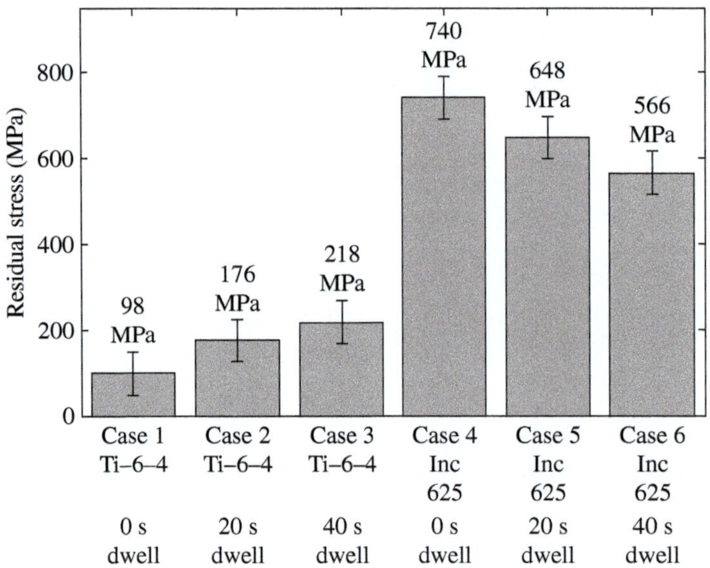

Figure 14.7 Contrasting behavior for increasing the interlayer dwell time during the deposition of Ti6Al4V vs. that for Inconel 625. The difference was attributed to differing microstructures. Source: Denlinger et al. [27]/with permission of Elsevier.

Figure 14.8 Blisk repair in process of a Ti6Al4V airfoil leading edge using the laser-engineered net shaping (LENS™) system. Source: Ref. [29]/Optomec Inc.

to an increase in residual stress from ~98 to ~218 MPa. These contrasting results are shown in Figure 14.7. The difference in behavior was attributed to the two-phase ($\alpha + \beta$) microstructure of titanium introducing complexities into the heating and cooling cycles that are not present in Inconel 625 that retains an fcc structure throughout the temperature cycles. The residual stress measurements in Figure 14.7 were done by the contour method using a CMM, as introduced in Section 14.2.

The surface residual stress in a high entropy CrMnFeCoNi alloy part produced by DED-LB using the **laser-engineered net shaping (LENS™)** system was measured to be ~182 MPa by XRD using the average of measurements at three points [28]. This relatively high value was associated with the high density of dislocations in the alloy. A post-processing surface treatment that successfully reduces the tensile stress and drives the stress into a compressive state will be discussed in Section 14.6.

As mentioned in the opening of this chapter, using AM to repair parts has advantages compared to conventional welding, including reduced residual stress due to lower heat input during the repair process. Figure 14.8 illustrates

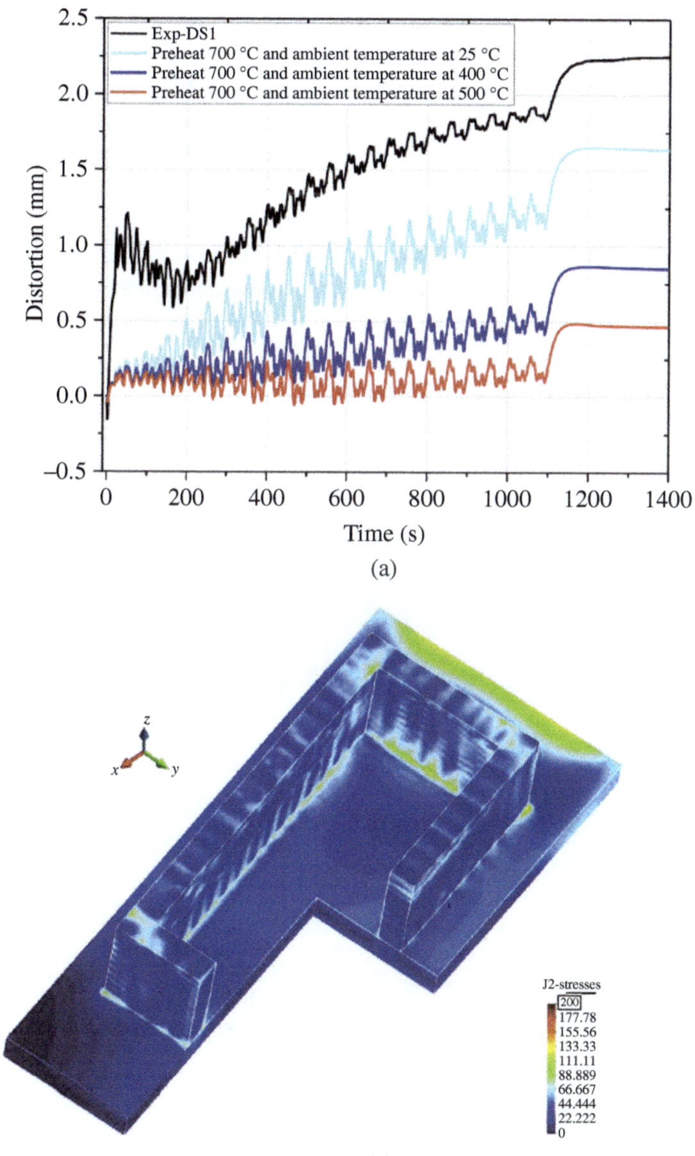

Figure 14.9 (a) A combination of substrate preheating and build chamber heating is seen to reduce the distortion in directed energy deposition produced Ti6Al4V parts by up to 90%, as modeled by thermo-mechanical FEM. Such distortion reduction corresponds to a residual stress reduction of up to 80%. (b) Residual von Mises stresses for the ambient temperature of 500 °C. Source: Lu et al. [31]/with permission of Elsevier.

such a repair using the LENS form of DED [29]. The part being repaired is a Ti6Al4V blisk, a high-value turbomachine component comprising both blades and a rotor disk. Such components are often subjected to harsh conditions (corrosive environments, impact events, thermal cycles, and high stresses). In addition to the ability to avoid the residual stress and geometric distortions of conventional welding, the low heat input and small heat-affected zone of the leading-edge repair shown in Figure 14.8 resulted in overall material properties equal to or better than the base material. A similar repair program for the T700 blisk made from AM355 steel passed Army-mandated spin testing (60 000 rpm proof test and 50 000 cycle low-cycle fatigue test) [29].

Design simulations are being used to optimize AM processing by determining the optimal orientation of the part on the build platform, as well as minimizing the support structures [30]. While potentially helpful, such simulations require various mechanical and thermophysical properties that are not available for many systems [1]. In a similar way, thermo-mechanical finite element method (FEM) has been used to model distortions and residual stresses in DED-produced Ti6Al4V parts. Figure 14.9 shows that a combination of substrate preheating and build chamber heating can reduce the distortions by up to 90% which corresponds to a residual stress reduction of up to 80% [31].

As mentioned in Section 14.2, the EBME technique is a means of monitoring interatomic dimensions and the corresponding residual stress in AM parts. Figure 14.10 shows the effective dynamic bulk modulus (K_d) and effective density (ρ) maps for PBF-LB produced Ti6Al4V alloy at 10 MHz (parts a and b) and at 20 MHz (parts c and d) [13]. The spatial resolution of the maps increases significantly for the 20 MHz scans, corresponding to the smaller wavelength of the higher frequency beam and a greater sensitivity to crystallographic defects in the alloy. In the 20 MHz scans, the values of K_d and ρ in the central core of the scanned region are relatively uniform (represented by large contour islands) in contrast to relatively small contour islands near the edges of the block. This result is associated with the bidirectional scanning pattern and affects the stress distribution in the part. The U-turns in the scans near the edges

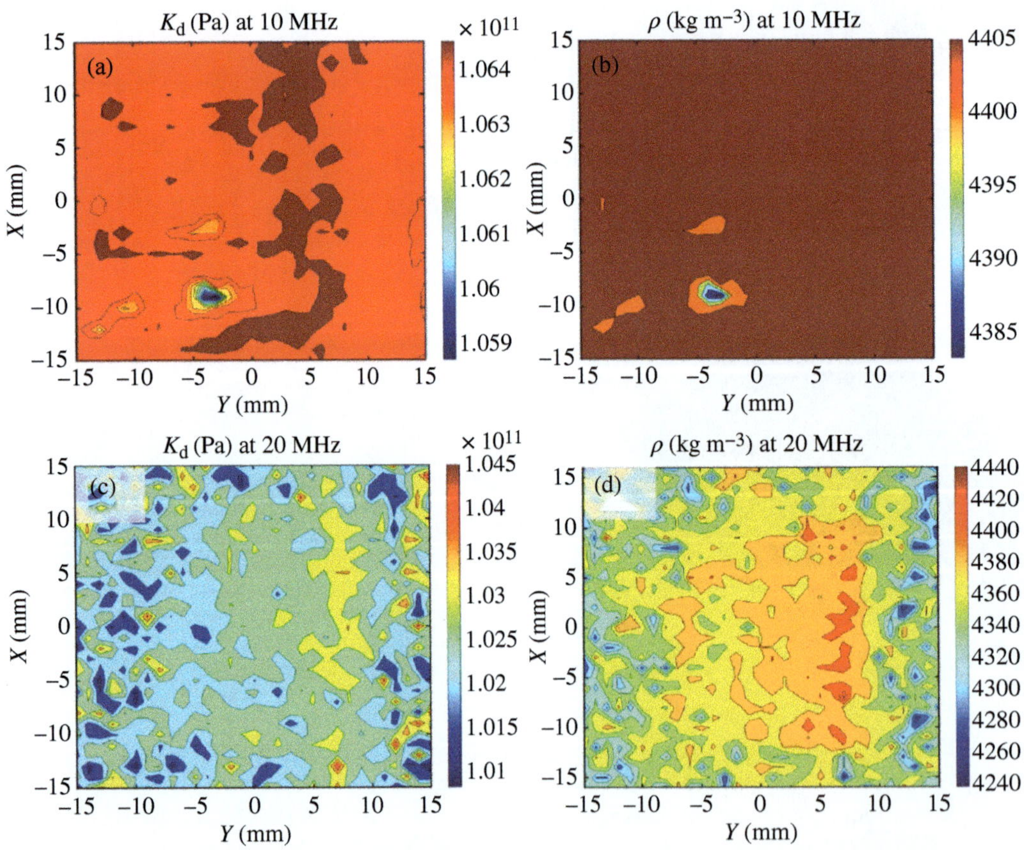

Figure 14.10 Effective dynamic bulk modulus (K_d) and effective density (ρ) maps for a Ti6Al4V alloy produced by powder bed fusion using a laser beam at 10 MHz (a, b) and at 20 MHz (c, d). Source: Pantawane et al. [13]/with permission of Elsevier.

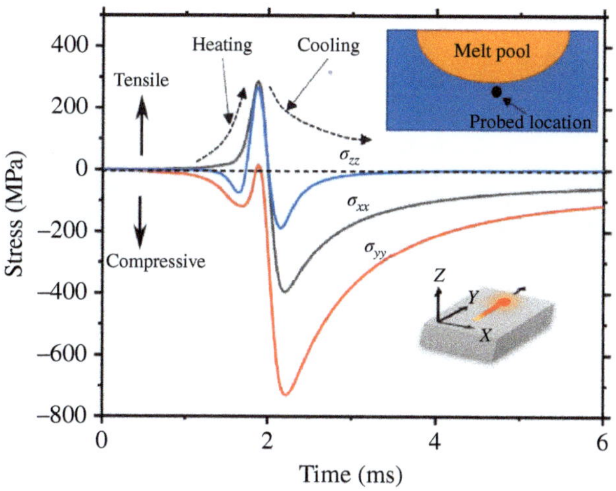

Figure 14.11 A numerical simulation of the residual stresses in the Ti6Al4V sample of Figure 14.10 along the building direction that peaks at a level of 720 MPa compressive. Source: Pantawane et al. [13]/with permission of Elsevier.

correspond to a decreased rate of heating and cooling leading to slower melting/solidifying and lower compressive residual stresses [12]. Figure 14.11 shows a numerical simulation of the residual stresses in the Ti6Al4V sample along the building direction that peaks at a level of 720 MPa compressive [13].

It is useful to recall the discussion in Section 13.3 of the NiCr intermediate transition layer used in preventing the production of brittle intermetallic phases in **functionally graded materials (FGM)** such as TiFe and $TiFe_2$ when joining Ti6Al4V to 410 SS [32]. The NiCr layer reduced the cracking, delamination, and associated residual stresses due to the brittle intermetallics. Similarly, a few transition layers of 100% V were used to produce a sufficient barrier between the Ti6Al4V and 304L SS alloys to prevent such intermetallics [33]. Overall, the tailored properties of FGM are beneficial in reducing sharp interfaces and, in turn, residual stresses [16].

14.6 Post-processing Treatments

Early in Chapter 13, the role of **hot isostatic pressing (HIP)** as a post-processing treatment was introduced as a means of reducing defects such as porosity. The same is generally true for residual stress minimization as well [1, 2]. While the DED AM technique has several advantages in comparison to the more widely used PBF-LB approach, a disadvantage is that residual stress levels tend to be higher. Nonetheless, post-processing HIPing is also effective in reducing stress levels in those AM parts [2]. For example, *in situ* compression tests were performed on Inconel 625 to determine the stress relaxation induced by post-processing heat treatment [34]. In this study, a conventionally produced alloy was compared to one produced by DED-LB in compression tests up to 9% engineering strain, with the stresses measured by *in situ* **neutron diffraction**. The overall stress relaxation was similar in the two cases, but the stress relaxation rate at a given temperature (600 or 700 °C) was twice as great for the AM-produced sample compared to the conventional sample, and the stress plateau was slightly lower in the AM sample as well, due to different texture and grain sizes in the two materials. In both cases, the mechanism of stress relaxation was deduced to be controlled by dislocation and glide, with dislocations interacting with grain boundaries, solute atoms, and secondary phases. Figure 14.12 summarizes the stress decay data. In a related study, neutron diffraction measurements indicated that residual stresses in Inconel 625 can be relieved by a heat treatment at 870 °C for one hour in an argon atmosphere [35]. However, the heat treatment also produced small amounts of carbide precipitates that reduced the reference strain-free lattice spacing leading to potential errors in the calculation of stress values.

While post-processing heat treatments have been effective, surface treatments have also shown positive results [2]. **Magnetic field-assisted finishing (MAF)** reduced the surface residual stress on PBF-LB-produced 316L stainless steel parts from ~200 MPa tensile to ~ −70 MPa compressive [36]. The MAF technique involves using a magnetic particle brush rotating along the surface in which either (i) a mixture of magnetic particles and an abrasive slurry is used

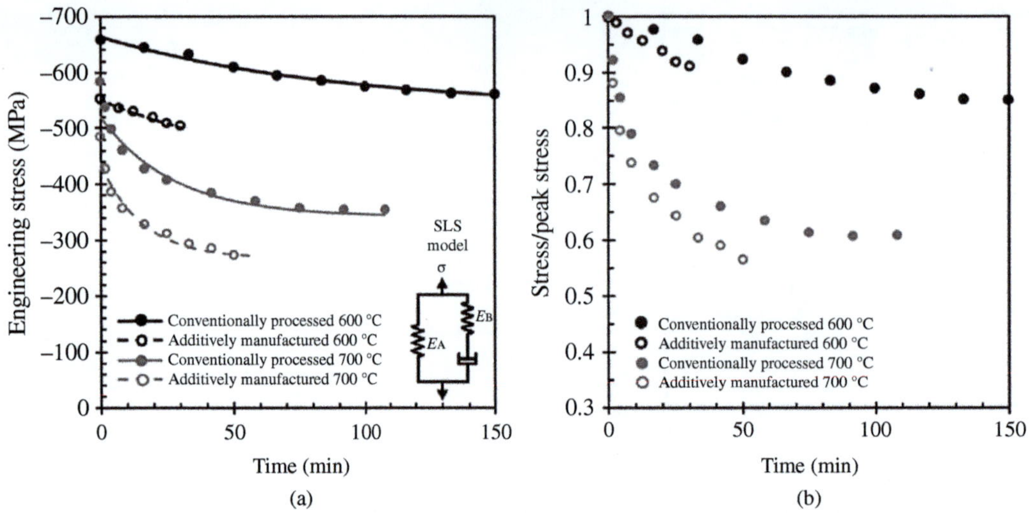

Figure 14.12 (a) Engineering stress relaxation and (b) engineering stress relaxation normalized by the peak stress are shown for Inconel 625, comparing a conventionally produced alloy and one produced by directed energy deposition AM at 600 and 700 °C. The experimental data are compared with plots of the standard linear solid (SLS) model illustrated in the lower right corner of (a). Source: Wang et al. [34]/with permission of Elsevier.

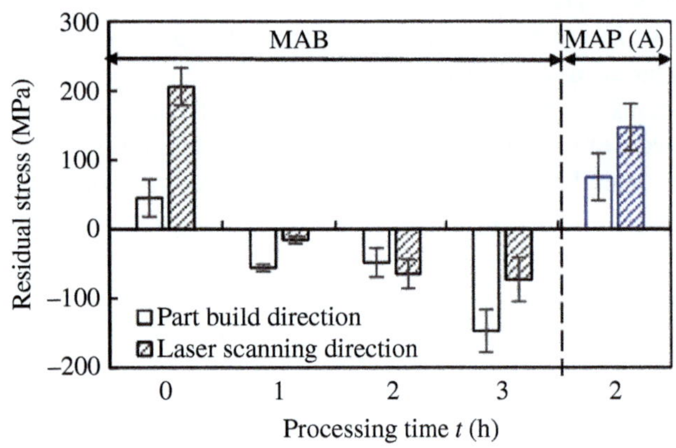

Figure 14.13 The significant reduction in residual stress of 316L stainless steel parts produced by powder bed fusion using a laser beam is shown for the **magnetic field-assisted burnishing (MAB)** technique for a period of one, two, and three hours. The data for the **magnetic field-assisted polishing (MAP)** method after two hours is shown as a reference. Source: Yamaguchi et al. [36]/with permission of Elsevier.

(designated as **magnetic field-assisted polishing or MAP**) or (ii) magnetic balls are used (designated as **magnetic field-assisted burnishing** or MAB). Figure 14.13 summarizes the effects of the burnishing (MAB) method for one, two, and three hours. The data for the MAP method after two hours is shown as a reference. In that case, the more aggressive polishing method did not produce a compressive stress state within the same time frame.

Laser shock peening (LSP) was discussed in Section 13.6 as a means of improving tensile properties and decreasing porosity in a high entropy CrMnFeCoNi alloy, at the expense of increased surface roughness [37]. In that same study, LSP was shown to decrease the surface residual stress from ~350 MPa tensile to ~ –350 MPa compressive after five LSP cycles. Figure 14.14 summarizes the XRD data that produced surface residual stress profiles for one, three, and five LSP cycles.

LSP has also been used as a post-processing treatment for the purpose of reducing residual stress in Ti6Al4V parts produced by DED. As shown in Figure 14.15, LSP reduced the surface residual stress from ~100 MPa tensile to ~ –200 MPa compressive, with the depth of the tensile-to-compressive effect extending to ~700 μm [38]. The residual stress improvement was accompanied by improvements in microhardness and elongation at failure. The use of LSP has also been shown to reduce residual stresses in WAAM processing [23].

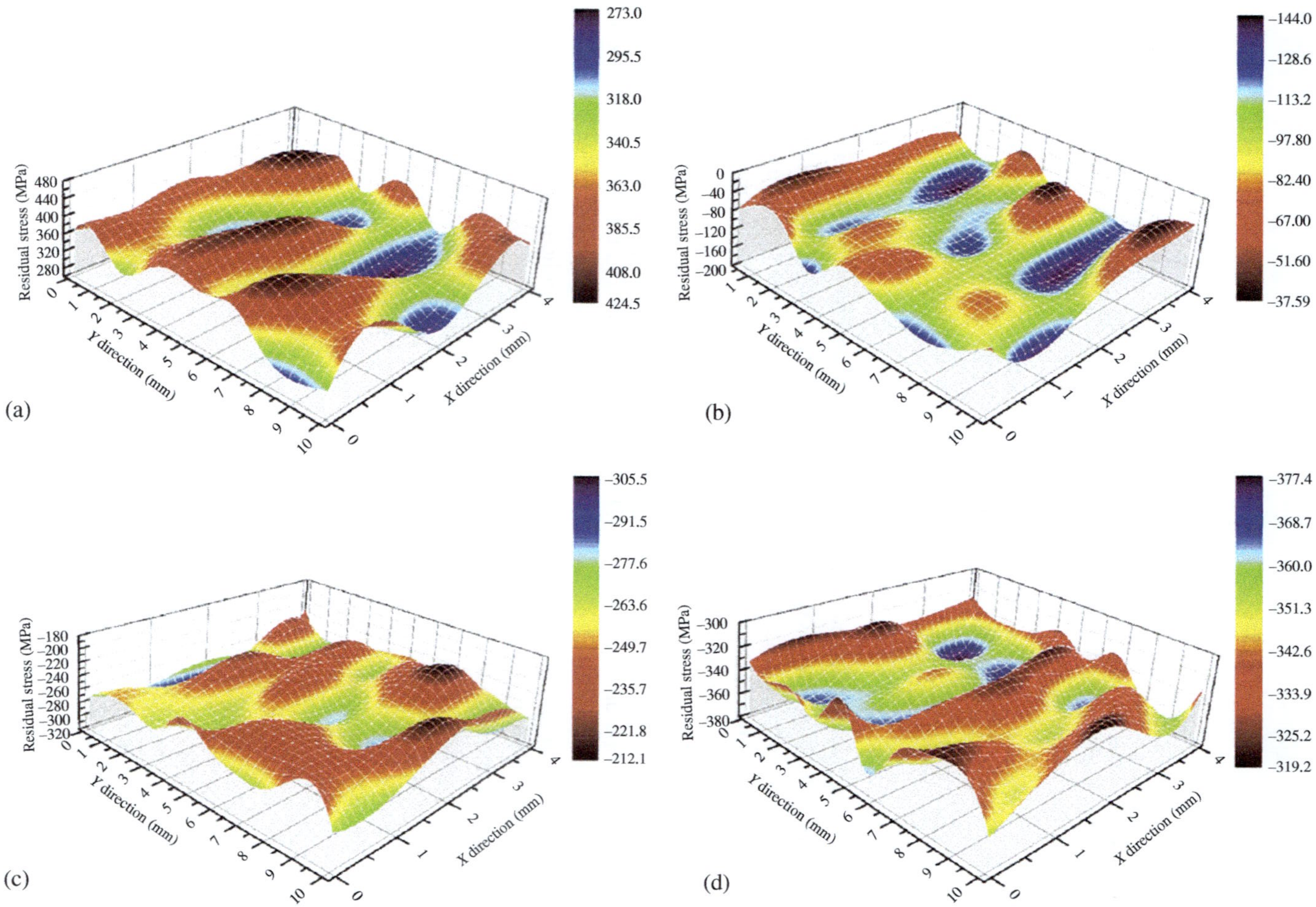

Figure 14.14 Surface residual stress profiles based on X-ray diffraction data for a high entropy CrMnFeCoNi alloy comparing (a) the as-built specimen with (b) one, (c) three, and (d) five laser shock peening cycles, showing a substantial decrease in the surface residual stress. Source: Tong et al. [37]/with permission of Elsevier.

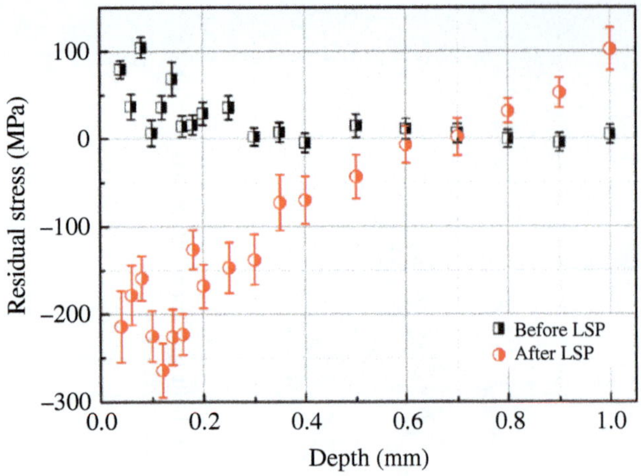

Figure 14.15 Residual stress is shown to be sharply reduced near the surface of a Ti6Al4V alloy produced by directed energy deposition as a result of laser shock peening. Source: [38]/with permission of Elsevier.

14.7 Summary

The rapid thermal cycling of each volume element during AM of metal parts puts residual stress at the core of the assessment of this technology. Residual stress levels in AM products are routinely evaluated in comparison to those in conventionally manufactured parts. A large portfolio of residual stress measuring technologies is available for such comparisons. Destructive methods such as the contour method and nondestructive methods such as X-ray and neutron diffraction and ultrasonic testing have been used to compare AM-produced parts with conventional ones as well as monitor various methods for residual stress mitigation. The nature of the metal powder used in AM is of fundamental importance. The distribution of powder size and its morphology along with the physical and chemical properties of the powder composition determine the flowability of the powder and the eventual success of the part construction. The preheating of the substrate, building chamber, and the printed part during deposition is effective in mitigating residual stress because of the reduction of the overall thermal gradients in the part during printing. As a primary source of residual stress is the rapid thermal cycling of each volume element in the structure during the various AM technologies, considerable benefit has been found in optimizing the scan strategy, including the laser scan speed and the scan pattern. Such process parameter adjustments are, however, material sensitive, and processing strategies must reflect factors such as phase changes and microstructural features including dislocation densities. Finally, HIP and various surface treatments have been found to be important post-processing methods for minimizing residual stresses in the final part.

Nomenclature

AM	additive manufacturing
BNA	Barkhausen noise analysis
CMM	coordinate measuring machine
CTE	coefficient of thermal expansion
DED	directed energy deposition
DED-LB	directed energy deposition using a laser beam
E	elastic modulus
EBME	effective bulk modulus elastography
EDM	electric discharge machine
FEM	finite element method
FGM	functionally graded materials
HIP	hot isostatic pressing

K_d	effective dynamic bulk modulus
LENS™	laser engineered net shaping
LSP	laser shock peening
MAB	magnetic field-assisted burnishing
MAF	magnetic field-assisted finishing
MAP	magnetic field-assisted polishing
NDT	nondestructive testing
PBF-EB	powder bed fusion using an electron beam
PBF-LB	powder bed fusion using a laser beam
PSD	particle size distribution
SAW	surface acoustic waves
SLS	standard linear solid
SMARTS	spectrometer for materials research at temperature and stress
UT	ultrasonic testing
WAAM	wire arc additive manufacturing
XRD	X-ray diffraction
θ_i	Bragg angle for diffraction relative to an angle inclined to the surface
θ_n	Bragg angle for diffraction relative to the surface normal
ν	Poisson's ratio
ρ	effective density
σ	residual stress
ϕ	angle relative to the principal stress direction
ψ	angle of inspection inclined to the surface normal

References

1 Blakey-Milner, B., Gradl, P., Snedden, G. et al. (2021). *Mater. Des.* 209: 110008. https://doi.org/10.1016/j.matdes.2021.110008.
2 Svetlisky, D., Das, M., Zheng, B. et al. (2021). *Mater. Today* 49: 271. https://doi.org/10.1016/j.mattod.2021.03.020.
3 Rossini, N.S., Dassisti, M., Benyounis, K.Y. et al. (2012). *Mater. Des.* 35: 572. https://doi.org/10.1016/j.matdes.2011.08.022.
4 Tebedge, N., Alpsten, G., and Tall, L. (1973). *Exp. Mech.* 13: 88. https://doi.org/10.1007/BF02322389.
5 Prime, M.B. and DeWald, A.T. (2013). The contour method. In: *Practical Residual Stress Measurement Methods* (ed. G.S. Schajer), 109. Hoboken, NJ: Wiley Blackwell.
6 DeWald, A.T., Legzdina, D., Clausen, B. et al. (2013). A comparison of residual stress measurements on a linear friction weld using the contour method and neutron diffraction. In: *Experimental and Applied Mechanics*, vol. 4 (ed. C.E. Ventura, W.C. Crone, and C. Furlong), 183. New York: Springer. https://doi.org/10.1007/978-1-4614-4226-4_22.
7 Olabi, A.G. and Hashmi, M.S.J. (1996). *J. Mater. Process. Technol.* 56: 552.
8 Ruud, C. (2002). Measurement of residual stresses. In: *Handbook of Residual Stress and Deformation of Steel* (ed. G. Totten, M. Howes, and T. Inoue), 99. Materials Park, OH: ASM International.
9 Cullity, B.D. and Stock, S.R. (2001). *Elements of X-Ray Diffraction*, 3e. Upper Saddle River, NJ: Prentice-Hall.
10 Withers, P.J. (2004). *J. Appl. Crystallogr.* 37: 607.
11 Hutchings, M.T., Withers, P.J., Holden, T.M. et al. (2005). *Introduction to the Characterization of Residual Stress by Neutron Diffraction*. Boca Raton, FL: Taylor & Francis.
12 Yang, T., Mazumder, S., Jin, Y. et al. (2021). *Materials* 14: 4929. https://doi.org/10.3390/ma14174929.
13 Pantawane, M.V., Yang, T., Jin, Y. et al. (2021). *Mater. Sci. Eng., A* 811: 140990. https://doi.org/10.1016/j.msea.2021.140990.
14 Chakrapani, S.K., Bond, L.J., and Edwards, R.S. (2018). Rayleigh wave nondestructive evaluation for defect detection and materials characterization. In: *ASM Handbook, Vol. 17 Nondestructive Evaluation of Materials* (ed. A. Ahmad and L.J. Bond), 266. Materials Park OH: ASM International.

15 DebRoy, T., Wei, H.L., Zuback, J.S. et al. (2018). *Prog. Mater Sci.* 92: 112. https://doi.org/10.1016/j.pmatsci.2017.10.001.

16 Svetlizky, D., Zheng, B., Vyatskikh, A. et al. (2022). *Mater. Sci. Eng., A* 840: 142967. https://doi.org/10.1016/j.msea.2022.142967.

17 Shi, Q., Gu, D., Xia, M. et al. (2016). *Opt. Laser Technol.* 84: 9. http://dx.doi.org/10.1016/j.optlastec.2016.04.009.

18 Zhang, J., Song, B., Wei, Q. et al. (2019). *J. Mater. Sci. Technol.* 35: 270. https://doi.org/10.1016/j.jmst.2018.09.004.

19 Ding, Y., Muniz-Lerma, J.A., Trask, M. et al. (2016). *MRS Bull.* 41: 745. https://doi.org/10.1557/mrs.2016.214.

20 Olakanmi, E.O. (2013). *J. Mater. Process. Technol.* 213: 1387. https://doi.org/10.1016/j.jmatprotec.2013.03.009.

21 Ye, H. (2003). *J. Mater. Eng. Perform.* 12: 288. https://doi.org/10.1361/105994903770343132.

22 Rolink, G., Vogt, S., Sencekova, L. et al. (2014). *J. Mater. Res.* 29: 2036. https://doi.org/10.1557/jmr.2014.131.

23 Rodrigues, T.A., Duarte, V., Miranda, R.M. et al. (2019). *Materials* 12: 1121. https://doi.org/10.3390/ma12071121.

24 Xiong, J., Lei, Y., and Li, R. (2017). *Appl. Therm. Eng.* 126: 43. https://doi.org/10.1016/j.applthermaleng.2017.07.168.

25 Gibson, I., Rosen, D., and Stucker, B. (2015). *Additive Manufacturing Technologies: 3D Printing, Rapid Prototyping, and Direct Digital Manufacturing*, 2e, 245. New York: Springer. https://doi-org.libproxy.mit.edu/10.1007/978-1-4939-2113-3_10.

26 Yu, J., Lin, X., Ma, L. et al. (2011). *Mater. Sci. Eng., A* 528: 1094. https://doi.org/10.1016/j.msea.2010.09.078.

27 Denlinger, E.R., Heigel, J.C., Michaleris, P. et al. (2015). *J. Mater. Process. Technol.* 215: 123. https://doi.org/10.1016/j.jmatprotec.2014.07.030.

28 Guan, S., Wan, D., Solberg, K. et al. (2019). *Mater. Sci. Eng., A* 761: 138056. https://doi.org/10.1016/j.msea.2019.138056.

29 Optomec (2016). LENS™ component repair solution. https://optomec.com/wp-content/uploads/2014/04/Optomec_LENS_Blisk_Repair_Datasheet.pdf (accessed 3 April 2023).

30 Leary, M. (2020). *Design for Additive Manufacturing*. Elsevier. https://doi.org/10.1016/C2017-0-04238-6.

31 Lu, X., Lin, X., Chiumenti, M. et al. (2019). *Addit. Manuf.* 26: 166. https://doi.org/10.1016/j.addma.2019.02.001.

32 Sahasrabudhe, H., Harrison, R., Carpenter, C. et al. (2015). *Addit. Manuf.* 5: 1. https://doi.org/10.1016/j.addma.2014.10.002.

33 Reichhardt, A., Dillon, R.P., Borgonia, J.P. et al. (2016). *Mater. Des.* 104: 404. https://doi.org/10.1016/j.matdes.2016.05.016.

34 Wang, Z., Stoica, A., Ma, D. et al. (2018). *Mater. Sci. Eng., A* 714: 75. https://doi.org/10.1016/j.msea.2017.12.058.

35 Wang, Z., Denlinger, E., Michaleris, P. et al. (2017). *Mater. Des.* 113: 169. https://doi.org/10.1016/j.matdes.2016.10.003.

36 Yamaguchi, H., Fergani, O., and Wu, P.Y. (2017). *CIRP Ann.* 66: 305. https://doi.org/10.1016/j.cirp.2017.04.084.

37 Tong, Z., Liu, H., Jiao, J. et al. (2020). *Addit. Manuf.* 35: 101417. https://doi.org/10.1016/j.addma.2020.101417.

38 Guo, W., Sun, R., Song, B. et al. (2018). *Surf. Coat. Technol.* 349: 503. https://doi.org/10.1016/j.surfcoat.2018.06.020.

15

Physical and Chemical Behavior and Powder Effects

15.1 Introduction

The point has been made in the opening paragraph of Chapters 13 and 14 that **additive manufacturing (AM)** can be a disruptive technology and replace conventional methods only if the resulting parts produced by AM do not have significantly greater levels of **defects** and **residual stress** as compared to those produced by conventional methods. The same argument can be made regarding the physical and chemical behavior of AM-produced parts. In fact, **physical behavior** is strongly affected by the defect structures discussed in Chapter 13 and the residual stresses discussed in Chapter 14. In this chapter, we discuss a wide range of physical properties including **density**, **surface appearance**, **deformation** (elastic and plastic), **hardness**, and the relationship between **fracture** and **fatigue**. We will focus on the interaction between defects and residual stress with these properties, as well as the role of powder characteristics in these interrelationships. We will also cover the topics of **corrosion** and **wear**, properties that are routinely discussed together, and provide a bridge between physical and **chemical properties**. An additional **chemical property** of fundamental importance to AM is **oxidation**. We also focus on the role of powder characteristics in determining these chemical properties. As with physical behavior, minimizing **porosity** is imperative to avoid the adverse effects of **oxidation** and corrosion of printed parts. The non-equilibrium nature of the **directed energy deposition (DED)** technique can lead to complex phase transformations, while microstructural changes, porosity, and cracking are especially challenging in this regard [1]. Nonetheless, considerable success has been shown in producing components with chemical behavior competitive with conventionally produced parts.

15.2 Density

As noted in Section 13.2, density and porosity defects are intimately related. Achieving a density of 99.5% is a primary goal of AM [2]. Both **powder bed fusion (PBF)** and DED, the two most mature AM methods, can reach such density levels [3]. Numerous methods for density measurement are available including the Archimedes method, ultrasonic pulse-echo velocity measurements, image analysis of metallographic cross-sections, X-ray micro-computed tomography (μ-CT), X-ray synchrotron measurements, and gas pycnometry [1].

Building on the introduction of the powder **particle size distribution (PSD)** in Section 2.5, the role of powder characteristics in affecting final part quality has been discussed numerous times throughout Part II of this book. Some general observations are available from a comprehensive review of powders in PBF technology [4]. First, a PSD skewed toward finer particle sizes allows those fine-sized particles to fill the interstices among the coarser ones leading to higher part densities. The widely observed lognormal distribution for powders introduced in Section 2.5 provides this concentration of finer particle sizes. Also, a wider distribution (i.e. a larger standard deviation for the lognormal distribution) has been found to provide a higher packing density of the powder and a subsequent higher part density for steel parts, although wider distributions were not found to produce greater densities in some other systems. The complexities that prevent a simple correlation between distribution width and final part density were demonstrated in the case discussed in Section 13.2 [5]. In those AlSi12 alloys produced by powder bed fusion using a laser beam (PBF-LB),

Metallic Powders for Additive Manufacturing: Science and Applications, First Edition.
Enrique J. Lavernia, Kaka Ma, Julie M. Schoenung, James F. Shackelford, and Baolong Zheng.

the smallest width (or span, as defined by Eq. (13.1)) had the lowest porosity and thus the highest final part density. In that study, it was shown that factors beyond the PSD are also important, including powder morphology as well as slight variations in the Si content. In a similar way, the correlation of other properties with the PSD width was also found to vary, as will be discussed in Sections 15.3–15.5 of this chapter.

15.3 Surface Appearance

The comprehensive review of powders in PBF technology introduced in Section 15.2 also provided some observations on surface appearance [4]. A PSD, such as the lognormal, that skewed toward finer particle sizes led to improved surface qualities, as well as higher part densities. Also, for steel parts, a wider PSD led to smoother surfaces parallel to the build direction, as well as higher powder bed and final part densities.

The **surface roughness** of the final part was discussed in some detail in Section 13.6 as one of the important defects associated with AM. Numerous methods for the measurement of surface roughness are available, including contact and non-contact, i.e. nondestructive, ones. A study of the effect of PBF-LB processing parameters on the surface roughness of Inconel 718 used two nondestructive tests, viz structured light (SL) scanning and **X-ray computed tomography (CT)** [6]. The surfaces of cylindrical fatigue bars were analyzed using ISO 25178 [7], with the surface

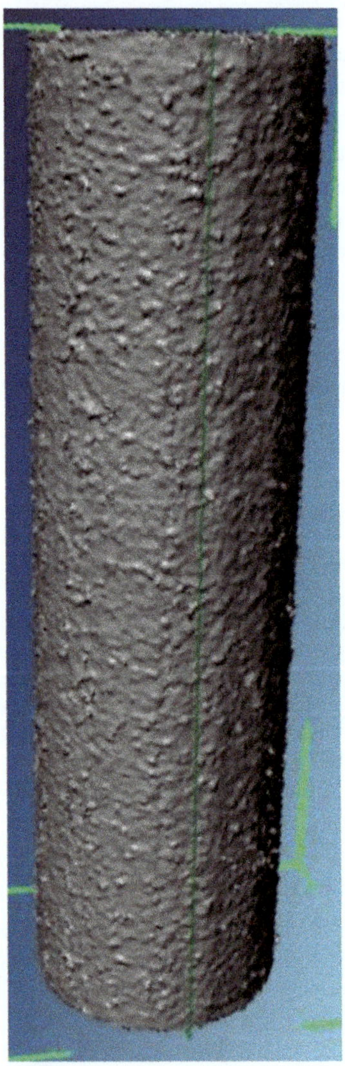

Figure 15.1 An X-ray computed tomography (CT) scan of the surface roughness of a fatigue bar gage section. Source: Gockel et al. [6]/with permission from Elsevier.

measured in the x–y plane and $z(x, y)$ representing the surface heights that are stored in a matrix. The mean plane is defined relative to all measured x, y points. The arithmetic mean height, S_a, and maximum pit height, S_v, are then defined relative to the mean plane by

$$S_a = (1/A) \int \int_A |z(x, y)| \, dxdy \tag{15.1}$$

$$S_v = |\min z(x, y)| \tag{15.2}$$

where A is the surface area and $|z(x, y)|$ is the matrix of surface heights.

Figure 15.1 shows the CT scan of the surface roughness of the fatigue bar gage section without any post-processing treatments. While the SL and CT inspections gave similar results, the CT data provide a truer representation of the surface topography, because, in SL scanning, underlying surface data are obscured by powder particles [6]. Figure 15.2 shows the CT surface parameters as a function of **process parameters**, laser speed and power. The maximum valley, S_v, is seen to (a) increase with the increasing speed at a fixed power and (b) decrease with increasing power at a fixed speed. No clear trends are shown for the average surface roughness, S_a. The trends in (a) and (b) are consistent with the size of the melt pool. Increasing the laser speed reduces the size of the melt pool and increasing the laser power increases it. The variation in S_v is related to the smaller layer thickness and pool instabilities associated with increasing melt pool size. As fracture is associated with the deepest surface notches (or valleys), the variation of S_v can be expected to be related to fracture behavior. That relationship is discussed further in Section 15.6.

While surface roughness is a substantial concern for AM processing, the aesthetics of the finished part can be affected as well. Excessive laser power can affect the color of the melt pool. For example, in the printing of stainless steel, the color of the final parts changed from silver-like (when the oxygen level was below ≈10 ppm) to bronze-like (when the oxygen level was above ≈150 ppm) (Kaka Ma, private communication).

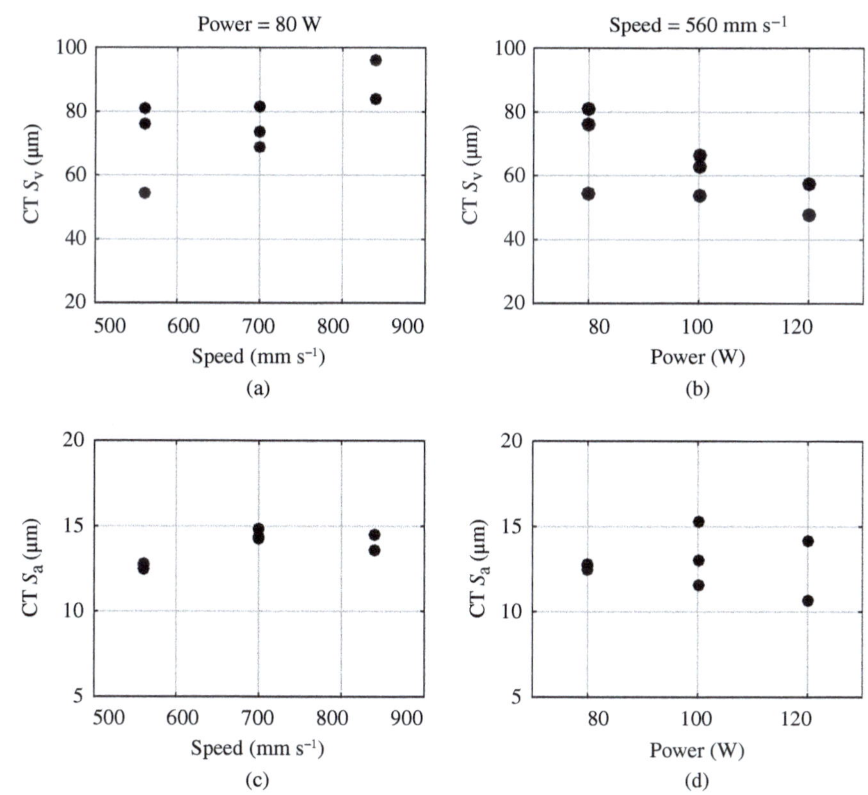

Figure 15.2 Comparison of the CT surface roughness parameters to the process parameters (a) S_v to speed, (b) S_v to power, (c) S_a to speed, (d) S_a to power [6].

15.4 Elastic and Plastic Deformation

The fundamental mechanical performance of engineered materials provided by a tensile test is another basis for comparing the AM-produced alloys with conventional ones. Figure 15.3 provides a graphical summary of this mechanical behavior (yield strength vs. elongation at failure) for some of the most common alloys produced by AM compared to those produced by conventional means [8]. While the stress-strain behavior of particular alloys produced by AM can be comparable to conventional counterparts, Figure 15.3 shows that the region of strength-elongation space in these Ashby charts [9] differs markedly for AM vs. conventional methods. This difference can be ascribed to the different bulk and surface structures characteristic of the two methods. As discussed previously in Part II of this book, bulk differences found in AM include fine grains, anisotropic microstructures with elongated grains, non-equilibrium microstructures, and the various defects discussed in Chapter 13. Surface differences include weld tracks and the various features associated with surface roughness.

It is useful to recall, as noted in Chapter 13, that **hot isostatic pressing (HIP)** is very useful for homogenizing the microstructure and closing pores, leading to improved ductility, and that this **post-processing** treatment has led to the wide use of AM in aerospace industries [3, 10]. As noted in Section 13.4, a useful application of DED AM is the

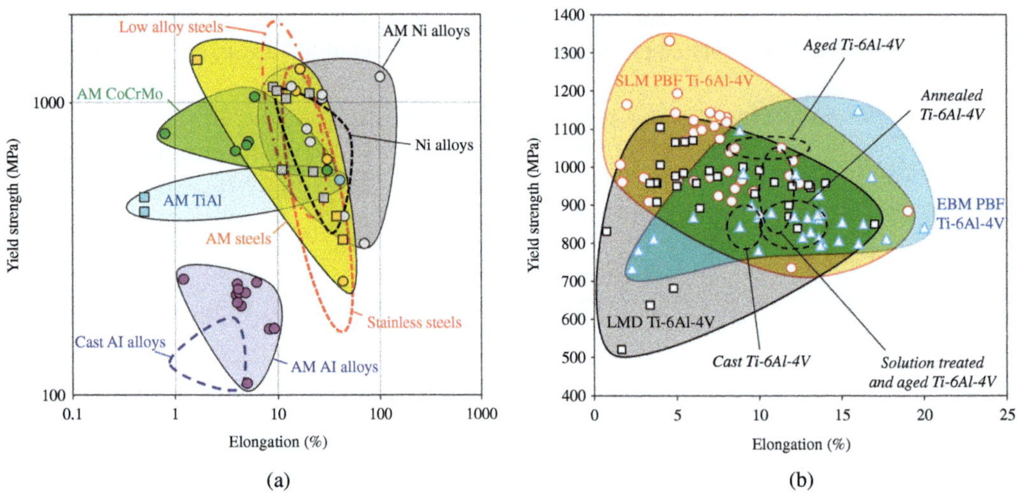

Figure 15.3 Ashby charts of the material property space for room temperature yield strength vs. elongation at failure of AM-produced alloys including (a) steels, Ni alloys, Al alloys, TiAl, and CoCrMo and (b) Ti6Al4V alloys [8]. Conventionally manufactured alloys are indicated by dashed lines. In part (b), the terms SLM PBF, EBM PBF, and LMD correspond to PBF-LB, PBF-EB, and DED-LB respectively in this chapter.

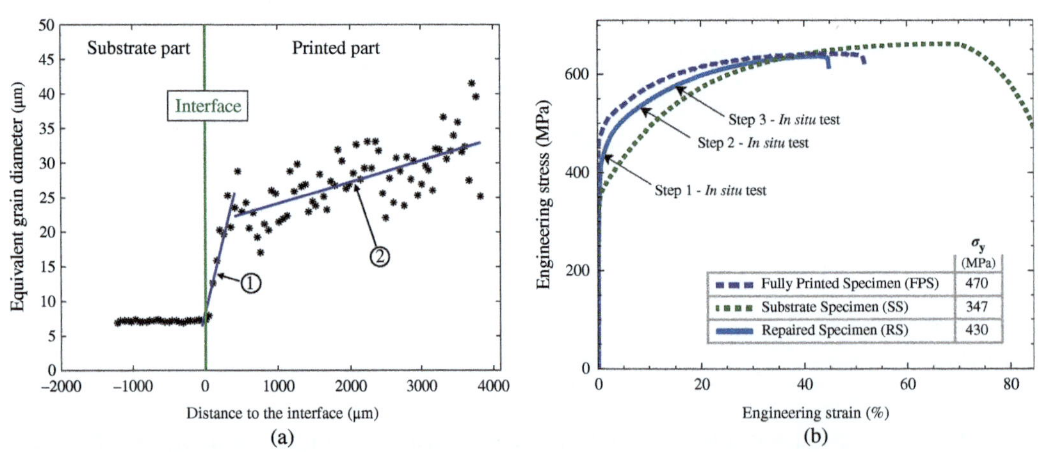

Figure 15.4 For a surface repair on top of a hot rolled 316L SS sheet, (a) the *in-situ* grain size measurements made in the repaired specimen as shown in Figure 13.20 are plotted along with (b) the *ex-situ* tensile test results for all three tensile test specimens [11].

ability to repair high-value parts, such as a surface layer repair on top of a hot rolled 316L SS sheet. The relationship of grain size in the original part to the build layer was studied in model systems and shown in Figure 13.20 [11]. As illustrated in Figure 13.20a, three dog bone-shaped test specimens were produced, viz a fully printed specimen (FPS) within the build layer, a repaired specimen (RS) straddling the interface, and a substrate specimen (SS) within the substrate. Figure 15.4 summarizes (a) the *in-situ* grain size measurements made in the RS and (b) the *ex-situ* tensile test results for all three specimens. The RS exhibits an intermediate yield strength with a lower ductility.

Returning to the issue of PSD raised in Sections 15.2 and 15.3, it is worth noting that, while finer-sized particles were associated with higher part densities and improved surface qualities, larger particles were associated with higher values of elongation at failure [4]. Also, a narrower PSD was associated with higher values of ultimate tensile strength for steel parts.

15.5 Hardness

Hardness closely correlates with the strength measurements from the tensile test discussed in Section 15.4 and is another basis for comparing the AM-produced alloys with conventional ones. Figure 15.5 shows an Ashby chart for the material's property space of room temperature hardness vs. density allowing complex concentrated alloy (CCA), more commonly referred to as **high entropy alloy (HEA)**, coatings to be compared with other conventional systems [8].

The HEA coatings produced by laser cladding are compared with conventional surface coatings (Co- and Ni-based alloys) and alloys compatible with cladding and thermal spraying (ceramic particle-reinforced [CPR] Mg-, Al-, and Ti-matrix, metal-bonded WC, and refractory alloys). For comparison, three families of bulk HEAs are shown as well (light-metal, 3D transition metal, and refractory metal HEAs). The bulk HEAs were processed by conventional methods, primarily arc-melting. The laser-cladding HEA coatings are represented by white circles. The corresponding materials property landscape (shown in yellow) overlaps the area for the conventional Co- and Ni-based coatings, as well as the 3D transition metal and refractory metal bulk HEAs, with substantial data showing higher hardness

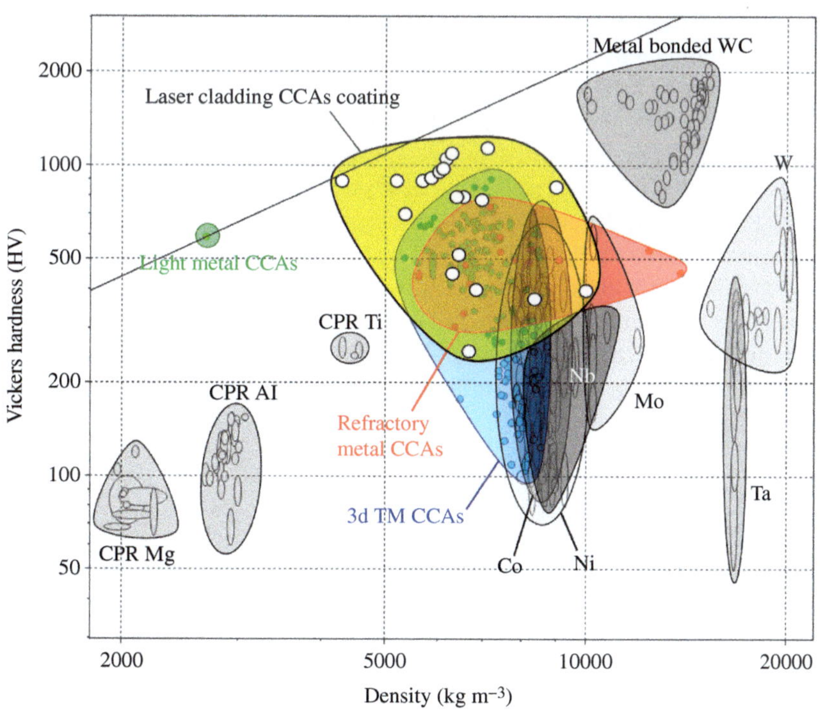

Figure 15.5 An Ashby chart of room temperature hardness vs. density of complex concentrated alloy (CCA), more commonly referred to as high entropy alloy (HEA), coatings, along with conventional metal alloy coatings, ceramic particle-reinforced (CPR) metal matrices, and bulk HEAs [8]. The laser-cladding HEA coatings are represented by the white circles within the materials property landscape shown in yellow.

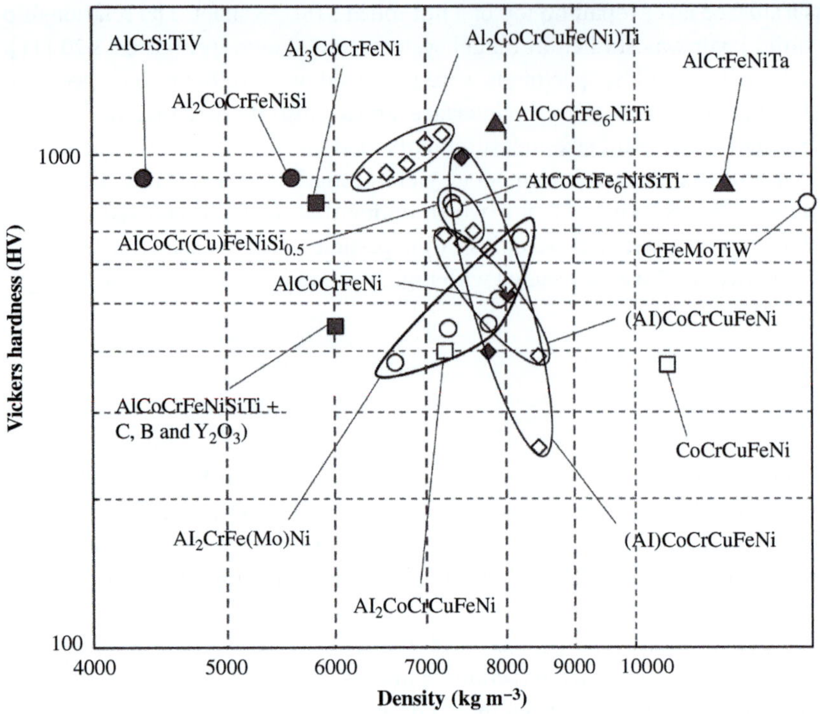

Figure 15.6 A detailed view of the Ashby chart in Figure 15.5 showing the laser cladding HEA coatings [8]. Squares represent fcc phases, and circles represent bcc phases. Diamonds represent duplex (fcc + bcc) phases, and triangles represent other phases. Solid solution and (solid solution + intermetallic phases) are shown by open and closed symbols respectfully.

values than both the conventional coatings and those bulk HEAs. The laser-cladding HEA coatings generally display higher hardness values than CPR systems. They also display superior hardness values to many of the bulk HEAs due to the prevalence of bcc phase along with grain refinement and nano-sized precipitates that result from the rapid cooling rate. While the laser-cladding HEA coatings do not match the best values for metal-bonded WC, they can provide superior hardness/density performance, as indicated by the upward-sloping solid line. Harder and lighter coatings correspond to data above that line. Figure 15.6 gives a more detailed view of the various HEA clads [8].

As with the issue of PSD raised in Sections 15.2–15.4 of this chapter, it is worth noting that a narrower PSD was associated with higher values of hardness in steel parts [4]. Regarding the analysis of the stainless-steel repair introduced in Figure 13.20 and tensile test results summarized in Figure 15.4, Figure 15.7a–c shows the evolution of the axial strain in the deformation steps 1, 2, and 3 of Figure 15.4b along with the corresponding microstructures obtained by electron backscatter diffraction. Microhardness measurements shown in Figure 15.7d indicate a low deformation zone within about 200 μm on both sides of the interface and correspondingly higher hardness, consistent with the presence of a heat-affected zone as well as the finer grain sizes shown in Figure 15.4a [11].

15.6 Fracture and Fatigue

Throughout the discussions of defects in AM processing in Chapter 13, a common goal was to minimize those defects in order to produce parts with fracture and fatigue behavior comparable to conventionally produced products. For example, in Section 13.2, it was pointed out that density has an especially adverse effect on fatigue leading to the goal of achieving 99.5% density. Avoiding irregular shaped pores was especially important. Similarly, in Section 13.6, we saw that increasing the surface roughness significantly decreases the fatigue life in a Ti6Al4V alloy being developed as a dental material. A primary justification for **post-processing** treatments was often to minimize porosity, surface roughness, and other defects to ensure acceptable fatigue behavior.

Fatigue behavior is especially important in dynamic loading conditions in aerospace applications. The presence of porosity is critical in fatigue loading as pores can act as stress concentrators and crack initiation locations. HIP

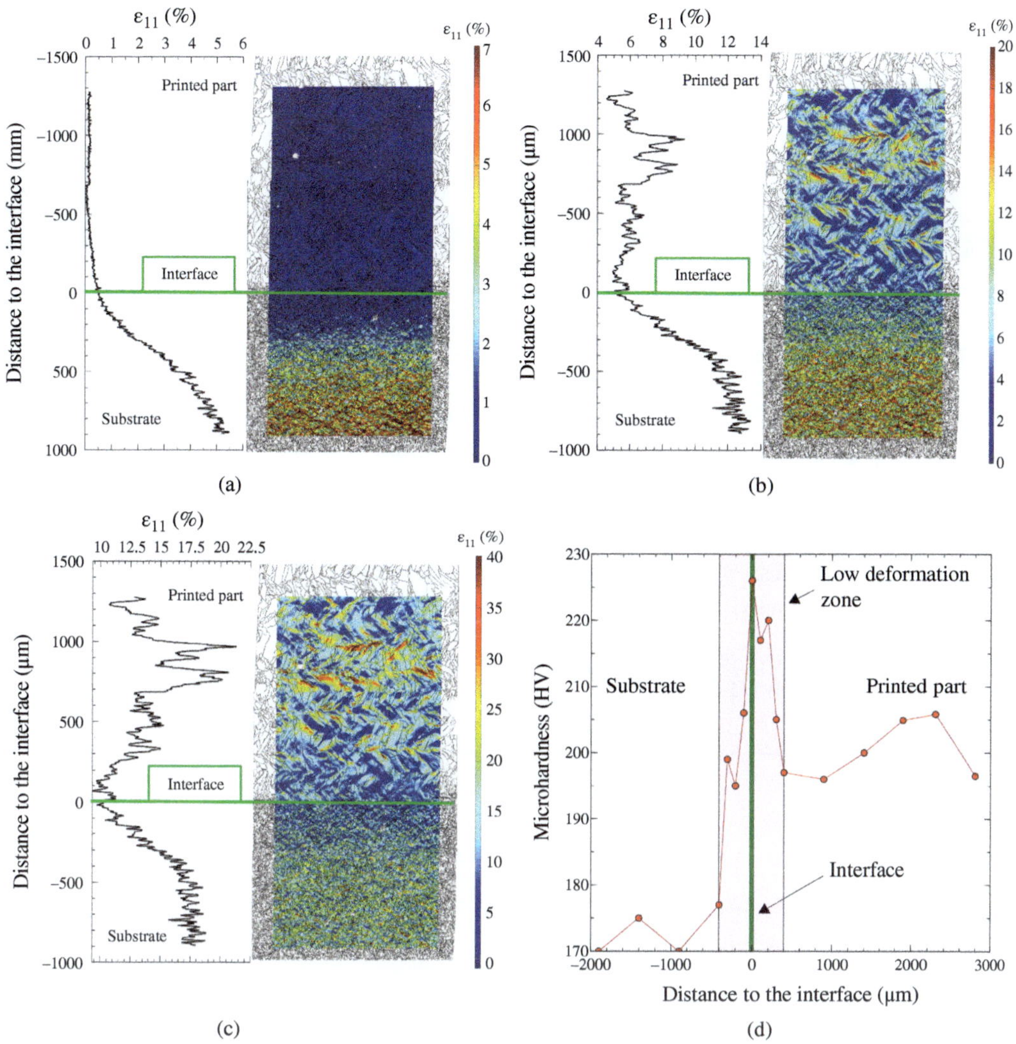

Figure 15.7 The evolution of axial strain is shown respectively in (a), (b), and (c) for steps 1,2, and 3 in Figure 15.4b. (d) Microhardness measurements indicate a low deformation zone within about 200 μm on both sides of the interface and a correspondingly higher hardness, consistent with the finer grain sizes shown in Figure 15.4a. Source: Balit et al. [11]/with permission from Elsevier.

is very useful for homogenizing the microstructure and closing pores with the resulting improvement in ductility also improving fatigue performance [3]. Surface treatments that reduce roughness are also important. For example, Figure 15.8 shows the dramatic improvement in the *S-N* curves for a Ti6Al4V alloy produced by PBF-LB when fatigue specimens were both HIPed at 920 °C and 100 MPa for two hours and surface polished with #600 emery paper [12]. Figure 15.9 shows similar results for the same alloy produced by the powder bed fusion using an electron beam (PBF-EB) technique [12]. Figure 15.10 shows *S-N* curves for another Ti6Al4V alloy produced by (a) PBF-LB and (b) PBF-EB methods [13]. The PBF-LB samples showed a fatigue limit of 550 MPa, a value higher than the value of 430 MPa for cast and annealed Ti6Al4V [14]. The PBF-EB value of 340 MPa was, however, inferior to both the PBF-LB sample and the conventional alloy. The better fatigue performance of the PBF-LB was attributed to its α′ martensitic phase that impedes dislocation motion leading to a strengthening effect. By contrast, the conventional α (hcp) phase and a small amount of β (bcc) phase was found in the PBF-EB samples.

In a study on Ti6Al4V alloy produced by DED using a laser beam (DED-LB) known as the **laser-engineered net shaping (LENS™)** system, fatigue life was found to be primarily affected by pores, especially densely grouped larger ones near the part's surface [15]. In another study on Ti6Al4V alloy also produced by the LENS system, the AM samples were compared directly to a wrought alloy. After a one hour anneal at 600 °C for the purpose of relieving residual

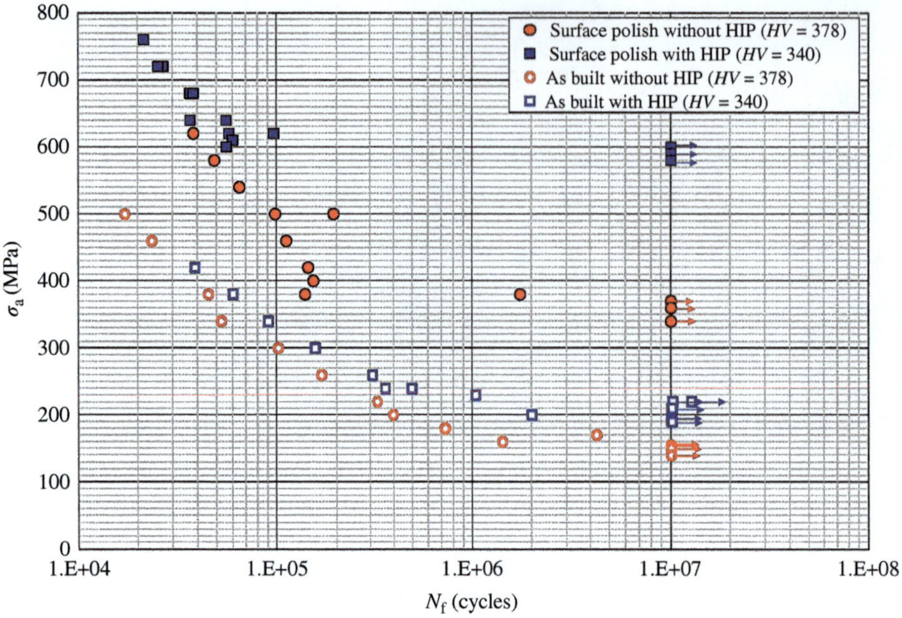

Figure 15.8 Fatigue behavior for a Ti6Al4V alloy produced by PBF-LB showing breaking stress (*S*) vs. the number of cycles to failure (N_f), along with Vickers hardness values (*HV*) [12].

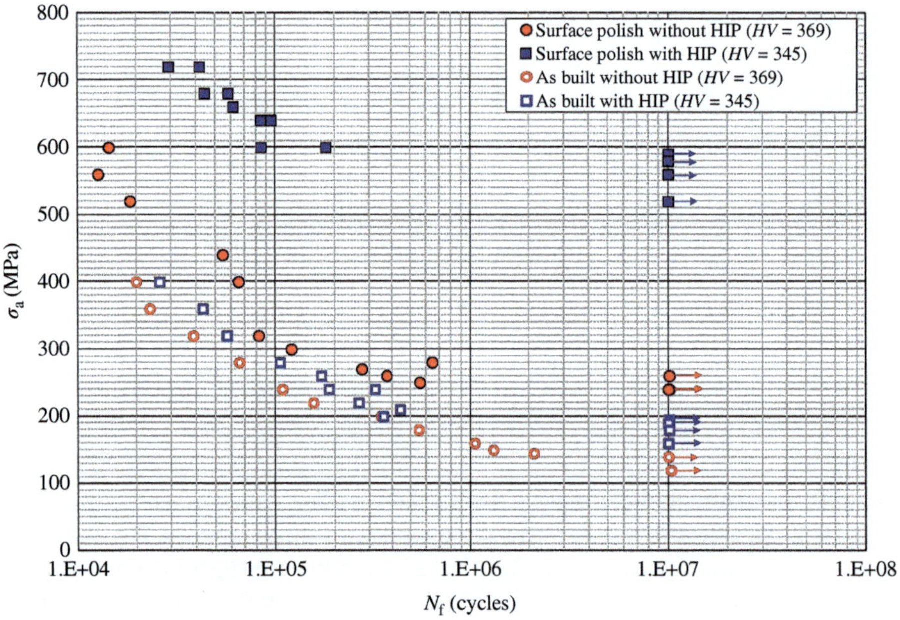

Figure 15.9 Fatigue behavior for a Ti6Al4V alloy produced by PBF-EB showing breaking stress (*S*) vs. the number of cycles to failure (N_f), along with Vickers hardness values (*HV*) [12].

stresses and surface machining to produce a mean surface roughness of 0.25 μm, the LENS samples demonstrated higher fatigue life and fatigue strength compared to wrought alloy samples machined to the same shape and surface roughness [16]. For example, a summary of the fatigue strength data at 1×10^6 cycles is given in Table 15.1. This superior performance was attributed to a finer and relatively complex microstructure, including lamellar α, basket-weave α + β, and very fine α′ (martensitic) grains within columnar β grains. By contrast, the wrought samples had relatively coarse, equiaxed α grains decorated with β grains. Fractographic analyses indicated a cross-colony slip-band crack initiation mechanism in the LENS samples in contrast to fatigue crack initiation from the edge of the wrought samples [16].

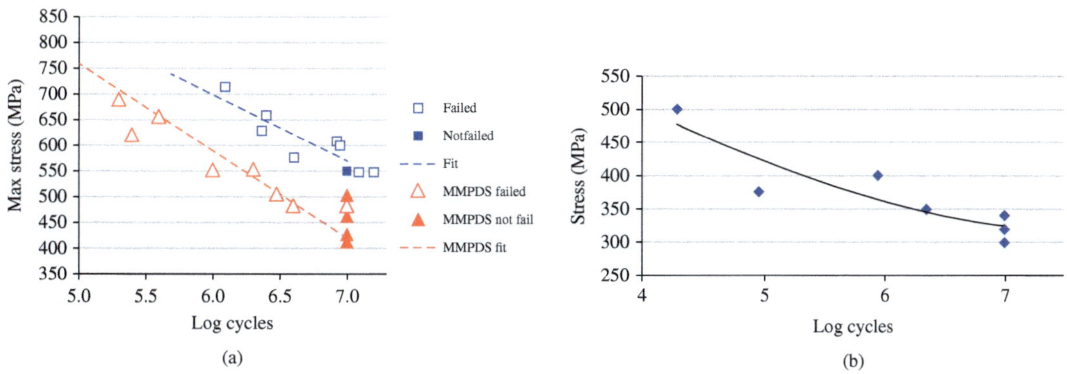

Figure 15.10 *S-N* curves for a Ti6Al4V alloy produced by (a) PBF-LB and (b) PBF-EB methods [13].

Table 15.1 Fatigue strength data for Ti6Al4V alloy at 1×10^6 cycles [16].

Sample geometry	LENS™	Wrought
Smooth	482	345
Semi-circular notch	477	344
V-notch	293	258

An AM technique that uses metal wire in place of powders is the wire arc additive manufacturing (WAAM) form of DED processing. WAAM for 304L has proven successful in producing fatigue behavior comparable to the wrought alloy [17]. While the data indicated that the build direction affects fatigue life, the as-built WAAM samples generally performed as well or better than the wrought sample and better than the LENS samples. In a study of high-cycle fatigue in 316L stainless steel produced by PBF-LB, parts directly produced with fatigue sample geometry without any post-printing treatments exhibited a fatigue strength of ~290 MPa at 10^6 cycles, comparing favorably with the fatigue strength of 250–290 MPa for wrought 316L under the sample cycling conditions [8]. Nonetheless, the authors noted that one might expect the as-built fatigue performance to be even better, as tensile test specimens were much stronger than the wrought counterpart and fatigue strength usually scales with tensile strength. They suggested that the explanation for the discrepancy could be some evidence of residual porosity and the inherent surface roughness of the PBF-LB-produced samples.

The relationship between processing parameters and surface roughness for Inconel 718 was discussed in Section 15.3, with a reference to the fact that surface roughness (specifically, maximum notch values) has implications for fracture and fatigue behavior [6]. Figure 15.11 shows how (a) fatigue life decreases with increasing laser speed at a fixed value of power and (b) increases with increasing laser power at a fixed value of speed. These trends are consistent with the discussion of maximum valley (or notch) size from CT measurements in Figure 15.2 and its relationship to melt pool size. Overall, the authors found a clear trend of greater fatigue life with decreasing values of S_v over the complete set of process parameters used in this study.

In a study of fatigue behavior in DED-LB produced Inconel 718 using the LENS technique, tests were carried out at 650 °C and room temperature [18]. A distinctive feature of the as-built microstructure was a dendritic structure that was not eliminated by a **post-processing** heat treatment. Needle-like δ phases were also observed on the grain boundaries. The LENS samples exhibited a slightly lower fatigue resistance than a wrought counterpart at room temperature but were comparable at 650 °C. In the low-cycle fatigue regime, the fatigue resistance of the LENS-produced sample was slightly inferior at the elevated temperature in comparison to room temperature due to the formation and growth of the brittle needle-like δ and Laves phases on grain boundaries during fatigue testing at the elevated temperature. Those brittle phases increased the fatigue crack growth rate. In the high-cycle fatigue regime, however, LENS samples tested at the elevated temperature were comparable to those at room temperature, attributed to surface oxidation at 650 °C retarding crack initiation and growth. In a related study using DED-LB, a suite of **post-processing** heat treatments (solution + aging + homogenization) was found to provide a toughening effect in

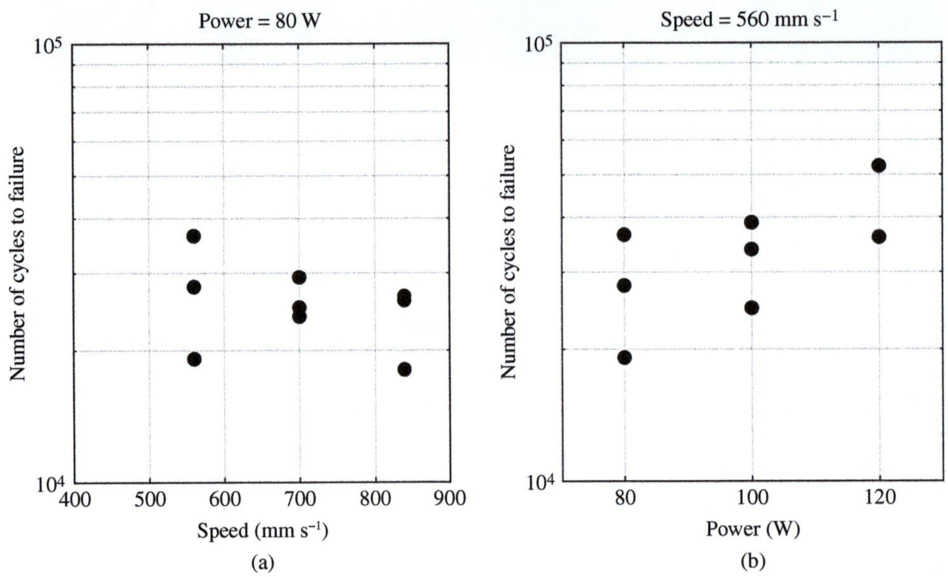

Figure 15.11 Comparison of fatigue life for Inconel 718 to the process parameters (a) laser speed and (b) laser power [6].

optimizing the fracture toughness values of Inconel 718 [19]. The K_{Ic} for the heat-treated sample increased by ~91% compared to the as-printed part and was ~60 MPa$\sqrt{m}$ higher than the lower limit of a comparable wrought alloy, reaching a high value of ~164 MPa$\sqrt{m}$.

15.7 Corrosion and Wear

Some of the alloys most widely produced by AM methods, including titanium alloys, stainless steels, HEA containing cobalt and chrome, and aluminum alloys, are often selected for engineering applications due to their corrosion resistance. As a result, comparing the corrosion resistance of AM-produced parts against conventionally produced counterparts is important. Also, the study of corrosion is often done in conjunction with wear. While wear is a physical property, it is appropriate to discuss corrosion and wear together as a broader picture of environmental degradation.

While aluminum alloys in general exhibit good corrosion resistance, it has been noted in Chapters 13 and 14 that residual stress and a high **coefficient of thermal expansion** (CTE) make them susceptible to shrinkage, cracking, and deformation during the severe thermal cycles involved in AM [20]. It was also noted that aluminum's low density leads to poor powder flowability affecting the quality of the build [21]. Aluminum's high moisture absorption and surface oxidation are also detrimental to laser-based deposition [20, 22]. As a result, PBF-LB and DED technology are generally focused on various AlSi alloys that exhibit excellent laser absorption, high corrosion resistance, good weldability, and good mechanical performance [21, 23].

The corrosion and wear behavior of an Al-12Si alloy produced by PFB-LB was characterized and compared to the alloy produced by conventional casting technology [24]. Figure 15.12 shows the weight loss with time due to corrosion for Al-12Si samples produced by both AM and casting. The corrosion was due to immersion in various HNO_3 concentrations over a span of several days. Figure 15.12a shows the increasing rate of corrosion with increasing HNO_3 concentrations for PBF-LB-produced samples. Figure 15.12b shows that the rate of corrosion for the as-prepared PBF-LB produced sample is comparable to that of the cast sample, but that additional heat treatment of the PBF-LB samples for six hours under an argon atmosphere at various temperatures increases the rate of corrosion significantly. While heat treatments are important **post-processing** techniques for AM processing, as noted in Section 14.6 with examples throughout Part II of this book, the microstructure of this Al-12Si alloy leads to a relatively complex corrosion process. (It should be noted that the term SLM in part (b) stands for "selective laser melting," an alternative descriptor for PBF-LB that is used in this chapter.) The nature of the corrosion is shown in greater detail in Figure 15.12c which shows the weight-loss curve for the as-prepared PBF-LB sample in the 0.1 M HNO_3 solution. In this case, we see two distinct time regions. The cast sample and heat-treated PBF-LB samples at 473 and 573 K in

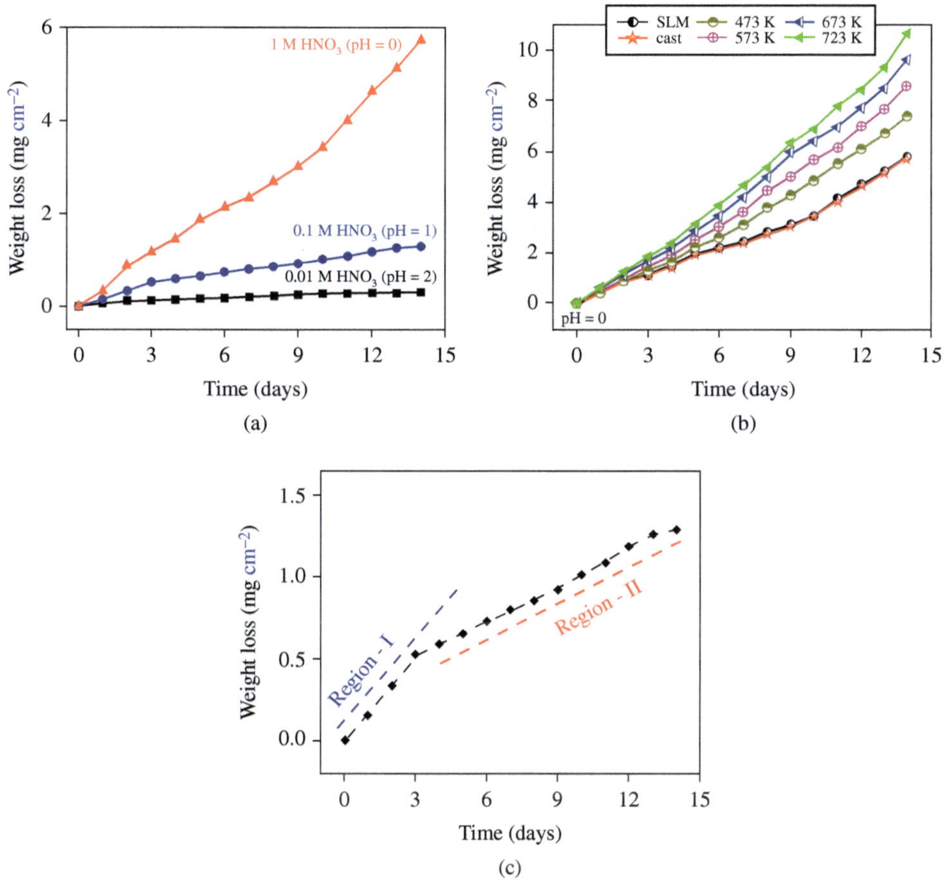

Figure 15.12 (a) Weight loss curves for as-prepared PBF-LB produced samples of Al-12Si as a function of immersion time in three different HNO_3 concentrations. (b) Weight loss curves for Al-12Si samples as-prepared by PBF-LB, cast, and as-prepared by PBF-LB with heat treatment for six hours under an argon atmosphere at various temperatures as a function of immersion time in the 1 M HNO_3 solution. (c) Weight loss curves for the as-prepared PBF-LB produced samples of Al-12Si as a function of immersion time in the 0.1 M HNO_3 solution showing two distinct regions [24]. (The term SLM in part (b) corresponds to PBF-LB in this chapter.)

Figure 15.12b showed similar behavior, but the samples heat treated at 673 and 723 K did not. This behavior can be explained with the aid of Figure 15.13 [24]. The rapid solidification during PBF-LB produces a cellular microstructure composed of a super-saturated primary Al-rich phase with residual Si segregated at the cellular boundaries. The Si surface oxidizes to a stable SiO_2 phase, while the Al dissolves in the electrolyte as Al^{3+} ions. As a result, the initial (Region I) corrosion rate in the first three days is relatively high. Once the Al-rich phase is removed from the surface, the continuous Si network hinders the access of HNO_3 electrolyte resulting in the reduced (Region II) corrosion rate. For the higher temperature heat treatments (673 and 723 K), the Si particles are still passivated but not interconnected, allowing more consistent access of the electrolyte to the Al-rich phase and surface pitting. Finally, the cast Al-12Si behaves in a manner similar to that described for the as-prepared PBF-LB sample.

The wear behavior of this Al-12Si alloy was measured by sliding wear tests carried out according to the ASTM G 99-05 standard and fretting wear tests carried out according to the ASTM-D5706-97 and ASTM-D5707-97 standards [24]. The as-prepared PBF-LB sample exhibited a lower sliding wear rate than the cast sample (0.92×10^{-12} m^3 m^{-1} compared to 1.05×10^{-12} m^3 m^{-1}), while heat-treated PBF-LB samples exhibited a sharp rise in sliding wear rate with increasing heat treatment temperatures reaching a value of 1.15×10^{-12} m^3 m^{-1} for the heat treatment at 723 K. These results were tied directly to the nature of the microstructures. Specifically, the Vickers hardness was found to decrease with the increasing size of the Si particles, and the size of those particles is higher in the cast sample than in the PBF-LB one and Si particle size increases with increasing heat treatment temperatures. The fretting wear results were consistent, as shown in Figure 15.14 as a plot of wear volume against Si particle size for both sliding wear and fretting wear, along with the behavior of the cast alloy [24].

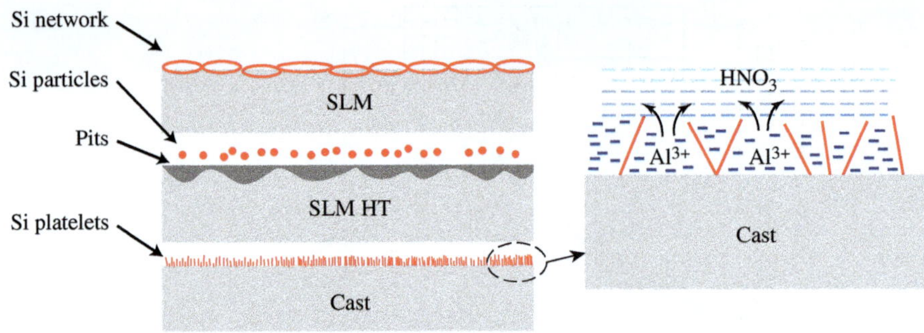

Figure 15.13 Schematic illustration of the corrosion behavior PBF-LB with and without heat treatments (HT) and produced by conventional casting [24]. (The term SLM corresponds to PBF-LB in this chapter.)

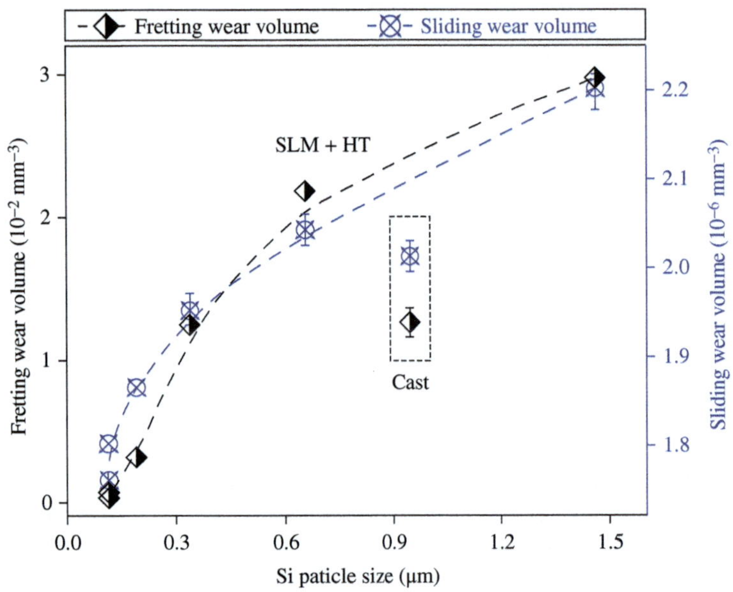

Figure 15.14 Fretting and sliding wear volumes for cast Al-12Si and PBF-LB produced samples as a function of Si particle size [24]. The increase in particle size is a function of the heat treatment temperature for the PBF-LB samples. (The term SLM corresponds to PBF-LB in this chapter.)

The AM processing of titanium alloys has been widely applied and increasingly so in recent years. The production of these parts by the DED technique can be especially challenging due to the non-equilibrium structures and complex microstructures mentioned above [25]. For example, challenges include the quality of the native oxide layer in Ti alloys and a significant dependence of corrosion resistance on the deposition orientation in those alloys. Nonetheless, many studies have shown that the generally good corrosion resistance of wrought forms of these alloys can be retained in DED-formed parts. A study of corrosion and wear in a γ-TiAl intermetallic alloy (33–35 w/o Al, 2.2–2.7 w/o Cr, 4.5–5.1 w/o Nb, balance Ti) was carried out after DED processing using the LENS system [26]. TiAl compounds are considered prime candidates for a variety of high-temperature applications such as aeronautical engines due to their attractive physical properties along with high oxidation resistance. Their inherent brittleness however makes conventional welding and forging processes difficult. While they can be produced by conventional casting, the resulting coarse microstructures can result in poor mechanical properties. Traditional powder metallurgy (PM) processing has been an attractive alternative but is limited in its ability to form complex shapes. AM processing has the potential to successfully process more complex structures. A wide range of laser power (in W)/laser scan speed (in mm s^{-1}) ratios were studied to identify optimal LENS processing conditions. Four such combinations produced visually sound parts (350/15, 300/15, 250/10, and 200/10). The compositions of these specific alloys are shown in Table 15.2, along with the percentages of porosity (≈1–2%) and pore size (≈0.5 μm), hardness data, and corrosion data [26]. The corrosion results are mixed with the 200/10 samples showing the lowest corrosion current density, I_{corr}, of 0.002 μA and

Table 15.2 Composition, porosity, hardness, and corrosion properties of LENS™ processed γ-TiAl parts [26]. The ID represents the laser power (in W)/laser scan speed (in mm s^{-1}) ratios.

				Gray relational analysis					
				Corrosion		Corrosion only		Overall	
ID	Al, Cr, Nb, Ti (wt%)	Porosity (%), pore size (μm)	Hardness ($HV_{0.5}$)	E_{corr} (mV vs. SCE)	I_{corr} (μA)	Grade	Order	Grade	Order
350/15	35.1 ± 0.3 3.2 ± 0.5 4.9 ± 0.3 56.8 ± 0.5	1.7 ± 0.03 0.5 ± 0.19	407 ± 7	−43	0.8	0.94	1	0.67	2
300/15	35.1 ± 0.8 2.9 ± 0.7 5.2 ± 0.3 56.8 ± 1.5	1.1 ± 0.04 0.5 ± 0.19	404 ± 10	−237	14.2	0.41	4	0.44	4
250/10	35.4 ± 0.7 3.0 ± 1.0 4.7 ± 0.2 56.9 ± 0.4	2.3 ± 0.29 0.5 ± 0.14	421 ± 15	−208	7.8	0.50	3	0.58	3
200/10	35.0 ± 0.1 3.3 ± 0.4 5.0 ± 0.3 56.7 ± 0.9	1.2 ± 0.03 0.5 ± 0.18	409 ± 6	−411	0.002	0.67	2	0.69	1

the 350/15 showing the most noble corrosion potential, $E_{corr,}$ of −43 mV vs. SCE. The Gray relational grade analysis attempts to indicate the best performance behavior for corrosion only and corrosion and wear resistance combined. For the samples in Table 15.2, the 350/15 and 200/10 samples demonstrated the best performance with 350/15 having the highest grade for corrosion and 200/10 having the highest grade for corrosion and wear combined.

The wear rate (in $mm^3(N\,m)^{-1}$) for the various parts is summarized in Figure 15.15 [26]. The wear performance in air (="dry" in Figure 15.15) and in a 3.5% NaCl solution is similar for each sample but varies significantly for different processing parameters. The superior performance of the 200/10 samples was attributed to relatively finer microstructural features (γ-TiAl with a small amount of α_2-Ti_3Al) that resulted from higher cooling rates than in the other samples. Inspection of the wear track morphology indicated that the dominant wear mechanism was abrasive in both air and the NaCl solution, while no corrosion was observed in the wear tracks.

Corrosion and wear performance in CoCrMo impeller samples produced by the LENS system have been reported [27]. The commercially available powder was fabricated at a laser power of 350 W, a 5 g min^{-1} feed rate,

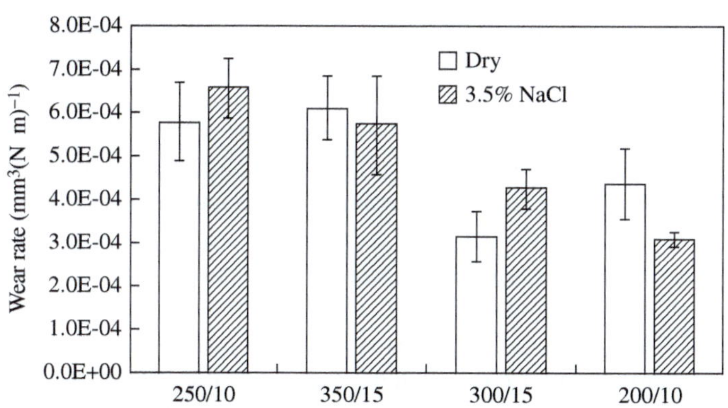

Figure 15.15 The wear rate of LENS™ processed γ-TiAl parts in dry air and 3.5% NaCl solution [26]. The horizontal axis indicates laser power (in W)/laser scan speed (in mm s^{-1}) ratios.

Table 15.3 Post-processing parameters for LENS™-produced CoCrMo alloys [27].

Sample no.	Solution treatment time (min)	Aging temperature (°C)	Aging time (h)
S1	30	No aging	
S2	30	815	4
S3	30	830	6
S4	45	No aging	
S5	45	815	6
S6	45	830	2
S7	60	No aging	
S8	60	815	2
S9	60	830	4

Table 15.4 Hardness, wear rate, and corrosion parameters for LENS™-produced CoCrMo alloys along with the Gray relational analysis to indicate the best overall combination of these parameters [27].

Sample ID	Hardness (HV)	Wear rate ($\times 10^{-4}$ mm^3 (N m)$^{-1}$)	Corrosion		Gray relational analysis	
			I_{corr} (μA cm^{-2})	E_{corr} (mV vs. SCE)	Grade	Order
S1	491 ± 51	1.44 ± 0.28	0.11	−368	0.563	7
S2	382 ± 25	1.18 ± 0.24	0.20	−300	0.463	9
S3	378 ± 11	1.19 ± 0.25	0.22	−61	0.514	8
S4	372 ± 44	0.96 ± 0.28	0.42	+64	0.624	3
S5	386 ± 25	1.17 ± 0.12	0.44	+85	0.568	6
S6	512 ± 58	1.08 ± 0.18	0.28	−116	0.670	2
S7	478 ± 81	0.96 ± 0.25	0.05	−178	0.739	1
S8	484 ± 37	0.97 ± 0.13	0.61	−86	0.603	5
S9	380 ± 35	0.90 ± 0.14	0.39	−22	0.619	4

and a 20 mm s^{-1} scan speed for the purpose of evaluating post-processing heat treatments on corrosion and wear behavior. The processing parameters had been optimized in an earlier study [28], and the sample heat treatments are summarized in Table 15.3. All samples were water quenched after the solution treatment at 1200 °C. Samples S_1, S_4, and S_7 were not subjected to any additional aging treatments after the solution treatment. Table 15.4 shows the hardness, wear rate, and corrosion parameters for all samples [27]. While sample 6 (1200 °C solution treatment for 45 minutes plus 2 hours of aging at 830 °C) had the highest hardness and sample 9 (1200 °C solution treatment for 60 minutes plus 4 hours of aging at 830 °C) had the lowest wear rate, the Gray relational analysis indicated that the CoCrMo alloy with the best overall combination of hardness, wear resistance, and corrosion resistance was sample 7 (1200 °C solution treatment without additional aging treatment). Microstructural analyses indicated that increasing solution times decreased the concentration of carbide precipitates at grain boundaries and within equiaxed Co grains while aging treatments increased the concentration of $Cr_{23}C_6$ precipitates. Corrosion was found to be primarily along the grain boundaries near the carbide precipitates, consistent with the local decrease in Cr concentration.

While the ability of AM-produced parts to provide good corrosion and wear resistance comparable to conventionally produced parts is desirable, an additional feature of laser deposition technology is the ability to produce coatings and claddings that can substantially improve the corrosion and wear performance of substrate materials [25].

Table 15.5 Bio-tribological testing results for Ti6Al4V Alloys Coated by LENS™ Laser Deposition [29]. Percentage differences are in comparison to uncoated Ti6Al4V.

Sample	Wear rate reduction	Coefficient of friction reduction	Hardness increase
Ti6Al4V + 5 wt% Si	~17%	~34%	61%
Ti6Al4V + 10 wt% Si	~71%	~38%	115%
Ti6Al4V + (3 wt% Si + 2 wt% HA)	~51%	~39%	104%

One example is a proposed modification to widely used total hip replacement (THR) designs in which a femoral stem of Ti6Al4V is attached to a spherical CoCrMo femoral head which provides a bearing surface into an acetabular cup of ultrahigh-molecular-weight polyethylene [9]. The problem with this traditional orthopedic device design is the potential for taper corrosion failure due to galvanic corrosion between the Ti6Al4V and CoCrMo alloys. The proposed solution was to make the femoral head from a material similar to the Ti6Al4V stem [29]. This alternative has not been used in the past because of the inferior wear performance of Ti6Al4V compared to CoCrMo. In this study, a Ti6Al4V substrate was coated with various composites of Ti6Al4V and 5–10 wt% Si or (3 wt% Si + 2 wt% hydroxyapatite [HA]) by laser deposition using the LENS system. Chemical analyses indicated the formation of Ti_5Si_3, V_5Si_3, and $TiSi_2$ silicides in the surface coatings. Bio-tribological testing results are summarized in Table 15.5 [29] and show significant reductions in wear rate and **coefficient of friction** (COF) along with significant increases in hardness. Microscopic analysis indicated that the wear mode evolved from abrasive-gouging to non-abrasive sliding wear for the 10 wt% Si and 3 wt% Si + 2 wt% HA samples. In addition, a positive shift in open circuit potential (OCP) was observed under linear wear conditions indicating a decreased tendency for corrosion. The authors concluded that the 3 wt% Si + 2 wt% HA coating was the most promising given its positive performance and evidence of surface protection by an *in situ* tribofilm [29]. Figure 15.16 shows a flowchart of the overall process from fabrication methodology to key results.

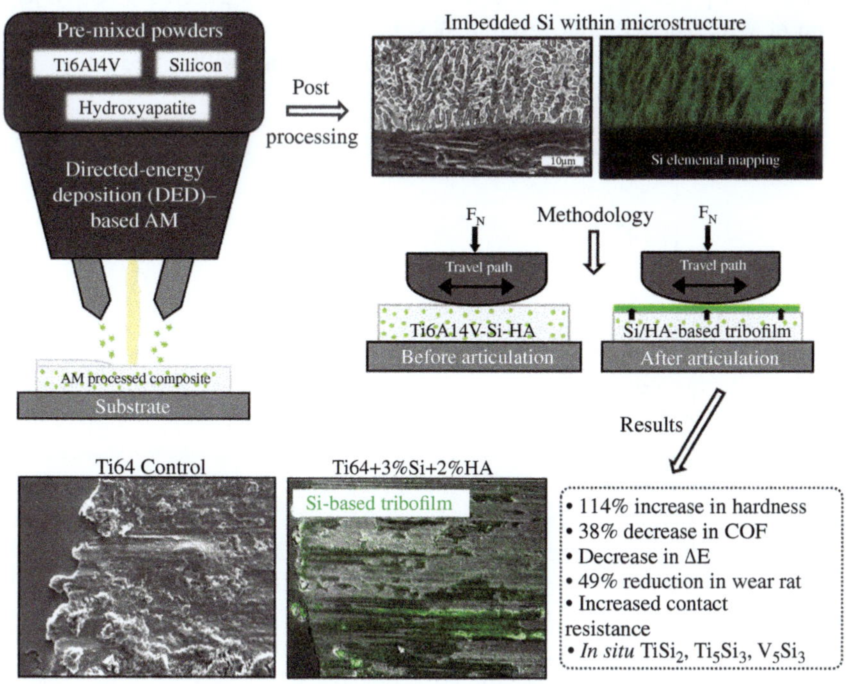

Figure 15.16 Flowchart of the overall process in which a Ti6Al4V substrate was coated with various composites of Ti6Al4V and 5–10 wt% Si or (3 wt% Si + 2 wt% hydroxyapatite [HA]) by laser deposition using the LENS™ system. The purpose of the study was to improve the design of a total hip replacement (THR) prosthesis. The 3 wt% Si + 2 wt% HA coating was the most promising given its positive performance and evidence of surface protection by an *in situ* tribofilm. Source: Avila et al. [29]/with permission from Elsevier.

Table 15.6 Chemical compositions of H13 steel and Stellite® 6 [30].

Material	Element (wt%)							
	C	Si	Mn	Cr	Mo	V	Co	Fe
H13	0.40	1.00	0.30	5.15	1.35	1.10	—	Bal.
Stellite 6	1.15	1.10	0.05	29.00	1.00	—	Bal.	3.00

Table 15.7 Compositions of three metal matrix composites (MMC) laser deposited onto an H13 substrate [30].

No.	Specimen 1	Specimen 2	Specimen 3
Substrate	H13	H13	H13
Coating	90% Stellite 6–10% Cr_3C_2 composite coatings	85% Stellite 6–10% Cr_3C_2-5% WS_2 composite coatings	80% Stellite 6–10% Cr_3C_2-10% WS_2 composite coatings

Laser cladding of significantly different materials onto a substrate for the purpose of improving corrosion and wear performance is an active area of research [25]. For example, Stellite® 6-Cr_3C_2-WS_2 composites have been laser deposited onto H13 tool steel for the purpose of significantly improving wear resistance [30]. Table 15.6 gives the chemical compositions of H13 and Stellite 6, and Table 15.7 gives the compositions of three **metal matrix composites (MMCs)** that were laser deposited onto the H13. Figure 15.17 shows the significant decrease in the extent of wear with the various claddings in a multi-functional friction and wear machine at 25 and 200 °C (to represent hot die temperatures). The improved wear resistance was attributed to the solid solution strengthening effect of γ-(Fe, Co) and the formation of a hard phase Cr_7C_3. In addition, samples 2 and 3 generated (Cr, W)C and CrS phases due to the decomposition of WS_2. The CrS and residual WS_2 phases contribute a self-lubricating effect, providing superior wear performance in Figure 15.17, especially at 200 °C along with a reduction in the COF.

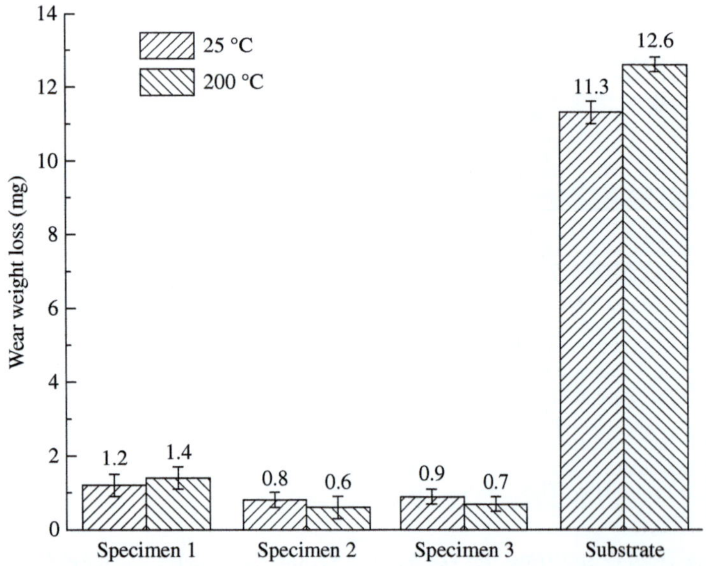

Figure 15.17 Weight loss due to wear for the three composite cladded H13 steels in Table 15.7 compared to the H13 substrate [30].

15.8 Oxidation

Along with corrosion, oxidation is a fundamental chemical property that plays a central role in the AM of metal alloys. While, as noted in Section 15.7, alloys manufactured by AM are often chosen for their corrosion resistance, the oxidation of those alloys can sometimes be an issue. As noted in Section 15.2, achieving a density of 99.5% is a primary goal of AM to achieve optimal performance. This goal is equally important regarding corrosion and oxidation, as porosity can be a significant contributor to such chemical reactions [1]. The focus on a limited number of aluminum alloys for AM was also noted in Section 15.7 to be the result of aluminum's high moisture absorption and surface oxidation. The high oxygen affinity of Cr in stainless steels leads to significant oxidation during laser-based deposition and requires controlling the oxygen concentration to a few ppm in the deposition chamber during the DED process [25]. There has also been considerable qualitative concern reported in the literature about the possible deleterious effect of the oxidation of powders during long-term storage prior to AM. This effect was quantified by *in situ* observation of the fluid dynamic behavior of Invar 36 feedstock during PBF-LB in a synchrotron X-ray beamline [31]. The fluid dynamics of virgin powder was compared with that oxidized by storage for a period of approximately one year. The oxygen content of the oxidized powder (0.343 wt%) was approximately six times greater than that of the virgin powder (0.057 wt%). Figure 15.18 shows the comparison between time series x-radiographs of melt features for PBF-LB of the first layer Invar 36 melt track for (a) virgin powder and (b) oxidized powder. No pores are evident in part (a) during melt track extension, but the spatter continues to eject from the powder bed in the direction of the laser scan and argon gas flow. Two distinct pore evolution mechanisms seen in (b) are shown in more detail as (c) centripetal Marangoni convection (white arrows) and (d) pore growth promoted by an oxide film (orange circles). The observed fluid flow was associated with changes in the surface tension of the molten pool due to the higher oxygen content. The origin of gas pores is shown in (c) and larger pores emerge due to Ostwald ripening in (d).

Oxidation has also played a role in the microstructural development and related mechanical behavior of two novel DED-LB high-carbon, high-speed steels, Fe_{bal}–C–Cr–Mo–V and $Fe_{bal\text{-}x}$–C–Cr–Mo–V–W_x [32]. High-temperature (500 °C) pin-on-disc tests of friction and wear were conducted. The microstructures of both alloys consisted of a martensitic matrix along with primary and eutectic carbides. Both alloy compositions gave microhardness values greater than 700 HV with no detrimental effect of repetitive laser thermal cycling. While both alloys demonstrated good wear resistance, the addition of W provided a superior performance due to *in situ* formation of a homogeneous

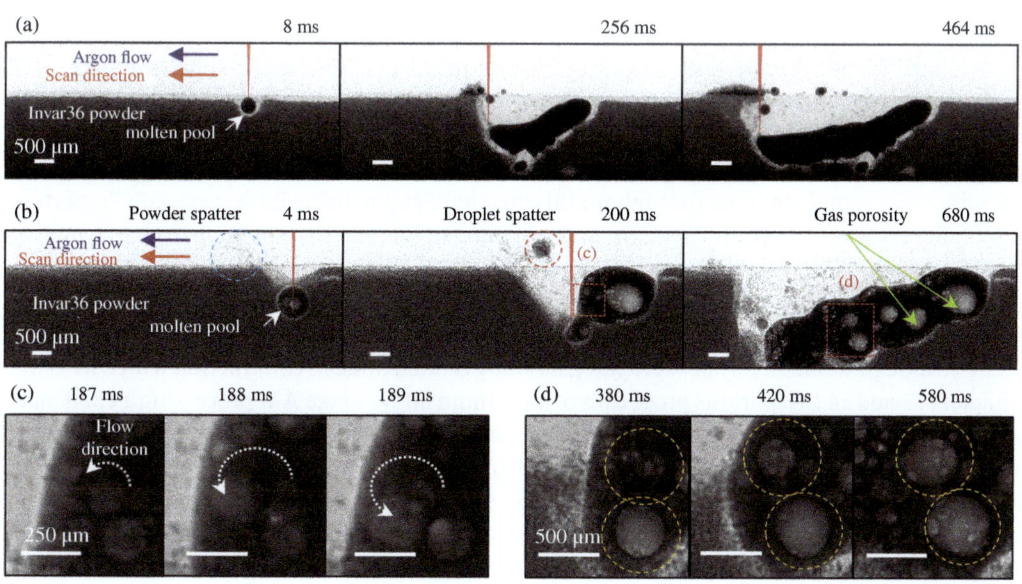

Figure 15.18 Comparison of the time series x-radiographs of melt features for PBF-LB of the first layer Invar 36 melt track for (a) virgin powder and (b) powder oxidized by storage for approximately one year. Two distinct pore evolution mechanisms seen in (b) are shown in more detail as (c) centripetal Marangoni convection (white arrows) and (d) pore growth promoted by oxide film (orange circles). Source: Leung et al. [31]/with permission from ELSEVIER.

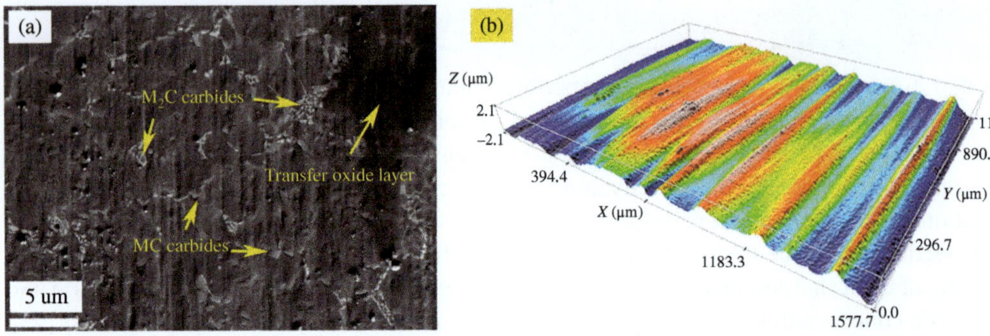

Figure 15.19 (a) The microstructure of an Fe_{bal-x}–C–Cr–Mo–V–W_x alloy is shown along with (b) the surface roughness Ra (μm) profile of the worn surface. This W-containing alloy exhibited superior wear performance due to the formation of a homogeneous oxide layer along with a larger phase fraction of metal carbides in comparison to a similar alloy that did not contain W. Source: Rahman et al. [32]/with permission from Elsevier.

oxide layer on the martensitic matrix, along with a larger phase fraction of metal carbides. Figure 15.19 shows the microstructure of the W-containing alloy along with the surface roughness profile of the worn surface.

15.9 Summary

The physical behavior of AM-produced parts has been reviewed from the perspective that, if AM is to be a disruptive technology and replace conventional methods, the resulting parts must display properties at least comparable to conventionally produced parts in the same way the level of defects and residual stresses covered in Chapters 13 and 14 needed to be comparable. During this review, we found that the various physical properties have been significantly affected by those defects and residual stresses. For example, density and porosity defects are intimately related, corresponding to the goal of achieving a density of 99.5%. Powder characteristics, including the PSD, were seen to play a significant role in optimizing density. The nature of the PSD, along with other processing parameters, plays a central role in the surface roughness of AM parts. In addition, excessive laser power can affect the color of the melt pool, and the color of the final parts can be affected by the atmospheric chemistry during processing. The elastic and plastic behavior of AM-produced parts, as measured by the traditional tensile test can be comparable to conventional counterparts, with differences related to the different bulk and surface structures characteristic of the two methods. HIP and other post-processing treatments are useful for addressing those differences. Hardness which closely correlates with the strength measurements of the tensile test has been used to monitor the quality of HEA coatings and the role of the PSD in producing steel parts by AM. Producing fracture and fatigue behavior comparable to conventionally produced products was raised as a significant goal throughout this chapter. Fatigue behavior is especially important in the dynamic loading conditions in aerospace applications, and a significant literature is emerging in characterizing fracture and fatigue performance in Ti6Al4V, stainless steels, and Inconel 718 alloys.

Alloys widely produced by AM methods are often selected for engineering applications due to their corrosion resistance. As a result, comparing the corrosion resistance of AM-produced parts against conventionally produced counterparts is important. Although wear is a physical property, its study is often done in conjunction with corrosion. It has been noted in Chapters 13 and 14 that various properties of aluminum alloys make AM processing a challenge; so, PBF-LB and DED technology are generally focused on a limited range of AlSi compositions such as Al-12Si that can be produced with both high corrosion resistance and good mechanical performance. The corrosion and wear of an Al-12Si alloy produced by PFB-LB were found to be comparable to a conventionally cast alloy but post-processing heat treatments routinely used for AM parts must be carried out with care to avoid excessive corrosion. In this system, the oxidation of the Si can eventually lead to a stable, corrosion-resistant SiO_2 phase at the surface. The corrosion and wear of various Ti and CoCrMo alloys produced by AM have been shown to be comparable to conventionally produced parts and, in some cases, can be formed in configurations not possible by conventional routes. An additional feature of laser deposition technology is the ability to produce coatings and claddings that can substantially improve the corrosion and wear performance of substrate materials. For example, a Ti6Al4V substrate coated with various composites of Ti6Al4V and Si and HA exhibits good corrosion and wear performance that could allow its use as the femoral head in a potential hip prosthesis design to avoid the galvanic corrosion failure between the Ti6Al4V femoral stem and the CoCrMo femoral head in traditional designs. Also, the desirability of high-density parts for maximizing mechanical performance is also important for corrosion and oxidation resistance, and oxygen concentrations often must be minimized for powder storage and during laser deposition to avoid excessive oxidation of the final parts. On the other hand, surface oxidation of some alloys can play a positive role in wear performance.

Nomenclature

(AM)	additive manufacturing
(S_a)	arithmetic mean height
(S)	breaking stress
(CPR)	ceramic particle-reinforced metal matrix
(COF)	coefficient of friction
(CTE)	coefficient of thermal expansion
(CCA)	complex concentrated alloy
(DED)	directed energy deposition
(FPS)	fully printed specimen
(HEA)	high entropy alloy
(HIP)	hot isostatic pressing
(LENS™)	laser engineered net shaping
($\lvert z(x, y)\rvert$)	matrix of surface heights
(S_v)	maximum pit height,
(MMC)	metal matrix composite
(N_f)	number of cycles to failure
(OCP)	open circuit potential
(PSD)	particle size distribution
(PBF)	powder bed fusion
(PBF-LB)	powder bed fusion using a laser beam
(PBF-EB)	powder bed fusion using an electron beam
(PM)	powder metallurgy
(RS)	repaired specimen
(SL)	structured light scanning
(SS)	substrate specimen
(A)	surface area
(THR)	total hip replacement
(HV)	Vickers hardness values
(WAAM)	wire arc additive manufacturing
(CT)	X-ray computed tomography
(μ-CT)	X-ray micro-computed tomography

References

1 Svetlisky, D., Das, M., Zheng, B. et al. (2021). *Mater. Today* 49: 271. https://doi.org/10.1016/j.mattod.2021.03.020.

2 Herzog, D., Seyda, V., Wycisk, E. et al. (2016). *Acta Mater.* 117: 371. http://dx.doi.org/10.1016/j.actamat.2016.07.019.

3 Blakey-Milner, B., Gradl, P., Snedden, G. et al. (2021). *Mater. Des.* 209: 110008. https://doi.org/10.1016/j.matdes.2021.110008.

4 Vock, S., Kloden, B., Kirchner, A. et al. (2019). *Prog. Addit. Manuf.* 4: 383. https://doi.org/10.1007/s40964-019-00078-6.

5 Baitimerov, R., Lykov, P., Zherebtsav, D. et al. (2018). *Materials* 11: 742. https://doi.org/10.3390/ma11050742.

6 Gockel, J., Sheridan, L., Koerper, B. et al. (2019). *Int. J. Fatigue* 124: 380. https://doi.org/10.1016/j.ijfatigue.2019.03.025.

7 ISO 25178-2 (2012). *Geometrical Product Specifications (GPS) – Surface Texture: Areal – Part 2: Terms, Definitions, and Surface Texture Parameters*. Geneva, Switzerland: International Organization for Standardization (ISO).

8 Gorsse, S., Hutchinson, C., Goune, M. et al. (2017). *Sci. Technol. Adv. Mater.* 18: 584. https://doi.org/10.1080/14686996.2017.1361305.

9 Shackelford, J.F. (2021). *Introduction to Materials Science for Engineers*, 9e, 565. Hoboken, NJ: Pearson.

10 du Plessis, A. and Macdonald, E. (2020). *Addit. Manuf.* 34: 101191. https://doi.org/10.1016/j.addma.2020.101191.

11 Balit, Y., Guevenoux, C., Tanguy, A. et al. (2020). *Mater. Lett.* 270: 127632. https://doi.org/10.1016/j.matlet.2020.127632.

12 Masuo, H., Tanaka, Y., Morokoshi, S. et al. (2018). *Int. J. Fatigue* 117: 163. https://doi.org/10.1016/j.ijfatigue.2018.07.020.

13 Rafi, H.K., Karthik, N.V., Gong, H. et al. (2013). *J. Mater. Eng. Perform.* 22: 3872. https://doi.org/10.1007/s11665-013-0658-0.

14 (2010). *Metallic Materials Properties Development and Standardization Handbook, MMPDS-05*. Columbus, OH: Battelle Memorial Institute.

15 Sterling, A.J., Torries, B., Shamsaei, N. et al. (2016). *Mater. Sci. Eng. A* 665: 100. http://dx.doi.org/10.1016/j.msea.2015.12.026.

16 Rajavi, S.M.J. and Berto, F. (2019). *Adv. Eng. Mater.* 21: 1900220. https://doi.org/10.1002/adem.201900220.

17 Gordon, J., Hochhalter, J., Haden, C. et al. (2019). *Mater. Des.* 168: 107630. https://doi.org/10.1016/j.matdes.2019.107630.

18 Nezhadfar, P.D., Johnson, A.S., and Shamsaei, N. (2020). *Int. J. Fatigue* 136: 105598. https://doi.org/10.1016/j.ijfatigue.2020.105598.

19 Yu, X., Lin, X., Liu, F. et al. (2020). *Mater. Sci. Eng. A* 798: 140092. https://doi.org/10.1016/j.msea.2020.140092.

20 Zhang, J., Song, B., Wei, Q. et al. (2019). *J. Mater. Sci. Technol.* 35: 270. https://doi.org/10.1016/j.jmst.2018.09.004.

21 Ding, Y., Muniz-Lerma, J.A., Trask, M. et al. (2016). *MRS Bull.* 41: 745. https://doi.org/10.1557/mrs.2016.214.

22 Olakanmi, E.O. (2013). *J. Mater. Process. Technol.* 213: 1387. https://doi.org/10.1016/j.jmatprotec.2013.03.009.

23 Ye, H. (2003). *J. Mater. Eng. Perform.* 12: 288. https://doi.org/10.1361/105994903770343132.

24 Prashanth, K.G., Debalina, B., Wang, Z. et al. (2014). *J. Mater. Res.* 29: 2044. https://doi.org/10.1557/jmr.2014.133.

25 Svetlizky, D., Zheng, B., Vyatskikh, A. et al. (2022). *Mater. Sci. Eng. A* 840: 142967. https://doi.org/10.1016/j.msea.2022.142967.

26 Balla, V.K., Das, M., Mohammad, A. et al. (2016). *Adv. Eng. Mater.* 18: 1208. https://doi.org/10.1002/adem.201500588.

27 Mantrala, K.M., Das, M., Balla, V.K. et al. (2015). *Front. Mech. Eng.* 1: 1. https://doi.org/10.3389/fmech.2015.00002.

28 Mantrala, K.M., Das, M., Balla, V.K. et al. (2014). *J. Mater. Res.* 29: 2021. https://doi.org/10.1557/jmr.2014.163.

29 Avila, J.D., Alrawahi, Z., Bose, S. et al. (2020). *Addit. Manuf.* 34: 101241. https://doi.org/10.1016/j.addma.2020.101241.

30 Chen, W., Liu, B., Chen, L. et al. (2020). *Metals* 10: 785. https://doi.org/10.3390/met10060785.

31 Leung, C.L.A., Marussi, S., Towrie, M. et al. (2019). *Acta Mater.* 166: 294. https://doi.org/10.1016/j.actamat.2018.12.027.

32 Rahman, N.U., Capuano, L., Cabeza, S. et al. (2019). *Addit. Manuf.* 30: 100838. https://doi.org/10.1016/j.addma.2019.100838.

16

Economic and Sustainability Assessments of Powder Production and Additive Manufacturing

16.1 Introduction

Metal additive manufacturing (AM) technologies are growing at an exponential rate, stepping into industrialization. In addition to technical performance requirements for metal powder and additively manufactured parts, as described elsewhere in this book, other aspects need to be considered including, **resource utilization**, **economic feasibility**, and **sustainability**-related **human health** and **environmental impacts**. These additional attributes are important in implementing and adopting metal AM in more industrial sectors. As a new era of Industry 4.0 is fostered by emerging technologies, marketplace competition calls for an increase in product value and a decrease in production time and costs for metal AM. To justify the large-scale implementation of metal AM systems and the realization of their potential cost and performance efficiencies, economic and sustainability assessments are in demand. Engineers and other decision-makers must be cognizant of the costs, both economic cost and environmental cost, of producing a given component via metal AM in addition to other performance metrics (e.g. part quality, material utilization, and process time). Therefore, the scope of this chapter covers **techno-economic analysis** (**TEA**) and **environmental impact analysis** such as **life cycle assessment** (**LCA**) of metal AM. TEA is an approach for evaluating the economic feasibility of a known technical structure, and economic feasibility refers to a measure of the economic attractiveness of a technology, typically determined by comparing the costs and the benefits of a technology for a certain stakeholder [1].

This chapter is structured as follows. Resource utilization is discussed first in Section 16.2, as the processes used for both powder production and AM require energy, materials, and water. Energy is of particular interest due to the energy intensity of the gas atomization process and of the high-energy sources needed for AM, including **laser beams** (**LB**) and **electron beams** (**EB**). Material flows are also important because of the extensive **waste** that results from both selecting a narrow **particle size distribution** of powder and unmolten or unfused powder particles wasted during AM. As noted previously in Sections 11.4 and 11.6, powder recycling has recently been explored for many AM systems, which is further discussed in this chapter from a resource utilization perspective. Section 16.3 is focused on economics. The cost breakdown commonly used in **economic assessment** of AM is presented, along with the models to estimate individual cost contributors, such as **machine costs**, **material costs**, and **labor costs**. Economic **tradeoffs** of AM compared to conventional materials processing with subtractive machining approaches are discussed. **Supply chain** effects are also addressed. In Section 16.4, sustainability-related attributes and analyses are presented. Potential **human health** and **occupational exposure risks** associated with the metallic elements of interest in the alloy powders described in Chapter 7 are presented. Environmental impact, such as **global warming potential** (**GWP**) and **embodied carbon**, are reviewed based on published LCA studies, for powder production and especially for comparisons of various AM technologies, relative to conventional processing.

16.2 Resource Utilization

Resource utilization in AM refers to the use of materials, energy, time, and other resources throughout the entire AM process, from design to production. Maximizing resource utilization is crucial for improving sustainability, cost effectiveness, and overall efficiency of AM processes. The specific resources can vary depending on the specific AM process and the equipment being employed, as different AM technologies, such as **powder bed fusion** (**PBF**),

Metallic Powders for Additive Manufacturing: Science and Applications, First Edition.
Enrique J. Lavernia, Kaka Ma, Julie M. Schoenung, James F. Shackelford, and Baolong Zheng.

directed energy deposition (DED), and **binder jetting**, have unique requirements and considerations, which are described in detail in Chapter 8. This section focuses on the common scheme of resource utilization among powder-based AM technologies. The two primary resources are materials and energy, regardless of what specific AM processes are used. The materials resources cover a wide range, including **feedstock** powders, **substrates** or **support structures**, gases for machine operation and maintenance, tools, **capital equipment**, and so forth. While the major energy source in various AM technologies is either laser or electron beam, the energy resources required to operate and maintain AM machines is electricity. As powder is the key feedstock for most of the metal AM technologies, the **upstream supply chain** for powder production, i.e. sourcing of raw materials, should also be addressed.

16.2.1 Materials Utilization

Metal powders, as the primary feedstock material for metal AM, are typically produced from various metals and alloy ingots via **atomization**. The upstream supply chain for powder production involves the mining, extraction, and refining of metal and mineral raw materials (e.g. naturally occurring ores). The **circular** material flow in AM, using a turbine blade as an example, is illustrated in Figure 16.1 [2], which also includes energy flow, as further discussed in Section 16.2.2. At each stage of the life cycle of an AM product, there are materials inputs, outputs, waste, and **emissions**. The simplest way to define the materials utilization is the ratio of the mass of useful product output to the mass of the feedstock or raw materials input. Nominally, AM technologies can significantly reduce material waste as they are designed to fuse together exclusively the powder necessary for the final parts. However, the practical material utilization factor, particularly the feedstock powder utilization, is fairly low, as more than 60% of the feedstock powder is left on the sample stage or inside the chamber [3]. For example, the feedstock utilization factor measured in optimized experimental systems was only 12.3% in the study by Lee on laser cladding (a DED-LB machine) [4], and up to 14% in the study by Unocic and DuPont on the **laser-engineered net shaping** (**LENS**™) process, also a DED-LB machine [5].

To improve **material efficiency**, powder can be reused or recycled. Recycling and reusing excess or unused metal powders can significantly reduce material waste. There is an essential difference between the terms **powder recycling** and **powder reuse**. Recycling relates to certain treatments, such as remelting or grinding, of powder or scrap materials that can be in-process recovered metal or post-consumer scrap, in order to reform them into new atomization feedstock. In contrast, reuse means repetitive use of a single (initially as-received and then reused) powder batch during multiple cycles in an AM machine, until the quality of the powder fails to meet morphology and/or chemical specifications, or other pre-defined limits. Reuse often requires little or limited processing. Unmolten or excess powder can be collected, sieved, and reused in subsequent builds. However, it is worth noting that the quality of the powder may degrade, e.g. an increase in powder particle size or rising oxygen content, after multiple reuse cycles [6–8], leading to a concern over the assumption of a high reusability rate of feedstock powders.

As discussed in Chapters 8 and 11, the dominant powder production method for metal AM is atomization, including gas atomization, centrifugal atomization, water atomization, and plasma atomization. Atomization accounts for a large fraction of metal powder manufacturing (over 60% of all metal powders produced in North America in 2015), as it provides good control over the production rate and desired physical characteristics of the resulting powders when compared to other powder production methods such as mechanical or chemical methods [7]. Different feedstock materials are utilized in different atomization methods. For instance, pre-alloyed wires are used as the raw materials in plasma atomization, while elemental or pre-alloyed ingots or billets are used as the feedstock for the molten materials in gas and water atomization. Further upstream sourcing of the ingots or billets is the natural resource of metals and minerals, i.e. ores, which are extracted from our Earth through mining and then processed via casting, as shown in Figure 16.1. Therefore, when the topic of materials utilization for metal AM is discussed in the context of a life cycle, the material efficiency of powder production cannot be overlooked.

In addition to metal powders, there are other **consumables**, referring to all the non-print materials and non-reusable components required by the AM process. Inert gas is a common example of consumables for most metal AM processes. An inert gas atmosphere, typically argon or nitrogen, is often used to purge the build chamber to prevent oxidation of the metal during the printing process. It helps maintain a controlled environment. In DED-type AM equipment, inert gas is also used as the carrier to deliver the feedstock powder from the powder feeder to the nozzle and then to inject the powder from the nozzle to the molten pool, thus building the parts. Forming gas, a mixture of hydrogen (mole fraction varies) and nitrogen, is used to regenerate the oxygen getter of a build

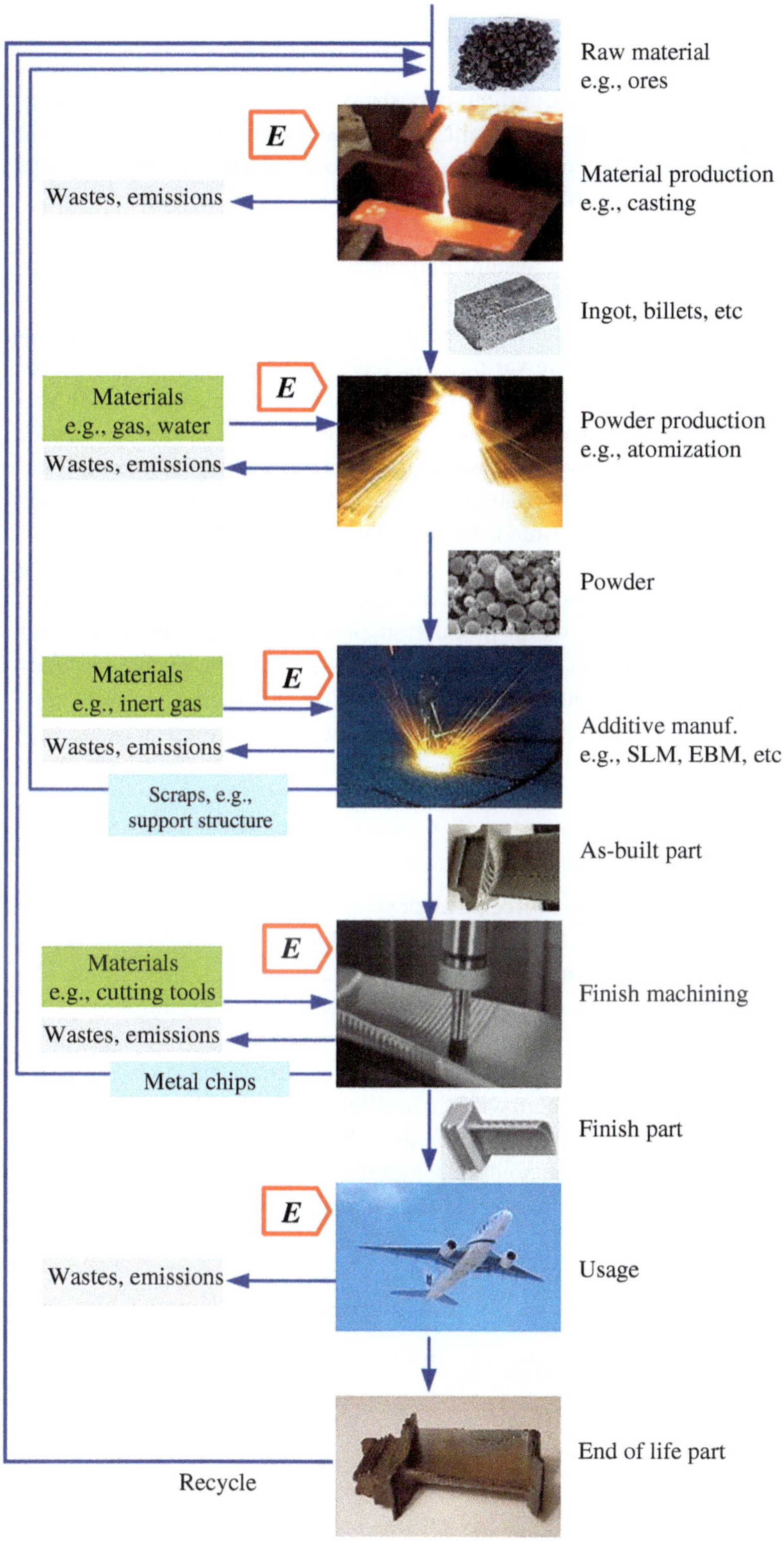

Figure 16.1 Circular material flow and energy flow in additive manufacturing. *E* represents energy inputs. Source: Liu et al. [2]/ Reproduced with permission from ELSEVIER.

chamber in PBF processes or catalysts in glove boxes that are often used in DED machines. Metal binder jetting utilizes a large quantity of organic binders as consumables to assist the printing of the parts. Another category of indispensable consumable in metal AM processes is sacrificial metal plates on which the parts are built, usually referred to as substrates or support structures. Some support structures are required to hold up overhanging features and prevent distortion during the AM process. These substrates and support structures are typically made from the same metallic material (chemical composition) as the part to be built, so as to keep compatibility and avoid thermal expansion mismatch during the layer-by-layer thermal cycling. These substrates and support structures are subsequently removed during post-processing. Other consumables include filters, tubes for powder delivery, optics components (focusing lenses or protective lenses), etc.

The scope of an LCA can be set by the evaluator, such as to a System Boundary A, which might encompass a single phase of the life cycle, or System Boundary B, which might encompass the direct inputs and emissions of the entire life [9]. When the latter is the case, capital equipment is regarded as an essential resource for the product manufacturing phase [9]. The equipment or machines for the various AM technologies can be considered as system-level material utilization. A specialized metal 3D printer is the **capital equipment** required for the AM process. For instance, a LENS™ system (Optomec, Inc.) is a type of capital equipment for the DED-LB AM process, and a Concept Laser M3 liner is a type of capital equipment for **selective laser melting** (**SLM**) or the PBF-LB process. One may argue that since ores and feedstock materials are included in System Boundary B, why not consider the material flows, energy demands, and emissions associated with mining and refining the materials needed to manufacture the equipment? These "infinite recession" questions lead to a vague system boundary, which is difficult to evaluate. Therefore, most studies limit the assessment to materials resources and utilization associated with the primary material flows directly required for the four main life stages – materials production, product manufacturing, use, and disposal. The secondary or indirect material flows required to make the primary ores are frequently excluded [9]. In addition to the AM capital equipment, there are also powder handling and storage equipment, quality control equipment, and post-processing capital equipment, such as cutting and machining equipment, furnaces for heat treatment, etc., which are system-level materials inputs. These additional pieces of equipment are discussed in Section 16.2.3.

16.2.2 Energy Utilization

In PBF or DED metal AM processes, a high-power-density energy source (laser or electron beam) is used to selectively melt and fuse or deposit the metal powder, making these processes energy-intensive. A measure called **cumulative energy demand** (**CED**) is commonly used as a proxy for other environmental impacts, such as greenhouse gas emissions that can lead to climate change. Liu et al. assessed the energy consumption and flow distribution in metal AM at two levels – machine level and process level [2]. The energy is consumed by subsystems of a given AM machine, such as laser or electron beam generator, cooling system, feedstock delivery system, gas circulation system, control system, and so on. Kellens et al. provided an example of energy breakdown by subsystems of an SLM machine [10, 11]. Their studies revealed that the peak power coincided with the time of powder spreading, and the power was slightly lower during the scan when the laser fired continuously. The laser generator (including cooling) consumed 2.24 kW for an output power of 100 W, which is not efficient energy utilization. Table 16.1 shows the power consumption and timeshare at different modes during the SLM process [11].

The energy utilization efficiency can be evaluated as the energy consumed to produce a unit mass of material (MJ kg^{-1}), often referred to as **specific energy consumption** (**SEC**). Morrow et al. reported an extremely high SEC

Table 16.1 Power consumption and timeshare at different modes of SLM process.

Modes	Power (kW)	Timeshare percentage (%)
Preheating and creating inert atmosphere	2.25	12
Laser scanning	3.25	68
Powder spreading	3.45	5
Cooling down and cleaning	0.7	15

Source: Kellens et al. [11].

Table 16.2 Specific energy consumption of additive manufacturing for selected metallic materials via various processes/machines.

Process	Machine	Materials	Energy usage (MJ kg^{-1})	Data sources (references)
DED	Optomec LENS™	Nistelle 625	1052	[14]
	—	H13 tool steel	7708	[12]
	EOSINT M 270	SS 17-4 PH	241–339	[15]
	—	Carbon steel	61.1	[16]
SLM	MTT SLM250	SS 316L	83–106	[15]
	MTT SLM250	SS 316L	111.6–139.5	[17]
	Concept laser	SS 316L	96.8	[18]
	M3 liner		423–588	[15]
	Concept laser Mlab	Aluminum	309–533	[10]
	Renishaw AM250	AlSi10Mg	566.2	[19]
EBM	Arcam AB A1	Ti-6Al-4V	61.2–176.7	[17]
			60	[19]
	Arcam	Ti-6Al-4V	94	[20]

Note: "—" means the machine was not specified in the references.

for the DED-LB process (1000–8000 MJ kg^{-1}), which was due to the low **deposition rate** (0.01 g s^{-1}) used in the lab-scale experiment [12], while a much higher deposition rate is often used in industrial production. In contrast, **electron beam melting (EBM)** has the lowest energy consumption resulting from its high process rate. The SEC for SLM ranges from 80 to 600 MJ kg^{-1}. Since the process/deposition rate significantly affects the SEC of AM, one may conclude that the energy utilization efficiency can be improved by increasing the process/deposition rates [13]. Nevertheless, the deposition or process rate is constrained by the heat transfer mechanisms that are required to deliver or fuse the melt stream to build a part. Increasing the process rate may sacrifice the quality of the build parts [13]. The SEC values also depend on the material being processed. Table 16.2 provides the energy consumption of AM for various metallic materials via various processes and/or machines. Another typical metric used for energy utilization analysis in metal AM is energy per part (MJ/part). For example, Morrow et al. reported that the manufacturing energy use for H13 tool steel made by **direct metal deposition (DMD)** is 3000 MJ/part [12].

At the process level, the energy input for AM is typically quantified by the term "**energy density**" (**ED**), which is defined as a function of the energy source beam power P, beam diameter D, scanning speed v, layer thickness λ, and hatch spacing H, shown in Eq. (16.1) [21].

$$\mathrm{ED} = \frac{P}{f(D, v, H, \lambda)} \tag{16.1}$$

The measure of ED can be quantified as linear, areal, or volumetric, with units of J mm^{-1}, J mm^{-2}, and J mm^{-3}, respectively, using Eqs. (16.2)–(16.6) [2].

Linear [22]:

$$\mathrm{ED}_{\mathrm{linear}} = \frac{P}{v} \tag{16.2}$$

Areal with optimized process parameters using hatch spacing [23]:

$$\mathrm{ED}_{\mathrm{areal}\ H} = \frac{P}{vH} \tag{16.3}$$

Areal actual heat flux of energy beam using beam diameter [24]:

$$\mathrm{ED}_{\mathrm{areal}\ D} = \frac{P}{vD} \tag{16.4}$$

Volumetric with optimized process parameters using hatch spacing [25]:

$$\mathrm{ED}_{\mathrm{vol}\ H} = \frac{P}{vH\lambda} \tag{16.5}$$

Table 16.3 Atomization energy and specific embodied energy for select metallic powders.

Material	Atomization energy (MJ kg^{-1})	Specific embodied energy (MJ kg^{-1})	Data source (references)
Carbon steel powder	9.7	56.1	[16]
Stainless steel powder	3.6–6	78–97	[29, 30]
Tool steel powder	1	15–26	[12]
Al alloy powder	8.1	160–230	[19]
Ni alloy powder	55.6	134–259	[14, 31]
Cu alloy powder	<1	33.0–64.5	[31]
Ti-6Al-4V powder	7–31.7	560	[20, 32, 33]

Volumetric actual heat flux of energy beam using beam diameter [26]:

$$ED_{\mathrm{vol}\,D} = \frac{P}{vD\lambda} \tag{16.6}$$

The energy flow distribution is shown in Figure 16.1. From the standpoint of the AM system, energy flow does not only include the direct energy consumption by the AM equipment but also the indirect embodied energy consumption resulting from material usage. Hence, the energy consumed during powder production needs to be analyzed to fully understand the energy-related environmental impact of metal AM processes. Following a unit process life cycle inventory methodology developed by Duflou et al., the LCA of energy utilization in AM should cover cradle-to-gate including material production, powder production, powder transportation, AM, finish machining, transportation, use phase, and end-of-life disposal (e.g. recycling) [27]. One critical concept for energy consumption analysis is **embodied energy**, defined as the energy required to extract raw materials from natural resources plus the energy consumed in manufacturing activities [28]. **Specific embodied energy** refers to the embodied energy of a material or a product per unit mass, typically having a unit of MJ kg^{-1}. All materials or products have inherent embodied energy. A material that is closer to its natural state at the time of use has a lower embodied energy. Atomization is the dominant approach for metal powder production for AM. All atomization methods use a high-energy source, e.g. an induction coil, a plasma arc, or plasma torches, to melt raw material into a molten stream, which is subsequently struck by a high-velocity jet of gas (e.g. nitrogen, argon, or helium), plasma or water to pulverize it into a cloud of fine particles. The specific embodied energy of feedstock powders for AM depends on the materials and the atomization machine being employed. Table 16.3 provides values for the specific embodied energy of some common metal powders for AM, noting that the data are not an exhaustive list of all the current literature.

16.2.3 Other Resources

Water. In addition to materials and energy, water is the third resource on which metal manufacturing depends [34]. The water demands associated with AM can be measured directly as inputs and outputs. Water is widely utilized for cooling the system and for cleaning typically required during routine operation of various metal AM machines. Water is also used in powder production, particularly being employed as the atomizing medium in water atomization [35]. Water consumption is commonly measured in units of volume (liters, L) of water per kilogram (kg) of material or parts produced, L kg^{-1}, which is equivalent to kg/kg, given that the density of water is approximately 1 kg L^{-1}.

Computational resources. Computer-aided design (CAD) software is used to create digital 3D models of the parts to be produced via AM. This digital model serves as the blueprint for the AM process. Before printing, pre-processing software is often used to optimize the 3D model for the specific metal AM technology being used. This software may involve slicing the model into thin layers and generating toolpaths for the printer. Simulation software is also often utilized to select and optimize the printing process parameters, reducing the need for trial and error, which can waste materials and time, and improving the quality of the final AM products.

Labor/workforce resources. Although AM technologies leverage automation and digital manufacturing, workers remain an indispensable resource to develop and implement AM processes. To utilize labor resources efficiently, education and training, and the relevant resources required for developing a qualified workforce, are in demand and must keep pace with the advances in AM technologies, which require interdisciplinary knowledge and collaborative effort.

Equipment resources other than the AM capital machines. In addition to the various AM machines themselves, namely "the printer," other equipment serves as a vital resource to achieve high-quality AM products and safety of operators or workers in the AM industry. Below are some typical examples of other vital equipment resources.

(a) **Materials Handling Equipment**: Handling equipment is used for loading and unloading metal powder into the 3D printer. It may include powder sieving and recycling equipment to ensure the quality of the powder used in the process.
(b) **Post-processing Equipment**: After the part is 3D printed, post-processing may be required, depending on the application. This can include heat treatment, surface finishing, and machining to achieve the desired mechanical properties and surface quality.
(c) **Quality Control Equipment**: Various inspection and testing facilities are used to ensure the final parts meet the required quality and dimensional standards. This may include non-destructive testing (e.g. X-ray or ultrasound) and dimensional measurement tools. In-process real-time monitoring, such as thermal imagers and high-speed cameras, also contributes to the process optimization and quality control of AM parts.
(d) **Safety Equipment:** Due to the high-energy sources involved and the potential for exposure to metal powders, safety equipment such as personal protective gear (e.g. gloves, safety glasses) and ventilation systems are essential for the safety of operators.
(e) **Materials Storage and Handling**: Proper storage and handling of metal powders are critical to maintaining their quality and preventing contamination. This may involve storage in controlled environments and specialized containers.
(f) **Maintenance Tools**: Like any industrial equipment, metal 3D printers require regular maintenance to ensure they operate correctly. Maintenance tools and spare parts are essential resources for keeping the printers in working condition.

16.3 Economic Assessment

Most conventional **subtractive metal manufacturing** technologies represent a **linear economy**, i.e. raw materials are extracted, transformed into parts or products during which significant wastes (e.g. metal scraps and contaminated machining lubricant or other cleaning fluid) are generated, and finally discarded. In contrast, AM can drive and accelerate a **circular economy** that links material, design, manufacturing, product, and end-of-life in a continuous, sustainable loop [36]. Specifically, AM processes are promising in reducing waste generated during production and saving input materials, thus saving material costs [37]. AM's capability to produce highly complex parts without tools also makes a decrease in production costs possible. Since conventional manufacturing often requires the need to produce a high quantity of parts to distribute the high cost of tooling, parts that are not produced in such high volumes are generally not cost-effective. AM techniques, on the other hand, are well suited for low-volume production. Hence, affordable and high-complexity individual products can be manufactured. Furthermore, reuse and recycling strategies are also increasingly adopted in AM processes, to use recycled materials as inputs for the powder and part manufacturing processes rather than using as-received materials (relating to extracting more natural resources) [38]. AM technologies, particularly DED processes, also promote the **repair** and **refurbishment** processes and extend the product lifecycle by 3D printing the parts required for broken/dysfunctional products onsite. Figure 16.2 displays these unique features of metal AM processes to support circular economy initiatives.

Cost-effectiveness and economic viability need to be assessed to promote wide adoption of different AM technologies in various industries. One major motivation for evaluating AM costs is to compare AM processes to other conventional processes to determine under what circumstances AM is more cost effective. Two common themes and metrics covered in economic assessments of AM are: (i) **total cost of ownership**, which includes not only the initial acquisition cost of the AM capital equipment but also operational costs, maintenance, material costs, and labor costs over the machine's lifespan, and (ii) **total cost per part** produced using AM, which is often compared to conventional manufacturing methods, such as casting and machining, to determine the cost competitiveness of AM for

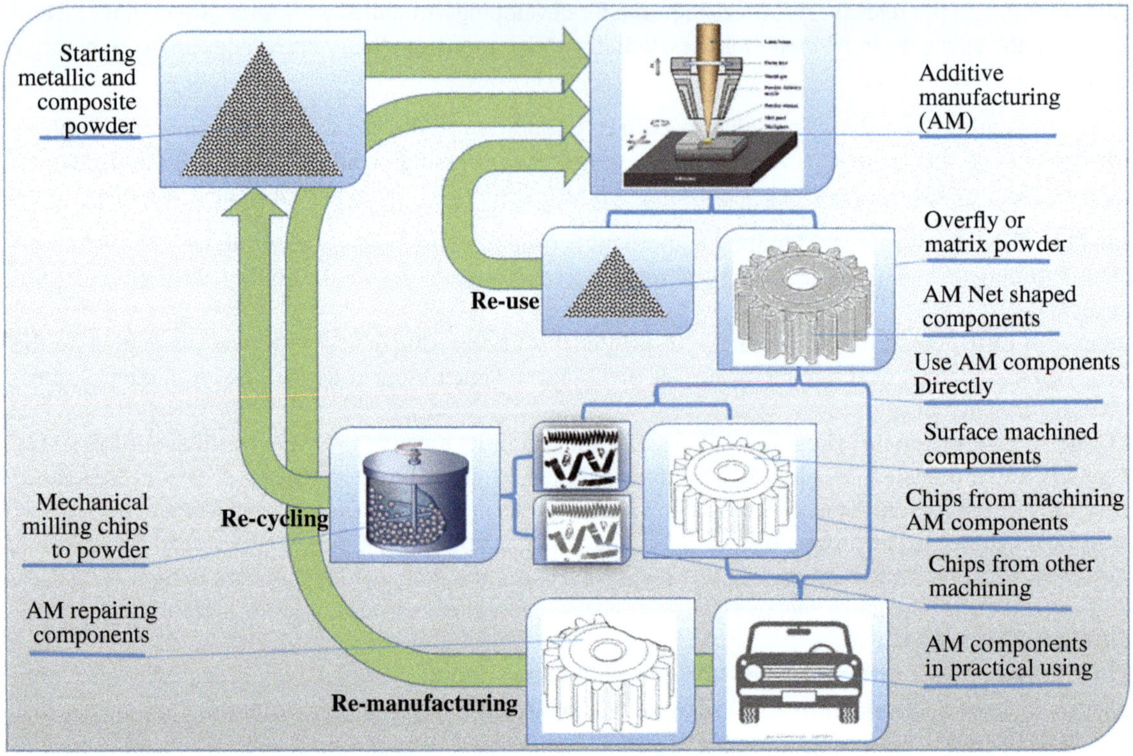

Figure 16.2 Circular economy as facilitated by additive manufacturing techniques.

specific applications. The cost of the product depends on the resources utilized for manufacturing the product. These resources are cost components of the products such as material, machine, labor, and tooling. To estimate the product cost, the cost associated with these components should be estimated. The total cost per part changes as production volumes increase. Jung et al. reported that AM has lower production costs when production volumes are relatively low, or when the parts are small and would have high material waste if conventionally manufactured [39]. For instance, the production volume needed to be below 42 parts per year for AlSi10Mg landing gear to make a PBF-LB process less costly than die casting [40]. Economies of scale may make AM more cost-effective for larger production runs. To understand the economic assessment of AM, methods to evaluate, breakdown, and model costs are described.

16.3.1 Cost Breakdown and Models

According to early work by Son, the costs of production can be categorized as "well-structured" costs and "ill-structured costs" [41]. The former refers to **material costs**, **machine costs**, and **labor costs**, while the latter is related to build failure, machine setup, and inventory. The metric "total cost per part," C_{tot}, is commonly used to compare AM to conventional manufacturing processes to assess its economic feasibility. C_{tot} is typically defined as the sum of material costs per part, machine costs per part, labor costs per part, and post-processing costs per part, as shown in Eq. (16.7) [42].

$$C_{tot} = C_{mc} + C_{mt} + C_{lb} + C_{pp} \tag{16.7}$$

where C_{tot} is the total cost per part, C_{mc} is machine cost, C_{mt} is the material cost, C_{lb} is the labor cost, and C_{pp} is the post-processing costs.

Machine costs are often calculated based on a time-driven model, comprising machine hourly operating costs and some fixed costs accounting for initial gas purging of the machine and labor costs, assuming the processes are constant for each building cycle [43, 44], which can be expressed as Eq. (16.8). The machine's hourly operating cost, C_{mr}, consisting of an indirect cost rate and a direct cost rate, is multiplied by the build time (t_{build}) to obtain the machine operating cost to build a part. Figure 16.3 shows the breakout of machine costs in detail. The capital equipment-relevant costs (e.g. machine purchase, depreciation time, maintenance, etc.) are included in the indirect

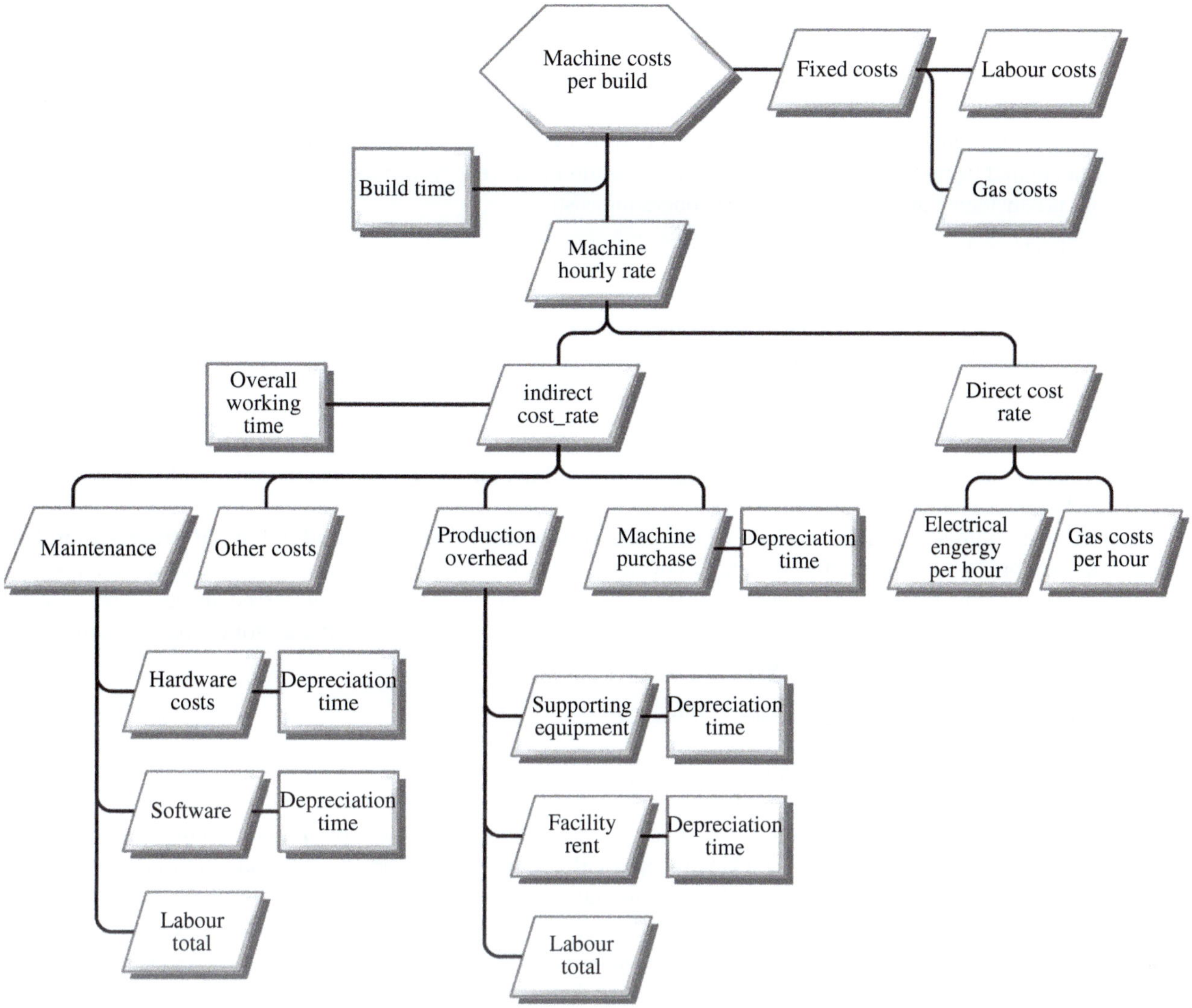

Figure 16.3 Breakdown of machine costs per part in detail. Source: Lindemann et al. [44].

cost rate. Energy costs, e.g. electrical energy consumed to operate the machine, and some material costs (inert gas), are regarded as direct cost rates.

$$C_{\text{mc}} = C_{\text{fixed}} + C_{\text{mr}} \cdot t_{\text{build}} \tag{16.8}$$

The build time is one of the most important factors that needs to be estimated. The build time is dependent on the build rate, the volume and design of the part, and some AM processing parameters such as hatch pattern, layer thickness, dwell time between layers, and so on. Based on the software packages currently available to estimate build time, there are two primary approaches to estimate the build time [45–47]. One approach is to prepare a detailed analysis that uses knowledge about the inner working mechanism of an AM machine, and the other is a parametric analysis that utilizes information on process time and parameters. It is important to note that the build time can be specific to a given AM machine and the material being built. Other factors significantly affecting machine costs include production rate, production volume (number of parts made in a year), and AM machine utilization time (i.e. hours per year when machine is in operation). Although it is counted as an indirect cost in machine costs per part, AM capital equipment is generally considered an expensive investment. Entry-level metal 3D printers cost approximately \$115 000 for a relatively small and basic configuration and can go as high as \$1.9 million for higher-end models that can produce parts as large as a full-scale V6 engine block [48]. It is not easy to obtain and get permission to publish a list of prices for commercial AM machines. Wohlers et al. provided base prices for AM machines in their report in

2017 [48, 49]. With more companies now offering their own systems, the pricing and number of options have become extensive. Market competition has started to force prices downward.

In a more recent study, Mahadik and Masel implemented an AM cost estimation tool using a breakdown approach, in which the estimation of the build time considered the activities undertaken by the machine for preparing a single layer and then multiplying it by the total number of layers [42]. In that work, the machine operation cost per part C_{op} was calculated using Eq. (16.9). Also, machine utilization time (e.g. how many hours per day the machine is in operation) is considered in the machine hourly operating cost:

$$C_{\text{op}} = C_{\text{mr}}(t_s + t_l + t_p + t_m + t_r + t_c) \tag{16.9}$$

where C_{mr} is the machine hourly operating cost as expressed in Eq. (16.10); and the time variables t_s, t_l, t_p, t_m, t_r, and t_c denote machine set up time, material loading time, parameter setting time, part manufacturing time, part removal time and machine cleaning time, respectively:

$$C_{\text{mr}} = \frac{C_{\text{EAC}}}{12 \cdot N_{\text{days}} \cdot E[H]} \tag{16.10}$$

C_{EAC} is the equivalent annual purchase cost, N_{days} is the number of working days in a month, and $E[H]$ is expected operating hours in a day.

Material costs, nominally, should include all the material resources that are utilized in the AM processes, as discussed in Section 16.2.1, such as the feedstock powders, substrates, support materials, etc. However, since some material costs, like the inert gas, have been counted in the machine operation costs discussed above, material costs primarily refer to the feedstock materials for the part and the substrate or support materials for the part. The material cost per part, C_{mt}, is estimated by Eq. (16.11):

$$C_{\text{mt}} = m_{\text{support}} \cdot C_{\text{support}} + \sum_i (C_{0\ \text{mater}\ i}) \cdot (m_{\text{mater}\ i}) \tag{16.11}$$

where m_{support} is the mass of the support material; C_{support} is the unit price of the support material per kilogram; $C_{0\ \text{mater}\ i}$ is the unit price of material i per kilogram, and $m_{\text{mater}\ i}$ is the mass of the material i in a part.

Depending on the natural reserve and the demand-supply relationship, prices of different metals and alloys are subject to fluctuations and **volatility**. The **U.S. Geological Survey** (**USGS**) provides statistics and information on the worldwide supply of, demand for, and flow of minerals and materials essential to the U.S. economy, national security, and protection of the environment [50]. Readers can find commodity prices for metals at the USGS website or other sources, such as the Daily Metal Price webpage [51]. Material criticality and supply risks need to be considered when assessing material costs. The metals used in AM processes vary significantly in cost due to format and quality requirements. Titanium (Ti) can range from \$100/kg for a production volume of thick wire to over \$500/kg for high-quality powder [52]. PBF-LB processes require very fine and highly spherical powders, which are expensive to produce, typically ranging from \$300/kg to \$500/kg for titanium. In contrast, PBF-EB, DED-LB, and binder jetting processes have lower requirements on powder quality, reducing powder costs by about 50%. Due to significant variances in the material price depending on powder characteristics, it is neither easy nor very meaningful to compile a price list for different metals and alloys. Rather, it is more important to understand the factors that affect material prices and how the volatility of materials prices influences the economic feasibility of AM technologies. The material price volatility is closely related to the material criticality. The "Critical Materials Assessment Report" provided by U.S. Department of Energy in 2023 list six metallic elements (Dy, Co, Ga, Ir, Nd, and Tb) as critical materials in the short term (2020–2025) and 11 metals (Li, Ni, Pt, Mg, Pr, plus the 6 elements above) as medium term (2025–2035) critical materials to the continued worldwide deployment of clean energy technologies [53]. Cu, Al, and electrical steel are also listed as near-critical materials in medium term. In other words, these critical metals or alloys are facing a high supply risk. Thus, prices of the raw materials for these critical metals and alloys are anticipated to increase over time, leading to an increase in material costs for metal AM. Besides metal powders, consumables, as described in Section 16.2.1, can also add significantly to the costs, yet they are often overlooked in economic models. Assessments of material costs often account for the quality and yield of parts produced through AM. Higher-quality parts and lower scrap rates can lead to savings in materials costs. Designing parts specifically for AM can also lead to cost reductions by reducing material usage and printing time.

Labor costs need to be considered in each stage of the AM process, e.g. pre-building (preparation for make a build), during the printing (considered as an indirect cost in machine costs), and post processing. Labor is associated with setting up, operating, and maintaining AM machines. Labor costs, C_{lb}, are typically estimated based on time requirements, using Eq. (16.12). Notably, labor hourly rates are highly affected by the location of the manufacturing plant.

$$C_{lb} = \text{labor hourly rate} \cdot \text{labor working time} \tag{16.12}$$

Post-processing costs sum up the costs of all necessary processes after printing, including removal of substrate and support material, surface finishing or polishing, heat treatment, and quality control. Post-processing costs can be estimated by multiplying the average of the processing time by the labor cost rate per hour.

Several studies have pointed out that machine costs are the largest cost driver for the total cost per part made by AM, and materials costs contribute as the second dominant cost driver [40, 44, 54]. Atzeni and Salmi compared the total cost per assembly between **selective laser sintering (SLS)** and high-pressure die-casting in a case study on an airplane landing gear assembly made from an aluminum alloy (AlSi10Mg) [40]. The total cost per assembly is defined as the sum of the single part cost plus the assembly cost for die casting, while for SLS, there was no assembly cost since the entire landing gear was printed as one part. Their findings showed that the total cost per assembly for SLS was €526 (for printing four parts at the same time) while that for high pressure die casting was € (21.29 + 21 000/N), where €21 000 corresponds to the die casting equipment cost, and N is the lot size (i.e. number of parts being cast). Thus, it was concluded in their study that AM could be economically convenient and competitive to conventional processes for small to medium-batch production of metal parts. The breakdown of the costs highlighted that machine cost per part is the largest factor contributing to the total cost per assembly. When comparing the material cost per part, using SLS (€25.81 per part) was nearly ten times more expensive than die casting (€2.59 per part). Based on their results, there is a breakeven point in production volume: when the number of parts is equal to or less than 42, the total cost per assembly for SLS becomes more economically competitive compared to die casting.

The cost assessment conducted by Lindemann et al. concurred that machine costs contributed the most to the costs of producing stainless-steel AM parts [44]. Figure 16.4 presents the cost distribution for stainless-steel AM parts as a function of four varying factors – build rates, machine utilization time, material costs, and machine investment costs (relevant to purchasing, housing, using, and maintaining the machine. The build rates were measured in units of cubic centimeters per hour ($cm^3\ h^{-1}$). Machine utilization was measured as the number of hours per year that the machine is operated. The production volume is a little less than 200 parts for the base case. Variations in key factors show that the machine costs per part can be reduced by increasing the build rate, increasing the machine utilization fraction, or reducing the machine investment costs. But no matter how these factors were modified, machine costs

	Building rate in ccm h^{-1}: 6.3	12	20	Machine utilization h yr^{-1}: 4500	6000	7800	Material costs (€/kg): 40	89	600	Machine investment costs (€ thousands): 200	320	500
Building process fix	2%	2%	3%	2%	2%	2%	2%	2%	1%	2%	2%	2%
Preparation costs	4%	7%	9%	4%	5%	6%	5%	4%	3%	6%	5%	4%
Oven	3%	4%	5%	3%	3%	4%	3%	3%	2%	4%	3%	3%
Material costs	11%	17%	23%	11%	14%	16%	5%	11%	46%	16%	14%	11%
Postprocessing costs	7%	10%	13%	7%	8%	9%	7%	7%	4%	9%	8%	7%
Machine costs	74%	59%	47%	74%	69%	64%	78%	74%	45%	63%	68%	74%

Figure 16.4 Cost distribution of additive manufacturing of metal parts by four varying factors. Source: Lindemann et al. [44].

remained one of the dominant cost drivers in the AM production process. These studies also reveal that material costs constitute the second largest proportion of the total cost of an AM product. The high material costs primarily result from the high cost associated with metal powder production, particularly the high energy consumption during atomization.

To sum up, machine costs and material costs constitute a primary proportion of the cost of metallic parts made by AM. As technologies can often be complementary, where two technologies, e.g. powder production and metal AM in the context of this book, are adopted alongside each other, the benefits can be greater than the situation when they were adopted individually. Increasing adoption of AM may reduce the cost of raw materials through economies of scale due to promoting recycling, reuse, and remanufacturing (repair), as displayed in Figure 16.2, instead of producing completely new parts using as-received materials. In turn, the reduced cost of raw materials may further expand the adoption of AM. There may also be economies of scale in raw material costs if particular materials become more commonly used rather than expanding into a wide range of different materials. It was reported that AM technologies represented 0.02% of the relevant manufacturing industries in the United States in 2016 [55]. As the machine costs and material costs are potentially declining over time, the adoption of AM will likely increase and consequently change the supply chain dynamics, i.e. the distribution of the suppliers, manufacturers, and consumers, and the interactions among them. Previous studies were limited in their scope to examine the costs of producing single parts or assembly of several parts produced by metal AM, due to data limitations and the complexity of measuring AM costs. Several aspects of AM that are not readily apparent when comparing final products are the roadblocks in determining whether AM is an environmentally and economically preferred alternative to conventional manufacturing. Those hidden factors include energy-intensive and expensive material feedstocks, excessive material waste during production, high machine costs, and slow rates of production. While the cost assessment models for the well-structured costs in AM (i.e. machine, materials, labor, and post-processing costs) are relatively mature, the supply chain effects, particularly those associated with the ill-structured costs, such as build failure, inventory, and transportation costs, along with AM's potentially decreased risk to supply disruption, have not been fully considered in the economic assessment of metal AM. Therefore, Section 16.3.2 addresses the supply chain effects.

16.3.2 Supply Chain Effects

A supply chain is a system through which products flow from supplier to customer. Many ill-structured costs are hidden in the supply chain of AM products. In turn, development and implementation of AM technologies have significant impacts on the design and size of the supply chain system, potentially reducing its associated costs [56, 57]. Figure 16.5 displays the supply chain for AM systems and the relevant costs and revenues in 2011 [58]. The supply chain for AM consists primarily of four constructs, namely AM system vendor, customers, material suppliers, and location of manufacture. The AM system vendor is a crucial stakeholder in the AM value chain, which is limited to a few other stakeholders, such as material and software suppliers, distributors, and resellers [59]. Approximately 62.8% of all commercial/industrial units sold in 2011 were made by the top three producers of additive manufacturing systems: Stratasys, Z Corporation, and 3D Systems based out of the United States [58]. Approximately 64.4% of all systems were made by companies based in the United States. The total global revenue from system sales was $502.5 million, with U.S. revenue estimated at $323.6 million in 2011, as seen in Figure 16.5. The global revenue of AM was estimated to be approximately $967 million in 2013, out of which the U.S. contributed approximately $367 million.

To elucidate the supply chain effects, the **inventory costs** in manufacturing must be considered. Inventory of products such as spare parts require resources (physical space, buildings, and land) to be consumed to produce and store those parts. Utilization of these resources leads to rent, utility costs, insurance, and taxes. Suppliers often bear these high inventory and distribution costs. In conventional manufacturing, it is too expensive and time-consuming to produce a specific type of product on demand when it is ordered, as idle machinery and equipment waiting for parts orders is too costly. Therefore, it is a more common practice that spare parts are produced in advance and stored while waiting for consumers' orders, which results in high inventory costs for infrequently ordered parts [57]. In addition, the stored products tend to deteriorate and become obsolete, leading to wasted inventory costs. In contrast, AM's capability to manufacture such parts on demand can reduce the need to maintain a large inventory and thus reduce the associated costs.

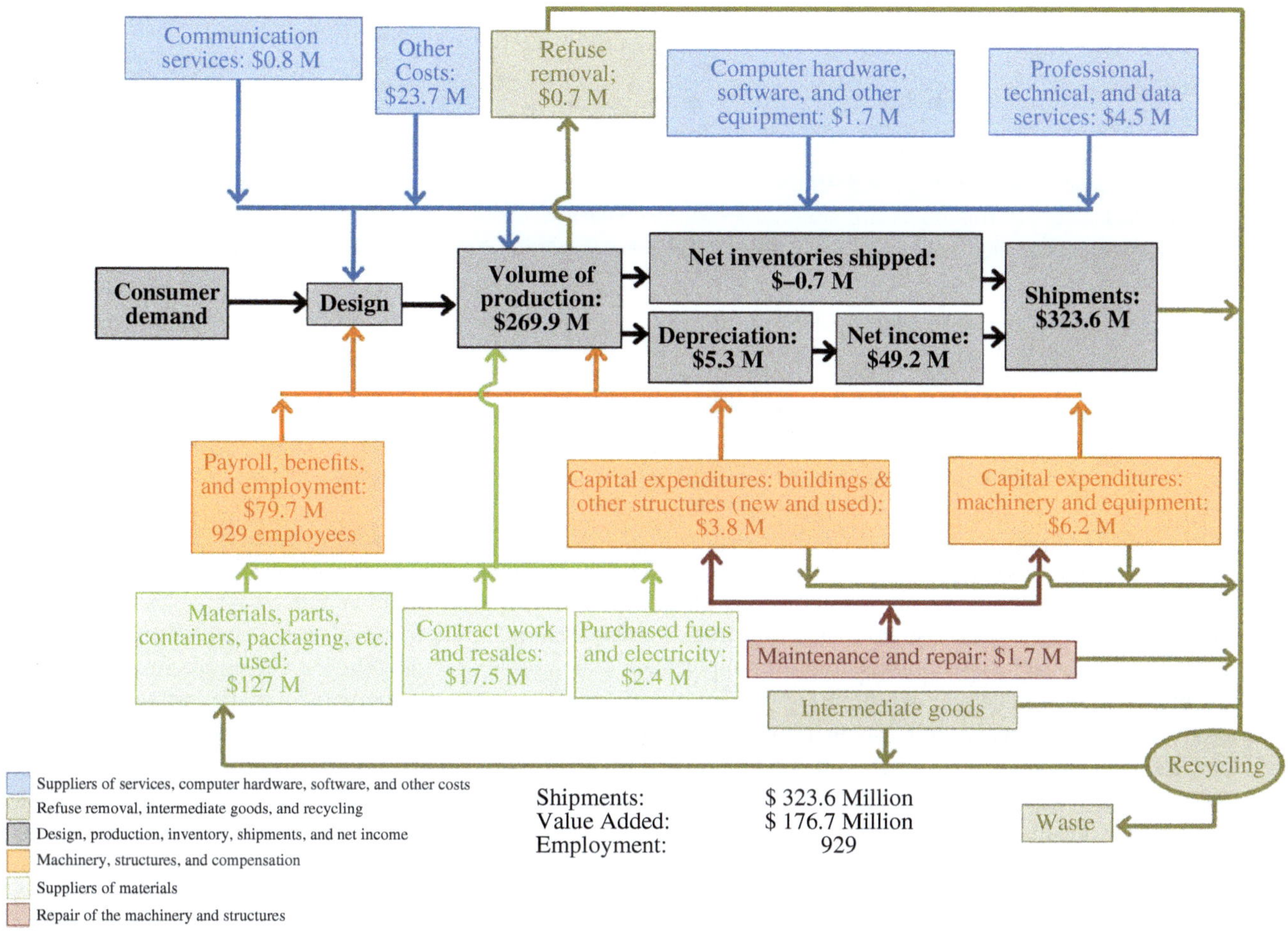

Figure 16.5 Supply chain and the relevant costs/revenues for additive manufacturing systems. Source: Wholers et al. [58].

Transportation cost is another ill-structured cost component. Conventional manufacturing commonly involves the production of multiple parts at multiple locations, and an inventory of each type of part is stored in those different locations. When an assembly product is ordered, those parts need to be shipped to an assembly facility from various locations, leading to high transportation costs. As AM processes enable the production of multiple parts simultaneously to obtain the entire assembled product in one build, it can eliminate transportation costs caused by producing parts at varying locations [60]. Since AM can produce a final product in one build, the exposure to hazardous conditions is limited, and little hazardous waste is generated [61]. Thus, it is potentially possible to bring production closer to the consumer for some products, i.e. distributed manufacturing or localized production, as opposed to centralized manufacturing. For instance, when using the conventional manufacturing supply chain, a remote geographic area may order automotive parts on demand, which may take multiple days to be delivered. In contrast, AM may enable the production of some of these parts or products near the point of use or even onsite [57].

AM disrupts the conventional manufacturing supply chain, allowing for local sourcing and shortening the distance between manufacturer and customer while involving fewer stakeholders to bring a product to the market [62]. Figure 16.6 displays the difference in supply chain flow between conventional manufacturing and AM: centralized manufacturing vs. localized production. Every manufacturing factory, transportation company, and distribution retailer in the supply chain for a product is a potential point where a disaster can stop, or a disruption can hinder the production and delivery of a product. A shorter supply chain that has fewer links gives rise to fewer points for potential disruption or disasters. AM can potentially reduce the number of links in the supply chain and bring production closer to consumers. As a result, AM reduces the vulnerability of supply chains to disasters and disruptions. In addition, when production is located closer to consumers, it gives rise to more decentralized production where many facilities are producing a few products rather than a few facilities producing a large number of products. Hence, when

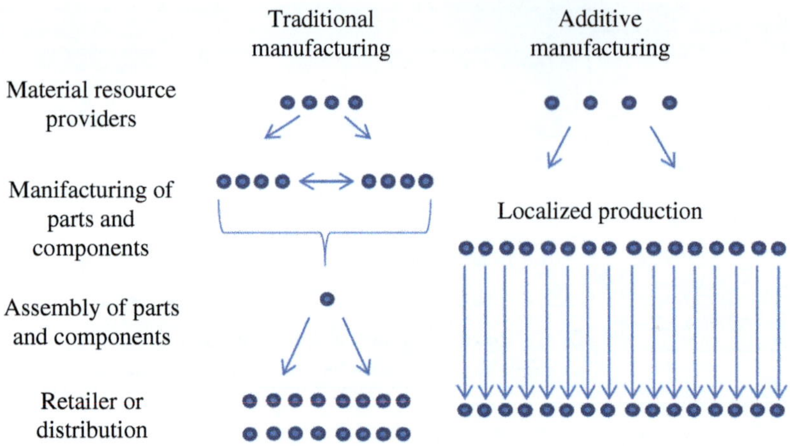

Figure 16.6 Schematic diagram comparing the supply chain for traditional manufacturing with that for additive manufacturing with localized production. Source: Thomas et al. [54].

disruptions occur in the supply chain, the impact would be localized or small in scale, instead of regional, national, or even international impacts. Moreover, localized production combined with simplified processes enabled by AM may begin to blur the boundaries between manufacturers, wholesalers, and retailers as each may potentially manufacture products in their facilities. Consequently, the supply chain dynamics is altered by the increasing adoption of AM.

Khajavi et al. estimated the operating cost of centralized AM production and compared it to the case of distributed production, where production was in proximity to the consumer [63]. Their findings suggest that the strategy of distributed production through AM reduced inventory obsolescence cost, initial inventory production costs, inventory carrying costs, and spare parts transportation costs. Nevertheless, the localized distributed production required significant increases in labor costs and the initial investment in AM capital equipment and associated machines, which made the total cost of distributed production higher than centralized production. To make the localized production more cost effective, it is necessary to increase automation (so as to reduce labor costs) and reduce machine costs. Although centralized production of spare parts is still the most common approach, distributed production will become more viable if AM processes become widely adopted, particularly for products that have complex designs.

To take into account those supply chain effects discussed above, Douglas proposed a cost model for AM production from a supply chain perspective, which included the potential reduction in inventory and transportation costs by distributed/localized production promoted by AM [55]. The model is represented as Eq. (16.13):

$$\begin{aligned}C_{\mathrm{AM}} &= (\mathrm{MI}_{\mathrm{R,AM}} + \mathrm{MI}_{\mathrm{M,AM}}) + (\mathrm{P}_{\mathrm{E,AM}} + P_{\mathrm{R,AM}} + P_{\mathrm{M,AM}})\\ &\quad + (\mathrm{FGI}_{\mathrm{E,AM}} + \mathrm{FGI}_{\mathrm{R,AM}} + \mathrm{FGI}_{\mathrm{M,AM}}) + \mathrm{WT}_{\mathrm{AM}} + \mathrm{RT}_{\mathrm{AM}} + T_{\mathrm{AM}}\end{aligned} \tag{16.13}$$

where C_{AM} is the total cost of producing an AM product; MI is the cost of material inventory, P stands for the processing costs, FGI corresponds to the inventory cost for finished goods, where the subscripts E, R and M refer to extraction of raw materials, refining of raw materials, and manufacturing, respectively. $\mathrm{WT}_{\mathrm{AM}}$ is the cost of wholesale trade for the AM product; $\mathrm{RT}_{\mathrm{AM}}$ is the cost of retail trade for the AM product; and T_{AM} is transportation cost throughout the supply chain for an AM product.

It should be noted that Eq. (16.13) and the discussion on the supply chain effect of economic assessment above is not limited to metal AM. Rather, it is applicable to any general AM processes. While the adoption and implementation of AM have expanded significantly in recent years, the total cost per part or per assembly are still higher for AM products relative to the counterparts made by conventional manufacturing. Up to this date, AM is more cost-effective when the production volume is small and/or when the products involve complex designs and/or complicated assembly of multiple components. From the standpoint of an entire supply chain, there is a tradeoff between flexibility and controllability. As the prices for materials and AM machines go down, the controllability of AM technologies can be increased, making them more attractive to various industrial sectors. At the basic level, improving economic feasibility of AM requires new knowledge, new designs, and enhanced educational and training effort of personnel.

16.4 Sustainability Assessments

Evaluating the sustainability of materials and processes is complex. Numerous metrics can be used to assess and compare alternatives on the basis of sustainability-related issues. These are often aggregated into three broad topics: **safer chemicals**, energy efficiency, and circular economy [64], as depicted in Figure 16.7.

The concept of safer chemicals focuses on assessing and evaluating the **intrinsic hazard** associated with the use, emission, and disposal of chemicals. Hazards include **physical hazards** such as **flammability** and **corrosivity**, **human health hazards** such as **carcinogenicity**, and **environmental hazards** such as water **toxicity** [65]. Concerns include **exposure** to workers (i.e. occupational) and to the general population (i.e. public health). In some circles, the terminology of "**chemical footprint**" is used to represent the impact of chemicals used in industry and by consumers.

The concept of energy efficiency entails the embodied energy in materials themselves due to synthesis and processing steps (e.g. mineral extraction and refining and chemical synthesis), and the energy associated with converting the materials into usable goods (e.g. energy for melting, casting, sintering, machining). Energy creation requires either **nonrenewable** or **renewable** sources, such as fossil fuel and coal-based energy plants, and solar and wind energy, respectively. Production of energy from any of these sources requires resources, and emissions into the air, and sometimes water or soil, are well-known by-products. Carbon dioxide is a common emission; thus, this topical area is often referred to as evaluating the "**carbon footprint**."

The concept of a circular economy is focused on minimizing waste, especially at **end-of-life**, and promoting reuse, remanufacturing, recycling, and recovery. This is of particular importance especially for critical, strategic, and rare materials, as well as for materials that have high embodied energy, which require extensive energy inputs to extract, refine and process into usable material feedstocks. In the context of metal powder production and AM, these concepts require that we address the intrinsic hazards of metals, and especially metallic powders, both to the general public, but especially for workers who handle them as part of their occupations. We must also consider the energy needed to melt the metal and produce gas atomized powder and the application of high-energy sources such as lasers and EB needed to create a melt pool during AM. The benefit of three-dimensional printing to avoid the need for machining can potentially provide an energy benefit, as well as a benefit in terms of waste reduction. Yet, AM techniques also generate waste powder, as noted elsewhere in this chapter and in this book, so maximizing the circularity of material needs to be further evaluated. In this Section 16.4, data on intrinsic hazard and potential occupational exposure

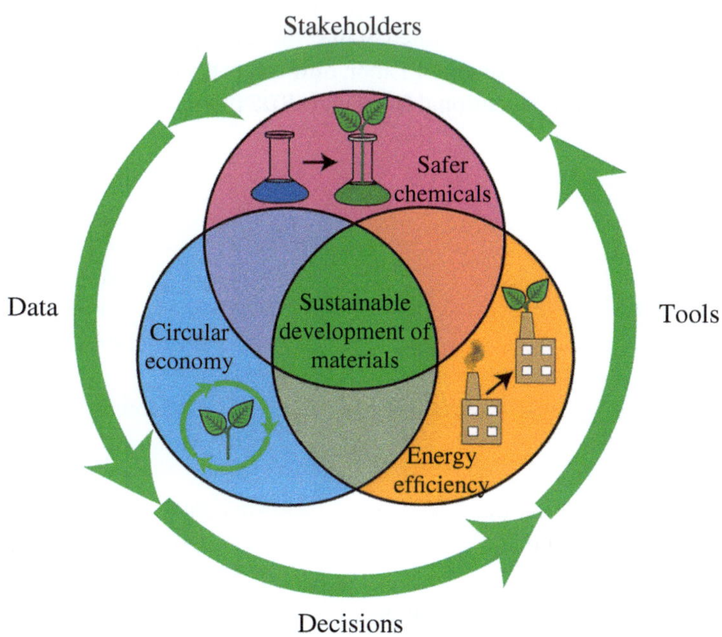

Figure 16.7 Schematic diagram presenting the concept of "sustainable development of materials." Source: Schoenung and Olivetti [64].

are described, followed by a review of recent literature on energy tradeoffs and associated environmental impact, as evaluated through the application of LCA.

16.4.1 Hazard Traits of Metals and Occupational Exposure Potential

In this section, data on the intrinsic **hazard traits** of metals and metallic powders are presented, followed by a summary of recent studies in the literature on **occupational exposure potential** during AM of metals.

The most commonly cited chemical hazard is the potential to cause cancer in humans, i.e. carcinogenicity. The most authoritative source of data on carcinogenicity is the **International Center for Research on Cancer (IARC)**, operated by the **World Health Organization (WHO)** [66]. The objective of the IARC is to promote international collaboration in cancer research, emphasizing interdisciplinary research that promotes cancer prevention [67]. One aspect of their efforts includes the publication of monographs that provide extensive data on cancer research and ultimately lead to the classification of substances that have been studied relative to their cancer-causing potential [68]. This list classifies substances into the following "groups":

- Group 1: Carcinogenic to humans (includes 127 substances)
- Group 2A: Probably carcinogenic to humans (includes 95 substances)
- Group 2B: Possibly carcinogenic to humans (includes 323 substances)
- Group 3: Not classifiable as to its carcinogenicity to humans (includes 500 substances)

While this list of ~1000 substances is very small in comparison to the hundreds of thousands of chemicals used in commerce today, it is a strategic list of chemicals that should be screened and closely evaluated prior to being selected for use within a product. For the purposes of this book, the metallic elements, identified in Chapter 7 as either primary or secondary alloying elements, have been tabulated in Table 16.4 together with their IARC classification, if any. Notably, nickel (Ni) and chromium (Cr) have been identified as Group 1 substances, which are known to be human carcinogens. Cobalt (Co) is in Group 2B, as are the oxides of titanium (Ti) and vanadium (V).

Beyond carcinogenicity, additional intrinsic hazard traits are frequently reported. In an effort to harmonize the classification of chemicals relative to such hazard traits, in the year 2003, the United Nations adopted the **Globally Harmonized System (GHS)** of Classification and Labelling of Chemicals [71]. The most recent version (10th edition) of the GHS guidance was published in 2023 by the United Nations Economic Commission for Europe [72]. While a driving force for establishing the GHS guidance was to harmonize hazard labeling around the globe (thereby avoiding the need for extensive language translation), the program has also resulted in harmonized classification of chemicals relative to nearly 30 intrinsic hazards, as detailed below. In the United States, the **Occupational Safety and Health Administration (OSHA)**, which is part of the U.S. Department of Labor, now refers to GHS data as the basis for **Safety Data Sheets (SDS)**, which have now replaced the conventional **Materials SDS** provided by material suppliers [73]. The GHS hazard traits include:

A. Physical Hazards

(1) Explosives
(2) Flammable gases
(3) Aerosols
(4) Oxidizing gases
(5) Gases under pressure
(6) Flammable liquids
(7) Flammable solids
(8) Self-reactive substances and mixtures
(9) Pyrophoric liquids
(10) Pyrophoric solids
(11) Self-heating substances and mixtures
(12) Substances and mixtures which, in contact with water, emit flammable gases
(13) Oxidizing liquids
(14) Oxidizing solids
(15) Organic peroxides

Table 16.4 Metallic alloying elements and their hazard attributes.

	CASRN[a)]	Carcinogenicity classification	NIOSH recommended exposure limit (REL)		Included in GHS-Japan
			Total (mg m^{-3})	Notes	
Source:		IARC [68]	NIOSH pocket guide [69]		NITE [70]
Primary alloying elements					
Al	7429-90-5	—	10	—	x
Co	7440-48-4	Group 2B	0.05	—	x
Fe	7439-89-6	—	5	for oxide	
Ni	7440-02-0	Group 1	0.015	—	x
Ti	7440-32-6	Group 2B (oxide)	—	—	x
Secondary alloying elements					
Ag	7440-22-4	—	0.01	—	x
B	7440-42-8	—	—	—	x
C	1333-86-4	—	—	—	
Ce	7440-45-1	—	—	—	x
Cr	7440-47-3	Group 1	0.5	—	x
Cu	7440-50-8	—	1	—	x
Ga	7440-55-3	—	—	—	
Hf	7440-58-6	—	0.5	—	x
Li	7439-93-2	—	0.025	For monohydride	x
Mg	7439-95-4	—	—	—	x
Mn	7439-96-5	—	1	—	x
Mo	7439-98-7	—	—	—	x
Nb	7440-03-1	—	—	—	
Pt	7440-06-4	—	—	—	x
Re	7440-15-5	—	—	—	
Ru	7440-18-8	—	—	—	
Sc	7440-20-2	—	—	—	
Si	7440-21-3	—	10	—	x
Sn	7440-31-5	—	2	—	x
Ta	7440-25-7	—	5	—	x
V	7440-62-2	Group 2B (oxide)	1	For FeV dust	
W	7440-33-7	—	—	—	x
Y	7440-65-5	—	1	—	x
Zn	7440-66-6	—	5	For ZnO	x
Zr	7440-67-7	—	5	—	x

a) Chemical Abstract Services Registry Number.

(16) Corrosive to metals
(17) Desensitized explosives

B. Health Hazards

(1) Acute toxicity
(2) Skin corrosion/irritation

(3) Serious eye damage/eye irritation
(4) Respiratory or skin sensitization
(5) Germ cell mutagenicity
(6) Carcinogenicity
(7) Reproductive toxicity
(8) Specific target organ toxicity – Single exposure
(9) Specific target organ toxicity – Repeated exposure
(10) Aspiration hazard

C. Environmental Hazards

(1) Hazardous to the aquatic environment
(2) Hazardous to the ozone layer

The most comprehensive, transparent, and easy-to-access and use GHS online database is provided by the **National Institute of Technology and Evaluation** (**NITE**), an accredited institute in Japan [70]. This database is often simply referred to as GHS-Japan. As illustrated for the example of aluminum (Al) in Tables 16.5A–16.5C, the GHS-Japan database includes not only the classification results for each GHS hazard trait ("Hazard Class" in Tables 16.5A–16.5C), but also the "Symbol" for the harmonized labeling, the "Signal Word" that accompanies the Symbol, the "Hazard Statement," the "Precautionary Statement," the "Rationale for the Classification," the "Classification Year (FY)," and the "GHS Classification Guidance for the Japanese Government" used as the basis for the classification. This GHS Classification Guidance for the Japanese Government [74] is similar to that published by the United Nations, mentioned above. The inclusion of the rationale for the classification within this simple, highly structured database, adds significantly to the transparency of the evaluation, especially in contrast to other data sources such as the lengthy monographs generated by IARC. Furthermore, the consistent structure of the database for each substance evaluated, even though many data cells are left blank or noted as "not classified," allows for easy comparison of materials, unlike other data sources on chemical hazards such as the **National Institute for Occupational Safety and Health** (**NIOSH**) Pocket Guide, published online by the U.S. **Centers for Disease Control and Prevention** (**CDC**) [69]. While the NIOSH Pocket Guide is also an excellent, independent source of data on hazard as well as occupational exposure prevention and treatment for a given substance of interest, the lack of standardization in format and content makes it less useful for direct material comparisons. Highlighting the ability to use the GHS-Japan database for rapid material comparisons, the "Classification" columns for this book's metallic elements of interest have been extracted and combined into Tables 16.6A–16.6C. Those metals with entries in the GHS-Japan database are also noted in Table 16.4. Note that Category 1 hazards are the most severe, followed by Category 2 hazards, Category 3 hazards, and Category 4 hazards, in an order of decreasing degree of severity. Notably, all 23 metallic elements presented in Tables 16.6A–16.6C exhibit at least one hazard; cobalt exhibits the most with 11. The majority of the metals exhibit two to five hazard traits. Most of the hazard traits relate to human health effects, with at least one metal exhibiting each of the ten listed human health hazard trait classes. Specific target organ toxicity is the most common, with 17 of the 23 metals exhibiting some type of target organ toxicity. Serious eye damage/eye irritation, respiratory sensitization, and aquatic toxicity are also very common.

One disadvantage of the GHS classifications is the potential difficulty in establishing priorities among the ~30 hazard traits when selecting materials. To address this challenge, various **chemical hazard assessment** (**CHA**) and **alternative assessment** decision frameworks have been developed. One well-established CHA decision framework is the **GreenScreen**® for Safer Chemicals [75]. This methodology uses a decision logic vetted by technical experts to consistently prioritize the hazard traits so that chemicals can be compared systematically and transparently. The method considers 18 hazard endpoints, shown in Table 16.7, and the evaluation criteria rely heavily on GHS classifications [76]. The method leads to giving each substance a "benchmark score" (see Figure 16.8), where Benchmark 1 corresponds to chemicals of high concern that should be avoided, Benchmark 2 corresponds to chemicals that can continue to be used but safer substitutes should be pursued, Benchmark 3 corresponds to chemicals that can be used, but there is still room for improvement, and Benchmark 4 corresponds to chemicals that are deemed preferred, safer chemicals. For situations where there is not sufficient data to complete the evaluation, the substance is labeled with a Benchmark "U" for "unknown." This benchmarking approach allows for simplified screening out of chemicals to not use (Benchmark 1) or to avoid using (Benchmark 2). One challenge with applying GreenScreen® to metals is that

Table 16.5A GHS Classification results for aluminum metal, CASRN: 7429-90-5, *physical hazards*.

	Hazard class	Classification	Symbol	Signal word	Hazard statement	Precautionary statement	Rationale for the classification	Classification year (FY)	GHS classification guidance for the Japanese government
1	Explosives	Not classified (not applicable)	—	—	—	—	There are no chemical groups present in the molecule associated with explosive properties	FY2015	FY2013 revised edition (Ver. 1.1)
2	Flammable gases	Not classified (not applicable)	—	—	—	—	"Solids" according to GHS definition	FY2015	FY2013 revised edition (Ver. 1.1)
3	Aerosols	Not classified (not applicable)	—	—	—	—	Not an aerosol product	FY2015	FY2013 revised edition (Ver. 1.1)
4	Oxidizing gases	Not classified (not applicable)	—	—	—	—	"Solids" according to GHS definition	FY2015	FY2013 revised edition (Ver. 1.1)
5	Gases under pressure	Not classified (not applicable)	—	—	—	—	"Solids" according to GHS definition	FY2015	FY2013 revised edition (Ver. 1.1)
6	Flammable liquids	Not classified (not applicable)	—	—	—	—	"Solids" according to GHS definition	FY2015	FY2013 revised edition (Ver. 1.1)
7	Flammable solids	Not classified	—	—	—	—	a)	FY2015	FY2013 revised edition (Ver. 1.1)
8	Self-reactive substances and mixtures	Not classified (not applicable)	—	—	—	—	There are no chemical groups present in the molecule associated with explosive or self-reactive properties	FY2015	FY2013 revised edition (Ver. 1.1)
9	Pyrophoric liquids	Not classified (not applicable)	—	—	—	—	"Solids" according to GHS definition	FY2015	FY2013 revised edition (Ver. 1.1)
10	Pyrophoric solids	Not classified	—	—	—	—	It is estimated that it does not ignite at normal temperatures from an ignition point of 590 degrees C (ICSC [2000])	FY2015	FY2013 revised edition (Ver. 1.1)
11	Self-heating substances and mixtures	Not classified	—	—	—	—	Blockish aluminum is non-flammable (GESTIS [Access on July 2015])	FY2015	FY2013 revised edition (Ver. 1.1)

(*continued*)

Table 16.5A (Continued)

Hazard class	Classification	Symbol	Signal word	Hazard statement	Precautionary statement	Rationale for the classification	Classification year (FY)	GHS classification guidance for the Japanese government
12 Substances and mixtures which, in contact with water, emit flammable gases	Category 2	Flame	Danger	H261: In contact with water releases flammable gas	b)	c)	FY2015	FY2013 revised edition (Ver. 1.1)
13 Oxidizing liquids	Not classified (not applicable)	—	—	—	—	"Solids" according to GHS definition	FY2015	FY2013 revised edition (Ver. 1.1)
14 Oxidizing solids	Not classified (not applicable)	—	—	—	—	It is an inorganic substance	FY2015	FY2013 revised edition (Ver. 1.1)
15 Organic peroxides	Not classified (not applicable)	—	—	—	—	It is an inorganic substance	FY2015	FY2013 revised edition (Ver. 1.1)
16 Corrosive to metals	Classification not possible	—	—	—	—	No established test method suitable for solid substances	FY2015	FY2013 revised edition (Ver. 1.1)
17 Desensitized explosives	—	—	—	—	—	—	—	—

a) Blockish aluminum is non-flammable (GESTIS [Access in July 2015]). Besides, it is written that finely dispersed particles are explosive in the air (ICSC [2000]).

b) P302 + P335 + P334: IF ON SKIN: Brush off loose particles from skin. Immerse in cool water [or wrap in wet bandages]; P231 + P232: Handle and store contents under inert gas/... Protect from moisture. P370 + P378: In case of fire: Use … to extinguish. P402 + P404: Store in a dry place. Store in a closed container. P223: Do not allow contact with water. P280: Wear protective gloves/protective clothing/eye protection/face protection. P501: Dispose of contents/container to ...

c) Blockish aluminum forms an oxide coating (alumite) in the air, and the inside does not corrode (Encyclopedic Dictionary of Chemistry). There is information that powder has a risk of fire and explosion in contact with water (ICSC [2000]). Aluminum powder (surface uncoated) is classified in class 4.3, PGII, III in UNRTDG (UN1396), corresponding to Category 2 or 3.

Table 16.5B GHS classification results for aluminum metal, CASRN: 7429-90-5, *health hazards*.

Hazard class	Classification	Symbol	Signal word	Hazard statement	Precautionary statement	Rationale for the classification	Classification year (FY)	GHS classification guidance for the Japanese government
1 Acute toxicity (oral)	Classification not possible	—	—	—	—	Due to lack of data, the classification is not possible	FY2015	FY2013 revised edition (Ver. 1.1)
1 Acute toxicity (dermal)	Classification not possible	—	—	—	—	Due to lack of data, the classification is not possible	FY2015	FY2013 revised edition (Ver. 1.1)
1 Acute toxicity (inhalation: gases)	Not classified (not applicable)	—	—	—	—	"Solids" according to GHS definition	FY2015	FY2013 revised edition (Ver. 1.1)
1 Acute toxicity (inhalation: vapors)	Not classified (not applicable)	—	—	—	—	"Solids" according to GHS definition	FY2015	FY2013 revised edition (Ver. 1.1)
1 Acute toxicity (inhalation: dusts and mists)	Classification not possible	—	—	—	—	Due to lack of data, the classification is not possible	FY2015	FY2013 revised edition (Ver. 1.1)
2 Skin corrosion/irritation	Classification not possible	—	—	—	—	Due to lack of data, the classification is not possible	FY2015	FY2013 revised edition (Ver. 1.1)
3 Serious eye damage/eye irritation	Classification not possible	—	—	—	—	a)	FY2015	FY2013 revised edition (Ver. 1.1)
4 Respiratory sensitization	Classification not possible	—	—	—	—	Due to lack of data, the classification is not possible	FY2015	FY2013 revised edition (Ver. 1.1)
4 Skin sensitization	Classification not possible	—	—	—	—	Due to lack of data, the classification is not possible	FY2015	FY2013 revised edition (Ver. 1.1)
5 Germ cell mutagenicity	Classification not possible	—	—	—	—	The classification is not possible due to lack of data. There are no *in vivo* and *in vitro* data for metallic aluminum	FY2015	FY2013 revised edition (Ver. 1.1)
6 Carcinogenicity	Classification not possible	—	—	—	—	b)	FY2015	FY2013 revised edition (Ver. 1.1)
7 Reproductive toxicity	Classification not possible	—	—	—	—	c)	FY2015	FY2013 revised edition (Ver. 1.1)
8 Specific target organ toxicity – Single exposure	Category 1 (respiratory organs)	Health Hazard	Danger	H370: Causes damage to organs (respiratory organs)	d)	e)	FY2015	FY2013 revised edition (Ver. 1.1)

(*continued*)

Table 16.5B (Continued)

Hazard class	Classification	Symbol	Signal word	Hazard statement	Precautionary statement	Rationale for the classification	Classification year (FY)	GHS classification guidance for the Japanese government
9 Specific target organ toxicity – Repeated exposure	Category 1 (respiratory organs)	Health Hazard	Danger	f)	g)	h)	FY2015	FY2013 revised edition (Ver. 1.1)
10 Aspiration hazard	Classification not possible	—	—	—	—	Due to lack of data, the classification is not possible	FY2015	FY2013 revised edition (Ver. 1.1)

a) The classification is not possible due to lack of data. Besides, it is written that dust of this substance may cause mild inflammation in eyes (HSDB (Access on June 2015)).

b) In 1970s, increased incidences of bladder and lung cancers were reported in workers at electrolysis tank of aluminum reduction plants. In IARC, working in the aluminum refining process is related to increased risk of cancers in workers, but it is estimated that the causative substance of cancers is not aluminum itself, but polycyclic aromatic hydrocarbon compounds, known carcinogens, or coal tar pitch volatiles, which are by-products derived from the decomposition of electrodes in the electrolysis tank. (IARC Suppl. 7 [1984], ACGIH [7th, 2008], ATSDR [2008]) As results of following epidemiological studies, some reports suggested a functional disorder of respiratory organs, or non-tumor lesions in respiratory organs by exposure to this substance, but there is no report on carcinogenicity (ACGIH [7th, 2008], ATSDR [2008]). Therefore, ACGIH classified in A4 (ACGIH [7th, 2008]). From the above, the substance was classified as "Classification not possible" in this hazard class.

c) Due to no information on reproductive toxicity by exposure to this substance, the classification is not possible due to lack of data. Besides, it is reported that in a test in which aluminum chloride or aluminum lactate was dosed in an oral route to pregnant rats during a gestation period, or to neonatal rats on days 5–17 after birth, delayed neurobehavioral development or hypoactivity was observed in newborns (ACGIH [7th, 2008], ATSDR [2008]).

d) P308 + P311: IF exposed or concerned: Call a POISON CENTER/doctor/... P260: Do not breathe dust/fume/gas/mist/vapors/spray. P264: Wash ... thoroughly after handling. P270: Do not eat, drink or smoke when using this product. P321: Specific treatment (see … on this label). P405: Store locked up. P501: Dispose of contents/container to ...

e) This substance (dust, powder) is irritating to the respiratory tract (HSDB [Access on June 2015]). As for humans, inhalation of this substance (dust) could cause lung disorders such as pneumoconiosis (aluminosis) (HSDB [Access on June 2015]). As for experimental animals, it is reported that in single inhalation exposure to this substance (dust) in rats, no change in pulmonary function but enzymatic and cytological changes in bronchoalveolar lavage fluid were observed at 0.05 $mg\,L^{-1}$, and microgranuloma development occurred in the lung and hilum lymph nodes (microscopic observation) at 0.2 $mg\,L^{-1}$. These were observed at doses corresponding to Category 1 (ACGIH (7th, 2008), PATTY (6th, 2012)). From the above, because this substance has effects on the respiratory organs in inhalation exposure, it was classified in Category 1 (respiratory system). Besides, respiratory tract irritation was included in effects on the respiratory organs. By adding new information, the previous classification was revised.

f) H372: Causes damage to organs through prolonged or repeated exposure (respiratory organs)

g) P260: Do not breathe dust/fume/gas/mist/vapors/spray. P264: Wash … thoroughly after handling. P270: Do not eat, drink or smoke when using this product. P314: Get medical advice/attention if you feel unwell. P501: Dispose of contents/container to …

h) As for humans, in an epidemiological survey of 1142 workers related to manufacturing aluminum and its compounds (1975–1981), effects on lung function were observed in exposure to dust of high concentration (>100 $mg\,m^{-3}$-year as total dust), and in chest X-ray examination, small and irregular nodes were reported in a lower part of the lung in 7–8% (ACGIH (7th, 2008)). As for experimental animals, there is no test report using this substance. Therefore, the substance was classified in Category 1 (respiratory system). Besides, as for humans, it is written that decreased cognitive test results were observed, but a clear conclusion was not drawn (ATSDR (2008)). Moreover, it is written that aluminum is not a factor causing Alzheimer's disease at this time, and that in many studies, there is no consistent relationship between aluminum and nervous disorder (ACGIH (7th, 2008)). Therefore, the central nervous system was not included in a target organ.

Table 16.5C GHS classification results for aluminum metal, CASRN: 7429-90-5, *environmental hazards.*

Hazard class		Classification	Symbol	Signal word	Hazard statement	Precautionary statement	Rationale for the classification	Classification year (FY)	GHS classification guidance for the Japanese government
11	Hazardous to the aquatic environment short term (acute)	Classification not possible	—	—	—	—	The classification is not possible because appropriate data are not obtained	FY2015	FY2013 revised edition (Ver. 1.1)
11	Hazardous to the aquatic environment Long term (chronic)	Classification not possible	—	—	—	—	No data	FY2015	FY2013 revised edition (Ver. 1.1)
12	Hazardous to the ozone layer	Classification not possible	—	—	—	—	No data.	FY2015	FY2013 revised edition (Ver. 1.1)

(1) GHS Classification Result by the Japanese Government is intended to provide a reference for preparing a GHS label or SDS for users. To include the same classification result in a label or SDS for Japan is NOT mandatory.
(2) Users can cite or copy this classification result when preparing a GHS label or SDS. Please be aware, however, that the responsibility for a label or SDS prepared by citing or copying this classification result lies with users.
(3) This GHS classification was conducted based on the information sources and the guidance for classification and judgment which are described in the GHS Classification Guidance for the Japanese Government etc. Using other literature, test results etc. as evidence and including different content from this classification result in a label or SDS are allowed.
(4) A blank or "—" in the column of "Classification" denotes that a classification for the hazard class was not conducted in the year.
(5) An asterisk "*" in the column of "Classification" denotes that "Not classified (or No applicable)" and/or "Classification not possible" is applicable. Details are described in the column of "Rationale for the classification." If not English translation for "Rationale for the classification" available, please refer to the Japanese version of the results.
Source: NITE [70].

Table 16.6A GHS classification results for select metals – *physical hazards.*

	CASRN:	7429-90-5	7440-48-4	7440-47-3	7440-50-8	7439-95-4	7439-96-5	7439-98-7	7440-02-0	7439-92-1	7440-21-3	7440-32-6	7440-66-6
	Chemical name:	**Al**	**Co**	**Cr**	**Cu**	**Mg**	**Mn**	**Mo**	**Ni**	**Pb**	**Si**	**Ti**	**Zn**
1	Explosives	NC (NA)	NC (NA)	NC (NA)	NC (NA)	NC (NA)	NC (NA)	NC (NA)	NC (NA)	NC (NA)	NC (NA)	NC (NA)	NC (NA)
2	Flammable gases	NC (NA)	NC (NA)	NC (NA)	NC (NA)	NC (NA)	NC (NA)	NC (NA)	NC (NA)	NC (NA)	NC (NA)	NC (NA)	NC (NA)
3	Aerosols	NC (NA)	NC (NA)	NC (NA)	NC (NA)	NC (NA)	NC (NA)	NC (NA)	NC (NA)	NC (NA)	NC (NA)	NC (NA)	NC (NA)
4	Oxidizing gases	NC (NA)	NC (NA)	NC (NA)	NC (NA)	NC (NA)	NC (NA)	NC (NA)	NC (NA)	NC (NA)	NC (NA)	NC (NA)	NC (NA)
5	Gases under pressure	NC (NA)	NC (NA)	NC (NA)	NC (NA)	NC (NA)	NC (NA)	NC (NA)	NC (NA)	NC (NA)	NC (NA)	NC (NA)	NC (NA)
6	Flammable liquids	NC (NA)	NC (NA)	NC (NA)	NC (NA)	NC (NA)	NC (NA)	NC (NA)	NC (NA)	NC (NA)	NC (NA)	NC (NA)	NC (NA)
7	Flammable solids	NC	CNP	CNP	CNP	NC	CNP	CNP	CNP	NC	Cat 2	Cat 1	CNP
8	Self-reactive substances and mixtures	NC (NA)	NC (NA)	NC (NA)	NC (NA)	NC (NA)	NC (NA)	NC (NA)	NC (NA)	NC (NA)	NC (NA)	NC (NA)	NC (NA)
9	Pyrophoric liquids	NC (NA)	NC (NA)	NC (NA)	NC (NA)	NC (NA)	NC (NA)	NC (NA)	NC (NA)	NC (NA)	NC (NA)	NC (NA)	NC (NA)
10	Pyrophoric solids	NC	NC	CNP	CNP	CNP	NC	CNP	NC	NC	NC	Cat 1	NC
11	Self-heating substances and mixtures	NC	CNP	CNP	CNP	CNP	CNP	CNP	CNP	NC	CNP	Cat 1	CNP
12	Substances and mixtures which, in contact with water, emit flammable gases	Cat 2	NC	NC	NC	CNP	CNP	NC	NC	NC	NC	NC	CNP
13	Oxidizing liquids	NC (NA)	NC (NA)	NC (NA)	NC (NA)	NC (NA)	NC (NA)	NC (NA)	NC (NA)	NC (NA)	NC (NA)	NC (NA)	NC (NA)
14	Oxidizing solids	NC (NA)	NC (NA)	NC (NA)	NC (NA)	NC (NA)	NC (NA)	NC (NA)	NC (NA)	NC (NA)	NC (NA)	NC (NA)	NC (NA)
15	Organic peroxides	NC (NA)	NC (NA)	NC (NA)	NC (NA)	NC (NA)	NC (NA)	NC (NA)	NC (NA)	NC (NA)	NC (NA)	NC (NA)	NC (NA)
16	Corrosive to metals	CNP	CNP	CNP	CNP	NC	CNP	CNP	CNP	CNP	CNP	CNP	CNP
17	Desensitized explosives	—	NC (NA)	NC	—	—	NC (NA)	—	—	—	—	—	—

(*continued*)

Table 16.6A (Continued)

	Category:	7440-22-4	7440-42-8	7440-45-1	12604-58-9	7440-58-6	7439-93-2	7440-06-4	7440-31-5	7440-25-7	7440-33-7	7440-67-7
	Chemical name:	**Ag**	**B**	**Ce**	**FeV**	**Hf**	**Li**	**Pt**	**Sn**	**Ta**	**W**	**Zr**
1	Explosives	NC (NA)	NC (NA)	NC (NA)	NC (NA)	NC (NA)	NC (NA)	NC (NA)	NC (NA)	NC (NA)	NC (NA)	NC (NA)
2	Flammable gases	NC (NA)	NC (NA)	NC (NA)	NC (NA)	NC (NA)	NC (NA)	NC (NA)	NC (NA)	NC (NA)	NC (NA)	NC (NA)
3	Aerosols	NC (NA)	NC (NA)	NC (NA)	NC (NA)	NC (NA)	NC (NA)	NC (NA)	NC (NA)	NC (NA)	NC (NA)	NC (NA)
4	Oxidizing gases	NC (NA)	NC (NA)	NC (NA)	NC (NA)	NC (NA)	NC (NA)	NC (NA)	NC (NA)	NC (NA)	NC (NA)	NC (NA)
5	Gases under pressure	NC (NA)	NC (NA)	NC (NA)	NC (NA)	NC (NA)	NC (NA)	NC (NA)	NC (NA)	NC (NA)	NC (NA)	NC (NA)
6	Flammable liquids	NC (NA)	NC (NA)	NC (NA)	NC (NA)	NC (NA)	NC (NA)	NC (NA)	NC (NA)	NC (NA)	NC (NA)	NC (NA)
7	Flammable solids	NC	CNP	Cat 1	CNP	CNP	NC	CNP	CNP	CNP	CNP	Cat 1
8	Self-reactive substances and mixtures	NC (NA)	NC (NA)	NC (NA)	NC (NA)	NC (NA)	NC (NA)	NC (NA)	NC (NA)	NC (NA)	NC (NA)	NC (NA)
9	Pyrophoric liquids	NC (NA)	NC (NA)	NC (NA)	NC (NA)	NC (NA)	NC (NA)	NC (NA)	NC (NA)	NC (NA)	NC (NA)	NC (NA)
10	Pyrophoric solids	NC	NC	CNP	NC	CNP	NC	CNP	NC	CNP	CNP	CNP
11	Self-heating substances and mixtures	NC	CNP	CNP	CNP	CNP	NC	CNP	CNP	CNP	CNP	CNP
12	Substances and mixtures which, in contact with water, emit flammable gases	NC	NC	Cat 2	CNP	CNP	Cat 1	CNP	NC	NC	NC	NC
13	Oxidizing liquids	NC (NA)	NC (NA)	NC (NA)	NC (NA)	NC (NA)	NC (NA)	NC (NA)	NC (NA)	NC (NA)	NC (NA)	NC (NA)
14	Oxidizing solids	NC (NA)	NC (NA)	NC (NA)	NC (NA)	NC (NA)	NC (NA)	NC (NA)	NC (NA)	NC (NA)	NC (NA)	NC (NA)
15	Organic peroxides	NC (NA)	NC (NA)	NC (NA)	NC (NA)	NC (NA)	NC (NA)	NC (NA)	NC (NA)	NC (NA)	NC (NA)	NC (NA)
16	Corrosive to metals	NC	CNP	CNP	CNP	CNP	NC	CNP	CNP	CNP	CNP	CNP
17	Desensitized explosives	—	NC	—	NC	—	—	—	NC	—	—	—

CASRN, Chemical Abstract Services Registry Number; NC, Not classified; NA, Not applicable; CNP, Classification not possible; Cat, Category; Category 1 is the most severe.

Table 16.6B GHS classification results for select metals – *health hazards*.

	CASRN:	7429-90-5	7440-48-4	7440-47-3	7440-50-8	7439-95-4	7439-96-5	7439-98-7	7440-02-0	7439-92-1	7440-21-3	7440-32-6	7440-66-6
	Chemical name:	**Al**	**Co**	**Cr**	**Cu**	**Mg**	**Mn**	**Mo**	**Ni**	**Pb**	**Si**	**Ti**	**Zn**
1	Acute toxicity (oral)	CNP	Cat 4	CNP	CNP	CNP	NC	CNP	NC	CNP	NC	CNP	NC
1	Acute toxicity (dermal)	CNP	CNP	CNP	CNP	CNP	CNP	CNP	CNP	CNP	CNP	CNP	CNP
1	Acute toxicity (inhalation: gases)	NC (NA)	NC	NC (NA)	NC (NA)	NC (NA)	NC (NA)	NC (NA)	NC (NA)	NC (NA)	NC (NA)	NC (NA)	NC (NA)
1	Acute toxicity (inhalation: vapors)	NC (NA)	CNP	CNP	NC (NA)	CNP	CNP	NC (NA)	CNP	CNP	CNP	CNP	CNP
1	Acute toxicity (inhalation: dusts and mists)	CNP	Cat 1	CNP	CNP	CNP	NC	NC	CNP	CNP	CNP	CNP	NC
2	Skin corrosion/irritation	CNP	NC	CNP	CNP	Cat 2	NC	Cat 2	CNP	CNP	CNP	CNP	NC
3	Serious eye damage/eye irritation	CNP	Cat 2B	Cat 2	CNP	Cat 2A	NC	Cat 2	CNP	CNP	Cat 2B	CNP	Cat 2B
4	Respiratory sensitization	CNP	Cat 1A	Cat 1A	CNP	CNP	CNP	CNP	Cat 1	CNP	CNP	CNP	CNP
4	Skin sensitization	CNP	Cat 1A	Cat 1A	Cat 1A	CNP	NC	CNP	Cat 1	CNP	CNP	CNP	NC
5	Germ cell mutagenicity	CNP	CNP	CNP	CNP	CNP	CNP	CNP	CNP	Cat 2	CNP	CNP	CNP
6	Carcinogenicity	CNP	Cat 2	CNP	CNP	CNP	CNP	CNP	Cat 2	Cat 2	CNP	CNP	CNP
7	Reproductive toxicity	CNP	Cat 1B	CNP	CNP	CNP	Cat 1B	CNP	CNP	Cat 1A	CNP	CNP	CNP
8	Specific target organ toxicity – single exposure	Cat 1	Cat 1	Cat 3	Cat 1, 3	Cat 3	CNP	Cat 3	Cat 1	CNP	CNP	CNP	CNP
9	Specific target organ toxicity – repeated exposure	Cat 1	Cat 1	CNP	CNP	CNP	Cat 1	CNP	Cat 1	Cat 1	CNP	CNP	CNP
10	Aspiration hazard	CNP	CNP	CNP	CNP	CNP	CNP	CNP	CNP	CNP	CNP	CNP	CNP

(*continued*)

Table 16.6B (Continued)

	Category:	7440-22-4	7440-42-8	7440-45-1	12604-58-9	7440-58-6	7439-93-2	7440-06-4	7440-31-5	7440-25-7	7440-33-7	7440-67-7
	Chemical name:	**Ag**	**B**	**Ce**	**FeV**	**Hf**	**Li**	**Pt**	**Sn**	**Ta**	**W**	**Zr**
1	Acute toxicity (oral)	NC	NC	CNP	CNP	CNP	CNP	CNP	CNP	CNP	CNP	CNP
1	Acute toxicity (dermal)	NC	CNP	CNP	CNP	CNP	CNP	CNP	CNP	CNP	CNP	CNP
1	Acute toxicity (inhalation: gases)	NC (NA)	NC (NA)	NC (NA)	NC (NA)	NC (NA)	NC (NA)	NC (NA)	NC (NA)	NC (NA)	NC (NA)	NC (NA)
1	Acute toxicity (inhalation: vapors)	CNP	CNP	CNP	CNP	CNP	CNP	CNP	CNP	NC (NA)	NC (NA)	NC (NA)
1	Acute toxicity (inhalation: dusts and mists)	CNP	NC	CNP	CNP	CNP	CNP	CNP	CNP	CNP	CNP	CNP
2	Skin corrosion/irritation	NC	NC	CNP	Cat 2	CNP	Cat 1	Cat 2	NC	CNP	CNP	CNP
3	Serious eye damage/eye irritation	Cat 2B	NC	CNP	Cat 2	CNP	Cat 1	Cat 2A	Cat 2	CNP	Cat 2B	CNP
4	Respiratory sensitization	CNP	CNP	CNP	CNP	CNP	CNP	Cat 1	CNP	CNP	CNP	CNP
4	Skin sensitization	Cat 1	NC	CNP	CAP	CNP	CNP	NC	CNP	CNP	CNP	CNP
5	Germ cell mutagenicity	CNP	CNP	CNP	CNP	CNP	CNP	CNP	CNP	CNP	CNP	CNP
6	Carcinogenicity	CNP	CNP	CNP	CNP	CNP	CNP	CNP	CNP	CNP	CNP	CNP
7	Reproductive toxicity	Cat 1B	CNP	CNP	CNP	CNP	CNP	CNP	CNP	CNP	CNP	CNP
8	Specific target organ toxicity – single exposure	Cat 1	Cat 2	Cat 1	Cat 1	CNP	Cat 2	Cat 3	Cat 3	Cat 3	Cat 3	Cat 3
9	Specific target organ toxicity – repeated exposure	Cat 1	CNP	CNP	Cat 1	Cat 2	CNP	NC	Cat 1	CNP	CNP	CNP
10	Aspiration hazard	NC (NA)	CNP	CNP	CNP	CNP	CNP	CNP	CNP	CNP	CNP	CNP

Table 16.6C GHS classification results for select metals – *environmental hazards.*

CASRN:	7429-90-5	7440-48-4	7440-47-3	7440-50-8	7439-95-4	7439-96-5	7439-98-7	7440-02-0	7439-92-1	7440-21-3	7440-32-6	7440-66-6
Chemical name:	**Al**	**Co**	**Cr**	**Cu**	**Mg**	**Mn**	**Mo**	**Ni**	**Pb**	**Si**	**Ti**	**Zn**
11 Hazardous to the aquatic environment short term (acute)	CNP	Cat 1	CNP	CNP	CNP	Cat 2	CNP	CNP	CNP	CNP	CNP	Cat 1
11 Hazardous to the aquatic environment long term (chronic)	CNP	Cat 1	CNP	CNP	CNP	Cat 2	CNP	CNP	CNP	CNP	CNP	Cat 1
12 Hazardous to the ozone layer	CNP	CNP	CNP	CNP	CNP	CNP	CNP	CNP	—	—	—	—

CASRN:	7440-22-4	7440-42-8	7440-45-1	12604-58-9	7440-58-6	7439-93-2	7440-06-4	7440-31-5	7440-25-7	7440-33-7	7440-67-7
Chemical name:	**Ag**	**B**	**Ce**	**FeV**	**Hf**	**Li**	**Pt**	**Sn**	**Ta**	**W**	**Zr**
11 Hazardous to the aquatic environment short term (acute)	CNP	CNP	Cat 1	CNP	CNP	CNP	CNP	CNP	CNP	CNP	CNP
11 Hazardous to the aquatic environment long term (chronic)	CNP	CNP	Cat 1	CNP	CNP	CNP	CNP	CNP	CNP	CNP	CNP
12 Hazardous to the ozone layer	CNP	CNP	—	CNP	—	—	—	CNP	CNP	CNP	CNP

CASRN, Chemical Abstract Services Registry Number; NC, Not classified; NA, Not applicable; CNP, Classification not possible; Cat, Category; Category 1 is the most severe.
(1) GHS Classification Result by the Japanese Government is intended to provide a reference for preparing a GHS label or SDS for users. To include the same classification result in a label or SDS for Japan is NOT mandatory.
(2) Users can cite or copy this classification result when preparing a GHS label or SDS. Please be aware, however, that the responsibility for a label or SDS prepared by citing or copying this classification result lies with users.
(3) This GHS classification was conducted based on the information sources and the guidance for classification and judgment which are described in the GHS Classification Guidance for the Japanese Government. Using other literature, test results, etc. as evidence and including different content from this classification results in a label or SDS are allowed.
(4) A blank or "—" in the column of "Classification" denotes that a classification for the hazard class was not conducted in the year.
(1) An asterisk "*" in the column of "Classification" denotes that "Not classified (or No applicable)" and/or "Classification not possible" is applicable. Details are described in the column "Rationale for the classification." If no English translation for "Rationale for the classification" available, please refer to the Japanese version of the results.
Source: NITE [70].

Table 16.7 Hazard endpoints evaluated in the GreenScreen® for Safer Chemicals decision logic.

Environmental fate	Environmental health	Human health group I	Human health group II	Physical hazards
Persistence (P)	Acute aquatic toxicity (AA)	Carcinogenicity (C)	Acute mammalian toxicity (AT)	Reactivity (Rx)
Bioaccumulation (B)	Chronic aquatic toxicity (CA)	Mutagenicity and genotoxicity (M)	Systemic toxicity and organ effects (incl. immunotoxicity) (ST)	Flammability (F)
		Reproductive toxicity (R)	Neurotoxicity (N)	
		Developmental toxicity (incl. developmental neurotoxicity) (D)	Sensitization (SnS)	
		Endocrine activity (E)	Respiratory sensitization (SnR)	
			Skin irritation (IrS)	
			Eye irritation (IrE)	

Source: GreenScreen Method [76].

two of the hazard endpoints are related to **environmental fate**, including "**persistence**." Since, unlike for chemical formulations, it is desirable for metals to be persistent and not degrade during use, it is logical to remove this hazard endpoint when evaluating a metal. Notably, nickel and chromium, because of their carcinogenicity, would be identified as Benchmark 1 substances and therefore their use should be avoided. Select studies in the published literature have conducted in-depth GreenScreen®-based assessments of metals and other substances, including **high entropy alloys** [77–82]. According to the analysis by Wang et al. [77], Al, Cr, and V were determined to be BM1 chemicals, and Hf, Mo, Nb, Ta, Ti, W, and Zr were determined to be BM2 chemicals.

From an occupational exposure perspective, intrinsic hazard should be considered, but also the potential for exposure, via inhalation, ingestion, or dermal, thus leading to a potential "risk" of causing disease [83]. An important indicator of exposure potential and concerns over occupational risk is the **Permissible Exposure Limits** (**PEL**), such as those mandated by OSHA [84], and the **Recommended Exposure Limits** (**REL**), such as those published by NIOSH [69]. Since the NIOSH REL values are updated more frequently, current values are provided for the metals of interest in this book, as shown in Table 16.4. High values are preferred. Notably, many of the metals do not have REL values. Of those that do, there is a wide range of values, with cobalt (Co), nickel (Ni), silver (Ag), chromium (Cr), and hafnium (Hf) all having REL values less than 1.0 $mg\,m^{-3}$.

Given the intrinsic hazard traits of metals and metallic powders, it is important to directly evaluate the potential for exposure to metal powders during AM. A recent review article on this topic was published by Arrizubieta et al. [85], in which they summarized various hazard classifications, LCA studies, preventive actions in the workplace, management of waste material, and recycling of powders. They highlight that metallic powders are indeed hazardous to humans and can cause disease, therefore, safety measures should be put in place to protect workers, including proper ventilation, protective masks, and machine enclosures. Two studies in the literature conduct direct measurements of particles in the workplace and **biomonitoring** of the workers, respectively [86, 87]. The study by Graff et al. focused on measuring particle numbers, masses, sizes, and identities present in the air during AM with metallic powder consisting essentially of chromium, nickel, and cobalt (specifically, Inconel 939) [86]. Results showed that there is a risk of particle exposure at certain operations and that particle sizes tended to be smaller in recycled metal powder compared to new. Specifically, nanosized particles were present in the AM work environment, and the operators were exposed to potential inhalation of these particles while handling the metallic powder. A subsequent study used biomonitoring of urine and dermal contamination among AM operators, office personnel, and welders when handling Hastelloy® X, which also contains nickel, cobalt, and chromium [87]. Results revealed elevated exposure during the first year the AM facility was in operation, which was reduced by control measures put into place for the second year. Overall, AM operators experienced a significant increase in exposure to cobalt compared with welders. Exposure to these metals is of concern to potential carcinogenicity, as mentioned above, and potential impact on lung, kidney, and liver function. Given this potential for occupational exposure, **robotic materials handling systems** are recommended.

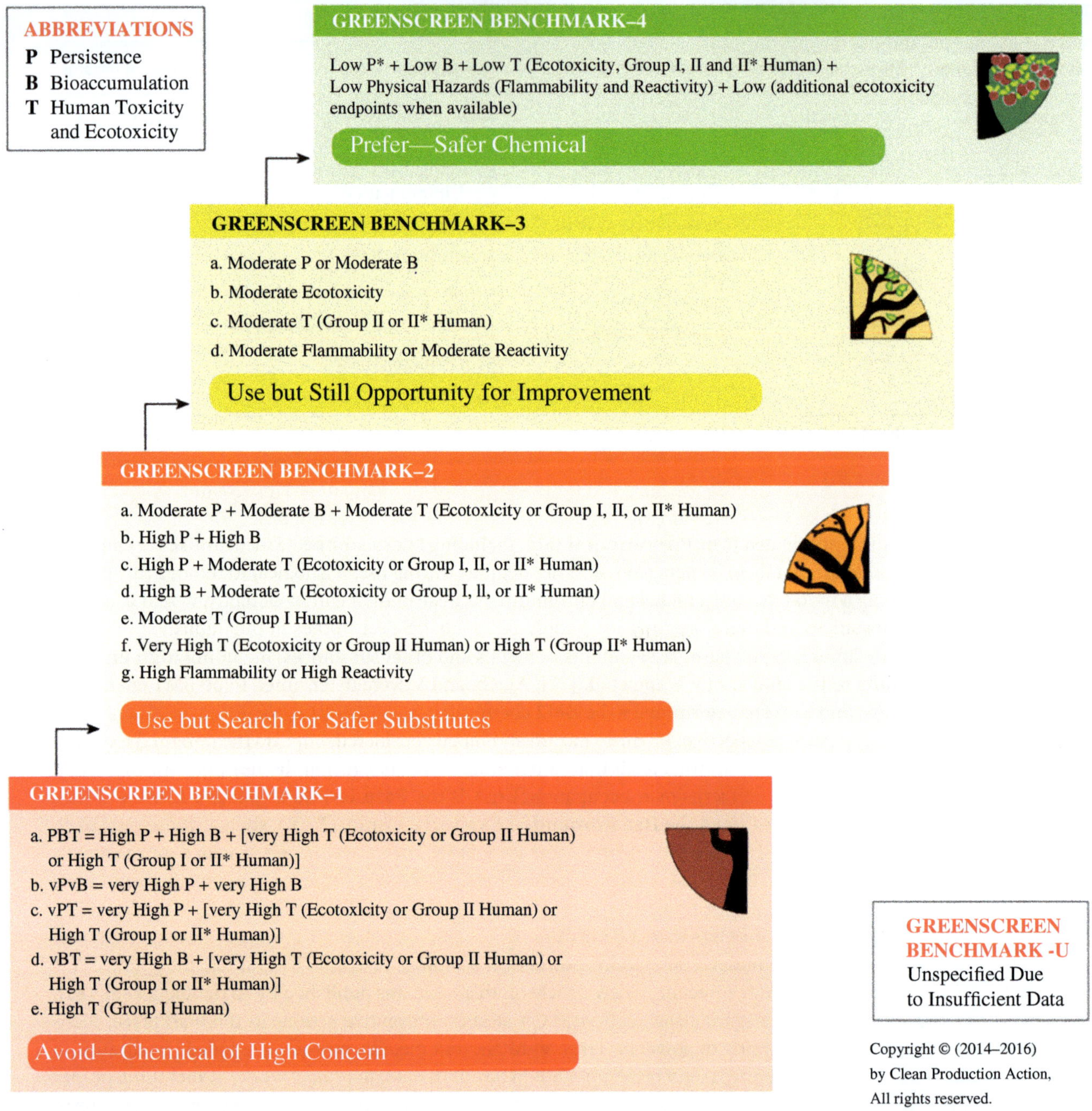

Figure 16.8 Details on the GreenScreen® for Safer Chemicals decision logic leading to benchmark scores. Source: GreenScreen Method [76].

16.4.2 Life Cycle Assessment of Environmental Impact

As described above in Section 16.3.1, it has become well-established that there are clear tradeoffs between conventional manufacturing and AM, such as the potential benefit of materials reduction coupled with the potential negative consequence of higher energy demand for powder production and machine operation in AM. The exact tradeoffs depend on the processing details, the material used, product design, and many other variables. In addition to these direct tradeoffs, there are also indirect effects, such as human health effects due to the use of hazardous substances or the emission of toxic pollutants from energy production. LCA has become a broadly recognized approach to evaluate these more complex impacts on the environment and human health, in a comprehensive, quantitative, analytical way. Key to this approach is to consider not just the manufacturing stage but also upstream processes such as raw material extraction and chemical synthesis, as well as downstream processes such as the use stage and end-of-life disposition (e.g. landfilling, recycling), as depicted in Figure 16.1. Transportation between these general life cycle stages is also considered. Although commonly simplified into four life cycle stages (raw materials, manufacturing, use, disposal/recycling), clearly, there could be dozens to hundreds of actual processing, use, and disposal steps along the way. For each step, inputs including materials, energy, water, and outputs such as air, soil, and water emissions, are quantified.

The **International Organization for Standardization (ISO)** formally describes the LCA methodology in their publication ISO 14001 (**environmental management**), specifically section ISO 14040 "Environmental management – Life cycle assessment – Principles and Framework" [88]. This method includes four steps: (i) Definition of Goal and Scope, (ii) **Life Cycle Inventory** (**LCI**), (iii) **Life Cycle Impact Assessment** (**LCIA**), and (iv) Interpretation, as described below. The definition of goal and scope requires the definition of the purpose of the study and the system boundaries. Most LCAs are comparative, so a functional unit needs to be defined as the basis of equivalency for the comparative studies. A commonly applied functional unit is one unit of product or one unit mass of material. System boundaries and scope definition include articulating which life cycle stages and process/use/disposal steps are to be included in the analysis, as well as which inputs and outputs. Geographic and temporal boundaries are also often defined. It is uncommon to do comprehensive LCAs covering all life cycle stages and all inputs and outputs; as for comparative studies, often many aspects are identical for the alternatives being compared. This streamlining of LCAs allows for less data to be collected, as data requirements and availability can often also limit the scope of a study. The LCI step in an LCA is the data-intensive step. Here, data are collected for all of the inputs and outputs associated with each process/use/disposal step within the defined scope. Numerous databases, such as **Ecoinvent** [89], now exist to facilitate this step in the LCA. The LCIA step requires converting the LCI data into environmental and human health impact categories that people value and can therefore make reasonable judgments when making tradeoff decisions. This LCIA step relies on environmental fate and transport modeling, combined with toxicology and exposure studies, to develop characterization factors that serve as multipliers to convert raw LCI values into LCIA impact values. The LCIA methodologies have also, over recent years, become relatively standardized. Commonly used LCIA methods include **ReCiPe** and **USEtox** [90, 91]. The LCI databases and LCIA characterization factors are commonly combined in commercially available software packages such as GaBi and SimaPro [92, 93]. The most complex step in LCA is the LCIA. Life cycle impact categories vary between methods, but commonly used categories are listed below [94]:

- Agricultural land occupation
- Climate change
- Fossil depletion
- Freshwater ecotoxicity
- Human toxicity
- Ionizing radiation
- Marine ecotoxicity
- Marine eutrophication
- Metal depletion
- Natural land transformation
- Ozone depletion
- Particulate matter formation
- Photochemical oxidant formation
- Terrestrial acidification
- Terrestrial ecotoxicity
- Urban land ecotoxicity
- Water depletion

These life cycle impact categories are frequently aggregated into damage categories, such as ecosystem health, human health, and resource scarcity, for ease of comparison. Some methods further aggregate these into a single-value indicator.

Given the inherent tradeoffs associated with switching from conventional manufacturing to AM, various researchers have evaluated these tradeoffs using LCA methods. While early studies were fairly limited in scope, often focusing only on the variable of part complexity [95, 96], the significant increased interest in AM in recent years has led to a similar increase in related LCA studies. While many of these still are fairly limited in scope, usually focusing on specific component case studies made of one material, with one AM technology, they still provide useful insights into key drivers that influence the overall environmental and human health impact of AM compared to conventional manufacturing technologies. Some of the studies also evaluate the impacts of metallic powder production. A few more recent studies are more comprehensive and highlight additional considerations that derive from the versatility of the AM techniques, as well as noting the refinement of the LCA methodologies themselves. Several of these LCA studies are briefly summarized below.

Lavery et al. conducted an LCA of sponge nickel produced by gas atomization for use as a catalyst [97]. Although not related to metal powder for AM, this study provides one of the few LCA studies on gas atomization of metallic powder. This study focused on Ni-Al powder production and provides a detailed inventory comparing gas atomization to two alternative processes: (i) cast and crush, and (ii) caustic leaching. Recycling and regeneration of the powder catalyst is considered. Their findings show that the gas atomization process requires more energy than the alternatives, increasing GWP. Notably, this increase is offset by the improved catalytic behavior during the use stage, ultimately leading to gas-atomized powder being preferred from a life cycle perspective. Bourhis et al. also evaluated gas atomization of powder, specifically that of metallic glass [98]. They provide LCI data on gas, water, and electricity consumption during atomization, and process efficiency, and they evaluated LCI using the **Eco-indicator** 99 approach. This was not a comparative study. Peng et al. [94] incorporated LCI data from a powder supplier to include gas atomization of stainless-steel powder as part their LCA study comparing SLM to conventional manufacturing (as described further below); their data included: yield, productivity, power, water consumption, and argon consumption. Their results indicate that the powder preparation stage represents the highest life cycle impact per part.

For the LCA case studies comparing conventional manufacturing to AM technologies, many of the components being evaluated are to be used in either the automotive industry or the aircraft industry, as for both applications, lightweighting of components can lead to significant energy savings during the "use stage" (i.e. while driving an automobile or truck, or flying an airplane), thus offsetting some of the additional energy required for the AM process. Paris et al. [20], for instance, evaluated the life cycle impacts of producing a Ti6Al4V aeronautic turbine, comparing EBM to conventional subtractive manufacturing. They considered not only energy tradeoffs but also 10 environmental and human health impact categories using the CML 2001 approach: abiotic depletion, acidification, global warming, freshwater aquatic ecotoxicity, marine aquatic ecotoxicity, terrestrial ecotoxicity, non-renewable fossil fuel-based energy, non-renewable nuclear energy, renewable energy, and renewable water. Their results indicate that EBM is preferred only when sufficient complexity in the part requires extensive energy-intensive machining. Tradeoffs with the use stage were not directly considered. Torres-Carillo et al. [99] conducted a similar LCA study on superalloy aircraft engine turbine blades, comparing the investment casting of Inconel 100 to SLM of Inconel 718. They used the CML 2001 and USEtox approaches to evaluate GWP, acidification potential, ozone layer depletion potential, human toxicity potential, and human toxicity with cancer and non-cancer effects. While they found the AM approach to be slightly preferred from the perspective of life cycle impact, they acknowledge that any change in the surface roughness of the preform to be machined in the conventional process will potentially offset this benefit, and therefore, the SLM process should be properly calibrated. This study also did not account for the use stage considerations.

In contrast, three recent studies consider both the manufacturing and use stages of the life cycle. Ingarao et al. [100] conducted an LCA on aluminum alloy components, comparing two conventional manufacturing process streams to SLM with gas-atomized powder. Overall, the latter was demonstrated to be more energy intensive than the conventional routes, for the manufacturing life cycle stage only. To consider the use stage, they applied a breakeven analysis assuming a **weight reduction** of 50%, which indicated that AM production was not environmentally beneficial for automotive (gas-powered) applications, but for both short-distance and long-distance aircraft applications, the benefits were significant. Böckin and Tillman conducted a more robust and detailed LCA of both the manufacturing and use life cycle stages [101]. Their study focused on an entire light-duty truck engine, manufactured by Volvo in Sweden. Three engine designs are considered with a range of materials (aluminum, cast iron, low-alloy steel, stainless steel) and various select components being made with AM techniques (gas atomization + PBF), resulting in weight reductions of up to ~20%. For the use stage, they considered life cycle impacts due to maintenance activities and fuel use for the truck, including impacts from fossil fuel extraction, refinement, and combustion. Their results indicate that the benefits from weight reduction do not offset the negative impact of the high energy demand in the AM process. Interestingly, they also extend their analysis to consider a "**clean energy scenario**" that would use

Table 16.8 Midpoints of different valve bodies.

Impact category	Abbreviation	Unit	V_{SLMopt}	V_{SLM}	V_{CM}
Agricultural land occupation	ALOP	m2a	1.906	2.217	3.102
Climate change	GWP100	kg CO2-Eq	25.624	30.651	36.610
Fossil depletion	FDP	kg oil-Eq	7.454	8.826	9.869
Freshwater ecotoxicity	FETPinf	kg 1,4-DCB-Eq	0.797	0.948	1.301
Freshwater eutrophication	FEP	kg P-Eq	0.012	0.015	0.017
Human toxicity	HTPinf	kg 1,4-DCB-Eq	10.407	12.534	15.475
Ionizing radiation	IRP_HE	kg U235-Eq	2.346	2.920	3.724
Marine ecotoxicity	METPinf	kg 1,4-DCB-Eq	0.757	0.904	1.247
Marine eutrophication	MEP	kg N-Eq	0.024	0.029	0.041
Metal depletion	MDP	kg Fe-Eq	8.271	10.421	14.917
Natural land transformation	NLTP	m^2	0.004	0.005	0.004
Ozone depletion	ODPinf	kg CFC-11-Eq	0.000	0.000	0.000
Particulate matter formation	PMFP	kg PM10-Eq	0.081	0.099	0.123
Photochemical oxidant formation	POFP	kg NMVOC	0.073	0.088	0.109
Terrestrial acidification	TAP100	kg SO2-Eq	0.127	0.151	0.170
Terrestrial ecotoxicity	TETPinf	kg 1,4-DCB-Eq	0.002	0.002	0.002
Urban land occupation	ULOP	m2a	0.201	0.244	0.363
Water depletion	WDP	m3	0.574	0.720	0.277

Source: Peng et al. [94].

electricity production with a low-carbon intensity and to consider a "**low-impact feedstock material**" replacing the nickel-based superalloys with steel. In both cases, the changes in the life cycle impact results are substantial, suggesting that AM technologies should avoid the application of nickel-based superalloys and support the application of low-carbon electricity production. Returning to the LCA study by Peng et al. [94], a comprehensive, detailed analysis and evaluation is presented for a steel hydraulic valve body, to be used in aircraft. They compare a conventionally manufactured part to a part with a sandwich structure made by SLM, yielding more than a 20% reduction in weight. As mentioned above, their analysis considers all 18 life cycle impact categories included in the ReCiPe method, as shown in Table 16.8 and Figure 16.9. Unlike several of the other studies cited here, they find that the SLM approach yields a lower impact than the conventional process overall, specifically in all 18 impact categories except water depletion, ozone layer depletion, and natural land transformation. Notably, the benefits in the human health category are particularly large (see Figure 16.10). To evaluate the weight reduction benefit in the use stage, they provide a quantitative analysis of the reduction in CO_2 emissions that result from the weight reduction in aircraft.

Two other LCA studies deserve mention here. In the study by Walachowicz et al. [102], the evaluation focuses on the application of AM to repair, rather than producing a new part of the component. This study compares conventional machining and **welding** to PBF for nickel-based superalloys and stainless steel as applied to gas turbine burners. Results indicate that the AM technique provides a significant reduction in material footprint (specifically abiotic depletion potential), primary energy consumption, and carbon footprint (GWP). Recycling of scrap to recover raw material was assumed in the study. In the study by Faludi et al. [19], an interesting expansion of the system boundary considered within the scope of the study was considered, specifically to also account for the life cycle impact of the materials in the machines and hardware themselves, which they refer to as "**machine embodied impacts**"; transportation of the machines from the location of manufacture to the location of use was also included. Notably, impacts due to actual manufacturing were not accounted for, only the embodied energy associated with the materials in the various components. For the aluminum alloy turbine wheel evaluated, they found that the machine-embodied impacts dominated in scenarios when machine utilization is low, and that certain impact categories (water depletion, metal depletion, and human toxicity) derive primarily from machine-embodied impacts, whereas other categories such as climate change, acidification, and land use, do not.

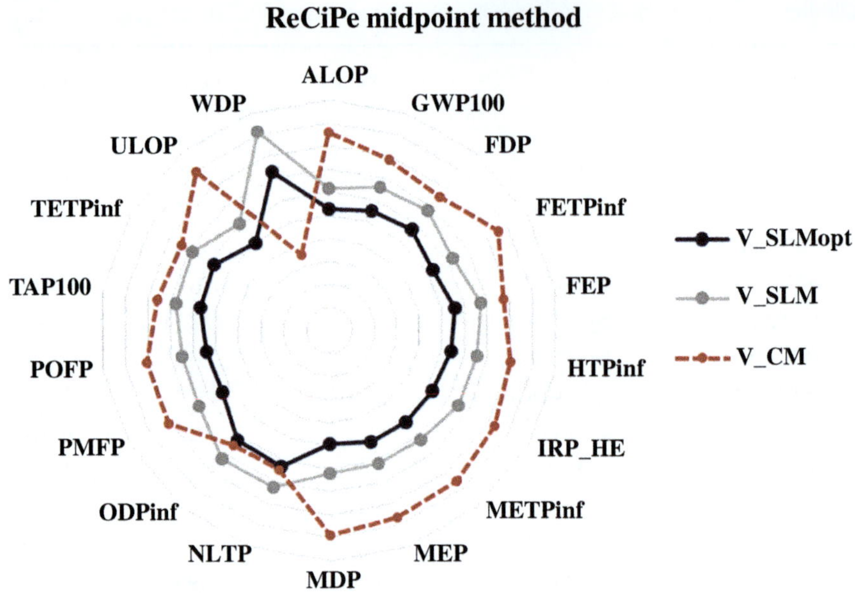

Figure 16.9 ReCiPe Midpoint H normalized impacts for different valve bodies. Source: Peng et al. [94].

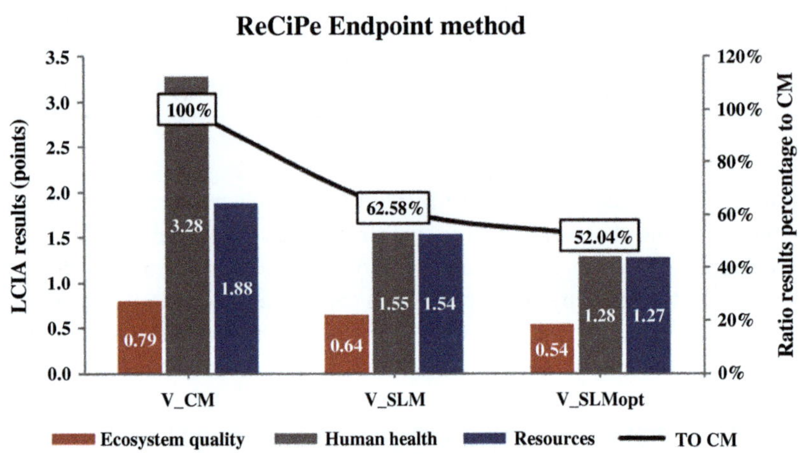

Figure 16.10 ReCiPe Endpoint H weighted impacts for different valve bodies. Source: Peng et al. [94].

While the LCA studies presented here provide insight into the life cycle impact tradeoffs of using metal powders and AM technologies to produce, and use, various industrial products, the scope of these studies is notably extremely limited. Further investigations are clearly needed that systematically, transparently, and robustly evaluate tradeoffs across the life cycle stages, materials, AM technologies, part designs, and applications. With the recent refinement of LCA methods themselves, such comprehensive yet detailed evaluations should be forthcoming.

16.5 Summary

This chapter provides an overview of resource utilization, economic, and sustainability assessment of metallic powder production and AM. As a rapidly evolving field, the specific resources used in these processes can vary depending on the technology and equipment being employed. As AM technologies advance, new materials and equipment may continue to be developed and utilized in the entire life cycle of AM products. Efficient utilization of resources including raw materials, consumables, energy, and water, not only reduces costs but also contributes to environmental sustainability and promotes responsible manufacturing practices. Continuous improvement and innovation in AM technologies and processes are essential for maximizing resource efficiency in the industry. Implementing sustainability initiatives and policies within organizations can encourage resource-conscious behavior and practices. The distributed production and consumption models enabled by AM technologies open a link between circular economics and AM. The prevailing material value chains are disrupted and shift toward distributed production from recycled materials. Consequently, deploying AM in the circular economic context comprises sourcing local materials and analyzing economic system conditions. While AM is promising in reducing environmental impacts and production costs, it can also lead to increases due to energy-intensive and costly material inputs, material waste during production, high machine costs, and slow rates of production. Future innovations will enable improved energy efficiency and materials utilization. Notably, for some applications, beneficial tradeoffs already exist due to potential energy savings during the use phase from weight reduction with AM. Yet the inherent physical, environmental, and human health hazards associated with metals and, especially, metallic powders need to be considered when selecting alloy compositions and designing manufacturing facilities to protect occupational workers from exposure and potential disease. As distributed production becomes more prevalent, vigilance is needed to ensure these protections. The inherent energy embodied in metal powders due to extraction, refining, and powder production provides incentives for powder reuse and recycling. From the perspective of LCA, standardized methods and databases are needed to allow detailed comparisons of the various materials, processes, and products of interest in this area of study and to account for not only the manufacturing stage but also the upstream materials production and downstream use, disposal and potential recycling and materials recovery. Robust consideration of system boundaries, including geographical boundaries and global energy sources, is needed. The dynamic nature of the ongoing transition from fossil-fuel-based non-renewable energy sources to renewable clean energy sources will further influence the materials – energy – use tradeoffs and therefore demand consideration when designing the metallic powder production and AM processes of the future.

Nomenclature

Variable	Description	Units
AM	Additive manufacturing	—
CASRN	Chemical Abstract Services Registry Number	—
CDC	Centers for Disease Control and Prevention	—
CED	Cumulative energy demand	MJ
CHA	Chemical hazard assessment	—
CNP	Classification not possible	—
DED	Directed energy deposition	—
DMD	Direct metal deposition	—
ED	Energy density	—
EB	Electron beams	—
EBM	Electron beam melting	—
GHS	Globally Harmonized System of Classification and Labeling of Chemicals	—
GWP	Global warming potential	—
IARC	International Center for Research on Cancer	—
ISO	International Organization for Standardization	—
LB	Laser beams	—
LCA	Life cycle assessment	—
LCI	Life Cycle Inventory	—
LCIA	Life Cycle Impact Assessment	—
LENS	Laser engineered net shaping	—
NA	Not applicable	—

Variable	Description	Units
NC	Not classified	—
NIOSH	National Institute for Occupational Safety and Health	—
NITE	National Institute of Technology and Evaluation	—
OSHA	Occupational Safety and Health Administration	—
PBF	Powder bed fusion	—
PEL	Permissible Exposure Limit	$mg\,m^{-3}$
REL	Recommended Exposure Limit	$mg\,m^{-3}$
SDS	Safety Data Sheets	—
SEC	Specific energy consumption	$MJ\,kg^{-1}$
SLM	Selective laser melting	—
SLS	Selective laser sintering	—
TEA	Techno-economic analysis	—
USGS	The United States Geological Survey	—
WHO	World Health Organization	—
C_{AM}	Total cost of producing an AM product	$ or €
C_{fixed}	Fixed costs	$ or €
C_{tot}	Total cost per part	$ or €
C_{mc}	Machine cost per part	$ or €
C_{mt}	Material cost per part	$ or €
C_{mr}	Machine hourly operating cost	$ or €
C_{lb}	Labor cost per part	$ or €
C_{op}	Machine operation cost per part	$ or €
C_{pp}	Post-processing cost per part	$ or €
C_{EAC}	Equivalent annual purchase cost	$ or €
$C_{support}$	Unit price of the support material per kilogram (kg)	$/kg
$C_{0\ mater\ i}$	Unit price of material *i* per kg	$/kg
D	Beam diameter	mm
ED_{linear}	Linear energy density	$J\,mm^{-1}$
ED_{areal}	Areal energy density	$J\,mm^{-2}$
ED_{vol}	Volumetric energy density	$J\,mm^{-3}$
$E[H]$	Expected operating hours in a day	hour
FGI	The inventory cost for finished goods	$ or €
H	Hatch spacing	mm
$m_{support}$	Mass of the support material	kg
$m_{mater\ i}$	Mass of the material *i* in a part	kg
MI	The cost of material inventory	$ or €

Variable	Description	Units
N_{days}	The number of working days in a month	—
P	Energy source beam power	Watt
RT_{AM}	The cost of retail trade for the AM product	
T_{AM}	Transportation cost throughout the supply chain for an AM product	$ or €
t_{build}	Build time	hour
t_s	Machine set up time	hour
t_l	Material loading time	hour
t_p	Parameter setting time	hour
t_m	Part manufacturing time	hour
t_r	Part removal time	hour
t_c	Machine cleaning time	hour
λ	Layer thickness	mm
v	Scanning speed	$mm\ s^{-1}$
WT_{AM}	Cost of wholesale trade for the AM product	$ or €

References

1 Oyesola, M., Mpofu, K., and Mathe, N. (2019). A techno-economic analytical approach of laser-based additive manufacturing processes for aerospace application. *Procedia Manuf.* 35: 155–163.

2 Liu, Z.Y., Li, C., Fang, X.Y., and Guo, Y.B. (2018). Energy consumption in additive manufacturing of metal parts. *Procedia Manuf.* 26: 834–845.

3 Ma, K., Smith, T., Lavernia, E.J., and Schoenung, J.M. (2017). Environmental sustainability of laser metal deposition: the role of feedstock powder and feedstock utilization factor. *Procedia Manuf.* 7: 198–204.

4 Lee, H.-K. (2008). Effects of the cladding parameters on the deposition efficiency in pulsed Nd:YAG laser cladding. *J. Mater. Process. Technol.* 202 (1): 321–327.

5 Unocic, R.R. and DuPont, J.N. (2004). Process efficiency measurements in the laser engineered net shaping process. *Metall. Mater. Trans. B* 35 (1): 143–152.

6 Delacroix, T., Lomello, F., Schuster, F. et al. (2022). Influence of powder recycling on 316L stainless steel feedstocks and printed parts in laser powder bed fusion. *Addit. Manuf.* 50: 102553.

7 Moghimian, P., Poirié, T., Habibnejad-Korayem, M. et al. (2021). Metal powders in additive manufacturing: a review on reusability and recyclability of common titanium, nickel and aluminum alloys. *Addit. Manuf.* 43: 102017.

8 Douglas, R., Lancaster, R., Jones, T. et al. (2022). The influence of powder reuse on the properties of laser powder bed-fused stainless steel 316L: a review. *Adv. Eng. Mater.* 24 (9): 2200596.

9 Ashby, M.F. (2021). The materials life cycle. In: *Materials and the Environment*, 3e (ed. M.F. Ashby), 41–64. Butterworth-Heinemann.

10 Kellens, K., Mertens, R., Paraskevas, D. et al. (2017). Environmental impact of additive manufacturing processes: does AM contribute to a more sustainable way of part manufacturing? *Procedia CIRP* 61: 582–587.

11 Kellens, K., Yasa, E., Renaldi, D. et al. (2011). Energy and resource efficiency of SLS/SLM processes. *22nd Annual International Solid Freeform Fabrication Symposium - An Additive Manufacturing Conference, Proceedings for the International Solid Freeform Fabrication Symposium 2011*, Austin, TX, USA (8–10 August 2011). pp. 1-16.

12 Morrow, W.R., Qi, H., Kim, I. et al. (2007). Environmental aspects of laser-based and conventional tool and die manufacturing. *J. Clean. Prod.* 15 (10): 932–943.

13 Gutowski, T., Jiang, S., Cooper, D. et al. (2017). Note on the rate and energy efficiency limits for additive manufacturing. *J. Ind. Ecol.* 21 (S1): S69–S79.

14 Wilson, J.M., Piya, C., Shin, Y.C. et al. (2014). Remanufacturing of turbine blades by laser direct deposition with its energy and environmental impact analysis. *J. Clean. Prod.* 80: 170–178.

15 Baumers, M., Tuck, C., Wildman, R. et al. (2011). Energy inputs to additive manufacturing: does capacity utilization matter? *2011 International Solid Freeform Fabrication Symposium*, Austin, TX, USA (8–10 August 2011).

16 Jackson, M.A., Van Asten, A., Morrow, J.D. et al. (2016). A comparison of energy consumption in wire-based and powder-based additive-subtractive manufacturing. *Procedia Manuf.* 5: 989–1005.

17 Baumers, M., Tuck, C., Hague, R. et al. (2010). A comparative study of metallic additive manufacturing power consumption. *2010 International Solid Freeform Fabrication Symposium*, Austin, TX, USA (9–11 August 2010).

18 Sachs, E., Cima, M., and Cornie, J. (1990). Three-dimensional printing: rapid tooling and prototypes directly from a CAD model. *CIRP Ann.* 39 (1): 201–204.

19 Faludi, J., Baumers, M., Maskery, I., and Hague, R. (2017). Environmental impacts of selective laser melting: do printer, powder, or power dominate? *J. Ind. Ecol.* 21 (S1): S144–S156.

20 Paris, H., Mokhtarian, H., Coatanéa, E. et al. (2016). Comparative environmental impacts of additive and subtractive manufacturing technologies. *CIRP Ann.* 65 (1): 29–32.

21 Bourell, D., Coholich, J., Chalancon, A., and Bhat, A. (2017). Evaluation of energy density measures and validation for powder bed fusion of polyamide. *CIRP Ann.* 66 (1): 217–220.

22 Gu, D. and Shen, Y. (2009). Effects of processing parameters on consolidation and microstructure of W–Cu components by DMLS. *J. Alloys Compd.* 473 (1): 107–115.

23 Ho, H.C.H., Gibson, I., and Cheung, W.L. (1999). Effects of energy density on morphology and properties of selective laser sintered polycarbonate. *J. Mater. Process. Technol.* 89–90: 204–210.

24 Campanelli, S.L., Casalino, G., Contuzzi, N. et al. (2014). Analysis of the molten/solidified zone in selective laser melted parts. *SPIE Proc.* 8963: 896311.

25 Gong, H., Rafi, K., Gu, H. et al. (2015). Influence of defects on mechanical properties of Ti–6Al–4V components produced by selective laser melting and electron beam melting. *Mater. Des.* 86: 545–554.

26 Ciurana, J., Hernandez, L., and Delgado, J. (2013). Energy density analysis on single tracks formed by selective laser melting with CoCrMo powder material. *Int. J. Adv. Manuf. Technol.* 68 (5): 1103–1110.

27 Duflou, J.R., Kellens, K., and Dewulf, W. (2011). Unit process impact assessment for discrete part manufacturing: a state of the art. *CIRP J. Manuf. Sci. Technol.* 4 (2): 129–135.

28 Kapur, A. and Graedel, T.E. (2004). Industrial ecology. In: *Encyclopedia of Energy* (ed. C.J. Cleveland), 373–382. New York: Elsevier.

29 Huang, R., Riddle, M., Graziano, D. et al. (2016). Energy and emissions saving potential of additive manufacturing: the case of lightweight aircraft components. *J. Clean. Prod.* 135: 1559–1570.

30 Azevedo, J.M.C., CabreraSerrenho, A., and Allwood, J.M. (2018). Energy and material efficiency of steel powder metallurgy. *Powder Technol.* 328: 329–336.

31 Rankin, J. (2012). Energy use in metal production. *Proceedings of the High Temperature Processing Symposium*, Austin, TX, USA (6–8 August 2012). pp. 6–7.

32 Serres, N., Tidu, D., Sankare, S., and Hlawka, F. (2011). Environmental comparison of MESO-CLAD® process and conventional machining implementing life cycle assessment. *J. Clean. Prod.* 19 (9): 1117–1124.

33 Baumers, M., Tuck, C., Wildman, R. et al. (2017). Shape complexity and process energy consumption in electron beam melting: a case of something for nothing in additive manufacturing? *J. Ind. Ecol.* 21 (S1): S157–S167.

34 Ashby, M.F. (2021). Resource consumption and its drivers. In: *Materials and the Environment*, 3e (ed. M.F. Ashby), 21–39. Butterworth-Heinemann.

35 Hoeges, S., Zwiren, A., and Schade, C. (2017). Additive manufacturing using water atomized steel powders. *Met. Powder Rep.* 72 (2): 111–117.

36 Kirchherr, J., Reike, D., and Hekkert, M. (2017). Conceptualizing the circular economy: an analysis of 114 definitions. *Resour. Conserv. Recycl.* 127: 221–232.

37 Ferreira, I.A., Godina, R., and Carvalho, H. (2021). Waste valorization through additive manufacturing in an industrial symbiosis setting. *Sustainability* 13 (1): 234.

38 Rosa, P., Sassanelli, C., Urbinati, A. et al. (2020). Assessing relations between circular economy and industry 4.0: a systematic literature review. *Int. J. Prod. Res.* 58 (6): 1662–1687.

39 Jung, S., Kara, L.B., Nie, Z. et al. (2023). Is additive manufacturing an environmentally and economically preferred alternative for mass production? *Environ. Sci. Technol.* 57 (16): 6373–6386.

40 Atzeni, E. and Salmi, A. (2012). Economics of additive manufacturing for end-usable metal parts. *Int. J. Adv. Manuf. Technol.* 62 (9): 1147–1155.

41 Son, Y.K. (1991). A cost estimation model for advanced manufacturing systems. *Int. J. Prod. Res.* 29 (3): 441–452.

42 Mahadik, A. and Masel, D. (2018). Implementation of additive manufacturing cost estimation tool (AMCET) using break-down approach. *Procedia Manuf.* 17: 70–77.

43 Dietrich, D. (2010). Emerging technology supply chain model for additive manufacturing. PhD dissertation. Missouri University of Science and Technology.

44 Lindemann, C., Jahnke, U., Habdank, M. et al. (2012). Analyzing product lifecycle costs for a better understanding of cost drivers in additive manufacturing. *23rd Annual International Solid Freeform Fabrication Symposium - An Additive Manufacturing Conference*, Austin, TX, USA (6–8 August 2012).

45 Ruffo, M., Tuck, C., and Hague, R. (2006). Empirical laser sintering time estimator for Duraform PA. *Int. J. Prod. Res.* 44 (23): 5131–5146.

46 Campbell, I., Combrinck, J., de Beer, D., and Barnard, L. (2008). Stereolithography build time estimation based on volumetric calculations. *Rapid Prototyp. J.* 14 (5): 271–279.

47 Di Angelo, L. and Di Stefano, P. (2011). A neural network-based build time estimator for layer manufactured objects. *Int. J. Adv. Manuf. Technol.* 57 (1): 215–224.

48 Wohlers, T. and Diegel, O. (2017). Investing in metal additive manufacturing: costs and considerations. *Metal AM Magazine* 3 (4): 93–98.

49 Campbell, I., Diegel, O., Kowen, J. et al. (2017). Wohlers Report 2017: 3D Printing and Additive Manufacturing State of the Industry Annual Worldwide Progress Report.

50 U.S. Geological Survey. (2023). Commodity Statistics and Information. https://www.usgs.gov/centers/national-minerals-information-center/commodity-statistics-and-information (accessed 15 November 2023).

51 Daily Metal Price: Free Metal Price Tables and Charts. https://www.dailymetalprice.com (accessed 15 November 2023).

52 Digital Alloys' Guide to Metal Additive Manufacturing - Part 5 Economics of Metal Additive Manufacturing. (2019). *Digital Alloys*. https://www.digitalalloys.com/blog/economics-metal-additive-manufacturing/ (accessed 15 November 2023).

53 Bauer, D.J., Nguyen, R.T., and Smith, B.J. (2023). Critical Materials Assessment. U.S. Department of Energy report. https://www.energy.gov/sites/default/files/2023-07/doe-critical-material-assessment_07312023.pdf (accessed 15 November 2023).

54 Thomas, D. and Gilbert, S. (2014). *Costs and cost effectiveness of additive manufacturing*, NIST Special Publication 1176. NIST https://doi.org/10.6028/NIST.SP.1176.

55 Thomas, D. (2016). Costs, benefits, and adoption of additive manufacturing: a supply chain perspective. *Int. J. Adv. Manuf. Technol.* 85 (5): 1857–1876.

56 Reeves, P. (2008). How the socioeconomic benefits of rapid manufacturing can offset technological limitations. *RAPID 2008 Conference and Exposition*, Lake Buena Vista, FL, USA (20–22 May 2008). pp. 1–12.

57 Holmström, J., Partanen, J., Tuomi, J., and Walter, M. (2010). Rapid manufacturing in the spare parts supply chain. *J. Manuf. Technol. Manag.* 21 (6): 687–697.

58 Wohlers, T. (2012). Wohlers Report 2012: Additive Manufacturing and 3D Printing State of the Industry.

59 Chan, H.K., Griffin, J., Lim, J.J. et al. (2018). The impact of 3D printing technology on the supply chain: manufacturing and legal perspectives. *Int. J. Prod. Econ.* 205: 156–162.

60 Thomas, D. (2013). *Economics of the U.S. Additive Manufacturing Industry*, NIST Special Publication 1163. NIST https://nvlpubs.nist.gov/nistpubs/SpecialPublications/NIST.SP.1163.pdf (accessed 15 November 2023).

61 Huang, S.H., Liu, P., Mokasdar, A., and Hou, L. (2013). Additive manufacturing and its societal impact: a literature review. *Int. J. Adv. Manuf. Technol.* 67 (5): 1191–1203.

62 Kunovjanek, M., Knofius, N., and Reiner, G. (2022). Additive manufacturing and supply chains – a systematic review. *Prod. Plan. Control* 33 (13): 1231–1251.

63 Khajavi, S.H., Partanen, J., and Holmström, J. (2014). Additive manufacturing in the spare parts supply chain. *Comput. Ind.* 65 (1): 50–63.

64 Schoenung, J.M., Olivetti, E.A., and Guest Editors (ed.) (2023). Sustainable development of materials: Broadening stakeholder engagement. *MRS Bull.* 48 (4): 362–367.

65 Lam, C.W., He, H., and Schoenung, J.M. (2017). Hazardous materials characterization and assessment. In: *Encyclopedia of Sustainability Science and Technology* (ed. R.A. Meyers), 1–21. New York: Springer New York.

66 International Agency for Research on Cancer (IARC). https://www.iarc.who.int/ (accessed 15 November 2023).

67 IARC. IARC's Mission: Cancer research for cancer prevention. https://www.iarc.who.int/about-iarc-mission/ (accessed 15 November 2023).

68 Agents Classified by the IARC Monographs, Volumes 1–134 – IARC Monographs on the Identification of Carcinogenic Hazards to Humans (2023). https://monographs.iarc.who.int/agents-classified-by-the-iarc/ (accessed 15 November 2023).

69 (2022). Pocket Guide to Chemical Hazards | NIOSH | CDC. https://www.cdc.gov/niosh/npg/default.html (accessed 15 November 2023).

70 NITE. (2023). NITE Japan-GHS | Chemical Management | National Institute of Technology and Evaluation (NITE). https://www.nite.go.jp/chem/english/ghs/ghs_nite_download_e.html (accessed 15 November 2023).

71 UNECE. About the GHS. https://unece.org/about-ghs (accessed 15 November 2023).

72 UNECE. Globally Harmonized System of Classification and Labelling of Chemicals (GHS Rev. 10, 2023). https://unece.org/transport/dangerous-goods/ghs-rev10-2023 (accessed 15 November 2023).

73 OSHA3514 Hazard Communication Standard: Safety Data Sheets. (2012). https://www.osha.gov/sites/default/files/publications/OSHA3514.pdf (accessed 15 November 2023).

74 (2013). GHS Classification Guidance for the Japanese Government 2013 Revised Edition. https://www.meti.go.jp/policy/chemical_management/int/files/ghs/h25jgov_en.pdf (accessed 15 November 2023).

75 GreenScreen® For Safer Chemicals. https://www.greenscreenchemicals.org (accessed 15 November 2023).

76 GreenScreen Method. (2018). *GreenScreen® For Safer Chemicals.* https://www.greenscreenchemicals.org/assess/method (accessed 15 November 2023).

77 Wang, X., Kramer, A., Glaubensklee, C. et al. (2022). Sustainability-based selection of materials for refractory high entropy alloys. *REWAS 2022: Developing Tomorrow's Technical Cycles (Volume I)*, 377–385.

78 Eisenberg, D.A., Yu, M., Lam, C.W. et al. (2013). Comparative alternative materials assessment to screen toxicity hazards in the life cycle of CIGS thin film photovoltaics. *J. Hazard. Mater.* 260: 534–542.

79 He, H., Gutierrez, Y., Young, T.M., and Schoenung, J.M. (2019). The role of data source selection in chemical hazard assessment: a case study on organic photovoltaics. *J. Hazard. Mater.* 365: 227–236.

80 He, H., Malloy, T.F., and Schoenung, J.M. (2019). Multicriteria decision analysis characterization of chemical hazard assessment data sources. *Integr. Environ. Assess. Manag.* 15 (6): 895–908.

81 Alcaraz Ochoa, M.L., He, H., Schoenung, J.M. et al. (2019). Design parameters and environmental impact of printed wiring board manufacture. *J. Clean. Prod.* 238: 117807.

82 He, H., Tian, S., Glaubensklee, C. et al. (2022). Advancing chemical hazard assessment with decision analysis: a case study on lithium-ion and redox flow batteries used for energy storage. *J. Hazard. Mater.* 437: 129301.

83 Davis, G., Hickox, W.H., and Denton, J. (2001). *A Guide to Health Risk Assessment*. California Environment Protection Agency, Office of Environmental Health Hazard Assessment https://oehha.ca.gov/media/downloads/risk-assessment/document/hrsguide2001.pdf (accessed 15 November 2023).

84 Occupational Safety and Health Administration. (2023). Permissible Exposure Limits – Annotated Tables. https://www.osha.gov/annotated-pels (accessed 15 November 2023).

85 Arrizubieta, J.I., Ukar, O., Ostolaza, M., and Mugica, A. (2020). Study of the environmental implications of using metal powder in additive manufacturing and its handling. *Metals* 10 (2): 261.

86 Graff, P., Ståhlbom, B., Nordenberg, E. et al. (2017). Evaluating measuring techniques for occupational exposure during additive manufacturing of metals: a pilot study. *J. Ind. Ecol.* 21 (S1): S120–S129.

87 Ljunggren, S.A., Karlsson, H., Ståhlbom, B. et al. (2019). Biomonitoring of metal exposure during additive manufacturing (3D printing). *Saf. Health Work* 10 (4): 518–526.

88 ISO (2014). ISO 14040:2006 Environmental management – Life cycle assessment – Principles and framework. *ISO*. https://www.iso.org/standard/37456.html (accessed 15 November 2023).

89 (2020). Ecoinvent Database - ecoinvent. https://ecoinvent.org/the-ecoinvent-database/ (accessed 15 November 2023).

90 RIVM. LCIA: the ReCiPe model. https://www.rivm.nl/en/life-cycle-assessment-lca/recipe (accessed 15 November 2023).

91 USEtox®. The USEtox Model. https://usetox.org/model (accessed 15 November 2023).

92 Sphera. Sustainability, Operational Risk Management & EHS Software. https://sphera.com/ (accessed 15 November 2023).

93 SimaPro. LCA software for informed changemakers. *SimaPro.* https://simapro.com/ (accessed 15 November 2023).

94 Peng, T., Wang, Y., Zhu, Y. et al. (2020). Life cycle assessment of selective-laser-melting-produced hydraulic valve body with integrated design and manufacturing optimization: a cradle-to-gate study. *Addit. Manuf.* 36: 101530.

95 Xiong, Y. and Schoenung, J.M. (2010). Process cost comparison for conventional and near-net-shape cermet fabrication. *Adv. Eng. Mater.* 12 (3): 235–241.

96 Xiong, Y., Lau, K., Zhou, X., and Schoenung, J.M. (2008). A streamlined life cycle assessment on the fabrication of WC–Co cermets. *J. Clean. Prod.* 16 (10): 1118–1126.

97 Lavery, N.P., Jarvis, D.J., Brown, S.G.R. et al. (2013). Life cycle assessment of sponge nickel produced by gas atomisation for use in industrial hydrogenation catalysis applications. *Int. J. Life Cycle Assess.* 18 (2): 362–376.

98 Bourhis, F.L., Kerbrat, O., Hascoet, J.-Y., and Mognol, P. (2013). Sustainable manufacturing: evaluation and modeling of environmental impacts in additive manufacturing. *Int. J. Adv. Manuf. Technol.* 69 (9): 1927–1939.

99 Torres-Carrillo, S., Siller, H.R., Vila, C. et al. (2020). Environmental analysis of selective laser melting in the manufacturing of aeronautical turbine blades. *J. Clean. Prod.* 246: 119068.

100 Ingarao, G., Priarone, P.C., Deng, Y., and Paraskevas, D. (2018). Environmental modelling of aluminium based components manufacturing routes: additive manufacturing versus machining versus forming. *J. Clean. Prod.* 176: 261–275.

101 Böckin, D. and Tillman, A.-M. (2019). Environmental assessment of additive manufacturing in the automotive industry. *J. Clean. Prod.* 226: 977–987.

102 Walachowicz, F., Bernsdorf, I., Papenfuss, U. et al. (2017). Comparative energy, resource and recycling lifecycle analysis of the industrial repair process of gas turbine burners using conventional machining and additive manufacturing. *J. Ind. Ecol.* 21 (S1): S203–S215.

17

Future Directions and Challenges

17.1 Introduction

In this concluding chapter, we explore the intersections of metal powders and additive manufacturing (AM) and identify future directions and challenges. First, we describe the goal and organization of the book and provide insights gleaned from each preceding chapter, followed by a discussion of challenges and promising opportunities that converge in this dynamic realm.

We present a comprehensive survey of the historical evolution and contemporary advancements in the production of metallic powders for AM, alongside an examination of the influence of powder chemistry and characteristics on AM processes and final component performance. The book is organized into distinct chapters as follows. Chapters 1 and 2 provide an overview of atomization techniques, highlighting their historical significance and applications in various technological fields. They discuss the complexity of atomizer designs, using gas jets to break down metals into fine powders, and the importance of controlling the atomization environment for optimal powder quality. Chapter 3 delves into the physical processes involved in atomization, such as the disintegration of molten metal into droplets and the subsequent heat transfer and solidification of these droplets. Various theoretical models and correlations are explored to gain insights into liquid breakup and droplet formation. Chapter 4 on composite powders for AM discusses the methods used to manufacture composite powders, including co-injection of ceramic particles with metal droplets and in-situ generation of oxide phases. The interactions between ceramic particles and metal droplets are complex and involve thermal, fluid, and mass transfer phenomena. Other fabrication methods like mechanical milling and surface coatings are also mentioned. Diagnostic and characterization techniques are reviewed in Chapter 5, focusing on their role in monitoring and controlling the atomization process. Different methods, such as laser light diffraction, thermal imaging, and surface ionization, are discussed, each with its own advantages and limitations. The progress made in monitoring and controlling atomization processes and the potential for further optimization are highlighted. Chapters 6 and 7 specifically address the atomization of alloys, providing insights into the production of metallic alloy powders using gas atomization. The importance of powder preparation processes and controlling physical characteristics and chemical composition are emphasized. Different atomization methods, including gas atomization, plasma atomization, and plasma rotating electrode process, are discussed in detail, along with the challenges of satellite particles and gas pores. An overview of metal AM technologies is provided in Chapter 8, focusing on the working principles of various AM processes and their requirements in terms of feedstock powder and post-processing treatments. This chapter highlights the significance of powder characteristics in controlling the process and ensuring the quality of the final parts. Chapter 9 describes the complex interactions among the powder, laser, and melt pool in AM processes. This chapter delves into the interactions between laser, powder, and melt pools within both directed energy deposition (DED) and powder-bed fusion (PBF) AM processes. Available information shows that an understanding of porosity and residual stresses hinges upon a comprehensive understanding of the dynamics of the melt pool. Advanced approaches like X-ray radiography offer the potential for deeper insights into intricate laser-material interplays. Subsequent research should center on validating prevailing theories and numerical simulations concerning melt pool behavior. Chapters 10 through 12 cover the influence of powder chemistry, physical powder characteristics, and microstructure evolution on AM. Chapter 10 discusses the point that the printability and final quality of AM parts are influenced by thermophysical properties like melting point, specific heat capacity, enthalpy of fusion, and reflectivity, which, in turn, affect the AM process. Alloys with diverse laser absorption properties exhibit differing local energy absorption, often modulated by chemical macro-segregation in the powders, subsequently impacting the

Metallic Powders for Additive Manufacturing: Science and Applications, First Edition.
Enrique J. Lavernia, Kaka Ma, Julie M. Schoenung, James F. Shackelford, and Baolong Zheng.

melting and solidification modes during AM. Chapters 11 and 12 provide an overview of pertinent aspects of metal powder characteristics in AM. They cover topics including powder characterization methods, production techniques, reuse and recycling, the impact of production methods and process variables on powder traits, and post-production treatment effects. These chapters also describe the pivotal factors shaping the microstructure of metallic parts in AM: process parameters and alloy compositions (powder effect) are often interdependent. Chapters 13 through 15 discuss the role of powders in defect formation, residual stress, and the physical and chemical behavior of AM-produced parts. These chapters emphasize the impact of powder properties, process optimization, and post-processing treatments on the quality, properties, and performance of AM parts. These chapters highlight key aspects of AM and its potential to replace conventional methods, emphasizing that AM-produced parts must exhibit properties comparable to traditionally manufactured ones. The analysis encompasses defects like porosity, cracking, delamination, and residual stresses. A range of strategies, including process optimization, feedstock control, and post-treatments, is explored for addressing these issues. The significance of powder characteristics, such as particle size distribution and morphology, is underlined for successful part construction. Consideration is given to factors like thermal cycling and microstructural features, which impact residual stress. Additionally, various physical properties such as density, surface roughness, elastic/plastic behavior, corrosion resistance, and wear performance are assessed in the context of AM-produced parts compared to conventional counterparts. The potential of AM to achieve equivalent or improved attributes in these areas is discussed. Chapter 16 discusses the economics of powder production for AM and strategies for improving sustainability, such as carefully selecting and designing metallic alloys and powder production techniques, using alternative feedstocks, implementing powder recycling, and minimizing waste and energy consumption.

In the subsequent sections, we leverage the information presented in Chapters 1–16 to explore the future prospects, hurdles, and upcoming trajectories surrounding atomized metallic powders in the context of AM. We dissect the potential avenues that lie ahead, offering opportunities for innovation and advancement in the realm of AM. Simultaneously, we describe the challenges that must be surmounted to fully harness the capabilities of atomized metallic powders in this transformative field. Moreover, we summarize future directions in the AM landscape, charting a course for the evolution of metallic component fabrication.

17.2 Future Directions in the Atomization of Powders

Building on the comprehensive review of metal powder atomization technologies provided in Chapters 1–7, as briefly summarized in Section 17.1, one can envision future advances in three areas, as outlined below. Metal powders play a crucial role in metallic AM and have become paramount in the field of 3D printing. The utilization of these metal powders in industries to fabricate AM components with intricate geometries has witnessed significant growth in recent decades. While several powder production methods for metallic powders have been developed, gas atomization remains the primary manufacturing process for AM powders. Researchers and manufacturers are actively working toward optimizing atomization processes and enhancing product quality to meet the surging demand for powder in AM applications. In addition, there is the potential for a role of atomization in producing nano-scale powders that could significantly contribute to the future of nanotechnology, and atomization processes could be integrated into Industry 4.0 frameworks, allowing for smarter and more automated production lines with real-time data monitoring and analysis. In the spirit of the concepts raised in Chapter 16, researchers must also find more eco-friendly and sustainable ways to atomize metals, produce feedstock powders for AM, and recycle and reuse waste powders, to reduce energy consumption, waste generation, and environmental impacts.

17.2.1 Technology Improvements

Enhanced atomization techniques and equipment may be developed to improve powder quality, size distribution, and efficiency. Advanced atomization methods, such as gas atomization, water atomization, or plasma atomization, might be further refined. These technological enhancements will be partly driven by the enhanced demand for metal powders, as shown in Figure 17.1 [1] and Figure 17.2 [2]. Growth will increase as market value increases in various sectors, especially defense and transportation.

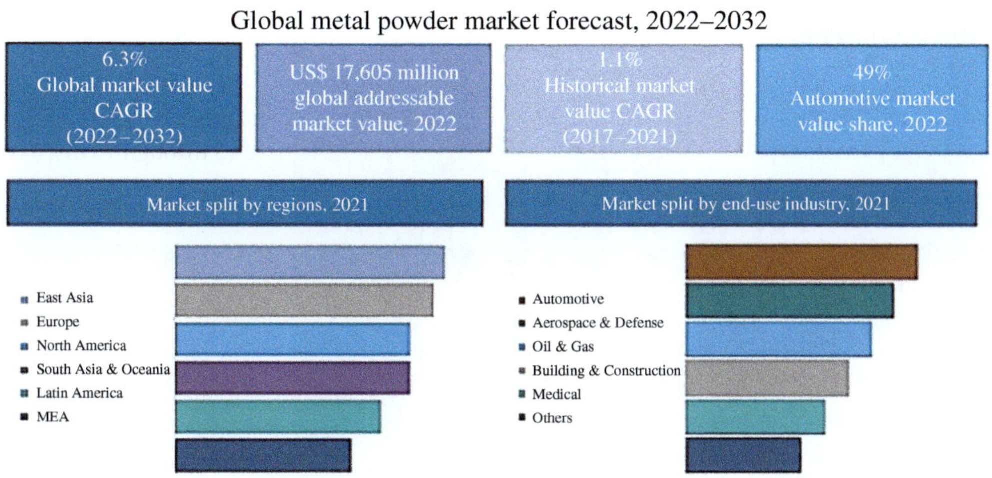

Figure 17.1 Metal powder market size, share, growth analysis forecast 2022–2032. Source: Fact.MR. https://www.factmr.com/report/3202/metal-powder-market/ last accessed Aug 31, 2023.

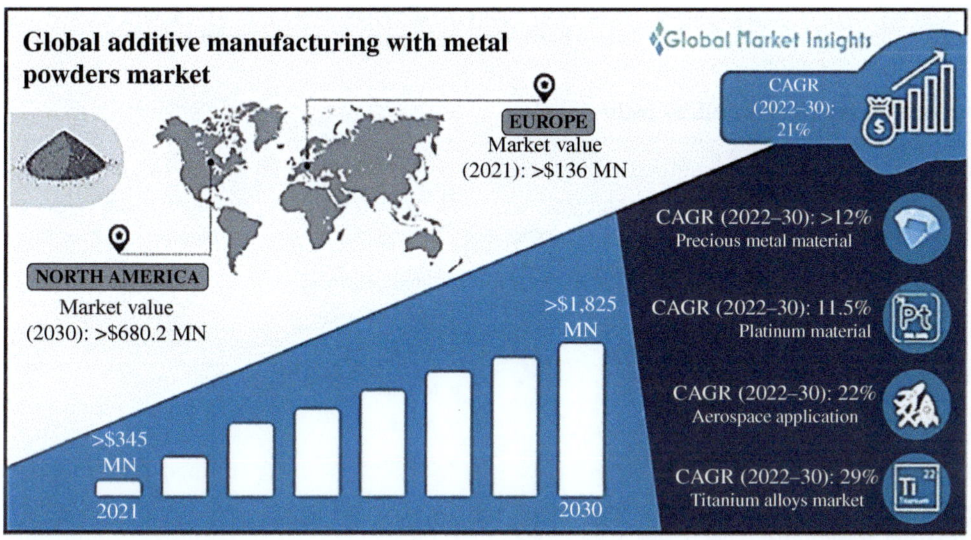

Figure 17.2 Projected additive manufacturing with metal powders market statistics, 2021–2030. Source: Ref. [2] Global Market Insights. https://www.gminsights.com/industry-analysis/additive-manufacturing-with-metal-powders-market last accessed Aug 31, 2023.

17.2.2 Custom Alloys and Composites

Atomization may enable the production of new and unique metal alloys and composites with tailored properties. This could lead to the development of materials with superior strength, corrosion resistance, or other specialized characteristics. Exponential growth in demand for powder is anticipated, especially for alloys of titanium, nickel, steel, and aluminum, as illustrated in Figure 17.3 [3] and described in Chapter 7.

17.2.3 Additive Manufacturing

As AM continues to grow, metal powders produced through atomization could be optimized for specific 3D printing processes. This could lead to the increased adoption of 3D printing in industries where metal parts are crucial. As shown in Figure 17.4 [4], the consumption of metal powder in AM has been projected to reach approximately $520 million in 2019, with a notable growth anticipated to reach $930 million by 2023.

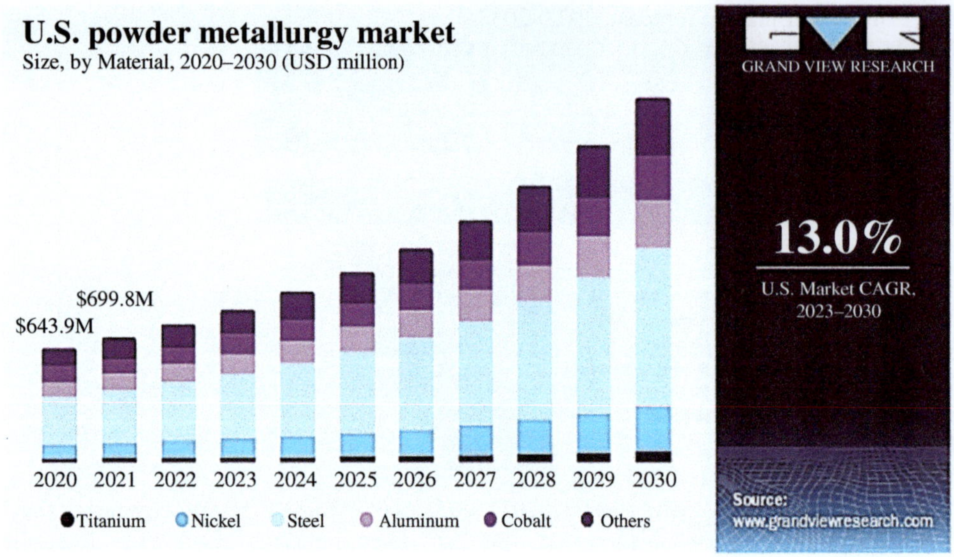

Figure 17.3 Projected powder metallurgy market size, share, and growth report, 2020–2030. Source: Ref. [3] Grand View Research. https://www.grandviewresearch.com/industry-analysis/powder-metallurgy-market# last accessed Aug 31, 2023.

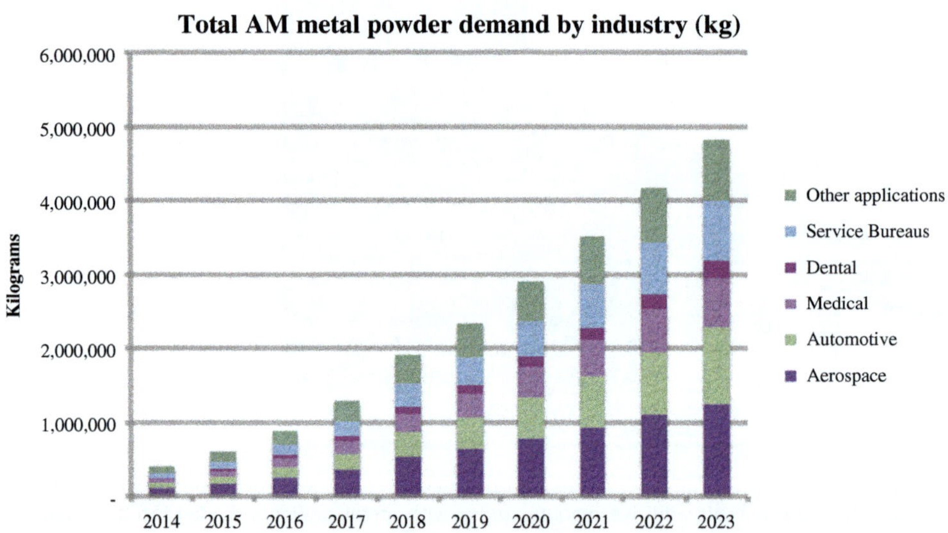

Figure 17.4 Projected metal powder shipments for additive manufacturing. Source: Ref. [4] 3DR HOLDINGS. https://www.smartechanalysis.com/news/metal-powder-shipments-am/ last accessed Aug 31, 2023.

17.3 Future Directions and Challenges in the Additive Manufacturing of Metal Alloys

The comprehensive review of the effects of atomized powders on AM in Part 2 of this book (Chapters 8–16) provides insights into the future directions of this technology. As shown in Figure 17.5 [5], the potential adoption rate for AM in various alloy groups is highly promising. This section identifies five directions and challenges along the path of this projected growth.

17.3.1 Machine Learning and Artificial Intelligence

Machine learning, modeling, and advanced control algorithms will be crucial in optimizing AM processes. By dynamically controlling process parameters during printing, the performance and quality of printed parts can be

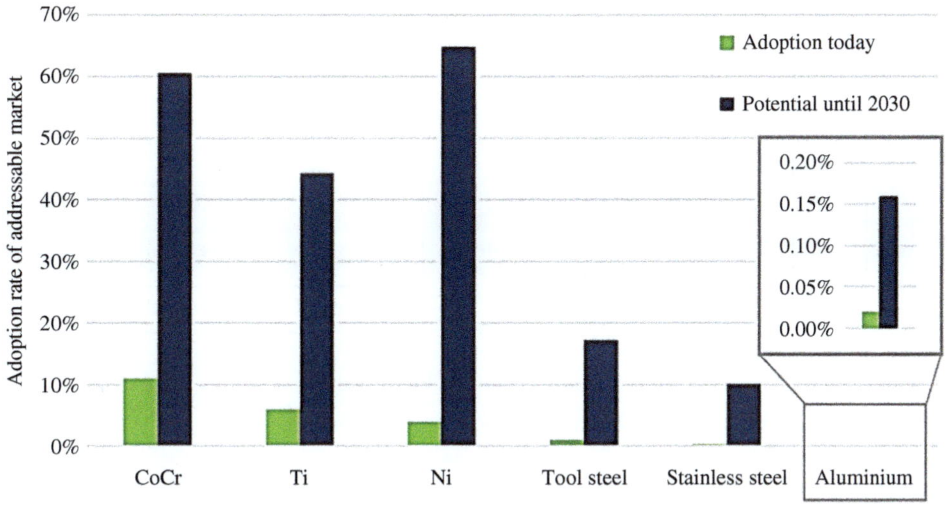

Figure 17.5 Strong increases in the adoption rate of additive manufacturing for various alloy groups as predicted by market analysts. (Courtesy of Dr. Munsch and Dr. Wycisk (AMPOWER, ampower.eu), published in Metal AM magazine, metal-am.com). Source: Ref. [5] Metal Additive Manufacturing. https://www.metal-am.com/metal-additive-manufacturing-magazine-archive/metal-additive-manufacturing-vol-9-no-2-summer-2023/ last accessed Aug 31, 2023.

maximized. For example, machine learning algorithms can be employed to analyze real-time sensor data and adjust parameters like laser power, scan speed, or powder flow rate to achieve optimal printing outcomes. This integration of artificial intelligence (AI)-based technologies with AM will lead to more efficient and consistent production.

As applied to AM, ongoing and future developments in AI hold great potential for revolutionizing the industry. Here are some key areas where AI is expected to make significant advancements: (i) Process Optimization: AI algorithms will play a vital role in optimizing the AM process parameters to achieve enhanced part quality, reduced production time, and minimized material waste. By analyzing vast amounts of data, including sensor readings, historical process data, and simulation results, AI models can identify optimal parameter settings for specific materials, geometries, and desired part characteristics. This will improve efficiency, reduce trial and error, and increase productivity in AM. Figure 17.6 illustrates a schematic diagram for a modeling framework comprising physics-based simulation and physics-model-based machine learning modeling to provide spreading predictions in powder-bed AM to offer layerwise control over spreading defects [6]. (ii) Real-time Monitoring and Quality Control: AI-based monitoring systems will provide real-time feedback on the printing process, allowing for early detection of anomalies or defects. By integrating sensors, computer vision, and machine learning techniques, AI algorithms can continuously analyze data to identify potential issues, such as layer defects, warping, or irregularities in the printing process. This will enable proactive intervention and quality control measures, leading to higher precision and reliability in AM production. (iii) Autonomous Printing: AI-driven autonomous systems will have the capability to autonomously control and adapt the printing process based on real-time feedback and predefined objectives. These systems will make decisions on the fly, adjusting parameters, optimizing toolpaths, or addressing unexpected challenges during the printing process. Autonomous printers will enhance efficiency, reduce human intervention, and improve AM's overall reliability and repeatability. (iv) Design Optimization: AI algorithms will assist in generating optimized designs for AM. By leveraging generative design techniques and machine learning, AI can explore a vast design space and identify novel geometries or lattice structures that maximize part performance while minimizing weight and material usage. AI-driven design optimization will enable the creation of highly efficient, lightweight, and complex parts that are tailored for specific applications, unlocking new levels of functionality and performance. (v) Material Development: AI will contribute to accelerating material development for AM. AI models can predict and recommend optimal material compositions or combinations for specific applications by analyzing material properties, performance requirements, and desired characteristics. This will enable the creation of new materials with enhanced properties, improved compatibility with AM processes, and tailored functionalities. (vi) Supply Chain Optimization: AI can optimize the AM supply chain by leveraging predictive analytics to anticipate material requirements, demand fluctuations, and production scheduling. AI algorithms can analyze historical data,

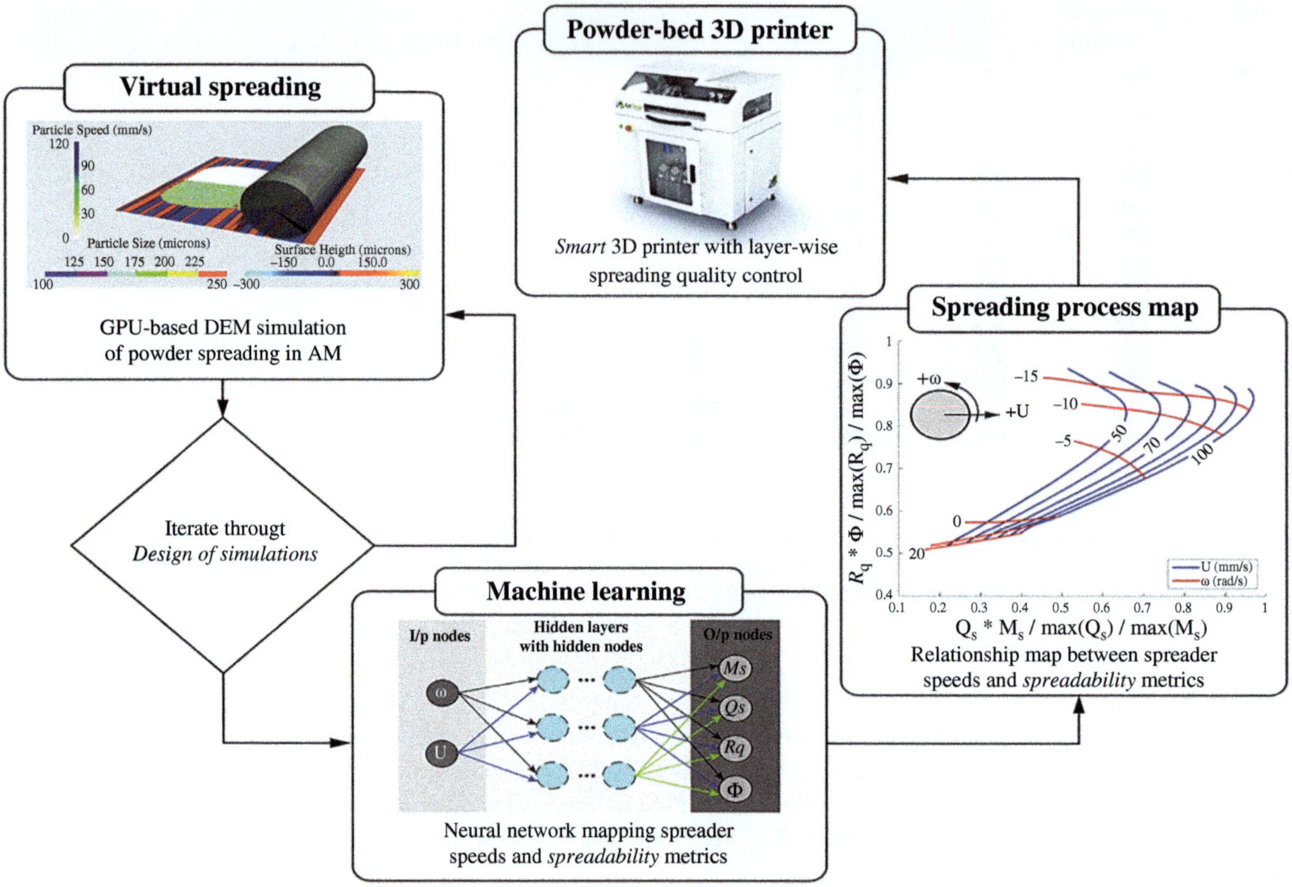

Figure 17.6 Schematic diagram of powder spreading process maps for powder-bed AM derived from physics model-based machine learning. Source: Ref. [6], MDPI/CC BY 4.0.

market trends, and other factors to optimize inventory levels, reduce lead times, and streamline logistics. This will improve cost efficiency, reduce waste, and enhance responsiveness to market demand.

17.3.2 Novel Structures

The advancements mentioned above will enable the fabrication of increasingly complex and hierarchical microstructures with unprecedented physical and mechanical properties combinations. For example, lattice structures with three-dimensional topography optimized for weight reduction or bone-like structures with enhanced strength and flexibility for medical implants can be fabricated using AM. Furthermore, multi-material printing or gradient structures can offer tailored material properties, such as varying stiffness or conductivity within a single component.

With the continuous advancements in AM, the size and complexity of parts that can be manufactured using these technologies will continue to increase. This opens new opportunities for engineering applications where traditional manufacturing methods may be limited. For example, large-scale structures, such as aircraft components or infrastructure elements, can be efficiently produced using AM, reducing the need for complex assembly, or joining processes. Additionally, the ability to print intricate geometries or internal features enables the creation of customized parts with improved functionality and performance.

17.3.3 Hybrid Manufacturing

AM technologies will be integrated with traditional fabrication processes, such as machining, welding, coating, joining, and cutting, to produce complex components efficiently and economically. This hybrid manufacturing approach combines the advantages of AM, such as geometric freedom and customization, with the high precision, surface finish, or material properties achievable through traditional processes. For instance, an AM-printed part can be post-processed using machining techniques to achieve tight tolerances or obtain specific surface finishes. In addition, the feedstock for AM can be expanded from powders and wires to larger, basic blocks such as cubes and cylinders, given the development of robotic technology and hybrid fast sintering methods with AM. When large, relatively defect-free blocks can be well controlled by robotic arms to be joined or assembled additively according to computer-aided designs, the production time of scaled-up parts will be significantly reduced, as some unnecessary powder particle-to-particle bonding/fusion at regular, noncomplex geometrical locations can be avoided. Therefore, hybrids of current AM technologies with the production of large but basic blocks for AM feedstock are also a future direction worthy of significant research effort.

17.3.4 Diagnostic Methods

One of the primary challenges in controlling the AM process is effectively managing both dimensions and properties, particularly in DED processing. Precise dimensional control can lead to significant cost savings in post-machining for surface finishing. Desired material properties can be achieved and costs reduced through meticulous process control and minimizing the need for post-deposition heat treatment.

Maintaining consistent molten pool size and deposition distance during AM deposition is crucial for ensuring microstructural homogeneity and dimensional accuracy of the components. To achieve this, proper thermal management of the molten pool and deposited layer thickness is essential during the deposition process. Two approaches can be employed to accomplish this [7]:

1. Utilizing appropriate process parameters in the traditional open-loop laser process, which heavily relies on the operator's expertise to establish a production process.
2. Implementing external sensors and control strategies in closed-loop laser processing to ensure the stability of molten pool size and deposited thickness throughout the deposition process. This approach enhances control and accuracy during AM.

To address the inherent stochastic nature of AM processes, novel diagnostic methods will be employed, particularly those that can be used in situ during the manufacturing process. These diagnostic techniques can provide real-time feedback on process conditions, allowing for timely adjustments and quality control. For instance, in-situ monitoring using sensors, thermal imaging, or acoustic emissions can help detect and mitigate defects, such as porosity or cracks, before they compromise the integrity of the final printed part.

17.3.5 Future Challenges

Future challenges with AM include: (i) Standards and Certification: Establishing consistent standards for material properties, process parameters, and part quality is crucial for the widespread adoption of AM. Developing certification processes and regulatory frameworks that ensure the reliability and safety of AM-produced components remain important challenges. (ii) Sustainability and Recycling: As discussed in Chapter 16, addressing the environmental impact of AM processes, including energy consumption and waste generation, is a growing concern. Researchers are exploring sustainable materials, recycling methods, and more energy-efficient printing techniques to minimize the ecological footprint of AM.

17.4 Summary

In summary, the future directions and challenges for the use of atomized metal powders in AM indicate optimistic horizons. As an example, Figure 17.7 summarizes a vision for both short-term (evolutionary) and long-term (revolutionary) horizons for AM that has been provided in a collaboration between General Motors (GM) and Lockheed Martin [8]. These "blueprints" known as "AM Dream Machine Requirements," indicate manufacturing advances moving from the current high-value/low-volume applications in the aerospace and medical industries to broader applications in the large-volume automotive industry. Early success with AM at GM led to the Additive Industrialization Center (AIC) opening in 2020 to accelerate the adoption of AM across the organization. This commitment was followed by the GM/Lockheed Martin collaboration, in which "evolutionary" requirements include upgrading their suite of PBF-LB machines to provide improvements in operations such as automated

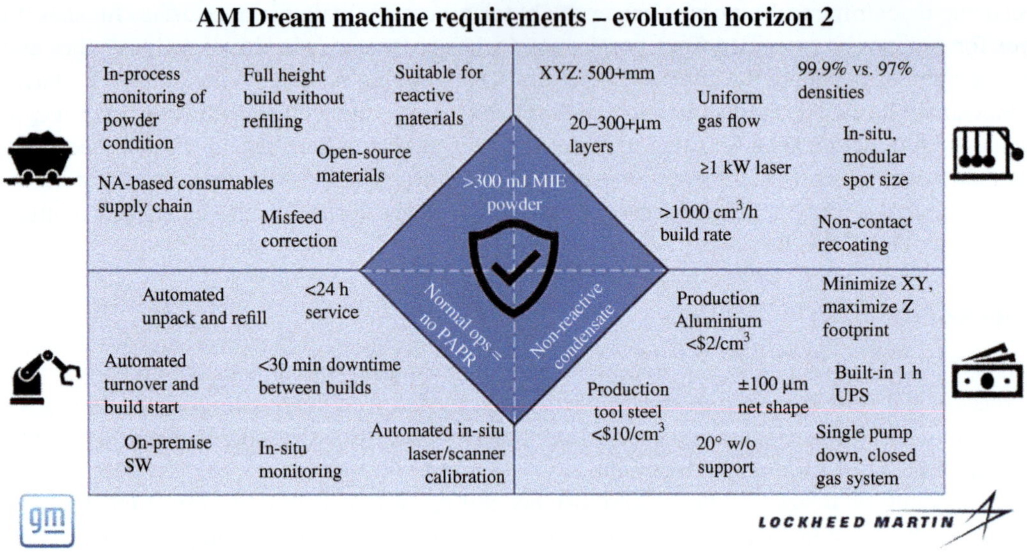

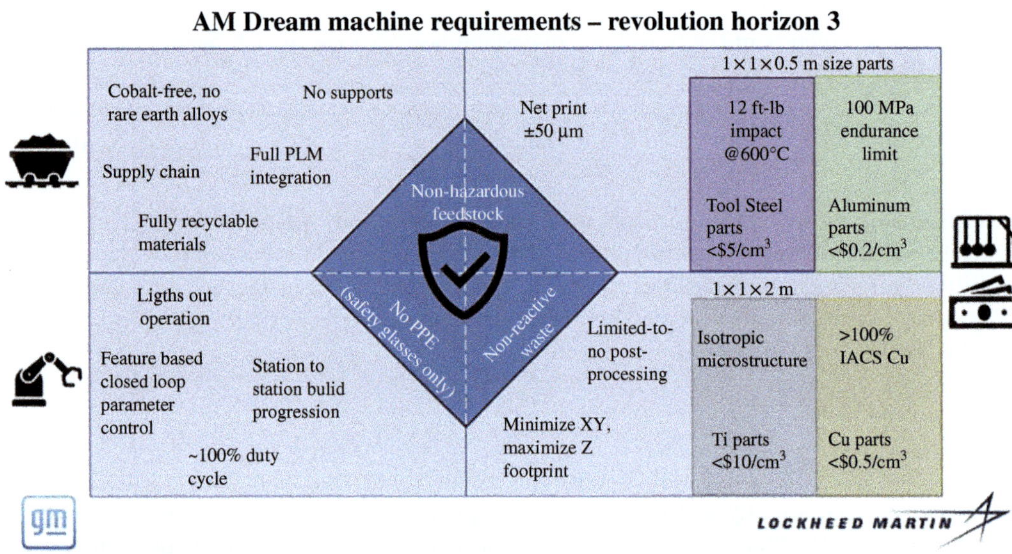

Figure 17.7 The future of additive manufacturing as envisioned by a major industrial collaboration. Goals are specified in terms of evolutionary and eventually revolutionary progress. (Courtesy of General Motors and Lockheed Martin) [8].

loading and unloading of build trays to provide realistic production rates necessary for the automotive industry, as well as acknowledging less stringent technical specifications than those required in aerospace and medical parts. The "revolutionary" requirements are envisioned to involve future AM machine designs to be informed by the large volume needs of the automotive industry, rather than simply adapting production to machines designed for aerospace and medical industries. Related goals would include nonhazardous feedstock, nonreactive waste, and dramatically reduced costs per part.

References

1 Jhn, S.N. (2022). Metal powder market growth from 2022 to 2032, *Fact.MR's research and consulting*, https://www.factmr.com/report/3202/metal-powder-market.

2 GMIPulse (2022). Additive manufacturing with metal powders market 2022–2030, GMIPulse, https://www.gminsights.com/industry-analysis/additive-manufacturing-with-metal-powders-market.

3 Grand View Research (2023). Powder metallurgy market size, share & trends analysis report 2023–2030, Grand View Research, https://www.grandviewresearch.com/industry-analysis/powder-metallurgy-market#. Report ID: GVR-4-68038-660-8.

4 SmarTech/Additive Manufacturing Research (2020). SmarTech Publishing announces metal powder shipments for additive manufacturing to reach $520 million by 2019, SmarTech, https://www.smartechanalysis.com/news/metal-powder-shipments-am/.

5 Munsch, M., and Wycisk, E. (2023). Metal AM's journey to industrialisation: Are we there yet? And what does the destination even look like?, Vol. 9, 141–147. AMPOWER: Metal AM's journey, https://www.metal-am.com/metal-additive-manufacturing-magazine-archive/metal-additive-manufacturing-magazine-archive-2023/.

6 Desai, P.S. and Higgs, C.F. (2019). *Metals* 9: 1176.

7 Smugeresky, J. E., Zheng, B., Ajdelsztajn, L. et al. (2005). 134th TMS annual meeting & exhibition, San Francisco, CA, Feb. 13–17, (2005).

8 Kerns, J. (2022) Revolution, not evolution: general motors on building an AM culture and the AM dream machine, Vol. 8, 135–145. MetalAM, https://www.metal-am.com/articles/revolution-not-evolution-general-motors-on-building-an-am-culture-and-the-am-dream-machine/.

Index

Metallic Powders for Additive Manufacturing: Science and Applications, First Edition.
Enrique J. Lavernia, Kaka Ma, Julie M. Schoenung, James F. Shackelford, and Baolong Zheng.